Modern Analytical Chemistry

David Harvey
DePauw University

Boston Burr Ridge, IL Dubuque, IA Madison, WI New York San Francisco St. Louis
Bangkok Bogotá Caracas Lisbon London Madrid
Mexico City Milan New Delhi Seoul Singapore Sydney Taipei Toronto

McGraw-Hill Higher Education

A Division of The **McGraw-Hill** *Companies*

MODERN ANALYTICAL CHEMISTRY

This book is printed on acid-free paper.

2 3 4 5 6 7 8 9 0 KGP/KGP 0 9 8 7 6 5 4 3 2 1 0

ISBN 0–07–237547–7

Vice president and editorial director: *Kevin T. Kane*
Publisher: *James M. Smith*
Sponsoring editor: *Kent A. Peterson*
Editorial assistant: *Jennifer L. Bensink*
Developmental editor: *Shirley R. Oberbroeckling*
Senior marketing manager: *Martin J. Lange*
Senior project manager: *Jayne Klein*
Production supervisor: *Laura Fuller*
Coordinator of freelance design: *Michelle D. Whitaker*
Senior photo research coordinator: *Lori Hancock*
Senior supplement coordinator: *Audrey A. Reiter*
Compositor: *Shepherd, Inc.*
Typeface: *10/12 Minion*
Printer: *Quebecor Printing Book Group/Kingsport*

Freelance cover/interior designer: *Elise Lansdon*
Cover image: *© George Diebold/The Stock Market*
Photo research: *Roberta Spieckerman Associates*

Colorplates: Colorplates 1–6, 8, 10: © David Harvey/Marilyn E. Culler, photographer; Colorplate 7: Richard Megna/Fundamental Photographs; Colorplate 9: © Alfred Pasieka/Science Photo Library/Photo Researchers, Inc.; Colorplate 11: From H. Black, *Environ. Sci. Technol.*, **1996,** *30,* 124A. Photos courtesy D. Pesiri and W. Tumas, Los Alamos National Laboratory; Colorplate 12: Courtesy of Hewlett-Packard Company; Colorplate 13: © David Harvey.

Library of Congress Cataloging-in-Publication Data

Harvey, David, 1956–
Modern analytical chemistry / David Harvey. — 1st ed.
p. cm.
Includes bibliographical references and index.
ISBN 0–07–237547–7
1. Chemistry, Analytic. I. Title.
QD75.2.H374 2000
543—dc21 99–15120
CIP

INTERNATIONAL EDITION ISBN 0–07–116953–9

www.mhhe.com

Contents

Chapter 6
Equilibrium Chemistry 135

Chapter 7
Obtaining and Preparing Samples for Analysis 179

Chapter 8

Gravimetric Methods of Analysis 232

Chapter 9

Titrimetric Methods of Analysis 273

Chapter 10

Spectroscopic Methods of Analysis 368

A Guide to Using This Text

. . . in Chapter

Representative Methods

Annotated methods of typical analytical procedures link theory with practice. The format encourages students to think about the design of the procedure and why it works.

246 Modern Analytical Chemistry

An additional problem is encountered when the isolated solid is nonstoichiometric. For example, precipitating Mn^{2+} as $Mn(OH)_2$, followed by heating to produce the oxide, frequently produces a solid with a stoichiometry of MnO_x, where x varies between 1 and 2. In this case the nonstoichiometric product results from the formation of a mixture of several oxides that differ in the oxidation state of manganese. Other nonstoichiometric compounds form as a result of lattice defects in the crystal structure.[6]

Representative Method The best way to appreciate the importance of the theoretical and practical details discussed in the previous section is to carefully examine the procedure for a typical precipitation gravimetric method. Although each method has its own unique considerations, the determination of Mg^{2+} in water and wastewater by precipitating $MgNH_4PO_4 \cdot 6H_2O$ and isolating $Mg_2P_2O_7$ provides an instructive example of a typical procedure.

Representative Methods

Method 8.1 Determination of Mg^{2+} in Water and Wastewater[7]

Description of Method. Magnesium is precipitated as $MgNH_4PO_4 \cdot 6H_2O$ using $(NH_4)_2HPO_4$ as the precipitant. The precipitate's solubility in neutral solutions (0.0065 g/100 mL in pure water at 10 °C) is relatively high, but it is much less soluble in the presence of dilute ammonia (0.0003 g/100 mL in 0.6 M NH_3). The precipitant is not very selective, so a preliminary separation of Mg^{2+} from potential interferents is necessary. Calcium, which is the most significant interferent, is usually removed by its prior precipitation as the oxalate. The presence of excess ammonium salts from the precipitant or the addition of too much ammonia can lead to the formation of $Mg(NH_4)_4(PO_4)_2$, which is subsequently isolated as $Mg(PO_3)_2$ after drying. The precipitate is isolated by filtration using a rinse solution of dilute ammonia. After filtering, the precipitate is converted to $Mg_2P_2O_7$ and weighed.

Procedure. Transfer a sample containing no more than 60 mg of Mg^{2+} into a 600-mL beaker. Add 2–3 drops of methyl red indicator, and, if necessary, adjust the volume to 150 mL. Acidify the solution with 6 M HCl, and add 10 mL of 30% w/v $(NH_4)_2HPO_4$. After cooling, add concentrated NH_3 dropwise, and while constantly stirring, until the methyl red indicator turns yellow (pH > 6.3). After stirring for 5 min, add 5 mL of concentrated NH_3, and continue stirring for an additional 10 min. Allow the solution and precipitate to stand overnight. Isolate the precipitate by filtration, rinsing with 5% w/v NH_3. Dissolve the precipitate in 50 mL of 10% w/v HCl, and precipitate a second time following the same procedure. After filtering, carefully remove the filter paper by charring. Heat the precipitate at 500 °C until the residue is white, and then bring the precipitate to constant weight at 1100 °C.

Questions

Margin Notes

Margin notes direct students to colorplates located toward the middle of the book

110 Modern Analytical Chemistry

either case, the calibration curve provides a means for relating S_{samp} to the analyte's concentration.

Color plate 1 shows an example of a set of external standards and their corresponding normal calibration curve.

Example

EXAMPLE 5.3

A second spectrophotometric method for the quantitative determination of Pb^{2+} levels in blood gives a linear normal calibration curve for which

$$S_{stand} = (0.296\ \text{ppb}^{-1}) \times C_S + 0.003$$

What is the Pb^{2+} level (in ppb) in a sample of blood if S_{samp} is 0.397?

SOLUTION

To determine the concentration of Pb^{2+} in the sample of blood, we replace S_{stand} in the calibration equation with S_{samp} and solve for C_A

$$C_A = \frac{S_{samp} - 0.003}{0.296\ \text{ppb}^{-1}} = \frac{0.397 - 0.003}{0.296\ \text{ppb}^{-1}} = 1.33\ \text{ppb}$$

It is worth noting that the calibration equation in this problem includes an extra term that is not in equation 5.3. Ideally, we expect the calibration curve to give a signal of zero when C_S is zero. This is the purpose of using a reagent blank to correct the measured signal. The extra term of +0.003 in our calibration equation results from uncertainty in measuring the signal for the reagent blank and the standards.

An external standardization allows a related series of samples to be analyzed using a single calibration curve. This is an important advantage in laboratories where many samples are to be analyzed or when the need for a rapid throughput of

Examples of Typical Problems

Each example problem includes a detailed solution that helps students in applying the chapter's material to practical problems.

There is a serious limitation, however, to an external standardization. The relationship between S_{stand} and C_S in equation 5.3 is determined when the analyte is present in the external standard's matrix. In using an external standardization, we assume that any difference between the matrix of the standards and the sample's matrix has no effect on the value of k. A proportional determinate error is introduced when differences between the two matrices cannot be ignored. This is shown in Figure 5.4, where the relationship between the signal and the amount of analyte is shown for both the sample's matrix and the standard's matrix. In this example, using a normal calibration curve results in a negative determinate error. When matrix problems are expected, an effort is made to match the matrix of the standards to that of the sample. This is known as **matrix matching.** When the sample's matrix is unknown, the matrix effect must be shown to be negligible, or an alternative method of standardization must be used. Both approaches are discussed in the following sections.

matrix matching
Adjusting the matrix of an external standard so that it is the same as the matrix of the samples to be analyzed.

method of standard additions
A standardization in which aliquots of a standard solution are added to the sample.

5B.4 Standard Additions

The complication of matching the matrix of the standards to that of the sample can be avoided by conducting the standardization in the sample. This is known as the **method of standard additions.** The simplest version of a standard addition is shown in Figure 5.5. A volume, V_o, of sample is diluted to a final volume, V_f, and the signal, S_{samp} is measured. A second identical **aliquot** of sample is

Bold-faced Key Terms with Margin Definitions

Key words appear in boldface when they are introduced within the text. The term and its definition appear in the margin for quick review by the student. All key words are also defined in the glossary.

. . . End of Chapter

5E KEY TERMS

aliquot (*p. 111*)
external standard (*p. 109*)
internal standard (*p. 116*)
linear regression (*p. 118*)
matrix matching (*p. 110*)
method of standard additions (*p. 110*)
multiple-point standardization (*p. 109*)
normal calibration curve (*p. 109*)
primary reagent (*p. 106*)
reagent grade (*p. 107*)
residual error (*p. 118*)
secondary reagent (*p. 107*)
single-point standardization (*p. 108*)
standard deviation about the regression (*p. 121*)
total Youden blank (*p. 129*)

List of Key Terms

The key terms introduced within the chapter are listed at the end of each chapter. Page references direct the student to the definitions in the text.

5F SUMMARY

In a quantitative analysis, we measure a signal and calculate the amount of analyte using one of the following equations.

$$S_{meas} = kn_A + S_{reag}$$

$$S_{meas} = kC_A + S_{reag}$$

To obtain accurate results we must eliminate determinate errors affecting the measured signal, S_{meas}, the method's sensitivity, k, and any signal due to the reagents, S_{reag}.

To ensure that S_{meas} is determined accurately, we calibrate the equipment or instrument used to obtain the signal. Balances are calibrated using standard weights. When necessary, we can also correct for the buoyancy of air. Volumetric glassware can be calibrated by measuring the mass of water contained or delivered and using the density of water to calculate the true volume. Most instruments have calibration standards suggested by the manufacturer.

An analytical method is standardized by determining its sensitivity. There are several approaches to standardization, including the use of external standards, the method of standard addition, and the use of an internal standard. The most desirable standardization strategy is an external standardization. The method of standard additions, in which known amounts of analyte are added to the sample, is used when the sample's matrix complicates the analysis. An internal standard, which is a species (not analyte) added to all samples and standards, is used when the procedure does not allow for the reproducible handling of samples and standards.

Standardizations using a single standard are common, but also are subject to greater uncertainty. Whenever possible, a multiple-point standardization is preferred. The results of a multiple-point standardization are graphed as a calibration curve. A linear regression analysis can provide an equation for the standardization.

A reagent blank corrects the measured signal for signals due to reagents other than the sample that are used in an analysis. The most common reagent blank is prepared by omitting the sample. When a simple reagent blank does not compensate for all constant sources of determinate error, other types of blanks, such as the total Youden blank, can be used.

Summary

The summary provides the student with a brief review of the important concepts within the chapter.

5G *Suggested* EXPERIMENTS

Experiments

The following exercises and experiments help connect the material in this chapter to the analytical laboratory.

Calibration—Volumetric glassware (burets, pipets, and volumetric flasks) can be calibrated in the manner described in Example 5.1. Most instruments have a calibration sample that can be prepared to verify the instrument's accuracy and precision. For example, as described in this chapter, a solution of 60.06 ppm $K_2Cr_2O_7$ in 0.0050 M H_2SO_4 should give an absorbance of 0.640 ± 0.010 at a wavelength of 350.0 nm when using 0.0050 M H_2SO_4 as a reagent blank. These exercises also provide practice with using volumetric glassware, weighing samples, and preparing solutions.

Standardization—External standards, standard additions, and internal standards are a common feature of many quantitative analyses. Suggested experiments using these standardization methods are found in later chapters. A good project experiment for introducing external standardization, standard additions, and the importance of the sample's matrix is to explore the effect of pH ... analysis of an acid–base indicator. U... as an example, external standards c... buffer and used to analyze samples ... in the range of 6–10. Results can be ... obtained using a standard addition.

Suggested Experiments

An annotated list of representative experiments is provided from the *Journal of Chemical Education.*

Suggested Readings

Suggested readings give the student access to more comprehensive discussion of the topics introduced within the chapter.

1G SUGGESTED READINGS

The role of analytical chemistry within the broader discipline of chemistry has been discussed by many prominent analytical chemists. Several notable examples follow.

Baiulescu, G. E.; Patroescu, C.; Chalmers, R. A. *Education and Teaching in Analytical Chemistry.* Ellis Horwood: Chichester, 1982.

Hieftje, G. M. "The Two Sides of Analytical Chemistry," *Anal. Chem.* **1985,** *57,* 256A–267A.

Kissinger, P. T. "Analytical Chemistry—What is It? Who Needs It? Why Teach It?" *Trends Anal. Chem.* **1992,** *11,* 54–57.

Laitinen, H. A. "Analytical Chemistry in a Changing World," *Anal. Chem.* **1980,** *52,* 605A–609A.

Laitinen, H. A. "History of Analytical Chemistry in the U.S.A.," *Talanta* **1989,** *36,* 1–9.

Laitinen, H. A.; Ewing, G. (eds). *A History of Analytical Chemistry.* The Division of Analytical Chemistry of the American Chemical Society: Washington, D.C., 1972.

McLafferty, F. W. "Analytical Chemistry: Historic and Modern," *Acc. Chem. Res.* **1990,** *23,* 63–64.

References

The references cited in the chapter are provided so the student can access them for further information.

1H REFERENCES

1. Ravey, M. *Spectroscopy* **1990,** *5(7),* 11.
2. de Haseth, J. *Spectroscopy* **1990,** *5(7),* 11.
3. Fresenius, C. R. *A System of Instruction in Quantitative Chemical Analysis.* John Wiley and Sons: New York, 1881.
4. Hillebrand, W. F.; Lundell, G. E. F. *Applied Inorganic Analysis,* John Wiley and Sons: New York, 1953.
5. Van Loon, J. C. *Analytical Atomic Absorption Spectroscopy.* Academic Press: New York, 1980.
6. Murray, R. W. *Anal. Chem.* **1991,** *63,* 271A.
7. For several different viewpoints see (a) Beilby, A. L. *J. Chem. Educ.* **1970,** *47,* 237–238; (b) Lucchesi, C. A. *Am. Lab.* **1980,** *October,* 113–119; (c) Atkinson, G. F. *J. Chem. Educ.* **1982,** *59,* 201–202; (d) Pardue, H. L.; Woo, J. *J. Chem. Educ.* **1984,** *61,* 409–412; (e) Guarnieri, M. *J. Chem. Educ.* **1988,** *65,* 201–203; (f) de Haseth, J. *Spectroscopy* **1990,** *5,* 20–21; (g) Strobel, H. A. *Am. Lab.* **1990,** *October,* 17–24.
8. Hieftje, G. M. *Am. Lab.* **1993,** *October,* 53–61.
9. See, for example, the following laboratory texts: (a) Sorum, C. H.; Lagowski, J. J. *Introduction to Semimicro Qualitative Analysis,* 5th ed. Prentice-Hall: Englewood Cliffs, NJ, 1977.; (b) Shriner, R. L.; Fuson, R. C.; Curtin, D. Y. *The Systematic Identification of Organic Compounds,* 5th ed. John Wiley and Sons: New York, 1964.

3J PROBLEMS

1. When working with a solid sample, it often is necessary to bring the analyte into solution by dissolving the sample in a suitable solvent. Any solid impurities that remain are removed by filtration before continuing with the analysis. In a typical total analysis method, the procedure might read

   After dissolving the sample in a beaker, remove any solid impurities by passing the solution containing the analyte through filter paper, collecting the solution in a clean Erlenmeyer flask. Rinse the beaker with several small portions of solvent, passing these rinsings through the filter paper, and collecting them in the same Erlenmeyer flask. Finally, rinse the filter paper with several portions of solvent, collecting the rinsings in the same Erlenmeyer flask.

   For a typical concentration method, however, the procedure ...

4. A sample was analyzed to determine the concentration of an analyte. Under the conditions of the analysis, the sensitivity is 17.2 ppm^{-1}. What is the analyte's concentration if S_{meas} is 35.2 and S_{reag} is 0.6?
5. A method for the analysis of Ca^{2+} in water suffers from an interference in the presence of Zn^{2+}. When the concentration of Ca^{2+} is 50 times greater than that of Zn^{2+}, an analysis for Ca^{2+} gives a relative error of –2.0%. What is the value of the selectivity coefficient for this method?
6. The quantitative analysis for reduced glutathione in blood is complicated by the presence of many potential interferents. In one study, when analyzing a solution of 10-ppb glutathione and 1.5-ppb ascorbic acid, the signal was 5.43 times greater than that obtained for the analysis of 10-ppb glutathione.[12] What is the selectivity coefficient for this analysis? The same study found that when analyzing a solution of 350-ppb methionine and 10-ppb glutathione the ...

Problems

A variety of problems, many based on data from the analytical literature, provide the student with practical examples of current research.

Preface

Preface

As currently taught, the introductory course in analytical chemistry emphasizes quantitative (and sometimes qualitative) methods of analysis coupled with a heavy dose of equilibrium chemistry. Analytical chemistry, however, is more than equilibrium chemistry and a collection of analytical methods; it is an approach to solving chemical problems. Although discussing different methods is important, that discussion should not come at the expense of other equally important topics. The introductory analytical course is the ideal place in the chemistry curriculum to explore topics such as experimental design, sampling, calibration strategies, standardization, optimization, statistics, and the validation of experimental results. These topics are important in developing good experimental protocols, and in interpreting experimental results. If chemistry is truly an experimental science, then it is essential that all chemistry students understand how these topics relate to the experiments they conduct in other chemistry courses.

Currently available textbooks do a good job of covering the diverse range of wet and instrumental analysis techniques available to chemists. Although there is some disagreement about the proper balance between wet analytical techniques, such as gravimetry and titrimetry, and instrumental analysis techniques, such as spectrophotometry, all currently available textbooks cover a reasonable variety of techniques. These textbooks, however, neglect, or give only brief consideration to, obtaining representative samples, handling interferents, optimizing methods, analyzing data, validating data, and ensuring that data are collected under a state of statistical control.

In preparing this textbook, I have tried to find a more appropriate balance between theory and practice, between "classical" and "modern" methods of analysis, between analyzing samples and collecting and preparing samples for analysis, and between analytical methods and data analysis. Clearly, the amount of material in this textbook exceeds what can be covered in a single semester; it's my hope, however, that the diversity of topics will meet the needs of different instructors, while, perhaps, suggesting some new topics to cover.

The anticipated audience for this textbook includes students majoring in chemistry, and students majoring in other science disciplines (biology, biochemistry, environmental science, engineering, and geology, to name a few), interested in obtaining a stronger background in chemical analysis. It is particularly appropriate for chemistry majors who are not planning to attend graduate school, and who often do not enroll in those advanced courses in analytical chemistry that require physical chemistry as a pre-requisite. Prior coursework of a year of general chemistry is assumed. Competence in algebra is essential; calculus is used on occasion, however, its presence is not essential to the material's treatment.

Key Features of This Textbook

Key features set this textbook apart from others currently available.

- *A stronger emphasis on the evaluation of data.* Methods for characterizing chemical measurements, results, and errors (including the propagation of errors) are included. Both the binomial distribution and normal distribution are presented, and the idea of a confidence interval is developed. Statistical methods for evaluating data include the *t*-test (both for paired and unpaired data), the *F*-test, and the treatment of outliers. Detection limits also are discussed from a statistical perspective. Other statistical methods, such as ANOVA and ruggedness testing, are presented in later chapters.
- *Standardizations and calibrations are treated in a single chapter.* Selecting the most appropriate calibration method is important and, for this reason, the methods of external standards, standard additions, and internal standards are gathered together in a single chapter. A discussion of curve-fitting, including the statistical basis for linear regression (with and without weighting) also is included in this chapter.
- *More attention to selecting and obtaining a representative sample.* The design of a statistically based sampling plan and its implementation are discussed earlier, and in more detail than in other textbooks. Topics that are covered include how to obtain a representative sample, how much sample to collect, how many samples to collect, how to minimize the overall variance for an analytical method, tools for collecting samples, and sample preservation.
- *The importance of minimizing interferents is emphasized.* Commonly used methods for separating interferents from analytes, such as distillation, masking, and solvent extraction, are gathered together in a single chapter.
- *Balanced coverage of analytical techniques.* The six areas of analytical techniques—gravimetry, titrimetry, spectroscopy, electrochemistry, chromatography, and kinetics—receive roughly equivalent coverage, meeting the needs of instructors wishing to emphasize wet methods and those emphasizing instrumental methods. Related methods are gathered together in a single chapter encouraging students to see the similarities between methods, rather than focusing on their differences.
- *An emphasis on practical applications.* Throughout the text applications from organic chemistry, inorganic chemistry, environmental chemistry, clinical chemistry, and biochemistry are used in worked examples, representative methods, and end-of-chapter problems.
- *Representative methods link theory with practice.* An important feature of this text is the presentation of representative methods. These boxed features present typical analytical procedures in a format that encourages students to think about why the procedure is designed as it is.
- *Separate chapters on developing a standard method and quality assurance.* Two chapters provide coverage of methods used in developing a standard method of analysis, and quality assurance. The chapter on developing a standard method includes topics such as optimizing experimental conditions using response surfaces, verifying the method through the blind analysis of standard samples and ruggedness testing, and collaborative testing using Youden's two-sample approach and ANOVA. The chapter on quality assurance covers quality control and internal and external techniques for quality assessment, including the use of duplicate samples, blanks, spike recoveries, and control charts.

- *Problems adapted from the literature.* Many of the in-chapter examples and end-of-chapter problems are based on data from the analytical literature, providing students with practical examples of current research in analytical chemistry.
- *An emphasis on critical thinking.* Critical thinking is encouraged through problems in which students are asked to explain why certain steps in an analytical procedure are included, or to determine the effect of an experimental error on the results of an analysis.
- *Suggested experiments from the Journal of Chemical Education.* Rather than including a short collection of experiments emphasizing the analysis of standard unknowns, an annotated list of representative experiments from the *Journal of Chemical Education* is included at the conclusion of most chapters. These experiments may serve as stand alone experiments, or as starting points for individual or group projects.

The Role of Equilibrium Chemistry in Analytical Chemistry

Equilibrium chemistry often receives a significant emphasis in the introductory analytical chemistry course. While an important topic, its overemphasis can cause students to confuse analytical chemistry with equilibrium chemistry. Although attention to solving equilibrium problems is important, it is equally important for students to recognize when such calculations are impractical, or when a simpler, more qualitative approach is all that is needed. For example, in discussing the gravimetric analysis of Ag^+ as AgCl, there is little point in calculating the equilibrium solubility of AgCl since the concentration of Cl^- at equilibrium is rarely known. It is important, however, to qualitatively understand that a large excess of Cl^- increases the solubility of AgCl due to the formation of soluble silver-chloro complexes. Balancing the presentation of a rigorous approach to solving equilibrium problems, this text also introduces the use of ladder diagrams as a means for providing a qualitative picture of a system at equilibrium. Students are encouraged to use the approach best suited to the problem at hand.

Computer Software

Many of the topics covered in analytical chemistry benefit from the availability of appropriate computer software. In preparing this text, however, I made a conscious decision to avoid a presentation tied to a single computer platform or software package. Students and faculty are increasingly experienced in the use of computers, spreadsheets, and data analysis software; their use is, I think, best left to the personal choice of each student and instructor.

Organization

The textbook's organization can be divided into four parts. Chapters 1–3 serve as an introduction, providing an overview of analytical chemistry (Chapter 1); a review of the basic tools of analytical chemistry, including significant figures, units, and stoichiometry (Chapter 2); and an introduction to the terminology used by analytical chemists (Chapter 3). Familiarity with the material in these chapters is assumed throughout the remainder of the text.

Chapters 4–7 cover a number of topics that are important in understanding how a particular analytical method works. Later chapters are mostly independent of the material in these chapters. Instructors may pick and choose from among the topics of these chapters, as needed, to support individual course goals. The statistical analysis of data is covered in Chapter 4 at a level that is more complete than that found in other introductory analytical textbooks. Methods for calibrating equipment, standardizing methods, and linear regression are gathered together in Chapter 5. Chapter 6 provides an introduction to equilibrium chemistry, stressing both the rigorous solution to equilibrium problems, and the use of semi-quantitative approaches, such as ladder diagrams. The importance of collecting the right sample, and methods for separating analytes and interferents are covered in Chapter 7.

Chapters 8–13 cover the major areas of analysis, including gravimetry (Chapter 8), titrimetry (Chapter 9), spectroscopy (Chapter 10), electrochemistry (Chapter 11), chromatography and electrophoresis (Chapter 12), and kinetic methods (Chapter 13). Related techniques, such as acid–base titrimetry and redox titrimetry, or potentiometry and voltammetry, are gathered together in single chapters. Combining related techniques together encourages students to see the similarities between methods, rather than focusing on their differences. The first technique presented in each chapter is generally that which is most commonly covered in the introductory course.

Finally, the textbook concludes with two chapters discussing the design and maintenance of analytical methods, two topics of importance to analytical chemists. Chapter 14 considers the development of an analytical method, including its optimization, verification, and validation. Quality control and quality assessment are discussed in Chapter 15.

Acknowledgments

Before beginning an academic career I was, of course, a student. My interest in chemistry and teaching was nurtured by many fine teachers at Westtown Friends School, Knox College, and the University of North Carolina at Chapel Hill; their collective influence continues to bear fruit. In particular, I wish to recognize David MacInnes, Alan Hiebert, Robert Kooser, and Richard Linton.

I have been fortunate to work with many fine colleagues during my nearly 17 years of teaching undergraduate chemistry at Stockton State College and DePauw University. I am particularly grateful for the friendship and guidance provided by Jon Griffiths and Ed Paul during my four years at Stockton State College. At DePauw University, Jim George and Bryan Hanson have willingly shared their ideas about teaching, while patiently listening to mine.

Approximately 300 students have joined me in thinking and learning about analytical chemistry; their questions and comments helped guide the development of this textbook. I realize that working without a formal textbook has been frustrating and awkward; all the more reason why I appreciate their effort and hard work.

The following individuals reviewed portions of this textbook at various stages during its development.

David Ballantine
Northern Illinois University

John E. Bauer
Illinois State University

Ali Bazzi
University of Michigan–Dearborn

Steven D. Brown
University of Delaware

Wendy Clevenger
University of Tennessee–Chattanooga

Cathy Cobb
Augusta State University

Paul Flowers
University of North Carolina–Pembroke

Nancy Gordon
University of Southern Maine

Virginia M. Indivero
Swarthmore College

Michael Janusa
Nicholls State University

J. David Jenkins
Georgia Southern University

Richard S. Mitchell
Arkansas State University

George A. Pearse, Jr.
Le Moyne College

Gary Rayson
New Mexico State University

David Redfield
NW Nazarene University

Vincent Remcho
West Virginia University

Jeanette K. Rice
Georgia Southern University

Martin W. Rowe
Texas A&M University

Alexander Scheeline
University of Illinois

James D. Stuart
University of Connecticut

Thomas J. Wenzel
Bates College

David Zax
Cornell University

I am particularly grateful for their detailed written comments and suggestions for improving the manuscript. Much of what is good in the final manuscript is the result of their interest and ideas. George Foy (*York College of Pennsylvania*), John McBride (*Hofstra University*), and David Karpovich (*Saginaw Valley State University*) checked the accuracy of problems in the textbook. Gary Kinsel (*University of Texas at Arlington*) reviewed the page proofs and provided additional suggestions.

This project began in the summer of 1992 with the support of a course development grant from DePauw University's Faculty Development Fund. Additional financial support from DePauw University's Presidential Discretionary Fund also is acknowledged. Portions of the first draft were written during a sabbatical leave in the Fall semester of the 1993/94 academic year. A Fisher Fellowship provided release time during the Fall 1995 semester to complete the manuscript's second draft.

Alltech and Associates (Deerfield, IL) graciously provided permission to use the chromatograms in Chapter 12; the assistance of Jim Anderson, Vice-President, and Julia Poncher, Publications Director, is greatly appreciated. Fred Soster and Marilyn Culler, both of DePauw University, provided assistance with some of the photographs.

The editorial staff at McGraw-Hill has helped guide a novice through the process of developing this text. I am particularly thankful for the encouragement and confidence shown by Jim Smith, Publisher for Chemistry, and Kent Peterson, Sponsoring Editor for Chemistry. Shirley Oberbroeckling, Developmental Editor for Chemistry, and Jayne Klein, Senior Project Manager, patiently answered my questions and successfully guided me through the publishing process.

Finally, I would be remiss if I did not recognize the importance of my family's support and encouragement, particularly that of my parents. A very special thanks to my daughter, Devon, for gifts too numerous to detail.

How to Contact the Author

Writing this textbook has been an interesting (and exhausting) challenge. Despite my efforts, I am sure there are a few glitches, better examples, more interesting end-of-chapter problems, and better ways to think about some of the topics. I welcome your comments, suggestions, and data for interesting problems, which may be addressed to me at DePauw University, 602 S. College St., Greencastle, IN 46135, or electronically at harvey@depauw.edu.

Chapter 1

Introduction

Chemistry is the study of matter, including its composition, structure, physical properties, and reactivity. There are many approaches to studying chemistry, but, for convenience, we traditionally divide it into five fields: organic, inorganic, physical, biochemical, and analytical. Although this division is historical and arbitrary, as witnessed by the current interest in interdisciplinary areas such as bioanalytical and organometallic chemistry, these five fields remain the simplest division spanning the discipline of chemistry.

Training in each of these fields provides a unique perspective to the study of chemistry. Undergraduate chemistry courses and textbooks are more than a collection of facts; they are a kind of apprenticeship. In keeping with this spirit, this text introduces the field of analytical chemistry and the unique perspectives that analytical chemists bring to the study of chemistry.

1A What Is Analytical Chemistry?

*"Analytical chemistry is what analytical chemists do."**

We begin this section with a deceptively simple question. What is analytical chemistry? Like all fields of chemistry, analytical chemistry is too broad and active a discipline for us to easily or completely define in an introductory textbook. Instead, we will try to say a little about what analytical chemistry is, as well as a little about what analytical chemistry is not.

Analytical chemistry is often described as the area of chemistry responsible for characterizing the composition of matter, both qualitatively (what is present) and quantitatively (how much is present). This description is misleading. After all, almost all chemists routinely make qualitative or quantitative measurements. The argument has been made that analytical chemistry is not a separate branch of chemistry, but simply the application of chemical knowledge.[1] In fact, you probably have performed quantitative and qualitative analyses in other chemistry courses. For example, many introductory courses in chemistry include qualitative schemes for identifying inorganic ions and quantitative analyses involving titrations.

Unfortunately, this description ignores the unique perspective that analytical chemists bring to the study of chemistry. The craft of analytical chemistry is not in performing a routine analysis on a routine sample (which is more appropriately called chemical analysis), but in improving established methods, extending existing methods to new types of samples, and developing new methods for measuring chemical phenomena.[2]

Here's one example of this distinction between analytical chemistry and chemical analysis. Mining engineers evaluate the economic feasibility of extracting an ore by comparing the cost of removing the ore with the value of its contents. To estimate its value they analyze a sample of the ore. The challenge of developing and validating the method providing this information is the analytical chemist's responsibility. Once developed, the routine, daily application of the method becomes the job of the chemical analyst.

Another distinction between analytical chemistry and chemical analysis is that analytical chemists work to improve established methods. For example, several factors complicate the quantitative analysis of Ni^{2+} in ores, including the presence of a complex heterogeneous mixture of silicates and oxides, the low concentration of Ni^{2+} in ores, and the presence of other metals that may interfere in the analysis. Figure 1.1 is a schematic outline of one standard method in use during the late nineteenth century.[3] After dissolving a sample of the ore in a mixture of H_2SO_4 and HNO_3, trace metals that interfere with the analysis, such as Pb^{2+}, Cu^{2+} and Fe^{3+}, are removed by precipitation. Any cobalt and nickel in the sample are reduced to Co and Ni, isolated by filtration and weighed (point A). After dissolving the mixed solid, Co is isolated and weighed (point B). The amount of nickel in the ore sample is determined from the difference in the masses at points A and B.

$$\%\text{Ni} = \frac{\text{mass point A} - \text{mass point B}}{\text{mass sample}} \times 100$$

*Attributed to C. N. Reilley (1925–1981) on receipt of the 1965 Fisher Award in Analytical Chemistry. Reilley, who was a professor of chemistry at the University of North Carolina at Chapel Hill, was one of the most influential analytical chemists of the last half of the twentieth century.

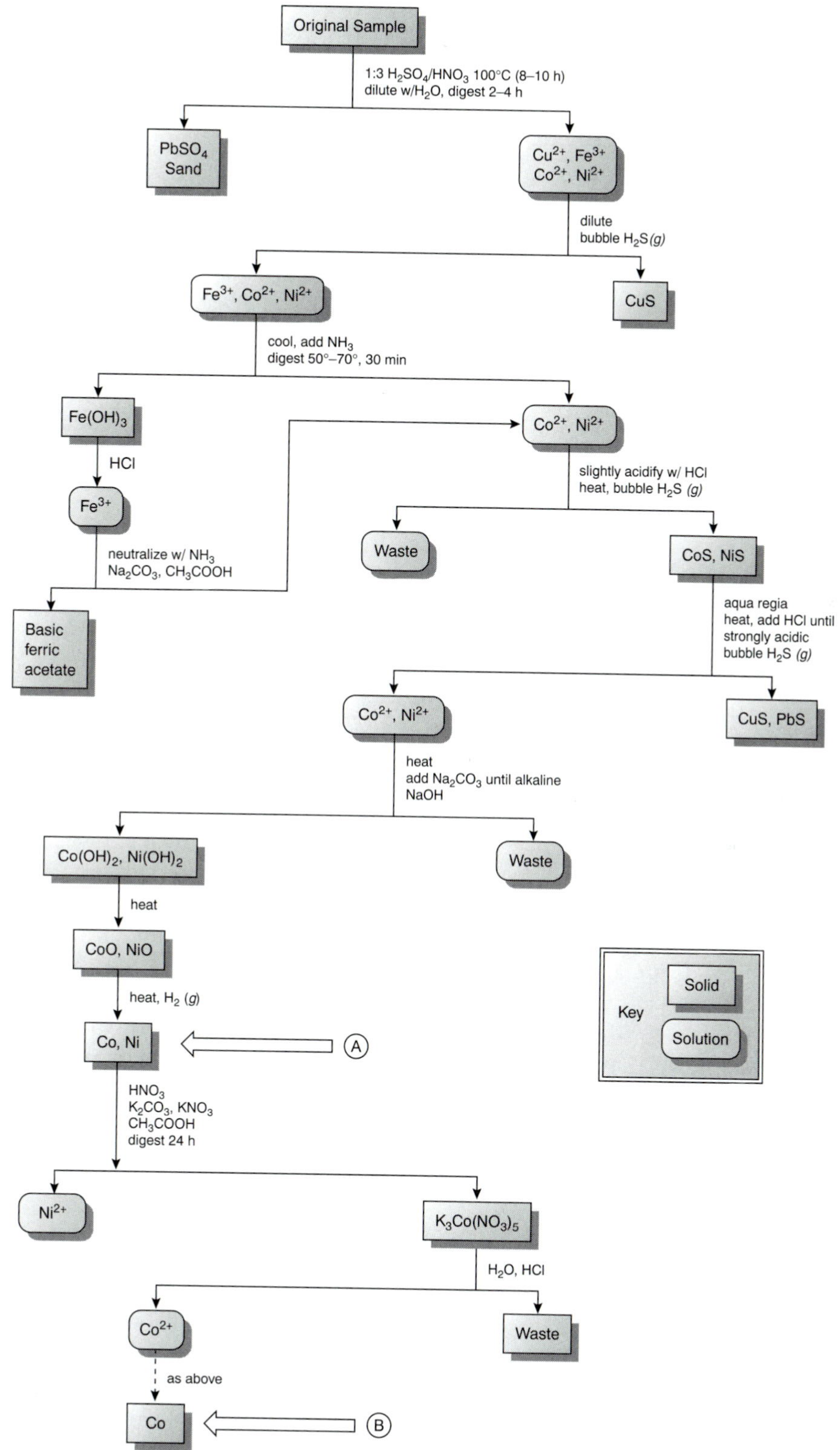

Figure 1.1
Analytical scheme outlined by Fresenius[3] for the gravimetric analysis of Ni in ores.

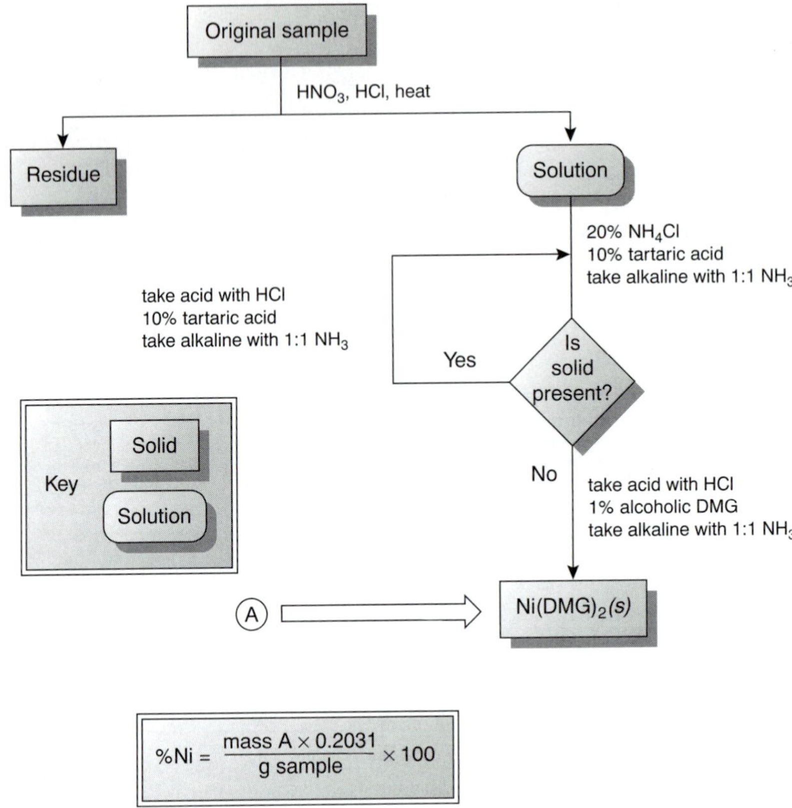

Figure 1.2
Analytical scheme outlined by Hillebrand and Lundell[4] for the gravimetric analysis of Ni in ores (DMG = dimethylgloxime). The factor of 0.2031 in the equation for %Ni accounts for the difference in the formula weights of $Ni(DMG)_2$ and Ni; see Chapter 8 for more details.

The combination of determining the mass of Ni^{2+} by difference, coupled with the need for many reactions and filtrations makes this procedure both time-consuming and difficult to perform accurately.

The development, in 1905, of dimethylgloxime (DMG), a reagent that selectively precipitates Ni^{2+} and Pd^{2+}, led to an improved analytical method for determining Ni^{2+} in ores.[4] As shown in Figure 1.2, the mass of Ni^{2+} is measured directly, requiring fewer manipulations and less time. By the 1970s, the standard method for the analysis of Ni^{2+} in ores progressed from precipitating $Ni(DMG)_2$ to flame atomic absorption spectrophotometry,[5] resulting in an even more rapid analysis. Current interest is directed toward using inductively coupled plasmas for determining trace metals in ores.

In summary, a more appropriate description of analytical chemistry is ". . . the science of inventing and applying the concepts, principles, and . . . strategies for measuring the characteristics of chemical systems and species."[6] Analytical chemists typically operate at the extreme edges of analysis, extending and improving the ability of all chemists to make meaningful measurements on smaller samples, on more complex samples, on shorter time scales, and on species present at lower concentrations. Throughout its history, analytical chemistry has provided many of the tools and methods necessary for research in the other four traditional areas of chemistry, as well as fostering multidisciplinary research in, to name a few, medicinal chemistry, clinical chemistry, toxicology, forensic chemistry, material science, geochemistry, and environmental chemistry.

You will come across numerous examples of qualitative and quantitative methods in this text, most of which are routine examples of chemical analysis. It is important to remember, however, that nonroutine problems prompted analytical chemists to develop these methods. Whenever possible, we will try to place these methods in their appropriate historical context. In addition, examples of current research problems in analytical chemistry are scattered throughout the text.

The next time you are in the library, look through a recent issue of an analytically oriented journal, such as *Analytical Chemistry*. Focus on the titles and abstracts of the research articles. Although you will not recognize all the terms and methods, you will begin to answer for yourself the question "What is analytical chemistry"?

1B The Analytical Perspective

Having noted that each field of chemistry brings a unique perspective to the study of chemistry, we now ask a second deceptively simple question. What is the "analytical perspective"? Many analytical chemists describe this perspective as an analytical approach to solving problems.[7] Although there are probably as many descriptions of the analytical approach as there are analytical chemists, it is convenient for our purposes to treat it as a five-step process:

1. Identify and define the problem.
2. Design the experimental procedure.
3. Conduct an experiment, and gather data.
4. Analyze the experimental data.
5. Propose a solution to the problem.

Figure 1.3 shows an outline of the analytical approach along with some important considerations at each step. Three general features of this approach deserve attention. First, steps 1 and 5 provide opportunities for analytical chemists to collaborate with individuals outside the realm of analytical chemistry. In fact, many problems on which analytical chemists work originate in other fields. Second, the analytical approach is not linear, but incorporates a "feedback loop" consisting of steps 2, 3, and 4, in which the outcome of one step may cause a reevaluation of the other two steps. Finally, the solution to one problem often suggests a new problem.

Analytical chemistry begins with a problem, examples of which include evaluating the amount of dust and soil ingested by children as an indicator of environmental exposure to particulate based pollutants, resolving contradictory evidence regarding the toxicity of perfluoro polymers during combustion, or developing rapid and sensitive detectors for chemical warfare agents.* At this point the analytical approach involves a collaboration between the analytical chemist and the individuals responsible for the problem. Together they decide what information is needed. It is also necessary for the analytical chemist to understand how the problem relates to broader research goals. The type of information needed and the problem's context are essential to designing an appropriate experimental procedure.

Designing an experimental procedure involves selecting an appropriate method of analysis based on established criteria, such as accuracy, precision, sensitivity, and detection limit; the urgency with which results are needed; the cost of a single analysis; the number of samples to be analyzed; and the amount of sample available for

*These examples are taken from a series of articles, entitled the "Analytical Approach," which has appeared as a regular feature in the journal *Analytical Chemistry* since 1974.

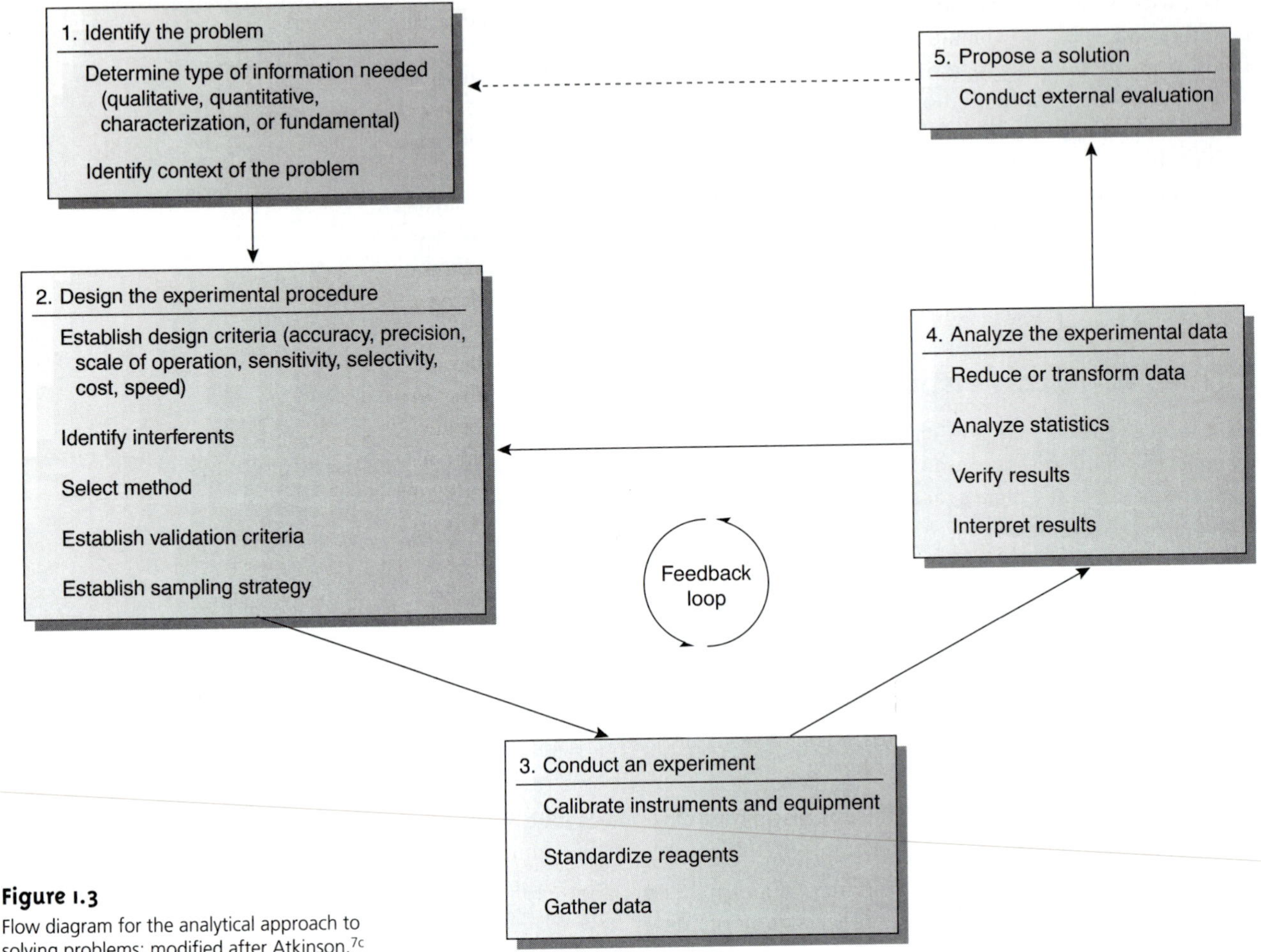

Figure 1.3
Flow diagram for the analytical approach to solving problems; modified after Atkinson.[7c]

analysis. Finding an appropriate balance between these parameters is frequently complicated by their interdependence. For example, improving the precision of an analysis may require a larger sample. Consideration is also given to collecting, storing, and preparing samples, and to whether chemical or physical interferences will affect the analysis. Finally, a good experimental procedure may still yield useless information if there is no method for validating the results.

The most visible part of the analytical approach occurs in the laboratory. As part of the validation process, appropriate chemical or physical standards are used to calibrate any equipment being used and any solutions whose concentrations must be known. The selected samples are then analyzed and the raw data recorded.

The raw data collected during the experiment are then analyzed. Frequently the data must be reduced or transformed to a more readily analyzable form. A statistical treatment of the data is used to evaluate the accuracy and precision of the analysis and to validate the procedure. These results are compared with the criteria established during the design of the experiment, and then the design is reconsidered, additional experimental trials are run, or a solution to the problem is proposed. When a solution is proposed, the results are subject to an external evaluation that may result in a new problem and the beginning of a new analytical cycle.

As an exercise, let's adapt this model of the analytical approach to a real problem. For our example, we will use the determination of the sources of airborne pollutant particles. A description of the problem can be found in the following article:

> "Tracing Aerosol Pollutants with Rare Earth Isotopes" by
> Ondov, J. M.; Kelly, W. R. *Anal. Chem.* **1991,** *63,* 691A–697A.

Before continuing, take some time to read the article, locating the discussions pertaining to each of the five steps outlined in Figure 1.3. In addition, consider the following questions:

1. What is the analytical problem?
2. What type of information is needed to solve the problem?
3. How will the solution to this problem be used?
4. What criteria were considered in designing the experimental procedure?
5. Were there any potential interferences that had to be eliminated? If so, how were they treated?
6. Is there a plan for validating the experimental method?
7. How were the samples collected?
8. Is there evidence that steps 2, 3, and 4 of the analytical approach are repeated more than once?
9. Was there a successful conclusion to the problem?

According to our model, the analytical approach begins with a problem. The motivation for this research was to develop a method for monitoring the transport of solid aerosol particulates following their release from a high-temperature combustion source. Because these particulates contain significant concentrations of toxic heavy metals and carcinogenic organic compounds, they represent a significant environmental hazard.

An aerosol is a suspension of either a solid or a liquid in a gas. Fog, for example, is a suspension of small liquid water droplets in air, and smoke is a suspension of small solid particulates in combustion gases. In both cases the liquid or solid particulates must be small enough to remain suspended in the gas for an extended time. Solid aerosol particulates, which are the focus of this problem, usually have micrometer or submicrometer diameters. Over time, solid particulates settle out from the gas, falling to the Earth's surface as dry deposition.

Existing methods for monitoring the transport of gases were inadequate for studying aerosols. To solve the problem, qualitative and quantitative information were needed to determine the sources of pollutants and their net contribution to the total dry deposition at a given location. Eventually the methods developed in this study could be used to evaluate models that estimate the contributions of point sources of pollution to the level of pollution at designated locations.

Following the movement of airborne pollutants requires a natural or artificial tracer (a species specific to the source of the airborne pollutants) that can be experimentally measured at sites distant from the source. Limitations placed on the tracer, therefore, governed the design of the experimental procedure. These limitations included cost, the need to detect small quantities of the tracer, and the absence of the tracer from other natural sources. In addition, aerosols are emitted from high-temperature combustion sources that produce an abundance of very reactive species. The tracer, therefore, had to be both thermally and chemically stable. On the basis of these criteria, rare earth isotopes, such as those of Nd, were selected as tracers. The choice of tracer, in turn, dictated the analytical method (thermal ionization mass spectrometry, or TIMS) for measuring the isotopic abundances of

Nd in samples. Unfortunately, mass spectrometry is not a selective technique. A mass spectrum provides information about the abundance of ions with a given mass. It cannot distinguish, however, between different ions with the same mass. Consequently, the choice of TIMS required developing a procedure for separating the tracer from the aerosol particulates.

Validating the final experimental protocol was accomplished by running a model study in which ^{148}Nd was released into the atmosphere from a 100-MW coal utility boiler. Samples were collected at 13 locations, all of which were 20 km from the source. Experimental results were compared with predictions determined by the rate at which the tracer was released and the known dispersion of the emissions.

Finally, the development of this procedure did not occur in a single, linear pass through the analytical approach. As research progressed, problems were encountered and modifications made, representing a cycle through steps 2, 3, and 4 of the analytical approach.

Others have pointed out, with justification, that the analytical approach outlined here is not unique to analytical chemistry, but is common to any aspect of science involving analysis.[8] Here, again, it helps to distinguish between a chemical analysis and analytical chemistry. For other analytically oriented scientists, such as physical chemists and physical organic chemists, the primary emphasis is on the problem, with the results of an analysis supporting larger research goals involving fundamental studies of chemical or physical processes. The essence of analytical chemistry, however, is in the second, third, and fourth steps of the analytical approach. Besides supporting broader research goals by developing and validating analytical methods, these methods also define the type and quality of information available to other research scientists. In some cases, the success of an analytical method may even suggest new research problems.

1C Common Analytical Problems

In Section 1A we indicated that analytical chemistry is more than a collection of qualitative and quantitative methods of analysis. Nevertheless, many problems on which analytical chemists work ultimately involve either a qualitative or quantitative measurement. Other problems may involve characterizing a sample's chemical or physical properties. Finally, many analytical chemists engage in fundamental studies of analytical methods. In this section we briefly discuss each of these four areas of analysis.

Many problems in analytical chemistry begin with the need to identify what is present in a sample. This is the scope of a **qualitative analysis,** examples of which include identifying the products of a chemical reaction, screening an athlete's urine for the presence of a performance-enhancing drug, or determining the spatial distribution of Pb on the surface of an airborne particulate. Much of the early work in analytical chemistry involved the development of simple chemical tests to identify the presence of inorganic ions and organic functional groups. The classical laboratory courses in inorganic and organic qualitative analysis,[9] still taught at some schools, are based on this work. Currently, most qualitative analyses use methods such as infrared spectroscopy, nuclear magnetic resonance, and mass spectrometry. These qualitative applications of identifying organic and inorganic compounds are covered adequately elsewhere in the undergraduate curriculum and, so, will receive no further consideration in this text.

qualitative analysis
An analysis in which we determine the identity of the constituent species in a sample.

Perhaps the most common type of problem encountered in the analytical lab is a **quantitative analysis.** Examples of typical quantitative analyses include the elemental analysis of a newly synthesized compound, measuring the concentration of glucose in blood, or determining the difference between the bulk and surface concentrations of Cr in steel. Much of the analytical work in clinical, pharmaceutical, environmental, and industrial labs involves developing new methods for determining the concentration of targeted species in complex samples. Most of the examples in this text come from the area of quantitative analysis.

quantitative analysis
An analysis in which we determine how much of a constituent species is present in a sample.

Another important area of analytical chemistry, which receives some attention in this text, is the development of new methods for characterizing physical and chemical properties. Determinations of chemical structure, equilibrium constants, particle size, and surface structure are examples of a **characterization analysis.**

characterization analysis
An analysis in which we evaluate a sample's chemical or physical properties.

The purpose of a qualitative, quantitative, and characterization analysis is to solve a problem associated with a sample. A **fundamental analysis,** on the other hand, is directed toward improving the experimental methods used in the other areas of analytical chemistry. Extending and improving the theory on which a method is based, studying a method's limitations, and designing new and modifying old methods are examples of fundamental studies in analytical chemistry.

fundamental analysis
An analysis whose purpose is to improve an analytical method's capabilities.

1D KEY TERMS

characterization analysis (*p. 9*)
fundamental analysis (*p. 9*)
qualitative analysis (*p. 8*)
quantitative analysis (*p. 9*)

1E SUMMARY

Analytical chemists work to improve the ability of all chemists to make meaningful measurements. Chemists working in medicinal chemistry, clinical chemistry, forensic chemistry, and environmental chemistry, as well as the more traditional areas of chemistry, need better tools for analyzing materials. The need to work with smaller quantities of material, with more complex materials, with processes occurring on shorter time scales, and with species present at lower concentrations challenges analytical chemists to improve existing analytical methods and to develop new analytical techniques.

Typical problems on which analytical chemists work include qualitative analyses (what is present?), quantitative analyses (how much is present?), characterization analyses (what are the material's chemical and physical properties?), and fundamental analyses (how does this method work and how can it be improved?).

1F PROBLEMS

1. For each of the following problems indicate whether its solution requires a qualitative, quantitative, characterization, or fundamental study. More than one type of analysis may be appropriate for some problems.

a. A hazardous-waste disposal site is believed to be leaking contaminants into the local groundwater.
b. An art museum is concerned that a recent acquisition is a forgery.
c. A more reliable method is needed by airport security for detecting the presence of explosive materials in luggage.
d. The structure of a newly discovered virus needs to be determined.
e. A new visual indicator is needed for an acid–base titration.
f. A new law requires a method for evaluating whether automobiles are emitting too much carbon monoxide.

2. Read a recent article from the column "Analytical Approach," published in *Analytical Chemistry,* or an article assigned by your instructor, and write an essay summarizing the nature of the problem and how it was solved. As a guide, refer back to Figure 1.3 for one model of the analytical approach.

1G SUGGESTED READINGS

The role of analytical chemistry within the broader discipline of chemistry has been discussed by many prominent analytical chemists. Several notable examples follow.

Baiulescu, G. E.; Patroescu, C.; Chalmers, R. A. *Education and Teaching in Analytical Chemistry.* Ellis Horwood: Chichester, 1982.

Hieftje, G. M. "The Two Sides of Analytical Chemistry," *Anal. Chem.* **1985,** *57,* 256A–267A.

Kissinger, P. T. "Analytical Chemistry—What is It? Who Needs It? Why Teach It?" *Trends Anal. Chem.* **1992,** *11,* 54–57.

Laitinen, H. A. "Analytical Chemistry in a Changing World," *Anal. Chem.* **1980,** *52,* 605A–609A.

Laitinen, H. A. "History of Analytical Chemistry in the U.S.A.," *Talanta* **1989,** *36,* 1–9.

Laitinen, H. A.; Ewing, G. (eds). *A History of Analytical Chemistry.* The Division of Analytical Chemistry of the American Chemical Society: Washington, D.C., 1972.

McLafferty, F. W. "Analytical Chemistry: Historic and Modern," *Acc. Chem. Res.* **1990,** *23,* 63–64.

Mottola, H. A. "The Interdisciplinary and Multidisciplinary Nature of Contemporary Analytical Chemistry and Its Core Components," *Anal. Chim. Acta* **1991,** *242,* 1–3.

Tyson, J. *Analysis: What Analytical Chemists Do.* Royal Society of Chemistry: Cambridge, England, 1988.

Several journals are dedicated to publishing broadly in the field of analytical chemistry, including *Analytical Chemistry, Analytica Chimica Acta, Analyst,* and *Talanta.* Other journals, too numerous to list, are dedicated to single areas of analytical chemistry.

Current research in the areas of quantitative analysis, qualitative analysis, and characterization analysis are reviewed biannually (odd-numbered years) in *Analytical Chemistry's* "Application Reviews."

Current research on fundamental developments in analytical chemistry are reviewed biannually (even-numbered years) in *Analytical Chemistry's* "Fundamental Reviews."

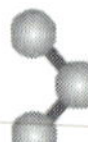

1H REFERENCES

1. Ravey, M. *Spectroscopy* **1990,** *5(7),* 11.
2. de Haseth, J. *Spectroscopy* **1990,** *5(7),* 11.
3. Fresenius, C. R. *A System of Instruction in Quantitative Chemical Analysis.* John Wiley and Sons: New York, 1881.
4. Hillebrand, W. F.; Lundell, G. E. F. *Applied Inorganic Analysis,* John Wiley and Sons: New York, 1953.
5. Van Loon, J. C. *Analytical Atomic Absorption Spectroscopy.* Academic Press: New York, 1980.
6. Murray, R. W. *Anal. Chem.* **1991,** *63,* 271A.
7. For several different viewpoints see (a) Beilby, A. L. *J. Chem. Educ.* **1970,** *47,* 237–238; (b) Lucchesi, C. A. *Am. Lab.* **1980,** *October,* 113–119; (c) Atkinson, G. F. *J. Chem. Educ.* **1982,** *59,* 201–202; (d) Pardue, H. L.; Woo, J. *J. Chem. Educ.* **1984,** *61,* 409–412; (e) Guarnieri, M. *J. Chem. Educ.* **1988,** *65,* 201–203; (f) de Haseth, J. *Spectroscopy* **1990,** *5,* 20–21; (g) Strobel, H. A. *Am. Lab.* **1990,** *October,* 17–24.
8. Hieftje, G. M. *Am. Lab.* **1993,** *October,* 53–61.
9. See, for example, the following laboratory texts: (a) Sorum, C. H.; Lagowski, J. J. *Introduction to Semimicro Qualitative Analysis,* 5th ed. Prentice-Hall: Englewood Cliffs, NJ, 1977.; (b) Shriner, R. L.; Fuson, R. C.; Curtin, D. Y. *The Systematic Identification of Organic Compounds,* 5th ed. John Wiley and Sons: New York, 1964.

Chapter 2

Basic Tools of Analytical Chemistry

In the chapters that follow we will learn about the specifics of analytical chemistry. In the process we will ask and answer questions such as "How do we treat experimental data?" "How do we ensure that our results are accurate?" "How do we obtain a representative sample?" and "How do we select an appropriate analytical technique?" Before we look more closely at these and other questions, we will first review some basic numerical and experimental tools of importance to analytical chemists.

2A Numbers in Analytical Chemistry

Analytical chemistry is inherently a quantitative science. Whether determining the concentration of a species in a solution, evaluating an equilibrium constant, measuring a reaction rate, or drawing a correlation between a compound's structure and its reactivity, analytical chemists make measurements and perform calculations. In this section we briefly review several important topics involving the use of numbers in analytical chemistry.

2A.1 Fundamental Units of Measure

Imagine that you find the following instructions in a laboratory procedure: "Transfer 1.5 of your sample to a 100 volumetric flask, and dilute to volume." How do you do this? Clearly these instructions are incomplete since the units of measurement are not stated. Compare this with a complete instruction: "Transfer 1.5 g of your sample to a 100-mL volumetric flask, and dilute to volume." This is an instruction that you can easily follow.

Measurements usually consist of a unit and a number expressing the quantity of that unit. Unfortunately, many different units may be used to express the same physical measurement. For example, the mass of a sample weighing 1.5 g also may be expressed as 0.0033 lb or 0.053 oz. For consistency, and to avoid confusion, scientists use a common set of fundamental units, several of which are listed in Table 2.1. These units are called **SI units** after the *Système International d'Unités.* Other measurements are defined using these fundamental SI units. For example, we measure the quantity of heat produced during a chemical reaction in joules, (J), where

SI units
Stands for *Système International d'Unités.* These are the internationally agreed on units for measurements.

$$1\ \text{J} = 1\frac{\text{m}^2\text{kg}}{\text{s}^2}$$

Table 2.2 provides a list of other important derived SI units, as well as a few commonly used non-SI units.

Chemists frequently work with measurements that are very large or very small. A mole, for example, contains 602,213,670,000,000,000,000,000 particles, and some analytical techniques can detect as little as 0.000000000000001 g of a compound. For simplicity, we express these measurements using **scientific notation;** thus, a mole contains 6.0221367×10^{23} particles, and the stated mass is 1×10^{-15} g. Sometimes it is preferable to express measurements without the exponential term, replacing it with a prefix. A mass of 1×10^{-15} g is the same as 1 femtogram. Table 2.3 lists other common prefixes.

scientific notation
A shorthand method for expressing very large or very small numbers by indicating powers of ten; for example, 1000 is 1×10^3.

Table 2.1 Fundamental SI Units

Measurement	Unit	Symbol
mass	kilogram	kg
volume	liter	L
distance	meter	m
temperature	kelvin	K
time	second	s
current	ampere	A
amount of substance	mole	mol

Table 2.2 Other SI and Non-SI Units

Measurement	Unit	Symbol	Equivalent SI units
length	angstrom	Å	1 Å = 1×10^{-10} m
force	newton	N	1 N = 1 m • kg/s^2
pressure	pascal	Pa	1 Pa = 1 N/m^2 = 1 kg/(m • s^2)
	atmosphere	atm	1 atm = 101,325 Pa
energy, work, heat	joule	J	1 J = 1 N • m = 1 m^2 • kg/s^2
power	watt	W	1 W = 1 J/s = 1 m^2 • kg/s^3
charge	coulomb	C	1 C = 1 A • s
potential	volt	V	1 V = 1 W/A = 1 m^2 • kg/(s^3 • A)
temperature	degree Celsius	°C	°C = K – 273.15
	degree Fahrenheit	°F	°F = 1.8(K – 273.15) + 32

Table 2.3 Common Prefixes for Exponential Notation

Exponential	Prefix	Symbol
10^{12}	tera	T
10^{9}	giga	G
10^{6}	mega	M
10^{3}	kilo	k
10^{-1}	deci	d
10^{-2}	centi	c
10^{-3}	milli	m
10^{-6}	micro	μ
10^{-9}	nano	n
10^{-12}	pico	p
10^{-15}	femto	f
10^{-18}	atto	a

2A.2 Significant Figures

Recording a measurement provides information about both its magnitude and uncertainty. For example, if we weigh a sample on a balance and record its mass as 1.2637 g, we assume that all digits, except the last, are known exactly. We assume that the last digit has an uncertainty of at least ±1, giving an absolute uncertainty of at least ±0.0001 g, or a relative uncertainty of at least

$$\frac{\pm 0.0001 \text{ g}}{1.2637 \text{ g}} \times 100 = \pm 0.0079\%$$

Significant figures are a reflection of a measurement's uncertainty. The number of significant figures is equal to the number of digits in the measurement, with the exception that a zero (0) used to fix the location of a decimal point is not considered significant. This definition can be ambiguous. For example, how many significant figures are in the number 100? If measured to the nearest hundred, then there is one significant figure. If measured to the nearest ten, however, then two

significant figures
The digits in a measured quantity, including all digits known exactly and one digit (the last) whose quantity is uncertain.

significant figures are included. To avoid ambiguity we use scientific notation. Thus, 1×10^2 has one significant figure, whereas 1.0×10^2 has two significant figures.

For measurements using logarithms, such as pH, the number of significant figures is equal to the number of digits to the right of the decimal, including all zeros. Digits to the left of the decimal are not included as significant figures since they only indicate the power of 10. A pH of 2.45, therefore, contains two significant figures.

Exact numbers, such as the stoichiometric coefficients in a chemical formula or reaction, and unit conversion factors, have an infinite number of significant figures. A mole of $CaCl_2$, for example, contains exactly two moles of chloride and one mole of calcium. In the equality

$$1000 \text{ mL} = 1 \text{ L}$$

both numbers have an infinite number of significant figures.

Recording a measurement to the correct number of significant figures is important because it tells others about how precisely you made your measurement. For example, suppose you weigh an object on a balance capable of measuring mass to the nearest ±0.1 mg, but record its mass as 1.762 g instead of 1.7620 g. By failing to record the trailing zero, which is a significant figure, you suggest to others that the mass was determined using a balance capable of weighing to only the nearest ±1 mg. Similarly, a buret with scale markings every 0.1 mL can be read to the nearest ±0.01 mL. The digit in the hundredth's place is the least significant figure since we must estimate its value. Reporting a volume of 12.241 mL implies that your buret's scale is more precise than it actually is, with divisions every 0.01 mL.

Significant figures are also important because they guide us in reporting the result of an analysis. When using a measurement in a calculation, the result of that calculation can never be more certain than that measurement's uncertainty. Simply put, the result of an analysis can never be more certain than the least certain measurement included in the analysis.

As a general rule, mathematical operations involving addition and subtraction are carried out to the last digit that is significant for all numbers included in the calculation. Thus, the sum of 135.621, 0.33, and 21.2163 is 157.17 since the last digit that is significant for all three numbers is in the hundredth's place.

$$135.6\underline{2}1 + 0.3\underline{3} + 21.2\underline{1}63 = 157.1673 = 157.1\underline{7}$$

When multiplying and dividing, the general rule is that the answer contains the same number of significant figures as that number in the calculation having the fewest significant figures. Thus,

$$\frac{22.91 \times 0.\underline{\underline{152}}}{16.302} = 0.21361 = 0.\underline{\underline{214}}$$

It is important to remember, however, that these rules are generalizations. What is conserved is not the number of significant figures, but absolute uncertainty when adding or subtracting, and relative uncertainty when multiplying or dividing. For example, the following calculation reports the answer to the correct number of significant figures, even though it violates the general rules outlined earlier.

$$\frac{101}{99} = 1.02$$

Since the relative uncertainty in both measurements is roughly 1% (101 ±1, 99 ±1), the relative uncertainty in the final answer also must be roughly 1%. Reporting the answer to only two significant figures (1.0), as required by the general rules, implies a relative uncertainty of 10%. The correct answer, with three significant figures, yields the expected relative uncertainty. Chapter 4 presents a more thorough treatment of uncertainty and its importance in reporting the results of an analysis.

Finally, to avoid "round-off" errors in calculations, it is a good idea to retain at least one extra significant figure throughout the calculation. This is the practice adopted in this textbook. Better yet, invest in a good scientific calculator that allows you to perform lengthy calculations without recording intermediate values. When the calculation is complete, the final answer can be rounded to the correct number of significant figures using the following simple rules.

1. Retain the least significant figure if it and the digits that follow are less than halfway to the next higher digit; thus, rounding 12.442 to the nearest tenth gives 12.4 since 0.442 is less than halfway between 0.400 and 0.500.
2. Increase the least significant figure by 1 if it and the digits that follow are more than halfway to the next higher digit; thus, rounding 12.476 to the nearest tenth gives 12.5 since 0.476 is more than halfway between 0.400 and 0.500.
3. If the least significant figure and the digits that follow are exactly halfway to the next higher digit, then round the least significant figure to the nearest even number; thus, rounding 12.450 to the nearest tenth gives 12.4, but rounding 12.550 to the nearest tenth gives 12.6. Rounding in this manner prevents us from introducing a bias by always rounding up or down.

2B Units for Expressing Concentration

Concentration is a general measurement unit stating the amount of solute present in a known amount of solution

$$\text{Concentration} = \frac{\text{amount of solute}}{\text{amount of solution}} \qquad 2.1$$

concentration
An expression stating the relative amount of solute per unit volume or unit mass of solution.

Although the terms "solute" and "solution" are often associated with liquid samples, they can be extended to gas-phase and solid-phase samples as well. The actual units for reporting concentration depend on how the amounts of solute and solution are measured. Table 2.4 lists the most common units of concentration.

2B.1 Molarity and Formality

Both molarity and formality express concentration as moles of solute per liter of solution. There is, however, a subtle difference between molarity and formality. **Molarity** is the concentration of a particular chemical species in solution. **Formality,** on the other hand, is a substance's total concentration in solution without regard to its specific chemical form. There is no difference between a substance's molarity and formality if it dissolves without dissociating into ions. The molar concentration of a solution of glucose, for example, is the same as its formality.

molarity
The number of moles of solute per liter of solution (M).

formality
The number of moles of solute, regardless of chemical form, per liter of solution (F).

For substances that ionize in solution, such as NaCl, molarity and formality are different. For example, dissolving 0.1 mol of NaCl in 1 L of water gives a solution containing 0.1 mol of Na^+ and 0.1 mol of Cl^-. The molarity of NaCl, therefore, is zero since there is essentially no undissociated NaCl in solution. The solution,

Table 2.4 Common Units for Reporting Concentration

Name	Units[a]	Symbol
molarity	$\frac{\text{moles solute}}{\text{liters solution}}$	M
formality	$\frac{\text{number FWs solute}}{\text{liters solution}}$	F
normality	$\frac{\text{number EWs solute}}{\text{liters solution}}$	N
molality	$\frac{\text{moles solute}}{\text{kg solvent}}$	m
weight %	$\frac{\text{g solute}}{\text{100 g solution}}$	% w/w
volume %	$\frac{\text{mL solute}}{\text{100 mL solution}}$	% v/v
weight-to-volume %	$\frac{\text{g solute}}{\text{100 mL solution}}$	% w/v
parts per million	$\frac{\text{g solute}}{10^6\ \text{g solution}}$	ppm
parts per billion	$\frac{\text{g solute}}{10^9\ \text{g solution}}$	ppb

[a]FW = formula weight; EW = equivalent weight.

instead, is 0.1 M in Na^+ and 0.1 M in Cl^-. The formality of NaCl, however, is 0.1 F because it represents the total amount of NaCl in solution. The rigorous definition of molarity, for better or worse, is largely ignored in the current literature, as it is in this text. When we state that a solution is 0.1 M NaCl we understand it to consist of Na^+ and Cl^- ions. The unit of formality is used only when it provides a clearer description of solution chemistry.

Molar concentrations are used so frequently that a symbolic notation is often used to simplify its expression in equations and writing. The use of square brackets around a species indicates that we are referring to that species' molar concentration. Thus, $[Na^+]$ is read as the "molar concentration of sodium ions."

2B.2 Normality

Normality is an older unit of concentration that, although once commonly used, is frequently ignored in today's laboratories. Normality is still used in some handbooks of analytical methods, and, for this reason, it is helpful to understand its meaning. For example, normality is the concentration unit used in *Standard Methods for the Examination of Water and Wastewater,*[1] a commonly used source of analytical methods for environmental laboratories.

normality
The number of equivalents of solute per liter of solution (N).

Normality makes use of the chemical equivalent, which is the amount of one chemical species reacting stoichiometrically with another chemical species. Note that this definition makes an equivalent, and thus normality, a function of the chemical reaction in which the species participates. Although a solution of H_2SO_4 has a fixed molarity, its normality depends on how it reacts.

The number of **equivalents,** n, is based on a reaction unit, which is that part of a chemical species involved in a reaction. In a precipitation reaction, for example, the reaction unit is the charge of the cation or anion involved in the reaction; thus for the reaction

equivalent
The moles of a species that can donate one reaction unit.

$$Pb^{2+}(aq) + 2I^-(aq) \rightleftharpoons PbI_2(s)$$

$n = 2$ for Pb^{2+} and $n = 1$ for I^-. In an acid–base reaction, the reaction unit is the number of H^+ ions donated by an acid or accepted by a base. For the reaction between sulfuric acid and ammonia

$$H_2SO_4(aq) + 2NH_3(aq) \rightleftharpoons 2NH_4^+(aq) + SO_4^{2-}(aq)$$

we find that $n = 2$ for H_2SO_4 and $n = 1$ for NH_3. For a complexation reaction, the reaction unit is the number of electron pairs that can be accepted by the metal or donated by the ligand. In the reaction between Ag^+ and NH_3

$$Ag^+(aq) + 2NH_3(aq) \rightleftharpoons Ag(NH_3)_2^+(aq)$$

the value of n for Ag^+ is 2 and that for NH_3 is 1. Finally, in an oxidation–reduction reaction the reaction unit is the number of electrons released by the reducing agent or accepted by the oxidizing agent; thus, for the reaction

$$2Fe^{3+}(aq) + Sn^{2+}(aq) \rightleftharpoons Sn^{4+}(aq) + 2Fe^{2+}(aq)$$

$n = 1$ for Fe^{3+} and $n = 2$ for Sn^{2+}. Clearly, determining the number of equivalents for a chemical species requires an understanding of how it reacts.

Normality is the number of **equivalent weights** (EW) per unit volume and, like formality, is independent of speciation. An equivalent weight is defined as the ratio of a chemical species' **formula weight** (FW) to the number of its equivalents

equivalent weight
The mass of a compound containing one equivalent (EW).

formula weight
The mass of a compound containing one mole (FW).

$$EW = \frac{FW}{n}$$

Consequently, the following simple relationship exists between normality and molarity.

$$N = n \times M$$

Example 2.1 illustrates the relationship among chemical reactivity, equivalent weight, and normality.

Example

EXAMPLE 2.1

Calculate the equivalent weight and normality for a solution of 6.0 M H_3PO_4 given the following reactions:

(a) $H_3PO_4(aq) + 3OH^-(aq) \rightleftharpoons PO_4^{3-}(aq) + 3H_2O(\ell)$

(b) $H_3PO_4(aq) + 2NH_3(aq) \rightleftharpoons HPO_4^{2-}(aq) + 2NH_4^+(aq)$

(c) $H_3PO_4(aq) + F^-(aq) \rightleftharpoons H_2PO_4^-(aq) + HF(aq)$

SOLUTION

For phosphoric acid, the number of equivalents is the number of H^+ ions donated to the base. For the reactions in (a), (b), and (c) the number of equivalents are 3, 2, and 1, respectively. Thus, the calculated equivalent weights and normalities are

$$\text{(a) EW} = \frac{\text{FW}}{n} = \frac{97.994}{3} = 32.665 \qquad \text{N} = n \times \text{M} = 3 \times 6.0 = 18\ \text{N}$$

$$\text{(b) EW} = \frac{\text{FW}}{n} = \frac{97.994}{2} = 48.997 \qquad \text{N} = n \times \text{M} = 2 \times 6.0 = 12\ \text{N}$$

$$\text{(c) EW} = \frac{\text{FW}}{n} = \frac{97.994}{1} = 97.994 \qquad \text{N} = n \times \text{M} = 1 \times 6.0 = 6.0\ \text{N}$$

2B.3 Molality

molality
The number of moles of solute per kilogram of solvent (m).

Molality is used in thermodynamic calculations where a temperature independent unit of concentration is needed. Molarity, formality and normality are based on the volume of solution in which the solute is dissolved. Since density is a temperature dependent property a solution's volume, and thus its molar, formal and normal concentrations, will change as a function of its temperature. By using the solvent's mass in place of its volume, the resulting concentration becomes independent of temperature.

2B.4 Weight, Volume, and Weight-to-Volume Ratios

weight percent
Grams of solute per 100 g of solution. (% w/w).

volume percent
Milliliters of solute per 100 mL of solution (% v/v).

weight-to-volume percent
Grams of solute per 100 mL of solution (% w/v).

parts per million
Micrograms of solute per gram of solution; for aqueous solutions the units are often expressed as milligrams of solute per liter of solution (ppm).

parts per billion
Nanograms of solute per gram of solution; for aqueous solutions the units are often expressed as micrograms of solute per liter of solution (ppb).

Weight percent (% w/w), **volume percent** (% v/v) and **weight-to-volume percent** (% w/v) express concentration as units of solute per 100 units of sample. A solution in which a solute has a concentration of 23% w/v contains 23 g of solute per 100 mL of solution.

Parts per million (ppm) and **parts per billion** (ppb) are mass ratios of grams of solute to one million or one billion grams of sample, respectively. For example, a steel that is 450 ppm in Mn contains 450 μg of Mn for every gram of steel. If we approximate the density of an aqueous solution as 1.00 g/mL, then solution concentrations can be expressed in parts per million or parts per billion using the following relationships.

$$\text{ppm} = \frac{\text{mg}}{\text{liter}} = \frac{\mu\text{g}}{\text{mL}}$$

$$\text{ppb} = \frac{\mu\text{g}}{\text{liter}} = \frac{\text{ng}}{\text{mL}}$$

For gases a part per million usually is a volume ratio. Thus, a helium concentration of 6.3 ppm means that one liter of air contains 6.3 μL of He.

2B.5 Converting Between Concentration Units

The units of concentration most frequently encountered in analytical chemistry are molarity, weight percent, volume percent, weight-to-volume percent, parts per million, and parts per billion. By recognizing the general definition of concentration given in equation 2.1, it is easy to convert between concentration units.

Example

EXAMPLE 2.2

A concentrated solution of aqueous ammonia is 28.0% w/w NH_3 and has a density of 0.899 g/mL. What is the molar concentration of NH_3 in this solution?

SOLUTION

$$\frac{28.0\ \text{g NH}_3}{100\ \text{g solution}} \times \frac{0.899\ \text{g solution}}{\text{mL solution}} \times \frac{1\ \text{mole NH}_3}{17.04\ \text{g NH}_3} \times \frac{1000\ \text{mL}}{\text{liter}} = 14.8\ \text{M}$$

Example

EXAMPLE 2.3

The maximum allowed concentration of chloride in a municipal drinking water supply is 2.50×10^2 ppm Cl^-. When the supply of water exceeds this limit, it often has a distinctive salty taste. What is this concentration in moles Cl^-/liter?

SOLUTION

$$\frac{2.50 \times 10^2 \text{ mg Cl}^-}{\text{L}} \times \frac{1 \text{ g}}{1000 \text{ mg}} \times \frac{1 \text{ mole Cl}^-}{35.453 \text{ g Cl}^-} = 7.05 \times 10^{-3} \text{ M}$$

2B.6 p-Functions

Sometimes it is inconvenient to use the concentration units in Table 2.4. For example, during a reaction a reactant's concentration may change by many orders of magnitude. If we are interested in viewing the progress of the reaction graphically, we might wish to plot the reactant's concentration as a function of time or as a function of the volume of a reagent being added to the reaction. Such is the case in Figure 2.1, where the molar concentration of H^+ is plotted (y-axis on left side of figure) as a function of the volume of NaOH added to a solution of HCl. The initial $[H^+]$ is 0.10 M, and its concentration after adding 75 mL of NaOH is 5.0×10^{-13} M. We can easily follow changes in the $[H^+]$ over the first 14 additions of NaOH. For the last ten additions of NaOH, however, changes in the $[H^+]$ are too small to be seen.

When working with concentrations that span many orders of magnitude, it is often more convenient to express the concentration as a **p-function.** The p-function of a number X is written as pX and is defined as

$$\text{p}X = -\log(X)$$

Thus, the pH of a solution that is 0.10 M H^+ is

$$\text{pH} = -\log[\text{H}^+] = -\log(0.10) = 1.00$$

and the pH of 5.0×10^{-13} M H^+ is

$$\text{pH} = -\log[\text{H}^+] = -\log(5.0 \times 10^{-13}) = 12.30$$

Figure 2.1 shows how plotting pH in place of $[H^+]$ provides more detail about how the concentration of H^+ changes following the addition of NaOH.

p-function
A function of the form pX, where pX = -log(X).

Example

EXAMPLE 2.4

What is pNa for a solution of 1.76×10^{-3} M Na_3PO_4?

SOLUTION

Since each mole of Na_3PO_4 contains three moles of Na^+, the concentration of Na^+ is

$$[\text{Na}^+] = \frac{3 \text{ mol Na}^+}{\text{mol Na}_3\text{PO}_4} \times 1.76 \times 10^{-3} \text{ M} = 5.28 \times 10^{-3} \text{ M}$$

and pNa is

$$\text{pNa} = -\log[\text{Na}^+] = -\log(5.28 \times 10^{-3}) = 2.277$$

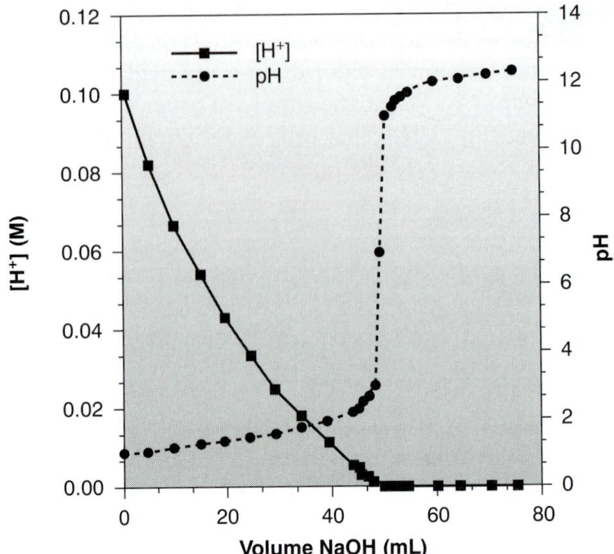

Figure 2.1
Graph of $[H^+]$ versus volume of NaOH and pH versus volume of NaOH for the reaction of 0.10 M HCl with 0.10 M NaOH.

Example

EXAMPLE 2.5

What is the $[H^+]$ in a solution that has a pH of 5.16?

SOLUTION

The concentration of H^+ is

$$\text{pH} = -\log[H^+] = 5.16$$

$$\log[H^+] = -5.16$$

$$[H^+] = \text{antilog}(-5.16) = 10^{-5.16} = 6.9 \times 10^{-6}\ \text{M}$$

2C Stoichiometric Calculations

A balanced chemical reaction indicates the quantitative relationships between the moles of reactants and products. These stoichiometric relationships provide the basis for many analytical calculations. Consider, for example, the problem of determining the amount of oxalic acid, $H_2C_2O_4$, in rhubarb. One method for this analysis uses the following reaction in which we oxidize oxalic acid to CO_2.

$$2Fe^{3+}(aq) + H_2C_2O_4(aq) + 2H_2O(\ell) \rightarrow 2Fe^{2+}(aq) + 2CO_2(g) + 2H_3O^+(aq) \quad \textbf{2.2}$$

The balanced chemical reaction provides the stoichiometric relationship between the moles of Fe^{3+} used and the moles of oxalic acid in the sample being analyzed—specifically, one mole of oxalic acid reacts with two moles of Fe^{3+}. As shown in Example 2.6, the balanced chemical reaction can be used to determine the amount of oxalic acid in a sample, provided that information about the number of moles of Fe^{3+} is known.

Example

EXAMPLE 2.6

The amount of oxalic acid in a sample of rhubarb was determined by reacting with Fe^{3+} as outlined in reaction 2.2. In a typical analysis, the oxalic acid in 10.62 g of rhubarb was extracted with a suitable solvent. The complete oxidation of the oxalic acid to CO_2 required 36.44 mL of 0.0130 M Fe^{3+}. What is the weight percent of oxalic acid in the sample of rhubarb?

SOLUTION

We begin by calculating the moles of Fe^{3+} used in the reaction

$$\frac{0.0130 \text{ mol Fe}^{3+}}{\text{L}} \times 0.03644 \text{ L} = 4.737 \times 10^{-4} \text{ mol Fe}^{3+}$$

The moles of oxalic acid reacting with the Fe^{3+}, therefore, is

$$4.737 \times 10^{-4} \text{ mol Fe}^{3+} \times \frac{1 \text{ mol C}_2\text{H}_2\text{O}_4}{2 \text{ mol Fe}^{3+}} = 2.369 \times 10^{-4} \text{ mol C}_2\text{H}_2\text{O}_4$$

Converting moles of oxalic acid to grams of oxalic acid

$$2.369 \times 10^{-4} \text{ mol C}_2\text{H}_2\text{O}_4 \times \frac{90.03 \text{ g C}_2\text{H}_2\text{O}_4}{\text{mol C}_2\text{H}_2\text{O}_4} = 2.132 \times 10^{-2} \text{ g oxalic acid}$$

and converting to weight percent gives the concentration of oxalic acid in the sample of rhubarb as

$$\frac{2.132 \times 10^{-2} \text{ g C}_2\text{H}_2\text{O}_4}{10.62 \text{ g rhubarb}} \times 100 = 0.201\% \text{ w/w C}_2\text{H}_2\text{O}_4$$

In the analysis described in Example 2.6 oxalic acid already was present in the desired form. In many analytical methods the compound to be determined must be converted to another form prior to analysis. For example, one method for the quantitative analysis of tetraethylthiuram disulfide ($C_{10}H_{20}N_2S_4$), the active ingredient in the drug Antabuse (disulfiram), requires oxidizing the S to SO_2, bubbling the SO_2 through H_2O_2 to produce H_2SO_4, followed by an acid–base titration of the H_2SO_4 with NaOH. Although we can write and balance chemical reactions for each of these steps, it often is easier to apply the principle of the conservation of reaction units.

A reaction unit is that part of a chemical species involved in a reaction. Consider, for example, the general unbalanced chemical reaction

$$\text{A} + \text{B} \rightarrow \text{Products}$$

Conservation of reaction units requires that the number of reaction units associated with the reactant A equal the number of reaction units associated with the reactant B. Translating the previous statement into mathematical form gives

$$\begin{aligned} &\text{Number of reaction units per A} \times \text{moles A} \\ &= \text{number of reaction units per B} \times \text{moles B} \end{aligned} \qquad \textbf{2.3}$$

If we know the moles of A and the number of reaction units associated with A and B, then we can calculate the moles of B. Note that a conservation of reaction units, as defined by equation 2.3, can only be applied between two species. There are five important principles involving a conservation of reaction units: mass, charge, protons, electron pairs, and electrons.

2C.1 Conservation of Mass

The easiest principle to appreciate is conservation of mass. Except for nuclear reactions, an element's total mass at the end of a reaction must be the same as that present at the beginning of the reaction; thus, an element serves as the most fundamental reaction unit. Consider, for example, the combustion of butane to produce CO_2 and H_2O, for which the unbalanced reaction is

$$C_4H_{10}(g) + O_2(g) \rightarrow CO_2(g) + H_2O(g)$$

All the carbon in CO_2 comes from the butane, thus we can select carbon as a reaction unit. Since there are four carbon atoms in butane, and one carbon atom in CO_2, we write

$$4 \times \text{moles } C_4H_{10} = 1 \times \text{moles } CO_2$$

Hydrogen also can be selected as a reaction unit since all the hydrogen in butane ends up in the H_2O produced during combustion. Thus, we can write

$$10 \times \text{moles } C_4H_{10} = 2 \times \text{moles } H_2O$$

Although the mass of oxygen is conserved during the reaction, we cannot apply equation 2.3 because the O_2 used during combustion does not end up in a single product.

Conservation of mass also can, with care, be applied to groups of atoms. For example, the ammonium ion, NH_4^+, can be precipitated as $Fe(NH_4)_2(SO_4)_2 \cdot 6H_2O$. Selecting NH_4^+ as the reaction unit gives

$$2 \times \text{moles } Fe(NH_4)_2(SO_4)_2 \cdot 6H_2O = 1 \times \text{moles } NH_4^+$$

2C.2 Conservation of Charge

The stoichiometry between two reactants in a precipitation reaction is governed by a conservation of charge, requiring that the total cation charge and the total anion charge in the precipitate be equal. The reaction units in a precipitation reaction, therefore, are the absolute values of the charges on the cation and anion that make up the precipitate. Applying equation 2.3 to a precipitate of $Ca_3(PO_4)_2$ formed from the reaction of Ca^{2+} and PO_4^{3-}, we write

$$2 \times \text{moles } Ca^{2+} = 3 \times \text{moles } PO_4^{3-}$$

2C.3 Conservation of Protons

In an acid–base reaction, the reaction unit is the proton. For an acid, the number of reaction units is given by the number of protons that can be donated to the base; and for a base, the number of reaction units is the number of protons that the base can accept from the acid. In the reaction between H_3PO_4 and NaOH, for example, the weak acid H_3PO_4 can donate all three of its protons to NaOH, whereas the strong base NaOH can accept one proton. Thus, we write

$$3 \times \text{moles } H_3PO_4 = 1 \times \text{moles NaOH}$$

Care must be exercised in determining the number of reaction units associated with the acid and base. The number of reaction units for an acid, for instance, depends not on how many acidic protons are present, but on how many of the protons are capable of reacting with the chosen base. In the reaction between H_3PO_4 and NH_3

$$H_3PO_4(aq) + 2NH_3(aq) \rightleftharpoons HPO_4^{-}(aq) + 2NH_4^{+}(aq)$$

a conservation of protons requires that

$$2 \times \text{moles } H_3PO_4 = \text{moles of } NH_3$$

2C.4 Conservation of Electron Pairs

In a complexation reaction, the reaction unit is an electron pair. For the metal, the number of reaction units is the number of coordination sites available for binding ligands. For the ligand, the number of reaction units is equivalent to the number of electron pairs that can be donated to the metal. One of the most important analytical complexation reactions is that between the ligand ethylenediaminetetracetic acid (EDTA), which can donate 6 electron pairs and 6 coordinate metal ions, such as Cu^{2+}; thus

$$6 \times \text{mole } Cu^{2+} = 6 \times \text{moles EDTA}$$

2C.5 Conservation of Electrons

In a redox reaction, the reaction unit is an electron transferred from a reducing agent to an oxidizing agent. The number of reaction units for a reducing agent is equal to the number of electrons released during its oxidation. For an oxidizing agent, the number of reaction units is given by the number of electrons needed to cause its reduction. In the reaction between Fe^{3+} and oxalic acid (reaction 2.2), for example, Fe^{3+} undergoes a 1-electron reduction. Each carbon atom in oxalic acid is initially present in a +3 oxidation state, whereas the carbon atom in CO_2 is in a +4 oxidation state. Thus, we can write

$$1 \times \text{moles } Fe^{3+} = 2 \times \text{moles of } H_2C_2O_4$$

Note that the moles of oxalic acid are multiplied by 2 since there are two carbon atoms, each of which undergoes a 1-electron oxidation.

2C.6 Using Conservation Principles in Stoichiometry Problems

As shown in the following examples, the application of conservation principles simplifies stoichiometric calculations.

Example

EXAMPLE 2.7

Rework Example 2.6 using conservation principles.

SOLUTION

Conservation of electrons for this redox reaction requires that

$$\text{moles } Fe^{3+} = 2 \times \text{moles } H_2C_2O_4$$

which can be transformed by writing moles as the product of molarity and volume or as grams per formula weight.

$$M_{Fe^{3+}} \times V_{Fe^{3+}} = \frac{2 \times \text{g } H_2C_2O_4}{\text{FW } H_2C_2O_4}$$

Solving for g $H_2C_2O_4$ gives

$$\frac{M_{Fe^{3+}} \times V_{Fe^{3+}} \times \text{FW } H_2C_2O_4}{2} = \frac{(0.0130\text{ M})(0.03644\text{ L})(90.03\text{ g/mole})}{2}$$
$$= 2.132 \times 10^{-2}\text{ g } H_2C_2O_4$$

and the weight percent oxalic acid is

$$\frac{2.132 \times 10^{-2}\text{ g } C_2H_2O_4}{10.62\text{ g rhubarb}} \times 100 = 0.201\%\text{ w/w } C_2H_2O_4$$

Example

EXAMPLE 2.8

One quantitative analytical method for tetraethylthiuram disulfide, $C_{10}H_{20}N_2S_4$ (Antabuse), requires oxidizing the sulfur to SO_2, and bubbling the resulting SO_2 through H_2O_2 to produce H_2SO_4. The H_2SO_4 is then reacted with NaOH according to the reaction

$$H_2SO_4(aq) + 2NaOH(aq) \rightarrow Na_2SO_4(aq) + 2H_2O(\ell)$$

Using appropriate conservation principles, derive an equation relating the moles of $C_{10}H_{20}N_2S_4$ to the moles of NaOH. What is the weight percent $C_{10}H_{20}N_2S_4$ in a sample of Antabuse if the H_2SO_4 produced from a 0.4613-g portion reacts with 34.85 mL of 0.02500 M NaOH?

SOLUTION

The unbalanced reactions converting $C_{10}H_{20}N_2S_4$ to H_2SO_4 are

$$C_{10}H_{20}N_2S_4 \rightarrow SO_2$$
$$SO_2 \rightarrow H_2SO_4$$

Using a conservation of mass we have

$$4 \times \text{moles } C_{10}H_{20}N_2S_4 = \text{moles } SO_2 = \text{moles } H_2SO_4$$

A conservation of protons for the reaction of H_2SO_4 with NaOH gives

$$2 \times \text{moles } H_2SO_4 = \text{moles of NaOH}$$

Combining the two conservation equations gives the following stoichiometric equation between $C_{10}H_{20}N_2S_4$ and NaOH

$$8 \times \text{moles } C_{10}H_{20}N_2S_4 = \text{moles NaOH}$$

Now we are ready to finish the problem. Making appropriate substitutions for moles of $C_{10}H_{20}N_2S_4$ and moles of NaOH gives

$$\frac{8 \times \text{g } C_{10}H_{20}N_2S_4}{\text{FW } C_{10}H_{20}N_2S_4} = M_{NaOH} \times V_{NaOH}$$

Solving for g $C_{10}H_{20}N_2S_4$ gives

$$\text{g } C_{10}H_{20}N_2S_4 = \frac{1}{8} \times M_{NaOH} \times V_{NaOH} \times \text{FW } C_{10}H_{20}N_2S_4$$

$$\frac{1}{8}(0.02500\text{ M})(0.03485\text{ L})(296.54\text{ g/mol}) = 0.032295\text{ g } C_{10}H_{20}N_2S_4$$

The weight percent $C_{10}H_{20}N_2S_4$ in the sample, therefore, is

$$\frac{0.32295 \text{ g } C_{10}H_{20}N_2S_4}{0.4613 \text{ g sample}} \times 100 = 7.001\% \text{ w/w } C_{10}H_{20}N_2S_4$$

2D Basic Equipment and Instrumentation

Measurements are made using appropriate equipment or instruments. The array of equipment and instrumentation used in analytical chemistry is impressive, ranging from the simple and inexpensive, to the complex and costly. With two exceptions, we will postpone the discussion of equipment and instrumentation to those chapters where they are used. The instrumentation used to measure mass and much of the equipment used to measure volume are important to all analytical techniques and are therefore discussed in this section.

2D.1 Instrumentation for Measuring Mass

An object's mass is measured using a **balance.** The most common type of balance is an electronic balance in which the balance pan is placed over an electromagnet (Figure 2.2). The sample to be weighed is placed on the sample pan, displacing the pan downward by a force equal to the product of the sample's mass and the acceleration due to gravity. The balance detects this downward movement and generates a counterbalancing force using an electromagnet. The current needed to produce this force is proportional to the object's mass. A typical electronic balance has a capacity of 100–200 g and can measure mass to the nearest ±0.01 to ±1 mg.

balance
An apparatus used to measure mass.

Another type of balance is the single-pan, unequal arm balance (Figure 2.3). In this mechanical balance the balance pan and a set of removable standard weights on one side of a beam are balanced against a fixed counterweight on the beam's other side. The beam itself is balanced on a fulcrum consisting of a sharp knife edge. Adding a sample to the balance pan tilts the beam away from its balance point. Selected standard weights are then removed until the beam is brought back into balance. The combined mass of the removed weights equals the sample's mass. The capacities and measurement limits of these balances are comparable to an electronic balance.

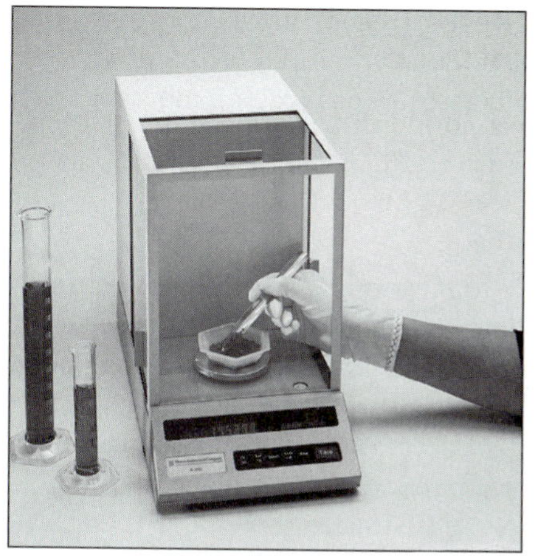

(a)

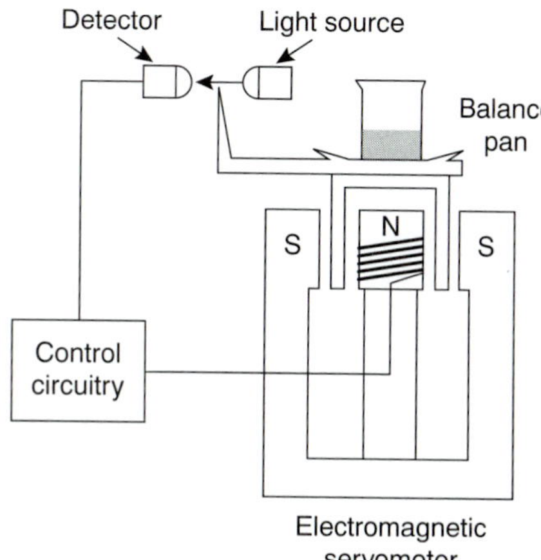

(b)

Figure 2.2
(a) Photo of a typical electronic balance.
(b) Schematic diagram of electronic balance; adding a sample moves the balance pan down, allowing more light to reach the detector. The control circuitry directs the electromagnetic servomotor to generate an opposing force, raising the sample up until the original intensity of light at the detector is restored.
Photo courtesy of Fisher Scientific.

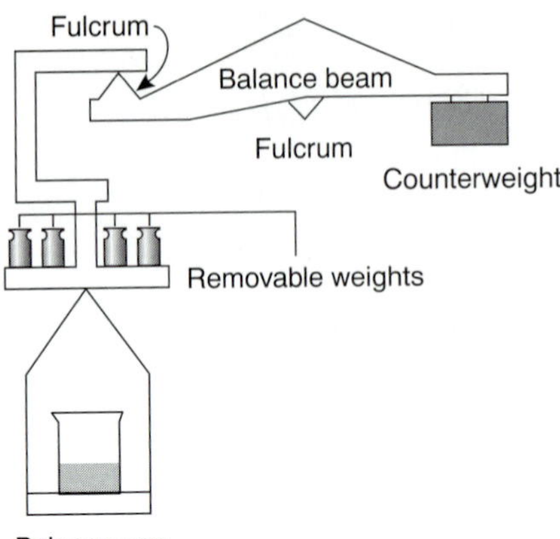

Figure 2.3
Schematic diagram of single-arm mechanical balance.

The mass of a sample is determined by difference. If the material being weighed is not moisture-sensitive, a clean and dry container is placed on the balance. The mass of this container is called the tare. Most balances allow the tare to be automatically adjusted to read a mass of zero. The sample is then transferred to the container, the new mass is measured and the sample's mass determined by subtracting the tare. Samples that absorb moisture from the air are weighed differently. The sample is placed in a covered weighing bottle and their combined mass is determined. A portion of the sample is removed, and the weighing bottle and remaining sample are reweighed. The difference between the two masses gives the mass of the transferred sample.

Several important precautions help to minimize errors in measuring an object's mass. Balances should be placed on heavy surfaces to minimize the effect of vibrations in the surrounding environment and should be maintained in a level position. Analytical balances are sensitive enough that they can measure the mass of a fingerprint. For this reason, materials placed on a balance should normally be handled using tongs or laboratory tissues. Volatile liquid samples should be weighed in a covered container to avoid the loss of sample by evaporation. Air currents can significantly affect a sample's mass. To avoid air currents, the balance's glass doors should be closed, or the balance's wind shield should be in place. A sample that is cooler or warmer than the surrounding air will create convective air currents that adversely affect the measurement of its mass. Finally, samples dried in an oven should be stored in a desiccator to prevent them from reabsorbing moisture from the atmosphere.

2D.2 Equipment for Measuring Volume

Analytical chemists use a variety of glassware to measure volume, several examples of which are shown in Figure 2.4. The type of glassware used depends on how exact the volume needs to be. Beakers, dropping pipets, and graduated cylinders are used to measure volumes approximately, typically with errors of several percent.

volumetric flask
Glassware designed to contain a specific volume of solution when filled to its calibration mark.

Pipets and volumetric flasks provide a more accurate means for measuring volume. When filled to its calibration mark, a **volumetric flask** is designed to contain a specified volume of solution at a stated temperature, usually 20 °C. The actual vol-

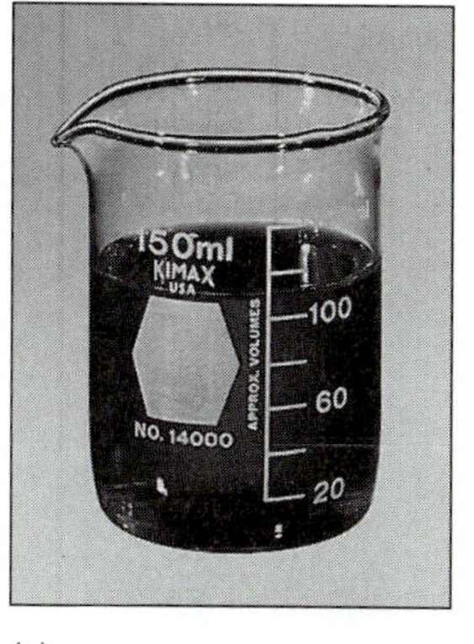

(a)

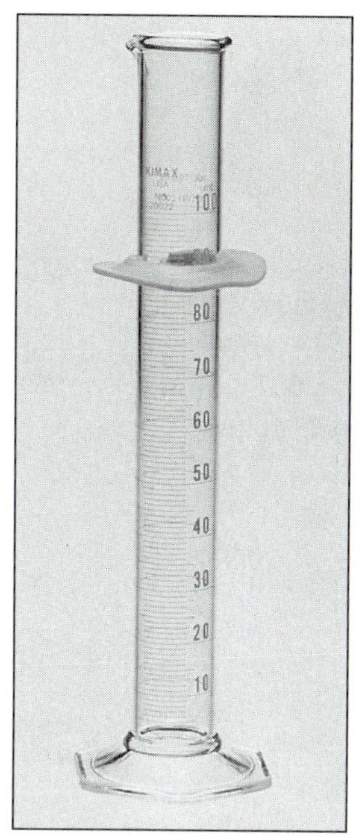

(b)

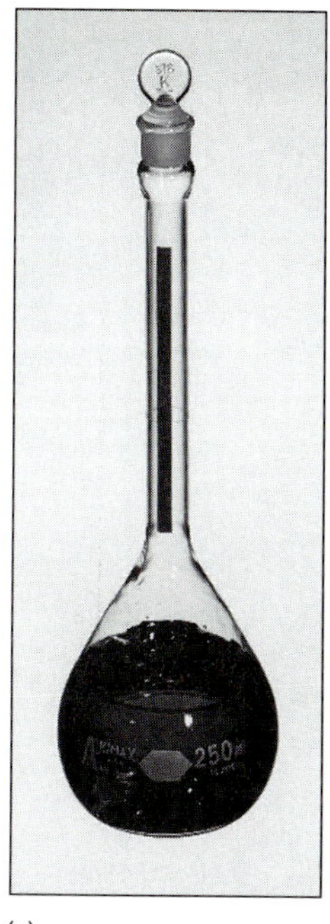

(c)

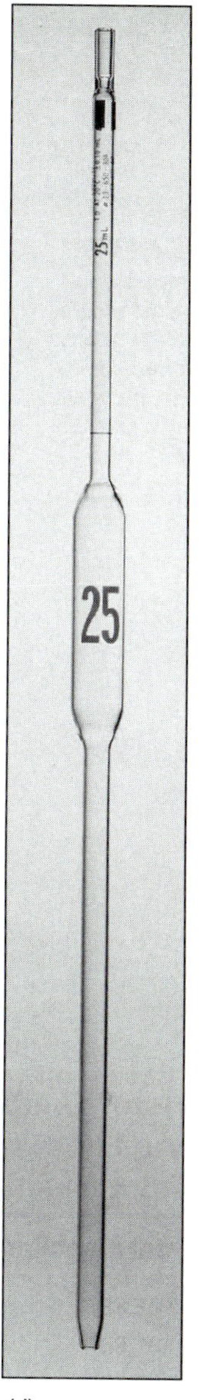

(d)

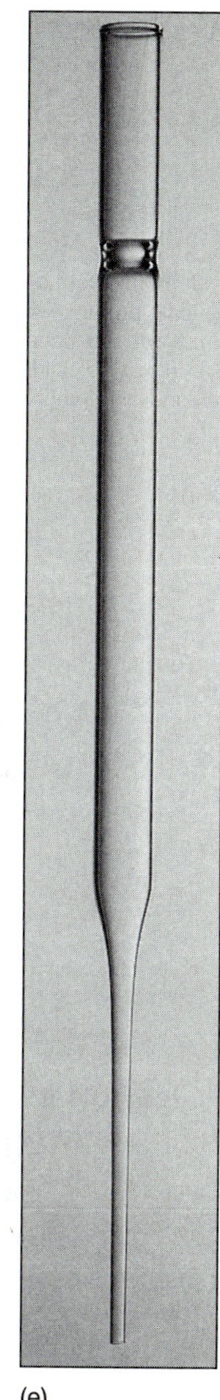
(e)

Figure 2.4
Common examples of glassware used to measure volume:
(a) beaker; (b) graduated cylinder; (c) volumetric flask; (d) pipet; (e) dropping pipet.
Photos courtesy of Fisher Scientific.

ume contained by the volumetric flask is usually within 0.03–0.2% of the stated value. Volumetric flasks containing less than 100 mL generally measure volumes to the hundredth of a milliliter, whereas larger volumetric flasks measure volumes to the tenth of a milliliter. For example, a 10-mL volumetric flask contains 10.00 mL, but a 250-mL volumetric flask holds 250.0 mL (this is important when keeping track of significant figures).

Because a volumetric flask contains a solution, it is useful in preparing solutions with exact concentrations. The reagent is transferred to the volumetric flask, and enough solvent is added to dissolve the reagent. After the reagent is dissolved, additional solvent is added in several portions, mixing the solution after each addition. The final adjustment of volume to the flask's calibration mark is made using a dropping pipet. To complete the mixing process, the volumetric flask should be inverted at least ten times.

A **pipet** is used to deliver a specified volume of solution. Several different styles of pipets are available (Figure 2.5). Transfer pipets provide the most accurate means for delivering a known volume of solution; their volume error is similar to that from an equivalent volumetric flask. A 250-mL transfer pipet, for instance, will deliver 250.0 mL. To fill a transfer pipet, suction from a rubber bulb is used to pull the liquid up past the calibration mark (*never* use your mouth to suck a solution into a pipet). After replacing the bulb with your finger, the liquid's level is adjusted to the calibration mark, and the outside of the pipet is wiped dry. The pipet's contents are allowed to drain into the receiving container with the tip of the pipet touching the container walls. A small portion of the liquid remains in the pipet's tip and should not be blown out. Measuring pipets are used to deliver variable volumes, but with less accuracy than transfer pipets. With some measuring

pipet
Glassware designed to deliver a specific volume of solution when filled to its calibration mark.

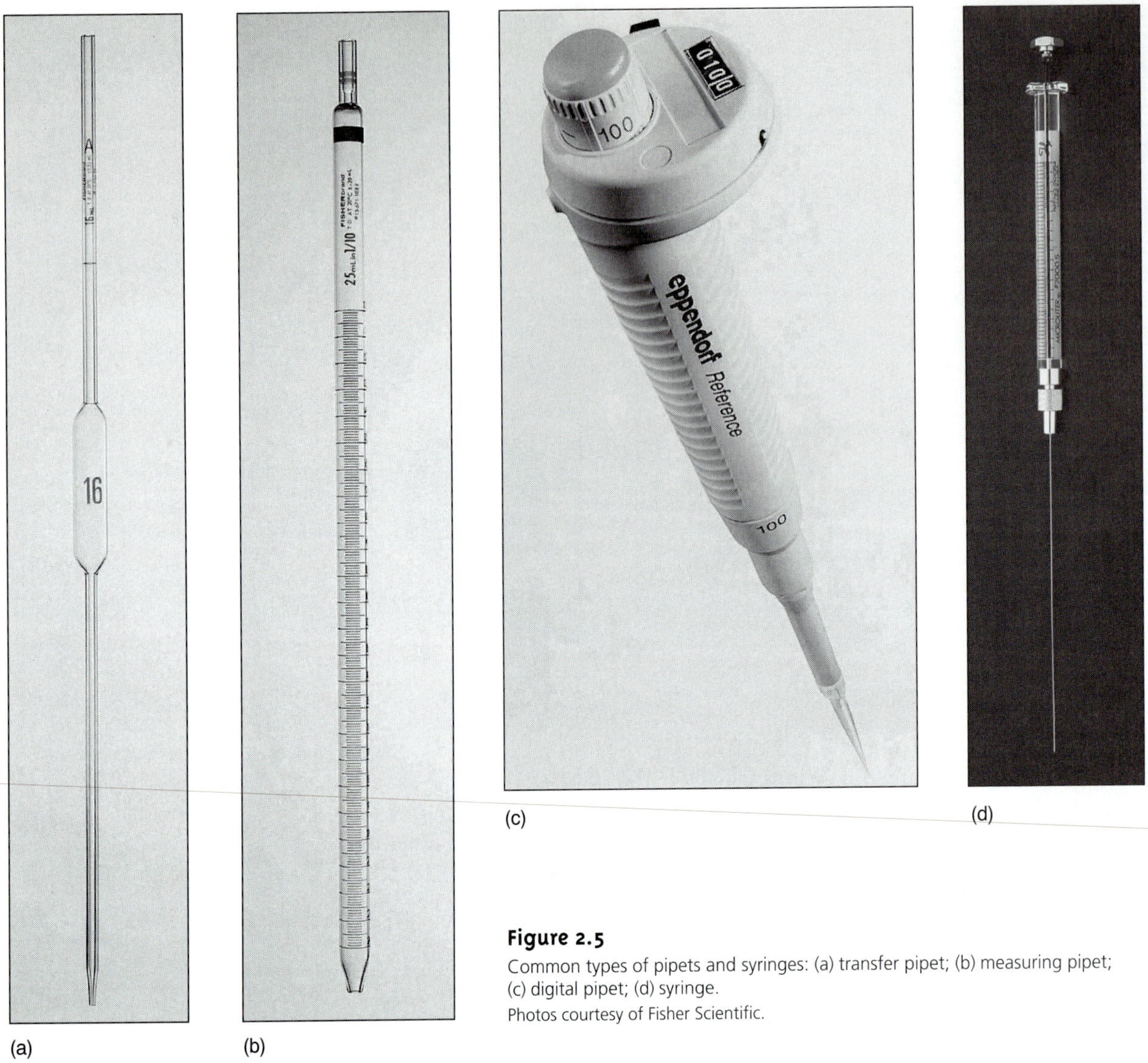

(a) (b) (c) (d)

Figure 2.5
Common types of pipets and syringes: (a) transfer pipet; (b) measuring pipet; (c) digital pipet; (d) syringe.
Photos courtesy of Fisher Scientific.

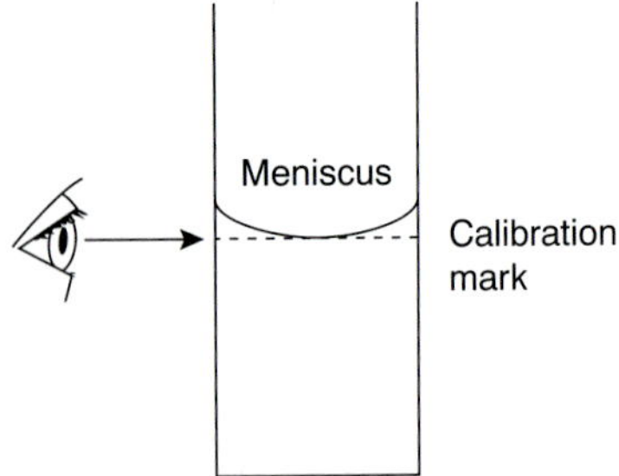

Figure 2.6
Proper means of reading the meniscus on a volumetric flask or pipet.

pipets, delivery of the calibrated volume requires that any solution remaining in the tip be blown out. Digital pipets and syringes can be used to deliver volumes as small as a microliter.

Three important precautions are needed when working with pipets and volumetric flasks. First, the volume delivered by a pipet or contained by a volumetric flask assumes that the glassware is clean. Dirt and grease on the inner glass surface prevents liquids from draining evenly, leaving droplets of the liquid on the container's walls. For a pipet this means that the delivered volume is less than the calibrated volume, whereas drops of liquid above the calibration mark mean that a volumetric flask contains more than its calibrated volume. Commercially available cleaning solutions can be used to clean pipets and volumetric flasks.

Second, when filling a pipet or volumetric flask, set the liquid's level exactly at the calibration mark. The liquid's top surface is curved into a **meniscus,** the bottom of which should be exactly even with the glassware's calibration mark (Figure 2.6). The meniscus should be adjusted with the calibration mark at eye level to avoid parallax errors. If your eye level is above the calibration mark the pipet or volumetric flask will be overfilled. The pipet or volumetric flask will be underfilled if your eye level is below the calibration mark.

Finally, before using a pipet or volumetric flask you should rinse it with several small portions of the solution whose volume is being measured. This ensures that any residual liquid remaining in the pipet or volumetric flask is removed.

meniscus
The curved surface of a liquid contained in a tube.

2D.3 Equipment for Drying Samples

Many materials need to be dried prior to their analysis to remove residual moisture. Depending on the material, heating to a temperature of 110–140 °C is usually sufficient. Other materials need to be heated to much higher temperatures to initiate thermal decomposition. Both processes can be accomplished using a laboratory oven capable of providing the required temperature.

Commercial laboratory ovens (Figure 2.7) are used when the maximum desired temperature is 160–325 °C (depending on the model). Some ovens include the ability to circulate heated air, allowing for a more efficient removal of moisture and shorter drying times. Other ovens provide a tight seal for the door, allowing the oven to be evacuated. In some situations a conventional laboratory oven can be replaced with a microwave oven. Higher temperatures, up to 1700° C, can be achieved using a muffle furnace (Figure 2.8).

After drying or decomposing a sample, it should be cooled to room temperature in a desiccator to avoid the readsorption of moisture. A **desiccator** (Figure 2.9) is a closed container that isolates the sample from the atmosphere. A drying agent, called a **desiccant,** is placed in the bottom of the container. Typical desiccants include calcium chloride and silica gel. A perforated plate sits above the desiccant, providing a shelf for storing samples. Some desiccators are equipped with stopcocks that allow them to be evacuated.

Figure 2.7
Conventional laboratory oven used for drying materials.

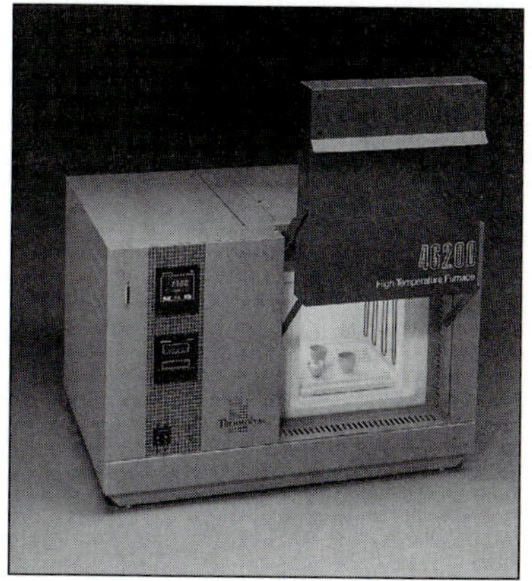

Figure 2.8
Example of a muffle furnace used for heating samples to maximum temperatures of 1100–1700 °C.
Courtesy of Fisher Scientific.

desiccator
A closed container containing a desiccant; used to store samples in a moisture-free environment.

desiccant
A drying agent.

(a)

(b)

Figure 2.9
(a) Desiccator. (b) Desiccator with stopcock for evacuating the desiccator.
Photos courtesy of Fisher Scientific.

2E Preparing Solutions

Preparing a solution of known concentration is perhaps the most common activity in any analytical lab. The method for measuring out the solute and solvent depend on the desired concentration units, and how exact the solution's concentration needs to be known. Pipets and volumetric flasks are used when a solution's concentration must be exact; graduated cylinders, beakers, and reagent bottles suffice when concentrations need only be approximate. Two methods for preparing solutions are described in this section.

2E.1 Preparing Stock Solutions

stock solution
A solution of known concentration from which other solutions are prepared.

A **stock solution** is prepared by weighing out an appropriate portion of a pure solid or by measuring out an appropriate volume of a pure liquid and diluting to a known volume. Exactly how this is done depends on the required concentration units. For example, to prepare a solution with a desired molarity you would weigh out an appropriate mass of the reagent, dissolve it in a portion of solvent, and bring to the desired volume. To prepare a solution where the solute's concentration is given as a volume percent, you would measure out an appropriate volume of solute and add sufficient solvent to obtain the desired total volume.

Example

EXAMPLE 2.9

Describe how you would prepare the following three solutions: (a) 500 mL of approximately 0.20 M NaOH using solid NaOH; (b) 1 L of 150.0 ppm Cu^{2+} using Cu metal; and (c) 2 L of 4% v/v acetic acid using concentrated glacial acetic acid.

SOLUTION

(a) Since the concentration only needs to be known to two significant figures, the mass of NaOH and volume of solution do not need to be measured exactly. The desired mass of NaOH is

$$\frac{0.20 \text{ mol}}{\text{L}} \times \frac{40.0 \text{ g}}{\text{mol}} \times 0.50 \text{ L} = 4.0 \text{ g}$$

To prepare the solution we place 4.0 g of NaOH, weighed to the nearest tenth of a gram, in a bottle or beaker and add approximately 500 mL of water.

(b) Since the concentration of Cu^{2+} needs to be exact, the mass of Cu metal and the final solution volume must be measured exactly. The desired mass of Cu metal is

$$\frac{150.0 \text{ mg}}{\text{L}} \times 1.000 \text{ L} = 150.0 \text{ mg} = 0.1500 \text{ g}$$

To prepare the solution we measure out exactly 0.1500 g of Cu into a small beaker. To dissolve the Cu we add a small portion of concentrated HNO_3 and gently heat until it completely dissolves. The resulting solution is poured into a 1-L volumetric flask. The beaker is rinsed repeatedly with small portions of water, which are added to the volumetric flask. This process, which is called a **quantitative transfer,** ensures that the Cu^{2+} is completely transferred to the volumetric flask. Finally, additional water is added to the volumetric flask's calibration mark.

quantitative transfer
The process of moving a sample from one container to another in a manner that ensures all material is transferred.

(c) The concentration of this solution is only approximate, so volumes do not need to be measured exactly. The necessary volume of glacial acetic acid is

$$\frac{4 \text{ mL } CH_3COOH}{100 \text{ mL}} \times 2000 \text{ mL} = 80 \text{ mL } CH_3COOH$$

To prepare the solution we use a graduated cylinder to transfer 80 mL of glacial acetic acid to a container that holds approximately 2 L, and we then add sufficient water to bring the solution to the desired volume.

2E.2 Preparing Solutions by Dilution

Solutions with small concentrations are often prepared by diluting a more concentrated stock solution. A known volume of the stock solution is transferred to a new container and brought to a new volume. Since the total amount of solute is the same before and after **dilution,** we know that

$$C_o \times V_o = C_d \times V_d \qquad \textbf{2.4}$$

where C_o is the concentration of the stock solution, V_o is the volume of the stock solution being diluted, C_d is the concentration of the dilute solution, and V_d is the volume of the dilute solution. Again, the type of glassware used to measure V_o and V_d depends on how exact the solution's concentration must be known.

dilution
The process of preparing a less concentrated solution from a more concentrated solution.

Example

EXAMPLE 2.10

A laboratory procedure calls for 250 mL of an approximately 0.10 M solution of NH_3. Describe how you would prepare this solution using a stock solution of concentrated NH_3 (14.8 M).

SOLUTION

Substituting known volumes in equation 2.4

$$14.8 \text{ M} \times V_o = 0.10 \text{ M} \times 0.25 \text{ L}$$

and solving for V_o gives 1.69×10^{-3} L, or 1.7 mL. Since we are trying to make a solution that is approximately 0.10 M NH_3, we can measure the appropriate amount of concentrated NH_3 using a graduated cylinder, transfer the NH_3 to a beaker, and add sufficient water to bring the total solution volume to approximately 250 mL.

As shown in the following example, equation 2.4 also can be used to calculate a solution's original concentration using its known concentration after dilution.

Example

EXAMPLE 2.11

A sample of an ore was analyzed for Cu^{2+} as follows. A 1.25-g sample of the ore was dissolved in acid and diluted to volume in a 250-mL volumetric flask. A 20-mL portion of the resulting solution was transferred by pipet to a 50-mL volumetric flask and diluted to volume. An analysis showed that the concentration of Cu^{2+} in the final solution was 4.62 ppm. What is the weight percent of Cu in the original ore?

SOLUTION

Substituting known volumes (with significant figures appropriate for pipets and volumetric flasks) into equation 2.4

$$(\text{ppm Cu}^{2+})_o \times 20.00 \text{ mL} = 4.62 \text{ ppm} \times 50.00 \text{ mL}$$

and solving for $(\text{ppm Cu}^{2+})_o$ gives the original solution concentration as 11.55 ppm. To calculate the grams of Cu^{2+} we multiply this concentration by the total volume

$$\frac{11.55 \text{ µg Cu}^{2+}}{\text{mL}} \times 250.0 \text{ mL} \times \frac{1 \text{ g}}{10^6 \text{ µg}} = 2.888 \times 10^{-3} \text{ g Cu}^{2+}$$

The weight percent Cu is then given by

$$\frac{2.888 \times 10^{-3} \text{ g Cu}^{2+}}{1.25 \text{ g sample}} \times 100 = 0.231\% \text{ w/w Cu}$$

2F The Laboratory Notebook

Finally, we cannot end a chapter on the basic tools of analytical chemistry without mentioning the laboratory notebook. Your laboratory notebook is your most important tool when working in the lab, providing a complete record of all your work. If kept properly, you should be able to look back at your laboratory notebook several years from now and reconstruct the experiments on which you worked.

Your instructor will probably provide you with detailed instructions on how he or she wants you to maintain your notebook. Of course, you should expect to bring your notebook to the lab. Everything you do, measure, or observe while working in the lab should be recorded in your notebook as it takes place. Preparing data tables to organize your data will help ensure that you record the data you need and that you can find the data when it is time to calculate and analyze your results. Writing a narrative to accompany your data will help you remember what you did, why you did it, and why you thought it was significant. Reserve space for your calculations, for analyzing your data, and for interpreting your results. Take your notebook with you when you do research in the library.

Maintaining a laboratory notebook may seem like a great deal of effort, but if you do it well you have a permanent record of your work. Scientists working in academic, industrial, and governmental research labs rely on their notebooks to provide a written record of their work. Questions about research carried out at some time in the past can be answered by finding the appropriate pages in the laboratory notebook. A laboratory notebook is also a legal document that helps establish patent rights and proof of discovery.

2G KEY TERMS

balance *(p. 25)*
concentration *(p. 15)*
desiccant *(p. 29)*
desiccator *(p. 29)*
dilution *(p. 31)*
equivalent *(p. 17)*
equivalent weight *(p. 17)*
formality *(p. 15)*
formula weight *(p. 17)*
meniscus *(p. 29)*
molality *(p. 18)*
molarity *(p. 15)*

2H SUMMARY

There are a few basic numerical and experimental tools with which you must be familiar. Fundamental measurements in analytical chemistry, such as mass and volume, use base SI units, such as the kilogram (kg) and the liter (L). Other units, such as power, are defined in terms of these base units. When reporting measurements, we must be careful to include only those digits that are significant and to maintain the uncertainty implied by these significant figures when transforming measurements into results.

The relative amount of a constituent in a sample is expressed as its concentration. There are many ways to express concentration, the most common of which are molarity, weight percent, volume percent, weight-to-volume percent, parts per million, and parts per billion. Concentrations also can be expressed using p-functions.

Stoichiometric relationships and calculations are important in many quantitative analyses. The stoichiometry between the reactants and products of a chemical reaction is given by the coefficients of a balanced chemical reaction. When it is inconvenient to balance reactions, conservation principles can be used to establish the stoichiometric relationships.

Balances, volumetric flasks, pipets, and ovens are standard pieces of laboratory instrumentation and equipment that are routinely used in almost all analytical work. You should be familiar with the proper use of this equipment. You also should be familiar with how to prepare a stock solution of known concentration, and how to prepare a dilute solution from a stock solution.

2I PROBLEMS

1. Indicate how many significant figures are in each of the following numbers.
a. 903 b. 0.903 c. 1.0903
d. 0.0903 e. 0.09030 f. 9.03×10^2

2. Round each of the following to three significant figures.
a. 0.89377 b. 0.89328 c. 0.89350
d. 0.8997 e. 0.08907

3. Round each of the following to the stated number of significant figures.
a. The atomic weight of carbon to four significant figures
b. The atomic weight of oxygen to three significant figures
c. Avogadro's number to four significant figures
d. Faraday's constant to three significant figures

4. Report results for the following calculations to the correct number of significant figures.
a. $4.591 + 0.2309 + 67.1 =$
b. $313 - 273.15 =$
c. $712 \times 8.6 =$
d. $1.43/0.026 =$
e. $(8.314 \times 298)/96485 =$
f. $\log(6.53 \times 10^{-5}) =$
g. $10^{-7.14} =$
h. $(6.51 \times 10^{-5})\,(8.14 \times 10^{-9}) =$

5. A 12.1374-g sample of an ore containing Ni and Co was carried through Fresenius' analytical scheme shown in Figure 1.1. At point A the combined mass of Ni and Co was found to be 0.2306 g, and at point B the mass of Co was found to be 0.0813 g. Report the weight percent Ni in the ore to the correct number of significant figures.

6. Hillebrand and Lundell's analytical scheme (see Figure 1.2) for the analysis of Ni in ores involves precipitating Ni^{2+} using dimethylgloxime. The formula for the precipitate is $Ni(C_4H_7N_2O_2)_2$. Calculate the precipitate's formula weight to the correct number of significant figures.

7. An analyst wishes to add 256 mg of Cl^- to a reaction mixture. How many milliliters of 0.217 M $BaCl_2$ should be added?

8. A solution of 0.10 M SO_4^{2-} is available. What is the normality of this solution when used in the following reactions?
a. $Pb^{2+}(aq) + SO_4^{2-}(aq) \rightleftharpoons PbSO_4(s)$
b. $HCl(aq) + SO_4^{2-}(aq) \rightleftharpoons HSO_4^-(aq) + Cl^-(aq)$
c. $SO_4^{2-} + 4H_3O^+(aq) + 2e^- \rightleftharpoons H_2SO_3(aq) + 5H_2O(\ell)$

9. The concentration of lead in an industrial waste stream is 0.28 ppm. What is its molar concentration?

10. Commercially available concentrated hydrochloric acid is 37.0% w/w HCl. Its density is 1.18 g/mL. Using this information calculate (a) the molarity of concentrated HCl, and (b) the mass and volume (in milliliters) of solution containing 0.315 mol of HCl.

11. The density of concentrated ammonia, which is 28.0% w/w NH_3, is 0.899 g/mL. What volume of this reagent should be diluted to 1.0×10^3 mL to make a solution that is 0.036 M in NH_3?

12. A 250.0-mL aqueous solution contains 45.1 μg of a pesticide. Express the pesticide's concentration in weight percent, parts per million, and parts per billion.

13. A city's water supply is fluoridated by adding NaF. The desired concentration of F^- is 1.6 ppm. How many milligrams of NaF should be added per gallon of treated water if the water supply already is 0.2 ppm in F^-?

14. What is the pH of a solution for which the concentration of H^+ is 6.92×10^{-6} M? What is the $[H^+]$ in a solution whose pH is 8.923?

15. Using conservation principles, write stoichiometric relationships for the following

a. The precipitation of Mg^{2+} as $Mg_2P_2O_7$

b. The acid–base reaction between $CaCO_3$ and HCl in which H_2CO_3 is formed

c. The reaction between AgCl and NH_3 to form $Ag(NH_3)_2^+$

d. The redox reaction between $Cr_2O_7^{2-}$ and Fe^{2+} to form Cr^{3+} and Fe^{3+}

16. Calculate the molarity of a potassium dichromate solution prepared by placing 9.67 g of $K_2Cr_2O_7$ in a 100-mL volumetric flask, dissolving, and diluting to the calibration mark.

17. For each of the following, explain how you would prepare 1.0 L of a solution that is 0.10 M in K^+. Repeat for concentrations of 1.0×10^2 ppm K^+ and 1.0% w/v K^+.

a. KCl b. K_2SO_4 c. $K_3Fe(CN)_6$

18. A series of dilute NaCl solutions is prepared, starting with an initial stock solution of 0.100 M NaCl. Solution A is prepared by pipeting 10 mL of the stock solution into a 250-mL volumetric flask and diluting to volume. Solution B is prepared by pipeting 25 mL of solution A into a 100-mL volumetric flask and diluting to volume. Solution C is prepared by pipeting 20 mL of solution B into a 500-mL volumetric flask and diluting to volume. What is the molar concentration of NaCl in solutions A, B, and C?

19. Calculate the molar concentration of NaCl, to the correct number of significant figures, if 1.917 g of NaCl is placed in a beaker and dissolved in 50 mL of water measured with a graduated cylinder. This solution is quantitatively transferred to a 250-mL volumetric flask and diluted to volume. Calculate the concentration of this second solution to the correct number of significant figures.

20. What is the molar concentration of NO_3^- in a solution prepared by mixing 50.0 mL of 0.050 M KNO_3 with 40.0 mL of 0.075 M $NaNO_3$? What is pNO_3 for the mixture?

21. What is the molar concentration of Cl^- in a solution prepared by mixing 25.0 mL of 0.025 M NaCl with 35.0 mL of 0.050 M $BaCl_2$? What is pCl for the mixture?

22. To determine the concentration of ethanol in cognac a 5.00-mL sample of cognac is diluted to 0.500 L. Analysis of the diluted cognac gives an ethanol concentration of 0.0844 M. What is the molar concentration of ethanol in the undiluted cognac?

2J SUGGESTED READINGS

Two useful articles providing additional information on topics covered in this chapter are

MacCarthy, P. "A Novel Classification of Concentration Units," *J. Chem. Educ.* **1983,** *60,* 187–189.

Schwartz, L. M. "Propagation of Significant Figures," *J. Chem. Educ.* **1985,** *62,* 693–697.

A useful resource for information on maintaining a useful laboratory notebook is

Kanare, H. M. *Writing the Laboratory Notebook,* American Chemical Society: Washington, DC; 1985.

2K REFERENCES

1. American Public Health Association. *Standard Methods for the Analysis of Waters and Wastewaters,* 19th ed., Washington, DC. 1995.

Chapter 3

The Language of Analytical Chemistry

Analytical chemists converse using terminology that conveys specific meaning to other analytical chemists. To discuss and learn analytical chemistry you must first understand its language. You are probably already familiar with some analytical terms, such as "accuracy" and "precision," but you may not have placed them in their appropriate analytical context. Other terms, such as "analyte" and "matrix," may be less familiar. This chapter introduces many important terms routinely used by analytical chemists. Becoming comfortable with these terms will make the material in the chapters that follow easier to read and understand.

3A Analysis, Determination, and Measurement

analysis
A process that provides chemical or physical information about the constituents in the sample or the sample itself.

analytes
The constituents of interest in a sample.

matrix
All other constituents in a sample except for the analytes.

determination
An analysis of a sample to find the identity, concentration, or properties of the analyte.

measurement
An experimental determination of an analyte's chemical or physical properties.

The first important distinction we will make is among the terms "analysis," "determination," and "measurement." An **analysis** provides chemical or physical information about a sample. The components of interest in the sample are called **analytes,** and the remainder of the sample is the **matrix.** In an analysis we determine the identity, concentration, or properties of the analytes. To make this **determination** we measure one or more of the analyte's chemical or physical properties.

An example helps clarify the differences among an analysis, a determination, and a **measurement.** In 1974, the federal government enacted the Safe Drinking Water Act to ensure the safety of public drinking water supplies. To comply with this act municipalities regularly monitor their drinking water supply for potentially harmful substances. One such substance is coliform bacteria. Municipal water departments collect and analyze samples from their water supply. To determine the concentration of coliform bacteria, a portion of water is passed through a membrane filter. The filter is placed in a dish containing a nutrient broth and incubated. At the end of the incubation period the number of coliform bacterial colonies in the dish is measured by counting (Figure 3.1). Thus, municipal water departments analyze samples of water to determine the concentration of coliform bacteria by measuring the number of bacterial colonies that form during a specified period of incubation.

3B Techniques, Methods, Procedures, and Protocols

Suppose you are asked to develop a way to determine the concentration of lead in drinking water. How would you approach this problem? To answer this question it helps to distinguish among four levels of analytical methodology: techniques, methods, procedures, and protocols.[1]

technique
A chemical or physical principle that can be used to analyze a sample.

A **technique** is any chemical or physical principle that can be used to study an analyte. Many techniques have been used to determine lead levels.[2] For example, in graphite furnace atomic absorption spectroscopy lead is atomized, and the ability of the free atoms to absorb light is measured; thus, both a chemical principle (atomization) and a physical principle (absorption of light) are used in this technique. Chapters 8–13 of this text cover techniques commonly used to analyze samples.

method
A means for analyzing a sample for a specific analyte in a specific matrix.

A **method** is the application of a technique for the determination of a specific analyte in a specific matrix. As shown in Figure 3.2, the graphite furnace atomic absorption spectroscopic method for determining lead levels in water is different from that for the determination of lead in soil or blood. Choosing a method for determining lead in water depends on how the information is to be used and the established design criteria (Figure 3.3). For some analytical problems the best method might use graphite furnace atomic absorption spectroscopy, whereas other problems might be more easily solved by using another technique, such as anodic stripping voltammetry or potentiometry with a lead ion-selective electrode.

procedure
Written directions outlining how to analyze a sample.

A **procedure** is a set of written directions detailing how to apply a method to a particular sample, including information on proper sampling, handling of interferents, and validating results. A method does not necessarily lead to a single procedure, as different analysts or agencies will adapt the method to their specific needs. As shown in Figure 3.2, the American Public Health Agency and the American Society for Testing Materials publish separate procedures for the determination of lead levels in water.

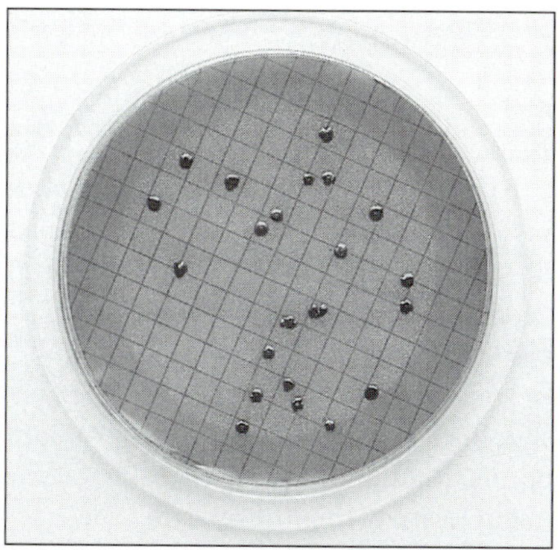

Figure 3.1
Membrane filter showing colonies of coliform bacteria. The number of colonies are counted and reported as colonies/100 mL of sample.
PourRite™ is a trademark of Hach Company/photo courtesy of Hach Company.

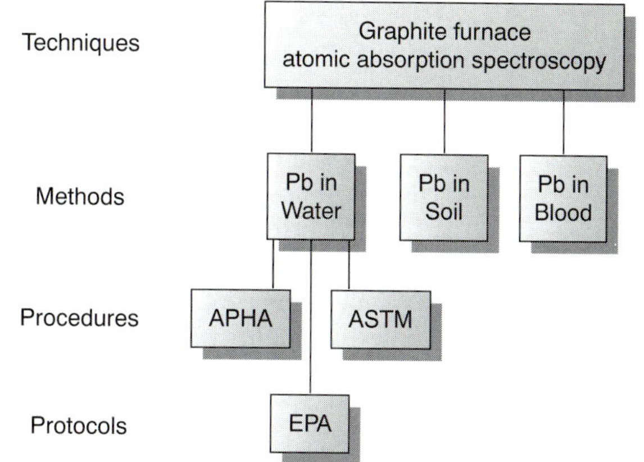

Figure 3.2
Chart showing hierarchical relationship among a technique, methods using that technique, and procedures and protocols for one method. (*Abbreviations:* APHA = American Public Health Association, ASTM = American Society for Testing Materials, EPA = Environmental Protection Agency)

Finally, a **protocol** is a set of stringent written guidelines detailing the procedure that must be followed if the agency specifying the protocol is to accept the results of the analysis. Protocols are commonly encountered when analytical chemistry is used to support or define public policy. For purposes of determining lead levels in water under the Safe Drinking Water Act, labs follow a protocol specified by the Environmental Protection Agency.

There is an obvious order to these four facets of analytical methodology. Ideally, a protocol uses a previously validated procedure. Before developing and validating a procedure, a method of analysis must be selected. This requires, in turn, an initial screening of available techniques to determine those that have the potential for monitoring the analyte. We begin by considering a useful way to classify analytical techniques.

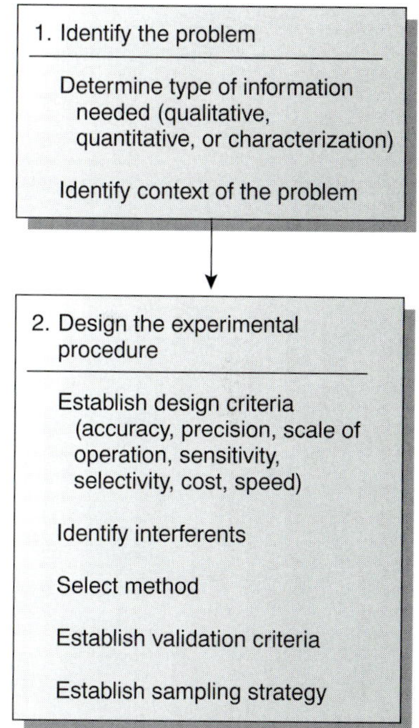

Figure 3.3
Subsection of the analytical approach to problem solving (see Figure 1.3), of relevance to the selection of a method and the design of an analytical procedure.

3C Classifying Analytical Techniques

Analyzing a sample generates a chemical or physical **signal** whose magnitude is proportional to the amount of analyte in the sample. The signal may be anything we can measure; common examples are mass, volume, and absorbance. For our purposes it is convenient to divide analytical techniques into two general classes based on whether this signal is proportional to an absolute amount of analyte or a relative amount of analyte.

Consider two graduated cylinders, each containing 0.01 M $Cu(NO_3)_2$ (Figure 3.4). Cylinder 1 contains 10 mL, or 0.0001 mol, of Cu^{2+}; cylinder 2 contains 20 mL, or 0.0002 mol, of Cu^{2+}. If a technique responds to the absolute amount of analyte in the sample, then the signal due to the analyte, S_A, can be expressed as

$$S_A = kn_A \qquad 3.1$$

where n_A is the moles or grams of analyte in the sample, and k is a proportionality constant. Since cylinder 2 contains twice as many moles of Cu^{2+} as cylinder 1, analyzing the contents of cylinder 2 gives a signal that is twice that of cylinder 1.

protocol
A set of written guidelines for analyzing a sample specified by an agency.

signal
An experimental measurement that is proportional to the amount of analyte (S).

A second class of analytical techniques are those that respond to the relative amount of analyte; thus

$$S_A = kC_A \qquad 3.2$$

where C_A is the concentration of analyte in the sample. Since the solutions in both cylinders have the same concentration of Cu^{2+}, their analysis yields identical signals.

total analysis techniques
A technique in which the signal is proportional to the absolute amount of analyte; also called "classical" techniques.

Techniques responding to the absolute amount of analyte are called **total analysis techniques.** Historically, most early analytical methods used total analysis techniques, hence they are often referred to as "classical" techniques. Mass, volume, and charge are the most common signals for total analysis techniques, and the corresponding techniques are gravimetry (Chapter 8), titrimetry (Chapter 9), and coulometry (Chapter 11). With a few exceptions, the signal in a total analysis technique results from one or more chemical reactions involving the analyte. These reactions may involve any combination of precipitation, acid–base, complexation, or redox chemistry. The stoichiometry of each reaction, however, must be known to solve equation 3.1 for the moles of analyte.

concentration techniques
A technique in which the signal is proportional to the analyte's concentration; also called "instrumental" techniques.

Techniques, such as spectroscopy (Chapter 10), potentiometry (Chapter 11), and voltammetry (Chapter 11), in which the signal is proportional to the relative amount of analyte in a sample are called **concentration techniques.** Since most concentration techniques rely on measuring an optical or electrical signal, they also are known as "instrumental" techniques. For a concentration technique, the relationship between the signal and the analyte is a theoretical function that depends on experimental conditions and the instrumentation used to measure the signal. For this reason the value of k in equation 3.2 must be determined experimentally.

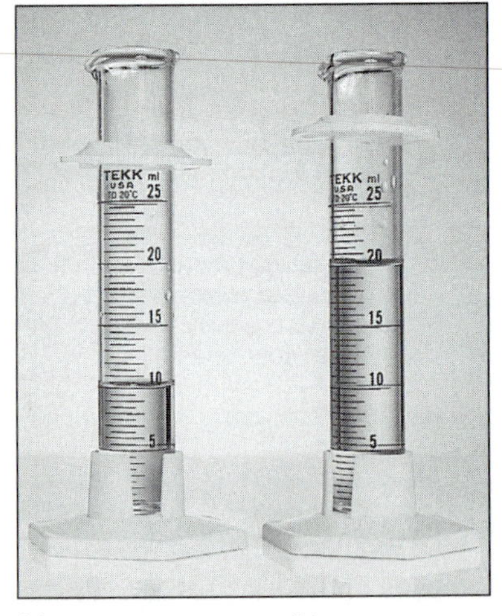

(a) (b)

Figure 3.4
Graduated cylinders containing 0.01 M $Cu(NO_3)_2$. (a) Cylinder 1 contains 10 mL, or 0.0001 mol, of Cu^{2+}. (b) Cylinder 2 contains 20 mL, or 0.0002 mol, of Cu^{2+}.

3D Selecting an Analytical Method

A method is the application of a technique to a specific analyte in a specific matrix. Methods for determining the concentration of lead in drinking water can be developed using any of the techniques mentioned in the previous section. Insoluble lead salts such as $PbSO_4$ and $PbCrO_4$ can form the basis for a gravimetric method. Lead forms several soluble complexes that can be used in a complexation titrimetric method or, if the complexes are highly absorbing, in a spectrophotometric method. Lead in the gaseous free-atom state can be measured by an atomic absorption spectroscopic method. Finally, the availability of multiple oxidation states (Pb, Pb^{2+}, Pb^{4+}) makes coulometric, potentiometric, and voltammetric methods feasible.

The requirements of the analysis determine the best method. In choosing a method, consideration is given to some or all the following design criteria: accuracy, precision, sensitivity, selectivity, robustness, ruggedness, scale of operation, analysis time, availability of equipment, and cost. Each of these criteria is considered in more detail in the following sections.

3D.1 Accuracy

accuracy
A measure of the agreement between an experimental result and its expected value.

Accuracy is a measure of how closely the result of an experiment agrees with the expected result. The difference between the obtained result and the expected result is usually divided by the expected result and reported as a percent relative error

$$\%\ \text{Error} = \frac{\text{obtained result} - \text{expected result}}{\text{expected result}} \times 100$$

Analytical methods may be divided into three groups based on the magnitude of their relative errors.[3] When an experimental result is within 1% of the correct result, the analytical method is highly accurate. Methods resulting in relative errors between 1% and 5% are moderately accurate, but methods of low accuracy produce relative errors greater than 5%.

The magnitude of a method's relative error depends on how accurately the signal is measured, how accurately the value of k in equations 3.1 or 3.2 is known, and the ease of handling the sample without loss or contamination. In general, total analysis methods produce results of high accuracy, and concentration methods range from high to low accuracy. A more detailed discussion of accuracy is presented in Chapter 4.

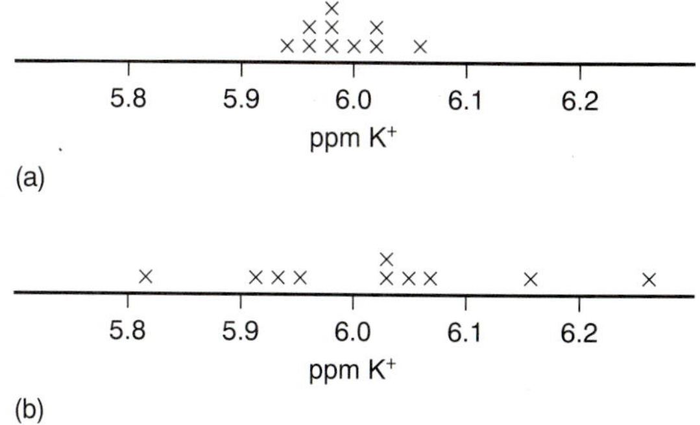

Figure 3.5
Two determinations of the concentration of K^+ in serum, showing the effect of precision. The data in (a) are less scattered and, therefore, more precise than the data in (b).

3D.2 Precision

When a sample is analyzed several times, the individual results are rarely the same. Instead, the results are randomly scattered. **Precision** is a measure of this variability. The closer the agreement between individual analyses, the more precise the results. For example, in determining the concentration of K^+ in serum, the results shown in Figure 3.5(a) are more precise than those in Figure 3.5(b). It is important to realize that precision does not imply accuracy. That the data in Figure 3.5(a) are more precise does not mean that the first set of results is more accurate. In fact, both sets of results may be very inaccurate.

precision
An indication of the reproducibility of a measurement or result.

As with accuracy, precision depends on those factors affecting the relationship between the signal and the analyte (equations 3.1 and 3.2). Of particular importance are the uncertainty in measuring the signal and the ease of handling samples reproducibly. In most cases the signal for a total analysis method can be measured with a higher precision than the corresponding signal for a concentration method. Precision is covered in more detail in Chapter 4.

3D.3 Sensitivity

The ability to demonstrate that two samples have different amounts of analyte is an essential part of many analyses. A method's **sensitivity** is a measure of its ability to establish that such differences are significant. Sensitivity is often confused with a method's detection limit.[4] The **detection limit** is the smallest amount of analyte that can be determined with confidence. The detection limit, therefore, is a statistical parameter and is discussed in Chapter 4.

sensitivity
A measure of a method's ability to distinguish between two samples; reported as the change in signal per unit change in the amount of analyte (k).

detection limit
A statistical statement about the smallest amount of analyte that can be determined with confidence.

Sensitivity is the change in signal per unit change in the amount of analyte and is equivalent to the proportionality constant, k, in equations 3.1 and 3.2. If ΔS_A is the smallest increment in signal that can be measured, then the smallest difference in the amount of analyte that can be detected is

$$\Delta n_A = \frac{\Delta S_A}{k} \quad \text{(total analysis method)}$$

$$\Delta C_A = \frac{\Delta S_A}{k} \quad \text{(concentration method)}$$

Suppose that for a particular total analysis method the signal is a measurement of mass using a balance whose smallest increment is ±0.0001 g. If the method's sensitivity is 0.200, then the method can conceivably detect a difference of as little as

$$\Delta n_A = \frac{\pm 0.0001 \text{ g}}{0.200} = \pm 0.0005 \text{ g}$$

in the absolute amount of analyte in two samples. For methods with the same ΔS_A, the method with the greatest sensitivity is best able to discriminate among smaller amounts of analyte.

3D.4 Selectivity

An analytical method is selective if its signal is a function of only the amount of analyte present in the sample. In the presence of an interferent, equations 3.1 and 3.2 can be expanded to include a term corresponding to the interferent's contribution to the signal, S_I,

$$S_{samp} = S_A + S_I = k_A n_A + k_I n_I \qquad \text{(total analysis method)} \qquad 3.3$$

$$S_{samp} = S_A + S_I = k_A C_A + k_I C_I \qquad \text{(concentration method)} \qquad 3.4$$

where S_{samp} is the total signal due to constituents in the sample; k_A and k_I are the sensitivities for the analyte and the interferent, respectively; and n_I and C_I are the moles (or grams) and concentration of the interferent in the sample.

selectivity
A measure of a method's freedom from interferences as defined by the method's selectivity coefficient.

selectivity coefficient
A measure of a method's sensitivity for an interferent relative to that for the analyte ($K_{A,I}$).

The **selectivity** of the method for the interferent relative to the analyte is defined by a **selectivity coefficient,** $K_{A,I}$

$$K_{A,I} = \frac{k_I}{k_A} \qquad 3.5$$

which may be positive or negative depending on whether the interferent's effect on the signal is opposite that of the analyte.* A selectivity coefficient greater than +1 or less than –1 indicates that the method is more selective for the interferent than for the analyte. Solving equation 3.5 for k_I

$$k_I = K_{A,I} \times k_A \qquad 3.6$$

substituting into equations 3.3 and 3.4, and simplifying gives

$$S_{samp} = k_A(n_A + K_{A,I} \times n_I) \qquad \text{(total analysis method)} \qquad 3.7$$

$$S_{samp} = k_A(C_A + K_{A,I} \times C_I) \qquad \text{(concentration method)} \qquad 3.8$$

The selectivity coefficient is easy to calculate if k_A and k_I can be independently determined. It is also possible to calculate $K_{A,I}$ by measuring S_{samp} in the presence and absence of known amounts of analyte and interferent.

Example

EXAMPLE 3.1

A method for the analysis of Ca^{2+} in water suffers from an interference in the presence of Zn^{2+}. When the concentration of Ca^{2+} is 100 times greater than that of Zn^{2+}, an analysis for Ca^{2+} gives a relative error of +0.5%. What is the selectivity coefficient for this method?

*Although k_A and k_I are usually positive, they also may be negative. For example, some analytical methods work by measuring the concentration of a species that reacts with the analyte. As the analyte's concentration increases, the concentration of the species producing the signal decreases, and the signal becomes smaller. If the signal in the absence of analyte is assigned a value of zero, then the subsequent signals are negative.

SOLUTION

Since only relative concentrations are reported, we can arbitrarily assign absolute concentrations. To make the calculations easy, let $C_{Ca} = 100$ (arbitrary units) and $C_{Zn} = 1$. A relative error of +0.5% means that the signal in the presence of Zn^{2+} is 0.5% greater than the signal in the absence of zinc. Again, we can assign values to make the calculation easier. If the signal in the absence of zinc is 100 (arbitrary units), then the signal in the presence of zinc is 100.5.

The value of k_{Ca} is determined using equation 3.2

$$k_{Ca} = \frac{S_{Ca}}{C_{Ca}} = \frac{100}{100} = 1$$

In the presence of zinc the signal is

$$S_{samp} = 100.5 = k_{Ca}C_{Ca} + k_{Zn}C_{Zn} = (1)(100) + k_{Zn}(1)$$

Solving for k_{Zn} gives a value of 0.5. The selectivity coefficient, therefore, is

$$K_{Ca/Zn} = \frac{k_{Zn}}{k_{Ca}} = \frac{0.5}{1} = 0.5$$

Knowing the selectivity coefficient provides a useful way to evaluate an interferent's potential effect on an analysis. An interferent will not pose a problem as long as the term $K_{A,I} \times n_I$ in equation 3.7 is significantly smaller than n_A, or $K_{A,I} \times C_I$ in equation 3.8 is significantly smaller than C_A.

Example

EXAMPLE 3.2

Barnett and colleagues[5] developed a new method for determining the concentration of codeine during its extraction from poppy plants. As part of their study they determined the method's response to codeine relative to that for several potential interferents. For example, the authors found that the method's signal for 6-methoxycodeine was 6 (arbitrary units) when that for an equimolar solution of codeine was 40.

(a) What is the value for the selectivity coefficient $K_{A,I}$ when 6-methoxycodeine is the interferent and codeine is the analyte?

(b) If the concentration of codeine is to be determined with an accuracy of ±0.50%, what is the maximum relative concentration of 6-methoxycodeine (i.e., [6-methoxycodeine]/[codeine]) that can be present?

SOLUTION

(a) The signals due to the analyte, S_A, and the interferent, S_I, are

$$S_A = k_A C_A \qquad S_I = k_I C_I$$

Solving these two expressions for k_A and k_I and substituting into equation 3.6 gives

$$K_{A,I} = \frac{S_I/C_I}{S_A/C_A}$$

Since equimolar concentrations of analyte and interferent were used ($C_A = C_I$), we have

$$K_{A,I} = \frac{S_I}{S_A} = \frac{6}{40} = 0.15$$

(b) To achieve an accuracy of better than ±0.50% the term $K_{A,I} \times C_I$ in equation 3.8 must be less than 0.50% of C_A; thus

$$0.0050 \times C_A \geq K_{A,I} \times C_I$$

Solving this inequality for the ratio C_I/C_A and substituting the value for $K_{A,I}$ determined in part (a) gives

$$\frac{C_I}{C_A} \leq \frac{0.0050}{K_{A,I}} = \frac{0.0050}{0.15} = 0.033$$

Therefore, the concentration of 6-methoxycodeine cannot exceed 3.3% of codeine's concentration.

Not surprisingly, methods whose signals depend on chemical reactivity are often less selective and, therefore, more susceptible to interferences. Problems with selectivity become even greater when the analyte is present at a very low concentration.[6]

3D.5 Robustness and Ruggedness

For a method to be useful it must provide reliable results. Unfortunately, methods are subject to a variety of chemical and physical interferences that contribute uncertainty to the analysis. When a method is relatively free from chemical interferences, it can be applied to the determination of analytes in a wide variety of sample matrices. Such methods are considered **robust.**

robust
A method that can be applied to analytes in a wide variety of matrices is considered robust.

rugged
A method that is insensitive to changes in experimental conditions is considered rugged.

Random variations in experimental conditions also introduce uncertainty. If a method's sensitivity is highly dependent on experimental conditions, such as temperature, acidity, or reaction time, then slight changes in those conditions may lead to significantly different results. A **rugged** method is relatively insensitive to changes in experimental conditions.

3D.6 Scale of Operation

Another way to narrow the choice of methods is to consider the scale on which the analysis must be conducted. Three limitations of particular importance are the amount of sample available for the analysis, the concentration of analyte in the sample, and the absolute amount of analyte needed to obtain a measurable signal. The first and second limitations define the scale of operations shown in Figure 3.6; the last limitation positions a method within the scale of operations.[7]

The scale of operations in Figure 3.6 shows the analyte's concentration in weight percent on the *y*-axis and the sample's size on the *x*-axis. For convenience, we divide analytes into major (>1% w/w), minor (0.01% w/w – 1% w/w), trace (10^{-7}% w/w – 0.01% w/w) and ultratrace (<10^{-7}% w/w) components, and we divide samples into macro (>0.1 g), meso (10 mg – 100 mg), micro (0.1 mg – 10 mg) and ultramicro (<0.1 mg) sample sizes. Note that both the *x*-axis and the *y*-axis use a logarithmic scale. The analyte's concentration and the amount of

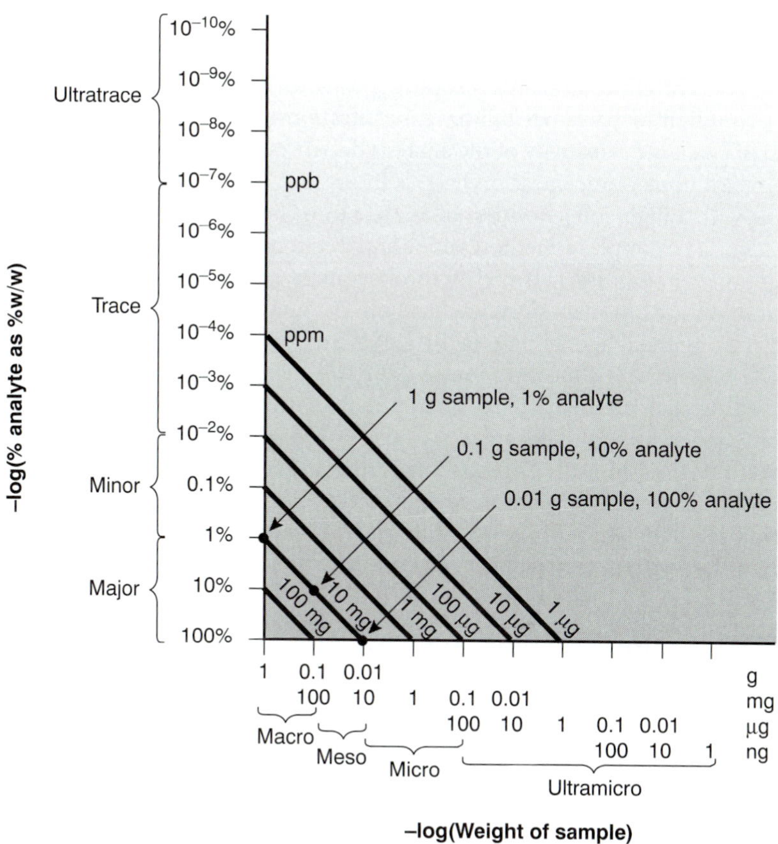

Figure 3.6
Scale of operation for analytical methods. Adapted from references 7*a* and 7*b*.

sample used provide a characteristic description for an analysis. For example, samples in a macro–major analysis weigh more than 0.1 g and contain more than 1% analyte.

Diagonal lines connecting the two axes show combinations of sample size and concentration of analyte containing the same absolute amount of analyte. As shown in Figure 3.6, for example, a 1-g sample containing 1% analyte has the same amount of analyte (0.010 g) as a 100-mg sample containing 10% analyte or a 10-mg sample containing 100% analyte.

Since total analysis methods respond to the absolute amount of analyte in a sample, the diagonal lines provide an easy way to define their limitations. Consider, for example, a hypothetical total analysis method for which the minimum detectable signal requires 100 mg of analyte. Using Figure 3.6, the diagonal line representing 100 mg suggests that this method is best suited for macro samples and major analytes. Applying the method to a minor analyte with a concentration of 0.1% w/w requires a sample of at least 100 g. Working with a sample of this size is rarely practical, however, due to the complications of carrying such a large amount of material through the analysis. Alternatively, the minimum amount of required analyte can be decreased by improving the limitations associated with measuring the signal. For example, if the signal is a measurement of mass, a decrease in the minimum amount of analyte can be accomplished by switching from a conventional analytical balance, which weighs samples to ±0.1 mg, to a semimicro (±0.01 mg) or microbalance (±0.001 mg).

Concentration methods frequently have both lower and upper limits for the amount of analyte that can be determined. The lower limit is dictated by the smallest concentration of analyte producing a useful signal and typically is in the parts per million or parts per billion concentration range. Upper concentration limits exist when the sensitivity of the analysis decreases at higher concentrations.

An upper concentration level is important because it determines how a sample with a high concentration of analyte must be treated before the analysis. Consider, for example, a method with an upper concentration limit of 1 ppm (micrograms per milliliter). If the method requires a sample of 1 mL, then the upper limit on the amount of analyte that can be handled is 1 μg. Using Figure 3.6, and following the diagonal line for 1 μg of analyte, we find that the analysis of an analyte present at a concentration of 10% w/w requires a sample of only 10 μg! Extending such an analysis to a major analyte, therefore, requires the ability to obtain and work with very small samples or the ability to dilute the original sample accurately. Using this example, analyzing a sample for an analyte whose concentration is 10% w/w requires a 10,000-fold dilution. Not surprisingly, concentration methods are most commonly used for minor, trace, and ultratrace analytes, in macro and meso samples.

3D.7 Equipment, Time, and Cost

Finally, analytical methods can be compared in terms of their need for equipment, the time required to complete an analysis, and the cost per sample. Methods relying on instrumentation are equipment-intensive and may require significant operator training. For example, the graphite furnace atomic absorption spectroscopic method for determining lead levels in water requires a significant capital investment in the instrument and an experienced operator to obtain reliable results. Other methods, such as titrimetry, require only simple equipment and reagents and can be learned quickly.

The time needed to complete an analysis for a single sample is often fairly similar from method to method. This is somewhat misleading, however, because much of this time is spent preparing the solutions and equipment needed for the analysis. Once the solutions and equipment are in place, the number of samples that can be analyzed per hour differs substantially from method to method. This is a significant factor in selecting a method for laboratories that handle a high volume of samples.

The cost of an analysis is determined by many factors, including the cost of necessary equipment and reagents, the cost of hiring analysts, and the number of samples that can be processed per hour. In general, methods relying on instruments cost more per sample than other methods.

3D.8 Making the Final Choice

Unfortunately, the design criteria discussed earlier are not mutually independent.[8] Working with smaller amounts of analyte or sample, or improving selectivity, often comes at the expense of precision. Attempts to minimize cost and analysis time may decrease accuracy. Selecting a specific method requires a careful balance among these design criteria. Usually, the most important design criterion is accuracy, and the best method is that capable of producing the most accurate results. When the need for results is urgent, as is often the case in clinical labs, analysis time may become the critical factor.

The best method is often dictated by the sample's properties. Analyzing a sample with a complex matrix may require a method with excellent selectivity to avoid interferences. Samples in which the analyte is present at a trace or ultratrace concentration usually must be analyzed by a concentration method. If the quantity of sample is limited, then the method must not require large amounts of sample.

Determining the concentration of lead in drinking water requires a method that can detect lead at the parts per billion concentrations. Selectivity is also important because other metal ions are present at significantly higher concentrations. Graphite furnace atomic absorption spectroscopy is a commonly used method for determining lead levels in drinking water because it meets these specifications. The same method is also used in determining lead levels in blood, where its ability to detect low concentrations of lead using a few microliters of sample are important considerations.

3E Developing the Procedure

After selecting a method, it is necessary to develop a procedure that will accomplish the goals of the analysis. In developing the procedure, attention is given to compensating for interferences, selecting and calibrating equipment, standardizing the method, acquiring a representative sample, and validating the method.

3E.1 Compensating for Interferences

The accuracy of a method depends on its selectivity for the analyte. Even the best methods, however, may not be free from interferents that contribute to the measured signal. Potential interferents may be present in the sample itself or the reagents used during the analysis. In this section we will briefly look at how to minimize these two sources of interference.

In the absence of an interferent, the total signal measured during an analysis, S_{meas}, is a sum of the signal due to the analyte, and the signal due to the reagents, S_{reag}

$$S_{meas} = S_A + S_{reag} = kn_A + S_{reag} \qquad \text{(total analysis method)} \qquad \textbf{3.9}$$

$$S_{meas} = S_A + S_{reag} = kC_A + S_{reag} \qquad \text{(concentration method)} \qquad \textbf{3.10}$$

Without an independent determination of S_{reag}, equation 3.9 or 3.10 cannot be solved for the moles or concentration of analyte. The contribution of S_{reag} is determined by measuring the signal for a reagent or **method blank** that does not contain the sample. Consider, for example, a procedure in which a 0.1-g sample is dissolved in 100 mL of solvent. After dissolving the sample, several reagents are added, and the signal is measured. The reagent blank is prepared by omitting the sample and adding the reagents to 100 mL of solvent. When the sample is a liquid, or is in solution, an equivalent volume of an inert solvent is substituted for the sample. Once S_{reag} is known, it is easy to correct S_{meas} for the reagent's contribution to the overall signal.

method blank
A sample that contains all components of the matrix except the analyte.

Compensating for an interference in the sample's matrix is more difficult. If the identity and concentration of the interferent are known, then it can be added to the reagent blank. In most analyses, however, the identity or concentration of matrix interferents is not known, and their contribution to S_{meas} is not included in S_{reag}. Instead, the signal from the interferent is included as an additional term

$$S_{meas} = k_A n_A + k_I n_I + S_{reag} \qquad \text{(total analysis method)} \qquad \textbf{3.11}$$

$$S_{meas} = k_A C_A + k_I C_I + S_{reag} \qquad \text{(concentration method)} \qquad \textbf{3.12}$$

Solving either equation 3.11 or 3.12 for the amount of analyte can be accomplished by separating the analyte and interferent before the analysis, thus eliminating the term for the interferent. Methods for effecting this separation are discussed in Chapter 7.

Alternatively, equations 3.11 or 3.12 can be solved for the amounts of both the analyte and the interferent. To do so, however, we must obtain two independent values for S_{meas}. Using a concentration method as an example, gives two equations

$$S_{meas,1} = k_{A,1}C_A + k_{I,1}C_I + S_{reag,1}$$

$$S_{meas,2} = k_{A,2}C_A + k_{I,2}C_I + S_{reag,2}$$

that can be solved simultaneously for C_A and C_I. This treatment is general. The composition of a solution with a total of n analytes and interferents can be determined by measuring n independent signals, and solving n independent simultaneous equations of the general form of equation 3.11 or 3.12.

Example

EXAMPLE 3.3

A sample was analyzed for the concentration of two analytes, A and B, under two sets of conditions. Under condition 1, the calibration sensitivities are

$$k_{A,1} = 76 \text{ ppm}^{-1} \qquad k_{B,1} = 186 \text{ ppm}^{-1}$$

and for condition 2

$$k_{A,2} = 33 \text{ ppm}^{-1} \qquad k_{B,2} = 243 \text{ ppm}^{-1}$$

The signals under the two sets of conditions are

$$S_{meas,1} = 33.4 \qquad S_{meas,2} = 29.7$$

Determine the concentration of A and B. You may assume that S_{reag} is zero under both conditions.

SOLUTION

Using equation 3.12, we write the following simultaneous equations

$$33.4 = (76 \text{ ppm}^{-1})C_A + (186 \text{ ppm}^{-1})C_B$$

$$29.7 = (33 \text{ ppm}^{-1})C_A + (243 \text{ ppm}^{-1})C_B$$

Multiplying the first equation by the ratio 33/76 gives the two equations as

$$14.5 = (33 \text{ ppm}^{-1})C_A + (80.8 \text{ ppm}^{-1})C_B$$

$$29.7 = (33 \text{ ppm}^{-1})C_A + (243 \text{ ppm}^{-1})C_B$$

Subtracting the first equation from the second gives

$$15.2 = (162.2 \text{ ppm}^{-1})C_B$$

Solving for C_B gives the concentration of B as 0.094 ppm. Substituting this concentration back into either of the two original equations gives the concentration of A, C_A, as 0.21 ppm.

3E.2 Calibration and Standardization

Analytical chemists make a distinction between calibration and standardization.[9] **Calibration** ensures that the equipment or instrument used to measure the signal is operating correctly by using a standard known to produce an exact signal. Balances, for example, are calibrated using a standard weight whose mass can be traced to the internationally accepted platinum–iridium prototype kilogram.

calibration
The process of ensuring that the signal measured by a piece of equipment or an instrument is correct.

Standardization is the process of experimentally determining the relationship between the signal and the amount of analyte (the value of k in equations 3.1 and 3.2). For a total analysis method, standardization is usually defined by the stoichiometry of the chemical reactions responsible for the signal. For a concentration method, however, the relationship between the signal and the analyte's concentration is a theoretical function that cannot be calculated without experimental measurements. To standardize a method, the value of k is determined by measuring the signal for one or more standards, each containing a known concentration of analyte. When several standards with different concentrations of analyte are used, the result is best viewed visually by plotting S_{meas} versus the concentration of analyte in the standards. Such a plot is known as a **calibration curve.** A more detailed discussion of calibration and standardization is found in Chapter 5.

standardization
The process of establishing the relationship between the amount of analtye and a method's signal.

calibration curve
The result of a standardization showing graphically how a method's signal changes with respect to the amount of analyte.

3E.3 Sampling

Selecting an appropriate method helps ensure that an analysis is accurate. It does not guarantee, however, that the result of the analysis will be sufficient to solve the problem under investigation or that a proposed answer will be correct. These latter concerns are addressed by carefully collecting the samples to be analyzed.

A proper sampling strategy ensures that samples are representative of the material from which they are taken. Biased or nonrepresentative sampling and contamination of samples during or after their collection are two sources of sampling error that can lead to significant errors. It is important to realize that sampling errors are completely independent of analysis errors. As a result, sampling errors cannot be corrected by evaluating a reagent blank. A more detailed discussion of sampling is found in Chapter 7.

3E.4 Validation

Before a procedure can provide useful analytical information, it is necessary to demonstrate that it is capable of providing acceptable results. **Validation** is an evaluation of whether the precision and accuracy obtained by following the procedure are appropriate for the problem. In addition, validation ensures that the written procedure has sufficient detail so that different analysts or laboratories following the same procedure obtain comparable results. Ideally, validation uses a standard sample whose composition closely matches the samples for which the procedure was developed. The comparison of replicate analyses can be used to evaluate the procedure's precision and accuracy. Intralaboratory and interlaboratory differences in the procedure also can be evaluated. In the absence of appropriate standards, accuracy can be evaluated by comparing results obtained with a new method to those obtained using a method of known accuracy. Chapter 14 provides a more detailed discussion of validation techniques.

validation
The process of verifying that a procedure yields acceptable results.

3F Protocols

Earlier we noted that a protocol is a set of stringent written guidelines, specifying an exact procedure that must be followed if results are to be accepted by the agency specifying the protocol. Besides all the considerations taken into account when designing the procedure, a protocol also contains very explicit instructions regarding internal and external **quality assurance and quality control** (QA/QC) procedures.[10] Internal QA/QC includes steps taken to ensure that the analytical work in a given laboratory is both accurate and precise. External QA/QC usually involves a process in which the laboratory is certified by an external agency.

quality assurance and quality control
Those steps taken to ensure that the work conducted in an analytical lab is capable of producing acceptable results; also known as QA/QC.

As an example, we will briefly outline some of the requirements in the Environmental Protection Agency's Contract Laboratory Program (CLP) protocol for the analysis of trace metals in aqueous samples by graphite furnace atomic absorption spectrophotometry. The CLP protocol (Figure 3.7) calls for daily standardization with a reagent blank and three standards, one of which is at the laboratory's contract required detection limit. The resulting calibration curve is then verified by analyzing initial calibration verification (ICV) and initial calibration blank (ICB) samples. The reported concentration of the ICV sample must fall within ±10% of the expected concentration. If the concentration falls outside this limit, the analysis must be stopped and the problem identified and corrected before continuing.

After a successful analysis of the ICV and ICB samples, standardization is reverified by analyzing a continuing calibration verification (CCV) sample and a continuing calibration blank (CCB). Results for the CCV also must be within ±10% of the expected concentration. Again, if the concentration of the CCV falls outside the established limits, the analysis must be stopped, the problem identified and corrected, and the system standardized as described earlier. The CCV and the CCB are analyzed before the first and after the last sample, and after every set of ten samples. Whenever the CCV or the CCB is unacceptable, the results for the most recent set of ten samples are discarded, the system is standardized, and the samples are reanalyzed. By following this protocol, every result is bound by successful checks on the standardization. Although not shown in Figure 3.7, the CLP also contains detailed instructions regarding the analysis of duplicate or split samples and the use of spike testing for accuracy.

3G The Importance of Analytical Methodology

The importance of analytical methodology is evident when examining the results of environmental monitoring programs. The purpose of a monitoring program is to determine the present status of an environmental system and to assess long-term trends in the quality of the system. These are broad and poorly defined goals. In many cases, such studies are initiated with little thought to the questions the data will be used to answer. This is not surprising since it can be hard to formulate questions in the absence of initial information about the system. Without careful planning, however, a poor experimental design may result in data that has little value.

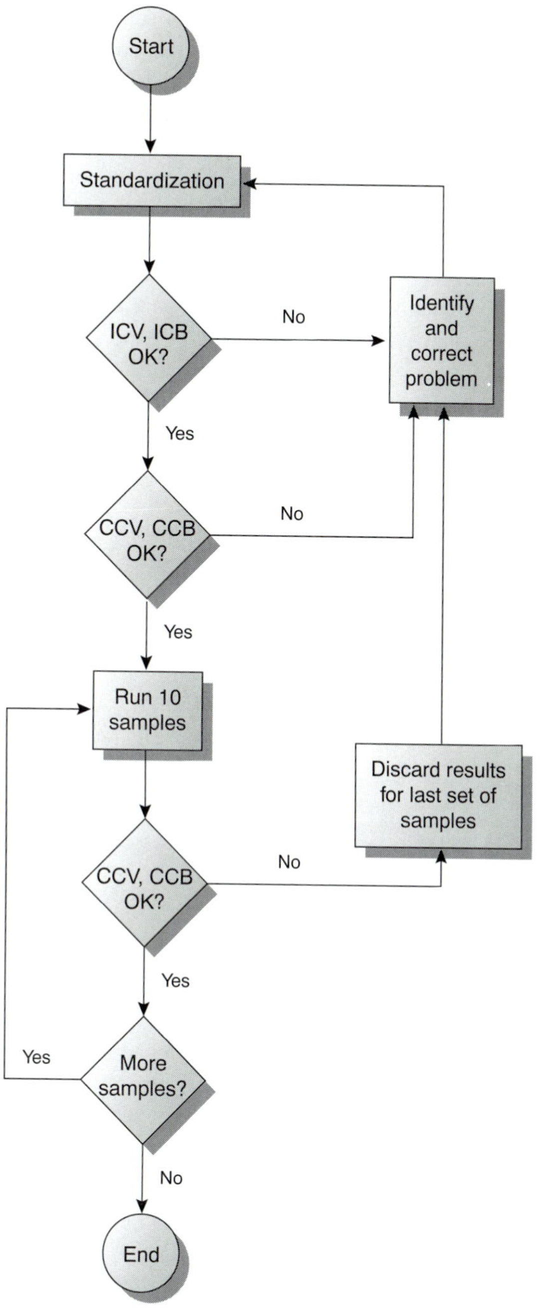

Figure 3.7
Schematic diagram of a portion of the Contract Laboratory Program protocol for the analysis of trace metals by graphite furnace atomic spectrophotometry, as specified by the Environmental Protection Agency. (*Abbreviations:* ICV = initial calibration verification; ICB = initial calibration blank, CCV = continuing calibration verification, CCB = continuing calibration blank)

These concerns are illustrated by the Chesapeake Bay monitoring program. This research program, designed to study nutrients and toxic pollutants in the Chesapeake Bay, was initiated in 1984 as a cooperative venture between the federal government, the state governments of Maryland, Virginia, and Pennsylvania, and the District of Columbia. A 1989 review of some of the problems with this program highlights the difficulties common to many monitoring programs.[11]

At the beginning of the Chesapeake Bay monitoring program, little attention was given to the proper choice of analytical methods, in large part because the intended uses of the monitoring data were not specified. The analytical methods initially chosen were those standard methods already approved by the EPA. In many cases these methods proved to be of little value for this monitoring project. Most of the EPA-approved methods were designed to detect pollutants at their legally mandated maximum allowed concentrations. The concentrations of these contaminants in natural waters, however, are often well below the detection limit of the EPA methods. For example, the EPA-approved standard method for phosphate had a detection limit of 7.5 ppb. Since actual phosphate concentrations in Chesapeake Bay usually were below the EPA detection limit, the EPA method provided no useful information. On the other hand, a nonapproved variant of the EPA method commonly used in chemical oceanography had a detection limit of 0.06 ppb. In other cases, such as the elemental analysis for particulate forms of carbon, nitrogen, and phosphorus, EPA-approved procedures provided poorer reproducibility than nonapproved methods.

3H KEY TERMS

accuracy (*p. 38*)
analysis (*p. 36*)
analytes (*p. 36*)
calibration (*p. 47*)
calibration curve (*p. 47*)
concentration techniques (*p. 38*)
detection limit (*p. 39*)
determination (*p. 36*)
matrix (*p. 36*)
measurement (*p. 36*)
method (*p. 36*)
method blank (*p. 45*)
precision (*p. 39*)
procedure (*p. 36*)
protocol (*p. 37*)
quality assurance and quality control (*p. 48*)
robust (*p. 42*)
rugged (*p. 42*)
selectivity (*p. 40*)
selectivity coefficient (*p. 40*)
sensitivity (*p. 39*)
signal (*p. 37*)
standardization (*p. 47*)
technique (*p. 36*)
total analysis techniques (*p. 38*)
validation (*p. 47*)

3I SUMMARY

Every discipline has its own terminology. Your success in studying analytical chemistry will improve if you master the language used by analytical chemists. Be sure that you understand the difference between an analyte and its matrix, a technique and a method, a procedure and a protocol, and a total analysis technique and a concentration technique.

An analytical method is selected on the basis of criteria such as accuracy, precision, sensitivity, selectivity, robustness, ruggedness, the amount of available sample, the amount of analyte in the sample, time, cost, and the availability of equipment. These criteria are not mutually independent, and it often is necessary to find an acceptable balance among them.

In developing a procedure or protocol, consideration is given to compensating for interferences, calibrating equipment and standardizing the method, obtaining an appropriate sample, and validating the analysis. Poorly designed procedures and protocols produce results that are insufficient to meet the needs of the analysis.

3J PROBLEMS

1. When working with a solid sample, it often is necessary to bring the analyte into solution by dissolving the sample in a suitable solvent. Any solid impurities that remain are removed by filtration before continuing with the analysis. In a typical total analysis method, the procedure might read

> After dissolving the sample in a beaker, remove any solid impurities by passing the solution containing the analyte through filter paper, collecting the solution in a clean Erlenmeyer flask. Rinse the beaker with several small portions of solvent, passing these rinsings through the filter paper, and collecting them in the same Erlenmeyer flask. Finally, rinse the filter paper with several portions of solvent, collecting the rinsings in the same Erlenmeyer flask.

For a typical concentration method, however, the procedure might state

> After dissolving the sample in a beaker, remove any solid impurities by filtering a portion of the solution containing the analyte. Collect and discard the first several milliliters of solution before collecting a sample of approximately 5 mL for further analysis.

Explain why these two procedures are different.

2. A certain concentration method works best when the analyte's concentration is approximately 10 ppb.
   a. If the sampling volume for the method is 0.5 mL, about what mass of analyte is being measured?
   b. If the analyte is present at 10% w/v, how would you prepare the sample for analysis?
   c. Repeat for the case in which the analyte is present at 10% w/w.
   d. Based on your results, comment on the suitability of this method for the analysis of a major analyte.

3. An analyst needs to evaluate the potential effect of an interferent, I, on the quantitative analysis for an analyte, A. She begins by measuring the signal for a sample in which the interferent is absent and the analyte is present with a concentration of 15 ppm, obtaining an average signal of 23.3 (arbitrary units). When analyzing a sample in which the analyte is absent and the interferent is present with a concentration of 25 ppm, she obtains an average signal of 13.7.
   a. What is the analyte's sensitivity?
   b. What is the interferent's sensitivity?
   c. What is the value of the selectivity coefficient?
   d. Is the method more selective for the analyte or the interferent?
   e. What is the maximum concentration of interferent relative to that of the analyte (i.e., [interferent]/[analyte]), if the error in the analysis is to be less than 1%?

4. A sample was analyzed to determine the concentration of an analyte. Under the conditions of the analysis, the sensitivity is 17.2 ppm^{-1}. What is the analyte's concentration if S_{meas} is 35.2 and S_{reag} is 0.6?

5. A method for the analysis of Ca^{2+} in water suffers from an interference in the presence of Zn^{2+}. When the concentration of Ca^{2+} is 50 times greater than that of Zn^{2+}, an analysis for Ca^{2+} gives a relative error of –2.0%. What is the value of the selectivity coefficient for this method?

6. The quantitative analysis for reduced glutathione in blood is complicated by the presence of many potential interferents. In one study, when analyzing a solution of 10-ppb glutathione and 1.5-ppb ascorbic acid, the signal was 5.43 times greater than that obtained for the analysis of 10-ppb glutathione.[12] What is the selectivity coefficient for this analysis? The same study found that when analyzing a solution of 350-ppb methionine and 10-ppb glutathione the signal was 0.906 times less than that obtained for the analysis of 10 ppb-glutathione. What is the selectivity coefficient for this analysis? In what way do these interferents behave differently?

7. Oungpipat and Alexander described a new method for determining the concentration of glycolic acid (GA) in a variety of samples, including physiological fluids such as urine.[13] In the presence of only GA, the signal is given as

$$S_{samp,1} = k_{GA}C_{GA}$$

and in the presence of both glycolic acid and ascorbic acid (AA), the signal is

$$S_{samp,2} = k_{GA}C_{GA} + k_{AA}C_{AA}$$

When the concentration of glycolic acid is 1.0×10^{-4} M and the concentration of ascorbic acid is 1.0×10^{-5} M, the ratio of the two signals was found to be

$$\frac{S_{samp,2}}{S_{samp,1}} = 1.44$$

   a. Using the ratio of the two signals, determine the value of the selectivity ratio

$$K_{GA,AA} = \frac{k_{AA}}{k_{GA}}$$

   b. Is the method more selective toward glycolic acid or ascorbic acid?
   c. If the concentration of ascorbic acid is 1.0×10^{-5} M, what is the smallest concentration of glycolic acid that can be determined such that the error introduced by failing to account for the signal from ascorbic acid is less than 1%?

8. Ibrahim and co-workers developed a new method for the quantitative analysis of hypoxanthine, a natural compound of some nucleic acids.[14] As part of their study they evaluated the method's selectivity for hypoxanthine in the presence of several possible interferents, including ascorbic acid.

a. When analyzing a solution of 1.12×10^{-6} M hypoxanthine, the authors obtained a signal of 7.45×10^{-5} amperes (A). What is the sensitivity for hypoxanthine? You may assume that the signal has been corrected for the method blank.

b. When a solution containing 1.12×10^{-6} M hypoxanthine and 6.5×10^{-5} M ascorbic acid was analyzed a signal of 4.04×10^{-5} A was obtained. What is the selectivity coefficient for this method?

c. Is the method more selective for hypoxanthine or for ascorbic acid?

d. What is the largest concentration of ascorbic acid that may be present if a concentration of 1.12×10^{-6} M hypoxanthine is to be determined within ±1%?

9. A sample was analyzed for the concentration of two analytes, C and D, under two sets of conditions. Under condition 1 the calibration sensitivities are

$$k_{C,1} = 23 \text{ ppm}^{-1} \qquad k_{D,1} = 415 \text{ ppm}^{-1}$$

and for condition 2

$$k_{C,2} = 115 \text{ ppm}^{-1} \qquad k_{D,2} = 45 \text{ ppm}^{-1}$$

The signals under the two sets of conditions are

$$S_{meas,1} = 78.6 \qquad S_{meas,2} = 47.9$$

Determine the concentration of C and D. You may assume that S_{reag} is zero under both conditions.

10. Examine a procedure from *Standard Methods for the Analysis of Waters and Wastewaters* (or another manual of standard analytical methods), and identify the steps taken to compensate for interferences, to calibrate equipment and instruments, to standardize the method, and to acquire a representative sample.

3K SUGGESTED READINGS

The following papers provide alternative schemes for classifying analytical methods

Booksh, K. S.; Kowalski, B. R. "Theory of Analytical Chemistry," *Anal. Chem.* **1994,** *66,* 782A–791A.

Phillips, J. B. "Classification of Analytical Methods," *Anal. Chem.* **1981,** *53,* 1463A–1470A.

Valcárcel, M.; Luque de Castro, M. D. "A Hierarchical Approach to Analytical Chemistry," *Trends Anal. Chem.,* **1995,** *14,* 242–250.

Further details on evaluating analytical methods may be found in

Wilson, A. L. "The Performance-Characteristics of Analytical Methods," Part I-*Talanta,* **1970,** *17,* 21–29; Part II-*Talanta,* **1970,** *17,* 31–44; Part III-*Talanta,* **1973,** *20,* 725–732; Part IV-*Talanta,* **1974,** *21,* 1109–1121.

Several texts provide numerous examples of analytical procedures for specific analytes in well-defined matrices.

Basset, J.; Denney, R. C.; Jeffery, G. H.; et al. *Vogel's Textbook of Quantitative Inorganic Analysis,* 4th ed. Longman: London, 1981.

Csuros, M. *Environmental Sampling and Analysis for Technicians,* Lewis: Boca Raton, 1994.

Keith, L. H., ed. *Compilation of EPA's Sampling and Analysis Methods,* Lewis: Boca Raton, 1996.

Rump, H. H.; Krist, H. *Laboratory Methods for the Examination of Water, Wastewater and Soil.* VCH Publishers: New York, 1988.

Standard Methods for the Analysis of Waters and Wastewaters, 19th ed. American Public Health Association: Washington, DC, 1995.

3L REFERENCES

1. Taylor, J. K. *Anal. Chem.* **1983,** *55,* 600A–608A.
2. Fitch, A.; Wang, Y.; Mellican, S.; et al. *Anal. Chem.* **1996,** *68,* 727A–731A.
3. Basset, J.; Denney, R. C.; Jeffery, G. H.; et al. *Vogel's Textbook of Quantitative Inorganic Analysis,* 4th ed. Longman: London, 1981, p. 8.
4. Ingle, J. D.; Crouch, S. R. *Spectrochemical Analysis.* Prentice-Hall: Englewood, NJ, 1988, pp. 171–172.
5. Barnett, N. W.; Bowser, T. A.; Gerardi, R. D.; et al. *Anal. Chim. Acta* **1996,** *318,* 309–317.
6. Rogers, L. B. *J. Chem. Ed.* **1986,** *63,* 3–6.
7. (a) Sandell, E. B.; Elving, P. J. In Kolthoff, I. M.; Elving, P. J., eds. *Treatise on Analytical Chemistry,* Interscience: New York; Part 1, Vol. 1, Chapter 1, pp. 3–6; (b) Potts, L. W. *Quantitative Analysis—Theory and Practice.* Harper and Row: New York, 1987, p. 12.
8. Valcárcel, M.; Ríos, A. *Anal. Chem.* **1993,** *65,* 781A–787A.
9. Valcárcel, M.; Ríos, A. *Analyst,* **1995,** *120,* 2291–2297.
10. (a) Amore, F. *Anal. Chem.* **1979,** *51,* 1105A–1110A; (b) Taylor, J. K. *Anal. Chem.* **1981,** *53,* 1588A–1593A.
11. D'Elia, C. F.; Sanders, J. G.; Capone, D. G. *Environ. Sci. Technol.* **1989,** *23,* 768–774.
12. Jiménez-Prieto, R.; Velasco, A.; Silva, M.; et al. *Anal. Chim. Acta* **1992,** *269,* 273–279.
13. Oungpipat, W.; Alexander, P. W. *Anal. Chim. Acta* **1994,** *295,* 36–46.
14. Ibrahim, M. S.; Ahmad, M. E.; Temerk, Y. M.; et al. *Anal. Chim. Acta* **1996,** *328,* 47–52.

Chapter 4

Evaluating Analytical Data

A problem dictates the requirements we place on our measurements and results. Regulatory agencies, for example, place stringent requirements on the reliability of measurements and results reported to them. This is the rationale for creating a protocol for regulatory problems. Screening the products of an organic synthesis, on the other hand, places fewer demands on the reliability of measurements, allowing chemists to customize their procedures.

When designing and evaluating an analytical method, we usually make three separate considerations of experimental error.[1] First, before beginning an analysis, errors associated with each measurement are evaluated to ensure that their cumulative effect will not limit the utility of the analysis. Errors known or believed to affect the result can then be minimized. Second, during the analysis the measurement process is monitored, ensuring that it remains under control. Finally, at the end of the analysis the quality of the measurements and the result are evaluated and compared with the original design criteria. This chapter is an introduction to the sources and evaluation of errors in analytical measurements, the effect of measurement error on the result of an analysis, and the statistical analysis of data.

4A Characterizing Measurements and Results

Let's begin by choosing a simple quantitative problem requiring a single measurement. The question to be answered is—What is the mass of a penny? If you think about how we might answer this question experimentally, you will realize that this problem is too broad. Are we interested in the mass of United State pennies or Canadian pennies, or is the difference in country of importance? Since the composition of a penny probably differs from country to country, let's limit our problem to pennies minted in the United States. There are other considerations. Pennies are minted at several locations in the United States (this is the meaning of the letter, or absence of a letter, below the date stamped on the lower right corner of the face of the coin). Since there is no reason to expect a difference between where the penny was minted, we will choose to ignore this consideration. Is there a reason to expect a difference between a newly minted penny not yet in circulation, and a penny that has been in circulation? The answer to this is not obvious. Let's simplify the problem by narrowing the question to—What is the mass of an average United States penny in circulation? This is a problem that we might expect to be able to answer experimentally.

A good way to begin the analysis is to acquire some preliminary data. Table 4.1 shows experimentally measured masses for seven pennies from my change jar at home. Looking at these data, it is immediately apparent that our question has no simple answer. That is, we cannot use the mass of a single penny to draw a specific conclusion about the mass of any other penny (although we might conclude that all pennies weigh at least 3 g). We can, however, characterize these data by providing a measure of the spread of the individual measurements around a central value.

4A.1 Measures of Central Tendency

One way to characterize the data in Table 4.1 is to assume that the masses of individual pennies are scattered around a central value that provides the best estimate of a penny's true mass. Two common ways to report this estimate of central tendency are the mean and the median.

mean
The average value of a set of data ($\overline{X}$).

Mean The **mean,** $\overline{X}$, is the numerical average obtained by dividing the sum of the individual measurements by the number of measurements

$$\overline{X} = \frac{\sum_{i=1}^{n} X_i}{n}$$

where X_i is the ith measurement, and n is the number of independent measurements.

Table 4.1 Masses of Seven United States Pennies in Circulation

Penny	Mass (g)
1	3.080
2	3.094
3	3.107
4	3.056
5	3.112
6	3.174
7	3.198

EXAMPLE 4.1

What is the mean for the data in Table 4.1?

SOLUTION

To calculate the mean, we add the results for all measurements

$$3.080 + 3.094 + 3.107 + 3.056 + 3.112 + 3.174 + 3.198 = 21.821$$

and divide by the number of measurements

$$\overline{X} = \frac{21.821}{7} = 3.117 \text{ g}$$

The mean is the most common estimator of central tendency. It is not considered a robust estimator, however, because extreme measurements, those much larger or smaller than the remainder of the data, strongly influence the mean's value.[2] For example, mistakenly recording the mass of the fourth penny as 31.07 g instead of 3.107 g, changes the mean from 3.117 g to 7.112 g!

Median The **median,** X_{med}, is the middle value when data are ordered from the smallest to the largest value. When the data include an odd number of measurements, the median is the middle value. For an even number of measurements, the median is the average of the $n/2$ and the $(n/2) + 1$ measurements, where n is the number of measurements.

median
That value for a set of ordered data, for which half of the data is larger in value and half is smaller in value ($\overline{X}_{\text{med}}$).

EXAMPLE 4.2

What is the median for the data in Table 4.1?

SOLUTION

To determine the median, we order the data from the smallest to the largest value

3.056 3.080 3.094 3.107 3.112 3.174 3.198

Since there is a total of seven measurements, the median is the fourth value in the ordered data set; thus, the median is 3.107.

As shown by Examples 4.1 and 4.2, the mean and median provide similar estimates of central tendency when all data are similar in magnitude. The median, however, provides a more robust estimate of central tendency since it is less sensitive to measurements with extreme values. For example, introducing the transcription error discussed earlier for the mean only changes the median's value from 3.107 g to 3.112 g.

4A.2 Measures of Spread

If the mean or median provides an estimate of a penny's true mass, then the spread of the individual measurements must provide an estimate of the variability in the masses of individual pennies. Although spread is often defined relative to a specific measure of central tendency, its magnitude is independent of the central value. Changing all measurements in the same direction, by adding or subtracting a constant value, changes the mean or median, but will not change the magnitude of the spread. Three common measures of spread are range, standard deviation, and variance.

range
The numerical difference between the largest and smallest values in a data set (w).

Range The **range,** w, is the difference between the largest and smallest values in the data set.

$$\text{Range} = w = X_{\text{largest}} - X_{\text{smallest}}$$

The range provides information about the total variability in the data set, but does not provide any information about the distribution of individual measurements. The range for the data in Table 4.1 is the difference between 3.198 g and 3.056 g; thus

$$w = 3.198 \text{ g} - 3.056 \text{ g} = 0.142 \text{ g}$$

standard deviation
A statistical measure of the "average" deviation of data from the data's mean value (s).

Standard Deviation The absolute **standard deviation,** s, describes the spread of individual measurements about the mean and is given as

$$s = \sqrt{\frac{\sum_{i=1}^{n} (X_i - \overline{X})^2}{n-1}} \qquad \textbf{4.1}$$

where X_i is one of n individual measurements, and $\overline{X}$ is the mean. Frequently, the relative standard deviation, s_r, is reported.

$$s_r = \frac{s}{\overline{X}}$$

The percent relative standard deviation is obtained by multiplying s_r by 100%.

Example

EXAMPLE 4.3

What are the standard deviation, the relative standard deviation, and the percent relative standard deviation for the data in Table 4.1?

SOLUTION

To calculate the standard deviation, we obtain the difference between the mean value (3.117; see Example 4.1) and each measurement, square the resulting differences, and add them to determine the sum of the squares (the numerator of equation 4.1)

$$
\begin{aligned}
(3.080 - 3.117)^2 &= (-0.037)^2 = 0.00137 \\
(3.094 - 3.117)^2 &= (-0.023)^2 = 0.00053 \\
(3.107 - 3.117)^2 &= (-0.010)^2 = 0.00010 \\
(3.056 - 3.117)^2 &= (-0.061)^2 = 0.00372 \\
(3.112 - 3.117)^2 &= (-0.005)^2 = 0.00003 \\
(3.174 - 3.117)^2 &= (+0.057)^2 = 0.00325 \\
(3.198 - 3.117)^2 &= (+0.081)^2 = \underline{0.00656} \\
& \qquad\qquad\qquad\; 0.01556
\end{aligned}
$$

The standard deviation is calculated by dividing the sum of the squares by $n - 1$, where n is the number of measurements, and taking the square root.

$$s = \sqrt{\frac{0.01556}{7-1}} = 0.051$$

The relative standard deviation and percent relative standard deviation are

$$s_r = \frac{0.051}{3.117} = 0.016$$

$$s_r\ (\%) = 0.016 \times 100\% = 1.6\%$$

It is much easier to determine the standard deviation using a scientific calculator with built-in statistical functions.*

Variance Another common measure of spread is the square of the standard deviation, or the **variance.** The standard deviation, rather than the variance, is usually reported because the units for standard deviation are the same as that for the mean value.

variance
The square of the standard deviation (s^2).

Example

EXAMPLE 4.4

What is the variance for the data in Table 4.1?

SOLUTION

The variance is just the square of the absolute standard deviation. Using the standard deviation found in Example 4.3 gives the variance as

$$\text{Variance} = s^2 = (0.051)^2 = 0.0026$$

4B Characterizing Experimental Errors

Realizing that our data for the mass of a penny can be characterized by a measure of central tendency and a measure of spread suggests two questions. First, does our measure of central tendency agree with the true, or expected value? Second, why are our data scattered around the central value? Errors associated with central tendency reflect the accuracy of the analysis, but the precision of the analysis is determined by those errors associated with the spread.

4B.1 Accuracy

Accuracy is a measure of how close a measure of central tendency is to the true, or expected value, μ.† Accuracy is usually expressed as either an absolute error

$$E = \overline{X} - \mu \qquad 4.2$$

or a percent relative error, E_r.

$$E_r = \frac{\overline{X} - \mu}{\mu} \times 100 \qquad 4.3$$

*Many scientific calculators include two keys for calculating the standard deviation, only one of which corresponds to equation 4.3. Your calculator's manual will help you determine the appropriate key to use.

†The standard convention for representing experimental parameters is to use a Roman letter for a value calculated from experimental data, and a Greek letter for the corresponding true value. For example, the experimentally determined mean is $\overline{X}$, and its underlying true value is μ. Likewise, the standard deviation by experiment is given the symbol s, and its underlying true value is identified as σ.

Although the mean is used as the measure of central tendency in equations 4.2 and 4.3, the median could also be used.

determinate error
Any systematic error that causes a measurement or result to always be too high or too small; can be traced to an identifiable source.

Errors affecting the accuracy of an analysis are called determinate and are characterized by a systematic deviation from the true value; that is, all the individual measurements are either too large or too small. A positive **determinate error** results in a central value that is larger than the true value, and a negative determinate error leads to a central value that is smaller than the true value. Both positive and negative determinate errors may affect the result of an analysis, with their cumulative effect leading to a net positive or negative determinate error. It is possible, although not likely, that positive and negative determinate errors may be equal, resulting in a central value with no net determinate error.

Determinate errors may be divided into four categories: sampling errors, method errors, measurement errors, and personal errors.

sampling error
An error introduced during the process of collecting a sample for analysis.

heterogeneous
Not uniform in composition.

Sampling Errors We introduce determinate **sampling errors** when our sampling strategy fails to provide a representative sample. This is especially important when sampling **heterogeneous** materials. For example, determining the environmental quality of a lake by sampling a single location near a point source of pollution, such as an outlet for industrial effluent, gives misleading results. In determining the mass of a U.S. penny, the strategy for selecting pennies must ensure that pennies from other countries are not inadvertently included in the sample. Determinate errors associated with selecting a sample can be minimized with a proper sampling strategy, a topic that is considered in more detail in Chapter 7.

method error
An error due to limitations in the analytical method used to analyze a sample.

Method Errors Determinate **method errors** are introduced when assumptions about the relationship between the signal and the analyte are invalid. In terms of the general relationships between the measured signal and the amount of analyte

$$S_{meas} = kn_A + S_{reag} \qquad \text{(total analysis method)} \tag{4.4}$$

$$S_{meas} = kC_A + S_{reag} \qquad \text{(concentration method)} \tag{4.5}$$

method errors exist when the sensitivity, k, and the signal due to the reagent blank, S_{reag}, are incorrectly determined. For example, methods in which S_{meas} is the mass of a precipitate containing the analyte (gravimetric method) assume that the sensitivity is defined by a pure precipitate of known stoichiometry. When this assumption fails, a determinate error will exist. Method errors involving sensitivity are minimized by standardizing the method, whereas method errors due to interferents present in reagents are minimized by using a proper reagent blank. Both are discussed in more detail in Chapter 5. Method errors due to interferents in the sample cannot be minimized by a reagent blank. Instead, such interferents must be separated from the analyte or their concentrations determined independently.

measurement error
An error due to limitations in the equipment and instruments used to make measurements.

tolerance
The maximum determinate measurement error for equipment or instrument as reported by the manufacturer.

Measurement Errors Analytical instruments and equipment, such as glassware and balances, are usually supplied by the manufacturer with a statement of the item's maximum **measurement error,** or **tolerance.** For example, a 25-mL volumetric flask might have a maximum error of ±0.03 mL, meaning that the actual volume contained by the flask lies within the range of 24.97–25.03 mL. Although expressed as a range, the error is determinate; thus, the flask's true volume is a fixed value within the stated range. A summary of typical measurement errors for a variety of analytical equipment is given in Tables 4.2–4.4.

Table 4.2 Measurement Errors for Selected Glassware[a]

Glassware	Volume (mL)	Measurement Errors for Class A Glassware (±mL)	Measurement Errors for Class B Glassware (±mL)
Transfer Pipets	1	0.006	0.012
	2	0.006	0.012
	5	0.01	0.02
	10	0.02	0.04
	20	0.03	0.06
	25	0.03	0.06
	50	0.05	0.10
Volumetric Flasks	5	0.02	0.04
	10	0.02	0.04
	25	0.03	0.06
	50	0.05	0.10
	100	0.08	0.16
	250	0.12	0.24
	500	0.20	0.40
	1000	0.30	0.60
	2000	0.50	1.0
Burets	10	0.02	0.04
	25	0.03	0.06
	50	0.05	0.10

[a]Specifications for class A and class B glassware are taken from American Society for Testing and Materials E288, E542 and E694 standards.

Table 4.3 Measurement Errors for Selected Balances

Balance	Capacity (g)	Measurement Error
Precisa 160M	160	±1 mg
A & D ER 120M	120	±0.1 mg
Metler H54	160	±0.01 mg

Table 4.4 Measurement Errors for Selected Digital Pipets

Pipet Range	Volume (mL or μL)[a]	Measurement Error (±%)
10–100 μL[b]	10	1.0
	50	0.6
	100	0.6
200–1000 μL[c]	200	1.5
	1000	0.8
1–10 mL[d]	1	0.6
	5	0.4
	10	0.3

[a]Units for volume same as for pipet range.

[b]Data for Eppendorf Digital Pipet 4710.

[c]Data for Oxford Benchmate.

[d]Data for Eppendorf Maxipetter 4720 with Maxitip P.

Volumetric glassware is categorized by class. Class A glassware is manufactured to comply with tolerances specified by agencies such as the National Institute of Standards and Technology. Tolerance levels for class A glassware are small enough that such glassware normally can be used without calibration. The tolerance levels for class B glassware are usually twice those for class A glassware. Other types of volumetric glassware, such as beakers and graduated cylinders, are unsuitable for accurately measuring volumes.

Determinate measurement errors can be minimized by calibration. A pipet can be calibrated, for example, by determining the mass of water that it delivers and using the density of water to calculate the actual volume delivered by the pipet. Although glassware and instrumentation can be calibrated, it is never safe to assume that the calibration will remain unchanged during an analysis. Many instruments, in particular, drift out of calibration over time. This complication can be minimized by frequent recalibration.

personal error
An error due to biases introduced by the analyst.

Personal Errors Finally, analytical work is always subject to a variety of **personal errors,** which can include the ability to see a change in the color of an indicator used to signal the end point of a titration; biases, such as consistently overestimating or underestimating the value on an instrument's readout scale; failing to calibrate glassware and instrumentation; and misinterpreting procedural directions. Personal errors can be minimized with proper care.

Identifying Determinate Errors Determinate errors can be difficult to detect. Without knowing the true value for an analysis, the usual situation in any analysis with meaning, there is no accepted value with which the experimental result can be compared. Nevertheless, a few strategies can be used to discover the presence of a determinate error.

constant determinate error
A determinate error whose value is the same for all samples.

Some determinate errors can be detected experimentally by analyzing several samples of different size. The magnitude of a **constant determinate error** is the same for all samples and, therefore, is more significant when analyzing smaller samples. The presence of a constant determinate error can be detected by running several analyses using different amounts of sample, and looking for a systematic change in the property being measured. For example, consider a quantitative analysis in which we separate the analyte from its matrix and determine the analyte's mass. Let's assume that the sample is 50.0% w/w analyte; thus, if we analyze a 0.100-g sample, the analyte's true mass is 0.050 g. The first two columns of Table 4.5 give the true mass of analyte for several additional samples. If the analysis has a positive constant determinate error of 0.010 g, then the experimentally determined mass for

Table 4.5 Effect of Constant Positive Determinate Error on Analysis of Sample Containing 50% Analyte (%w/w)

Mass Sample (g)	True Mass of Analyte (g)	Constant Error (g)	Mass of Analyte Determined (g)	Percent Analyte Reported (%w/w)
0.100	0.050	0.010	0.060	60.0
0.200	0.100	0.010	0.110	55.0
0.400	0.200	0.010	0.210	52.5
0.800	0.400	0.010	0.410	51.2
1.000	0.500	0.010	0.510	51.0

any sample will always be 0.010 g, larger than its true mass (column four of Table 4.5). The analyte's reported weight percent, which is shown in the last column of Table 4.5, becomes larger when we analyze smaller samples. A graph of % w/w analyte versus amount of sample shows a distinct upward trend for small amounts of sample (Figure 4.1). A smaller concentration of analyte is obtained when analyzing smaller samples in the presence of a constant negative determinate error.

A **proportional determinate error,** in which the error's magnitude depends on the amount of sample, is more difficult to detect since the result of an analysis is independent of the amount of sample. Table 4.6 outlines an example showing the effect of a positive proportional error of 1.0% on the analysis of a sample that is 50.0% w/w in analyte. In terms of equations 4.4 and 4.5, the reagent blank, S_{reag}, is an example of a constant determinate error, and the sensitivity, *k*, may be affected by proportional errors.

proportional determinate error
A determinate error whose value depends on the amount of sample analyzed.

Potential determinate errors also can be identified by analyzing a standard sample containing a known amount of analyte in a matrix similar to that of the samples being analyzed. Standard samples are available from a variety of sources, such as the National Institute of Standards and Technology (where they are called **standard reference materials**) or the American Society for Testing and Materials. For example, Figure 4.2 shows an analysis sheet for a typical reference material. Alternatively, the sample can be analyzed by an independent method known to give accurate results, and the results of the two methods can be compared. Once identified, the source of a determinate error can be corrected. The best prevention against errors affecting accuracy, however, is a well-designed procedure that identifies likely sources of determinate errors, coupled with careful laboratory work.

standard reference material
A material available from the National Institute of Standards and Technology certified to contain known concentrations of analytes.

The data in Table 4.1 were obtained using a calibrated balance, certified by the manufacturer to have a tolerance of less than ±0.002 g. Suppose the Treasury Department reports that the mass of a 1998 U.S. penny is approximately 2.5 g. Since the mass of every penny in Table 4.1 exceeds the reported mass by an amount significantly greater than the balance's tolerance, we can safely conclude that the error in this analysis is not due to equipment error. The actual source of the error is revealed later in this chapter.

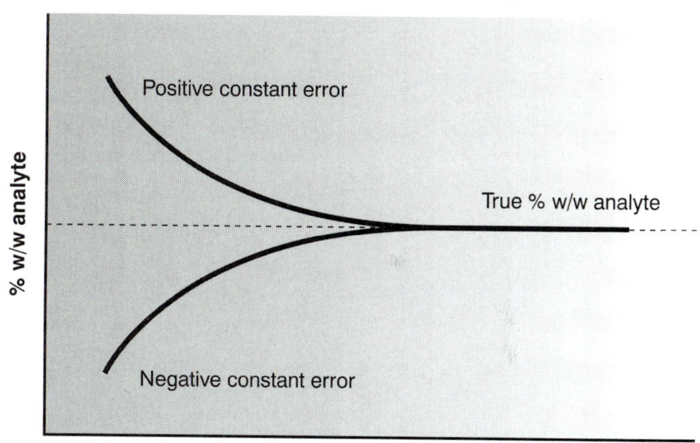

Figure 4.1
Effect of a constant determinate error on the reported concentration of analyte.

Table 4.6 Effect of Proportional Positive Determinate Error on Analysis of Sample Containing 50% Analyte (%w/w)

Mass Sample (g)	True Mass of Analyte (g)	Proportional Error (%)	Mass of Analyte Determined (g)	Percent Analyte Reported (%w/w)
0.200	0.100	1.00	0.101	50.5
0.400	0.200	1.00	0.202	50.5
0.600	0.300	1.00	0.303	50.5
0.800	0.400	1.00	0.404	50.5
1.000	0.500	1.00	0.505	50.5

Simulated Rainwater (liquid form)

This SRM was developed to aid in the analysis of acidic rainwater by providing a stable, homogeneous material at two levels of acidity.

SRM	Type	Unit of issue	
2694a	Simulated rainwater	Set of 4: 2 of 50 mL at each of 2 levels	
	Constituent element parameter	**2694a-I**	**2694a-II**
	pH, 25°C	4.30	3.60
	Electrolytic Conductivity (S/cm, 25°C)	25.4	129.3
	Acidity, meq/L	0.0544	0.283
	Fluoride, mg/L	0.057	0.108
	Chloride, mg/L	(0.23)*	(0.94)*
	Nitrate, mg/L	(0.53)*	7.19
	Sulfate, mg/L	2.69	10.6
	Sodium, mg/L	0.208	0.423
	Potassium, mg/L	0.056	0.108
	Ammonium, mg/L	(0.12)*	(1.06)*
	Calcium, mg/L	0.0126	0.0364
	Magnesium, mg/L	0.0242	0.0484

* Values in parentheses are not certified and are given for information only.

Figure 4.2
Analysis sheet for Simulated Rainwater (SRM 2694a). Adapted from NIST Special Publication 260: *Standard Reference Materials Catalog 1995–96,* p. 64; U.S. Department of Commerce, Technology Administration, National Institute of Standards and Technology.

4B.2 Precision

Precision is a measure of the spread of data about a central value and may be expressed as the range, the standard deviation, or the variance. Precision is commonly divided into two categories: repeatability and reproducibility. **Repeatability** is the precision obtained when all measurements are made by the same analyst during a single period of laboratory work, using the same solutions and equipment. **Reproducibility,** on the other hand, is the precision obtained under any other set of conditions, including that between analysts, or between laboratory sessions for a single analyst. Since reproducibility includes additional sources of variability, the reproducibility of an analysis can be no better than its repeatability.

repeatability
The precision for an analysis in which the only source of variability is the analysis of replicate samples.

reproducibility
The precision when comparing results for several samples, for several analysts or several methods.

indeterminate error
Any random error that causes some measurements or results to be too high while others are too low.

Errors affecting the distribution of measurements around a central value are called indeterminate and are characterized by a random variation in both magnitude and direction. **Indeterminate errors** need not affect the accuracy of an analysis. Since indeterminate errors are randomly scattered around a central value, positive and negative errors tend to cancel, provided that enough measurements are made. In such situations the mean or median is largely unaffected by the precision of the analysis.

Sources of Indeterminate Error Indeterminate errors can be traced to several sources, including the collection of samples, the manipulation of samples during the analysis, and the making of measurements.

When collecting a sample, for instance, only a small portion of the available material is taken, increasing the likelihood that small-scale inhomogeneities in the sample will affect the repeatability of the analysis. Individual pennies, for example, are expected to show variation from several sources, including the manufacturing process, and the loss of small amounts of metal or the addition of dirt during circulation. These variations are sources of indeterminate error associated with the sampling process.

During the analysis numerous opportunities arise for random variations in the way individual samples are treated. In determining the mass of a penny, for example, each penny should be handled in the same manner. Cleaning some pennies but not cleaning others introduces an indeterminate error.

Finally, any measuring device is subject to an indeterminate error in reading its scale, with the last digit always being an estimate subject to random fluctuations, or background noise. For example, a buret with scale divisions every 0.1 mL has an inherent indeterminate error of ±0.01 – 0.03 mL when estimating the volume to the hundredth of a milliliter (Figure 4.3). Background noise in an electrical meter (Figure 4.4) can be evaluated by recording the signal without analyte and observing the fluctuations in the signal over time.

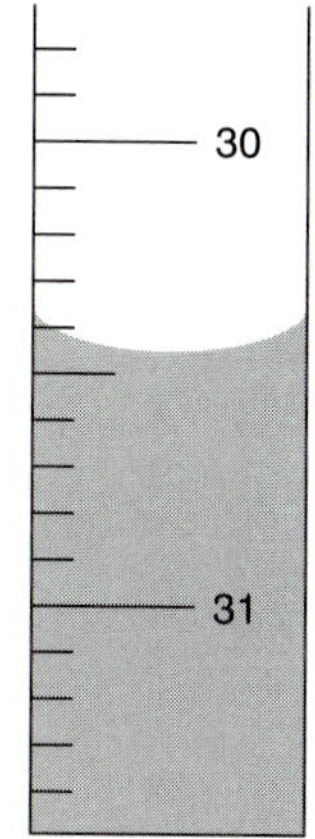

Figure 4.3
Close-up of buret, showing difficulty in estimating volume. With scale divisions every 0.1 mL it is difficult to read the actual volume to better than ±0.01 – 0.03 mL.

Evaluating Indeterminate Error Although it is impossible to eliminate indeterminate error, its effect can be minimized if the sources and relative magnitudes of the indeterminate error are known. Indeterminate errors may be estimated by an appropriate measure of spread. Typically, a standard deviation is used, although in some cases estimated values are used. The contribution from analytical instruments and equipment are easily measured or estimated. Indeterminate errors introduced by the analyst, such as inconsistencies in the treatment of individual samples, are more difficult to estimate.

To evaluate the effect of indeterminate error on the data in Table 4.1, ten replicate determinations of the mass of a single penny were made, with results shown in Table 4.7. The standard deviation for the data in Table 4.1 is 0.051, and it is 0.0024 for the data in Table 4.7. The significantly better precision when determining the mass of a single penny suggests that the precision of this analysis is not limited by the balance used to measure mass, but is due to a significant variability in the masses of individual pennies.

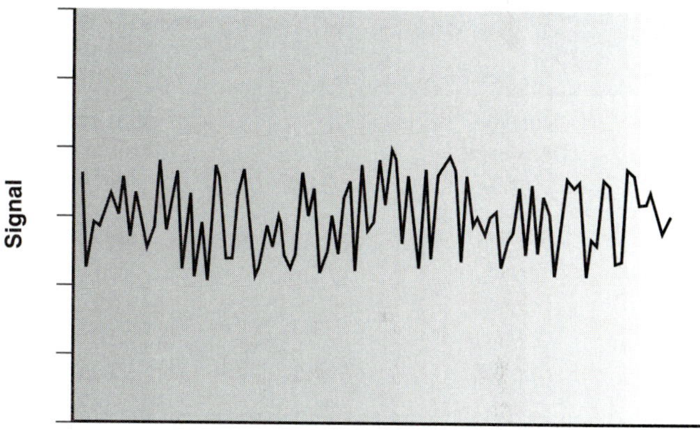

Figure 4.4
Background noise in a meter obtained by measuring signal over time in the absence of analyte.

Table 4.7 Replicate Determinations of the Mass of a Single United States Penny in Circulation

Replicate Number	Mass (g)
1	3.025
2	3.024
3	3.028
4	3.027
5	3.028
6	3.023
7	3.022
8	3.021
9	3.026
10	3.024

4B.3 Error and Uncertainty

error
A measure of bias in a result or measurement.

Analytical chemists make a distinction between error and uncertainty.[3] **Error** is the difference between a single measurement or result and its true value. In other words, error is a measure of bias. As discussed earlier, error can be divided into determinate and indeterminate sources. Although we can correct for determinate error, the indeterminate portion of the error remains. Statistical significance testing, which is discussed later in this chapter, provides a way to determine whether a bias resulting from determinate error might be present.

uncertainty
The range of possible values for a measurement.

Uncertainty expresses the range of possible values that a measurement or result might reasonably be expected to have. Note that this definition of uncertainty is not the same as that for precision. The precision of an analysis, whether reported as a range or a standard deviation, is calculated from experimental data and provides an estimation of indeterminate error affecting measurements. Uncertainty accounts for all errors, both determinate and indeterminate, that might affect our result. Although we always try to correct determinate errors, the correction itself is subject to random effects or indeterminate errors.

Table 4.8 Experimentally Determined Volumes Delivered by a 10-mL Class A Pipet

Trial	Volume Delivered (mL)	Trial	Volume Delivered (mL)
1	10.002	6	9.983
2	9.993	7	9.991
3	9.984	8	9.990
4	9.996	9	9.988
5	9.989	10	9.999

To illustrate the difference between precision and uncertainty, consider the use of a class A 10-mL pipet for delivering solutions. A pipet's uncertainty is the range of volumes in which its true volume is expected to lie. Suppose you purchase a 10-mL class A pipet from a laboratory supply company and use it without calibration. The pipet's tolerance value of ±0.02 mL (see Table 4.2) represents your uncertainty since your best estimate of its volume is 10.00 mL ±0.02 mL. Precision is determined experimentally by using the pipet several times, measuring the volume of solution delivered each time. Table 4.8 shows results for ten such trials that have a mean of 9.992 mL and a standard deviation of 0.006. This standard deviation represents the precision with which we expect to be able to deliver a given solution using any class A 10-mL pipet. In this case the uncertainty in using a pipet is worse than its precision. Interestingly, the data in Table 4.8 allow us to calibrate this specific pipet's delivery volume as 9.992 mL. If we use this volume as a better estimate of this pipet's true volume, then the uncertainty is ±0.006. As expected, calibrating the pipet allows us to lower its uncertainty.

4C Propagation of Uncertainty

Suppose that you need to add a reagent to a flask by several successive transfers using a class A 10-mL pipet. By calibrating the pipet (see Table 4.8), you know that it delivers a volume of 9.992 mL with a standard deviation of 0.006 mL. Since the pipet is calibrated, we can use the standard deviation as a measure of uncertainty. This uncertainty tells us that when we use the pipet to repetitively deliver 10 mL of solution, the volumes actually delivered are randomly scattered around the mean of 9.992 mL.

If the uncertainty in using the pipet once is 9.992 ± 0.006 mL, what is the uncertainty when the pipet is used twice? As a first guess, we might simply add the uncertainties for each delivery; thus

$$(9.992\text{ mL} + 9.992\text{ mL}) \pm (0.006\text{ mL} + 0.006\text{ mL}) = 19.984 \pm 0.012\text{ mL}$$

It is easy to see that combining uncertainties in this way overestimates the total uncertainty. Adding the uncertainty for the first delivery to that of the second delivery assumes that both volumes are either greater than 9.992 mL or less than 9.992 mL. At the other extreme, we might assume that the two deliveries will always be on opposite sides of the pipet's mean volume. In this case we subtract the uncertainties for the two deliveries,

$$(9.992 \text{ mL} + 9.992 \text{ mL}) \pm (0.006 \text{ mL} - 0.006 \text{ mL}) = 19.984 \pm 0.000 \text{ mL}$$

underestimating the total uncertainty.

So what is the total uncertainty when using this pipet to deliver two successive volumes of solution? From the previous discussion we know that the total uncertainty is greater than ±0.000 mL and less than ±0.012 mL. To estimate the cumulative effect of multiple uncertainties, we use a mathematical technique known as the propagation of uncertainty. Our treatment of the propagation of uncertainty is based on a few simple rules that we will not derive. A more thorough treatment can be found elsewhere.[4]

4C.1 A Few Symbols

Propagation of uncertainty allows us to estimate the uncertainty in a calculated result from the uncertainties of the measurements used to calculate the result. In the equations presented in this section the result is represented by the symbol R and the measurements by the symbols A, B, and C. The corresponding uncertainties are s_R, s_A, s_B, and s_C. The uncertainties for A, B, and C can be reported in several ways, including calculated standard deviations or estimated ranges, as long as the same form is used for all measurements.

4C.2 Uncertainty When Adding or Subtracting

When measurements are added or subtracted, the absolute uncertainty in the result is the square root of the sum of the squares of the absolute uncertainties for the individual measurements. Thus, for the equations $R = A + B + C$ or $R = A + B - C$, or any other combination of adding and subtracting A, B, and C, the absolute uncertainty in R is

$$s_R = \sqrt{s_A^2 + s_B^2 + s_C^2} \qquad 4.6$$

Example

EXAMPLE 4.5

The class A 10-mL pipet characterized in Table 4.8 is used to deliver two successive volumes. Calculate the absolute and relative uncertainties for the total delivered volume.

SOLUTION

The total delivered volume is obtained by adding the volumes of each delivery; thus

$$V_{\text{tot}} = 9.992 \text{ mL} + 9.992 \text{ mL} = 19.984 \text{ mL}$$

Using the standard deviation as an estimate of uncertainty, the uncertainty in the total delivered volume is

$$s_R = \sqrt{(0.006)^2 + (0.006)^2} = 0.0085$$

Thus, we report the volume and its absolute uncertainty as 19.984 ± 0.008 mL. The relative uncertainty in the total delivered volume is

$$\frac{0.0085}{19.984} \times 100 = 0.043\%$$

4C.3 Uncertainty When Multiplying or Dividing

When measurements are multiplied or divided, the relative uncertainty in the result is the square root of the sum of the squares of the relative uncertainties for the individual measurements. Thus, for the equations $R = A \times B \times C$ or $R = A \times B/C$, or any other combination of multiplying and dividing A, B, and C, the relative uncertainty in R is

$$\frac{s_R}{R} = \sqrt{\left(\frac{s_A}{A}\right)^2 + \left(\frac{s_B}{B}\right)^2 + \left(\frac{s_C}{C}\right)^2} \qquad 4.7$$

Example

EXAMPLE 4.6

The quantity of charge, Q, in coulombs passing through an electrical circuit is

$$Q = I \times t$$

where I is the current in amperes and t is the time in seconds. When a current of 0.15 ± 0.01 A passes through the circuit for 120 ± 1 s, the total charge is

$$Q = (0.15 \text{ A}) \times (120 \text{ s}) = 18 \text{ C}$$

Calculate the absolute and relative uncertainties for the total charge.

SOLUTION

Since charge is the product of current and time, its relative uncertainty is

$$\frac{s_R}{R} = \sqrt{\left(\frac{0.01}{0.15}\right)^2 + \left(\frac{1}{120}\right)^2} = \pm 0.0672$$

or ±6.7%. The absolute uncertainty in the charge is

$$s_R = R \times 0.0672 = (18) \times (\pm 0.0672) = \pm 1.2$$

Thus, we report the total charge as 18 C ± 1 C.

4C.4 Uncertainty for Mixed Operations

Many chemical calculations involve a combination of adding and subtracting, and multiply and dividing. As shown in the following example, the propagation of uncertainty is easily calculated by treating each operation separately using equations 4.6 and 4.7 as needed.

Example

EXAMPLE 4.7

For a concentration technique the relationship between the measured signal and an analyte's concentration is given by equation 4.5

$$S_{meas} = kC_A + S_{reag}$$

Calculate the absolute and relative uncertainties for the analyte's concentration if S_{meas} is 24.37 ± 0.02, S_{reag} is 0.96 ± 0.02, and k is 0.186 ± 0.003 ppm^{-1}.

SOLUTION

Rearranging equation 4.5 and solving for C_A

$$C_A = \frac{S_{meas} - S_{reag}}{k} = \frac{24.37 - 0.96}{0.186 \text{ ppm}^{-1}} = 125.9 \text{ ppm}$$

gives the analyte's concentration as 126 ppm. To estimate the uncertainty in C_A, we first determine the uncertainty for the numerator, $S_{meas} - S_{reag}$, using equation 4.6

$$s_R = \sqrt{(0.02)^2 + (0.02)^2} = 0.028$$

The numerator, therefore, is 23.41 ± 0.028 (note that we retain an extra significant figure since we will use this uncertainty in further calculations). To complete the calculation, we estimate the relative uncertainty in C_A using equation 4.7, giving

$$\frac{s_R}{R} = \sqrt{\left(\frac{0.028}{23.41}\right)^2 + \left(\frac{0.003}{0.186}\right)^2} = 0.0162$$

or a percent relative uncertainty of 1.6%. The absolute uncertainty in the analyte's concentration is

$$s_R = (125.9 \text{ ppm}) \times (0.0162) = \pm 2.0 \text{ ppm}$$

giving the analyte's concentration as 126 ± 2 ppm.

4C.5 Uncertainty for Other Mathematical Functions

Many other mathematical operations are commonly used in analytical chemistry, including powers, roots, and logarithms. Equations for the propagation of uncertainty for some of these functions are shown in Table 4.9.

Example

EXAMPLE 4.8

The pH of a solution is defined as

$$\text{pH} = -\log[\text{H}^+]$$

where [H^+] is the molar concentration of H^+. If the pH of a solution is 3.72 with an absolute uncertainty of ±0.03, what is the [H^+] and its absolute uncertainty?

SOLUTION

The molar concentration of H^+ for this pH is

$$[H^+] = 10^{-pH} = 10^{-3.72} = 1.91 \times 10^{-4} \text{ M}$$

or 1.9×10^{-4} M to two significant figures. From Table 4.9 the relative uncertainty in $[H^+]$ is

$$\frac{s_R}{R} = 2.303 \times s_A = 2.303 \times 0.03 = 0.069$$

and the absolute uncertainty is

$$(1.91 \times 10^{-4} \text{ M}) \times (0.069) = 1.3 \times 10^{-5} \text{ M}$$

We report the $[H^+]$ and its absolute uncertainty as $1.9 (\pm 0.1) \times 10^{-4}$ M.

Table 4.9 Propagation of Uncertainty for Selected Functions[a]

Function	s_R
$R = kA$	$s_R = ks_A$
$R = A + B$	$s_R = \sqrt{s_A^2 + s_B^2}$
$R = A - B$	$s_R = \sqrt{s_A^2 + s_B^2}$
$R = A \times B$	$\frac{s_R}{R} = \sqrt{\left(\frac{s_A}{A}\right)^2 + \left(\frac{s_B}{B}\right)^2}$
$R = \frac{A}{B}$	$\frac{s_R}{R} = \sqrt{\left(\frac{s_A}{A}\right)^2 + \left(\frac{s_B}{B}\right)^2}$
$R = \ln(A)$	$s_R = \frac{s_A}{A}$
$R = \log(A)$	$s_R = 0.4343 \times \frac{s_A}{A}$
$R = e^A$	$\frac{s_R}{R} = s_A$
$R = 10^A$	$\frac{s_R}{R} = 2.303 s_A$
$R = A^k$	$\frac{s_R}{R} = \left[k \frac{s_A}{A}\right]$

[a]These equations assume that the measurements A and B are uncorrelated; that is, s_A is independent of s_B.

4C.6 Is Calculating Uncertainty Actually Useful?

Given the complexity of determining a result's uncertainty when several measurements are involved, it is worth examining some of the reasons why such calculations are useful. A propagation of uncertainty allows us to estimate an expected uncertainty for an analysis. Comparing the expected uncertainty to that which is actually obtained can provide useful information. For example, in determining the mass of a penny, we estimated the uncertainty in measuring mass as ±0.002 g based on the balance's tolerance. If we measure a single penny's mass several times and obtain a standard deviation of ±0.020 g, we would have reason to believe that our measurement process is out of control. We would then try to identify and correct the problem.

A propagation of uncertainty also helps in deciding how to improve the uncertainty in an analysis. In Example 4.7, for instance, we calculated the concentration of an analyte, obtaining a value of 126 ppm with an absolute uncertainty of ±2 ppm and a relative uncertainty of 1.6%. How might we improve the analysis so that the absolute uncertainty is only ±1 ppm (a relative uncertainty of 0.8%)? Looking back on the calculation, we find that the relative uncertainty is determined by the relative uncertainty in the measured signal (corrected for the reagent blank)

$$\frac{0.028}{23.41} = \pm 0.0012, \text{ or } \pm 0.12\%$$

and the relative uncertainty in the method's sensitivity, k,

$$\frac{0.003}{0.186} = \pm 0.016, \text{ or } \pm 1.6\%$$

Of these two terms, the sensitivity's uncertainty dominates the total uncertainty. Measuring the signal more carefully will not improve the overall uncertainty of the analysis. On the other hand, the desired improvement in uncertainty can be achieved if the sensitivity's absolute uncertainty can be decreased to $\pm 0.0015 \text{ ppm}^{-1}$.

As a final example, a propagation of uncertainty can be used to decide which of several procedures provides the smallest overall uncertainty. Preparing a solution by diluting a stock solution can be done using several different combinations of volumetric glassware. For instance, we can dilute a solution by a factor of 10 using a 10-mL pipet and a 100-mL volumetric flask, or by using a 25-mL pipet and a 250-mL volumetric flask. The same dilution also can be accomplished in two steps using a 50-mL pipet and a 100-mL volumetric flask for the first dilution, and a 10-mL pipet and a 50-mL volumetric flask for the second dilution. The overall uncertainty, of course, depends on the uncertainty of the glassware used in the dilutions. As shown in the following example, we can use the tolerance values for volumetric glassware to determine the optimum dilution strategy.[5]

Example

EXAMPLE 4.9

Which of the following methods for preparing a 0.0010 M solution from a 1.0 M stock solution provides the smallest overall uncertainty?

(a) A one-step dilution using a 1-mL pipet and a 1000-mL volumetric flask.

(b) A two-step dilution using a 20-mL pipet and a 1000-mL volumetric flask for the first dilution and a 25-mL pipet and a 500-mL volumetric flask for the second dilution.

SOLUTION

Letting M_a and M_b represent the molarity of the final solutions from method (a) and method (b), we can write the following equations

$$M_a = 0.0010 \text{ M} = \frac{(1.0 \text{ M})(1.000 \text{ mL})}{1000.0 \text{ mL}}$$

$$M_b = 0.0010 \text{ M} = \frac{(1.0 \text{ M})(20.00 \text{ mL})(25.00 \text{ mL})}{(1000.0 \text{ mL})(500.0 \text{ mL})}$$

Using the tolerance values for pipets and volumetric flasks given in Table 4.2, the overall uncertainties in M_a and M_b are

$$\left(\frac{s_R}{R}\right)_{M_a} = \sqrt{\left(\frac{0.006}{1.000}\right)^2 + \left(\frac{0.3}{1000.0}\right)^2} = 0.006$$

$$\left(\frac{s_R}{R}\right)_{M_b} = \sqrt{\left(\frac{0.03}{20.00}\right)^2 + \left(\frac{0.03}{25.00}\right)^2 + \left(\frac{0.2}{500.0}\right)^2 + \left(\frac{0.3}{1000.0}\right)^2} = 0.002$$

Since the relative uncertainty for M_b is less than that for M_a, we find that the two-step dilution provides the smaller overall uncertainty.

4D The Distribution of Measurements and Results

An analysis, particularly a quantitative analysis, is usually performed on several replicate samples. How do we report the result for such an experiment when results for the replicates are scattered around a central value? To complicate matters further, the analysis of each replicate usually requires multiple measurements that, themselves, are scattered around a central value.

Consider, for example, the data in Table 4.1 for the mass of a penny. Reporting only the mean is insufficient because it fails to indicate the uncertainty in measuring a penny's mass. Including the standard deviation, or other measure of spread, provides the necessary information about the uncertainty in measuring mass. Nevertheless, the central tendency and spread together do not provide a definitive statement about a penny's true mass. If you are not convinced that this is true, ask yourself how obtaining the mass of an additional penny will change the mean and standard deviation.

How we report the result of an experiment is further complicated by the need to compare the results of different experiments. For example, Table 4.10 shows results for a second, independent experiment to determine the mass of a U.S. penny in circulation. Although the results shown in Tables 4.1 and 4.10 are similar, they are not identical; thus, we are justified in asking whether the results are in agreement. Unfortunately, a definitive comparison between these two sets of data is not possible based solely on their respective means and standard deviations.

Developing a meaningful method for reporting an experiment's result requires the ability to predict the true central value and true spread of the population under investigation from a limited sampling of that population. In this section we will take a quantitative look at how individual measurements and results are distributed around a central value.

Table 4.10 Results for a Second Determination of the Mass of a United States Penny in Circulation

Penny	Mass (g)
1	3.052
2	3.141
3	3.083
4	3.083
5	3.048
$\overline{X}$	3.081
s	0.037

4D.1 Populations and Samples

In the previous section we introduced the terms "population" and "sample" in the context of reporting the result of an experiment. Before continuing, we need to understand the difference between a population and a sample. A **population** is the set of all objects in the system being investigated. These objects, which also are members of the population, possess qualitative or quantitative characteristics, or values, that can be measured. If we analyze every member of a population, we can determine the population's true central value, μ, and spread, σ.

population
All members of a system.

The probability of occurrence for a particular value, $P(V)$, is given as

$$P(V) = \frac{M}{N}$$

where V is the value of interest, M is the value's frequency of occurrence in the population, and N is the size of the population. In determining the mass of a circulating United States penny, for instance, the members of the population are all United States pennies currently in circulation, while the values are the possible masses that a penny may have.

In most circumstances, populations are so large that it is not feasible to analyze every member of the population. This is certainly true for the population of circulating U.S. pennies. Instead, we select and analyze a limited subset, or **sample,** of the population. The data in Tables 4.1 and 4.10, for example, give results for two samples drawn at random from the larger population of all U.S. pennies currently in circulation.

sample
Those members of a population that we actually collect and analyze.

4D.2 Probability Distributions for Populations

To predict the properties of a population on the basis of a sample, it is necessary to know something about the population's expected distribution around its central value. The distribution of a population can be represented by plotting the frequency of occurrence of individual values as a function of the values themselves. Such plots are called **probability distributions.** Unfortunately, we are rarely able to calculate the exact probability distribution for a chemical system. In fact, the probability distribution can take any shape, depending on the nature of the chemical system being investigated. Fortunately many chemical systems display one of several common probability distributions. Two of these distributions, the binomial distribution and the normal distribution, are discussed next.

probability distribution
Plot showing frequency of occurrence for members of a population.

binomial distribution
Probability distribution showing chance of obtaining one of two specific outcomes in a fixed number of trials.

Binomial Distribution The **binomial distribution** describes a population in which the values are the number of times a particular outcome occurs during a fixed number of trials. Mathematically, the binomial distribution is given as

$$P(X, N) = \frac{N!}{X!(N-X)!} \times p^X \times (1-p)^{N-X}$$

where $P(X,N)$ is the probability that a given outcome will occur X times during N trials, and p is the probability that the outcome will occur in a single trial.* If you flip a coin five times, $P(2,5)$ gives the probability that two of the five trials will turn up "heads."

A binomial distribution has well-defined measures of central tendency and spread. The true mean value, for example, is given as

$$\mu = Np$$

and the true spread is given by the variance

$$\sigma^2 = Np(1-p)$$

or the standard deviation

$$\sigma = \sqrt{Np(1-p)}$$

The binomial distribution describes a population whose members have only certain, discrete values. A good example of a population obeying the binomial distribution is the sampling of **homogeneous** materials. As shown in Example 4.10, the binomial distribution can be used to calculate the probability of finding a particular isotope in a molecule.

homogeneous
Uniform in composition.

Example

EXAMPLE 4.10

Carbon has two common isotopes, ^{12}C and ^{13}C, with relative isotopic abundances of, respectively, 98.89% and 1.11%. (a) What are the mean and standard deviation for the number of ^{13}C atoms in a molecule of cholesterol? (b) What is the probability of finding a molecule of cholesterol ($C_{27}H_{44}O$) containing no atoms of ^{13}C?

SOLUTION

The probability of finding an atom of ^{13}C in cholesterol follows a binomial distribution, where X is the sought for frequency of occurrence of ^{13}C atoms, N is the number of C atoms in a molecule of cholesterol, and p is the probability of finding an atom of ^{13}C.

(a) The mean number of ^{13}C atoms in a molecule of cholesterol is

$$\mu = Np = 27 \times 0.0111 = 0.300$$

with a standard deviation of

$$\sigma = \sqrt{(27)(0.0111)(1-0.0111)} = 0.172$$

(b) Since the mean is less than one atom of ^{13}C per molecule, most molecules of cholesterol will not have any ^{13}C. To calculate

*$N!$ is read as N-factorial and is the product $N \times (N-1) \times (N-2) \times \cdots \times 1$. For example, 4! is $4 \times 3 \times 2 \times 1$, or 24. Your calculator probably has a key for calculating factorials.

the probability, we substitute appropriate values into the binomial equation

$$P(0, 27) = \frac{27!}{0!\,(27-0)!} \times (0.0111)^0 \times (1-0.0111)^{27-0} = 0.740$$

There is therefore a 74.0% probability that a molecule of cholesterol will not have an atom of ^{13}C.

A portion of the binomial distribution for atoms of ^{13}C in cholesterol is shown in Figure 4.5. Note in particular that there is little probability of finding more than two atoms of ^{13}C in any molecule of cholesterol.

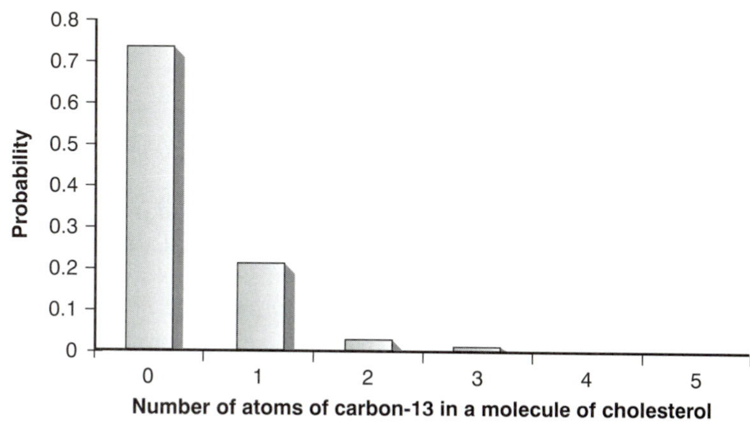

Figure 4.5
Portion of the binomial distribution for the number of naturally occurring ^{13}C atoms in a molecule of cholesterol.

Normal Distribution The binomial distribution describes a population whose members have only certain, discrete values. This is the case with the number of ^{13}C atoms in a molecule, which must be an integer number no greater then the number of carbon atoms in the molecule. A molecule, for example, cannot have 2.5 atoms of ^{13}C. Other populations are considered continuous, in that members of the population may take on any value.

The most commonly encountered continuous distribution is the Gaussian, or **normal distribution,** where the frequency of occurrence for a value, X, is given by

normal distribution
"Bell-shaped" probability distribution curve for measurements and results showing the effect of random error.

$$f(X) = \frac{1}{\sqrt{2\pi\sigma^2}} \exp\left[\frac{-(X-\mu)^2}{2\sigma^2}\right]$$

The shape of a normal distribution is determined by two parameters, the first of which is the population's central, or true mean value, μ, given as

$$\mu = \frac{\sum_{i=1}^{N} X_i}{n}$$

where n is the number of members in the population. The second parameter is the population's variance, σ^2, which is calculated using the following equation*

$$\sigma^2 = \frac{\sum_{i=1}^{N} (X_i - \mu)^2}{n} \qquad \textbf{4.8}$$

*Note the difference between the equation for a population's variance, which includes the term n in the denominator, and the similar equation for the variance of a sample (the square of equation 4.3), which includes the term $n - 1$ in the denominator. The reason for this difference is discussed later in the chapter.

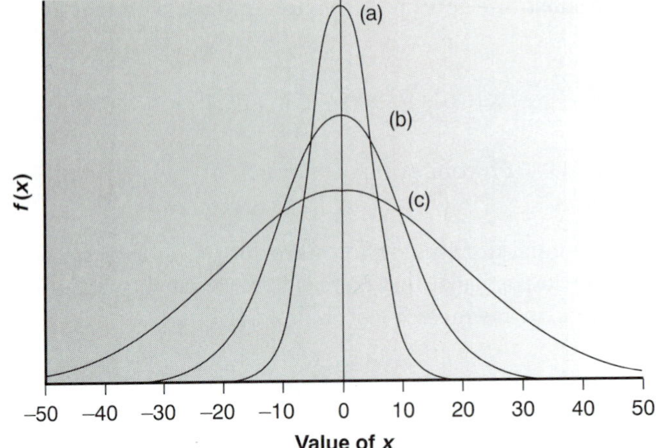

Figure 4.6
Normal distributions for (a) $\mu = 0$ and $\sigma^2 = 25$; (b) $\mu = 0$ and $\sigma^2 = 100$; and (c) $\mu = 0$ and $\sigma^2 = 400$.

Examples of normal distributions with $\mu = 0$ and $\sigma^2 = 25$, 100 or 400, are shown in Figure 4.6. Several features of these normal distributions deserve attention. First, note that each normal distribution contains a single maximum corresponding to μ and that the distribution is symmetrical about this value. Second, increasing the population's variance increases the distribution's spread while decreasing its height. Finally, because the normal distribution depends solely on μ and σ^2, the area, or probability of occurrence between any two limits defined in terms of these parameters is the same for all normal distribution curves. For example, 68.26% of the members in a normally distributed population have values within the range $\mu \pm 1\sigma$, regardless of the actual values of μ and σ. As shown in Example 4.11, probability tables (Appendix 1A) can be used to determine the probability of occurrence between any defined limits.

EXAMPLE 4.11

The amount of aspirin in the analgesic tablets from a particular manufacturer is known to follow a normal distribution, with $\mu = 250$ mg and $\sigma^2 = 25$. In a random sampling of tablets from the production line, what percentage are expected to contain between 243 and 262 mg of aspirin?

SOLUTION

The normal distribution for this example is shown in Figure 4.7, with the shaded area representing the percentage of tablets containing between 243 and 262 mg of aspirin. To determine the percentage of tablets between these limits, we first determine the percentage of tablets with less than 243 mg of aspirin, and the percentage of tablets having more than 262 mg of aspirin. This is accomplished by calculating the deviation, z, of each limit from μ, using the following equation

$$z = \frac{X - \mu}{\sigma}$$

where X is the limit in question, and σ, the population standard deviation, is 5. Thus, the deviation for the lower limit is

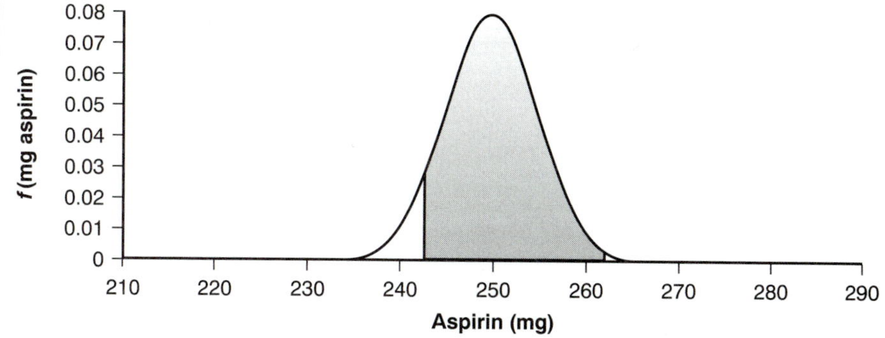

Figure 4.7
Normal distribution for population of aspirin tablets with μ = 250 mg aspirin and σ^2 = 25. The shaded area shows the percentage of tablets containing between 243 and 262 mg of aspirin.

$$z_{\text{low}} = \frac{243 - 250}{5} = -1.4$$

and the deviation for the upper limit is

$$z_{\text{up}} = \frac{262 - 250}{5} = +2.4$$

Using the table in Appendix 1A, we find that the percentage of tablets with less than 243 mg of aspirin is 8.08%, and the percentage of tablets with more than 262 mg of aspirin is 0.82%. The percentage of tablets containing between 243 and 262 mg of aspirin is therefore

$$100.00\% - 8.08\% - 0.82\ \% = 91.10\%$$

4D.3 Confidence Intervals for Populations

If we randomly select a single member from a population, what will be its most likely value? This is an important question, and, in one form or another, it is the fundamental problem for any analysis. One of the most important features of a population's probability distribution is that it provides a way to answer this question.

Earlier we noted that 68.26% of a normally distributed population is found within the range of $\mu \pm 1\sigma$. Stating this another way, there is a 68.26% probability that a member selected at random from a normally distributed population will have a value in the interval of $\mu \pm 1\sigma$. In general, we can write

$$X_i = \mu \pm z\sigma \qquad \textbf{4.9}$$

where the factor z accounts for the desired level of confidence. Values reported in this fashion are called **confidence intervals.** Equation 4.9, for example, is the confidence interval for a single member of a population. Confidence intervals can be quoted for any desired probability level, several examples of which are shown in Table 4.11. For reasons that will be discussed later in the chapter, a 95% confidence interval frequently is reported.

Table 4.11 Confidence Intervals for Normal Distribution Curves Between the Limits $\mu \pm z\sigma$

z	Confidence Interval (%)
0.50	38.30
1.00	68.26
1.50	86.64
1.96	95.00
2.00	95.44
2.50	98.76
3.00	99.73
3.50	99.95

confidence interval
Range of results around a mean value that could be explained by random error.

Example

EXAMPLE 4.12

What is the 95% confidence interval for the amount of aspirin in a single analgesic tablet drawn from a population where μ is 250 mg and σ^2 is 25?

SOLUTION

According to Table 4.11, the 95% confidence interval for a single member of a normally distributed population is

$$X_i = \mu \pm 1.96\sigma = 250 \text{ mg} \pm (1.96)(5) = 250 \text{ mg} \pm 10 \text{ mg}$$

Thus, we expect that 95% of the tablets in the population contain between 240 and 260 mg of aspirin.

Alternatively, a confidence interval can be expressed in terms of the population's standard deviation and the value of a single member drawn from the population. Thus, equation 4.9 can be rewritten as a confidence interval for the population mean

$$\mu = X_i \pm z\sigma \qquad \textbf{4.10}$$

Example

EXAMPLE 4.13

The population standard deviation for the amount of aspirin in a batch of analgesic tablets is known to be 7 mg of aspirin. A single tablet is randomly selected, analyzed, and found to contain 245 mg of aspirin. What is the 95% confidence interval for the population mean?

SOLUTION

The 95% confidence interval for the population mean is given as

$$\mu = X_i \pm z\sigma = 245 \pm (1.96)(7) = 245 \text{ mg} \pm 14 \text{ mg}$$

There is, therefore, a 95% probability that the population's mean, μ, lies within the range of 231–259 mg of aspirin.

Confidence intervals also can be reported using the mean for a sample of size n, drawn from a population of known σ. The standard deviation for the mean value, $\sigma_{\overline{X}}$, which also is known as the standard error of the mean, is

$$\sigma_{\overline{X}} = \frac{\sigma}{\sqrt{n}}$$

The confidence interval for the population's mean, therefore, is

$$\mu = \overline{X} \pm \frac{z\sigma}{\sqrt{n}} \qquad \textbf{4.11}$$

Example

EXAMPLE 4.14

What is the 95% confidence interval for the analgesic tablets described in Example 4.13, if an analysis of five tablets yields a mean of 245 mg of aspirin?

SOLUTION

In this case the confidence interval is given as

$$\mu = 245 \pm \frac{(1.96)(7)}{\sqrt{5}} = 245 \text{ mg} \pm 6 \text{ mg}$$

Thus, there is a 95% probability that the population's mean is between 239 and 251 mg of aspirin. As expected, the confidence interval based on the mean of five members of the population is smaller than that based on a single member.

4D.4 Probability Distributions for Samples

In Section 4D.2 we introduced two probability distributions commonly encountered when studying populations. The construction of confidence intervals for a normally distributed population was the subject of Section 4D.3. We have yet to address, however, how we can identify the probability distribution for a given population. In Examples 4.11–4.14 we assumed that the amount of aspirin in analgesic tablets is normally distributed. We are justified in asking how this can be determined without analyzing every member of the population. When we cannot study the whole population, or when we cannot predict the mathematical form of a population's probability distribution, we must deduce the distribution from a limited sampling of its members.

Sample Distributions and the Central Limit Theorem Let's return to the problem of determining a penny's mass to explore the relationship between a population's distribution and the distribution of samples drawn from that population. The data shown in Tables 4.1 and 4.10 are insufficient for our purpose because they are not large enough to give a useful picture of their respective probability distributions. A better picture of the probability distribution requires a larger sample, such as that shown in Table 4.12, for which $\overline{X}$ is 3.095 and s^2 is 0.0012.

The data in Table 4.12 are best displayed as a **histogram,** in which the frequency of occurrence for equal intervals of data is plotted versus the midpoint of each interval. Table 4.13 and Figure 4.8 show a frequency table and histogram for the data in Table 4.12. Note that the histogram was constructed such that the mean value for the data set is centered within its interval. In addition, a normal distribution curve using $\overline{X}$ and s^2 to estimate μ and σ^2 is superimposed on the histogram.

histogram
A plot showing the number of times an observation occurs as a function of the range of observed values.

It is noteworthy that the histogram in Figure 4.8 approximates the normal distribution curve. Although the histogram for the mass of pennies is not perfectly symmetrical, it is roughly symmetrical about the interval containing the greatest number of pennies. In addition, we know from Table 4.11 that 68.26%, 95.44%, and 99.73% of the members of a normally distributed population are within, respectively, $\pm 1\sigma$, $\pm 2\sigma$, and $\pm 3\sigma$. If we assume that the mean value, 3.095 g, and the sample variance, 0.0012, are good approximations for μ and σ^2, we find that 73%,

Table 4.12 Individual Masses for a Large Sample of U.S. Pennies in Circulation[a]

Penny	Weight (g)	Penny	Weight (g)	Penny	Weight (g)	Penny	Weight (g)
1	3.126	26	3.073	51	3.101	76	3.086
2	3.140	27	3.084	52	3.049	77	3.123
3	3.092	28	3.148	53	3.082	78	3.115
4	3.095	29	3.047	54	3.142	79	3.055
5	3.080	30	3.121	55	3.082	80	3.057
6	3.065	31	3.116	56	3.066	81	3.097
7	3.117	32	3.005	57	3.128	82	3.066
8	3.034	33	3.115	58	3.112	83	3.113
9	3.126	34	3.103	59	3.085	84	3.102
10	3.057	35	3.086	60	3.086	85	3.033
11	3.053	36	3.103	61	3.084	86	3.112
12	3.099	37	3.049	62	3.104	87	3.103
13	3.065	38	2.998	63	3.107	88	3.198
14	3.059	39	3.063	64	3.093	89	3.103
15	3.068	40	3.055	65	3.126	90	3.126
16	3.060	41	3.181	66	3.138	91	3.111
17	3.078	42	3.108	67	3.131	92	3.126
18	3.125	43	3.114	68	3.120	93	3.052
19	3.090	44	3.121	69	3.100	94	3.113
20	3.100	45	3.105	70	3.099	95	3.085
21	3.055	46	3.078	71	3.097	96	3.117
22	3.105	47	3.147	72	3.091	97	3.142
23	3.063	48	3.104	73	3.077	98	3.031
24	3.083	49	3.146	74	3.178	99	3.083
25	3.065	50	3.095	75	3.054	100	3.104

[a]Pennies are identified in the order in which they were sampled and weighed.

Table 4.13 Frequency Distribution for the Data in Table 4.12

Interval	Frequency
2.991–3.009	2
3.010–3.028	0
3.029–3.047	4
3.048–3.066	19
3.067–3.085	15
3.086–3.104	23
3.105–3.123	19
3.124–3.142	12
3.143–3.161	13
3.162–3.180	1
3.181–3.199	2

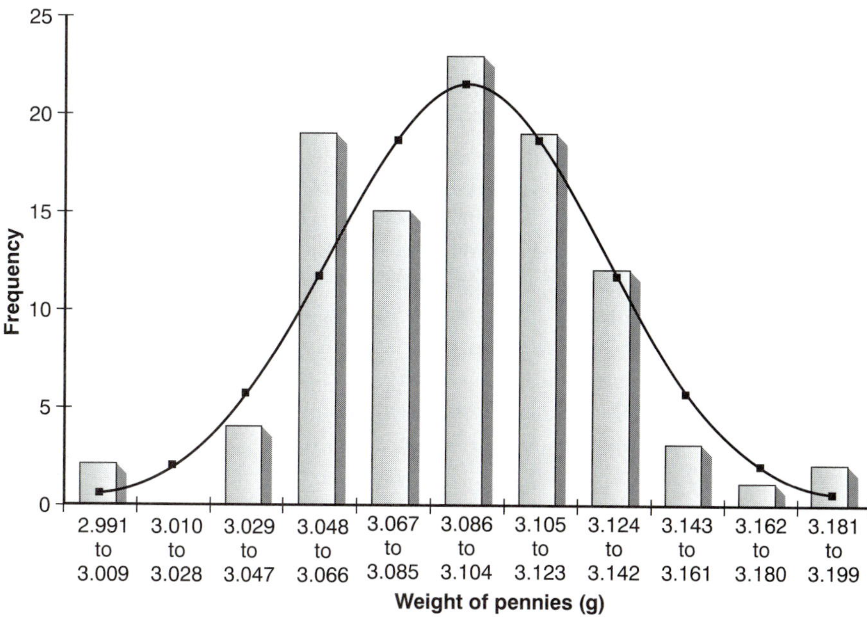

Figure 4.8
Histogram for data in Table 4.12. A normal distribution curve for the data, based on $\overline{X}$ and s^2, is superimposed on the histogram.

95%, and 100% of the pennies fall within these limits. It is easy to imagine that increasing the number of pennies in the sample will result in a histogram that even more closely approximates a normal distribution.

We will not offer a formal proof that the sample of pennies in Table 4.12 and the population from which they were drawn are normally distributed; however, the evidence we have seen strongly suggests that this is true. Although we cannot claim that the results for all analytical experiments are normally distributed, in most cases the data we collect in the laboratory are, in fact, drawn from a normally distributed population. That this is generally true is a consequence of the **central limit theorem.**[6] According to this theorem, in systems subject to a variety of indeterminate errors, the distribution of results will be approximately normal. Furthermore, as the number of contributing sources of indeterminate error increases, the results come even closer to approximating a normal distribution. The central limit theorem holds true even if the individual sources of indeterminate error are not normally distributed. The chief limitation to the central limit theorem is that the sources of indeterminate error must be independent and of similar magnitude so that no one source of error dominates the final distribution.

central limit theorem
The distribution of measurements subject to indeterminate errors is often a normal distribution.

Estimating μ and σ^2 Our comparison of the histogram for the data in Table 4.12 to a normal distribution assumes that the sample's mean, $\overline{X}$, and variance, s^2, are appropriate estimators of the population's mean, μ, and variance, σ^2. Why did we select $\overline{X}$ and s^2, as opposed to other possible measures of central tendency and spread? The explanation is simple; $\overline{X}$ and s^2 are considered unbiased estimators of μ and σ^2.[7,8] If we could analyze every possible sample of equal size for a given population (e.g., every possible sample of five pennies), calculating their respective means and variances, the average mean and the average variance would equal μ and σ^2. Although $\overline{X}$ and s^2 for any single sample probably will not be the same as μ or σ^2, they provide a reasonable estimate for these values.

Degrees of Freedom Unlike the population's variance, the variance of a sample includes the term $n - 1$ in the denominator, where n is the size of the sample

$$s^2 = \frac{\sum_{i=1}^{n} (X_i - \overline{X})^2}{n - 1} \qquad \textbf{4.12}$$

degrees of freedom
The number of independent values on which a result is based (ν).

Defining the sample's variance with a denominator of n, as in the case of the population's variance leads to a biased estimation of σ^2. The denominators of the variance equations 4.8 and 4.12 are commonly called the **degrees of freedom** for the population and the sample, respectively. In the case of a population, the degrees of freedom is always equal to the total number of members, n, in the population. For the sample's variance, however, substituting $\overline{X}$ for μ removes a degree of freedom from the calculation. That is, if there are n members in the sample, the value of the nth member can always be deduced from the remaining $n - 1$ members and $\overline{X}$. For example, if we have a sample with five members, and we know that four of the members are 1, 2, 3, and 4, and that the mean is 3, then the fifth member of the sample must be

$$(\overline{X} \times n) - X_1 - X_2 - X_3 - X_4 = (3 \times 5) - 1 - 2 - 3 - 4 = 5$$

4D.5 Confidence Intervals for Samples

Earlier we introduced the confidence interval as a way to report the most probable value for a population's mean, μ, when the population's standard deviation, σ, is known. Since s^2 is an unbiased estimator of σ^2, it should be possible to construct confidence intervals for samples by replacing σ in equations 4.10 and 4.11 with s. Two complications arise, however. The first is that we cannot define s^2 for a single member of a population. Consequently, equation 4.10 cannot be extended to situations in which s^2 is used as an estimator of σ^2. In other words, when σ is unknown, we cannot construct a confidence interval for μ by sampling only a single member of the population.

The second complication is that the values of z shown in Table 4.11 are derived for a normal distribution curve that is a function of σ^2, not s^2. Although s^2 is an unbiased estimator of σ^2, the value of s^2 for any randomly selected sample may differ significantly from σ^2. To account for the uncertainty in estimating σ^2, the term z in equation 4.11 is replaced with the variable t, where t is defined such that $t \geq z$ at all confidence levels. Thus, equation 4.11 becomes

$$\mu = \overline{X} \pm \frac{ts}{\sqrt{n}} \qquad \textbf{4.13}$$

Values for t at the 95% confidence level are shown in Table 4.14. Note that t becomes smaller as the number of the samples (or degrees of freedom) increase, approaching z as n approaches infinity. Additional values of t for other confidence levels can be found in Appendix 1B.

Example

EXAMPLE 4.15

What is the 95% confidence interval for the data in Table 4.1?

SOLUTION

The mean and standard deviation for this sample are, respectively, 3.117 g and 0.051 g. Since the sample consists of seven measurements, there are six degrees of freedom. The value of t from Table 4.14, is 2.45. Substituting into equation 4.13 gives

$$\mu = \overline{X} \pm \frac{ts}{\sqrt{n}} = 3.117 \pm \frac{(2.45)(0.051)}{\sqrt{7}} = 3.117 \pm 0.047 \text{ g}$$

Thus, there is a 95% probability that the population's mean is between 3.070 and 3.164 g.

Table 4.14 Values of t for the 95% Confidence Interval

Degrees of Freedom	t
1	12.71
2	4.30
3	3.18
4	2.78
5	2.57
6	2.45
7	2.36
8	2.31
9	2.26
10	2.23
12	2.18
14	2.14
16	2.12
18	2.10
20	2.09
30	2.04
50	2.01
∞	1.96

4D.6 A Cautionary Statement

There is a temptation when analyzing data to plug numbers into an equation, carry out the calculation, and report the result. This is never a good idea, and you should develop the habit of constantly reviewing and evaluating your data. For example, if analyzing five samples gives an analyte's mean concentration as 0.67 ppm with a standard deviation of 0.64 ppm, then the 95% confidence interval is

$$\mu = 0.67 \pm \frac{(2.78)(0.64)}{\sqrt{5}} = 0.64 \pm 0.80 \text{ ppm}$$

This confidence interval states that the analyte's true concentration lies within the range of –0.16 ppm to 1.44 ppm. Including a negative concentration within the confidence interval should lead you to reevaluate your data or conclusions. On further investigation your data may show that the standard deviation is larger than expected, making the confidence interval too broad, or you may conclude that the analyte's concentration is too small to detect accurately.*

A second example is also informative. When samples are obtained from a normally distributed population, their values must be random. If results for several samples show a regular pattern or trend, then the samples cannot be normally distributed. This may reflect the fact that the underlying population is not normally distributed, or it may indicate the presence of a time-dependent determinate error. For example, if we randomly select 20 pennies and find that the mass of each penny exceeds that of the preceding penny, we might suspect that the balance on which the pennies are being weighed is drifting out of calibration.

4E Statistical Analysis of Data

In the previous section we noted that the result of an analysis is best expressed as a confidence interval. For example, a 95% confidence interval for the mean of five results gives the range in which we expect to find the mean for 95% of all samples of equal size, drawn from the same population. Alternatively, and in the absence of determinate errors, the 95% confidence interval indicates the range of values in which we expect to find the population's true mean.

The probabilistic nature of a confidence interval provides an opportunity to ask and answer questions comparing a sample's mean or variance to either the accepted values for its population or similar values obtained for other samples. For example, confidence intervals can be used to answer questions such as "Does a newly developed method for the analysis of cholesterol in blood give results that are significantly different from those obtained when using a standard method?" or "Is there a significant variation in the chemical composition of rainwater collected at different sites downwind from a coalburning utility plant?" In this section we introduce a general approach to the statistical analysis of data. Specific statistical methods of analysis are covered in Section 4F.

4E.1 Significance Testing

Let's consider the following problem. Two sets of blood samples have been collected from a patient receiving medication to lower her concentration of blood glucose. One set of samples was drawn immediately before the medication was administered; the second set was taken several hours later. The samples are analyzed and their respective means and variances reported. How do we decide if the medication was successful in lowering the patient's concentration of blood glucose?

One way to answer this question is to construct probability distribution curves for each sample and to compare the curves with each other. Three possible outcomes are shown in Figure 4.9. In Figure 4.9a, the probability distribution curves are completely separated, strongly suggesting that the samples are significantly different. In Figure 4.9b, the probability distributions for the two samples are highly overlapped, suggesting that any difference between the samples is insignificant. Figure 4.9c, however, presents a dilemma. Although the means for the two samples appear to be different, the probability distributions overlap to an extent that a significant number of possible outcomes could belong to either distribution. In this case we can, at best, only make a statement about the probability that the samples are significantly different.

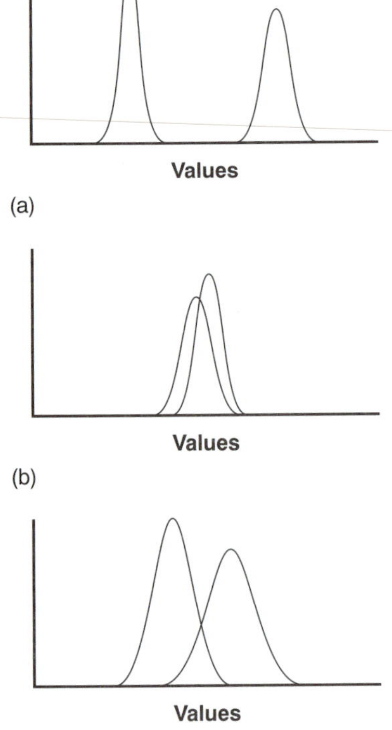

Figure 4.9
Three examples of possible relationships between the probability distributions for two populations. (a) Completely separate distributions; (b) Distributions with a great deal of overlap; (c) Distributions with some overlap.

*The topic of detection limits is discussed at the end of this chapter.

The process by which we determine the probability that there is a significant difference between two samples is called significance testing or hypothesis testing. Before turning to a discussion of specific examples, however, we will first establish a general approach to conducting and interpreting significance tests.

4E.2 Constructing a Significance Test

A **significance test** is designed to determine whether the difference between two or more values is too large to be explained by indeterminate error. The first step in constructing a significance test is to state the experimental problem as a yes-or-no question, two examples of which were given at the beginning of this section. A null hypothesis and an alternative hypothesis provide answers to the question. The **null hypothesis,** H_0, is that indeterminate error is sufficient to explain any difference in the values being compared. The **alternative hypothesis,** H_A, is that the difference between the values is too great to be explained by random error and, therefore, must be real. A significance test is conducted on the null hypothesis, which is either retained or rejected. If the null hypothesis is rejected, then the alternative hypothesis must be accepted. When a null hypothesis is not rejected, it is said to be retained rather than accepted. A null hypothesis is retained whenever the evidence is insufficient to prove it is incorrect. Because of the way in which significance tests are conducted, it is impossible to prove that a null hypothesis is true.

significance test
A statistical test to determine if the difference between two values is significant.

null hypothesis
A statement that the difference between two values can be explained by indeterminate error; retained if the significance test does not fail (H_0).

alternative hypothesis
A statement that the difference between two values is too great to be explained by indeterminate error; accepted if the significance test shows that null hypothesis should be rejected (H_A).

The difference between retaining a null hypothesis and proving the null hypothesis is important. To appreciate this point, let us return to our example on determining the mass of a penny. After looking at the data in Table 4.12, you might pose the following null and alternative hypotheses

H_0: Any U.S. penny in circulation has a mass that falls in the range of 2.900–3.200 g

H_A: Some U.S. pennies in circulation have masses that are less than 2.900 g or more than 3.200 g.

To test the null hypothesis, you reach into your pocket, retrieve a penny, and determine its mass. If the mass of this penny is 2.512 g, then you have proved that the null hypothesis is incorrect. Finding that the mass of your penny is 3.162 g, however, does not prove that the null hypothesis is correct because the mass of the next penny you sample might fall outside the limits set by the null hypothesis.

After stating the null and alternative hypotheses, a significance level for the analysis is chosen. The significance level is the confidence level for retaining the null hypothesis or, in other words, the probability that the null hypothesis will be incorrectly rejected. In the former case the significance level is given as a percentage (e.g., 95%), whereas in the latter case, it is given as α, where α is defined as

$$\alpha = 1 - \frac{\text{confidence level}}{100}$$

Thus, for a 95% confidence level, α is 0.05.

Next, an equation for a test statistic is written, and the test statistic's critical value is found from an appropriate table. This critical value defines the breakpoint between values of the test statistic for which the null hypothesis will be retained or rejected. The test statistic is calculated from the data, compared with the critical value, and the null hypothesis is either rejected or retained. Finally, the result of the significance test is used to answer the original question.

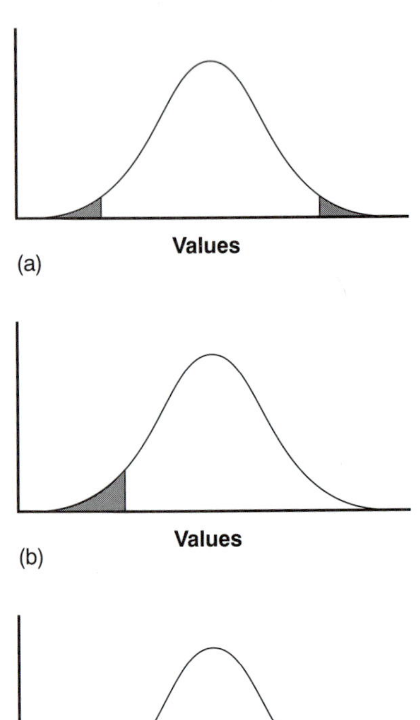

Figure 4.10
Examples of (a) two-tailed, (b) and (c) one-tailed, significance tests. The shaded areas in each curve represent the values for which the null hypothesis is rejected.

4E.3 One-Tailed and Two-Tailed Significance Tests

Consider the situation when the accuracy of a new analytical method is evaluated by analyzing a standard reference material with a known μ. A sample of the standard is analyzed, and the sample's mean is determined. The null hypothesis is that the sample's mean is equal to μ

$$H_0\colon\ \overline{X} = \mu$$

If the significance test is conducted at the 95% confidence level ($\alpha = 0.05$), then the null hypothesis will be retained if a 95% confidence interval around $\overline{X}$ contains μ. If the alternative hypothesis is

$$H_A\colon\ \overline{X} \neq \mu$$

then the null hypothesis will be rejected, and the alternative hypothesis accepted if μ lies in either of the shaded areas at the tails of the sample's probability distribution (Figure 4.10a). Each of the shaded areas accounts for 2.5% of the area under the probability distribution curve. This is called a **two-tailed significance test** because the null hypothesis is rejected for values of μ at either extreme of the sample's probability distribution.

The alternative hypothesis also can be stated in one of two additional ways

$$H_A\colon\ \overline{X} > \mu$$

$$H_A\colon\ \overline{X} < \mu$$

for which the null hypothesis is rejected if μ falls within the shaded areas shown in Figure 4.10(b) and Figure 4.10(c), respectively. In each case the shaded area represents 5% of the area under the probability distribution curve. These are examples of **one-tailed significance tests.**

For a fixed confidence level, a two-tailed test is always the more conservative test because it requires a larger difference between $\overline{X}$ and μ to reject the null hypothesis. Most significance tests are applied when there is no a priori expectation about the relative magnitudes of the parameters being compared. A two-tailed significance test, therefore, is usually the appropriate choice. One-tailed significance tests are reserved for situations when we have reason to expect one parameter to be larger or smaller than the other. For example, a one-tailed significance test would be appropriate for our earlier example regarding a medication's effect on blood glucose levels since we believe that the medication will lower the concentration of glucose.

two-tailed significance test
Significance test in which the null hypothesis is rejected for values at either end of the normal distribution.

one-tailed significance test
Significance test in which the null hypothesis is rejected for values at only one end of the normal distribution.

4E.4 Errors in Significance Testing

Since significance tests are based on probabilities, their interpretation is naturally subject to error. As we have already seen, significance tests are carried out at a significance level, α, that defines the probability of rejecting a null hypothesis that is true. For example, when a significance test is conducted at $\alpha = 0.05$, there is a 5% probability that the null hypothesis will be incorrectly rejected. This is known as a **type 1 error,** and its risk is always equivalent to α. Type 1 errors in two-tailed and one-tailed significance tests are represented by the shaded areas under the probability distribution curves in Figure 4.10.

The second type of error occurs when the null hypothesis is retained even though it is false and should be rejected. This is known as a **type 2 error,** and its probability of occurrence is β. Unfortunately, in most cases β cannot be easily calculated or estimated.

type 1 error
The risk of falsely rejecting the null hypothesis (α).

type 2 error
The risk of falsely retaining the null hypothesis (β).

The probability of a type 1 error is inversely related to the probability of a type 2 error. Minimizing a type 1 error by decreasing α, for example, increases the likelihood of a type 2 error. The value of α chosen for a particular significance test, therefore, represents a compromise between these two types of error. Most of the examples in this text use a 95% confidence level, or $\alpha = 0.05$, since this is the most frequently used confidence level for the majority of analytical work. It is not unusual, however, for more stringent (e.g. $\alpha = 0.01$) or for more lenient (e.g. $\alpha = 0.10$) confidence levels to be used.

4F Statistical Methods for Normal Distributions

The most commonly encountered probability distribution is the normal, or Gaussian, distribution. A normal distribution is characterized by a true mean, μ, and variance, σ^2, which are estimated using $\overline{X}$ and s^2. Since the area between any two limits of a normal distribution is well defined, the construction and evaluation of significance tests are straightforward.

4F.1 Comparing $\overline{X}$ to μ

One approach for validating a new analytical method is to analyze a standard sample containing a known amount of analyte, μ. The method's accuracy is judged by determining the average amount of analyte in several samples, $\overline{X}$, and using a significance test to compare it with μ. The null hypothesis is that $\overline{X}$ and μ are the same and that any difference between the two values can be explained by indeterminate errors affecting the determination of $\overline{X}$. The alternative hypothesis is that the difference between $\overline{X}$ and μ is too large to be explained by indeterminate error.

The equation for the test (experimental) statistic, t_{exp}, is derived from the confidence interval for μ

$$\mu = \overline{X} \pm \frac{t_{exp} s}{\sqrt{n}} \qquad 4.14$$

Rearranging equation 4.14

$$t_{exp} = \frac{|\mu - \overline{X}| \times \sqrt{n}}{s} \qquad 4.15$$

gives the value of t_{exp} when μ is at either the right or left edge of the sample's apparent confidence interval (Figure 4.11a). The value of t_{exp} is compared with a critical value, $t(\alpha,\nu)$, which is determined by the chosen significance level, α, the degrees of freedom for the sample, ν, and whether the significance test is one-tailed or two-tailed. Values for $t(\alpha,\nu)$ are found in Appendix 1B. The critical value $t(\alpha,\nu)$ defines the confidence interval that can be explained by indeterminate errors. If t_{exp} is greater than $t(\alpha,\nu)$, then the confidence interval for the data is wider than that expected from indeterminate errors (Figure 4.11b). In this case, the null hypothesis is rejected and the alternative hypothesis is accepted. If t_{exp} is less than or equal to $t(\alpha,\nu)$, then the confidence interval for the data could be attributed to indeterminate error, and the null hypothesis is retained at the stated significance level (Figure 4.11c).

A typical application of this significance test, which is known as a ***t*-test** of $\overline{X}$ to μ, is outlined in the following example.

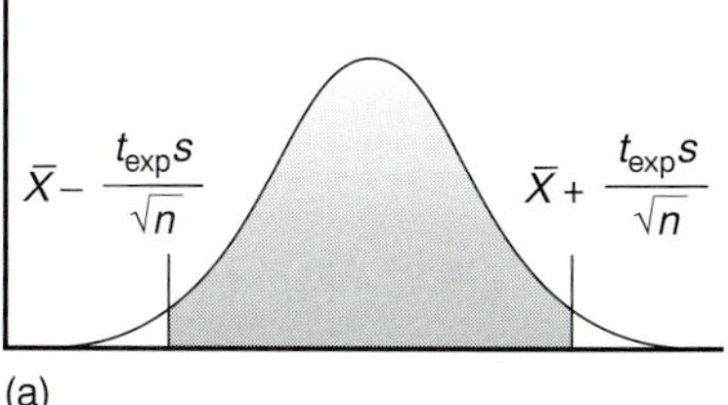

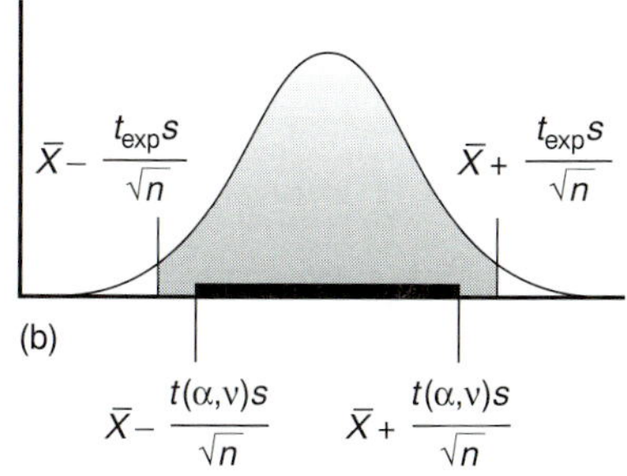

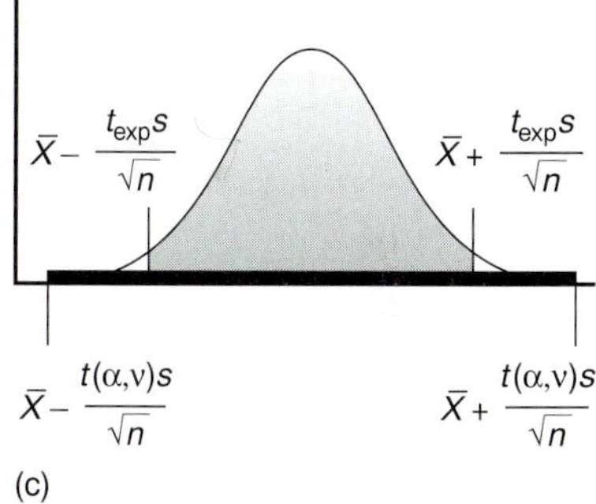

Figure 4.11
Relationship between confidence intervals and results of a significance test. (a) The shaded area under the normal distribution curves shows the apparent confidence intervals for the sample based on t_{exp}. The solid bars in (b) and (c) show the actual confidence intervals that can be explained by indeterminate error using the critical value of (α,ν). In part (b) the null hypothesis is rejected and the alternative hypothesis is accepted. In part (c) the null hypothesis is retained.

***t*-test**
Statistical test for comparing two mean values to see if their difference is too large to be explained by indeterminate error.

Example

EXAMPLE 4.16

Before determining the amount of Na_2CO_3 in an unknown sample, a student decides to check her procedure by analyzing a sample known to contain 98.76% w/w Na_2CO_3. Five replicate determinations of the %w/w Na_2CO_3 in the standard were made with the following results

98.71% 98.59% 98.62% 98.44% 98.58%

Is the mean for these five trials significantly different from the accepted value at the 95% confidence level ($\alpha = 0.05$)?

SOLUTION

The mean and standard deviation for the five trials are

$$\overline{X} = 98.59 \qquad s = 0.0973$$

Since there is no reason to believe that $\overline{X}$ must be either larger or smaller than μ, the use of a two-tailed significance test is appropriate. The null and alternative hypotheses are

$$H_0\text{:} \quad \overline{X} = \mu \qquad H_A\text{:} \quad \overline{X} \neq \mu$$

The test statistic is

$$t_{\text{exp}} = \frac{|\mu - \overline{X}| \times \sqrt{n}}{s} = \frac{|98.76 - 98.59| \times \sqrt{5}}{0.0973} = 3.91$$

The critical value for $t(0.05,4)$, as found in Appendix 1B, is 2.78. Since t_{exp} is greater than $t(0.05, 4)$, we must reject the null hypothesis and accept the alternative hypothesis. At the 95% confidence level the difference between $\overline{X}$ and μ is significant and cannot be explained by indeterminate sources of error. There is evidence, therefore, that the results are affected by a determinate source of error.

If evidence for a determinate error is found, as in Example 4.16, its source should be identified and corrected before analyzing additional samples. Failing to reject the null hypothesis, however, does not imply that the method is accurate, but only indicates that there is insufficient evidence to prove the method inaccurate at the stated confidence level.

The utility of the t-test for $\overline{X}$ and μ is improved by optimizing the conditions used in determining $\overline{X}$. Examining equation 4.15 shows that increasing the number of replicate determinations, n, or improving the precision of the analysis enhances the utility of this significance test. A t-test can only give useful results, however, if the standard deviation for the analysis is reasonable. If the standard deviation is substantially larger than the expected standard deviation, σ, the confidence interval around $\overline{X}$ will be so large that a significant difference between $\overline{X}$ and μ may be difficult to prove. On the other hand, if the standard deviation is significantly smaller than expected, the confidence interval around $\overline{X}$ will be too small, and a significant difference between $\overline{X}$ and μ may be found when none exists. A significance test that can be used to evaluate the standard deviation is the subject of the next section.

4F.2 Comparing s^2 to σ^2

When a particular type of sample is analyzed on a regular basis, it may be possible to determine the expected, or true variance, σ^2, for the analysis. This often is the case in clinical labs where hundreds of blood samples are analyzed each day. Replicate analyses of any single sample, however, results in a sample variance, s^2. A statistical comparison of s^2 to σ^2 provides useful information about whether the analysis is in a state of "statistical control." The null hypothesis is that s^2 and σ^2 are identical, and the alternative hypothesis is that they are not identical.

The test statistic for evaluating the null hypothesis is called an ***F*-test,** and is given as either

$$F_{\text{exp}} = \frac{s^2}{\sigma^2} \quad (s^2 > \sigma^2) \qquad \text{or} \qquad F_{\text{exp}} = \frac{\sigma^2}{s^2} \quad (\sigma^2 > s^2) \qquad \textbf{4.16}$$

depending on whether s^2 is larger or smaller than σ^2. Note that F_{exp} is defined such that its value is always greater than or equal to 1.

If the null hypothesis is true, then F_{exp} should equal 1. Due to indeterminate errors, however, the value for F_{exp} usually is greater than 1. A critical value, $F(\alpha, \nu_{\text{num}}, \nu_{\text{den}})$, gives the largest value of F that can be explained by indeterminate error. It is chosen for a specified significance level, α, and the degrees of freedom for the variances in the numerator, ν_{num}, and denominator, ν_{den}. The degrees of freedom for s^2 is $n - 1$, where n is the number of replicates used in determining the sample's variance. Critical values of F for $\alpha = 0.05$ are listed in Appendix 1C for both one-tailed and two-tailed significance tests.

***F*-test**
Statistical test for comparing two variances to see if their difference is too large to be explained by indeterminate error.

Example

EXAMPLE 4.17

A manufacturer's process for analyzing aspirin tablets has a known variance of 25. A sample of ten aspirin tablets is selected and analyzed for the amount of aspirin, yielding the following results

254 249 252 252 249 249 250 247 251 252

Determine whether there is any evidence that the measurement process is not under statistical control at $\alpha = 0.05$.

SOLUTION

The variance for the sample of ten tablets is 4.3. A two-tailed significance test is used since the measurement process is considered out of statistical control if the sample's variance is either too good or too poor. The null hypothesis and alternative hypotheses are

$$H_0:\ s^2 = \sigma^2 \qquad H_A:\ s^2 \neq \sigma^2$$

The test statistic is

$$F_{\text{exp}} = \frac{\sigma^2}{s^2} = \frac{25}{4.3} = 5.8$$

The critical value for $F(0.05, \infty, 9)$ from Appendix 1C is 3.33. Since F is greater than $F(0.05,\infty, 9)$, we reject the null hypothesis and accept the alternative hypothesis that the analysis is not under statistical control. One explanation for the unreasonably small variance could be that the aspirin tablets were not selected randomly.

4F.3 Comparing Two Sample Variances

The F-test can be extended to the comparison of variances for two samples, A and B, by rewriting equation 4.16 as

$$F_{exp} = \frac{s_A^2}{s_B^2}$$

where A and B are defined such that s_A^2 is greater than or equal to s_B^2. An example of this application of the F-test is shown in the following example.

Example

EXAMPLE 4.18

Tables 4.1 and 4.8 show results for two separate experiments to determine the mass of a circulating U.S. penny. Determine whether there is a difference in the precisions of these analyses at $\alpha = 0.05$.

SOLUTION

Letting A represent the results in Table 4.1 and B represent the results in Table 4.8, we find that the variances are $s_A^2 = 0.00259$ and $s_B^2 = 0.00138$. A two-tailed significance test is used since there is no reason to suspect that the results for one analysis will be more precise than that of the other. The null and alternative hypotheses are

$$H_0: \; s_A^2 = s_B^2 \qquad H_A: \; s_A^2 \neq s_B^2$$

and the test statistic is

$$F_{exp} = \frac{s_A^2}{s_B^2} = \frac{0.00259}{0.00138} = 1.88$$

The critical value for $F(0.05, 6, 4)$ is 9.197. Since F_{exp} is less than $F(0.05, 6, 4)$, the null hypothesis is retained. There is no evidence at the chosen significance level to suggest that the difference in precisions is significant.

4F.4 Comparing Two Sample Means

The result of an analysis is influenced by three factors: the method, the sample, and the analyst. The influence of these factors can be studied by conducting a pair of experiments in which only one factor is changed. For example, two methods can be compared by having the same analyst apply both methods to the same sample and examining the resulting means. In a similar fashion, it is possible to compare two analysts or two samples.

Significance testing for comparing two mean values is divided into two categories depending on the source of the data. Data are said to be **unpaired** when each mean is derived from the analysis of several samples drawn from the same source. **Paired data** are encountered when analyzing a series of samples drawn from different sources.

unpaired data
Two sets of data consisting of results obtained using several samples drawn from a single source.

paired data
Two sets of data consisting of results obtained using several samples drawn from different sources.

Unpaired Data Consider two samples, A and B, for which mean values, $\overline{X}_A$ and $\overline{X}_B$, and standard deviations, s_A and s_B, have been measured. Confidence intervals for μ_A and μ_B can be written for both samples

$$\mu_A = \overline{X}_A \pm \frac{ts_A}{\sqrt{n_A}} \qquad 4.17$$

$$\mu_B = \overline{X}_B \pm \frac{ts_B}{\sqrt{n_B}} \qquad 4.18$$

where n_A and n_B are the number of replicate trials conducted on samples A and B. A comparison of the mean values is based on the null hypothesis that $\overline{X}_A$ and $\overline{X}_B$ are identical, and an alternative hypothesis that the means are significantly different.

A test statistic is derived by letting μ_A equal μ_B, and combining equations 4.17 and 4.18 to give

$$\overline{X}_A \pm \frac{ts_A}{\sqrt{n_A}} = \overline{X}_B \pm \frac{ts_B}{\sqrt{n_B}}$$

Solving for $|\overline{X}_A - \overline{X}_B|$ and using a propagation of uncertainty, gives

$$|\overline{X}_A - \overline{X}_B| = t \times \sqrt{\frac{s_A^2}{n_A} + \frac{s_B^2}{n_B}}$$

Finally, solving for t, which we replace with t_{exp}, leaves us with

$$t_{exp} = \frac{|\overline{X}_A - \overline{X}_B|}{\sqrt{(s_A^2/n_A) + (s_B^2/n_B)}} \qquad 4.19$$

The value of t_{exp} is compared with a critical value, $t(\alpha, \nu)$, as determined by the chosen significance level, α, the degrees of freedom for the sample, ν, and whether the significance test is one-tailed or two-tailed.

It is unclear, however, how many degrees of freedom are associated with $t(\alpha, \nu)$ since there are two sets of independent measurements. If the variances s_A^2 and s_B^2 estimate the same σ^2, then the two standard deviations can be factored out of equation 4.19 and replaced by a pooled standard deviation, s_{pool}, which provides a better estimate for the precision of the analysis. Thus, equation 4.19 becomes

$$t_{exp} = \frac{|\overline{X}_A - \overline{X}_B|}{s_{pool}\sqrt{(1/n_A) + (1/n_B)}} \qquad 4.20$$

with the pooled standard deviation given as

$$s_{pool} = \sqrt{\frac{(n_A - 1)s_A^2 + (n_B - 1)s_B^2}{n_A + n_B - 2}} \qquad 4.21$$

As indicated by the denominator of equation 4.21, the degrees of freedom for the pooled standard deviation is $n_A + n_B - 2$.

If s_A and s_B are significantly different, however, then t_{exp} must be calculated using equation 4.19. In this case, the degrees of freedom is calculated using the following imposing equation.

$$\nu = \frac{[(s_A^2/n_A) + (s_B^2/n_B)]^2}{[(s_A^2/n_A)^2/(n_A + 1)] + [(s_B^2/n_B)^2/(n_B + 1)]} - 2 \qquad 4.22$$

Since the degrees of freedom must be an integer, the value of ν obtained using equation 4.22 is rounded to the nearest integer.

Regardless of whether equation 4.19 or 4.20 is used to calculate t_{exp}, the null hypothesis is rejected if t_{exp} is greater than $t(\alpha, \nu)$, and retained if t_{exp} is less than or equal to $t(\alpha, \nu)$.

Example

EXAMPLE 4.19

Tables 4.1 and 4.8 show results for two separate experiments to determine the mass of a circulating U.S. penny. Determine whether there is a difference in the means of these analyses at $\alpha = 0.05$.

SOLUTION

To begin with, we must determine whether the variances for the two analyses are significantly different. This is done using an *F*-test as outlined in Example 4.18. Since no significant difference was found, a pooled standard deviation with 10 degrees of freedom is calculated

$$s_{pool} = \sqrt{\frac{(n_A - 1)s_A^2 + (n_B - 1)s_B^2}{n_A + n_B - 2}}$$

$$= \sqrt{\frac{(7-1)(0.00259) + (5-1)(0.00138)}{7 + 5 - 2}}$$

$$= 0.0459$$

where the subscript A indicates the data in Table 4.1, and the subscript B indicates the data in Table 4.8. The comparison of the means for the two analyses is based on the null hypothesis

$$H_0: \quad \overline{X}_A = \overline{X}_B$$

and a two-tailed alternative hypothesis

$$H_A: \quad \overline{X}_A \neq \overline{X}_B$$

Since the standard deviations can be pooled, the test statistic is calculated using equation 4.20

$$t_{exp} = \frac{|\overline{X}_A - \overline{X}_B|}{s_{pool}\sqrt{(1/n_A + 1/n_B)}} = \frac{|3.117 - 3.081|}{0.0459\sqrt{(1/7 + 1/5)}} = 1.34$$

The critical value for $t(0.05, 10)$, from Appendix 1B, is 2.23. Since t_{exp} is less than $t(0.05, 10)$ the null hypothesis is retained, and there is no evidence that the two sets of pennies are significantly different at the chosen significance level.

Example

EXAMPLE 4.20

The %w/w Na_2CO_3 in soda ash can be determined by an acid–base titration. The results obtained by two analysts are shown here. Determine whether the difference in their mean values is significant at $\alpha = 0.05$.

Analyst A	Analyst B
86.82	81.01
87.04	86.15
86.93	81.73
87.01	83.19
86.20	80.27
87.00	83.94

SOLUTION

We begin by summarizing the mean and standard deviation for the data reported by each analyst. These values are

$$\begin{aligned} \overline{X}_A &= 86.83\% \\ s_A &= 0.32 \\ \overline{X}_B &= 82.71\% \\ s_B &= 2.16 \end{aligned}$$

A two-tailed *F*-test of the following null and alternative hypotheses

$$H_0: \; s_A^2 = s_B^2 \qquad H_A: \; s_A^2 \neq s_B^2$$

is used to determine whether a pooled standard deviation can be calculated. The test statistic is

$$F_{exp} = \frac{s_B^2}{s_A^2} = \frac{(2.16)^2}{(0.32)^2} = 45.6$$

Since F_{exp} is larger than the critical value of 7.15 for $F(0.05, 5, 5)$, the null hypothesis is rejected and the alternative hypothesis that the variances are significantly different is accepted. As a result, a pooled standard deviation cannot be calculated.

The mean values obtained by the two analysts are compared using a two-tailed *t*-test. The null and alternative hypotheses are

$$H_0: \; \overline{X}_A = \overline{X}_B \qquad H_A: \; \overline{X}_A \neq \overline{X}_B$$

Since a pooled standard deviation could not be calculated, the test statistic, t_{exp}, is calculated using equation 4.19

$$t_{exp} = \frac{\left|\overline{X}_A - \overline{X}_B\right|}{\sqrt{(s_A^2/n_A) + (s_B^2/n_B)}} = \frac{\left|86.83 - 82.71\right|}{\sqrt{[(0.32)^2/6] + [(2.16)^2/6]}} = 4.62$$

and the degrees of freedom are calculated using equation 4.22

$$\nu = \frac{[(0.32^2/6) + (2.16^2/6)]^2}{\{(0.32^2/6)^2/(6+1)\} + \{(2.16^2/6)^2/(6+1)\}} - 2 = 5.3 \approx 5$$

The critical value for $t(0.05, 5)$ is 2.57. Since the calculated value of t_{exp} is greater than $t(0.05, 5)$ we reject the null hypothesis and accept the alternative hypothesis that the mean values for %w/w Na_2CO_3 reported by the two analysts are significantly different at the chosen significance level.

Paired Data In some situations the variation within the data sets being compared is more significant than the difference between the means of the two data sets. This is commonly encountered in clinical and environmental studies, where the data being compared usually consist of a set of samples drawn from several populations. For example, a study designed to investigate two procedures for monitoring the concentration of glucose in blood might involve blood samples drawn from ten patients. If the variation in the blood glucose levels among the patients is significantly larger than the anticipated variation between the methods, then an analysis in which the data are treated as unpaired will fail to find a significant difference between the methods. In general, paired data sets are used whenever the variation being investigated is smaller than other potential sources of variation.

In a study involving paired data the difference, d_i, between the paired values for each sample is calculated. The average difference, $\overline{d}$, and standard deviation of the differences, s_d, are then calculated. The null hypothesis is that $\overline{d}$ is 0, and that there is no difference in the results for the two data sets. The alternative hypothesis is that the results for the two sets of data are significantly different, and, therefore, $\overline{d}$ is not equal to 0.

The test statistic, t_{exp}, is derived from a confidence interval around $\overline{d}$

$$0 = \overline{d} \pm \frac{ts_d}{\sqrt{n}}$$

where n is the number of paired samples. Replacing t with t_{exp} and rearranging gives

$$t_{exp} = \frac{|\overline{d}|\sqrt{n}}{s_d}$$

The value of t_{exp} is then compared with a critical value, $t(\alpha, \nu)$, which is determined by the chosen significance level, α, the degrees of freedom for the sample, ν, and whether the significance test is one-tailed or two-tailed. For paired data, the degrees of freedom is $n - 1$. If t_{exp} is greater than $t(\alpha, \nu)$, then the null hypothesis is rejected and the alternative hypothesis is accepted. If t_{exp} is less than or equal to $t(\alpha, \nu)$, then the null hypothesis is retained, and a significant difference has not been demonstrated at the stated significance level. This is known as the **paired *t*-test.**

paired *t*-test
Statistical test for comparing paired data to determine if their difference is too large to be explained by indeterminate error.

Example

EXAMPLE 4.21

Marecek and colleagues developed a new electrochemical method for the rapid quantitative analysis of the antibiotic monensin in the fermentation vats used during its production.[9] The standard method for the analysis, which is based on a test for microbiological activity, is both difficult and time-consuming. As part of the study, samples taken at different times from a fermentation production vat were analyzed for the concentration of monensin using both the electrochemical and microbiological procedures. The results, in parts per thousand (ppt),* are reported in the following table.

Sample	Microbiological	Electrochemical
1	129.5	132.3
2	89.6	91.0
3	76.6	73.6
4	52.2	58.2
5	110.8	104.2
6	50.4	49.9
7	72.4	82.1
8	141.4	154.1
9	75.0	73.4
10	34.1	38.1
11	60.3	60.1

Determine whether there is a significant difference between the methods at $\alpha = 0.05$.

*1 ppt is equivalent to 0.1%.

SOLUTION

This is an example of a paired data set since the acquisition of samples over an extended period introduces a substantial time-dependent change in the concentration of monensin. The comparison of the two methods must be done with the paired *t*-test, using the following null and two-tailed alternative hypotheses

$$H_0\colon\ \bar{d} = 0 \qquad H_A\colon\ \bar{d} \neq 0$$

Defining the difference between the methods as

$$d = X_{\text{elect}} - X_{\text{micro}}$$

we can calculate the difference for each sample

Sample	1	2	3	4	5	6	7	8	9	10	11
d	2.8	1.4	−3.0	6.0	−6.6	−0.5	9.7	12.7	−1.6	4.0	−0.2

The mean and standard deviation for the differences are 2.25 and 5.63, respectively. The test statistic is

$$t_{\text{exp}} = \frac{|\bar{d}|\sqrt{n}}{s_d} = \frac{|2.25|\sqrt{11}}{5.63} = 1.33$$

which is smaller than the critical value of 2.23 for $t(0.05, 10)$. Thus, the null hypothesis is retained, and there is no evidence that the two methods yield different results at the stated significance level.

A paired *t*-test can only be applied when the individual differences, d_i, belong to the same population. This will only be true if the determinate and indeterminate errors affecting the results are independent of the concentration of analyte in the samples. If this is not the case, a single sample with a larger error could result in a value of d_i that is substantially larger than that for the remaining samples. Including this sample in the calculation of $\bar{d}$ and s_d leads to a biased estimate of the true mean and standard deviation. For samples that span a limited range of analyte concentrations, such as that in Example 4.21, this is rarely a problem. When paired data span a wide range of concentrations, however, the magnitude of the determinate and indeterminate sources of error may not be independent of the analyte's concentration. In such cases the paired *t*-test may give misleading results since the paired data with the largest absolute determinate and indeterminate errors will dominate $\bar{d}$. In this situation a comparison is best made using a linear regression, details of which are discussed in the next chapter.

4F.5 Outliers

On occasion, a data set appears to be skewed by the presence of one or more data points that are not consistent with the remaining data points. Such values are called **outliers.** The most commonly used significance test for identifying outliers is **Dixon's Q-test.** The null hypothesis is that the apparent outlier is taken from the same population as the remaining data. The alternative hypothesis is that the outlier comes from a different population, and, therefore, should be excluded from consideration.

The Q-test compares the difference between the suspected outlier and its nearest numerical neighbor to the range of the entire data set. Data are ranked from smallest to largest so that the suspected outlier is either the first or the last data

outlier
Data point whose value is much larger or smaller than the remaining data.

Dixon's Q-test
Statistical test for deciding if an outlier can be removed from a set of data.

point. The test statistic, Q_{exp}, is calculated using equation 4.23 if the suspected outlier is the smallest value (X_1)

$$Q_{exp} = \frac{X_2 - X_1}{X_n - X_1} \tag{4.23}$$

or using equation 4.24 if the suspected outlier is the largest value (X_n)

$$Q_{exp} = \frac{X_n - X_{n-1}}{X_n - X_1} \tag{4.24}$$

where n is the number of members in the data set, including the suspected outlier. It is important to note that equations 4.23 and 4.24 are valid only for the detection of a single outlier. Other forms of Dixon's Q-test allow its extension to the detection of multiple outliers.[10] The value of Q_{exp} is compared with a critical value, $Q(\alpha, n)$, at a significance level of α. The Q-test is usually applied as the more conservative two-tailed test, even though the outlier is the smallest or largest value in the data set. Values for $Q(\alpha, n)$ can be found in Appendix 1D. If Q_{exp} is greater than $Q(\alpha, n)$, then the null hypothesis is rejected and the outlier may be rejected. When Q_{exp} is less than or equal to $Q(\alpha, n)$ the suspected outlier must be retained.

Example

EXAMPLE 4.22

The following masses, in grams, were recorded in an experiment to determine the average mass of a U.S. penny.

3.067 3.049 3.039 2.514 3.048 3.079 3.094 3.109 3.102

Determine if the value of 2.514 g is an outlier at $\alpha = 0.05$.

SOLUTION

To begin with, place the masses in order from smallest to largest

2.514 3.039 3.048 3.049 3.067 3.079 3.094 3.102 3.109

and calculate Q_{exp}

$$Q_{exp} = \frac{X_2 - X_1}{X_9 - X_1} = \frac{3.039 - 2.514}{3.109 - 2.514} = 0.882$$

The critical value for $Q(0.05, 9)$ is 0.493. Since $Q_{exp} > Q(0.05, 9)$ the value is assumed to be an outlier, and can be rejected.

The Q-test should be applied with caution since there is a probability, equal to α, that an outlier identified by the Q-test actually is not an outlier. In addition, the Q-test should be avoided when rejecting an outlier leads to a precision that is unreasonably better than the expected precision determined by a propagation of uncertainty. Given these two concerns it is not surprising that some statisticians caution against the removal of outliers.[11] On the other hand, testing for outliers can provide useful information if we try to understand the source of the suspected outlier. For example, the outlier identified in Example 4.22 represents a significant change in the mass of a penny (an approximately 17% decrease in mass), due to a change in the composition of the U.S. penny. In 1982, the composition of a U.S. penny was changed from a brass alloy consisting of 95% w/w Cu and 5% w/w Zn, to a zinc core covered with copper.[12] The pennies in Example 4.22 were therefore drawn from different populations.

4G Detection Limits

The focus of this chapter has been the evaluation of analytical data, including the use of statistics. In this final section we consider how statistics may be used to characterize a method's ability to detect trace amounts of an analyte.

A method's **detection limit** is the smallest amount or concentration of analyte that can be detected with statistical confidence. The International Union of Pure and Applied Chemistry (IUPAC) defines the detection limit as the smallest concentration or absolute amount of analyte that has a signal significantly larger than the signal arising from a reagent blank. Mathematically, the analyte's signal at the detection limit, $(S_A)_{DL}$, is

detection limit
The smallest concentration or absolute amount of analyte that can be reliably detected.

$$(S_A)_{DL} = S_{reag} + z\sigma_{reag} \qquad 4.25$$

where S_{reag} is the signal for a reagent blank, σ_{reag} is the known standard deviation for the reagent blank's signal, and z is a factor accounting for the desired confidence level. The concentration, $(C_A)_{DL}$, or absolute amount of analyte, $(n_A)_{DL}$, at the detection limit can be determined from the signal at the detection limit.

$$(C_A)_{DL} = \frac{(S_A)_{DL}}{k}$$

$$(n_A)_{DL} = \frac{(S_A)_{DL}}{k}$$

The value for z depends on the desired significance level for reporting the detection limit. Typically, z is set to 3, which, from Appendix 1A, corresponds to a significance level of $\alpha = 0.00135$. Consequently, only 0.135% of measurements made on the blank will yield signals that fall outside this range (Figure 4.12a). When σ_{reag} is unknown, the term $z\sigma_{reag}$ may be replaced with ts_{reag}, where t is the appropriate value from a t-table for a one-tailed analysis.[13]

In analyzing a sample to determine whether an analyte is present, the signal for the sample is compared with the signal for the blank. The null hypothesis is that the sample does not contain any analyte, in which case $(S_A)_{DL}$ and S_{reag} are identical. The alternative hypothesis is that the analyte is present, and $(S_A)_{DL}$ is greater than S_{reag}. If $(S_A)_{DL}$ exceeds S_{reag} by $z\sigma$(or ts), then the null hypothesis is rejected and there is evidence for the analyte's presence in the sample. The probability that the null hypothesis will be falsely rejected, a type 1 error, is the same as the significance level. Selecting z to be 3 minimizes the probability of a type 1 error to 0.135%.

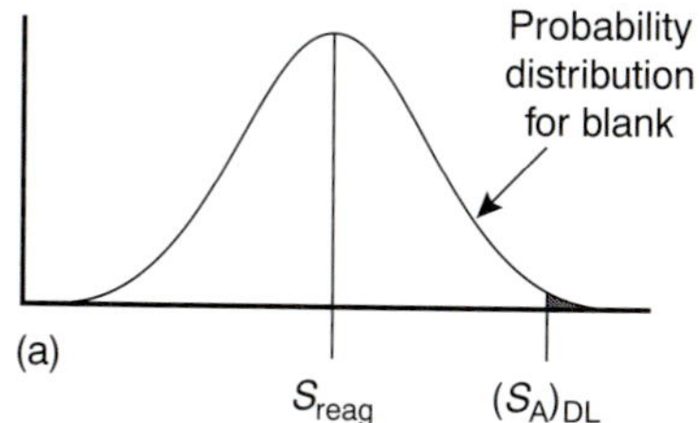

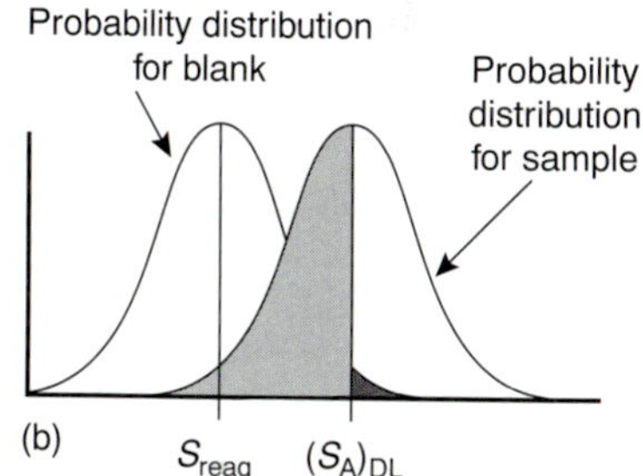

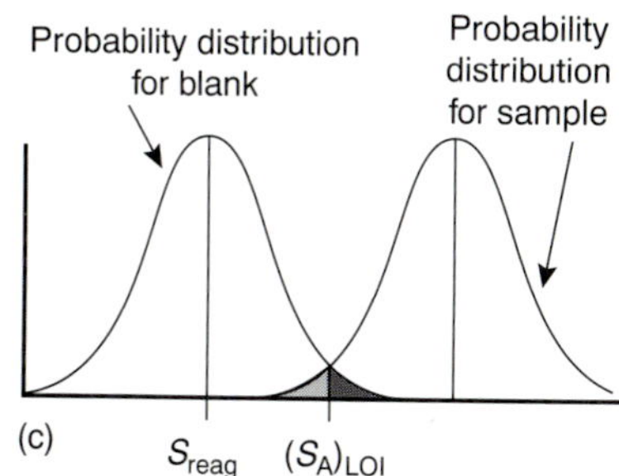

Figure 4.12
Normal distribution curves showing the definition of detection limit and limit of identification (LOI). The probability of a type 1 error is indicated by the dark shading, and the probability of a type 2 error is indicated by light shading.

Significance tests, however, also are subject to type 2 errors in which the null hypothesis is falsely retained. Consider, for example, the situation shown in Figure 4.12b, where S_A is exactly equal to $(S_A)_{DL}$. In this case the probability of a type 2 error is 50% since half of the signals arising from the sample's population fall below the detection limit. Thus, there is only a 50:50 probability that an analyte at the IUPAC detection limit will be detected. As defined, the IUPAC definition for the detection limit only indicates the smallest signal for which we can say, at a significance level of α, that an analyte is present in the sample. Failing to detect the analyte, however, does not imply that it is not present.

An alternative expression for the detection limit, which minimizes both type 1 and type 2 errors, is the **limit of identification,** $(S_A)_{LOI}$, which is defined as[14]

limit of identification
The smallest concentration or absolute amount of analyte such that the probability of type 1 and type 2 errors are equal (LOI).

$$(S_A)_{LOI} = S_{reag} + z\sigma_{reag} + z\sigma_{samp}$$

limit of quantitation
The smallest concentration or absolute amount of analyte that can be reliably determined (LOQ).

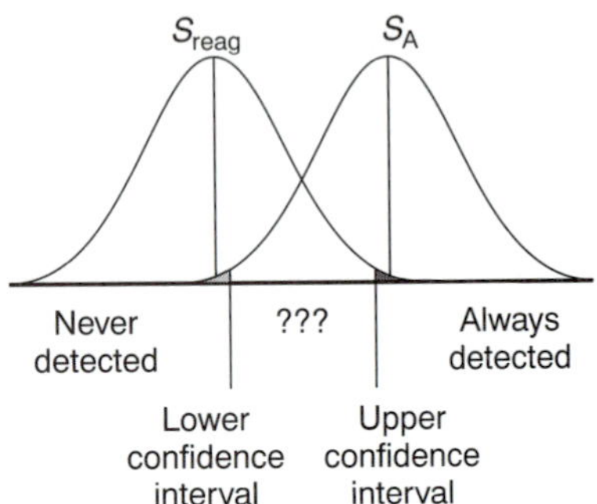

Figure 4.13
Establishment of areas where the signal is never detected, always detected, and where results are ambiguous. The upper and lower confidence limits are defined by the probability of a type 1 error (*dark shading*), and the probability of a type 2 error (*light shading*).

As shown in Figure 4.12c, the limit of identification is selected such that there is an equal probability of type 1 and type 2 errors. The American Chemical Society's Committee on Environmental Analytical Chemistry recommends the **limit of quantitation,** $(S_A)_{LOQ}$, which is defined as[15]

$$(S_A)_{LOQ} = S_{reag} + 10\sigma_{reag}$$

Other approaches for defining the detection limit have also been developed.[16]

The detection limit is often represented, particularly when used in debates over public policy issues, as a distinct line separating analytes that can be detected from those that cannot be detected.[17] This use of a detection limit is incorrect. Defining the detection limit in terms of statistical confidence levels implies that there may be a gray area where the analyte is sometimes detected and sometimes not detected. This is shown in Figure 4.13 where the upper and lower confidence limits are defined by the acceptable probabilities for type 1 and type 2 errors. Analytes producing signals greater than that defined by the upper confidence limit are always detected, and analytes giving signals smaller than the lower confidence limit are never detected. Signals falling between the upper and lower confidence limits, however, are ambiguous because they could belong to populations representing either the reagent blank or the analyte. Figure 4.12c represents the smallest value of S_A for which no such ambiguity exists.

4H KEY TERMS

alternative hypothesis (*p. 83*)
binomial distribution (*p. 72*)
central limit theorem (*p. 79*)
confidence interval (*p. 75*)
constant determinate error (*p. 60*)
degrees of freedom (*p. 80*)
detection limit (*p. 95*)
determinate error (*p. 58*)
Dixon's Q-test (*p. 93*)
error (*p. 64*)
F-test (*p. 87*)
heterogeneous (*p. 58*)
histogram (*p. 77*)
homogeneous (*p. 72*)
indeterminate error (*p. 62*)
limit of identification (*p. 95*)
limit of quantitation (*p. 96*)
mean (*p. 54*)
measurement error (*p. 58*)
median (*p. 55*)
method error (*p. 58*)
normal distribution (*p. 73*)
null hypothesis (*p. 83*)
one-tailed significance test (*p. 84*)
outlier (*p. 93*)
paired data (*p. 88*)
paired *t*-test (*p. 92*)
personal error (*p. 60*)
population (*p. 71*)
probability distribution (*p. 71*)
proportional determinate error (*p. 61*)
range (*p. 56*)
repeatability (*p. 62*)
reproducibility (*p. 62*)
sample (*p. 71*)
sampling error (*p. 58*)
significance test (*p. 83*)
standard deviation (*p. 56*)
standard reference material (*p. 61*)
tolerance (*p. 58*)
t-test (*p. 85*)
two-tailed significance test (*p. 84*)
type 1 error (*p. 84*)
type 2 error (*p. 84*)
uncertainty (*p. 64*)
unpaired data (*p. 88*)
variance (*p. 57*)

4I SUMMARY

The data we collect are characterized by their central tendency (where the values are clustered), and their spread (the variation of individual values around the central value). Central tendency is reported by stating the mean or median. The range, standard deviation, or variance may be used to report the data's spread. Data also are characterized by their errors, which include determinate errors affecting the data's accuracy, and indeterminate errors affecting the data's precision. A propagation of uncertainty allows us to estimate the affect of these determinate and indeterminate errors on results determined from our data.

The distribution of the results of an analysis around a central value is often described by a probability distribution, two examples of which are the binomial distribution and the normal distribution. Knowing the type of distribution allows us to determine the probability of obtaining results within a specified range. For a normal distribution this range is best expressed as a confidence interval.

A statistical analysis allows us to determine whether our results are significantly different from known values, or from values obtained by other analysts, by other methods of analysis, or for other samples. A *t*-test is used to compare mean values, and an *F*-test to compare precisions. Comparisons between two sets of data require an initial evaluation of whether the data is paired or unpaired. For unpaired data it is also necessary to decide if the standard deviations can be pooled. A decision about whether to retain an outlying value can be made using Dixon's *Q*-test.

Finally, we have seen that the detection limit is a statistical statement about the smallest amount of analyte that can be detected with confidence. A detection limit is not exact because its value depends on how willing we are to falsely report the analyte's presence or absence in a sample. When reporting a detection limit, you should clearly indicate how you arrived at its value.

4J *Suggested* EXPERIMENTS

Experiments

The following experiments may be used to introduce the statistical analysis of data in the analytical chemistry laboratory. Each experiment is annotated with a brief description of the data collected and the type of statistical analysis used in evaluating the data.

Cunningham, C. C.; Brown, G. R.; St Pierre, L. E. "Evaluation of Experimental Data," *J. Chem. Educ.* **1981,** *58,* 509–511.

In this experiment students determine the density of glass marbles and the radius of the bore of a glass capillary tube. Density is determined by measuring a marble's mass and volume, the latter by measuring a marble's diameter and assuming a spherical shape. Results are compared with those expected for a normal distribution. The radius of a glass capillary tube is determined using Poiseuille's equation by measuring the volume flow rate of water as a function of the hydrostatic head. In both experiments the experimentally obtained standard deviation is compared with that estimated by a propagation of uncertainty.

Gordus, A. A. "Statistical Evaluation of Class Data for Two Buret Readings," *J. Chem. Educ.* **1987,** *64,* 376–377.

The volumes of water in two burets are read, and the difference between the volumes are calculated. Students analyze the data by drawing histograms for each of the three volumes, comparing results with those predicted for a normal distribution.

Harvey, D. T. "Statistical Evaluation of Acid/Base Indicators," *J. Chem. Educ.* **1991,** *68,* 329–331.

In this experiment students standardize a solution of HCl by titration using several different indicators to signal the titration's end point. A statistical analysis of the data using *t*-tests and *F*-tests allows students to compare results obtained using the same indicator, with results obtained using different indicators. The results of this experiment can be used later when discussing the selection of appropriate indicators.

O'Reilley, J. E. "The Length of a Pestle," *J. Chem. Educ.* **1986,** *63,* 894–896.

In this experiment students measure the length of a pestle using a wooden meter stick, a stainless-steel ruler, and a vernier caliper. The data collected in this experiment provide an opportunity to discuss significant figures and sources of error. Statistical analysis includes the *Q*-test, *t*-test, and *F*-test.

Paselk, R. A. "An Experiment for Introducing Statistics to Students of Analytical and Clinical Chemistry," *J. Chem. Educ.* **1985,** *62,* 536.

Students use a commercial diluter to prepare five sets of dilutions of a stock dye solution (each set contains ten replicates) using two different diluters. Results are compared using *t*-tests and *F*-tests.

Richardson, T. H. "Reproducible Bad Data for Instruction in Statistical Methods," *J. Chem. Educ.* **1991,** *68,* 310–311.

This experiment uses the change in the mass of a U.S. penny to create data sets with outliers. Students are given a sample of ten pennies, nine of which are from one population. The *Q*-test is used to verify that the outlier can be rejected. Class data from each of the two populations of pennies are pooled and compared with results predicted for a normal distribution.

Sheeran, D. "Copper Content in Synthetic Copper Carbonate: A Statistical Comparison of Experimental and Expected Results," *J. Chem. Educ.* **1998,** *75,* 453–456.

In this experiment students synthesize basic copper(II) carbonate and determine the %w/w Cu by reducing the copper to Cu. A statistical analysis of the results shows that the synthesis does not produce $CuCO_3$, the compound that many predict to be the product (although it does not exist). Results are shown to be consistent with a hemihydrate of malachite, $Cu_2(OH)_2(CO_3) \cdot 1/2H_2O$, or azurite, $Cu_3(OH)_2(CO_3)_2$.

—*Continued*

Experiments

Continued from page 97

Spencer, R. D. "The Dependence of Strength in Plastics upon Polymer Chain Length and Chain Orientation," *J. Chem. Educ.* **1984,** *61,* 555–563.

The stretching properties of polymers are investigated by examining the effect of polymer orientation, polymer chain length, stretching rate, and temperature. Homogeneity of polymer films and consistency between lots of polymer films also are investigated. Statistical analysis of data includes Q-tests and *t*-tests.

Thomasson, K.; Lofthus-Merschman, S.; Humbert, M.; et al. "Applying Statistics in the Undergraduate Chemistry Laboratory: Experiments with Food Dyes," *J. Chem. Educ.* **1998,** *75,* 231–233.

The absorbance of solutions of food dyes is used to explore the treatment of outliers and the application of the *t*-test for comparing means.

Vitha, M. F.; Carr, P. W. "A Laboratory Exercise in Statistical Analysis of Data," *J. Chem. Educ.* **1997,** *74,* 998–1000.

Students determine the average weight of vitamin E pills using several different methods (one at a time, in sets of ten pills, and in sets of 100 pills). The data collected by the class are pooled together, plotted as histograms, and compared with results predicted by a normal distribution. The histograms and standard deviations for the pooled data also show the effect of sample size on the standard error of the mean.

4K PROBLEMS

1. The following masses were recorded for 12 different U.S. quarters (all given in grams):

5.683	5.549	5.548	5.552
5.620	5.536	5.539	5.684
5.551	5.552	5.554	5.632

Report the mean, median, range, standard deviation, and variance for these data.

2. Shown in the following rows are results for the determination of acetaminophen (in milligrams) in ten separate tablets of Excedrin Extra Strength Pain Reliever.[18]

224.3	240.4	246.3	239.4	253.1
261.7	229.4	255.5	235.5	249.7

(a) Report the mean, median, range, standard deviation, and variance for these data. (b) Assuming that $\overline{X}$ and s^2 are good approximations for μ and σ^2, and that the population is normally distributed, what percentage of tablets are expected to contain more than the standard amount of 250 mg acetaminophen per tablet?

3. Salem and Galan have developed a new method for determining the amount of morphine hydrochloride in tablets.[19] Results, in milligrams, for several tablets containing different nominal dosages follow

100-mg tablets	60-mg tablets	30-mg tablets	10-mg tablets
99.17	54.21	28.51	19.06
94.31	55.62	26.25	8.83
95.92	57.40	25.92	9.08
94.55	57.51	28.62	
93.83	52.59	24.93	

(a) For each dosage, calculate the mean and standard deviation for the milligrams of morphine hydrochloride per tablet. (b) Assuming that $\overline{X}$ and s^2 are good approximations for μ and σ^2, and that the population is normally distributed, what percentage of tablets at each dosage level are expected to contain more than the nominal amount of morphine hydrochloride per tablet?

4. Daskalakis and co-workers recently evaluated several procedures for digesting the tissues of oysters and mussels prior to analyzing the samples for silver.[20] One of the methods used to evaluate the procedure is a spike recovery in which a known amount of silver is added to the tissue sample and the percent of the added silver found on analysis is reported. Ideally, spike recoveries should fall within the range 100 ± 15%. The results for one method are

106% 108% 92% 99% 104% 101% 93% 93%

Assuming that the spike recoveries are normally distributed, what is the probability that any single spike recovery will be within the accepted range?

5. The formula weight (*FW*) of a gas can be determined using the following form of the ideal gas law

$$FW = \frac{gRT}{PV}$$

where *g* is the mass in grams, *R* is the gas constant, *T* is the temperature in kelvins, *P* is the pressure in atmospheres, and *V* is the volume in liters. In a typical analysis the following data are obtained (with estimated uncertainties in parentheses)

g = 0.118 (±0.002)
R = 0.082056 (±0.000001)
T = 298.2 (±0.1)
P = 0.724 (±0.005)
V = 0.250 (±0.005)

(a) What is the compound's formula weight and its estimated uncertainty? (b) To which variable(s) should you direct your attention if you wish to improve the uncertainty in the compound's molecular weight?

6. A standard solution of Mn^{2+} was prepared by dissolving 0.250 g of Mn in 10 mL of concentrated HNO_3 (measured with a graduated cylinder). The resulting solution was quantitatively transferred to a 100-mL volumetric flask and diluted to volume with distilled water. A 10-mL aliquot of the solution was pipeted into a 500-mL volumetric flask and diluted to volume. (a) Express the concentration of Mn in parts per million, and estimate uncertainty by a propagation of uncertainty calculation. (b) Would the uncertainty in the solution's concentration be improved by using a pipet to measure the HNO_3, instead of a graduated cylinder?

7. Hydroscopic materials often are measured by the technique of weighing by difference. In this technique the material is placed in a sealed container and weighed. A portion of the material is removed, and the container and the remaining material are reweighed. The difference between the two masses gives the amount of material that was sampled. A solution of a hydroscopic material with a gram formula weight of 121.34 (±0.01) was prepared in the following manner. A sample of the compound and its container has a mass of 23.5811 g. A portion of the compound was transferred to a 100-mL volumetric flask and diluted to volume. The mass of the compound and container after the transfer is 22.1559 g. Calculate the molarity of the solution, and estimate its uncertainty by a propagation of uncertainty calculation.

8. Show by a propagation of uncertainty calculation that the standard error of the mean for n determinations is given as $s/\sqrt{n}$.

9. What is the smallest mass that can be measured on an analytical balance with a tolerance of ±0.1 mg, such that the relative error is less than 0.1%?

10. Which of the following is the best way to dispense 100.0 mL of a reagent: (a) use a 50-mL pipet twice; (b) use a 25-mL pipet four times; or (c) use a 10-mL pipet ten times?

11. A solution can be diluted by a factor of 200 using readily available pipets (1-mL to 100-mL) and volumetric flasks (10-mL to 1000-mL) in either one, two, or three steps. Limiting yourself to glassware listed in Table 4.2, determine the proper combination of glassware to accomplish each dilution, and rank them in order of their most probable uncertainties.

12. Explain why changing all values in a data set by a constant amount will change $\overline{X}$ but will have no effect on s.

13. Obtain a sample of a metal from your instructor, and determine its density by one or both of the following methods:

Method A: Obtain the sample's mass with a balance. Calculate the sample's volume using appropriate linear dimensions.

Method B: Obtain the sample's mass with a balance. Calculate the sample's volume by measuring the amount of water that it displaces. This can be done by adding water to a graduated cylinder, reading the volume, adding the object, and reading the new volume. The difference in volumes is equal to the object's volume.

Determine the density at least five times. (a) Report the mean, the standard deviation, and the 95% confidence interval for your results. (b) Find the accepted value for the density of your metal, and determine the absolute and relative error for your experimentally determined density. (c) Use the propagation of uncertainty to determine the uncertainty for your chosen method. Are the results of this calculation consistent with your experimental results? If not, suggest some possible reasons for this disagreement.

14. How many carbon atoms must a molecule have if the mean number of ^{13}C atoms per molecule is 1.00? What percent of such molecules will have no atoms of ^{13}C?

15. In Example 4.10 we determined the probability that a molecule of cholesterol, $C_{27}H_{44}O$, had no atoms of ^{13}C. (a) Calculate the probability that a molecule of cholesterol, has one atom of ^{13}C. (b) What is the probability that a molecule of cholesterol will have two or more atoms of ^{13}C?

16. Berglund and Wichart investigated the quantitative determination of Cr in high-alloy steels by a potentiometric titration of Cr^{6+}.[21] Before titrating the steel was dissolved in acid and the chromium oxidized to Cr^{6+} by peroxydisulfate. Following are their results (%w/w Cr) for the analysis of a single reference steel.

16.968	16.922	16.840	16.883
16.887	16.977	16.857	16.728

Calculate the mean, the standard deviation, and the 95% confidence interval about the mean. What does this confidence interval mean?

17. Ketkar and co-workers developed a new analytical method for measuring trace levels of atmospheric gases.[22] The analysis of a sample containing 40.0 parts per thousand (ppt) 2-chloroethylsulfide yielded the following results

43.3	34.8	31.9	37.8	34.4	31.9	42.1	33.6	35.3

(a)Determine whether there is a significant difference between the experimental mean and the expected value at $\alpha = 0.05$. (b) As part of this study a reagent blank was analyzed 12 times, giving a mean of 0.16 ppt and a standard deviation of 1.20 ppt. What are the IUPAC detection limit, the limit of identification, and limit of quantitation for this method assuming $\alpha = 0.05$?

18. To test a spectrophotometer for its accuracy, a solution of 60.06 ppm $K_2Cr_2O_7$ in 5.0 mM H_2SO_4 is prepared and analyzed. This solution has a known absorbance of 0.640 at 350.0 nm in a 1.0-cm cell when using 5.0 mM H_2SO_4 as a reagent blank. Several aliquots of the solution are analyzed with the following results

0.639 0.638 0.640 0.639 0.640 0.639 0.638

Determine whether there is a significant difference between the experimental mean and the expected value at $\alpha = 0.01$.

19. Monna and co-workers studied the use of radioactive isotopes as a means of dating sediments collected from the bottom of lakes and estuaries.[23] To verify this method they analyzed a ^{208}Po standard known to have an activity of 77.5 decays/min, obtaining the following results

77.09	75.37	72.42	76.84	77.84	76.69
78.03	74.96	77.54	76.09	81.12	75.75

Determine whether there is a significant difference between the mean and the expected value at $\alpha = 0.05$.

20. A 2.6540-g sample of an iron ore known to contain 53.51% w/w Fe is dissolved in a small portion of concentrated HCl and diluted to volume in a 250-mL volumetric flask. A spectrophotometric method is used to determine the concentration of Fe in this solution, yielding results of 5840, 5770, 5650, and 5660 ppm. Determine whether there is a significant difference between the experimental mean and the expected value at $\alpha = 0.05$.

21. Horvat and colleagues investigated the application of atomic absorption spectroscopy to the analysis of Hg in coal fly ash.[24] Of particular interest was the development of an appropriate procedure for digesting the samples in order to release the Hg for analysis. As part of their study they tested several reagents for digesting samples. Results obtained with HNO_3 and with a 1 + 3 mixture of HNO_3 and HCl are shown here. All concentrations are given as nanograms of Hg per gram of sample.

HNO_3:	161	165	160	167	166	
1 + 3 HNO_3–HCl:	159	145	140	147	143	156

Determine whether there is a significant difference between these methods at $\alpha = 0.05$.

22. Lord Rayleigh, John William Strutt (1842–1919) was one of the most well-known scientists of the late nineteenth and early twentieth centuries, publishing over 440 papers and receiving the Nobel Prize in chemistry in 1904 for the discovery of argon. An important turning point in the discovery of Ar was Rayleigh's experimental measurements of the density of N_2. Rayleigh approached this experiment in two ways: first by taking atmospheric air and removing any O_2 and H_2 that was present; and second, by chemically producing N_2 by decomposing nitrogen-containing compounds (NO, N_2O, and NH_4NO_3) and again removing any O_2 and H_2. His results for the density of N_2, published in *Proc. Roy. Soc.* **1894,** *LV*, 340 (publication 210), follow (all values are for grams of gas at equivalent volume, pressure, and temperature).

Atmospheric Origin:	2.31017	2.30986	2.31010	2.31001
	2.31024	2.31010	2.31028	
Chemical Origin:	2.30143	2.29890	2.29816	2.30182
	2.29869	2.29940	2.29849	2.29889

Explain why these data led Rayleigh to look for and discover Ar.

23. Gács and Ferraroli reported a new method for monitoring the concentration of SO_2 in air.[25] They compared their method with the standard method by sampling and analyzing urban air from a single location. Air samples were collected by drawing air through a collection solution for 6 min. Following is a summary of their results with SO_2 concentrations reported in microliters per cubic meter.

Standard method:	21.62	22.20	24.27	23.54	24.25	23.09	21.02
New method:	21.54	20.51	22.31	21.30	24.62	25.72	21.54

Using an appropriate statistical test, determine whether there is any significant difference between the standard and new methods at $\alpha = 0.05$.

24. The accuracy of a spectrophotometer can be checked by measuring absorbances for a series of standard dichromate solutions that can be obtained in sealed cuvettes from the National Institute of Standards and Technology. Absorbances are measured at 257 nm and compared with the accepted values. The results obtained when testing a newly purchased spectrophotometer are shown here. Determine if the tested spectrophotometer is accurate at $\alpha = 0.05$.

Standard:	1	2	3	4	5
Measured absorbance:	0.2872	0.5773	0.8674	1.1623	1.4559
Accepted absorbance:	0.2871	0.5760	0.8677	1.1608	1.4565

25. Maskarinec and associates investigated the stability of volatile organics in environmental water samples.[26] Of particular interest was establishing proper conditions for maintaining the sample's integrity between its collection and analysis. Two preservatives were investigated (ascorbic acid and sodium bisulfate), and maximum holding times were determined for a number of volatile organics and water matrices. Results (in days) for surface waters follow.

	Ascorbic acid	Sodium bisulfate
methylene chloride	77	62
carbon disulfide	23	54
trichloroethane	52	51
benzene	62	42
1,1,2-trichloroethane	57	53
1,1,2,2-tetrachloroethane	33	85
tetrachloroethene	41	63
toluene	32	94
chlorobenzene	36	86

Determine whether there is a significant difference in the effectiveness of the two preservatives at $\alpha = 0.10$.

26. Using X-ray diffraction, Karstang and Kvalhein reported a new method for determining the weight percent of kalonite in complex clay minerals.[27] To test the method, nine samples containing known amounts of kalonite were prepared and analyzed. The results (as %w/w kalonite) are shown.

Actual:	5.0	10.0	20.0	40.0	50.0	60.0	80.0	90.0	95.0
Found:	6.8	11.7	19.8	40.5	53.6	61.7	78.9	91.7	94.7

Evaluate the accuracy of the method at $\alpha = 0.05$.

27. Mizutani and colleagues reported the development of a new method for the analysis of l-malate.[28] As part of their study they analyzed a series of beverages using both their method and a standard spectrophotometric procedure based on a clinical kit purchased from Boerhinger Scientific. A summary follows of their results (in parts per million).

Sample	Electrode	Spectrophotometric
Apple juice 1	34.0	33.4
Apple juice 2	22.6	28.4
Apple juice 3	29.7	29.5
Apple juice 4	24.9	24.8
Grape juice 1	17.8	18.3
Grape juice 2	14.8	15.4
Mixed fruit juice 1	8.6	8.5
Mixed fruit juice 2	31.4	31.9
White wine 1	10.8	11.5
White wine 2	17.3	17.6
White wine 3	15.7	15.4
White wine 4	18.4	18.3

Determine whether there is a significant difference between the methods at $\alpha = 0.05$.

28. Alexiev and associates describe an improved photometric method for the determination of Fe^{3+} based on its catalytic effect on the oxidation of sulphanilic acid by KIO_4.[29] As part of their study the concentration of Fe^{3+} in human serum samples was determined by the proposed method and the standard method. Following are the results, with concentrations in micromoles/L.

Sample	Proposed Method	Standard Method
1	8.25	8.06
2	9.75	8.84
3	9.75	8.36
4	9.75	8.73
5	10.75	13.13
6	11.25	13.65
7	13.88	13.85
8	14.25	13.43

Determine whether there is a significant difference between the two methods at $\alpha = 0.05$.

29. The following data were collected during a study of the concentration of Zn in samples drawn from several locations in Lake Erie (all concentrations in parts per million).

Location	$[Zn^{2+}]$ at the air–water interface	$[Zn^{2+}]$ at the sediment–water interface
1	0.430	0.415
2	0.266	0.238
3	0.567	0.390
4	0.531	0.410
5	0.707	0.605
6	0.716	0.609

Determine whether there is a significant difference between the concentration of Zn^{2+} at the air–water interface and the sediment–water interface at $\alpha = 0.05$.

30. Ten laboratories were asked to determine the concentration of an analyte A in three standard test samples. Following are the results, in parts per million.[30]

Laboratory	Sample 1	Sample 2	Sample 3
1	22.6	13.6	16.0
2	23.0	14.2	15.9
3	21.5	13.9	16.3
4	21.9	13.9	16.9
5	21.3	13.5	16.7
6	22.1	13.5	17.4
7	23.1	13.9	17.5
8	21.7	13.5	16.8
9	22.2	12.9	17.2
10	21.7	13.8	16.7

Determine if there are any potential outliers in Sample 1, Sample 2, or Sample 3 at a significance level of $\alpha = 0.05$.

31. When copper metal and powdered sulfur are placed in a crucible and ignited, the product is a sulfide with an empirical formula of Cu_xS. The value of x can be determined by weighing the Cu and S before ignition, and finding the mass of Cu_xS when the reaction is complete. Following are the Cu/S ratios from 62 such experiments.

1.764	1.838	1.865	1.866	1.872	1.877	1.890	1.891	1.891
1.897	1.899	1.900	1.906	1.908	1.910	1.911	1.916	1.919
1.920	1.922	1.927	1.931	1.935	1.936	1.936	1.937	1.939
1.939	1.940	1.941	1.941	1.942	1.943	1.948	1.953	1.955
1.957	1.957	1.957	1.959	1.962	1.963	1.963	1.963	1.966
1.968	1.969	1.973	1.975	1.976	1.977	1.981	1.981	1.988
1.993	1.993	1.995	1.995	1.995	2.017	2.029	2.042	

(a) Calculate the mean and standard deviation for these data. (b) Construct a histogram for this data set. From a visual inspection of your histogram, do the data appear to be normally distributed? (c) In a normally distributed population, 68.26% of all members lie within the range $\mu \pm 1\sigma$. What percentage of the data lies within the range $\overline{X} \pm 1s$? Does this support your answer to the previous question? (d) Assuming that $\overline{X}$ and σ^2 are good approximations for μ and σ^2, what percentage of all experimentally determined Cu/S ratios will be greater than 2? How does this compare with the experimental data? Does this support your conclusion about whether the data are normally distributed? (e) It has been reported that this method for preparing copper sulfide results in a nonstoichiometric compound with a Cu/S ratio of less than 2. Determine if the mean value for these data is significantly less than 2 at a significance level of $\alpha = 0.01$.

4L SUGGESTED READINGS

A more comprehensive discussion of the analysis of data, covering all topics considered in this chapter as well as additional material, can be found in any textbook on statistics or data analysis; following are several such texts.

Anderson, R. L. *Practical Statistics for Analytical Chemists.* Van Nostrand Reinhold: New York, 1987.

Graham, R. C. *Data Analysis for the Chemical Sciences.* VCH Publishers: New York, 1993.

Mark, H.; Workman, J. *Statistics in Spectroscopy.* Academic Press: Boston, 1991.

Mason, R. L.; Gunst, R. F.; Hess, J. L. *Statistical Design and Analysis of Experiments.* Wiley: New York, 1989.

Miller, J. C.; Miller, J. N. *Statistics for Analytical Chemistry,* 3rd ed. Ellis Horwood PTR Prentice-Hall: New York, 1993.

Sharaf, M. H.; Illman, D. L.; Kowalski, B. R. *Chemometrics.* Wiley-Interscience: New York, 1986.

The difference between precision and accuracy and a discussion of indeterminate and determinate sources of error is covered in the following paper.

Treptow, R. S. "Precision and Accuracy in Measurements," *J. Chem. Educ.* **1998,** *75,* 992–995.

The detection of outliers, particularly when working with a small number of samples, is discussed in the following papers.

Efstathiou, C. "Stochastic Calculation of Critical Q-Test Values for the Detection of Outliers in Measurements," *J. Chem. Educ.* **1992,** *69,* 773–736.

Kelly, P. C. "Outlier Detection in Collaborative Studies," *Anal. Chem.* **1990,** *73,* 58–64.

Mitschele, J. "Small Sample Statistics," *J. Chem. Educ.* **1991,** *68,* 470–473.

The following papers provide additional information on error and uncertainty, including the propagation of uncertainty.

Andraos, J. "On the Propagation of Statistical Errors for a Function of Several Variables," *J. Chem. Educ.* **1996,** *73,* 150–154.

Donato, H.; Metz, C. "A Direct Method for the Propagation of Error Using a Personal Computer Spreadsheet Program," *J. Chem. Educ.* **1988,** *65,* 867–868.

Gordon, R.; Pickering, M.; Bisson, D. "Uncertainty Analysis by the 'Worst Case' Method," *J. Chem. Educ.* **1984,** *61,* 780–781.

Guare, C. J. "Error, Precision and Uncertainty," *J. Chem. Educ.* **1991,** *68,* 649–652.

Guedens, W. J.; Yperman, J.; Mullens, J.; et al. "Statistical Analysis of Errors: A Practical Approach for an Undergraduate Chemistry Lab," Part 1. The Concept, *J. Chem. Educ.* **1993,** *70,* 776–779; Part 2. Some Worked Examples, *J. Chem. Educ.* **1993,** *70,* 838–841.

Heydorn, K. "Detecting Errors in Micro and Trace Analysis by Using Statistics," *Anal. Chim. Acta* **1993,** *283,* 494–499.

Taylor, B. N.; Kuyatt, C. E. "Guidelines for Evaluating and Expressing the Uncertainty of NIST Measurement Results," *NIST Technical Note* 1297, **1994.**

A further discussion of detection limits is found in

Currie, L. A., ed. *Detection in Analytical Chemistry: Importance, Theory and Practice.* American Chemical Society: Washington, DC, 1988.

4M REFERENCES

1. Goedhart, M. J.; Verdonk, A. H. *J. Chem. Educ.* **1991,** *68,* 1005–1009.
2. Rousseeuw, P. J. *J. Chemom.* **1991,** *5,* 1–20.
3. Ellison, S.; Wegscheider, W.; Williams, A. *Anal. Chem.* **1997,** *69,* 607A–613A.
4. Shoemaker, D. P.; Garland, C. W.; Nibler, J. W. *Experiments in Physical Chemistry,* 5th ed. McGraw-Hill: New York, **1989,** pp. 55–63.
5. Lam, R. B.; Isenhour, T. L. *Anal. Chem.* **1980,** *52,* 1158–1161.
6. Mark, H.; Workman, J. *Spectroscopy,* **1988,** *3(1),* 44–48.
7. Winn, R. L. *Statistics for Scientists and Engineers,* Prentice-Hall: Englewood Cliffs, NJ, 1964; pp. 165–174.
8. Mark, H.; Workman, J. *Spectroscopy,* **1989,** *4(3),* 56–58.
9. Marecek, V.; Janchenova, H.; Brezina, M.; et al. *Anal. Chim. Acta* **1991,** *244,* 15–19.
10. Rorabacher, D. B. *Anal. Chem.* **1991,** *63,* 139–146.
11. Deming, W. E. *Statistical Adjustment of Data.* Wiley: New York, 1943 (republished by Dover: New York, 1961); p. 171.
12. Richardson, T. H. *J. Chem. Educ.* **1991,** *68,* 310–311.
13. Kirchner, C. J. "Estimation of Detection Limits for Environmental Analytical Procedures," In Currie, L. A., ed. *Detection in Analytical Chemistry: Importance, Theory and Practice.* American Chemical Society: Washington, DC, 1988.
14. Long, G. L.; Winefordner, J. D. *Anal. Chem.* **1983,** *55,* 712A–724A.

15. "Guidelines for Data Acquisition and Data Quality Control Evaluation in Environmental Chemistry," *Anal. Chem.* **1980,** *52,* 2242–2249.
16. (a) Ferrus, R.; Egea, M. R. *Anal. Chim. Acta* **1994,** *287,* 119–145; (b) Glaser, J. A.; Foerst, D. L.; McKee, G. D.; et al. *Environ. Sci. Technol.* **1981,** *15,* 1426–1435; (c) Boumans, P. W. J. M. *Anal. Chem.* **1994,** *66,* 459A–467A; (d) Kimbrough, D. E.; Wakakuwa, J. *Environ. Sci. Technol.* **1994,** *28,* 338–345; (e) Currie, L. A. *Anal. Chem.* **1968,** *40,* 586–593.
17. (a) Rogers, L. B. *J. Chem. Educ.* **1986,** *63,* 3–6; (b) Mark, H.; Workman, J. *Spectroscopy,* **1989,** *4(1),* 52–55.
18. Simonian, M. H.; Dinh, S.; Fray, L. A. *Spectroscopy* **1993,** *8(6),* 37–47.
19. Salem, I. I.; Galan, A. C. *Anal. Chim. Acta* **1993,** *283,* 334–337.
20. Daskalakis, K. D.; O'Connor, T. P.; Crecelius, E. A. *Environ. Sci. Technol.* **1997,** *31,* 2303–2306.
21. Berglund, B.; Wichardt, C. *Anal. Chim. Acta* **1990,** *236,* 399–410.
22. Ketkar, S. N.; Dulak, J. G.; Dheandhanou, S.; et al. *Anal. Chim. Acta.* **1991,** *245,* 267–270.
23. Monna, F.; Mathieu, D.; Marques Jr., et al. *Anal. Chim. Acta* **1996,** *330,* 107–116.
24. Horvat, M.; Lupsina, V.; Pihlar, B. *Anal. Chim. Acta.* **1991,** *243,* 71–79.
25. Gács, I.; Ferraroli, R. *Anal. Chim. Acta* **1992,** *269,* 177–185.
26. Maskarinec, M. P.; Johnson, L. H.; Holladay, S. K.; et al. *Environ. Sci. Technol.* **1990,** *24,* 1665–1670.
27. Karstang, T. V.; Kvalhein, O. M. *Anal. Chem.* **1991,** *63,* 767–772.
28. Mizutani, F.; Yabuki, S.; Asai, M. *Anal. Chim. Acta.* **1991,** *245,* 145–150.
29. Alexiev, A.; Rubino, S.; Deyanova, M.; et al. *Anal. Chem. Acta* **1994,** *295,* 211–219.
30. These data are adapted from Steiner, E. H. "Planning and Analysis of Results of Collaborative Tests" published in *Statistical Manual of the Association of Official Analytical Chemists,* Association of Official Analytical Chemists: Washington, DC, 1975.

Chapter 5

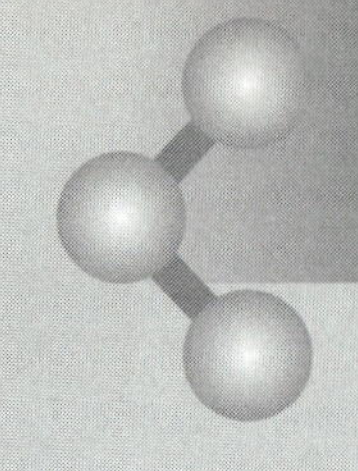

Calibrations, Standardizations, and Blank Corrections

In Chapter 3 we introduced a relationship between the measured signal, S_{meas}, and the absolute amount of analyte

$$S_{meas} = kn_A + S_{reag} \qquad 5.1$$

or the relative amount of analyte in a sample

$$S_{meas} = kC_A + S_{reag} \qquad 5.2$$

where n_A is the moles of analyte, C_A is the analyte's concentration, k is the method's sensitivity, and S_{reag} is the contribution to S_{meas} from constant errors introduced by the reagents used in the analysis. To obtain an accurate value for n_A or C_A it is necessary to avoid determinate errors affecting S_{meas}, k, and S_{reag}. This is accomplished by a combination of calibrations, standardizations, and reagent blanks.

5A Calibrating Signals

Signals are measured using equipment or instruments that must be properly calibrated if S_{meas} is to be free of determinate errors. Calibration is accomplished against a standard, adjusting S_{meas} until it agrees with the standard's known signal. Several common examples of calibration are discussed here.

When the signal is a measurement of mass, S_{meas} is determined with an analytical balance. Before a balance can be used, it must be calibrated against a reference weight meeting standards established by either the National Institute for Standards and Technology or the American Society for Testing and Materials. With an electronic balance the sample's mass is determined by the current required to generate an upward electromagnetic force counteracting the sample's downward gravitational force. The balance's calibration procedure invokes an internally programmed calibration routine specifying the reference weight to be used. The reference weight is placed on the balance's weighing pan, and the relationship between the displacement of the weighing pan and the counteracting current is automatically adjusted.

Calibrating a balance, however, does not eliminate all sources of determinate error. Due to the buoyancy of air, an object's weight in air is always lighter than its weight in vacuum. If there is a difference between the density of the object being weighed and the density of the weights used to calibrate the balance, then a correction to the object's weight must be made.[1] An object's true weight in vacuo, W_v, is related to its weight in air, W_a, by the equation

$$W_v = W_a \times \left[1 + \left(\frac{1}{D_o} - \frac{1}{D_w}\right) \times 0.0012\right]$$

where D_o is the object's density, D_w is the density of the calibration weight, and 0.0012 is the density of air under normal laboratory conditions (all densities are in units of g/cm^3). Clearly the greater the difference between D_o and D_w the more serious the error in the object's measured weight.

The buoyancy correction for a solid is small, and frequently ignored. It may be significant, however, for liquids and gases of low density. This is particularly important when calibrating glassware. For example, a volumetric pipet is calibrated by carefully filling the pipet with water to its calibration mark, dispensing the water into a tared beaker and determining the mass of water transferred. After correcting for the buoyancy of air, the density of water is used to calculate the volume of water dispensed by the pipet.

Example

EXAMPLE 5.1

A 10-mL volumetric pipet was calibrated following the procedure just outlined, using a balance calibrated with brass weights having a density of 8.40 g/cm^3. At 25 °C the pipet was found to dispense 9.9736 g of water. What is the actual volume dispensed by the pipet?

SOLUTION

At 25 °C the density of water is 0.99705 g/cm^3. The water's true weight, therefore, is

$$W_v = 9.9736 \text{ g} \times \left[1 + \left(\frac{1}{0.99705} - \frac{1}{8.40}\right) \times 0.0012\right] = 9.9842 \text{ g}$$

and the actual volume of water dispensed by the pipet is

$$\frac{9.9842 \text{ g}}{0.99705 \text{ g/cm}^3} = 10.014 \text{ cm}^3 = 10.014 \text{ mL}$$

If the buoyancy correction is ignored, the pipet's volume is reported as

$$\frac{9.9736 \text{ g}}{0.99705 \text{ g/cm}^3} = 10.003 \text{ cm}^3 = 10.003 \text{ mL}$$

introducing a negative determinate error of –0.11%.

Balances and volumetric glassware are examples of laboratory equipment. Laboratory instrumentation also must be calibrated using a standard providing a known response. For example, a spectrophotometer's accuracy can be evaluated by measuring the absorbance of a carefully prepared solution of 60.06 ppm $K_2Cr_2O_7$ in 0.0050 M H_2SO_4, using 0.0050 M H_2SO_4 as a reagent blank.[2] The spectrophotometer is considered calibrated if the resulting absorbance at a wavelength of 350.0 nm is 0.640 ± 0.010 absorbance units. Be sure to read and carefully follow the calibration instructions provided with any instrument you use.

5B Standardizing Methods

The American Chemical Society's Committee on Environmental Improvement defines standardization as the process of determining the relationship between the measured signal and the amount of analyte.[3] A method is considered standardized when the value of k in equation 5.1 or 5.2 is known.

In principle, it should be possible to derive the value of k for any method by considering the chemical and physical processes responsible for the signal. Unfortunately, such calculations are often of limited utility due either to an insufficiently developed theoretical model of the physical processes or to nonideal chemical behavior. In such situations the value of k must be determined experimentally by analyzing one or more standard solutions containing known amounts of analyte. In this section we consider several approaches for determining the value of k. For simplicity we will assume that S_{reag} has been accounted for by a proper reagent blank, allowing us to replace S_{meas} in equations 5.1 and 5.2 with the signal for the species being measured.

5B.1 Reagents Used as Standards

The accuracy of a standardization depends on the quality of the reagents and glassware used to prepare standards. For example, in an acid–base titration, the amount of analyte is related to the absolute amount of titrant used in the analysis by the stoichiometry of the chemical reaction between the analyte and the titrant. The amount of titrant used is the product of the signal (which is the volume of titrant) and the titrant's concentration. Thus, the accuracy of a titrimetric analysis can be no better than the accuracy to which the titrant's concentration is known.

primary reagent
A reagent of known purity that can be used to make a solution of known concentration.

Primary Reagents Reagents used as standards are divided into primary reagents and secondary reagents. A **primary reagent** can be used to prepare a standard containing an accurately known amount of analyte. For example, an accurately weighed sample of 0.1250 g $K_2Cr_2O_7$ contains exactly 4.249×10^{-4} mol of $K_2Cr_2O_7$. If this same sample is placed in a 250-mL volumetric flask and diluted to volume, the concentration of the resulting solution is exactly 1.700×10^{-3} M. A primary reagent must have a known stoichiometry, a known purity (or assay), and be stable during long-term storage both in solid and solution form. Because of the difficulty in establishing the degree of hydration, even after drying, hydrated materials usually are not considered primary reagents. Reagents not meeting these criteria are called **secondary reagents.** The purity of a secondary reagent in solid form or the concentration of a standard prepared from a secondary reagent must be determined relative to a primary reagent. Lists of acceptable primary reagents are available.[4] Appendix 2 contains a selected listing of primary standards.

secondary reagent
A reagent whose purity must be established relative to a primary reagent.

Other Reagents Preparing a standard often requires additional substances that are not primary or secondary reagents. When a standard is prepared in solution, for example, a suitable solvent and solution matrix must be used. Each of these solvents and reagents is a potential source of additional analyte that, if unaccounted for, leads to a determinate error. If available, **reagent grade** chemicals conforming to standards set by the American Chemical Society should be used.[5] The packaging label included with a reagent grade chemical (Figure 5.1) lists either the maximum allowed limit for specific impurities or provides the actual assayed values for the impurities as reported by the manufacturer. The purity of a reagent grade chemical can be improved by purification or by conducting a more accurate assay. As discussed later in the chapter, contributions to S_{meas} from impurities in the sample matrix can be compensated for by including an appropriate blank determination in the analytical procedure.

reagent grade
Reagents conforming to standards set by the American Chemical Society.

(a)

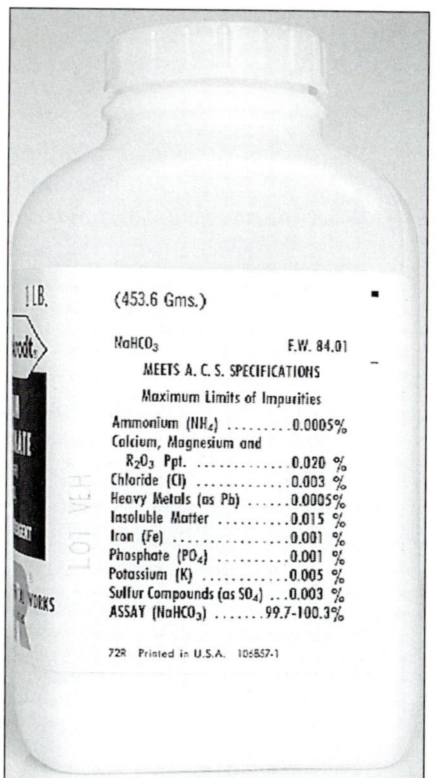

(b)

Figure 5.1
Examples of typical packaging labels from reagent grade chemicals. Label (a) provides the actual lot assay for the reagent as determined by the manufacturer. Note that potassium has been flagged with an asterisk (*) because its assay exceeds the maximum limit established by the American Chemical Society (ACS). Label (b) does not provide assayed values, but indicates that the reagent meets the specifications of the ACS for the listed impurities. An assay for the reagent also is provided.

Preparing Standard Solutions Solutions of primary standards generally are prepared in class A volumetric glassware to minimize determinate errors. Even so, the relative error in preparing a primary standard is typically ±0.1%. The relative error can be improved if the glassware is first calibrated as described in Example 5.1. It also is possible to prepare standards gravimetrically by taking a known mass of standard, dissolving it in a solvent, and weighing the resulting solution. Relative errors of ±0.01% can typically be achieved in this fashion.

It is often necessary to prepare a series of standard solutions, each with a different concentration of analyte. Such solutions may be prepared in two ways. If the range of concentrations is limited to only one or two orders of magnitude, the solutions are best prepared by transferring a known mass or volume of the pure standard to a volumetric flask and diluting to volume. When working with larger concentration ranges, particularly those extending over more than three orders of magnitude, standards are best prepared by a serial dilution from a single stock solution. In a serial dilution a volume of a concentrated stock solution, which is the first standard, is diluted to prepare a second standard. A portion of the second standard is then diluted to prepare a third standard, and the process is repeated until all necessary standards have been prepared. Serial dilutions must be prepared with extra care because a determinate error in the preparation of any single standard is passed on to all succeeding standards.

5B.2 Single-Point versus Multiple-Point Standardizations*

single-point standardization
Any standardization using a single standard containing a known amount of analyte.

The simplest way to determine the value of k in equation 5.2 is by a **single-point standardization.** A single standard containing a known concentration of analyte, C_S, is prepared and its signal, S_{stand}, is measured. The value of k is calculated as

$$k = \frac{S_{stand}}{C_S} \qquad 5.3$$

A single-point standardization is the least desirable way to standardize a method. When using a single standard, all experimental errors, both determinate and indeterminate, are carried over into the calculated value for k. Any uncertainty in the value of k increases the uncertainty in the analyte's concentration. In addition, equation 5.3 establishes the standardization relationship for only a single concentration of analyte. Extending equation 5.3 to samples containing concentrations of analyte different from that in the standard assumes that the value of k is constant, an assumption that is often not true.[6] Figure 5.2 shows how assuming a constant value of k may lead to a determinate error. Despite these limitations, single-point standardizations are routinely used in many laboratories when the analyte's range of expected concentrations is limited. Under these conditions it is often safe to assume that k is constant (although this assumption should be verified experimentally). This is the case, for example, in clinical laboratories where many automated analyzers use only a single standard.

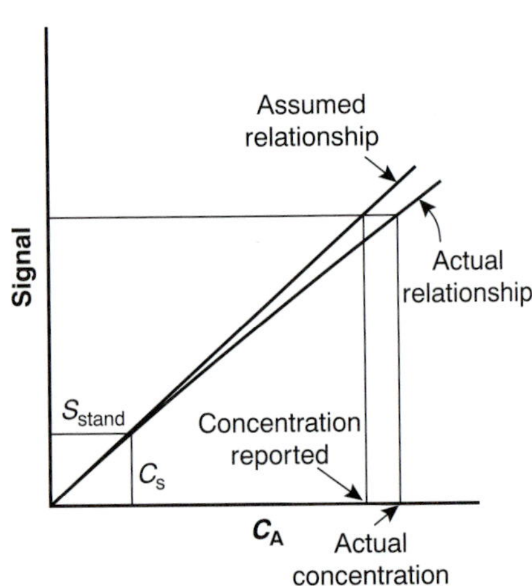

Figure 5.2
Example showing how an improper use of a single-point standardization can lead to a determinate error in the reported concentration of analyte.

The preferred approach to standardizing a method is to prepare a series of standards, each containing the analyte at a different concentration. Standards are chosen such that they bracket the expected range for the

*The following discussion of standardizations assumes that the amount of analyte is expressed as a concentration. It also applies, however, when the absolute amount of analyte is given in grams or moles.

analyte's concentration. Thus, a **multiple-point standardization** should use at least three standards, although more are preferable. A plot of S_{stand} versus C_S is known as a calibration curve. The exact standardization, or calibration relationship, is determined by an appropriate curve-fitting algorithm.* Several approaches to standardization are discussed in the following sections.

multiple-point standardization
Any standardization using two or more standards containing known amounts of analyte.

5B.3 External Standards

The most commonly employed standardization method uses one or more **external standards** containing known concentrations of analyte. These standards are identified as external standards because they are prepared and analyzed separately from the samples.

external standard
A standard solution containing a known amount of analyte, prepared separately from samples containing the analyte.

A quantitative determination using a single external standard was described at the beginning of this section, with k given by equation 5.3. Once standardized, the concentration of analyte, C_A, is given as

$$C_A = \frac{S_{samp}}{k} \qquad 5.4$$

EXAMPLE 5.2

A spectrophotometric method for the quantitative determination of Pb^{2+} levels in blood yields an S_{stand} of 0.474 for a standard whose concentration of lead is 1.75 ppb. How many parts per billion of Pb^{2+} occur in a sample of blood if S_{samp} is 0.361?

SOLUTION

Equation 5.3 allows us to calculate the value of k for this method using the data for the standard

$$k = \frac{S_{stand}}{C_S} = \frac{0.474}{1.75 \text{ ppb}} = 0.2709 \text{ ppb}^{-1}$$

Once k is known, the concentration of Pb^{2+} in the sample of blood can be calculated using equation 5.4

$$C_A = \frac{S_{samp}}{k} = \frac{0.361}{0.2709 \text{ ppb}^{-1}} = 1.33 \text{ ppb}$$

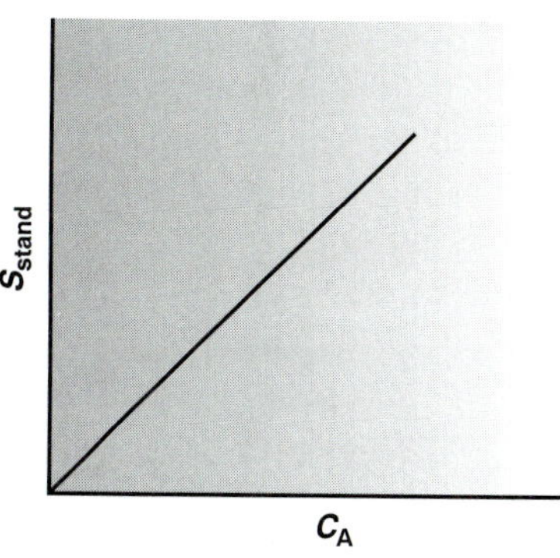

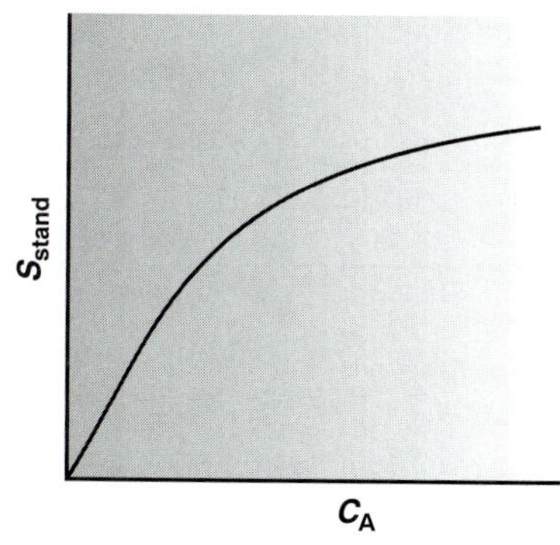

Figure 5.3
Examples of (a) straight-line and (b) curved normal calibration curves.

A multiple-point external standardization is accomplished by constructing a calibration curve, two examples of which are shown in Figure 5.3. Since this is the most frequently employed method of standardization, the resulting relationship often is called a **normal calibration curve.** When the calibration curve is a linear (Figure 5.3a), the slope of the line gives the value of k. This is the most desirable situation since the method's sensitivity remains constant throughout the standard's concentration range. When the calibration curve is nonlinear, the method's sensitivity is a function of the analyte's concentration. In Figure 5.3b, for example, the value of k is greatest when the analyte's concentration is small and decreases continuously as the amount of analyte is increased. The value of k at any point along the calibration curve is given by the slope at that point. In

normal calibration curve
A calibration curve prepared using several external standards.

*Linear regression, also known as the method of least squares, is covered in Section 5C.

either case, the calibration curve provides a means for relating S_{samp} to the analyte's concentration.

Colorplate 1 shows an example of a set of external standards and their corresponding normal calibration curve.

Example

EXAMPLE 5.3

A second spectrophotometric method for the quantitative determination of Pb^{2+} levels in blood gives a linear normal calibration curve for which

$$S_{stand} = (0.296 \text{ ppb}^{-1}) \times C_S + 0.003$$

What is the Pb^{2+} level (in ppb) in a sample of blood if S_{samp} is 0.397?

SOLUTION

To determine the concentration of Pb^{2+} in the sample of blood, we replace S_{stand} in the calibration equation with S_{samp} and solve for C_A

$$C_A = \frac{S_{samp} - 0.003}{0.296 \text{ ppb}^{-1}} = \frac{0.397 - 0.003}{0.296 \text{ ppb}^{-1}} = 1.33 \text{ ppb}$$

It is worth noting that the calibration equation in this problem includes an extra term that is not in equation 5.3. Ideally, we expect the calibration curve to give a signal of zero when C_S is zero. This is the purpose of using a reagent blank to correct the measured signal. The extra term of +0.003 in our calibration equation results from uncertainty in measuring the signal for the reagent blank and the standards.

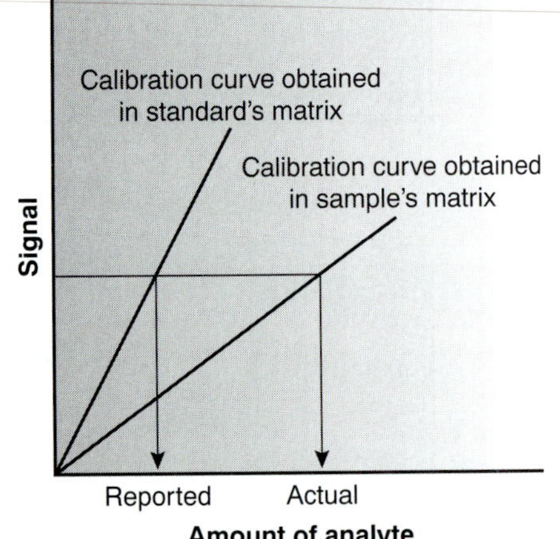

Figure 5.4
Effect of the sample's matrix on a normal calibration curve.

An external standardization allows a related series of samples to be analyzed using a single calibration curve. This is an important advantage in laboratories where many samples are to be analyzed or when the need for a rapid throughput of samples is critical. Not surprisingly, many of the most commonly encountered quantitative analytical methods are based on an external standardization.

There is a serious limitation, however, to an external standardization. The relationship between S_{stand} and C_S in equation 5.3 is determined when the analyte is present in the external standard's matrix. In using an external standardization, we assume that any difference between the matrix of the standards and the sample's matrix has no effect on the value of k. A proportional determinate error is introduced when differences between the two matrices cannot be ignored. This is shown in Figure 5.4, where the relationship between the signal and the amount of analyte is shown for both the sample's matrix and the standard's matrix. In this example, using a normal calibration curve results in a negative determinate error. When matrix problems are expected, an effort is made to match the matrix of the standards to that of the sample. This is known as **matrix matching.** When the sample's matrix is unknown, the matrix effect must be shown to be negligible, or an alternative method of standardization must be used. Both approaches are discussed in the following sections.

matrix matching
Adjusting the matrix of an external standard so that it is the same as the matrix of the samples to be analyzed.

method of standard additions
A standardization in which aliquots of a standard solution are added to the sample.

5B.4 Standard Additions

The complication of matching the matrix of the standards to that of the sample can be avoided by conducting the standardization in the sample. This is known as the **method of standard additions.** The simplest version of a standard addi-

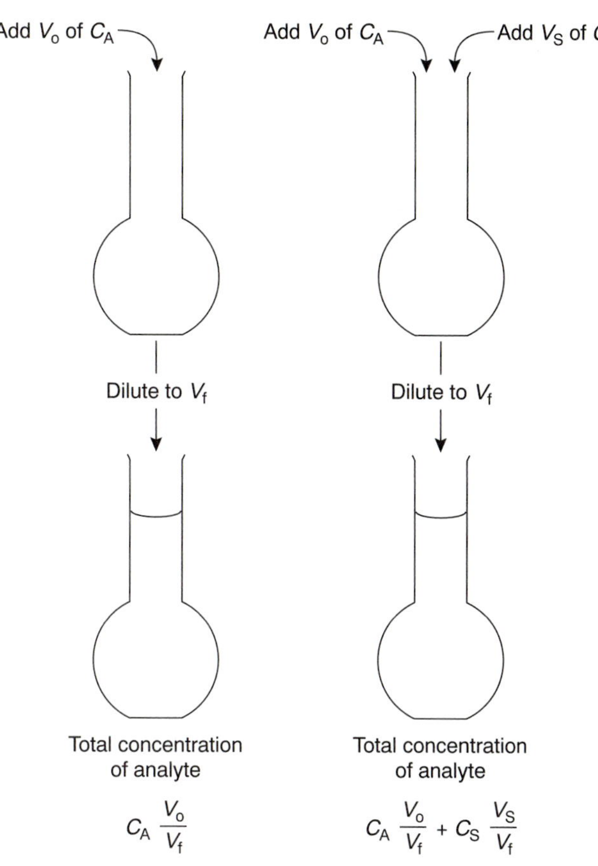

Figure 5.5
Illustration showing the method of standard additions in which separate aliquots of sample are diluted to the same final volume. One aliquot of sample is spiked with a known volume of a standard solution of analyte before diluting to the final volume.

tion is shown in Figure 5.5. A volume, V_o, of sample is diluted to a final volume, V_f, and the signal, S_{samp} is measured. A second identical **aliquot** of sample is spiked with a volume, V_s, of a standard solution for which the analyte's concentration, C_S, is known. The spiked sample is diluted to the same final volume and its signal, S_{spike}, is recorded. The following two equations relate S_{samp} and S_{spike} to the concentration of analyte, C_A, in the original sample

aliquot
A portion of a solution.

$$S_{samp} = kC_A \frac{V_o}{V_f} \qquad 5.5$$

$$S_{spike} = k\left(C_A \frac{V_o}{V_f} + C_S \frac{V_s}{V_f}\right) \qquad 5.6$$

where the ratios V_o/V_f and V_s/V_f account for the dilution. As long as V_s is small relative to V_o, the effect of adding the standard to the sample's matrix is insignificant, and the matrices of the sample and the spiked sample may be considered identical. Under these conditions the value of k is the same in equations 5.5 and 5.6. Solving both equations for k and equating gives

$$\frac{S_{samp}}{C_A(V_o/V_f)} = \frac{S_{spike}}{C_A(V_o/V_f) + C_S(V_s/V_f)} \qquad 5.7$$

Equation 5.7 can be solved for the concentration of analyte in the original sample.

Example

EXAMPLE 5.4

A third spectrophotometric method for the quantitative determination of the concentration of Pb^{2+} in blood yields an S_{samp} of 0.193 for a 1.00-mL sample of blood that has been diluted to 5.00 mL. A second 1.00-mL sample is spiked with 1.00 μL of a 1560-ppb Pb^{2+} standard and diluted to 5.00 mL, yielding an S_{spike} of 0.419. Determine the concentration of Pb^{2+} in the original sample of blood.

SOLUTION

The concentration of Pb^{2+} in the original sample of blood can be determined by making appropriate substitutions into equation 5.7 and solving for C_A. Note that all volumes must be in the same units, thus V_s is converted from 1.00 μL to 1.00×10^{-3} mL.

$$\frac{0.193}{C_A\left(\dfrac{1.00 \text{ mL}}{5.00 \text{ mL}}\right)} = \frac{0.419}{C_A\left(\dfrac{1.00 \text{ mL}}{5.00 \text{ mL}}\right) + 1560 \text{ ppb}\left(\dfrac{1.00 \times 10^{-3} \text{ mL}}{5.00 \text{ mL}}\right)}$$

$$\frac{0.193}{0.200C_A} = \frac{0.419}{0.200C_A + 0.312 \text{ ppb}}$$

$$0.0386C_A + 0.0602 \text{ ppb} = 0.0838C_A$$

$$0.0452C_A = 0.0602 \text{ ppb}$$

$$C_A = 1.33 \text{ ppb}$$

Thus, the concentration of Pb^{2+} in the original sample of blood is 1.33 ppb.

It also is possible to make a standard addition directly to the sample after measuring S_{samp} (Figure 5.6). In this case, the final volume after the standard addition is $V_o + V_s$ and equations 5.5–5.7 become

$$S_{samp} = kC_A$$

$$S_{spike} = k\left(C_A \frac{V_o}{V_o + V_s} + C_S \frac{V_s}{V_o + V_s}\right) \qquad 5.8$$

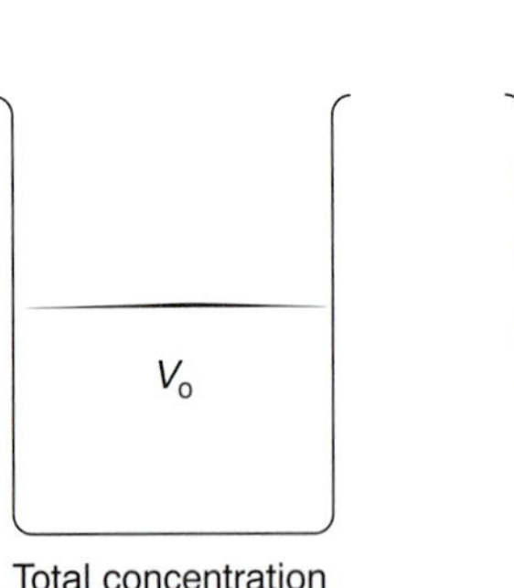

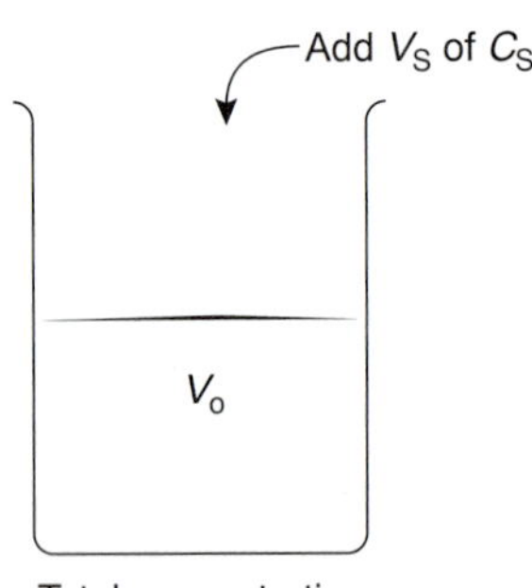

Figure 5.6
Illustration showing an alternative form of the method of standard additions. In this case a sample containing the analyte is spiked with a known volume of a standard solution of analyte without further diluting either the sample or the spiked sample.

$$\frac{S_{samp}}{C_A} = \frac{S_{spike}}{C_A[V_o/(V_o + V_s)] + C_S[V_s/(V_o + V_s)]} \qquad 5.9$$

Example

EXAMPLE 5.5

A fourth spectrophotometric method for the quantitative determination of the concentration of Pb^{2+} in blood yields an S_{samp} of 0.712 for a 5.00-mL sample of blood. After spiking the blood sample with 5.00 μL of a 1560-ppb Pb^{2+} standard, an S_{spike} of 1.546 is measured. Determine the concentration of Pb^{2+} in the original sample of blood.

SOLUTION

The concentration of Pb^{2+} in the original sample of blood can be determined by making appropriate substitutions into equation 5.9 and solving for C_A.

$$\frac{0.712}{C_A} = \frac{1.546}{C_A\left[\dfrac{5.00\text{ mL}}{(5.00\text{ mL} + 5.00 \times 10^{-3}\text{ mL})}\right] + 1560\text{ ppb}\left[\dfrac{5.00 \times 10^{-3}\text{ mL}}{(5.00\text{ mL} + 5.00 \times 10^{-3}\text{ mL})}\right]}$$

$$\frac{0.712}{C_A} = \frac{1.546}{0.9990C_A + 1.558\text{ ppb}}$$

$$0.7113C_A + 1.109\text{ ppb} = 1.546C_A$$

$$C_A = 1.33\text{ ppb}$$

Thus, the concentration of Pb^{2+} in the original sample of blood is 1.33 ppb.

The single-point standard additions outlined in Examples 5.4 and 5.5 are easily adapted to a multiple-point standard addition by preparing a series of spiked samples containing increasing amounts of the standard. A calibration curve is prepared by plotting S_{spike} versus an appropriate measure of the amount of added standard. Figure 5.7 shows two examples of a standard addition calibration curve based on equation 5.6. In Figure 5.7(a) S_{spike} is plotted versus the volume of the standard solution spikes, V_s. When k is constant, the calibration curve is linear, and it is easy to show that the x-intercept's absolute value is C_AV_o/C_S.

Colorplate 2 shows an example of a set of standard additions and their corresponding standard additions calibration curve.

Example

EXAMPLE 5.6

Starting with equation 5.6, show that the equations for the slope, y-intercept, and x-intercept in Figure 5.7(a) are correct.

SOLUTION

We begin by rewriting equation 5.6 as

$$S_{spike} = \frac{kC_AV_o}{V_f} + \frac{kC_S}{V_f} \times V_s$$

which is in the form of the linear equation

$$Y = y\text{-intercept} + \text{slope} \times X$$

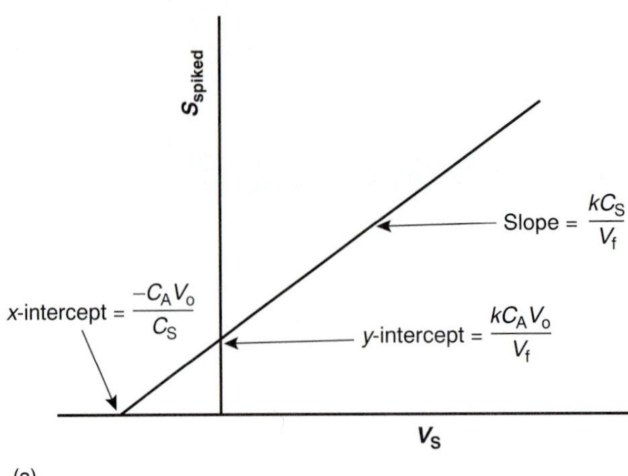

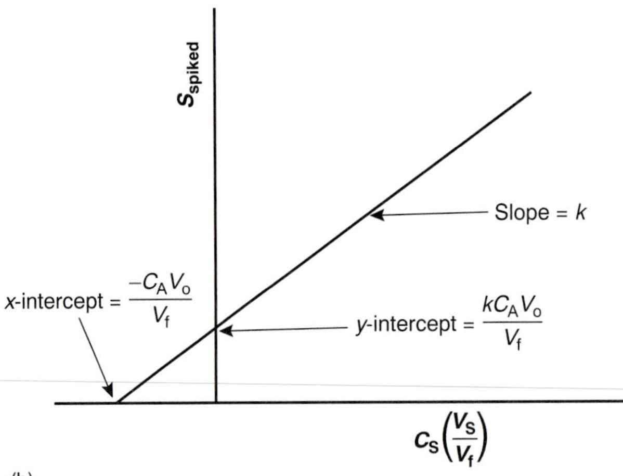

Figure 5.7
Examples of calibration curves for the method of standard additions. In (a) the signal is plotted versus the volume of the added standard, and in (b) the signal is plotted versus the concentration of the added standard after dilution.

where Y is S_{spike} and X is V_s. The slope of the line, therefore, is kC_S/V_f, and the y-intercept is kC_AV_o/V_f. The x-intercept is the value of X when Y is 0, or

$$0 = \frac{kC_AV_o}{V_f} + \frac{kC_S}{V_f} \times (x\text{-intercept})$$

$$x\text{-intercept} = -\frac{(kC_AV_o/V_f)}{(kC_S/V_f)} = -\frac{C_AV_o}{C_S}$$

Thus, the absolute value of the x-intercept is C_AV_o/C_S.

Since both V_o and C_S are known, the x-intercept can be used to calculate the analyte's concentration.

EXAMPLE 5.7

A fifth spectrophotometric method for the quantitative determination of the concentration of Pb^{2+} in blood uses a multiple-point standard addition based on equation 5.6. The original blood sample has a volume of 1.00 mL, and the standard used for spiking the sample has a concentration of 1560 ppb Pb^{2+}. All samples were diluted to 5.00 mL before measuring the signal. A calibration curve of S_{spike} versus V_s is described by

$$S_{spike} = 0.266 + 312\ \text{mL}^{-1} \times V_s$$

Determine the concentration of Pb^{2+} in the original sample of blood.

SOLUTION

To find the x-intercept we let S_{spike} equal 0

$$0 = 0.266 + 312\ \text{mL}^{-1} \times (x\text{-intercept})$$

and solve for the x-intercept's absolute value, giving a value of 8.526×10^{-4} mL. Thus

$$x\text{-intercept} = 8.526 \times 10^{-4}\ \text{mL} = \frac{C_A V_o}{C_S} = \frac{C_A \times (1.00\ \text{mL})}{1560\ \text{ppb}}$$

and the concentration of Pb^{2+} in the blood sample, C_A, is 1.33 ppb.

Figure 5.7(b) shows the relevant relationships when S_{spike} is plotted versus the concentrations of the spiked standards after dilution. Standard addition calibration curves based on equation 5.8 are also possible.

Since a standard additions calibration curve is constructed in the sample, it cannot be extended to the analysis of another sample. Each sample, therefore, requires its own standard additions calibration curve. This is a serious drawback to the routine application of the method of standard additions, particularly in laboratories that must handle many samples or that require a quick turnaround time. For example, suppose you need to analyze ten samples using a three-point calibration curve. For a normal calibration curve using external standards, only 13 solutions need to be analyzed (3 standards and 10 samples). Using the method of standard additions, however, requires the analysis of 30 solutions, since each of the 10 samples must be analyzed three times (once before spiking and two times after adding successive spikes).

The method of standard additions can be used to check the validity of an external standardization when matrix matching is not feasible. To do this, a normal calibration curve of S_{stand} versus C_S is constructed, and the value of k is determined from its slope. A standard additions calibration curve is then constructed using equation 5.6, plotting the data as shown in Figure 5.7(b). The slope of this standard additions calibration curve gives an independent determination of k. If the two values of k are identical, then any difference between the sample's matrix and that of the external standards can be ignored. When the values of k are different, a proportional determinate error is introduced if the normal calibration curve is used.

5B.5 Internal Standards

The successful application of an external standardization or the method of standard additions, depends on the analyst's ability to handle samples and standards reproducibly. When a procedure cannot be controlled to the extent that all samples and standards are treated equally, the accuracy and precision of the standardization may suffer. For example, if an analyte is present in a volatile solvent, its concentration will increase if some solvent is lost to evaporation. Suppose that you have a sample and a standard with identical concentrations of analyte and identical signals. If both experience the same loss of solvent their concentrations of analyte and signals will continue to be identical. In effect, we can ignore changes in concentration due to evaporation provided that the samples and standards experience an equivalent loss of solvent. If an identical standard and sample experience different losses of solvent, however, their concentrations and signals will no longer be equal. In this case, an external standardization or standard addition results in a determinate error.

A standardization is still possible if the analyte's signal is referenced to a signal generated by another species that has been added at a fixed concentration to all samples and standards. The added species, which must be different from the analyte, is called an **internal standard.**

internal standard
A standard, whose identity is different from the analyte's, that is added to all samples and standards containing the analyte.

Since the analyte and internal standard in any sample or standard receive the same treatment, the ratio of their signals will be unaffected by any lack of reproducibility in the procedure. If a solution contains an analyte of concentration C_A, and an internal standard of concentration, C_{IS}, then the signals due to the analyte, S_A, and the internal standard, S_{IS}, are

$$S_A = k_A C_A$$

$$S_{IS} = k_{IS} C_{IS}$$

where k_A and k_{IS} are the sensitivities for the analyte and internal standard, respectively. Taking the ratio of the two signals gives

$$\frac{S_A}{S_{IS}} = \frac{k_A}{k_{IS}} \times \frac{C_A}{C_{IS}} = K \times \frac{C_A}{C_{IS}} \qquad 5.10$$

Because equation 5.10 is defined in terms of a ratio, K, of the analyte's sensitivity and the internal standard's sensitivity, it is not necessary to independently determine values for either k_A or k_{IS}.

In a single-point internal standardization, a single standard is prepared, and K is determined by solving equation 5.10

$$K = \left(\frac{C_{IS}}{C_A}\right)\left(\frac{S_A}{S_{IS}}\right)_{\text{stand}} \qquad 5.11$$

Once the method is standardized, the analyte's concentration is given by

$$C_A = \left(\frac{C_{IS}}{K}\right)\left(\frac{S_A}{S_{IS}}\right)_{\text{samp}}$$

Example

EXAMPLE 5.8

A sixth spectrophotometric method for the quantitative determination of Pb^{2+} levels in blood uses Cu^{2+} as an internal standard. A standard containing 1.75 ppb Pb^{2+} and 2.25 ppb Cu^{2+} yields a ratio of S_A/S_{IS} of 2.37. A sample of blood is spiked with the same concentration of Cu^{2+}, giving a signal ratio of 1.80. Determine the concentration of Pb^{2+} in the sample of blood.

SOLUTION

Equation 5.11 allows us to calculate the value of K using the data for the standard

$$K = \left(\frac{C_{IS}}{C_A}\right)\left(\frac{S_A}{S_{IS}}\right)_{\text{stand}} = \frac{2.25}{1.75} \times 2.37 = 3.05$$

The concentration of Pb^{2+}, therefore, is

$$C_A = \left(\frac{C_{IS}}{K}\right)\left(\frac{S_A}{S_{IS}}\right)_{\text{samp}} = \frac{2.25}{3.05} \times 1.80 = 1.33 \text{ ppb } Pb^{2+}$$

A single-point internal standardization has the same limitations as a single-point normal calibration. To construct an internal standard calibration curve, it is necessary to prepare several standards containing different concentrations of analyte. These standards are usually prepared such that the internal standard's concentration is constant. Under these conditions a calibration curve of $(S_A/S_{IS})_{stand}$ versus C_A is linear with a slope of K/C_{IS}.

Example

EXAMPLE 5.9

A seventh spectrophotometric method for the quantitative determination of Pb^{2+} levels in blood gives a linear internal standards calibration curve for which

$$\left(\frac{S_A}{S_{IS}}\right)_{stand} = (2.11\,\text{ppb}^{-1}) \times C_A - 0.006$$

What is the concentration (in ppb) of Pb^{2+} in a sample of blood if $(S_A/S_{IS})_{samp}$ is 2.80?

SOLUTION

To determine the concentration of Pb^{2+} in the sample of blood, we replace $(S_A/S_{IS})_{stand}$ in the calibration equation with $(S_A/S_{IS})_{samp}$ and solve for C_A

$$C_A = \frac{(S_A/S_{IS})_{samp} + 0.006}{2.11\text{ ppb}^{-1}} = \frac{2.80 + 0.006}{2.11\text{ ppb}^{-1}} = 1.33\text{ ppb}$$

The concentration of Pb^{2+} in the sample of blood is 1.33 ppb.

When the internal standard's concentration cannot be held constant the data must be plotted as $(S_A/S_{IS})_{stand}$ versus C_A/C_{IS}, giving a linear calibration curve with a slope of K.

5C Linear Regression and Calibration Curves

In a single-point external standardization, we first determine the value of k by measuring the signal for a single standard containing a known concentration of analyte. This value of k and the signal for the sample are then used to calculate the concentration of analyte in the sample (see Example 5.2). With only a single determination of k, a quantitative analysis using a single-point external standardization is straightforward. This is also true for a single-point standard addition (see Examples 5.4 and 5.5) and a single-point internal standardization (see Example 5.8).

A multiple-point standardization presents a more difficult problem. Consider the data in Table 5.1 for a multiple-point external standardization. What is the best estimate of the relationship between S_{meas} and C_S? It is tempting to treat this data as five separate single-point standardizations, determining k for each standard and reporting the mean value. Despite its simplicity, this is not an appropriate way to treat a multiple-point standardization.

In a single-point standardization, we assume that the reagent blank (the first row in Table 5.1) corrects for all constant sources of determinate error. If this is not the case, then the value of k determined by a single-point standardization will have a determinate error.

Table 5.1 Data for Hypothetical Multiple-Point External Standardization

C_S	S_{meas}
0.000	0.00
0.100	12.36
0.200	24.83
0.300	35.91
0.400	48.79
0.500	60.42

Table 5.2 Effect of a Constant Determinate Error on the Value of k Calculated Using a Single-Point Standardization

C_A	S_{meas} (true)	k (true)	S_{meas} (with constant error)	k (apparent)
1.00	1.00	1.00	1.50	1.50
2.00	2.00	1.00	2.50	1.25
3.00	3.00	1.00	3.50	1.17
4.00	4.00	1.00	4.50	1.13
5.00	5.00	1.00	5.50	1.10
	mean k(true) =	1.00	mean k (apparent) =	1.23

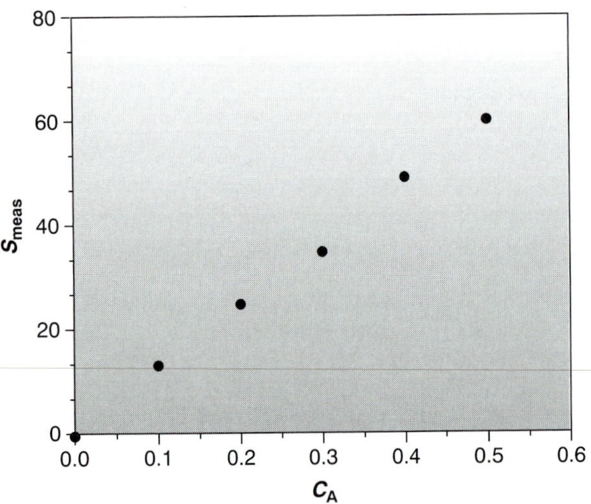

Figure 5.8
Normal calibration plot of hypothetical data from Table 5.1.

Table 5.2 demonstrates how an uncorrected constant error affects our determination of k. The first three columns show the concentration of analyte, the true measured signal (no constant error) and the true value of k for five standards. As expected, the value of k is the same for each standard. In the fourth column a constant determinate error of +0.50 has been added to the measured signals. The corresponding values of k are shown in the last column. Note that a different value of k is obtained for each standard and that all values are greater than the true value. As we noted in Section 5B.2, this is a significant limitation to any single-point standardization.

How do we find the best estimate for the relationship between the measured signal and the concentration of analyte in a multiple-point standardization? Figure 5.8 shows the data in Table 5.1 plotted as a normal calibration curve. Although the data appear to fall along a straight line, the actual calibration curve is not intuitively obvious. The process of mathematically determining the best equation for the calibration curve is called regression.

5C.1 Linear Regression of Straight-Line Calibration Curves

A calibration curve shows us the relationship between the measured signal and the analyte's concentration in a series of standards. The most useful calibration curve is a straight line since the method's sensitivity is the same for all concentrations of analyte. The equation for a linear calibration curve is

$$y = \beta_0 + \beta_1 x \qquad 5.12$$

linear regression
A mathematical technique for fitting an equation, such as that for a straight line, to experimental data.

residual error
The difference between an experimental value and the value predicted by a regression equation.

where y is the signal and x is the amount of analyte. The constants β_0 and β_1 are the true y-intercept and the true slope, respectively. The goal of **linear regression** is to determine the best estimates for the slope, b_1, and y-intercept, b_0. This is accomplished by minimizing the **residual error** between the experimental values, y_i, and those values, $\hat{y}_i$, predicted by equation 5.12 (Figure 5.9). For obvious reasons, a regression analysis is also called a least-squares treatment. Several approaches to the linear regression of equation 5.12 are discussed in the following sections.

5C.2 Unweighted Linear Regression with Errors in y

The most commonly used form of linear regression is based on three assumptions: (1) that any difference between the experimental data and the calculated regression line is due to indeterminate errors affecting the values of y, (2) that these indeterminate errors are normally distributed, and (3) that the indeterminate errors in y do not depend on the value of x. Because we assume that indeterminate errors are the same for all standards, each standard contributes equally in estimating the slope and y-intercept. For this reason the result is considered an unweighted linear regression.

The second assumption is generally true because of the central limit theorem outlined in Chapter 4. The validity of the two remaining assumptions is less certain and should be evaluated before accepting the results of a linear regression. In particular, the first assumption is always suspect since there will certainly be some indeterminate errors affecting the values of x. In preparing a calibration curve, however, it is not unusual for the relative standard deviation of the measured signal (y) to be significantly larger than that for the concentration of analyte in the standards (x). In such circumstances, the first assumption is usually reasonable.

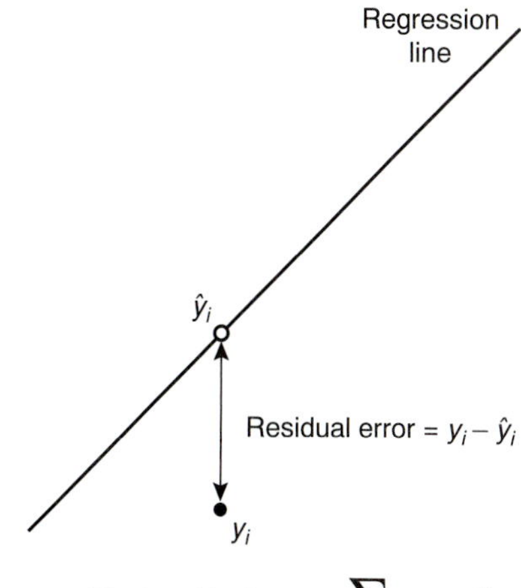

Figure 5.9
Residual error in linear regression, where the filled circle shows the experimental value y_i, and the open circle shows the predicted value $\hat{y}_i$.

Finding the Estimated Slope and y-Intercept The derivation of equations for calculating the estimated slope and y-intercept can be found in standard statistical texts[7] and is not developed here. The resulting equation for the slope is given as

$$b_1 = \frac{n\sum x_i y_i - \sum x_i \sum y_i}{n\sum x_i^2 - (\sum x_i)^2} \qquad 5.13$$

and the equation for the y-intercept is

$$b_0 = \frac{\sum y_i - b_1 \sum x_i}{n} \qquad 5.14$$

Although equations 5.13 and 5.14 appear formidable, it is only necessary to evaluate four summation terms. In addition, many calculators, spreadsheets, and other computer software packages are capable of performing a linear regression analysis based on this model. To save time and to avoid tedious calculations, learn how to use one of these tools. For illustrative purposes, the necessary calculations are shown in detail in the following example.

Example

EXAMPLE 5.10

Using the data from Table 5.1, determine the relationship between S_{meas} and C_S by an unweighted linear regression.

SOLUTION

Equations 5.13 and 5.14 are written in terms of the general variables x and y. As you work through this example, remember that x represents the concentration of analyte in the standards (C_S), and that y corresponds to the signal (S_{meas}). We begin by setting up a table to help in the calculation of the summation terms Σx_i, Σy_i, Σx_i^2, and $\Sigma x_i y_i$ which are needed for the calculation of b_0 and b_1

x_i	y_i	x_i^2	$x_i y_i$
0.000	0.00	0.000	0.000
0.100	12.36	0.010	1.236
0.200	24.83	0.040	4.966
0.300	35.91	0.090	10.773
0.400	48.79	0.160	19.516
0.500	60.42	0.250	30.210

Adding the values in each column gives

$$\Sigma x_i = 1.500 \qquad \Sigma y_i = 182.31 \qquad \Sigma x_i^2 = 0.550 \qquad \Sigma x_i y_i = 66.701$$

Substituting these values into equations 5.12 and 5.13 gives the estimated slope

$$b_1 = \frac{(6)(66.701) - (1.500)(182.31)}{(6)(0.550) - (1.500)^2} = 120.706$$

and the estimated y-intercept

$$b_0 = \frac{182.31 - (120.706)(1.500)}{6} = 0.209$$

The relationship between the signal and the analyte, therefore, is

$$S_{\text{meas}} = 120.70 \times C_S + 0.21$$

Note that for now we keep enough significant figures to match the number of decimal places to which the signal was measured. The resulting calibration curve is shown in Figure 5.10.

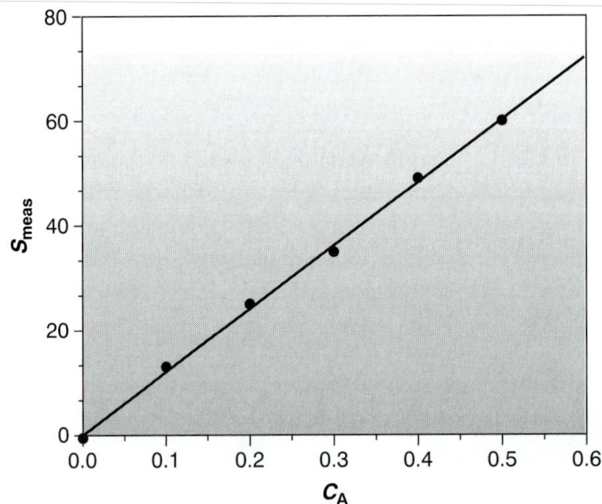

Figure 5.10
Normal calibration curve for the hypothetical data in Table 5.1, showing the regression line.

Uncertainty in the Regression Analysis As shown in Figure 5.10, the regression line need not pass through the data points (this is the consequence of indeterminate errors affecting the signal). The cumulative deviation of the data from the regression line is used to calculate the uncertainty in the regression due to indeterminate error. This is called the **standard deviation about the regression,** s_r, and is given as

standard deviation about the regression The uncertainty in a regression analysis due to indeterminate error (s_r).

$$s_r = \sqrt{\frac{\sum_{i=1}^{n}(y_i - \hat{y}_i)^2}{n-2}} \qquad \textbf{5.15}$$

where y_i is the ith experimental value, and $\hat{y}_i$ is the corresponding value predicted by the regression line

$$\hat{y}_i = b_0 + b_1 x_i$$

There is an obvious similarity between equation 5.15 and the standard deviation introduced in Chapter 4, except that the sum of squares term for s_r is determined relative to $\hat{y}_i$ instead of $\bar{y}$, and the denominator is $n - 2$ instead of $n - 1$; $n - 2$ indicates that the linear regression analysis has only $n - 2$ degrees of freedom since two parameters, the slope and the intercept, are used to calculate the values of $\hat{y}_i$.

A more useful representation of uncertainty is to consider the effect of indeterminate errors on the predicted slope and intercept. The standard deviation of the slope and intercept are given as

$$s_{b_1} = \sqrt{\frac{ns_r^2}{n\sum x_i^2 - (\sum x_i)^2}} = \sqrt{\frac{s_r^2}{\sum (x_i - \bar{x})^2}} \qquad \textbf{5.16}$$

$$s_{b_0} = \sqrt{\frac{s_r^2 \sum x_i^2}{n\sum x_i^2 - (\sum x_i)^2}} = \sqrt{\frac{s_r^2 \sum x_i^2}{n\sum (x_i - \bar{x})^2}} \qquad \textbf{5.17}$$

These standard deviations can be used to establish confidence intervals for the true slope and the true y-intercept

$$\beta_1 = b_1 \pm ts_{b_1} \qquad \textbf{5.18}$$

$$\beta_o = b_o \pm ts_{b_0} \qquad \textbf{5.19}$$

where t is selected for a significance level of α and for $n - 2$ degrees of freedom. Note that the terms ts_{b_1} and ts_{b_0} do not contain a factor of $(\sqrt{n})^{-1}$ because the confidence interval is based on a single regression line. Again, many calculators, spreadsheets, and computer software packages can handle the calculation of s_{b_0} and s_{b_1} and the corresponding confidence intervals for β_0 and β_1. Example 5.11 illustrates the calculations.

Example

EXAMPLE 5.11

Calculate the 95% confidence intervals for the slope and y-intercept determined in Example 5.10.

SOLUTION

Again, as you work through this example, remember that x represents the concentration of analyte in the standards (C_S), and y corresponds to the signal (S_{meas}). To begin with, it is necessary to calculate the standard deviation about the regression. This requires that we first calculate the predicted signals, $\hat{y}_i$, using the slope and y-intercept determined in Example 5.10. Taking the first standard as an example, the predicted signal is

$$\hat{y}_i = b_0 + b_1 x = 0.209 + (120.706)(0.100) = 12.280$$

The results for all six solutions are shown in the following table.

x_i	y_i	$\hat{y}_i$	$(y_i - \hat{y}_i)^2$
0.000	0.00	0.209	0.0437
0.100	12.36	12.280	0.0064
0.200	24.83	24.350	0.2304
0.300	35.91	36.421	0.2611
0.400	48.79	48.491	0.0894
0.500	60.42	60.562	0.0202

Adding together the data in the last column gives the numerator of equation 5.15, $\Sigma(y_i - \hat{y}_i)^2$, as 0.6512. The standard deviation about the regression, therefore, is

$$s_r = \sqrt{\frac{0.6512}{6-2}} = 0.4035$$

Next we calculate s_{b_1} and s_{b_0} using equations 5.16 and 5.17. Values for the summation terms Σx_i^2 and Σx_i are found in Example 5.10.

$$s_{b_1} = \sqrt{\frac{ns_r^2}{n\sum x_i^2 - (\sum x_i)^2}} = \sqrt{\frac{(6)(0.4035)^2}{(6)(0.550) - (1.500)^2}} = 0.965$$

$$s_{b_0} = \sqrt{\frac{s_r^2 \sum x_i^2}{n\sum x_i^2 - (\sum x_i)^2}} = \sqrt{\frac{(0.4035)^2 (0.550)}{(6)(0.550) - (1.500)^2}} = 0.292$$

Finally, the 95% confidence intervals ($\alpha = 0.05$, 4 degrees of freedom) for the slope and y-intercept are

$$\beta_1 = b_1 \pm ts_{b_1} = 120.706 \pm (2.78)(0.965) = 120.7 \pm 2.7$$

$$\beta_0 = b_0 \pm ts_{b_0} = 0.209 \pm (2.78)(0.292) = 0.2 \pm 0.8$$

The standard deviation about the regression, s_r, suggests that the measured signals are precise to only the first decimal place. For this reason, we report the slope and intercept to only a single decimal place.

To minimize the uncertainty in the predicted slope and y-intercept, calibration curves are best prepared by selecting standards that are evenly spaced over a wide range of concentrations or amounts of analyte. The reason for this can be rationalized by examining equations 5.16 and 5.17. For example, both s_{b_0} and s_{b_1} can be minimized by increasing the value of the term $\Sigma(x_i - \overline{x})^2$, which is present in the denominators of both equations. Thus, increasing the range of concentrations used in preparing standards decreases the uncertainty in the slope and the y-intercept. Furthermore, to minimize the uncertainty in the y-intercept, it also is necessary to decrease the value of the term Σx_i^2 in equation 5.17. This is accomplished by spreading the calibration standards evenly over their range.

Using the Regression Equation Once the regression equation is known, we can use it to determine the concentration of analyte in a sample. When using a normal calibration curve with external standards or an internal standards calibration curve, we measure an average signal for our sample, $\overline{Y}_X$, and use it to calculate the value of X

$$X = \frac{\overline{Y}_X - b_0}{b_1} \tag{5.20}$$

The standard deviation for the calculated value of X is given by the following equation

$$s_X = \frac{s_r}{b_1}\left\{\frac{1}{m} + \frac{1}{n} + \frac{(\overline{Y}_X - \overline{y})^2}{b_1^2 \sum (x_i - \overline{x})^2}\right\}^{1/2} \tag{5.21}$$

where m is the number of replicate samples used to establish $\overline{Y}_X$, n is the number of calibration standards, $\overline{y}$ is the average signal for the standards, and x_i and $\overline{x}$ are the individual and mean concentrations of the standards.[8] Once s_X is known the confidence interval for the analyte's concentration can be calculated as

$$\mu_X = X \pm t s_X$$

where μ_X is the expected value of X in the absence of determinate errors, and the value of t is determined by the desired level of confidence and for $n - 2$ degrees of freedom. The following example illustrates the use of these equations for an analysis using a normal calibration curve with external standards.

Example

EXAMPLE 5.12

Three replicate determinations are made of the signal for a sample containing an unknown concentration of analyte, yielding values of 29.32, 29.16, and 29.51. Using the regression line from Examples 5.10 and 5.11, determine the analyte's concentration, C_A, and its 95% confidence interval.

SOLUTION

The equation for a normal calibration curve using external standards is

$$S_{meas} = b_0 + b_1 \times C_A$$

thus, $\overline{Y}_X$ is the average signal of 29.33, and X is the analyte's concentration. Substituting the value of $\overline{Y}_X$ into equation 5.20 along with the estimated slope and the y-intercept for the regression line gives the analyte's concentration as

$$C_A = X = \frac{\overline{Y}_X - b_0}{b_1} = \frac{29.33 - 0.209}{120.706} = 0.241$$

To calculate the standard deviation for the analyte's concentration, we must determine the values for $\overline{y}$ and $\Sigma(x_i - \overline{x})^2$. The former is just the average signal for the standards used to construct the calibration curve. From the data in Table 5.1, we easily calculate that $\overline{y}$ is 30.385. Calculating $\Sigma(x_i - \overline{x})^2$ looks formidable, but we can simplify the calculation by recognizing that this sum of squares term is simply the numerator in a standard deviation equation; thus,

$$\sum(x_i - \overline{x})^2 = s^2(n - 1)$$

where s is the standard deviation for the concentration of analyte in the standards used to construct the calibration curve. Using the data in Table 5.1, we find that s is 0.1871 and

$$\sum(x_i - \overline{x})^2 = (0.1871)^2(6 - 1) = 0.175$$

Substituting known values into equation 5.21 gives

$$s_A = s_X = \frac{0.4035}{120.706}\left\{\frac{1}{3} + \frac{1}{6} + \frac{(29.33 - 30.385)^2}{(120.706)^2 (0.175)}\right\}^{1/2} = 0.0024$$

Finally, the 95% confidence interval for 4 degrees of freedom is

$$\mu_A = C_A \pm ts_A = 0.241 \pm (2.78)(0.0024) = 0.241 \pm 0.007$$

In a standard addition the analyte's concentration is determined by extrapolating the calibration curve to find the x-intercept. In this case the value of X is

$$X = x\text{-intercept} = \frac{-b_0}{b_1}$$

and the standard deviation in X is

$$s_X = \frac{s_r}{b_1}\left\{\frac{1}{n} + \frac{(\overline{y})^2}{b_1^2 \sum (x_i - \overline{x})^2}\right\}^{1/2}$$

where n is the number of standards used in preparing the standard additions calibration curve (including the sample with no added standard), and $\overline{y}$ is the average signal for the n standards. Because the analyte's concentration is determined by extrapolation, rather than by interpolation, s_X for the method of standard additions generally is larger than for a normal calibration curve.

A linear regression analysis should not be accepted without evaluating the validity of the model on which the calculations were based. Perhaps the simplest way to evaluate a regression analysis is to calculate and plot the residual error for each value of x. The residual error for a single calibration standard, r_i, is given as

$$r_i = y_i - \hat{y}_i$$

If the regression model is valid, then the residual errors should be randomly distributed about an average residual error of 0, with no apparent trend toward either smaller or larger residual errors (Figure 5.11a). Trends such as those shown in Figures 5.11b and 5.11c provide evidence that at least one of the assumptions on which the regression model is based are incorrect. For example, the trend toward larger residual errors in Figure 5.11b suggests that the indeterminate errors affecting y are not independent of the value of x. In Figure 5.11c the residual errors are not randomly distributed, suggesting that the data cannot be modeled with a straight-line relationship. Regression methods for these two cases are discussed in the following sections.

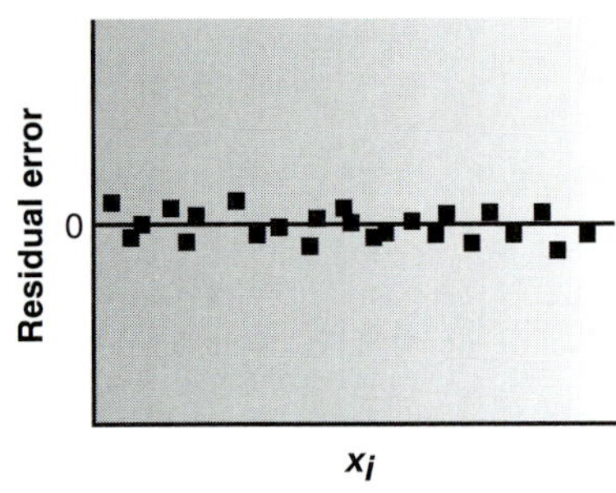

(a)

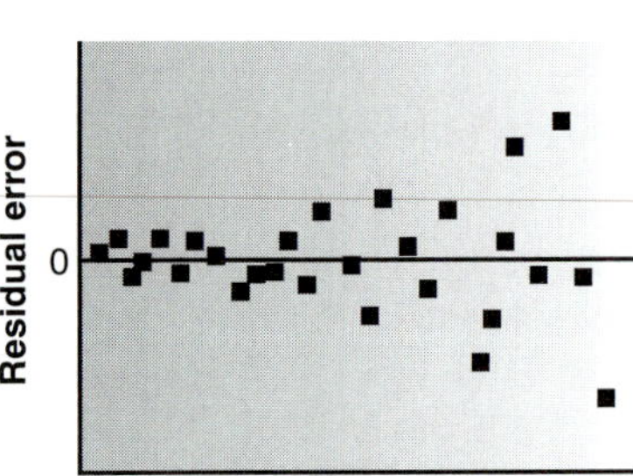

(b)

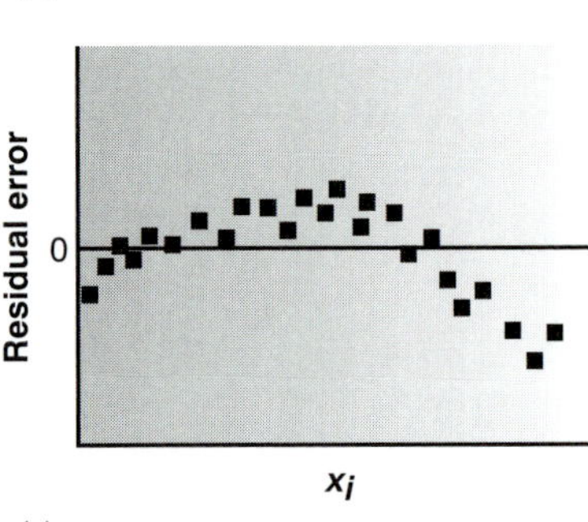

(c)

Figure 5.11
Plot of the residual error in y as a function of x. The distribution of the residuals in (a) indicates that the regression model was appropriate for the data, and the distributions in (b) and (c) indicate that the model does not provide a good fit for the data.

5C.3 Weighted Linear Regression with Errors in *y*

Equations 5.13 for the slope, b_1, and 5.14 for the y-intercept, b_0, assume that indeterminate errors equally affect each value of y. When this assumption is false, as shown in Figure 5.11b, the variance associated with each value of y must be included when estimating β_0 and β_1. In this case the predicted slope and intercept are

$$b_1 = \frac{n \sum w_i x_i y_i - \sum w_i x_i \sum w_i y_i}{n \sum w_i x_i^2 - (\sum w_i x_i)^2} \qquad 5.22$$

and

$$b_0 = \frac{\sum w_i y_i - b_1 \sum w_i x_i}{n} \qquad \textbf{5.23}$$

where w_i is a weighting factor accounting for the variance in measuring y_i. Values of w_i are calculated using equation 5.24.

$$w_i = \frac{n s_i^{-2}}{\sum s_i^{-2}} \qquad \textbf{5.24}$$

where s_i is the standard deviation associated with y_i. The use of a weighting factor ensures that the contribution of each pair of xy values to the regression line is proportional to the precision with which y_i is measured.

Example

EXAMPLE 5.13

The following data were recorded during the preparation of a calibration curve, where $\overline{S}_{meas}$ and s are the mean and standard deviation, respectively, for three replicate measurements of the signal.

C_A	$\overline{S}_{meas}$	s
0.000	0.00	0.02
0.100	12.36	0.02
0.200	24.83	0.07
0.300	35.91	0.13
0.400	48.79	0.22
0.500	60.42	0.33

Determine the relationship between $\overline{S}_{meas}$ and C_A using a weighted linear regression model.

SOLUTION

Once again, as you work through this example, remember that x represents the concentration of analyte in the standards (C_S), and y corresponds to the average signal ($\overline{S}_{meas}$). We begin by setting up a table to aid in the calculation of the weighting factor.

x_i	y_i	s_i	s_i^{-2}	w_i
0.000	0.00	0.02	2500.00	2.8339
0.100	12.36	0.02	2500.00	2.8339
0.200	24.83	0.07	204.08	0.2313
0.300	35.91	0.13	59.17	0.0671
0.400	48.79	0.22	20.66	0.0234
0.500	60.42	0.33	9.18	0.0104

Adding together the values in the forth column gives

$$\sum s_i^{-2} = 5293.09$$

which is used to calculate the weights in the last column. As a check on the calculation, the sum of the weights in the last column should equal the number of calibration standards, n. In this case

$$\Sigma w_i = 6.0000$$

After the individual weights have been calculated, a second table is used to aid in calculating the four summation terms in equations 5.22 and 5.23.

x_i	y_i	w_i	w_ix_i	w_iy_i	$w_ix_i^2$	$w_ix_iy_i$
0.000	0.00	2.8339	0.0000	0.0000	0.0000	0.0000
0.100	12.36	2.8339	0.2834	35.0270	0.0283	3.5027
0.200	24.83	0.2313	0.0463	5.7432	0.0093	1.1486
0.300	35.91	0.0671	0.0201	2.4096	0.0060	0.7229
0.400	48.79	0.0234	0.0094	1.1417	0.0037	0.4567
0.500	60.42	0.0104	0.0052	0.6284	0.0026	0.3142

Adding the values in the last four columns gives

$$\sum w_ix_i = 0.3644 \quad \sum w_iy_i = 44.9499 \quad \sum w_ix_i^2 = 0.0499 \quad \sum w_ix_iy_i = 6.1451$$

Substituting these values into the equations 5.22 and 5.23 gives the estimated slope

$$b_1 = \frac{(6)(6.1451) - (0.3644)(44.9499)}{(6)(0.0499) - (0.3644)^2} = 122.985$$

and the estimated y-intercept

$$b_0 = \frac{44.9499 - (122.985)(0.3644)}{6} = 0.0224$$

The relationship between the signal and the concentration of the analyte, therefore, is

$$\bar{S}_{meas} = 122.98 \times C_A + 0.02$$

with the calibration curve shown in Figure 5.12.

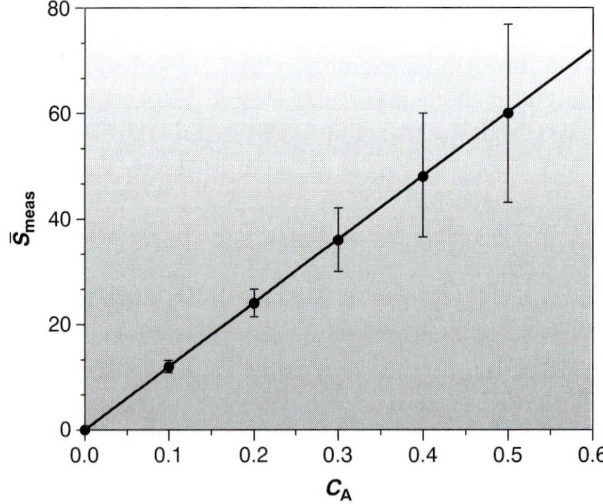

Figure 5.12
Weighted normal calibration curve for the data in Example 5.13. The lines through the data points show the standard deviation of the signal for the standards. These lines have been scaled by a factor of 50 so that they can be seen on the same scale as the calibration curve.

Equations for calculating confidence intervals for the slope, the y-intercept, and the concentration of analyte when using a weighted linear regression are not as easily defined as for an unweighted linear regression.[9] The confidence interval for the concentration of an analyte, however, will be at its optimum value when the analyte's signal is near the weighted centroid, $\bar{y}$, of the calibration curve

$$\bar{y} = \frac{1}{n}\sum w_i y_i$$

5C.4 Weighted Linear Regression with Errors in Both x and y

If we remove the assumption that the indeterminate errors affecting a calibration curve are found only in the signal (y), then indeterminate errors affecting the preparation of standards containing known amounts of analyte (x) must be factored into the regression model. The solution for the resulting regression line is computationally more involved than that for either the unweighted or weighted regression lines, and is not presented in this text. The suggested readings at the end of the chapter list several papers discussing algorithms for this regression method.

5C.5 Curvilinear and Multivariate Regression

Regression models based on a straight line, despite their apparent complexity, use the simplest functional relationship between two variables. In many cases, calibration curves show a pronounced curvature at high concentrations of analyte (see Figure 5.3b). One approach to constructing a calibration curve when curvature exists is to seek a transformation function that will make the data linear. Logarithms, exponentials, reciprocals, square roots, and trigonometric functions have all been used in this capacity. A plot of y versus log x is a typical example. Such transformations are not without complications. Perhaps the most obvious is that data that originally has a uniform variance for the y values will not maintain that uniform variance when the variable is transformed.

A more rigorous approach to developing a regression model for a nonlinear calibration curve is to fit a polynomial equation such as $y = a + bx + cx^2$ to the data. Equations for calculating the parameters a, b, and c are derived in the same manner as that described earlier for the straight-line model.[10] When a single polynomial equation cannot be fitted to the calibration data, it may be possible to fit separate polynomial equations to short segments of the calibration curve. The result is a single continuous calibration curve known as a spline function.

The regression models considered earlier apply only to functions containing a single independent variable. Analytical methods, however, are frequently subject to determinate sources of error due to interferents that contribute to the measured signal. In the presence of a single interferent, equations 5.1 and 5.2 become

$$S_{meas} = k_A n_A + k_I n_I + S_{reag}$$

$$S_{meas} = k_A C_A + k_I C_I + S_{reag}$$

where k_I is the interferent's sensitivity, n_I is the moles of interferent, and C_I is the interferent's concentration. Multivariate calibration curves can be prepared using standards that contain known amounts of analyte and interferent.[11]

5D Blank Corrections

In discussing ways to standardize a method, we assumed that an appropriate reagent blank had been used to correct S_{meas} for signals originating from sources other than the analyte. At that time we did not ask an important question—"What constitutes an appropriate reagent blank?" Surprisingly, the answer is not intuitively obvious.

In one study,[12] analytical chemists were asked to evaluate a data set consisting of a normal calibration curve, three samples of different size but drawn from the same source, and an analyte-free sample (Table 5.3). At least four different approaches for correcting the signals were used by the participants: (1) ignore the correction entirely, which clearly is incorrect; (2) use the *y*-intercept of the calibration curve as a calibration blank, CB; (3) use the analyte-free sample as a reagent blank, RB; and (4) use both the calibration and reagent blanks. Equations for calculating the concentration of analyte using each approach are shown in Table 5.4, along with the resulting concentration for the analyte in each of the three samples.

That all four methods give a different result for the concentration of analyte underscores the importance of choosing a proper blank but does not tell us which of the methods is correct. In fact, the variation within each method for the reported concentration of analyte indicates that none of these four methods has adequately corrected for the blank. Since the three samples were drawn from the same source, they must have the same true concentration of analyte. Since all four methods predict concentrations of analyte that are dependent on the size of the sample, we can conclude that none of these blank corrections has accounted for an underlying constant source of determinate error.

To correct for all constant method errors, a blank must account for signals due to the reagents and solvent used in the analysis and any bias due to interac-

Table 5.3 Hypothetical Data Used to Study Procedures for Method Blanks

W_s[a]	S_{stand}	Sample Number	W_x[b]	S_{samp}
1.6667	0.2500	1	62.4746	0.8000
5.0000	0.5000	2	82.7915	1.0000
8.3333	0.7500	3	103.1085	1.2000
9.5507	0.8413			
11.6667	1.0000		analyte-free[c]	0.1000
18.1600	1.4870			
19.9333	1.6200			

Calibration equation: $S_{stand} = 0.0750 \times W_s + 0.1250$

Source: Modified from Cardone, M. J. *Anal. Chem.* **1986,** *58,* 433–438.

[a]W_s = weight of analyte used to prepare standard solution by diluting to a fixed volume, *V*.

[b]W_x = weight of sample used to prepare sample solution by diluting to a fixed volume, *V*.

[c]Analyte-free sample prepared in the same fashion as samples, but without the analyte being present.

tions between the analyte and the sample matrix. Both the calibration blank and the reagent blank correct for signals due to the reagents and solvents. Any difference in their values is due to the number and composition of samples contributing to the determination of the blank.

Unfortunately, neither the calibration blank nor the reagent blank can correct for bias due to analyte–matrix interactions because the analyte is missing in the reagent blank, and the sample's matrix is missing from the calibration blank. The true method blank must include both the matrix and the analyte and, consequently, can only be determined using the sample itself. One approach is to measure the signal for samples of different size and determine the regression line from a plot of signal versus the amount of sample. The resulting y-intercept gives the signal for the condition of no sample and is known as the **total Youden blank.**[13] This is the true blank correction. The regression line for the sample data in Table 5.3 is

total Youden blank
A blank that corrects the signal for analyte–matrix interactions.

$$S_{samp} = 0.009844 \times W_x + 0.185$$

giving a true blank correction of 0.185. Using this value to correct the signals gives identical values for the concentration of analyte in all three samples (see Table 5.4, bottom row).

The total Youden blank is not encountered frequently in analytical work, because most chemists rely on a calibration blank when using calibration curves and rely on reagent blanks when using a single-point standardization. As long as any constant bias due to analyte–matrix interactions can be ignored, which is often the case, the accuracy of the method will not suffer. It is always a good idea, however, to check for constant sources of error, by analyzing samples of different sizes, before relying on either a calibration or reagent blank.

Table 5.4 Equations and Resulting Concentrations for Different Approaches to Correcting for the Method Blank

		Concentration of Analyte in		
Approach for Correcting Method Blank	**Equation**[a]	**Sample 1**	**Sample 2**	**Sample 3**
Ignore blank corrections	$C_A = \frac{W_a}{W_x} = \frac{S_{samp}}{kW_x}$	0.1707	0.1610	0.1552
Use calibration blank	$C_A = \frac{W_a}{W_x} = \frac{S_{samp} - CB}{kW_x}$	0.1441	0.1409	0.1390
Use reagent blank	$C_A = \frac{W_a}{W_x} = \frac{S_{samp} - RB}{kW_x}$	0.1494	0.1449	0.1422
Use both calibration and reagent blank	$C_A = \frac{W_a}{W_x} = \frac{S_{samp} - CB - RB}{kW_x}$	0.1227	0.1248	0.1261
Use total Youden blank	$C_A = \frac{W_a}{W_x} = \frac{S_{samp} - TYB}{kW_x}$	0.1313	0.1313	0.1313

[a]C_A = concentration of analyte; W_a = weight of analyte; W_x = weight of sample; k = slope of calibration curve = 0.075 (see Table 5.3).
Abbreviations: CB = calibration blank = 0.125 (see Table 5.3); RB = reagent blank = 0.100 (see Table 5.3); TYB = total Youden blank = 0.185 (see text).

5E KEY TERMS

aliquot (p. 111)
external standard (p. 109)
internal standard (p. 116)
linear regression (p. 118)
matrix matching (p. 110)
method of standard additions (p. 110)
multiple-point standardization (p. 109)
normal calibration curve (p. 109)
primary reagent (p. 106)
reagent grade (p. 107)
residual error (p. 118)
secondary reagent (p. 107)
single-point standardization (p. 108)
standard deviation about the regression (p. 121)
total Youden blank (p. 129)

5F SUMMARY

In a quantitative analysis, we measure a signal and calculate the amount of analyte using one of the following equations.

$$S_{meas} = kn_A + S_{reag}$$

$$S_{meas} = kC_A + S_{reag}$$

To obtain accurate results we must eliminate determinate errors affecting the measured signal, S_{meas}, the method's sensitivity, k, and any signal due to the reagents, S_{reag}.

To ensure that S_{meas} is determined accurately, we calibrate the equipment or instrument used to obtain the signal. Balances are calibrated using standard weights. When necessary, we can also correct for the buoyancy of air. Volumetric glassware can be calibrated by measuring the mass of water contained or delivered and using the density of water to calculate the true volume. Most instruments have calibration standards suggested by the manufacturer.

An analytical method is standardized by determining its sensitivity. There are several approaches to standardization, including the use of external standards, the method of standard addition, and the use of an internal standard. The most desirable standardization strategy is an external standardization. The method of standard additions, in which known amounts of analyte are added to the sample, is used when the sample's matrix complicates the analysis. An internal standard, which is a species (not analyte) added to all samples and standards, is used when the procedure does not allow for the reproducible handling of samples and standards.

Standardizations using a single standard are common, but also are subject to greater uncertainty. Whenever possible, a multiple-point standardization is preferred. The results of a multiple-point standardization are graphed as a calibration curve. A linear regression analysis can provide an equation for the standardization.

A reagent blank corrects the measured signal for signals due to reagents other than the sample that are used in an analysis. The most common reagent blank is prepared by omitting the sample. When a simple reagent blank does not compensate for all constant sources of determinate error, other types of blanks, such as the total Youden blank, can be used.

5G *Suggested* EXPERIMENTS

Experiments

The following exercises and experiments help connect the material in this chapter to the analytical laboratory.

Calibration—Volumetric glassware (burets, pipets, and volumetric flasks) can be calibrated in the manner described in Example 5.1. Most instruments have a calibration sample that can be prepared to verify the instrument's accuracy and precision. For example, as described in this chapter, a solution of 60.06 ppm $K_2Cr_2O_7$ in 0.0050 M H_2SO_4 should give an absorbance of 0.640 ± 0.010 at a wavelength of 350.0 nm when using 0.0050 M H_2SO_4 as a reagent blank. These exercises also provide practice with using volumetric glassware, weighing samples, and preparing solutions.

Standardization—External standards, standard additions, and internal standards are a common feature of many quantitative analyses. Suggested experiments using these standardization methods are found in later chapters. A good project experiment for introducing external standardization, standard additions, and the importance of the sample's matrix is to explore the effect of pH on the quantitative analysis of an acid–base indicator. Using bromothymol blue as an example, external standards can be prepared in a pH 9 buffer and used to analyze samples buffered to different pHs in the range of 6–10. Results can be compared with those obtained using a standard addition.

5H PROBLEMS

1. In calibrating a 10-mL pipet, a measured volume of water was transferred to a tared flask and weighed, yielding a mass of 9.9814 g. (a) Calculate, with and without correcting for buoyancy, the volume of water delivered by the pipet. Assume that the density of water is 0.99707 g/cm^3 and that the density of the weights is 8.40 g/cm^3. (b) What are the absolute and relative errors introduced by failing to account for the effect of buoyancy? Is this a significant source of determinate error for the calibration of a pipet? Explain.

2. Repeat the questions in problem 1 for the case when a mass of 0.2500 g is measured for a solid that has a density of 2.50 g/cm^3.

3. Is the failure to correct for buoyancy a constant or proportional source of determinate error?

4. What is the minimum density of a substance necessary to keep the buoyancy correction to less than 0.01% when using brass calibration weights with a density of 8.40 g/cm^3?

5. Describe how you would use a serial dilution to prepare 100 mL each of a series of standards with concentrations of 1.000×10^{-5}, 1.000×10^{-4}, 1.000×10^{-3}, and 1.000×10^{-2} M from a 0.1000 M stock solution. Calculate the uncertainty for each solution using a propagation of uncertainty, and compare to the uncertainty if each solution was prepared by a single dilution of the stock solution. Tolerances for different types of volumetric glassware and digital pipets are found in Tables 4.2 and 4.4. Assume that the uncertainty in the molarity of the stock solution is ± 0.0002.

6. Three replicate determinations of the signal for a standard solution of an analyte at a concentration of 10.0 ppm give values of 0.163, 0.157, and 0.161 (arbitrary units), respectively. The signal for a method blank was found to be 0.002. Calculate the concentration of analyte in a sample that gives a signal of 0.118.

7. A 10.00-g sample containing an analyte was transferred to a 250-mL volumetric flask and diluted to volume. When a 10.00-mL aliquot of the resulting solution was diluted to 25.00 mL it was found to give a signal of 0.235 (arbitrary units). A second 10.00-mL aliquot was spiked with 10.00 mL of a 1.00-ppm standard solution of the analyte and diluted to 25.00 mL. The signal for the spiked sample was found to be 0.502. Calculate the weight percent of analyte in the original sample.

8. A 50.00-mL sample containing an analyte gives a signal of 11.5 (arbitrary units). A second 50-mL aliquot of the sample, which is spiked with 1.00-mL of a 10.0-ppm standard solution of the analyte, gives a signal of 23.1. What is the concentration of analyte in the original sample?

9. An appropriate standard additions calibration curve based on equation 5.8 plots $S_{spike}(V_o + V_s)$ on the y-axis and $C_s V_s$ on the x-axis. Clearly explain why you cannot plot S_{spike} on the y-axis and $C_s[V_s/(V_o + V_s)]$ on the x-axis. Derive equations for the slope and y-intercept, and explain how the amount of analyte in a sample can be determined from the calibration curve.

10. A standard sample was prepared containing 10.0 ppm of an analyte and 15.0 ppm of an internal standard. Analysis of the sample gave signals for the analyte and internal standard of 0.155 and 0.233 (arbitrary units), respectively. Sufficient internal standard was added to a sample to make it 15.0 ppm in the internal standard. Analysis of the sample yielded signals for the analyte and internal standard of 0.274 and 0.198, respectively. Report the concentration of analyte in the sample.

11. For each of the pairs of calibration curves in Figure 5.13 on page 132, select the calibration curve with the better set of standards. Briefly explain the reasons for your selections. The scales for the x-axes and y-axes are the same for each pair.

12. The following standardization data were provided for a series of external standards of Cd^{2+} that had been buffered to a pH of 4.6.[14]

$[Cd^{2+}]$ (nM)	15.4	30.4	44.9	59.0	72.7	86.0
S_{meas} (nA)	4.8	11.4	18.2	26.6	32.3	37.7

(a) Determine the standardization relationship by a linear regression analysis, and report the confidence intervals for the slope and y-intercept. (b) Construct a plot of the residuals, and comment on their significance.

At a pH of 3.7 the following data were recorded

$[Cd^{2+}]$ (nM)	15.4	30.4	44.9	59.0	72.7	86.0
S_{meas} (nA)	15.0	42.7	58.5	77.0	101	118

(c) How much more or less sensitive is this method at the lower pH? (d) A single sample is buffered to a pH of 3.7 and analyzed for cadmium, yielding a signal of 66.3. Report the concentration of Cd^{2+} in the sample and its 95% confidence interval.

13. To determine the concentration of analyte in a sample, a standard additions was performed. A 5.00-mL portion of the sample was analyzed and then successive 0.10-mL spikes of a 600.0-ppb standard of the analyte

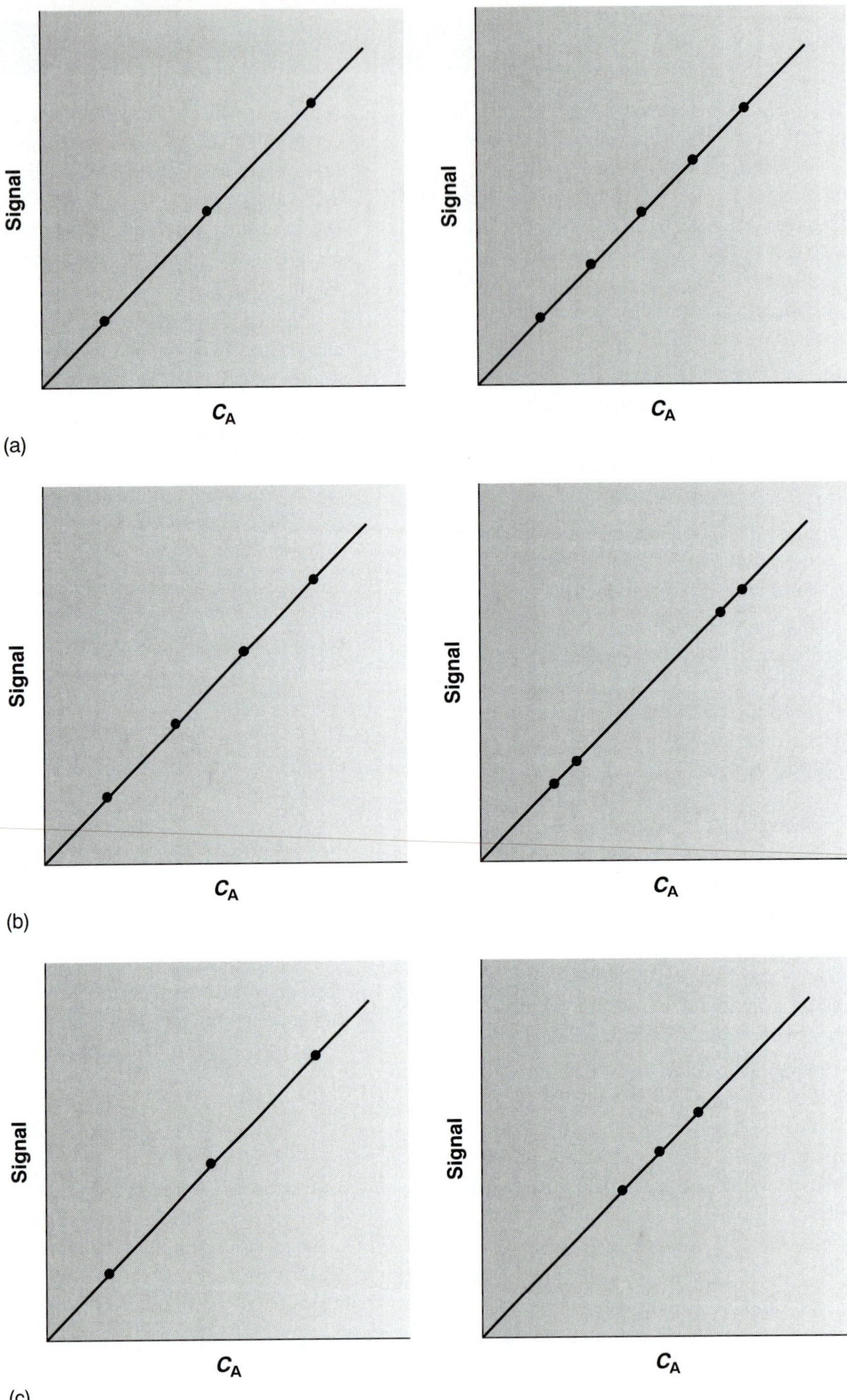

Figure 5.13

were added, analyzing after each spike. The following results were obtained

Volume of Spike (mL)	Signal (arbitrary units)
0.00	0.119
0.10	0.231
0.20	0.339
0.30	0.442

Construct an appropriate standard additions calibration curve, and use a linear regression analysis to determine the concentration of analyte in the original sample and its 95% confidence interval.

14. Troost and Olavesen investigated the application of an internal standardization to the quantitative analysis of polynuclear aromatic hydrocarbons.[15] The following results were obtained for the analysis of the analyte phenanthrene using isotopically labeled phenanthrene as an internal standard

	S_A/S_{IS}	
C_A/C_{IS}	Replicate 1	Replicate 2
0.50	0.514	0.522
1.25	0.993	1.024
2.00	1.486	1.471
3.00	2.044	2.080
4.00	2.342	2.550

(a) Determine the standardization relationship by a linear regression, and report the confidence intervals for the slope and y-intercept. (b) Based on your results, explain why the authors of this paper concluded that the internal standardization was inappropriate.

15. In Chapter 4 we used a paired t-test to compare two methods that had been used to independently analyze a series of samples of variable composition. An alternative approach is to plot the results for one method versus those for the other. If the two methods yield identical results, then the plot should have a true slope (β_1) of 1.00 and a true y-intercept (β_0) of 0.0. A t-test can be used to compare the actual slope and y-intercept with these ideal values. The appropriate test statistic for the y-intercept is found by rearranging equation 5.18

$$t_{\text{exp}} = \frac{|\beta_0 - b_0|}{s_{b_0}} = \frac{|b_0|}{s_{b_0}}$$

Rearranging equation 5.17 gives the test statistic for the slope

$$t_{\text{exp}} = \frac{|\beta_1 - b_1|}{s_{b_1}} = \frac{|1.00 - b_1|}{s_{b_1}}$$

Reevaluate the data in problem 24 in Chapter 4 using the same significance level as in the original problem.*

16. Franke and co-workers evaluated a standard additions method for a voltammetric determination of Tl.[16] A summary of their results is tabulated here.

ppm Tl added	Instrument Response for Replicates (μA)						
0.000	2.53	2.50	2.70	2.63	2.70	2.80	2.52
0.387	8.42	7.96	8.54	8.18	7.70	8.34	7.98
1.851	29.65	28.70	29.05	28.30	29.20	29.95	28.95
5.734	84.8	85.6	86.0	85.2	84.2	86.4	87.8

Determine the standardization relationship using a weighted linear regression.

5I SUGGESTED READINGS

In addition to the texts listed as suggested readings in Chapter 4, the following text provides additional details on regression

Draper, N. R.; Smith, H. *Applied Regression Analysis*, 2nd. ed. Wiley: New York, 1981.

Several articles providing more details about linear regression follow.

Boqué, R.; Rius, F. X.; Massart, D. L. "Straight Line Calibration: Something More Than Slopes, Intercepts, and Correlation Coefficients," *J. Chem. Educ.* **1993,** *70,* 230–232.

Henderson, G. "Lecture Graphic Aids for Least-Squares Analysis," *J. Chem. Educ.* **1988,** *65,* 1001–1003.

Renman, L., Jagner, D. "Asymmetric Distribution of Results in Calibration Curve and Standard Addition Evaluations," *Anal. Chim. Acta* **1997,** *357,* 157–166.

Two useful papers providing additional details on the method of standard additions are

Bader, M. "A Systematic Approach to Standard Addition Methods in Instrumental Analysis," *J. Chem. Educ.* **1980,** *57,* 703–706.

*Although this is a commonly used procedure for comparing two methods, it does violate one of the assumptions of an ordinary linear regression. Since both methods are expected to have indeterminate errors, an unweighted regression with errors in y may produce a biased result, with the slope being underestimated and the y-intercept being overestimated. This limitation can be minimized by placing the more precise method on the x-axis, using ten or more samples to increase the degrees of freedom in the analysis, and by using samples that uniformly cover the range of concentrations. For more information see Miller, J. C.; Miller, J. N. *Statistics for Analytical Chemistry*, 3rd ed. Ellis Horwood PTR Prentice-Hall: New York, 1993. Alternative approaches are discussed in Hartman, C.; Smeyers-Verbeke, J.; Penninckx, W.; Massart, D. L. *Anal. Chim. Acta* **1997,** *338,* 19–40 and Zwanziger, H. W.; Sârbu, C. *Anal. Chem.* **1998,** *70,* 1277–1280.

Nimura, Y.; Carr, M. R. "Reduction of the Relative Error in the Standard Additions Method," *Analyst* **1990,** *115,* 1589–1595.

The following paper discusses the importance of weighting experimental data when using linear regression

Karolczak, M. "To Weight or Not to Weight? An Analyst's Dilemma," *Curr. Separations* **1995,** *13,* 98–104.

Algorithms for performing a linear regression with errors in both *x* and *y* are discussed in

Irvin, J. A.; Quickenden, T. L. "Linear Least Squares Treatment When There Are Errors in Both *x* and *y*," *J. Chem. Educ.* **1983,** *60,* 711–712.

Kalantar, A. H. "Kerrich's Method for $y = \alpha x$ Data When Both *y* and *x* Are Uncertain," *J. Chem. Educ.* **1991,** *68,* 368–370.

Macdonald, J. R.; Thompson, W. J. "Least-Squares Fitting When Both Variables Contain Errors: Pitfalls and Possibilities," *Am. J. Phys.* **1992,** *60,* 66–73.

Ogren, P. J.; Norton, J. R. "Applying a Simple Linear Least-Squares Algorithm to Data with Uncertainties in Both Variables," *J. Chem. Educ.* **1992,** *69,* A130–A131.

The following paper discusses some of the problems that may be encountered when using linear regression to model data that have been mathematically transformed into a linear form.

Chong, D. P. "On the Use of Least Squares to Fit Data in Linear Form," *J. Chem. Educ.* **1994,** *71,* 489–490.

The analysis of nonlinear data is covered in the following papers.

Harris, D. C. "Nonlinear Least-Squares Curve Fitting with Microsoft Excel Solver," *J. Chem. Educ.* **1998,** *75,* 119–121.

Lieb, S. G. "Simplex Method of Nonlinear Least-Squares—A Logical Complementary Method to Linear Least-Squares Analysis of Data," *J. Chem. Educ.* **1997,** *74,* 1008–1011.

Machuca-Herrera, J. G. "Nonlinear Curve Fitting with Spreadsheets," *J. Chem. Educ.* **1997,** *74,* 448–449.

Zielinski, T. J.; Allendoerfer, R. D. "Least Squares Fitting of Nonlinear Data in the Undergraduate Laboratory," *J. Chem. Educ.* **1997,** *74,* 1001–1007.

More information on multivariate regression can be found in

Lang, P. M.; Kalivas, J. H. "A Global Perspective on Multivariate Calibration Methods," *J. Chemometrics* **1993,** *7,* 153–164.

Kowalski, B. R.; Seasholtz, M. B. "Recent Developments in Multivariate Calibration," *J. Chemometrics* **1991,** *5* 129–145.

An additional discussion on method blanks is found in the following two papers.

Cardone, M. J. "Detection and Determination of Error in Analytical Methodology. Part II. Correction for Corrigible Systematic Error in the Course of Real Sample Analysis," *J. Assoc. Off. Anal. Chem.* **1983,** *66,* 1283–1294.

Cardone, M. J. "Detection and Determination of Error in Analytical Methodology. Part IIB. Direct Calculational Technique for Making Corrigible Systematic Error Corrections," *J. Assoc. Off. Anal. Chem.* **1985,** *68,* 199–202.

5J REFERENCES

1. Battino, R.; Williamson, A. G. *J. Chem. Educ.* **1984,** *61,* 51–52.
2. Ebel, S. *Fresenius J. Anal. Chem.* **1992,** *342,* 769.
3. ACS Committee on Environmental Improvement "Guidelines for Data Acquisition and Data Quality Evaluation in Environmental Chemistry," *Anal. Chem.* **1980,** *52,* 2242–2249.
4. Moody, J. R.; Greenburg, P. R.; Pratt, K. W.; et al. *Anal. Chem.* **1988,** *60,* 1203A–1218A.
5. Committee on Analytical Reagents, *Reagent Chemicals,* 8th ed., American Chemical Society: Washington, DC, 1993.
6. Cardone, M. J.; Palmero, P. J.; Sybrandt, L. B. *Anal. Chem.* **1980,** *52,* 1187–1191.
7. Draper, N. R.; Smith, H. *Applied Regression Analysis,* 2nd ed. Wiley: New York, 1981.
8. (a) Miller, J. N. *Analyst* **1991,** *116,* 3–14; and (b) Sharaf, M. A.; Illman, D. L.; Kowalski, B. R. *Chemometrics,* Wiley-Interscience: New York, 1986; pp. 126–127.
9. Bonate, P. *J. Anal. Chem.* **1993,** *65,* 1367–1372.
10. (a) Sharaf, M. A.; Illman, D. L.; Kowalski, B. R. *Chemometrics,* Wiley-Interscience: New York, 1986; (b) Deming, S. N.; Morgan, S. L. *Experimental Design: A Chemometric Approach,* Elsevier: Amsterdam, 1987.
11. Beebe, K. R.; Kowalski, B. R. *Anal. Chem.* **1987,** *59,* 1007A–1017A.
12. Cardone, M. *J. Anal. Chem.* **1986,** *58,* 433–438.
13. Cardone, M. *J. Anal. Chem.* **1986,** *58,* 438–445.
14. Wojciechowski, M; Balcerzak, *J. Anal. Chim. Acta* **1991,** *249,* 433–445.
15. Troost, J. R.; Olavesen, E. Y. *Anal. Chem.* **1996,** *68,* 708–711.
16. Franke, J. P.; de Zeeuw, R. A.; Hakkert, R. *Anal. Chem.* **1978,** *50,* 1374–1380.

Chapter 6

Equilibrium Chemistry

Regardless of the problem on which an analytical chemist is working, its solution ultimately requires a knowledge of chemistry and the ability to reason with that knowledge. For example, an analytical chemist developing a method for studying the effect of pollution on spruce trees needs to know, or know where to find, the structural and chemical differences between *p*-hydroxybenzoic acid and *p*-hydroxyacetophenone, two common phenols found in the needles of spruce trees (Figure 6.1). Chemical reasoning is a product of experience and is constructed from knowledge acquired in the classroom, the laboratory, and the chemical literature.

The material in this text assumes familiarity with topics covered in the courses and laboratory work you have already completed. This chapter provides a review of equilibrium chemistry. Much of the material in this chapter should be familiar to you, but other ideas are natural extensions of familiar topics.

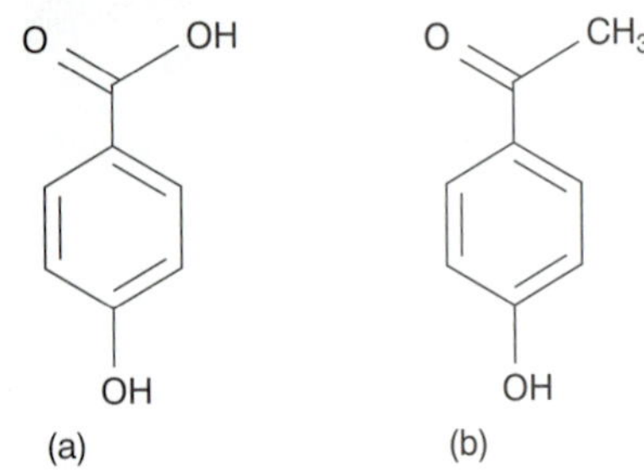

Figure 6.1
Structures of (a) *p*-hydroxybenzoic acid and (b) *p*-hydroxyacetophenone.

6A Reversible Reactions and Chemical Equilibria

In 1798, the chemist Claude Berthollet (1748–1822) accompanied a French military expedition to Egypt. While visiting the Natron Lakes, a series of salt water lakes carved from limestone, Berthollet made an observation that contributed to an important discovery. Upon analyzing water from the Natron Lakes, Berthollet found large quantities of common salt, NaCl, and soda ash, Na_2CO_3, a result he found surprising. Why would Berthollet find this result surprising and how did it contribute to an important discovery? Answering these questions provides an example of chemical reasoning and introduces the topic of this chapter.

Berthollet "knew" that a reaction between Na_2CO_3 and $CaCl_2$ goes to completion, forming NaCl and a precipitate of $CaCO_3$ as products.

$$Na_2CO_3 + CaCl_2 \rightarrow 2NaCl + CaCO_3$$

Understanding this, Berthollet expected that large quantities of NaCl and Na_2CO_3 could not coexist in the presence of $CaCO_3$. Since the reaction goes to completion, adding a large quantity of $CaCl_2$ to a solution of Na_2CO_3 should produce NaCl and $CaCO_3$, leaving behind no unreacted Na_2CO_3. In fact, this result is what he observed in the laboratory. The evidence from Natron Lakes, where the coexistence of NaCl and Na_2CO_3 suggests that the reaction has not gone to completion, ran counter to Berthollet's expectations. Berthollet's important insight was recognizing that the chemistry occurring in the Natron Lakes is the reverse of what occurs in the laboratory.

$$CaCO_3 + 2NaCl \rightarrow Na_2CO_3 + CaCl_2$$

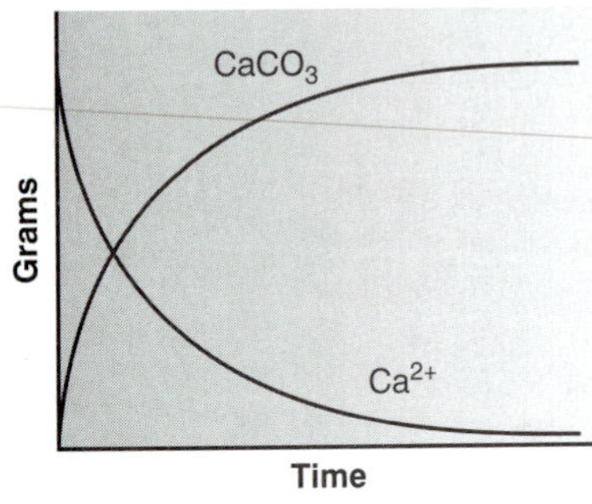

Figure 6.2
Change in mass of undissolved Ca^{2+} and solid $CaCO_3$ over time during the precipitation of $CaCO_3$.

Using this insight Berthollet reasoned that the reaction is reversible, and that the relative amounts of "reactants" and "products" determine the direction in which the reaction occurs, and the final composition of the reaction mixture. We recognize a reaction's ability to move in both directions by using a double arrow when writing the reaction.

$$Na_2CO_3 + CaCl_2 \rightleftharpoons 2NaCl + CaCO_3$$

Berthollet's reasoning that reactions are reversible was an important step in understanding chemical reactivity. When we mix together solutions of Na_2CO_3 and $CaCl_2$, they react to produce NaCl and $CaCO_3$. If we monitor the mass of dissolved Ca^{2+} remaining and the mass of $CaCO_3$ produced as a function of time, the result will look something like the graph in Figure 6.2. At the start of the reaction the mass of dissolved Ca^{2+} decreases and the mass of $CaCO_3$ increases. Eventually, however, the reaction reaches a point after which no further changes occur in the amounts of these species. Such a condition is called a state of **equilibrium.**

equilibrium
A system is at equilibrium when the concentrations of reactants and products remain constant.

Although a system at equilibrium appears static on a macroscopic level, it is important to remember that the forward and reverse reactions still occur. A reaction at equilibrium exists in a "steady state," in which the rate at which any species forms equals the rate at which it is consumed.

6B Thermodynamics and Equilibrium Chemistry

Thermodynamics is the study of thermal, electrical, chemical, and mechanical forms of energy. The study of thermodynamics crosses many disciplines, including physics, engineering, and chemistry. Of the various branches of thermodynamics, the most important to chemistry is the study of the changes in energy occurring during a chemical reaction.

Consider, for example, the general equilibrium reaction shown in equation 6.1, involving the solutes A, B, C, and D, with stoichiometric coefficients *a, b, c,* and *d.*

$$aA + bB \rightleftharpoons cC + dD \qquad 6.1$$

By convention, species to the left of the arrows are called reactants, and those on the right side of the arrows are called products. As Berthollet discovered, writing a reaction in this fashion does not guarantee that the reaction of A and B to produce C and D is favorable. Depending on initial conditions, the reaction may move to the left, to the right, or be in a state of equilibrium. Understanding the factors that determine the final position of a reaction is one of the goals of chemical thermodynamics.

Chemical systems spontaneously react in a fashion that lowers their overall free energy. At a constant temperature and pressure, typical of many bench-top chemical reactions, the free energy of a chemical reaction is given by the **Gibb's free energy** function

Gibb's free energy
A thermodynamic function for systems at constant temperature and pressure that indicates whether or not a reaction is favorable ($\Delta G < 0$), unfavorable ($\Delta G > 0$), or at equilibrium ($\Delta G = 0$).

$$\Delta G = \Delta H - T\,\Delta S \qquad 6.2$$

where T is the temperature in kelvins, and ΔG, ΔH, and ΔS are the differences in the Gibb's free energy, the enthalpy, and the entropy between the products and reactants.

Enthalpy is a measure of the net flow of energy, as heat, during a chemical reaction. Reactions in which heat is produced have a negative ΔH and are called exothermic. Endothermic reactions absorb heat from their surroundings and have a positive ΔH. **Entropy** is a measure of randomness, or disorder. The entropy of an individual species is always positive and tends to be larger for gases than for solids and for more complex rather than simpler molecules. Reactions that result in a large number of simple, gaseous products usually have a positive ΔS.

enthalpy
A change in enthalpy indicates the heat absorbed or released during a chemical reaction at constant pressure.

entropy
A measure of disorder.

The sign of ΔG can be used to predict the direction in which a reaction moves to reach its equilibrium position. A reaction is always thermodynamically favored when enthalpy decreases and entropy increases. Substituting the inequalities $\Delta H < 0$ and $\Delta S > 0$ into equation 6.2 shows that ΔG is negative when a reaction is thermodynamically favored. When ΔG is positive, the reaction is unfavorable as written (although the reverse reaction is favorable). Systems at equilibrium have a ΔG of zero.

As a system moves from a nonequilibrium to an equilibrium position, ΔG must change from its initial value to zero. At the same time, the species involved in the reaction undergo a change in their concentrations. The Gibb's free energy, therefore, must be a function of the concentrations of reactants and products.

As shown in equation 6.3, the Gibb's free energy can be divided into two terms.

$$\Delta G = \Delta G^\circ + RT \ln Q \qquad 6.3$$

The first term, ΔG°, is the change in Gibb's free energy under **standard-state** conditions; defined as a temperature of 298 K, all gases with partial pressures of 1 atm, all solids and liquids pure, and all solutes present with 1 M concentrations. The second term, which includes the reaction quotient, Q, accounts for nonstandard-state pressures or concentrations. For reaction 6.1 the reaction quotient is

standard state
Condition in which solids and liquids are in pure form, gases have partial pressures of 1 atm, solutes have concentrations of 1 M, and the temperature is 298 K.

$$Q = \frac{[C]^c[D]^d}{[A]^a[B]^b} \qquad 6.4$$

where the terms in brackets are the molar concentrations of the solutes. Note that the reaction quotient is defined such that the concentrations of products are placed in the numerator, and the concentrations of reactants are placed in the denominator. In addition, each concentration term is raised to a power equal to its stoichiometric coefficient in the balanced chemical reaction. Partial pressures are substituted for concentrations when the reactant or product is a gas. The concentrations of pure solids and pure liquids do not change during a chemical reaction and are excluded from the reaction quotient.

At equilibrium the Gibb's free energy is zero, and equation 6.3 simplifies to

$$\Delta G^\circ = -RT \ln K$$

equilibrium constant
For a reaction at equilibrium, the equilibrium constant determines the relative concentrations of products and reactants.

where K is an **equilibrium constant** that defines the reaction's equilibrium position. The equilibrium constant is just the numerical value obtained when substituting the concentrations of reactants and products at equilibrium into equation 6.4; thus,

$$K = \frac{[\mathrm{C}]_{\mathrm{eq}}^c[\mathrm{D}]_{\mathrm{eq}}^d}{[\mathrm{A}]_{\mathrm{eq}}^a[\mathrm{B}]_{\mathrm{eq}}^b} \qquad \mathbf{6.5}$$

where the subscript "eq" indicates a concentration at equilibrium. Although the subscript "eq" is usually omitted, it is important to remember that the value of K is determined by the concentrations of solutes at equilibrium.

As written, equation 6.5 is a limiting law that applies only to infinitely dilute solutions, in which the chemical behavior of any species in the system is unaffected by all other species. Corrections to equation 6.5 are possible and are discussed in more detail at the end of the chapter.

6C Manipulating Equilibrium Constants

We will use two useful relationships when working with equilibrium constants. First, if we reverse a reaction's direction, the equilibrium constant for the new reaction is simply the inverse of that for the original reaction. For example, the equilibrium constant for the reaction

$$\mathrm{A} + 2\mathrm{B} \rightleftharpoons \mathrm{AB_2} \qquad K_1 = \frac{[\mathrm{AB_2}]}{[\mathrm{A}][\mathrm{B}]^2}$$

is the inverse of that for the reaction

$$\mathrm{AB_2} \rightleftharpoons \mathrm{A} + 2\mathrm{B} \qquad K_2 = \frac{1}{K_1} = \frac{[\mathrm{A}][\mathrm{B}]^2}{[\mathrm{AB_2}]}$$

Second, if we add together two reactions to obtain a new reaction, the equilibrium constant for the new reaction is the product of the equilibrium constants for the original reactions.

$$\mathrm{A} + \mathrm{C} \rightleftharpoons \mathrm{AC} \qquad K_1 = \frac{[\mathrm{AC}]}{[\mathrm{A}][\mathrm{C}]}$$

$$\mathrm{AC} + \mathrm{C} \rightleftharpoons \mathrm{AC_2} \qquad K_2 = \frac{[\mathrm{AC_2}]}{[\mathrm{AC}][\mathrm{C}]}$$

$$\mathrm{A} + 2\mathrm{C} \rightleftharpoons \mathrm{AC_2} \qquad K_3 = K_1K_2 = \frac{[\mathrm{AC}]}{[\mathrm{A}][\mathrm{C}]} \times \frac{[\mathrm{AC_2}]}{[\mathrm{AC}][\mathrm{C}]} = \frac{[\mathrm{AC_2}]}{[\mathrm{A}][\mathrm{C}]^2}$$

Example

EXAMPLE 6.1

Calculate the equilibrium constant for the reaction

$$2A + B \rightleftharpoons C + 3D$$

given the following information

Rxn 1:	$A + B \rightleftharpoons D$	$K_1 = 0.40$
Rxn 2:	$A + E \rightleftharpoons C + D + F$	$K_2 = 0.10$
Rxn 3:	$C + E \rightleftharpoons B$	$K_3 = 2.0$
Rxn 4:	$F + C \rightleftharpoons D + B$	$K_4 = 5.0$

SOLUTION

The overall reaction is given as

$$\text{Rxn 1} + \text{Rxn 2} - \text{Rxn 3} + \text{Rxn 4}$$

If Rxn 3 is reversed, giving

$$\text{Rxn 5: } B \rightleftharpoons C + E \qquad K_5 = \frac{1}{K_3} = \frac{1}{2.0} = 0.50$$

then the overall reaction is

$$\text{Rxn 1} + \text{Rxn 2} + \text{Rxn 5} + \text{Rxn 4}$$

and the overall equilibrium constant is

$$K_{\text{overall}} = K_1 \times K_2 \times K_5 \times K_4 = 0.40 \times 0.10 \times 0.50 \times 5.0 = 0.10$$

6D Equilibrium Constants for Chemical Reactions

Several types of reactions are commonly used in analytical procedures, either in preparing samples for analysis or during the analysis itself. The most important of these are precipitation reactions, acid–base reactions, complexation reactions, and oxidation–reduction reactions. In this section we review these reactions and their equilibrium constant expressions.

6D.1 Precipitation Reactions

A precipitation reaction occurs when two or more soluble species combine to form an insoluble product that we call a **precipitate.** The most common precipitation reaction is a metathesis reaction, in which two soluble ionic compounds exchange parts. When a solution of lead nitrate is added to a solution of potassium chloride, for example, a precipitate of lead chloride forms. We usually write the balanced reaction as a net ionic equation, in which only the precipitate and those ions involved in the reaction are included. Thus, the precipitation of $PbCl_2$ is written as

precipitate
An insoluble solid that forms when two or more soluble reagents are combined.

$$Pb^{2+}(aq) + 2Cl^-(aq) \rightleftharpoons PbCl_2(s)$$

In the equilibrium treatment of precipitation, however, the reverse reaction describing the dissolution of the precipitate is more frequently encountered.

$$PbCl_2(s) \rightleftharpoons Pb^{2+}(aq) + 2Cl^-(aq)$$

solubility product
The equilibrium constant for a reaction in which a solid dissociates into its ions (K_{sp}).

The equilibrium constant for this reaction is called the **solubility product,** K_{sp}, and is given as

$$K_{sp} = [Pb^{2+}][Cl^-]^2 = 1.7 \times 10^{-5} \qquad 6.6$$

Note that the precipitate, which is a solid, does not appear in the K_{sp} expression. It is important to remember, however, that equation 6.6 is valid only if $PbCl_2(s)$ is present and in equilibrium with the dissolved Pb^{2+} and Cl^-. Values for selected solubility products can be found in Appendix 3A.

6D.2 Acid–Base Reactions

acid
A proton donor.

base
A proton acceptor.

A useful definition of acids and bases is that independently introduced by Johannes Brønsted (1879–1947) and Thomas Lowry (1874–1936) in 1923. In the Brønsted-Lowry definition, **acids** are proton donors, and **bases** are proton acceptors. Note that these definitions are interrelated. Defining a base as a proton acceptor means an acid must be available to provide the proton. For example, in reaction 6.7 acetic acid, CH_3COOH, donates a proton to ammonia, NH_3, which serves as the base.

$$CH_3COOH(aq) + NH_3(aq) \rightleftharpoons CH_3COO^-(aq) + NH_4^+(aq) \qquad 6.7$$

When an acid and a base react, the products are a new acid and base. For example, the acetate ion, CH_3COO^-, in reaction 6.7 is a base that reacts with the acidic ammonium ion, NH_4^+, to produce acetic acid and ammonia. We call the acetate ion the conjugate base of acetic acid, and the ammonium ion is the conjugate acid of ammonia.

Strong and Weak Acids The reaction of an acid with its solvent (typically water) is called an acid dissociation reaction. Acids are divided into two categories based on the ease with which they can donate protons to the solvent. Strong acids, such as HCl, almost completely transfer their protons to the solvent molecules.

$$HCl(aq) + H_2O(\ell) \rightarrow H_3O^+(aq) + Cl^-(aq)$$

In this reaction H_2O serves as the base. The hydronium ion, H_3O^+, is the conjugate acid of H_2O, and the chloride ion is the conjugate base of HCl. It is the hydronium ion that is the acidic species in solution, and its concentration determines the acidity of the resulting solution. We have chosen to use a single arrow ($\rightarrow$) in place of the double arrows ($\rightleftharpoons$) to indicate that we treat HCl as if it were completely dissociated in aqueous solutions. A solution of 0.10 M HCl is effectively 0.10 M in H_3O^+ and 0.10 M in Cl^-. In aqueous solutions, the common strong acids are hydrochloric acid (HCl), hydroiodic acid (HI), hydrobromic acid (HBr), nitric acid (HNO_3), perchloric acid ($HClO_4$), and the first proton of sulfuric acid (H_2SO_4).

Weak acids, of which aqueous acetic acid is one example, cannot completely donate their acidic protons to the solvent. Instead, most of the acid remains undissociated, with only a small fraction present as the conjugate base.

acid dissociation constant
The equilibrium constant for a reaction in which an acid donates a proton to the solvent (K_a).

$$CH_3COOH(aq) + H_2O(\ell) \rightleftharpoons H_3O^+(aq) + CH_3COO^-(aq)$$

The equilibrium constant for this reaction is called an **acid dissociation constant,** K_a, and is written as

$$K_a = \frac{[H_3O^+][CH_3COO^-]}{[CH_3COOH]} = 1.75 \times 10^{-5}$$

Note that the concentration of H_2O is omitted from the K_a expression because its value is so large that it is unaffected by the dissociation reaction.* The magnitude of K_a provides information about the relative strength of a weak acid, with a smaller K_a corresponding to a weaker acid. The ammonium ion, for example, with a K_a of 5.70×10^{-10}, is a weaker acid than acetic acid.

Monoprotic weak acids, such as acetic acid, have only a single acidic proton and a single acid dissociation constant. Some acids, such as phosphoric acid, can donate more than one proton and are called polyprotic weak acids. Polyprotic acids are described by a series of acid dissociation steps, each characterized by it own acid dissociation constant. Phosphoric acid, for example, has three acid dissociation reactions and acid dissociation constants.

$$H_3PO_4(aq) + H_2O(\ell) \rightleftharpoons H_3O^+(aq) + H_2PO_4^-(aq)$$

$$K_{a1} = \frac{[H_2PO_4^-][H_3O^+]}{[H_3PO_4]} = 7.11 \times 10^{-3}$$

$$H_2PO_4^-(aq) + H_2O(\ell) \rightleftharpoons H_3O^+(aq) + HPO_4^{2-}(aq)$$

$$K_{a2} = \frac{[HPO_4^{2-}][H_3O^+]}{[H_2PO_4^-]} = 6.32 \times 10^{-8}$$

$$HPO_4^{2-}(aq) + H_2O(\ell) \rightleftharpoons H_3O^+(aq) + PO_4^{3-}(aq)$$

$$K_{a3} = \frac{[PO_4^{3-}][H_3O^+]}{[HPO_4^{2-}]} = 4.5 \times 10^{-13}$$

The decrease in the acid dissociation constant from K_{a1} to K_{a3} tells us that each successive proton is harder to remove. Consequently, H_3PO_4 is a stronger acid than $H_2PO_4^-$, and $H_2PO_4^-$ is a stronger acid than HPO_4^{2-}.

Strong and Weak Bases Just as the acidity of an aqueous solution is a measure of the concentration of the hydronium ion, H_3O^+, the basicity of an aqueous solution is a measure of the concentration of the hydroxide ion, OH^-. The most common example of a strong base is an alkali metal hydroxide, such as sodium hydroxide, which completely dissociates to produce the hydroxide ion.

$$NaOH(aq) \rightarrow Na^+(aq) + OH^-(aq)$$

Weak bases only partially accept protons from the solvent and are characterized by a **base dissociation constant,** K_b. For example, the base dissociation reaction and base dissociation constant for the acetate ion are

base dissociation constant
The equilibrium constant for a reaction in which a base accepts a proton from the solvent (K_b).

$$CH_3COO^-(aq) + H_2O(\ell) \rightleftharpoons OH^-(aq) + CH_3COOH(aq)$$

$$K_b = \frac{[CH_3COOH][OH^-]}{[CH_3COO^-]} = 5.71 \times 10^{-10}$$

Polyprotic bases, like polyprotic acids, also have more than one base dissociation reaction and base dissociation constant.

Amphiprotic Species Some species can behave as either an acid or a base. For example, the following two reactions show the chemical reactivity of the bicarbonate ion, HCO_3^-, in water.

*The concentration of pure water is approximately 55.5 M

$$HCO_3^-(aq) + H_2O(\ell) \rightleftharpoons H_3O^+(aq) + CO_3^{2-}(aq) \qquad 6.8$$

$$HCO_3^-(aq) + H_2O(\ell) \rightleftharpoons OH^-(aq) + H_2CO_3(aq) \qquad 6.9$$

amphiprotic
A species capable of acting as both an acid and a base.

A species that can serve as both a proton donor and a proton acceptor is called **amphiprotic.** Whether an amphiprotic species behaves as an acid or as a base depends on the equilibrium constants for the two competing reactions. For bicarbonate, the acid dissociation constant for reaction 6.8

$$K_{a2} = 4.69 \times 10^{-11}$$

is smaller than the base dissociation constant for reaction 6.9.

$$K_{b2} = 2.25 \times 10^{-8}$$

Since bicarbonate is a stronger base than it is an acid ($k_{b2} > k_{a2}$), we expect that aqueous solutions of HCO_3^- will be basic.

Dissociation of Water Water is an amphiprotic solvent in that it can serve as an acid or a base. An interesting feature of an amphiprotic solvent is that it is capable of reacting with itself as an acid and a base.

$$H_2O(\ell) + H_2O(\ell) \rightleftharpoons H_3O^+(aq) + OH^-(aq)$$

The equilibrium constant for this reaction is called water's dissociation constant, K_w,

$$K_w = [H_3O^+][OH^-] \qquad 6.10$$

which has a value of 1.0000×10^{-14} at a temperature of 24 °C. The value of K_w varies substantially with temperature. For example, at 20 °C, K_w is 6.809×10^{-15}, but at 30 °C K_w is 1.469×10^{-14}. At the standard state temperature of 25 °C, K_w is 1.008×10^{-14}, which is sufficiently close to 1.00×10^{-14} that the latter value can be used with negligible error.

The pH Scale An important consequence of equation 6.10 is that the concentrations of H_3O^+ and OH^- are related. If we know $[H_3O^+]$ for a solution, then $[OH^-]$ can be calculated using equation 6.10.

Example

EXAMPLE 6.2

What is the $[OH^-]$ if the $[H_3O^+]$ is 6.12×10^{-5} M?

SOLUTION

$$[OH^-] = \frac{K_w}{[H_3O^+]} = \frac{1.00 \times 10^{-14}}{6.12 \times 10^{-5}} = 1.63 \times 10^{-10}$$

pH
Defined as $pH = -\log[H_3O^+]$.

Equation 6.10 also allows us to develop a **pH** scale that indicates the acidity of a solution. When the concentrations of H_3O^+ and OH^- are equal, a solution is neither acidic nor basic; that is, the solution is neutral. Letting

$$[H_3O^+] = [OH^-]$$

and substituting into equation 6.10 leaves us with

$$K_w = [H_3O^+]^2 = 1.00 \times 10^{-14}$$

Solving for $[H_3O^+]$ gives

$$[H_3O^+] = \sqrt{1.00 \times 10^{-14}} = 1.00 \times 10^{-7}$$

A neutral solution has a hydronium ion concentration of 1.00×10^{-7} M and a pH of 7.00.* For a solution to be acidic, the concentration of H_3O^+ must be greater than that for OH^-, or

$$[H_3O^+] > 1.00 \times 10^{-7}\ \text{M}$$

The pH of an acidic solution, therefore, must be less than 7.00. A basic solution, on the other hand, will have a pH greater than 7.00. Figure 6.3 shows the pH scale along with pH values for some representative solutions.

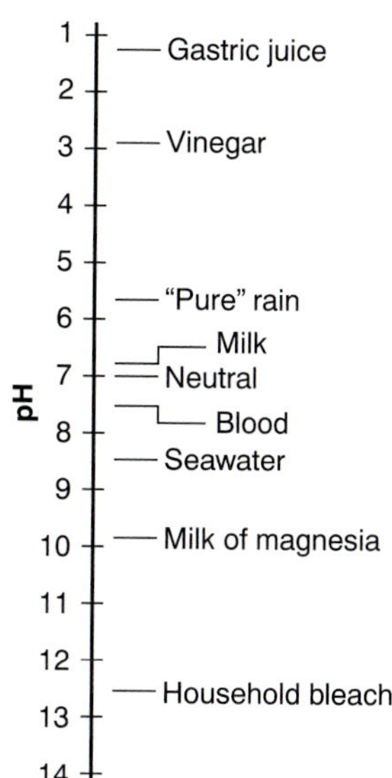

Figure 6.3
pH scale showing values for representative solutions.

Tabulating Values for K_a and K_b A useful observation about acids and bases is that the strength of a base is inversely proportional to the strength of its conjugate acid. Consider, for example, the dissociation reactions of acetic acid and acetate.

$$CH_3COOH(aq) + H_2O(\ell) \rightleftharpoons H_3O^+(aq) + CH_3COO^-(aq) \qquad \mathbf{6.11}$$

$$CH_3COO^-(aq) + H_2O(\ell) \rightleftharpoons CH_3COOH(aq) + OH^-(aq) \qquad \mathbf{6.12}$$

Adding together these two reactions gives

$$2H_2O(\ell) \rightleftharpoons H_3O^+(aq) + OH^-(aq) \qquad \mathbf{6.13}$$

The equilibrium constant for equation 6.13 is K_w. Since equation 6.13 is obtained by adding together reactions 6.11 and 6.12, K_w may also be expressed as the product of K_a for CH_3COOH and K_b for CH_3COO^-. Thus, for a weak acid, HA, and its conjugate weak base, A^-,

$$K_w = K_a \times K_b \qquad \mathbf{6.14}$$

This relationship between K_a and K_b simplifies the tabulation of acid and base dissociation constants. Acid dissociation constants for a variety of weak acids are listed in Appendix 3B. The corresponding values of K_b for their conjugate weak bases are determined using equation 6.14.

Example

EXAMPLE 6.3

Using Appendix 3B, calculate the following equilibrium constants

(a) K_b for pyridine, C_5H_5N
(b) K_b for dihydrogen phosphate, $H_2PO_4^-$

SOLUTION

$$\text{(a)}\ K_{b,C_5H_5N} = \frac{K_w}{K_{a,C_5H_5NH^+}} = \frac{1.00 \times 10^{-14}}{5.90 \times 10^{-6}} = 1.69 \times 10^{-9}$$

$$\text{(b)}\ K_{b,H_2PO_4^-} = \frac{K_w}{K_{a,H_3PO_4}} = \frac{1.00 \times 10^{-14}}{7.11 \times 10^{-3}} = 1.41 \times 10^{-12}$$

*The use of a p-function to express a concentration is covered in Chapter 2.

6D.3 Complexation Reactions

A more general definition of acids and bases was proposed by G. N. Lewis (1875–1946) in 1923. The Brønsted–Lowry definition of acids and bases focuses on an acid's proton-donating ability and a base's proton-accepting ability. Lewis theory, on the other hand, uses the breaking and forming of covalent bonds to describe acid–base characteristics. In this treatment, an acid is an electron pair acceptor, and a base is an electron pair donor. Although Lewis theory can be applied to the treatment of acid–base reactions, it is more useful for treating complexation reactions between metal ions and ligands.

ligand
A Lewis base that binds with a metal ion.

The following reaction between the metal ion Cd^{2+} and the **ligand** NH_3 is typical of a complexation reaction.

$$Cd^{2+}(aq) + 4(:NH_3)(aq) \rightleftharpoons Cd(:NH_3)_4^{2+}(aq) \qquad \textbf{6.15}$$

The product of this reaction is called a metal–ligand complex. In writing the equation for this reaction, we have shown ammonia as $:NH_3$ to emphasize the pair of electrons it donates to Cd^{2+}. In subsequent reactions we will omit this notation.

formation constant
The equilibrium constant for a reaction in which a metal and a ligand bind to form a metal–ligand complex (K_f).

The formation of a metal–ligand complex is described by a **formation constant,** K_f. The complexation reaction between Cd^{2+} and NH_3, for example, has the following equilibrium constant

$$K_f = \frac{[Cd(NH_3)_4^{2+}]}{[Cd^{2+}][NH_3]^4} = 5.5 \times 10^7 \qquad \textbf{6.16}$$

dissociation constant
The equilibrium constant for a reaction in which a metal–ligand complex dissociates to form uncomplexed metal ion and ligand (K_d).

The reverse of reaction 6.15 is called a dissociation reaction and is characterized by a **dissociation constant,** K_d, which is the reciprocal of K_f.

Many complexation reactions occur in a stepwise fashion. For example, the reaction between Cd^{2+} and NH_3 involves four successive reactions

$$Cd^{2+}(aq) + NH_3(aq) \rightleftharpoons Cd(NH_3)^{2+}(aq) \qquad \textbf{6.17}$$

$$Cd(NH_3)^{2+}(aq) + NH_3(aq) \rightleftharpoons Cd(NH_3)_2^{2+}(aq) \qquad \textbf{6.18}$$

$$Cd(NH_3)_2^{2+}(aq) + NH_3(aq) \rightleftharpoons Cd(NH_3)_3^{2+}(aq) \qquad \textbf{6.19}$$

$$Cd(NH_3)_3^{2+}(aq) + NH_3(aq) \rightleftharpoons Cd(NH_3)_4^{2+}(aq) \qquad \textbf{6.20}$$

stepwise formation constant
The formation constant for a metal–ligand complex in which only one ligand is added to the metal ion or to a metal–ligand complex (K_i).

cumulative formation constant
The formation constant for a metal–ligand complex in which two or more ligands are simultaneously added to a metal ion or to a metal–ligand complex (β_i).

This creates a problem since it no longer is clear what reaction is described by a formation constant. To avoid ambiguity, formation constants are divided into two categories. **Stepwise formation constants,** which are designated as K_i for the ith step, describe the successive addition of a ligand to the metal–ligand complex formed in the previous step. Thus, the equilibrium constants for reactions 6.17–6.20 are, respectively, K_1, K_2, K_3, and K_4. Overall, or **cumulative formation constants,** which are designated as β_i, describe the addition of i ligands to the free metal ion. The equilibrium constant expression given in equation 6.16, therefore, is correctly identified as β_4, where

$$\beta_4 = K_1 \times K_2 \times K_3 \times K_4$$

In general

$$\beta_i = K_1 \times K_2 \times \cdots \times K_i$$

Stepwise and cumulative formation constants for selected metal–ligand complexes are given in Appendix 3C.

Equilibrium constants for complexation reactions involving solids are defined by combining appropriate K_{sp} and K_f expressions. For example, the solubility of AgCl increases in the presence of excess chloride as the result of the following complexation reaction

$$AgCl(s) + Cl^-(aq) \rightleftharpoons AgCl_2^-(aq) \qquad \textbf{6.21}$$

This reaction can be separated into three reactions for which equilibrium constants are known—the solubility of AgCl, described by its K_{sp}

$$AgCl(s) \rightleftharpoons Ag^+(aq) + Cl^-(aq)$$

and the stepwise formation of $AgCl_2^-$, described by K_1 and K_2

$$Ag^+(aq) + Cl^-(aq) \rightleftharpoons AgCl(aq)$$

$$AgCl(aq) + Cl^-(aq) \rightleftharpoons AgCl_2^-(aq)$$

The equilibrium constant for reaction 6.21, therefore, is equal to $K_{sp} \times K_1 \times K_2$.

Example

EXAMPLE 6.4

Determine the value of the equilibrium constant for the reaction

$$PbCl_2(s) \rightleftharpoons PbCl_2(aq)$$

SOLUTION

This reaction can be broken down into three reactions. The first of these reactions is the solubility of $PbCl_2$, described by its K_{sp}

$$PbCl_2(s) \rightleftharpoons Pb^{2+}(aq) + 2Cl^-(aq)$$

and the second and third are the stepwise formation of $PbCl_2$ (aq), described by K_1 and K_2

$$Pb^{2+}(aq) + Cl^-(aq) \rightleftharpoons PbCl^+(aq)$$

$$PbCl^+(aq) + Cl^-(aq) \rightleftharpoons PbCl_2(aq)$$

Using values for K_{sp}, K_1, and K_2 from Appendices 3A and 3C, we find the equilibrium constant to be

$$K = K_{sp} \times K_1 \times K_2 = (1.7 \times 10^{-5})(38.9)(1.62) = 1.1 \times 10^{-3}$$

6D.4 Oxidation–Reduction Reactions

In a complexation reaction, a Lewis base donates a pair of electrons to a Lewis acid. In an oxidation–reduction reaction, also known as a **redox reaction,** electrons are not shared, but are transferred from one reactant to another. As a result of this electron transfer, some of the elements involved in the reaction undergo a change in oxidation state. Those species experiencing an increase in their oxidation state are oxidized, while those experiencing a decrease in their oxidation state are reduced. For example, in the following redox reaction between Fe^{3+} and oxalic acid, $H_2C_2O_4$, iron is reduced since its oxidation state changes from +3 to +2.

redox reaction
An electron-transfer reaction.

$$2Fe^{3+}(aq) + H_2C_2O_4(aq) + 2H_2O(\ell) \rightleftharpoons 2Fe^{2+}(aq) + 2CO_2(g) + 2H_3O^+(aq) \qquad \textbf{6.22}$$

Oxalic acid, on the other hand, is oxidized since the oxidation state for carbon increases from +3 in $H_2C_2O_4$ to +4 in CO_2.

Redox reactions, such as that shown in equation 6.22, can be divided into separate half-reactions that individually describe the **oxidation** and the **reduction** processes.

oxidation
A loss of electrons.

reduction
A gain of electrons.

$$H_2C_2O_4(aq) + 2H_2O(\ell) \rightarrow 2CO_2(g) + 2H_3O^+(aq) + 2e^-$$

$$Fe^{3+}(aq) + e^- \rightarrow Fe^{2+}(aq)$$

It is important to remember, however, that oxidation and reduction reactions always occur in pairs.* This relationship is formalized by the convention of calling the species being oxidized a **reducing agent,** because it provides the electrons for the reduction half-reaction. Conversely, the species being reduced is called an **oxidizing agent.** Thus, in reaction 6.22, Fe^{3+} is the oxidizing agent and $H_2C_2O_4$ is the reducing agent.

reducing agent
A species that donates electrons to another species.

oxidizing agent
A species that accepts electrons from another species.

The products of a redox reaction also have redox properties. For example, the Fe^{2+} in reaction 6.22 can be oxidized to Fe^{3+}, while CO_2 can be reduced to $H_2C_2O_4$. Borrowing some terminology from acid–base chemistry, we call Fe^{2+} the conjugate reducing agent of the oxidizing agent Fe^{3+} and CO_2 the conjugate oxidizing agent of the reducing agent $H_2C_2O_4$.

Unlike the reactions that we have already considered, the equilibrium position of a redox reaction is rarely expressed by an equilibrium constant. Since redox reactions involve the transfer of electrons from a reducing agent to an oxidizing agent, it is convenient to consider the thermodynamics of the reaction in terms of the electron.

The free energy, ΔG, associated with moving a charge, Q, under a potential, E, is given by

$$\Delta G = EQ$$

Charge is proportional to the number of electrons that must be moved. For a reaction in which one mole of reactant is oxidized or reduced, the charge, in coulombs, is

$$Q = nF$$

where n is the number of moles of electrons per mole of reactant, and F is Faraday's constant (96,485 $C \cdot mol^{-1}$). The change in free energy (in joules per mole; J/mol) for a redox reaction, therefore, is

$$\Delta G = -nFE \tag{6.23}$$

where ΔG has units of joules per mole. The appearance of a minus sign in equation 6.23 is due to a difference in the conventions for assigning the favored direction for reactions. In thermodynamics, reactions are favored when ΔG is negative, and redox reactions are favored when E is positive.

The relationship between electrochemical potential and the concentrations of reactants and products can be determined by substituting equation 6.23 into equation 6.3

$$-nFE = -nFE^\circ + RT \ln Q$$

where E° is the electrochemical potential under standard-state conditions. Dividing through by $-nF$ leads to the well-known **Nernst equation.**

Nernst equation
An equation relating electrochemical potential to the concentrations of products and reactants.

*Separating a redox reaction into its half-reactions is useful if you need to balance the reaction. One method for balancing redox reactions is reviewed in Appendix 4.

$$E = E^{\circ} - \frac{RT}{nF}\ln Q$$

Substituting appropriate values for R and F, assuming a temperature of 25 °C (298 K), and switching from ln to log* gives the potential in volts as

$$E = E^{\circ} - \frac{0.05916}{n}\log Q \qquad \textbf{6.24}$$

The standard-state electrochemical potential, E°, provides an alternative way of expressing the equilibrium constant for a redox reaction. Since a reaction at equilibrium has a ΔG of zero, the electrochemical potential, E, also must be zero. Substituting into equation 6.24 and rearranging shows that

$$E^{\circ} = \frac{RT}{nF}\log K \qquad \textbf{6.25}$$

Standard-state potentials are generally not tabulated for chemical reactions, but are calculated using the standard-state potentials for the oxidation, E°_{ox}, and reduction half-reactions, E°_{red}. By convention, standard-state potentials are only listed for reduction half-reactions, and E° for a reaction is calculated as

$$E^{\circ}_{reac} = E^{\circ}_{red} - E^{\circ}_{ox}$$

where both E°_{red} and E°_{ox} are standard-state reduction potentials.

Since the potential for a single half-reaction cannot be measured, a reference half-reaction is arbitrarily assigned a standard-state potential of zero. All other reduction potentials are reported relative to this reference. The standard half-reaction is

$$2H_3O^+(aq) + 2e^- \rightleftharpoons 2H_2O(\ell) + H_2(g)$$

Appendix 3D contains a listing of the standard-state reduction potentials for selected species. The more positive the standard-state reduction potential, the more favorable the reduction reaction will be under standard-state conditions. Thus, under standard-state conditions, the reduction of Cu^{2+} to Cu ($E^{\circ} = +0.3419$) is more favorable than the reduction of Zn^{2+} to Zn ($E^{\circ} = -0.7618$).

Example

EXAMPLE 6.5

Calculate (a) the standard-state potential, (b) the equilibrium constant, and (c) the potential when $[Ag^+] = 0.020$ M and $[Cd^{2+}] = 0.050$ M, for the following reaction taking place at 25 °C.

$$Cd(s) + 2Ag^+(aq) \rightleftharpoons Cd^{2+}(aq) + 2Ag(s)$$

SOLUTION

(a) In this reaction Cd is undergoing oxidation, and Ag^+ is undergoing reduction. The standard-state cell potential, therefore, is

$$E^{\circ} = E^{\circ}_{Ag^+/Ag} - E^{\circ}_{Cd^{2+}/Cd} = 0.7996\text{ V} - (-0.4030\text{ V}) = 1.2026\text{ V}$$

(b) To calculate the equilibrium constant, we substitute the values for the standard-state potential and number of electrons into equation 6.25.

$$1.2026 = \frac{0.05916}{2}\log K$$

*$\ln(x) = 2.303\log(x)$

Solving for K gives the equilibrium constant as

$$\log K = 40.6558$$

$$K = 4.527 \times 10^{40}$$

(c) The potential when the $[Ag^+]$ is 0.020 M and the $[Cd^{2+}]$ is 0.050 M is calculated using equation 6.24 employing the appropriate relationship for the reaction quotient Q.

$$E = E^\circ - \frac{0.05916}{n} \log \frac{[Cd^{2+}]}{[Ag^+]^2}$$

$$= 1.2026 - \frac{0.05916}{2} \log \frac{(0.050)}{(0.020)^2}$$

$$= 1.14\text{V}$$

6E Le Châtelier's Principle

The equilibrium position for any reaction is defined by a fixed equilibrium constant, not by a fixed combination of concentrations for the reactants and products. This is easily appreciated by examining the equilibrium constant expression for the dissociation of acetic acid.

$$K_a = \frac{[H_3O^+][CH_3COO^-]}{[CH_3COOH]} = 1.75 \times 10^{-5} \qquad 6.26$$

As a single equation with three variables, equation 6.26 does not have a unique solution for the concentrations of CH_3COOH, CH_3COO^-, and H_3O^+. At constant temperature, different solutions of acetic acid may have different values for $[H_3O^+]$, $[CH_3COO^-]$ and $[CH_3COOH]$, but will always have the same value of K_a.

If a solution of acetic acid at equilibrium is disturbed by adding sodium acetate, the $[CH_3COO^-]$ increases, suggesting an apparent increase in the value of K_a. Since K_a must remain constant, however, the concentration of all three species in equation 6.26 must change in a fashion that restores K_a to its original value. In this case, equilibrium is reestablished by the partial reaction of CH_3COO^- and H_3O^+ to produce additional CH_3COOH.

Le Châtelier's principle
When stressed, a system that was at equilibrium returns to its equilibrium state by reacting in a manner that relieves the stress.

The observation that a system at equilibrium responds to a stress by reequilibrating in a manner that diminishes the stress, is formalized as **Le Châtelier's principle.** One of the most common stresses that we can apply to a reaction at equilibrium is to change the concentration of a reactant or product. We already have seen, in the case of sodium acetate and acetic acid, that adding a product to a reaction mixture at equilibrium converts a portion of the products to reactants. In this instance, we disturb the equilibrium by adding a product, and the stress is diminished by partially reacting the excess product. Adding acetic acid has the opposite effect, partially converting the excess acetic acid to acetate.

In our first example, the stress to the equilibrium was applied directly. It is also possible to apply a concentration stress indirectly. Consider, for example, the following solubility equilibrium involving AgCl

$$AgCl(s) \rightleftharpoons Ag^+(aq) + Cl^-(aq) \qquad 6.27$$

The effect on the solubility of AgCl of adding $AgNO_3$ is obvious,* but what is the effect of adding a ligand that forms a stable, soluble complex with Ag^+? Ammonia, for example, reacts with Ag^+ as follows

$$Ag^+(aq) + 2NH_3(aq) \rightleftharpoons Ag(NH_3)_2^+(aq) \qquad \textbf{6.28}$$

Adding ammonia decreases the concentration of Ag^+ as the $Ag(NH_3)_2^+$ complex forms. In turn, decreasing the concentration of Ag^+ increases the solubility of AgCl as reaction 6.27 reestablishes its equilibrium position. Adding together reactions 6.27 and 6.28 clarifies the effect of ammonia on the solubility of AgCl, by showing that ammonia is a reactant.

$$AgCl(s) + 2NH_3(aq) \rightleftharpoons Ag(NH_3)_2^+(aq) + Cl^-(aq) \qquad \textbf{6.29}$$

Example

EXAMPLE 6.6

What is the effect on the solubility of AgCl if HNO_3 is added to the equilibrium solution defined by reaction 6.29?

SOLUTION

Nitric acid is a strong acid that reacts with ammonia as shown here

$$HNO_3(aq) + NH_3(aq) \rightleftharpoons NH_4^+(aq) + NO_3^-(aq)$$

Adding nitric acid lowers the concentration of ammonia. Decreasing ammonia's concentration causes reaction 6.29 to move from products to reactants, decreasing the solubility of AgCl.

Increasing or decreasing the partial pressure of a gas is the same as increasing or decreasing its concentration.† The effect on a reaction's equilibrium position can be analyzed as described in the preceding example for aqueous solutes. Since the concentration of a gas depends on its partial pressure, and not on the total pressure of the system, adding or removing an inert gas has no effect on the equilibrium position of a gas-phase reaction.

Most reactions involve reactants and products that are dispersed in a solvent. If the amount of solvent is changed, either by diluting or concentrating the solution, the concentrations of all reactants and products either decrease or increase. The effect of these changes in concentration is not as intuitively obvious as when the concentration of a single reactant or product is changed. As an example, let's consider how dilution affects the equilibrium position for the formation of the aqueous silver-amine complex (reaction 6.28). The equilibrium constant for this reaction is

$$\beta_2 = \frac{[Ag(NH_3)_2^+]_{eq}}{[Ag^+]_{eq}[NH_3]_{eq}^2} \qquad \textbf{6.30}$$

*Adding $AgNO_3$ decreases the solubility of AgCl.

†The relationship between pressure and concentration can be deduced from the ideal gas law. Starting with $PV = nRT$, we solve for the molar concentration

$$\text{Molar concentration} = \frac{n}{V} = \frac{P}{RT}$$

Of course, this assumes an ideal gas (which is usually a reasonable assumption under normal laboratory conditions).

where the subscript "eq" is included for clarification. If a portion of this solution is diluted with an equal volume of water, each of the concentration terms in equation 6.30 is cut in half. Thus, the reaction quotient becomes

$$Q = \frac{(0.5)[Ag(NH_3)_2^+]_{eq}}{(0.5)[Ag^+]_{eq}(0.5)^2[NH_3]_{eq}^2}$$

which we can rewrite as

$$Q = \left(\frac{0.5}{(0.5)^3}\right)\left(\frac{[Ag(NH_3)_2^+]_{eq}}{[Ag^+][NH_3]_{eq}^2}\right) = 4 \times \beta_2$$

Since Q is greater than β_2, equilibrium must be reestablished by shifting the reaction to the left, decreasing the concentration of $Ag(NH_3)_2^+$. Furthermore, this new equilibrium position lies toward the side of the equilibrium reaction with the greatest number of solutes (one Ag^+ ion and two molecules of NH_3 versus the single metal–ligand complex). If the solution of $Ag(NH_3)_2^+$ is concentrated, by evaporating some of the solvent, equilibrium is reestablished in the opposite direction. This is a general conclusion that can be applied to any reaction, whether gas-phase, liquid-phase, or solid-phase. Increasing volume always favors the direction producing the greatest number of particles, and decreasing volume always favors the direction producing the fewest particles. If the number of particles is the same on both sides of the equilibrium, then the equilibrium position is unaffected by a change in volume.

6F Ladder Diagrams

When developing or evaluating an analytical method, we often need to understand how the chemistry taking place affects our results. We have already seen, for example, that adding NH_3 to a solution of Ag^+ is a poor idea if we intend to isolate the Ag^+ as a precipitate of AgCl (reaction 6.29). One of the primary sources of determinate method errors is a failure to account for potential chemical interferences.

ladder diagram
A visual tool for evaluating systems at equilibrium.

In this section we introduce the **ladder diagram** as a simple graphical tool for evaluating the chemistry taking place during an analysis.[1] Using ladder diagrams, we will be able to determine what reactions occur when several reagents are combined, estimate the approximate composition of a system at equilibrium, and evaluate how a change in solution conditions might affect our results.

6F.1 Ladder Diagrams for Acid–Base Equilibria

To see how a ladder diagram is constructed, we will use the acid–base equilibrium between HF and F^-

$$HF(aq) + H_2O(\ell) \rightleftharpoons H_3O^+(aq) + F^-(aq)$$

for which the acid dissociation constant is

$$K_{a,HF} = \frac{[H_3O^+][F^-]}{[HF]}$$

Taking the log of both sides and multiplying through by –1 gives

$$-\log K_{a,HF} = -\log[H_3O^+] - \log\frac{[F^-]}{[HF]}$$

Finally, replacing the negative log terms with p-functions and rearranging leaves us with

$$pH = pK_a + \log\frac{[F^-]}{[HF]} \qquad \textbf{6.31}$$

Examining equation 6.31 tells us a great deal about the relationship between pH and the relative amounts of F^- and HF at equilibrium. If the concentrations of F^- and HF are equal, then equation 6.31 reduces to

$$pH = pK_{a,HF} = -\log(K_{a,HF}) = -\log(6.8 \times 10^{-4}) = 3.17$$

For concentrations of F^- greater than that of HF, the log term in equation 6.31 is positive and

$$pH > pK_{a,HF} \qquad \text{or} \qquad pH > 3.17$$

This is a reasonable result since we expect the concentration of hydrofluoric acid's conjugate base, F^-, to increase as the pH increases. Similar reasoning shows that the concentration of HF exceeds that of F^- when

$$pH < pK_{a,HF} \qquad \text{or} \qquad pH < 3.17$$

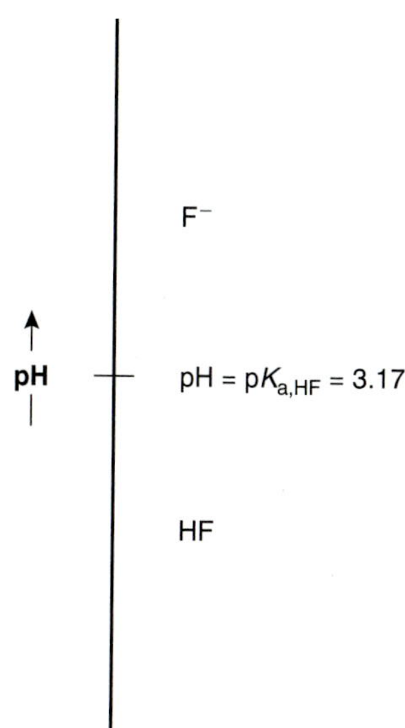

Figure 6.4
Ladder diagram for HF, showing areas of predominance for HF and F^-.

Now we are ready to construct the ladder diagram for HF (Figure 6.4). The ladder diagram consists of a vertical scale of pH values oriented so that smaller (more acidic) pH levels are at the bottom and larger (more basic) pH levels are at the top. A horizontal line is drawn at a pH equal to $pK_{a,HF}$. This line, or step, separates the solution into regions where each of the two conjugate forms of HF predominate. By referring to the ladder diagram, we see that at a pH of 2.5 hydrofluoric acid will exist predominately as HF. If we add sufficient base to the solution such that the pH increases to 4.5, the predominate form becomes F^-.

Figure 6.5 shows a second ladder diagram containing information about HF/F^- and NH_4^+/NH_3. From this ladder diagram we see that if the pH is less than 3.17, the predominate species are HF and NH_4^+. For pH's between 3.17 and 9.24 the predominate species are F^- and NH_4^+, whereas above a pH of 9.24 the predominate species are F^- and NH_3.

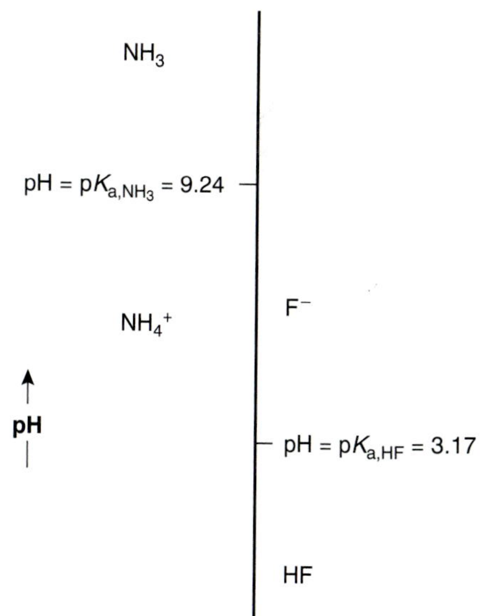

Figure 6.5
Ladder diagram for HF and NH_3.

Ladder diagrams are particularly useful for evaluating the reactivity of acids and bases. An acid and a base cannot coexist if their respective areas of predominance do not overlap. If we mix together solutions of NH_3 and HF, the reaction

$$HF(aq) + NH_3(aq) \rightleftharpoons NH_4^+(aq) + F^-(aq) \qquad \textbf{6.32}$$

occurs because the predominance areas for HF and NH_3 do not overlap. Before continuing, let us show that this conclusion is reasonable by calculating the equilibrium constant for reaction 6.32. To do so we need the following three reactions and their equilibrium constants.

$$HF(aq) + H_2O(\ell) \rightleftharpoons H_3O^+(aq) + F^-(aq) \qquad K_a = 6.8 \times 10^{-4}$$

$$NH_3(aq) + H_2O(\ell) \rightleftharpoons OH^-(aq) + NH_4^+(aq) \qquad K_b = 1.75 \times 10^{-5}$$

$$H_3O^+(aq) + OH^-(aq) \rightleftharpoons 2H_2O(\ell) \qquad K = \frac{1}{K_w} = \frac{1}{1.00 \times 10^{-14}}$$

Adding together these reactions gives us reaction 6.32, for which the equilibrium constant is

$$K = \frac{K_a K_b}{K_w} = \frac{(6.8 \times 10^{-4})(1.75 \times 10^{-5})}{(1.00 \times 10^{-14})} = 1.19 \times 10^6$$

Since the equilibrium constant is significantly greater than 1, the reaction's equilibrium position lies far to the right. This conclusion is general and applies to all ladder diagrams. The following example shows how we can use the ladder diagram in Figure 6.5 to evaluate the composition of any solution prepared by mixing together solutions of HF and NH_3.

Example

EXAMPLE 6.7

Predict the pH and composition of a solution prepared by adding 0.090 mol of HF to 0.040 mol of NH_3.

SOLUTION

Since HF is present in excess and the reaction between HF and NH_3 is favorable, the NH_3 will react to form NH_4^+. At equilibrium, essentially no NH_3 remains and

$$\text{Moles } NH_4^+ = 0.040 \text{ mol}$$

Converting NH_3 to NH_4^+ consumes 0.040 mol of HF; thus

$$\text{Moles HF} = 0.090 - 0.040 = 0.050 \text{ mol}$$

$$\text{Moles } F^- = 0.040 \text{ mol}$$

According to the ladder diagram for this system (see Figure 6.5), a pH of 3.17 results when there is an equal amount of HF and F^-. Since we have more HF than F^-, the pH will be slightly less than 3.17. Similar reasoning will show you that mixing together 0.090 mol of NH_3 and 0.040 mol of HF will result in a solution whose pH is slightly larger than 9.24.

If the areas of predominance for an acid and a base overlap each other, then practically no reaction occurs. For example, if we mix together solutions of NaF and NH_4Cl, we expect that there will be no significant change in the moles of F^- and NH_4^+. Furthermore, the pH of the mixture must be between 3.17 and 9.24. Because F^- and NH_4^+ can coexist over a range of pHs we cannot be more specific in estimating the solution's pH.

The ladder diagram for HF/F^- also can be used to evaluate the effect of pH on other equilibria that include either HF or F^-. For example, the solubility of CaF_2

$$CaF_2(s) \rightleftharpoons Ca^{2+}(aq) + 2F^-(aq)$$

is affected by pH because F^- is a weak base. Using Le Châtelier's principle, if F^- is converted to HF, the solubility of CaF_2 will increase. To minimize the solubility of CaF_2 we want to control the solution's pH so that F^- is the predominate species. From the ladder diagram we see that maintaining a pH of more than 3.17 ensures that solubility losses are minimal.

6F.2 Ladder Diagrams for Complexation Equilibria

The same principles used in constructing and interpreting ladder diagrams for acid–base equilibria can be applied to equilibria involving metal–ligand complexes. For complexation reactions the ladder diagram's scale is defined by the concentration of uncomplexed, or free ligand, pL. Using the formation of $Cd(NH_3)^{2+}$ as an example

$$Cd^{2+}(aq) + NH_3(aq) \rightleftharpoons Cd(NH_3)^{2+}(aq)$$

we can easily show that the dividing line between the predominance regions for Cd^{2+} and $Cd(NH_3)^{2+}$ is $\log(K_1)$.

$$K_1 = \frac{[Cd(NH_3)^{2+}]}{[Cd^{2+}][NH_3]}$$

$$\log(K_1) = \log\frac{[Cd(NH_3)^{2+}]}{[Cd^{2+}]} - \log[NH_3]$$

$$\log(K_1) = \log\frac{[Cd(NH_3)^{2+}]}{[Cd^{2+}]} + pNH_3$$

$$pNH_3 = \log(K_1) + \log\frac{[Cd^{2+}]}{[Cd(NH_3)^{2+}]}$$

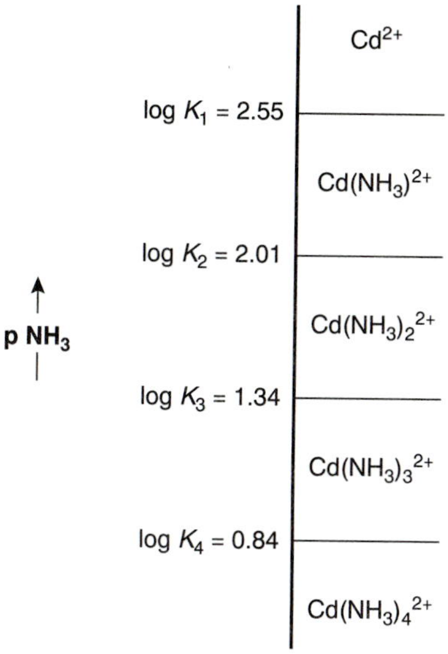

Figure 6.6
Ladder diagram for metal–ligand complexes of Cd^{2+} and NH_3.

Since K_1 for $Cd(NH_3)^{2+}$ is 3.55×10^2, $\log(K_1)$ is 2.55. Thus, for a pNH_3 greater than 2.55 (concentrations of NH_3 less than 2.8×10^{-3} M), Cd^{2+} is the predominate species. A complete ladder diagram for the metal–ligand complexes of Cd^{2+} and NH_3 is shown in Figure 6.6.

Example

EXAMPLE 6.8

Using the ladder diagram in Figure 6.7, predict the result of adding 0.080 mol of Ca^{2+} to 0.060 mol of $Mg(EDTA)^{2-}$. EDTA is an abbreviation for the ligand ethylenediaminetetraacetic acid.

SOLUTION

The predominance regions for Ca^{2+} and $Mg(EDTA)^{2-}$ do not overlap, therefore, the reaction

$$Ca^{2+} + Mg(EDTA)^{2-} \rightleftharpoons Mg^{2+} + Ca(EDTA)^{2-}$$

will take place. Since there is an excess of Ca^{2+}, the composition of the final solution is approximately

$$\text{Moles } Ca^{2+} = 0.080 - 0.060 = 0.020 \text{ mol}$$

$$\text{Moles } Ca(EDTA)^{2-} = 0.060 \text{ mol}$$

$$\text{Moles } Mg^{2+} = 0.060 \text{ mol}$$

$$\text{Moles } Mg(EDTA)^{2-} = 0 \text{ mol}$$

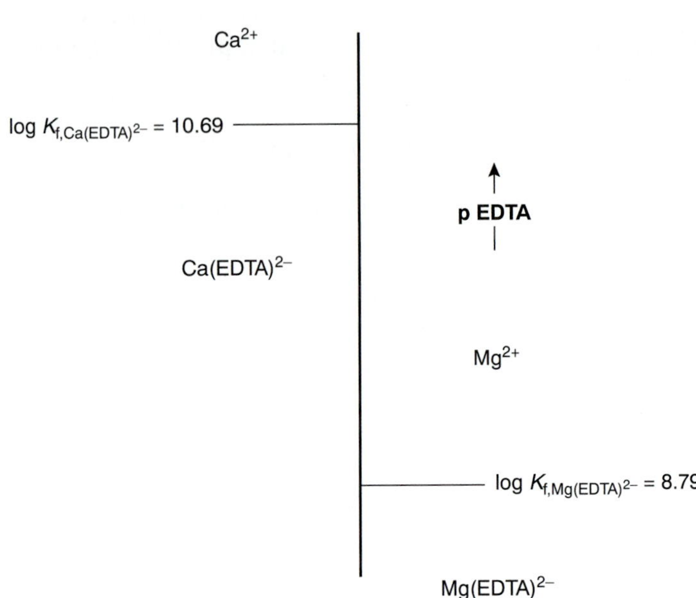

Figure 6.7
Ladder diagram for metal–ligand complexes of ethylenediaminetetraacetic acid (EDTA) with Ca^{2+} and Mg^{2+}.

We can also construct ladder diagrams using cumulative formation constants in place of stepwise formation constants. The first three stepwise formation constants for the reaction of Zn^{2+} with NH_3

$$Zn^{2+}(aq) + NH_3(aq) \rightleftharpoons Zn(NH_3)^{2+}(aq) \qquad K_1 = 1.6 \times 10^2$$

$$Zn(NH_3)^{2+}(aq) + NH_3(aq) \rightleftharpoons Zn(NH_3)_2^{2+}(aq) \qquad K_2 = 1.95 \times 10^2$$

$$Zn(NH_3)_2^{2+}(aq) + NH_3(aq) \rightleftharpoons Zn(NH_3)_3^{2+}(aq) \qquad K_3 = 2.3 \times 10^2$$

show that the formation of $Zn(NH_3)_3^{2+}$ is more favorable than the formation of $Zn(NH_3)^{2+}$ or $Zn(NH_3)_2^{2+}$. The equilibrium, therefore, is best represented by the cumulative formation reaction

$$Zn^{2+}(aq) + 3NH_3(aq) \rightleftharpoons Zn(NH_3)_3^{2+}(aq)$$

for which

$$\beta_3 = \frac{[Zn(NH_3)_3^{2+}]}{[Zn^{2+}][NH_3]^3} = 7.2 \times 10^6$$

Taking the log of each side gives

$$\log \beta_3 = \log \frac{[Zn(NH_3)_3^{2+}]}{[Zn^{2+}]} - 3 \log [NH_3]$$

or

$$pNH_3 = \frac{1}{3}\log\beta_3 + \frac{1}{3}\log\frac{[Zn^{2+}]}{[Zn(NH_3)_3^{2+}]}$$

The concentrations of Zn^{2+} and $Zn(NH_3)_3^{2+}$, therefore, are equal when

$$pNH_3 = \frac{1}{3}\log\beta_3 = \frac{1}{3}\log(7.2 \times 10^6) = 2.29$$

A complete ladder diagram for the Zn^{2+}–NH_3 system is shown in Figure 6.8.

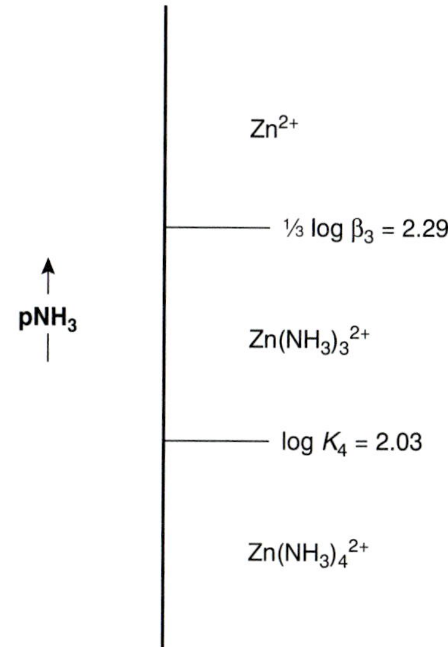

Figure 6.8
Ladder diagram for Zn^{2+}, $Zn(NH_3)_3^{2+}$, and $Zn(NH_3)_4^{2+}$, showing how cumulative formation constants are included.

6F.3 Ladder Diagram for Oxidation–Reduction Equilibria

Ladder diagrams can also be used to evaluate equilibrium reactions in redox systems. Figure 6.9 shows a typical ladder diagram for two half-reactions in which the scale is the electrochemical potential, *E*. Areas of predominance are defined by the Nernst equation. Using the Fe^{3+}/Fe^{2+} half-reaction as an example, we write

$$E = E^\circ - 0.05916\log\frac{[Fe^{2+}]}{[Fe^{3+}]} = +0.771\text{V} - 0.05916\log\frac{[Fe^{2+}]}{[Fe^{3+}]}$$

For potentials more positive than the standard-state potential, the predominate species is Fe^{3+}, whereas Fe^{2+} predominates for potentials more negative than E°. When coupled with the step for the Sn^{4+}/Sn^{2+} half-reaction, we see that Sn^{2+} can be used to reduce Fe^{3+}. If an excess of Sn^{2+} is added, the potential of the resulting solution will be near +0.154 V.

Using standard-state potentials to construct a ladder diagram can present problems if solutes are not at their standard-state concentrations. Because the concentrations of the reduced and oxidized species are in a logarithmic term, deviations from standard-state concentrations can usually be ignored if the steps being compared are separated by at least 0.3 V.[1b] A trickier problem occurs when a half-reaction's potential is affected by the concentration of another species. For example, the potential for the following half-reaction

$$UO_2^{2+}(aq) + 4H_3O^+(aq) + 2e^- \rightleftharpoons U^{4+}(aq) + 6H_2O(\ell)$$

depends on the pH of the solution. To define areas of predominance in this case, we begin with the Nernst equation

$$E = 0.327 - \frac{0.05916}{2}\log\frac{[U^{4+}]}{[UO_2^{2+}][H_3O^+]^4}$$

and factor out the concentration of H_3O^+.

$$E = 0.327 + \frac{0.05916}{2}\log[H_3O^+]^4 - \frac{0.05916}{2}\log\frac{[U^{4+}]}{[UO_2^{2+}]}$$

From this equation we see that the areas of predominance for UO_2^{2+} and U^{4+} are defined by a step whose potential is

$$E = 0.327 + \frac{0.05916}{2}\log[H_3O^+]^4 = 0.327 - 0.1183\,\text{pH}$$

Figure 6.10 shows how a change in pH affects the step for the UO_2^{2+}/U^{4+} half-reaction.

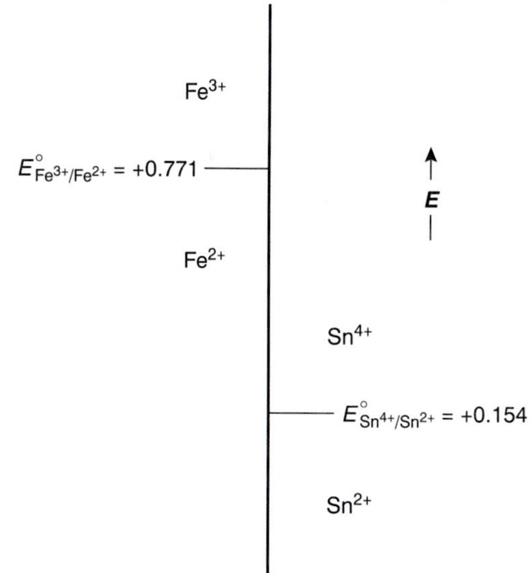

Figure 6.9
Ladder diagram for the Fe^{3+}/Fe^{2+} and Sn^{4+}/SN^{2+} half-reactions.

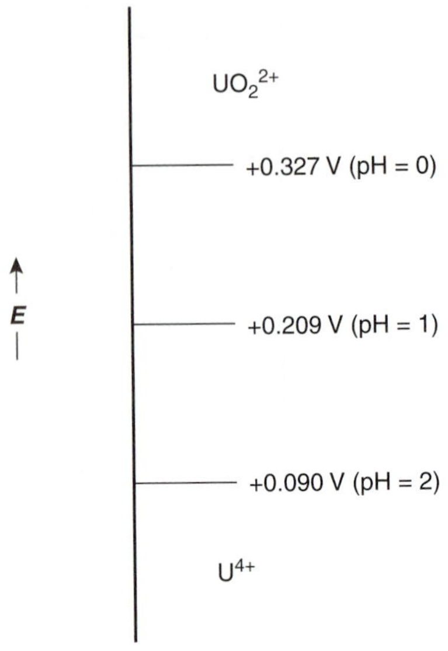

Figure 6.10
Ladder diagram showing the effect of a change in pH on the areas of predominance for the UO_2^{2+}/U^{4+} half-reaction.

6G Solving Equilibrium Problems

Ladder diagrams are a useful tool for evaluating chemical reactivity, usually providing a reasonable approximation of a chemical system's composition at equilibrium. When we need a more exact quantitative description of the equilibrium condition, a ladder diagram may not be sufficient. In this case we can find an algebraic solution. Perhaps you recall solving equilibrium problems in your earlier coursework in chemistry. In this section we will learn how to set up and solve equilibrium problems. We will start with a simple problem and work toward more complex ones.

6G.1 A Simple Problem: Solubility of $Pb(IO_3)_2$ in Water

When an insoluble compound such as $Pb(IO_3)_2$ is added to a solution a small portion of the solid dissolves. Equilibrium is achieved when the concentrations of Pb^{2+} and IO_3^- are sufficient to satisfy the solubility product for $Pb(IO_3)_2$. At equilibrium the solution is saturated with $Pb(IO_3)_2$. How can we determine the concentrations of Pb^{2+} and IO_3^-, and the solubility of $Pb(IO_3)_2$ in a saturated solution prepared by adding $Pb(IO_3)_2$ to distilled water?

We begin by writing the equilibrium reaction

$$Pb(IO_3)_2(s) \rightleftharpoons Pb^{2+}(aq) + 2IO_3^-(aq)$$

and its equilibrium constant

$$K_{sp} = [Pb^{2+}][IO_3^-]^2 = 2.5 \times 10^{-13} \qquad 6.33$$

As equilibrium is established, two IO_3^- ions are produced for each ion of Pb^{2+}. If we assume that the molar concentration of Pb^{2+} at equilibrium is x then the molar concentration of IO_3^- is $2x$. To help keep track of these relationships, we can use the following table.

	$Pb(IO_3)_2(s)$	$\rightleftharpoons$	$Pb^{2+}(aq)$	+	$2IO_3^-(aq)$
Initial concentration	solid		0		0
Change in concentration	solid		$+x$		$+2x$
Equilibrium concentration	solid		$0 + x = x$		$0 + 2x = 2x$

Substituting the equilibrium concentrations into equation 6.33

$$(x)(2x)^2 = 2.5 \times 10^{-13}$$

and solving gives

$$4x^3 = 2.5 \times 10^{-13}$$

$$x = 3.97 \times 10^{-5}$$

The equilibrium concentrations of Pb^{2+} and IO_3^-, therefore, are

$$[Pb^{2+}] = x = 4.0 \times 10^{-5} \text{ M}$$

$$[IO_3^-] = 2x = 7.9 \times 10^{-5} \text{ M}$$

Since one mole of $Pb(IO_3)_2$ contains one mole of Pb^{2+}, the solubility of $Pb(IO_3)_2$ is the same as the concentration of Pb^{2+}; thus, the solubility of $Pb(IO_3)_2$ is 4.0×10^{-5} M.

6G.2 A More Complex Problem: The Common Ion Effect

Calculating the solubility of $Pb(IO_3)_2$ in distilled water is a straightforward problem since the dissolution of the solid is the only source of Pb^{2+} or IO_3^-. How is the solubility of $Pb(IO_3)_2$ affected if we add $Pb(IO_3)_2$ to a solution of 0.10 M $Pb(NO_3)_2$? Before we set up and solve the problem algebraically, think about the chemistry occurring in this system, and decide whether the solubility of $Pb(IO_3)_2$ will increase, decrease, or remain the same. This is a good habit to develop. Knowing what answers are reasonable will help you spot errors in your calculations and give you more confidence that your solution to a problem is correct.

We begin by setting up a table to help us keep track of the concentrations of Pb^{2+} and IO_3^- in this system.

	$Pb(IO_3)_2(s)$	$\rightleftharpoons$	$Pb^{2+}(aq)$	+	$2IO_3^-(aq)$
Initial concentration	solid		0.10		0
Change in concentration	solid		$+x$		$+2x$
Equilibrium concentration	solid		$0.10 + x$		$0 + 2x = 2x$

Substituting the equilibrium concentrations into the solubility product expression (equation 6.33)

$$(0.10 + x)(2x)^2 = 2.5 \times 10^{-13}$$

and multiplying out the terms on the left leaves us with

$$4x^3 + 0.40x^2 = 2.5 \times 10^{-13} \qquad \mathbf{6.34}$$

This is a more difficult equation to solve than that for the solubility of $Pb(IO_3)_2$ in distilled water, and its solution is not immediately obvious. A rigorous solution to equation 6.34 can be found using available computer software packages and spreadsheets.

How might we solve equation 6.34 if we do not have access to a computer? One possibility is that we can apply our understanding of chemistry to simplify the algebra. From Le Châtelier's principle, we expect that the large initial concentration of Pb^{2+} will significantly decrease the solubility of $Pb(IO_3)_2$. In this case we can reasonably expect the equilibrium concentration of Pb^{2+} to be very close to its initial concentration; thus, the following approximation for the equilibrium concentration of Pb^{2+} seems reasonable

$$[Pb^{2+}] = 0.10 + x \approx 0.10 \text{ M}$$

Substituting into equation 6.34

$$(0.10)(2x)^2 = 2.5 \times 10^{-13}$$

and solving for x gives

$$0.40x^2 = 2.5 \times 10^{-13}$$

$$x = 7.91 \times 10^{-7}$$

Before accepting this answer, we check to see if our approximation was reasonable. In this case the approximation $0.10 + x \approx 0.10$ seems reasonable since the difference between the two values is negligible. The equilibrium concentrations of Pb^{2+} and IO_3^-, therefore, are

$$[Pb^{2+}] = 0.10 + x \approx 0.10 \text{ M}$$

$$[IO_3^-] = 2x = 1.6 \times 10^{-6} \text{ M}$$

The solubility of $Pb(IO_3)_2$ is equal to the additional concentration of Pb^{2+} in solution, or 7.9×10^{-7} mol/L. As expected, the solubility of $Pb(IO_3)_2$ decreases in the presence of a solution that already contains one of its ions. This is known as the **common ion effect.**

common ion effect
The solubility of an insoluble salt decreases when it is placed in a solution already containing one of the salt's ions.

As outlined in the following example, the process of making and evaluating approximations can be extended if the first approximation leads to an unacceptably large error.

Example

EXAMPLE 6.9

Calculate the solubility of $Pb(IO_3)_2$ in 1.0×10^{-4} M $Pb(NO_3)_2$.

SOLUTION

Letting x equal the change in the concentration of Pb^{2+}, the equilibrium concentrations are

$$[Pb^{2+}] = 1.0 \times 10^{-4} + x \qquad [IO_3^-] = 2x$$

and

$$(1.0 \times 10^{-4} + x)(2x)^2 = 2.5 \times 10^{-13}$$

We start by assuming that

$$[Pb^{2+}] = 1.0 \times 10^{-4} + x \approx 1.0 \times 10^{-4} \text{ M}$$

and solve for x, obtaining a value of 2.50×10^{-5}. Substituting back gives the calculated concentration of Pb^{2+} at equilibrium as

$$[Pb^{2+}] = 1.0 \times 10^{-4} + 2.50 \times 10^{-5} = 1.25 \times 10^{-4} \text{ M}$$

a value that differs by 25% from our approximation that the equilibrium concentration is 1.0×10^{-4} M. This error seems unreasonably large. Rather than shouting in frustration, we make a new assumption. Our first assumption that the concentration of Pb^{2+} is 1.0×10^{-4} M was too small. The calculated concentration of 1.25×10^{-4} M, therefore, is probably a little too large. Let us assume that

$$[Pb^{2+}] = 1.0 \times 10^{-4} + x \approx 1.2 \times 10^{-4} \text{ M}$$

Substituting into the solubility product equation and solving for x gives us

$$x = 2.28 \times 10^{-5}$$

or a concentration of Pb^{2+} at equilibrium of

$$[Pb^{2+}] = 1.0 \times 10^{-4} + (2.28 \times 10^{-5}) = 1.23 \times 10^{-4} \text{ M}$$

which differs from our assumed concentration of 1.2×10^{-4} M by 2.5%. This seems to be a reasonable error since the original concentration of Pb^{2+} is given to only two significant figures. Our final solution, to two significant figures, is

$$[Pb^{2+}] = 1.2 \times 10^{-4} \text{ M} \qquad [IO_3^-] = 4.6 \times 10^{-5} \text{ M}$$

and the solubility of $Pb(IO_3)_2$ is 2.3×10^{-5} mol/L. This iterative approach to solving an equation is known as the method of successive approximations.

6G.3 Systematic Approach to Solving Equilibrium Problems

Calculating the solubility of $Pb(IO_3)_2$ in a solution of $Pb(NO_3)_2$ was more complicated than calculating its solubility in distilled water. The necessary calculations, however, were still relatively easy to organize, and the assumption used to simplify the problem was fairly obvious. This problem was reasonably straightforward because it involved only a single equilibrium reaction, the solubility of $Pb(IO_3)_2$. Calculating the equilibrium composition of a system with multiple equilibrium reactions can become quite complicated. In this section we will learn how to use a systematic approach to setting up and solving equilibrium problems.

As its name implies, a systematic approach involves a series of steps:

1. Write all relevant equilibrium reactions and their equilibrium constant expressions.
2. Count the number of species whose concentrations appear in the equilibrium constant expressions; these are your unknowns. If the number of unknowns equals the number of equilibrium constant expressions, then you have enough information to solve the problem. If not, additional equations based on the conservation of mass and charge must be written. Continue to add equations until you have the same number of equations as you have unknowns.
3. Decide how accurate your final answer needs to be. This decision will influence your evaluation of any assumptions you use to simplify the problem.
4. Combine your equations to solve for one unknown (usually the one you are most interested in knowing). Whenever possible, simplify the algebra by making appropriate assumptions.
5. When you obtain your final answer, be sure to check your assumptions. If any of your assumptions prove invalid, then return to the previous step and continue solving. The problem is complete when you have an answer that does not violate any of your assumptions.

Besides equilibrium constant equations, two other types of equations are used in the systematic approach to solving equilibrium problems. The first of these is a **mass balance equation,** which is simply a statement of the conservation of matter. In a solution of a monoprotic weak acid, for example, the combined concentrations of the conjugate weak acid, HA, and the conjugate weak base, A^-, must equal the weak acid's initial concentration, C_{HA}.*

mass balance equation
An equation stating that matter is conserved, and that the total amount of a species added to a solution must equal the sum of the amount of each of its possible forms present in solution.

The second type of equation is a **charge balance equation.** A charge balance equation is a statement of solution electroneutrality.

charge balance equation
An equation stating that the total concentration of positive charge in a solution must equal the total concentration of negative charge.

Total positive charge from cations = total negative charge from anions

Mathematically, the charge balance expression is expressed as

$$\sum_{i=1}^{n} \left|(z^+)_i\right| \times [M^{z+}]_i = \sum_{j=1}^{m} \left|(z^-)_j\right| \times [A^{z-}]_j$$

where $[M^{z+}]_i$ and $[A^{z-}]_j$ are, respectively, the concentrations of the ith cation and the jth anion, and $(z^+)_i$ and $(z^-)_j$ are the charges of the ith cation and the jth anion. Note that the concentration terms are multiplied by the absolute values of each ion's charge, since electroneutrality is a conservation of charge, not concentration. Every ion in solution, even those not involved in any equilibrium

*You may recall that this is the difference between a formal concentration and a molar concentration.

reactions, must be included in the charge balance equation. The charge balance equation for an aqueous solution of $Ca(NO_3)_2$ is

$$2 \times [Ca^{2+}] + [H_3O^+] = [OH^-] + [NO_3^-]$$

Note that the concentration of Ca^{2+} is multiplied by 2, and that the concentrations of H_3O^+ and OH^- are also included. Charge balance equations must be written carefully since every ion in solution must be included. This presents a problem when the concentration of one ion in solution is held constant by a reagent of unspecified composition. For example, in many situations pH is held constant using a buffer. If the composition of the buffer is not specified, then a charge balance equation cannot be written.

Example

EXAMPLE 6.10

Write a mass balance and charge balance equations for a 0.10 M solution of $NaHCO_3$.

SOLUTION

It is easier to keep track of what species are in solution if we write down the reactions that control the solution's composition. These reactions are the dissolution of a soluble salt

$$NaHCO_3(s) \rightarrow Na^+(aq) + HCO_3^-(aq)$$

and the acid–base dissociation reactions of HCO_3^- and H_2O

$$HCO_3^-(aq) + H_2O(\ell) \rightleftharpoons H_3O^+(aq) + CO_3^{2-}(aq)$$

$$HCO_3^-(aq) + H_2O(\ell) \rightleftharpoons OH^-(aq) + H_2CO_3(aq)$$

$$2H_2O(\ell) \rightleftharpoons H_3O^+(aq) + OH^-(aq)$$

The mass balance equations are

$$0.10 \text{ M} = [H_2CO_3] + [HCO_3^-] + [CO_3^{2-}]$$

$$0.10 \text{ M} = [Na^+]$$

The charge balance equation is

$$[Na^+] + [H_3O^+] = [OH^-] + [HCO_3^-] + 2 \times [CO_3^{2-}]$$

6G.4 pH of a Monoprotic Weak Acid

To illustrate the systematic approach, let us calculate the pH of 1.0 M HF. Two equilbria affect the pH of this system. The first, and most obvious, is the acid dissociation reaction for HF

$$HF(aq) + H_2O(\ell) \rightleftharpoons H_3O^+(aq) + F^-(aq)$$

for which the equilibrium constant expression is

$$K_a = \frac{[H_3O^+][F^-]}{[HF]} = 6.8 \times 10^{-4} \qquad 6.35$$

The second equilibrium reaction is the dissociation of water, which is an obvious yet easily disregarded reaction

$$2H_2O(\ell) \rightleftharpoons H_3O^+(aq) + OH^-(aq)$$

$$K_w = [H_3O^+][OH^-] = 1.00 \times 10^{-14} \qquad \textbf{6.36}$$

Counting unknowns, we find four ([HF], [F^-], [H_3O^+], and [OH^-]). To solve this problem, therefore, we need to write two additional equations involving these unknowns. These equations are a mass balance equation

$$C_{HF} = [HF] + [F^-] \qquad \textbf{6.37}$$

and a charge balance equation

$$[H_3O^+] = [F^-] + [OH^-] \qquad \textbf{6.38}$$

We now have four equations (6.35, 6.36, 6.37, and 6.38) and four unknowns ([HF], [F^-], [H_3O^+], and [OH^-]) and are ready to solve the problem. Before doing so, however, we will simplify the algebra by making two reasonable assumptions. First, since HF is a weak acid, we expect the solution to be acidic; thus it is reasonable to assume that

$$[H_3O^+] >> [OH^-]$$

simplifying the charge balance equation (6.38) to

$$[H_3O^+] = [F^-] \qquad \textbf{6.39}$$

Second, since HF is a weak acid we expect that very little dissociation occurs, and

$$[HF] >> [F^-]$$

Thus, the mass balance equation (6.36) simplifies to

$$C_{HF} = [HF] \qquad \textbf{6.40}$$

For this exercise we will accept our assumptions if the error introduced by each assumption is less than ±5%.

Substituting equations 6.39 and 6.40 into the equilibrium constant expression for the dissociation of HF (equation 6.35) and solving for the concentration of H_3O^+ gives us

$$K_a = \frac{[H_3O^+][H_3O^+]}{C_{HF}}$$

$$[H_3O^+] = \sqrt{K_aC_{HF}} = \sqrt{(6.8 \times 10^{-4})(1.0)} = 2.6 \times 10^{-2}\ M$$

Before accepting this answer, we must verify that our assumptions are acceptable. The first assumption was that the [OH^-] is significantly smaller than the [H_3O^+]. To calculate the concentration of OH^- we use the K_w expression (6.36)

$$[OH^-] = \frac{K_w}{[H_3O^+]} = \frac{1.00 \times 10^{-14}}{2.6 \times 10^{-2}} = 3.8 \times 10^{-13}\ M$$

Clearly this assumption is reasonable. The second assumption was that the [F^-] is significantly smaller than the [HF]. From equation 6.39 we have

$$[F^-] = 2.6 \times 10^{-2}\ M$$

Since the [F^-] is 2.6% of C_{HF}, this assumption is also within our limit that the error be no more than ±5%. Accepting our solution for the concentration of H_3O^+, we find that the pH of 1.0 M HF is 1.59.

How does the result of this calculation change if we require our assumptions to have an error of less than ±1%. In this case we can no longer assume that [HF] >> [F^-]. Solving the mass balance equation (6.37) for [HF]

$$[HF] = C_{HF} - [F^-]$$

and substituting into the K_a expression along with equation 6.39 gives

$$K_a = \frac{[H_3O^+]^2}{C_{HF} - [H_3O^+]}$$

Rearranging leaves us with a quadratic equation

$$[H_3O^+]^2 = K_aC_{HF} - K_a[H_3O^+]$$

$$[H_3O^+]^2 + K_a[H_3O^+] - K_aC_{HF} = 0$$

which we solve using the quadratic formula

$$x = \frac{-b \pm \sqrt{b^2 - 4ac}}{2a}$$

where *a*, *b*, and *c* are the coefficients in the quadratic equation $ax^2 + bx + c = 0$. Solving the quadratic formula gives two roots, only one of which has any chemical significance. For our problem the quadratic formula gives roots of

$$x = \frac{-6.8 \times 10^{-4} \pm \sqrt{(6.8 \times 10^{-4})^2 - (4)(1)(-6.8 \times 10^{-4})(1.0)}}{2(1)}$$

$$= \frac{-6.8 \times 10^{-4} \pm 5.22 \times 10^{-2}}{2}$$

$$= 2.57 \times 10^{-2} \quad \text{or} \quad -2.63 \times 10^{-2}$$

Only the positive root has any chemical significance since the negative root implies that the concentration of H_3O^+ is negative. Thus, the [H_3O^+] is 2.6×10^{-2} M, and the pH to two significant figures is still 1.59.

This same approach can be extended to find the pH of a monoprotic weak base, replacing K_a with K_b, C_{HF} with the weak base's concentration, and solving for the [OH^-] in place of [H_3O^+].

Example

EXAMPLE 6.11

Calculate the pH of 0.050 M NH_3. State any assumptions made in simplifying the calculation, and verify that the error is less than 5%.

SOLUTION

Since NH_3 is a weak base ($K_b = 1.75 \times 10^{-5}$), we assume that

$$[OH^-] >> [H_3O^+] \qquad \text{and} \qquad C_{NH_3} = 0.050 \text{ M}$$

With these assumptions, we find (be sure to check the derivation)

$$[OH^-] = \sqrt{K_bC_{NH_3}} = \sqrt{(1.75 \times 10^{-5})(0.050)} = 9.35 \times 10^{-4} \text{ M}$$

Both assumptions are acceptable (again, verify that this is true). The concentration of H_3O^+ is calculated using K_w

$$[H_3O^+] = \frac{K_w}{[OH^-]} = \frac{1.00 \times 10^{-14}}{9.35 \times 10^{-4}} = 1.07 \times 10^{-11}$$

giving a pH of 10.97.

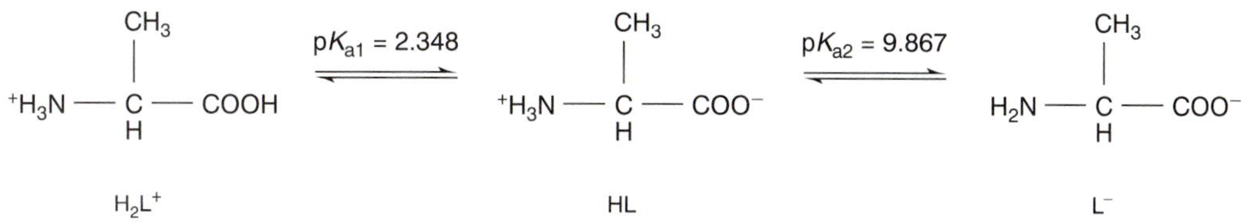

$$\underset{H_2L^+}{{}^+H_3N-\underset{H}{\overset{CH_3}{C}}-COOH} \;\underset{}{\overset{pK_{a1} = 2.348}{\rightleftharpoons}}\; \underset{HL}{{}^+H_3N-\underset{H}{\overset{CH_3}{C}}-COO^-} \;\overset{pK_{a2} = 9.867}{\rightleftharpoons}\; \underset{L^-}{H_2N-\underset{H}{\overset{CH_3}{C}}-COO^-}$$

Figure 6.11
Acid–base equilibria for the amino acid alanine.

6G.5 pH of a Polyprotic Acid or Base

A more challenging problem is to find the pH of a solution prepared from a polyprotic acid or one of its conjugate species. As an example, we will use the amino acid alanine whose structure and acid dissociation constants are shown in Figure 6.11.

pH of 0.10 M H_2L^+ Alanine hydrochloride is a salt consisting of the diprotic weak acid H_2L^+ and Cl^-. Because H_2L^+ has two acid dissociation reactions, a complete systematic solution to this problem will be more complicated than that for a monoprotic weak acid. Using a ladder diagram (Figure 6.12) can help us simplify the problem. Since the areas of predominance for H_2L^+ and L^- are widely separated, we can assume that any solution containing an appreciable quantity of H_2L^+ will contain essentially no L^-. In this case, HL is such a weak acid that H_2L^+ behaves as if it were a monoprotic weak acid.

To find the pH of 0.10 M H_2L^+, we assume that

$$[H_3O^+] >> [OH^-]$$

Because H_2L^+ is a relatively strong weak acid, we cannot simplify the problem further, leaving us with

$$K_a = \frac{[H_3O^+]^2}{C_{H_2L^+} - [H_3O^+]}$$

Solving the resulting quadratic equation gives the $[H_3O^+]$ as 1.91×10^{-2} M or a pH of 1.72. Our assumption that $[H_3O^+]$ is significantly greater than $[OH^-]$ is acceptable.

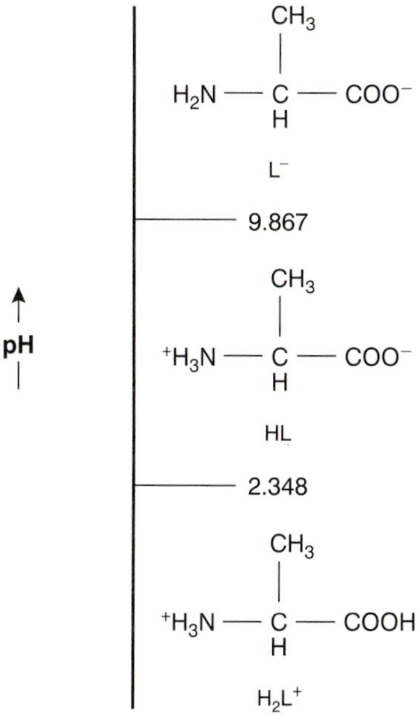

Figure 6.12
Ladder diagram for the amino acid alanine.

pH of 0.10 M L^- The alaninate ion is a diprotic weak base, but using the ladder diagram as a guide shows us that we can treat it as if it were a monoprotic weak base. Following the steps in Example 6.11 (which is left as an exercise), we find that the pH of 0.10 M alaninate is 11.42.

pH of 0.1 M HL Finding the pH of a solution of alanine is more complicated than that for H_2L^+ or L^- because we must consider two equilibrium reactions involving HL. Alanine is an amphiprotic species, behaving as an acid

$$HL(aq) + H_2O(\ell) \rightleftharpoons H_3O^+(aq) + L^-(aq)$$

and a base

$$HL(aq) + H_2O(\ell) \rightleftharpoons OH^-(aq) + H_2L^+(aq)$$

As always, we must also consider the dissociation of water

$$2H_2O(\ell) \rightleftharpoons H_3O^+(aq) + OH^-(aq)$$

This leaves us with five unknowns ($[H_2L^+]$, $[HL]$, $[L^-]$, $[H_3O^+]$, and $[OH^-]$), for which we need five equations. These equations are K_{a2} and K_{b2} for HL,

$$K_{a2} = \frac{[H_3O^+][L^-]}{[HL]}$$

$$K_{b2} = \frac{K_w}{K_{a1}} = \frac{[OH^-][H_2L^+]}{[HL]}$$

the K_w equation,

$$K_w = [H_3O^+][OH^-]$$

a mass balance equation on HL,

$$C_{HL} = [H_2L^+] + [HL] + [L^-]$$

and a charge balance equation

$$[H_2L^+] + [H_3O^+] = [OH^-] + [L^-]$$

From the ladder diagram it appears that we may safely assume that the concentrations of H_2L^+ and L^- are significantly smaller than that for HL, allowing us to simplify the mass balance equation to

$$C_{HL} = [HL]$$

Next we solve K_{b2} for $[H_2L^+]$

$$[H_2L^+] = \frac{K_w[HL]}{K_{a1}[OH^-]} = \frac{[HL][H_3O^+]}{K_{a1}} = \frac{C_{HL}[H_3O^+]}{K_{a1}}$$

and K_{a2} for $[L^-]$

$$[L^-] = \frac{K_{a2}[HL]}{[H_3O^+]} = \frac{K_{a2}C_{HL}}{[H_3O^+]}$$

Substituting these equations, along with the equation for K_w, into the charge balance equation gives us

$$\frac{C_{HL}[H_3O^+]}{K_{a1}} + [H_3O^+] = \frac{K_w}{[H_3O^+]} + \frac{K_{a2}C_{HL}}{[H_3O^+]}$$

which simplifies to

$$[H_3O^+]\left(\frac{C_{HL}}{K_{a1}} + 1\right) = \frac{1}{[H_3O^+]}(K_w + K_{a2}C_{HL})$$

$$[H_3O^+]^2 = \frac{K_{a2}C_{HL} + K_w}{(C_{HL}/K_{a1}) + 1}$$

$$[H_3O^+] = \sqrt{\frac{K_{a1}K_{a2}C_{HL} + K_{a1}K_w}{C_{HL} + K_{a1}}} \quad \textbf{6.41}$$

We can simplify this equation further if $K_{a1}K_w << K_{a1}K_{a2}C_{HL}$, and if $K_{a1} << C_{HL}$, giving

$$[H_3O^+] = \sqrt{K_{a1}K_{a2}} \qquad \textbf{6.42}$$

For a solution of 0.10 M alanine, the $[H_3O^+]$ is

$$[H_3O^+] = \sqrt{(4.487 \times 10^{-3})(1.358 \times 10^{-10})} = 7.807 \times 10^{-7}\ M$$

or a pH of 6.11. Verifying that the assumptions are acceptable is left as an exercise.

Triprotic Acids and Bases, and Beyond The treatment of a diprotic acid or base is easily extended to acids and bases having three or more acid–base sites. For a triprotic weak acid such as H_3PO_4, for example, we can treat H_3PO_4 as if it was a monoprotic weak acid, $H_2PO_4^-$ and HPO_4^{2-} as if they were intermediate forms of diprotic weak acids, and PO_4^{3-} as if it was a monoprotic weak base.

Example

EXAMPLE 6.12

Calculate the pH of 0.10 M Na_2HPO_4.

SOLUTION

We treat HPO_4^{2-} as the intermediate form of a diprotic weak acid

$$H_2PO_4^-(aq) \rightleftharpoons HPO_4^{2-}(aq) \rightleftharpoons PO_4^{3-}(aq)$$

where the equilibrium constants are $K_{a2} = 6.32 \times 10^{-8}$ and $K_{a3} = 4.5 \times 10^{-13}$. Since the value of K_{a3} is so small, we use equation 6.41 instead of equation 6.42.

$$[H_3O^+] = \sqrt{\frac{(6.32 \times 10^{-8})(4.5 \times 10^{-13})(0.10) + (6.32 \times 10^{-8})(1.00 \times 10^{-14})}{0.10 + 6.32 \times 10^{-8}}}$$

$$= 1.86 \times 10^{-10}$$

or a pH of 9.73.

6G.6 Effect of Complexation on Solubility

The solubility of a precipitate can be improved by adding a ligand capable of forming a soluble complex with one of the precipitate's ions. For example, the solubility of AgI increases in the presence of NH_3 due to the formation of the soluble $Ag(NH_3)_2^+$ complex. As a final illustration of the systematic approach to solving equilibrium problems, let us find the solubility of AgI in 0.10 M NH_3.

We begin by writing the equilibria that we need to consider

$$AgI(s) \rightleftharpoons Ag^+(aq) + I^-(aq)$$

$$Ag^+(aq) + 2NH_3(aq) \rightleftharpoons Ag(NH_3)_2^+(aq)$$

$$NH_3(aq) + H_2O(\ell) \rightleftharpoons OH^-(aq) + NH_4^+(aq)$$

$$2H_2O(\ell) \rightleftharpoons H_3O^+(aq) + OH^-(aq)$$

Counting unknowns, we find that there are seven—$[Ag^+]$, $[I^-]$, $[Ag(NH_3)_2{}^+]$, $[NH_3]$, $[NH_4{}^+]$, $[OH^-]$, and $[H_3O^+]$. Four of the equations needed to solve this problem are given by the equilibrium constant expressions

$$K_{sp} = [Ag^+][I^-] = 8.3 \times 10^{-17}$$

$$\beta_2 = \frac{[Ag(NH_3)_2{}^+]}{[Ag^+][NH_3]^2} = 1.7 \times 10^7$$

$$K_b = \frac{[NH_4{}^+][OH^-]}{[NH_3]} = 1.75 \times 10^{-5}$$

$$K_w = [H_3O^+][OH^-] = 1.00 \times 10^{-14}$$

Three additional equations are needed. The first of these equations is a mass balance for NH_3.

$$C_{NH_3} = [NH_3] + [NH_4{}^+] + 2 \times [Ag(NH_3)_2{}^+]$$

Note that in writing this mass balance equation, the concentration of $Ag(NH_3)_2{}^+$ must be multiplied by 2 since two moles of NH_3 occurs per mole of $Ag(NH_3)_2{}^+$. The second additional equation is a mass balance on iodide and silver. Since AgI is the only source of I^- and Ag^+, every iodide in solution must have an associated silver ion; thus

$$[I^-] = [Ag^+] + [Ag(NH_3)_2{}^+]$$

Finally, the last equation is a charge balance equation

$$[Ag^+] + [Ag(NH_3)_2{}^+] + [NH_4{}^+] + [H_3O^+] = [I^-] + [OH^-]$$

Our problem looks challenging, but several assumptions greatly simplify the algebra. First, since the formation of the $Ag(NH_3)_2{}^+$ complex is favorable, we will assume that

$$[Ag^+] << [Ag(NH_3)_2{}^+]$$

Second, since NH_3 is a base, we will assume that

$$[H_3O^+] << [OH^-]$$

$$[NH_4{}^+] << [NH_3] + [Ag(NH_3)_2{}^+]$$

Finally, since K_{sp} is significantly smaller than β_2, it seems likely that the solubility of AgI is small and

$$[Ag(NH_3)_2{}^+] << [NH_3]$$

Using these assumptions allows us to simplify several equations. The mass balance for NH_3 is now

$$C_{NH_3} = [NH_3]$$

and the mass balance for I^- is

$$[I^-] = [Ag(NH_3)_2{}^+]$$

Simplifying the charge balance expression by dropping $[H_3O^+]$ and $[Ag^+]$, and replacing $[Ag(NH_3)_2{}^+]$ with $[I^-]$ gives

$$[NH_4^+] = [OH^-]$$

We continue by multiplying together the equations for K_{sp} and β_2, giving

$$K_{sp}\beta_2 = \frac{[Ag(NH_3)_2^+][I^-]}{[NH_3]^2} = 1.4 \times 10^{-9}$$

Substituting in the new mass balance equations for NH_3 and I^-

$$\frac{[I^-]^2}{(C_{NH_3})^2} = 1.4 \times 10^{-9}$$

and solving for the $[I^-]$ gives

$$\frac{[I^-]^2}{(0.10\text{ M})^2} = 1.4 \times 10^{-9}$$

$$[I^-] = 3.7 \times 10^{-6}\text{ M}$$

Before accepting this answer, we first check our assumptions. Using the K_{sp} equation we calculate the $[Ag^+]$ to be

$$[Ag^+] = \frac{K_{sp}}{[I^-]} = \frac{8.3 \times 10^{-17}}{3.7 \times 10^{-6}} = 2.2 \times 10^{-11}\text{ M}$$

From the simplified mass balance equation for I^-, we have

$$[Ag(NH_3)_2^+] = [I^-] = 3.7 \times 10^{-6}\text{ M}$$

Our first assumption that the $[Ag^+]$ is significantly smaller than the $[Ag(NH_3)_2^+]$, therefore, is reasonable. Furthermore, our third assumption that the $[Ag(NH_3)_2^+]$ is significantly less than the $[NH_3]$ also is reasonable. Our second assumption was

$$[NH_4^+] << [NH_3] + [Ag(NH_3)_2^+]$$

To verify this assumption, we solve the K_b equation for $[NH_4^+]$

$$\frac{[NH_4^+][OH^-]}{[NH_3]} = \frac{[NH_4^+]^2}{0.10\text{ M}} = 1.75 \times 10^{-5}$$

giving

$$[NH_4^+] = 1.3 \times 10^{-3}\text{ M}$$

Although the $[NH_4^+]$ is not significantly smaller than the combined concentrations of NH_3 and $Ag(NH_3)_2^+$, the error is only about 1%. Since this is not an excessively large error, we will accept this approximation as reasonable.

Since one mole of AgI produces one mole of I^-, the solubility of AgI is the same as the concentration of iodide, or 3.7×10^{-6} mol/L.

6H Buffer Solutions

Adding as little as 0.1 mL of concentrated HCl to a liter of H_2O shifts the pH from 7.0 to 3.0. The same addition of HCl to a liter solution that is 0.1 M in both a weak acid and its conjugate weak base, however, results in only a negligible change in pH. Such solutions are called **buffers,** and their buffering action is a consequence of the relationship between pH and the relative concentrations of the conjugate weak acid/weak base pair.

buffer
A solution containing a conjugate weak acid/weak base pair that is resistant to a change in pH when a strong acid or strong base is added.

A mixture of acetic acid and sodium acetate is one example of an acid/base buffer. The equilibrium position of the buffer is governed by the reaction

$$CH_3COOH(aq) + H_2O(\ell) \rightleftharpoons H_3O^+(aq) + CH_3COO^-(aq)$$

and its acid dissociation constant

$$K_a = \frac{[H_3O^+][CH_3COO^-]}{[CH_3COOH]} = 1.75 \times 10^{-5} \qquad 6.43$$

The relationship between the pH of an acid–base buffer and the relative amounts of CH_3COOH and CH_3COO^- is derived by taking the negative log of both sides of equation 6.43 and solving for the pH

$$pH = pK_a + \log\frac{[CH_3COO^-]}{[CH_3COOH]} = 4.76 + \log\frac{[CH_3COO^-]}{[CH_3COOH]} \qquad 6.44$$

Buffering occurs because of the logarithmic relationship between pH and the ratio of the weak base and weak acid concentrations. For example, if the equilibrium concentrations of CH_3COOH and CH_3COO^- are equal, the pH of the buffer is 4.76. If sufficient strong acid is added such that 10% of the acetate ion is converted to acetic acid, the concentration ratio $[CH_3COO^-]/[CH_3COOH]$ changes to 0.818, and the pH decreases to 4.67.

6H.1 Systematic Solution to Buffer Problems

Equation 6.44 is written in terms of the concentrations of CH_3COOH and CH_3COO^- at equilibrium. A more useful relationship relates the buffer's pH to the initial concentrations of weak acid and weak base. A general buffer equation can be derived by considering the following reactions for a weak acid, HA, and the salt of its conjugate weak base, NaA.

$$NaA(aq) \rightarrow Na^+(aq) + A^-(aq)$$

$$HA(aq) + H_2O(\ell) \rightleftharpoons H_3O^+(aq) + A^-(aq)$$

$$2H_2O(\ell) \rightleftharpoons H_3O^+(aq) + OH^-(aq)$$

Since the concentrations of Na^+, A^-, HA, H_3O^+, and OH^- are unknown, five equations are needed to uniquely define the solution's composition. Two of these equations are given by the equilibrium constant expressions

$$K_a = \frac{[H_3O^+][A^-]}{[HA]}$$

$$K_w = [H_3O^+][OH^-]$$

The remaining three equations are given by mass balance equations on HA and Na^+

$$C_{HA} + C_{NaA} = [HA] + [A^-] \qquad 6.45$$

$$C_{NaA} = [Na^+] \qquad 6.46$$

and a charge balance equation

$$[H_3O^+] + [Na^+] = [OH^-] + [A^-]$$

Substituting equation 6.46 into the charge balance equation and solving for $[A^-]$ gives

$$[A^-] = C_{NaA} - [OH^-] + [H_3O^+] \qquad \textbf{6.47}$$

which is substituted into equation 6.45 to give the concentration of HA

$$[HA] = C_{HA} + [OH^-] - [H_3O^+] \qquad \textbf{6.48}$$

Finally, substituting equations 6.47 and 6.48 into the K_a equation for HA and solving for pH gives the general buffer equation

$$pH = pK_a + \log\frac{C_{NaA} - [OH^-] + [H_3O^+]}{C_{HA} + [OH^-] - [H_3O^+]}$$

If the initial concentrations of weak acid and weak base are greater than $[H_3O^+]$ and $[OH^-]$, the general equation simplifies to the **Henderson–Hasselbalch equation.**

Henderson–Hasselbalch equation Equation showing the relationship between a buffer's pH and the relative amounts of the buffer's conjugate weak acid and weak base.

$$pH = pK_a + \log\frac{C_{NaA}}{C_{HA}} \qquad \textbf{6.49}$$

As in Example 6.13, the Henderson–Hasselbalch equation provides a simple way to calculate the pH of a buffer and to determine the change in pH upon adding a strong acid or strong base.

Example

EXAMPLE 6.13

Calculate the pH of a buffer that is 0.020 M in NH_3 and 0.030 M in NH_4Cl. What is the pH after adding 1.00 mL of 0.10 M NaOH to 0.10 L of this buffer?

SOLUTION

The acid dissociation constant for NH_4^+ is 5.70×10^{-10}; thus the initial pH of the buffer is

$$pH = 9.24 + \log\frac{C_{NH_3}}{C_{NH_4^+}} = 9.24 + \log\frac{0.020}{0.030} = 9.06$$

Adding NaOH converts a portion of the NH_4^+ to NH_3 due to the following reaction

$$NH_4^+(aq) + OH^-(aq) \rightleftharpoons NH_3(aq) + H_2O(\ell)$$

Since the equilibrium constant for this reaction is large, we may treat the reaction as if it went to completion. The new concentrations of NH_4^+ and NH_3 are therefore

$$C_{NH_4^+} = \frac{\text{moles } NH_4^+ - \text{moles } OH^-}{V_{tot}}$$

$$= \frac{(0.030\,M)(0.10\,L) - (0.10\,M)(1.00 \times 10^{-3}\,L)}{0.101\,L} = 0.029\,M$$

$$C_{NH_3} = \frac{\text{moles } NH_3 + \text{moles } OH^-}{V_{tot}}$$

$$= \frac{(0.020\,M)(0.10\,L) + (0.10\,M)(1.00 \times 10^{-3}\,L)}{0.101\,L} = 0.021\,M$$

Substituting the new concentrations into the Henderson–Hasselbalch equation gives a pH of

$$pH = 9.24 + \log\frac{0.021}{0.029} = 9.10$$

Multiprotic weak acids can be used to prepare buffers at as many different pH's as there are acidic protons. For example, a diprotic weak acid can be used to prepare buffers at two pH's and a triprotic weak acid can be used to prepare three different buffers. The Henderson–Hasselbalch equation applies in each case. Thus, buffers of malonic acid ($pK_{a1} = 2.85$ and $pK_{a2} = 5.70$) can be prepared for which

$$pH = 2.85 + \log\frac{C_{HM^-}}{C_{H_2M}}$$

$$pH = 5.70 + \log\frac{C_{M^{2-}}}{C_{HM^-}}$$

where H_2M, HM^-, and M^{2-} are the different forms of malonic acid.

The capacity of a buffer to resist a change in pH is a function of the absolute concentration of the weak acid and the weak base, as well as their relative proportions. The importance of the weak acid's concentration and the weak base's concentration is obvious. The more moles of weak acid and weak base that a buffer has, the more strong base or strong acid it can neutralize without significantly changing the buffer's pH. The relative proportions of weak acid and weak base affect the magnitude of the change in pH when adding a strong acid or strong base. Buffers that are equimolar in weak acid and weak base require a greater amount of strong acid or strong base to effect a change in pH of one unit.[2] Consequently, buffers are most effective to the addition of either acid or base at pH values near the pK_a of the weak acid.

Buffer solutions are often prepared using standard "recipes" found in the chemical literature.[3] In addition, computer programs have been developed to aid in the preparation of other buffers.[4] Perhaps the simplest means of preparing a buffer, however, is to prepare a solution containing an appropriate conjugate weak acid and weak base and measure its pH. The pH is easily adjusted to the desired pH by adding small portions of either a strong acid or a strong base.

Although this treatment of buffers was based on acid–base chemistry, the idea of a buffer is general and can be extended to equilibria involving complexation or redox reactions. For example, the Nernst equation for a solution containing Fe^{2+} and Fe^{3+} is similar in form to the Henderson–Hasselbalch equation.

$$E = E^{\circ}_{Fe^{3+}/Fe^{2+}} - 0.05916\log\frac{[Fe^{2+}]}{[Fe^{3+}]}$$

Consequently, solutions of Fe^{2+} and Fe^{3+} are buffered to a potential near the standard-state reduction potential for Fe^{3+}.

6H.2 Representing Buffer Solutions with Ladder Diagrams

Ladder diagrams provide a simple graphical description of a solution's predominate species as a function of solution conditions. They also provide a convenient way to show the range of solution conditions over which a buffer is most effective. For ex-

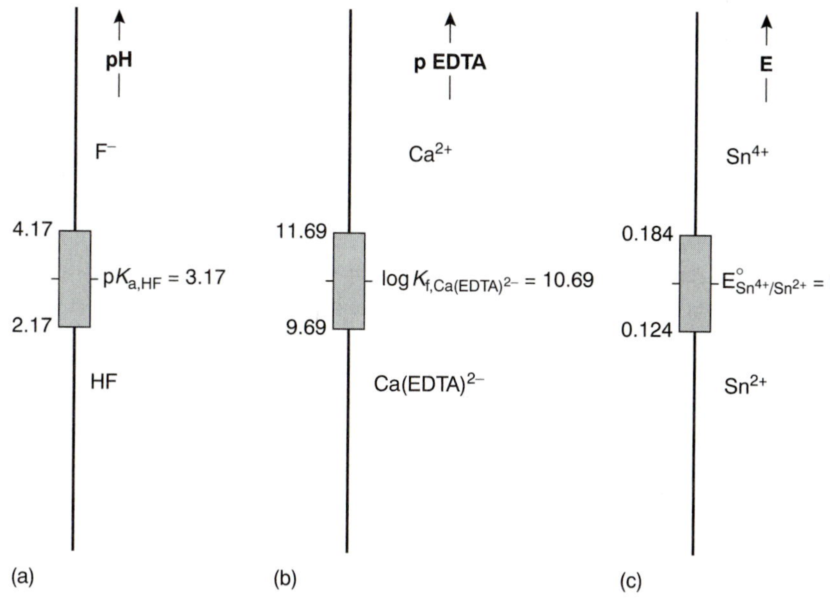

Figure 6.13
Ladder diagrams showing buffer regions for (a) HF/F^- acid–base buffer; (b) $Ca^{2+}/Ca(EDTA)^{2-}$ metal–ligand complexation buffer; and (c) SN^{4+}/Sn^{2+} oxidation–reduction buffer.

ample, an acid–base buffer can only exist when the relative abundance of the weak acid and its conjugate weak base are similar. For convenience, we will assume that an acid–base buffer exists when the concentration ratio of weak base to weak acid is between 0.1 and 10. Applying the Henderson–Hasselbalch equation

$$\text{pH} = \text{p}K_a + \log\frac{1}{10} = \text{p}K_a - 1$$

$$\text{pH} = \text{p}K_a + \log\frac{10}{1} = \text{p}K_a + 1$$

shows that acid–base buffer exists within the range of $\text{pH} = \text{p}K_a \pm 1$. In the same manner, it is easy to show that a complexation buffer for the metal–ligand complex ML_n exists when $\text{pL} = \log K_f \pm 1$, and that a redox buffer exists for $E = E^\circ \pm (0.05916/n)$. Ladder diagrams showing buffer regions for several equilibria are shown in Figure 6.13.

6I Activity Effects

Suppose you need to prepare a buffer with a pH of 9.36. Using the Henderson–Hasselbalch equation, you calculate the amounts of acetic acid and sodium acetate needed and prepare the buffer. When you measure the pH, however, you find that it is 9.25. If you have been careful in your calculations and measurements, what can account for the difference between the obtained and expected pHs? In this section, we will examine an important limitation to our use of equilibrium constants and learn how this limitation can be corrected.

Careful measurements of the solubility of $AgIO_3$ show that it increases in the presence of KNO_3, even though neither K^+ or NO_3^- participates in the solubility reaction.[5] Clearly the equilibrium position for the reaction

$$AgIO_3(s) \rightleftharpoons Ag^+(aq) + IO_3^-(aq)$$

depends on the composition of the solution. When the solubility product for $AgIO_3$ is calculated using the equilibrium concentrations of Ag^+ and IO_3^-

$$K_{sp} = [Ag^+][IO_3^-]$$

its apparent value increases when an inert electrolyte such as KNO_3 is added.

Why should adding an inert electrolyte affect the equilibrium position of a chemical reaction? We can explain the effect of KNO_3 on the solubility of $AgIO_3$ by considering the reaction on a microscopic scale. The solution in which equilibrium is established contains a variety of cations and anions—K^+, Ag^+, H_3O^+, NO_3^-, IO_3^- and OH^-. Although the solution is homogeneous, on the average, there are more anions in regions near Ag^+ ions, and more cations in regions near IO_3^- ions. Thus, Ag^+ and IO_3^- are surrounded by charged ionic atmospheres that partially screen the ions from each other. The formation of $AgIO_3$ requires the disruption of the ionic atmospheres surrounding the Ag^+ and IO_3^- ions. Increasing the concentrations of ions in solution, by adding KNO_3, increases the size of these ionic atmospheres. Since more energy is now required to disrupt the ionic atmospheres, there is a decrease in the formation of $AgIO_3$, and an apparent increase in the equilibrium constant.

ionic strength
A quantitative method for reporting the ionic composition of a solution that takes into account the greater effect of more highly charged ions (μ).

The ionic composition of a solution frequently is expressed by its **ionic strength,** μ

$$\mu = \frac{1}{2}\sum_i c_i z_i^2$$

where c_i and z_i are the concentration and charge of the *i*th ion.

EXAMPLE 6.14

Calculate the ionic strength of 0.10 M NaCl. Repeat the calculation for a solution of 0.10 M Na_2SO_4.

SOLUTION

The ionic strength for 0.10 M NaCl is

$$\mu = \frac{1}{2}([Na^+](+1)^2 + [Cl^-](-1)^2) = \frac{1}{2}[(0.10)(+1)^2 + (0.10)(-1)^2] = 0.10\text{ M}$$

For 0.10 M Na_2SO_4, the ionic strength is

$$\mu = \frac{1}{2}([Na^+](+1)^2 + [SO_4^{\ 2-}](-2)^2) = \frac{1}{2}[(0.20)(+1)^2 + (0.10)(-2)^2] = 0.30\text{ M}$$

Note that the unit for ionic strength is molarity, but that the molar ionic strength need not match the molar concentration of the electrolyte. For a 1:1 electrolyte, such as NaCl, ionic strength and molar concentration are identical. The ionic strength of a 2:1 electrolyte, such as Na_2SO_4, is three times larger than the electrolyte's molar concentration.

activity
True thermodynamic constants use a species activity in place of its molar concentration (a).

activity coefficient
The number that when multiplied by a species' concentration gives that species' activity (γ).

The true thermodynamic equilibrium constant is a function of **activity** rather than concentration. The activity of a species, a_A, is defined as the product of its molar concentration, [A], and a solution-dependent **activity coefficient,** γ_A.

$$a_A = [A]\gamma_A$$

The true thermodynamic equilibrium constant, K_{sp}, for the solubility of $AgIO_3$, therefore, is

$$K_{sp} = (a_{Ag^+})(a_{IO_3^-}) = [Ag^+][IO_3^-](\gamma_{Ag^+})(\gamma_{IO_3^-})$$

To accurately calculate the solubility of $AgIO_3$, we must know the activity coefficients for Ag^+ and IO_3^-.

For gases, pure solids, pure liquids, and nonionic solutes, activity coefficients are approximately unity under most reasonable experimental conditions. For reactions involving only these species, differences between activity and concentration are negligible. Activity coefficients for ionic solutes, however, depend on the ionic composition of the solution. It is possible, using the extended Debye–Hückel theory,* to calculate activity coefficients using equation 6.50

$$-\log \gamma_A = \frac{0.51 \times z_A^2 \times \sqrt{\mu}}{1 + 3.3 \times \alpha_A \times \sqrt{\mu}} \qquad \textbf{6.50}$$

where Z_A is the charge of the ion, α_A is the effective diameter of the hydrated ion in nanometers (Table 6.1), μ is the solution's ionic strength, and 0.51 and 3.3 are constants appropriate for aqueous solutions at 25 °C.

Several features of equation 6.50 deserve mention. First, as the ionic strength approaches zero, the activity coefficient approaches a value of one. Thus, in a solution where the ionic strength is zero, an ion's activity and concentration are identical. We can take advantage of this fact to determine a reaction's thermodynamic equilibrium constant. The equilibrium constant based on concentrations is measured for several increasingly smaller ionic strengths and the results extrapolated

Table 6.1 Effective Diameters (α) for Selected Inorganic Cations and Anions

Ion	Effective Diameter (nm)
H_3O^+	0.9
Li^+	0.6
Na^+, IO_3^-, HSO_3^-, HCO_3^-, $H_2PO_4^-$	0.45
OH^-, F^-, SCN^-, HS^-, ClO_3^-, ClO_4^-, MnO_4^-	0.35
K^+, Cl^-, Br^-, I^-, CN^-, NO_2^-, NO_3^-	0.3
Cs^+, Tl^+, Ag^+, NH_4^+	0.25
Mg^{2+}, Be^{2+}	0.8
Ca^{2+}, Cu^{2+}, Zn^{2+}, Sn^{2+}, Mn^{2+}, Fe^{2+}, Ni^{2+}, Co^{2+}	0.6
Sr^{2+}, Ba^{2+}, Cd^{2+}, Hg^{2+}, S^{2-}	0.5
Pb^{2+}, CO_3^{2-}, SO_3^{2-}	0.45
Hg_2^{2+}, SO_4^{2-}, $S_2O_3^{2-}$, CrO_4^{2-}, HPO_4^{2-}	0.40
Al^{3+}, Fe^{3+}, Cr^{3+}	0.9
PO_4^{3-}, $Fe(CN)_6^{3-}$	0.4
Zr^{4+}, Ce^{4+}, Sn^{4+}	1.1
$Fe(CN)_6^{4-}$	0.5

Source: Values from Kielland, J. *J. Am. Chem. Soc.* **1937**, *59*, 1675.

*See any standard textbook on physical chemistry for more information on the Debye–Hückel theory and its application to solution equilibrium

back to zero ionic strength to give the thermodynamic equilibrium constant. Second, activity coefficients are smaller, and thus activity effects are more important, for ions with higher charges and smaller effective diameters. Finally, the extended Debye–Hückel equation provides reasonable activity coefficients for ionic strengths of less than 0.1. Modifications to the extended Debye–Hückel equation, which extend the calculation of activity coefficients to higher ionic strength, have been proposed.[6]

Example

EXAMPLE 6.15

Calculate the solubility of $Pb(IO_3)_2$ in a matrix of 0.020 M $Mg(NO_3)_2$.

SOLUTION

We begin by calculating the ionic strength of the solution. Since $Pb(IO_3)_2$ is only sparingly soluble, we will assume that its contribution to the ionic strength can be ignored; thus

$$\mu = \frac{1}{2}[(0.20\ \text{M})(+2)^2 + (0.040\ \text{M})(-1)^2] = 0.060\ \text{M}$$

Activity coefficients for Pb^{2+} and I^- are calculated using equation 6.50

$$-\log \gamma_{Pb^{2+}} = \frac{0.51 \times (+2)^2 \times \sqrt{0.060}}{1 + 3.3 \times 0.45 \times \sqrt{0.060}} = 0.366$$

giving an activity coefficient for Pb^{2+} of 0.43. A similar calculation for IO_3^- gives its activity coefficient as 0.81. The equilibrium constant expression for the solubility of PbI_2 is

$$K_{sp} = [Pb^{2+}][IO_3^-]^2\gamma_{Pb^{2+}}\gamma_{IO_3^-} = 2.5 \times 10^{-13}$$

Letting

$$[Pb^{2+}] = x \qquad \text{and} \qquad [IO_3^-] = 2x$$

we have

$$(x)(2x)^2(0.45)(0.81)^2 = 2.5 \times 10^{-13}$$

Solving for x gives a value of 6.0×10^{-5} or a solubility of 6.0×10^{-5} mol/L. This compares to a value of 4.0×10^{-5} mol/L when activity is ignored. Failing to correct for activity effects underestimates the solubility of PbI_2 in this case by 33%.

As this example shows, failing to correct for the effect of ionic strength can lead to significant differences between calculated and actual concentrations. Nevertheless, it is not unusual to ignore activities and assume that the equilibrium constant is expressed in terms of concentrations. There is a practical reason for this—in an analysis one rarely knows the composition, much less the ionic strength of a sample solution. Equilibrium calculations are often used as a guide when developing an analytical method. Only by conducting the analysis and evaluating the results can we judge whether our theory matches reality.

Colorplate 3 provides a visual demonstration of the effect of ionic strength on the equilibrium reaction $Fe^{3+}(aq) + SCN^-(aq) \rightleftharpoons Fe(SCN)^{2+}(aq)$

6J Two Final Thoughts About Equilibrium Chemistry

In this chapter we have reviewed and extended our understanding of equilibrium chemistry. We also have developed several tools for evaluating the composition of a system at equilibrium. These tools differ in how accurately they allow us to answer questions involving equilibrium chemistry. They also differ in their ease of use. An important part of having several tools available to you is knowing when to use them. If you need to know whether a reaction is favorable, or the approximate pH of a solution, a ladder diagram may be sufficient to meet your needs. On the other hand, if you require an accurate estimate of a compound's solubility, a rigorous calculation using the systematic approach and activity coefficients is necessary.

Finally, a consideration of equilibrium chemistry can only help us decide what reactions are favorable. Knowing that a reaction is favorable does not guarantee that the reaction will occur. How fast a reaction approaches its equilibrium position does not depend on the magnitude of the equilibrium constant. The rate of a chemical reaction is a kinetic, not a thermodynamic, phenomenon. Kinetic effects and their application in analytical chemistry are discussed in Chapter 13.

6K KEY TERMS

acid (*p. 140*)
acid dissociation constant (*p. 140*)
activity (*p. 172*)
activity coefficient (*p. 172*)
amphiprotic (*p. 142*)
base (*p. 140*)
base dissociation constant (*p. 141*)
buffer (*p. 167*)
charge balance equation (*p. 159*)
common ion effect (*p. 158*)
cumulative formation constant (*p. 144*)
dissociation constant (*p. 144*)
enthalpy (*p. 137*)
entropy (*p. 137*)
equilibrium (*p. 136*)
equilibrium constant (*p. 138*)
formation constant (*p. 144*)
Gibb's free energy (*p. 137*)
Henderson–Hasselbalch equation (*p. 169*)
ionic strength (*p. 172*)
ladder diagram (*p. 150*)
Le Châtelier's principle (*p. 148*)
ligand (*p. 144*)
mass balance equation (*p. 159*)
Nernst equation (*p. 146*)
oxidation (*p. 146*)
oxidizing agent (*p. 146*)
pH (*p. 142*)
precipitate (*p. 139*)
redox reaction (*p. 145*)
reducing agent (*p. 146*)
reduction (*p. 146*)
solubility product (*p. 140*)
standard state (*p. 137*)
stepwise formation constant (*p. 144*)

6L SUMMARY

Analytical chemistry is more than a collection of techniques; it is the application of chemistry to the analysis of samples. As you will see in later chapters, almost all analytical methods use chemical reactivity to accomplish one or more of the following—dissolve the sample, separate analytes and interferents, transform the analyte to a more useful form, or provide a signal. Equilibrium chemistry and thermodynamics provide us with a means for predicting which reactions are likely to be favorable.

The most important types of reactions are precipitation reactions, acid–base reactions, metal–ligand complexation reactions, and redox reactions. In a precipitation reaction two or more soluble species combine to produce an insoluble product called a precipitate. The equilibrium properties of a precipitation reaction are described by a solubility product.

Acid–base reactions occur when an acid donates a proton to a base. The equilibrium position of an acid–base reaction is described using either the dissociation constant for the acid, K_a, or the dissociation constant for the base, K_b. The product of K_a and K_b for an acid and its conjugate base is K_w (water's dissociation constant).

Ligands have electron pairs that they can donate to a metal ion, forming a metal–ligand complex. The formation of the metal–ligand complex ML_2, for example, may be described by a stepwise formation constant in which each ligand is added one at a time; thus, K_1 represents the addition of the first ligand to M, and K_2 represents the addition of the second ligand to ML. Alternatively, the formation of ML_2 can be described by a cumulative, or overall formation constant, β_2, in which both ligands are added to M.

In a redox reaction, one of the reactants is oxidized while another reactant is reduced. Equilibrium constants are rarely used when characterizing redox reactions. Instead, we use the electrochemical potential, positive values of which indicate a favorable reaction. The Nernst equation relates this potential to the concentrations of reactants and products.

Le Châtelier's principle provides a means for predicting how systems at equilibrium respond to a change in conditions. When a stress is applied to an equilibrium by adding a reactant or product, by adding a reagent that reacts with one of the reactants or products, or by changing the volume, the system responds by moving in the direction that relieves the stress.

You should be able to describe a system at equilibrium both qualitatively and quantitatively. Rigorous solutions to equilibrium problems can be developed by combining equilibrium constant expressions with appropriate mass balance and charge balance equations. Using this systematic approach, you can solve some quite complicated equilibrium problems. When a less rigorous answer is needed, a ladder diagram may help you decide the equilibrium system's composition.

Solutions containing a weak acid and its conjugate base show only a small change in pH upon the addition of small amounts of strong acid or strong base. Such solutions are called buffers. Buffers can also be formed using a metal and its metal–ligand complex, or an oxidizing agent and its conjugate reducing agent. Both the systematic approach to solving equilibrium problems and ladder diagrams can be used to characterize a buffer.

A quantitative solution to an equilibrium problem may give an answer that does not agree with the value measured experimentally. This result occurs when the equilibrium constant based on concentrations is matrix-dependent. The true, thermodynamic equilibrium constant is based on the activities, *a*, of the reactants and products. A species' activity is related to its molar concentration by an activity coefficient, γ, where $a_i = \gamma_i[\ \ ]_i$. Activity coefficients often can be calculated, making possible a more rigorous treatment of equilibria.

6M *Suggested* EXPERIMENTS

Experiments

The following experiments involve the experimental determination of equilibrium constants and, in some cases, demonstrate the importance of activity effects.

"The Effect of Ionic Strength on an Equilibrium Constant (A Class Study)." In J. A. Bell, ed. *Chemical Principles in Practice.* Addison-Wesley: Reading, MA, 1967.

In this experiment the equilibrium constant for the dissociation of bromocresol green is measured at several ionic strengths. Results are extrapolated to zero ionic strength to find the thermodynamic equilibrium constant.

"Equilibrium Constants for Calcium Iodate Solubility and Iodic Acid Dissociation." In J. A. Bell, ed. *Chemical Principles in Practice.* Addison-Wesley: Reading, MA, 1967.

The effect of pH on the solubility of $Ca(IO_3)_2$ is studied in this experiment.

"The Solubility of Silver Acetate." In J. A. Bell, ed. *Chemical Principles in Practice.* Addison-Wesley: Reading, MA, 1967.

In this experiment the importance of the soluble silver acetate complexes $AgCH_3COO(aq)$ and $Ag(CH_3COO)_2^-(aq)$ in describing the solubility of $AgCH_3COO(s)$ is investigated.

Green, D. B.; Rechtsteiner, G.; Honodel, A. "Determination of the Thermodynamic Solubility Product, K_{sp}, of PbI_2 Assuming Nonideal Behavior," *J. Chem. Educ.* **1996,** *73,* 789–792.

The thermodynamic solubility product for PbI_2 is determined in this experiment by measuring its solubility at several ionic strengths.

6N PROBLEMS

1. Write equilibrium constant expressions for the following reactions. Determine the value for the equilibrium constant for each reaction using appropriate equilibrium constants from Appendix 3.

a. $NH_3(aq) + HCl(aq) \rightleftharpoons NH_4^+(aq) + Cl^-(aq)$

b. $PbI_2(s) + S^{2-}(aq) \rightleftharpoons PbS(s) + 2I^-(aq)$

c. $CdY^{2-}(aq) + 4CN^-(aq) \rightleftharpoons Cd(CN)_4^{2-}(aq) + Y^{4-}(aq)$ [Y^{4-} is EDTA]

d. $AgCl(s) + 2NH_3(aq) \rightleftharpoons Ag(NH_3)_2^+(aq) + Cl^-(aq)$

e. $BaCO_3(s) + 2H_3O^+(aq) \rightleftharpoons Ba^{2+}(aq) + H_2CO_3(aq) + 2H_2O(\ell)$

2. Using a ladder diagram, explain why the following reaction

$$H_3PO_4(aq) + F^-(aq) \rightleftharpoons HF(aq) + H_2PO_4^-(aq)$$

is favorable, whereas

$$H_3PO_4(aq) + 2F^-(aq) \rightleftharpoons 2HF(aq) + H_2PO_4^{2-}(aq)$$

is unfavorable. Determine the equilibrium constant for these reactions, and verify that they are consistent with your ladder diagram.

3. Calculate the potential for the following redox reaction when the $[Fe^{3+}] = 0.050$ M, $[Fe^{2+}] = 0.030$ M, $[Sn^{2+}] = 0.015$ M and $[Sn^{4+}] = 0.020$ M

$$2Fe^{3+}(aq) + Sn^{2+}(aq) \rightleftharpoons Sn^{4+}(aq) + 2Fe^{2+}(aq)$$

4. Balance the following redox reactions, and calculate the standard-state potential and the equilibrium constant for each. Assume that the $[H_3O^+]$ is 1 M for acidic solutions, and that the $[OH^-]$ is 1 M for basic solutions.

a. $MnO_4^-(aq) + H_2SO_3(aq) \rightleftharpoons Mn^{2+}(aq) + SO_4^{2-}(aq)$ (acidic solution)
b. $IO_3^-(aq) + I^-(aq) \rightleftharpoons I_2(s)$ (acidic solution)
c. $ClO^-(aq) + I^- \rightleftharpoons IO_3^-(aq) + Cl^-(aq)$ (basic solution)

5. Sulfur can be determined quantitatively by oxidizing to SO_4^{2-} and precipitating as $BaSO_4$. The solubility reaction for $BaSO_4$ is

$$BaSO_4(s) \rightleftharpoons Ba^{2+}(aq) + SO_4^{2-}(aq)$$

How will the solubility of $BaSO_4$ be affected by (a) decreasing the pH of the solution; (b) adding $BaCl_2$; (c) decreasing the volume of the solution?

6. Write charge balance and mass balance equations for the following solutions

a. 0.1 M NaCl
b. 0.1 M HCl
c. 0.1 M HF
d. 0.1 M NaH_2PO_4
e. $MgCO_3$ (saturated solution)
f. 0.1 M $Ag(CN)_2^-$ (from $AgNO_3$ and KCN)
g. 0.1 M HCl and 0.050 M $NaNO_2$

7. Using the systematic approach, calculate the pH of the following solutions

a. 0.050 M $HClO_4$
b. 1.00×10^{-7} M HCl
c. 0.025 M HClO
d. 0.010 M HCOOH
e. 0.050 M $Ba(OH)_2$
f. 0.010 M C_5H_5N

8. Construct ladder diagrams for the following diprotic weak acids (H_2L), and estimate the pH of 0.10 M solutions of H_2L, HL^-, and L^{2-}. Using the systematic approach, calculate the pH of each of these solutions.

a. maleic acid
b. malonic acid
c. succinic acid

9. Ignoring activity effects, calculate the solubility of Hg_2Cl_2 in the following

a. A saturated solution of Hg_2Cl_2
b. 0.025 M $Hg_2(NO_3)_2$ saturated with Hg_2Cl_2
c. 0.050 M NaCl saturated with Hg_2Cl_2

10. The solubility of CaF_2 is controlled by the following two reactions

$$CaF_2(s) \rightleftharpoons Ca^{2+}(aq) + F^-(aq)$$

$$HF(aq) + H_2O(\ell) \rightleftharpoons H_3O^+(aq) + F^-(aq)$$

Calculate the solubility of CaF_2 in a solution buffered to a pH of 7.00. Use a ladder diagram to help simplify the calculations. How would your approach to this problem change if the pH is buffered to 2.00? What is the solubility of CaF_2 at this pH?

11. Calculate the solubility of $Mg(OH)_2$ in a solution buffered to a pH of 7.00. How does this compare with its solubility in unbuffered water?

12. Calculate the solubility of Ag_3PO_4 in a solution buffered to a pH of 9.00.

13. Determine the equilibrium composition of saturated solution of AgCl. Assume that the solubility of AgCl is influenced by the following reactions.

$$AgCl(s) \rightleftharpoons Ag^+(aq) + Cl^-(aq)$$

$$Ag^+(aq) + Cl^-(aq) \rightleftharpoons AgCl(aq)$$

$$AgCl(aq) + Cl^-(aq) \rightleftharpoons AgCl_2^-(aq)$$

14. Calculate the ionic strength of the following solutions

a. 0.050 M NaCl
b. 0.025 M $CuCl_2$
c. 0.10 M Na_2SO_4

15. Repeat the calculations in problem 9, this time correcting for activity effects.

16. With the permission of your instructor, carry out the following experiment. In a beaker, mix equal volumes of 0.001 M NH_4SCN and 0.001 M $FeCl_3$ (the latter solution must be acidified with concentrated HNO_3 at a ratio of 4 drops/L to prevent the precipitation of $Fe(OH)_3$). Divide solution in half, and add solid KNO_3 to one portion at a ratio of 4 g per 100 mL. Compare the colors of the two solutions (see Color Plate 3), and explain why they are different. The relevant reaction is

$$Fe^{3+}(aq) + SCN^-(aq) \rightleftharpoons Fe(SCN)^{2+}(aq)$$

17. Over what pH range do you expect $Ca_3(PO_4)_2$ to have its minimum solubility?

18. Construct ladder diagrams for the following systems, and describe the information that can be obtained from each

a. HF and H_3PO_4
b. $Ag(CN)_2^-$, $Ni(CN)_4^{2-}$ and $Fe(CN)_6^{4-}$
c. $Cr_2O_7^{2-}/Cr^{3+}$ and Fe^{3+}/Fe^{2+}

19. Calculate the pH of the following acid–base buffers

a. 100 mL of 0.025 M formic acid and 0.015 M sodium formate
b. 50.00 mL of 0.12 M NH_3 and 3.50 mL of 1.0 M HCl
c. 5.00 g of Na_2CO_3 and 5.00 g of $NaHCO_3$ in 0.100 L

20. Calculate the pH of the buffers in problem 19 after adding 5.0×10^{-4} mol of HCl.

21. Calculate the pH of the buffers in problem 19 after adding 5.0×10^{-4} mol of NaOH.

22. Consider the following hypothetical complexation reaction between a metal, M, and a ligand, L

$$M(aq) + L(aq) \rightleftharpoons ML(aq)$$

with a formation constant of 1.5×10^8. Derive an equation, similar to the Henderson–Hasselbalch equation, which relates pM to the concentrations of L and ML. What will be the pM for a solution containing 0.010 mol of M and 0.020 mol of L? What will the pM be if 0.002 mol of M are added?

23. A redox buffer contains an oxidizing agent and its conjugate reducing agent. Calculate the potential of a solution containing 0.010 mol of Fe^{3+} and 0.015 mol of Fe^{2+}. What is the potential if sufficient oxidizing agent is added such that 0.002 mol of Fe^{2+} is converted to Fe^{3+}?

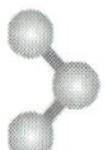

6O SUGGESTED READINGS

A lucid discussion of Berthollet's discovery of the reversibility of reactions is found in

Roots-Bernstein, R. S. *Discovering.* Harvard University Press: Cambridge, MA, 1989.

The following texts and articles provide additional coverage of equilibrium chemistry and the systematic approach to solving equilibrium problems.

Butler, J. N. *Ionic Equilibria: A Mathematical Approach.* Addison-Wesley: Reading, MA, 1964.

Butler, J. N. *Solubility and pH Calculations.* Addison-Wesley: Reading, MA, 1973.

Chaston, S. "Calculating Complex Equilibrium Concentrations by a Next Guess Factor Method," *J. Chem. Educ.* **1993,** *70,* 622–624.

Fernando, Q.; Ryan, M. D. *Calculations in Analytical Chemistry,* Harcourt Brace Jovanovich: New York, 1982.

Freiser, H. *Concepts and Calculations in Analytical Chemistry,* CRC Press: Boca Raton, 1992.

Freiser, H.; Fernando, Q. *Ionic Equilibria in Analytical Chemistry,* Wiley: New York, 1963.

Gordus, A. A. "Chemical Equilibrium I. The Thermodynamic Equilibrium Concept," *J. Chem. Educ.* **1991,** *68,* 138–140.

Gordus, A. A. "Chemical Equilibrum II. Deriving an Exact Equilibrium Equation," *J. Chem. Educ.* **1991,** *68,* 215–217.

Gordus, A. A. "Chemical Equilibrium III. A Few Math Tricks," *J. Chem. Educ.* **1991,** *68,* 291–293.

Gordus, A. A. "Chemical Equilibrium IV. Weak Acids and Bases," *J. Chem. Educ.* **1991,** *68,* 397–399.

Gordus, A. A. "Chemical Equilibrium VI. Buffer Solutions," *J. Chem. Educ.* **1991,** *68,* 656–658.

Gordus, A. A. "Chemical Equilibrium VII. Precipitates," *J. Chem. Educ.* **1991,** *68,* 927–930.

Gordus, A. A. *Schaum's Outline of Analytical Chemistry.* McGraw-Hill: New York, 1985.

Olivieri, A. C. "Solution of Acid–Base Equilibria by Successive Approximations," *J. Chem. Educ.* **1990,** *67,* 229–231.

Ramette, R. W. *Chemical Equilibrium and Analysis.* Addison-Wesley: Reading, MA, 1981.

Thomson, B. M.; Kessick, M. A. "On the Preparation of Buffer Solutions," *J. Chem. Educ.* **1981,** *58,* 743–746.

Weltin, E. "Are the Equilibrium Concentrations for a Chemical Reaction Always Uniquely Determined by the Initial Concentrations?" *J. Chem. Educ.* **1990,** *67,* 548.

Weltin, E. "A Numerical Method to Calculate Equilibrium Concentrations for Single-Equation Systems," *J. Chem. Educ.* **1991,** *68,* 486–487.

Weltin, E. "Calculating Equilibrium Concentrations," *J. Chem. Educ.* **1992,** *69,* 393–396.

Weltin, E. "Calculating Equilibrium Concentrations for Stepwise Binding of Ligands and Polyprotic Acid-Base Systems," *J. Chem. Educ.* **1993,** *70,* 568–571.

Weltin, E. "Equilibrium Calculations are Easier Than You Think— But You Do Have to Think!" *J. Chem. Educ.* **1993,** *70,* 571–573.

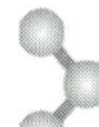

6P REFERENCES

1. (a) Runo, J. R.; Peters, D. G. *J. Chem. Educ.* **1993,** *70,* 708–713; (b) Vale, J.; Fernandez-Pereira, C.; Alcalde, M. *J. Chem. Educ.* **1993,** *70,* 790–795.
2. Van Slyke, D. D. *J. Biol. Chem.* **1922,** *52,* 525–570.
3. (a) Bower, V. E.; Bates, R. G. *J. Res. Natl. Bur. Stand. (U. S.)* **1955,** *55,* 197–200; (b) Bates, R. G. *Ann. N. Y. Acad. Sci.* **1961,** *92,* 341–356; (c) Bates, R. G. *Determination of pH,* 2nd ed. Wiley-Interscience: New York, **1973,** p. 73.
4. Lambert, W. J. *J. Chem. Educ.* **1990,** *67,* 150–153.
5. Kolthoff, I. M.; Lingane, J. *J. Phys. Chem.* **1938,** *42,* 133–140.
6. Davies, C. W. *Ion Association.* Butterworth: London, 1962.

Chapter 7

Obtaining and Preparing Samples for Analysis

When we first use an analytical method to solve a problem, it is not unusual to find that our results are of questionable accuracy or so imprecise as to be meaningless. Looking back we may find that nothing in the method seems amiss. In designing the method we considered sources of determinate and indeterminate error and took appropriate steps, such as including a reagent blank and calibrating our instruments, to minimize their effect. Why, then, might a carefully designed method give such poor results? One explanation is that we may not have accounted for errors associated with the sample. When we collect the wrong sample or lose analyte while preparing the sample for analysis, we introduce a determinate source of error. If we do not collect enough samples or collect samples of the wrong size, the precision of the analysis may suffer. In this chapter we consider how collecting samples and preparing them for analysis can affect the accuracy and precision of our results.

7A The Importance of Sampling

When a manufacturer produces a chemical they wish to list as ACS Reagent Grade, they must demonstrate that it conforms to specifications established by the American Chemical Society (ACS). For example, ACS specifications for $NaHCO_3$ require that the concentration of iron be less than or equal to 0.001% w/w. To verify that a production lot meets this standard, the manufacturer performs a quantitative analysis, reporting the result on the product's label. Because it is impractical to analyze the entire production lot, its properties are estimated from a limited sampling. Several samples are collected and analyzed, and the resulting mean, $\overline{X}$, and standard deviation, s, are used to establish a confidence interval for the production lot's true mean, μ

$$\mu = \overline{X} \pm \frac{ts}{\sqrt{n}} \qquad 7.1$$

where n is the number of samples, and t is a statistical factor whose value is determined by the number of samples and the desired confidence level.*

Selecting a sample introduces a source of determinate error that cannot be corrected during the analysis. If a sample does not accurately represent the population from which it is drawn, then an analysis that is otherwise carefully conducted will yield inaccurate results. Sampling errors are introduced whenever we extrapolate from a sample to its target population. To minimize sampling errors we must collect the right sample.

Even when collecting the right sample, indeterminate or random errors in sampling may limit the usefulness of our results. Equation 7.1 shows that the width of a confidence interval is directly proportional to the standard deviation. The overall standard deviation for an analysis, s_o, is determined by random errors affecting each step of the analysis. For convenience, we divide the analysis into two steps. Random errors introduced when collecting samples are characterized by a standard deviation for sampling, s_s. The standard deviation for the analytical method, s_m, accounts for random errors introduced when executing the method's procedure. The relationship among s_o, s_s, and s_m is given by a propagation of random error

$$s_o^2 = s_m^2 + s_s^2 \qquad 7.2$$

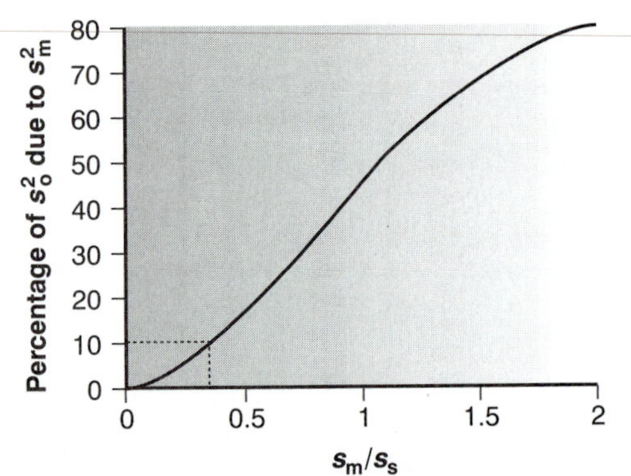

Figure 7.1
Percent of overall variance (s_o^2) due to the method as a function of the relative magnitudes of the standard deviation of the method and the standard deviation of sampling (s_m/s_s). The dotted lines show that the variance due to the method accounts for 10% of the overall variance when $s_s = 3 \times s_m$.

Equation 7.2 shows that an analysis' overall variance may be limited by either the analytical method or sample collection. Unfortunately, analysts often attempt to minimize overall variance by improving only the method's precision. This is futile, however, if the standard deviation for sampling is more than three times greater than that for the method.[1] Figure 7.1 shows how the ratio s_m/s_s affects the percentage of overall variance attributable to the method. When the method's standard deviation is one third of that for sampling, indeterminate method errors explain only 10% of the overall variance. Attempting to improve the analysis by decreasing s_m provides only a nominal change in the overall variance.

*Values for t can be found in Appendix 1B.

EXAMPLE 7.1

A quantitative analysis for an analyte gives a mean concentration of 12.6 ppm. The standard deviation for the method is found to be 1.1 ppm, and that due to sampling is 2.1 ppm. (a) What is the overall variance for the analysis? (b) By how much does the overall variance change if s_m is improved by 10% to 0.99 ppm? (c) By how much does the overall variance change if s_s is improved by 10% to 1.9 ppm?

SOLUTION

(a) The overall variance is

$$s_o^2 = s_m^2 + s_s^2 = (1.1)^2 + (2.1)^2 = 1.21 + 4.41 = 5.62 \approx 5.6$$

(b) Improving the method's standard deviation changes the overall variance to

$$s_o^2 = (0.99)^2 + (2.1)^2 = 0.98 + 4.41 = 5.39 \approx 5.4$$

Thus, a 10% improvement in the method's standard deviation changes the overall variance by approximately 4%.

(c) Changing the standard deviation for sampling

$$s_o^2 = (1.1)^2 + (1.9)^2 = 1.21 + 3.61 = 4.82 \approx 4.8$$

improves the overall variance by almost 15%. As expected, since s_s is larger than s_m, a more significant improvement in the overall variance is realized when we focus our attention on sampling problems.

To determine which step has the greatest effect on the overall variance, both s_m^2 and s_s^2 must be known. The analysis of replicate samples can be used to estimate the overall variance. The variance due to the method is determined by analyzing a standard sample, for which we may assume a negligible sampling variance. The variance due to sampling is then determined by difference.

EXAMPLE 7.2

The following data were collected as part of a study to determine the effect of sampling variance on the analysis of drug animal-feed formulations.[2]

% Drug (w/w)		
0.0114	0.0099	0.0105
0.0102	0.0106	0.0087
0.0100	0.0095	0.0098
0.0105	0.0095	0.0097

% Drug (w/w)		
0.0105	0.0109	0.0107
0.0103	0.0103	0.0104
0.0101	0.0101	0.0103

The data on the left were obtained under conditions in which random errors in sampling and the analytical method contribute to the overall variance. The data on the right were obtained in circumstances in which the sampling variance is known to be insignificant. Determine the overall variance and the contributions from sampling and the analytical method.

SOLUTION

The overall variance, s_o^2, is determined using the data on the left and is equal to 4.71×10^{-7}. The method's contribution to the overall variance, s_m^2, is determined using the data on the right and is equal to 7.00×10^{-8}. The variance due to sampling, s_s^2, is therefore

$$s_s^2 = s_o^2 - s_m^2 = 4.71 \times 10^{-7} - 7.00 \times 10^{-8} = 4.01 \times 10^{-7}$$

7B Designing A Sampling Plan

sampling plan
A plan that ensures that a representative sample is collected.

A **sampling plan** must support the goals of an analysis. In characterization studies a sample's purity is often the most important parameter. For example, a material scientist interested in the surface chemistry of a metal is more likely to select a freshly exposed surface, created by fracturing the sample under vacuum, than a surface that has been exposed to the atmosphere for an extended time. In a qualitative analysis the sample's composition does not need to be identical to that of the substance being analyzed, provided that enough sample is taken to ensure that all components can be detected. In fact, when the goal of an analysis is to identify components present at trace levels, it may be desirable to discriminate against major components when sampling. In a quantitative analysis, however, the sample's composition must accurately represent the target population. The focus of this section, therefore, is on designing a sampling plan for a quantitative analysis.

Five questions should be considered when designing a sampling plan:

1. From where within the target population should samples be collected?
2. What type of samples should be collected?
3. What is the minimum amount of sample needed for each analysis?
4. How many samples should be analyzed?
5. How can the overall variance be minimized?

Each of these questions is considered below in more detail.

7B.1 Where to Sample the Target Population

Sampling errors occur when a sample's composition is not identical to that of the population from which it is drawn. When the material being sampled is homogeneous, individual samples can be taken without regard to possible sampling errors. Unfortunately, in most situations the target population is heterogeneous in either time or space. As a result of settling, for example, medications available as oral suspensions may have a higher concentration of their active ingredients at the bottom of the container. Before removing a dose (sample), the suspension is shaken to minimize the effect of this spatial heterogeneity. Clinical samples, such as blood or urine, frequently show a temporal heterogeneity. A patient's blood glucose level, for instance, will change in response to eating, medication, or exercise. Other systems show both spatial and temporal heterogeneities. The concentration of dissolved O_2 in a lake shows a temporal heterogeneity due to the change in seasons, whereas point sources of pollution may produce a spatial heterogeneity.

When the target population's heterogeneity is of concern, samples must be acquired in a manner that ensures that determinate sampling errors are insignificant. If the target population can be thoroughly homogenized, then samples can be taken without introducing sampling errors. In most cases, however, homogenizing the target population is impracticable. Even more important, homogenization destroys information about the analyte's spatial or temporal distribution within the target population.

Random Sampling The ideal sampling plan provides an unbiased estimate of the target population's properties. This requirement is satisfied if the sample is collected at random from the target population.[3] Despite its apparent simplicity, a true **random sample** is difficult to obtain. Haphazard sampling, in which samples are collected without a sampling plan, is not random and may reflect an analyst's unintentional biases. The best method for ensuring the collection of a random sample is to divide the target population into equal units, assign a unique number to each unit, and use a random number table (Appendix 1E) to select the units from which to sample. Example 7.3 shows how this is accomplished.

random sample
A sample collected at random from the target population.

Example

EXAMPLE 7.3

To analyze the properties of a 100 cm × 100 cm polymer sheet, ten 1 cm × 1 cm samples are to be selected at random and removed for analysis. Explain how a random number table can be used to ensure that samples are drawn at random.

SOLUTION

As shown in the following grid, we divide the polymer sheet into 10,000 1 cm × 1 cm squares, each of which can be identified by its row number and its column number.

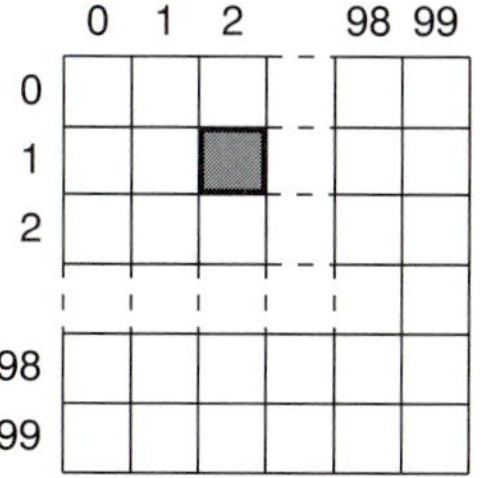

For example, the highlighted square is in row 1 and column 2. To pick ten squares at random, we enter the random number table at an arbitrary point, and let that number represent the row for the first sample. We then move through the table in a predetermined fashion, selecting random numbers for the column of the first sample, the row of the second sample, and so on until all ten samples have been selected. Since our random number table (Appendix 1E) uses five-digit numbers we will use only the last two digits. Let's begin with the fifth entry and use every other entry after that. The fifth entry is 65423 making the first row number 23. The next entry we use is 41812, giving the first column number as 12. Continuing in this manner, the ten samples are as follows:

Sample	Row	Column	Sample	Row	Column
1	23	12	6	93	83
2	45	80	7	91	17
3	81	12	8	45	13
4	66	17	9	12	92
5	46	01	10	97	52

A randomly collected sample makes no assumptions about the target population, making it the least biased approach to sampling. On the other hand, random sampling requires more time and expense than other sampling methods since a greater number of samples are needed to characterize the target population.

judgmental sampling
Samples collected from the target population using available information about the analyte's distribution within the population.

Judgmental Sampling The opposite of random sampling is selective, or **judgmental sampling,** in which we use available information about the target population to help select samples. Because assumptions about the target population are included in the sampling plan, judgmental sampling is more biased than random sampling; however, fewer samples are required. Judgmental sampling is common when we wish to limit the number of independent variables influencing the results of an analysis. For example, a researcher studying the bioaccumulation of polychlorinated biphenyls (PCBs) in fish may choose to exclude fish that are too small or that appear diseased. Judgmental sampling is also encountered in many protocols in which the sample to be collected is specifically defined by the regulatory agency.

systematic sampling
Samples collected from the target population at regular intervals in time or space.

Systematic Sampling Random sampling and judgmental sampling represent extremes in bias and the number of samples needed to accurately characterize the target population. **Systematic sampling** falls in between these extremes. In systematic sampling the target population is sampled at regular intervals in space or time. For a system exhibiting a spatial heterogeneity, such as the distribution of dissolved O_2 in a lake, samples can be systematically collected by dividing the system into discrete units using a two- or three-dimensional grid pattern (Figure 7.2). Samples are collected from the center of each unit, or at the intersection of grid lines. When a heterogeneity is time-dependent, as is common in clinical studies, samples are drawn at regular intervals.

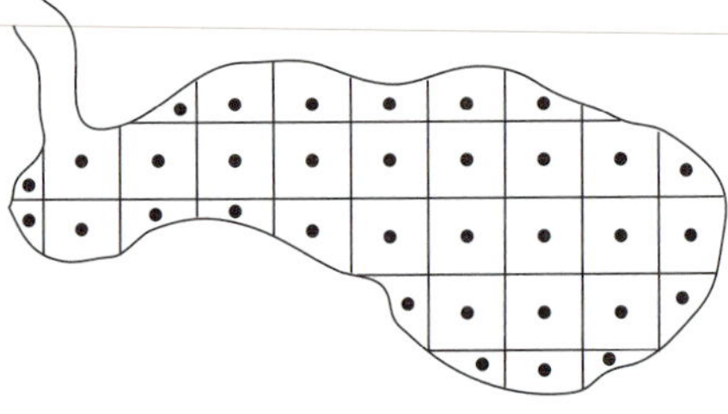

Figure 7.2
Example of a systematic sampling plan for collecting samples from a lake. Each solid dot represents a sample collected from within the sampling grid.

When a target population's spatial or temporal heterogeneity shows a periodic trend, a systematic sampling leads to a significant bias if samples are not collected frequently enough. This is a common problem when sampling electronic signals, in which case the problem is known as aliasing. Consider, for example, a signal consisting of a simple sine wave. Figure 7.3a shows how an insufficient sampling frequency underestimates the signal's true frequency.

Nyquist theorem
Statement that a periodic signal must be sampled at least twice each period to avoid a determinate error in measuring its frequency.

According to the **Nyquist theorem,** to determine a periodic signal's true frequency, we must sample the signal at a rate that is at least twice its frequency (Figure 7.3b); that is, the signal must be sampled at least twice during a single cycle or period. When samples are collected at an interval of Δt, the highest frequency that can be accurately monitored has a frequency of $(2\ \Delta t)^{-1}$. For example, if samples are collected every hour, the highest frequency that we can monitor is 0.5 h^{-1}, or a periodic cycle lasting 2 h. A signal with a cycling period of less than 2 h (a frequency of more than 0.5 h^{-1}) cannot be monitored. Ideally, the sampling frequency should be at least three to four times that of the highest frequency signal of interest. Thus, if an hourly periodic cycle is of interest, samples should be collected at least every 15–20 min.

systematic–judgmental sampling
A sampling plan that combines judgmental sampling with systematic sampling.

Systematic–Judgmental Sampling Combinations of the three primary approaches to sampling are also possible.[4] One such combination is **systematic–judgmental sampling,** which is encountered in environmental studies when a spatial or temporal distribution of pollutants is anticipated. For example, a plume of waste leaching from a landfill can reasonably be expected to move in the same direction as the flow of groundwater. The systematic–judgmental sampling plan shown in Figure 7.4 includes a rectangular grid for systematic sampling and linear transects extending the sampling along the plume's suspected major and minor axes.[5]

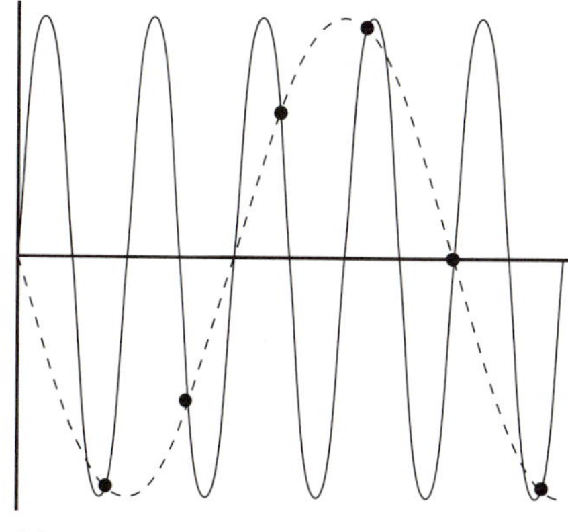

(a)

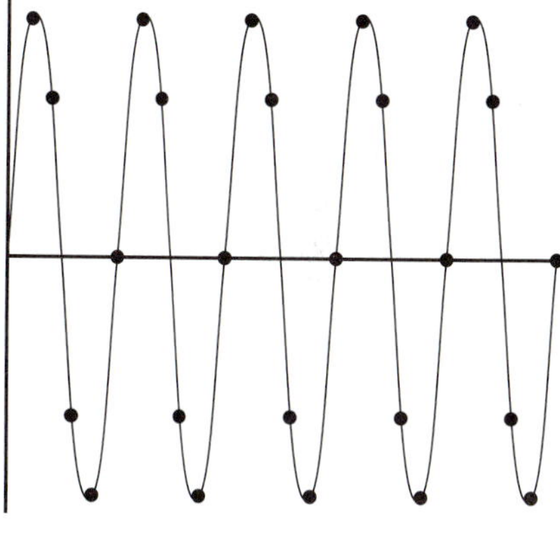

(b)

Figure 7.3
Effect of sampling frequency when monitoring periodic signals. In (a) the sampling frequency is 1.2 samples per period. The dashed line shows the apparent signal, while the solid line shows the true signal. In (b) a sampling frequency of five samples per period is sufficient to give an accurate estimation of the true signal.

Stratified Sampling Another combination of the three primary approaches to sampling is judgmental–random, or **stratified sampling.** Many target populations are conveniently subdivided into distinct units, or strata. For example, in determining the concentration of particulate Pb in urban air, the target population can be subdivided by particle size. In this case samples can be collected in two ways. In a random sampling, differences in the strata are ignored, and individual samples are collected at random from the entire target population. In a stratified sampling the target population is divided into strata, and random samples are collected from within each stratum. Strata are analyzed separately, and their respective means are pooled to give an overall mean for the target population.

The advantage of stratified sampling is that the composition of each stratum is often more homogeneous than that of the entire target population. When true, the sampling variance for each stratum is less than that when the target population is treated as a single unit. As a result, the overall sampling variance for stratified sampling is always at least as good as, and often better than, that obtained by simple random sampling.

Convenience Sampling One additional method of sampling deserves brief mention. In **convenience sampling,** sample sites are selected using criteria other than minimizing sampling error and sampling variance. In a survey of groundwater quality, for example, samples can be collected by drilling wells at randomly selected sites, or by making use of existing wells. The latter method is usually the preferred choice. In this case, cost, expedience, and accessibility are the primary factors used in selecting sampling sites.

stratified sampling
A sampling plan that divides the population into distinct strata from which random samples are collected.

convenience sampling
A sampling plan in which samples are collected because they are easily obtained.

grab sample
A single sample removed from the target population.

7B.2 What Type of Sample to Collect

After determining where to collect samples, the next step in designing a sampling plan is to decide what type of sample to collect. Three methods are commonly used to obtain samples: grab sampling, composite sampling, and in situ sampling. The most common type of sample is a **grab sample,** in which a portion of the target population is removed at a given time and location in space. A grab sample, therefore, provides a "snapshot" of the target population. Grab sampling is easily adapted to any of the sampling schemes discussed in the previous section. If the target population is fairly uniform in time and space, a set of grab samples collected at random can be used to establish its properties. A systematic sampling using grab samples can be used to characterize a target population whose composition varies over time or space.

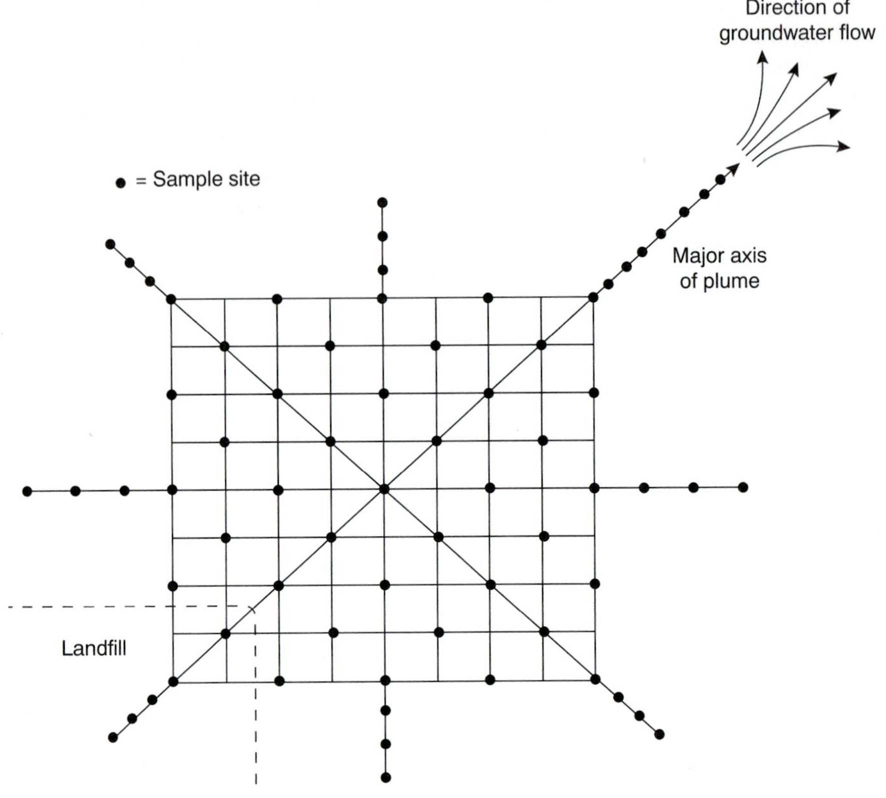

Figure 7.4
Systematic–judgmental sampling scheme for monitoring the leaching of pollutants from a landfill. Sites where samples are collected are represented by the solid dots.

composite sample
Several grab samples combined to form a single sample.

A **composite sample** consists of a set of grab samples that are combined to form a single sample. After thoroughly mixing, the composite sample is analyzed. Because information is lost when individual samples are combined, it is normally desirable to analyze each grab sample separately. In some situations, however, there are advantages to working with composite samples. One such situation is in determining a target population's average composition over time or space. For example, wastewater treatment plants are required to monitor and report the average composition of treated water released to the environment. One approach is to analyze a series of individual grab samples, collected using a systematic sampling plan, and average the results. Alternatively, the individual grab samples can be combined to form a single composite sample. Analyzing a single composite sample instead of many individual grab samples, provides an appreciable savings in time and cost. Composite sampling is also useful when a single sample cannot supply sufficient material for an analysis. For example, methods for determining PCBs in fish often require as much as 50 g of tissue, an amount that may be difficult to obtain from a single fish. Tissue samples from several fish can be combined and homogenized, and a 50-g portion of the composite sample taken for analysis.

A significant disadvantage of grab samples and composite samples is the need to remove a portion of the target population for analysis. As a result, neither type of sample can be used to continuously monitor a time-dependent change in the target population. **In situ sampling,** in which an analytical sensor is placed directly in the target population, allows continuous monitoring without removing individual grab samples. For example, the pH of a solution moving through an industrial production line can be continually monitored by immersing a pH electrode within the solution's flow.

in situ sampling
Sampling done within the population without physically removing the sample.

7B.3 How Much Sample to Collect

To minimize sampling errors, a randomly collected grab sample must be of an appropriate size. If the sample is too small its composition may differ substantially from that of the target population, resulting in a significant sampling error. Samples that are too large, however, may require more time and money to collect and analyze, without providing a significant improvement in sampling error.

As a starting point, let's assume that our target population consists of two types of particles. Particles of type A contain analyte at a fixed concentration, and type B particles contain no analyte. If the two types of particles are randomly distributed, then a sample drawn from the population will follow the binomial distribution.* If we collect a sample containing n particles, the expected number of particles containing analyte, n_A, is

$$n_A = np$$

where p is the probability of selecting a particle of type A. The sampling standard deviation is

$$s_s = \sqrt{np(1-p)} \qquad \textbf{7.3}$$

The relative standard deviation for sampling, $s_{s,r}$, is obtained by dividing equation 7.3 by n_A.

$$s_{s,r} = \frac{\sqrt{np(1-p)}}{np}$$

Solving for n allows us to calculate the number of particles that must be sampled to obtain a desired sampling variance.

$$n = \frac{1-p}{p} \times \frac{1}{s_{s,r}^2} \qquad \textbf{7.4}$$

Note that the relative sampling variance is inversely proportional to the number of particles sampled. Increasing the number of particles in a sample, therefore, improves the sampling variance.

Example

EXAMPLE 7.4

Suppose you are to analyze a solid where the particles containing analyte represent only $1 \times 10^{-7}\%$ of the population. How many particles must be collected to give a relative sampling variance of 1%?

SOLUTION

Since the particles of interest account for $1 \times 10^{-7}\%$ of all particles in the population, the probability of selecting one of these particles is only 1×10^{-9}. Substituting into equation 7.4 gives

$$n = \frac{1-(1\times 10^{-9})}{1\times 10^{-9}} \times \frac{1}{(0.01)^2} = 1 \times 10^{13}$$

Thus, to obtain the desired sampling variance we need to collect 1×10^{13} particles.

*See Chapter 4 to review the properties of a binomial distribution.

A sample containing 10^{13} particles can be fairly large. Suppose this is equivalent to a mass of 80 g. Working with a sample this large is not practical; but does this mean we must work with a smaller sample and accept a larger relative sampling variance? Fortunately the answer is no. An important feature of equation 7.4 is that the relative sampling variance is a function of the number of particles but not their combined mass. We can reduce the needed mass by crushing and grinding the particles to make them smaller. Our sample must still contain 10^{13} particles, but since each particle is smaller their combined mass also is smaller. If we assume that a particle is spherical, then its mass is proportional to the cube of its radius.

$$\text{Mass} \propto r^3$$

Decreasing a particle's radius by a factor of 2, for example, decreases its mass by a factor of 2^3, or 8. Instead of an 80-g sample, a 10-g sample will now contain 10^{13} particles.

Example

EXAMPLE 7.5

Assume that the sample of 10^{13} particles from Example 7.4 weighs 80 g. By how much must you reduce the radius of the particles if you wish to work with a sample of 0.6 g?

SOLUTION

To reduce the sample from 80 g to 0.6 g you must change its mass by a factor of

$$\frac{80}{0.6} = 133 \text{ times}$$

This can be accomplished by decreasing the radius of the particles by a factor of

$$x^3 = 133$$

$$x = 5.1$$

Decreasing the radius by a factor of approximately 5 allows you to decrease the sample's mass from 80 g to 0.6 g.

Treating a population as though it contains only two types of particles is a useful exercise because it shows us that the relative sampling variance can be improved by collecting more particles of sample. Furthermore, we learned that the mass of sample needed can be reduced by decreasing particle size without affecting the relative sampling variance. Both are important conclusions.

Few populations, however, meet the conditions for a true binomial distribution. Real populations normally contain more than two types of particles, with the analyte present at several levels of concentration. Nevertheless, many well-mixed populations, in which the population's composition is homogeneous on the scale at which we sample, approximate binomial sampling statistics. Under these conditions the following relationship between the mass of a randomly collected grab sample, m, and the percent relative standard deviation for sampling, R, is often valid.[6]

$$mR^2 = K_s \qquad \textbf{7.5}$$

where K_s is a sampling constant equal to the mass of sample producing a percent relative standard deviation for sampling of ±1%.* The sampling constant is evalu-

*Problem 8 in the end-of-chapter problem set asks you to consider the relationship between equations 7.4 and 7.5.

ated by determining R using several samples of similar mass. Once K_s is known, the mass of sample needed to achieve a desired relative standard deviation for sampling can be calculated.

Example

EXAMPLE 7.6

The following data were obtained in a preliminary determination of the amount of inorganic ash in a breakfast cereal.

Mass of Cereal (g)	Percentage Ash (w/w)
0.9956	1.34
0.9981	1.29
1.0036	1.32
0.9994	1.26
1.0067	1.28

Determine K_s and the amount of sample needed to give a relative standard deviation for sampling of ±2.0%. Predict the percent relative standard deviation and the absolute standard deviation if samples of 5 g are collected.

SOLUTION

To determine K_s we need to know the average mass of the cereal samples and the relative standard deviation for the %(w/w) ash. The average mass of the five cereal samples is 1.0007 g. The average %(w/w) ash and the absolute standard deviation are, respectively, 1.298% and 0.03194. The percent relative standard deviation, therefore, is

$$R = \frac{s_s}{\overline{X}} \times 100 = \frac{0.03194}{1.298} \times 100 = 2.46\%$$

Thus

$$K_s = mR^2 = (1.0007 \text{ g})(2.46)^2 = 6.06 \text{ g}$$

The amount of sample needed to give a relative standard deviation of ±2%, therefore, is

$$m = \frac{K_s}{R^2} = \frac{6.06 \text{ g}}{(2.0)^2} = 1.5 \text{ g}$$

If we use 5.00-g samples, then the expected percent relative standard deviation is

$$R = \sqrt{\frac{K_s}{m}} = \sqrt{\frac{6.06 \text{ g}}{5.00 \text{ g}}} = 1.10\%$$

and the expected absolute standard deviation is

$$R = \frac{s_s}{\overline{X}} \times 100$$

$$s_s = \frac{R\overline{X}}{100} = \frac{(1.10)(1.298)}{100} = 0.0143$$

When the target population is segregated, or stratified, equation 7.5 provides a poor estimate of the amount of sample needed to achieve a desired relative standard deviation for sampling. A more appropriate relationship, which can be applied to both segregated and nonsegregated samples, has been proposed.[7]

$$s_s^2 = \frac{A}{mn_s} + \frac{B}{n_s} \qquad 7.6$$

where n_s is the number of samples to be analyzed, m is the mass of each sample, A is a homogeneity constant accounting for the random distribution of analyte in the target population, and B is a segregation constant accounting for the nonrandom distribution of analyte in the target population. Equation 7.6 shows that sampling variance due to the random distribution of analyte can be minimized by increasing either the mass of each sample or the total number of samples. Sampling errors due to the nonrandom distribution of the analyte, however, can only be minimized by increasing the total number of samples. Values for the homogeneity constant and heterogeneity constant are determined using two sets of samples that differ significantly in mass.

Example

EXAMPLE 7.7

To develop a sampling plan for the determination of PCBs in lake sediments, the following two experiments are conducted. First, 15 samples, each with a mass of 1.00 g, are analyzed, giving a sampling variance of 0.0183. In a second experiment, ten samples, each with a mass of 10.0 g, are analyzed, giving a sampling variance of 0.0069. If samples weighing 5.00 g are to be collected, how many are needed to give a sampling variance of 0.0100? If five samples are to be collected, how much should each sample weigh?

SOLUTION

Substituting known values for the two experiments into equation 7.6 gives the following pair of simultaneous equations

$$0.0183 = \frac{A}{(1.00)(15)} + \frac{B}{15}$$

$$0.0069 = \frac{A}{(10.0)(10)} + \frac{B}{10}$$

Solving for A and B gives values of 0.228 and 0.0462, respectively. The number of 5.00-g samples is determined by solving

$$0.0100 = \frac{0.228}{(5.00)n} + \frac{0.0462}{n}$$

for n, giving $n = 9.2 \approx 9$ samples. When using five samples, the mass of each is given by the equation

$$0.0100 = \frac{0.228}{m(5)} + \frac{0.0462}{5}$$

for which m is 60.0 g.

7B.4 How Many Samples to Collect

In the previous section we considered the amount of sample needed to minimize the sampling variance. Another important consideration is the number of samples required to achieve a desired maximum sampling error. If samples drawn from the target population are normally distributed, then the following equation describes the confidence interval for the sampling error

$$\mu = \overline{X} \pm \frac{ts_s}{\sqrt{n_s}}$$

where n_s is the number of samples and s_s is the sampling standard deviation. Rearranging and substituting e for the quantity $(\mu - \overline{X})$, gives the number of samples as

$$n_s = \frac{t^2 s_s^2}{e^2} \tag{7.7}$$

where s_s^2 and e^2 are both expressed as absolute uncertainties or as relative uncertainties. Finding a solution to equation 7.7 is complicated by the fact that the value of t depends on n_s. As shown in Example 7.8, equation 7.7 is solved iteratively.

Example

EXAMPLE 7.8

In Example 7.6 we found that an analysis for the inorganic ash content of a breakfast cereal required a sample of 1.5 g to establish a relative standard deviation for sampling of ±2.0%. How many samples are needed to obtain a relative sampling error of no more than 0.80% at the 95% confidence level?

SOLUTION

Because the value of t depends on n_s, and the value of n_s is not yet known, we begin by letting $n_s = \infty$ and use the associated value of t. From Appendix 1B, the value for t is 1.96. Substituting known values into equation 7.7

$$n_s = \frac{(1.96)^2(2.0)^2}{(0.80)^2} = 24$$

Letting $n_s = 24$, the value of t from Appendix 1B is 2.075. Recalculating n_s gives

$$n_s = \frac{(2.075)^2(2.0)^2}{(0.80)^2} = 26.9 \approx 27$$

When $n_s = 27$, the value of t is 2.066. Recalculating n_s, we find

$$n_s = \frac{(2.066)^2(2.0)^2}{(0.80)^2} = 26.7 \approx 27$$

Since two successive calculations give the same value for n_s, an iterative solution has been found. Thus, 27 samples are needed to achieve the desired sampling error.

Equation 7.7 only provides an estimate for the smallest number of samples expected to produce the desired sampling error. The actual sampling error may be substantially higher if the standard deviation for the samples that are collected is significantly greater than the standard deviation due to sampling used to calculate n_s.

This is not an uncommon problem. For a target population with a relative sampling variance of 50 and a desired relative sampling error of ±5%, equation 7.7 predicts that ten samples are sufficient. In a simulation in which 1000 samples of size 10 were collected, however, only 57% of the samples resulted in sampling errors of less than ±5%.[8] By increasing the number of samples to 17 it was possible to ensure that the desired sampling error was achieved 95% of the time.

7B.5 Minimizing the Overall Variance

A final consideration in developing a sampling plan is to minimize the overall variance for the analysis. Equation 7.2 shows that the overall variance is a function of the variance due to the method and the variance due to sampling. As we have seen, we can improve the variance due to sampling by collecting more samples of proper size. Increasing the number of times we analyze each sample improves the variance due to the method. If s_s^2 is significantly greater than s_m^2, then the method's variance can be ignored and equation 7.7 used to estimate the number of samples to analyze. Analyzing any sample more than once will not improve the overall variance, since the variance due to the method is insignificant.

If s_m^2 is significantly greater than s_s^2, then we only need to collect and analyze a single sample. The number of replicate analyses, n_r, needed to minimize the error due to the method is given by an equation similar to equation 7.7

$$n_r = \frac{t^2 s_m^2}{e^2}$$

Unfortunately, the simple situations just described are often the exception. In many cases, both the sampling variance and method variance are significant, and both multiple samples and replicate analyses of each sample are required. The overall error in this circumstance is given by

$$e = t\left(\frac{s_s^2}{n_s} + \frac{s_m^2}{n_s n_r}\right)^{1/2} \qquad 7.8$$

Equation 7.8 does not have a unique solution because different combinations of n_s and n_r give the same overall error. The choice of how many samples to collect and how many times each sample should be analyzed is determined by other concerns, such as the cost of collecting and analyzing samples, and the amount of available sample.

Example

EXAMPLE 7.9

A certain analytical method has a relative sampling variance of 0.40% and a relative method variance of 0.070%. Evaluate the relative error ($\alpha = 0.05$) if (a) you collect five samples, analyzing each twice; and, (b) you collect two samples, analyzing each five times.

SOLUTION

Both sampling strategies require a total of ten determinations. Using Appendix 1B, we find that the value of t is 2.26. Substituting into equation 7.8, we find that the relative error for the first sampling strategy is

$$e = 2.26\left(\frac{0.40}{5} + \frac{0.070}{(5)(2)}\right)^{1/2} = 0.67\%$$

and that for the second sampling strategy is

$$e = 2.26\left(\frac{0.40}{2} + \frac{0.070}{(2)(5)}\right)^{1/2} = 1.0\%$$

As expected, since the relative method variance is better than the relative sampling variance, a sampling strategy that favors the collection of more samples and few replicate analyses gives the better relative error.

7C Implementing the Sampling Plan

After a sampling plan has been developed, it is put into action. Implementing a sampling plan normally involves three steps: physically removing the sample from its target population, preserving the sample, and preparing the sample for analysis. Except for in situ sampling, the analysis of a sample occurs after removing it from the target population. Since sampling exposes the target population to potential contamination, the sampling device must be inert and clean.

Once a sample is withdrawn from a target population, there is a danger that it may undergo a chemical or physical change. This is a serious problem since the properties of the sample will no longer be representative of the target population. For this reason, samples are often preserved before transporting them to the laboratory for analysis. Even when samples are analyzed in the field, preservation may still be necessary.

The initial sample is called the primary, or **gross sample** and may be a single increment drawn from the target population, or a composite of several increments. In many cases the gross sample cannot be analyzed without further treatment. Processing the gross sample may be used to reduce the sample's particle size, to transfer the sample into a more readily analyzable form, or to improve its homogeneity.

gross sample
The initial sample, collected from the target population without any processing.

In the sections that follow, these three steps are considered for the sampling of liquids (including solutions), gases, and solids.

7C.1 Solutions

Typical examples of liquid samples include those drawn from containers of commercial solvents; beverages, such as milk or fruit juice; natural waters, including from lakes, streams, seawater, and rain; bodily fluids, such as blood and urine; and, suspensions, such as those found in many oral medications.

Sample Collection Homogeneous solutions are easily sampled by siphoning, decanting, or by using a pipet or syringe. Unfortunately, few solutions are truly homogeneous. When the material to be sampled is of manageable size, manual shaking is often sufficient to ensure homogeneity. Samples may then be collected with a pipet, a syringe, or a bottle. The majority of solutions, however, cannot be sampled in this manner. To minimize the effect of heterogeneity, the method for collecting the gross sample must be adapted to the material being sampled.

The environmental sampling of waters and wastewaters provides a good illustration of many of the methods used to sample solutions. The chemical composition of surface waters, such as streams, rivers, lakes, estuaries, and oceans, is influenced by flow rate and depth. Rapidly flowing shallow streams and rivers, and shallow (<5 m) lakes are usually well mixed and show little stratification with depth. Grab samples are conveniently collected by submerging a capped bottle below the surface and removing the cap. The air–water interface, which may be enriched with heavy metals[9] or contaminated with oil, is avoided when collecting the sample. After the sample bottle is filled, the cap is replaced and the bottle removed. Slowly moving streams and rivers, lakes deeper than 5 m, estuaries, and oceans may show substantial stratification. Grab samples from near the surface can be collected as described earlier, whereas samples at greater depths are collected with a weighted sample bottle that is lowered to the desired depth. Once it has reached the desired depth, the sample bottle is opened, allowed to fill, and closed before retrieving. Grab samples can be analyzed individually, giving information about changes in the analyte's concentration with depth. Alternatively, the grab samples may be pooled to form a composite sample.

Wells used for collecting groundwater samples must be purged before the sample is collected, since the chemical composition of water in the well-casing and in the adjacent matrix may be significantly different from that of the surrounding groundwater. These differences may result from contaminants introduced when drilling the well, or differences in the groundwater's redox potential when exposed to atmospheric oxygen. In general, wells are purged by pumping out a volume of water equivalent to several well-casing volumes, or until the water's temperature, pH, or specific conductance are constant. Samples collected from municipal water supplies must also be purged since the chemical composition of water left standing in pipes may differ significantly from the treated water supply. Samples are collected at faucets after flushing the pipes for 2–3 min.

Samples from municipal wastewater treatment plants and samples of industrial discharges often are collected as 24-h composites. Samples are obtained using an automatic sampler that periodically removes individual grab samples. The volume of each sample increment and the frequency of sampling may be constant or may vary in response to changes in flow rate.

Sample containers for collecting solutions are made from glass or plastic. Containers made from Kimax or Pyrex brand borosilicate glass have the advantage of being sterilizable, easy to clean, and inert to all solutions except those that are strongly alkaline. The disadvantages of glass containers are cost, weight, and the likelihood of breakage. Plastic containers are made from a variety of polymers, including polyethylene, polypropylene, polycarbonate, polyvinyl chloride, and Teflon (polytetrafluoroethylene). Plastic containers are lightweight, durable, and, except for those manufactured from Teflon, inexpensive. In most cases glass or plastic bottles may be used, although polyethylene bottles are generally preferred because of their lower cost. Glass containers are always used when collecting samples for the analysis of pesticides, oil and grease, and organics because these species often interact with plastic surfaces. Since glass surfaces easily adsorb metal ions, plastic bottles are preferred when collecting samples for the analysis of trace metals.

In most cases the sample bottle has a wide mouth, making it easy to fill and remove the sample. A narrow-mouth sample bottle is used when exposing the sample to the container cap or to the outside environment is undesirable. Unless exposure to plastic is a problem, caps for sample bottles are manufactured from polyethylene. When polyethylene must be avoided, the container cap includes an inert interior liner of neoprene or Teflon.

Sample Preservation Once removed from its target population, a liquid sample's chemical composition may change as a result of chemical, biological, or physical processes. Following its collection, samples are preserved by controlling the solu-

Table 7.1 Preservation Methods and Maximum Holding Times for Selected Water and Wastewater Parameters

Parameter	Preservation	Maximum Holding Time
ammonia	cool to 4 °C; H_2SO_4 to pH < 2	28 days
chloride	none required	28 days
metals—Cr(VI)	cool to 4 °C	24 h
metals—Hg	HNO_3 to pH < 2	28 days
metals—all others	HNO_3 to pH < 2	6 months
nitrate	none required	48 h
organochlorine pesticides	1 mL 10 mg/mL $HgCl_2$; or addition of extracting solvent	7 days without extraction 40 days with extraction
pH	none required	analyze immediately

tion's pH and temperature, limiting its exposure to light or to the atmosphere, or by adding a chemical preservative. After preserving, samples may be safely stored for later analysis. The maximum holding time between preservation and analysis depends on the analyte's stability and the effectiveness of sample preservation. Table 7.1 provides a list of sample preservation methods and maximum holding times for several analytes of importance in the analysis of water and wastewater.

Sample Preparation Most analytical methods can be applied to analytes in a liquid or solution state. For this reason a gross sample of a liquid or solution does not need additional processing to bring it into a more suitable form for analysis.

7C.2 Gases

Typical examples of gaseous samples include automobile exhaust, emissions from industrial smokestacks, atmospheric gases, and compressed gases. Also included with gaseous samples are solid aerosol particulates.

Sample Collection The simplest approach for collecting a gas sample is to fill a container, such as a stainless steel canister or a Tedlar/Teflon bag, with a portion of the gas. A pump is used to pull the gas into the container, and, after flushing the container for a predetermined time, the container is sealed. This method has the advantage of collecting a more representative sample of the gas than other collection techniques. Disadvantages include the tendency for some gases to adsorb to the container's walls, the presence of analytes at concentrations too low to detect with accuracy and precision, and the presence of reactive gases, such as ozone and nitrogen oxides, that may change the sample's chemical composition with time, or react with the container. When using a stainless steel canister many of these disadvantages can be overcome with cryogenic cooling, which changes the sample from a gaseous to a liquid state.

Due to the difficulty of storing gases, most gas samples are collected using either a trap containing a solid sorbent or by filtering. Solid sorbents are used to collect volatile gases (vapor pressures more than approximately 10^{-6} atm) and semivolatile gases (vapor pressures between approximately 10^{-6} atm and 10^{-12} atm), and filtration is used to collect nonvolatile gases.

Solid sorbent sampling is accomplished by passing the gas through a canister packed with sorbent particles. Typically 2–100 L of gas is sampled when collecting volatile compounds, and 2–500 m^3 when collecting semivolatile gases.* A variety of inorganic, organic polymer and carbon sorbents have been used. Inorganic sorbents, such as silica gel, alumina, magnesium aluminum silicate, and molecular sieves, are efficient collectors for polar compounds. Their efficiency for collecting water, however, limits their sorption capacity for many organic compounds.

Organic polymeric sorbents are manufactured using polymeric resins of 2,4-diphenyl-*p*-phenylene oxide or styrene-divinylbenzene for volatile compounds, or polyurethane foam for semivolatile compounds. These materials have a low affinity for water and are efficient collectors for all but the most highly volatile organic compounds and some low-molecular-weight alcohols and ketones. The adsorbing ability of carbon sorbents is superior to that of organic polymer resins. Thus, carbon sorbents can be used to collect those highly volatile organic compounds that cannot be collected by polymeric resins. The adsorbing ability of carbon sorbents may be a disadvantage, however, since the adsorbed compounds may be difficult to desorb.

Nonvolatile compounds are normally present either as solid particulates or bound to solid particulates. Samples are collected by pulling large volumes of gas through a filtering unit where the particulates are collected on glass fiber filters.

One of the most significant problems with sorbent sampling is the limited capacity of the sorbent to retain gaseous analytes. If a sorbent's capacity is exceeded before sampling is complete then a portion of the analyte will pass through the canister without being retained, making an accurate determination of its concentration impossible. For this reason it is not uncommon to place a second sorbent canister downstream from the first. If the analyte is not detected in the second canister, then it is safe to assume that the first canister's capacity has not been exceeded. The volume of gas that can be sampled before exceeding the sorbent's capacity is called the **breakthrough volume** and is normally reported with units of $m^3/(g_{pack})$, where g_{pack} is the grams of sorbent.

breakthrough volume
The volume of sample that can be passed through a solid sorbent before the analytes are no longer retained.

The short-term exposure of humans, animals, and plants to gaseous pollutants is more severe than that for pollutants in other matrices. Since the composition of atmospheric gases can show a substantial variation over a time, the continuous monitoring of atmospheric gases such as O_3, CO, SO_2, NH_3, H_2O_2, and NO_2 by in situ sampling is important.[10]

Sample Preservation and Preparation After collecting the gross sample there is generally little need for sample preservation or preparation. The chemical composition of a gas sample is usually stable when it is collected using a solid sorbent, a filter, or by cryogenic cooling. When using a solid sorbent, gaseous compounds may be removed before analysis by thermal desorption or by extracting with a suitable solvent. Alternatively, when the sorbent is selective for a single analyte, the increase in the sorbent's mass can be used to determine the analyte's concentration in the sample.

7C.3 Solids

Typical examples of solid samples include large particulates, such as those found in ores; smaller particulates, such as soils and sediments; tablets, pellets, and capsules used in dispensing pharmaceutical products and animal feeds; sheet materials, such as polymers and rolled metals; and tissue samples from biological specimens.

*1 m^3 is equivalent to 1000 L.

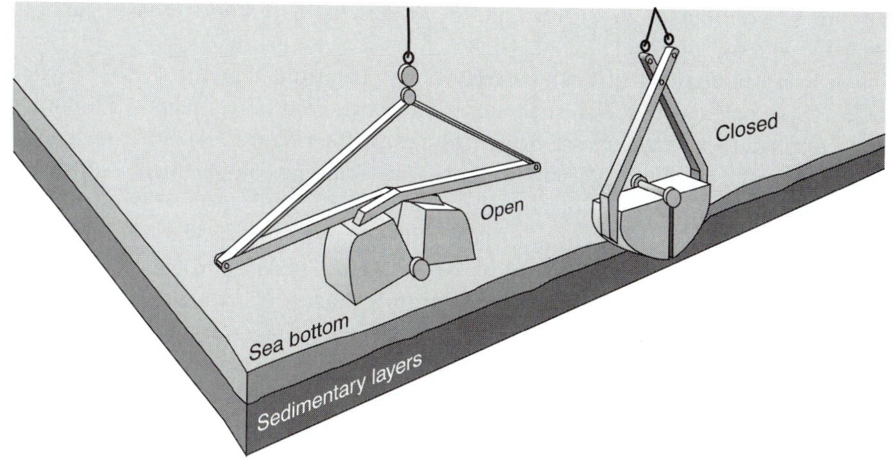

Figure 7.5
Schematic diagram of grab sampler. When the sampler reaches the sediment, the jaws of the grab sampler are closed, collecting a sample of the sediment.

Sample Collection Solids are usually heterogeneous, and samples must be collected carefully if they are to be representative of the target population. As noted earlier, solids come in a variety of forms, each of which is sampled differently.

Sediments from the bottom of streams, rivers, lakes, estuaries, and oceans are collected with a bottom grab sampler or with a corer. Grab samplers are equipped with a pair of "jaws" that close when they contact the sediment, scooping up sediment in the process (Figure 7.5). Their principal advantages are ease of use and the ability to collect a large sample. Disadvantages include the tendency to lose finer-grained sediment particles as water flows out of the sampler and the loss of spatial information, both laterally and with depth, due to mixing of the sample.

An alternative method for sampling sediments uses a cylindrical coring device (Figure 7.6). The corer is dropped into the sediment, collecting a column of sediment and the water in contact with the sediment. With the possible exception of sediment at the surface, which may experience mixing, samples collected with a corer maintain their vertical profile. As a result, changes in the sediment's composition with depth are preserved. The main disadvantage to a corer is that only a small surface area is sampled. For this reason sampling with a corer usually requires more samples.

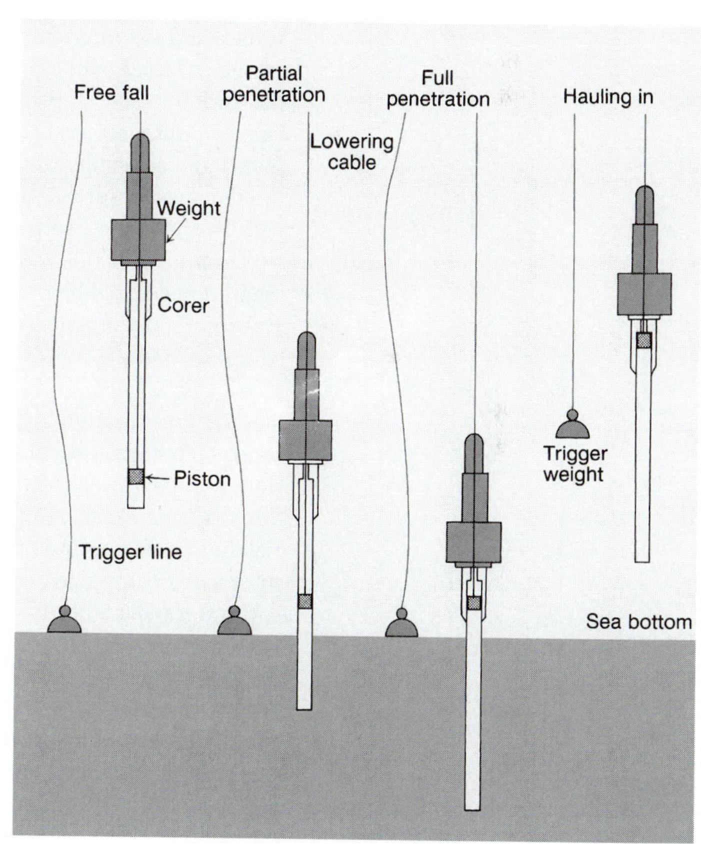

Figure 7.6
Schematic diagram of a piston corer in operation. The weight of the corer is sufficient to cause its penetration into the sediment, while the upward motion of the piston allows water pressure to help force the sediment column into the barrel of the corer.

Soil samples collected at depths of up to 30 cm are easily collected with scoops or shovels, although the sampling variance is generally high. A better method for obtaining soil samples near the surface is to use a soil punch. This thin-walled steel tube retains a core sample when it is pushed into the soil and removed. Soil samples collected at depths greater than 30 cm are obtained by digging a trench and collecting lateral samples with a soil punch. Alternatively,

Figure 7.7
Example of a four-unit riffle. A sample added through the top is divided into four piles, two on each of the riffle's sides.

an auger may be used to drill a hole to the desired depth and the sample collected with a soil punch.

The sampling of particulate material is often determined by the size of the particles. Large particulate solids, such as coals and ores, can be sampled by randomly collecting samples with a shovel or by riffling. A riffle (Figure 7.7) is a trough containing an even number of compartments with adjacent compartments emptying on opposite sides of the riffle. Particulate material dumped into a riffle is divided into two parts. By repeatedly passing half of the separated material back through the riffle, a sample of any desired size may be collected. Smaller particulate materials, such as powders, are best collected with a sample thief, which allows material to be collected simultaneously from several locations (Figure 7.8). A typical sample thief consists of two tubes that are nestled together. Each tube has an identical set of slots aligned down their length. Before the sample thief is inserted into the material being sampled, the inner tube is rotated so that slots are closed. When the sample thief is in place, the inner tube is rotated to open the slots, allowing the powder to enter the sample thief through each slot. The inner tube is then rotated to the closed position and the sample thief withdrawn.

When sampling a metal, it usually is necessary to obtain material from both the surface and the interior. When the metal is in the form of a sheet, random samples can be collected with a metal punch. Samples can be obtained from a metal wire by randomly cutting off pieces of an appropriate length. Larger pieces of metal, such as bars or bricks, are best sampled by sawing through the metal at randomly selected points and collecting the "sawdust" or by drilling through the metal and collecting the shavings. A surface coating can be sampled in situ or by dissolving the coating with an appropriate solvent.

Sampling of biological tissue is done by removing the entire organ, which is then homogenized before smaller portions are taken for analysis. Alternatively, several small portions of tissue may be combined to form a composite sample. The composite sample is then homogenized and analyzed.

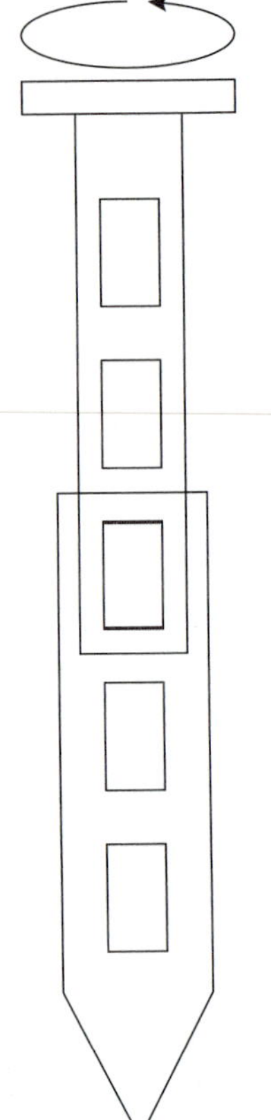

Figure 7.8
Schematic diagram of a sample thief. Rotating the inner cylinder opens and closes the openings along the outer cylinder's shaft.

Sample Preservation Without preservation, many solid samples are subject to changes in chemical composition due to the loss of volatile material, biodegradation, and chemical reactivity (particularly redox reactions). Samples stored at reduced temperatures are less prone to biodegradation and the loss of volatile material, but fracturing and phase separations may present problems. The loss of volatile material is minimized by ensuring that the sample completely fills its container without leaving a headspace where gases can collect. Samples collected from materials that have not been exposed to O_2 are particularly susceptible to oxidation reactions. For example, the contact of air with anaerobic sediments must be prevented.

Sample Preparation Unlike gases and liquids, which generally require little sample preparation, solid samples usually need some processing before analysis. There are two reasons for this. First, as discussed in Section 7B.3, sampling variance is a function of the number of particles sampled, not their combined mass. For extremely heterogeneous populations consisting of large particulates, the gross sample may be too large to analyze. For example, a boxcar containing a load of a Ni-bearing ore with an average particle size of 5 mm may require a sample weighing one ton to obtain a reasonable sampling variance. Reducing the sample's average particle size allows the same number of particles to be sampled with a smaller, more manageable combined mass.

Second, the majority of analytical techniques, particularly those used for a quantitative analysis, require that the analyte be in solution. Solid samples, or at least the analytes in a solid sample, must be brought into solution.

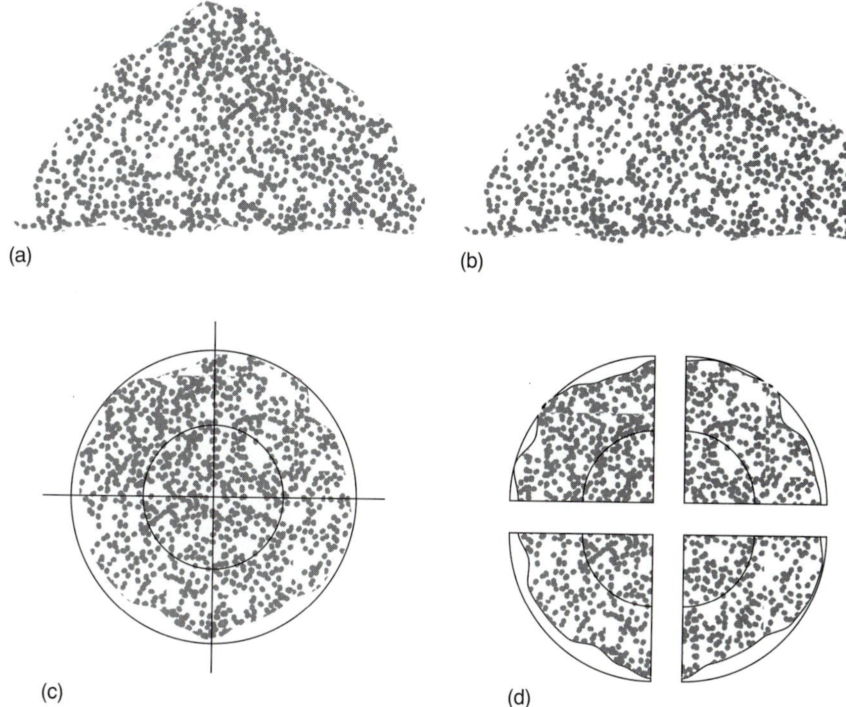

Figure 7.9
Illustration showing the method of coning and quartering as a means of reducing a gross sample for subsampling. (a) The gross sample is first piled into a cone and (b) flattened. Looking down from above, (c) the cone is divided into four quarters, (d) which are then separated.

Reducing Particle Size A reduction in particle size is accomplished by a combination of crushing and grinding the gross sample. The resulting particulates are then thoroughly mixed and divided into samples of smaller mass containing the appropriate number of particles. The process seldom occurs in a single step. Instead, samples are cycled through the process several times until a **laboratory sample** of desired mass is obtained.

laboratory sample
Sample taken into the lab for analysis after processing the gross sample.

Crushing and grinding uses mechanical force to break larger particles into smaller ones. A variety of tools are used depending on the particle's size and hardness. Large particles are crushed using jaw crushers capable of reducing particles to diameters of a few millimeters. Ball mills, disk mills, and mortars and pestles are used to further reduce particle size.

Significant changes in composition may occur during crushing and grinding. Decreasing particle size increases available surface area. With more surface area there is a greater risk of losing volatile components, a problem made worse by the frictional heat accompanying the crushing and grinding. An increase in surface area also means that portions of the sample are freshly exposed to the atmosphere where oxidation may alter the sample's composition. Other problems include contamination from the mechanical abrasion of the materials used to crush and grind the sample, and differences in the ease with which particles are reduced in size. Softer particles are reduced in size more easily and may be lost as dust before the rest of the sample has been processed. This is a problem since the analyte's distribution may not be uniform between particles of different size.

To ensure that all particles are reduced to a uniform size, the sample is intermittently passed through a sieve. Processing of those particles not passing through the sieve continues until the entire sample is of uniform size. The sample is then mixed thoroughly to ensure homogeneity, and a secondary sample obtained with a riffle or by **coning and quartering.** The latter approach is outlined in Figure 7.9. The gross sample is piled into a cone, flattened, divided into four quarters, and two diagonally

coning and quartering
A process for reducing the size of a gross sample.

Table 7.2 Acids and Bases Used for Sample Digestion

Solution (≈ %w/w)	Uses and Properties
HCl (37%)	• dissolves metals more easily reduced than H_2 ($E^o < 0$) • dissolves insoluble carbonates, sulfides, phosphates, fluorides, sulfates, and many oxides
HNO_3 (70%)	• strong oxidizing agent • dissolves most common metals except Al and Cr • decomposes organics and biological samples (wet ashing)
H_2SO_4 (98%)	• dissolves many metals and alloys • decomposes organics by oxidation and dehydration
HF (50%)	• dissolves silicates forming volatile SiF_4
$HClO_4$ (70%)	• hot, concentrated solutions are strong oxidizing agents • dissolves many metals and alloys • decomposes organics (*reactions with organics are often explosive, use only in specially equipped hoods with a blast shield and after prior decomposition with HNO_3*)
HCl:HNO_3 (3:1 v/v)	• also known as aqua regia • dissolves Au and Pt
NaOH	• dissolves Al and amphoteric oxides of Sn, Pb, Zn, and Cr

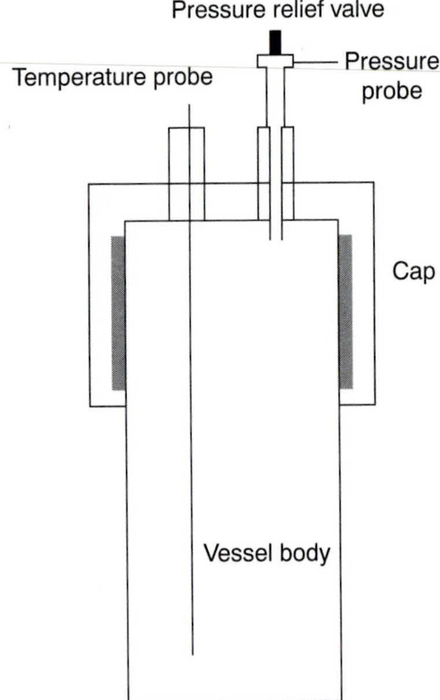

Figure 7.10
Schematic diagram of a microwave digestion vessel.

opposed quarters are discarded. The remaining material is cycled through the process of coning and quartering until the desired amount of sample remains.

Bringing Solid Samples into Solution If you are fortunate, the sample with which you are working will easily dissolve in a suitable solvent, requiring no more effort than gentle swirling and heating. Distilled water is usually the solvent of choice for inorganic salts, but an organic solvent, such as methanol, chloroform, or toluene, is used for organic materials. More often, one or more of the sample's components resist simple dissolution.

With samples that are difficult to dissolve, the first approach is usually to try digesting the sample with an acid or base. Table 7.2 lists the most commonly used acids and bases and summarizes their use. Digestion is commonly carried out in an open container, such as a beaker, using a hot plate as a source of heat. The chief advantage of this approach is its low cost as it requires no special equipment. Volatile reaction products, however, are lost, leading to a determinate error if analyte is included among the volatile substances.

Many digestions are now carried out in closed containers using microwave radiation as a source of energy for heating the solution. Vessels for microwave digestion are manufactured using Teflon (or some other fluoropolymer) or fused silica. Both materials are thermally stable, chemically resistant, transparent to microwave radiation, and capable of withstanding elevated pressures. A typical microwave digestion vessel is shown in Figure 7.10 and consists of the vessel body and cap, a temperature probe, and a pressure relief valve. Vessels are placed in a microwave oven (typically 6–12 vessels can be accommodated), and microwave energy is controlled by monitoring the temperature or pressure within the vessels. A microwave digestion has several important advantages over an open container digestion, including higher temperatures (200–300 °C) and pressures (40–100 bar). As a result, digestions requiring several hours in an open container may be accomplished in less than

Table 7.3 Common Fluxes for Decomposing Inorganic Samples

Flux	Melting Temperature (°C)	Crucible	Typical Samples
Na_2CO_3	851	Pt	silicates, oxides, phosphates, sulfides
$Li_2B_4O_7$	930	Pt, graphite	aluminosilicates, carbonates
$LiBO_2$	845		
NaOH	318	Au, Ag	silicates, silicon carbide
KOH	380		
Na_2O_2	—	Ni	silicates, chromium steel, Pt alloys
$K_2S_2O_7$	300	Pt, porcelain	oxides
B_2O_3	577	Pt	silicates, oxides

30 min using microwave digestion. In addition, the closed container prevents the loss of volatile gases. Disadvantages include the inability to add reagents during digestion, limitations on the amount of sample that can be used (typically 1 g or less), and safety concerns due to the use of high pressures and corrosive reagents. Applications include environmental and biological samples.

Inorganic samples that resist decomposition by digestion with acids or bases often can be brought into solution by fusing with a large excess of an alkali metal salt, called a flux. The sample and flux are mixed together in a crucible and heated till the substances fuse together in a molten state. The resulting melt is allowed to cool slowly to room temperature. Typically the melt dissolves readily in distilled water or dilute acid. Several common fluxes and their uses are listed in Table 7.3. Fusion works when other methods of decomposition do not because of the higher temperatures obtained and the high concentration of the reactive flux in the molten liquid. Disadvantages include a greater risk of contamination from the large quantity of flux and the crucible and the loss of volatile materials.

Finally, organic materials may be decomposed by dry ashing. In this method the sample is placed in a suitable crucible and heated over a flame or in a furnace. Any carbon present in the sample is oxidized to CO_2, and hydrogen, sulfur, and nitrogen are removed as H_2O, SO_2 and N_2. These gases can be trapped and weighed to determine their content in the organic material. Often the goal of dry ashing is the removal of organic material, leaving behind an inorganic residue, or ash, that can be further analyzed.

7D Separating the Analyte from Interferents

When a method shows a high degree of selectivity for the analyte, the task of performing a quantitative, qualitative, or characterization analysis is simplified. For example, a quantitative analysis for glucose in honey is easier to accomplish if the method is selective for glucose, even in the presence of other reducing sugars, such as fructose. Unfortunately, analytical methods are rarely selective toward a single species.

In the absence of interferents, the relationship between the sample's signal, S_{samp}, and the concentration of analyte, C_A, is

$$S_{samp} = k_A C_A \qquad 7.9$$

where k_A is the analyte's sensitivity.* In the presence of an interferent, equation 7.9 becomes

$$S_{samp} = k_A C_A + k_I C_I \qquad 7.10$$

where k_I and C_I are the interferent's sensitivity and concentration, respectively. A method's selectivity is determined by the relative difference in its sensitivity toward the analyte and interferent. If k_A is greater than k_I, then the method is more selective for the analyte. The method is more selective for the interferent if k_I is greater than k_A.

Even if a method is more selective for an interferent, it can be used to determine an analyte's concentration if the interferent's contribution to S_{samp} is insignificant. The selectivity coefficient, $K_{A,I}$, was introduced in Chapter 3 as a means of characterizing a method's selectivity.

$$K_{A,I} = \frac{k_I}{k_A} \qquad 7.11$$

Solving equation 7.11 for k_I and substituting into equation 7.10 gives, after simplifying

$$S_{samp} = k_A(C_A + K_{A,I} \times C_I) \qquad 7.12$$

An interferent, therefore, will not pose a problem as long as the product of its concentration and the selectivity coefficient is significantly smaller than the analyte's concentration.

$$K_{A,I} \times C_I << C_A$$

When an interferent cannot be ignored, an accurate analysis must begin by separating the analyte and interferent.

7E General Theory of Separation Efficiency

The goal of an analytical separation is to remove either the analyte or the interferent from the sample matrix. To achieve a separation there must be at least one significant difference between the chemical or physical properties of the analyte and interferent. Relying on chemical or physical properties, however, presents a fundamental problem—a separation also requires selectivity. A separation that completely removes an interferent may result in the partial loss of analyte. Altering the separation to minimize the loss of analyte, however, may leave behind some of the interferent.

A separation's efficiency is influenced both by the failure to recover all the analyte and the failure to remove all the interferent. We define the analyte's **recovery,** R_A, as

recovery
The fraction of analyte or interferent remaining after a separation (R).

$$R_A = \frac{C_A}{(C_A)_o}$$

where C_A is the concentration of analyte remaining after the separation, and $(C_A)_o$ is the analyte's initial concentration. A recovery of 1.00 means that none of the analyte is lost during the separation. The recovery of the interferent, R_I, is defined in the same manner

$$R_I = \frac{C_I}{(C_I)_o} \qquad 7.13$$

*In equation 7.9, and the equations that follow, the concentration of analyte, C_A, can be replaced by the moles of analyte, n_A, when considering a total analysis technique.

where C_I is the concentration of interferent remaining after the separation, and $(C_I)_o$ is the interferent's initial concentration. The degree of separation is given by a **separation factor,** $S_{I,A}$, which is the change in the ratio of interferent to analyte caused by the separation.[11]

separation factor
A measure of the effectiveness of a separation at separating an analyte from an interferent ($S_{I,A}$).

$$S_{I,A} = \frac{C_I / C_A}{(C_I)_o / (C_A)_o} = \frac{R_I}{R_A}$$

Example

EXAMPLE 7.10

An analysis to determine the concentration of Cu in an industrial plating bath uses a procedure for which Zn is an interferent. When a sample containing 128.6 ppm Cu is carried through a separation to remove Zn, the concentration of Cu remaining is 127.2 ppm. When a 134.9-ppm solution of Zn is carried through the separation, a concentration of 4.3 ppm remains. Calculate the recoveries for Cu and Zn and the separation factor.

SOLUTION

The recoveries for the analyte and interferent are

$$R_{Cu} = \frac{127.2 \text{ ppm}}{128.6 \text{ ppm}} = 0.9891, \text{ or } 98.91\%$$

and

$$R_{Zn} = \frac{4.3 \text{ ppm}}{134.9 \text{ ppm}} = 0.032, \text{ or } 3.2\%$$

The separation factor is

$$S_{Zn,Cu} = \frac{R_{Zn}}{R_{Cu}} = \frac{0.032}{0.9891} = 0.032$$

In an ideal separation $R_A = 1$, $R_I = 0$, and $S_{I,A} = 0$. In general, the separation factor should be approximately 10^{-7} for the quantitative analysis of a trace analyte in the presence of a macro interferent, and 10^{-3} when the analyte and interferent are present in approximately equal amounts.

Recoveries and separation factors are useful ways to evaluate the effectiveness of a separation. They do not, however, give a direct indication of the relative error introduced by failing to remove all interferents or failing to recover all the analyte. The relative error introduced by the separation, E, is defined as

$$E = \frac{S_{samp} - S^*_{samp}}{S^*_{samp}} \qquad \textbf{7.14}$$

where S^*_{samp} is the expected signal for an ideal separation when all the analyte is recovered.

$$S^*_{samp} = k_A(C_A)_o \qquad \textbf{7.15}$$

Substituting equations 7.12 and 7.15 into 7.14 gives

$$E = \frac{k_A(C_A + K_{A,I} \times C_I) - k_A(C_A)_o}{k_A(C_A)_o}$$

which simplifies to

$$E = \frac{C_A + K_{A,I} \times C_I - (C_A)_o}{(C_A)_o}$$

$$= \frac{C_A}{(C_A)_o} - \frac{(C_A)_o}{(C_A)_o} + \frac{K_{A,I} \times C_I}{(C_A)_o}$$

$$= (R_A - 1) + \frac{K_{A,I} \times C_I}{(C_A)_o} \qquad \textbf{7.16}$$

A more useful equation for the relative error is obtained by solving equation 7.13 for C_I and substituting back into equation 7.16

$$E = (R_A - 1) + \left[\frac{K_{A,I} \times (C_I)_o}{(C_A)_o} \times R_I\right] \qquad \textbf{7.17}$$

The first term of equation 7.17 accounts for the incomplete recovery of analyte, and the second term accounts for the failure to remove all the interferent.

Example

EXAMPLE 7.11

Following the separation outlined in Example 7.10, an analysis is to be carried out for the concentration of Cu in an industrial plating bath. The concentration ratio of Cu to Zn in the plating bath is 7:1. Analysis of standard solutions containing only Cu or Zn give the following standardization equations

$$S_{Cu} = 1250 \times (\text{ppm Cu})$$

$$S_{Zn} = 2310 \times (\text{ppm Zn})$$

(a) What error is expected if no attempt is made to remove Zn before analyzing for Cu? (b) What is the error if the separation is carried out? (c) What is the maximum acceptable recovery for Zn if Cu is completely recovered and the error due to the separation must be no greater than 0.10%?

SOLUTION

(a) If the analysis is carried out without a separation, then R_{Cu} and R_{Zn} are equal to 1.000, and equation 7.17 simplifies to

$$E = \frac{K_{Cu,Zn} \times (\text{ppm Zn})_o}{(\text{ppm Cu})_o}$$

From equation 7.11 the selectivity coefficient is

$$K_{Cu,Zn} = \frac{k_{Zn}}{k_{Cu}} = \frac{2310}{1250} = 1.85$$

Although we do not know the actual concentrations of Zn or Cu in the sample, we do know that the concentration ratio (ppm Zn)$_o$/(ppm Cu)$_o$ is 1/7. Thus

$$E = \frac{(1.85)(1)}{(7)} = 0.264, \text{ or } 26.4\%$$

(b) To calculate the error, we substitute the recoveries calculated in Example 7.10 into equation 7.17

$$E = (0.9891 - 1) + \left[\frac{(1.85)(1)}{(7)} \times 0.032\right]$$

$$= (-0.0109) + (0.0085)$$

$$= -0.0024, \text{ or } -0.24\%$$

Note that a negative determinate error introduced by failing to recover all the analyte is partially offset by a positive determinate error due to a failure to remove all the interferent.

(c) To determine the maximum allowed recovery for Zn, we make appropriate substitutions into equation 7.17

$$0.0010 = (1.000 - 1) + \left[\frac{(1.85)(1)}{(7)} \times R_{Zn}\right]$$

and solve for R_{Zn}, obtaining a recovery of 0.0038, or 0.38%. Thus, at least 99.62% of the Zn must be removed by the separation.

7F Classifying Separation Techniques

An analyte and an interferent can be separated if there is a significant difference in at least one of their chemical or physical properties. Table 7.4 provides a partial list of several separation techniques, classified by the chemical or physical property that is exploited.

7F.1 Separations Based on Size

The simplest physical property that can be exploited in a separation is size. The separation is accomplished using a porous medium through which only the analyte or interferent can pass. Filtration, in which gravity, suction, or pressure is used to pass a sample through a porous filter is the most commonly encountered separation technique based on size.

Particulate interferents can be separated from dissolved analytes by filtration, using a filter whose pore size retains the interferent. This separation technique is important in the analysis of many natural waters, for which the presence of suspended solids may interfere in the analysis. Filtration also can be used to isolate analytes present as solid particulates from dissolved ions in the sample matrix. For example, this is a necessary step in gravimetry, in which the analyte is isolated as a precipitate. A more detailed description of the types of available filters is found in the discussion of precipitation gravimetry and particulate gravimetry in Chapter 8.

Table 7.4 Classification of Separation Techniques

Basis of Separation	Separation Technique
size	filtration dialysis size-exclusion chromatography
mass and density	centrifugation
complex formation	masking
change in physical state	distillation sublimation recrystallization
change in chemical state	precipitation ion exchange electrodeposition volatilization
partitioning between phases	extraction chromatography

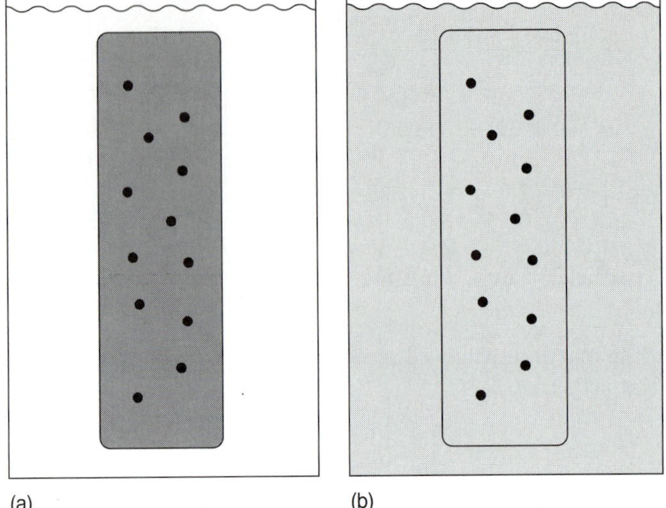

Figure 7.11
Illustration of a dialysis membrane in action. In (a) the sample solution is placed in the dialysis tube and submerged in the solvent. (b) Smaller particles pass through the membrane, but larger particles remain within the dialysis tube.

dialysis
A method of separation that uses a semipermeable membrane.

Another example of a separation technique based on size is **dialysis,** in which a semipermeable membrane is used to separate the analyte and interferent. Dialysis membranes are usually constructed from cellulose, with pore sizes of 1–5 nm. The sample is placed inside a bag or tube constructed from the membrane. The dialysis membrane and sample are then placed in a container filled with a solution whose composition differs from that of the sample. If the concentration of a particular species is not the same on the two sides of the membrane, the resulting concentration gradient provides a driving force for its diffusion across the membrane. Although small particles may freely pass through the membrane, larger particles are unable to pass (Figure 7.11). Dialysis is frequently used to purify proteins, hormones, and enzymes. During kidney dialysis, metabolic waste products, such as urea, uric acid, and creatinine, are removed from blood by passing it over a dialysis membrane.

size-exclusion chromatography
A separation method in which a mixture passes through a bed of porous particles, with smaller particles taking longer to pass through the bed due to their ability to move into the porous structure.

Size-exclusion chromatography, which also is called gel permeation or molecular-exclusion chromatography, is a third example of a separation technique based on size. In this technique a column is packed with small, approximately 10-μm, porous particles of cross-linked dextrin or polyacrylamide. The pore size of the particles is controlled by the degree of cross-linking, with greater cross-linking resulting in smaller pore sizes. The sample to be separated is placed into a stream of solvent that is pumped through the column at a fixed flow rate. Particles too large to enter the pores are not retained and pass through the column at the same rate as the solvent. Those particles capable of entering into the pore structure take longer to pass through the column. Smaller particles, which penetrate more deeply into the pore structure, take the longest time to pass through the column. Size-exclusion chromatography is widely used in the analysis of polymers and in biochemistry, where it is used for the separation of proteins.

7F.2 Separations Based on Mass or Density

If there is a difference in the mass or density of the analyte and interferent, then a separation using centrifugation may be possible. The sample, as a suspension, is placed in a centrifuge tube and spun at a high angular velocity (high numbers of revolutions per minute, rpm). Particles experiencing a greater centrifugal force have faster sedimentation rates and are preferentially pulled toward the bottom of the

Table 7.5 Conditions for the Separation of Selected Cellular Components by Centrifugation

Components	Centrifugal Force ($\times g$)	Time (min)
eukaryotic cell	1000	5
cell membranes, nuclei	4000	10
mitochondria, bacterial cells	15,000	20
lysosomes, bacterial membranes	30,000	30
ribosomes	100,000	180

Source: Adapted from Zubay G. *Biochemistry,* 2nd ed. Macmillan: New York, 1988, p. 120.

centrifuge tube. For particles of equal density the separation is based on mass, with heavier particles having greater sedimentation rates. When the particles are of equal mass, those with the highest density have the greatest sedimentation rate.

Centrifugation is of particular importance as a separation technique in biochemistry. As shown in Table 7.5, cellular components can be separated by centrifugation.[12] For example, lysosomes can be separated from other cellular components by repeated differential centrifugation, in which the sample is divided into a solid residue and a solution called the supernatant. After destroying the cell membranes, the solution is centrifuged at 15,000 $\times g$ (a centrifugal field strength that is 15,000 times that of the Earth's gravitational field) for 20 min, leaving a residue of cell membranes and mitochondria. The supernatant is isolated by decanting from the residue and is centrifuged at 30,000 $\times g$ for 30 min, leaving a residue of lysosomes.

An alternative approach to differential centrifugation is equilibrium–density–gradient centrifugation. The sample is either placed in a solution with a preformed density gradient or in a solution that, when centrifuged, forms a density gradient. For example, density gradients can be established with solutions of sucrose or CsCl. During centrifugation, the sample's components undergo sedimentation at a rate determined by their centrifugal force. Because the solution's density increases toward the bottom of the centrifuge tube, the sedimentation rate for each component decreases as it moves toward the bottom of the centrifuge tube. When a component reaches a position where its density is equal to that of the solution, the centrifugal force drops to zero and sedimentation stops. Each component, therefore, is isolated as a separate band positioned where the density of the component is equal to the density of the solution. For example, a mixture of proteins, RNA, and DNA can be separated in this way since their densities are different. A density gradient from 1.65 g/cm^3 to 1.80 g/cm^3 is established using CsCl. Proteins, with a density of less than 1.3 g/cm^3 experience no sedimentation, whereas RNA, with a density of greater than 1.8 g/cm^3 collects as a residue at the bottom of the centrifuge tube. The DNA, which has a density of approximately 1.7 g/cm^3 separates as a band near the middle of the centrifuge tube (Figure 7.12).

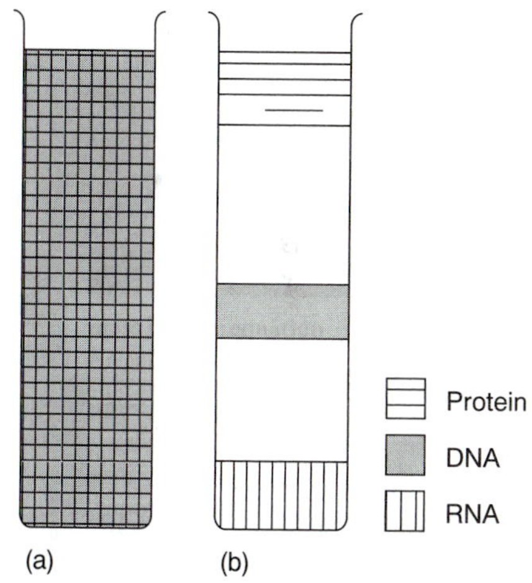

Figure 7.12
Illustration showing separation by equilibrium–density–gradient centrifugation. The homogeneous mixture in (a) separates into three bands (b) after applying centrifugal force.

7F.3 Separations Based on Complexation Reactions (Masking)

One of the most widely used techniques for preventing an interference is to bind the interferent as a soluble complex, preventing it from interfering in the analyte's determination. This process is known as **masking.** Technically, masking is not a separation

masking
A pseudo-separation method in which a species is prevented from participating in a chemical reaction by binding it with a masking agent in an unreactive complex.

Table 7.6 Selected Masking Agents

Masking Agent	Species Which Can Be Masked
CN^-	Ag, Au, Cd, Co, Cu, Fe, Hg, Mn, Ni, Pd, Pt, Zn
SCN^-	Ag, Cd, Co, Cu, Fe, Ni, Pd, Pt, Zn
NH_3	Ag, Co, Cu, Fe, Pd, Pt
F^-	Al, Co, Cr, Mg, Mn, Sn, Zn
$S_2O_3^{2-}$	Au, Cd, Co, Cu, Fe, Pb, Pd, Pt, Sb
tartrate	Al, Ba, Bi, Ca, Ce, Co, Cr, Cu, Fe, Hg, Mn, Pb, Pd, Pt, Sb, Sn, Zn
oxalate	Al, Fe, Mg, Mn, Sn
thioglycolic acid	Cu, Fe, Sn

Source: Adapted from Meites, L. *Handbook of Analytical Chemistry,* McGraw-Hill: New York, 1963.

technique because the analyte and interferent are never physically separated from each other. Masking can, however, be considered a pseudo-separation technique, and is included here for that reason. A wide variety of ions and molecules have been used as **masking agents** (Table 7.6), and, as a result, selectivity is usually not a problem.[13]

masking agent
The reagent used to bind the species to be masked in an unreactive complex.

EXAMPLE 7.12

Suggest a masking agent for the analysis of Al in the presence of Fe. Repeat for the analysis of Fe when Al is an interferent.

SOLUTION

To find a suitable masking agent, we look for a species that binds with the interferent but does not bind with the analyte. Oxalate, for example, is an inappropriate choice because it binds with both Al and Fe. From Table 7.6 we find that thioglycolic acid is a selective masking agent for Fe in the presence of Al and that F^- is a selective masking agent for Al in the presence of Fe.

As shown in Example 7.13, a masking agent's effectiveness can be judged by considering the equilibrium constants for the analytical and masking reactions.

EXAMPLE 7.13

Show that CN^- is an appropriate masking agent for Ni^{2+} in a method in which nickel's complexation with EDTA is an interference.

SOLUTION

The relevant reactions and equilibrium constants from Appendix 3C are

$$Ni^{2+} + Y^{4-} \rightarrow NiY^{2-} \qquad K_f = 4.2 \times 10^{18}$$

$$Ni^{2+} + 4CN^- \rightarrow Ni(CN)_4^{2-} \qquad \beta_4 = 1.7 \times 10^{30}$$

in which Y^{4-} is an abbreviation for ethylenediaminetetraacetic acid (EDTA). Cyanide is an appropriate masking agent because the formation constant for

$Ni(CN)_4^{2-}$ is greater than that for the Ni–EDTA complex. In fact, the equilibrium constant for the reaction in which EDTA displaces the masking agent

$$Ni(CN)_4^{2-} + Y^{4-} \rightleftharpoons NiY^{2-} + 4CN^-$$

$$K = \frac{K_f}{\beta_4} = \frac{4.2 \times 10^{18}}{1.7 \times 10^{30}} = 2.5 \times 10^{-12}$$

is very small, indicating that $Ni(CN)_4^{2-}$ is relatively inert in the presence of EDTA.

7F.4 Separations Based on a Change of State

Since an analyte and interferent are usually in the same phase, a separation often can be effected by inducing a change in one of their physical or chemical states. Changes in physical state that have been exploited for the purpose of a separation include liquid-to-gas and solid-to-gas phase transitions. Changes in chemical state involve one or more chemical reactions.

Changes in Physical State When the analyte and interferent are miscible liquids, a separation based on distillation may be possible if their boiling points are significantly different. The progress of a distillation is outlined in Figure 7.13, which shows a plot of temperature versus the vapor-phase and liquid-phase composition of a mixture consisting of a low-boiling analyte and a high-boiling interferent. The initial mixture is indicated by the point labeled A. When this solution is brought to its boiling point, a vapor phase with the composition indicated by the point labeled B is in equilibrium with the original liquid phase. This equilibrium is indicated by the horizontal tie-line between points A and B. When the vapor phase at point B condenses, a new liquid phase with the same composition as the vapor phase (point C) results. The liquid phase at point C boils at a lower temperature, with an equilibrium established with the vapor-phase composition indicated by point D. This process of repeated vaporization and condensation gradually separates the analyte and interferent.

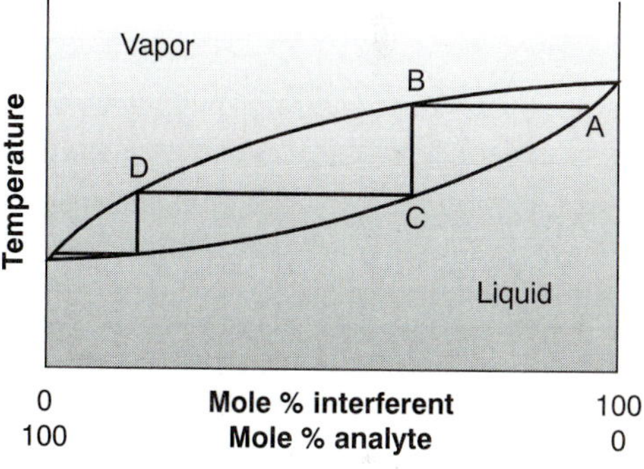

Figure 7.13
Boiling points versus composition diagram for a near-ideal solution, showing the progress of a distillation.

Two examples of the equipment used for distillations are shown in Figure 7.14. The simple distillation apparatus shown in Figure 7.14a does not produce a very efficient separation and is useful only for separating a volatile liquid from nonvolatile liquids or for separating liquids with boiling points that differ by more than 150 °C. A more efficient separation is achieved by a fractional distillation (Figure 7.14b). Packing the distillation column with a high-surface-area material, such as a steel sponge or glass beads, provides more opportunity for the repeated process of vaporization and condensation necessary to effect a complete separation.

When the sample is a solid, a separation of the analyte and interferent by sublimation may be possible. The sample is heated at a temperature and pressure below its triple point where the solid vaporizes without passing through the liquid state. The vapor is then condensed to recover the purified solid. A good example of the use of sublimation is in the isolation of amino acids from fossil mollusk shells and deep-sea sediments.[14]

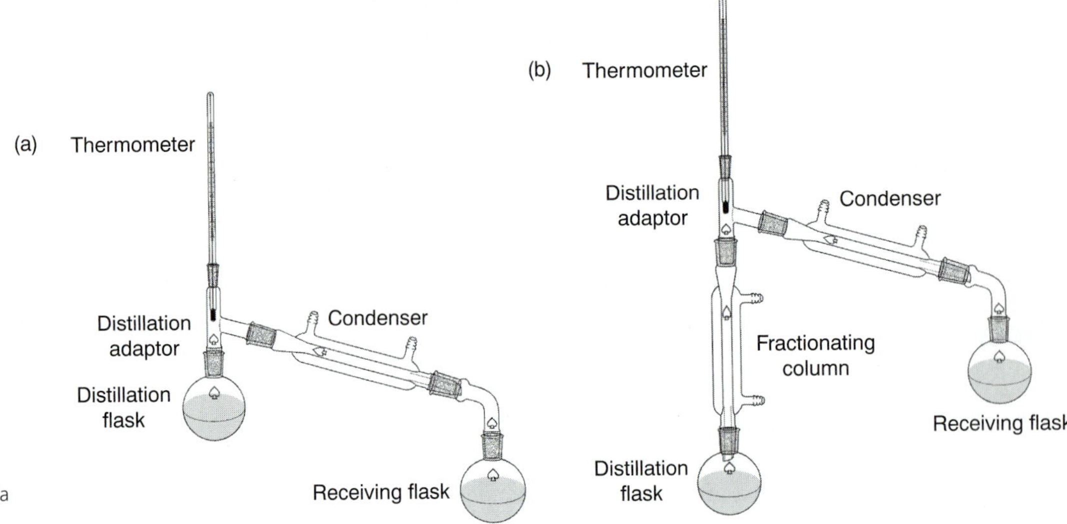

Figure 7.14
Typical equipment for a (a) simple distillation; and a (b) fractional distillation.

Another approach for purifying solids is recrystallization. The solid is dissolved in a minimum volume of solvent, for which the analyte's solubility is significant when the solvent is hot, and minimal when the solvent is cold. The interferents must be less soluble in the hot solvent than the analyte, or present in much smaller amounts. A portion of the solvent is heated in an Erlenmeyer flask, and small amounts of sample are added until undissolved sample is visible. Additional heated solvent is added until the sample is again dissolved or until only insoluble impurities remain. The process of adding sample and solvent is repeated until the entire sample has been added to the Erlenmeyer flask. If necessary, insoluble impurities are removed by filtering the heated solution. The solution is allowed to cool slowly, promoting the growth of large, pure crystals, and then cooled in an ice bath to minimize solubility losses. The purified sample is isolated by filtration and rinsed to remove soluble impurities. Finally, the sample is dried to remove any remaining traces of the solvent. Further purification, if necessary, can be accomplished by additional recrystallizations.

Changes in Chemical State Distillation, sublimation, and recrystallization use a change in physical state as a means of separation. Chemical reactivity also can be used in a separation by effecting a change in the chemical state of the analyte or interferent. For example, SiO_2 can be separated from a sample by reacting with HF. The volatile SiF_4 that forms is easily removed by evaporation. In other cases distillation may be used to remove a nonvolatile inorganic ion after chemically converting it to a more volatile form. For example, NH_4^+ can be separated from a sample by making the solution basic, resulting in the formation of NH_3. The ammonia that is produced can then be removed by distillation. Other examples are listed in Table 7.7.

Table 7.7 Selected Examples of the Application of Distillation to the Separation of Inorganic Ions

Analyte or Interferent	Treatment[a]
CO_3^{2-}	$CO_3^{2-} + 2H_3O^+ \rightarrow \underline{CO_2} + 3H_2O$
NH_4^+	$NH_4^+ + OH^- \rightarrow \underline{NH_3} + H_2O$
SO_3^{2-}	$SO_3^{2-} + 2H_3O^+ \rightarrow \underline{SO_2} + 3H_2O$
S^{2-}	$S^{2-} + 2H_3O^+ \rightarrow \underline{H_2S} + 2H_2O$

[a]Underlined species is removed by distillation.

Other types of reactions can be used to chemically separate an analyte and interferent, including precipitation, electrodeposition, and ion exchange. Two important examples of the application of precipitation are the pH-dependent solubility of metal oxides and hydroxides, and the solubility of metal sulfides.

Separations based on the pH-dependent solubility of oxides and hydroxides are usually accomplished using strong acids, strong bases, or NH_3/NH_4Cl buffers. Most metal oxides and hydroxides are soluble in hot concentrated HNO_3, although a few oxides, such as WO_3, SiO_2, and SnO_2 remain insoluble even under these harsh conditions. In determining the amount of Cu in brass, for example, an interference from Sn is avoided by dissolving the sample with a strong acid. An insoluble residue of SnO_2 remains that can then be removed by filtration.

Most metals will precipitate as the hydroxide in the presence of concentrated NaOH. Metals forming amphoteric hydroxides, however, remain soluble in concentrated NaOH due to the formation of higher-order hydroxo-complexes. For example, Zn^{2+} and Al^{3+} will not precipitate in concentrated NaOH due to the formation of $Zn(OH)_3^-$ and $Al(OH)_4^-$. The solubility of Al^{3+} in concentrated NaOH is used to isolate aluminum from impure bauxite, an ore of Al_2O_3. The ore is powdered and placed in a solution of concentrated NaOH where the Al_2O_3 dissolves to form $Al(OH)_4^-$. Other oxides that may be present in the ore, such as Fe_2O_3 and SiO_2, remain insoluble. After filtering, the filtrate is acidified to recover the aluminum as a precipitate of $Al(OH)_3$.

The pH of an NH_3/NH_4Cl buffer ($pK_a = 9.24$) is sufficient to ensure the precipitation of most metals as the hydroxide. The alkaline earths and alkaline metals, however, will not precipitate at this pH. In addition, metal ions that form soluble complexes with NH_3, such as Cu^{2+}, Zn^{2+}, Ni^{2+}, and Co^{2+}, also will not precipitate under these conditions.

Historically, the use of S^{2-} as a precipitating reagent is one of the earliest examples of a separation technique. In Fresenius's 1881 text, *A System of Instruction in Quantitative Chemical Analysis,*[15] sulfide is frequently used as a means for separating metal ions from the remainder of the sample matrix. The importance of sulfide as a precipitating reagent for separations is due to two factors: most metal ions, except for the alkaline earths and alkaline metals, form insoluble sulfides; and the solubilities of these metal sulfides show a substantial variation. Since the concentration of S^{2-} is pH-dependent, control of pH was used to determine which metal ions would precipitate. For example, in Fresenius's gravimetric procedure for the determination of Ni in ore samples (see Figure 1.1 in Chapter 1 for a schematic diagram of this procedure), sulfide is used three times as a means of separating Co^{2+} and Ni^{2+} from Cu^{2+} and, to a lesser extent from Pb^{2+}.

7F.5 Separations Based on a Partitioning Between Phases

The most important class of separation techniques is based on the selective partitioning of the analyte or interferent between two immiscible phases. When a phase containing a solute, *S*, is brought into contact with a second phase, the solute partitions itself between the two phases.

$$S_{\text{phase 1}} \rightleftharpoons S_{\text{phase 2}} \qquad \textbf{7.18}$$

The equilibrium constant for reaction 7.18

$$K_D = \frac{[S_{\text{phase 2}}]}{[S_{\text{phase 1}}]}$$

is called the distribution constant, or **partition coefficient.** If K_D is sufficiently large, then the solute will move from phase 1 to phase 2. The solute will remain in phase 1,

partition coefficient
An equilibrium constant describing the distribution of a solute between two phases; only one form of the solute is used in defining the partition coefficient (K_D).

extraction
The process by which a solute is transferred from one phase to a new phase.

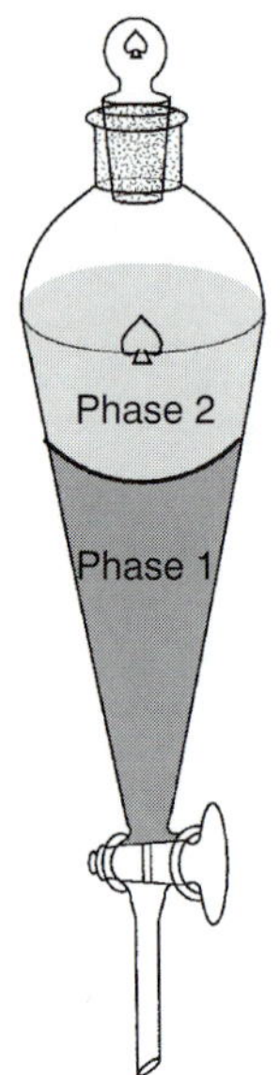

Figure 7.15
Separatory funnel for use in a liquid–liquid extraction.

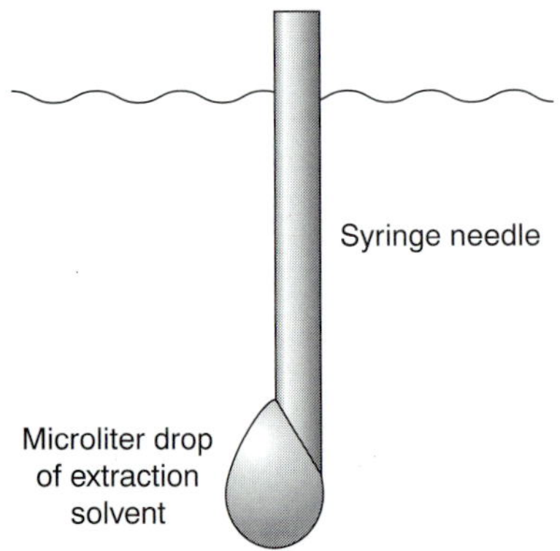

Figure 7.16
Schematic of a liquid–liquid microextraction showing syringe needle with attached 1-μL droplet.

however, if the partition coefficient is sufficiently small. If a phase containing two solutes is brought into contact with a second phase, and K_D is favorable for only one of the solutes, then a separation of the solutes may be possible. The physical states of the two phases are identified when describing the separation process, with the phase containing the sample listed first. For example, when the sample is in a liquid phase and the second phase is a solid, the separation involves liquid–solid partitioning.

Extraction Between Two Phases When the sample is initially present in one of the phases, the separation is known as an **extraction.** In a simple extraction the sample is extracted one or more times with portions of the second phase. Simple extractions are particularly useful for separations in which only one component has a favorable distribution ratio. Several important separation techniques are based on simple extractions, including liquid–liquid, liquid–solid, solid–liquid, and gas–solid extractions.

Liquid–Liquid Extractions Liquid–liquid extractions are usually accomplished with a separatory funnel (Figure 7.15). The two liquids are placed in the separatory funnel and shaken to increase the surface area between the phases. When the extraction is complete, the liquids are allowed to separate, with the denser phase settling to the bottom of the separatory funnel. Liquid–liquid extractions also may be carried out in the sample container by adding the extracting solvent when the sample is collected. Pesticides in water, for example, may be preserved for longer periods by extracting into a small volume of hexane added to the sample in the field. Liquid–liquid microextractions, in which the extracting phase is a 1-μL drop suspended from a microsyringe (Figure 7.16) also have been described.[16] Because of its importance, a more thorough discussion of liquid–liquid extraction is given in Section 7G.

Solid-Phase Extractions In a solid-phase extraction the sample is passed through a cartridge containing solid particulates that serve as the adsorbent material. For liquid samples the solid adsorbent is isolated in either a disk cartridge or a column (Figure 7.17). The choice of adsorbent is determined by the properties of the species being retained and the matrix in which it is found. Representative solid adsorbents

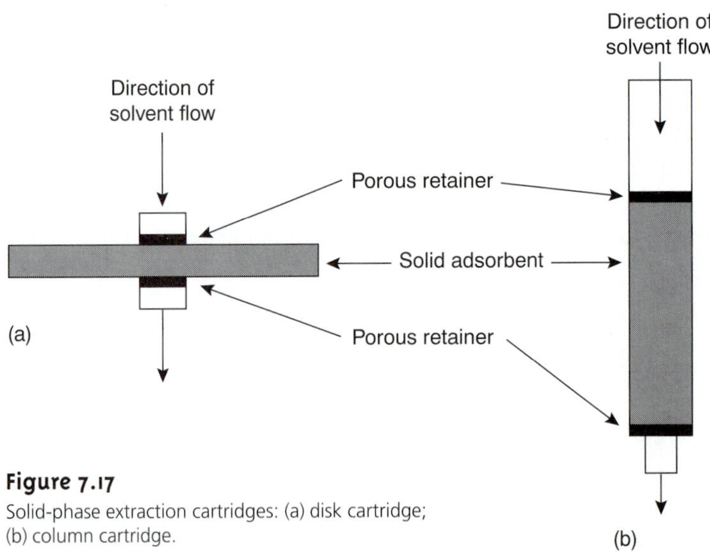

Figure 7.17
Solid-phase extraction cartridges: (a) disk cartridge; (b) column cartridge.

are listed in Table 7.8. For example, sedatives, such as secobarbital and phenobarbital, can be isolated from serum by a solid-phase extraction using a C-18 solid adsorbent. Typically a 500-μL sample of serum is passed through the cartridge, with the sedatives being retained by a liquid–solid extraction. The cartridge is then washed with distilled water to remove any residual traces of the serum's matrix. Finally, the retained sedatives are eluted from the cartridge by a solid–liquid extraction using 500 μL of acetone. For many analyses, solid–phase extractions are replacing liquid–liquid extractions due to their ease of use, faster extraction times, decreased volumes of solvent, and their superior ability to concentrate the analytes. The last advantage is discussed in more detail in the final section of this chapter.

Solid-phase microextractions also have been developed. In one approach, a fused silica fiber is placed inside a syringe needle. The fiber, which is coated with a thin organic film, such as poly(dimethyl siloxane), is lowered into the sample by depressing a plunger and exposed to the sample for a predetermined time. The fiber is then withdrawn into the needle and transferred to a gas chromatograph for analysis.[17]

In gas–solid extractions the sample is passed through a container packed with a solid adsorbent. One example of the application of gas–solid extraction is in the analysis of organic compounds for carbon and hydrogen. The sample is combusted in a flowing stream of O_2, and the gaseous combustion products are passed through a series of solid-phase adsorbents that remove the CO_2 and H_2O.

Continuous Extractions An extraction is still feasible even when the component of interest has an unfavorable partition coefficient, provided that all other components in the sample have significantly smaller partition coefficients. Because the partition

Table 7.8 Selected Adsorbents for Solid-Phase Extraction of Liquid Samples

Adsorbent	Surface Structure	Properties and Uses
silica	—O—Si(OH)—O—Si(OH)—O—	• retains low-to-moderate polarity species from organic matrices • fat-soluble vitamins, steroids
alumina	—O—Al(OH)—O—Al(OH)—O—	• retains hydrophilic species from organic matrices
cyanopropyl	$—C_3H_6CN$	• retains wide variety of species from aqueous and organic matrices • pesticides, hydrophobic peptides
diol	—CH(OH)—CH(OH)—	• retains wide variety of species from aqueous and organic matrices • proteins, peptides, fungicides
octadecyl (C-18)	$—C_{18}H_{37}$	• retains hydrophobic species from aqueous matrices • caffeine, sedatives, polyaromatic hydrocarbons, carbohydrates, pesticides
octyl (C-8)	$—C_8H_{17}$	• similar to C-18
styrene divinylbenzene		• wide variety of organic species from aqueous matrices • polyaromatic hydrocarbons

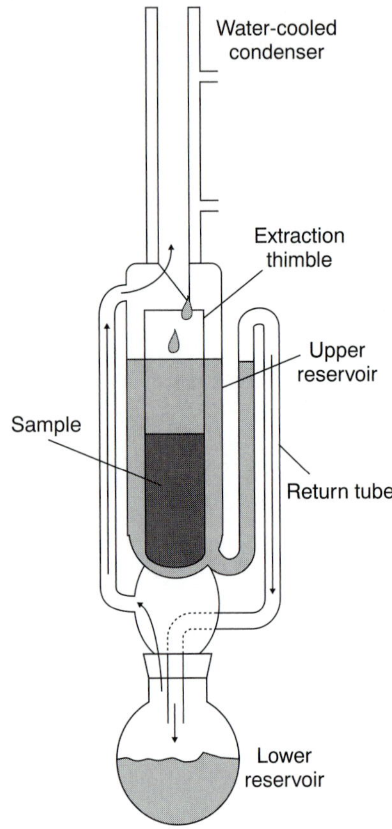

Figure 7.18
Schematic diagram of a Soxhlet extractor.

coefficient is unfavorable, a simple extraction will not be quantitative. Instead, the extraction is accomplished by continuously passing the extracting phase through the sample until a quantitative extraction is achieved.

Many continuous extractions involving solid samples are carried out with a Soxhlet extractor (Figure 7.18). The extracting solvent is placed in the lower reservoir and heated to its boiling point. Solvent in the vapor phase moves upward through the tube on the left side of the apparatus to the condenser where it condenses back to the liquid state. The solvent then passes through the sample, which is held in a porous cellulose filter thimble, collecting in the upper reservoir. When the volume of solvent in the upper reservoir reaches the upper bend of the return tube, the solvent and any extracted components are siphoned back to the lower reservoir. Over time, the concentration of the extracted component in the lower reservoir increases.

Soxhlet extractions have been replaced in some applications by microwave-assisted extractions.[18] The process is the same as that described earlier for microwave digestion. The sample is placed in a sealed digestion vessel along with the liquid extraction phase, and a microwave oven is used to heat the extraction mixture. Using a sealed digestion vessel allows the extraction to take place at a higher temperature and pressure, thereby reducing the amount of time needed for a quantitative extraction. In a Soxhlet extraction the temperature is limited by the solvent's boiling point at atmospheric pressure. For example, when acetone is the solvent, a Soxhlet extraction is limited to 56 °C. With a microwave-assisted extraction, however, a temperature of over 150 °C can be obtained when using acetone as the solvent.

Two other examples of a continuous extraction deserve mention. Volatile organic compounds (VOCs) can be quantitatively removed from liquid samples by a liquid–gas extraction. As shown in Figure 7.19, the VOCs are removed by passing an inert purging gas, such as He, through the sample. The He removes the VOCs, which are then carried by the He to a tube where they are collected on a solid adsorbent. When the extraction is complete, the VOCs can then be removed from the trap for analysis by rapidly heating the tube while flushing with He. This technique is known as a **purge and trap.** Recoveries for analytes using a purge and trap may not be reproducible, requiring the use of internal standards for quantitative work.

purge and trap
A technique for separating volatile analytes from liquid samples in which the analytes are subsequently trapped on a solid adsorbent.

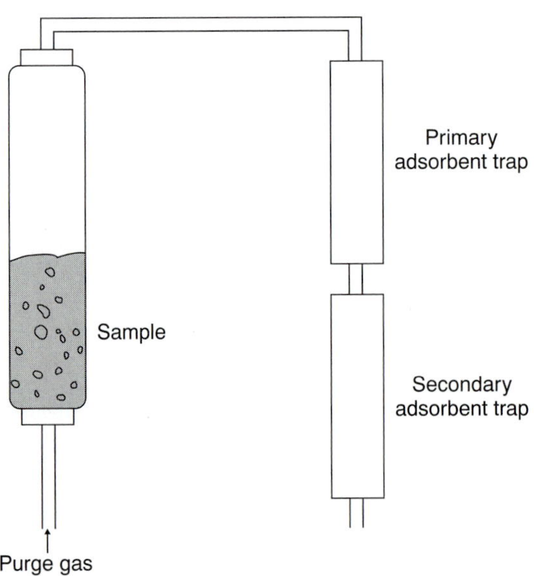

Figure 7.19
Schematic diagram of a purge-and-trap system. Analyte is collected in the primary adsorption trap. The secondary adsorption trap is monitored for evidence of breakthrough.

Continuous extractions also can be accomplished with **supercritical fluids.**[19] When a substance is heated above its critical temperature and pressure, it forms a supercritical fluid whose properties are between those of a gas and a liquid. Supercritical fluids are better solvents than gases, making them a better reagent for extractions. In addition, the viscosity of a supercritical fluid is significantly less than that of a liquid solvent, allowing it to pass more readily through particulate samples. One example of a supercritical extraction is the determination of total petroleum hydrocarbons (TPHs) in soils, sediments, and sludges with supercritical CO_2. Approximately 3 g of sample is placed in a 10-mL stainless steel cartridge, and supercritical CO_2, at a pressure of 340 atm and a temperature of 80 °C, is passed through the cartridge for 30 min at flow rate of 1–2 mL/min. The petroleum hydrocarbons are collected by passing the effluent from the cartridge through 3 mL of tetrachloroethylene at room temperature. At this temperature the CO_2 reverts to the gas phase and is released to the atmosphere.[20]

supercritical fluid
A state of matter where a substance is held at a temperature and pressure that exceeds its critical temperature and pressure.

Chromatographic Separations In an extraction, the sample is initially present in one phase, and the component of interest is extracted into a second phase. Separations can also be accomplished by continuously passing one sample-free phase, called the mobile phase, over a second sample-free phase that remains fixed or stationary. The sample is then injected or placed into the mobile phase. As the sample's components move with the mobile phase, they partition themselves between the mobile and stationary phases. Those components having the largest partition coefficients are more likely to move into the stationary phase, taking longer to pass through the system. This is the basis of all chromatographic separation techniques. As currently practiced, modern chromatography provides a means both of separating analytes and interferents and of performing a qualitative or quantitative analysis of the analyte. For this reason a more thorough treatment of chromatography is found in Chapter 12.

7G Liquid–Liquid Extractions

A liquid–liquid extraction is one of the most important separation techniques used in environmental, clinical, and industrial laboratories. Two examples from environmental analysis serve to illustrate its importance. Public drinking water supplies are routinely monitored for trihalomethanes ($CHCl_3$, $CHBrCl_2$, $CHBr_2Cl$, and $CHBr_3$) because of their known or suspected carcinogeneity. Before their analysis by gas chromatography, trihalomethanes are separated from their aqueous matrix by a liquid–liquid extraction using pentane.[21] A liquid–liquid extraction is also used in screening orange juice for the presence of organophosphorous pesticides. A sample of orange juice is mixed with acetonitrite and filtered. Any organophosphorous pesticides that might be present in the filtrate are extracted with petroleum ether before a gas chromatographic analysis.[22]

In a simple liquid–liquid extraction the solute is partitioned between two immiscible phases. In most cases one of the phases is aqueous, and the other phase is an organic solvent such as diethyl ether or chloroform. Because the phases are immiscible, they form two layers, with the denser phase on the bottom. The solute is initially present in one phase, but after extraction it is present in both phases. The efficiency of a liquid–liquid extraction is determined by the equilibrium constant for the solute's partitioning between the two phases. Extraction efficiency is also influenced by any secondary reactions involving the solute. Examples of secondary reactions include acid–base and complexation equilibria.

7G.1 Partition Coefficients and Distribution Ratios

Earlier we learned that the partitioning of a solute between two phases is described by a partition coefficient. If the solute is initially in an aqueous phase and is extracted into an organic phase*

$$S_{aq} \rightleftharpoons S_{org}$$

the partition coefficient is

$$K_D = \frac{[S_{org}]}{[S_{aq}]}$$

A large value for K_D indicates that the extraction of the solute into the organic phase is favorable.

distribution ratio
A ratio expressing the total concentration of solute in one phase relative to a second phase; all forms of the solute are considered in defining the distribution ratio (*D*).

In evaluating the efficiency of an extraction, however, we must consider the solute's total concentration in each phase. We define the **distribution ratio,** *D,* to be the ratio of the solute's total concentration in each phase.

$$D = \frac{[S_{org}]_{tot}}{[S_{aq}]_{tot}}$$

When the solute exists in only one form in each phase, then the partition coefficient and the distribution ratio are identical. If, however, the solute exists in more than one form in either phase, then K_D and D usually have different values. For example, if the solute exists in two forms in the aqueous phase, A and B, only one of which, A, partitions itself between the two phases, then

$$D = \frac{[S_{org}]_A}{[S_{aq}]_A + [S_{aq}]_B} \leq K_D = \frac{[S_{org}]_A}{[S_{aq}]_A}$$

This distinction between K_D and D is important. The partition coefficient is an equilibrium constant and has a fixed value for the solute's partitioning between the two phases. The value of the distribution ratio, however, changes with solution conditions if the relative amounts of forms A and B change. If we know the equilibrium reactions taking place within each phase and between the phases, we can derive an algebraic relationship between K_D and D.

7G.2 Liquid–Liquid Extraction with No Secondary Reactions

In the simplest form of liquid–liquid extraction, the only reaction affecting extraction efficiency, is the partitioning of the solute between the two phases (Figure 7.20). In this case the distribution ratio and the partition coefficient are equal.

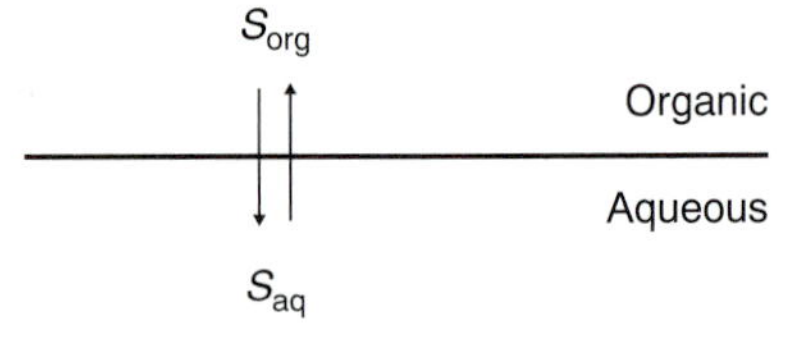

Figure 7.20
Scheme for a simple liquid–liquid extraction without any secondary reactions.

$$D = \frac{[S_{org}]_{tot}}{[S_{aq}]_{tot}} = \frac{[S_{org}]}{[S_{aq}]} \qquad 7.19$$

Conservation of mass requires that the moles of solute initially present in one phase equal the combined moles of solute in the aqueous and organic phases after the extraction; thus

$$(\text{Moles aq})_0 = (\text{moles aq})_1 + (\text{moles org})_1 \qquad 7.20$$

*Although the following treatment assumes that the solute is initially present in the aqueous phase, the resulting equations for the distribution of the solute between the two phases are independent of which phase originally contains the solute.

where the subscript indicates the extraction number. The concentration of S in the aqueous phase after the extraction is

$$[S_{aq}] = \frac{(\text{moles aq})_1}{V_{aq}} \tag{7.21}$$

whereas the solute's concentration in the organic phase is

$$[S_{org}] = \frac{(\text{moles org})_1}{V_{org}} \tag{7.22}$$

where V_{aq} and V_{org} are the volumes of the aqueous and organic phases. Solving equation 7.20 for $(\text{moles org})_1$ and substituting into equation 7.22 leave us with

$$[S_{org}] = \frac{(\text{moles aq})_0 - (\text{moles aq})_1}{V_{org}} \tag{7.23}$$

Substituting equations 7.21 and 7.23 into equation 7.19, we obtain

$$D = \frac{[(\text{moles aq})_0 - (\text{moles aq})_1]/V_{org}}{(\text{moles aq})_1/V_{aq}} = \frac{(\text{moles aq})_0 V_{aq} - (\text{moles aq})_1 V_{aq}}{(\text{moles aq})_1 V_{org}}$$

Rearranging and solving for the fraction of solute remaining in the aqueous phase after one extraction, $(q_{aq})_1$, gives

$$(q_{aq})_1 = \frac{(\text{moles aq})_1}{(\text{moles aq})_0} = \frac{V_{aq}}{DV_{org} + V_{aq}} \tag{7.24}$$

The fraction present in the organic phase after one extraction, $(q_{org})_1$, is

$$(q_{org})_1 = \frac{(\text{moles org})_1}{(\text{moles org})_0} = 1 - (q_{aq})_1 = \frac{DV_{org}}{DV_{org} + V_{aq}}$$

Example 7.14 shows how equation 7.24 is used to calculate the efficiency of a simple liquid–liquid extraction.

Example

EXAMPLE 7.14

A solute, S, has a K_D between water and chloroform of 5.00. A 50.00-mL sample of a 0.050 M aqueous solution of the solute is extracted with 15.00 mL of chloroform. (a) What is the extraction efficiency for this separation? (b) What is the solute's final concentration in each phase? (c) What volume of chloroform is needed to extract 99.9% of the solute?

SOLUTION

For a simple liquid–liquid extraction, the distribution ratio, D, and the partition coefficient, K_D, are identical.

(a) The fraction of solute remaining in the aqueous phase after the extraction is given by equation 7.24

$$(q_{aq})_1 = \frac{50.00 \text{ mL}}{(5.00)(15.00 \text{ mL}) + 50.00 \text{ mL}} = 0.400$$

The fraction of solute present in the organic phase is, therefore, 0.600. Extraction efficiency is the percentage of solute successfully transferred from its initial phase to the extracting phase. The extraction efficiency is, therefore, 60.0%.

(b) The moles of solute present in the aqueous phase before the extraction is

$$(\text{Moles aq})_0 = [S_{aq}]_0 \times V_{aq} = \frac{0.050 \text{ mol}}{\text{L}} \times 0.05000 \text{ L} = 0.0025 \text{ mol}$$

Since 40.0% of the solute remains in the aqueous phase, and 60.0% has been extracted into the organic phase, the moles of solute in the two phases after extraction are

$$(\text{Moles aq})_1 = (\text{moles aq})_0 \times (q_{aq})_1 = 0.0025 \text{ mol} \times (0.400) = 0.0010 \text{ mol}$$

$$(\text{Moles org})_1 = (\text{moles aq})_0 - (\text{moles aq})_1 = 0.0025 \text{ mol} - 0.0010 \text{ mol} = 0.0015 \text{ mol}$$

The solute's concentration in each phase is

$$[S_{aq}]_1 = \frac{(\text{moles aq})_1}{V_{aq}} = \frac{0.0010 \text{ mol}}{0.05000 \text{ L}} = 0.020 \text{ M}$$

$$[S_{org}]_1 = \frac{(\text{moles org})_1}{V_{org}} = \frac{0.0015 \text{ mol}}{0.01500 \text{ L}} = 0.10 \text{ M}$$

(c) To extract 99.9% of the solute $(q_{aq})_1$ must be 0.001. Solving equation 7.24 for V_{org}, and making appropriate substitutions for $(q_{aq})_1$ and V_{aq} gives

$$V_{org} = \frac{V_{aq} - (q_{aq})_1 V_{aq}}{(q_{aq})_1 D} = \frac{50.00 \text{ mL} - (0.001)(50.00 \text{ mL})}{(0.001)(5.00)} = 9990 \text{ mL}$$

Clearly, a single extraction is not reasonable under these conditions.

In Example 7.14 a single extraction results in an extraction efficiency of only 60%. If a second extraction is carried out, the fraction of solute remaining in the aqueous phase, $(q_{aq})_2$, is given by

$$(q_{aq})_2 = \frac{(\text{moles aq})_2}{(\text{moles aq})_1} = \frac{V_{aq}}{DV_{org} + V_{aq}}$$

Colorplate 4 shows an example of a liquid–liquid extraction.

If the volumes of the aqueous and organic layers are the same for both extractions, then the cumulative fraction of solute remaining in the aqueous layer after two extractions, $(Q_{aq})_2$, is

$$(Q_{aq})_2 = \frac{(\text{moles aq})_2}{(\text{moles aq})_0} = (q_{aq})_1(q_{aq})_2 = \left(\frac{V_{aq}}{DV_{org} + V_{aq}}\right)^2$$

In general, for a series of n identical extractions, the fraction of analyte remaining in the aqueous phase after the last extraction is

$$(Q_{aq})_n = \left(\frac{V_{aq}}{DV_{org} + V_{aq}}\right)^n \qquad 7.25$$

Example

EXAMPLE 7.15

For the extraction described in Example 7.14, determine (a) the extraction efficiency for two extractions and for three extractions; and (b) the number of extractions required to ensure that 99.9% of the solute is extracted.

SOLUTION

(a) The fraction of solute remaining in the aqueous phase after two and three extractions is

$$(Q_{aq})_2 = \left(\frac{50.00 \text{ mL}}{(5.00)(15.00 \text{ mL}) + 50.00 \text{ mL}}\right)^2 = 0.160$$

$$(Q_{aq})_3 = \left(\frac{50.00 \text{ mL}}{(5.00)(15.00 \text{ mL}) + 50.00 \text{ mL}}\right)^3 = 0.064$$

Thus, the extraction efficiencies are 84.0% with two extractions and 93.6% with three extractions.

(b) To determine the minimum number of extractions for an efficiency of 99.9%, we set $(Q_{aq})_n$ to 0.001 and solve for n in equation 7.25

$$0.001 = \left(\frac{50.00 \text{ mL}}{(5.00)(15.00 \text{ mL}) + 50.00 \text{ mL}}\right)^n = (0.400)^n$$

Taking the log of both sides

$$\log(0.001) = n\log(0.400)$$

and solving for n gives

$$n = 7.54$$

Thus, a minimum of eight extractions is necessary.

An important observation from Examples 7.14 and 7.15 is that an extraction efficiency of 99.9% can be obtained with less solvent when using multiple extractions. Obtaining this extraction efficiency with one extraction requires 9990 mL of the organic solvent. Eight extractions using separate 15-mL portions of the organic solvent, however, requires only 120 mL. Although extraction efficiency increases dramatically with the first few multiple extractions, the effect quickly diminishes as the number of extractions is increased (Figure 7.21). In most cases there is little gain in extraction efficiency after five or six extractions. In Example 7.15 five extractions are needed to reach an extraction efficiency of 99%, and an additional three extractions are required to obtain the extra 0.9% increase in extraction efficiency.

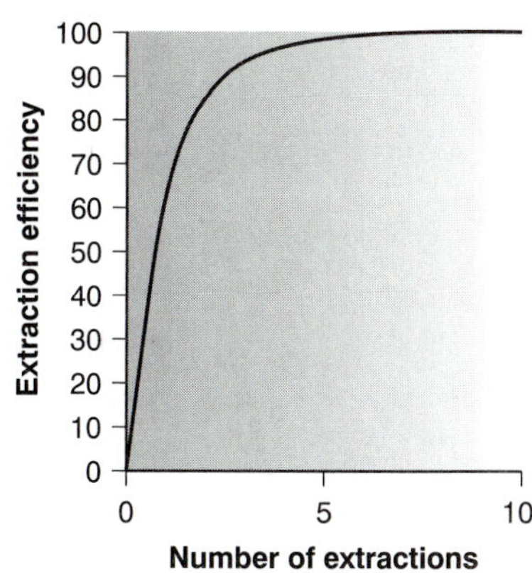

Figure 7.21
Plot of extraction efficiency versus number of extractions for the liquid–liquid extraction scheme in Figure 7.20.

7G.3 Liquid–Liquid Extractions Involving Acid–Base Equilibria

In a simple liquid–liquid extraction the distribution ratio and the partition coefficient are identical. As a result, the distribution ratio is unaffected by any change in the composition of the aqueous or organic phase. If the solute also participates in a single-phase equilibrium reaction, then the distribution ratio and the partition coefficient may not be the same. For example, Figure 7.22 shows the equilibria occurring when extracting an aqueous solution containing a molecular weak acid, HA, with an organic phase in which ionic species are not soluble. In this case the partition coefficient and the distribution ratio are

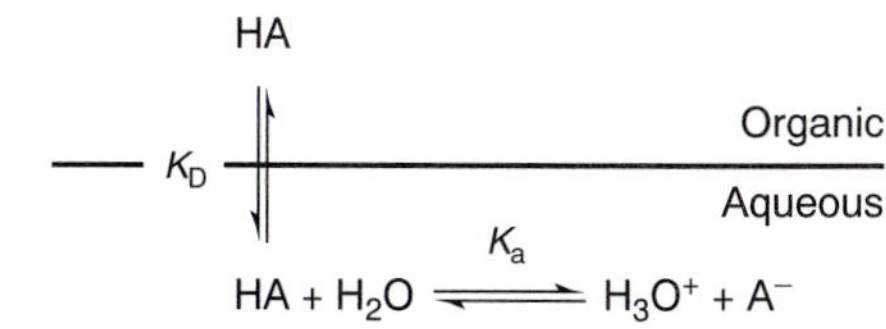

Figure 7.22
Scheme for the liquid–liquid extraction of a molecular weak acid.

$$K_D = \frac{[HA_{org}]}{[HA_{aq}]} \qquad 7.26$$

$$D = \frac{[\text{HA}_{\text{org}}]_{\text{tot}}}{[\text{HA}_{\text{aq}}]_{\text{tot}}} = \frac{[\text{HA}_{\text{org}}]}{[\text{HA}_{\text{aq}}] + [\text{A}^-_{\text{aq}}]} \qquad 7.27$$

Since the position of an acid–base equilibrium depends on the pH, the distribution ratio must also be pH-dependent. To derive an equation for D showing this dependency, we begin with the acid dissociation constant for HA.

$$K_{\text{a}} = \frac{[\text{H}_3\text{O}^+_{\text{aq}}][\text{A}^-_{\text{aq}}]}{[\text{HA}_{\text{aq}}]} \qquad 7.28$$

Solving equation 7.28 for $[\text{A}^-_{\text{aq}}]$

$$[A^-_{\text{aq}}] = \frac{K_{\text{a}}[\text{HA}_{\text{aq}}]}{[\text{H}_3\text{O}^+_{\text{aq}}]}$$

and substituting into equation 7.27 gives

$$D = \frac{[\text{HA}_{\text{org}}]}{[\text{HA}_{\text{aq}}] + (K_{\text{a}}[\text{HA}_{\text{aq}}]/[\text{H}_3\text{O}^+_{\text{aq}}])}$$

Factoring $[\text{HA}_{\text{aq}}]$ from the denominator

$$D = \frac{[\text{HA}_{\text{org}}]}{[\text{HA}_{\text{aq}}]\{1 + (K_{\text{a}}/[\text{H}_3\text{O}^+_{\text{aq}}])\}}$$

and substituting equation 7.26

$$D = \frac{K_{\text{D}}}{1 + (K_{\text{a}}/[\text{H}_3\text{O}^+_{\text{aq}}])}$$

gives, after simplifying, the sought-after relationship between the distribution ratio and the pH of the aqueous solution

$$D = \frac{K_{\text{D}}[\text{H}_3\text{O}^+_{\text{aq}}]}{[\text{H}_3\text{O}^+_{\text{aq}}] + K_{\text{a}}} \qquad 7.29$$

The value for D given by equation 7.29 can be used in equation 7.25 to determine extraction efficiency.

Example

EXAMPLE 7.16

An acidic solute, HA, has an acid dissociation constant of 1.00×10^{-5}, and a partition coefficient between water and benzene of 3.00. Calculate the extraction efficiency when 50.00 mL of a 0.025 M aqueous solution of HA buffered to a pH of 3.00, is extracted with 50.00 mL of benzene. Repeat for cases in which the pH of the aqueous solution is buffered to 5.00 and 7.00.

SOLUTION

When the pH is 3.00, the $[\text{H}_3\text{O}^+_{\text{aq}}]$ is 1.00×10^{-3}, and the distribution ratio for the extraction is

$$D = \frac{(3.00)(1.00 \times 10^{-3})}{1.00 \times 10^{-3} + 1.00 \times 10^{-5}} = 2.97$$

The fraction of solute remaining in the aqueous phase is

$$(Q_{aq})_1 = \frac{50.00 \text{ mL}}{(2.97)(50.00 \text{ mL}) + 50.00 \text{ mL}} = 0.252$$

The extraction efficiency, therefore, is almost 75%. When the same calculation is carried out at a pH of 5.00, the extraction efficiency is 60%, but the extraction efficiency is only 3% at a pH of 7.00. As expected, extraction efficiency is better at more acidic pHs when HA is the predominate species in the aqueous phase. A graph of extraction efficiency versus pH for this system is shown in Figure 7.23. Note that the extraction efficiency is greatest for pHs more acidic than the weak acid's pK_a and decreases substantially at pHs more basic than the pK_A. A ladder diagram for HA is superimposed on the graph to help illustrate this effect.

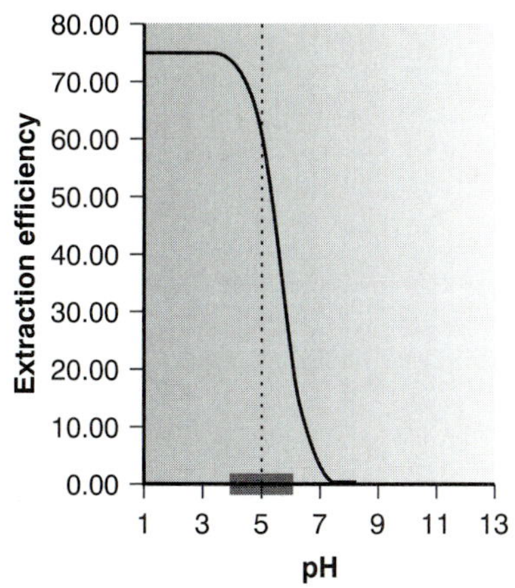

Figure 7.23
Plot of extraction efficiency versus pH of the aqueous phase for the liquid–liquid extraction of the molecular weak acid in Example 7.16.

The same approach can be used to derive an equation for the distribution ratio when the solute is a molecular weak base, B, (Figure 7.24). The resulting distribution ratio is

$$D = \frac{K_D K_a}{K_a + [H_3O^+_{aq}]}$$

where K_a is the acid dissociation constant for the weak base's conjugate weak acid.

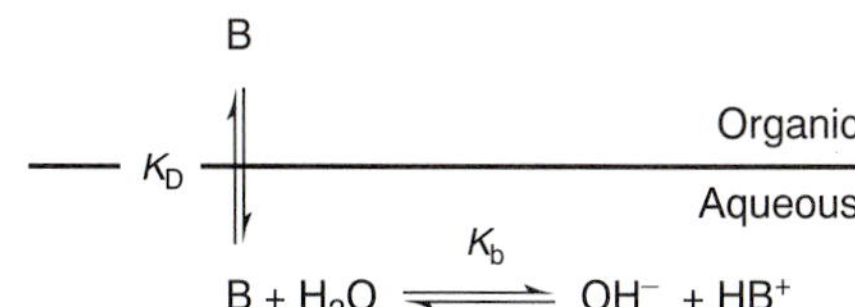

Figure 7.24
Scheme for the liquid–liquid extraction of a molecular weak base.

7G.4 Liquid–Liquid Extractions Involving Metal Chelators

One of the most common applications of a liquid–liquid extraction is the selective extraction of metal ions using a chelating agent. Unfortunately, many chelating agents have a limited solubility in water or are subject to hydrolysis or air oxidation in aqueous solutions. For these reasons the chelating agent is added to the organic solvent instead of the aqueous phase. The chelating agent is extracted into the aqueous phase, where it reacts to form a stable metal–ligand complex with the metal ion. The metal–ligand complex is then extracted into the organic phase. A summary of the relevant equilibria is shown in Figure 7.25.

If the ligand's concentration is much greater than the metal ion's concentration, the distribution ratio is given as*

$$D = \frac{\beta K_{D,c} K_a^n C_L^n}{K_{D,L}^n [H_3O^+_{aq}]^n + \beta K_a^n C_L^n} \qquad 7.30$$

where C_L is the initial concentration of ligand in the organic phase before the extraction. The distribution ratio calculated using equation 7.30 can be substituted back into equation 7.25 to determine the extraction efficiency. As shown in Example 7.17, the extraction efficiency for metal ions shows a marked pH dependency.

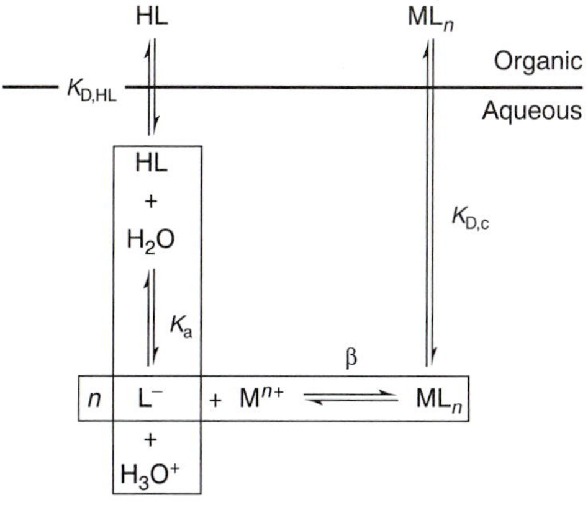

Figure 7.25
Scheme for the liquid–liquid extraction of a metal ion by a metal chelator.

*The derivation of equation 7.31 is considered in problem 33 in the end-of-chapter problem set.

Example

EXAMPLE 7.17

A divalent metal ion, M^{2+}, is to be extracted from an aqueous solution into an organic solvent using a chelating agent, HL, dissolved in the organic solvent. The partition coefficients for the chelating agent, $K_{D,L}$, and the metal–ligand complex, $K_{D,c}$, are 1.0×10^4 and 7.0×10^4, respectively. The acid dissociation constant for the chelating agent, K_a, is 5.0×10^{-5}, and the formation constant for the metal–ligand complex, β, is 2.5×10^{16}. Calculate the extraction efficiency when 100.0 mL of a 1.0×10^{-6} M aqueous solution of M^{2+}, buffered to a pH of 1.00, is extracted with 10.00 mL of an organic solvent that is 0.1 mM in the chelating agent. Repeat the calculation at a pH of 3.00.

SOLUTION

At a pH of 1.00 ($[H_3O^+_{aq}] = 0.10$), the distribution ratio for the extraction is

$$D = \frac{(2.5 \times 10^{16})(7.0 \times 10^4)(5.0 \times 10^{-5})^2(1.0 \times 10^{-4})^2}{(1.0 \times 10^4)^2(0.10)^2 + (2.5 \times 10^{16})(5.0 \times 10^{-5})^2(1.0 \times 10^{-4})^2} = 0.0438$$

and the fraction of metal ion remaining in the aqueous phase is

$$(Q_{aq})_1 = \frac{100.0 \text{ mL}}{(0.0438)(10.00 \text{ mL}) + 100.0 \text{ mL}} = 0.996$$

Thus, at a pH of 1.00, only 0.40% of the metal is extracted. Changing the pH to 3.00, however, gives an extraction efficiency of 97.8%. A plot of extraction efficiency versus the pH of the aqueous phase is shown in Figure 7.26.

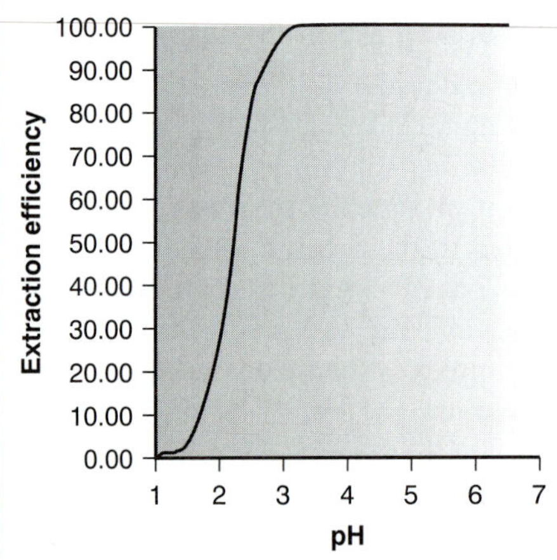

Figure 7.26
Typical plot of extraction efficiency versus pH for the liquid–liquid extraction of a metal ion by metal chelator.

One of the advantages of using a chelating agent is the high degree of selectivity that it brings to the extraction of metal ions. As shown in Example 7.17 and Figure 7.26, the extraction efficiency for a divalent cation increases from approximately 0%–100% over a range of only 2 pH units. Furthermore, a chelating agent's metal–ligand formation constant varies substantially between metal ions. As a result, significant differences arise in the pH range over which different metal ions experience an increase in extraction efficiency from 0% to 100% (Figure 7.27).

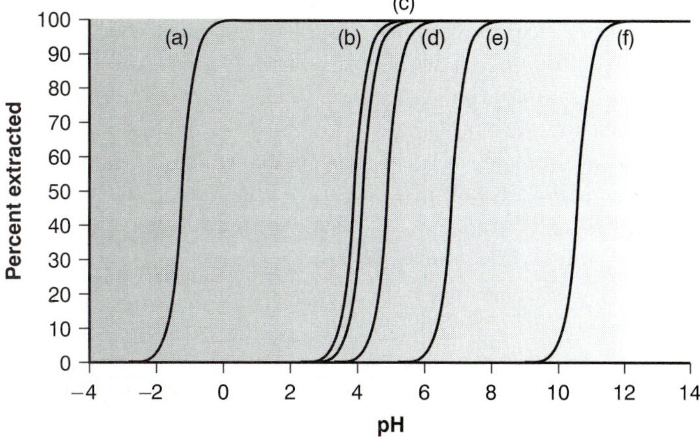

Figure 7.27
Plot of extraction efficiency for selected metals using dithizone in CCl_4. The metal ions are: (a) Cu^{2+}; (b) Co^{2+}; (c) Ni^{2+}; (d) Sn^{2+}; (e) Pb^{2+}; and (f) Cd^{2+}.

Example

EXAMPLE 7.18

Using Figure 7.27, explain how an aqueous mixture of Cu^{2+}, Pb^{2+}, and Cd^{2+} can be separated by extraction with dithizone in CCl_4.

SOLUTION

From Figure 7.27 we see that a quantitative separation of Cu^{2+} from Pb^{2+} and Cd^{2+} can be accomplished if the aqueous phase is buffered to a pH of less than 5.5. After the extraction is complete, the pH can be buffered to approximately 9.5, allowing the selective extraction of Pb^{2+}.

Liquid–liquid extractions using ammonium pyrrolidine dithiocarbamate (APDC) as a metal chelating agent are commonly encountered in the analysis of metal ions in aqueous samples. The sample and APDC are mixed together, and the resulting metal–ligand complexes are extracted into methyl isobutyl ketone before analysis.

7H Separation Versus Preconcentration

Two frequently encountered analytical problems are: (1) the presence of matrix components interfering with the analysis of the analyte; and (2) the presence of analytes at concentrations too small to analyze accurately. We have seen how a separation can be used to solve the former problem. Interestingly, separation techniques can often be used to solve the second problem as well. For separations in which a complete recovery of the analyte is desired, it may be possible to transfer the analyte in a manner that increases its concentration. This step in an analytical procedure is known as a **preconcentration.**

preconcentration
The process of increasing an analyte's concentration before its analysis.

Two examples from the analysis of water samples illustrate how a separation and preconcentration can be accomplished simultaneously. In the gas chromatographic analysis for organophosphorous pesticides in environmental waters, the analytes in a 1000-mL sample may be separated from their aqueous matrix by a solid-phase extraction using 15 mL of ethyl acetate.[23] After the extraction, the analytes are present in the ethyl acetate at a concentration that is 67 times greater than that in the original sample (if the extraction is 100% efficient). The preconcentration of metal ions is accomplished by a liquid–liquid extraction with a metal chelator. For example, before their analysis by atomic absorption spectrophotometry, metal ions in aqueous samples can be concentrated by extraction into methyl isobutyl ketone (MIBK) using ammonium pyrrolidine dithiocarbamate (APDC) as a chelating agent. Typically, a 100-mL sample is treated with 1 mL of APDC, and extracted with ten mL of MIBK. The result is a ten-fold increase in the concentration of the metal ions. This procedure can be adjusted to increase the concentrations of the metal ions by as much as a factor of 40.

7I KEY TERMS

breakthrough volume (*p. 196*)
composite sample (*p. 186*)
coning and quartering (*p. 199*)
convenience sampling (*p. 185*)
dialysis (*p. 206*)
distribution ratio (*p. 216*)
extraction (*p. 212*)
grab sample (*p. 185*)
gross sample (*p. 193*)
in situ sampling (*p. 186*)
judgmental sampling (*p. 184*)
laboratory sample (*p. 199*)
masking (*p. 207*)
masking agent (*p. 208*)
Nyquist theorem (*p. 184*)
partition coefficient (*p. 211*)
preconcentration (*p. 223*)
purge and trap (*p. 214*)
random sample (*p. 183*)
recovery (*p. 202*)
sampling plan (*p. 182*)
separation factor (*p. 203*)
size-exclusion chromatography (*p. 206*)
stratified sampling (*p. 185*)
systematic–judgmental sampling (*p. 184*)
systematic sampling (*p. 184*)
supercritical fluid (*p. 215*)

7J SUMMARY

An analysis requires a sample, and how we acquire the sample is critical. To be useful, the samples we collect must accurately represent their target population. Just as important, our sampling plan must provide a sufficient number of samples of appropriate size so that the variance due to sampling does not limit the precision of our analysis.

A complete sampling plan requires several considerations, including the type of sampling (random, judgmental, systematic, systematic–judgmental, stratified, or convenience); whether to collect grab samples, composite samples, or in situ samples; whether the population is homogeneous or heterogeneous; the appropriate size for each sample; and, the number of samples to collect.

Removing a sample from its population may induce a change in its composition due to a chemical or physical process. For this reason, samples are collected in inert containers and are often preserved at the time of collection.

When the analytical method's selectivity is insufficient, it may be necessary to separate the analyte from potential interferents. Such separations can take advantage of physical properties, such as size, mass or density, or chemical properties. Important examples of chemical separations include masking, distillation, and extractions.

In a solid-phase extraction the analytes are first extracted from their solution matrix into a solid adsorbent. After washing to remove impurities, the analytes are removed from the adsorbent with a suitable solvent. Alternatively, the extraction can be carried out using a Soxhlet extractor.

In a liquid–liquid extraction, the analyte (or interferent) is extracted from one liquid phase into a second, immiscible liquid phase. When the analyte is involved in secondary equilibrium reactions, it is often possible to improve selectivity by carefully adjusting the composition of one or both phases.

7K *Suggested* EXPERIMENTS

Experiments

The following set of experiments introduce students to the important effect of sampling on the quality of analytical results. Each experiment is annotated with a brief description of the principles that it emphasizes.

Bauer, C. F. "Sampling Error Lecture Demonstration," *J. Chem. Educ.* **1985,** *62,* 253.

This short paper describes a demonstration suitable for use in the classroom. Two populations of corks are sampled to determine the concentration of labeled corks. The exercise demonstrates how increasing the number of particles sampled improves the standard deviation due to sampling.

Clement, R. E. "Environmental Sampling for Trace Analysis," *Anal. Chem.* **1992,** *64,* 1076A–1081A.

Sampling of a large population ($n = 900$) of colored candies (M&M's work well) is used to demonstrate the importance of sample size in determining the concentration of species at several different concentration levels. This experiment is similar to the preceding one described by Bauer but incorporates several analytes.

Guy, R. D.; Ramaley, L.; Wentzell, P. D. "An Experiment in the Sampling of Solids for Chemical Analysis," *J. Chem. Educ.* **1998,** *75,* 1028–1033.

This experiment uses the molybdenum-blue method to determine the concentration of phosphate in a phosphate/sodium chloride mixture. Flow-injection analysis is used to increase the speed of analysis, allowing students to collect a sufficient amount of data in a single laboratory period. The overall variance of the analysis is partitioned into components due to the method, to the preparation of samples, and to sample collection. The validity of equation 7.5 is also evaluated.

Kratochvil, B.; Reid, R. S.; Harris, W. E. "Sampling Error in a Particulate Mixture," *J. Chem. Educ.* **1980,** *57,* 518–520.

In this experiment the overall variance for the analysis of potassium hydrogen phthalate (KHP) in a mixture of KHP and sucrose is partitioned into that due to sampling and that due to the analytical method (an acid–base titration). By having individuals analyze samples with different % w/w KHP, the relationship between sampling error and concentration of analyte can be explored.

Lochmuler, C. "Atomic Spectroscopy—Determination of Calcium and Magnesium in Sand with a Statistical Treatment of Measurements" published on the web at http://www.chem.duke.edu/~clochmul/exp4/exp4.html.

This experiment introduces random sampling. The experiment's overall variance is divided into that due to the instrument, that due to sample preparation, and that due to sampling.

The following experiments describe homemade sampling devices for collecting samples in the field.

Delumyea, R. D.; McCleary, D. L. "A Device to Collect Sediment Cores," *J. Chem. Educ.* **1993,** *70,* 172–173.

Directions are provided for preparing and using a simple coring device using PVC pipe. This experiment also details a procedure for determining the weight percent of organic material in sediments as a function of depth.

Rockwell, D. M.; Hansen, T. "Sampling and Analyzing Air Pollution," *J. Chem. Educ.* **1994,** *71,* 318–322.

Two simple air samplers are described as well as their use for determining particulates in air.

Saxena, S.; Upadhyay, R.; Upadhyay, P. "A Simple and Low-Cost Air Sampler," *J. Chem. Educ.* **1996,** *73,* 787–788.

This experiment describes the construction of an air sampler using an aquarium pump, a flow meter, a filter holder, and bottles that serve as traps for analytes. Applications include the determinations of SO_2, NO_2, HCHO, and suspended particulate matter.

Shooter, D. "Nitrogen Dioxide and Its Determination in the Atmosphere," *J. Chem. Educ.* **1993,** *70,* A133–A140.

This paper describes the construction and use of a diffusion tube for sampling NO_2 from the atmosphere. Examples of its use include the determination of NO_2 concentrations at various heights above ground level in an urban environment and through a tree's leaf canopy.

—Continued

Experiments

Continued from page 225

The following experiments introduce students to the importance of sample preparation and methods for extracting analytes from their matrix. Each experiment includes a brief description of the sample and analyte, as well as the method of analysis used to measure the analyte's concentration.

Dunn, J. G.; Phillips, D. N.; von Bronswijk, W. "An Exercise to Illustrate the Importance of Sample Preparation in Chemical Analysis," *J. Chem. Educ.* **1997,** *74,* 1188–1190.

Ore samples are analyzed for %w/w Ni. A jaw crusher is used to break the original ore sample into smaller pieces that are then sieved into 5 size fractions. A portion of each fraction is reduced in size using a disk mill and samples taken for analysis by coning and quartering. The effect of particle size on the determination of %w/w Ni is evaluated.

"Extract-Clean SPE Sample Preparation Guide Volume 1," Bulletin No. 83, Alltech Associates, Inc. Deerfield, IL.

This publication provides several examples of the use of solid-phase extractions for separating analytes from their matrices. Some of the examples included are caffeine from coffee, polyaromatic hydrocarbons from water, parabens from cosmetics, chlorinated pesticides from water, and steroids from hydrocortisone creams. Extracted analytes may be determined quantitatively by gas (GC) or liquid chromatography (LC).

Freeman, R. G.; McCurdy, D. L. "Using Microwave Sample Decomposition in Undergraduate Analytical Chemistry," *J. Chem. Educ.* **1998,** *75,* 1033–1032.

Although experimental details are not included, the application of microwave digestion is discussed for the determination of Na in food samples by flame atomic emission spectroscopy, and the Kjeldahl analysis of nitrogen is detailed.

Snow, N. H.; Dunn, M.; Patel, S. "Determination of Crude Fat in Food Products by Supercritical Fluid Extraction and Gravimetric Analysis," *J. Chem. Educ.* **1997,** *74,* 1108–1111.

Supercritical CO_2 is used to extract fat from candy bars. Samples are placed in an extraction vessel and its weight determined before and after extraction. The %w/w fat content is determined by difference. The volume of CO_2 needed to effect a complete extraction is determined experimentally. Variations in results for different samples illustrate the importance of sampling.

Yang, M. J.; Orton, M. L.; Pawliszyn, J. "Quantitative Determination of Caffeine in Beverages Using a Combined SPME-GC/MS Method," *J. Chem. Educ.* **1997,** *74,* 1130–1132.

Caffeine is extracted from beverages by a solid-phase microextraction using an uncoated fused silica fiber. The fiber is suspended in the sample for 5 min and the sample stirred to assist the mass transfer of analyte to the fiber. Immediately after removing the fiber from the sample it is transferred to the gas chromatograph's injection port where the analyte is thermally desorbed. Quantitation is accomplished by using a $^{13}C_3$ caffeine solution as an internal standard.

7L PROBLEMS

1. Because of the risk of lead poisoning, the exposure of children to lead-based paint is a significant public health concern. The first step in the quantitative analysis of lead in dried paint chips is to dissolve the sample. Corl evaluated several dissolution techniques.[24] In this study, samples of paint were collected and pulverized with a Pyrex mortar and pestle. Replicate portions of the powdered paint were then taken for analysis. Results for an unknown paint sample and for a standard reference material, in which dissolution was accomplished by a 4–6-h digestion with HNO_3 on a hot plate, are shown in the following table.

Replicate	%w/w Pb in Unknown	%w/w Pb in Standard
1	5.09	11.48
2	6.29	11.62
3	6.64	11.47
4	4.63	11.86

(a) Determine the overall variance, the variance due to the method, and the variance due to sampling. (b) What percentage of the overall variance is due to sampling? How might the variance due to sampling be decreased?

2. A shipment of 100 barrels of an organic solvent is to be evaluated by collecting and analyzing single samples from 10 of the barrels. A random number table is used to determine the barrels to be sampled. From which barrels should the samples be drawn if the first barrel is given by the twelfth entry in the random number table in Appendix 1E, with subsequent barrels given by every third entry?

3. The concentration of dissolved O_2 in a lake shows a daily cycle due to the effect of photosynthesis and a yearly cycle due to seasonal changes in temperature. Suggest an appropriate systematic sampling plan for monitoring the daily changes in dissolved O_2. Suggest an appropriate systematic sampling plan for monitoring the yearly changes in dissolved O_2.

4. The following data were collected during a preliminary study of the pH of an industrial wastewater stream

Time (h)	pH	Time (h)	pH
0.5	4.4	9.0	5.7
1.0	4.8	9.5	5.5
1.5	5.2	10.0	6.5
2.0	5.2	10.5	6.0
2.5	5.6	11.0	5.8
3.0	5.4	11.5	6.0
3.5	5.4	12.0	5.6
4.0	4.4	12.5	5.6
4.5	4.8	13.0	5.4
5.0	4.8	13.5	4.9
5.5	4.2	14.0	5.2
6.0	4.2	14.5	4.4
6.5	3.8	15.0	4.0
7.0	4.0	15.5	4.5
7.5	4.0	16.0	4.0
8.0	3.9	16.5	5.0
8.5	4.7	17.0	5.0

Construct a graph of pH as a function of time, and suggest an appropriate sampling frequency for a long-term monitoring program.

5. Suppose you have been asked to monitor the daily fluctuations in atmospheric ozone levels in the downtown area of a city to determine the relationship between daily traffic patterns and ozone levels. (a) Describe the sampling plan you would choose (random, systematic, judgmental, systematic–judgmental, or stratified). (b) Would you choose to collect and analyze a series of grab samples or form a single composite sample? (c) How would your answers to these questions change if the purpose of your work is to determine if the average daily ozone level exceeds a threshold value?

6. The best sampling plan for collecting samples depends on whether the population is homogeneous or heterogeneous. (a) Define homogeneous and heterogeneous. (b) If you collect and analyze a single sample, can you determine if the population is homogeneous or heterogeneous?

7. Examine equation 7.6 for sampling a heterogeneous population. (a) Explain why the contribution of heterogeneity to the overall sampling variance can be minimized by increasing the number of samples, but not the mass of the individual samples. (b) Explain why the contribution of homogeneity to the overall sampling variance can be minimized by increasing both the number of samples and the mass of individual samples.

8. Show that equation 7.5 can be derived from equation 7.4. Assume that the particles are spherical with a radius of r and a density of d.

9. The sampling constant for the radioisotope ^{24}Na in a sample of homogenized human liver has been reported as approximately 35 g.[25] (a) What is the expected relative standard deviation for sampling if 1.0-g samples are analyzed? (b) How many 1.0-g samples need to be analyzed to obtain a maximum sampling error of ±5% at the 95% confidence level?

10. Engels and Ingamells reported the following results for the %w/w K_2O in a mixture of amphibolite and orthoclase.[26]

0.247	0.300	0.236	0.258	0.304	0.330
0.247	0.275	0.212	0.311	0.258	0.187

Each of the 12 samples had a nominal weight of 0.1 g. Determine the approximate value for K_s, and the mass of sample needed to achieve a percent relative standard deviation of 2%.

11. The following data have been reported for the determination of KH_2PO_4 in a mixture of KH_2PO_4 and NaCl.[27]

Nominal Mass (g)	Actual Mass (g)	%w/w KH_2PO_4
0.10	0.1039	0.085
	0.1015	1.078
	0.1012	0.413
	0.1010	1.248
	0.1060	0.654
	0.0997	0.507
0.25	0.2515	0.847
	0.2465	0.598
	0.2770	0.431
	0.2460	0.842
	0.2485	0.964
	0.2590	1.178
0.50	0.5084	1.009
	0.4954	0.947
	0.5286	0.618
	0.5232	0.744
	0.4965	0.572
	0.4995	0.709

1.00	1.027	0.696
	0.987	0.843
	0.991	0.535
	0.998	0.750
	0.997	0.711
	1.001	0.639
2.50	2.496	0.766
	2.504	0.769
	2.496	0.682
	2.496	0.609
	2.557	0.589
	2.509	0.617

(a) Prepare a graph of %w/w KH_2PO_4 versus actual sample mass, and discuss how this graph is consistent with your understanding of factors affecting sampling variance. (b) For each nominal mass, calculate the percent relative standard deviation for the analysis. The value of K_s for this analysis has been estimated as 350. For each nominal mass, use K_s to determine the percent relative standard deviation due to sampling. Considering these two calculations, what conclusion can you make about the importance of indeterminate sampling errors for this analysis? (c) For each nominal mass, convert the percent relative standard deviation to an absolute standard deviation. Plot points on your graph corresponding to ±1 absolute standard deviation about the overall average %w/w KH_2PO_4. Draw smooth curves through these two sets of points. Considering these results, does the sample appear to be homogeneous on the scale at which it is sampled?

12. In this problem you will collect and analyze data in a simulation of the sampling process. Obtain a pack of M&M's or other similar candy. Obtain a sample of five candies, and count the number that are red. Report the result of your analysis as % red. Return the candies to the bag, mix thoroughly, and repeat the analysis for a total of 20 determinations. Calculate the mean and standard deviation for your data. Remove all candies, and determine the true % red for the population. Sampling in this exercise should follow binomial statistics. Calculate the expected mean value and expected standard deviation, and compare to your experimental results.

13. The following two experiments were conducted in developing a sampling plan for determining the concentration of the herbicide diaquat in soil. First, eight samples, each with a mass of 1.50 g, were analyzed, giving a sampling variance of 0.013. In the second experiment, eight samples, each with a mass of 3.00 g, were analyzed, giving a sampling variance of 0.0071. (a) If samples of 1.00 g are to be collected, how many are needed to give a sampling variance of 0.0100? (b) If five samples are to be collected, how much should each sample weigh?

14. Determine the error ($\alpha = 0.05$) for the following situations. In each case, assume that the variance for a single determination is 0.0025 and that the variance for collecting a single sample is 0.050. (a) Nine samples are collected, each of which is analyzed once. (b) One sample is collected and analyzed nine times. (c) Five samples are collected, each of which is analyzed three times.

15. Which of the sampling schemes in problem 14 is best if you wish to have an overall error of less than ±0.25, and the cost of collecting a single sample is 1 (arbitrary units), and the cost of analyzing a single sample is 10? Which is best if the cost of collecting a single sample is 7, and the cost of analyzing a single sample is 3?

16. Maw, Witry, and Emond evaluated a microwave digestion method for Hg against the standard open-vessel digestion method.[28] The standard method requires a 2-h digestion and is operator-intensive, whereas the microwave digestion is complete in approximately 0.5 h and requires little monitoring by the operator. Samples of baghouse dust from air-pollution-control equipment were collected from a hazardous waste incinerator and digested in triplicate by each method before determining the concentration of Hg in parts per million (ppm). Results are summarized in the following table.

Parts per Million Hg Following Microwave Digestion

Sample	Replicate 1	Replicate 2	Replicate 3
1	7.12	7.66	7.17
2	16.1	15.7	15.6
3	4.89	4.62	4.28
4	9.64	9.03	8.44
5	6.76	7.22	7.50
6	6.19	6.61	7.61
7	9.44	9.56	10.7
8	30.8	29.0	26.2

Parts per million Hg Following Standard Digestion

Sample	Replicate 1	Replicate 2	Replicate 3
1	5.50	5.54	5.40
2	13.1	12.8	13.0
3	5.39	5.12	5.36
4	6.59	6.52	7.20
5	6.20	6.03	5.77
6	6.25	5.65	5.61
7	15.0	13.9	14.0
8	20.4	16.1	20.0

Evaluate whether the microwave digestion method yields acceptable results in comparison with the standard digestion method.

17. Simpson, Apte, and Batley investigated methods for preserving water samples collected from anoxic (O_2-poor) environments containing high concentrations of dissolved sulfide.[29] They found that preserving the water samples by adding HNO_3 (which is a common preservation method for aerobic samples) gave significant negative determinate errors when analyzing for Cu^{2+}. When samples were preserved by adding H_2O_2 followed by HNO_3, the concentration of Cu^{2+} was determined without a determinate error. Explain their observations.

18. In a particular analysis the selectivity coefficient, $K_{A,I}$, is 0.816. When a standard sample known to contain an analyte-to-interferent ratio of 5:1 is carried through the analysis, the error in determining the analyte is +6.3%. (a) Determine the apparent recovery for the analyte if $R_I = 0$. (b) Determine the apparent recovery for the interferent if $R_A = 1$.

19. The amount of Co in an ore sample is to be determined using a procedure for which Fe is an interferent. To evaluate the procedure's accuracy, a standard sample of ore known to have a Co/Fe ratio of 10.2:1 is analyzed. When pure samples of Co and Fe are taken through the procedure, the following calibration relationships are obtained

$$S_{Co} = 0.786 \times \text{number of grams Co}$$

$$S_{Fe} = 0.699 \times \text{number of grams Fe}$$

When 278.3 mg of Co is taken through the separation step, 275.9 mg is recovered. Only 3.6 mg of Fe is recovered when a 184.9-mg sample of Fe is carried through the separation step. Calculate (a) the recoveries for Co and Fe; (b) the separation factor; (c) the error if no attempt is made to separate the Co and Fe before the analysis; (d) the error if the separation step is carried out; and (e) the maximum recovery for Fe if all the Co is recovered and the maximum allowed error is 0.05%.

20. The amount of calcium in a sample of urine was determined by a method for which magnesium is an interferent. The selectivity coefficient, $K_{Ca,Mg}$, for the method is 0.843. When a sample with a Mg/Ca ratio of 0.50 was carried through the procedure, an error of –3.7% was obtained. The error was +5.5% when a sample with a Mg/Ca ratio of 2.0 was used. (a) Determine the recoveries for Ca and Mg. (b) What is the expected error for a urine sample in which the Mg/Ca ratio is 10.0?

21. Show that F^- is an effective masking agent in preventing a reaction of Al^{3+} with EDTA. Assume that the only significant forms of fluoride and EDTA are F^- and Y^{4-}.

22. Cyanide is frequently used as a masking agent for metal ions. The effectiveness of CN^- as a masking agent is generally better in more basic solutions. Explain the reason for this pH dependency.

23. Explain how an aqueous sample consisting of Cu^{2+}, Sn^{4+}, Pb^{2+}, and Zn^{2+} can be separated into its component parts by adjusting the pH of the solution.

24. A solute, S, has a distribution ratio between water and ether of 7.5. Calculate the extraction efficiency if a 50.0-mL aqueous sample of S is extracted using 50.0 mL of ether as (a) a single portion of 50.0 mL; (b) two portions, each of 25.0 mL; (c) four portions, each of 12.5 mL; and (d) five portions, each of 10.0 mL. Assume that the solute is not involved in any secondary equilibria.

25. What volume of ether is needed to extract 99.9% of the solute in problem 24 when using (a) one extraction; (b) two extractions; (c) four extractions; and (d) five extractions.

26. What must a solute's distribution ratio be if 99% of the solute in a 50.0-mL sample is to be extracted with a single 50.0-mL portion of an organic solvent? Repeat for the case where two 25.0-mL portions of the organic solvent are used.

27. A weak acid, HA, with a K_a of 1.0×10^{-5}, has a partition coefficient, K_D, between water and an organic solvent of 1200. What restrictions on the sample's pH are necessary to ensure that a minimum of 99.9% of the weak acid is extracted in a single step from 50.0 mL of water using 50.0 mL of the organic solvent?

28. For problem 27, how many extractions will be necessary if the pH of the sample cannot be decreased below 7.0?

29. A weak base, with a K_b of 1.0×10^{-3}, has a partition coefficient, K_D, between water and an organic solvent of 500. What restrictions on the pH of the sample are necessary to ensure that a minimum of 99.9% of the weak base is extracted from 50.0 mL of water using two 25.00-mL portions of the organic solvent?

30. A sample contains a weak acid analyte, HA, and a weak acid interferent, HB. The acid dissociation constants and partition coefficients for the weak acids are as follows: $K_{a,HA} = 1.0 \times 10^{-3}$, $K_{a,HB} = 1.0 \times 10^{-7}$, $K_{D,HA} = K_{D,HB} = 500$. (a) Calculate the extraction efficiency for HA and HB when 50.0 mL of sample, buffered to a pH of 7.0, is extracted with 50.0 mL of the organic solvent. (b) Which phase is enriched in the analyte? (c) What are the recoveries for the analyte and interferent in this phase? (d) What is the separation factor? (e) A quantitative analysis is conducted on the contents of the phase enriched in analyte. What is the expected relative error if the selectivity coefficient, $K_{HA,HB}$, is 0.500 and the initial ratio of HB/HA was 10.0?

31. The relevant equilibria for the extraction of I_2 from an aqueous solution of KI into an organic phase are shown in the following diagram.

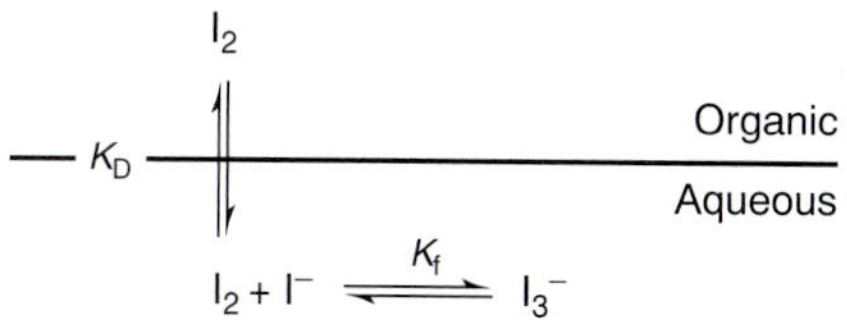

(a) Will the extraction efficiency for I_2 be greater for higher or lower concentrations of I^-? (b) Derive an expression for the distribution ratio for this extraction.

32. The relevant equilibria for extracting a neutral metal–ligand complex from an aqueous solution into an organic phase are shown in the following diagram.

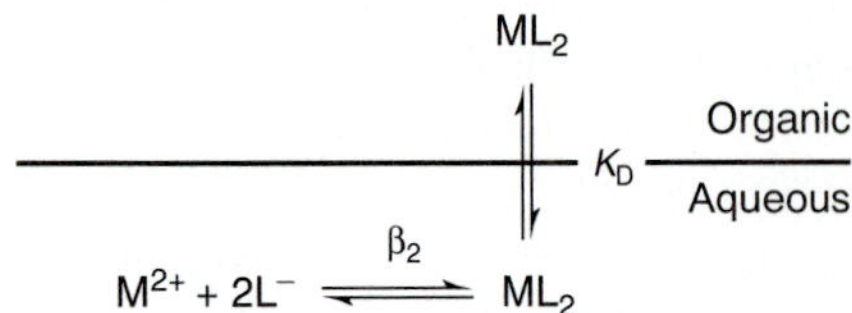

(a) Derive an expression for the distribution ratio for this extraction. (b) Calculate the extraction efficiency when 50.0 mL of an aqueous solution that is 0.15 mM in M^{2+} and 0.12 M in L^- is extracted with 25.0 mL of the organic phase. Assume that K_D is 10.3 and β_2 is 560.

33. Derive equation 7.30.

34. The following information is available for the extraction of Cu^{2+} into CCl_4 with dithizone: $K_{D,c} = 7 \times 10^4$; $\beta = 5 \times 10^{22}$; $K_{a,HL} = 3 \times 10^{-5}$; $K_{D,HL} = 1.1 \times 10^4$; $n = 2$. What is the extraction efficiency if 100 mL of an aqueous solution of 1.0×10^{-7} M Cu^{2+} that is 1 M in HCl is extracted with 10 mL of CCl_4 containing 4.0×10^{-4} M HL?

35. Cupferron is a ligand whose strong affinity for metal ions makes it useful as a chelating agent in liquid–liquid extractions. The following distribution ratios are known for the extraction of Hg^{2+}, Pb^{2+}, and Zn^{2+} from aqueous solutions to an organic solvent.

	Distribution Ratio for		
pH	Hg^{2+}	Pb^{2+}	Zn^{2+}
1	3.3	0.0	0.0
2	10.0	0.43	0.0
3	32.3	999	0.0
4	32.3	9999	0.0
5	19.0	9999	0.18
6	4.0	9999	0.33
7	1.0	9999	0.82
8	0.54	9999	1.50
9	0.15	9999	2.57
10	0.05	9999	2.57

(a) Suppose that you have 50.0 mL of an aqueous solution containing Hg^{2+}, Pb^{2+}, and Zn^{2+}. Describe how you would go about effecting a separation in which these metal ions are extracted into separate portions of the organic solvent. (b) Under the conditions you have selected for extracting Hg^{2+}, what percent of the Hg^{2+} remains in the aqueous phase after three extractions with 50.0 mL each of the organic solvent? (c) Under the conditions you have chosen for extracting Pb^{2+}, what is the minimum volume of organic solvent needed to extract 99.5% of the Pb^{2+} in a single extraction? (d) Under the conditions you have chosen for extracting Zn^{2+}, how many extractions are needed to remove 99.5% of the Zn^{2+} if each extraction is to use 25.0 mL of organic solvent?

7M SUGGESTED READINGS

The following paper provides a general introduction to the terminology used in describing sampling.

Majors, R. E. "Nomenclature for Sampling in Analytical Chemistry." *LC•GC* **1992,** *10,* 500–506.

Further information on the statistics of sampling is covered in the following papers.

Kratochvil, B.; Goewie, C. E.; Taylor, J. K. "Sampling Theory for Environmental Analysis," *Trends Anal. Chem.* **1986,** *5,* 253–256.

Kratochvil, B.; Taylor, J. K. "Sampling for Chemical Analysis," *Anal. Chem.* **1981,** *53,* 924A–938A.

The following sources may be consulted for further details regarding the collection of environmental samples. The paper by Benoit and colleagues provides a good discussion of how easily samples can be contaminated during collection and preservation.

Barceló, D.; Hennion, M. C. "Sampling of Polar Pesticides from Water Matrices," *Anal. Chim. Acta* **1997,** *338,* 3–18.

Batley, G. E.; Gardner, D. "Sampling and Storage of Natural Waters for Trace Metal Analysis", *Wat. Res.* **1977,** *11,* 745-756.

Benoit, G.; Hunter, K. S.; Rozan, T. F. "Sources of Trace Metal Contamination Artifacts During Collection, Handling, and Analysis of Freshwaters," *Anal. Chem.* **1997,** *69,* 1006–1011.

Keith, L. H., ed. *Principles of Environmental Sampling,* American Chemical Society: Washington, DC, 1988.

Keith, L. H. *Environmental Sampling and Analysis—A Practical Guide,* Lewis Publishers, Boca Raton, FL, 1991.

The following sources provide additional information on preparing samples for analysis, including the separation of analytes and interferents.

Anderson, R. *Sample Pretreatment and Separation,* Wiley: Chichester, 1987.

Baiulescu, G. E.; Dumitrescu, P.; Zuaravescu, P. G. *Sampling.* Ellis Horwood: New York, 1991.

Compton, T. R. *Direct Preconcentration Techniques.* Oxford Science Publications: Oxford, 1993.

Compton, T. R. *Complex-Formation Preconcentration Techniques.* Oxford Science Publications: Oxford, 1993.

Gy, P. M. *Sampling of Particulate Materials: Theory and Practice;* Elsevier: Amsterdam, 1979.

Gy, P. M. *Sampling of Heterogeneous and Dynamic Materials: Theories of Heterogeneity, Sampling and Homogenizing.* Elsevier: Amsterdam, 1992.

Gy, P. ed. *Sampling for Analytical Purposes.* Wiley: New York, 1998.

Karger, B. L.; Snyder, L. R.; Harvath, C. *An Introduction to Separation Science,* Wiley-Interscience: New York, 1973.

Kingston, H. M.; Haswell, S. J., eds. *Microwave-Enhanced Chemistry: Fundamentals, Sample Preparation, and Applications.* American Chemical Society: Washington, DC, 1997.

Majors, R. E.; Raynie, D. E. "Sample Preparation and Solid-Phase Extraction," *LC•GC* **1997,** *15,* 1106–1117.

Miller, J. M. *Separation Methods in Chemical Analysis,* Wiley-Interscience: New York, 1975.

Morrison, G. H.; Freiser, H. *Solvent Extraction in Analytical Chemistry,* John Wiley and Sons: New York, 1957.

Pawliszyn, J. *Solid-Phase Microextraction: Theory and Practice,* Wiley: New York, 1997.

Smith, R.; James, G. V. *The Sampling of Bulk Materials.* Royal Society of Chemistry: London, 1981.

Sulcek, Z.; Povondra, P. *Methods of Decomposition in Inorganic Analysis.* CRC Press: Boca Raton, FL, 1989.

Thurman, E. M.; Mills, M. S. *Solid-Phase Extraction: Principles and Practice,* Wiley: New York, 1998.

A web-site dedicated to sample preparation, which contains useful information about acid digestion and microwave digestion, is found at

http://www.sampleprep.duq.edu/sampleprep/

7N REFERENCES

1. Youden, Y. J. *J. Assoc. Off. Anal. Chem.* **1981,** *50,* 1007–1013.
2. Fricke, G. H.; Mischler, P. G.; Staffieri, F. P.; et al. *Anal. Chem.* **1987,** *59,* 1213–1217.
3. (a) Cohen, R. D. *J. Chem. Educ.* **1991,** *68,* 902–903; (b) Cohen, R. D. *J. Chem. Educ.* **1992,** *69,* 200–203.
4. Keith, L. H. *Environ. Sci. Technol.* **1990,** *24,* 610–617.
5. Flatman, G. T.; Englund, E. J.; Yfantis, A. A. In Keith, L. H., ed. *Principles of Environmental Sampling.* American Chemical Society: Washington, DC, 1988, 73–84.
6. Ingamells; C. O.; Switzer, P. *Talanta* **1973,** *20,* 547–568.
7. Viseman, J. *Mat. Res. Stds.* **1969,** *9(11),* 8–13.
8. Blackwood, L. G. *Environ. Sci. Technol.* **1991,** *25,* 1366–1367.
9. Duce, R. A.; Quinn, J. G.; Olney, C. E. et al. *Science,* **1972,** *176,* 161–163.
10. Tanner, R. L. In Keith, L. H., ed. *Principles of Environmental Sampling.* American Chemical Society: Washington, DC, 1988, 275–286.
11. (a) Sandell, E. B. *Colorimetric Determination of Traces of Metals,* Interscience Publishers: New York, 1950, pp. 19–20; (b) Sandell, E. B. *Anal. Chem.* **1968,** *40,* 834–835.
12. Zubay, G. *Biochemistry,* 2nd ed. Macmillan: New York, 1988, p. 120.
13. Meites, L. *Handbook of Analytical Chemistry,* McGraw-Hill: New York, 1963.
14. Glavin, D. P.; Bada, J. L. *Anal. Chem.* **1998,** *70,* 3119–3122.
15. Fresenius, C. R. *A System of Instruction in Quantitative Chemical Analysis.* John Wiley and Sons: New York, 1881.
16. Jeannot, M. A.; Cantwell, F. F. *Anal. Chem.* **1997,** *69,* 235–239.
17. Zhang, A.; Yang, M. J.; Pawliszyn, J. *Anal. Chem.* **1994,** *66,* 844A–853A.
18. Renoe, B. W. *Am. Lab.* August **1994,** 34–40.
19. McNally, M. E. *Anal. Chem.* **1995,** *67,* 308A–315A.
20. "TPH Extraction by SFE", ISCO, Inc., Lincoln, NE, Revised Nov. 1992.
21. "The Analysis of Trihalomethanes in Drinking Water by Liquid Extraction"; US Environmental Protection Agency, Environmental Monitoring and Support Laboratory, Cincinnati, OH, 9 Sept. 1977.
22. *Official Methods of Analysis,* 11th ed., Association of Official Analytical Chemists, Washington, DC, 1970, p. 475.
23. Aguilar, C.; Borrull, F.; Marcé, R. M. *LC•GC* **1996,** *14,* 1048–1054.
24. Corl, W. E. *Spectroscopy* **1991,** 6(8), 40–43.
25. Kratochvil, B.; Taylor, J. K. *Anal. Chem.* **1981,** *53,* 924A–938A.
26. Engels, J. C.; Ingamells, C. O. *Geochim. Cosmochim. Acta* **1970,** *34,* 1007–1017.
27. Guy, R. D.; Ramaley, L.; Wentzell, P. D. *J. Chem. Educ.* **1998,** *75,* 1028–1033.
28. Maw, R.; Witry, L.; Emond, T. *Spectroscopy* **1994,** *9,* 39–41.
29. Simpson, S. L.; Apte, S. C.; Batley, G. E. *Anal. Chem.* **1998,** *70,* 4202–4205.

Chapter 8

Gravimetric Methods of Analysis

Gravimetry encompasses all techniques in which we measure mass or a change in mass. When you step on a scale after exercising you are making, in a sense, a gravimetric determination of your mass. Measuring mass is the most fundamental of all analytical measurements, and gravimetry is unquestionably the oldest analytical technique.

8A Overview of Gravimetry

Before we look more closely at specific gravimetric methods and their applications, let's take a moment to develop a broad survey of **gravimetry.** Later, as you read through the sections of this chapter discussing different gravimetric methods, this survey will help you focus on their similarities. It is usually easier to understand a new method of analysis when you can see its relationship to other similar methods.

gravimetry
Any method in which the signal is a mass or change in mass.

8A.1 Using Mass as a Signal

At the beginning of this chapter we indicated that in gravimetry we measure mass or a change in mass. This suggests that there are at least two ways to use mass as an analytical signal. We can, of course, measure an analyte's mass directly by placing it on a balance and recording its mass. For example, suppose you are to determine the total suspended solids in water released from a sewage-treatment facility. Suspended solids are just that; solid matter that has yet to settle out of its solution matrix. The analysis is easy. You collect a sample and pass it through a preweighed filter that retains the suspended solids. After drying to remove any residual moisture, you weigh the filter. The difference between the filter's original mass and final mass gives the mass of suspended solids. We call this a direct analysis because the analyte itself is the object being weighed.

What if the analyte is an aqueous ion, such as Pb^{2+}? In this case we cannot isolate the analyte by filtration because the Pb^{2+} is dissolved in the solution's matrix. We can still measure the analyte's mass, however, by chemically converting it to a solid form. If we suspend a pair of Pt electrodes in our solution and apply a sufficiently positive potential between them for a long enough time, we can force the reaction

$$Pb^{2+}(aq) + 4H_2O(\ell) \rightleftharpoons PbO_2(s) + H_2(g) + 2H_3O^+(aq)$$

to go to completion. The Pb^{2+} ion in solution oxidizes to PbO_2 and deposits on the Pt electrode serving as the anode. If we weigh the Pt anode before and after applying the potential, the difference in the two measurements gives the mass of PbO_2 and, from the reaction's stoichiometry, the mass of Pb^{2+}. This also is a direct analysis because the material being weighed contains the analyte.

Sometimes it is easier to remove the analyte and use a change in mass as the analytical signal. Imagine how you would determine a food's moisture content by a direct analysis. One possibility is to heat a sample of the food to a temperature at which the water in the sample vaporizes. If we capture the vapor in a preweighed absorbent trap, then the change in the absorbent's mass provides a direct determination of the amount of water in the sample. An easier approach, however, is to weigh the sample of food before and after heating, using the change in its mass as an indication of the amount of water originally present. We call this an indirect analysis since we determine the analyte by a signal representing its disappearance.

The indirect determination of moisture content in foods is done by difference. The sample's initial mass includes the water, whereas the final mass is measured after removing the water. We can also determine an analyte indirectly without its ever being weighed. Again, as with the determination of Pb^{2+} as $PbO_2(s)$, we take advantage of the analyte's chemistry. For example, phosphite, PO_3^{3-}, reduces Hg^{2+} to Hg_2^{2+}. In the presence of Cl^- a solid precipitate of Hg_2Cl_2 forms.

$$2HgCl_2(aq) + PO_3^{3-}(aq) + 3H_2O(\ell) \rightleftharpoons Hg_2Cl_2(s) + 2H_3O^+(aq) + 2Cl^-(aq) + PO_4^{3-}(aq)$$

If $HgCl_2$ is added in excess, each mole of PO_3^{3-} produces one mole of Hg_2Cl_2. The precipitate's mass, therefore, provides an indirect measurement of the mass of PO_3^{3-} present in the original sample.

Summarizing, we can determine an analyte gravimetrically by directly determining its mass, or the mass of a compound containing the analyte. Alternatively, we can determine an analyte indirectly by measuring a change in mass due to its loss, or the mass of a compound formed as the result of a reaction involving the analyte.

8A.2 Types of Gravimetric Methods

In the previous section we used four examples to illustrate the different ways that mass can serve as an analytical signal. These examples also illustrate the four gravimetric methods of analysis. When the signal is the mass of a precipitate, we call the method **precipitation gravimetry.** The indirect determination of PO_3^{3-} by precipitating Hg_2Cl_2 is a representative example, as is the direct determination of Cl^- by precipitating AgCl.

precipitation gravimetry
A gravimetric method in which the signal is the mass of a precipitate.

In **electrogravimetry** the analyte is deposited as a solid film on one electrode in an electrochemical cell. The oxidation of Pb^{2+}, and its deposition as PbO_2 on a Pt anode is one example of electrogravimetry. Reduction also may be used in electrogravimetry. The electrodeposition of Cu on a Pt cathode, for example, provides a direct analysis for Cu^{2+}.

electrogravimetry
A gravimetric method in which the signal is the mass of an electrodeposit on the cathode or anode in an electrochemical cell.

When thermal or chemical energy is used to remove a volatile species, we call the method **volatilization gravimetry.** In determining the moisture content of food, thermal energy vaporizes the H_2O. The amount of carbon in an organic compound may be determined by using the chemical energy of combustion to convert C to CO_2.

volatilization gravimetry
A gravimetric method in which the loss of a volatile species gives rise to the signal.

Finally, in **particulate gravimetry** the analyte is determined following its removal from the sample matrix by filtration or extraction. The determination of suspended solids is one example of particulate gravimetry.

particulate gravimetry
A gravimetric method in which the mass of a particulate analyte is determined following its separation from its matrix.

8A.3 Conservation of Mass

An accurate gravimetric analysis requires that the mass of analyte present in a sample be proportional to the mass or change in mass serving as the analytical signal. For all gravimetric methods this proportionality involves a conservation of mass. For gravimetric methods involving a chemical reaction, the analyte should participate in only one set of reactions, the stoichiometry of which indicates how the precipitate's mass relates to the analyte's mass. Thus, for the analysis of Pb^{2+} and PO_3^{3-} described earlier, we can write the following conservation equations

$$\text{Moles } Pb^{2+} = \text{moles } PbO_2$$

$$\text{Moles } PO_3^{3-} = \text{moles } Hg_2Cl_2$$

Removing the analyte from its matrix by filtration or extraction must be complete. When true, the analyte's mass can always be found from the analytical signal; thus, for the determination of suspended solids we know that

$$\text{Filter's final mass} - \text{filter's initial mass} = \text{g suspended solid}$$

whereas for the determination of the moisture content we have

$$\text{Sample's initial mass} - \text{sample's final mass} = \text{g } H_2O$$

Specific details, including worked examples, are found in the sections of this chapter covering individual gravimetric methods.

8A.4 Why Gravimetry Is Important

Except for particulate gravimetry, which is the most trivial form of gravimetry, it is entirely possible that you will never use gravimetry after you are finished with this course. Why, then, is familiarity with gravimetry still important? The answer is that gravimetry is one of only a small number of techniques whose measurements require only base SI units, such as mass and moles, and defined constants, such as Avogadro's number and the mass of ^{12}C.* The result of an analysis must ultimately be traceable to methods, such as gravimetry, that can be related to fundamental physical properties.[1] Most analysts never use gravimetry to validate their methods. Verifying a method by analyzing a standard reference material, however, is common. Estimating the composition of these materials often involves a gravimetric analysis.[2]

8B Precipitation Gravimetry

Precipitation gravimetry is based on the formation of an insoluble compound following the addition of a precipitating reagent, or **precipitant,** to a solution of the analyte. In most methods the precipitate is the product of a simple metathesis reaction between the analyte and precipitant; however, any reaction generating a precipitate can potentially serve as a gravimetric method. Most precipitation gravimetric methods were developed in the nineteenth century as a means for analyzing ores. Many of these methods continue to serve as standard methods of analysis.

precipitant
A reagent that causes the precipitation of a soluble species.

8B.1 Theory and Practice

A precipitation gravimetric analysis must have several important attributes. First, the precipitate must be of low solubility, high purity, and of known composition if its mass is to accurately reflect the analyte's mass. Second, the precipitate must be in a form that is easy to separate from the reaction mixture. The theoretical and experimental details of precipitation gravimetry are reviewed in this section.

Solubility Considerations An accurate precipitation gravimetric method requires that the precipitate's solubility be minimal. Many total analysis techniques can routinely be performed with an accuracy of better than ±0.1%. To obtain this level of accuracy, the isolated precipitate must account for at least 99.9% of the analyte. By extending this requirement to 99.99% we ensure that accuracy is not limited by the precipitate's solubility.

Solubility losses are minimized by carefully controlling the composition of the solution in which the precipitate forms. This, in turn, requires an understanding of the relevant equilibrium reactions affecting the precipitate's solubility. For example, Ag^+ can be determined gravimetrically by adding Cl^- as a precipitant, forming a precipitate of AgCl.

$$Ag^+(aq) + Cl^-(aq) \rightleftharpoons AgCl(s) \qquad 8.1$$

*Two other techniques that depend only on base SI units are coulometry and isotope-dilution mass spectrometry. Coulometry is discussed in Chapter 11. Isotope-dilution mass spectroscopy is beyond the scope of an introductory text, however, the list of suggested readings includes a useful reference.

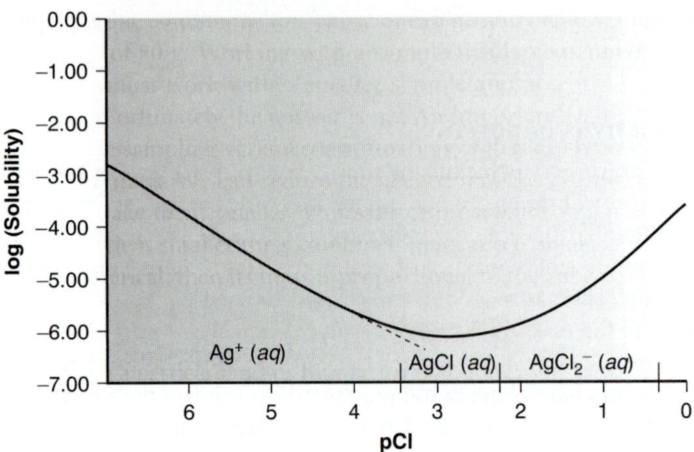

Figure 8.1
Solubility of AgCl as a function of pCl. The dashed line shows the predicted S_{AgCl}, assuming that only reaction 8.1 and equation 8.2 affect the solubility of AgCl. The solid line is calculated using equation 8.7, and includes the effect of reactions 8.3–8.5. A ladder diagram for the AgCl complexation equilibria is superimposed on the pCl axis.

If this is the only reaction considered, we would falsely conclude that the precipitate's solubility, S_{AgCl}, is given by

$$S_{AgCl} = [Ag^+] = \frac{K_{sp}}{[Cl^-]} \qquad \textbf{8.2}$$

and that solubility losses may be minimized by adding a large excess of Cl^-. In fact, as shown in Figure 8.1, adding a large excess of Cl^- eventually increases the precipitate's solubility.

To understand why AgCl shows a more complex solubility relationship than that suggested by equation 8.2, we must recognize that Ag^+ also forms a series of soluble chloro-complexes

$$Ag^+(aq) + Cl^-(aq) \overset{K_1}{\rightleftharpoons} AgCl(aq) \qquad \textbf{8.3}$$

$$Ag^+(aq) + 2Cl^-(aq) \overset{\beta_2}{\rightleftharpoons} AgCl_2^-(aq) \qquad \textbf{8.4}$$

$$Ag^+(aq) + 3Cl^-(aq) \overset{\beta_3}{\rightleftharpoons} AgCl_3^{2-}(aq) \qquad \textbf{8.5}$$

The solubility of AgCl, therefore, is the sum of the equilibrium concentrations for all soluble forms of Ag^+.

$$S_{AgCl} = [Ag^+] + [AgCl(aq)] + [AgCl_2^-] + [AgCl_3^{2-}] \qquad \textbf{8.6}$$

Substituting the equilibrium constant expressions for reactions 8.3–8.5 into equation 8.6 defines the solubility of AgCl in terms of the equilibrium concentration of Cl^-.

$$S_{AgCl} = \frac{K_{sp}}{[Cl^-]} + K_1K_{sp} + \beta_2K_{sp}[Cl^-] + \beta_3K_{sp}[Cl^-]^2 \qquad \textbf{8.7}$$

Equation 8.7 explains the solubility curve for AgCl shown in Figure 8.1. As Cl^- is added to a solution of Ag^+, the solubility of AgCl initially decreases because of reaction 8.1. Note that under these conditions, the final three terms in equation 8.7 are small, and that equation 8.1 is sufficient to describe the solubility of AgCl. Increasing the concentration of chloride, however, leads to an increase in the solubility of AgCl due to the soluble chloro-complexes formed in reactions 8.3–8.5.*

*Also shown in Figure 8.1 is a ladder diagram for this system. Note that the increase in solubility begins when the higher-order soluble complexes, $AgCl_2^-$ and $AgCl_3^{2-}$, become the dominant species.

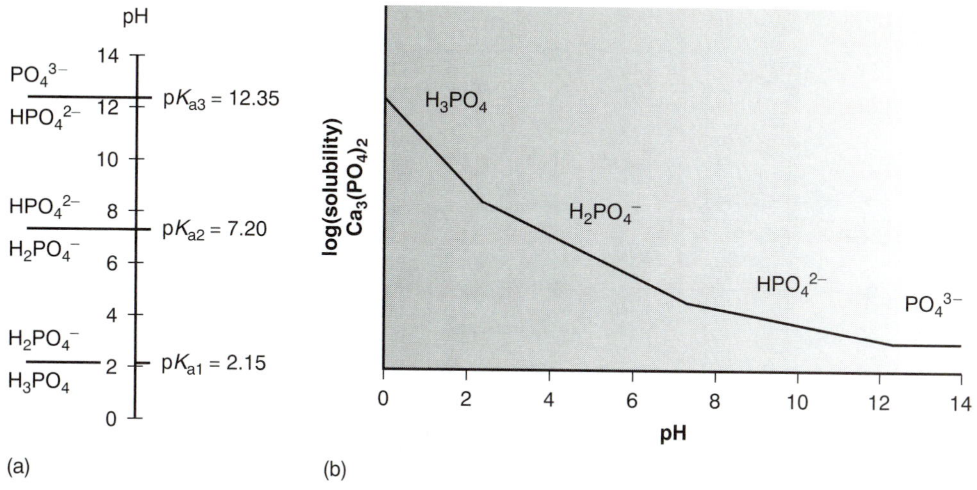

Figure 8.2
(a) Ladder diagram for phosphate; (b) Solubility diagram for $Ca_3(PO_4)_2$ showing the predominate form of phosphate for each segment of the solubility curve.

Clearly the equilibrium concentration of chloride is an important parameter if the concentration of silver is to be determined gravimetrically by precipitating AgCl. In particular, a large excess of chloride must be avoided.

Another important parameter that may affect a precipitate's solubility is the pH of the solution in which the precipitate forms. For example, hydroxide precipitates, such as $Fe(OH)_3$, are more soluble at lower pH levels at which the concentration of OH^- is small. The effect of pH on solubility is not limited to hydroxide precipitates, but also affects precipitates containing basic or acidic ions. The solubility of $Ca_3(PO_4)_2$ is pH-dependent because phosphate is a weak base. The following four reactions, therefore, govern the solubility of $Ca_3(PO_4)_2$.

$$Ca_3(PO_4)_2(s) \overset{K_{sp}}{\rightleftharpoons} 3Ca^{2+}(aq) + 2PO_4^{3-}(aq) \qquad \textbf{8.8}$$

$$PO_4^{3-}(aq) + H_2O(\ell) \overset{K_{b1}}{\rightleftharpoons} HPO_4^{2-}(aq) + OH^-(aq) \qquad \textbf{8.9}$$

$$HPO_4^{2-}(aq) + H_2O(\ell) \overset{K_{b2}}{\rightleftharpoons} H_2PO_4^-(aq) + OH^-(aq) \qquad \textbf{8.10}$$

$$H_2PO_4^-(aq) + H_2O(\ell) \overset{K_{b3}}{\rightleftharpoons} H_3PO_4(aq) + OH^-(aq) \qquad \textbf{8.11}$$

Depending on the solution's pH, the predominate phosphate species is either PO_4^{3-}, HPO_4^{2-}, $H_2PO_4^-$, or H_3PO_4. The ladder diagram for phosphate, shown in Figure 8.2a, provides a convenient way to evaluate the pH-dependent solubility of phosphate precipitates. When the pH is greater than 12.4, the predominate phosphate species is PO_4^{3-}, and the solubility of $Ca_3(PO_4)_2$ will be at its minimum because only reaction 8.8 occurs to an appreciable extent (see Figure 8.2b). As the solution becomes more acidic, the solubility of $Ca_3(PO_4)_2$ increases due to the contributions of reactions 8.9–8.11.

Solubility can often be decreased by using a nonaqueous solvent. A precipitate's solubility is generally greater in aqueous solutions because of the ability of water molecules to stabilize ions through solvation. The poorer solvating ability of nonaqueous solvents, even those that are polar, leads to a smaller solubility product. For example, $PbSO_4$ has a K_{sp} of 1.6×10^{-8} in H_2O, whereas in a 50:50 mixture of H_2O/ethanol the K_{sp} at 2.6×10^{-12} is four orders of magnitude smaller.

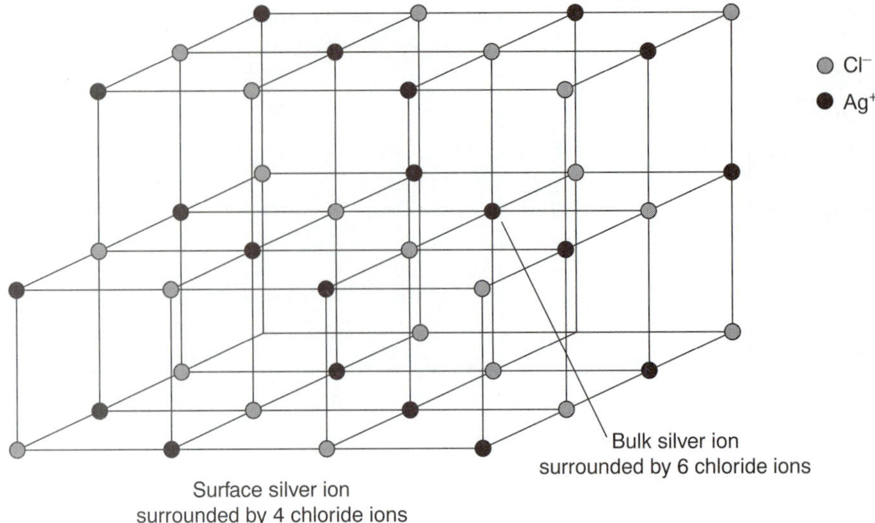

Figure 8.3
Schematic model of AgCl showing difference between bulk and surface atoms of silver. Silver and chloride ions are not shown to scale.

Avoiding Impurities Precipitation gravimetry is based on a known stoichiometry between the analyte's mass and the mass of a precipitate. It follows, therefore, that the precipitate must be free from impurities. Since precipitation typically occurs in a solution rich in dissolved solids, the initial precipitate is often impure. Any impurities present in the precipitate's matrix must be removed before obtaining its weight.

The greatest source of impurities results from chemical and physical interactions occurring at the precipitate's surface. A precipitate is generally crystalline, even if only on a microscopic scale, with a well-defined lattice structure of cations and anions. Those cations and anions at the surface of the precipitate carry, respectively, a positive or a negative charge as a result of their incomplete coordination spheres. In a precipitate of AgCl, for example, each Ag^+ ion in the bulk of the precipitate is bound to six Cl^- ions. Silver ions at the surface, however, are bound to no more than five Cl^- ions, and carry a partial positive charge (Figure 8.3).

Precipitate particles grow in size because of the electrostatic attraction between charged ions on the surface of the precipitate and oppositely charged ions in solution. Ions common to the precipitate are chemically adsorbed, extending the crystal lattice. Other ions may be physically adsorbed and, unless displaced, are incorporated into the crystal lattice as a coprecipitated impurity. Physically adsorbed ions are less strongly attracted to the surface and can be displaced by chemically adsorbed ions.

inclusion
A coprecipitated impurity in which the interfering ion occupies a lattice site in the precipitate.

One common type of impurity is an **inclusion.** Potential interfering ions whose size and charge are similar to a lattice ion may substitute into the lattice structure by chemical adsorption, provided that the interferent precipitates with the same crystal structure (Figure 8.4a). The probability of forming an inclusion is greatest when the interfering ion is present at substantially higher concentrations than the dissolved lattice ion. The presence of an inclusion does not decrease the amount of analyte that precipitates, provided that the precipitant is added in sufficient excess. Thus, the precipitate's mass is always larger than expected.

Inclusions are difficult to remove since the included material is chemically part of the crystal lattice. The only way to remove included material is through reprecipitation. After isolating the precipitate from the supernatant solution, it is dissolved in a small portion of a suitable solvent at an elevated temperature. The solution is then cooled to re-form the precipitate. Since the concentration ratio of interferent to analyte is lower in the new solution than in the original supernatant solution, the mass percent of included material in the precipitate decreases. This process of reprecipitation is repeated as needed to completely remove the inclusion. Potential solubility losses of the analyte, however, cannot be ignored. Thus, reprecipitation requires a precipitate of low solubility, and a solvent for which there is a significant difference in the precipitate's solubility as a function of temperature.

occlusion
A coprecipitated impurity trapped within a precipitate as it forms.

Occlusions, which are a second type of coprecipitated impurity, occur when physically adsorbed interfering ions become trapped within the growing precipitate. Occlusions form in two ways. The most common mechanism occurs when physically adsorbed ions are surrounded by additional precipitate before they can be desorbed or displaced (see Figure 8.4a). In this case the precipitate's mass is always greater than expected. Occlusions also form when rapid precipitation traps a pocket of solution within the growing precipitate (Figure 8.4b). Since the trapped solution contains dissolved solids, the precipitate's mass normally increases. The mass of the precipitate may be less than expected, however, if the occluded material consists primarily of the analyte in a lower-molecular-weight form from that of the precipitate.

digestion
The process by which a precipitate is given time to form larger, purer particles.

Occlusions are minimized by maintaining the precipitate in equilibrium with its supernatant solution for an extended time. This process is called **digestion** and may be carried out at room temperature or at an elevated temperature. During digestion, the dynamic nature of the solubility–precipitation equilibrium, in which the precipitate dissolves and re-forms, ensures that occluded material is eventually exposed to the supernatant solution. Since the rate of dissolution and reprecipitation are slow, the chance of forming new occlusions is minimal.

adsorbate
A coprecipitated impurity that adsorbs to the surface of a precipitate.

After precipitation is complete the surface continues to attract ions from solution (Figure 8.4c). These surface **adsorbates,** which may be chemically or physically adsorbed, constitute a third type of coprecipitated impurity. Surface adsorption is minimized by decreasing the precipitate's available surface area. One benefit of digestion is that it also increases the average size of precipitate particles. This is not surprising since the probability that a particle will dissolve is inversely proportional to its size. During digestion larger particles of precipitate increase in size at the expense of smaller particles. One consequence of forming fewer particles of larger size is an overall decrease in the precipitate's surface area. Surface adsorbates also may be removed by washing the precipitate. Potential solubility losses, however, cannot be ignored.

Inclusions, occlusions, and surface adsorbates are called coprecipitates because they represent soluble species that are brought into solid form along with the desired precipitate. Another source of impurities occurs when other species in solution precipitate under the conditions of the analysis. Solution conditions necessary to minimize the solubility of a desired precipitate may lead to the formation of an additional precipitate that interferes in the analysis. For example, the precipitation of nickel dimethylgloxime requires a pH that is slightly basic. Under these conditions, however, any Fe^{3+} that might be present precipitates as $Fe(OH)_3$. Finally, since most precipitants are not selective toward a single analyte, there is always a risk that the precipitant will react, sequentially, with more than one species.

The formation of these additional precipitates can usually be minimized by carefully controlling solution conditions. Interferents forming precipitates that are less soluble than the analyte may be precipitated and removed by filtration, leaving the analyte behind in solution. Alternatively, either the analyte or the interferent can be masked using a suitable complexing agent, preventing its precipitation.

CACACACACACACACACACA
ACACACACACACACACACAC
CACACACAMACACACACACA
ACACACACACACAMACACAC
CACACACACACACACACACA
AMACACACACACACACACAC
CACACACACACACACACACA
(a)

CACACACACACACACACACA
ACACACACACACACACACAC
CACAC CACA
ACACA ACAC
CACACA CACA
ACACACACACACACACACAC
CACACACACACACACACACA
(b)

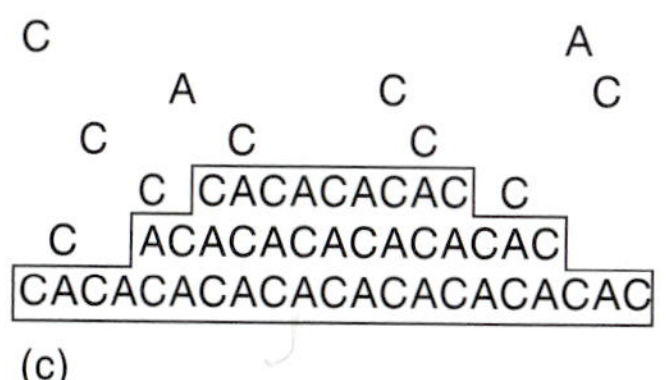

(c)

Figure 8.4
Example of coprecipitation: (a) schematic of a chemically adsorbed inclusion or a physically adsorbed occlusion in a crystal lattice, where C and A represent the cation–anion pair comprising the analyte and the precipitant, and M is the impurity; (b) schematic of an occlusion by entrapment of supernatant solution; (c) surface adsorption of excess C.

Both of the above-mentioned approaches are illustrated in Fresenius's analytical method for determining Ni and Co in ores containing Pb^{2+}, Cu^{2+}, and Fe^{3+} as potential interfering ions (see Figure 1.1 in Chapter 1). The ore is dissolved in a solution containing H_2SO_4, selectively precipitating Pb^{2+} as $PbSO_4$. After filtering, the supernatant solution is treated with H_2S. Because the solution is strongly acidic, however, only CuS precipitates. After removing the CuS by filtration, the solution is made basic with ammonia until $Fe(OH)_3$ precipitates. Cobalt and nickel, which form soluble amine complexes, remain in solution.

In some situations the rate at which a precipitate forms can be used to separate an analyte from a potential interferent. For example, due to similarities in their chemistry, a gravimetric analysis for Ca^{2+} may be adversely affected by the presence of Mg^{2+}. Precipitates of $Ca(OH)_2$, however, form more rapidly than precipitates of $Mg(OH)_2$. If $Ca(OH)_2$ is filtered before $Mg(OH)_2$ begins to precipitate, then a quantitative analysis for Ca^{2+} is feasible.

Finally, in some cases it is easier to isolate and weigh both the analyte and the interferent. After recording its weight, the mixed precipitate is treated to convert at least one of the two precipitates to a new chemical form. This new mixed precipitate is also isolated and weighed. For example, a mixture containing Ca^{2+} and Mg^{2+} can be analyzed for both cations by first isolating a mixed precipitate of $CaCO_3$ and $MgCO_3$. After weighing, the mixed precipitate is heated, converting it to a mixture of CaO and MgO. Thus

$$\text{Grams of mixed precipitate 1} = \text{grams } CaCO_3 + \text{grams } MgCO_3$$

$$\text{Grams of mixed precipitate 2} = \text{grams CaO} + \text{grams MgO}$$

Although these equations contain four unknowns (grams $CaCO_3$, grams $MgCO_3$, grams CaO, and grams MgO), the stoichiometric relationships between $CaCO_3$ and CaO

$$\text{Moles } CaCO_3 = \text{moles CaO}$$

and between $MgCO_3$ and MgO

$$\text{Moles } MgCO_3 = \text{moles MgO}$$

provide enough additional information to determine the amounts of both Ca^{2+} and Mg^{2+} in the sample.*

Controlling Particle Size Following precipitation and digestion, the precipitate must be separated from the supernatant solution and freed of any remaining impurities, including residual solvent. These tasks are accomplished by filtering, rinsing, and drying the precipitate. The size of the precipitate's particles determines the ease and success of filtration. Smaller, colloidal particles are difficult to filter because they may readily pass through the pores of the filtering device. Large, crystalline particles, however, are easily filtered.

By carefully controlling the precipitation reaction we can significantly increase a precipitate's average particle size. Precipitation consists of two distinct events: nucleation, or the initial formation of smaller stable particles of precipitate, and the subsequent growth of these particles. Larger particles form when the rate of particle growth exceeds the rate of nucleation.

*Example 8.2 shows how to solve this type of problem.

A solute's **relative supersaturation,** *RSS,* can be expressed as

relative supersaturation
A measure of the extent to which a solution, or a localized region of solution, contains more dissolved solute than that expected at equilibrium (*RSS*).

$$RSS = \frac{Q - S}{S} \qquad \textbf{8.12}$$

where Q is the solute's actual concentration, S is the solute's expected concentration at equilibrium, and $Q - S$ is a measure of the solute's supersaturation when precipitation begins.[3] A large, positive value of *RSS* indicates that a solution is highly supersaturated. Such solutions are unstable and show high rates of nucleation, producing a precipitate consisting of numerous small particles. When *RSS* is small, precipitation is more likely to occur by particle growth than by nucleation.

Examining equation 8.12 shows that we can minimize *RSS* by either decreasing the solute's concentration or increasing the precipitate's solubility. A precipitate's solubility usually increases at higher temperatures, and adjusting pH may affect a precipitate's solubility if it contains an acidic or basic anion. Temperature and pH, therefore, are useful ways to increase the value of S. Conducting the precipitation in a dilute solution of analyte, or adding the precipitant slowly and with vigorous stirring are ways to decrease the value of Q.

There are, however, practical limitations to minimizing *RSS.* Precipitates that are extremely insoluble, such as $Fe(OH)_3$ and PbS, have such small solubilities that a large *RSS* cannot be avoided. Such solutes inevitably form small particles. In addition, conditions that yield a small *RSS* may lead to a relatively stable supersaturated solution that requires a long time to fully precipitate. For example, almost a month is required to form a visible precipitate of $BaSO_4$ under conditions in which the initial *RSS* is 5.[4]

An increase in the time required to form a visible precipitate under conditions of low *RSS* is a consequence of both a slow rate of nucleation and a steady decrease in *RSS* as the precipitate forms. One solution to the latter problem is to chemically generate the precipitant in solution as the product of a slow chemical reaction. This maintains the *RSS* at an effectively constant level. The precipitate initially forms under conditions of low *RSS,* leading to the nucleation of a limited number of particles. As additional precipitant is created, nucleation is eventually superseded by particle growth. This process is called **homogeneous precipitation.**[5]

homogeneous precipitation
A precipitation in which the precipitant is generated in situ by a chemical reaction.

Two general methods are used for homogeneous precipitation. If the precipitate's solubility is pH-dependent, then the analyte and precipitant can be mixed under conditions in which precipitation does not occur. The pH is then raised or lowered as needed by chemically generating OH^- or H_3O^+. For example, the hydrolysis of urea can be used as a source of OH^-.

$$CO(NH_2)_2(aq) + H_2O(\ell) \rightleftharpoons CO_2(g) + 2NH_3(aq)$$

$$NH_3(aq) + H_2O(\ell) \rightleftharpoons NH_4^+(aq) + OH^-(aq)$$

The hydrolysis of urea is strongly temperature-dependent, with the rate being negligible at room temperature. The rate of hydrolysis, and thus the rate of precipitate formation, can be controlled by adjusting the solution's temperature. Precipitates of $BaCrO_4$, for example, have been produced in this manner.

Color Plate 5 shows the difference between a precipitate formed by direct precipitation and a precipitate formed by a homogeneous precipitation.

In the second method of homogeneous precipitation, the precipitant itself is generated by a chemical reaction. For example, Ba^{2+} can be homogeneously precipitated as $BaSO_4$ by hydrolyzing sulphamic acid to produce SO_4^{2-}.

$$NH_2SO_3H(aq) + 2H_2O(\ell) \rightleftharpoons NH_4^+(aq) + H_3O^+(aq) + SO_4^{2-}(aq)$$

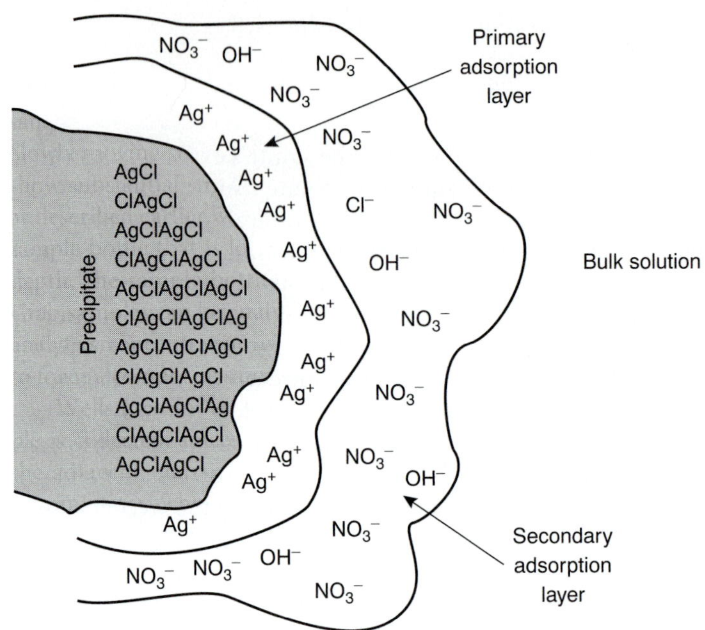

Figure 8.5
Schematic model of the solid–solution interface at a particle of AgCl in a solution containing excess $AgNO_3$.

Homogeneous precipitation affords the dual advantages of producing large particles of precipitate that are relatively free from impurities. These advantages, however, may be offset by increasing the time needed to produce the precipitate, and a tendency for the precipitate to deposit as a thin film on the container's walls. The latter problem is particularly severe for hydroxide precipitates generated using urea.

An additional method for increasing particle size deserves mention. When a precipitate's particles are electrically neutral, they tend to coagulate into larger particles. Surface adsorption of excess lattice ions, however, provides the precipitate's particles with a net positive or negative surface charge. Electrostatic repulsion between the particles prevents them from coagulating into larger particles.

Consider, for instance, the precipitation of AgCl from a solution of $AgNO_3$, using NaCl as a precipitant. Early in the precipitation, when NaCl is the limiting reagent, excess Ag^+ ions chemically adsorb to the AgCl particles, forming a positively charged primary adsorption layer (Figure 8.5). Anions in solution, in this case NO_3^- and OH^-, are attracted toward the surface, forming a negatively charged secondary adsorption layer that balances the surface's positive charge. The solution outside the secondary adsorption layer remains electrically neutral. **Coagulation** cannot occur if the secondary adsorption layer is too thick because the individual particles of AgCl are unable to approach one another closely enough.

coagulation
The process of smaller particles of precipitate clumping together to form larger particles.

Coagulation can be induced in two ways: by increasing the concentration of the ions responsible for the secondary adsorption layer or by heating the solution. One way to induce coagulation is to add an inert electrolyte, which increases the concentration of ions in the secondary adsorption layer. With more ions available, the thickness of the secondary absorption layer decreases. Particles of precipitate may now approach one another more closely, allowing the precipitate to coagulate. The amount of electrolyte needed to cause spontaneous coagulation is called the critical coagulation concentration.

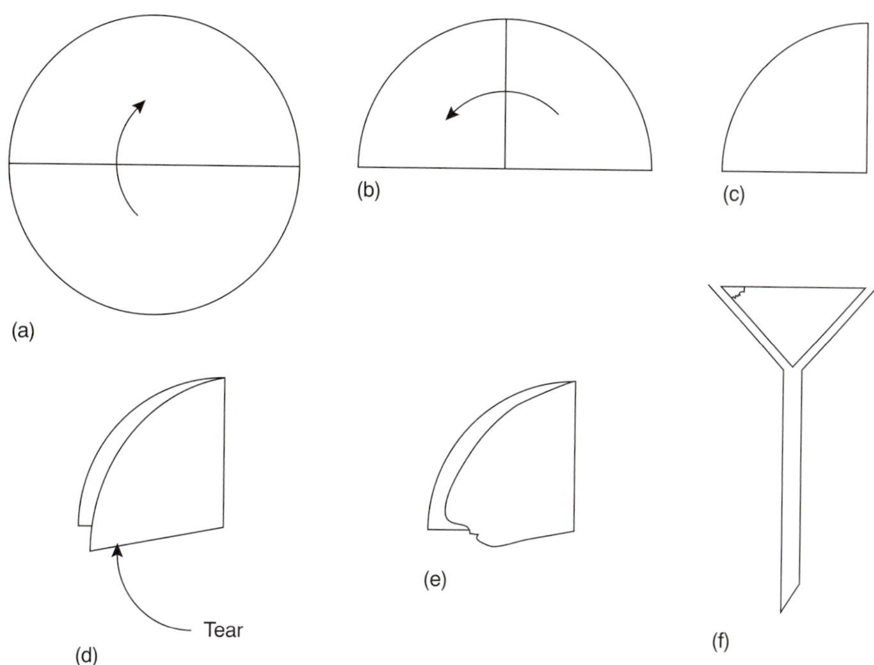

Figure 8.6
Proper procedure for filtering solids using filter paper. The filter paper circle in (a) is folded in half (b), and folded in half again (c). The filter paper is parted (d), and a small corner is torn off (e). The filter paper is opened up into a cone and placed in the funnel (f). Note that the torn corner is placed to the outside.

Heating the solution and precipitate provides a second way to induce coagulation. As the temperature increases, the number of ions in the primary adsorption layer decreases, lowering the precipitate's surface charge. In addition, increasing the particle's kinetic energy may be sufficient to overcome the electrostatic repulsion preventing coagulation at lower temperatures.

Filtering the Precipitate After precipitation and digestion are complete, the precipitate is separated from solution by filtration using either filter paper or a filtering crucible. The most common filtering medium is cellulose-based filter paper, which is classified according to its filtering speed, its size, and its ash content on ignition. Filtering speed is a function of the paper's pore size, which determines the particle sizes retained by the filter. Filter paper is rated as fast (retains particles $> 20–25$ μm), medium fast (retains particles > 16 μm), medium (retains particles > 8 μm), and slow (retains particles $> 2–3$ μm). The proper choice of filtering speed is important. If the filtering speed is too fast, the precipitate may pass through the filter paper resulting in a loss of precipitate. On the other hand, the filter paper can become clogged when using a filter paper that is too slow.

Filter paper is hygroscopic and is not easily dried to a constant weight. As a result, in a quantitative procedure the filter paper must be removed before weighing the precipitate. This is accomplished by carefully igniting the filter paper. Following ignition, a residue of noncombustible inorganic ash remains that contributes a positive determinate error to the precipitate's final mass. For quantitative analytical procedures a low-ash filter paper must be used. This grade of filter paper is pretreated by washing with a mixture of HCl and HF to remove inorganic materials. Filter paper classed as quantitative has an ash content of less than 0.010% w/w. Qualitative filter paper typically has a maximum ash content of 0.06% w/w.

Filtering is accomplished by folding the filter paper into a cone, which is then placed in a long-stem funnel (Figure 8.6). A seal between the filter cone and the funnel is formed by dampening the paper with water and pressing the paper to the wall of the funnel. When properly prepared, the stem of the funnel will fill with the solution being filtered, increasing the rate of filtration. Filtration is accomplished by the force of gravity.

supernatant
The solution that remains after a precipitate forms.

The precipitate is transferred to the filter in several steps (Figure 8.7). The first step is to decant the majority of the **supernatant** through the filter paper without transferring the precipitate. This is done to prevent the filter paper from becoming clogged at the beginning of the filtration process. Initial rinsing of the precipitate is done in the beaker in which the precipitation was performed. These rinsings are also decanted through the filter paper. Finally, the precipitate is transferred onto the filter paper using a stream of rinse solution. Any precipitate clinging to the walls of the beaker is transferred using a rubber policeman (which is simply a flexible rubber spatula attached to the end of a glass stirring rod).

Figure 8.7
Proper procedure for transferring the supernatant to the filter paper cone.

An alternative method for filtering the precipitate is a filtering crucible (Figure 8.8). The most common is a fritted glass crucible containing a porous glass disk filter. Fritted glass crucibles are classified by their porosity: coarse (retaining particles > 40–60 μm), medium (retaining particles > 10–15 μm), and fine (retaining particles > 4–5.5 μm). Another type of filtering crucible is the Gooch crucible, a porcelain crucible with a perforated bottom. A glass fiber mat is placed in the crucible to retain the precipitate, which is transferred to the crucible in the same manner described for filter paper. The supernatant is drawn through the crucible with the assistance of suction from a vacuum aspirator or pump.

Rinsing the Precipitate Filtering removes most of the supernatant solution. Residual traces of the supernatant, however, must be removed to avoid a source of determinate error. Rinsing the precipitate to remove this residual material must be done carefully to avoid significant losses of the precipitate. Of greatest concern is the potential for solubility losses. Usually the rinsing medium is selected to ensure that solubility losses are negligible. In many cases this simply involves the use of cold solvents or rinse solutions containing organic solvents such as ethanol. Precipitates containing acidic or basic ions may experience solubility losses if the rinse solution's pH is not appropriately adjusted. When coagulation plays an important role in de-

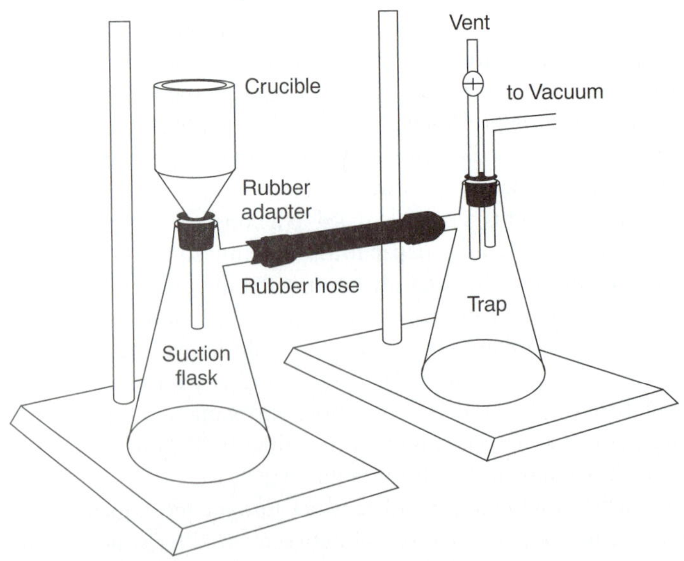

Figure 8.8
Procedure for filtering through a filtering crucible. The trap is used to prevent water from a water aspirator from backwashing into the suction flask.

termining particle size, a volatile inert electrolyte is often added to the rinse water to prevent the precipitate from reverting into smaller particles that may not be retained by the filtering device. This process of reverting to smaller particles is called **peptization.** The volatile electrolyte is removed when drying the precipitate.

peptization
The reverse of coagulation in which a coagulated precipitate reverts to smaller particles.

When rinsing a precipitate there is a trade-off between introducing positive determinate errors due to ionic impurities from the precipitating solution and introducing negative determinate errors from solubility losses. In general, solubility losses are minimized by using several small portions of the rinse solution instead of a single large volume. Testing the used rinse solution for the presence of impurities is another way to ensure that the precipitate is not overrinsed. This can be done by testing for the presence of a targeted solution ion and rinsing until the ion is no longer detected in a freshly collected sample of the rinse solution. For example, when Cl^- is known to be a residual impurity, its presence can be tested for by adding a small amount of $AgNO_3$ to the collected rinse solution. A white precipitate of AgCl indicates that Cl^- is present and additional rinsing is necessary. Additional rinsing is not needed, however, if adding $AgNO_3$ does not produce a precipitate.

Drying the Precipitate Finally, after separating the precipitate from its supernatant solution the precipitate is dried to remove any residual traces of rinse solution and any volatile impurities. The temperature and method of drying depend on the method of filtration, and the precipitate's desired chemical form. A temperature of 110 °C is usually sufficient when removing water and other easily volatilized impurities. A conventional laboratory oven is sufficient for this purpose. Higher temperatures require the use of a muffle furnace, or a Bunsen or Meker burner, and are necessary when the precipitate must be thermally decomposed before weighing or when using filter paper. To ensure that drying is complete the precipitate is repeatedly dried and weighed until a constant weight is obtained.

Filter paper's ability to absorb moisture makes its removal necessary before weighing the precipitate. This is accomplished by folding the filter paper over the precipitate and transferring both the filter paper and the precipitate to a porcelain or platinum crucible. Gentle heating is used to first dry and then to char the filter paper. Once the paper begins to char, the temperature is slowly increased. Although the paper will often show traces of smoke, it is not allowed to catch fire as any precipitate retained by soot particles will be lost. After the paper is completely charred the temperature is slowly raised to a higher temperature. At this stage any carbon left after charring is oxidized to CO_2.

Fritted glass crucibles cannot withstand high temperatures and, therefore, should only be dried in an oven at temperatures below 200 °C. The glass fiber mats used in Gooch crucibles can be heated to a maximum temperature of approximately 500 °C.

Composition of Final Precipitate The quantitative application of precipitation gravimetry, which is based on a conservation of mass, requires that the final precipitate have a well-defined composition. Precipitates containing volatile ions or substantial amounts of hydrated water are usually dried at a temperature that is sufficient to completely remove the volatile species. For example, one standard gravimetric method for the determination of magnesium involves the precipitation of $MgNH_4PO_4 \cdot 6H_2O$. Unfortunately, this precipitate is difficult to dry at lower temperatures without losing an inconsistent amount of hydrated water and ammonia. Instead, the precipitate is dried at temperatures above 1000 °C, where it decomposes to magnesium pyrophosphate, $Mg_2P_2O_7$.

An additional problem is encountered when the isolated solid is nonstoichiometric. For example, precipitating Mn^{2+} as $Mn(OH)_2$, followed by heating to produce the oxide, frequently produces a solid with a stoichiometry of MnO_x, where x varies between 1 and 2. In this case the nonstoichiometric product results from the formation of a mixture of several oxides that differ in the oxidation state of manganese. Other nonstoichiometric compounds form as a result of lattice defects in the crystal structure.[6]

Representative Method The best way to appreciate the importance of the theoretical and practical details discussed in the previous section is to carefully examine the procedure for a typical precipitation gravimetric method. Although each method has its own unique considerations, the determination of Mg^{2+} in water and wastewater by precipitating $MgNH_4PO_4 \cdot 6H_2O$ and isolating $Mg_2P_2O_7$ provides an instructive example of a typical procedure.

Representative Methods

Method 8.1 Determination of Mg^{2+} in Water and Wastewater[7]

Description of Method. Magnesium is precipitated as $MgNH_4PO_4 \cdot 6H_2O$ using $(NH_4)_2HPO_4$ as the precipitant. The precipitate's solubility in neutral solutions (0.0065 g/100 mL in pure water at 10 °C) is relatively high, but it is much less soluble in the presence of dilute ammonia (0.0003 g/100 mL in 0.6 M NH_3). The precipitant is not very selective, so a preliminary separation of Mg^{2+} from potential interferents is necessary. Calcium, which is the most significant interferent, is usually removed by its prior precipitation as the oxalate. The presence of excess ammonium salts from the precipitant or the addition of too much ammonia can lead to the formation of $Mg(NH_4)_4(PO_4)_2$, which is subsequently isolated as $Mg(PO_3)_2$ after drying. The precipitate is isolated by filtration using a rinse solution of dilute ammonia. After filtering, the precipitate is converted to $Mg_2P_2O_7$ and weighed.

Procedure. Transfer a sample containing no more than 60 mg of Mg^{2+} into a 600-mL beaker. Add 2–3 drops of methyl red indicator, and, if necessary, adjust the volume to 150 mL. Acidify the solution with 6 M HCl, and add 10 mL of 30% w/v $(NH_4)_2HPO_4$. After cooling, add concentrated NH_3 dropwise, and while constantly stirring, until the methyl red indicator turns yellow (pH > 6.3). After stirring for 5 min, add 5 mL of concentrated NH_3, and continue stirring for an additional 10 min. Allow the resulting solution and precipitate to stand overnight. Isolate the precipitate by filtration, rinsing with 5% v/v NH_3. Dissolve the precipitate in 50 mL of 10% v/v HCl, and precipitate a second time following the same procedure. After filtering, carefully remove the filter paper by charring. Heat the precipitate at 500 °C until the residue is white, and then bring the precipitate to constant weight at 1100 °C.

Questions

1. **Why does the procedure call for a sample containing no more than 60 mg of Mg^{2+}?**

   A sample containing 60 mg of Mg^{2+} will generate approximately 600 mg, or 0.6 g, of $MgNH_4PO_4 \cdot 6H_2O$. This is a substantial amount of precipitate to work with during the filtration step. Large quantities of precipitate may be difficult to filter and difficult to adequately rinse free of impurities.

—Continued

2. **Why is the solution acidified with HCl before the precipitant is added?**

   The HCl is added to ensure that $MgNH_4PO_4 \cdot 6H_2O$ does not precipitate when the precipitant is initially added. Because PO_4^{3-} is a weak base, the precipitate is soluble in a strongly acidic solution. If the precipitant is added under neutral or basic conditions (high *RSS*), the resulting precipitate will consist of smaller, less pure particles. Increasing the pH by adding base allows the precipitate of $MgNH_4PO_4 \cdot 6H_2O$ to form under more favorable (low *RSS*) conditions.

3. **Why is the acid–base indicator methyl red added to the solution?**

   The indicator's color change, which occurs at a pH of approximately 6.3, indicates when sufficient NH_3 has been added to neutralize the HCl added at the beginning of the procedure. The amount of NH_3 added is crucial to this procedure. If insufficient NH_3 is added, the precipitate's solubility increases, leading to a negative determinate error. If too much NH_3 is added, the precipitate may contain traces of $Mg(NH_4)_4(PO_4)_2$, which, on ignition, forms $Mg(PO_3)_2$. This increases the mass of the ignited precipitate, giving a positive determinate error. Once enough NH_3 has been added to neutralize the HCl, additional NH_3 is added to quantitatively precipitate $MgNH_4PO_4 \cdot 6H_2O$.

4. **Explain why the formation of $Mg(PO_3)_2$ in place of $Mg_2P_2O_7$ increases the mass of precipitate.**

   The desired final precipitate, $Mg_2P_2O_7$, contains two moles of Mg, and the impurity, $Mg(PO_3)_2$, contains only one mole of Mg. Conservation of mass, therefore, requires that two moles of $Mg(PO_3)_2$ must form in place of each mole of $Mg_2P_2O_7$. One mole of $Mg_2P_2O_7$ weights 222.6 g. Two moles of $Mg(PO_3)_2$ weigh 364.5 g. Any replacement of $Mg_2P_2O_7$ with $Mg(PO_3)_2$ must increase the precipitate's mass.

5. **What additional steps in the procedure, beyond those discussed in questions 2 and 3, are taken to improve the precipitate's purity?**

   Two additional steps in the procedure help form a precipitate that is free of impurities: digestion and reprecipitation.

6. **Why is the precipitate rinsed with a solution of 5% v/v NH_3?**

   This is done for the same reason that precipitation is carried out in an ammonical solution; using dilute ammonia minimizes solubility losses when rinsing the precipitate.

8B.2 Quantitative Applications

Although not in common use, precipitation gravimetry still provides a reliable means for assessing the accuracy of other methods of analysis or for verifying the composition of standard reference materials. In this section we review the general application of precipitation gravimetry to the analysis of inorganic and organic compounds.

Inorganic Analysis The most important precipitants for inorganic cations are chromate, the halides, hydroxide, oxalate, sulfate, sulfide, and phosphate. A summary of selected methods, grouped by precipitant, is shown in Table 8.1. Many inorganic anions can be determined using the same reactions by reversing the analyte

Table 8.1 Selected Gravimetric Method for Inorganic Cations Based on Precipitation

Analyte	Precipitant	Precipitate Formed	Precipitate Weighed
Ba^{2+}	$(NH_4)_2CrO_4$	$BaCrO_4$	$BaCrO_4$
Pb^{2+}	K_2CrO_4	$PbCrO_4$	$PbCrO_4$
Ag^+	HCl	AgCl	AgCl
Hg_2^{2+}	HCl	Hg_2Cl_2	Hg_2Cl_2
Al^{3+}	NH_3	$Al(OH)_3$	Al_2O_3
Be^{2+}	NH_3	$Be(OH)_2$	BeO
Fe^{3+}	NH_3	$Fe(OH)_3$	Fe_2O_3
Ca^{2+}	$(NH_4)_2C_2O_4$	CaC_2O_4	$CaCO_3$ or CaO
Sb^{3+}	H_2S	Sb_2S_3	Sb_2S_3
As^{3+}	H_2S	As_2S_3	As_2S_3
Hg^{2+}	H_2S	HgS	HgS
Ba^{2+}	H_2SO_4	$BaSO_4$	$BaSO_4$
Pb^{2+}	H_2SO_4	$PbSO_4$	$PbSO_4$
Sr^{2+}	H_2SO_4	$SrSO_4$	$SrSO_4$
Be^{2+}	$(NH_4)_2HPO_4$	NH_4BePO_4	$Be_2P_2O_7$
Mg^{2+}	$(NH_4)_2HPO_4$	NH_4MgPO_4	$Mg_2P_2O_7$
Sr^{2+}	KH_2PO_4	$SrHPO_4$	$Sr_2P_2O_7$
Zn^{2+}	$(NH_4)_2HPO_4$	NH_4ZnPO_4	$Zn_2P_2O_7$

Table 8.2 Selected Gravimetric Methods for Inorganic Anions Based on Precipitation

Analyte	Precipitant	Precipitate Formed	Precipitate Weighed
CN^-	$AgNO_3$	AgCN	AgCN
I^-	$AgNO_3$	AgI	AgI
Br^-	$AgNO_3$	AgBr	AgBr
Cl^-	$AgNO_3$	AgCl	AgCl
ClO_3^-	$FeSO_4/AgNO_3$	AgCl	AgCl
SCN^-	$SO_2/CuSO_4$	CuSCN	CuSCN
SO_4^{2-}	$BaCl_2$	$BaSO_4$	$BaSO_4$

and precipitant. For example, chromate can be determined by adding $BaCl_2$ and precipitating $BaCrO_4$. Methods for other selected inorganic anions are summarized in Table 8.2. Methods for the homogeneous generation of precipitants are shown in Table 8.3.

The majority of inorganic precipitants show poor selectivity. Most organic precipitants, however, are selective for one or two inorganic ions. Several common organic precipitants are listed in Table 8.4.

Precipitation gravimetry continues to be listed as a standard method for the analysis of Mg^{2+} and SO_4^{2-} in water and wastewater analysis. A description of the procedure for Mg^{2+} was discussed earlier in Method 8.1. Sulfate is analyzed by precipitating $BaSO_4$, using $BaCl_2$ as the precipitant. Precipitation is carried out in an

Table 8.3 Reactions for the Homogeneous Preparation of Selected Inorganic Precipitants

Precipitant	Reaction
OH^-	$(NH_2)_2CO + 3H_2O \rightleftharpoons 2NH_4^+ + CO_2 + 2OH^-$
SO_4^{2-}	$NH_2HSO_3 + 2H_2O \rightleftharpoons NH_4^+ + H_3O^+ + SO_4^{2-}$
S^{2-}	$CH_3CSNH_2 + H_2O \rightleftharpoons CH_3CONH_2 + H_2S$
IO_3^-	$HOCH_2CH_2OH + IO_4^- \rightleftharpoons 2HCHO + H_2O + IO_3^-$
PO_4^{2-}	$(CH_3O)_3PO + 3H_2O \rightleftharpoons 3CH_3OH + H_3PO_4$
$C_2O_4^{2-}$	$(C_2H_5)_2C_2O_4 + 2H_2O \rightleftharpoons 2C_2H_5OH + H_2C_2O_4$
CO_3^{2-}	$Cl_3CCOOH + 2OH^- \rightleftharpoons CHCl_3 + CO_3^{2-} + H_2O$

Table 8.4 Selected Gravimetric Methods for Inorganic Cations Based on Precipitation with Organic Precipitants

Analyte	Precipitant	Structure	Precipitate Formed	Precipitate Weighed
Ni^{2+}	dimethylgloxime		$Ni(C_4H_7O_2N_2)_2$	$Ni(C_4H_7O_2N_2)_2$
Fe^{3+}	cupferron		$Fe(C_6H_5N_2O_2)_3$	Fe_2O_3
Cu^{2+}	cupron		$CuC_{14}H_{11}O_2N$	$CuC_{14}H_{11}O_2N$
Co^{2+}	1-nitroso-2-naphthol		$Co(C_{10}H_6O_2N)_3$	Co or $CoSO_4$
K^+	sodium tetraphenylborate	$Na[B(C_6H_5)_4]$	$K[B(C_6H_5)_4]$	$K[B(C_6H_5)_4]$
NO_3^-	nitron		$C_{20}H_{16}N_4HNO_3$	$C_{20}H_{16}N_4HNO_3$

Table 8.5 Selected Gravimetric Methods for the Analysis of Organic Functional Groups and Heteroatoms Based on Precipitation

Analyte	Treatment	Precipitant	Precipitate
Organic halides R-X X = Cl, Br, I	Oxidation with HNO_3 in presence of Ag^+	$AgNO_3$	AgX
Organic Halides R-X X = Cl, Br, I	Combustion in O_2 (with Pt catalyst) in presence of Ag^+	$AgNO_3$	AgX
Organic sulfur	Oxidation with HNO_3 in presence of Ba^{2+}	$BaCl_2$	$BaSO_4$
Organic sulfur	Combustion in O_2 (with Pt catalyst) to produce SO_2 and SO_3, which are collected in dilute H_2O_2	$BaCl_2$	$BaSO_4$
Alkoxy groups R′-OR R = CH_3 or C_2H_5 or R′—C(=O)—OR	Reaction with HI to produce RI	$AgNO_3$	AgI
Alkimide group >N—R R = CH_3, C_2H_5 N may be 1°, 2°, or 3°	Reaction with HI to produce RI	$AgNO_3$	AgI

acidic solution (acidified to pH 4.5–5.0 with HCl) to prevent the possible precipitation of $BaCO_3$ or $Ba_3(PO_4)_2$ and performed near the solution's boiling point. The precipitate is digested at 80–90 °C for at least 2 h. Ashless filter paper pulp is added to the precipitate to aid in filtration. After filtering, the precipitate is ignited to constant weight at 800 °C. Alternatively, the precipitate can be filtered through a fine-porosity fritted glass crucible (without adding filter paper pulp) and dried to constant weight at 105 °C. This procedure is subject to a variety of errors, including occlusions of $Ba(NO_3)_2$, $BaCl_2$, and alkali sulfates.

Organic Analysis Several organic functional groups or heteroatoms can be determined using gravimetric precipitation methods; examples are outlined in Table 8.5. Note that the procedures for the alkoxy and alkimide functional groups are examples of indirect analyses.

Quantitative Calculations In precipitation gravimetry the relationship between the analyte and the precipitate is determined by the stoichiometry of the relevant reactions. As discussed in Section 2C, gravimetric calculations can be simplified by applying the principle of conservation of mass. The following example demonstrates the application of this approach to the direct analysis of a single analyte.

Example

EXAMPLE 8.1

An ore containing magnetite, Fe_3O_4, was analyzed by dissolving a 1.5419-g sample in concentrated HCl, giving a mixture of Fe^{2+} and Fe^{3+}. After adding HNO_3 to oxidize any Fe^{2+} to Fe^{3+}, the resulting solution was diluted with water and the Fe^{3+} precipitated as $Fe(OH)_3$ by adding NH_3. After filtering and rinsing, the residue was ignited, giving 0.8525 g of pure Fe_2O_3. Calculate the %w/w Fe_3O_4 in the sample.

SOLUTION

This is an example of a direct analysis since the iron in the analyte, Fe_3O_4, is part of the isolated precipitate, Fe_2O_3. Applying a conservation of mass to Fe, we write

$$3 \times \text{moles } Fe_3O_4 = 2 \times \text{moles } Fe_2O_3$$

Using formula weights, FW, to convert from moles to grams in the preceding equation leaves us with

$$\frac{3 \times \text{g } Fe_3O_4}{\text{FW } Fe_3O_4} = \frac{2 \times \text{g } Fe_2O_3}{\text{FW } Fe_2O_3}$$

which can be solved for grams of Fe_3O_4 and %w/w Fe_3O_4 in the sample.

$$\frac{2 \times \text{g } Fe_2O_3 \times \text{FW } Fe_3O_4}{3 \times \text{FW } Fe_2O_3} = \frac{2 \times 0.8525 \text{ g} \times 231.54 \text{ g/mol}}{3 \times 159.69 \text{ g/mol}} = 0.82405 \text{ g } Fe_3O_4$$

$$\frac{\text{g } Fe_3O_4}{\text{g sample}} \times 100 = \frac{0.82405 \text{ g}}{1.5419 \text{ g}} \times 100 = 53.44\% \text{ w/w Fe}$$

As discussed earlier, the simultaneous analysis of samples containing two analytes requires the isolation of two precipitates. As shown in Example 8.2, conservation of mass can be used to write separate stoichiometric equations for each precipitate. These equations can then be solved simultaneously for both analytes.

Example

EXAMPLE 8.2

A 0.6113-g sample of Dow metal, containing aluminum, magnesium, and other metals, was dissolved and treated to prevent interferences by the other metals. The aluminum and magnesium were precipitated with 8-hydroxyquinoline. After filtering and drying, the mixture of $Al(C_9H_6NO)_3$ and $Mg(C_9H_6NO)_2$ was found to weigh 7.8154 g. The mixture of dried precipitates was then ignited, converting the precipitate to a mixture of Al_2O_3 and MgO. The weight of this mixed solid was found to be 1.0022 g. Calculate the %w/w Al and %w/w Mg in the alloy.

SOLUTION

This is an example of a direct analysis in which the two analytes are determined without a prior separation. The weight of the original precipitate and the ignited precipitate are given by the following two equations

$$\text{g } Al(C_9H_6NO)_3 + \text{g } Mg(C_9H_6NO)_2 = 7.8154$$

$$\text{g } Al_2O_3 + \text{g MgO} = 1.0022$$

containing four unknown terms. To solve this pair of equations, we must find two additional equations relating the four unknowns to one another. These additional equations describe the stoichiometric relationships between the two compounds containing aluminum and the two compounds containing magnesium and are based on the conservation of Al and Mg. Thus, for Al we have

$$2 \times \text{moles } Al_2O_3 = \text{moles } Al(C_9H_6NO)_3$$

Converting from moles to grams and solving yields an equation relating the grams of Al_2O_3 to the grams of $Al(C_9H_6NO)_3$

$$\frac{2 \times \text{g } Al_2O_3}{\text{FW } Al_2O_3} = \frac{\text{g } Al(C_9H_6NO)_3}{\text{FW } Al(C_9H_6NO)_3}$$

$$\text{g } Al_2O_3 = \frac{\text{g } Al(C_9H_6NO)_3 \times \text{FW } Al_2O_3}{2 \times \text{FW } Al(C_9H_6NO)_3} = \frac{\text{g } Al(C_9H_6NO)_3 \times 101.96 \text{ g/mol}}{2 \times 459.45 \text{ g/mol}}$$

$$= 0.11096 \times \text{g } Al(C_9H_6NO)_3$$

For Mg we have

$$\text{Moles MgO} = \text{moles } Mg(C_9H_6NO)_2$$

$$\frac{\text{g MgO}}{\text{FW MgO}} = \frac{\text{g } Mg(C_9H_6NO)_2}{\text{FW } Mg(C_9H_6NO)_2}$$

$$\text{g MgO} = \frac{\text{g } Mg(C_9H_6NO)_2 \times \text{FW MgO}}{\text{FW } Mg(C_9H_6NO)_2} = \frac{\text{g } Mg(C_9H_6NO)_2 \times 40.304 \text{ g/mol}}{312.61 \text{ g/mol}}$$

$$= 0.12893 \times \text{g } Mg(C_9H_6NO)_2$$

Substituting the equations for g MgO and g Al_2O_3 into the equation for the combined weights of MgO and Al_2O_3 leaves us with two equations and two unknowns.

$$\text{g } Al(C_9H_6NO)_3 + \text{g } Mg(C_9H_6NO)_2 = 7.8154$$

$$0.11096 \times \text{g } Al(C_9H_6NO)_3 + 0.12893 \times \text{g } Mg(C_9H_6NO)_2 = 1.0022$$

Multiplying the first equation by 0.11096 and subtracting the second equation gives

$$-0.01797 \times \text{g } Mg(C_9H_6NO)_2 = -0.1350$$

which can be solved for the mass of $Mg(C_9H_6NO)_2$.

$$\text{g } Mg(C_9H_6NO)_2 = 7.5125 \text{ g}$$

The mass of $Al(C_9H_6NO)_3$ can then be calculated using the known combined mass of the two original precipitates.

$$7.8154 \text{ g} - \text{g } Mg(C_9H_6NO)_2 = 7.8154 \text{ g} - 7.5125 \text{ g} = 0.3029 \text{ g } Al(C_9H_6NO)_3$$

Using the conservation of Mg and Al, the %w/w Mg and %w/w Al in the sample can now be determined as in Example 8.1, where AW is an atomic weight.

Moles Mg = moles $Mg(C_9H_6NO)_2$

$$\frac{\text{g Mg}}{\text{AW Mg}} = \frac{\text{g Mg}(C_9H_6NO)_2}{\text{FW Mg}(C_9H_6NO)_2}$$

$$\text{g Mg} = \frac{\text{g Mg}(C_9H_6NO)_2 \times \text{AW Mg}}{\text{FW Mg}(C_9H_6NO)_2}$$

$$= \frac{7.5125\text{ g} \times 24.305\text{ g/mol}}{312.61\text{ g/mol}}$$

$$= 0.5841\text{ g}$$

$$\%\text{Mg} = \frac{\text{g Mg}}{\text{g sample}} \times 100$$

$$\frac{0.5841\text{ g}}{0.6113\text{ g}} \times 100 = 95.55\%\text{ w/w Mg}$$

Moles Al = moles $Al(C_9H_6NO)_3$

$$\frac{\text{g Al}}{\text{AW Al}} = \frac{\text{g Al}(C_9H_6NO)_3}{\text{FW Al}(C_9H_6NO)_3}$$

$$\text{g Al} = \frac{\text{g Al}(C_9H_6NO)_3 \times \text{AW Al}}{\text{FW Al}(C_9H_6NO)_3}$$

$$= \frac{0.3029\text{ g} \times 26.982\text{ g/mol}}{459.45\text{ g/mol}}$$

$$= 0.0178\text{ g}$$

$$\%\text{Al} = \frac{\text{g Al}}{\text{g sample}} \times 100$$

$$\frac{0.0178\text{ g}}{0.6113\text{ g}} \times 100 = 2.91\%\text{ w/w Al}$$

In an indirect analysis the precipitate does not contain the analyte, but is the product of a reaction involving the analyte. Despite the additional complexity, a stoichiometric relationship between the analyte and the precipitate can be written by applying the conservation principles discussed in Section 2C.

Example

EXAMPLE 8.3

An impure sample of Na_3PO_3 weighing 0.1392 g was dissolved in 25 mL of water. A solution containing 50 mL of 3% w/v mercury(II) chloride, 20 mL of 10% w/v sodium acetate and 5 mL of glacial acetic acid was then prepared. The solution containing the phosphite was added dropwise to the second solution, oxidizing PO_3^{3-} to PO_4^{3-} and precipitating Hg_2Cl_2. After digesting, filtering, and rinsing, the precipitated Hg_2Cl_2 was found to weigh 0.4320 g. Report the purity of the original sample as %w/w Na_3PO_3.

SOLUTION

This is an example of an indirect analysis since the isolated precipitate, Hg_2Cl_2, does not contain the analyte, Na_3PO_3. The stoichiometry of the redox reaction in which Na_3PO_3 is a reactant and Hg_2Cl_2 is a product was given earlier in the chapter. It also can be determined by using the conservation of electrons. Phosphorus has an oxidation state of +3 in PO_3^{3-} and +5 in PO_4^{3-}; thus, oxidizing PO_3^{3-} to PO_4^{3-} requires two electrons. The formation of Hg_2Cl_2 by reduction of $HgCl_2$ requires 2 electrons as the oxidation state of each mercury changes from +2 to +1. Since the oxidation of PO_3^{3-} and the formation of Hg_2Cl_2 both require two electrons, we have

$$\text{Moles } Na_3PO_3 = \text{moles } Hg_2Cl_2$$

Converting from moles to grams leaves us with

$$\frac{\text{g } Na_3PO_3}{\text{FW } Na_3PO_3} = \frac{\text{g } Hg_2Cl_2}{\text{FW } Hg_2Cl_2}$$

which can be solved for g Na_3PO_3 and %w/w Na_3PO_3 in the sample.

$$\text{g Na}_3\text{PO}_3 = \frac{\text{g Hg}_2\text{Cl}_2 \times \text{FW Na}_3\text{PO}_3}{\text{FW Hg}_2\text{Cl}_2} = \frac{0.4320 \text{ g} \times 147.94 \text{ g/mol}}{472.09 \text{ g/mol}} = 0.1354 \text{ g}$$

$$\frac{\text{g Na}_3\text{PO}_3}{\text{g sample}} \times 100 = \frac{0.1354 \text{ g}}{0.1392 \text{ g}} \times 100 = 97.27\% \text{ w/w Na}_3\text{PO}_3$$

8B.3 Qualitative Applications

Precipitation gravimetry can also be applied to the identification of inorganic and organic analytes, using precipitants such as those outlined in Tables 8.1, 8.2, 8.4, and 8.5. Since this does not require quantitative measurements, the analytical signal is simply the observation that a precipitate has formed. Although qualitative applications of precipitation gravimetry have been largely replaced by spectroscopic methods of analysis, they continue to find application in spot testing for the presence of specific analytes.[8]

8B.4 Evaluating Precipitation Gravimetry

Scale of Operation The scale of operation for precipitation gravimetry is governed by the sensitivity of the balance and the availability of sample. To achieve an accuracy of ±0.1% using an analytical balance with a sensitivity of ±0.1 mg, the precipitate must weigh at least 100 mg. As a consequence, precipitation gravimetry is usually limited to major or minor analytes, and macro or meso samples (see Figure 3.6 in Chapter 3). The analysis of trace level analytes or micro samples usually requires a microanalytical balance.

Accuracy For macro–major samples, relative errors of 0.1–0.2% are routinely achieved. The principal limitations are solubility losses, impurities in the precipitate, and the loss of precipitate during handling. When it is difficult to obtain a precipitate free from impurities, an empirical relationship between the precipitate's mass and the mass of the analyte can be determined by an appropriate standardization.

Precision The relative precision of precipitation gravimetry depends on the amount of sample and precipitate involved. For smaller amounts of sample or precipitate, relative precisions of 1–2 ppt are routinely obtained. When working with larger amounts of sample or precipitate, the relative precision can be extended to several parts per million. Few quantitative techniques can achieve this level of precision.

Sensitivity For any precipitation gravimetric method, we can write the following general equation relating the signal (grams of precipitate) to the absolute amount of analyte in the sample

$$\text{Grams precipitate} = k \times \text{grams of analyte} \qquad \mathbf{8.13}$$

where k, the method's sensitivity, is determined by the stoichiometry between the precipitate and the analyte. Note that equation 8.13 assumes that a blank has been used to correct the signal for the reagent's contribution to the precipitate's mass.

Consider, for example, the determination of Fe as Fe_2O_3. Using a conservation of mass for Fe we write

$$2 \times \text{moles Fe}_2\text{O}_3 = \text{moles Fe}$$

Converting moles to grams and rearranging yields an equation in the form of 8.13

$$\text{g Fe}_2\text{O}_3 = \frac{1}{2} \times \frac{\text{FW Fe}_2\text{O}_3}{\text{AW Fe}} \times \text{g Fe}$$

where k is equal to

$$k = \frac{1}{2} \times \frac{\text{FW Fe}_2\text{O}_3}{\text{AW Fe}} \qquad \textbf{8.14}$$

As can be seen from equation 8.14, we may improve a method's sensitivity in two ways. The most obvious way is to increase the ratio of the precipitate's molar mass to that of the analyte. In other words, it is desirable to form a precipitate with as large a formula weight as possible. A less obvious way to improve the calibration sensitivity is indicated by the term of 1/2 in equation 8.14, which accounts for the stoichiometry between the analyte and precipitate. Sensitivity also may be improved by forming precipitates containing fewer units of the analyte.

Selectivity Due to the chemical nature of the precipitation process, precipitants are usually not selective for a single analyte. For example, silver is not a selective precipitant for chloride because it also forms precipitates with bromide and iodide. Consequently, interferents are often a serious problem that must be considered if accurate results are to be obtained.

Time, Cost, and Equipment Precipitation gravimetric procedures are time-intensive and rarely practical when analyzing a large number of samples. However, since much of the time invested in precipitation gravimetry does not require an analyst's immediate supervision, it may be a practical alternative when working with only a few samples. Equipment needs are few (beakers, filtering devices, ovens or burners, and balances), inexpensive, routinely available in most laboratories, and easy to maintain.

8C Volatilization Gravimetry

A second approach to gravimetry is to thermally or chemically decompose a solid sample. The volatile products of the decomposition reaction may be trapped and weighed to provide quantitative information. Alternatively, the residue remaining when decomposition is complete may be weighed. In **thermogravimetry,** which is one form of volatilization gravimetry, the sample's mass is continuously monitored while the applied temperature is slowly increased.

thermogravimetry
A form of volatilization gravimetry in which the change in a sample's mass is monitored while it is heated.

8C.1 Theory and Practice

Whether the analysis is direct or indirect, volatilization gravimetry requires that the products of the decomposition reaction be known. This requirement is rarely a problem for organic compounds for which volatilization is usually accomplished by combustion and the products are gases such as CO_2, H_2O, and N_2. For inorganic compounds, however, the identity of the volatilization products may depend on the temperature at which the decomposition is conducted.

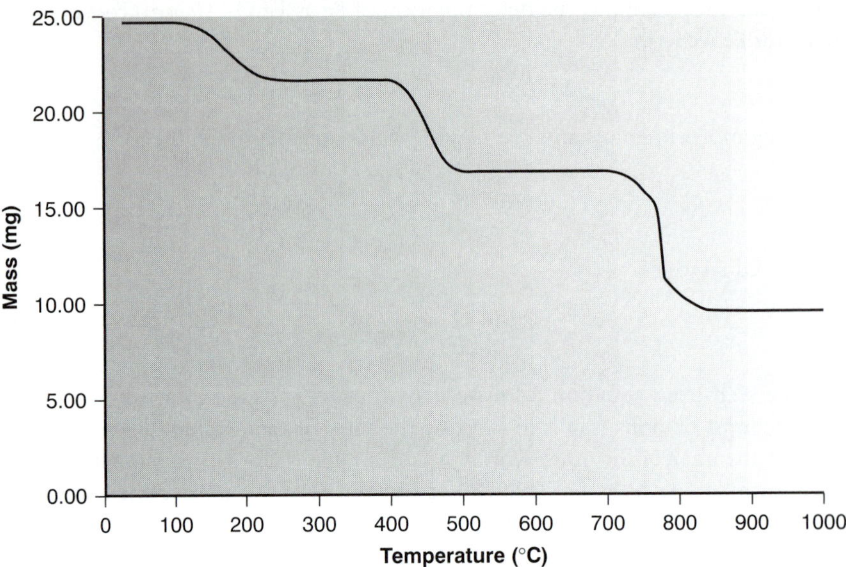

Figure 8.9
Thermogram for $CaC_2O_4 \cdot H_2O$.

thermogram
A graph showing change in mass as a function of applied temperature.

Thermogravimetry The products of a thermal decomposition can be deduced by monitoring the sample's mass as a function of applied temperature. (Figure 8.9). The loss of a volatile gas on thermal decomposition is indicated by a step in the **thermogram.** As shown in Example 8.4, the change in mass at each step in a thermogram can be used to identify both the volatilized species and the solid residue.

Example

EXAMPLE 8.4

The thermogram in Figure 8.9 shows the change in mass for a sample of calcium oxalate monohydrate, $CaC_2O_4 \cdot H_2O$. The original sample weighed 24.60 mg and was heated from room temperature to 1000 °C at a rate of 5 °C min. The following changes in mass and corresponding temperature ranges were observed:

Loss of 3.03 mg from 100–250 °C

Loss of 4.72 mg from 400–500 °C

Loss of 7.41 mg from 700–850 °C

Determine the identities of the volatilization products and the solid residue at each step of the thermal decomposition.

SOLUTION

The loss of 3.03 mg from 100–250 °C corresponds to a 12.32% decrease in the original sample's mass.

$$\frac{3.03 \text{ mg}}{24.60 \text{ mg}} \times 100 = 12.32\%$$

In terms of $CaC_2O_4 \cdot H_2O$, this corresponds to a loss of 18.00 g/mol.

$$0.1232 \times 146.11 \text{ g/mol} = 18.00 \text{ g/mol}$$

The product's molar mass, coupled with the temperature range, suggests that this represents the loss of H_2O. The residue is CaC_2O_4.

The loss of 4.72 mg from 400–500 °C represents a 19.19% decrease in the original mass of 24.60 g, or a loss of

$$0.1919 \times 146.11 \text{ g/mol} = 28.04 \text{ g/mol}$$

This loss is consistent with CO as the volatile product, leaving a residue of $CaCO_3$.

Finally, the loss of 7.41 mg from 700–850 °C is a 30.12% decrease in the original mass of 24.60 g. This is equivalent to a loss of

$$0.3012 \times 146.11 \text{ g/mol} = 44.01 \text{ g/mol}$$

suggesting the loss of CO_2. The final residue is CaO.

Once the products of thermal decomposition have been determined, an analytical procedure can be developed. For example, the thermogram in Figure 8.9 shows that a precipitate of $CaC_2O_4 \cdot H_2O$ must be heated at temperatures above 250 °C, but below 400 °C if it is to be isolated as CaC_2O_4. Alternatively, by heating the sample to 1000 °C, the precipitate can be isolated as CaO. Knowing the identity of the volatilization products also makes it possible to design an analytical method in which one or more of the gases are trapped. Thus, a sample of $CaC_2O_4 \cdot H_2O$ could be analyzed by heating to 1000 °C and passing the volatilized gases through a trap that selectively retains H_2O, CO, or CO_2.

Equipment Depending on the method, the equipment for volatilization gravimetry may be simple or complex. In the simplest experimental design, the weight of a solid residue is determined following either thermal decomposition at a fixed temperature or combustion. Thermal decomposition or combustion is accomplished using a Bunsen or Meker burner, a laboratory oven or a muffle furnace, with the volatile products vented to the atmosphere. The weight of the sample and the solid residue are determined using an analytical balance.

Constant-temperature decomposition or combustion, followed by trapping and weighing the volatilized gases, requires more specialized equipment. Decomposition of the sample is conducted in a closed container, and the volatilized gases are carried by a purge-gas stream through one or more selective absorbent traps.

In a thermogravimetric analysis, the sample is placed in a small weighing boat attached to one arm of a specially designed electromagnetic balance and placed inside an electric furnace. The temperature of the electric furnace is slowly increased at a fixed rate of a few degrees per minute, and the sample's weight is monitored.

Representative Method Although each volatilization gravimetric procedure has its own unique characteristics, the following indirect method for the determination of Si in ores and alloys by formation of volatile SiF_4 provides an instructive example of a typical procedure.

Representative Methods

Method 8.2 Determination of Si in Ores and Alloys[9]

Description of Method. Silicon is determined by dissolving the sample in acid. Dehydration of the resulting solution precipitates silicon as SiO_2. Because a variety of other insoluble oxides also form, the precipitate's mass does not provide a direct measure of the amount of silicon in the sample. Treating the solid residue with HF results in the formation of volatile SiF_4. The decrease in mass following the loss of SiF_4 provides an indirect measure of the amount of silicon in the original sample.

Procedure. Transfer a sample of between 0.5 and 5 g to a platinum crucible along with an excess of Na_2CO_3, and heat until a melt is formed. After cooling, dissolve the residue in dilute HCl. Dehydrating silicon to SiO_2 is accomplished by evaporating the solution to dryness on a steam bath and heating the residue for 1 hour at 110 °C. Moisten the residue with HCl, and repeat the dehydration. Remove any acid-soluble materials from the residue by adding 50 mL of water and 5 mL of concentrated HCl. Bring to a boil, and filter through #40 filter paper. Wash the residue with hot 2% v/v HCl followed by hot water. Evaporate the filtrate to dryness twice, and, following the same procedure, treat to remove any acid–soluble materials. Combine the two precipitates, and dry and ignite to a constant weight at 1200 °C. After cooling, add 2 drops of 50% v/v H_2SO_4 and 10 mL of HF. Remove the volatile SiF_4 by evaporating to dryness on a hot plate. Finally, bring the residue to constant weight by igniting at 1200 °C.

Questions

1. **According to the procedure, the sample should weigh between 0.5 and 5 g. On what basis should a decision on the amount of sample be made?**

   In this procedure the critical measurement is the decrease in mass following the volatilization of SiF_4. The reaction that occurs is

   $$SiO_2(s) + 4HF(aq) \rightarrow SiF_4(g) + 2H_2O(\ell)$$

   The water and any excess HF are removed during the final ignition and do not contribute to the change in mass. The loss in weight, therefore, is equal to the grams of SiO_2 present after the dehydration step. For every 0.1 g of Si in the original sample, a weight loss of 0.21 g is expected. The amount of sample used is determined by how much Si is present. If a sample is 50% w/w Si, a 0.5-g sample will give a respectable weight loss of 0.53 g. A 0.5-g sample that is only 5% w/w Si, however, will give a weight loss of only 0.053 g. In this case, a larger sample is needed.

2. **Why are acid-soluble materials removed before the dehydrated residue is treated with HF?**

   Any acid-soluble materials present in the sample will react with HF or H_2SO_4. If the products of these reactions are volatile or decompose at the ignition temperature of 1200 °C, then the change in weight will not be due solely to the volatilization of SiF_4. The result is a positive determinate error.

3. **Why is H_2SO_4 added with the HF?**

   Many samples containing silicon also contain aluminum and iron. After dehydration, these metals are present as Al_2O_3 and Fe_2O_3. These oxides are potential interferents since they also are capable of forming volatile fluorides. In the presence of H_2SO_4, however, aluminum and iron form nonvolatile sulfates. These sulfates decompose back to their respective oxides when ignited to 1200 °C. As a result, the change in weight after treating with HF and H_2SO_4 is due only to the loss of SiF_4.

8C.2 Quantitative Applications

Unlike precipitation gravimetry, which is rarely used as a standard method of analysis, gravimetric methods based on volatilization reactions continue to play an important role in chemical analysis. Several important examples are discussed in the following sections.

Inorganic Analysis Determining the inorganic ash content of organic materials, such as polymers and paper, is an example of a direct volatilization gravimetric analysis. The sample is weighed, placed in an appropriate crucible, and the organic material is carefully removed by combustion. The crucible containing the residue is then heated to a constant weight using either a burner or an oven.

Another example of volatilization gravimetry is the determination of dissolved solids in water and wastewater. In this method a sample of the water is transferred to a weighed dish and dried to a constant weight at either 103–105 °C, or at 180 °C. Samples dried at the lower temperature retain some occluded water and lose some carbonate as CO_2. The loss of organic material, however, is minimal. At the higher temperature, the residue is free from occluded water, but losses of carbonate are greater. In addition, some chloride, nitrate, and organic material are lost through thermal decomposition. The residue remaining after drying at either temperature can be ignited to constant weight at 500 °C. The loss in weight on ignition provides an indirect measure of the amount of volatile solids in the sample, and the weight of the remaining residue gives the amount of fixed solids.

Indirect analyses based on the weight of the residue remaining after volatilization are commonly used in determining moisture in a variety of products and in determining silica in water, wastewater, and rocks. Moisture is determined by drying a preweighed sample with an infrared lamp or in a low-temperature oven. The difference between the original weight and the weight after drying equals the mass of water lost.

The determination of silicon is commonly encountered in metallurgical and mining laboratories responsible for the analysis of ores, slags, and alloys. The volatilization gravimetric method, which is appropriate for samples containing high concentrations of silicon, was described earlier in Method 8.2.

As a final example, the determination of carbon in steels and other metal alloys can be determined by heating the sample. The carbon is converted to CO_2, which is collected in an appropriate absorbent trap, providing a direct measure of the amount of C in the original sample.

Organic Analysis The most important application of volatilization gravimetry to the analysis of organic materials is an elemental analysis. When burned in a stream of pure O_2, many elements, such as carbon and hydrogen, are released as gaseous combustion products, such as CO_2 and H_2O. The combustion products are passed through preweighed tubes containing appropriate absorbents. The increase in the mass of these tubes provides a direct indication of the mass percent of carbon and hydrogen in the organic material.

Alkaline metals and earths in organic materials can be determined by adding H_2SO_4 to the sample before combustion. Following combustion, the metal remains behind as a solid residue of metal sulfate. Silver, gold, and platinum can be determined by burning the organic sample, leaving a metallic residue of Ag, Au, or Pt. Other metals are determined by adding HNO_3 before combustion, leaving a residue of the metal oxide.

Volatilization gravimetry is also used to determine biomass in water and wastewater. Biomass is a water quality index, providing an indication of the total mass of organisms contained within a sample of water. A known volume of the sample is passed through a preweighed 0.45-μm membrane filter or a glass-fiber filter and dried at 105 °C for 24 h. The residue's mass provides a direct measure of biomass. If samples are known to contain a substantial amount of dissolved inorganic solids, the residue can be ignited at 500 °C for 1 h, thereby removing all organic materials. The resulting residue is wetted with distilled water to rehydrate any clay minerals and dried to a constant weight at 105 °C. The difference in weight before and after ignition provides an indirect measure of biomass.

Quantitative Calculations When needed, the relationship between the analyte and the analytical signal is given by the stoichiometry of any relevant reactions. Calculations are simplified, however, by applying the principle of conservation of mass. The most frequently encountered example of a direct volatilization gravimetric analysis is the determination of a compound's elemental composition.

Example

EXAMPLE 8.5

A 101.3-mg sample of an organic compound known to contain Cl is burned in pure O_2 and the combustion gases collected in absorbent tubes. The tube used to trap CO_2 increases in mass by 167.6 mg, and the tube for trapping H_2O shows a 13.7-mg increase. A second sample of 121.8 mg is treated with concentrated HNO_3 producing Cl_2, which subsequently reacts with Ag^+, forming 262.7 mg of AgCl. Determine the compound's composition, as well as its empirical formula.

SOLUTION

Applying the principle of conservation of mass to carbon, we write

$$\text{Moles C} = \text{moles CO}_2$$

Converting from moles to grams, and rearranging to solve for milligrams of carbon gives

$$\frac{\text{g CO}_2 \times \text{AW C} \times 1000 \text{ mg/g}}{\text{FW CO}_2} = \frac{0.1676 \text{ g} \times 12.011 \text{ g/mol} \times 1000 \text{ mg/g}}{44.011 \text{ g/mol}}$$

$$= 45.74 \text{ mg C}$$

Thus, the %w/w C in the sample is

$$\frac{\text{mg C}}{\text{mg sample}} \times 100 = \frac{45.74 \text{ mg}}{101.3 \text{ mg}} \times 100 = 45.15\% \text{ w/w C}$$

The calculation is then repeated for hydrogen

$$\text{Moles H} = 2 \times \text{moles H}_2\text{O}$$

$$\frac{2 \times \text{g H}_2\text{O} \times \text{AW H} \times 1000 \text{ mg/g}}{\text{FW H}_2\text{O}} = \frac{2 \times 0.0137 \text{ g} \times 1.008 \text{ g/mol} \times 1000 \text{ mg/g}}{18.015 \text{ g/mol}}$$

$$= 1.533 \text{ mg H}$$

$$\frac{\text{mg H}}{\text{mg sample}} \times 100 = \frac{1.533 \text{ mg}}{101.3 \text{ mg}} \times 100 = 1.51\% \text{ w/w H}$$

and for chlorine

$$\text{Moles Cl} = \text{moles AgCl}$$

$$\frac{\text{g AgCl} \times \text{AW Cl} \times 1000 \text{ mg/g}}{\text{FW AgCl}} = \frac{0.2627 \text{ g} \times 35.453 \text{ g/mol} \times 1000 \text{ mg/g}}{143.32 \text{ g/mol}}$$

$$= 64.98 \text{ mg Cl}$$

$$\frac{\text{mg Cl}}{\text{mg sample}} \times 100 = \frac{64.98 \text{ mg}}{121.8 \text{ mg}} \times 100 = 53.35\% \text{ w/w Cl}$$

Adding together the weight percents for C, H, and Cl gives a total of 100.01%. The compound, therefore, contains only these three elements. To determine the compound's empirical formula, we assume that we have 1 g of the compound, giving 0.4515 g of C, 0.0151 g of H, and 0.5335 g of Cl. Expressing each element in moles gives 0.0376 mol C, 0.0150 mol H, and 0.0150 mol Cl. Hydrogen and chlorine are present in a 1:1 molar ratio. The molar ratio of C to moles of H or Cl is

$$\frac{\text{Moles C}}{\text{Moles H}} = \frac{\text{moles C}}{\text{moles Cl}} = \frac{0.0376}{0.0150} = 2.51 \approx 2.5$$

Thus, the simplest, or empirical, formula for the compound is $C_5H_2Cl_2$.

In an indirect volatilization gravimetric analysis, the change in the sample's weight is proportional to the amount of analyte. Note that in the following example it is not necessary to apply the conservation of mass to relate the analytical signal to the analyte.

Example

EXAMPLE 8.6

A sample of slag from a blast furnace is analyzed for SiO_2 by decomposing a 0.5003-g sample with HCl, leaving a residue with a mass of 0.1414 g. After treating with HF and H_2SO_4 and evaporating the volatile SiF_4, a residue with a mass of 0.0183 g remains. Determine the %w/w SiO_2 in the sample.

SOLUTION

In this procedure the difference in the residue's mass before and after volatilizing SiF_4 gives the mass of SiO_2 in the sample. Thus the sample contained

$$0.1414 \text{ g} - 0.0183 \text{ g} = 0.1231 \text{ g } SiO_2$$

The %w/w SiO_2, therefore, is

$$\frac{\text{g } SiO_2}{\text{g sample}} \times 100 = \frac{0.1231 \text{ g}}{0.5003 \text{ g}} \times 100 = 24.61\% \text{ w/w } SiO_2$$

Finally, in some quantitative applications it is necessary to compare the result for a sample with a similar result obtained using a standard.

Example

EXAMPLE 8.7

A 26.23-mg sample of $MgC_2O_4 \cdot H_2O$ and inert materials is heated to constant weight at 1200 °C, leaving a residue weighing 20.98 mg. A sample of pure $MgC_2O_4 \cdot H_2O$, when treated in the same fashion, undergoes a 69.08% change in its mass. Determine the %w/w $MgC_2O_4 \cdot H_2O$ in the sample.

SOLUTION

The change in mass when analyzing the mixture is 5.25 mg, thus the grams of $MgC_2O_4 \cdot H_2O$ in the sample is

$$5.25 \text{ mg lost} \times \frac{100 \text{ mg MgC}_2\text{O}_4 \cdot \text{H}_2\text{O}}{69.08 \text{ mg lost}} = 7.60 \text{ mg MgC}_2\text{O}_4 \cdot \text{H}_2\text{O}$$

The %w/w $MgC_2O_4 \cdot H_2O$, therefore, is

$$\frac{\text{mg MgC}_2\text{O}_4 \cdot \text{H}_2\text{O}}{\text{mg sample}} \times 100 = \frac{7.60 \text{ mg}}{26.23 \text{ mg}} \times 100 = 29.0\% \text{ w/w MgC}_2\text{O}_4 \cdot \text{H}_2\text{O}$$

8C.3 Evaluating Volatilization Gravimetry

The scale of operation, accuracy, and precision of gravimetric volatilization methods are similar to that described in Section 8B.4 for precipitation gravimetry. The sensitivity for a direct analysis is fixed by the analyte's chemical form following combustion or volatilization. For an indirect analysis, however, sensitivity can be improved by carefully choosing the conditions for combustion or volatilization so that the change in mass is as large as possible. For example, the thermogram in Figure 8.9 shows that an indirect analysis for $CaC_2O_4 \cdot H_2O$ based on the weight of the residue following ignition at 1000 °C will be more sensitive than if the ignition was done 300 °C. Selectivity does not present a problem for direct volatilization gravimetric methods in which the analyte is a gaseous product retained in an absorbent trap. A direct analysis based on the residue's weight following combustion or volatilization is possible when the residue only contains the analyte of interest. As noted earlier, indirect analyses are only feasible when the residue's change in mass results from the loss of a single volatile product containing the analyte.

Volatilization gravimetric methods are time- and labor-intensive. Equipment needs are few except when combustion gases must be trapped or for a thermogravimetric analysis, which requires specialized equipment.

8D Particulate Gravimetry

Gravimetric methods based on precipitation or volatilization reactions require that the analyte, or some other species in the sample, participate in a chemical reaction producing a change in physical state. For example, in direct precipitation gravimetry, a soluble analyte is converted to an insoluble form that precipitates from solution. In some situations, however, the analyte is already present in a form that may be readily separated from its liquid, gas, or solid matrix. When such a separation is possible, the analyte's mass can be directly determined with an appropriate balance. In this section the application of particulate gravimetry is briefly considered.

8D.1 Theory and Practice

Two approaches have been used to separate the analyte from its matrix in particulate gravimetry. The most common approach is filtration, in which solid particulates are separated from their gas, liquid, or solid matrix. A second approach uses a liquid-phase or solid-phase extraction.

Filtration Liquid samples are filtered by pulling the liquid through an appropriate filtering medium, either by gravity or by applying suction from a vacuum pump or aspirator. The choice of filtering medium is dictated primarily by the size of the solid particles and the sample's matrix. Filters are constructed from a variety of materials, including cellulose fibers, glass fibers, cellulose nitrate, and polytetrafluoroethylene (PTFE). Particle retention depends on the size of the filter's pores. Cellulose fiber filters, commonly referred to as filter paper, range in pore size from 30 μm to 2–3 μm. Glass fiber filters, constructed from chemically inert borosilicate glass, range in pore size from 2.5 μm to 0.3 μm. Membrane filters, which are made from a variety of materials, including cellulose nitrate and PTFE, are available with pore sizes from 5.0 μm to 0.1 μm.

Solid aerosol particulates in gas samples are filtered using either a single or multiple stage. In a single-stage system the gas is passed through a single filter, retaining particles larger than the filter's pore size. When sampling a gas line, the filter is placed directly in line. Atmospheric gases are sampled with a high-volume sampler that uses a vacuum pump to pull air through the filter at a rate of approximately 75 m^3/h. In either case, the filtering medium used for liquid samples also can be used for gas samples. In a multiple-stage system, a series of filtering units is used to separate the particles by size.

Solid samples are separated by particle size using one or more sieves. By selecting several sieves of different mesh size, particulates with a narrow size range can be isolated from the solid matrix. Sieves are available in a variety of mesh sizes, ranging from approximately 25 mm to 40 μm.

Extraction Filtering limits particulate gravimetry to solid particulate analytes that are easily separated from their matrix. Particulate gravimetry can be extended to the analysis of gas-phase analytes, solutes, and poorly filterable solids if the analyte can be extracted from its matrix with a suitable solvent. After extraction, the solvent can be evaporated and the mass of the extracted analyte determined. Alternatively, the analyte can be determined indirectly by measuring the change in a sample's mass after extracting the analyte. Solid-phase extractions, such as those described in Chapter 7, also may be used.

More recently, methods for particulate gravimetry have been developed in which the analyte is separated by adsorption onto a metal surface, by absorption into a thin polymer or chemical film coated on a solid support, or by chemically binding to a suitable receptor covalently bound to a solid support (Figure 8.10). Adsorption, absorption, and binding occur at the interface between the metal surface, the thin film, or the receptor, and the solution containing the analyte. Consequently, the amount of analyte extracted is minuscule, and the resulting change in mass is too small to detect with a conventional balance. This problem is overcome by using a quartz crystal microbalance as a support.

The measurement of mass using a quartz crystal microbalance is based on the piezoelectric effect.[10] When a piezoelectric material, such as a quartz crystal, experiences a mechanical stress, it generates an electrical potential whose magnitude is proportional to the applied stress. Conversely, when an alternating electrical field is

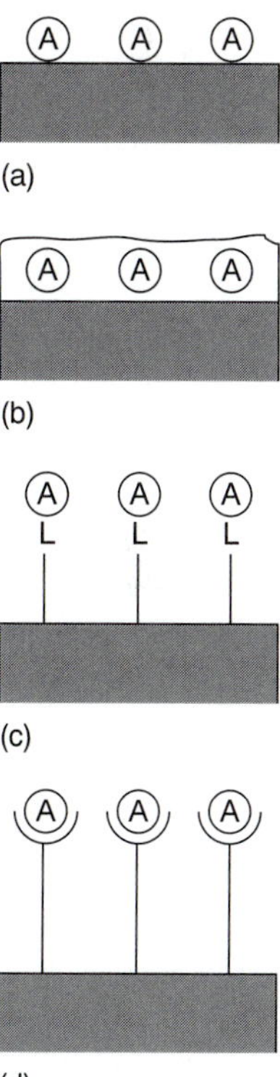

Figure 8.10
Four possible mechanisms for solid-state extraction: (a) adsorption onto a solid substrate; (b) absorption into a thin polymer or chemical film coated on a solid substrate; (c) metal–ligand complexation in which the ligand is covalently bound to the solid substrate; and (d) antibody–antigen binding in which the receptor is covalently bound to the solid substrate.

applied across a quartz crystal, an oscillatory vibrational motion is induced in the crystal. Every quartz crystal vibrates at a characteristic resonant frequency that is a function of the crystal's properties, including the mass per unit area of any material coated on the crystal's surface. The change in mass following adsorption, absorption, or binding of the analyte, therefore, can be determined by monitoring the change in the quartz crystal's characteristic resonant frequency. The exact relationship between the change in frequency and mass is determined by a calibration curve.

8D.2 Quantitative Applications

Particulate gravimetry is commonly encountered in the environmental analysis of water, air, and soil samples. The analysis for suspended solids in water samples, for example, is accomplished by filtering an appropriate volume of a well-mixed sample through a glass fiber filter and drying the filter to constant weight at 103–105 °C.

Microbiological testing of water also is accomplished by particulate gravimetry. For example, in the analysis for coliform bacteria an appropriate volume of sample is passed through a sterilized 0.45-μm membrane filter. The filter is then placed on a sterilized absorbent pad saturated with a culturing medium and incubated for 22–24 h at 35 ±0.5 °C. Coliform bacteria are identified by the presence of individual bacterial colonies that form during the incubation period. As with qualitative applications of precipitation gravimetry, the signal in this case is a visual observation rather than a measurement of mass.

Total airborne particulates are determined using a high-volume air sampler equipped with either cellulose fiber or glass fiber filters. Samples taken from urban environments require approximately 1 h of sampling time, but samples from rural environments require substantially longer times.

Grain size distributions for sediments and soils are used to determine the amount of sand, silt, and clay present in a sample. For example, a grain size of 2 mm serves as a boundary between gravel and sand. Grain size boundaries for sand–silt and silt–clay are given as 1/16 mm and 1/256 mm, respectively.

Several standard methods for the quantitative analysis of food samples are based on measuring the sample's mass following a selective solvent extraction. For example, the crude fat content in chocolate can be determined by extracting with ether for 16 h in a Soxhlet extractor. After the extraction is complete, the ether is allowed to evaporate, and the residue is weighed after drying at 100 °C. This analysis has also been accomplished indirectly by weighing a sample before and after extracting with supercritical CO_2.

Quartz crystal microbalances equipped with thin-film polymer or chemical coatings have found numerous quantitative applications in environmental analysis. Methods have been reported for the analysis of a variety of gaseous pollutants, including ammonia, hydrogen sulfide, ozone, sulfur dioxide, and mercury.[10] Biochemical particulate gravimetric sensors also have been developed. For example, a piezoelectric immunosensor has been developed that shows a high selectivity for human serum albumin and is capable of detecting microgram quantities.[11]

Quantitative Calculations The result of a quantitative analysis by particulate gravimetry is just the ratio, using appropriate units, of the amount of analyte to the amount of sample.

Example

EXAMPLE 8.8

A 200.0-mL sample of water was filtered through a preweighed glass fiber filter. After drying to constant weight at 105 °C, the filter was found to have increased in mass by 48.2 mg. Determine the total suspended solids for the sample in parts per million.

SOLUTION

Parts per million is the same as milligrams of analyte per liter of solution; thus, the total suspended solids for the sample is

$$\frac{48.2 \text{ mg}}{0.2000 \text{ L}} = 241 \text{ ppm}$$

8D.3 Evaluating Particulate Gravimetry

The scale of operation and detection limit for particulate gravimetry can be extended beyond that of other gravimetric methods by increasing the size of the sample taken for analysis. This is usually impossible for other gravimetric methods because of the difficulty of manipulating a larger sample through the individual steps of the analysis. With particulate gravimetry, however, the part of the sample that is not analyte is removed when filtering or extracting. Consequently, particulate gravimetry is easily extended to the analysis of trace-level analytes.

Except for methods relying on a quartz crystal microbalance, particulate gravimetry uses the same balances as other gravimetric methods and is capable of achieving similar levels of accuracy and precision. Since particulate gravimetry is defined in terms of the mass of the particle itself, the sensitivity of the analysis is given by the balance's sensitivity. Selectivity, on the other hand, is determined by either the filter's pore size or the properties of the extracting phase. Particulate gravimetric methods based on filtration are generally less time-, labor-, and capital-intensive than other gravimetric methods since they require only a filtration step.

8E KEY TERMS

adsorbate *(p. 239)*
coagulation *(p. 242)*
digestion *(p. 239)*
electrogravimetry *(p. 234)*
gravimetry *(p. 233)*
homogeneous precipitation *(p. 241)*
inclusion *(p. 238)*
occlusion *(p. 239)*
particulate gravimetry *(p. 234)*
peptization *(p. 245)*
precipitant *(p. 235)*
precipitation gravimetry *(p. 234)*
relative supersaturation *(p. 241)*
supernatant *(p. 244)*
thermogram *(p. 256)*
thermogravimetry *(p. 255)*
volatilization gravimetry *(p. 234)*

8F SUMMARY

In a gravimetric analysis a measurement of mass or change in mass provides quantitative information about the amount of analyte in a sample. The most common form of gravimetry uses a precipitation reaction to generate a product whose mass is proportional to the analyte. In many cases the precipitate includes the analyte; however, an indirect analysis in which the analyte causes the precipitation of another compound also is possible. Precipitation gravimetric procedures must be carefully controlled to produce precipitates that are easily filterable, free from impurities, and of known stoichiometry.

In volatilization gravimetry, thermal or chemical energy is used to decompose the sample containing the analyte. The mass of residue remaining after decomposition, the mass of volatile products collected with a suitable trap, or a change in mass due to the loss of volatile material are all gravimetric measurements.

When the analyte is already present in a particulate form that is easily separated from its matrix, then a particulate gravimetric analysis may be feasible. Examples include the determination of dissolved solids and the determination of fat in foods.

8G *Suggested* EXPERIMENTS

Experiments

A number of gravimetric methods, such as the determination of Cl^- in a soluble salt, have been part of the "standard" repertoire of experiments for introductory courses in analytical chemistry. Listed here are additional experiments that may be used to provide practical examples of gravimetry.

Burrows, H. D.; Ellis, H. A.; Odilora, C. A. "The Dehydrochlorination of PVC," *J. Chem. Educ.* **1995,** *72,* 448–450.

This experiment describes a simple gravimetric procedure for determining the %w/w Cl in samples of poly(vinyl chloride).

Carmosini, N.; Ghoreshy, S.; Koether, M. C. "The Gravimetric Analysis of Nickel Using a Microwave Oven," *J. Chem. Educ.* **1997,** *74,* 986–987.

A procedure for using a microwave oven to digest samples of Ni ore and to dry precipitates of nickel–dimethylgloxime is described in this experiment.

Harris, T. M. "Revitalizing the Gravimetric Determination in Quantitative Analysis Laboratory," *J. Chem. Educ.* **1995,** *72,* 355–356.

This experiment investigates the accuracy of a gravimetric analysis when modifying a standard procedure. The gravimetric procedure is the determination of Ba^{2+} as $BaSO_4$. Modifications that are investigated include the addition of a potential interferent (Ca^{2+}), changing the pH at which precipitation occurs, changing the rate at which the precipitant is added, changing the conditions for digesting the precipitate, and changing the procedure for filtering and drying the precipitate. Errors introduced by modifying the standard procedure can be explained by considering the process by which precipitation occurs.

Henrickson, C. H.; Robinson, P. R. "Gravimetric Determination of Calcium as $CaC_2O_4 \cdot H_2O$," *J. Chem. Educ.* **1979,** *56,* 341–342.

A procedure is provided for the analysis of calcium in samples of $CaCO_3$. Precipitation is done homogeneously using urea and acidified ammonium oxalate. By acidifying the ammonium oxalate, the oxalate is introduced as oxalic acid and does not precipitate the Ca^{2+}. Heating the solution hydrolyzes the urea, forming NH_3. As the NH_3 neutralizes the acid in solution, oxalate acid is converted to oxalate and $CaC_2O_4 \cdot H_2O$ precipitates. The acid–base indicator methyl red is used to signal the completion of the precipitation.

Snow, N. H.; Dunn, M.; Patel, S. "Determination of Crude Fat in Food Products by Supercritical Fluid Extraction and Gravimetric Analysis," *J. Chem. Educ.* **1997,** *74,* 1108–1111.

The %w/w fat in candy bars is determined by an indirect particulate gravimetric analysis. Supercritical CO_2 is used to extract the fat from the sample, and the change in the sample's weight is used to determine the fat content.

Thompson, R. Q.; Ghadiali, M. "Microwave Drying of Precipitates for Gravimetric Analysis," *J. Chem. Educ.* **1993,** *70,* 170–171.

This article describes conditions for using a household microwave oven to dry precipitates for the determination of Cl^- as AgCl, the determination of SO_4^{2-} as $BaSO_4$, and the determination of Ca^{2+} as $CaC_2O_4 \cdot H_2O$.

8H PROBLEMS

1. Starting with the equilibrium constant expressions for reactions 8.1, and 8.3–8.5, verify that equation 8.7 is correct.

2. Equation 8.7 shows how the solubility of AgCl varies as a function of the equilibrium concentration of Cl^-. Derive a similar equation to describe the solubility of AgCl as a function of the equilibrium concentration of Ag^+. Graph the resulting solubility function and compare it with that shown in Figure 8.1.

3. Derive a solubility diagram (solubility versus pH) for $Zn(OH)_2$ that takes into account the following soluble zinc hydroxide complexes: $Zn(OH)^+$, $Zn(OH)_3^-$, $Zn(OH)_4^{2-}$. What is the optimum pH for the quantitative precipitation of $Zn(OH)_2$?

4. For what pH range will the following precipitates have their lowest solubility?

a. CaC_2O_4
b. $PbCrO_4$
c. $BaSO_4$
d. $SrCO_3$
e. ZnS

5. When solutions of 1.5 M KNO_3 and 1.5 M $HClO_4$ are mixed, a white precipitate of $KClO_4$ is formed. If traces of MnO_4^- are present, an inclusion of $KMnO_4$ is possible. Impure precipitates of $KClO_4$ are colored purple by the included $KMnO_4$. Following are the descriptions and results for two experiments in which $KClO_4$ is precipitated in the presence of MnO_4^-. Explain why the two experiments lead to different results (see Color Plate 6).

Experiment 1. Place 1 mL of 1.5 M KNO_3 in a test tube, add 3 drops of 0.1 M $KMnO_4$, and swirl to mix. Add 1 mL of 1.5 M $HClO_4$ dropwise, agitating the solution between drops. Destroy the excess $KMnO_4$ by adding 0.1 M $NaHSO_3$ dropwise. The resulting precipitate of $KClO_4$ has an intense purple color.

Experiment 2. Place 1 mL of 1.5 M $HClO_4$ in a test tube, add 3 drops of 0.1 M $KMnO_4$, and swirl to mix. Add 1 mL of 1.5 M KNO_3 dropwise, agitating the solution between drops. Destroy the excess $KMnO_4$ by adding 0.1 M $NaHSO_3$ dropwise. The resulting precipitate of $KClO_4$ is pale purple or white in color.

6. When solutions of $Ba(SCN)_2$ and $MgSO_4$ are mixed, a precipitate of $BaSO_4$ forms. Following are the descriptions and results for several experiments in which only the concentrations of $Ba(SCN)_2$ and $MgSO_4$ are different. Explain why these experiments produce different results.

Experiment 1. When equal volumes of 3.5 M $Ba(SCN)_2$ and 3.5 M $MgSO_4$ are mixed, a gelatinous precipitate immediately forms.

Experiment 2. When equal volumes of 1.5 M $Ba(SCN)_2$ and 1.5 M $MgSO_4$ are mixed, a curdy precipitate immediately forms. Individual particles of $BaSO_4$ can be seen as points under magnification at 1500×.

Experiment 3. When equal volumes of 0.0005 M $Ba(SCN)_2$ and 0.0005 M $MgSO_4$ are mixed, the complete precipitation of $BaSO_4$ requires 2–3 h. Individual crystals of $BaSO_4$ obtain lengths of approximately 0.005 mm.

7. Aluminum can be determined gravimetrically by precipitating as $Al(OH)_3$ and isolating as Al_2O_3. A sample containing approximately 0.1 g of Al is dissolved in 200 mL of H_2O and 5 g of NH_4Cl and a few drops of methyl red indicator is added (methyl red is red at pH levels below 4 and yellow at pH levels above 6). The solution is heated to boiling, and 1:1 NH_3 is added dropwise till the indicator turns yellow, precipitating $Al(OH)_3$. The precipitate is held at the solution's boiling point for several minutes, filtered, and washed with a hot solution of 2%, w/v NH_4NO_3. The precipitate is then ignited at 1000–1100 °C, forming Al_2O_3.
(a) Cite two ways in which this procedure has been designed to encourage the formation of larger particles of precipitate. (b) The ignition step must be carried out carefully to ensure that $Al(OH)_3$ is quantitatively converted to Al_2O_3. What effect would an incomplete conversion have on the reported %w/w Al? (c) What role do NH_4Cl and methyl red indicator play in this procedure? (d) An alternative procedure involves isolating and weighing the precipitate as the 8-hydroxyquinolate, $Al(C_9H_6ON)_3$. Why might this be a more advantageous form of Al for a gravimetric analysis?

8. Calcium is determined gravimetrically by precipitating it as $CaC_2O_4 \cdot H_2O$, followed by isolating the precipitate as $CaCO_3$. The sample to be analyzed is dissolved in 10 mL of water and 15 mL of 6 M HCl. After dissolution, the resulting solution is heated to boiling, and a warm solution of excess ammonium oxalate is added. The solution is maintained at 80 °C, and 6 M NH_3 is added dropwise, with stirring, until the solution is faintly alkaline. The resulting precipitate and solution are removed from the heat and allowed to stand for at least 1 h. After testing the solution for completeness of precipitation, the sample is filtered, washed with 0.1% w/v ammonium oxalate, and dried at 100–120 °C for 1 h. The precipitate is then transferred to a muffle furnace where it is converted to $CaCO_3$ by drying at 500 ± 25 °C until a constant weight.

(a) Why is the precipitate of $CaC_2O_4 \cdot H_2O$ converted to $CaCO_3$? (b) In the final step, if the sample is heated at too high of a temperature, some $CaCO_3$ may be converted to CaO. What effect would this have on the reported %w/w Ca? (c) Why is the precipitant, $(NH_4)_2C_2O_4$, added to a hot, acidic solution rather than to a cold, alkaline solution?

9. Iron can be determined gravimetrically by precipitating as $Fe(OH)_3$ and igniting to Fe_2O_3. The sample to be analyzed is weighed and transferred to a 400-mL beaker where it is dissolved in 50 mL of H_2O and 10 mL of 6 M HCl. Any Fe^{2+} that is present is oxidized to Fe^{3+} with 1–2 mL of concentrated HNO_3. After boiling to remove the oxides of nitrogen, the solution is diluted to 200 mL, brought to boiling, and $Fe(OH)_3$ is precipitated by slowly adding 1:1 NH_3 until the odor of NH_3 is detected. The solution is boiled for an additional minute, and the precipitate is allowed to settle to the bottom of the beaker. The precipitate is then filtered and washed with several portions of hot 1% w/v NH_4NO_3 until no Cl^- is found in the wash water. Finally, the precipitate is ignited to constant weight at 500–550 °C, and weighed as Fe_2O_3.
(a) If the ignition is not carried out under oxidizing conditions (plenty of O_2 present), the final product will contain some Fe_3O_4. What effect would this have on the reported %w/w Fe? (b) The precipitate is washed with a dilute solution of NH_4NO_3. Why is NH_4NO_3 added to the wash water? (c) Why does the procedure call for adding NH_3 until the odor of ammonia is detected? (d) Describe how you might test the filtrate for Cl^-.

10. Sinha and Shome described a gravimetric method for molybdenum in which it is precipitated with *n*-benzoylphenylhydroxylamine, $C_{13}H_{11}NO_2$, giving a precipitate of $MoO_2(C_{13}H_{10}NO_2)_2$.[12] The precipitate is weighed after igniting to MoO_3. As part of their study they determined the optimum conditions for the analysis. Samples containing a known amount of Mo were taken through the procedure while varying the temperature, the amount of precipitant added, and the pH of the solution. The solution volume was held constant at 300 mL for all experiments. A summary of their results are shown in the following table

g Mo taken	Temperature (°C)	g precipitant	mL 10 M HCl	g MoO_3
0.0770	30	0.20	0.9	0.0675
0.0770	30	0.30	0.9	0.1014
0.0770	30	0.35	0.9	0.1140
0.0770	30	0.42	0.9	0.1155
0.0770	30	0.42	0.3	0.1150
0.0770	30	0.42	18.0	0.1152
0.0770	30	0.42	48.0	0.1160
0.0770	30	0.42	75.0	0.1159
0.0770	50	0.42	0.9	0.1156
0.0770	75	0.42	0.9	0.1158
0.0770	80	0.42	0.9	0.1129

Considering these results, discuss the optimum conditions for the determination of Mo by this method. Express your results for the precipitant as the minimum concentration in excess, as %w/v, needed to ensure a quantitative precipitation.

11. A sample of an impure iron ore is believed to be approximately 55% w/w Fe. The amount of Fe in the sample is to be determined gravimetrically by isolating it as Fe_2O_3. How many grams of sample should be taken to ensure that approximately 1 g of Fe_2O_3 will be isolated?

12. The concentration of arsenic in an insecticide can be determined gravimetrically by precipitating it as $MgNH_4AsO_4$. The precipitate is ignited and weighed as $Mg_2As_2O_7$. Determine the %w/w As_2O_3 in a 1.627-g sample of insecticide that yields 106.5 mg of $Mg_2As_2O_7$.

13. After preparing a sample of alum, $K_2SO_4 \cdot Al_2(SO_4)_3 \cdot 24H_2O$, a student determined its purity gravimetrically. A 1.2931-g sample was dissolved and the aluminum precipitated as $Al(OH)_3$. The precipitate was collected by filtration, washed, and ignited to Al_2O_3, yielding 0.1357 g. What is the purity of the alum preparation?

14. To determine the amount of iron in a dietary supplement, a random sample of 15 tablets weighing a total of 20.505 g was ground into a fine powder. A 3.116-g sample of the powdered tablets was dissolved and treated to precipitate the iron as $Fe(OH)_3$. The precipitate was collected, rinsed, and ignited to a constant weight as Fe_2O_3, yielding 0.355 g. Report the iron content of the dietary supplement as g $FeSO_4 \cdot 7H_2O$ per tablet.

15. A 1.4639-g sample of limestone was analyzed for Fe, Ca, and Mg. The iron was determined as Fe_2O_3, yielding 0.0357 g. Calcium was isolated as $CaSO_4$, yielding a precipitate of 1.4058 g, and Mg was isolated as 0.0672 g of $Mg_2P_2O_7$. Report the amount of Fe, Ca, and Mg in the limestone sample as %w/w Fe_2O_3, %w/w CaO, and %w/w MgO.

16. The number of ethoxy groups (CH_3CH_2O–) in an organic compound can be determined by the following sequence of reactions

$$R(OCH_2CH_3)_x + xHI \rightarrow R(OH)_x + xCH_3CH_2I$$

$$CH_3CH_2I + Ag^+ + H_2O \rightarrow AgI(s) + CH_3CH_2OH$$

A 36.92-mg sample of an organic compound with an approximate molecular weight of 176 was treated in this fashion, yielding 0.1478 g of AgI. How many ethoxy groups are there in each molecule?

17. A 516.7-mg sample containing a mixture of K_2SO_4 and $(NH_4)_2SO_4$ was dissolved in water and treated with $BaCl_2$, precipitating the SO_4^{2-} as $BaSO_4$. The resulting precipitate was isolated by filtration, rinsed free of impurities, and dried to a constant weight, yielding 863.5 mg of $BaSO_4$. What is the %w/w K_2SO_4 in the sample?

18. The amount of iron and manganese in an alloy can be determined by precipitating the metals with 8-hydroxyquinoline, C_9H_7NO. After weighing the mixed precipitate, the precipitate is dissolved and the amount of 8-hydroxyquinoline determined by another method. In a typical analysis, a 127.3-mg sample of an alloy containing iron, manganese, and other metals was dissolved in acid and treated with appropriate masking agents to prevent an interference from other metals. The iron and manganese were precipitated and isolated as $Fe(C_9H_6NO)_3$ and $Mn(C_9H_6NO)_2$, yielding a total mass of 867.8 mg. The amount of 8-hydroxyquinolate in the mixed precipitate was determined to be 5.276 mmol. Calculate the %w/w Fe and %w/w Mn in the alloy.

19. A 0.8612-g sample of a mixture consisting of NaBr, NaI, and $NaNO_3$ was analyzed by adding $AgNO_3$ to precipitate the Br^- and I^-, yielding a 1.0186-g mixture of AgBr and AgI. The precipitate was then heated in a stream of Cl_2, converting it to 0.7125 g of AgCl. Calculate the %w/w $NaNO_3$ in the sample.

20. The earliest determinations of elemental atomic weights were accomplished gravimetrically. In determining the atomic weight of manganese, a carefully purified sample of $MnBr_2$ weighing 7.16539 g was dissolved and the Br^- precipitated as AgBr, yielding 12.53112 g. What is the atomic weight for Mn if the atomic weights for Ag and Br are taken to be 107.868 and 79.904, respectively?

21. While working as a laboratory assistant you prepared 0.4 M solutions of $AgNO_3$, $Pb(NO_3)_2$, $BaCl_2$, KI, and Na_2SO_4. Unfortunately, you became distracted and forgot to label the solutions before leaving the laboratory. Realizing your error, you labeled the solutions A–E and performed all possible binary mixings of the five solutions. The following results were obtained where NR means no reaction was observed, W means a white precipitate formed, and Y means a yellow precipitate formed.

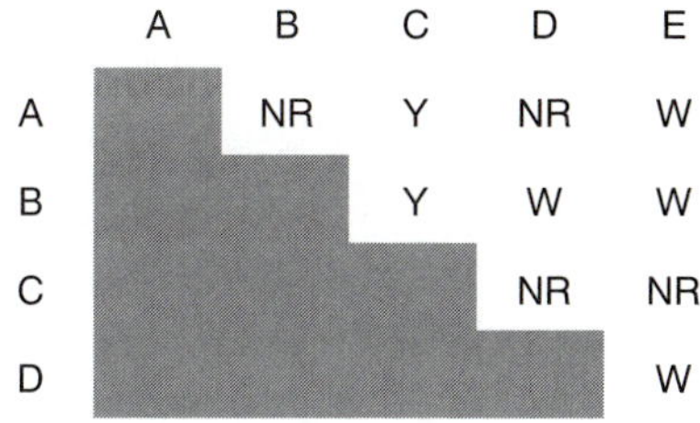

	A	B	C	D	E
A		NR	Y	NR	W
B			Y	W	W
C				NR	NR
D					W

Identify solutions A–E.

22. A solid sample is known to consist of approximately equal amounts of two or more of the following soluble salts.

$AgNO_3$ $ZnCl_2$ K_2CO_3 $MgSO_4$ $Ba(C_2H_3O_2)_2$ NH_4NO_3

A sample of the solid, sufficient to give at least 0.04 mol of any single salt, was added to 100 mL of water, yielding a white precipitate and a clear solution. The precipitate was collected and rinsed with water. When a portion of the precipitate was placed in dilute HNO_3, it completely dissolved, leaving a colorless solution. A second portion of the precipitate was placed in dilute HCl, yielding a precipitate and a clear solution. Finally, the filtrate from the original precipitate was treated with excess NH_3, yielding a white precipitate. Indicate which of the soluble salts must be present in the sample, which must be absent and those for which there is insufficient information to make this determination.[13]

23. Two methods have been proposed for the analysis of sulfur in impure samples of pyrite, FeS_2. Sulfur can be determined in a direct analysis by oxidizing it to SO_4^{2-} and precipitating as $BaSO_4$. An indirect analysis is also possible if the iron is precipitated as $Fe(OH)_3$ and isolated as Fe_2O_3. Which of these methods will provide a more sensitive determination for sulfur? What other factors should be considered in deciding between these methods?

24. A sample of impure pyrite known to be approximately 90–95% w/w FeS_2 is to be analyzed by oxidizing the sulfur to SO_4^{2-} and precipitating as $BaSO_4$. How many grams of the sample must be taken if a minimum of 1 g of $BaSO_4$ is desired?

25. A series of samples consisting of any possible combination of KCl, NaCl, and NH_4Cl is to be analyzed by adding $AgNO_3$ to precipitate AgCl. What is the minimum volume of 5% w/v $AgNO_3$ necessary to completely precipitate the chloride in any 0.5-g sample?

26. When precipitation does not result in a "stoichiometric" precipitate, a gravimetric analysis is still feasible if the stoichiometric ratio between analyte and precipitate can be established experimentally. Consider, for example, the precipitation gravimetric analysis of Pb as $PbCrO_4$. (a) For each gram of Pb in a sample, how many grams of $PbCrO_4$ are expected to form? (b) In a study of this procedure, Grote found that 1.568 g of $PbCrO_4$ formed per gram of Pb.[14] What is the apparent stoichiometry between Pb and $PbCrO_4$? (c) Does failing to account for the actual stoichiometry lead to a positive or negative determinate error for the analysis?

27. Determine the uncertainty for the gravimetric analysis described in Example 8.1. (a) How does your result compare with the expected accuracy of 0.1–0.2% for precipitation gravimetry? (b) What sources of error might account for any discrepancy between the most probable measurement error and the expected accuracy?

28. A 38.63-mg sample of potassium ozonide, KO_3, was heated to 70 °C for 1 h, undergoing a weight loss of 7.10 mg. Write a balanced chemical reaction describing this decomposition reaction. A 29.6-mg sample of impure KO_3 experiences a 4.86-mg weight loss when treated under similar condition. What is the %w/w KO_3 in the sample?

29. The water content of an 875.4-mg sample of cheese was determined with a moisture analyzer. What is the %w/w H_2O in the cheese if the final mass was found to be 545.8 mg?

30. Method 8.2 describes a procedure for determining Si in ores and alloys. In this analysis a weight loss of 0.21 g corresponds to 0.1 g of Si. Show that this relationship is correct.

31. The iron content of an organometallic compound was determined by treating a 0.4873-g sample with HNO_3 and heating to volatilize the organic material. After ignition, the residue of Fe_2O_3 was found to weigh 0.2091 g. (a) What is the %w/w Fe in this compound? The carbon and hydrogen in a second sample of the compound were determined by a combustion analysis. When a 0.5123-g sample was carried through the analysis, 1.2119 g of CO_2 and 0.2482 g of H_2O were collected. (b) What are the %w/w C and %w/w H in this compound? (c) What is the compound's empirical formula?

32. A polymer's ash content is determined by placing a weighed sample in a Pt crucible that has been previously brought to a constant weight. The polymer is melted under gentle heating from a Bunsen burner until the volatile vapor ignites. The polymer is allowed to burn until only a noncombustible residue remains. The residue is then brought to constant weight at 800 °C in a muffle furnace. The following data were collected during the analysis of two samples of a polymer resin:

Polymer A	g crucible	g crucible + polymer	g crucible + ash
replicate 1	19.1458	21.2287	19.7717
replicate 2	15.9193	17.9522	16.5310
replicate 3	15.6992	17.6660	16.2909

Polymer B	g crucible	g crucible + polymer	g crucible + ash
replicate 1	19.1457	21.0693	19.7187
replicate 2	15.6991	17.8273	16.3327
replicate 3	15.9196	17.9037	16.5110

(a) Determine the average and standard deviation for the %w/w ash of each polymer resin. (b) Is there any evidence at $\alpha = 0.05$ for a significant difference between the two polymer resins?

33. In the presence of water vapor the surface of zirconia, ZrO_2, chemically adsorbs H_2O, forming surface hydroxyls, ZrOH (additional water is physically adsorbed as H_2O). When heated above 200 °C the surface hydroxyls convert to $H_2O(g)$, with one molecule of water released for every two surface hydroxyls. Below 200 °C only physically absorbed water is lost. Nawrocki and coworkers[15] used thermogravimetry to determine the density of surface hydroxyls on a sample of zirconia that was heated to 700 °C and allowed to cool in a desiccator containing humid N_2, finding that 0.006 g H_2O was lost between 200 °C and 900 °C for every gram of dehydroxylated ZrO_2. Given that the zirconia sample had a surface area of 33 m^2/g and that one molecule of H_2O forms two surface hydroxyls, calculate the density of surface hydroxyls in micromoles per square meter.

34. Suppose that you have a mixture of CaC_2O_4, MgC_2O_4, and some inert impurities, and that you wish to determine the %w/w CaC_2O_4 in the sample. One way to accomplish this analysis is by volatilization gravimetry. The following information is available to you as the result of a thermogravimetric study: CaC_2O_4 converts to $CaCO_3$ at a temperature of 500 °C; $CaCO_3$ converts to CaO at a temperature of 900 °C; and MgC_2O_4 converts to MgO at a temperature of 500 °C. (a) Describe how you would carry out this analysis, paying particular attention to the measurements that you need to make. (b) How would you calculate the %w/w CaC_2O_4 in the sample?

35. The concentration of airborne particulates in an industrial workplace was determined by pulling the air through a single-stage air sampler equipped with a glass fiber filter. The air was sampled for 20 min at a rate of 75 m^3/h. At the end of the sampling period the glass fiber filter was found to have increased in mass by 345.2 mg. What is the concentration of particulates in the air sample in milligrams per cubic meter and in milligrams per liter?

36. The fat content of potato chips can be determined indirectly by weighing a sample before and after extracting the fat with supercritical CO_2. The following data were obtained for the analysis of one sample of potato chips.[16]

Sample Number	Initial Sample Weight (g)	Final Sample Weight (g)
1	1.1661	0.9253
2	1.1723	0.9252
3	1.2525	0.9850
4	1.2280	0.9562
5	1.2837	1.0119

(a) Determine the average fat content (%w/w) for this sample of potato chips. (b) This sample of potato chips is known to have a fat content of 22.7% w/w. Is there any evidence for a determinate error in the data at $\alpha = 0.05$?

37. Delumyea and McCleary report data for the determination of %w/w organic material in sediment samples collected at different depths from a cove on the St. Johns River in Jacksonville, FL.[17] After collecting the sediment core it was sectioned into a set of 2-cm increments. Each increment was placed in 50 mL of deionized water and the slurry filtered through a piece of preweighed filter paper. The filter paper and sediment increment were then placed in a preweighed evaporating dish and dried in an oven at

110 °C. After weighing, the filter paper and sediment were transferred to a muffle furnace where the filter paper and any organic material were removed by ashing. Finally, the inorganic residue remaining after ashing was weighed. Using the following data, determine the %w/w organic material as a function of the average depth for each increment.

Depth (cm)	Weight of Paper Filter (g)	Weight of Evaporating Dish (g)	Weight After Drying (g)	Weight After Ashing (g)
0–2	1.590	43.21	52.10	49.49
2–4	1.745	40.62	48.83	46.00
4–6	1.619	41.23	52.86	47.84
6–8	1.611	42.10	50.59	47.13
8–10	1.658	43.62	51.88	47.53
10–12	1.628	43.24	49.45	45.31
12–14	1.633	43.08	47.92	44.20
14–16	1.630	43.96	58.31	55.53
16–18	1.636	43.36	54.37	52.75

38. Yao and associates recently described a method for the quantitative analysis of thiourea based on its reaction with I_2.[18]

$$CS(NH_2)_2 + 4I_2 + 6H_2O \rightarrow (NH_4)_2SO_4 + 8\ HI + CO_2$$

The procedure calls for placing a 100-μL aqueous sample containing the thiourea in a 60-mL separatory funnel along with 10 mL of a pH 7 buffer and 10 mL of a 12 μM solution of I_2 in CCl_4. The contents of the separatory funnel are shaken, and the organic and aqueous layers are allowed to separate. The organic layer, containing the excess I_2, is transferred to the surface of a piezoelectric crystal on which a thin layer of Au has been deposited. After allowing the I_2 to adsorb to the Au, the CCl_4 is removed and the crystal's frequency shift is measured. The following data are reported for a series of thiourea standards.

[thiourea] (M)	Δf (Hz)
3.00×10^{-7}	74.6
5.00×10^{-7}	120
7.00×10^{-7}	159
9.00×10^{-7}	205
15.00×10^{-7}	327
25.00×10^{-7}	543
35.00×10^{-7}	789
50.00×10^{-7}	1089

(a) Characterize this method with respect to the scale of operation shown in Figure 3.6 of Chapter 3. (b) Using a regression analysis, determine the relationship between the crystal's frequency shift and the concentration of thiourea. (c) A sample containing an unknown amount of thiourea is taken through the procedure and gives a Δf of 176 Hz. What is the molar concentration of thiourea in the sample? (d) What is the 95% confidence interval for the concentration of thiourea in this sample assuming one replicate?

8I SUGGESTED READINGS

The following resources provide a general history of gravimetry.

Beck, C. M. "Classical Analysis: A Look at the Past, Present, and Future," *Anal. Chem.* **1991,** *63,* 993A–1003A.

Laitinen, H. A.; Ewing, G. W., eds. *A History of Analytical Chemistry.* The Division of Analytical Chemistry of the American Chemical Society: Washington, DC, 1977, pp. 10–24.

Sources providing additional examples of inorganic and organic gravimetric methods include the following texts.

Bassett, J.; Denney, R. C.; Jeffery, G. H.; et al. *Vogel's Textbook of Quantitative Inorganic Analysis,* 4th ed. Longman: London, 1981.

Erdey, L. *Gravimetric Analysis,* Pergamon: Oxford, 1965.

Steymark, A. *Quantitative Organic Microanalysis,* The Blakiston Co.: New York, 1951.

The following text provides more information on thermogravimetry.

Wendlandt, W. W. *Thermal Methods of Analysis,* 2nd ed. Wiley: New York, 1986.

For a review of isotope dilution mass spectrometry see the following article.

Fassett, J. D.; Paulsen, P. J. "Isotope Dilution Mass Spectrometry for Accurate Elemental Analysis," *Anal. Chem.* **1989,** *61,* 643A–649A.

8J REFERENCES

1. Valcárcel, M.; Rios, A. *Analyst* **1995,** *120,* 2291–2297.
2. (a) Moody, J. R.; Epstein, M. S. *Spectrochim. Acta* **1991,** *46B,* 1571–1575. (b) Epstein, M. S. *Spectrochim. Acta* **1991,** *46B,* 1583–1591.
3. Von Weimarn, P. P. *Chem. Revs.* **1925,** *2,* 217.
4. Bassett, J.; Denney, R. C.; Jeffery, G. H.; et al. *Vogel's Textbook of Quantitative Inorganic Analysis,* 4th ed. Longman: London, 1981, p. 408.
5. Gordon, L; Salutsky, M. L.; Willard, H. H. *Precipitation from Homogeneous Solution,* Wiley: New York, 1959.
6. Ward, R., ed. *Non-Stoichiometric Compounds* (*Ad. Chem. Ser. 39*); American Chemical Society: Washington, DC, 1963.
7. Method 3500-Mg D as published in *Standard Methods for the Examination of Water and Wastewater,* 18th ed. American Public Health Association: Washington, DC, 1992, pp. 3–73 to 3–74.
8. Jungreis, E. *Spot Test Analysis,* 2nd ed. Wiley: New York, 1997.
9. Young, R. S. *Chemical Analysis in Extractive Metallurgy,* Griffen: London, 1971, pp. 302–304.
10. (a) Ward, M. D.; Buttry, D. A. *Science* **1990,** *249,* 1000–1007; (b) Grate, J. W.; Martin, S. J.; White, R. M. *Anal. Chem.* **1993**, *65,* 940A–948A; (c) Grate, J. W.; Martin, S. J.; White, R. M. *Anal. Chem.* **1993,** *65,* 987A–996A.
11. Muratsugu, M.; Ohta, F.; Miya, Y. et al. *Anal. Chem.* **1993,** *65,* 2933–2937.
12. Sinha, S. K.; Shome, S. C. *Anal. Chim. Acta* **1960,** *24,* 33–36.
13. Adapted from Sorum, C. H.; Lagowski, J. J. *Introduction to Semimicro Qualitative Analysis,* 5th ed. Prentice-Hall: Englewood Cliffs, N. J., 1977, p. 285.
14. Grote, Z. *Anal. Chem.* **1941,** *122,* 395.
15. Nawrocki, J.; Carr, P. W.; Annen, M. J.; et al. *Anal. Chim. Acta* **1996,** *327,* 261–266.
16. Data taken from the pamphlet *Fat Determination by SFE,* ISCO, Inc.: Lincoln, NE.
17. Delumyea, R. D.; McCleary, D. L. *J. Chem. Educ.* **1993,** *70,* 172–173.
18. Yao, S. F.; He, F. J.; Nie, L. H. *Anal. Chim. Acta* **1992,** *268,* 311–314.

Colorplates

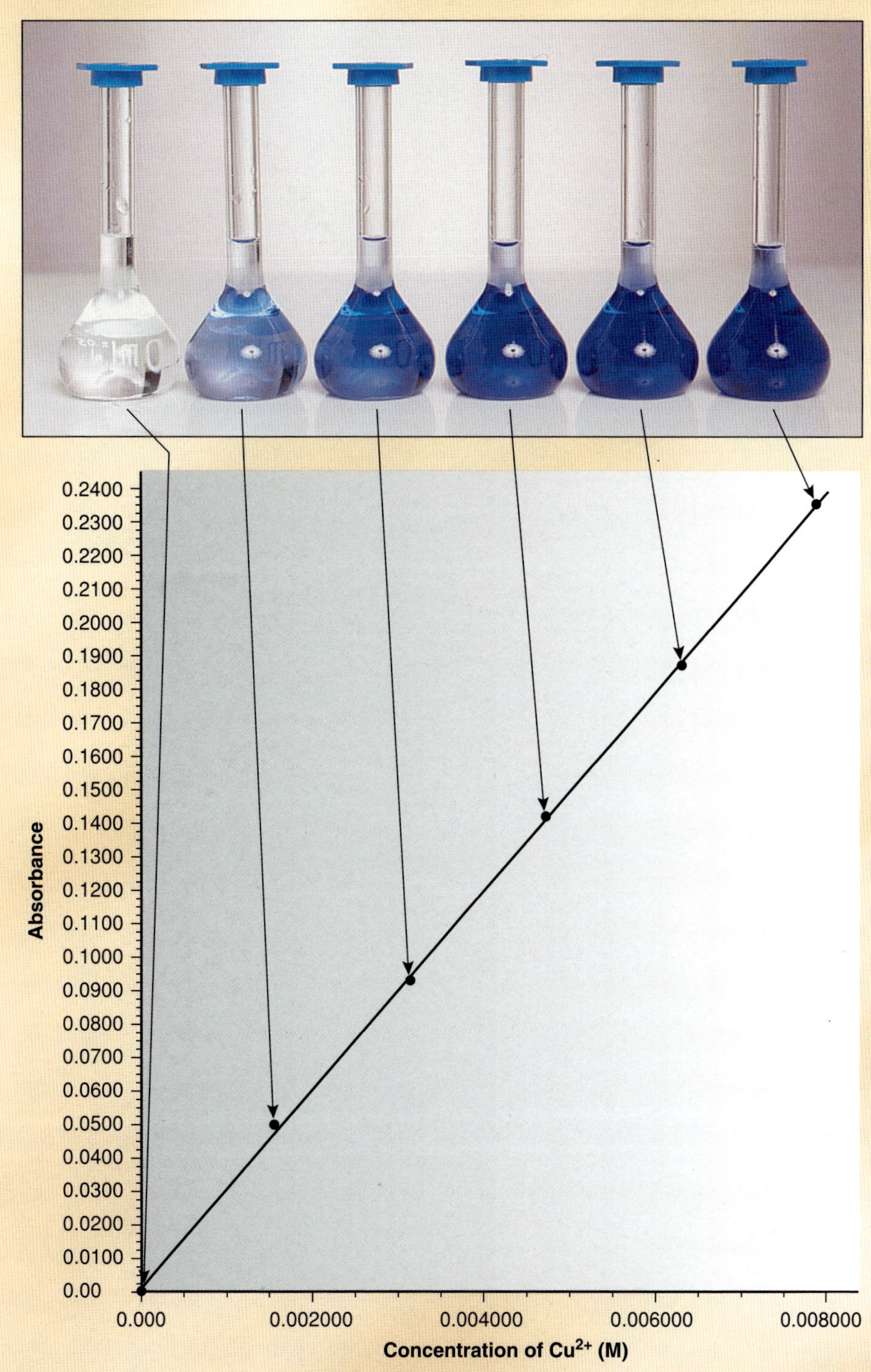

Colorplate 1
A set of external standards for Cu^{2+} showing the associated calibration curve.

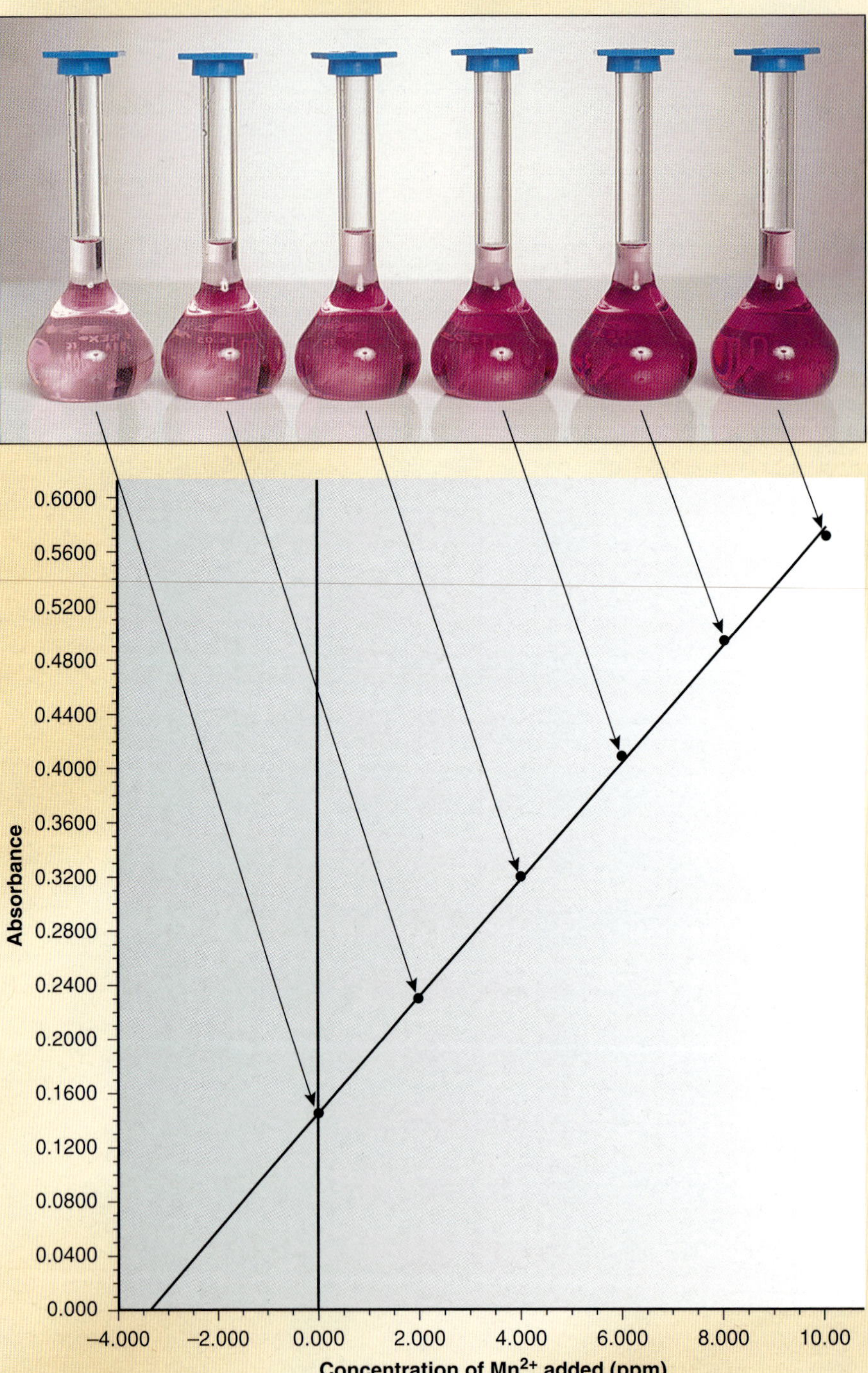

Colorplate 2
A set of standard additions for the determination of Mn^{2+} showing the associated calibration curve.

Colorplate 3

The effect of ionic strength on the formation of a complex ion is shown by these two solutions of $Fe(SCN)^{2+}$. Both solutions were prepared by mixing together 150 mL of 1.0 mM $FeCl_3$ and 150 mL of 1.0 mM NH_4SCN. Adding an inert salt (12 g of KNO_3) to the beaker on the right causes a portion of the $Fe(SCN)^{2+}$ to dissociate back to Fe^{3+} and SCN^-.

Colorplate 4

Extraction of I_2 from H_2O to CCl_4. The separatory funnel on the left contains 50 mL of an aqueous solution of 0.1 M I_2. Extracting with 50 mL of CCl_4 (bottom layer) shows a partial extraction of the I_2 into the CCl_4 (note the purple color of I_2 in CCl_4). A second extraction, as shown on the right, removes additional I_2 from the aqueous layer.

Colorplate 5
Two precipitates of $PbCrO_4$. The yellow precipitate on the left was formed by mixing together solutions of $Pb(NO_3)_2$ and K_2CrO_4 and consists of very small particles due to the relatively large relative supersaturation (RSS). When prepared homogeneously (small RSS), by the reaction $6BrO_3^- + 10Cr^{3+} + 22H_2O + 10Pb^{2+} \rightarrow 10PbCrO_4(s) + 3Br_2 + 44H^+$ the precipitate consists of much larger crystals and is orange in color.

Colorplate 6
Precipitates of $KClO_4$ formed from $HClO_4$ and KNO_3 in the presence of $KMnO_4$. The precipitate on the left, which was formed under conditions favoring the formation of an inclusion of $KMnO_4$, shows a deep purple color due to the inclusion. The precipitate on the right, which is a pale pink, was formed under conditions that minimize the formation of included $KMnO_4$. See problem 5 in Chapter 8 for further details.

Colorplate 7
Solutions of universal indicator over a pH range of 1–12 showing the variation in the indicator's color with pH.

(a)

(b)

(c)

Colorplate 8

(a) End point for the Kjeldahl titration using methyl red as an indicator (see Method 9.1). The flask on the left contains excess HCl and shows the indicator's initial faint red color. The excess HCl is titrated with a standard solution of NaOH to the indicator's orange end point as shown in the middle flask. In the presence of excess NaOH, the indicator is yellow.

(b) End point for the complexometric determination of water hardness by titration with EDTA using calmagite as an indicator (see Method 9.2). The flask on the left shows the red color of the Mg^{2+}–calmagite complex. Titrating with EDTA displaces the Mg^{2+}. Near the end point the solution shows a purple color due to the increasing presence of uncomplexed calmagite (middle flask). As shown by the flask on the right, the indicator has a blue color at the end point.

(c) End point for the determination of total chlorine residual (see Method 9.3). After acidifying the solution, excess I^- is added, forming a brown solution containing I_3^- (left-most flask). Titrating with a standard solution of $S_2O_3^{2-}$ reduces I_3^- to I^-, leaving the solution with a pale yellow color as the titration nears its end point (left center flask). Starch is added, forming a dark blue starch-I_3^- complex (right center flask), which is then titrated to the colorless end point (right-most flask).

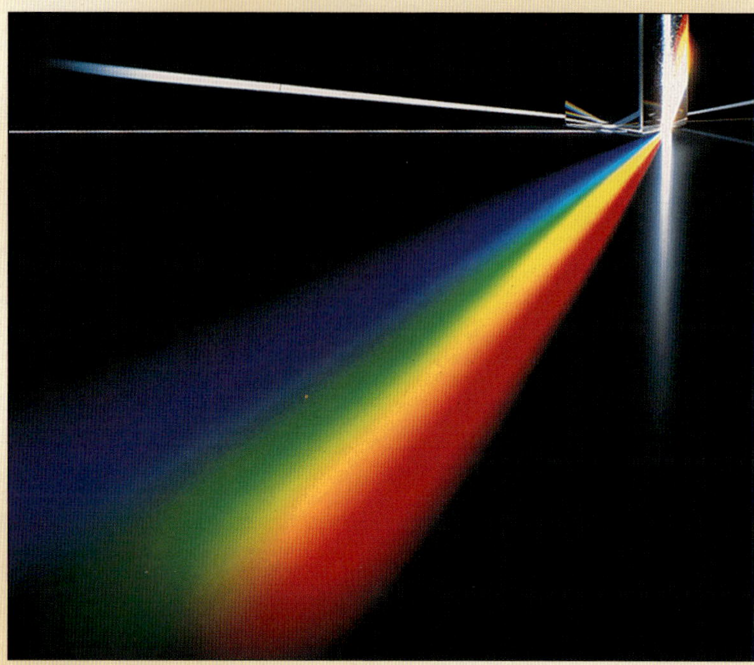

Colorplate 9

Visible spectrum of white light produced by passing the light through a glass prism.

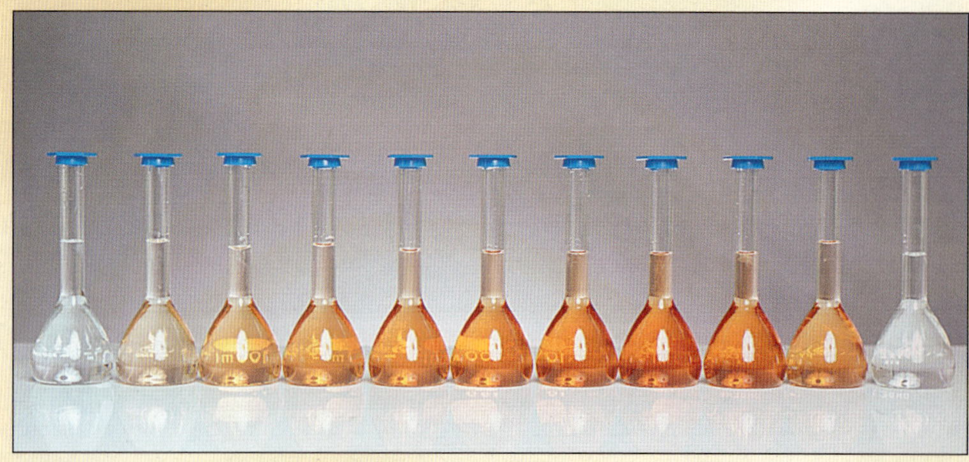

Colorplate 10

Solutions of Fe^{3+} and *o*-phenanthroline (total combined concentration of 3.15×10^{-4} M) showing the method of continuous variations. Flasks contain, from left to right, mole fractions of *o*-phenanthroline of 0, 0.1, 0.2, 0.3, 0.4, 0.5, 0.6, 0.7, 0.8, 0.9, and 1.0. See example 10.7 in Chapter 10 for additional details.

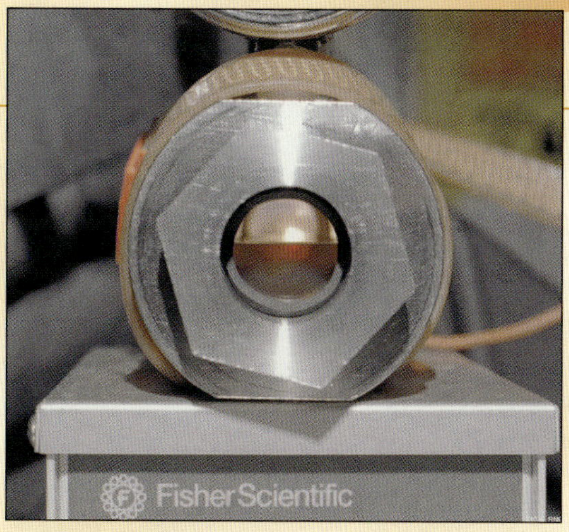

(b)

(a)

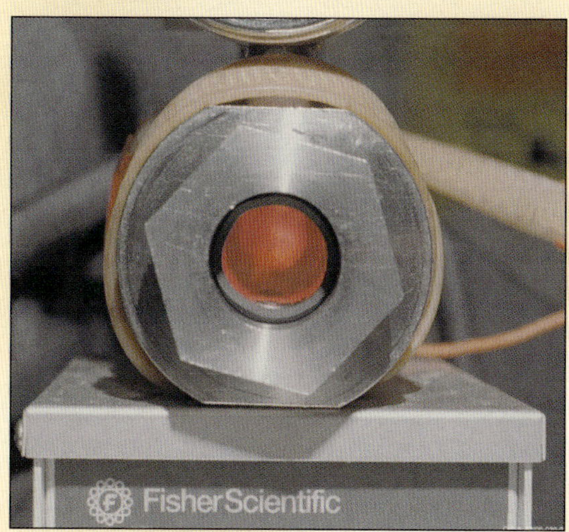

(c)

Colorplate 11

(a) Apparatus for studying the phase-transition of liquid CO_2 to supercritical CO_2, including a 60-mL stainless steel chamber with window and a pressure gauge. (b) Close-up showing the liquid CO_2. (c) Close-up showing supercritical CO_2 obtained by increasing the temperature.

Colorplate 12
HPLC equipped with a multichannel photodiode array detector. The image on the computer screen shows a 3-D chromatogram.

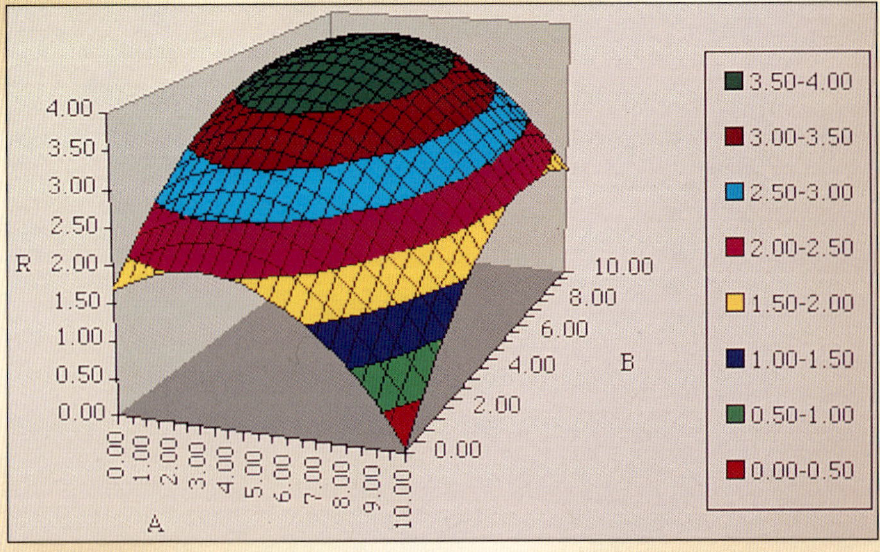

Colorplate 13
Response surface for problem 14.1a.

Chapter 9

Titrimetric Methods of Analysis

Titrimetry, in which we measure the volume of a reagent reacting stoichiometrically with the analyte, first appeared as an analytical method in the early eighteenth century. Unlike gravimetry, titrimetry initially did not receive wide acceptance as an analytical technique. Many prominent late-nineteenth century analytical chemists preferred gravimetry over titrimetry and few of the standard texts from that era include titrimetric methods. By the early twentieth century, however, titrimetry began to replace gravimetry as the most commonly used analytical method.

Interestingly, precipitation gravimetry developed in the absence of a theory of precipitation. The relationship between the precipitate's mass and the mass of analyte, called a gravimetric factor, was determined experimentally by taking known masses of analyte (an external standardization). Gravimetric factors could not be calculated using the precipitation reaction's stoichiometry because chemical formulas and atomic weights were not yet available! Unlike gravimetry, the growth and acceptance of titrimetry required a deeper understanding of stoichiometry, thermodynamics, and chemical equilibria. By the early twentieth century the accuracy and precision of titrimetric methods were comparable to that of gravimetry, establishing titrimetry as an accepted analytical technique.

9A Overview of Titrimetry

titrimetry
Any method in which volume is the signal.

titrant
The reagent added to a solution containing the analyte and whose volume is the signal.

Titrimetric methods are classified into four groups based on the type of reaction involved. These groups are acid–base titrations, in which an acidic or basic **titrant** reacts with an analyte that is a base or an acid; complexometric titrations involving a metal–ligand complexation reaction; redox titrations, where the titrant is an oxidizing or reducing agent; and precipitation titrations, in which the analyte and titrant react to form a precipitate. Despite the difference in chemistry, all titrations share several common features, providing the focus for this section.

9A.1 Equivalence Points and End Points

equivalence point
The point in a titration where stoichiometrically equivalent amounts of analyte and titrant react.

For a titration to be accurate we must add a stoichiometrically equivalent amount of titrant to a solution containing the analyte. We call this stoichiometric mixture the **equivalence point.** Unlike precipitation gravimetry, where the precipitant is added in excess, determining the exact volume of titrant needed to reach the equivalence point is essential. The product of the equivalence point volume, V_{eq}, and the titrant's concentration, C_T, gives the moles of titrant reacting with the analyte.

$$\text{Moles titrant} = V_{eq} \times C_T$$

Knowing the stoichiometry of the titration reaction(s), we can calculate the moles of analyte.

end point
The point in a titration where we stop adding titrant.

indicator
A colored compound whose change in color signals the end point of a titration.

titration error
The determinate error in a titration due to the difference between the end point and the equivalence point.

Unfortunately, in most titrations we usually have no obvious indication that the equivalence point has been reached. Instead, we stop adding titrant when we reach an **end point** of our choosing. Often this end point is indicated by a change in the color of a substance added to the solution containing the analyte. Such substances are known as **indicators.** The difference between the end point volume and the equivalence point volume is a determinate method error, often called the **titration error.** If the end point and equivalence point volumes coincide closely, then the titration error is insignificant and can be safely ignored. Clearly, selecting an appropriate end point is critical if a titrimetric method is to give accurate results.

9A.2 Volume as a Signal*

Almost any chemical reaction can serve as a titrimetric method provided that three conditions are met. The first condition is that all reactions involving the titrant and analyte must be of known stoichiometry. If this is not the case, then the moles of titrant used in reaching the end point cannot tell us how much analyte is in our sample. Second, the titration reaction must occur rapidly. If we add titrant at a rate that is faster than the reaction's rate, then the end point will exceed the equivalence point by a significant amount. Finally, a suitable method must be available for determining the end point with an acceptable level of accuracy. These are significant limitations and, for this reason, several titration strategies are commonly used.

A simple example of a titration is an analysis for Ag^+ using thiocyanate, SCN^-, as a titrant.

$$Ag^+(aq) + SCN^-(aq) \rightleftharpoons AgSCN(s)$$

*Instead of measuring the titrant's volume we also can measure its mass. Since the titrant's density is a measure of its mass per unit volume, the mass of titrant and volume of titrant are proportional.

This reaction occurs quickly and is of known stoichiometry. A titrant of SCN^- is easily prepared using KSCN. To indicate the titration's end point we add a small amount of Fe^{3+} to the solution containing the analyte. The formation of the red-colored $Fe(SCN)^{2+}$ complex signals the end point. This is an example of a direct titration since the titrant reacts with the analyte.

If the titration reaction is too slow, a suitable indicator is not available, or there is no useful direct titration reaction, then an indirect analysis may be possible. Suppose you wish to determine the concentration of formaldehyde, H_2CO, in an aqueous solution. The oxidation of H_2CO by I_3^-

$$H_2CO(aq) + 3OH^-(aq) + I_3^-(aq) \rightleftharpoons HCO_2^-(aq) + 3I^-(aq) + 2H_2O(\ell)$$

is a useful reaction, except that it is too slow for a direct titration. If we add a known amount of I_3^-, such that it is in excess, we can allow the reaction to go to completion. The I_3^- remaining can then be titrated with thiosulfate, $S_2O_3^{2-}$.

$$I_3^-(aq) + 2S_2O_3^{2-}(aq) \rightleftharpoons S_4O_6^{2-}(aq) + 3I^-(aq)$$

This type of titration is called a **back titration.**

back titration
A titration in which a reagent is added to a solution containing the analyte, and the excess reagent remaining after its reaction with the analyte is determined by a titration.

Calcium ion plays an important role in many aqueous environmental systems. A useful direct analysis takes advantage of its reaction with the ligand ethylenediaminetetraacetic acid (EDTA), which we will represent as Y^{4-}.

$$Ca^{2+}(aq) + Y^{4-}(aq) \rightleftharpoons CaY^{2-}(aq)$$

Unfortunately, it often happens that there is no suitable indicator for this direct titration. Reacting Ca^{2+} with an excess of the Mg^{2+}–EDTA complex

$$Ca^{2+}(aq) + MgY^{2-}(aq) \rightleftharpoons CaY^{2-}(aq) + Mg^{2+}(aq)$$

releases an equivalent amount of Mg^{2+}. Titrating the released Mg^{2+} with EDTA

$$Mg^{2+}(aq) + Y^{4-}(aq) \rightleftharpoons MgY^{2-}(aq)$$

gives a suitable end point. The amount of Mg^{2+} titrated provides an indirect measure of the amount of Ca^{2+} in the original sample. Since the analyte displaces a species that is then titrated, we call this a **displacement titration.**

displacement titration
A titration in which the analyte displaces a species, usually from a complex, and the amount of the displaced species is determined by a titration.

When a suitable reaction involving the analyte does not exist it may be possible to generate a species that is easily titrated. For example, the sulfur content of coal can be determined by using a combustion reaction to convert sulfur to sulfur dioxide.

$$S(s) + O_2(g) \rightarrow SO_2(g)$$

Passing the SO_2 through an aqueous solution of hydrogen peroxide, H_2O_2,

$$SO_2(g) + H_2O_2(aq) \rightarrow H_2SO_4(aq)$$

produces sulfuric acid, which we can titrate with NaOH,

$$H_2SO_4(aq) + 2OH^-(aq) \rightleftharpoons SO_4^{2-}(aq) + 2H_2O(\ell)$$

providing an indirect determination of sulfur.

9A.3 Titration Curves

To find the end point we monitor some property of the titration reaction that has a well-defined value at the equivalence point. For example, the equivalence point for a titration of HCl with NaOH occurs at a pH of 7.0. We can find the end point,

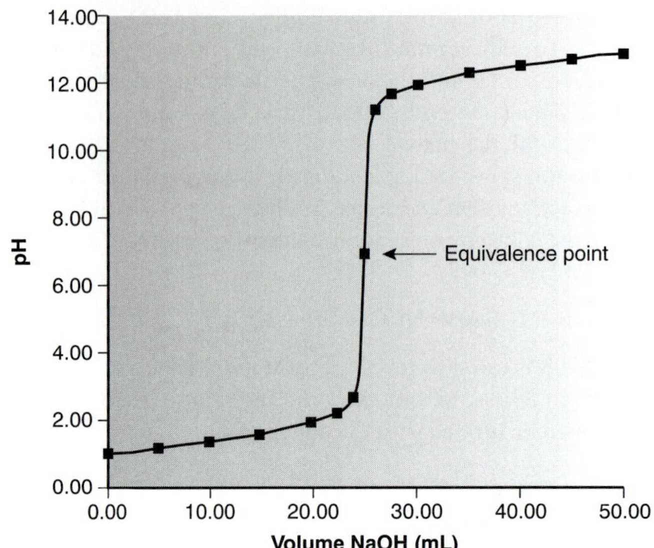

Figure 9.1
Acid–base titration curve for 25.0 mL of 0.100 M HCl with 0.100 M NaOH.

therefore, by monitoring the pH with a pH electrode or by adding an indicator that changes color at a pH of 7.0.

Suppose that the only available indicator changes color at a pH of 6.8. Is this end point close enough to the equivalence point that the titration error may be safely ignored? To answer this question we need to know how the pH changes during the titration.

titration curve
A graph showing the progress of a titration as a function of the volume of titrant added.

A **titration curve** provides us with a visual picture of how a property, such as pH, changes as we add titrant (Figure 9.1). We can measure this titration curve experimentally by suspending a pH electrode in the solution containing the analyte, monitoring the pH as titrant is added. As we will see later, we can also calculate the expected titration curve by considering the reactions responsible for the change in pH. However we arrive at the titration curve, we may use it to evaluate an indicator's likely titration error. For example, the titration curve in Figure 9.1 shows us that an end point pH of 6.8 produces a small titration error. Stopping the titration at an end point pH of 11.6, on the other hand, gives an unacceptably large titration error.

The titration curve in Figure 9.1 is not unique to an acid–base titration. Any titration curve that follows the change in concentration of a species in the titration reaction (plotted logarithmically) as a function of the volume of titrant has the same general sigmoidal shape. Several additional examples are shown in Figure 9.2.

Concentration is not the only property that may be used to construct a titration curve. Other parameters, such as temperature or the absorbance of light, may be used if they show a significant change in value at the equivalence point. Many titration reactions, for example, are exothermic. As the titrant and analyte react, the temperature of the system steadily increases. Once the titration is complete, further additions of titrant do not produce as exothermic a response, and the change in temperature levels off. A typical titration curve of temperature versus volume of titrant is shown in Figure 9.3. The titration curve contains two linear segments, the intersection of which marks the equivalence point.

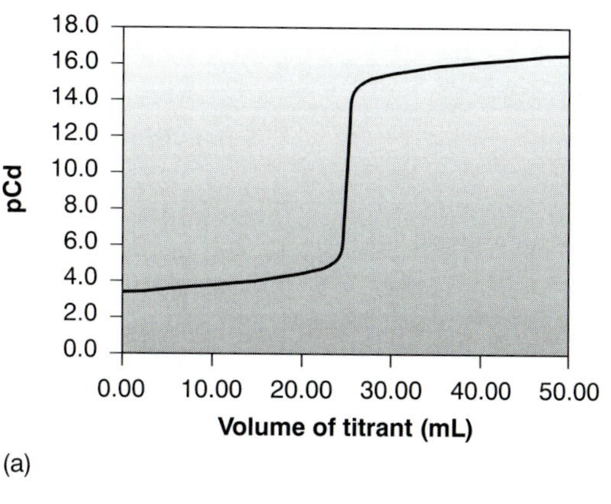

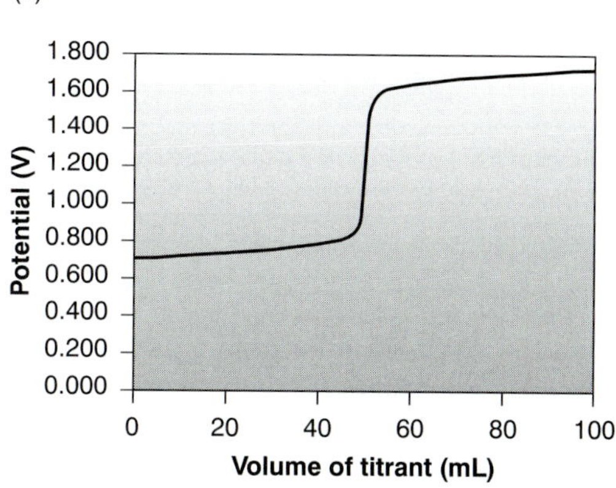

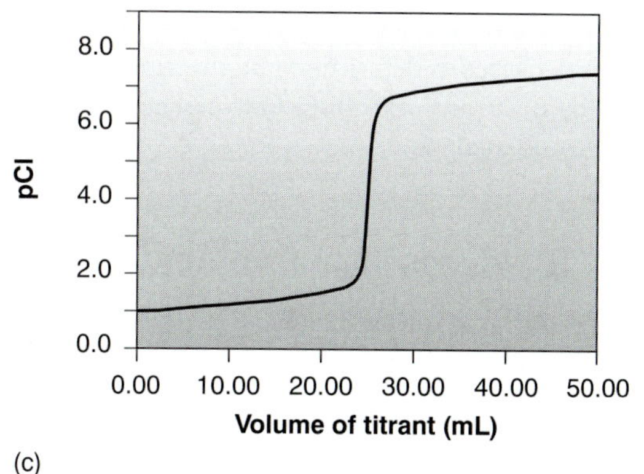

Figure 9.2
Examples of titration curves for (a) a complexation titration, (b) a redox titration, and (c) a precipitation titration.

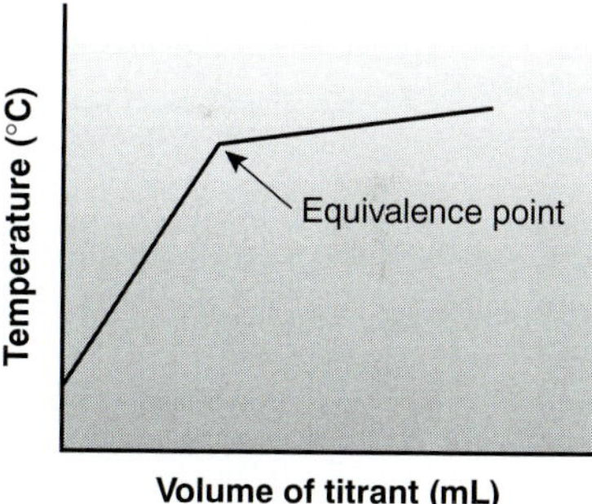

Figure 9.3
Example of a thermometric titration curve.

9A.4 The Buret

The only essential piece of equipment for an acid–base titration is a means for delivering the titrant to the solution containing the analyte. The most common method for delivering the titrant is a **buret** (Figure 9.4). A buret is a long, narrow tube with graduated markings, and a stopcock for dispensing the titrant. Using a buret with a small internal diameter provides a better defined meniscus, making it easier to read the buret's volume precisely. Burets are available in a variety of sizes and tolerances

buret
Volumetric glassware used to deliver variable, but known volumes of solution.

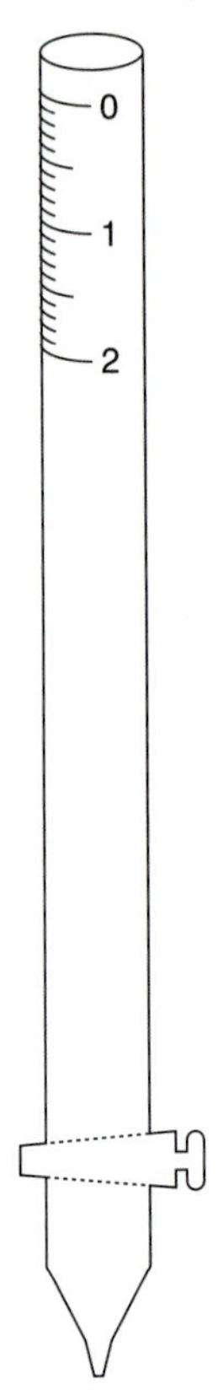

Figure 9.4
Volumetric buret showing a portion of its graduated scale.

Table 9.1 Specifications for Volumetric Burets

Volume (mL)	Class[a]	Subdivision (mL)	Tolerance (mL)
5	A	0.01	±0.01
	B	0.01	±0.02
10	A	0.02	±0.02
	B	0.02	±0.04
25	A	0.1	±0.03
	B	0.1	±0.06
50	A	0.1	±0.05
	B	0.1	±0.10
100	A	0.2	±0.10
	B	0.2	±0.20

[a]Specifications for class A and class B glassware are taken from American Society for Testing Materials (ASTM) E288, E542, and E694 standards.

(Table 9.1), with the choice of buret determined by the demands of the analysis. The accuracy obtainable with a buret can be improved by calibrating it over several intermediate ranges of volumes using the same method described in Chapter 5 for calibrating pipets. In this manner, the volume of titrant delivered can be corrected for any variations in the buret's internal diameter.

Titrations may be automated using a pump to deliver the titrant at a constant flow rate, and a solenoid valve to control the flow (Figure 9.5). The volume of titrant delivered is determined by multiplying the flow rate by the elapsed time. Automated titrations offer the additional advantage of using a microcomputer for data storage and analysis.

9B Titrations Based on Acid–Base Reactions

acid–base titration
A titration in which the reaction between the analyte and titrant is an acid–base reaction.

The earliest **acid–base titrations** involved the determination of the acidity or alkalinity of solutions, and the purity of carbonates and alkaline earth oxides. Before 1800, acid–base titrations were conducted using H_2SO_4, HCl, and HNO_3 as acidic titrants, and K_2CO_3 and Na_2CO_3 as basic titrants. End points were determined using visual indicators such as litmus, which is red in acidic solutions and blue in basic solutions, or by observing the cessation of CO_2 effervescence when neutralizing CO_3^{2-}. The accuracy of an acid–base titration was limited by the usefulness of the indicator and by the lack of a strong base titrant for the analysis of weak acids.

The utility of acid–base titrimetry improved when NaOH was first introduced as a strong base titrant in 1846. In addition, progress in synthesizing organic dyes led to the development of many new indicators. Phenolphthalein was first synthesized by Bayer in 1871 and used as a visual indicator for acid–base titrations in 1877. Other indicators, such as methyl orange, soon followed. Despite the increasing availability of indicators, the absence of a theory of acid–base reactivity made selecting a proper indicator difficult.

Developments in equilibrium theory in the late nineteenth century led to significant improvements in the theoretical understanding of acid–base chemistry and,

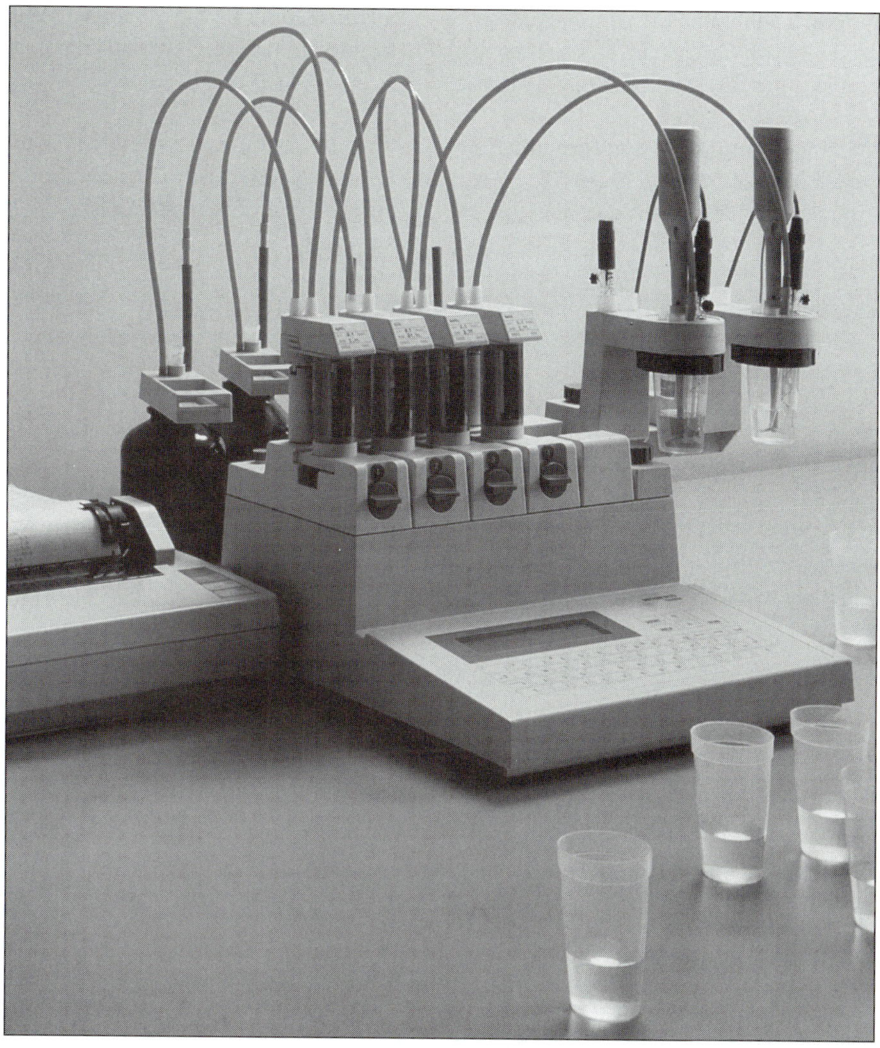

Figure 9.5
Typical instrumentation for performing an automatic titration.
Courtesy of Fisher Scientific.

in turn, of acid–base titrimetry. Sørenson's establishment of the pH scale in 1909 provided a rigorous means for comparing visual indicators. The determination of acid–base dissociation constants made the calculation of theoretical titration curves possible, as outlined by Bjerrum in 1914. For the first time a rational method existed for selecting visual indicators, establishing acid–base titrimetry as a useful alternative to gravimetry.

9B.1 Acid–Base Titration Curves

In the overview to this chapter we noted that the experimentally determined end point should coincide with the titration's equivalence point. For an acid–base titration, the equivalence point is characterized by a pH level that is a function of the acid–base strengths and concentrations of the analyte and titrant. The pH at the end point, however, may or may not correspond to the pH at the equivalence point. To understand the relationship between end points and equivalence points we must know how the pH changes during a titration. In this section we will learn how to construct titration curves for several important types of acid–base titrations. Our approach will make use of the equilibrium calculations described in Chapter 6. We also will learn how to sketch a good approximation to any titration curve using only a limited number of simple calculations.

Titrating Strong Acids and Strong Bases For our first titration curve let's consider the titration of 50.0 mL of 0.100 M HCl with 0.200 M NaOH. For the reaction of a strong base with a strong acid the only equilibrium reaction of importance is

$$H_3O^+(aq) + OH^-(aq) \rightleftharpoons 2H_2O(\ell) \qquad \mathbf{9.1}$$

The first task in constructing the titration curve is to calculate the volume of NaOH needed to reach the equivalence point. At the equivalence point we know from reaction 9.1 that

$$\text{Moles HCl} = \text{moles NaOH}$$

or

$$M_aV_a = M_bV_b$$

where the subscript 'a' indicates the acid, HCl, and the subscript 'b' indicates the base, NaOH. The volume of NaOH needed to reach the equivalence point, therefore, is

$$V_{eq} = V_b = \frac{M_aV_a}{M_b} = \frac{(0.100\text{ M})(50.0\text{ mL})}{(0.200\text{ M})} = 25.0\text{ mL}$$

Before the equivalence point, HCl is present in excess and the pH is determined by the concentration of excess HCl. Initially the solution is 0.100 M in HCl, which, since HCl is a strong acid, means that the pH is

$$\text{pH} = -\log[H_3O^+] = -\log[\text{HCl}] = -\log(0.100) = 1.00$$

The equilibrium constant for reaction 9.1 is $(K_w)^{-1}$, or 1.00×10^{14}. Since this is such a large value we can treat reaction 9.1 as though it goes to completion. After adding 10.0 mL of NaOH, therefore, the concentration of excess HCl is

$$[\text{HCl}] = \frac{\text{moles excess HCl}}{\text{total volume}} = \frac{M_aV_a - M_bV_b}{V_a + V_b}$$

$$= \frac{(0.100\text{ M})(50.0\text{ mL}) - (0.200\text{ M})(10.0\text{ mL})}{50.0\text{ mL} + 10.0\text{ mL}} = 0.050\text{ M}$$

giving a pH of 1.30.

At the equivalence point the moles of HCl and the moles of NaOH are equal. Since neither the acid nor the base is in excess, the pH is determined by the dissociation of water.

$$K_w = 1.00 \times 10^{-14} = [H_3O^+][OH^-] = [H_3O^+]^2$$

$$[H_3O^+] = 1.00 \times 10^{-7}\text{ M}$$

Thus, the pH at the equivalence point is 7.00.

Finally, for volumes of NaOH greater than the equivalence point volume, the pH is determined by the concentration of excess OH^-. For example, after adding 30.0 mL of titrant the concentration of OH^- is

Table 9.2 Data for Titration of 50.00 mL of 0.100 M HCl with 0.0500 M NaOH

Volume (mL) of Titrant	pH
0.00	1.00
5.00	1.14
10.00	1.30
15.00	1.51
20.00	1.85
22.00	2.08
24.00	2.57
25.00	7.00
26.00	11.42
28.00	11.89
30.00	12.50
35.00	12.37
40.00	12.52
45.00	12.62
50.00	12.70

$$[\mathrm{OH^-}] = \frac{\text{moles excess NaOH}}{\text{total volume}} = \frac{M_\mathrm{b}V_\mathrm{b} - M_\mathrm{a}V_\mathrm{a}}{V_\mathrm{a} + V_\mathrm{b}}$$

$$= \frac{(0.200\ \mathrm{M})(30.0\ \mathrm{mL}) - (0.100\ \mathrm{M})(50.0\ \mathrm{mL})}{50.0\ \mathrm{mL} + 30.0\ \mathrm{mL}} = 0.0125\ \mathrm{M}$$

To find the concentration of H_3O^+, we use the K_w expression

$$[\mathrm{H_3O^+}] = \frac{K_\mathrm{w}}{[\mathrm{OH^-}]} = \frac{1.00 \times 10^{-14}}{0.0125} = 8.00 \times 10^{-13}$$

giving a pH of 12.10. Table 9.2 and Figure 9.1 show additional results for this titration curve. Calculating the titration curve for the titration of a strong base with a strong acid is handled in the same manner, except that the strong base is in excess before the equivalence point and the strong acid is in excess after the equivalence point.

Titrating a Weak Acid with a Strong Base For this example let's consider the titration of 50.0 mL of 0.100 M acetic acid, CH_3COOH, with 0.100 M NaOH. Again, we start by calculating the volume of NaOH needed to reach the equivalence point; thus

$$\text{Moles } \mathrm{CH_3COOH} = \text{moles NaOH}$$

$$M_\mathrm{a}V_\mathrm{a} = M_\mathrm{b}V_\mathrm{b}$$

$$V_\mathrm{eq} = V_\mathrm{b} = \frac{M_\mathrm{a}V_\mathrm{a}}{M_\mathrm{b}} = \frac{(0.100\ \mathrm{M})(50.0\ \mathrm{mL})}{(0.100\ \mathrm{M})} = 50.0\ \mathrm{mL}$$

Before adding any NaOH the pH is that for a solution of 0.100 M acetic acid. Since acetic acid is a weak acid, we calculate the pH using the method outlined in Chapter 6.

$$CH_3COOH(aq) + H_2O(\ell) \rightleftharpoons H_3O^+(aq) + CH_3COO^-(aq)$$

$$K_a = \frac{[H_3O^+][CH_3COO^-]}{[CH_3COOH]} = \frac{(x)(x)}{0.100 - x} = 1.75 \times 10^{-5}$$

$$x = [H_3O^+] = 1.32 \times 10^{-3}$$

At the beginning of the titration the pH is 2.88.

Adding NaOH converts a portion of the acetic acid to its conjugate base.

$$CH_3COOH(aq) + OH^-(aq) \rightleftharpoons H_2O(\ell) + CH_3COO^-(aq) \qquad 9.2$$

Any solution containing comparable amounts of a weak acid, HA, and its conjugate weak base, A^-, is a buffer. As we learned in Chapter 6, we can calculate the pH of a buffer using the Henderson–Hasselbalch equation.

$$pH = pK_a + \log\frac{[A^-]}{[HA]}$$

The equilibrium constant for reaction 9.2 is large ($K = K_a/K_w = 1.75 \times 10^9$), so we can treat the reaction as one that goes to completion. Before the equivalence point, the concentration of unreacted acetic acid is

$$[CH_3COOH] = \frac{\text{moles unreacted } CH_3COOH}{\text{total volume}} = \frac{M_aV_a - M_bV_b}{V_a + V_b}$$

and the concentration of acetate is

$$[CH_3COO^-] = \frac{\text{moles NaOH added}}{\text{total volume}} = \frac{M_bV_b}{V_a + V_b}$$

For example, after adding 10.0 mL of NaOH the concentrations of CH_3COOH and CH_3COO^- are

$$[CH_3COOH] = \frac{(0.100\text{ M})(50.0\text{ mL}) - (0.100\text{ M})(10.0\text{ mL})}{50.0\text{ mL} + 10.0\text{ mL}} = 0.0667\text{ M}$$

$$[CH_3COO^-] = \frac{(0.100\text{ M})(10.0\text{ mL})}{50.0\text{ mL} + 10.0\text{ mL}} = 0.0167\text{ M}$$

giving a pH of

$$pH = 4.76 + \log\frac{0.0167}{0.0667} = 4.16$$

A similar calculation shows that the pH after adding 20.0 mL of NaOH is 4.58.

At the equivalence point, the moles of acetic acid initially present and the moles of NaOH added are identical. Since their reaction effectively proceeds to completion, the predominate ion in solution is CH_3COO^-, which is a weak base. To calculate the pH we first determine the concentration of CH_3COO^-.

$$[CH_3COO^-] = \frac{\text{moles } CH_3COOH}{\text{total volume}} = \frac{(0.100\text{ M})(50.0\text{ mL})}{50.0\text{ mL} + 50.0\text{ mL}} = 0.0500\text{ M}$$

The pH is then calculated as shown in Chapter 6 for a weak base.

$$CH_3COO^-(aq) + H_2O(\ell) \rightleftharpoons OH^-(aq) + CH_3COOH(aq)$$

$$K_b = \frac{[OH^-][CH_3COOH]}{[CH_3COO^-]} = \frac{(x)(x)}{0.0500 - x} = 5.71 \times 10^{-10}$$

$$x = [OH^-] = 5.34 \times 10^{-6}\ M$$

The concentration of H_3O^+, therefore, is 1.87×10^{-9}, or a pH of 8.73.

After the equivalence point NaOH is present in excess, and the pH is determined in the same manner as in the titration of a strong acid with a strong base. For example, after adding 60.0 mL of NaOH, the concentration of OH^- is

$$[OH^-] = \frac{(0.100\ M)(60.0\ mL) - (0.100\ M)(50.0\ mL)}{50.0\ mL + 60.0\ mL} = 0.00909\ M$$

giving a pH of 11.96. Table 9.3 and Figure 9.6 show additional results for this titration. The calculations for the titration of a weak base with a strong acid are handled in a similar manner except that the initial pH is determined by the weak base, the pH at the equivalence point by its conjugate weak acid, and the pH after the equivalence point by the concentration of excess strong acid.

Table 9.3 Data for Titration of 50.0 mL of 0.100 M Acetic Acid with 0.100 M NaOH

Volume of NaOH (mL)	pH
0.00	2.88
5.00	3.81
10.00	4.16
15.00	4.39
20.00	4.58
25.00	4.76
30.00	4.94
35.00	5.13
40.00	5.36
45.00	5.71
48.00	6.14
50.00	8.73
52.00	11.29
55.00	11.68
60.00	11.96
65.00	12.12
70.00	12.22
75.00	12.30
80.00	12.36
85.00	12.41
90.00	12.46
95.00	12.49
100.00	12.52

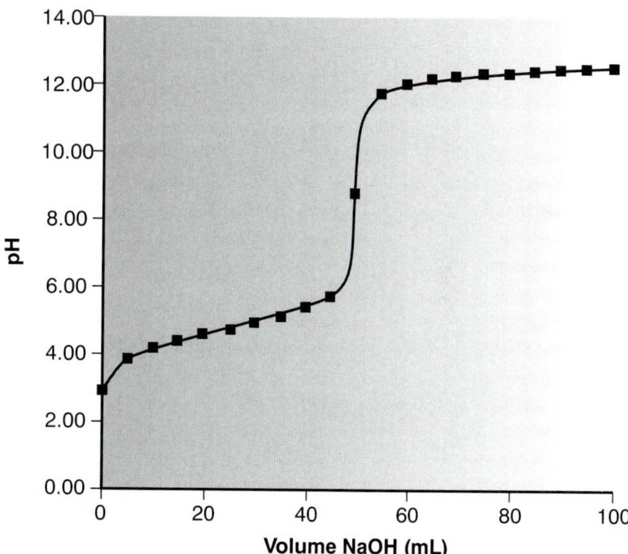

Figure 9.6
Titration curve for 50.0 mL of 0.100 M acetic acid ($pK_a = 4.76$) with 0.100 M NaOH.

The approach that we have worked out for the titration of a monoprotic weak acid with a strong base can be extended to reactions involving multiprotic acids or bases and mixtures of acids or bases. As the complexity of the titration increases, however, the necessary calculations become more time-consuming. Not surprisingly, a variety of algebraic[1] and computer spreadsheet[2] approaches have been described to aid in constructing titration curves.

Sketching an Acid–Base Titration Curve To evaluate the relationship between an equivalence point and an end point, we only need to construct a reasonable approximation to the titration curve. In this section we demonstrate a simple method for sketching any acid–base titration curve. Our goal is to sketch the titration curve quickly, using as few calculations as possible.

To quickly sketch a titration curve we take advantage of the following observation. Except for the initial pH and the pH at the equivalence point, the pH at any point of a titration curve is determined by either an excess of strong acid or strong base, or by a buffer consisting of a weak acid and its conjugate weak base. As we have seen in the preceding sections, calculating the pH of a solution containing excess strong acid or strong base is straightforward.

We can easily calculate the pH of a buffer using the Henderson–Hasselbalch equation. We can avoid this calculation, however, if we make the following assumption. You may recall that in Chapter 6 we stated that a buffer operates over a pH range extending approximately ±1 pH units on either side of the buffer's pK_a. The pH is at the lower end of this range, $pH = pK_a - 1$, when the weak acid's concentration is approximately ten times greater than that of its conjugate weak base. Conversely, the buffer's pH is at its upper limit, $pH = pK_a + 1$, when the concentration of weak acid is ten times less than that of its conjugate weak base. When titrating a weak acid or weak base, therefore, the buffer region spans a range of volumes from approximately 10% of the equivalence point volume to approximately 90% of the equivalence point volume.*

Our strategy for quickly sketching a titration curve is simple. We begin by drawing our axes, placing pH on the *y*-axis and volume of titrant on the *x*-axis. After calculating the volume of titrant needed to reach the equivalence point, we draw a vertical line that intersects the *x*-axis at this volume. Next, we determine the pH for two volumes before the equivalence point and for two volumes after the equivalence point. To save time we only calculate pH values when the pH is determined by excess strong acid or strong base. For weak acids or bases we use the limits of their buffer region to estimate the two points. Straight lines are drawn through each pair of points, with each line intersecting the vertical line representing the equivalence point volume. Finally, a smooth curve is drawn connecting the three straight-line segments. Example 9.1 illustrates this approach for the titration of a weak acid with a strong base.

Example

EXAMPLE 9.1

Sketch the titration curve for the titration of 50.0 mL of 0.100 M acetic acid with 0.100 M NaOH. This is the same titration for which we previously calculated the titration curve (Table 9.3 and Figure 9.6).

SOLUTION

We begin by drawing the axes for the titration curve (Figure 9.7a). We have already shown that the volume of NaOH needed to reach the equivalence point is 50 mL, so we draw a vertical line intersecting the *x*-axis at this volume (Figure 9.7b).

*Question 4 in the end-of-chapter problems asks you to consider why these pH limits correspond to approximately 10% and 90% of the equivalence point volume.

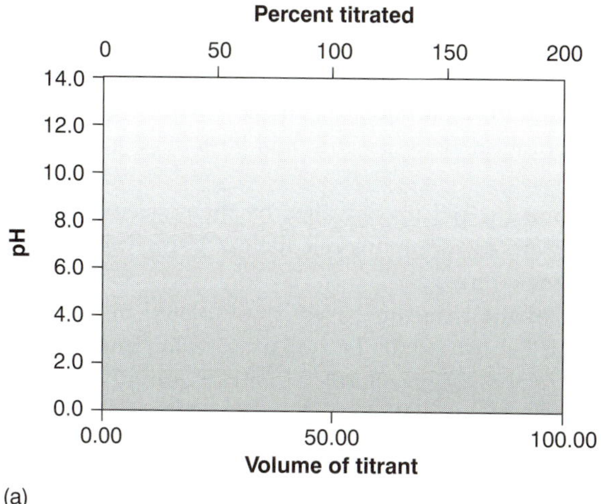

(a)

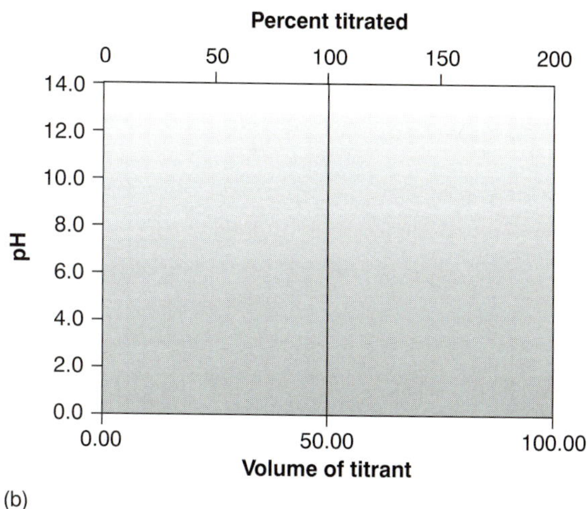

(b)

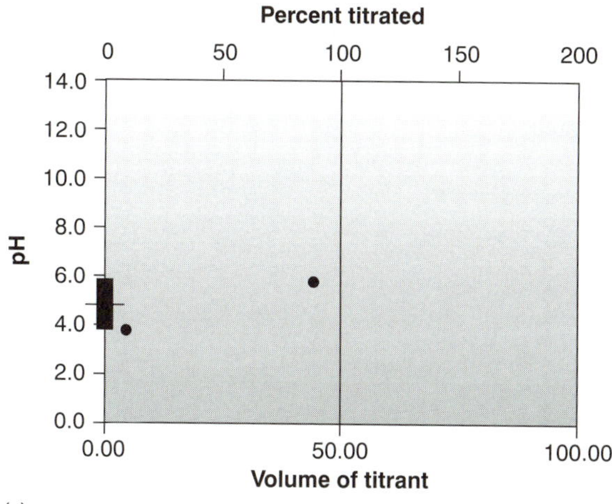

(c)

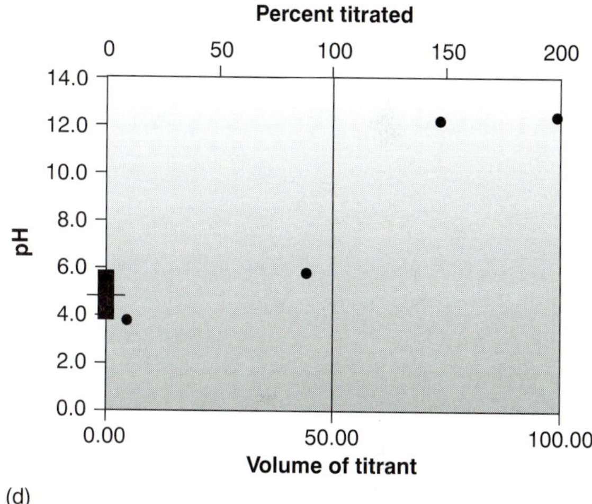

(d)

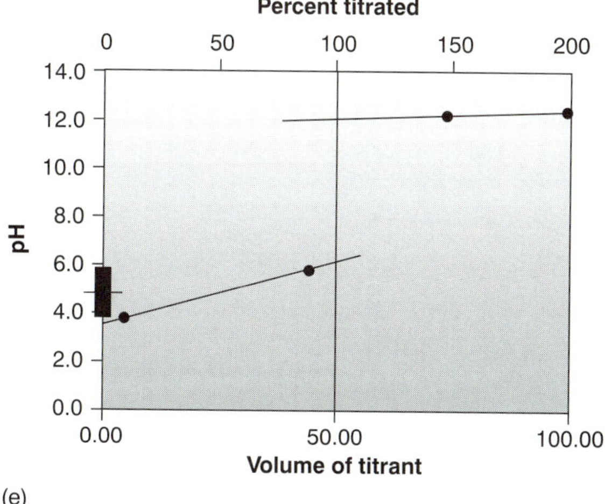

(e)

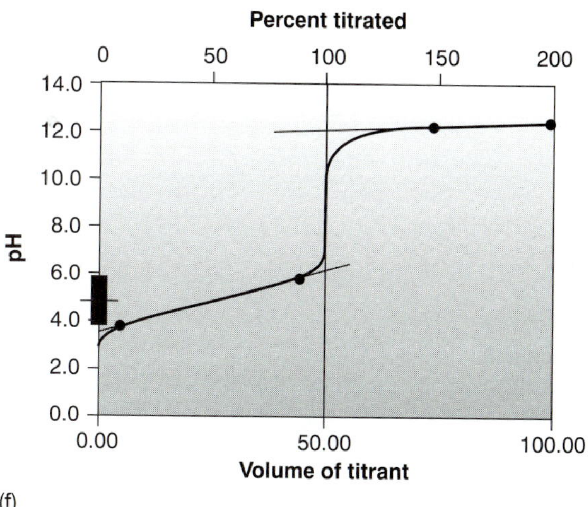

(f)

Figure 9.7
How to sketch an acid–base titration curve; see text for explanation.

Before the equivalence point the titrant is the limiting reagent, and the pH is controlled by a buffer consisting of unreacted acetic acid and its conjugate weak base, acetate. The pH limits for the buffer region are plotted by superimposing the ladder diagram for acetic acid on the *y*-axis (Figure 9.7c) and adding the appropriate points at 10% (5.0 mL) and 90% (45.0 mL) of the equivalence point volume.

After the equivalence point the pH is controlled by the concentration of excess NaOH. Again, we have already done this calculation. Using values from Table 9.3, we plot two additional points.

An approximate sketch of the titration curve is completed by drawing separate straight lines through the two points in the buffer region and the two points in the excess titrant region (Figure 9.7e). Finally, a smooth curve is drawn connecting the three straight-line segments (Figure 9.7f).

This approach can be used to sketch titration curves for other acid–base titrations including those involving polyprotic weak acids and bases or mixtures of weak acids and bases (Figure 9.8). Figure 9.8a, for example, shows the titration curve when titrating a diprotic weak acid, H_2A, with a strong base. Since the analyte is

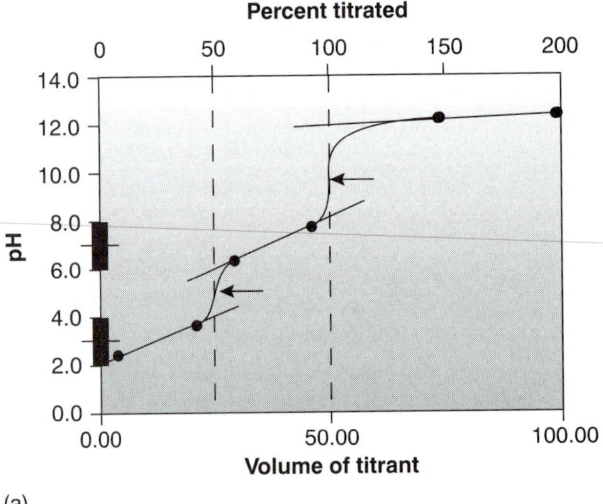

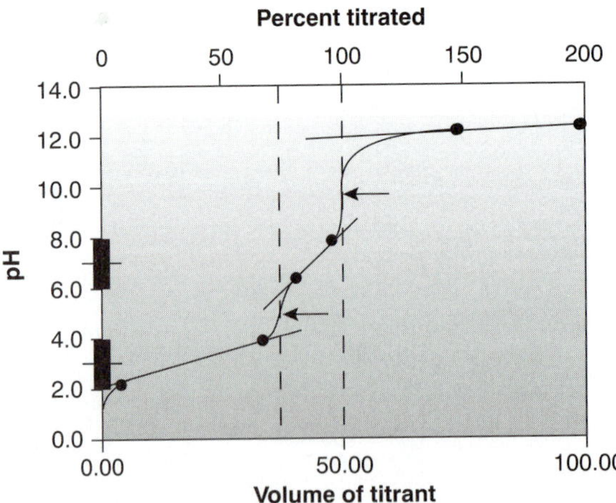

Figure 9.8
Sketches of titration curves for (a) 50.00 mL of 0.0500 M diprotic weak acid ($pK_{a1} = 3$, $pK_{a2} = 7$) with 0.100 M strong base; and (b) 50.00 mL of a mixture of weak acids consisting of 0.075 M HA ($pK_{a,HA} = 3$) and 0.025 M HB ($pK_{a,HB} = 7$) with 0.100 M strong base. The points used to sketch the titration curves are indicated by the dots (•). Equivalence points are indicated by the arrows.

diprotic there are two equivalence points, each requiring the same volume of titrant. Before the first equivalence point the pH is controlled by a buffer consisting of H_2A and HA^-, and the HA^-/A^{2-} buffer determines the pH between the two equivalence points. After the second equivalence point, the pH reflects the concentration of the excess strong base titrant.

Figure 9.8b shows a titration curve for a mixture consisting of two weak acids: HA and HB. Again, there are two equivalence points. In this case, however, the equivalence points do not require the same volume of titrant because the concentration of HA is greater than that for HB. Since HA is the stronger of the two weak acids, it reacts first; thus, the pH before the first equivalence point is controlled by the HA/A^- buffer. Between the two equivalence points the pH reflects the titration of HB and is determined by the HB/B^- buffer. Finally, after the second equivalence point, the excess strong base titrant is responsible for the pH.

9B.2 Selecting and Evaluating the End Point

Earlier we made an important distinction between an end point and an equivalence point. The difference between these two terms is important and deserves repeating. The equivalence point occurs when stoichiometrically equal amounts of analyte and titrant react. For example, if the analyte is a triprotic weak acid, a titration with NaOH will have three equivalence points corresponding to the addition of one, two, and three moles of OH^- for each mole of the weak acid. An equivalence point, therefore, is a theoretical not an experimental value.

An end point for a titration is determined experimentally and represents the analyst's best estimate of the corresponding equivalence point. Any difference between an equivalence point and its end point is a source of determinate error. As we shall see, it is even possible that an equivalence point will not have an associated end point.

Where Is the Equivalence Point? We have already learned how to calculate the equivalence point for the titration of a strong acid with a strong base, and for the titration of a weak acid with a strong base. We also have learned to sketch a titration curve with a minimum of calculations. Can we also locate the equivalence point without performing any calculations? The answer, as you may have guessed, is often yes!

It has been shown[3] that for most acid–base titrations the inflection point, which corresponds to the greatest slope in the titration curve, very nearly coincides with the equivalence point. The inflection point actually precedes the equivalence point, with the error approaching 0.1% for weak acids or weak bases with dissociation constants smaller than 10^{-9}, or for very dilute solutions. Equivalence points determined in this fashion are indicated on the titration curves in Figure 9.8.

The principal limitation to using a titration curve to locate the equivalence point is that an inflection point must be present. Sometimes, however, an inflection point may be missing or difficult to detect. Figure 9.9, for example, demonstrates the influence of the acid dissociation constant, K_a, on the titration curve for a weak acid with a strong base titrant. The inflection point is visible, even if barely so, for acid dissociation constants larger than 10^{-9}, but is missing when K_a is 10^{-11}.

Another situation in which an inflection point may be missing or difficult to detect occurs when the analyte is a multiprotic weak acid or base whose successive dissociation constants are similar in magnitude. To see why this is true let's consider the titration of a diprotic weak acid, H_2A, with NaOH. During the titration the following two reactions occur.

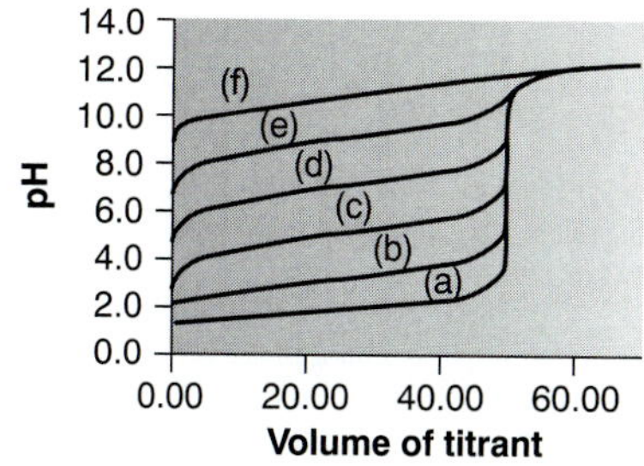

Figure 9.9
Titration curves for 50.00 mL of 0.100 M weak acid with 0.100 M strong base. The pK_as of the weak acids are (a) 1, (b) 3, (c) 5, (d) 7, (e) 9, (f) 11.

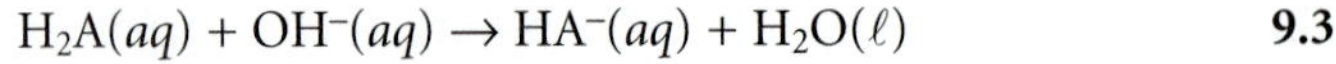

$$H_2A(aq) + OH^-(aq) \rightarrow HA^-(aq) + H_2O(\ell) \qquad 9.3$$

$$HA^-(aq) + OH^-(aq) \rightarrow A^{2-}(aq) + H_2O(\ell) \qquad 9.4$$

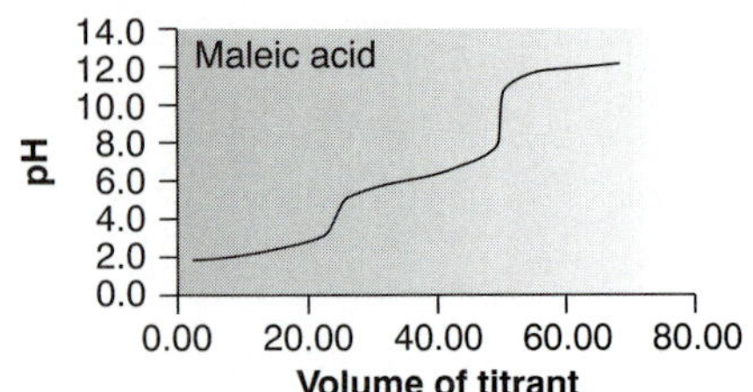

(a)

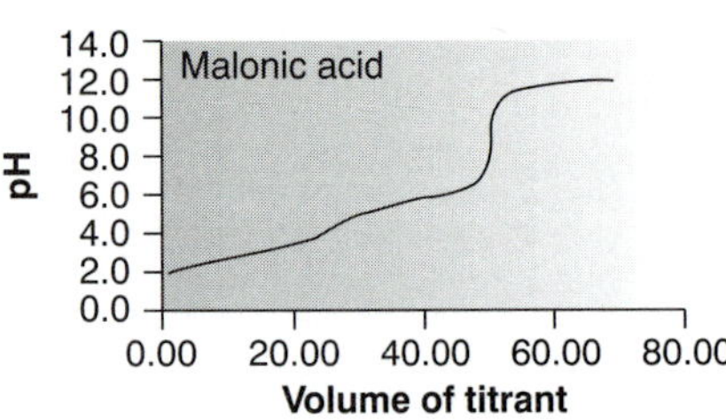

(b)

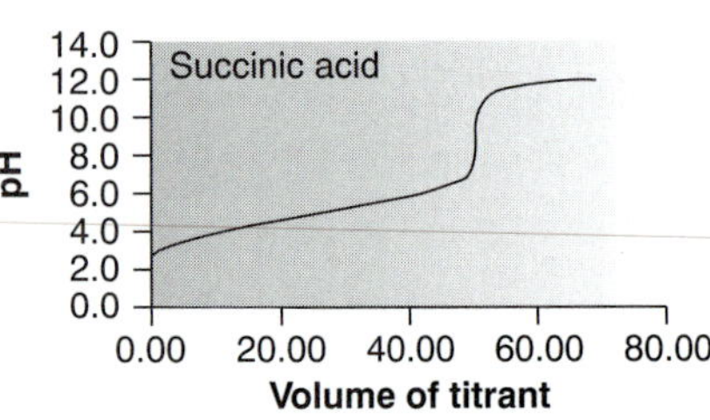

(c)

Figure 9.10

Titration curves for (a) maleic acid, $pK_{a1} = 1.91$, $pK_{a2} = 6.33$; (b) malonic acid, $pK_{a1} = 2.85$, $pK_{a2} = 5.70$; (c) succinic acid, $pK_{a1} = 4.21$, $pK_{a2} = 5.64$. Titration curves are for 50.00 mL of 0.0500 M acid with 0.100 M strong base. Equivalence points for all three titrations occur at 25.00 and 50.00 mL of titrant.

Two distinct inflection points are seen if reaction 9.3 is essentially complete before reaction 9.4 begins.

Figure 9.10 shows titration curves for three diprotic weak acids. The titration curve for maleic acid, for which K_{a1} is approximately 20,000 times larger than K_{a2}, shows two very distinct inflection points. Malonic acid, on the other hand, has acid dissociation constants that differ by a factor of approximately 690. Although malonic acid's titration curve shows two inflection points, the first is not as distinct as that for maleic acid. Finally, the titration curve for succinic acid, for which the two K_a values differ by a factor of only 27, has only a single inflection point corresponding to the neutralization of $HC_4H_4O_4^-$ to $C_4H_4O_4^{2-}$. In general, separate inflection points are seen when successive acid dissociation constants differ by a factor of at least 500 (a ΔpK_a of at least 2.7).

Finding the End Point with a Visual Indicator One interesting group of weak acids and bases are derivatives of organic dyes. Because such compounds have at least one conjugate acid–base species that is highly colored, their titration results in a change in both pH and color. This change in color can serve as a useful means for determining the end point of a titration, provided that it occurs at the titration's equivalence point.

The pH at which an acid–base indicator changes color is determined by its acid dissociation constant. For an indicator that is a monoprotic weak acid, HIn, the following dissociation reaction occurs

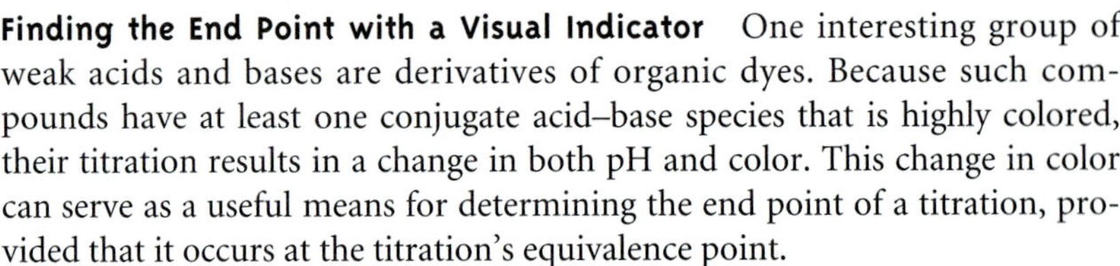

$$HIn(aq) + H_2O(\ell) \rightleftharpoons H_3O^+(aq) + In^-(aq)$$

for which the equilibrium constant is

$$K_a = \frac{[In^-][H_3O^+]}{[HIn]} \qquad 9.5$$

Taking the negative log of each side of equation 9.5, and rearranging to solve for pH gives a familiar equation.

$$pH = pK_a + \log\frac{[In^-]}{[HIn]} \qquad 9.6$$

The two forms of the indicator, HIn and In^-, have different colors. The color of a solution containing an indicator, therefore, continuously changes as the concentration of HIn decreases and the concentration of In^- increases. If we assume that both HIn and In^- can be detected with equal ease, then the transition between the two colors reaches its midpoint when their concentrations are identical or when the pH is equal to the indicator's pK_a. The equivalence point and the end point coincide, therefore, if an indicator is selected whose pK_a is equal to the pH at the equivalence point, and the titration is continued until the indicator's color is exactly halfway between that for HIn and In^-. Unfortunately, the exact pH at the equivalence point is rarely known. In addition, detecting the point where the concentrations of HIn and In^- are equal may be difficult if the change in color is subtle.

We can establish a range of pHs over which the average analyst will observe a change in color if we assume that a solution of the indicator is the color of HIn whenever its concentration is ten times more than that of In^-, and the color of In^-

whenever the concentration of HIn is ten times less than that of In^-. Substituting these inequalities into equation 9.6

$$pH = pK_a + \log\frac{1}{10} = pK_a - 1$$

$$pH = pK_a + \log\frac{10}{1} = pK_a + 1$$

shows that an indicator changes color over a pH range of ±1 units on either side of its pK_a (Figure 9.11). Thus, the indicator will be the color of HIn when the pH is less than $pK_a - 1$, and the color of In^- for pHs greater than $pK_a + 1$.

The pH range of an indicator does not have to be equally distributed on either side of the indicator's pK_a. For some indicators only the weak acid or weak base is colored. For other indicators both the weak acid and weak base are colored, but one form may be easier to see. In either case, the pH range is skewed toward those pH levels for which the less colored form of the indicator is present in higher concentration.

A list of several common acid–base indicators, along with their pK_as, color changes, and pH ranges, is provided in the top portion of Table 9.4. In some cases,

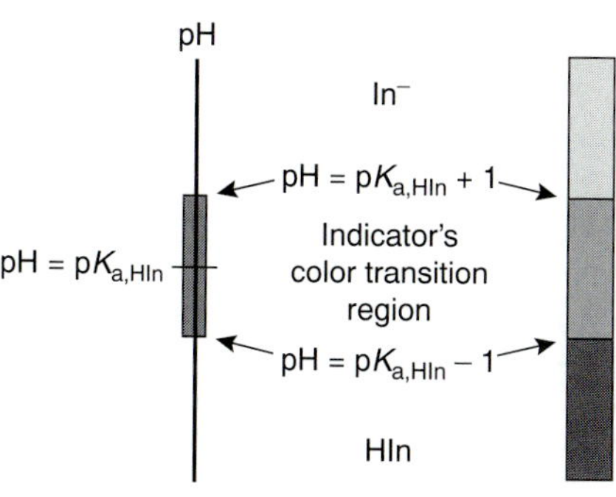

Figure 9.11
Ladder diagram showing the range of pH levels over which a typical acid–base indicator changes color.

Table 9.4 Properties of Selected Indicators, Mixed Indicators, and Screened Indicators for Acid–Base Titrations

Indicator	Acid Color	Base Color	pH Range	pK_a
cresol red	red	yellow	0.2–1.8	—
thymol blue	red	yellow	1.2–2.8	1.7
bromophenol blue	yellow	blue	3.0–4.6	4.1
methyl orange	red	orange	3.1–4.4	3.7
Congo red	blue	red	3.0–5.0	—
bromocresol green	yellow	blue	3.8–5.4	4.7
methyl red	red	yellow	4.2–6.3	5.0
bromocresol purple	yellow	purple	5.2–6.8	6.1
litmus	red	blue	5.0–8.0	—
bromothymol blue	yellow	blue	6.0–7.6	7.1
phenol red	yellow	red	6.8–8.4	7.8
cresol red	yellow	red	7.2–8.8	8.2
thymol blue	yellow	blue	8.0–9.6	8.9
phenolphthalein	colorless	red	8.3–10.0	9.6
alizarin yellow R	yellow	orange/red	10.1–12.0	—

Mixed Indicator	Acid Color	Base Color	pH Range
bromocresol green and methyl orange	orange	blue-green	3.5–4.3
bromocresol green and chlorophenol red	yellow-green	blue-violet	5.4–6.2
bromothymol blue and phenol red	yellow	violet	7.2–7.6

Screened Indicator	Acid Color	Base Color	pH Range
dimethyl yellow and methylene blue	blue-violet	green	3.2–3.4
methyl red and methylene blue	red-violet	green	5.2–5.6
neutral red and methylene blue	violet-blue	green	6.8–7.3

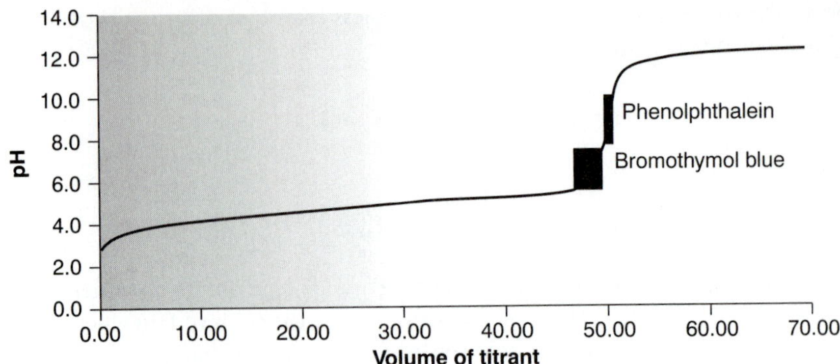

Figure 9.12
Titration curve for 50.00 mL of 0.100 M CH_3COOH with 0.100 M NaOH showing the range of pHs and volumes of titrant over which the indicators bromothymol blue and phenolphthalein are expected to change color.

mixed indicators, which are a mixture of two or more acid–base indicators, provide a narrower range of pHs over which the color change occurs. A few examples of such mixed indicators are included in the middle portion of Table 9.4. Adding a neutral screening dye, such as methylene blue, also has been found to narrow the pH range over which an indicator changes color (lower portion of Table 9.4). In this case, the neutral dye provides a gray color at the midpoint of the indicator's color transition.

The relatively broad range of pHs over which any indicator changes color places additional limitations on the feasibility of a titration. To minimize a determinate titration error, an indicator's entire color transition must lie within the sharp transition in pH occurring near the equivalence point. Thus, in Figure 9.12 we see that phenolphthalein is an appropriate indicator for the titration of 0.1 M acetic acid with 0.1 M NaOH. Bromothymol blue, on the other hand, is an inappropriate indicator since its change in color begins before the initial sharp rise in pH and, as a result, spans a relatively large range of volumes. The early change in color increases the probability of obtaining inaccurate results, and the range of possible end point volumes increases the probability of obtaining imprecise results.

The need for the indicator's color transition to occur in the sharply rising portion of the titration curve justifies our earlier statement that not every equivalence point has an end point. For example, trying to use a visual indicator to find the first equivalence point in the titration of succinic acid (see Figure 9.10c) is pointless since any difference between the equivalence point and the end point leads to a large titration error.

Finding the End Point by Monitoring pH An alternative approach to finding a titration's end point is to monitor the titration reaction with a suitable sensor whose signal changes as a function of the analyte's concentration. Plotting the data gives us the resulting titration curve. The end point may then be determined from the titration curve with only a minimal error.

The most obvious sensor for an acid–base titration is a pH electrode.* For example, Table 9.5 lists values for the pH and volume of titrant obtained during the titration of a weak acid with NaOH. The resulting titration curve, which is called a potentiometric titration curve, is shown in Figure 9.13a. The simplest method for finding the end point is to visually locate the inflection point of the titration curve. This is also the least accurate method, particularly if the titration curve's slope at the equivalence point is small.

*See Chapter 11 for more details about pH electrodes.

Table 9.5 Data for the Titration of a Weak Acid with 0.100 M NaOH

Normal Titration		First Derivative		Second Derivative	
Volume (mL)	pH	Volume (mL)	$\frac{\Delta pH}{\Delta V}$	Volume (mL)	$\frac{\Delta^2 pH}{\Delta V^2}$
0.00	2.89				
		1.00	0.815		
2.00	4.52			2.00	−0.273
		3.00	0.270		
4.00	5.06			5.00	−0.033
		7.00	0.138		
10.00	5.89			9.00	−0.002
		11.00	0.130		
12.00	6.15			12.00	0.055
		13.00	0.240		
14.00	6.63			13.75	0.140
		14.50	0.450		
15.00	7.08			14.88	1.19
		15.25	1.34		
15.50	7.75			15.40	17.2
		15.55	6.50		
15.60	8.40			15.60	24.0
		15.65	8.90		
15.70	9.29			15.70	−11.0
		15.75	7.80		
15.80	10.07			15.83	−32.8
		15.90	2.90		
16.00	10.65			16.20	−3.68
		16.50	0.690		
17.00	11.34			17.00	−0.400
		17.50	0.290		
18.00	11.63			18.25	−0.070
		19.00	0.185		
20.00	12.00			20.50	−0.027
		22.00	0.103		
24.00	12.41				

Another method for finding the end point is to plot the first or second derivative of the titration curve. The slope of a titration curve reaches its maximum value at the inflection point. The first derivative of a titration curve, therefore, shows a separate peak for each end point. The first derivative is approximated as $\Delta pH/\Delta V$, where ΔpH is the change in pH between successive additions of titrant. For example, the initial point in the first derivative titration curve for the data in Table 9.5 is

$$\frac{\Delta pH}{\Delta V} = \frac{4.52 - 2.89}{2.00 - 0.00} = 0.815$$

and is plotted at the average of the two volumes (1.00 mL). The remaining data for the first derivative titration curve are shown in Table 9.5 and plotted in Figure 9.13b.

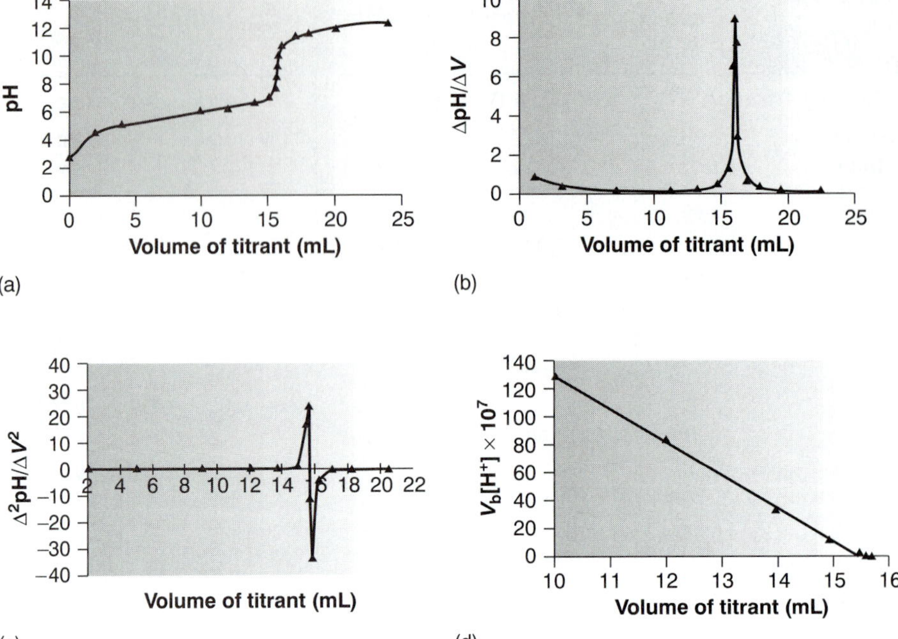

Figure 9.13
Titration curves for a weak acid with 0.100 M NaOH—(a) normal titration curve; (b) first derivative titration curve; (c) second derivative titration curve; (d) Gran plot.

The second derivative of a titration curve may be more useful than the first derivative, since the end point is indicated by its intersection with the volume axis. The second derivative is approximated as $\Delta(\Delta\text{pH}/\Delta V)/\Delta V$, or $\Delta^2\text{pH}/\Delta V^2$. For the titration data in Table 9.5, the initial point in the second derivative titration curve is

$$\frac{\Delta^2\text{pH}}{\Delta V^2} = \frac{0.270 - 0.815}{3.00 - 1.00} = -0.273$$

and is plotted as the average of the two volumes (2.00 mL). The remainder of the data for the second derivative titration curve are shown in Table 9.5 and plotted in Figure 9.13c.

Derivative methods are particularly well suited for locating end points in multiprotic and multicomponent systems, in which the use of separate visual indicators for each end point is impractical. The precision with which the end point may be located also makes derivative methods attractive for the analysis of samples with poorly defined normal titration curves.

Derivative methods work well only when sufficient data are recorded during the sharp rise in pH occurring near the equivalence point. This is usually not a problem when the titration is conducted with an automatic titrator, particularly when operated under computer control. Manual titrations, however, often contain only a few data points in the equivalence point region, due to the limited range of volumes over which the transition in pH occurs. Manual titrations are, however, information-rich during the more gently rising portions of the titration curve before and after the equivalence point.

Consider again the titration of a monoprotic weak acid, HA, with a strong base. At any point during the titration the weak acid is in equilibrium with H_3O^+ and A^-

$$\text{HA}(aq) + \text{H}_2\text{O}(\ell) \rightleftharpoons \text{H}_3\text{O}^+(aq) + \text{A}^-$$

for which

$$K_a = \frac{[H_3O^+][A^-]}{[HA]}$$

Before the equivalence point, and for volumes of titrant in the titration curve's buffer region, the concentrations of HA and A^- are given by the following equations.

$$[HA] = \frac{\text{moles HA} - \text{moles OH}^- \text{ added}}{\text{total volume}} = \frac{M_aV_a - M_bV_b}{V_a + V_b}$$

$$[A^-] = \frac{\text{moles OH}^- \text{ added}}{\text{total volume}} = \frac{M_bV_b}{V_a + V_b}$$

Substituting these equations into the K_a expression for HA, and rearranging gives

$$K_a = \frac{[H_3O^+](M_bV_b)/(V_a + V_b)}{(M_aV_a - M_bV_b)/(V_a + V_b)}$$

$$= \frac{[H_3O^+]M_bV_b}{M_aV_a - M_bV_b}$$

$$K_aM_aV_a - K_aM_bV_b = [H_3O^+]M_bV_b$$

$$\frac{K_aM_aV_a}{M_b} - K_aV_b = [H_3O^+]V_b$$

Finally, recognizing that the equivalence point volume is

$$V_{eq} = \frac{M_aV_a}{M_b}$$

leaves us with the following equation.

$$V_b \times [H_3O^+] = K_a \times V_{eq} - K_a \times V_b$$

For volumes of titrant before the equivalence point, a plot of $V_b \times [H_3O^+]$ versus V_b is a straight line with an *x*-intercept equal to the volume of titrant at the end point and a slope equal to $-K_a$.* Results for the data in Table 9.5 are shown in Table 9.6 and plotted in Figure 9.13d. Plots such as this, which convert a portion of a titration curve into a straight line, are called **Gran plots.**

Gran plot
A linearized form of a titration curve.

Finding the End Point by Monitoring Temperature The reaction between an acid and a base is exothermic. Heat generated by the reaction increases the temperature of the titration mixture. The progress of the titration, therefore, can be followed by monitoring the change in temperature.

An idealized thermometric titration curve (Figure 9.14a) consists of three distinct linear regions. Before adding titrant, any change in temperature is due to the cooling or warming of the solution containing the analyte. Adding titrant initiates the exothermic acid–base reaction, resulting in an increase in temperature. This part of a thermometric titration curve is called the titration branch. The

*Values of K_a determined by this method may be in substantial error if the effect of activity is not considered.

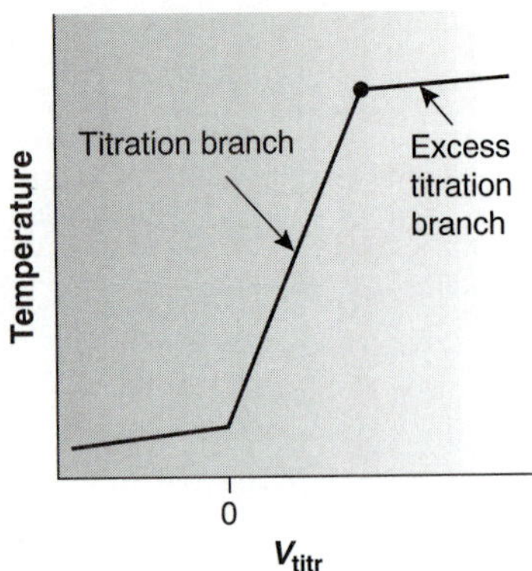

(a)

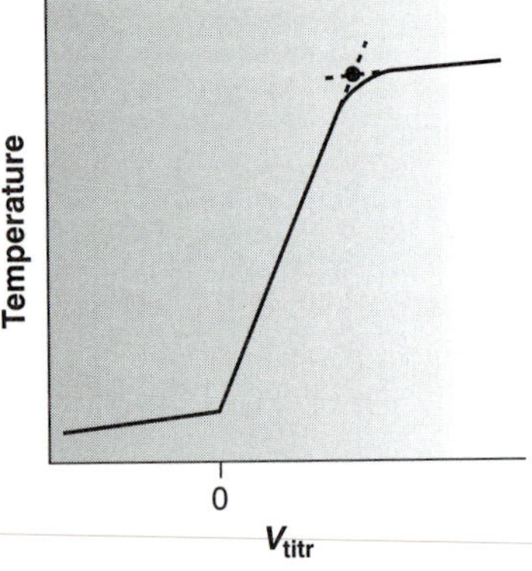

(b)

Figure 9.14
Thermometric titration curves—(a) ideal; (b) showing curvature at the intersection of the titration and excess titrant branches. Equivalence points are indicated by the dots (•).

Table 9.6 Gran Plot Treatment of the Data in Table 9.5

Volume (mL)	$V_b \times [H_3O^+] \times 10^7$
2.00	604
4.00	348
10.00	129
12.00	85.0
14.00	33.8
15.00	12.5
15.55	2.76
15.60	0.621
15.70	0.0805
15.80	0.0135

temperature continues to rise with each addition of titrant until the equivalence point is reached. After the equivalence point, any change in temperature is due to the difference between the temperatures of the analytical solution and the titrant, and the enthalpy of dilution for the excess titrant. Actual thermometric titration curves (Figure 9.14b) frequently show curvature at the intersection of the titration branch and the excess titrant branch due to the incompleteness of the neutralization reaction, or excessive dilution of the analyte during the titration. The latter problem is minimized by using a titrant that is 10–100 times more concentrated than the analyte, although this results in a very small end point volume and a larger relative error.

The end point is indicated by the intersection of the titration branch and the excess titrant branch. In the idealized thermometric titration curve (see Figure 9.14a) the end point is easily located. When the intersection between the two branches is curved, the end point can be found by extrapolation (Figure 9.14b).

Although not commonly used, thermometric titrations have one distinct advantage over methods based on the direct or indirect monitoring of pH. As discussed earlier, visual indicators and potentiometric titration curves are limited by the magnitude of the relevant equilibrium constants. For example, the titration of boric acid, H_3BO_3, for which K_a is 5.8×10^{-10}, yields a poorly defined equivalence point (Figure 9.15a). The enthalpy of neutralization for boric acid with NaOH, however, is only 23% less than that for a strong acid (–42.7 kJ/mol

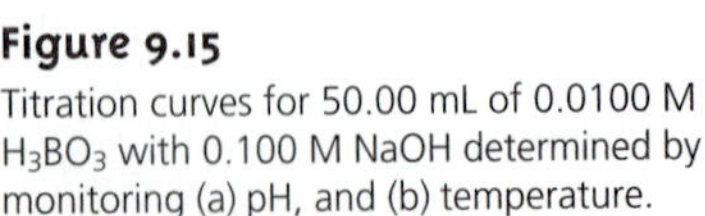

Figure 9.15
Titration curves for 50.00 mL of 0.0100 M H_3BO_3 with 0.100 M NaOH determined by monitoring (a) pH, and (b) temperature.

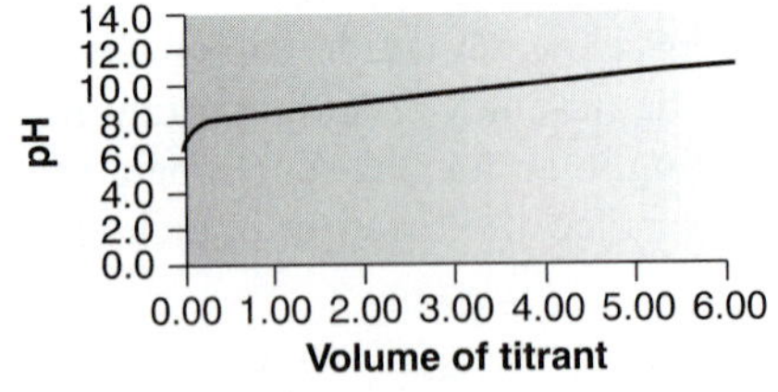

(a)

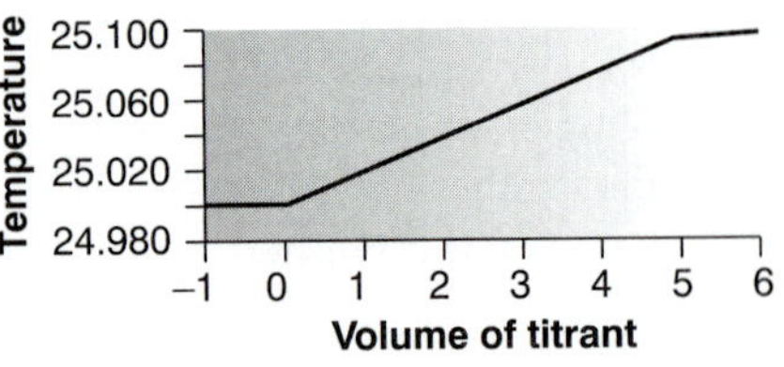

(b)

for H_3BO_3 versus –55.6 kJ/mol for HCl), resulting in a favorable thermometric titration curve (Figure 9.15b).

9B.3 Titrations in Nonaqueous Solvents

Thus far we have assumed that the acid and base are in an aqueous solution. Indeed, water is the most common solvent in acid–base titrimetry. When considering the utility of a titration, however, the solvent's influence cannot be ignored.

The dissociation, or autoprotolysis constant for a solvent, SH, relates the concentration of the protonated solvent, SH_2^+, to that of the deprotonated solvent, S^-. For amphoteric solvents, which can act as both proton donors and proton acceptors, the autoprotolysis reaction is

$$2SH \rightleftharpoons SH_2^+ + S^-$$

with an equilibrium constant of

$$K_s = [SH_2^+][S^-]$$

You should recognize that K_w is just the specific form of K_s for water. The pH of a solution is now seen to be a general statement about the relative abundance of protonated solvent

$$\text{pH} = -\log[SH_2^+]$$

where the pH of a neutral solvent is given as

$$\text{pH}_{\text{neut}} = \frac{1}{2}\text{p}K_s$$

Perhaps the most obvious limitation imposed by K_s is the change in pH during a titration. To see why this is so, let's consider the titration of a 50 mL solution of 10^{-4} M strong acid with equimolar strong base. Before the equivalence point, the pH is determined by the untitrated strong acid, whereas after the equivalence point the concentration of excess strong base determines the pH. In an aqueous solution the concentration of H_3O^+ when the titration is 90% complete is

$$[H_3O^+] = \frac{M_aV_a - M_bV_b}{V_a + V_b}$$

$$= \frac{(1 \times 10^{-4}\text{ M})(50\text{ mL}) - (1 \times 10^{-4}\text{ M})(45\text{ mL})}{50 + 45} = 5.3 \times 10^{-6}\text{ M}$$

corresponding to a pH of 5.3. When the titration is 110% complete, the concentration of OH^- is

$$[OH^-] = \frac{M_bV_b - M_aV_a}{V_a + V_b}$$

$$= \frac{(1 \times 10^{-4}\text{ M})(55\text{ mL}) - (1 \times 10^{-4}\text{ M})(50\text{ mL})}{50 + 55} = 4.8 \times 10^{-6}\text{ M}$$

or a pOH of 5.3. The pH, therefore, is

$$\text{pH} = \text{p}K_w - \text{pOH} = 14.0 - 5.3 = 8.7$$

The change in pH when the titration passes from 90% to 110% completion is

$$\Delta\text{pH} = 8.7 - 5.3 = 3.4$$

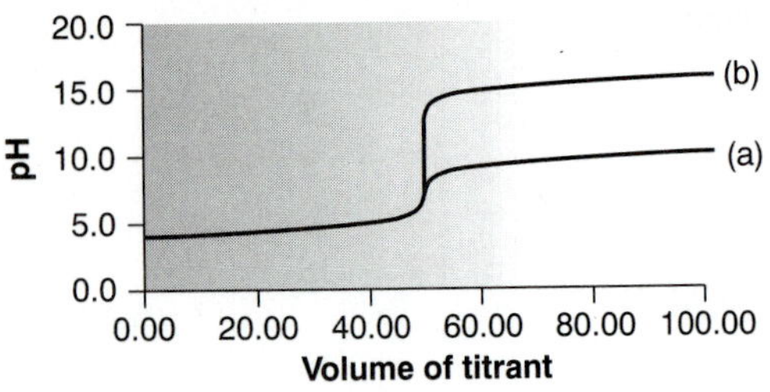

Figure 9.16
Titration curves for 50.00 mL of 10^{-4} M HCl with 10^{-4} M NaOH in (a) water, $K_w = 1 \times 10^{-14}$, and (b) nonaqueous solvent, $K_s = 1 \times 10^{-20}$.

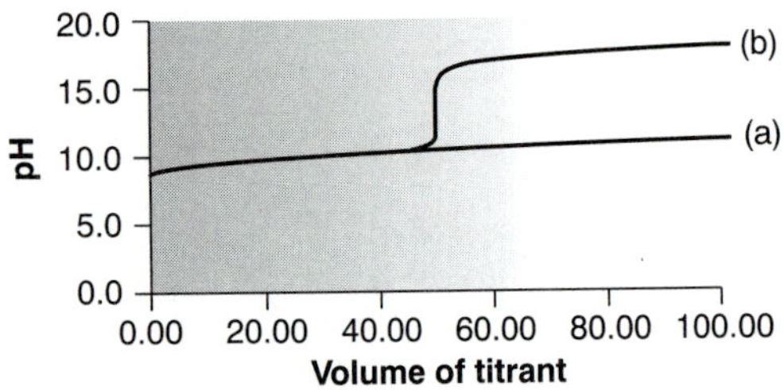

Figure 9.17
Titration curves for 50.00 mL of 0.100 M weak acid ($pK_a = 11$) with 0.100 M NaOH in (a) water, $K_w = 1 \times 10^{-14}$; and (b) nonaqueous solvent, $K_s = 1 \times 10^{-20}$. The titration curve in (b) assumes that the change in solvent has no effect on the acid dissociation constant of the weak acid.

leveling
Acids that are better proton donors than the solvent are leveled to the acid strength of the protonated solvent; bases that are better proton acceptors than the solvent are leveled to the base strength of the deprotonated solvent.

If the same titration is carried out in a nonaqueous solvent with a K_s of 1.0×10^{-20}, the pH when the titration is 90% complete is still 5.3. However, the pH when the titration is 110% complete is now

$$\text{pH} = \text{p}K_s - \text{pOH} = 20.0 - 5.3 = 14.7$$

In this case the change in pH of

$$\Delta\text{pH} = 14.7 - 5.3 = 9.4$$

is significantly greater than that obtained when the titration is carried out in water. Figure 9.16 shows the titration curves in both the aqueous and nonaqueous solvents. Nonaqueous solvents also may be used to increase the change in pH when titrating weak acids or bases (Figure 9.17).

Another parameter affecting the feasibility of a titration is the dissociation constant of the acid or base being titrated. Again, the solvent plays an important role. In the Brønsted–Lowry view of acid–base behavior, the strength of an acid or base is a relative measure of the ease with which a proton is transferred from the acid to the solvent, or from the solvent to the base. For example, the strongest acid that can exist in water is H_3O^+. The acids HCl and HNO_3 are considered strong because they are better proton donors than H_3O^+. Strong acids essentially donate all their protons to H_2O, "**leveling**" their acid strength to that of H_3O^+. In a different solvent HCl and HNO_3 may not behave as strong acids.

When acetic acid, which is a weak acid, is placed in water, the dissociation reaction

$$CH_3COOH(aq) + H_2O(\ell) \rightleftharpoons H_3O^+(aq) + CH_3COO^-(aq)$$

does not proceed to a significant extent because acetate is a stronger base than water and the hydronium ion is a stronger acid than acetic acid. If acetic acid is placed in a solvent that is a stronger base than water, such as ammonia, then the reaction

$$CH_3COOH + NH_3 \rightleftharpoons NH_4^+ + CH_3COO^-$$

proceeds to a greater extent. In fact, HCl and CH_3COOH are both strong acids in ammonia.

All other things being equal, the strength of a weak acid increases if it is placed in a solvent that is more basic than water, whereas the strength of a weak base increases if it is placed in a solvent that is more acidic than water. In some cases, however, the opposite effect is observed. For example, the pK_b for ammonia is 4.76 in water and 6.40 in the more acidic glacial acetic acid. In contradiction to our expectations, ammonia is a weaker base in the more acidic solvent. A full description of the solvent's effect on a weak acid's pK_a or on the pK_b of a weak base is beyond the scope of this text. You should be aware, however, that titrations that are not feasible in water may be feasible in a different solvent.

9B.4 Representative Method

Although each acid–base titrimetric method has its own unique considerations, the following description of the determination of protein in bread provides an instructive example of a typical procedure.

Method 9.1 Determination of Protein in Bread[4]

Description of the Method. This quantitative method of analysis for proteins is based on a determination of the %w/w N in the sample. Since different cereal proteins have similar amounts of nitrogen, the experimentally determined %w/w N is multiplied by a factor of 5.7 to give the %w/w protein in the sample (on average there are 5.7 g of cereal protein for every gram of nitrogen). As described here, nitrogen is determined by the Kjeldahl method. The protein in a sample of bread is oxidized in hot concentrated H_2SO_4, converting the nitrogen to NH_4^+. After making the solution alkaline, converting NH_4^+ to NH_3, the ammonia is distilled into a flask containing a known amount of standard strong acid. Finally, the excess strong acid is determined by a back titration with a standard strong base titrant.

Procedure. Transfer a 2.0-g sample of bread, which has previously been air dried and ground into a powder, to a suitable digestion flask, along with 0.7 g of HgO as a catalyst, 10 g of K_2SO_4, and 25 mL of concentrated H_2SO_4. Bring the solution to a boil, and continue boiling until the solution turns clear, and for at least an additional 30 min. After cooling to below room temperature, add 200 mL of H_2O and 25 mL of 4% w/v K_2S to remove the Hg^{2+} catalyst. Add a few Zn granules to serve as boiling stones, and 25 g of NaOH. Quickly connect the flask to a distillation apparatus, and distill the NH_3 into a collecting flask containing a known amount of standardized HCl. The tip of the condenser should be placed below the surface of the strong acid. After the distillation is complete, titrate the excess strong acid with a standard solution of NaOH, using methyl red as a visual indicator.

The photo in Colorplate 8a shows the indicator's color change for this titration.

Questions

1. **Oxidizing the protein converts the nitrogen to NH_4^+. Why is the amount of nitrogen not determined by titrating the NH_4^+ with a strong base?**

   There are two reasons for not titrating the ammonium ion. First, NH_4^+ is a very weak acid ($K_a = 5.7 \times 10^{-10}$) that yields a poorly defined end point when titrated with a strong base. Second, even if the end point can be determined with acceptable accuracy and precision, the procedure calls for adding a substantial amount of H_2SO_4. After the oxidation is complete, the amount of excess H_2SO_4 will be much greater than the amount of NH_4^+ that is produced. The presence of two acids that differ greatly in concentration makes for a difficult analysis. If the titrant's concentration is similar to that of H_2SO_4, then the equivalence point volume for the titration of NH_4^+ may be too small to measure reliably. On the other hand, if the concentration of the titrant is similar to that of NH_4^+, the volume needed to neutralize the H_2SO_4 will be unreasonably large.

2. **Ammonia is a volatile compound as evidenced by the strong smell of even dilute solutions. This volatility presents a possible source of determinate error. Will this determinate error be negative or positive?**

   The conversion of N to NH_3 follows the following pathway

$$N \rightarrow NH_4^+$$

$$NH_4^+ \rightarrow NH_3$$

   Any loss of NH_3 is loss of analyte and a negative determinate error.

—Continued

Continued from page 297

3. **Discuss the steps taken in this procedure to minimize this determinate error.**

   Three specific steps are taken to minimize the loss of ammonia: (1) the solution is cooled to below room temperature before adding NaOH; (2) the digestion flask is quickly connected to the distillation apparatus after adding NaOH; and (3) the condenser tip of the distillation apparatus is placed below the surface of the HCl to ensure that the ammonia will react with the HCl before it can be lost through volatilization.

4. **How does K_2S remove Hg^{2+}, and why is this important?**

   Adding sulfide precipitates the Hg^{2+} as HgS. This is important because NH_3 forms stable complexes with many metal ions, including Hg^{2+}. Any NH_3 that is complexed with Hg^{2+} will not be collected by distillation, providing another source of determinate error.

9B.5 Quantitative Applications

Although many quantitative applications of acid–base titrimetry have been replaced by other analytical methods, there are several important applications that continue to be listed as standard methods. In this section we review the general application of acid–base titrimetry to the analysis of inorganic and organic compounds, with an emphasis on selected applications in environmental and clinical analysis. First, however, we discuss the selection and standardization of acidic and basic titrants.

Selecting and Standardizing a Titrant Most common acid–base titrants are not readily available as primary standards and must be standardized before they can be used in a quantitative analysis. Standardization is accomplished by titrating a known amount of an appropriate acidic or basic primary standard.

The majority of titrations involving basic analytes, whether conducted in aqueous or nonaqueous solvents, use HCl, $HClO_4$, or H_2SO_4 as the titrant. Solutions of these titrants are usually prepared by diluting a commercially available concentrated stock solution and are stable for extended periods of time. Since the concentrations of concentrated acids are known only approximately,* the titrant's concentration is determined by standardizing against one of the primary standard weak bases listed in Table 9.7.

The most common strong base for titrating acidic analytes in aqueous solutions is NaOH. Sodium hydroxide is available both as a solid and as an approximately 50% w/v solution. Solutions of NaOH may be standardized against any of the primary weak acid standards listed in Table 9.7. The standardization of NaOH, however, is complicated by potential contamination from the following reaction between CO_2 and OH^-.

$$CO_2(g) + 2OH^-(aq) \rightarrow CO_3^{2-}(aq) + H_2O(\ell) \qquad 9.7$$

When CO_2 is present, the volume of NaOH used in the titration is greater than that needed to neutralize the primary standard because some OH^- reacts with the CO_2.

*The nominal concentrations are 12.1 M HCl, 11.7 M $HClO_4$, and 18.0 M H_2SO_4.

Table 9.7 Selected Primary Standards for the Standardization of Strong Acid and Strong Base Titrants

Standardization of Acidic Titrants

Primary Standard	Titration Reaction	Comment
Na_2CO_3	$Na_2CO_3 + 2H_3O^+ \rightarrow H_2CO_3 + 2Na^+ + 2H_2O$	a
TRIS	$(HOCH_2)_3CNH_2 + H_3O^+ \rightarrow (HOCH_2)_3CNH_3^+ + H_2O$	b
$Na_2B_4O_7$	$Na_2B_4O_7 + 2H_3O^+ + 3H_2O \rightarrow 2Na^+ + 4H_3BO_3$	

Standardization of Basic Titrants

Primary Standard	Titration Reaction	Comment
$KHC_8H_4O_4$	$KHC_8H_4O_4 + OH^- \rightarrow K^+ + C_8H_4O_4^{2-} + H_2O$	c
C_6H_5COOH	$C_6H_5COOH + OH^- \rightarrow C_6H_5COO^- + H_2O$	d
$KH(IO_3)_2$	$KH(IO_3)_2 + OH^- \rightarrow K^+ + 2IO_3^- + H_2O$	

[a]The end point for this titration is improved by titrating to the second equivalence point, boiling the solution to expel CO_2, and retitrating to the second equivalence point. In this case the reaction is

$$Na_2CO_3 + 2H_3O^+ \rightarrow CO_2 + 2Na^+ + 3H_2O$$

[b]TRIS stands for *tris*-(hydroxymethyl)aminomethane.

[c]$KHC_8H_4O_4$ is also known as potassium hydrogen phthalate, or KHP.

[d]Due to its poor solubility in water, benzoic acid is dissolved in a small amount of ethanol before being diluted with water.

The calculated concentration of OH^-, therefore, is too small. This is not a problem when titrations involving NaOH are restricted to an end point pH less than 6. Below this pH any CO_3^{2-} produced in reaction 9.7 reacts with H_3O^+ to form carbonic acid.

$$CO_3^{2-}(aq) + 2H_3O^+(aq) \rightarrow H_2CO_3(aq) + 2H_2O(\ell) \qquad \textbf{9.8}$$

Combining reactions 9.7 and 9.8 gives an overall reaction of

$$CO_2(g) + H_2O(\ell) \rightarrow H_2CO_3(aq)$$

which does not include OH^-. Under these conditions the presence of CO_2 does not affect the quantity of OH^- used in the titration and, therefore, is not a source of determinate error.

For pHs between 6 and 10, however, the neutralization of CO_3^{2-} requires only one proton

$$CO_3^{2-}(aq) + H_3O^+(aq) \rightarrow HCO_3^-(aq) + H_2O(\ell)$$

and the net reaction between CO_2 and OH^- is

$$CO_2(g) + OH^-(aq) \rightarrow HCO_3^-(aq)$$

Under these conditions some OH^- is consumed in neutralizing CO_2. The result is a determinate error in the titrant's concentration. If the titrant is used to analyze an analyte that has the same end point pH as the primary standard used during standardization, the determinate errors in the standardization and the analysis cancel, and accurate results may still be obtained.

Solid NaOH is always contaminated with carbonate due to its contact with the atmosphere and cannot be used to prepare carbonate-free solutions of NaOH. Solutions of carbonate-free NaOH can be prepared from 50% w/v NaOH since Na_2CO_3 is very insoluble in concentrated NaOH. When CO_2 is absorbed, Na_2CO_3

precipitates and settles to the bottom of the container, allowing access to the carbonate-free NaOH. Dilution must be done with water that is free from dissolved CO_2. Briefly boiling the water expels CO_2 and, after cooling, it may be used to prepare carbonate-free solutions of NaOH. Provided that contact with the atmosphere is minimized, solutions of carbonate-free NaOH are relatively stable when stored in polyethylene bottles. Standard solutions of sodium hydroxide should not be stored in glass bottles because NaOH reacts with glass to form silicate.

Inorganic Analysis Acid–base titrimetry is a standard method for the quantitative analysis of many inorganic acids and bases. Standard solutions of NaOH can be used in the analysis of inorganic acids such as H_3PO_4 or H_3AsO_4, whereas standard solutions of HCl can be used for the analysis of inorganic bases such as Na_2CO_3.

Inorganic acids and bases too weak to be analyzed by an aqueous acid–base titration can be analyzed by adjusting the solvent or by an indirect analysis. For example, the accuracy in titrating boric acid, H_3BO_3, with NaOH is limited by boric acid's small acid dissociation constant of 5.8×10^{-10}. The acid strength of boric acid, however, increases when mannitol is added to the solution because it forms a complex with the borate ion. The increase in K_a to approximately 1.5×10^{-4} results in a sharper end point and a more accurate titration. Similarly, the analysis of ammonium salts is limited by the small acid dissociation constant of 5.7×10^{-10} for NH_4^+. In this case, NH_4^+ can be converted to NH_3 by neutralizing with strong base. The NH_3, for which K_b is 1.8×10^{-5}, is then removed by distillation and titrated with a standard strong acid titrant.

Inorganic analytes that are neutral in aqueous solutions may still be analyzed if they can be converted to an acid or base. For example, NO_3^- can be quantitatively analyzed by reducing it to NH_3 in a strongly alkaline solution using Devarda's alloy, a mixture of 50% w/w Cu, 45% w/w Al, and 5% w/w Zn.

$$3NO_3^-(aq) + 8Al(s) + 5OH^-(aq) + 2H_2O(\ell) \rightarrow 8AlO_2^-(aq) + 3NH_3(aq)$$

The NH_3 is removed by distillation and titrated with HCl. Alternatively, NO_3^- can be titrated as a weak base in an acidic nonaqueous solvent such as anhydrous acetic acid, using $HClO_4$ as a titrant.

alkalinity
A measure of a water's ability to neutralize acid.

Acid–base titrimetry continues to be listed as the standard method for the determination of alkalinity, acidity, and free CO_2 in water and wastewater analysis. **Alkalinity** is a measure of the acid-neutralizing capacity of a water sample and is assumed to arise principally from OH^-, HCO_3^-, and CO_3^{2-}, although other weak bases, such as phosphate, may contribute to the overall alkalinity. Total alkalinity is determined by titrating with a standard solution of HCl or H_2SO_4 to a fixed end point at a pH of 4.5, or to the bromocresol green end point. Alkalinity is reported as milligrams $CaCO_3$ per liter.

When the sources of alkalinity are limited to OH^-, HCO_3^-, and CO_3^{2-}, titrations to both a pH of 4.5 (bromocresol green end point) and a pH of 8.3 (phenolphthalein or metacresol purple end point) can be used to determine which species are present, as well as their respective concentrations. Titration curves for OH^-, HCO_3^-, and CO_3^{2-} are shown in Figure 9.18. For a solution containing only OH^- alkalinity, the volumes of strong acid needed to reach the two end points are identical. If a solution contains only HCO_3^- alkalinity, the volume of strong acid needed to reach the end point at a pH of 8.3 is zero, whereas that for the pH 4.5 end point is greater than zero. When the only source of alkalinity is CO_3^{2-}, the volume of strong acid needed to reach the end point at a pH of 4.5 is exactly twice that needed to reach the end point at a pH of 8.3.

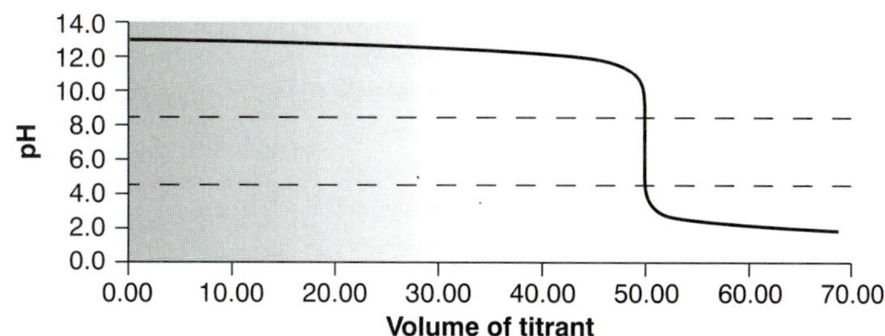

(a)

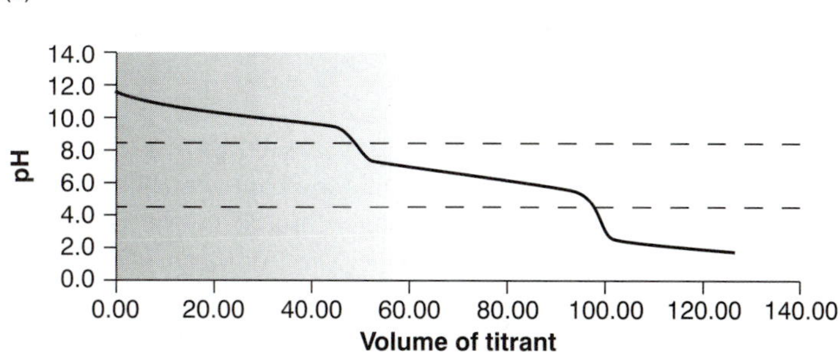

(b)

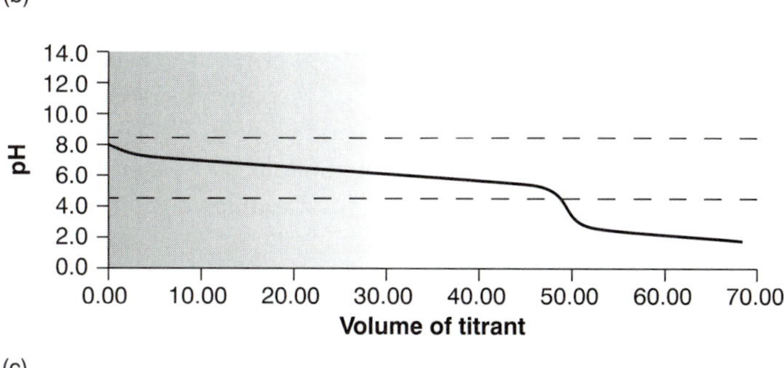

(c)

Figure 9.18

Titration curves for (a) 50.00 mL of 0.100 M NaOH with 0.100 M HCl; (b) 50.00 mL of 0.100 M Na_2CO_3 with 0.100 M HCl; and (c) 50.00 mL of 0.100 M $NaHCO_3$ with M HCl. The dashed lines indicate the pH 8.3 and pH 4.5 end points.

Mixtures of OH^- and CO_3^{2-}, or HCO_3^- and CO_3^{2-} alkalinities also are possible. Consider, for example, a mixture of OH^- and CO_3^{2-}. The volume of strong acid needed to titrate OH^- will be the same whether we titrate to the pH 8.3 or pH 4.5 end point. Titrating CO_3^{2-} to the end point at a pH of 4.5, however, requires twice as much strong acid as when titrating to the pH 8.3 end point. Consequently, when titrating a mixture of these two ions, the volume of strong acid needed to reach the pH 4.5 end point is less than twice that needed to reach the end point at a pH of 8.3. For a mixture of HCO_3^- and CO_3^{2-}, similar reasoning shows that the volume of strong acid needed to reach the end point at a pH of 4.5 is more than twice that need to reach the pH 8.3 end point. Solutions containing OH^- and HCO_3^- alkalinities are unstable with respect to the formation of CO_3^{2-} and do not exist. Table 9.8 summarizes the relationship between the sources of alkalinity and the volume of titrant needed to reach the two end points.

Acidity is a measure of a water sample's capacity for neutralizing base and is conveniently divided into strong acid and weak acid acidity. Strong acid acidity is due to the presence of inorganic acids, such as HCl, HNO_3, and H_2SO_4, and is commonly found in industrial effluents and acid mine drainage. Weak acid acidity is usually dominated by the formation of H_2CO_3 from dissolved CO_2, but also

acidity
A measure of a water's ability to neutralize base.

Table 9.8 Relationship Between End Point Volumes and Sources of Alkalinity

Source of Alkalinity	Relationship Between End Point Volumes
OH^-	$V_{pH\ 4.5} = V_{pH\ 8.3}$
CO_3^{2-}	$V_{pH\ 4.5} = 2 \times V_{pH\ 8.3}$
HCO_3^-	$V_{pH\ 8.3} = 0$; $V_{pH\ 4.5} > 0$
OH^- and CO_3^{2-}	$V_{pH\ 4.5} < 2 \times V_{pH\ 8.3}$
CO_3^{2-} and HCO_3^-	$V_{pH\ 4.5} > 2 \times V_{pH\ 8.3}$

includes contributions from hydrolyzable metal ions such as Fe^{3+}, Al^{3+}, and Mn^{2+}. In addition, weak acid acidity may include a contribution from organic acids.

Acidity is determined by titrating with a standard solution of NaOH to fixed end points at pH 3.7 and pH 8.3. These end points are located potentiometrically, using a pH meter, or by using an appropriate indicator (bromophenol blue for pH 3.7, and metacresol purple or phenolphthalein for pH 8.3). Titrating to a pH of 3.7 provides a measure of strong acid acidity,* and titrating to a pH of 8.3 provides a measure of total acidity. Weak acid acidity is given indirectly as the difference between the total and strong acid acidities. Results are expressed as the milligrams of $CaCO_3$ per liter that could be neutralized by the water sample's acidity. An alternative approach for determining strong and weak acidity is to obtain a potentiometric titration curve and use Gran plot methodology to determine the two equivalence points. This approach has been used, for example, in determining the forms of acidity in atmospheric aerosols.[5]

Water in contact with either the atmosphere or carbonate-bearing sediments contains dissolved or free CO_2 that exists in equilibrium with gaseous CO_2 and the aqueous carbonate species H_2CO_3, HCO_3^-, and CO_3^{2-}. The concentration of free CO_2 is determined by titrating with a standard solution of NaOH to the phenolphthalein end point, or to a pH of 8.3, with results reported as milligrams CO_2 per liter. This analysis is essentially the same as that for the determination of total acidity, and can only be applied to water samples that do not contain any strong acid acidity.

Organic Analysis The use of acid–base titrimetry for the analysis of organic compounds continues to play an important role in pharmaceutical, biochemical, agricultural, and environmental laboratories. Perhaps the most widely employed acid–base titration is the **Kjeldahl analysis** for organic nitrogen, described earlier in Method 9.1. This method continues to be used in the analysis of caffeine and saccharin in pharmaceutical products, as well as for the analysis of proteins, fertilizers, sludges, and sediments. Any nitrogen present in the –3 oxidation state is quantitatively oxidized to NH_4^+. Some aromatic heterocyclic compounds, such as pyridine, are difficult to oxidize. A catalyst, such as HgO, is used to ensure that oxidation is complete. Nitrogen in an oxidation state other than –3, such as nitro- and azo-nitrogens, is often oxidized to N_2, resulting in a negative determinate error. Adding a reducing agent, such as salicylic acid, reduces the nitrogen to a –3 oxidation state,

Kjeldahl analysis
An acid–base titrimetric method for determining the amount of nitrogen in organic compounds.

*This is sometimes referred to as "methyl orange acidity" since, at one time, methyl orange was the traditional indicator of choice.

Table 9.9 Selected Elemental Analyses Based on Acid–Base Titrimetry

Element	Liberated as	Reaction Producing Acid or Base to Be Titrated[a]	Titration
N	$NH_3(g)$	$NH_3(g) + \mathbf{H_3O^+}(aq) \rightarrow NH_4^+(aq) + H_2O(\ell)$	excess H_3O^+ with strong base
S	$SO_2(g)$	$SO_2(g) + H_2O_2(aq) \rightarrow \mathbf{H_2SO_4}(aq)$	H_2SO_4 with strong base
C	$CO_2(g)$	$CO_2(g) + \mathbf{Ba(OH)_2}(aq) \rightarrow BaCO_3(s) + H_2O(\ell)$	excess $Ba(OH)_2$ with strong acid
Cl	$HCl(g)$	$HCl(g) + H_2O(\ell) \rightarrow \mathbf{H_3O^+}(aq) + Cl^-(aq)$	H_3O^+ with strong base
F	$SiF_4(g)$	$3SiF_4(g) + 2H_2O(\ell) \rightarrow \mathbf{2H_2SiF_6}(aq) + SiO_2(s)$	H_2SiF_6 with strong base

[a]The acid or base that is eventually titrated is indicated in **bold**.

Table 9.10 Selected Acid–Base Titrimetric Procedures for Organic Functional Groups Based on the Production or Consumption of Acid or Base

Functional Group	Reaction Producing Acid or Base to Be Titrated[a]	Titration
ester	$RCOOR'(aq) + \mathbf{OH^-}(aq) \rightarrow RCOO^-(aq) + HOR'(aq)$	excess OH^- with strong acid
carbonyl	$R_2C=O(aq) + NH_2OH \cdot HCl(aq) \rightarrow R_2C=NOH(aq) + \mathbf{HCl}(aq) + H_2O(\ell)$	HCl with strong base
alcohol[b]	[1] $(CH_3CO)_2O + ROH \rightarrow CH_3COOR + \mathbf{CH_3COOH}$ [2] $(CH_3CO)_2O + H_2O \rightarrow \mathbf{2CH_3COOH}$	CH_3COOH with strong base; ROH is determined from the difference in the amount of titrant needed to react with a blank consisting only of acetic anhydride, and the amount reacting with the sample.

[a]The acid or base that is eventually titrated is indicated in **bold**.
[b]The acetylation reaction, [1], is carried out in pyridine to avoid the hydrolysis of acetic anhydride by water. After the acetylation is complete, water is added to convert the remaining acetic anhydride to acetic acid, [2].

eliminating this source of error. Other examples of elemental analyses based on the conversion of the element to an acid or base are outlined in Table 9.9.

Several organic functional groups have weak acid or weak base properties that allow their direct determination by an acid–base titration. Carboxylic (—COOH), sulfonic ($—SO_3H$), and phenolic ($—C_6H_5OH$) functional groups are weak acids that can be successfully titrated in either aqueous or nonaqueous solvents. Sodium hydroxide is the titrant of choice for aqueous solutions. Nonaqueous titrations are often carried out in a basic solvent, such as ethylenediamine, using tetrabutylammonium hydroxide, $(C_4H_9)_4NOH$, as the titrant. Aliphatic and aromatic amines are weak bases that can be titrated using HCl in aqueous solution or $HClO_4$ in glacial acetic acid. Other functional groups can be analyzed indirectly by use of a functional group reaction that produces or consumes an acid or base. Examples are shown in Table 9.10.

Many pharmaceutical compounds are weak acids or bases that can be analyzed by an aqueous or nonaqueous acid–base titration; examples include salicylic acid, phenobarbital, caffeine, and sulfanilamide. Amino acids and proteins can be analyzed in glacial acetic acid, using $HClO_4$ as the titrant. For example, a procedure for determining the amount of nutritionally available protein has been developed that is based on an acid–base titration of lysine residues.[6]

Quantitative Calculations In acid–base titrimetry the quantitative relationship between the analyte and the titrant is determined by the stoichiometry of the relevant reactions. As outlined in Section 2C, stoichiometric calculations may be simplified by focusing on appropriate conservation principles. In an acid–base reaction the number of protons transferred between the acid and base is conserved; thus

$$\frac{\text{moles of H}^+\text{ donated}}{\text{mole acid}} \times \text{moles acid} = \frac{\text{moles of H}^+\text{ accepted}}{\text{mole base}} \times \text{moles base}$$

The following example demonstrates the application of this approach in the direct analysis of a single analyte.

Example

EXAMPLE 9.2

A 50.00-mL sample of a citrus drink requires 17.62 mL of 0.04166 M NaOH to reach the phenolphthalein end point (Figure 9.19a). Express the sample's acidity in terms of grams of citric acid, $C_6H_8O_7$, per 100 mL.

SOLUTION

Since citric acid is a triprotic weak acid, we must first decide to which equivalence point the titration has been carried. The three acid dissociation constants are

$$pK_{a1} = 3.13 \qquad pK_{a2} = 4.76 \qquad pK_{a3} = 6.40$$

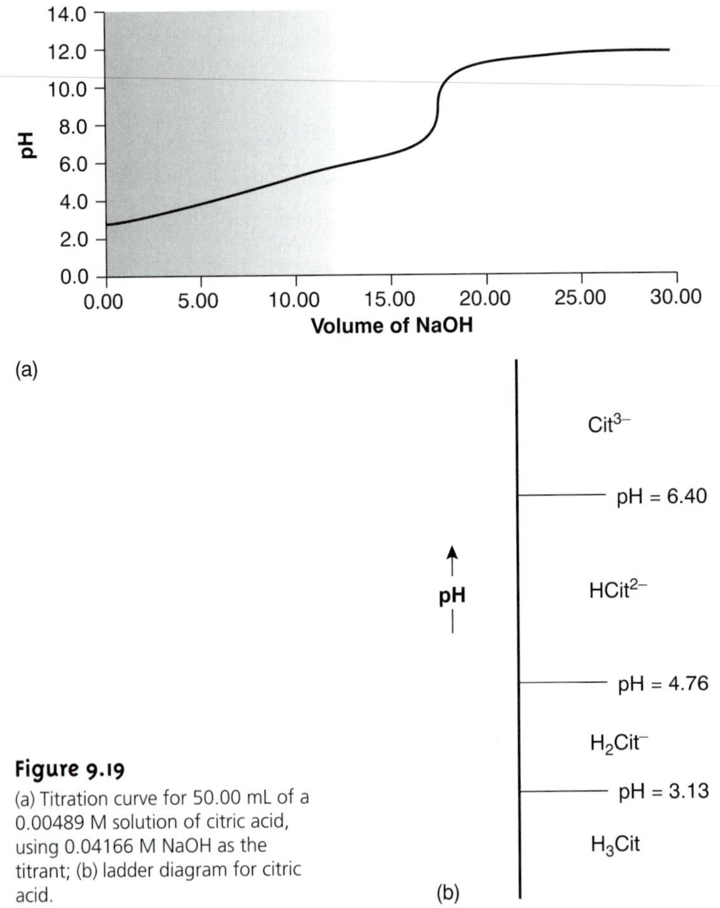

Figure 9.19
(a) Titration curve for 50.00 mL of a 0.00489 M solution of citric acid, using 0.04166 M NaOH as the titrant; (b) ladder diagram for citric acid.

The phenolphthalein end point is basic, occurring at a pH of approximately 8.3 and can be reached only if the titration proceeds to the third equivalence point (Figure 9.19b); thus, we write

$$3 \times \text{moles citric acid} = \text{moles NaOH}$$

Making appropriate substitutions for the moles of citric acid and moles of NaOH gives the following equation

$$\frac{3 \times \text{ g citric acid}}{\text{FW citric acid}} = M_b \times V_b$$

which can be solved for the grams of citric acid

$$\frac{M_b \times V_b \times \text{FW citric acid}}{3} = \frac{(0.04166 \text{ M})(0.01762 \text{ L})(192.13 \text{ g/mol})}{3}$$

$$= 0.04701 \text{ g citric acid}$$

Since this is the grams of citric acid in a 50.00-mL sample, the concentration of citric acid in the citrus drink is 0.09402 g/100 mL.

In an indirect analysis the analyte participates in one or more preliminary reactions that produce or consume acid or base. Despite the additional complexity, the stoichiometry between the analyte and the amount of acid or base produced or consumed may be established by applying the conservation principles outlined in Section 2C. Example 9.3 illustrates the application of an indirect analysis in which an acid is produced.

Example

EXAMPLE 9.3

The purity of a pharmaceutical preparation of sulfanilamide, $C_6H_4N_2O_2S$, can be determined by oxidizing the sulfur to SO_2 and bubbling the SO_2 through H_2O_2 to produce H_2SO_4. The acid is then titrated with a standard solution of NaOH to the bromothymol blue end point, where both of sulfuric acid's acidic protons have been neutralized. Calculate the purity of the preparation, given that a 0.5136-g sample required 48.13 mL of 0.1251 M NaOH.

SOLUTION

Conservation of protons for the titration reaction requires that

$$2 \times \text{moles } H_2SO_4 = \text{moles NaOH}$$

Since all the sulfur in H_2SO_4 comes from sulfanilamide, we use a conservation of mass on sulfur to establish the following stoichiometric relationship.

$$\text{Moles } C_6H_4N_2O_2S = \text{moles } H_2SO_4$$

Combining the two conservation equations gives a single equation relating the moles of analyte to the moles of titrant.

$$2 \times \text{moles } C_6H_4N_2O_2S = \text{moles NaOH}$$

Making appropriate substitutions for the moles of sulfanilamide and moles of NaOH gives

$$\frac{2 \times \text{g sulfanilamide}}{\text{FW sulfanilamide}} = M_b \times V_b$$

which can be solved for the grams of sulfanilamide

$$\frac{M_b \times V_b \times \text{FW sulfanilamide}}{2} = \frac{(0.1251\ \text{M})(0.04813\ \text{L})(168.18\ \text{g/mol})}{2}$$

$$= 0.5063\ \text{g sulfanilamide}$$

Thus, the purity of the preparation is

$$\frac{\text{g sulfanilamide}}{\text{g sample}} \times 100 = \frac{0.5063\ \text{g}}{0.5136\ \text{g}} \times 100 = 98.58\%\ \text{w/w sulfanilamide}$$

This approach is easily extended to back titrations, as shown in the following example.

Example

EXAMPLE 9.4

The amount of protein in a sample of cheese is determined by a Kjeldahl analysis for nitrogen. After digesting a 0.9814-g sample of cheese, the nitrogen is oxidized to NH_4^+, converted to NH_3 with NaOH, and distilled into a collection flask containing 50.00 mL of 0.1047 M HCl. The excess HCl is then back titrated with 0.1183 M NaOH, requiring 22.84 mL to reach the bromothymol blue end point. Report the %w/w protein in the cheese given that there is 6.38 g of protein for every gram of nitrogen in most dairy products.

SOLUTION

In this procedure, the HCl reacts with two different bases; thus

$$\text{Moles HCl} = \text{moles HCl reacting with } NH_3 + \text{moles HCl reacting with NaOH}$$

Conservation of protons requires that

$$\text{Moles HCl reacting with } NH_3 = \text{moles } NH_3$$

$$\text{Moles HCl reacting with NaOH} = \text{moles NaOH}$$

A conservation of mass on nitrogen gives the following equation.

$$\text{Moles } NH_3 = \text{moles N}$$

Combining all four equations gives a final stoichiometric equation of

$$\text{Moles HCl} = \text{moles N} + \text{moles NaOH}$$

Making appropriate substitutions for the moles of HCl, N, and NaOH gives

$$M_a \times V_a = \frac{\text{g N}}{\text{AW N}} + M_b \times V_b$$

which we solve for the grams of nitrogen.

$$\text{g N} = (M_a \times V_a - M_b \times V_b) \times \text{AW N}$$

$$(0.1047\ \text{M} \times 0.05000\ \text{L} - 0.1183\ \text{M} \times 0.02284\ \text{L}) \times 14.01\ \text{g/mol} = 0.03549\ \text{g N}$$

The mass of protein, therefore, is

$$\frac{6.38 \text{ g protein}}{\text{g N}} \times 0.03549 \text{ g N} = 0.2264 \text{ g protein}$$

and the %w/w protein is

$$\frac{\text{g protein}}{\text{g sample}} \times 100 = \frac{0.2264 \text{ g}}{0.9814 \text{ g}} \times 100 = 23.1\% \text{ w/w protein}$$

Earlier we noted that an acid–base titration may be used to analyze a mixture of acids or bases by titrating to more than one equivalence point. The concentration of each analyte is determined by accounting for its contribution to the volume of titrant needed to reach the equivalence points.

Example

EXAMPLE 9.5

The alkalinity of natural waters is usually controlled by OH^-, CO_3^{2-}, and HCO_3^-, which may be present singularly or in combination. Titrating a 100.0-mL sample to a pH of 8.3 requires 18.67 mL of a 0.02812 M solution of HCl. A second 100.0-mL aliquot requires 48.12 mL of the same titrant to reach a pH of 4.5. Identify the sources of alkalinity and their concentrations in parts per million.

SOLUTION

Since the volume of titrant needed to reach a pH of 4.5 is more than twice that needed to reach a pH of 8.3, we know, from Table 9.8, that the alkalinity of the sample is controlled by CO_3^{2-} and HCO_3^{2-}.

Titrating to a pH of 8.3 neutralizes CO_3^{2-} to HCO_3^-, but does not lead to a reaction of the titrant with HCO_3^- (see Figure 9.14b). Thus

$$\text{Moles HCl to pH 8.3} = \text{moles } CO_3^{2-}$$

or

$$M_a \times V_{a,\text{pH } 8.3} = \frac{\text{g } CO_3^{2-}}{\text{FW } CO_3^{2-}}$$

Solving for the grams of carbonate gives

$$\text{g } CO_3^{2-} = M_a \times V_{a,\text{ pH } 8.3} \times \text{FW } CO_3^{2-}$$

$$0.02812 \text{ M} \times 0.01867 \text{ L} \times 60.01 \text{ g/mol} = 0.03151 \text{ g } CO_3^{2-}$$

The concentration of CO_3^{2-}, therefore, is

$$\frac{\text{mg } CO_3^{2-}}{\text{liter}} = \frac{31.51 \text{ mg}}{0.1000 \text{ L}} = 315.1 \text{ ppm } CO_3^{2-}$$

Titrating to the second end point at pH 4.5 neutralizes CO_3^{2-} to H_2CO_3, and HCO_3^- to H_2CO_3 (see Figures 9.18b,c). The conservation of protons, therefore, requires that

$$\text{Moles HCl to pH 4.5} = 2 \times \text{moles } CO_3^{2-} + \text{moles } HCO_3^-$$

or

$$M_a \times V_{a,\ pH\ 4.5} = \frac{2 \times g\ CO_3^{2-}}{FW\ CO_3^{2-}} + \frac{g\ HCO_3^-}{FW\ HCO_3^-}$$

Solving for the grams of bicarbonate gives

$$g\ HCO_3^- = \left(M_a \times V_{a,\ pH\ 4.5} - \frac{2 \times g\ CO_3^{2-}}{FW\ CO_3^{2-}} \right) \times FW\ HCO_3^-$$

$$\left(0.02812\ M \times 0.04812\ L - \frac{2 \times 0.03151\ g}{60.01\ g/mol} \right) \times 61.02\ g/mol = 0.01849\ g\ HCO_3^{2-}$$

The concentration of HCO_3^-, therefore, is

$$\frac{mg\ HCO_3^-}{liter} = \frac{18.49\ mg}{0.1000\ L} = 184.9\ ppm\ HCO_3^-$$

9B.6 Qualitative Applications

We have already come across one example of the qualitative application of acid–base titrimetry in assigning the forms of alkalinity in waters (see Example 9.5). This approach is easily extended to other systems. For example, the composition of solutions containing one or two of the following species

$$H_3PO_4 \quad H_2PO_4^- \quad HPO_4^{2-} \quad PO_4^{3-} \quad NaOH \quad HCl$$

can be determined by titrating with either a strong acid or a strong base to the methyl orange and phenolphthalein end points. As outlined in Table 9.11, each species or mixture of species has a unique relationship between the volumes of titrant needed to reach these two end points.

Table 9.11 Relationship Between End Point Volumes for Solutions of Phosphate Species with HCl and NaOH

Solution Composition	Relationship Between End Point Volumes with Strong Base Titrant[a]	Relationship Between End Point Volumes with Strong Acid Titrant[a]
H_3PO_4	$V_{PH} = 2 \times V_{MO}$	—[b]
$H_2PO_4^-$	$V_{PH} > 0$; $V_{MO} = 0$	—
HPO_4^{2-}	—	$V_{MO} > 0$; $V_{PH} = 0$
PO_4^{3-}	—	$V_{MO} = 2 \times V_{PH}$
HCl	$V_{PH} = V_{MO}$	—
NaOH	—	$V_{MO} = V_{PH}$
HCl and H_3PO_4	$V_{PH} < 2 \times V_{MO}$	—
H_3PO_4 and $H_2PO_4^-$	$V_{PH} > 2 \times V_{MO}$	—
$H_2PO_4^-$ and HPO_4^{2-}	$V_{PH} > 0$; $V_{MO} = 0$	$V_{MO} > 0$; $V_{PH} = 0$
HPO_4^{2-} and PO_4^{3-}	—	$V_{MO} > 2 \times V_{PH}$
PO_4^{3-} and NaOH	—	$V_{MO} < 2 \times V_{PH}$

[a] V_{MO} and V_{PH} are, respectively, the volume of titrant needed to reach the methyl orange and phenolphthalein end points.
[b] When no information is given, the volume of titrant needed to reach either end point is zero.

9B.7 Characterization Applications

Two useful characterization applications involving acid–base titrimetry are the determination of equivalent weight, and the determination of acid–base dissociation constants.

Equivalent Weights Acid–base titrations can be used to characterize the chemical and physical properties of matter. One simple example is the determination of the equivalent weight* of acids and bases. In this method, an accurately weighed sample of a pure acid or base is titrated to a well-defined equivalence point using a monoprotic strong acid or strong base. If we assume that the titration involves the transfer of n protons, then the moles of titrant needed to reach the equivalence point is given as

$$\text{Moles titrant} = n \times \text{moles analyte}$$

and the formula weight is

$$\text{FW} = \frac{\text{g analyte}}{\text{moles analyte}} = n \times \frac{\text{g analyte}}{\text{moles titrant}}$$

Since the actual number of protons transferred between the analyte and titrant is uncertain, we define the analyte's equivalent weight (EW) as the apparent formula weight when $n = 1$. The true formula weight, therefore, is an integer multiple of the calculated equivalent weight.

$$\text{FW} = n \times \text{EW}$$

Thus, if we titrate a monoprotic weak acid with a strong base, the EW and FW are identical. If the weak acid is diprotic, however, and we titrate to its second equivalence point, the FW will be twice as large as the EW.

Example

EXAMPLE 9.6

A 0.2521-g sample of an unknown weak acid is titrated with a 0.1005 M solution of NaOH, requiring 42.68 mL to reach the phenolphthalein end point. Determine the compound's equivalent weight. Which of the following compounds is most likely to be the unknown weak acid?

ascorbic acid	$C_6H_8O_6$	FW = 176.1	monoprotic
malonic acid	$C_3H_4O_4$	FW = 104.1	diprotic
succinic acid	$C_4H_6O_4$	FW = 118.1	diprotic
citric acid	$C_6H_8O_7$	FW = 192.1	triprotic

SOLUTION

The moles of NaOH needed to reach the end point is

$$M_b \times V_b = 0.1005 \text{ M} \times 0.04268 \text{ L} = 4.289 \times 10^{-3} \text{ mol NaOH}$$

giving an equivalent weight of

$$\text{EW} = \frac{\text{g analyte}}{\text{moles titrant}} = \frac{0.2521 \text{ g}}{4.289 \times 10^{-3} \text{ mol}} = 58.78 \text{ g/mol}$$

*See Section 2B.2 for a review of chemical equivalents and equivalent weights.

The possible formula weights for the unknown weak acid are

$$\text{for } n = 1: \quad \text{FW} = \text{EW} = 58.78 \text{ g/mol}$$
$$\text{for } n = 2: \quad \text{FW} = 2 \times \text{EW} = 117.6 \text{ g/mol}$$
$$\text{for } n = 3: \quad \text{FW} = 3 \times \text{EW} = 176.3 \text{ g/mol}$$

If the weak acid is monoprotic, then the FW must be 58.78 g/mol, eliminating ascorbic acid as a possibility. If the weak acid is diprotic, then the FW may be either 58.78 g/mol or 117.6 g/mol, depending on whether the titration was to the first or second equivalence point. Succinic acid, with a formula weight of 118.1 g/mol is a possibility, but malonic acid is not. If the analyte is a triprotic weak acid, then its FW must be 58.78 g/mol, 117.6 g/mol, or 176.3 g/mol. None of these values is close to the formula weight for citric acid, eliminating it as a possibility. Only succinic acid provides a possible match.

Equilibrium Constants Another application of acid–base titrimetry is the determination of equilibrium constants. Consider, for example, the titration of a weak acid, HA, with a strong base. The dissociation constant for the weak acid is

$$K_a = \frac{[A^-][H_3O^+]}{[HA]} \qquad 9.9$$

When the concentrations of HA and A^- are equal, equation 9.9 reduces to $K_a = [H_3O^+]$, or pH = pK_a. Thus, the pK_a for a weak acid can be determined by measuring the pH for a solution in which half of the weak acid has been neutralized. On a titration curve, the point of half-neutralization is approximated by the volume of titrant that is half of that needed to reach the equivalence point. As shown in Figure 9.20, an estimate of the weak acid's pK_a can be obtained directly from the titration curve.

This method provides a reasonable estimate of the pK_a, provided that the weak acid is neither too strong nor too weak. These limitations are easily appreciated by considering two limiting cases. For the first case let's assume that the acid is strong enough that it is more than 50% dissociated before the titration begins. As a result the concentration of HA before the equivalence point is always less than the concentration of A^-, and there is no point along the titration curve where [HA] = [A^-]. At the other extreme, if the acid is too weak, the equilibrium constant for the titration reaction

$$HA(aq) + OH^-(aq) \rightleftharpoons H_2O(\ell) + A^-(aq)$$

may be so small that less than 50% of HA will have reacted at the equivalence point. In this case the concentration of HA before the equivalence point is always greater

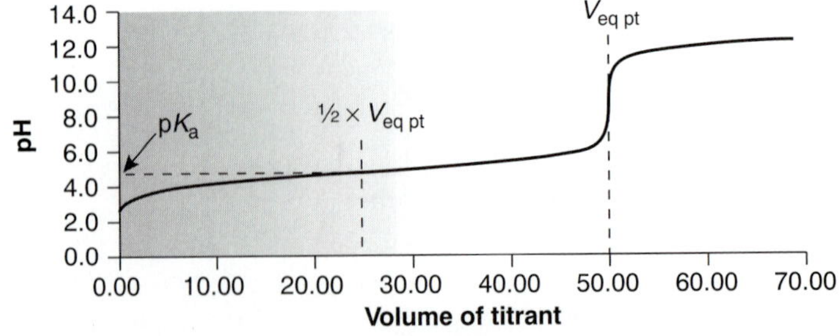

Figure 9.20
Estimating the pK_a for a weak acid from its titration curve with a strong base.

than that of A^-. Determining the pK_a by the half-equivalence point method overestimates its value if the acid is too strong and underestimates its value if the acid is too weak.

A second approach for determining the pK_a of an acid is to replot the titration curve in a linear form as a Gran plot. For example, earlier we learned that the titration of a weak acid with a strong base can be plotted in a linear form using the following equation

$$V_b \times [H_3O^+] = K_a \times V_{eq} - K_a \times V_b$$

Plotting $V_b \times [H_3O^+]$ versus V_b, for volumes less than the equivalence point volume yields a straight line with a slope of $-K_a$. Other linearizations have been developed that use all the points on a titration curve[7] or require no assumptions.[8] This approach to determining acidity constants has been used to study the acid–base properties of humic acids, which are naturally occurring, large-molecular-weight organic acids with multiple acidic sites. In one study, a sample of humic acid was found to have six titratable sites, three of which were identified as carboxylic acids, two of which were believed to be secondary or tertiary amines, and one of which was identified as a phenolic group.[9]

9B.8 Evaluation of Acid–Base Titrimetry

Scale of Operation In an acid–base titration the volume of titrant needed to reach the equivalence point is proportional to the absolute amount of analyte present in the analytical solution. Nevertheless, the change in pH at the equivalence point, and thus the utility of an acid–base titration, is a function of the analyte's concentration in the solution being titrated.

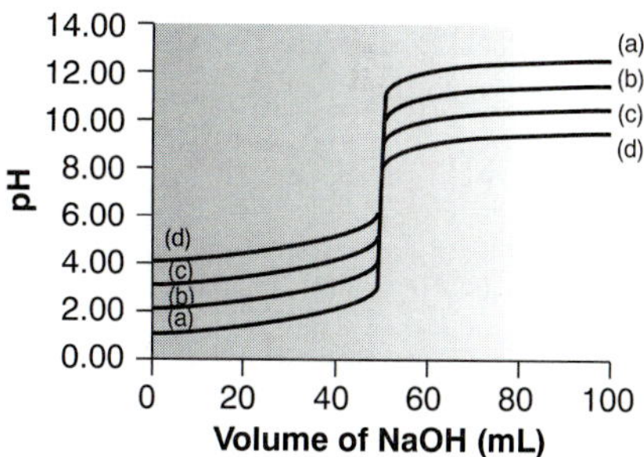

Figure 9.21
Titration curves for (a) 10^{-1} M HCl, (b) 10^{-2} M HCl, (c) 10^{-3} M HCl, and (d) 10^{-4} M HCl. In each case the titrant is an equimolar solution of NaOH.

When the sample is available as a solution, the smallest concentration of analyte that can be readily analyzed is approximately 10^{-3} M (Figure 9.21). If, for example, the analyte has a gram formula weight of 120 g/mol, then the lower concentration limit is 120 ppm. When the analyte is a solid, it must first be placed into solution, the volume of which must be sufficient to allow the titration's end point to be monitored using a visual indicator or a suitable probe. If we assume a minimum volume of 25 mL, and a lower concentration limit of 120 ppm, then a sample containing at least 3 mg of analyte is required. Acid–base titrations involving solid or solution samples, therefore, are generally limited to major and minor analytes (see Figure 3.6 in Chapter 3). The analysis of gases can be extended to trace analytes by pulling a large volume of the gas through a suitable collection solution.

Efforts have been made to develop methods for conducting acid–base titrations on a much smaller scale. In one experimental design, samples of 20–100 μL were held by capillary action between a flat-surface pH electrode and a stainless steel rod.[10] The titrant was added by using the oscillations of a piezoelectric ceramic device to move an angled glass rod in and out of a tube connected to a reservoir containing the titrant (see Figure 9.22). Each time the glass tube was withdrawn an approximately 2-nL microdroplet of titrant was released. The microdroplets were allowed to fall onto the steel rod containing the sample, with mixing accomplished by spinning the rod at 120 rpm. A total of 450 microdroplets, with a combined volume of 0.81–0.84 μL, was dispensed between each pH measurement. In this fashion a titration curve was constructed. This method was used to titrate solutions of 0.1 M HCl and 0.1 M CH_3COOH with 0.1 M NaOH. Absolute errors ranged from a minimum of +0.1% to a maximum of –4.1%, with relative

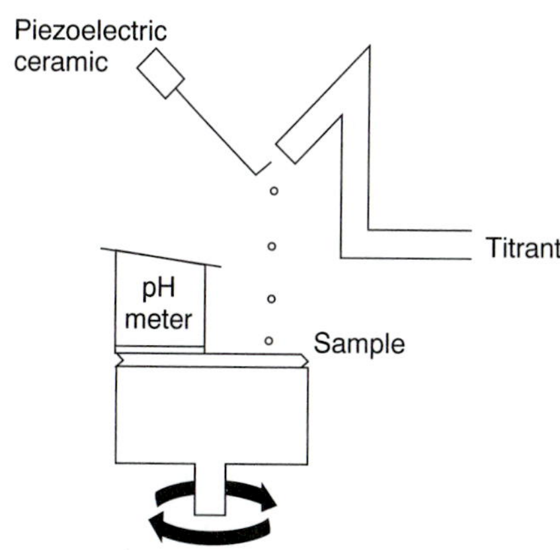

Figure 9.22
Experimental design for a microdroplet titration apparatus.

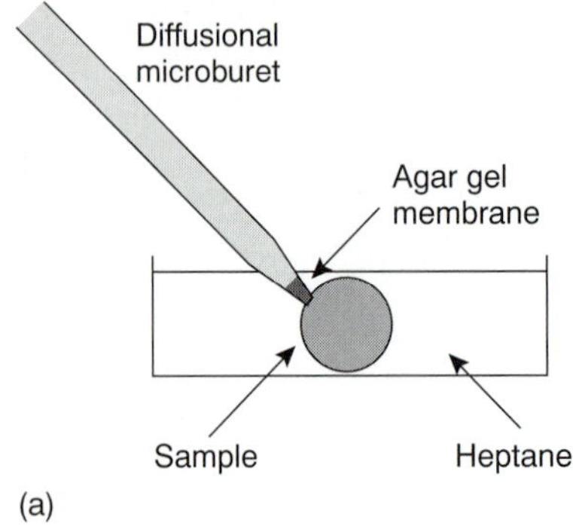

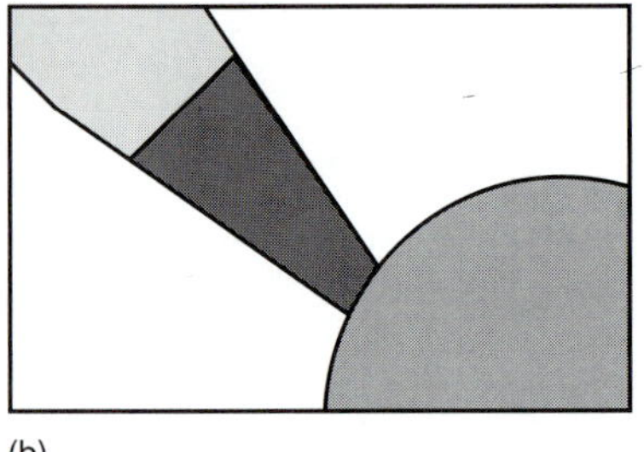

Figure 9.23
(a) Experimental set-up for a diffusional microtitration; (b) close-up showing the tip of the diffusional microburet in contact with the drop of sample.

standard deviations from 0.15% to 4.7%. The smallest volume of sample that was successfully titrated was 20 μL.

More recently, a method has been described in which the acid–base titration is conducted within a single drop of solution.[11] The titrant is added using a microburet fashioned from a glass capillary micropipet (Figure 9.23). The microburet has a 1–2 μm tip filled with an agar gel membrane. The tip of the microburet is placed within a drop of the sample solution, which is suspended in heptane, and the titrant is allowed to diffuse into the sample. The titration is followed visually using a colored indicator, and the time needed to reach the end point is measured. The rate of the titrant's diffusion from the microburet must be determined by calibration. Once calibrated, the end point time can be converted to an end point volume. Samples usually consisted of picoliter volumes (10^{-12} L), with the smallest sample being 0.7 pL. The precision of the titrations was usually about 2%.

Titrations conducted with microliter or picoliter sample volumes require a smaller absolute amount of analyte. For example, diffusional titrations have been successfully conducted on as little as 29 femtomoles (10^{-15} mol) of nitric acid. Nevertheless, the analyte must still be present in the sample at a major or minor level for the titration to be performed accurately and precisely.

Accuracy When working with macro–major and macro–minor samples, acid–base titrations can be accomplished with relative errors of 0.1–0.2%. The principal limitation to accuracy is the difference between the end point and the equivalence point.

Precision The relative precision of an acid–base titration depends primarily on the precision with which the end point volume can be measured and the precision of the end point signal. Under optimum conditions, an acid–base titration can be accomplished with a relative precision of 0.1–0.2%. The relative precision can be improved by using the largest volume buret that is feasible and ensuring that most of its capacity is used to reach the end point. Smaller volume burets are used when the cost of reagents or waste disposal is of concern or when the titration must be completed quickly to avoid competing chemical reactions. Automatic titrators are particularly useful for titrations requiring small volumes of titrant, since the precision with which the volume can be measured is significantly better (typically about ±0.05% of the buret's volume).

The precision of the end point signal depends on the method used to locate the end point and the shape of the titration curve. With a visual indicator, the precision of the end point signal is usually between ±0.03 mL and 0.10 mL. End points determined by direct monitoring often can be determined with a greater precision.

Sensitivity For an acid–base titration we can write the following general analytical equation

$$\text{Volume of titrant} = k \times \text{moles of analyte}$$

where *k*, the sensitivity, is determined by the stoichiometric relationship between analyte and titrant. Note that this equation assumes that a blank has been analyzed to correct the signal for the volume of titrant reacting with the reagents.

Consider, for example, the determination of sulfurous acid, H_2SO_3, by titrating with NaOH to the first equivalence point. Using the conservation of protons, we write

$$\text{Moles NaOH} = \text{moles } H_2SO_3$$

Substituting the molarity and volume of titrant for moles, and rearranging gives

$$V_b = \frac{1}{M_b} \times \text{moles } H_2SO_3 \qquad \textbf{9.10}$$

where k is equivalent to

$$k = \frac{1}{M_b}$$

There are two ways in which the sensitivity can be increased. The first, and most obvious, is to decrease the concentration of the titrant, since it is inversely proportional to the sensitivity, k. The second method, which only applies if the analyte is multiprotic, is to titrate to a later equivalence point. When H_2SO_3 is titrated to the second equivalence point, for instance, equation 9.10 becomes

$$V_b = 2 \times \frac{1}{M_b} \times \text{moles } H_2SO_3$$

where k is now equal to

$$k = \frac{2}{M_b}$$

In practice, however, any improvement in the sensitivity of an acid–base titration due to an increase in k is offset by a decrease in the precision of the equivalence point volume when the buret needs to be refilled. Consequently, standard analytical procedures for acid–base titrimetry are usually written to ensure that titrations require 60–100% of the buret's volume.

Selectivity Acid–base titrants are not selective. A strong base titrant, for example, will neutralize any acid, regardless of strength. Selectivity, therefore, is determined by the relative acid or base strengths of the analyte and the interferent. Two limiting situations must be considered. First, if the analyte is the stronger acid or base, then the titrant will begin reacting with the analyte before reacting with the interferent. The feasibility of the analysis depends on whether the titrant's reaction with the interferent affects the accurate location of the analyte's equivalence point. If the acid dissociation constants are substantially different, the end point for the analyte can be accurately determined (Figure 9.24a). Conversely, if the acid dissociation constants for the analyte and interferent are similar, then an accurate end point for the analyte may not be found (Figure 9.24b). In the latter case, a quantitative analysis for the analyte is not possible.

In the second limiting situation the analyte is a weaker acid or base than the interferent. In this case the volume of titrant needed to reach the analyte's equivalence point is determined by the concentration of both the analyte and the interferent. To account for the contribution from the interferent, an equivalence point for the interferent must be present. Again, if the acid dissociation constants for the analyte and interferent are significantly different, the analyte's determination is possible. If, however, the acid dissociation constants are similar, only a single equivalence point is found, and the analyte's and interferent's contributions to the equivalence point volume cannot be separated.

Time, Cost, and Equipment Acid–base titrations require less time than most gravimetric procedures, but more time than many instrumental methods of analysis, particularly when analyzing many samples. With the availability of instruments for

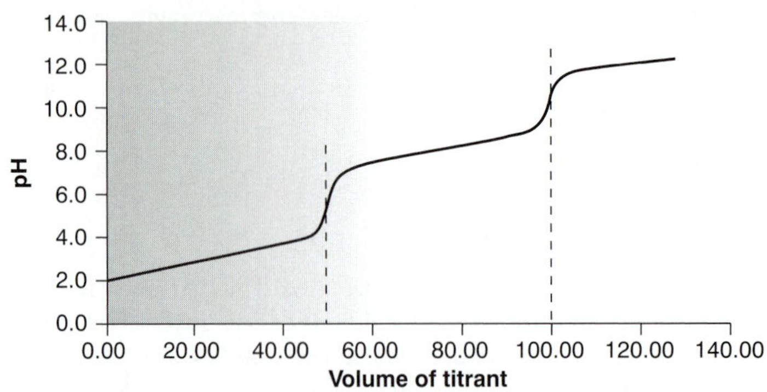

(a)

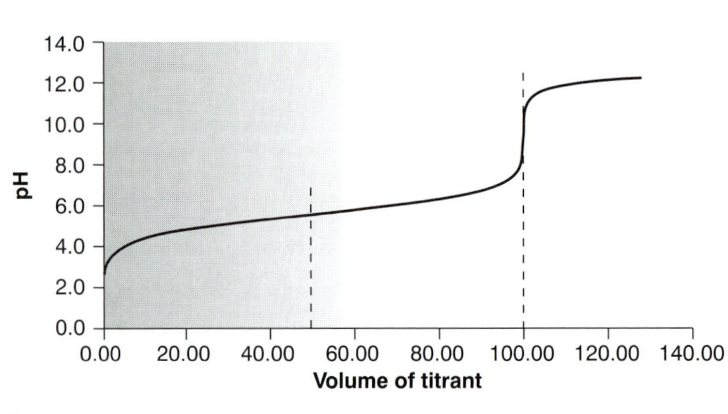

(b)

Figure 9.24
Titration curves for a 50.00 mL mixture of 0.100 M HA and 0.100 M HB with 0.100 M NaOH, where (a) $pK_{a,HA} = 3$ and $pK_{a,HB} = 8$; and (b) $pK_{a,HA} = 5$ and $pK_{a,HB} = 6$. The dashed lines indicate the location of the two equivalence points.

performing automated titrations, however, concerns about analysis time are less of a problem. When performing a titration manually the equipment needs are few (a buret and possibly a pH meter), inexpensive, routinely available in most laboratories, and easy to maintain. Instrumentation for automatic titrations can be purchased for around $3000.

9C Titrations Based on Complexation Reactions

complexation titration
A titration in which the reaction between the analyte and titrant is a complexation reaction.

The earliest titrimetric applications involving metal–ligand complexation were the determinations of cyanide and chloride using, respectively, Ag^+ and Hg^{2+} as titrants. Both methods were developed by Justus Liebig (1803–1873) in the 1850s. The use of a monodentate ligand, such as Cl^- and CN^-, however, limited the utility of **complexation titrations** to those metals that formed only a single stable complex, such as $Ag(CN)_2^-$ and $HgCl_2$. Other potential metal–ligand complexes, such as CdI_4^{2-}, were not analytically useful because the stepwise formation of a series of metal–ligand complexes (CdI^+, CdI_2, CdI_3^-, and CdI_4^{2-}) resulted in a poorly defined end point.

The utility of complexation titrations improved following the introduction by Schwarzenbach, in 1945, of aminocarboxylic acids as multidentate ligands capable of forming stable 1:1 complexes with metal ions. The most widely used of these new ligands was ethylenediaminetetraacetic acid, EDTA, which forms strong 1:1 complexes with many metal ions. The first use of EDTA as a titrant occurred in

1946, when Schwarzenbach introduced metallochromic dyes as visual indicators for signaling the end point of a complexation titration.

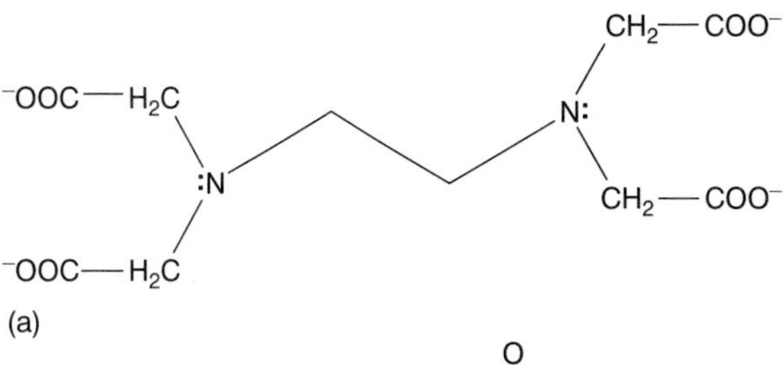

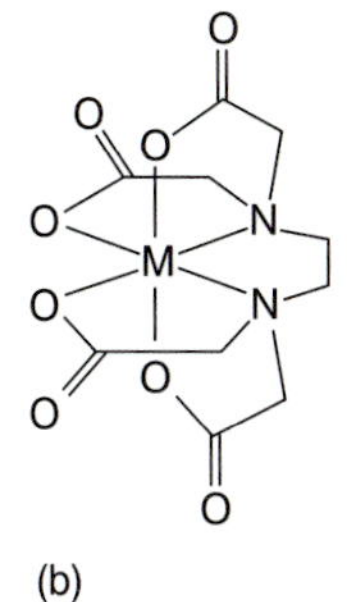

Figure 9.25
Structures of (a) EDTA, and (b) a six-coordinate metal–EDTA complex.

9C.1 Chemistry and Properties of EDTA

Ethylenediaminetetraacetic acid, or EDTA, is an aminocarboxylic acid. The structure of EDTA is shown in Figure 9.25a. EDTA, which is a Lewis acid, has six binding sites (the four carboxylate groups and the two amino groups), providing six pairs of electrons. The resulting metal–ligand complex, in which EDTA forms a cage-like structure around the metal ion (Figure 9.25b), is very stable. The actual number of coordination sites depends on the size of the metal ion; however, all metal–EDTA complexes have a 1:1 stoichiometry.

Metal–EDTA Formation Constants To illustrate the formation of a metal–EDTA complex consider the reaction between Cd^{2+} and EDTA

$$Cd^{2+}(aq) + Y^{4-}(aq) \rightleftharpoons CdY^{2-}(aq)$$

where Y^{4-} is a shorthand notation for the chemical form of EDTA shown in Figure 9.25. The formation constant for this reaction

$$K_f = \frac{[CdY^{2-}]}{[Cd^{2+}][Y^{4-}]} = 2.9 \times 10^{16} \tag{9.11}$$

is quite large, suggesting that the reaction's equilibrium position lies far to the right. Formation constants for other metal–EDTA complexes are found in Appendix 3C.

EDTA Is a Weak Acid Besides its properties as a ligand, EDTA is also a weak acid. The fully protonated form of EDTA, H_6Y^{2+}, is a hexaprotic weak acid with successive pK_a values of

$$pK_{a1} = 0.0 \quad pK_{a2} = 1.5 \quad pK_{a3} = 2.0 \quad pK_{a4} = 2.68 \quad pK_{a5} = 6.11 \quad pK_{a6} = 10.17$$

The first four values are for the carboxyl protons, and the remaining two values are for the ammonium protons. A ladder diagram for EDTA is shown in Figure 9.26. The species Y^{4-} becomes the predominate form of EDTA at pH levels greater than 10.17. It is only for pH levels greater than 12 that Y^{4-} becomes the only significant form of EDTA.

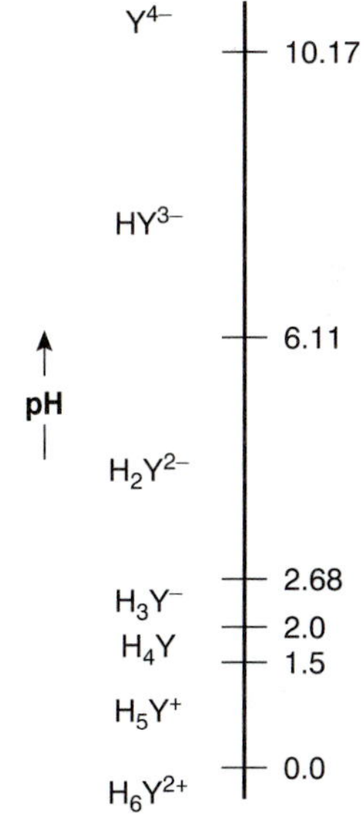

Figure 9.26
Ladder diagram for EDTA.

Conditional Metal–Ligand Formation Constants Recognizing EDTA's acid–base properties is important. The formation constant for CdY^{2-} in equation 9.11 assumes that EDTA is present as Y^{4-}. If we restrict the pH to levels greater than 12, then equation 9.11 provides an adequate description of the formation of CdY^{2-}. For pH levels less than 12, however, K_f overestimates the stability of the CdY^{2-} complex.

At any pH a mass balance requires that the total concentration of unbound EDTA equal the combined concentrations of each of its forms.

$$C_{EDTA} = [H_6Y^{2+}] + [H_5Y^+] + [H_4Y] + [H_3Y^-] + [H_2Y^{2-}] + [HY^{3-}] + [Y^{4-}]$$

To correct the formation constant for EDTA's acid–base properties, we must account for the fraction, $\alpha_{Y^{4-}}$, of EDTA present as Y^{4-}.

$$\alpha_{Y^{4-}} = \frac{[Y^{4-}]}{C_{EDTA}} \tag{9.12}$$

Table 9.12 Values of $\alpha_{Y^{4-}}$ for Selected pHs

pH	$\alpha_{Y^{4-}}$	pH	$\alpha_{Y^{4-}}$
2	3.7×10^{-14}	8	5.4×10^{-3}
3	2.5×10^{-11}	9	5.2×10^{-2}
4	3.6×10^{-9}	10	0.35
5	3.5×10^{-7}	11	0.85
6	2.2×10^{-5}	12	0.98
7	4.8×10^{-4}	13	1.00

Table 9.13 Conditional Formation Constants for CdY^{2-}

pH	K_f'	pH	K_f'
2	1.1×10^{3}	8	1.6×10^{14}
3	7.3×10^{5}	9	1.5×10^{15}
4	1.0×10^{8}	10	1.0×10^{16}
5	1.0×10^{10}	11	2.5×10^{16}
6	6.4×10^{11}	12	2.8×10^{16}
7	1.4×10^{13}	13	2.9×10^{16}

Values of $\alpha_{Y^{4-}}$ are shown in Table 9.12. Solving equation 9.12 for $[Y^{4-}]$ and substituting into the equation for the formation constant gives

$$K_f = \frac{[CdY^{2-}]}{[Cd^{2+}]\alpha_{Y^{4-}}C_{EDTA}}$$

If we fix the pH using a buffer, then $\alpha_{Y^{4-}}$ is a constant. Combining $\alpha_{Y^{4-}}$ with K_f gives

$$K_f' = \alpha_{Y^{4-}} \times K_f = \frac{[CdY^{2-}]}{[Cd^{2+}]C_{EDTA}} \qquad 9.13$$

conditional formation constant
The equilibrium formation constant for a metal–ligand complex for a specific set of solution conditions, such as pH.

where K_f' is a **conditional formation constant** whose value depends on the pH. As shown in Table 9.13 for CdY^{2-}, the conditional formation constant becomes smaller, and the complex becomes less stable at lower pH levels.

EDTA Must Compete with Other Ligands To maintain a constant pH, we must add a buffering agent. If one of the buffer's components forms a metal–ligand complex with Cd^{2+}, then EDTA must compete with the ligand for Cd^{2+}. For example, an NH_4^+/NH_3 buffer includes the ligand NH_3, which forms several stable Cd^{2+}–NH_3 complexes. EDTA forms a stronger complex with Cd^{2+} and will displace NH_3. The presence of NH_3, however, decreases the stability of the Cd^{2+}–EDTA complex.

auxiliary complexing agent
A second ligand in a complexation titration that initially binds with the analyte but is displaced by the titrant.

We can account for the effect of an **auxiliary complexing agent,** such as NH_3, in the same way we accounted for the effect of pH. Before adding EDTA, a mass balance on Cd^{2+} requires that the total concentration of Cd^{2+}, C_{Cd}, be

$$C_{Cd} = [Cd^{2+}] + [Cd(NH_3)^{2+}] + [Cd(NH_3)_2^{2+}] + [Cd(NH_3)_3^{2+}] + [Cd(NH_3)_4^{2+}]$$

Table 9.14 Values of $\alpha_{M^{n+}}$ for Selected Concentrations of Ammonia

$[NH_3]$ (M)	α_{Ag^+}	$\alpha_{Ca^{2+}}$	$\alpha_{Cd^{2+}}$	$\alpha_{Co^{2+}}$	$\alpha_{Cu^{2+}}$	$\alpha_{Mg^{2+}}$	$\alpha_{Ni^{2+}}$	$\alpha_{Zn^{2+}}$
1	1.00×10^{-7}	5.50×10^{-1}	6.09×10^{-8}	1.00×10^{-6}	3.79×10^{-14}	1.76×10^{-1}	9.20×10^{-10}	3.95×10^{-10}
0.5	4.00×10^{-7}	7.36×10^{-1}	1.05×10^{-6}	2.22×10^{-5}	6.86×10^{-13}	4.13×10^{-1}	3.44×10^{-8}	6.27×10^{-9}
0.1	9.98×10^{-6}	9.39×10^{-1}	3.51×10^{-4}	6.64×10^{-3}	4.63×10^{-10}	8.48×10^{-1}	5.12×10^{-5}	3.68×10^{-6}
0.05	3.99×10^{-5}	9.69×10^{-1}	2.72×10^{-3}	3.54×10^{-2}	7.17×10^{-9}	9.22×10^{-1}	6.37×10^{-4}	5.45×10^{-5}
0.01	9.83×10^{-4}	9.94×10^{-1}	8.81×10^{-2}	3.55×10^{-1}	3.22×10^{-6}	9.84×10^{-1}	4.32×10^{-2}	1.82×10^{-2}
0.005	3.86×10^{-3}	9.97×10^{-1}	2.27×10^{-1}	5.68×10^{-1}	3.62×10^{-5}	9.92×10^{-1}	1.36×10^{-1}	1.27×10^{-1}
0.001	7.95×10^{-2}	9.99×10^{-1}	6.90×10^{-1}	8.84×10^{-1}	4.15×10^{-3}	9.98×10^{-1}	5.76×10^{-1}	7.48×10^{-1}

The fraction, $\alpha_{Cd^{2+}}$, present as uncomplexed Cd^{2+} is

$$\alpha_{Cd^{2+}} = \frac{[Cd^{2+}]}{C_{Cd}} \qquad 9.14$$

Solving equation 9.14 for $[Cd^{2+}]$ and substituting into equation 9.13 gives

$$K'_f = \alpha_{Y^{4-}} \times K_f = \frac{[CdY^{2-}]}{\alpha_{Cd^{2+}} C_{Cd} C_{EDTA}}$$

If the concentration of NH_3 is held constant, as it usually is when using a buffer, then we can rewrite this equation as

$$K''_f = \alpha_{Cd^{2+}} \times \alpha_{Y^{4-}} \times K_f = \frac{[CdY^{2-}]}{C_{Cd} C_{EDTA}} \qquad 9.15$$

where K''_f is a new conditional formation constant accounting for both pH and the presence of an auxiliary complexing agent. Values of $\alpha_{M^{n+}}$ for several metal ions are provided in Table 9.14.

9C.2 Complexometric EDTA Titration Curves

Now that we know something about EDTA's chemical properties, we are ready to evaluate its utility as a titrant for the analysis of metal ions. To do so we need to know the shape of a complexometric EDTA titration curve. In Section 9B we saw that an acid–base titration curve shows the change in pH following the addition of titrant. The analogous result for a titration with EDTA shows the change in pM, where M is the metal ion, as a function of the volume of EDTA. In this section we learn how to calculate the titration curve. We then show how to quickly sketch the titration curve using a minimum number of calculations.

Calculating the Titration Curve As an example, let's calculate the titration curve for 50.0 mL of 5.00×10^{-3} M Cd^{2+} with 0.0100 M EDTA at a pH of 10 and in the presence of 0.0100 M NH_3. The formation constant for Cd^{2+}–EDTA is 2.9×10^{16}.

Since the titration is carried out at a pH of 10, some of the EDTA is present in forms other than Y^{4-}. In addition, the presence of NH_3 means that the EDTA must compete for the Cd^{2+}. To evaluate the titration curve, therefore, we must use the appropriate conditional formation constant. From Tables 9.12 and 9.14 we find that $\alpha_{Y^{4-}}$ is 0.35 at a pH of 10, and that $\alpha_{Cd^{2+}}$ is 0.0881 when the concentration of NH_3 is 0.0100 M. Using these values, we calculate that the conditional formation constant is

$$K_f'' = \alpha_{Y^{4-}} \times \alpha_{Cd^{2+}} \times K_f = (0.35)(0.0881)(2.9 \times 10^{16}) = 8.9 \times 10^{14}$$

Because K_f'' is so large, we treat the titration reaction as though it proceeds to completion.

The first task in calculating the titration curve is to determine the volume of EDTA needed to reach the equivalence point. At the equivalence point we know that

$$\text{Moles EDTA} = \text{moles } Cd^{2+}$$

or

$$M_{EDTA}V_{EDTA} = M_{Cd}V_{Cd}$$

Solving for the volume of EDTA

$$V_{EDTA} = \frac{M_{Cd}V_{Cd}}{M_{EDTA}} = \frac{(5.00 \times 10^{-3}\ \text{M})(50.0\ \text{mL})}{0.0100\ \text{M}} = 25.0\ \text{mL}$$

shows us that 25.0 mL of EDTA is needed to reach the equivalence point.

Before the equivalence point, Cd^{2+} is in excess, and pCd is determined by the concentration of free Cd^{2+} remaining in solution. Not all the untitrated Cd^{2+} is free (some is complexed with NH_3), so we will have to account for the presence of NH_3. For example, after adding 5.0 mL of EDTA, the total concentration of Cd^{2+} is

$$C_{Cd} = \frac{\text{moles excess } Cd^{2+}}{\text{total volume}} = \frac{M_{Cd}V_{Cd} - M_{EDTA}V_{EDTA}}{V_{Cd} + V_{EDTA}}$$

$$= \frac{(5.00 \times 10^{-3}\ \text{M})(50.0\ \text{mL}) - (0.0100\ \text{M})(5.0\ \text{mL})}{50.0\ \text{mL} + 5.0\ \text{mL}} = 3.64 \times 10^{-3}\ \text{M}$$

To calculate the concentration of free Cd^{2+} we use equation 9.14.

$$[Cd^{2+}] = \alpha_{Cd^{2+}} \times C_{Cd} = (0.0881)(3.64 \times 10^{-3}\ \text{M}) = 3.21 \times 10^{-4}\ \text{M}$$

Thus, pCd is

$$\text{pCd} = -\log[Cd^{2+}] = -\log(3.21 \times 10^{-4}) = 3.49$$

At the equivalence point, all the Cd^{2+} initially present is now present as CdY^{2-}. The concentration of Cd^{2+}, therefore, is determined by the dissociation of the CdY^{2-} complex. To find pCd we must first calculate the concentration of the complex.

$$[CdY^{2-}] = \frac{\text{initial moles } Cd^{2+}}{\text{total volume}} = \frac{M_{Cd}V_{Cd}}{V_{Cd} + V_{EDTA}}$$

$$= \frac{(5.00 \times 10^{-3}\ \text{M})(50.0\ \text{mL})}{50.0\ \text{mL} + 25.0\ \text{mL}} = 3.33 \times 10^{-3}\ \text{M}$$

Letting the variable x represent the concentration of Cd^{2+} due to the dissociation of the CdY^{2-} complex, we have

$$K_f'' = \frac{[CdY^{2-}]}{C_{Cd}C_{EDTA}} = \frac{3.33 \times 10^{-3} - x}{(x)(x)} = 8.94 \times 10^{14}$$

$$x = C_{Cd} = 1.93 \times 10^{-9}\ \text{M}$$

Once again, to find the $[Cd^{2+}]$ we must account for the presence of NH_3; thus

$$[Cd^{2+}] = \alpha_{Cd^{2+}} \times C_{Cd} = (0.0881)(1.93 \times 10^{-9}\ M) = 1.70 \times 10^{-10}\ M$$

giving pCd as 9.77.

After the equivalence point, EDTA is in excess, and the concentration of Cd^{2+} is determined by the dissociation of the CdY^{2-} complex. Examining the equation for the complex's conditional formation constant (equation 9.15), we see that to calculate C_{Cd} we must first calculate $[CdY^{2-}]$ and C_{EDTA}. After adding 30.0 mL of EDTA, these concentrations are

$$[CdY^{2-}] = \frac{\text{initial moles } Cd^{2+}}{\text{total volume}} = \frac{M_{Cd}V_{Cd}}{V_{Cd} + V_{EDTA}}$$

$$= \frac{(5.00 \times 10^{-3}\ M)(50.0\ mL)}{50.0\ mL + 30.0\ mL} = 3.13 \times 10^{-3}\ M$$

$$C_{EDTA} = \frac{\text{moles excess EDTA}}{\text{total volume}} = \frac{M_{EDTA}V_{EDTA} - M_{Cd}V_{Cd}}{V_{Cd} + V_{EDTA}}$$

$$= \frac{(0.0100\ M)(30.0\ mL) - (5.00 \times 10^{-3}\ M)(50.0\ mL)}{50.0\ mL + 30.0\ mL} = 6.25 \times 10^{-4}\ M$$

Substituting these concentrations into equation 9.15 and solving for C_{Cd} gives

$$\frac{[CdY^{2-}]}{C_{Cd}C_{EDTA}} = \frac{3.13 \times 10^{-3}\ M}{C_{Cd}(6.25 \times 10^{-4})} = 8.94 \times 10^{14}$$

$$C_{Cd} = 5.60 \times 10^{-15}\ M$$

Thus,

$$[Cd^{2+}] = \alpha_{Cd^{2+}} \times C_{Cd} = (0.0881)(5.60 \times 10^{-15}\ M) = 4.93 \times 10^{-16}\ M$$

and pCd is 15.31. Figure 9.27 and Table 9.15 show additional results for this titration.

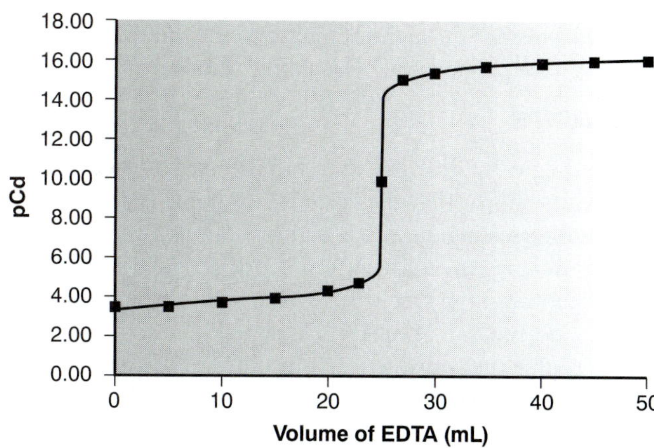

Figure 9.27
Complexometric titration curve for 50.0 mL of 5.00×10^{-3} M Cd^{2+} with 0.0100 M EDTA at a pH of 10.0 in the presence of 0.0100 M NH_3.

Table 9.15 Data for Titration of 5.00×10^{-3} M Cd^{2+} with 0.0100 M EDTA at a pH of 10.0 and in the Presence of 0.0100 M NH_3

Volume of EDTA (mL)	pCd
0.00	3.36
5.00	3.49
10.00	3.66
15.00	3.87
20.00	4.20
23.00	4.62
25.00	9.77
27.00	14.91
30.00	15.31
35.00	15.61
40.00	15.78
45.00	15.91
50.00	16.01

Sketching an EDTA Titration Curve Our strategy for sketching an EDTA titration curve is similar to that for sketching an acid–base titration curve. We begin by drawing axes, placing pM on the *y*-axis and volume of EDTA on the *x*-axis. After calculating the volume of EDTA needed to reach the equivalence point, we add a vertical line intersecting the *x*-axis at this volume. Next we calculate and plot two values of pM for volumes of EDTA before the equivalence point and two values of pM for volumes after the equivalence point. Straight lines are drawn through each pair of points. Finally, a smooth curve is drawn connecting the three straight-line segments.

Example

EXAMPLE 9.7

Sketch the titration curve for 50.0 mL of 5.00×10^{-3} M Cd^{2+} with 0.010 M EDTA at a pH of 10, and in the presence of an ammonia concentration that is held constant throughout the titration at 0.010 M. This is the same titration for which we previously calculated the titration curve (Table 9.15 and Figure 9.27).

SOLUTION

We begin by drawing axes for the titration curve (Figure 9.28a). We have already shown that the equivalence point is at 25.0 mL, so we draw a vertical line intersecting the *x*-axis at this volume (Figure 9.28b).

Before the equivalence point, pCd is determined by the excess concentration of free Cd^{2+}. Using values from Table 9.15, we plot pCd for 5.0 mL and 10.0 mL of EDTA (Figure 9.28c).

After the equivalence point, pCd is determined by the dissociation of the Cd^{2+}–EDTA complex. Using values from Table 9.15, we plot pCd for 30.0 mL and 40.0 mL of EDTA (Figure 9.28d).

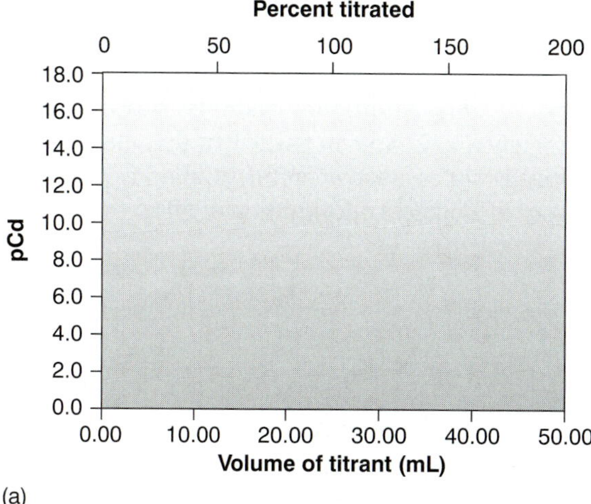

(a)

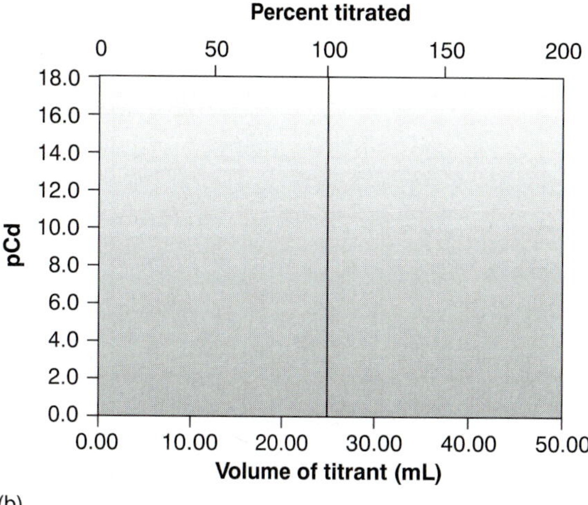

(b)

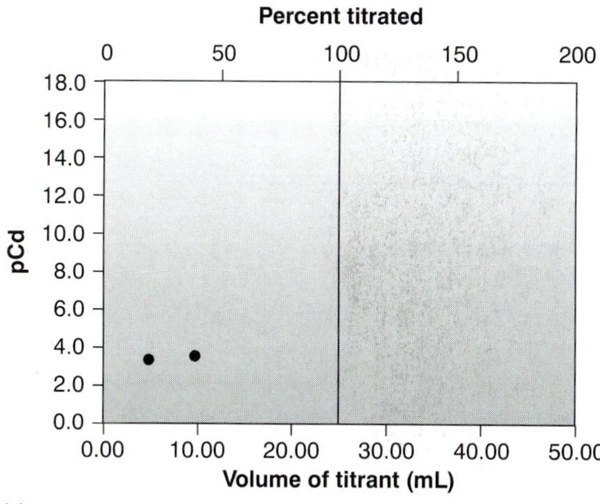

(c)

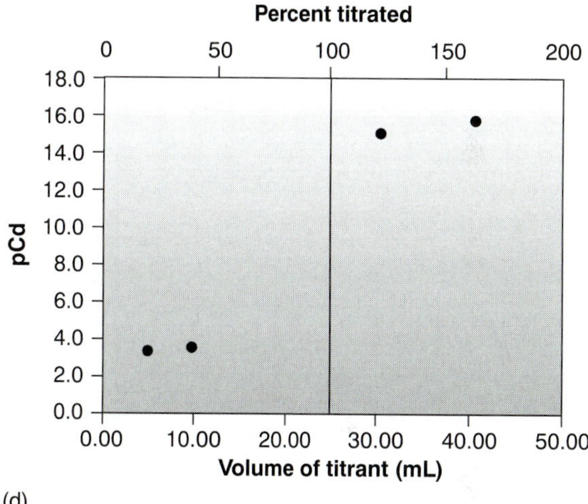

(d)

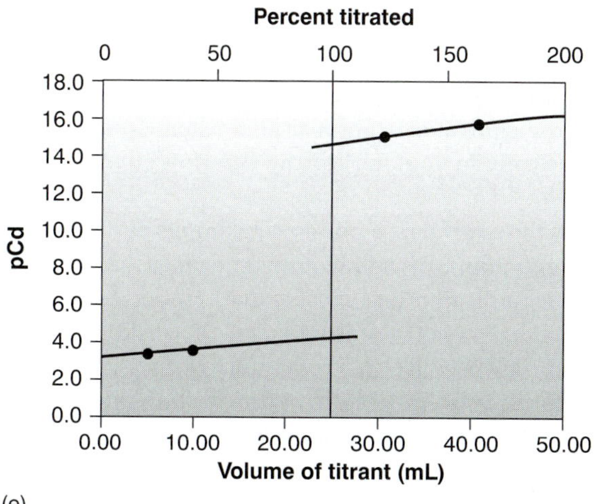

(e)

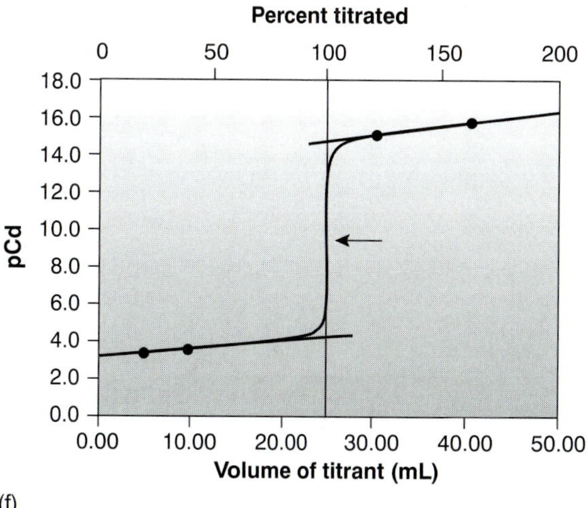

(f)

Figure 9.28
How to sketch an EDTA complexometric titration curve; see text for explanation.

An approximate sketch of the titration curve is completed by drawing separate straight lines through the two points before and after the equivalence point (Figure 9.28e). Finally, a smooth curve is drawn to connect the three straight-line segments (Figure 9.28f). Additional examples of Cd^{2+}–EDTA titration curves are shown in Figure 9.29. Note in particular, that the change in pCd near the equivalence point is not as great when the titration is carried out at a lower pH or in the presence of a higher concentration of NH_3.

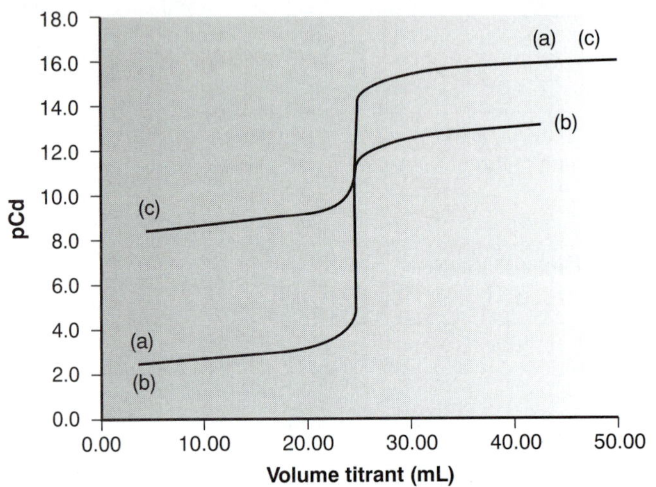

Figure 9.29
Effect of pH and $[NH_3]$ on the titration curve for 50.0 mL of 5.00×10^{-3} M Cd^{2+} with 0.0100 M EDTA. (a) pH = 10, $[NH_3]$ = 0; (b) pH = 7, $[NH_3]$ = 0; (c) pH = 10, $[NH_3]$ = 0.5 M.

9C.3 Selecting and Evaluating the End Point

The equivalence point of a complexation titration occurs when stoichiometrically equivalent amounts of analyte and titrant have reacted. For titrations involving metal ions and EDTA, the equivalence point occurs when C_M and C_{EDTA} are equal and may be located visually by looking for the titration curve's inflection point.

As with acid–base titrations, the equivalence point of a complexation titration is estimated by an experimental end point. A variety of methods have been used to find the end point, including visual indicators and sensors that respond to a change in the solution conditions. Typical examples of sensors include recording a potentiometric titration curve using an ion-selective electrode (analogous to measuring pH with a pH electrode),* monitoring the temperature of the titration mixture, and monitoring the absorbance of electromagnetic radiation by the titration mixture.

*See Chapter 11 for a further discussion of ion-selective electrodes.

The first two sensors were discussed in Section 9B.3 for acid–base titrations and are not considered further in this section.

Finding the End Point with a Visual Indicator Most indicators for complexation titrations are organic dyes that form stable complexes with metal ions. These dyes are known as **metallochromic indicators.** To function as an indicator for an EDTA titration, the metal–indicator complex must possess a color different from that of the uncomplexed indicator. Furthermore, the formation constant for the metal–indicator complex must be less favorable than that for the metal–EDTA complex.

metallochromic indicator
A visual indicator used to signal the end point in a complexation titration.

The indicator, In^{m-}, is added to the solution of analyte, forming a colored metal–indicator complex, MIn^{n-m}. As EDTA is added, it reacts first with the free analyte, and then displaces the analyte from the metal–indicator complex, affecting a change in the solution's color. The accuracy of the end point depends on the strength of the metal–indicator complex relative to that of the metal–EDTA complex. If the metal–indicator complex is too strong, the color change occurs after the equivalence point. If the metal–indicator complex is too weak, however, the end point is signaled before reaching the equivalence point.

Most metallochromic indicators also are weak acids or bases. The conditional formation constant for the metal–indicator complex, therefore, depends on the solution's pH. This provides some control over the indicator's titration error. The apparent strength of a metal–indicator complex can be adjusted by controlling the pH at which the titration is carried out. Unfortunately, because they also are acid–base indicators, the color of the uncomplexed indicator changes with pH. For example, calmagite, which we may represent as H_3In, undergoes a change in color from the red of H_2In^- to the blue of HIn^{2-} at a pH of approximately 8.1, and from the blue of HIn^{2-} to the red-orange of In^{3-} at a pH of approximately 12.4. Since the color of calmagite's metal–indicator complexes are red, it is only useful as a metallochromic indicator in the pH range of 9–11, at which almost all the indicator is present as HIn^{2-}.

A partial list of metallochromic indicators, and the metal ions and pH conditions for which they are useful, is given in Table 9.16. Even when a suitable indicator does not exist, it is often possible to conduct an EDTA titration by introducing a small amount of a secondary metal–EDTA complex, provided that the secondary metal ion forms a stronger complex with the indicator and a weaker complex with EDTA than the analyte. For example, calmagite can be used in the determination of

Table 9.16 Selected Metallochromic Indicators

Indicator	Useful pH Range	Useful for
calmagite	9–11	Ba, Ca, Mg, Zn
Eriochrome Black T	7.5–10.5	Ba, Ca, Mg, Zn
Eriochrome Blue Black R	8–12	Ca, Mg, Zn, Cu
murexide	6–13	Ca, Ni, Cu
PAN	2–11	Cd, Cu, Zn
salicylic acid	2–3	Fe

Ca^{2+} if a small amount of Mg^{2+}–EDTA is added to the solution containing the analyte. The Mg^{2+} is displaced from the EDTA by Ca^{2+}, freeing the Mg^{2+} to form the red Mg^{2+}–indicator complex. After all the Ca^{2+} has been titrated, Mg^{2+} is displaced from the Mg^{2+}–indicator complex by EDTA, signaling the end point by the presence of the uncomplexed indicator's blue form.

Finding the End Point by Monitoring Absorbance. An important limitation when using a visual indicator is the need to observe the change in color signaling the end point. This may be difficult when the solution is already colored. For example, ammonia is used to adjust the pH of solutions containing Cu^{2+} before its titration with EDTA. The presence of the intensely colored $Cu(NH_3)_4{}^{2+}$ complex obscures the indicator's color, making an accurate determination of the end point difficult. Other absorbing species present within the sample matrix may also interfere in a similar fashion. This is often a problem when analyzing clinical samples such as blood or environmental samples such as natural waters.

As long as at least one species in a complexation titration absorbs electromagnetic radiation, the equivalence point can be located by monitoring the absorbance of the analytical solution at a carefully selected wavelength.* For example, the equivalence point for the titration of Cu^{2+} with EDTA, in the presence of NH_3, can be located by monitoring the absorbance at a wavelength of 745 nm, where the $Cu(NH_3)_4{}^{2+}$ complex absorbs strongly. At the beginning of the titration the absorbance is at a maximum. As EDTA is added, however, the reaction

$$Cu(NH_3)_4{}^{2+}(aq) + Y^{4-}(aq) \rightarrow CuY^{2-}(aq) + 4NH_3(aq)$$

occurs, decreasing both the concentration of $Cu(NH_3)_4{}^{2+}$ and the absorbance. The absorbance reaches a minimum at the equivalence point and remains essentially unchanged as EDTA is added in excess. The resulting spectrophotometric titration curve is shown in Figure 9.30a. In order to keep the individual segments of the titration curve linear, the measured absorbance, A_{meas}, is corrected for dilution

$$A_{corr} = A_{meas} \times \frac{V_{EDTA} + V_{Cu}}{V_{Cu}}$$

where A_{corr} is the corrected absorbance, and V_{EDTA} and V_{Cu} are, respectively, the volumes of EDTA and Cu. The equivalence point is given by the intersection of the linear segments, which are extrapolated if necessary to correct for any curvature in the titration curve. Other common spectrophotometric titration curves are shown in Figures 9.30b–f.

9C.4 Representative Method

Although every complexation titrimetric method has its own unique considerations, the following description for determining the hardness of water provides an instructive example of a typical procedure.

*See Chapter 10 for a further discussion of absorbance and spectroscopy.

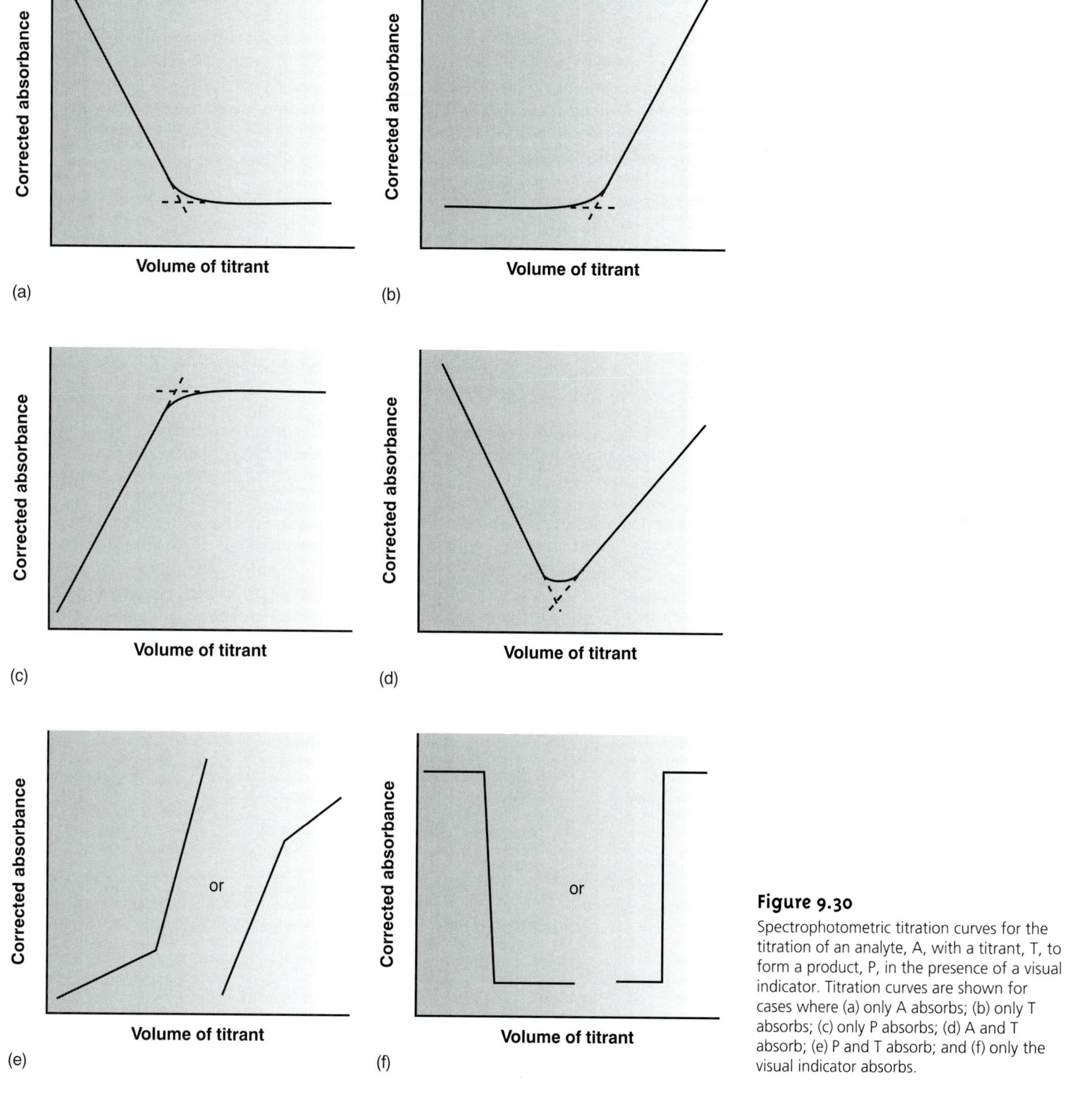

Figure 9.30

Spectrophotometric titration curves for the titration of an analyte, A, with a titrant, T, to form a product, P, in the presence of a visual indicator. Titration curves are shown for cases where (a) only A absorbs; (b) only T absorbs; (c) only P absorbs; (d) A and T absorb; (e) P and T absorb; and (f) only the visual indicator absorbs.

Representative Methods

Method 9.2 Determination of Hardness of Water and Wastewater[12]

Description of the Method. The operational definition of water hardness is the total concentration of cations in a sample capable of forming insoluble complexes with soap. Although most divalent and trivalent metal ions contribute to hardness, the most important are Ca^{2+} and Mg^{2+}. Hardness is determined by titrating with EDTA at a buffered pH of 10. Eriochrome Black T or calmagite is used as a visual indicator. Hardness is reported in parts per million $CaCO_3$.

The photo in Colorplate 8b shows the indicator's color change for this titration.

Procedure. Select a volume of sample requiring less than 15 mL of titrant to keep the analysis time under 5 min and, if necessary, dilute the sample to 50 mL with distilled water. Adjust the pH by adding 1–2 mL of a pH 10 buffer containing a small amount of Mg^{2+}–EDTA. Add 1–2 drops of indicator, and titrate with a standard solution of EDTA until the red-to-blue end point is reached.

Questions

1. **Why is the sample buffered to a pH of 10? What problems might be expected at higher or lower pHs?**

   Of the cations contributing to hardness, Mg^{2+} forms the weakest complex with EDTA and is the last cation to be titrated. Calmagite was selected as the indicator because it gives a distinct end point with Mg^{2+}. Because of calmagite's acid–base properties the indicator is only useful in the pH range of 9–11 (see Table 9.16). Figure 9.31 shows the titration curve for a solution of 10^{-3} M Mg^{2+} with 10^{-2} M EDTA at pHs of 9, 10, and 11. Superimposed on each titration curve is the range of conditions in which the average analyst will find the end point. At a pH of 9 an early end point is possible, leading to a negative determinate error, and at a pH of 11 there is a chance of a late end point and a positive determinate error.

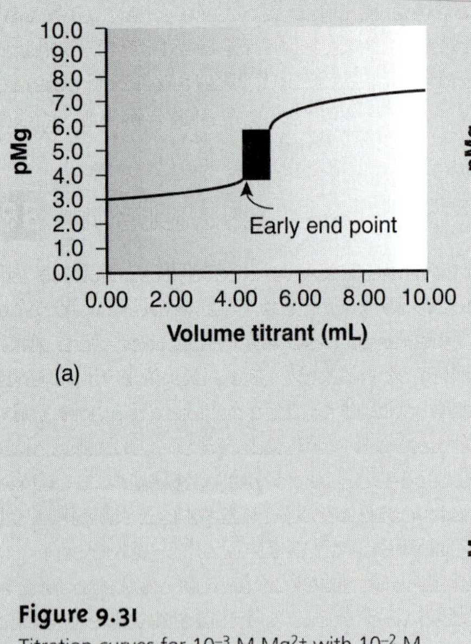

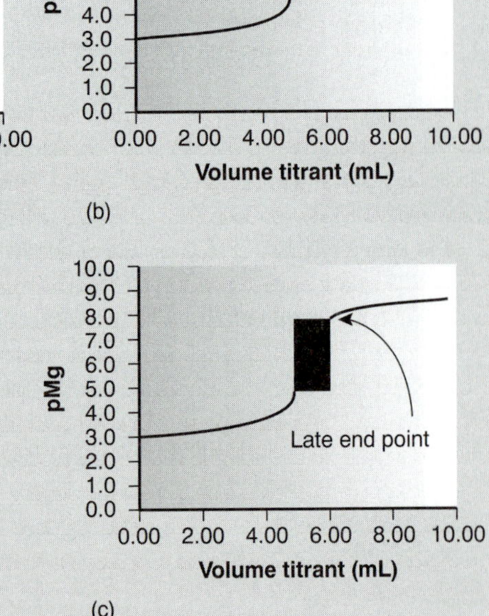

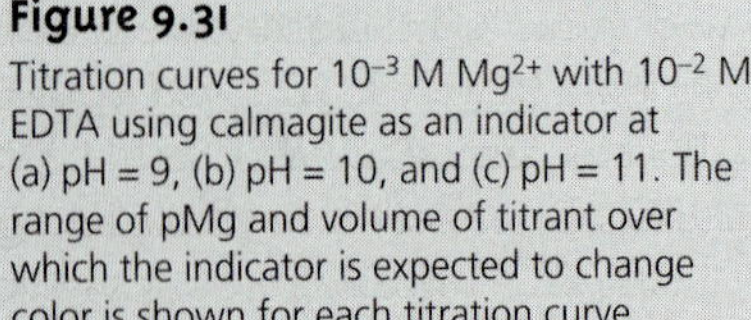

Figure 9.31
Titration curves for 10^{-3} M Mg^{2+} with 10^{-2} M EDTA using calmagite as an indicator at (a) pH = 9, (b) pH = 10, and (c) pH = 11. The range of pMg and volume of titrant over which the indicator is expected to change color is shown for each titration curve.

—Continued

2. **Why is a small amount of Mg^{2+}–EDTA complex added to the buffer?**

   The titration's end point is signaled by the indicator calmagite, which gives a good end point with magnesium, but a poor end point with other cations such as calcium. If the sample does not contain any Mg^{2+} as a source of hardness, then the titration will have a poorly defined end point and inaccurate results will be obtained. By adding a small amount of Mg^{2+}–EDTA to the buffer, a source of Mg^{2+} is ensured. When the buffer is added to the sample, the Mg^{2+} is displaced by Ca^{2+}, because Ca^{2+} forms a stronger complex with EDTA. Since the displacement is stoichiometric, the total concentration of hardness cations remains unchanged, and there is no change in the amount of EDTA needed to reach the equivalence point.

3. **Why does the procedure specify that the titration take no longer than 5 min?**

   The presence of a time limitation suggests that there must be a kinetically controlled interference, possibly arising from a competing chemical reaction. In this case the interference is the possible precipitation of $CaCO_3$.

9C.5 Quantitative Applications

With a few exceptions, most quantitative applications of complexation titrimetry have been replaced by other analytical methods. In this section we review the general application of complexation titrimetry with an emphasis on selected applications from the analysis of water and wastewater. We begin, however, with a discussion of the selection and standardization of complexation titrants.

Selection and Standardization of Titrants EDTA is a versatile titrant that can be used for the analysis of virtually all metal ions. Although EDTA is the most commonly employed titrant for complexation titrations involving metal ions, it cannot be used for the direct analysis of anions or neutral ligands. In the latter case, standard solutions of Ag^+ or Hg^{2+} are used as the titrant.

Solutions of EDTA are prepared from the soluble disodium salt, $Na_2H_2Y \cdot 2H_2O$. Concentrations can be determined directly from the known mass of EDTA; however, for more accurate work, standardization is accomplished by titrating against a solution made from the primary standard $CaCO_3$. Solutions of Ag^+ and Hg^{2+} are prepared from $AgNO_3$ and $Hg(NO_3)_2$, both of which are secondary standards. Standardization is accomplished by titrating against a solution prepared from primary standard grade NaCl.

Inorganic Analysis Complexation titrimetry continues to be listed as a standard method for the determination of hardness, Ca^{2+}, CN^-, and Cl^- in water and wastewater analysis. The evaluation of hardness was described earlier in Method 9.2. The determination of Ca^{2+} is complicated by the presence of Mg^{2+}, which also reacts with EDTA. To prevent an interference from Mg^{2+}, the pH is adjusted to 12–13, precipitating any Mg^{2+} as $Mg(OH)_2$. Titrating with EDTA using murexide or Eriochrome Blue Black R as a visual indicator gives the concentration of Ca^{2+}.

Cyanide is determined at concentrations greater than 1 ppm by making the sample alkaline with NaOH and titrating with a standard solution of $AgNO_3$, forming the soluble $Ag(CN)_2^-$ complex. The end point is determined using *p*-dimethylaminobenzalrhodamine as a visual indicator, with the solution turning from yellow to a salmon color in the presence of excess Ag^+.

Chloride is determined by titrating with $Hg(NO_3)_2$, forming soluble $HgCl_2$. The sample is acidified to within the pH range of 2.3–3.8 where diphenylcarbazone, which forms a colored complex with excess Hg^{2+}, serves as the visual indicator. Xylene cyanol FF is added as a pH indicator to ensure that the pH is within the desired range. The initial solution is a greenish blue, and the titration is carried out to a purple end point.

Quantitative Calculations The stoichiometry of complexation reactions is given by the conservation of electron pairs between the ligand, which is an electron-pair donor, and the metal, which is an electron-pair acceptor (see Section 2C); thus

$$\frac{\text{moles of electron pairs donated}}{\text{mole ligand}} \times \text{moles ligand} = \frac{\text{moles of electron pairs accepted}}{\text{mole metal}} \times \text{moles metal}$$

This is simplified for titrations involving EDTA where the stoichiometry is always 1:1 regardless of how many electron pairs are involved in the formation of the metal–ligand complex.

Example

EXAMPLE 9.8

The concentration of a solution of EDTA was determined by standardizing against a solution of Ca^{2+} prepared from the primary standard $CaCO_3$. A 0.4071-g sample of $CaCO_3$ was transferred to a 500-mL volumetric flask, dissolved using a minimum of 6 M HCl, and diluted to volume. A 50.00-mL portion of this solution was transferred into a 250-mL Erlenmeyer flask and the pH adjusted by adding 5 mL of a pH 10 NH_3–NH_4Cl buffer containing a small amount of Mg^{2+}–EDTA. After adding calmagite as a visual indicator, the solution was titrated with the EDTA, requiring 42.63 mL to reach the end point. Report the molar concentration of the titrant.

SOLUTION

Conservation of electron pairs for the titration reaction requires that

$$\text{Moles EDTA} = \text{moles Ca}^{2+}$$

Making appropriate substitution for the moles of EDTA and Ca^{2+} gives

$$M_{\text{EDTA}} \times V_{\text{EDTA}} = M_{\text{Ca}} \times V_{\text{Ca}}$$

The molarity of the Ca^{2+} solution is

$$\frac{\text{moles CaCO}_3}{V_{\text{flask}}} = \frac{\text{g CaCO}_3}{\text{FW CaCO}_3 \times V_{\text{flask}}} = \frac{0.4071\text{ g}}{100.09\text{ g/mol} \times 0.5000\text{ L}}$$
$$= 8.135 \times 10^{-3}\text{ M Ca}^{2+}$$

Substituting known values and solving for M_{EDTA} gives

$$\frac{M_{\text{Ca}}V_{\text{Ca}}}{V_{\text{EDTA}}} = \frac{(8.135 \times 10^{-3}\text{ M})(50.00\text{ mL})}{42.63\text{ mL}} = 9.541 \times 10^{-3}\text{ M EDTA}$$

The principle of the conservation of electron pairs is easily extended to other complexation reactions, as shown in the following example.

Example

EXAMPLE 9.9

The concentration of Cl^- in a 100.0-mL sample of water drawn from a fresh water acquifer suffering from encroachment of sea water, was determined by titrating with 0.0516 M $Hg(NO_3)_2$. The sample was acidified and titrated to the diphenylcarbazone end point, requiring 6.18 mL of the titrant. Report the concentration of Cl^- in parts per million.

SOLUTION

Conservation of electron pairs requires that

$$\text{Moles Cl}^- = 2 \times \text{moles Hg}^{2+}$$

Making appropriate substitutions for the moles of Cl^- and Hg^{2+}

$$\frac{\text{g Cl}^-}{\text{AW Cl}} = 2 \times M_{\text{Hg}} \times V_{\text{Hg}}$$

and rearranging leaves us with

$$\text{g Cl}^- = 2 \times M_{\text{Hg}} \times V_{\text{Hg}} \times \text{AW Cl}$$

Substituting known values and solving gives

$$2 \times 0.0516 \text{ M} \times 0.00618 \text{ L} \times 35.453 \text{ g/mol} = 0.0226 \text{ g Cl}^-$$

The concentration of Cl^- in parts per million, therefore, is

$$\frac{\text{mg Cl}^-}{\text{liter}} = \frac{22.6 \text{ mg}}{0.1000 \text{ L}} = 226 \text{ ppm}$$

Finally, quantitative problems involving multiple analytes and back titrations also can be solved by applying the principle of conservation of electron pairs.

Example

EXAMPLE 9.10

An alloy of chromel containing Ni, Fe, and Cr was analyzed by a complexation titration using EDTA as the titrant. A 0.7176-g sample of the alloy was dissolved in HNO_3 and diluted to 250 mL in a volumetric flask. A 50.00-mL aliquot of the sample, treated with pyrophosphate to mask the Fe and Cr, required 26.14 mL of 0.05831 M EDTA to reach the murexide end point. A second 50.00-mL aliquot was treated with hexamethylenetetramine to mask the Cr. Titrating with 0.05831 M EDTA required 35.43 mL to reach the murexide end point. Finally, a third 50.00-mL aliquot was treated with 50.00 mL of 0.05831 M EDTA, and back titrated to the murexide end point with 6.21 mL of 0.06316 M Cu^{2+}. Report the weight percents of Ni, Fe, and Cr in the alloy.

SOLUTION

Conservation of electron pairs for the three titrations requires that for

Titration 1: moles Ni = moles EDTA1 (Fe, Cr masked)

Titration 2: moles Ni + moles Fe = moles EDTA2 (Cr masked)

Titration 3: moles Ni + moles Fe + moles Cr + moles Cu = moles EDTA3

Note that the third titration is a back titration. Titration 1 can be used to determine the amount of Ni in the alloy. Once the amount of Ni is known, the amount of Fe can be determined from the results for titration 2. Finally, titration 3 can be solved for the amount of Cr.

Titration 1

$$\frac{\text{g Ni}}{\text{AW Ni}} = M_{\text{EDTA1}} \times V_{\text{EDTA1}}$$

$$\text{g Ni} = M_{\text{EDTA1}} \times V_{\text{EDTA1}} \times \text{AW Ni}$$

$$0.05831\ \text{M} \times 0.02614\ \text{L} \times 58.69\ \text{g / mol} = 0.08946\ \text{g Ni}$$

Titration 2

$$\text{Moles EDTA1} + \text{moles Fe} = \text{moles EDTA2}$$

$$\frac{\text{g Fe}}{\text{AW Fe}} = \text{moles EDTA2} - \text{moles EDTA1}$$

$$\text{g Fe} = (M_{\text{EDTA2}} \times V_{\text{EDTA2}} - M_{\text{EDTA1}} \times V_{\text{EDTA1}}) \times \text{AW Fe}$$

$$(0.05831\ \text{M} \times 0.03543\ \text{L} - 0.05831\ \text{M} \times 0.02614\ \text{L}) \times 55.847\ \text{g/mol} = 0.03025\ \text{g Fe}$$

Titration 3

$$\text{Moles EDTA2} + \text{moles Cr} + \text{moles Cu} = \text{moles EDTA3}$$

$$\frac{\text{g Cr}}{\text{AW Cr}} = \text{moles EDTA3} - \text{moles EDTA2} - \text{moles Cu}$$

$$\text{g Cr} = (M_{\text{EDTA3}} \times V_{\text{EDTA3}} - M_{\text{EDTA2}} \times V_{\text{EDTA2}} - M_{\text{Cu}} \times V_{\text{Cu}}) \times \text{AW Cr}$$

$$(0.05831\ \text{M} \times 0.05000\ \text{L} - 0.05831\ \text{M} \times 0.03543\ \text{L} - 0.06316\ \text{M} \times 0.00621\ \text{L}) \times 51.996\ \text{g/mol} = 0.02378\ \text{g Cr}$$

Each of these titrations was conducted on a 50.00-mL aliquot of the original 250.0-mL sample. The mass of each analyte, therefore, must be corrected by multiplying by a factor of 5. Thus, the grams of Ni, Fe, and Cr in the original sample are

$$0.08946 \text{ g} \times 5 = 0.4473 \text{ g Ni}$$

$$0.03025 \text{ g} \times 5 = 0.1513 \text{ g Fe}$$

$$0.02378 \text{ g} \times 5 = 0.1189 \text{ g Cr}$$

and the %w/w for each metal is

$$\frac{0.4473 \text{ g}}{0.7176 \text{ g}} \times 100 = 62.33\% \text{ w/w Ni}$$

$$\frac{0.1513 \text{ g}}{0.7176 \text{ g}} \times 100 = 21.08\% \text{ w/w Fe}$$

$$\frac{0.1189 \text{ g}}{0.7176 \text{ g}} \times 100 = 16.57\% \text{ w/w Cr}$$

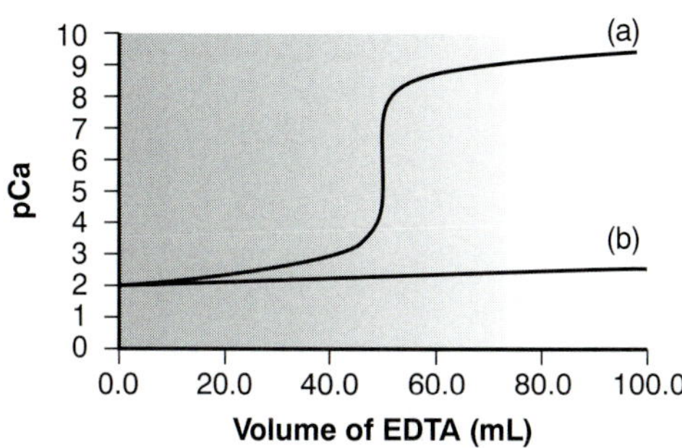

Figure 9.32
Titration curve for 10^{-2} M Ca^{2+} with 10^{-2} M EDTA at (a) pH = 9 and (b) pH = 3.

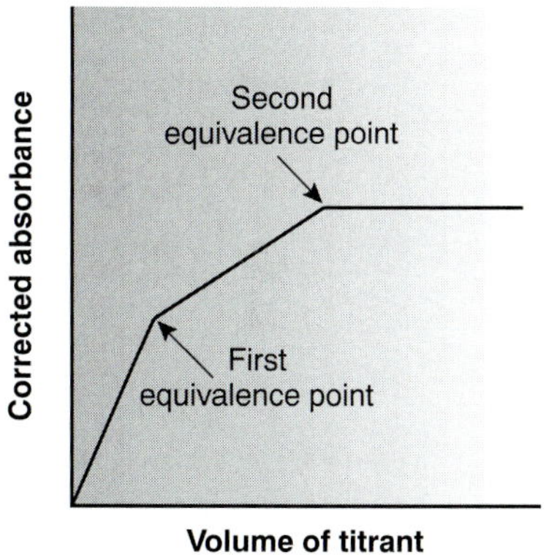

9C.6 Evaluation of Complexation Titrimetry

The scale of operations, accuracy, precision, sensitivity, time, and cost of methods involving complexation titrations are similar to those described earlier for acid–base titrimetric methods. Compared with acid–base titrations, however, complexation titrations are more selective. Despite the ability of EDTA to form strong complexes with virtually all metal ions, carefully controlling the pH at which the titration is carried out makes it possible to analyze samples containing two or more analytes (see Example 9.10). The reason that pH can be used to provide selectivity is easily appreciated by examining Figure 9.32. A titration of Ca^{2+} at a pH of 9 gives a distinct break in the titration curve because the conditional formation constant ($K_f' = 2.6 \times 10^9$) is large enough to ensure that the reaction of Ca^{2+} and EDTA goes to completion. At a pH of 3, however, the conditional formation constant ($K_f' = 1.23$) is so small that very little Ca^{2+} reacts with the EDTA.

Spectrophotometric titrations are particularly useful for the analysis of mixtures if a suitable difference in absorbance exists between the analytes and products, or titrant. For example, the analysis of a two-component mixture can be accomplished if there is a difference between the absorbance of the two metal–ligand complexes (Figure 9.33).

Corrected absorbance
Second equivalence point
First equivalence point
Volume of titrant

Figure 9.33
Spectrophotometric titration curve for the complexation titration of a mixture.

9D Titrations Based on Redox Reactions

Redox titrations were introduced shortly after the development of acid–base titrimetry. The earliest methods took advantage of the oxidizing power of chlorine. In 1787, Claude Berthollet introduced a method for the quantitative analysis of chlorine water (a mixture of Cl_2, HCl, and HOCl) based on its ability to oxidize solutions of the dye indigo (indigo is colorless in its oxidized state). In 1814, Joseph Louis Gay-Lussac (1778–1850), developed a similar method for chlorine in bleaching powder. In both methods the end point was signaled visually. Before the equivalence point, the solution remains clear due to the oxidation of indigo. After the equivalence point, however, unreacted indigo imparts a permanent color to the solution.

redox titration
A titration in which the reaction between the analyte and titrant is an oxidation/reduction reaction.

The number of redox titrimetric methods increased in the mid-1800s with the introduction of MnO_4^-, $Cr_2O_7^{2-}$ and I_2 as oxidizing titrants, and Fe^{2+} and $S_2O_3^{2-}$ as reducing titrants. Even with the availability of these new titrants, however, the routine application of redox titrimetry to a wide range of samples was limited by the lack of suitable indicators. Titrants whose oxidized and reduced forms differ significantly in color could be used as their own indicator. For example, the intensely purple MnO_4^- ion serves as its own indicator since its reduced form, Mn^{2+}, is almost colorless. The utility of other titrants, however, required a visual indicator that could be added to the solution. The first such indicator was diphenylamine, which was introduced in the 1920s. Other redox indicators soon followed, increasing the applicability of redox titrimetry.

9D.1 Redox Titration Curves

To evaluate a redox titration we must know the shape of its titration curve. In an acid–base titration or a complexation titration, a titration curve shows the change in concentration of H_3O^+ (as pH) or M^{n+} (as pM) as a function of the volume of titrant. For a redox titration, it is convenient to monitor electrochemical potential.

You will recall from Chapter 6 that the Nernst equation relates the electrochemical potential to the concentrations of reactants and products participating in a redox reaction. Consider, for example, a titration in which the analyte in a reduced state, A_{red}, is titrated with a titrant in an oxidized state, T_{ox}. The titration reaction is

$$A_{red} + T_{ox} \rightleftharpoons T_{red} + A_{ox}$$

The electrochemical potential for the reaction is the difference between the reduction potentials for the reduction and oxidation half-reactions; thus,

$$E_{rxn} = E_{T_{ox}/T_{red}} - E_{A_{ox}/A_{red}}$$

After each addition of titrant, the reaction between the analyte and titrant reaches a state of equilibrium. The reaction's electrochemical potential, E_{rxn}, therefore, is zero, and

$$E_{T_{ox}/T_{red}} = E_{A_{ox}/A_{red}}$$

Consequently, the potential for either half-reaction may be used to monitor the titration's progress.

Before the equivalence point the titration mixture consists of appreciable quantities of both the oxidized and reduced forms of the analyte, but very little unreacted titrant. The potential, therefore, is best calculated using the Nernst equation for the analyte's half-reaction

$$E_{A_{ox}/A_{red}} = E^{\circ}_{A_{ox}A_{red}} - \frac{RT}{nF}\ln\frac{[A_{red}]}{[A_{ox}]}$$

formal potential
The potential of a redox reaction for a specific set of solution conditions, such as pH and ionic composition.

Although $E^{\circ}_{A_{ox}/A_{red}}$ is the standard-state potential for the analyte's half-reaction, a matrix-dependent **formal potential** is used in its place. After the equivalence point, the potential is easiest to calculate using the Nernst equation for the titrant's half-reaction, since significant quantities of its oxidized and reduced forms are present.

$$E_{T_{ox}/T_{red}} = E^{\circ}_{T_{ox}T_{red}} - \frac{RT}{nF}\ln\frac{[T_{red}]}{[T_{ox}]}$$

Calculating the Titration Curve As an example, let's calculate the titration curve for the titration of 50.0 mL of 0.100 M Fe^{2+} with 0.100 M Ce^{4+} in a matrix of 1 M $HClO_4$. The reaction in this case is

$$Fe^{2+}(aq) + Ce^{4+}(aq) \rightleftharpoons Ce^{3+}(aq) + Fe^{3+}(aq) \qquad \textbf{9.16}$$

The equilibrium constant for this reaction is quite large (it is approximately 6×10^{15}), so we may assume that the analyte and titrant react completely.

The first task is to calculate the volume of Ce^{4+} needed to reach the equivalence point. From the stoichiometry of the reaction we know

$$\text{Moles } Fe^{2+} = \text{moles } Ce^{4+}$$

or

$$M_{Fe}V_{Fe} = M_{Ce}V_{Ce}$$

Solving for the volume of Ce^{4+}

$$V_{Ce} = \frac{M_{Fe}V_{Fe}}{M_{Ce}} = \frac{(0.100 \text{ M})(50.0 \text{ mL})}{(0.100 \text{ M})} = 50.0 \text{ mL}$$

gives the equivalence point volume as 50.0 mL.

Before the equivalence point the concentration of unreacted Fe^{2+} and the concentration of Fe^{3+} produced by reaction 9.16 are easy to calculate. For this reason we find the potential using the Nernst equation for the analyte's half-reaction

$$E = E^{\circ}_{Fe^{3+}/Fe^{2+}} - 0.05916 \ln \frac{[Fe^{2+}]}{[Fe^{3+}]} \qquad \textbf{9.17}$$

The concentrations of Fe^{2+} and Fe^{3+} after adding 5.0 mL of titrant are

$$[Fe^{2+}] = \frac{\text{moles unreacted } Fe^{2+}}{\text{total volume}} = \frac{M_{Fe}V_{Fe} - M_{Ce}V_{Ce}}{V_{Fe} + V_{Ce}}$$

$$= \frac{(0.100 \text{ M})(50.0 \text{ mL}) - (0.100 \text{ M})(5.0 \text{ mL})}{50.0 \text{ mL} + 5.0 \text{ mL}} = 8.18 \times 10^{-2} \text{ M}$$

$$[Fe^{3+}] = \frac{\text{moles } Ce^{4+} \text{ added}}{\text{total volume}} = \frac{M_{Ce}V_{Ce}}{V_{Fe} + V_{Ce}}$$

$$= \frac{(0.100 \text{ M})(5.0 \text{ mL})}{50.0 \text{ ml} + 5.0 \text{ mL}} = 9.09 \times 10^{-3} \text{ M}$$

Substituting these concentrations into equation 9.17 along with the formal potential for the Fe^{3+}/Fe^{2+} half-reaction from Appendix 3D, we find that the potential is

$$E = +0.767 \text{ V} - 0.05916 \log\left(\frac{8.18 \times 10^{-2}}{9.09 \times 10^{-3}}\right) = +0.711 \text{ V}$$

At the equivalence point, the moles of Fe^{2+} initially present and the moles of Ce^{4+} added are equal. Because the equilibrium constant for reaction 9.16 is large, the concentrations of Fe^{2+} and Ce^{4+} are exceedingly small and difficult to calculate without resorting to a complex equilibrium problem. Consequently, we cannot calculate the potential at the equivalence point, E_{eq}, using just the Nernst equation for the analyte's half-reaction or the titrant's half-reaction. We can, however, calculate

E_{eq} by combining the two Nernst equations. To do so we recognize that the potentials for the two half-reactions are the same; thus,

$$E_{eq} = E^{\circ}_{Fe^{3+}/Fe^{2+}} - 0.05916 \log \frac{[Fe^{2+}]}{[Fe^{3+}]}$$

$$E_{eq} = E^{\circ}_{Ce^{4+}/Ce^{3+}} - 0.05916 \log \frac{[Ce^{3+}]}{[Ce^{4+}]}$$

Adding together these two Nernst equations leaves us with

$$2E_{eq} = E^{\circ}_{Fe^{3+}/Fe^{2+}} + E^{\circ}_{Ce^{4+}/Ce^{3+}} - 0.05916 \log \frac{[Fe^{2+}][Ce^{3+}]}{[Fe^{3+}][Ce^{4+}]} \qquad 9.18$$

At the equivalence point, the titration reaction's stoichiometry requires that

$$[Fe^{2+}] = [Ce^{4+}]$$

$$[Fe^{3+}] = [Ce^{3+}]$$

The ratio in the log term of equation 9.18, therefore, equals one and the log term is zero. Equation 9.18 simplifies to

$$E_{eq} = \frac{E^{\circ}_{Fe^{3+}/Fe^{2+}} + E^{\circ}_{Ce^{4+}/Ce^{3+}}}{2} = \frac{0.767\text{ V} + 1.70\text{ V}}{2} = 1.23\text{ V}$$

After the equivalence point, the concentrations of Ce^{3+} and excess Ce^{4+} are easy to calculate. The potential, therefore, is best calculated using the Nernst equation for the titrant's half-reaction.

$$E = E^{\circ}_{Ce^{4+}/Ce^{3+}} - 0.05916 \log \frac{[Ce^{3+}]}{[Ce^{4+}]} \qquad 9.19$$

For example, after adding 60.0 mL of titrant, the concentrations of Ce^{3+} and Ce^{4+} are

$$[Ce^{3+}] = \frac{\text{intial moles Fe}^{2+}}{\text{total volume}} = \frac{M_{Fe}V_{Fe}}{V_{Fe} + V_{Ce}}$$

$$= \frac{(0.100\text{ M})(50.0\text{ mL})}{50.0\text{ mL} + 60.0\text{ mL}} = 4.55 \times 10^{-2}\text{ M}$$

$$[Ce^{4+}] = \frac{\text{moles excess Ce}^{4+}}{\text{total volume}} = \frac{M_{Ce}V_{Ce} - M_{Fe}V_{Fe}}{V_{Fe} + V_{Ce}}$$

$$= \frac{(0.100\text{ M})(60.0\text{ mL}) - (0.100\text{ M})(50.0\text{ mL})}{50.0\text{ mL} + 60.0\text{ mL}} = 9.09 \times 10^{-3}\text{ M}$$

Substituting these concentrations into equation 9.19 gives the potential as

$$E = +1.70\text{ V} - 0.05916 \log \frac{4.55 \times 10^{-2}}{9.09 \times 10^{-3}} = 1.66\text{ V}$$

Additional results for this titration curve are shown in Table 9.17 and Figure 9.34.

Table 9.17 Data for Titration of 50.0 mL of 0.100 M Fe^{2+} with 0.100 M Ce^{4+}

Volume Ce^{4+} (mL)	*E* (V)	Volume Ce^{4+} (mL)	*E* (V)
5.00	0.711	55.00	1.64
10.00	0.731	60.00	1.66
15.00	0.745	65.00	1.67
20.00	0.757	70.00	1.68
25.00	0.767	75.00	1.68
30.00	0.777	80.00	1.69
35.00	0.789	85.00	1.69
40.00	0.803	90.00	1.69
45.00	0.823	95.00	1.70
50.00	1.23	100.00	1.70

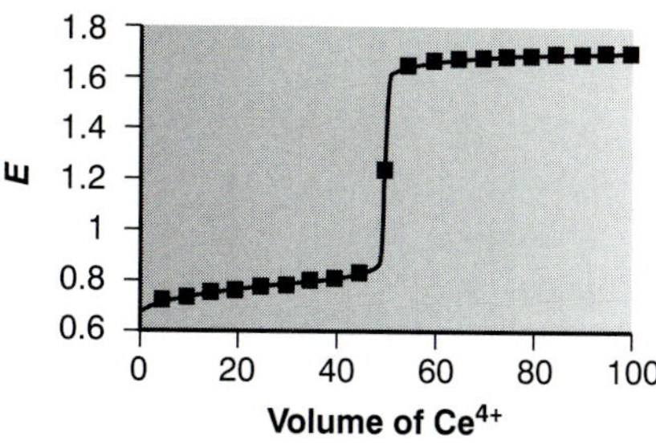

Figure 9.34
Redox titration curve for 50.0 mL of 0.100 M Fe^{2+} with 0.100 M Ce^{4+} in 1 M $HClO_4$.

Sketching a Redox Titration Curve As we have done for acid–base and complexometric titrations, we now show how to quickly sketch a redox titration curve using a minimum number of calculations.

Example

EXAMPLE 9.11

Sketch a titration curve for the titration of 50.0 mL of 0.100 M Fe^{2+} with 0.100 M Ce^{4+} in a matrix of 1 M $HClO_4$. This is the same titration for which we previously calculated the titration curve (Table 9.17 and Figure 9.34).

SOLUTION

We begin by drawing axes for the titration curve (Figure 9.35a). Having shown that the equivalence point volume is 50.0 mL, we draw a vertical line intersecting the *x*-axis at this volume (Figure 9.35b).

Before the equivalence point, the solution's electrochemical potential is determined by the concentration of excess Fe^{2+} and the concentration of Fe^{3+} produced by the titration reaction. Using values from Table 9.17, we plot *E* for 5.0 mL and 45.0 mL of titrant (Figure 9.35c).

After the equivalence point, the solution's electrochemical potential is determined by the concentration of excess Ce^{4+} and the concentration of Ce^{3+}. Using values from Table 9.17, we plot points for 60.0 mL and 80.0 mL of titrant (Figure 9.35d).

To complete an approximate sketch of the titration curve, we draw separate straight lines through the two points before and after the equivalence point (Figure 9.35e). Finally, a smooth curve is drawn to connect the three straight-line segments (Figure 9.35f).

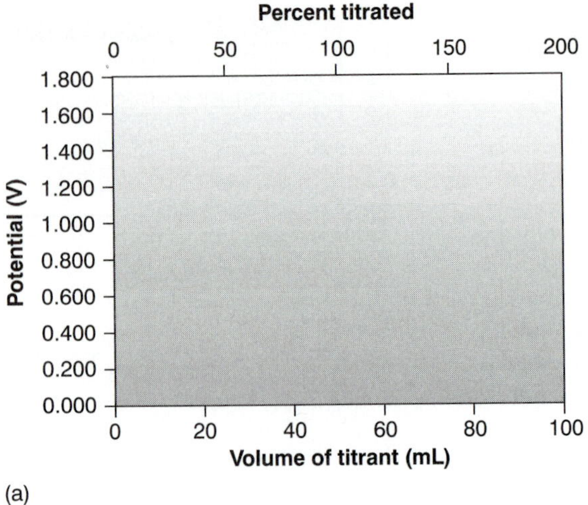

(a)

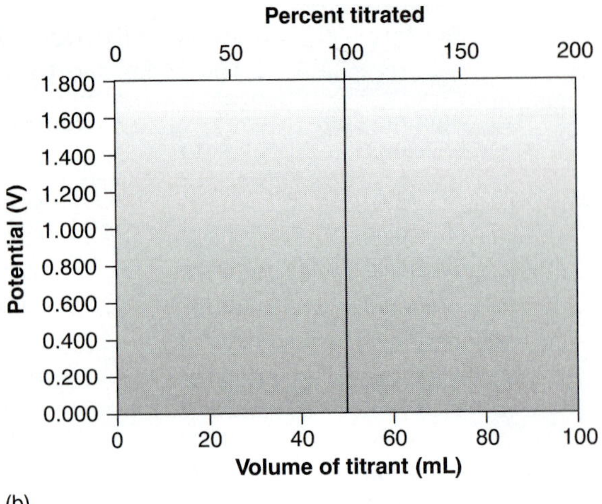

(b)

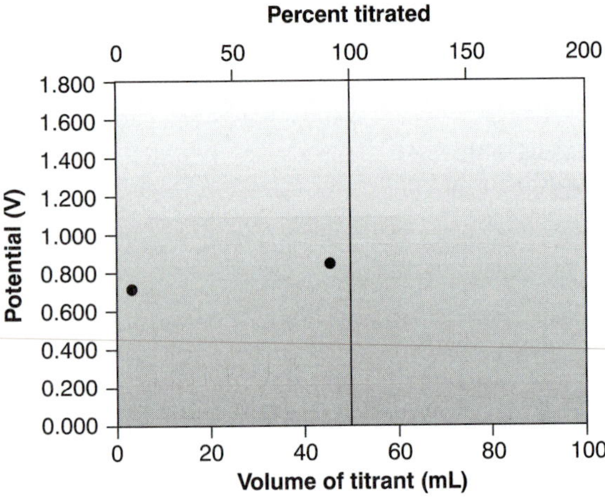

(c)

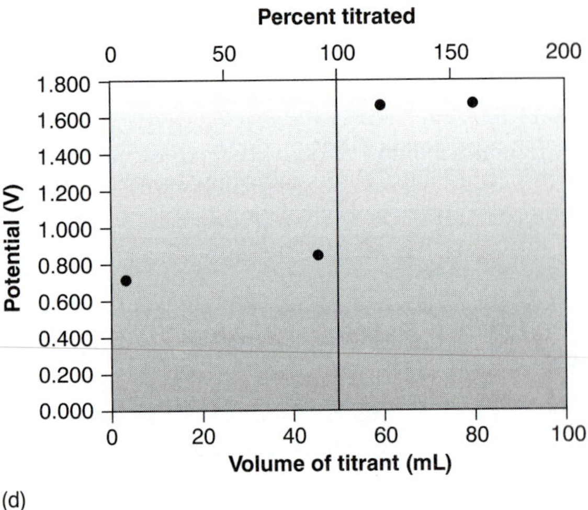

(d)

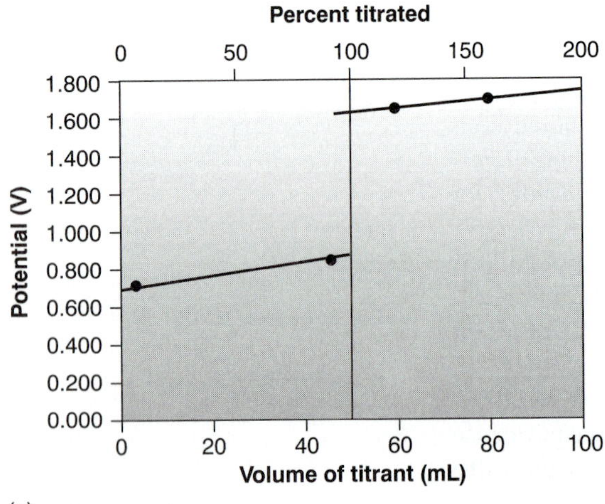

(e)

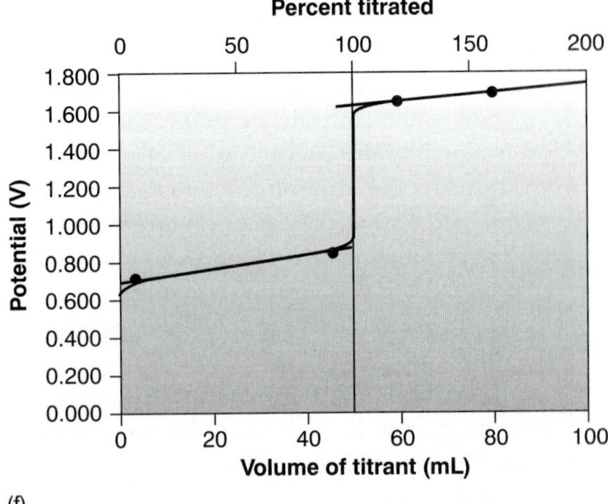

(f)

Figure 9.35
How to sketch a redox titration curve; see text for explanation.

9D.2 Selecting and Evaluating the End Point

The equivalence point of a redox titration occurs when stoichiometrically equivalent amounts of analyte and titrant react. As with other titrations, any difference between the equivalence point and the end point is a determinate source of error.

Where Is the Equivalence Point? In discussing acid–base titrations and complexometric titrations, we noted that the equivalence point is almost identical with the inflection point located in the sharply rising part of the titration curve. If you look back at Figures 9.8 and 9.28, you will see that for acid–base and complexometric titrations the inflection point is also in the middle of the titration curve's sharp rise (we call this a symmetrical equivalence point). This makes it relatively easy to find the equivalence point when you sketch these titration curves. When the stoichiometry of a redox titration is symmetrical (one mole analyte per mole of titrant), then the equivalence point also is symmetrical. If the stoichiometry is not symmetrical, then the equivalence point will lie closer to the top or bottom of the titration curve's sharp rise. In this case the equivalence point is said to be asymmetrical. Example 9.12 shows how to calculate the equivalence point potential in this situation.

Example

EXAMPLE 9.12

Derive a general equation for the electrochemical potential at the equivalence point for the titration of Fe^{2+} with MnO_4^-; the reaction is

$$5Fe^{2+}(aq) + MnO_4^-(aq) + 8H_3O^+(aq) \rightleftharpoons 5Fe^{3+}(aq) + Mn^{2+}(aq) + 12H_2O(\ell)$$

SOLUTION

The redox half-reactions for the analyte and the titrant are

$$Fe^{2+}(aq) \rightleftharpoons Fe^{3+}(aq) + e^-$$

$$MnO_4^-(aq) + 8H_3O^+(aq) + 5e^- \rightleftharpoons Mn^{2+}(aq) + 12H_2O(\ell)$$

for which the Nernst equations are

$$E_{eq} = E^\circ_{Fe^{3+}/Fe^{2+}} - 0.05916 \log \frac{[Fe^{2+}]}{[Fe^{3+}]}$$

$$E_{eq} = E^\circ_{MnO_4^-/Mn^{2+}} - \frac{0.05916}{5} \log \frac{[Mn^{2+}]}{[MnO_4^-][H_3O^+]^8}$$

Before adding together these two equations, the second equation must be multiplied by 5 so that the log terms can be combined; thus

$$6E_{eq} = E^\circ_{Fe^{3+}/Fe^{2+}} + 5E^\circ_{MnO_4^-/Mn^{2+}} - 0.05916 \log \frac{[Fe^{2+}][Mn^{2+}]}{[Fe^{3+}][MnO_4^-][H_3O^+]^8}$$

At the equivalence point, we know that

$$[Fe^{2+}] = 5 \times [MnO_4^-]$$

$$[Fe^{3+}] = 5 \times [Mn^{2+}]$$

Substituting these equalities into the equation for E_{eq} and rearranging gives

$$E_{eq} = \frac{E^{\circ}_{Fe^{3+}/Fe^{2+}} + 5E^{\circ}_{MnO_4^-/Mn^{2+}}}{6} - \frac{0.05916}{6}\log\frac{5[MnO_4^-][Mn^{2+}]}{5[Mn^{2+}][MnO_4^-][H_3O^+]^8}$$

$$= \frac{E^{\circ}_{Fe^{3+}/Fe^{2+}} + 5E^{\circ}_{MnO_4^-/Mn^{2+}}}{6} - \frac{0.05916}{6}\log\frac{1}{[H_3O^+]^8}$$

$$= \frac{E^{\circ}_{Fe^{3+}/Fe^{2+}} + 5E^{\circ}_{MnO_4^-/Mn^{2+}}}{6} + \frac{(0.05916)(8)}{6}\log[H_3O^+]$$

$$= \frac{E^{\circ}_{Fe^{3+}/Fe^{2+}} + 5E^{\circ}_{MnO_4^-/Mn^{2+}}}{6} - 0.07888\text{pH}$$

For this titration the electrochemical potential at the equivalence point consists of two terms. The first term is a weighted average of the standard state or formal potentials for the analyte and titrant, in which the weighting factors are the number of electrons in their respective redox half-reactions. The second term shows that E_{eq} is pH-dependent. Figure 9.36 shows a typical titration curve for the analysis of Fe^{2+} by titration with MnO_4^-, showing the asymmetrical equivalence point. Note that the change in potential near the equivalence point is sharp enough that selecting an end point near the middle of the titration curve's sharply rising portion does not introduce a significant titration error.

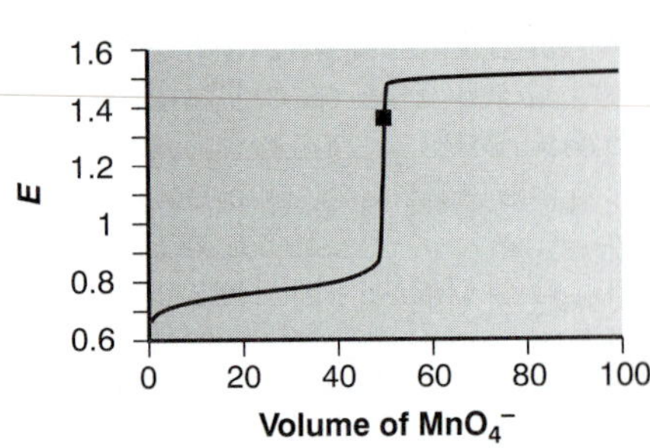

Figure 9.36
Titration curve for Fe^{2+} with MnO_4^- in 1 M H_2SO_4; equivalence point is shown by the symbol ■.

Finding the End Point with a Visual Indicator Three types of visual indicators are used to signal the end point in a redox titration. A few titrants, such as MnO_4^-, have oxidized and reduced forms whose colors in solution are significantly different. Solutions of MnO_4^- are intensely purple. In acidic solutions, however, permanganate's reduced form, Mn^{2+}, is nearly colorless. When MnO_4^- is used as an oxidizing titrant, the solution remains colorless until the first drop of excess MnO_4^- is added. The first permanent tinge of purple signals the end point.

A few substances indicate the presence of a specific oxidized or reduced species. Starch, for example, forms a dark blue complex with I_3^- and can be used to signal the presence of excess I_3^- (color change: colorless to blue), or the completion of a reaction in which I_3^- is consumed (color change: blue to colorless). Another example of a specific indicator is thiocyanate, which forms a soluble red-colored complex, $Fe(SCN)^{2+}$, with Fe^{3+}.

The most important class of redox indicators, however, are substances that do not participate in the redox titration, but whose oxidized and reduced forms differ in color. When added to a solution containing the analyte, the indicator imparts a color that depends on the solution's electrochemical potential. Since the indicator changes color in response to the electrochemical potential, and not to the presence or absence of a specific species, these compounds are called general **redox indicators.**

redox indicator
A visual indicator used to signal the end point in a redox titration.

The relationship between a redox indicator's change in color and the solution's electrochemical potential is easily derived by considering the half-reaction for the indicator

$$\text{In}_{\text{ox}} + ne^- \rightleftharpoons \text{In}_{\text{red}}$$

where In_{ox} and In_{red} are, respectively, the indicator's oxidized and reduced forms. The Nernst equation for this reaction is

$$E = E^\circ_{\text{In}_{\text{ox}}/\text{In}_{\text{red}}} - \frac{0.05916}{n}\log\frac{[\text{In}_{\text{red}}]}{[\text{In}_{\text{ox}}]}$$

If we assume that the indicator's color in solution changes from that of In_{ox} to that of In_{red} when the ratio $[\text{In}_{\text{red}}]/[\text{In}_{\text{ox}}]$ changes from 0.1 to 10, then the end point occurs when the solution's electrochemical potential is within the range

$$E = E^\circ_{\text{In}_{\text{ox}}/\text{In}_{\text{red}}} \pm \frac{0.05916}{n}$$

A partial list of general redox indicators is shown in Table 9.18. Examples of appropriate and inappropriate indicators for the titration of Fe^{2+} with Ce^{4+} are shown in Figure 9.37.

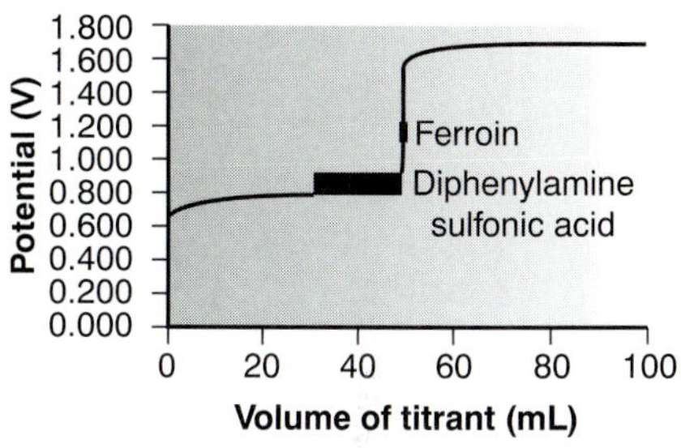

Figure 9.37
Titration curve for 50.00 mL of 0.0500 M Fe^{2+} with 0.0500 M Ce^{4+} showing the range of *E* and volume of titrant over which the indicators ferroin and diphenylamine sulfonic acid are expected to change color.

Finding the End Point Potentiometrically Another method for locating the end point of a redox titration is to use an appropriate electrode to monitor the change in electrochemical potential as titrant is added to a solution of analyte. The end point can then be found from a visual inspection of the titration curve. The simplest experimental design (Figure 9.38) consists of a Pt indicator electrode whose potential is governed by the analyte's or titrant's redox half-reaction, and a reference electrode that has a fixed potential. A further discussion of potentiometry is found in Chapter 11.

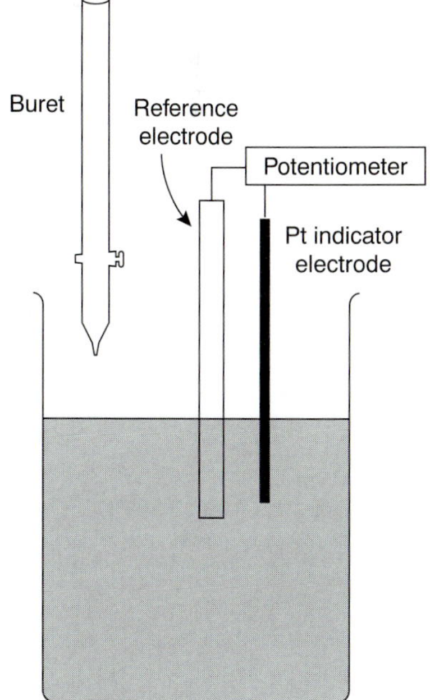

Figure 9.38
Experimental arrangement for recording a potentiometric redox titration curve.

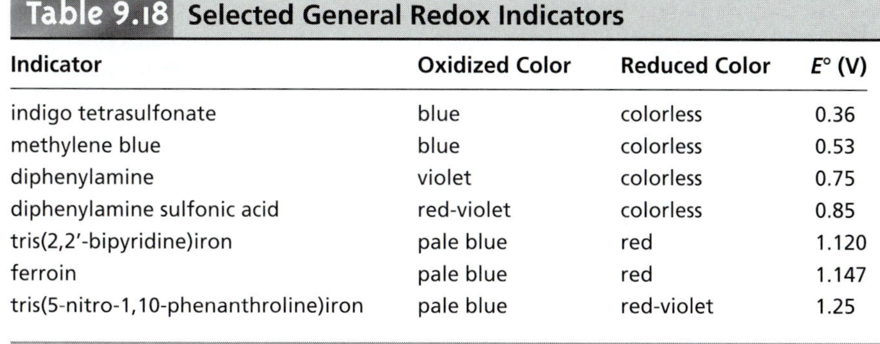

Table 9.18 Selected General Redox Indicators

Indicator	Oxidized Color	Reduced Color	*E*° (V)
indigo tetrasulfonate	blue	colorless	0.36
methylene blue	blue	colorless	0.53
diphenylamine	violet	colorless	0.75
diphenylamine sulfonic acid	red-violet	colorless	0.85
tris(2,2′-bipyridine)iron	pale blue	red	1.120
ferroin	pale blue	red	1.147
tris(5-nitro-1,10-phenanthroline)iron	pale blue	red-violet	1.25

9D.3 Representative Method

Although every redox titrimetric method has its own unique considerations, the following description of the determination of total residual chlorine in water provides an instructive example of a typical procedure.

Representative Methods

Method 9.3 Determination of Total Chlorine Residual[13]

Description of the Method. The chlorination of public water supplies results in the formation of several chlorine-containing species, the combined concentration of which is called the total chlorine residual. Chlorine may be present in a variety of states including free residual chlorine, consisting of Cl_2, HOCl, and OCl^-, and combined chlorine residual, consisting of NH_2Cl, $NHCl_2$, and NCl_3. The total chlorine residual is determined by using the oxidizing power of chlorine to convert I^- to I_3^-. The amount of I_3^- formed is then determined by a redox titration using $S_2O_3^{2-}$ as a titrant and starch as an indicator. Regardless of its form, the total chlorine residual is calculated as if all the chlorine were available as Cl_2, and is reported as parts per million of Cl.

The photo in Colorplate 8c shows the indicator's color change for this titration.

Procedure. Select a volume of sample requiring less than 20 mL of $S_2O_3^{2-}$ to reach the end point. Using glacial acetic acid, acidify the sample to a pH in the range of 3 to 4, and add about 1 g of KI. Titrate with $Na_2S_2O_3$ until the yellow color due to I_3^- begins to disappear. Add 1 mL of a starch indicator solution, and continue titrating until the blue color of the starch–I_3^- complex disappears. The volume of titrant needed to reach the end point should be corrected for reagent impurities by conducting a blank titration.

Questions

1. **Is this an example of a direct or an indirect analysis?**

   This is an indirect method of analysis because the chlorine-containing species do not react with the titrant. Instead the total chlorine residual oxidizes I^- to I_3^-, and the amount of I_3^- is determined by the redox titration with $Na_2S_2O_3$.

2. **Why is the procedure not carried out directly using KI as a titrant?**

   The redox half-reaction when I^- is used as a titrant is

   $$3I^- \rightleftharpoons I_3^- + 2e^-$$

   Because the product of the reaction, I_3^-, is itself colored, the color of the solution containing the analyte changes with each addition of titrant. For this reason it is difficult to find a suitable visual indicator for the titration's end point.

3. **Both oxidizing and reducing agents can interfere with this analysis. Explain what effect each of these interferents will have on the result of an analysis.**

   An interferent that is an oxidizing agent will convert additional I^- to I_3^-. This extra I_3^- requires an additional volume of $S_2O_3^{2-}$ to reach the end point, overestimating the total chlorine residual. If the interferent is a reducing agent, it reduces some of the I_3^-, produced by the reaction between the total chlorine residual and iodide, back to I^-. The result is an underestimation of the total chlorine residual.

9D.4 Quantitative Applications

As with acid–base and complexation titrations, redox titrations are not frequently used in modern analytical laboratories. Nevertheless, several important applications continue to find favor in environmental, pharmaceutical, and industrial laboratories. In this section we review the general application of redox titrimetry. We begin, however, with a brief discussion of selecting and characterizing redox titrants, and methods for controlling the analyte's oxidation state.

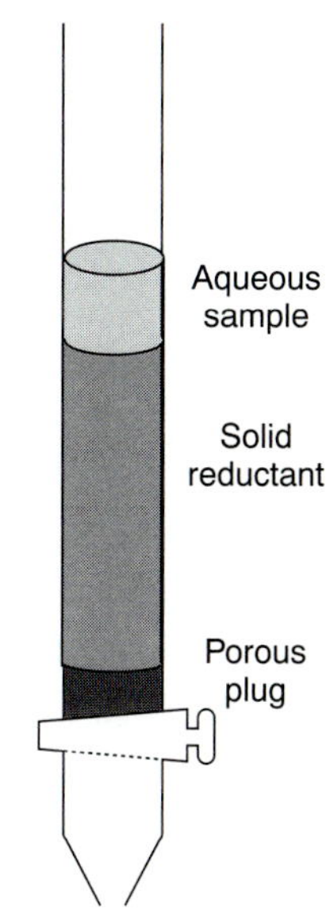

Figure 9.39
Schematic diagram of a reductor column.

Adjusting the Analyte's Oxidation State If a redox titration is to be used in a quantitative analysis, the analyte must initially be present in a single oxidation state. For example, the iron content of a sample can be determined by a redox titration in which Ce^{4+} oxidizes Fe^{2+} to Fe^{3+}. The process of preparing the sample for analysis must ensure that all iron is present as Fe^{2+}. Depending on the sample and the method of sample preparation, however, the iron may initially be present in both the +2 and +3 oxidation states. Before titrating, any Fe^{3+} that is present must be reduced to Fe^{2+}. This type of pretreatment can be accomplished with an auxiliary reducing or oxidizing agent.

Metals that are easily oxidized, such as Zn, Al, and Ag, can serve as **auxiliary reducing agents.** The metal, as a coiled wire or powder, is placed directly in the solution where it reduces the analyte. Of course any unreacted auxiliary reducing agent will interfere with the analysis by reacting with the titrant. The residual auxiliary reducing agent, therefore, must be removed once the analyte is completely reduced. This can be accomplished by simply removing the coiled wire or by filtering.

auxiliary reducing agent
A reagent used to reduce the analyte before its analysis by a redox titration.

An alternative approach to using an auxiliary reducing agent is to immobilize it in a column. To prepare a reduction column, an aqueous slurry of the finely divided metal is packed in a glass tube equipped with a porous plug at the bottom (Figure 9.39). The sample is placed at the top of the column and moves through the column under the influence of gravity or vacuum suction. The length of the reduction column and the flow rate are selected to ensure the analyte's complete reduction.

Two common reduction columns are used. In the **Jones reductor** the column is filled with amalgamated Zn prepared by briefly placing Zn granules in a solution of $HgCl_2$ to form Zn(Hg). Oxidation of the amalgamated Zn

Jones reductor
A reduction column using a Zn amalgam as a reducing agent.

$$Zn(Hg)(s) \rightleftharpoons Zn^{2+}(aq) + Hg(\ell) + 2e^-$$

provides the electrons for reducing the analyte. In the **Walden reductor** the column is filled with granular Ag metal. The solution containing the analyte is acidified with HCl and passed through the column where the oxidation of Ag

Walden reductor
A reduction column using granular Ag as a reducing agent.

$$Ag(s) + Cl^-(aq) \rightleftharpoons AgCl(s) + e^-$$

provides the necessary electrons for reducing the analyte. Examples of both reduction columns are shown in Table 9.19.

Several reagents are commonly used as **auxiliary oxidizing agents,** including ammonium peroxydisulfate, $(NH_4)_2S_2O_8$, and hydrogen peroxide, H_2O_2. Ammonium peroxydisulfate is a powerful oxidizing agent

auxiliary oxidizing agent
A reagent used to oxidize the analyte before its analysis by a redox titration.

$$S_2O_8^{2-}(aq) + 2e^- \rightleftharpoons 2SO_4^{2-}(aq)$$

capable of oxidizing Mn^{2+} to MnO_4^-, Cr^{3+} to $Cr_2O_7^{2-}$, and Ce^{3+} to Ce^{4+}. Excess peroxydisulfate is easily destroyed by briefly boiling the solution. The reduction of hydrogen peroxide in acidic solution

$$H_2O_2(aq) + 2H_3O^+(aq) + 2e^- \rightleftharpoons 4H_2O(\ell)$$

Table 9.19 Selected Reductions Using Metal Reductor Columns

Oxidized Species	Walden Reductor	Jones Reductor
Cr^{3+}	—[a]	$Cr^{3+} + e^- \rightarrow Cr^{2+}$
Cu^{2+}	$Cu^{2+} + e^- \rightarrow Cu^+$	$Cu^{2+} + 2e^- \rightarrow Cu$
Fe^{3+}	$Fe^{3+} + e^- \rightarrow Fe^{2+}$	$Fe^{3+} + e^- \rightarrow Fe^{2+}$
TiO^{2+}	—[a]	$TiO^{2+} + 2H_3O^+ + e^- \rightarrow Ti^{3+} + 3H_2O$
MoO_2^{2+}	$MoO_2^{2+} + e^- \rightarrow MoO_2^+$	$MoO_2^{2+} + 4H_3O^+ + 3e^- \rightarrow Mo^{3+} + 6H_2O$
VO_2^+	$VO_2^+ + 2H_3O^+ + e^- \rightarrow VO^{2+} + 3H_2O$	$VO_2^+ + 4H_3O^+ + 3e^- \rightarrow V^{2+} + 6H_2O$

[a]No reaction.

provides another method for oxidizing an analyte. Excess H_2O_2 also can be destroyed by briefly boiling the solution.

Selecting and Standardizing a Titrant In quantitative work the titrant's concentration must remain stable during the analysis. Since titrants in a reduced state are susceptible to air oxidation, most redox titrations are carried out using an oxidizing agent as the titrant. The choice of which of several common oxidizing titrants is best for a particular analysis depends on the ease with which the analyte can be oxidized. Analytes that are strong reducing agents can be successfully titrated with a relatively weak oxidizing titrant, whereas a strong oxidizing titrant is required for the analysis of analytes that are weak reducing agents.

The two strongest oxidizing titrants are MnO_4^- and Ce^{4+}, for which the reduction half-reactions are

$$MnO_4^-(aq) + 8H_3O^+(aq) + 5e^- \rightleftharpoons Mn^{2+}(aq) + 12H_2O(\ell)$$

$$Ce^{4+}(aq) + e^- \rightleftharpoons Ce^{3+}(aq)$$

Solutions of Ce^{4+} are prepared from the primary standard cerium ammonium nitrate, $Ce(NO_3)_4 \cdot 2NH_4NO_3$, in 1 M H_2SO_4. When prepared from reagent grade materials, such as $Ce(OH)_4$, the solution must be standardized against a primary standard reducing agent such as $Na_2C_2O_4$ or Fe^{2+} (prepared using Fe wire). Ferroin is a suitable indicator when standardizing against Fe^{2+} (Table 9.20). Despite its availability as a primary standard and its ease of preparation, Ce^{4+} is not as frequently used as MnO_4^- because of its greater expense.

Solutions of MnO_4^- are prepared from $KMnO_4$, which is not available as a primary standard. Aqueous solutions of permanganate are thermodynamically unstable due to its ability to oxidize water.

$$4MnO_4^-(aq) + 2H_2O(\ell) \rightleftharpoons 4MnO_2(s) + 3O_2(g) + 4OH^-(aq)$$

This reaction is catalyzed by the presence of MnO_2, Mn^{2+}, heat, light, and the presence of acids and bases. Moderately stable solutions of permanganate can be prepared by boiling for an hour and filtering through a sintered glass filter to remove any solid MnO_2 that precipitates. Solutions prepared in this fashion are stable for 1–2 weeks, although the standardization should be rechecked periodically. Standardization may be accomplished using the same primary standard reducing agents that are used with Ce^{4+}, using the pink color of MnO_4^- to signal the end point (Table 9.20).

Table 9.20 Standardization Reactions for Selected Redox Titrants

Titration Reaction
$Ce^{4+} + Fe^{2+} \rightarrow Ce^{3+} + Fe^{3+}$
$2Ce^{4+} + H_2C_2O_4 + 2H_2O \rightarrow 2Ce^{3+} + 2CO_2 + 2H_3O^+$
$MnO_4^- + 5Fe^{2+} + 8H_3O^+ \rightarrow Mn^{2+} + 5Fe^{3+} + 12H_2O$
$2MnO_4^- + 5H_2C_2O_4 + 6H_3O^+ \rightarrow 2Mn^{2+} + 10CO_2 + 14H_2O$
$I_3^- + 2S_2O_3^{2-} \rightarrow 3I^- + S_4O_6^{2-}$
$I_2 + 2S_2O_3^{2-} \rightarrow 2I^- + S_4O_6^{2-}$

Potassium dichromate is a relatively strong oxidizing agent whose principal advantages are its availability as a primary standard and the long-term stability of its solutions. It is not, however, as strong an oxidizing agent as MnO_4^- or Ce^{4+}, which prevents its application to the analysis of analytes that are weak reducing agents. Its reduction half-reaction is

$$Cr_2O_7^{2-}(aq) + 14H_3O^+(aq) + 6e^- \rightleftharpoons 2Cr^{3+}(aq) + 21H_2O(\ell)$$

Although solutions of $Cr_2O_7^{2-}$ are orange and those of Cr^{3+} are green, neither color is intense enough to serve as a useful indicator. Diphenylamine sulfonic acid, whose oxidized form is purple and reduced form is colorless, gives a very distinct end point signal with $Cr_2O_7^{2-}$.

Iodine is another commonly encountered oxidizing titrant. In comparison with MnO_4^-, Ce^{4+}, and $Cr_2O_7^{2-}$, it is a weak oxidizing agent and is useful only for the analysis of analytes that are strong reducing agents. This apparent limitation, however, makes I_2 a more selective titrant for the analysis of a strong reducing agent in the presence of weaker reducing agents. The reduction half-reaction for I_2 is

$$I_2(aq) + 2e^- \rightleftharpoons 2I^-(aq)$$

Because of iodine's poor solubility, solutions are prepared by adding an excess of I^-. The complexation reaction

$$I_2(aq) + I^-(aq) \rightleftharpoons I_3^-(aq)$$

increases the solubility of I_2 by forming the more soluble triiodide ion, I_3^-. Even though iodine is present as I_3^- instead of I_2, the number of electrons in the reduction half-reaction is unaffected.

$$I_3^-(aq) + 2e^- \rightleftharpoons 3I^-(aq)$$

Solutions of I_3^- are normally standardized against $Na_2S_2O_3$ (see Table 9.20) using starch as a specific indicator for I_3^-.

Oxidizing titrants such as MnO_4^-, Ce^{4+}, $Cr_2O_7^{2-}$ and I_3^-, are used to titrate analytes that are in a reduced state. When the analyte is in an oxidized state, it can be reduced with an auxiliary reducing agent and titrated with an oxidizing titrant. Alternatively, the analyte can be titrated with a suitable reducing titrant. Iodide is a relatively strong reducing agent that potentially could be used for the analysis of analytes in higher oxidation states. Unfortunately, solutions of I^- cannot be used as a direct titrant because they are subject to the air oxidation of I^- to I_3^-.

$$3I^-(aq) \rightleftharpoons I_3^-(aq) + 2e^-$$

Instead, an excess of KI is added, reducing the analyte and liberating a stoichiometric amount of I_3^-. The amount of I_3^- produced is then determined by a back titration using $Na_2S_2O_3$ as a reducing titrant.

$$2S_2O_3^{2-}(aq) \rightleftharpoons S_4O_6^{2-}(aq) + 2e^-$$

Solutions of $Na_2S_2O_3$ are prepared from the pentahydrate and must be standardized before use. Standardization is accomplished by dissolving a carefully weighed portion of the primary standard KIO_3 in an acidic solution containing an excess of KI. When acidified, the reaction between IO_3^- and I^-

$$IO_3^-(aq) + 8I^-(aq) + 6H_3O^+(aq) \rightleftharpoons 3I_3^-(aq) + 9H_2O(\ell)$$

liberates a stoichiometric amount of I_3^-. Titrating I_3^- using starch as a visual indicator allows the determination of the titrant's concentration.

Although thiosulfate is one of the few reducing titrants not readily oxidized by contact with air, it is subject to a slow decomposition to bisulfite and elemental sulfur. When used over a period of several weeks, a solution of thiosulfate should be restandardized periodically. Several forms of bacteria are able to metabolize thiosulfate, which also can lead to a change in its concentration. This problem can be minimized by adding a preservative such as HgI_2 to the solution.

Another reducing titrant is ferrous ammonium sulfate, $Fe(NH_4)_2(SO_4)_2 \cdot 6H_2O$, in which iron is present in the +2 oxidation state. Solutions of Fe^{2+} are normally very susceptible to air oxidation, but when prepared in 0.5 M H_2SO_4 the solution may remain stable for as long as a month. Periodic restandardization with $K_2Cr_2O_7$ is advisable. The titrant can be used in either a direct titration in which the Fe^{2+} is oxidized to Fe^{3+}, or an excess of the solution can be added and the quantity of Fe^{3+} produced determined by a back titration using a standard solution of Ce^{4+} or $Cr_2O_7^{2-}$.

Inorganic Analysis Redox titrimetry has been used for the analysis of a wide range of inorganic analytes. Although many of these methods have been replaced by newer methods, a few continue to be listed as standard methods of analysis. In this section we consider the application of redox titrimetry to several important environmental, public health, and industrial analyses. Other examples can be found in the suggested readings listed at the end of this chapter.

One of the most important applications of redox titrimetry is in evaluating the chlorination of public water supplies. In Method 9.3 an approach for determining the total chlorine residual was described in which the oxidizing power of chlorine is used to oxidize I^- to I_3^-. The amount of I_3^- formed is determined by a back titration with $S_2O_3^{2-}$.

The efficiency of chlorination depends on the form of the chlorinating species. For this reason it is important to distinguish between the free chlorine residual, due to Cl_2, HOCl, and OCl^-, and the combined chlorine residual. The latter form of chlorine results from the reaction of ammonia with the free chlorine residual, forming NH_2Cl, $NHCl_2$, and NCl_3. When a sample of iodide-free chlorinated water is mixed with an excess of the indicator *N*,*N*-diethyl-*p*-phenylenediamine (DPD), the free chlorine oxidizes a stoichiometric portion of DPD to its red-colored form. The oxidized DPD is then titrated back to its colorless form with ferrous ammonium sulfate, with the volume of titrant being proportional to the amount of free residual chlorine. Adding a small amount of KI reduces monochloramine, NH_2Cl, forming I_3^-. The I_3^- then oxidizes a portion of the DPD to its red-colored form. Titrating the oxidized DPD with ferrous ammonium sulfate yields the amount of NH_2Cl in the sample. The amount of dichloramine and trichloramine are determined in a similar fashion.

The methods described earlier for determining the total, free, or combined chlorine residual also are used in establishing the chlorine demand of a water supply. The chlorine demand is defined as the quantity of chlorine that must be added to a water supply to completely react with any substance that can be oxidized by chlorine while also maintaining the desired chlorine residual. It is determined by adding progressively greater amounts of chlorine to a set of samples drawn from the water supply and determining the total, free, or combined chlorine residual.

Another important example of redox titrimetry that finds applications in both public health and environmental analyses is the determination of dissolved oxygen. In natural waters the level of dissolved O_2 is important for two reasons: it is the most readily available oxidant for the biological oxidation of inorganic and organic pollutants; and it is necessary for the support of aquatic life. In wastewater treatment plants, the control of dissolved O_2 is essential for the aerobic oxidation of waste materials. If the level of dissolved O_2 falls below a critical value, aerobic bacteria are replaced by anaerobic bacteria, and the oxidation of organic waste produces undesirable gases such as CH_4 and H_2S.

One standard method for determining the dissolved O_2 content of natural waters and wastewaters is the Winkler method. A sample of water is collected in a fashion that prevents its exposure to the atmosphere (which might change the level of dissolved O_2). The sample is then treated with a solution of $MnSO_4$, and then with a solution of NaOH and KI. Under these alkaline conditions Mn^{2+} is oxidized to MnO_2 by the dissolved oxygen.

$$2Mn^{2+}(aq) + 4OH^-(aq) + O_2(aq) \rightarrow 2MnO_2(s) + 2H_2O(\ell)$$

After the reaction is complete, the solution is acidified with H_2SO_4. Under the now acidic conditions I^- is oxidized to I_3^- by MnO_2.

$$MnO_2(s) + 3I^-(aq) + 4H_3O^+(aq) \rightarrow Mn^{2+}(aq) + I_3^-(aq) + 6H_2O(\ell)$$

The amount of I_3^- formed is determined by titrating with $S_2O_3^{2-}$ using starch as an indicator. The Winkler method is subject to a variety of interferences, and several modifications to the original procedure have been proposed. For example, NO_2^- interferes because it can reduce I_3^- to I^- under acidic conditions. This interference is eliminated by adding sodium azide, NaN_3, reducing NO_2^- to N_2. Other reducing agents, such as Fe^{2+}, are eliminated by pretreating the sample with $KMnO_4$, and destroying the excess permanganate with $K_2C_2O_4$.

Another important example of a redox titration for inorganic analytes, which is important in industrial labs, is the determination of water in nonaqueous solvents. The titrant for this analysis is known as the Karl Fischer reagent and consists of a mixture of iodine, sulfur dioxide, pyridine, and methanol. The concentration of pyridine is sufficiently large so that I_2 and SO_2 are complexed with the pyridine (py) as py • I_2 and py • SO_2. When added to a sample containing water, I_2 is reduced to I^-, and SO_2 is oxidized to SO_3.

$$\text{py} \cdot I_2 + \text{py} \cdot SO_2 + \text{py} + H_2O \rightarrow 2\text{py} \cdot HI + \text{py} \cdot SO_3$$

Methanol is included to prevent the further reaction of py • SO_3 with water. The titration's end point is signaled when the solution changes from the yellow color of the products to the brown color of the Karl Fischer reagent.

Organic Analysis Redox titrimetric methods also are used for the analysis of organic analytes. One important example is the determination of the chemical oxygen demand (COD) in natural waters and wastewaters. The COD provides a measure of the quantity of oxygen necessary to completely oxidize all the organic matter in a sample to CO_2 and H_2O. No attempt is made to correct for organic matter that cannot be decomposed biologically or for which the decomposition kinetics are very slow. Thus, the COD always overestimates a sample's true oxygen demand. The determination of COD is particularly important in managing industrial wastewater treatment facilities where it is used to monitor the release of organic-rich wastes into municipal sewer systems or the environment.

The COD is determined by refluxing the sample in the presence of excess $K_2Cr_2O_7$, which serves as the oxidizing agent. The solution is acidified with H_2SO_4, and Ag_2SO_4 is added as a catalyst to speed the oxidation of low-molecular-weight fatty acids. Mercuric sulfate, $HgSO_4$, is added to complex any chloride that is present, thus preventing the precipitation of the Ag^+ catalyst as AgCl. Under these conditions, the efficiency for oxidizing organic matter is 95–100%. After refluxing for 2h, the solution is cooled to room temperature, and the excess $Cr_2O_7^{2-}$ is determined by a back titration, using ferrous ammonium sulfate as the titrant and ferroin as the indicator. Since it is difficult to completely remove all traces of organic matter from the reagents, a blank titration must be performed. The difference in the amount of ferrous ammonium sulfate needed to titrate the blank and the sample is proportional to the COD.

Iodine has been used as an oxidizing titrant for a number of compounds of pharmaceutical interest. Earlier we noted that the reaction of $S_2O_3^{2-}$ with I_3^- produces the tetrathionate ion, $S_4O_6^{2-}$. The tetrathionate ion is actually a dimer consisting of two thiosulfate ions connected through a disulfide (-S-S-) linkage. In the same fashion, I_3^- can be used to titrate mercaptans of the general formula RSH, forming the dimer RSSR as a product. The amino acid cysteine also can be titrated with I_3^-. The product of this titration is cystine, which is a dimer of cysteine. Triiodide also can be used for the analysis of ascorbic acid (vitamin C) by oxidizing the enediol functional group to an alpha diketone

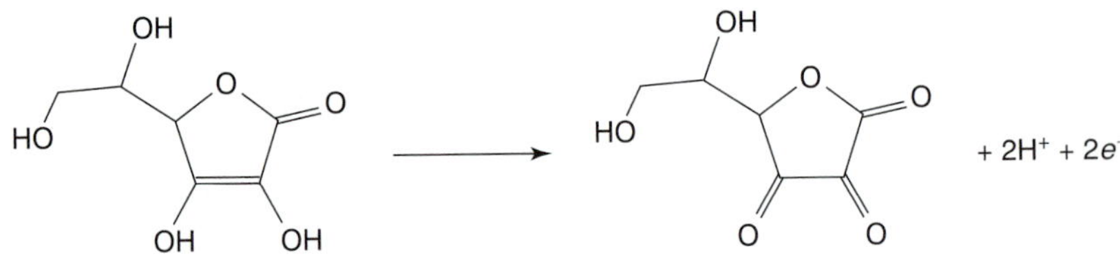

$$\text{ascorbic acid (enediol)} \longrightarrow \text{dehydroascorbic acid (diketone)} + 2H^+ + 2e^-$$

and for the analysis of reducing sugars, such as glucose, by oxidizing the aldehyde functional group to a carboxylate ion in a basic solution.

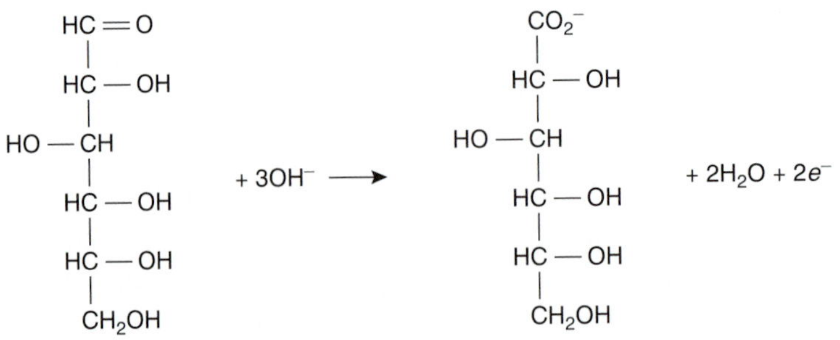

$$\begin{array}{c} HC{=}O \\ | \\ HC{-}OH \\ | \\ HO{-}CH \\ | \\ HC{-}OH \\ | \\ HC{-}OH \\ | \\ CH_2OH \end{array} + 3OH^- \longrightarrow \begin{array}{c} CO_2^- \\ | \\ HC{-}OH \\ | \\ HO{-}CH \\ | \\ HC{-}OH \\ | \\ HC{-}OH \\ | \\ CH_2OH \end{array} + 2H_2O + 2e^-$$

Organic compounds containing a hydroxyl, carbonyl, or amine functional group adjacent to a hydoxyl or carbonyl group can be oxidized using metaperiodate, IO_4^-, as an oxidizing titrant.

$$IO_4^-(aq) + H_2O(\ell) + 2e^- \rightarrow IO_3^-(aq) + 2OH^-(aq)$$

A two-electron oxidation cleaves the C—C bond between the two functional groups, with hydroxyl groups being oxidized to aldehydes or ketones, carbonyl functional groups being oxidized to carboxylic acids, and amines being oxidized to an aldehyde and an amine (ammonia if the original amine was primary). For example, treatment of serine with IO_4^- results in the following oxidation reaction

$$H_2C(OH)—CH(NH_3^+)—CO_2^- + 2OH^- \longrightarrow CH_2{=}O + O{=}HC—CO_2^- + NH_4^+ + H_2O + 2e^-$$

The analysis is conducted by adding a known excess of IO_4^- to the solution containing the analyte and allowing the oxidation to take place for approximately 1 h at room temperature. When the oxidation is complete, an excess of KI is added, which reacts with the unreacted IO_4^- to form IO_3^- and I_3^-.

$$IO_4^-(aq) + 3I^-(aq) + H_2O(\ell) \rightarrow IO_3^-(aq) + I_3^-(aq) + 2OH^-(aq)$$

The I_3^- is then determined by titrating with $S_2O_3^{2-}$ using starch as an indicator.

Quantitative Calculations The stoichiometry of a redox reaction is given by the conservation of electrons between the oxidizing and reducing agents (see Section 2C); thus

$$\frac{\text{Moles of electrons lost}}{\text{Mole reducing agent}} \times \text{moles reducing agent} = \frac{\text{moles of electrons gained}}{\text{mole oxidizing agent}} \times \text{moles oxidizing agent}$$

Example 9.13 shows how this equation is applied to an analysis based on a direct titration.

Example

EXAMPLE 9.13

The amount of Fe in a 0.4891-g sample of an ore was determined by a redox titration with $K_2Cr_2O_7$. The sample was dissolved in HCl and the iron brought into the +2 oxidation state using a Jones reductor. Titration to the diphenylamine sulfonic acid end point required 36.92 mL of 0.02153 M $K_2Cr_2O_7$. Report the iron content of the ore as %w/w Fe_2O_3.

SOLUTION

In this titration the analyte is oxidized from Fe^{2+} to Fe^{3+}, and the titrant is reduced from $Cr_2O_7^{2-}$ to Cr^{3+}. Oxidation of Fe^{2+} requires only a single electron. Reducing $Cr_2O_7^{2-}$, in which chromium is in the +6 oxidation state, requires a total of six electrons. Conservation of electrons for the redox reaction, therefore, requires that

$$\text{Moles } Fe^{2+} = 6 \times \text{moles } Cr_2O_7^{2-}$$

A conservation of mass relates the moles of Fe^{2+} to the moles of Fe_2O_3

$$\text{Moles } Fe^{2+} = 2 \times \text{moles } Fe_2O_3$$

Combining the two conservation equations gives

$$2 \times \text{moles Fe}_2\text{O}_3 = 6 \times \text{moles Cr}_2\text{O}_7^{2-}$$

Making appropriate substitutions for the moles of Fe_2O_3 and $Cr_2O_7^{2-}$ gives the following equation

$$\frac{2 \times \text{grams Fe}_2\text{O}_3}{\text{FW Fe}_2\text{O}_3} = 6 \times M_{\text{Cr}_2\text{O}_7^{2-}} \times V_{\text{Cr}_2\text{O}_7^{2-}}$$

which we solve for the grams of Fe_2O_3.

$$\frac{6 \times M_{\text{Cr}_2\text{O}_7^{2-}} \times V_{\text{Cr}_2\text{O}_7^{2-}} \times \text{FW Fe}_2\text{O}_3}{2} =$$

$$\frac{(6)(0.02153 \text{ M})(0.03692 \text{ L})(159.69 \text{ g/mol})}{2} = 0.3808 \text{ g Fe}_2\text{O}_3$$

Thus, the %w/w Fe_2O_3 in the sample of ore is

$$\frac{\text{grams Fe}_2\text{O}_3}{\text{grams sample}} \times 100 = \frac{0.3808 \text{ g}}{0.4891 \text{ g}} \times 100 = 77.86\% \text{ w/w Fe}_2\text{O}_3$$

As shown in the following two examples, this approach is easily extended to situations that require an indirect analysis or a back titration.

Example

EXAMPLE 9.14

A 25.00-mL sample of a liquid bleach was diluted to 1000 mL in a volumetric flask. A 25-mL portion of the diluted sample was transferred by pipet into an Erlenmeyer flask and treated with excess KI, oxidizing the OCl^- to Cl^-, and producing I_3^-. The liberated I_3^- was determined by titrating with 0.09892 M $Na_2S_2O_3$, requiring 8.96 mL to reach the starch indicator end point. Report the %w/v NaOCl in the sample of bleach.

SOLUTION

Reducing OCl^- to Cl^- requires two electrons, and oxidizing a single I^- to I_3^- requires 2/3 of an electron; thus, conservation of electrons requires that

$$2 \times \text{moles NaOCl} = 2/3 \times \text{moles I}^-$$

In the titration, two electrons are needed when I_3^- is reduced, but oxidizing $S_2O_3^{2-}$ releases one electron; thus

$$2 \times \text{moles I}_3^- = \text{moles S}_2\text{O}_3^{2-}$$

These two equations can be combined by recognizing that a conservation of mass for iodine requires that

$$\text{Moles I}^- = 3 \times \text{moles I}_3^-$$

Thus

$$2 \times \text{moles NaOCl} = \text{moles S}_2\text{O}_3^{2-}$$

Substituting for moles of NaOCl and $S_2O_3^{2-}$ leaves us with an equation

$$\frac{2 \times \text{grams NaOCl}}{\text{FW NaOCl}} = M_{\text{S}_2\text{O}_3^{2-}} \times V_{\text{S}_2\text{O}_3^{2-}}$$

that is solved for the grams of NaOCl.

$$\frac{M_{S_2O_3^{2-}} \times V_{S_2O_3^{2-}} \times \text{FW NaOCl}}{2} = \frac{(0.09892\ \text{M})(0.00896\ \text{L})(74.44\ \text{g/mol})}{2}$$

$$= 0.03299\ \text{g NaOCl}$$

Thus, the %w/v NaOCl in the diluted sample is

$$\frac{\text{Grams NaOCl}}{\text{mL sample}} \times 100 = \frac{0.03299\ \text{g}}{25.00\ \text{mL}} \times 100 = 0.132\%\ \text{w/v NaOCl}$$

Since the bleach was diluted by a factor of 40 (25 mL to 1000 mL), the concentration of NaOCl in the bleach is 5.28% (w/v).

Example

EXAMPLE 9.15

The amount of ascorbic acid, $C_6H_8O_6$, in orange juice was determined by oxidizing the ascorbic acid to dehydroascorbic acid, $C_6H_6O_6$, with a known excess of I_3^-, and back titrating the excess I_3^- with $Na_2S_2O_3$. A 5.00-mL sample of filtered orange juice was treated with 50.00 mL of excess 0.01023 M I_3^-. After the oxidation was complete, 13.82 mL of 0.07203 M $Na_2S_2O_3$ was needed to reach the starch indicator end point. Report the concentration of ascorbic acid in milligrams per 100 mL.

SOLUTION

Oxidizing ascorbic acid requires two electrons, and reducing I_3^- to I^- also requires two electrons. Thus

$$(\text{Moles } I_3^-)_{\text{ascorbic acid}} = \text{moles } C_6H_8O_6$$

For the back titration, the stoichiometric relationship between I_3^- and $S_2O_3^{2-}$ is (see Example 9.14)

$$(\text{Moles } I_3^-)_{\text{back titration}} = 0.5 \times \text{moles } S_2O_3^{2-}$$

The total moles of I_3^- used in the analysis is the sum of that reacting with ascorbic acid and $S_2O_3^{2-}$

$$(\text{Moles } I_3^-)_{\text{tot}} = (\text{moles } I_3^-)_{\text{ascorbic acid}} + (\text{moles } I_3^-)_{\text{back titration}}$$

or

$$\text{Moles } I_3^- = \text{moles } C_6H_8O_6 + 0.5 \times \text{moles } S_2O_3^{2-}$$

Making appropriate substitutions for the moles of I_3^-, $C_6H_8O_6$, and $S_2O_3^{2-}$

$$M_{I_3^-} \times V_{I_3^-} = \frac{\text{g } C_6H_8O_6}{\text{FW } C_6H_8O_6} + 0.5 \times M_{S_2O_3^{2-}} \times V_{S_2O_3^{2-}}$$

and solving for the grams of $C_6H_8O_6$ gives

$$(M_{I_3^-} \times V_{I_3^-} - 0.5 \times M_{S_2O_3^{2-}} \times V_{S_2O_3^{2-}}) \times \text{FW } C_6H_8O_6 =$$

$$[(0.01023\ \text{M})(0.0500\ \text{L}) - (0.5)(0.07203\ \text{M})(0.01382\ \text{L})](176.13\ \text{g/mol})$$

$$= 0.00243\ \text{g } C_6H_8O_6$$

Thus, there is 2.43 mg of ascorbic acid in the 5.00-mL sample, or 48.6 mg/100 mL of orange juice.

9D.5 Evaluation of Redox Titrimetry

The scale of operations, accuracy, precision, sensitivity, time, and cost of methods involving redox titrations are similar to those described earlier in the chapter for acid–base and complexometric titrimetric methods. As with acid–base titrations, redox titrations can be extended to the analysis of mixtures if there is a significant difference in the ease with which the analytes can be oxidized or reduced. Figure 9.40 shows an example of the titration curve for a mixture of Fe^{2+} and Sn^{2+}, using Ce^{4+} as the titrant. The titration of a mixture of analytes whose standard-state potentials or formal potentials differ by at least 200 mV will result in a separate equivalence point for each analyte.

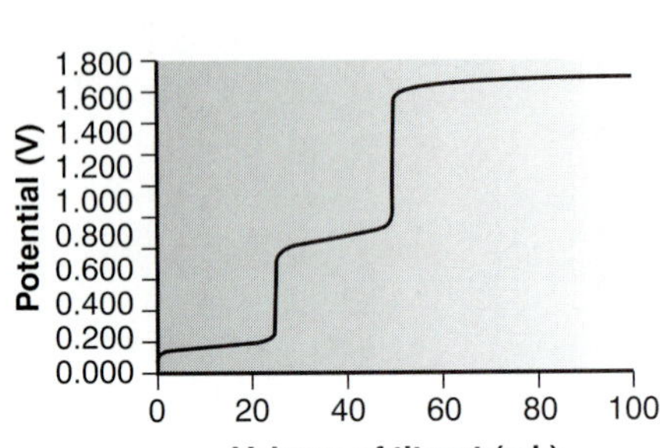

Figure 9.40
Titration curve for 50.00 mL of 0.0250 M Sn^{2+} and 0.0250 M Fe^{2+} with 0.055 M Ce^{4+}.

9E Precipitation Titrations

Thus far we have examined titrimetric methods based on acid–base, complexation, and redox reactions. A reaction in which the analyte and titrant form an insoluble precipitate also can form the basis for a titration. We call this type of titration a **precipitation titration.**

precipitation titration
A titration in which the reaction between the analyte and titrant involves a precipitation.

One of the earliest precipitation titrations, developed at the end of the eighteenth century, was for the analysis of K_2CO_3 and K_2SO_4 in potash. Calcium nitrate, $Ca(NO_3)_2$, was used as a titrant, forming a precipitate of $CaCO_3$ and $CaSO_4$. The end point was signaled by noting when the addition of titrant ceased to generate additional precipitate. The importance of precipitation titrimetry as an analytical method reached its zenith in the nineteenth century when several methods were developed for determining Ag^+ and halide ions.

9E.1 Titration Curves

The titration curve for a precipitation titration follows the change in either the analyte's or titrant's concentration as a function of the volume of titrant. For example, in an analysis for I^- using Ag^+ as a titrant

$$Ag^+(aq) + I^-(aq) \rightleftharpoons AgI(s)$$

the titration curve may be a plot of pAg or pI as a function of the titrant's volume. As we have done with previous titrations, we first show how to calculate the titration curve and then demonstrate how to quickly sketch the titration curve.

Calculating the Titration Curve As an example, let's calculate the titration curve for the titration of 50.0 mL of 0.0500 M Cl^- with 0.100 M Ag^+. The reaction in this case is

$$Ag^+(aq) + Cl^-(aq) \rightleftharpoons AgCl(s)$$

The equilibrium constant for the reaction is

$$K = (K_{sp})^{-1} = (1.8 \times 10^{-10})^{-1} = 5.6 \times 10^9$$

Since the equilibrium constant is large, we may assume that Ag^+ and Cl^- react completely.

By now you are familiar with our approach to calculating titration curves. The first task is to calculate the volume of Ag^+ needed to reach the equivalence point. The stoichiometry of the reaction requires that

$$\text{Moles Ag}^+ = \text{moles Cl}^-$$

or

$$M_{Ag}V_{Ag} = M_{Cl}V_{Cl}$$

Solving for the volume of Ag^+

$$V_{Ag} = \frac{M_{Cl}V_{Cl}}{M_{Ag}} = \frac{(0.0500\text{ M})(50.0\text{ mL})}{(0.100\text{ M})} = 25.0\text{ mL}$$

shows that we need 25.0 mL of Ag^+ to reach the equivalence point.

Before the equivalence point Cl^- is in excess. The concentration of unreacted Cl^- after adding 10.0 mL of Ag^+, for example, is

$$[\text{Cl}^-] = \frac{\text{moles excess Cl}^-}{\text{total volume}} = \frac{M_{Cl}V_{Cl} - M_{Ag}V_{Ag}}{V_{Cl} + V_{Ag}}$$

$$= \frac{(0.0500\text{ M})(50.0\text{ mL}) - (0.100\text{ M})(10.0\text{ mL})}{50.0\text{ mL} + 10.0\text{ mL}}$$

$$= 2.50 \times 10^{-2}\text{ M}$$

If the titration curve follows the change in concentration for Cl^-, then we calculate pCl as

$$\text{pCl} = -\log[\text{Cl}^-] = -\log(2.50 \times 10^{-2}) = 1.60$$

However, if we wish to follow the change in concentration for Ag^+ then we must first calculate its concentration. To do so we use the K_{sp} expression for AgCl

$$K_{sp} = [\text{Ag}^+][\text{Cl}^-] = 1.8 \times 10^{-10}$$

Solving for the concentration of Ag^+

$$[\text{Ag}^+] = \frac{K_{sp}}{[\text{Cl}^-]} = \frac{1.8 \times 10^{-10}}{2.50 \times 10^{-2}} = 7.2 \times 10^{-9}\text{ M}$$

gives a pAg of 8.14.

At the equivalence point, we know that the concentrations of Ag^+ and Cl^- are equal. Using the solubility product expression

$$K_{sp} = [\text{Ag}^+][\text{Cl}^-] = [\text{Ag}^+]^2 = 1.8 \times 10^{-10}$$

gives

$$[\text{Ag}^+] = [\text{Cl}^-] = 1.3 \times 10^{-5}\text{ M}$$

At the equivalence point, therefore, pAg and pCl are both 4.89.

After the equivalence point, the titration mixture contains excess Ag^+. The concentration of Ag^+ after adding 35.0 mL of titrant is

$$[\text{Ag}^+] = \frac{\text{moles excess Ag}^+}{\text{total volume}} = \frac{M_{Ag}V_{Ag} - M_{Cl}V_{Cl}}{V_{Cl} + V_{Ag}}$$

$$= \frac{(0.100\text{ M})(35.0\text{ mL}) - (0.0500\text{ M})(50.0\text{ mL})}{50.0\text{ mL} + 35.0\text{ mL}}$$

$$= 1.18 \times 10^{-2}\text{ M}$$

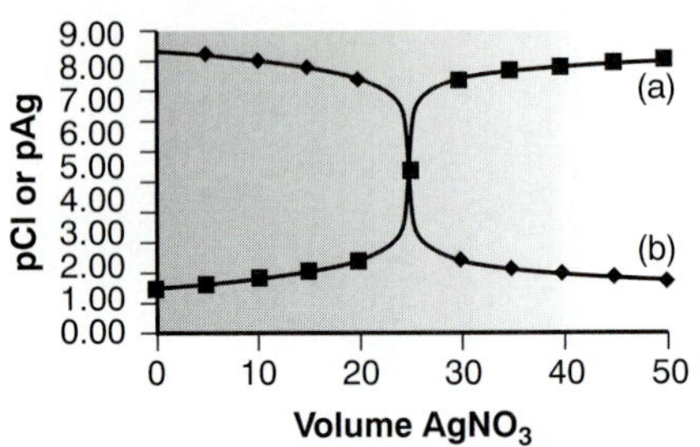

Figure 9.41
Precipitation titration curve for 50.0 mL of 0.0500 M Cl^- with 0.100 M Ag^+. (a) pCl versus volume of titrant; (b) pAg versus volume of titrant.

Table 9.21 Data for Titration of 50.0 mL of 0.0500 M Cl^- with 0.100 M Ag^+

Volume $AgNO_3$ (mL)	pCl	pAg
0.00	1.30	—
5.00	1.44	8.31
10.00	1.60	8.14
15.00	1.81	7.93
20.00	2.15	7.60
25.00	4.89	4.89
30.00	7.54	2.20
35.00	7.82	1.93
40.00	7.97	1.78
45.00	8.07	1.68
50.00	8.14	1.60

or a pAg of 1.93. The concentration of Cl^- is

$$[Cl^-] = \frac{K_{sp}}{[Ag^+]} = \frac{1.8 \times 10^{-10}}{1.18 \times 10^{-2}} = 1.5 \times 10^{-8} \text{ M}$$

or a pCl of 7.82. Additional results for the titration curve are shown in Table 9.21 and Figure 9.41.

Sketching the Titration Curve As we have done for acid–base, complexometric titrations, and redox titrations, we now show how to quickly sketch a precipitation titration curve using a minimum number of calculations.

Example

EXAMPLE 9.16

Sketch a titration curve for the titration of 50.0 mL of 0.0500 M Cl^- with 0.100 M Ag^+. This is the same titration for which we previously calculated the titration curve (Table 9.21 and Figure 9.41).

SOLUTION

We begin by drawing axes for the titration curve (Figure 9.42a). Having shown that the equivalence point volume is 25.0 mL, we draw a vertical line intersecting the *x*-axis at this volume (Figure 9.42b).

Before the equivalence point, pCl and pAg are determined by the concentration of excess Cl^-. Using values from Table 9.21, we plot either pAg or pCl for 10.0 mL and 20.0 mL of titrant (Figure 9.42c).

After the equivalence point, pCl and pAg are determined by the concentration of excess Ag^+. Using values from Table 9.21, we plot points for 30.0 mL and 40.0 mL of titrant (Figure 9.42d).

To complete an approximate sketch of the titration curve, we draw separate straight lines through the two points before and after the equivalence point (Figure 9.42e). Finally, a smooth curve is drawn to connect the three straight-line segments (Figure 9.42f).

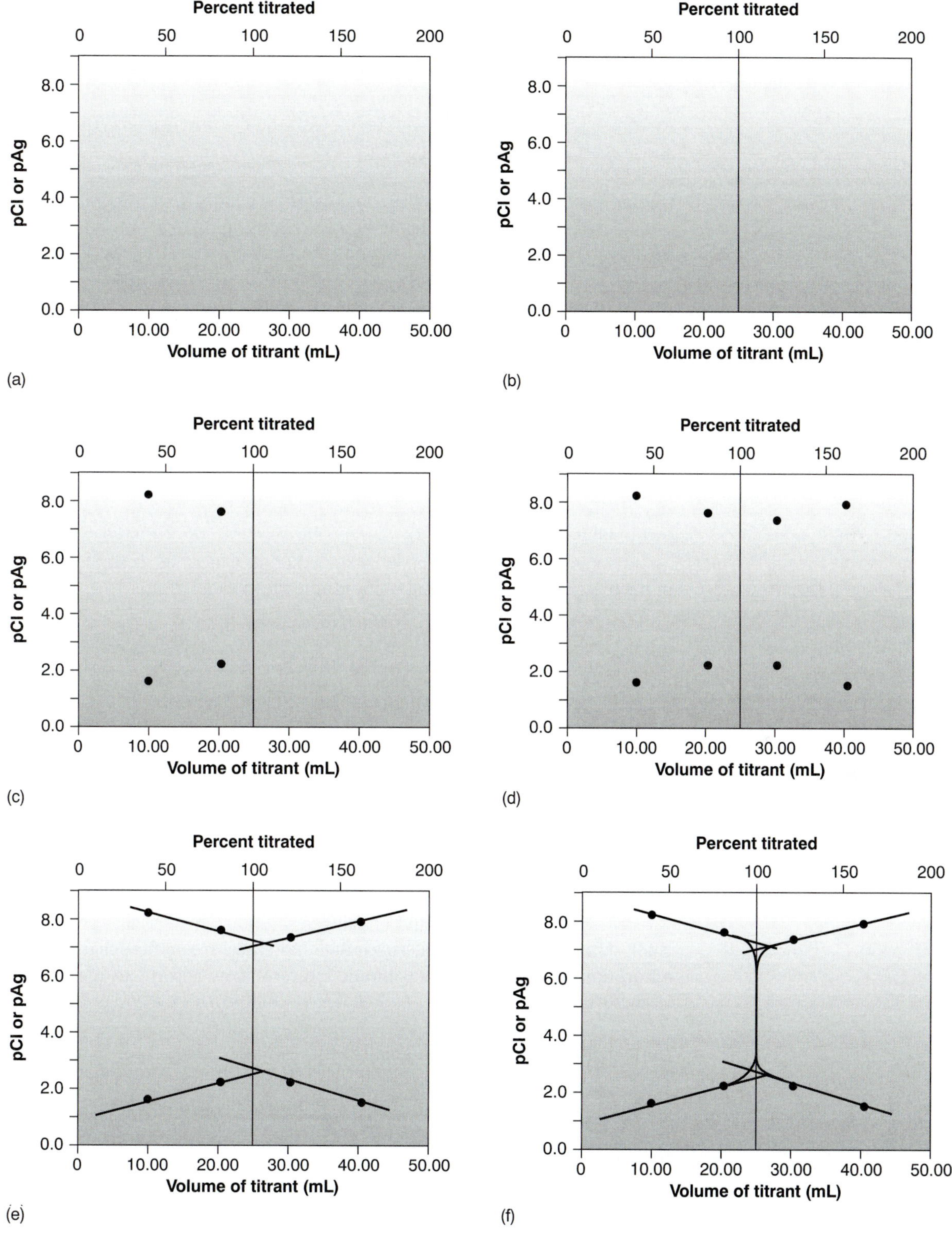

Figure 9.42
How to sketch a precipitation titration curve; see text for explanation.

9E.2 Selecting and Evaluating the End Point

Initial attempts at developing precipitation titration methods were limited by a poor end point signal. Finding the end point by looking for the first addition of titrant that does not yield additional precipitate is cumbersome at best. The feasibility of precipitation titrimetry improved with the development of visual indicators and potentiometric ion-selective electrodes.

Finding the End Point with a Visual Indicator The first important visual indicator to be developed was the Mohr method for Cl^- using Ag^+ as a titrant. By adding a small amount of K_2CrO_4 to the solution containing the analyte, the formation of a precipitate of reddish-brown Ag_2CrO_4 signals the end point. Because K_2CrO_4 imparts a yellow color to the solution, obscuring the end point, the amount of CrO_4^{2-} added is small enough that the end point is always later than the equivalence point. To compensate for this positive determinate error an analyte-free reagent blank is analyzed to determine the volume of titrant needed to effect a change in the indicator's color. The volume for the reagent blank is subsequently subtracted from the experimental end point to give the true end point. Because CrO_4^{2-} is a weak base, the solution usually is maintained at a slightly alkaline pH. If the pH is too acidic, chromate is present as $HCrO_4^-$, and the Ag_2CrO_4 end point will be in significant error. The pH also must be kept below a level of 10 to avoid precipitating silver hydroxide.

A second end point is the Volhard method in which Ag^+ is titrated with SCN^- in the presence of Fe^{3+}. The end point for the titration reaction

$$Ag^+(aq) + SCN^-(aq) \rightleftharpoons AgSCN(s)$$

is the formation of the reddish colored $Fe(SCN)^{2+}$ complex.

$$SCN^-(aq) + Fe^{3+}(aq) \rightleftharpoons Fe(SCN)^{2+}(aq)$$

The titration must be carried out in a strongly acidic solution to achieve the desired end point.

A third end point is evaluated with Fajans' method, which uses an adsorption indicator whose color when adsorbed to the precipitate is different from that when it is in solution. For example, when titrating Cl^- with Ag^+ the anionic dye dichlorofluoroscein is used as the indicator. Before the end point, the precipitate of AgCl has a negative surface charge due to the adsorption of excess Cl^-. The anionic indicator is repelled by the precipitate and remains in solution where it has a greenish yellow color. After the end point, the precipitate has a positive surface charge due to the adsorption of excess Ag^+. The anionic indicator now adsorbs to the precipitate's surface where its color is pink. This change in color signals the end point.

Finding the End Point Potentiometrically Another method for locating the end point of a precipitation titration is to monitor the change in concentration for the analyte or titrant using an ion-selective electrode. The end point can then be found from a visual inspection of the titration curve. A further discussion of potentiometry is found in Chapter 11.

9E.3 Quantitative Applications

Precipitation titrimetry is rarely listed as a standard method of analysis, but may still be useful as a secondary analytical method for verifying results obtained by other methods. Most precipitation titrations involve Ag^+ as either an analyte or

Table 9.22 Representative Examples of Precipitation Titrations

Analyte	Titrant[a]	End Point[b]
AsO_4^{3-}	$AgNO_3$, KSCN	Volhard
Br^-	$AgNO_3$	Mohr or Fajans
	$AgNO_3$, KSCN	Volhard
Cl^-	$AgNO_3$	Mohr or Fajans
	$AgNO_3$, KSCN	Volhard*
CO_3^{2-}	$AgNO_3$, KSCN	Volhard*
$C_2O_4^{2-}$	$AgNO_3$, KSCN	Volhard*
CrO_4^{2-}	$AgNO_3$, KSCN	Volhard*
I^-	$AgNO_3$	Fajans
	$AgNO_3$, KSCN	Volhard
PO_4^{3-}	$AgNO_3$, KSCN	Volhard*
S^{2-}	$AgNO_3$, KSCN	Volhard*
SCN^-	$AgNO_3$, KSCN	Volhard

[a]When two reagents are listed, the analysis is by a back titration. The first reagent is added in excess, and the second reagent is used to back titrate the excess.

[b]For Volhard methods identified by an asterisk (*) the precipitated silver salt must be removed before carrying out the back titration.

titrant. Those titrations in which Ag^+ is the titrant are called **argentometric titrations.** Table 9.22 provides a list of several typical precipitation titrations.

argentometric titration
A precipitation titration in which Ag^+ is the titrant.

Quantitative Calculations The stoichiometry of a precipitation reaction is given by the conservation of charge between the titrant and analyte (see Section 2C); thus

$$\frac{\text{Moles of charge}}{\text{mole titrant}} \times \text{moles titrant} = \frac{\text{moles of charge}}{\text{mole analyte}} \times \text{moles analyte}$$

Example 9.17 shows how this equation is applied to an analysis based on a direct titration.

Example

EXAMPLE 9.17

A mixture containing only KCl and NaBr is analyzed by the Mohr method. A 0.3172-g sample is dissolved in 50 mL of water and titrated to the Ag_2CrO_4 end point, requiring 36.85 mL of 0.1120 M $AgNO_3$. A blank titration requires 0.71 mL of titrant to reach the same end point. Report the %w/w KCl and NaBr in the sample.

SOLUTION

The volume of titrant reacting with the analytes is

$$V_{Ag} = 36.85 \text{ mL} - 0.71 \text{ mL} = 36.14 \text{ mL}$$

Conservation of charge for the titration requires that

$$\text{Moles } Ag^+ = \text{moles KCl} + \text{moles NaBr}$$

Making appropriate substitutions for the moles of Ag^+, KCl, and NaBr gives the following equation.

$$M_{Ag}V_{Ag} = \frac{\text{g KCl}}{\text{FW KCl}} + \frac{\text{g NaBr}}{\text{FW NaBr}}$$

Since the sample contains just KCl and NaBr, we know that

$$\text{g NaBr} = 0.3172 \text{ g} - \text{g KCl}$$

and

$$M_{Ag}V_{Ag} = \frac{\text{g KCl}}{\text{FW KCl}} + \frac{0.3172 \text{ g} - \text{g KCl}}{\text{FW NaBr}}$$

Solving, we find

$$(0.1120 \text{ M})(0.03614 \text{ L}) = \frac{\text{g KCl}}{74.551 \text{ g/mol}} + \frac{0.3172 \text{ g} - \text{g KCl}}{102.89 \text{ g/mol}}$$

$$4.048 \times 10^{-3} = 1.341 \times 10^{-2} \text{ (g KCl)} + 3.083 \times 10^{-3} - 9.719 \times 10^{-3} \text{ (g KCl)}$$

$$3.691 \times 10^{-3} \text{ (g KCl)} = 9.650 \times 10^{-4}$$

$$\text{g KCl} = 0.2614 \text{ g}$$

that there is 0.2614 g of KCl and

$$\text{g NaBr} = 0.3172 \text{ g} - 0.2614 \text{ g} = 0.0558 \text{ g}$$

0.0558 g of NaBr. The weight percents for the two analytes, therefore, are

$$\frac{0.2614 \text{ g KCl}}{0.3172 \text{ g}} \times 100 = 82.41\% \text{ w/w KCl}$$

$$\frac{0.0558 \text{ g NaBr}}{0.3172 \text{ g}} \times 100 = 17.59\% \text{ w/w NaBr}$$

The analysis for I^- using the Volhard method requires a back titration. A typical calculation is shown in the following example.

Example

EXAMPLE 9.18

The %w/w I^- in a 0.6712-g sample was determined by a Volhard titration. After adding 50.00 mL of 0.05619 M $AgNO_3$ and allowing the precipitate to form, the remaining silver was back titrated with 0.05322 M KSCN, requiring 35.14 mL to reach the end point. Report the %w/w I^- in the sample.

SOLUTION

Conservation of charge for this back titration requires that

$$\text{Moles Ag}^+ = \text{moles I}^- + \text{moles SCN}^-$$

Making appropriate substitutions for moles of Ag^+, I^-, and SCN^- leaves us with

$$M_{Ag}V_{Ag} = \frac{\text{g I}^-}{\text{AW I}^-} + M_{SCN}V_{SCN}$$

Solving for the grams of I^-, we find

$$g\ I^- = (AW\ I^-)(M_{Ag}V_{Ag} - M_{SCN}V_{SCN})$$

$$= (126.9\ g/mol)[(0.05619\ M)(0.05000\ L) - (0.05322\ M)(0.03514\ L)]$$

$$= 0.1192\ g$$

that there is 0.1192 g of iodide. The weight percent iodide, therefore, is

$$\frac{0.1192\ g}{0.6712\ g} \times 100 = 17.76\%\ w/w\ I^-$$

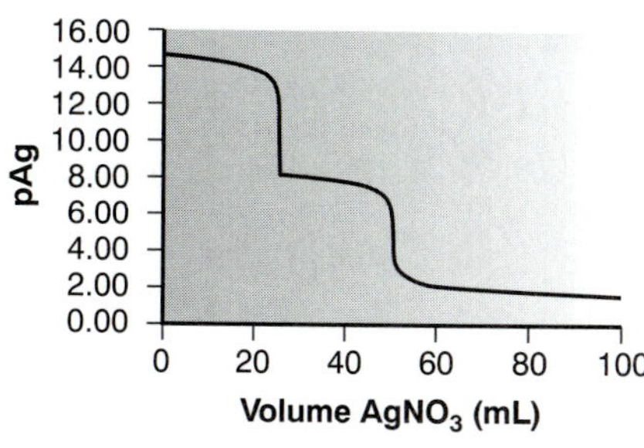

Figure 9.43
Titration curve for a mixture of I^- and Cl^- using $AgNO_3$ as a titrant.

9E.4 Evaluation of Precipitation Titrimetry

The scale of operations, accuracy, precision, sensitivity, time, and cost of methods involving precipitation titrations are similar to those described earlier in the chapter for other titrimetric methods. Precipitation titrations also can be extended to the analysis of mixtures, provided that there is a significant difference in the solubilities of the precipitates. Figure 9.43 shows an example of the titration curve for a mixture of I^- and Cl^- using Ag^+ as a titrant.

9F KEY TERMS

acid–base titration *(p. 278)*
acidity *(p. 301)*
alkalinity *(p. 300)*
argentometric titration *(p. 355)*
auxiliary complexing agent *(p. 316)*
auxiliary oxidizing agent *(p. 341)*
auxiliary reducing agent *(p. 341)*
back titration *(p. 275)*
buret *(p. 277)*
complexation titration *(p. 314)*
conditional formation constant *(p. 316)*
displacement titration *(p. 275)*
end point *(p. 274)*
equivalence point *(p. 274)*
formal potential *(p. 332)*
Gran plot *(p. 293)*
indicator *(p. 274)*
Jones reductor *(p. 341)*
Kjeldahl analysis *(p. 302)*
leveling *(p. 296)*
metallochromic indicator *(p. 323)*
precipitation titration *(p. 350)*
redox indicator *(p. 339)*
redox titration *(p. 331)*
titrant *(p. 274)*
titration curve *(p. 276)*
titration error *(p. 274)*
titrimetry *(p. 274)*
Walden reductor *(p. 341)*

9G SUMMARY

In a titrimetric method of analysis the volume of titrant reacting stoichiometrically with the analyte provides quantitative information about the amount of analyte in a sample. The volume of titrant required to achieve this stoichiometric reaction is called the equivalence point. Experimentally we determine the titration's end point using a visual indicator that changes color near the equivalence point. Alternatively, we can locate the end point by recording a titration curve showing the titration reaction's progress as a function of the titrant's volume. In either case, the end point must closely match the equivalence point if a titration is to be accurate. Knowing the shape of a titration curve is critical to evaluating the feasibility of a proposed titrimetric method.

Many titrations are direct, in which the titrant reacts with the analyte. Other titration strategies may be used when a direct reaction between the analyte and titrant is not feasible. In a back titration a reagent is added in excess to a solution containing the analyte. When the reaction between the reagent and the analyte is complete, the amount of excess reagent is determined titrimetrically. In a displacement titration the analyte displaces a reagent, usually from a complex, and the amount of the displaced reagent is determined by an appropriate titration.

Titrimetric methods have been developed using acid–base, complexation, redox, and precipitation reactions. Acid–base titrations use a strong acid or strong base as a titrant. The most common titrant for a complexation titration is EDTA. Because of their stability against air oxidation, most redox titrations use an oxidizing agent as a titrant. Titrations with reducing agents also are possible. Precipitation titrations usually involve Ag^+ as either the analyte or titrant.

9H *Suggested* EXPERIMENTS

Experiments

The following experiments may be used to illustrate the application of titrimetry to quantitative, qualitative, or characterization problems. Experiments are grouped into four categories based on the type of reaction (acid–base, complexation, redox, and precipitation). A brief description is included with each experiment providing details such as the type of sample analyzed, the method for locating end points, or the analysis of data. Additional experiments emphasizing potentiometric electrodes are found in Chapter 11.

The first block of experiments cover acid–base titrimetry.

Castillo, C. A.; Jaramillo, A. "An Alternative Procedure for Titration Curves of a Mixture of Acids of Different Strengths," *J. Chem. Educ.* **1989,** *66,* 341.

This short paper describes a modification to the traditional Gran plot for determining the concentration of a weak acid in the presence of a strong acid.

Clay, J. T.; Walters, E. A.; Brabson, G. D. "A Dibasic Acid Titration for the Physical Chemistry Laboratory," *J. Chem. Educ.* **1995,** *72,* 665–667.

Values for K_{a1} and K_{a2} for acids of the form H_2A are determined from a least-squares analysis of data from a potentiometric titration.

Crossno, S. K; Kalbus, L. H.; Kalbus, G. E. "Determinations of Carbon Dioxide by Titration," *J. Chem. Educ.* **1996,** *73,* 175–176.

Experiments are described for determining CO_2 in carbonated beverages, $NaHCO_3$ in Alka-Seltzer tablets, and the molecular weight of CO_2. Carbon dioxide is collected in NaOH and the concentrations of CO_3^{2-} are determined by titrating with a standard solution of HCl to the phenolphthalein and methyl orange end points.

Flowers, P. A. "Potentiometric Measurement of Transition Ranges and Titration Errors for Acid–Base Indicators," *J. Chem. Educ.* **1997,** *74,* 846–847.

Dilute solutions of nominally 0.001 M NaOH and HCl are used to demonstrate the effect of an indicator's color transition range on titration error. Potentiometric titration curves are measured, and the indicator's color transition range is noted. Titration errors are calculated using the volume of titrant needed to effect the first color change and for a complete color change.

Graham. R. C.; DePew, S. "Determination of Ammonia in Household Cleaners," *J. Chem. Educ.* **1983,** *60,* 765–766.

A quantitative analysis for NH_3 in several household cleaning products is carried out by titrating with a standard solution of HCl. The titration's progress is followed thermometrically by monitoring the temperature of the titration mixture as a function of the volume of added titrant. Household cleaning products may contain other basic components, such as sodium citrate or sodium carbonate, that will also be titrated by HCl. By comparing titration curves for prepared samples of NH_3 to titration curves for the samples, it is possible to determine that portion of the thermometric titration curve due to the neutralization of NH_3.

Kalbus, L. H.; Petrucci, R. H.; Forman, J. E.; et al. "Titration of Chromate–Dichromate Mixtures," *J. Chem. Educ.* **1991,** *68,* 677–678.

A known amount of HCl in excess is added to a mixture of CrO_4^- and $Cr_2O_7^{2-}$, with the HCl converting CrO_4^{2-} to $HCrO_4^-$. The resulting solution is then back titrated with a standard solution of NaOH. The titration is followed potentiometrically, showing two end points. Titration to the first end point provides a quantitative measure of the excess HCl, providing an indirect measure of the amount of CrO_4^{2-} in the sample. Titration to the second end point gives the total amount of $Cr_2O_7^{2-}$ and CrO_4^{2-} in the sample.

Ophardt, C. E. "Acid Rain Analysis by Standard Addition Titration," *J. Chem. Educ.* **1985,** *62,* 257–258.

This experiment describes a method for determining the acidity, reported as an equivalent molarity of H_2SO_4, of rain water. Because the volume of standard base needed to titrate a sample of rain water is small, the analysis is done by a standard addition. A 10.00-mL sample of nominally 0.005 M H_2SO_4 is diluted with 100.0 mL of distilled water and standardized by titrating with 0.0100 M NaOH. A second 10.00-mL sample of the sulfuric acid is mixed with 100.0 mL of rain water and titrated with the same solution of NaOH. The difference between the two equivalence point volumes

—Continued

Experiments

gives the volume of NaOH needed to neutralize the acidity in the rain water. Titration curves are measured potentiometrically and the equivalence point determined from a Gran plot.

Partanen, J. I.; Kärki, M. H. "Determination of the Thermodynamic Dissociation Constant of a Weak Acid by Potentiometric Acid–Base Titration," *J. Chem. Educ.* **1994,** *71,* A120–A122.

Directions are provided in this experiment for determining the dissociation constant for a weak acid. Potentiometric titration data are analyzed by a modified Gran plot. The experiment is carried out at a variety of ionic strengths and the thermodynamic dissociation constant determined by extrapolating to zero ionic strength.

Thompson, R. Q. "Identification of Weak Acids and Bases by Titration with Primary Standards," *J. Chem. Educ.* **1988,** *65,* 179–180.

Potentiometric titration curves are used to determine the molecular weight and K_a or K_b for weak acid or weak base analytes. The analysis is accomplished using a nonlinear least squares fit to the potentiometric curve. The appropriate master equation can be provided, or its derivation can be left as a challenge.

Tucker, S. A.; Acree, Jr., W. E. "A Student-Designed Analytical Laboratory Method," *J. Chem. Educ.* **1994,** *71,* 71–74.

In this experiment the effect of a mixed aqueous–organic solvent on the color transition range of common indicators is investigated. One goal of the experiment is to design an appropriate titrimetric method for analyzing sparingly soluble acids and bases.

Tucker, S. A.; Amszi, V. L.; Acree, Jr., W. E. "Studying Acid-Base Equilibria in Two-Phase Solvent Media," *J. Chem. Educ.* **1993,** *70,* 80–82.

This experiment shows how modifying the matrix of the solution containing the analyte can dramatically improve the shape of the titration curve. Trialkylammonium salts, such as lidocaine hydrochloride, are titrated in an aqueous solution containing a surfactant. The presence of the surfactant increases the trialkylammonium salt's K_a, giving a titration curve with a more pronounced break. The effect of adding an immiscible organic solvent, such as methylene chloride or toluene, also is demonstrated.

Werner, J. A.; Werner, T. C. "Multifunctional Base Unknowns in the Introductory Analytical Chemistry Lab," *J. Chem. Educ.* **1991,** *68,* 600–601.

A potentiometric titration is used to determine if an unknown sample is pure Na_2CO_3, a mixture of Na_2CO_3 and $NaHCO_3$, pure Na_3PO_4, or a mixture of Na_3PO_4 and Na_2HPO_4.

Three experiments involving complexometric titrations are described in this second block of experiments.

Fulton, R.; Ross, M.; Schroeder, K. "Spectrophotometric Titration of a Mixture of Calcium and Magnesium," *J. Chem. Educ.* **1986,** *63,* 721–723.

In this experiment the concentrations of Ca^{2+} and Mg^{2+} in aqueous solutions are determined by titrating with EDTA. The titration is followed spectrophotometrically by measuring the absorbance of a visual indicator. The effect of changing the indicator, the pH at which the titration is carried out, and the relative concentrations of Ca^{2+} and Mg^{2+} are also investigated.

Novick, S. G. "Complexometric Titration of Zinc," *J. Chem. Educ.* **1997,** *74,* 1463.

Most experiments involving EDTA as a titrant use Ca^{2+} and Mg^{2+} as an analyte. This experiment describes a quantitative analysis for Zn^{2+} in cold lozenges using EDTA as a titrant and xylenol orange as a visual indicator.

Smith, R. L.; Popham, R. E. "The Quantitative Resolution of a Mixture of Group II Metal Ions by Thermometric Titration with EDTA," *J. Chem. Educ.* **1983,** *60,* 1076–1077.

Binary mixtures of Ca^{2+} and Mg^{2+}, and ternary mixtures of Ca^{2+}, Mg^{2+}, and Ba^{2+} are determined by titrating with EDTA. The progress of the titration is followed thermometrically. Complexation of Ca^{2+} and Ba^{2+} with EDTA is exothermic, whereas complexation of Mg^{2+} with EDTA is endothermic. As EDTA is added, the temperature initially rises due to the complexation of Ca^{2+}. The temperature then falls as Mg^{2+} is titrated, rising again as Ba^{2+} is titrated.

The following five experiments provide examples of redox titrimetric methods.

Guenther, W. B. "Supertitrations: High-Precision Methods," *J. Chem. Educ.* **1988,** *65,* 1097–1098.

The purity of ferrous ammonium sulfate is determined by a redox titration with $K_2Cr_2O_7$, using the weight of the reagents as the signal in place of volume.

—Continued

Experiments

Continued from page 359

Haddad, P. "Vitamin C Content of Commercial Orange Juices," *J. Chem. Educ.* **1977,** *54,* 192–193.

The content of ascorbic acid, in milligrams per 100 mL, in orange juice is determined by a redox titration using either 2,6-dichlorophenolindephenol or *N*-bromosuccinimide as the titrant.

Harris, D. C.; Hills, M. E.; Hewston, T. A. "Preparation, Iodometric Analysis and Classroom Demonstration of Superconductivity in $YBa_2Cu_3O_{8-x}$," *J. Chem. Educ.* **1987,** *64,* 847–850.

The superconductor $YBa_2Cu_3O_{8-x}$ contains copper in both the +2 and +3 oxidation states. Procedures are described for synthesizing the superconductor, demonstrating the superconducting effect, and for determining the amount of Cu^{2+} and Cu^{3+} in the prepared material.

Phinyocheep, P.; Tang, I. M. "Determination of the Hole Concentration (Copper Valency) in the High T_c Superconductors," *J. Chem. Educ.* **1994,** *71,* A115–A118.

This experiment outlines a potentiometric titration for determining the valency of copper in superconductors in place of the visual end point used in the preceding experiment of Harris, Hill, and Hewston. The analysis of several different superconducting materials is described.

Powell, J. R.; Tucker, S. A.; Acree, Jr., et al. "A Student–Designed Potentiometric Titration: Quantitative Determination of Iron(II) by Caro's Acid Titration," *J. Chem. Educ.* **1996,** *73,* 984–986.

Caro's acid, H_2SO_5 is used as a titrant for determining Fe^{2+}. Directions are given for exploring the method of end point detection, the titrant's shelf-life, the method's accuracy and precision, and the susceptibility of the method to interference from other species.

Finally, the last experiment describes an unusual precipitation titration.

Ueno, K.; Kina, K. "Colloid Titration—A Rapid Method for the Determination of Charged Colloid," *J. Chem . Educ.* **1985,** *62,* 627–629.

The reaction of a positively charged polyelectrolyte with a negatively charged polyelectrolyte produces a precipitate, forming the basis for a precipitation titration. This paper provides an overview of colloid titrations, discussing compounds that can be used as positively charged titrants or negatively charged titrants. Several methods for detecting the end point, including the use of visual indicators or potentiometric measurements, are discussed. Procedures for titrating positive polyelectrolytes and negative polyelectrolytes also are given.

9I PROBLEMS

1. Calculate or sketch (or both) qualitatively correct titration curves for the following acid–base titrations.
   (a) 25.0 mL of 0.100 M NaOH with 0.0500 M HCl
   (b) 50.0 mL of 0.0500 M HCOOH with 0.100 M NaOH
   (c) 50.0 mL of 0.100 M NH_3 with 0.100 M HCl
   (d) 50.0 mL of 0.0500 M ethylenediamine with 0.100 M HCl
   (e) 50.0 mL of 0.0400 M citric acid with 0.120 M NaOH
   (f) 50.0 mL of 0.0400 M H_3PO_4 with 0.120 M NaOH
2. Locate the equivalence point for each of the titration curves in problem 1. What is the stoichiometric relationship between the moles of acid and moles of base at each of these equivalence points?
3. Suggest an appropriate visual indicator for each of the titrations in problem 1.
4. In sketching the titration curve for a weak acid, we approximated the pH at 10% of the equivalence point volume as $pK_a - 1$, and the pH at 90% of the equivalence point volume as $pK_a + 1$. Show that these assumptions are reasonable.
5. Tartaric acid, $H_2C_4H_4O_6$, is a diprotic weak acid with a pK_{a1} of 3.0 and a pK_{a2} of 4.4. Suppose you have a sample of impure tartaric acid (%purity > 80) and that you plan to determine its purity by titrating with a solution of 0.1 M NaOH using a visual indicator to signal the end point. Describe how you would carry out the analysis, paying particular attention to how much sample you would use, the desired pH range over which you would like the visual indicator to operate, and how you would calculate the %w/w tartaric acid.
6. The following data were collected with an automatic titrator during the titration of a monoprotic weak acid with a strong base. Prepare normal, first-derivative, second-derivative, and Gran plot titration curves for this data, and locate the equivalence point for each.

Volume NaOH (mL)	pH	Volume NaOH (mL)	pH
0.25	3.0	49.97	8.0
0.86	3.2	49.98	8.2
1.63	3.4	49.99	8.4
2.72	3.6	50.00	8.7
4.29	3.8	50.01	9.1
6.54	4.0	50.02	9.4
9.67	4.2	50.04	9.6
13.79	4.4	50.06	9.8
18.83	4.6	50.10	10.0
24.47	4.8	50.16	10.2
30.15	5.0	50.25	10.4
35.33	5.2	50.40	10.6
39.62	5.4	50.63	10.8
42.91	5.6	51.01	11.0
45.28	5.8	51.61	11.2
46.91	6.0	52.58	11.4
48.01	6.2	54.15	11.6
48.72	6.4	56.73	11.8
49.19	6.6	61.11	12.0
49.48	6.8	68.83	12.2
49.67	7.0	83.54	12.4
49.79	7.2	116.14	12.6
49.87	7.4		
49.92	7.6		
49.95	7.8		

7. Schwartz has published some hypothetical data for the titration of a 1.02×10^{-4} M solution of a monoprotic weak acid ($pK_a = 8.16$) with 1.004×10^{-3} M NaOH.[14] A 50-mL pipet is used to transfer a portion of the weak acid solution to the titration vessel. Calibration of the pipet, however, shows that it delivers a volume of only 49.94 mL. Prepare normal, first-derivative, second-derivative, and Gran plot titration curves for these data, and determine the equivalence point for each. How do these equivalence points compare with the expected equivalence point? Comment on the utility of each titration curve for the analysis of very dilute solutions of very weak acids.

Volume Strong Base (mL)	pH	Volume Strong Base (mL)	pH
0.03	6.212	4.79	8.858
0.09	6.504	4.99	8.926
0.29	6.936	5.21	8.994
0.72	7.367	5.41	9.056
1.06	7.567	5.61	9.118
1.32	7.685	5.85	9.180
1.53	7.776	6.05	9.231
1.76	7.863	6.28	9.283
1.97	7.938	6.47	9.327
2.18	8.009	6.71	9.374
2.38	8.077	6.92	9.414
2.60	8.146	7.15	9.451
2.79	8.208	7.36	9.484
3.01	8.273	7.56	9.514
3.19	8.332	7.79	9.545
3.41	8.398	7.99	9.572
3.60	8.458	8.21	9.599
3.80	8.521	8.44	9.624
3.99	8.584	8.64	9.645
4.18	8.650	8.84	9.666
4.40	8.720	9.07	9.688
4.57	8.784	9.27	9.706

8. Calculate or sketch (or both) the titration curves for 50.0 mL of a 0.100 M solution of a monoprotic weak acid ($pK_a = 8$) with 0.1 M strong base in (a) water; and (b) a non-aqueous solvent with $K_s = 10^{-20}$. You may assume that the change in solvent does not affect the weak acid's pK_a.

9. The titration of a mixture of *p*-nitrophenol ($pK_a = 7.0$) and *m*-nitrophenol ($pK_a = 8.3$) can be followed spectrophotometrically. Neither acid absorbs at a wavelength of 545 nm, but their respective conjugate bases do absorb at this wavelength. The *m*-nitrophenolate ion has a greater absorbance than an equimolar solution of the *p*-nitrophenolate ion. Sketch the spectrophotometric titration curve for a 50.00-mL mixture consisting of 0.0500 M *p*-nitrophenol and 0.0500 M *m*-nitrophenol with 0.100 M NaOH, and compare the curve with the expected potentiometric titration curves.

10. The quantitative analysis for aniline ($C_6H_5NH_2$; $K_b = 3.94 \times 10^{-10}$) can be carried out by an acid–base titration, using glacial acetic acid as the solvent and $HClO_4$ as the titrant. A known volume of sample containing 3–4 mmol of aniline is transferred to a 250-mL Erlenmeyer flask and diluted to approximately 75 mL with glacial acetic acid. Two drops of a methyl violet visual indicator are added, and the solution is titrated with previously standardized 0.1000 M $HClO_4$ (prepared in glacial acetic acid using anhydrous $HClO_4$) until the visual end point is reached. Results are reported as parts per million of aniline.
    (a) Explain why this titration is conducted using glacial acetic acid as the solvent instead of water. (b) One problem with using glacial acetic acid as solvent is its relatively high coefficient of thermal expansion of 0.11%/°C. Thus, for example, 100.00 mL of glacial acetic acid at 25 °C will occupy 100.22 mL at 27 °C. What is the effect on the reported parts per million of aniline if the standardization of $HClO_4$ was conducted at a lower temperature than the analysis of the unknown? (c) The procedure calls for a sample containing only 3–4 mmol of aniline. Why was this requirement made?

11. Using a ladder diagram, explain why the presence of dissolved CO_2 leads to a determinate error in the standardization of

NaOH when the end point's pH falls between 6 and 10, but no determinate error is observed when the end point's pH is less than 6.

12. The acidity of a water sample is determined by titrating to fixed end points of 3.7 and 8.3, with the former providing a measure of the concentration of strong acid, and the latter a measure of the combined concentrations of strong acid and weak acid. Sketch a titration curve for a mixture of 0.10 M HCl and 0.10 M H_2CO_3 with 0.20 M strong base, and use it to justify the choice of these end points.

13. Ethylenediaminetetraacetic acid, H_4Y, is a tetraprotic weak acid with successive acid dissociation constants of 0.010, 2.1×10^{-3}, 7.8×10^{-7}, and 6.8×10^{-11}. The titration curve shown here is for H_4Y with NaOH. What is the stoichiometric relationship between H_4Y and NaOH at the equivalence point marked with the arrow?

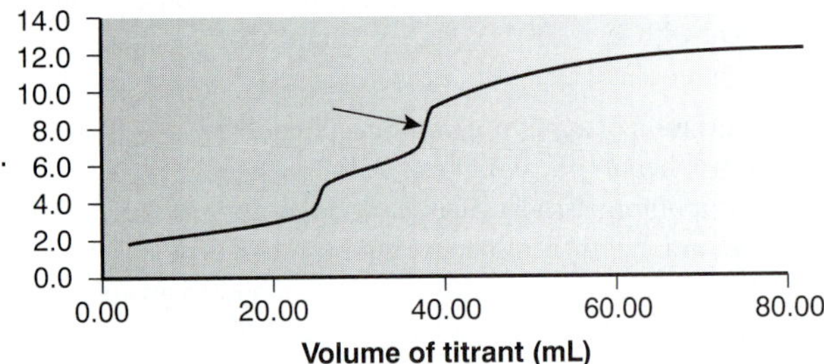

14. A Gran plot method has been described for determining the quantitative analysis of a mixture consisting of a strong acid and a monoprotic weak acid.[15] A 50.00-mL sample that is a mixture of HCl and CH_3COOH is transferred to an Erlenmeyer flask and titrated by using a pipet to add 1.00-mL aliquots of 0.09186 M NaOH. The progress of the titration is monitored by recording the pH after the addition of each aliquot of titrant. Using the two papers listed in reference 15 (p. 367), prepare a Gran plot for the following data, and determine the concentrations of HCl and CH_3COOH.

Volume of NaOH (mL)	pH	Volume of NaOH (mL)	pH	Volume of NaOH (mL)	pH
1	1.83	24	4.45	47	12.14
2	1.86	25	4.53	48	12.17
3	1.89	26	4.61	49	12.20
4	1.92	27	4.69	50	12.23
5	1.95	28	4.76	51	12.26
6	1.99	29	4.84	52	12.28
7	2.03	30	4.93	53	12.30
8	2.10	31	5.02	54	12.32
9	2.18	32	5.13	55	12.34
10	2.31	33	5.23	56	12.36
11	2.51	34	5.37	57	12.38
12	2.81	35	5.52	58	12.39
13	3.16	36	5.75	59	12.40
14	3.36	37	6.14	60	12.42
15	3.54	38	10.30	61	12.43
16	3.69	39	11.31	62	12.44
17	3.81	40	11.58	63	12.45
18	3.93	41	11.74	64	12.47
19	4.02	42	11.85	65	12.48
20	4.14	43	11.93	66	12.49
21	4.22	44	12.00	67	12.50
22	4.30	45	12.05	68	12.51
23	4.38	46	12.10	69	12.52

15. Explain why it is not possible for a sample of water to simultaneously have OH^- and HCO_3^- as sources of alkalinity.

16. For each of the following, determine the forms of alkalinity (OH^-, HCO_3^-, CO_3^{2-}) that are present, and their respective concentrations in parts per million. In each case, a 25.00-mL sample is titrated with 0.1198 M HCl to the bromocresol green and phenolphthalein end points.

	Volume of HCl (mL) to the Phenolphthalein End Point	Volume of HCl (mL) to the Bromocresol Green End Point
(a)	21.36	21.38
(b)	5.67	21.13
(c)	0.00	14.28
(d)	17.12	34.26
(e)	21.36	25.69

17. A sample may contain any of the following: HCl, NaOH, H_3PO_4, $H_2PO_4^-$, HPO_4^{2-}, or PO_4^{3-}. The composition of a sample is determined by titrating a 25.00-mL portion with 0.1198 M HCl or 0.1198 M NaOH to the phenolphthalein and methyl orange end points. For each of the following, determine which species are present in the sample, and their respective molar concentrations.

	Titrant	Volume (mL) to the Phenolphthalein End Point	Volume (mL) to the Methyl Orange End Point
(a)	HCl	11.54	35.29
(b)	NaOH	19.79	9.89
(c)	HCl	22.76	22.78
(d)	NaOH	39.42	17.48

18. The protein in a 1.2846-g sample of an oat cereal is determined by the Kjeldahl procedure for organic nitrogen. The sample is digested with H_2SO_4, the resulting solution made basic with NaOH, and the NH_3 distilled into 50.00 mL of 0.09552 M HCl. The excess HCl is then back titrated using 37.84 mL of 0.05992 M NaOH. Given that the protein in grains averages 17.54% w/w N, report the %w/w protein in the sample of cereal.

19. The concentration of SO_2 in atmospheric samples can be determined by bubbling a sample of air through a trap containing H_2O_2. Oxidation of SO_2 by H_2O_2 results in the

formation of H_2SO_4, the amount of which can be determined by titrating with NaOH. In a typical analysis, a sample of air was passed through the peroxide trap at a rate of 1.25 L/min for 60 min and required 10.08 mL of 0.0244 M NaOH to reach the phenolphthalein end point. Calculate the parts per million of SO_2 (μL/L) in the sample of air. The density of SO_2 at the temperature of the air sample is 2.86 mg/mL.

20. The concentration of CO_2 in air can be determined by an indirect acid–base titration. A sample of the air is bubbled through a solution containing an excess of $Ba(OH)_2$, precipitating $BaCO_3$. The excess $Ba(OH)_2$ is back titrated with HCl. In a typical analysis, a 3.5-L sample of air was bubbled through 50.00 mL of 0.0200 M $Ba(OH)_2$. Back titrating with 0.0316 M HCl requires 38.58 mL to reach the end point. Determine the parts per million of CO_2 in the sample of air, given that the density of CO_2 at the temperature of the sample is 1.98 g/L.

21. The purity of a synthetic preparation of methylethyl ketone (C_4H_8O) can be determined by reacting the ketone with hydroxylamine hydrochloride, liberating HCl (see Table 9.10). In a typical analysis, a 3.00-mL sample was diluted to 50.00 mL and treated with an excess of hydroxylamine hydrochloride. The liberated HCl was titrated with 0.9989 M NaOH, requiring 32.68 mL to reach the end point. Report the percent purity of the sample, given that the density of methylethyl ketone is 0.805 g/mL.

22. Animal fats and vegetable oils are triacylglycerols, or triesters, formed from the reaction of glycerol (1, 2, 3-propanetriol) with three long-chain fatty acids. One of the methods used to characterize a fat or an oil is a determination of its saponification number. When treated with boiling aqueous KOH, an ester is saponified into the parent alcohol and fatty acids (as carboxylate ions). The saponification number is the number of milligrams of KOH required to saponify 1.000 g of the fat or oil. In a typical analysis, a 2.085-g sample of butter is added to 25.00 mL of 0.5131 M KOH. After saponification is complete, the excess KOH is back titrated with 10.26 mL of 0.5000 M HCl. What is the saponification number for this sample of butter?

23. A 250.0-mg sample of an organic weak acid was dissolved in an appropriate solvent and titrated with 0.0556 M NaOH, requiring 32.58 mL to reach the end point. Determine the compound's equivalent weight.

24. The potentiometric titration curve shown here was recorded on a 0.4300-g sample of a purified amino acid that was dissolved in 50.00 mL of water and titrated with 0.1036 M NaOH. Identify the amino acid from the possibilities listed in the following table.

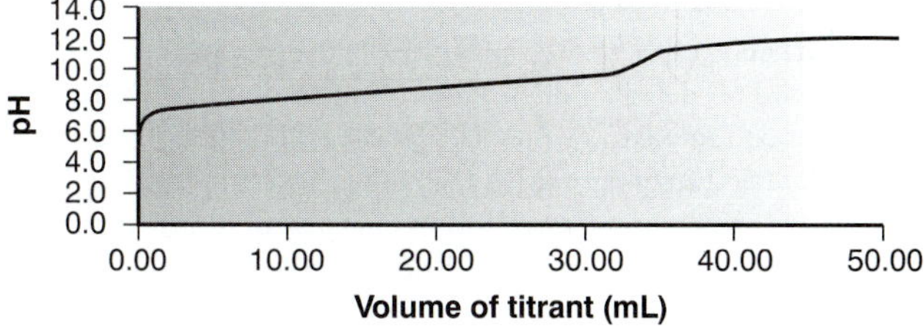

Amino Acid	FW	K_a	Amino Acid	FW	K_a
alanine	89.1	1.36×10^{-10}	asparagine	150	1.9×10^{-9}
glycine	75.1	1.67×10^{-10}	leucine	131.2	1.79×10^{-10}
methionine	149.2	8.9×10^{-10}	phenylalanine	166.2	4.9×10^{-10}
taurine	125.2	1.8×10^{-9}	valine	117.2	1.91×10^{-10}

Abbreviation: FW=formula weight.

25. Using its titration curve, determine the acid dissociation constant for the weak acid in problem 6.

26. Where in the scale of operations do the microtitration techniques discussed in Section 9B.8 belong?

27. An acid–base titration can be used to determine an analyte's equivalent weight, but cannot be used to determine its formula weight. Explain why.

28. Commercial washing soda contains approximately 30–40% w/w Na_2CO_3. One procedure for the quantitative analysis of washing soda contains the following instructions:

    Transfer an approximately 4-g sample of the washing soda to a 250-mL volumetric flask. Dissolve the sample in about 100 mL of H_2O, and then dilute to the mark. Using a pipet, transfer a 25-mL aliquot of this solution to a 125-mL Erlenmeyer flask, and add 25-mL of H_2O and 2 drops of bromocresol green indicator. Titrate the sample with 0.1 M HCl to the indicator's end point.

    What modifications, if any, would need to be made to this procedure if it were to be adapted for the evaluation of the purity of commercial Na_2CO_3 (>98+% pure)?

29. When standardizing a solution of NaOH against potassium hydrogen phthalate (KHP), a variety of systematic and random errors are possible. Identify, with justification, whether the following are systematic or random sources of error, or if they have no effect. If the error is systematic, then indicate whether the experimentally determined molarity for NaOH will be too high or too low. The standardization reaction is

$$C_8H_5O_4^- + OH^- \rightarrow C_8H_4O_4^{2-} + H_2O$$

    (a) The balance used to weigh the KHP is not properly calibrated and always reads 0.15 g too low. (b) The visual indicator selected for the standardization changes color between a pH of 3 and 4. (c) An air bubble is lodged in the tip of the buret at the beginning of the analysis, but is dislodged during the titration. (d) Samples of KHP are weighed into separate Erlenmeyer flasks, but the balance is only tared with the first flask. (e) The KHP was not dried before it was used. (f) The NaOH was not dried before it was used. (g) The procedure calls for the sample of KHP to be dissolved in 25 mL of water, but it is accidentally dissolved in 35 mL.

30. The concentration of *o*-phthalic acid in an organic solvent, such as *n*-butanol, may be determined by an acid–base titration using aqueous NaOH as the titrant. As the titrant is added, the *o*-phthalic acid is extracted into the aqueous

solution, where it reacts with the titrant. The titrant must be added slowly to allow sufficient time for the extraction to occur. (a) What type of error would you expect if the titration were to be carried out too quickly? (b) Propose an alternative acid–base titration method that would allow for the more rapid determination of the concentration of *o*-phthalic acid in *n*-butanol.

31. Calculate or sketch (or both) the titration curves for 50.00 mL of 0.0500 Mg^{2+} with 0.0500 M EDTA at a pH of 7 and 10. Locate the equivalence point for each titration curve.

32. Calculate or sketch (or both) the titration curves for 25.0 mL of 0.0500 M Cu^{2+} with 0.025 M EDTA at a pH of 10 and in the presence of 10^{-3} M and 10^{-1} M NH_3. Locate the equivalence point for each titration curve.

33. Sketch the spectrophotometric titration curve for the titration of a mixture of 5.00×10^{-3} M Bi^{3+} and 5.00×10^{-3} M Cu^{2+} with 0.0100 M EDTA. Assume that only the Cu^{2+}–EDTA complex absorbs at the selected wavelength.

34. The EDTA titration of mixtures of Ca^{2+} and Mg^{2+} can be followed thermometrically because the formation of the Ca^{2+}–EDTA complex is exothermic, but the formation of the Mg^{2+}–EDTA complex is endothermic. Sketch the thermometric titration curve for a mixture of 5.00×10^{-3} M Ca^{2+} with 5.00×10^{-3} M Mg^{+} with 0.0100 M EDTA. The heats of formation for CaY^{2-} and MgY^{2-} are, respectively, –23.9 kJ/mol and 23.0 kJ/mol.

35. EDTA is one member of a class of aminocarboxylate ligands that form very stable 1:1 complexes with metal ions. The following table shows log K_f values for several ligands with Ca^{2+} and Mg^{2+}. Which ligand is the best choice for the direct titration of Ca^{2+} in the presence of Mg^{2+}?

		Mg^{2+}	Ca^{2+}
EDTA	ethylenediaminetetraacetic acid	8.7	10.7
HEDTA	N-hydroxyethylenediaminetriacetic acid	7.0	8.0
EEDTA	ethyletherdiaminetetraacetic acid	8.3	10.0
EGTA	ethyleneglycol-bis(β-aminoethylether) -N,N′-tetraacetic acid	5.4	10.9
DTPA	diethylenetriaminepentaacetic acid	9.0	10.7
CyDTA	cyclohexanediaminetetraacetic acid	10.3	12.3

36. The amount of calcium in physiologic fluids can be determined by a complexometric titration with EDTA. In one such analysis, a 0.100-mL sample of blood serum was made basic by adding 2 drops of NaOH and titrated with 0.00119 M EDTA, requiring 0.268 mL to reach the end point. Report the concentration of calcium in the sample as miligrams of Ca per 100 mL.

37. After removing the membranes from an eggshell, the shell is dried and its mass recorded as 5.613 g. The eggshell is transferred to a 250-mL beaker and dissolved in 25 mL of 6 M HCl. After filtering, the solution containing the dissolved eggshell is diluted to 250 mL in a volumetric flask. A 10.00-mL aliquot is placed in a 125-mL Erlenmeyer flask and buffered to a pH of 10. Titrating with 0.04988 M EDTA requires 44.11 mL to reach the end point. Determine the amount of calcium in the eggshell as %w/w $CaCO_3$.

38. The concentration of cyanide, CN^-, in a copper electroplating bath can be determined by a complexometric titration with Ag^+, forming the soluble $Ag(CN)_2^-$ complex. In a typical analysis a 5.00-mL sample from an electroplating bath is transferred to a 250-mL Erlenmeyer flask, and treated with 100 mL of H_2O, 5 mL of 20% w/v NaOH, and 5 mL of 10% w/v KI. The sample is titrated with 0.1012 M $AgNO_3$, requiring 27.36 mL to reach the end point as signaled by the formation of a yellow precipitate of AgI. Report the concentration of cyanide as parts per million of NaCN.

39. Before the introduction of EDTA most complexation titrations used Ag^+ or CN^- as the titrant. The analysis for Cd^{2+}, for example, was accomplished indirectly by adding an excess of KCN to form $Cd(CN)_4^{2-}$, and back titrating the excess CN^- with Ag^+, forming $Ag(CN)_2^-$. In one such analysis, a 0.3000-g sample of an ore was dissolved and treated with 20.00 mL of 0.5000 M KCN. The excess CN^- required 13.98 mL of 0.1518 M $AgNO_3$ to reach the end point. Determine the %w/w Cd in the ore.

40. Solutions containing both Fe^{3+} and Al^{3+} can be selectively analyzed for Fe^{3+} by buffering to a pH of 2 and titrating with EDTA. The pH of the solution is then raised to 5 and an excess of EDTA added, resulting in the formation of the Al^{3+}–EDTA complex. The excess EDTA is back titrated using a standard solution of Fe^{3+}, providing an indirect analysis for Al^{3+}.
    (a) Show that at a pH of 2 the formation of the Fe^{3+}–EDTA complex is favorable, but the formation of the Al^{3+}–EDTA complex is not favorable. (b) A 50.00-mL aliquot of a sample containing Fe^{3+} and Al^{3+} is transferred to a 250-mL Erlenmeyer flask and buffered to a pH of 2. A small amount of salicylic acid is added, forming the soluble red-colored Fe^{3+}–salicylic acid complex. The solution is titrated with 0.05002 M EDTA, requiring 24.82 mL to reach the end point as signaled by the disappearance of the Fe^{3+}–salicylic acid complex's red color. The solution is buffered to a pH of 5, and 50.00 mL of 0.05002 M EDTA is added. After ensuring that the formation of the Al^{3+}–EDTA complex is complete, the excess EDTA was back titrated with 0.04109 M Fe^{3+}, requiring 17.84 mL to reach the end point as signaled by the reappearance of the red-colored Fe^{3+}–salicylic acid complex. Report the molar concentrations of Fe^{3+} and Al^{3+} in the sample.

41. Prada and colleagues recently described a new indirect method for determining sulfate in natural samples, such as sea water and industrial effluents.[16] The method is based on (1) precipitating the sulfate as $PbSO_4$; (2) dissolving the $PbSO_4$ in an ammonical solution of excess EDTA to form the soluble PbY^{2-} complex; and (3) titrating the excess EDTA

with a standard solution of Mg^{2+}. The following reactions and equilibrium constants are known

$PbSO_4(s) \rightleftharpoons Pb^{2+} + SO_4^{2-}$ $\quad K_{sp} = 1.6 \times 10^{-8}$

$Pb^{2+} + Y^{4-} \rightleftharpoons PbY^{2-}$ $\quad K_f = 1.1 \times 10^{18}$

$Mg^{2+} + Y^{4-} \rightleftharpoons MgY^{2-}$ $\quad K_f = 4.9 \times 10^{8}$

$Zn^{2+} + Y^{4-} \rightleftharpoons ZnY^{2-}$ $\quad K_f = 3.2 \times 10^{16}$

(a) Show that a precipitate of $PbSO_4$ will dissolve in a solution of Y^{4-}. (b) Sporek proposed a similar method using Zn^{2+} as a titrant and found that the accuracy was frequently poor.[17] One explanation is that Zn^{2+} might react with the PbY^{2-} complex, forming ZnY^{2-}. Using the preceding equilibrium constants, show that this might be a problem when using Zn^{2+} as a titrant, but will not be a problem when using Mg^{2+} as a titrant. Would such a displacement of Pb^{2+} by Zn^{2+} lead to the reporting of too much or too little sulfate? (c) In a typical analysis, a 25.00-mL sample of an industrial effluent was carried through the procedure using 50.00 mL of 0.05000 M EDTA. Titrating the excess EDTA required 12.42 mL of 0.1000 M Mg^{2+}. Report the molar concentration of SO_4^{2-} in the effluent sample.

42. Calculate or sketch (or both) titration curves for the following (unbalanced) redox titration reactions at 25 °C. Assume that the analyte is initially present at a concentration of 0.0100 M and that a 25.0-mL sample is taken for analysis. The titrant, which is the underlined species in each reaction, is 0.0100 M.

(a) $V^{2+} + \underline{Ce^{4+}} \rightarrow V^{3+} + Ce^{3+}$

(b) $Ti^{2+} + \underline{Fe^{3+}} \rightarrow Ti^{3+} + Fe^{2+}$

(c) $Fe^{2+} + \underline{MnO_4^-} \rightarrow Fe^{3+} + Mn^{2+}$ (pH = 1)

43. What are the equivalence points for the titrations listed in problem 42?

44. Suggest an appropriate indicator for each of the titrations in problem 42.

45. The iron content of an ore can be determined by a redox titration using $K_2Cr_2O_7$ as the titrant. The ore sample is dissolved in concentrated HCl using Sn^{2+} to speed the ore's dissolution by reducing the Fe^{3+} to Fe^{2+}. After the sample is dissolved, the Fe^{2+} and any excess Sn^{2+} are oxidized to Fe^{3+} and Sn^{4+}, using MnO_4^-. The iron is then carefully reduced to Fe^{2+} by adding a 2–3-drop excess of Sn^{2+}. A solution of $HgCl_2$ is added and, if a white precipitate of Hg_2Cl_2 forms, the analysis is continued by titrating with $K_2Cr_2O_7$. The sample is discarded without completing the analysis if no precipitate forms, or if a gray precipitate (due to Hg) forms.
(a) Explain the role of the $HgCl_2$ in the analysis. (b) Will a determinate error be introduced into the analysis if the analyst forgets to add Sn^{2+} in the step where the iron ore is dissolved? (c) Will a determinate error be introduced if the iron is not quantitatively oxidized back to Fe^{3+} by the MnO_4^-?

46. The amount of Cr^{3+} in inorganic salts can be determined by a redox titration. A portion of sample containing approximately 0.25 g of Cr^{3+} is accurately weighed and dissolved in <u>50 mL of H_2O</u>. The Cr^{3+} is oxidized to $Cr_2O_7^{2-}$ by adding <u>20 mL of 0.1 M $AgNO_3$</u>, which serves as a catalyst, and <u>50 mL of 10% w/v $(NH_4)_2S_2O_8$</u>, which serves as the oxidizing agent. After the reaction is complete, the resulting solution is boiled for 20 min to destroy the excess $S_2O_8^{2-}$, cooled to room temperature, and diluted to 250 mL in a volumetric flask. A <u>50-mL portion of the resulting solution</u> is transferred to an Erlenmeyer flask, treated with <u>50 mL of a standard solution of Fe^{2+}</u>, and acidified with <u>200 mL of 1 M H_2SO_4</u>, reducing the $Cr_2O_7^{2-}$ to Cr^{3+}. The excess Fe^{2+} is then determined by a back titration with a standard solution of $K_2Cr_2O_7$, using an appropriate visual indicator. The results are reported as %w/w Cr^{3+}.
(a) Several volume measurements are noted in this procedure by underlining. Which of these measurements must be made using a volumetric pipet? (b) The excess peroxydisulfate, $S_2O_8^{2-}$, which was used to oxidize the Cr^{3+} to $Cr_2O_7^{2-}$, is destroyed by boiling the solution. What would be the effect on the reported %w/w Cr^{3+} if some of the $S_2O_8^{2-}$ was not destroyed during this step?
(c) Solutions of Fe^{2+} undergo slow air oxidation to Fe^{3+}. What would be the effect on the reported %w/w Cr^{3+} if the standard solution of Fe^{2+} had been inadvertently allowed to partially oxidize?

47. The exact concentration of H_2O_2 in a solution that is nominally 6% w/v H_2O_2 can be determined by a redox titration with MnO_4^-. A 25-mL aliquot of the sample is transferred to a 250-mL volumetric flask and diluted to volume with distilled water. A 25 mL aliquot of the diluted sample is added to an Erlenmeyer flask, diluted with 200 mL of distilled water, and acidified with 20 mL of 25% v/v H_2SO_4. The resulting solution is then titrated with a standard solution of $KMnO_4$ until a faint pink color persists for 30 s. The results are reported as %w/v H_2O_2.
(a) Many commercially available solutions of H_2O_2 contain an inorganic or organic stabilizer to prevent the autodecomposition of the peroxide to H_2O and O_2. What effect would the presence of this stabilizer have on the reported %w/v H_2O_2 if it also reacts with MnO_4^-?
(b) Laboratory distilled water often contains traces of dissolved organic material that also can react with MnO_4^-. Describe a simple method to correct for this potential interference. (c) What modifications, if any, would need to be made to this procedure if the sample to be analyzed has a nominal concentration of 30% w/v H_2O_2.

48. The amount of iron in a meteorite was determined by a redox titration using $KMnO_4$ as the titrant. A 0.4185-g sample was dissolved in acid and the liberated Fe^{3+} quantitatively reduced to Fe^{2+}, using a reductor column. Titrating with 0.02500 M $KMnO_4$ requires 41.27 mL to reach the end point. Determine the %w/w Fe_2O_3 in the sample of meteorite.

49. Under basic conditions, MnO_4^- can be used as a titrant for the analysis of Mn^{2+}, with both the analyte and the titrant ending up as MnO_2. In the analysis of a mineral sample for

manganese, a 0.5165-g sample is dissolved, and the manganese reduced to Mn^{2+}. The solution is made basic and titrated with 0.03358 M $KMnO_4$, requiring 34.88 mL to reach the end point. Calculate the %w/w Mn in the mineral sample.

50. The amount of uranium in an ore sample can be determined by an indirect redox titration. The analysis is accomplished by dissolving the ore in sulfuric acid and reducing the resulting UO_2^{2+} to U^{4+} with a Walden reductor. The resulting solution is treated with an excess of Fe^{3+}, forming Fe^{2+} and U^{6+}. The Fe^{2+} is titrated with a standard solution of $K_2Cr_2O_7$ to a visual end point. In a typical analysis, a 0.315-g sample of ore is passed through the Walden reductor and treated with an excess of Fe^{3+}. Titrating with 0.00987 M $K_2Cr_2O_7$ requires 10.52 mL. What is the %w/w U in the sample?

51. The thickness of the chromium plate on an auto fender was determined by dissolving a 30.0-cm^2 section of the fender in acid and oxidizing the liberated Cr^{3+} to $Cr_2O_7^{2-}$ with peroxydisulfate. After removing the excess peroxydisulfate by boiling, 500.0 mg of $Fe(NH_4)_2(SO_4)_2 \cdot 6H_2O$ was added, reducing the $Cr_2O_7^{2-}$ to Cr^{3+}. The excess Fe^{2+} was back-titrated, requiring 18.29 mL of 0.00389 M $K_2Cr_2O_7$ to reach the end point. Determine the average thickness of the chromium plate given that the density of Cr is 7.20 g/cm^3.

52. The concentration of CO in air can be determined by passing a known volume of air through a tube containing I_2O_5, resulting in the formation of CO_2 and I_2. The I_2 is removed from the tube by distillation and is collected in a solution containing an excess of KI, producing I_3^-. The I_3^- is titrated with a standard solution of $Na_2S_2O_3$. In a typical analysis, a 4.79-L sample of air was sampled as described here, requiring 7.17 mL of 0.00329 M $Na_2S_2O_3$ to reach the end point. If the air has a density of 1.23×10^{-3} g/mL, determine the parts per million of CO in the air.

53. The level of dissolved oxygen in a water sample can be determined by the Winkler method. In a typical analysis, a 100.0-mL sample is made basic, and treated with a solution of $MnSO_4$, resulting in the formation of MnO_2. An excess of KI is added, and the solution is acidified, resulting in the formation of Mn^{2+} and I_2. The liberated I_2 is titrated with a solution of 0.00870 M $Na_2S_2O_3$, requiring 8.90 mL to reach the starch indicator end point. Calculate the concentration of dissolved oxygen as parts per million of O_2.

54. The analysis for Cl^- using the Volhard method requires a back titration. A known amount of $AgNO_3$ in excess is added, precipitating AgCl. The unreacted Ag^+ is determined by back titrating with KSCN. There is a complication, however, because AgCl is more soluble than AgSCN. (a) Why do the relative solubilities of AgCl and AgSCN lead to a titration error? (b) Is the resulting titration error a positive or negative determinate error? (c) How might you modify the procedure described here to prevent this source of determinate error? (d) Will this source of determinate error be of concern when using the Volhard method to determine Br^-?

55. Voncina and co-workers suggest that precipitation titrations can be monitored by measuring pH as a function of the volume of titrant if the titrant is a weak base.[18] For example, in titrating Pb^{2+} with CrO_4^{2-} the solution containing the analyte is initially acidified to a pH of 3.50 using HNO_3. Before the equivalence point, the concentration of CrO_4^{2-} is controlled by the solubility product of $PbCrO_4$. After the equivalence point, the concentration of CrO_4^{2-} is determined by the amount of excess titrant added. Considering the reactions controlling the concentration of CrO_4^{2-}, sketch the expected titration curve of pH versus volume of titrant.

56. Calculate or sketch (or both) the titration curve for the titration of 50.0 mL of 0.0250 M KI with 0.0500 M $AgNO_3$. Plot both pAg and pI.

57. Calculate or sketch (or both) the titration curve for the titration of 25.0 mL of 0.0500 M KI and 0.0500 M KSCN with 0.0500 M $AgNO_3$.

58. A 0.5131-g sample containing KBr is dissolved in 50 mL of distilled water. Titrating with 0.04614 M $AgNO_3$ requires 25.13 mL to reach the Mohr end point. A blank titration requires 0.65 mL to reach the same end point. Report the %w/w KBr in the sample.

59. A 0.1036-g sample containing only $BaCl_2$ and NaCl is dissolved in 50 mL of distilled water. Titrating with 0.07916 M $AgNO_3$ requires 19.46 mL to reach the Fajans end point. Report the %w/w $BaCl_2$ in the sample.

60. A 0.1093-g sample of impure Na_2CO_3 was analyzed by the Volhard method. After adding 50.00 mL of 0.06911 M $AgNO_3$, the sample was back titrated with 0.05781 M KSCN, requiring 27.36 mL to reach the end point. Report the purity of the Na_2CO_3 sample.

9J SUGGESTED READINGS

For a general history of titrimetry, see the following sources.

Kolthoff, I. M. "Analytical Chemistry in the USA in the First Quarter of This Century," *Anal. Chem.* **1994,** *66,* 241A–249A.

Laitinen, H. A.; Ewing, G. W., eds. *A History of Analytical Chemistry.* The Division of Analytical Chemistry of the American Chemical Society: Washington, DC, 1977, pp. 52–93.

The use of weight instead of volume as a signal for titrimetry is reviewed in the following paper.

Kratochvil, B.; Maitra, C. "Weight Titrations: Past and Present," *Am. Lab.* **1983,** January, 22–29.

A more thorough discussion of nonaqueous titrations, with numerous practical examples, is provided in the following text.

Fritz, J. S. *Acid–Base Titrations in Nonaqueous Solvents.* Allyn and Bacon: Boston, 1973.

The text listed below provides more details on how the potentiometric titration data may be used to calculate equilibrium constants. This text provides a number of examples and includes a discussion of several computer programs that have been developed to model equilibrium reactions.

Meloun, M.; Havel, J.; Högfeldt, E. *Computation of Solution Equilibria.* Ellis Horwood Limited: Chichester, England, 1988.

The following paper provides additional information about the use of Gran plots.

Schwartz, L. M. "Advances in Acid–Base Gran Plot Methodology," *J. Chem. Educ.* **1987,** *64,* 947–950.

For a complete discussion of the application of complexation titrimetry, see the following texts.

Pribil, R. *Applied Complexometry.* Pergamon Press: Oxford, 1982.

Ringbom, A. *Complexation in Analytical Chemistry.* John Wiley and Sons, Inc.: New York, 1963.

Schwarzenbach, G. *Complexometric Titrations.* Methuen & Co. Ltd: London, 1957.

A good source for additional examples of the application of all forms of titrimetry is

Vogel's Textbook of Quantitative Inorganic Analysis, 4th ed. Longman: London, 1981.

9K REFERENCES

1. (a) Willis, C. J. *J. Chem. Educ.* **1981,** *58,* 659–663; (b) Nakagawa, K. *J. Chem. Educ.* **1990,** *67,* 673–676; (c) Gordus, A. A. *J. Chem. Educ.* **1991,** *68,* 759–761; (d) de Levie, R. *J. Electroanal. Chem.* **1992,** *323,* 347–355; (e) de Levie, R. *J. Chem. Educ.* **1993,** *70,* 209–217; (f) Chaston, S. *J. Chem. Educ.* **1993,** *70,* 878–880; (g) de Levie, R. *Anal. Chem.* **1996,** *68,* 585–590.
2. (a) Currie, J. O.; Whiteley, R. V. *J. Chem. Educ.* **1991,** *68,* 923–926; (b) Breneman, G. L.; Parker, O. J. *J. Chem. Educ.* **1992,** *69,* 46–47; (c) Carter, D. R.; Frye, M. S.; Mattson, W. A. *J. Chem. Educ.* **1993,** *70,* 67–71; (d) Freiser, H. *Concepts and Calculations in Analytical Chemistry,* CRC Press: Boca Raton, FL, 1992.
3. Meites, L.; Goldman, J. A. *Anal. Chim. Acta* **1963,** *29,* 472–479.
4. Method 13.86 as published in *Official Methods of Analysis,* 8th Ed., Association of Official Agricultural Chemists: Washington, DC, 1955, p. 226.
5. Ferek, R. J.; Lazrus, A. L.; Haagenson, P. L.; et al. *Environ. Sci. Technol.* **1983,** *17,* 315–324.
6. (a) Molnár-Perl, I.; Pintée-Szakás, M. *Anal. Chim. Acta* **1987,** *202,* 159; (b) Barbosa, J.; Bosch, E.; Cortina, J. L., et al. *Anal. Chim. Acta* **1992,** *256,* 177–181.
7. Gonzalez, A. G.; Asuero, A. G. *Anal. Chim. Acta* **1992,** *257,* 29–33.
8. Papanastasiou, G.; Ziogas, I.; Kokkinidis, G. *Anal. Chim. Acta* **1993,** *277,* 119–135.
9. Alexio, L. M.; Godinho, O. E. S.; da Costa, W. F. *Anal. Chim. Acta* **1992,** *257,* 35–39.
10. Steele, A. W.; Hieftje, G. M. *Anal. Chem.* **1984,** *56,* 2884–2888.
11. (a) Gratzl, M.; Yi, C. *Anal. Chem.* **1993,** *65,* 2085–2088; (b) Yi, C.; Gratzl, M. *Anal. Chem.* **1994,** *66,* 1976–1982; (c) Hui, K. Y.; Gratzl. M. *Anal. Chem.* **1997,** *69,* 695–698; (d) Yi, C.; Huang, D.; Gratzl, M. *Anal. Chem.* **1996,** *68,* 1580–1584; (e) Xie, H.; Gratzl, M. *Anal. Chem.* **1996,** *68,* 3665–3669.
12. Method 2340 C as published in *Standard Methods for the Examination of Water and Wastewater,* 20th ed. American Public Health Association: Washington, DC, 1998, pp. 2–27 to 2–29.
13. Method 4500-ClA as published in *Standard Methods for the Examination of Water and Wastewater,* 20th ed. American Public Health Association: Washington, DC, 1998, pp. 4–53 to 4–55.
14. Schwartz, L. M. *J. Chem. Educ.* **1992,** *69,* 879–883.
15. (a) Boiani, J. A. *J. Chem. Educ.* **1986,** *63,* 724–726; (b) Castillo, C. A.; Jaramillo, A. *J. Chem. Educ.* **1989,** *66,* 341.
16. Prada, S.; Guekezian, M.; Suarez-Iha, M. E. V. *Anal. Chim. Acta* **1996,** *329,* 197–202.
17. Sporek, K. F. *Anal. Chem.* **1958,** *30,* 1032.
18. Voncina, D. B.; Dobcnik, D.; Gomiscek, S. *Anal. Chim. Acta* **1992,** *263,* 147–153.

Chapter 10

Spectroscopic Methods of Analysis

Before the beginning of the twentieth century most quantitative chemical analyses used gravimetry or titrimetry as the analytical method. With these methods, analysts achieved highly accurate results, but were usually limited to the analysis of major and minor analytes. Other methods developed during this period extended quantitative analysis to include trace level analytes. One such method was colorimetry.

One example of an early colorimetric analysis is Nessler's method for ammonia, which was first proposed in 1856. Nessler found that adding an alkaline solution of HgI_2 and KI to a dilute solution of ammonia produced a yellow to reddish brown colloid with the color determined by the concentration of ammonia. A comparison of the sample's color to that for a series of standards was used to determine the concentration of ammonia. Equal volumes of the sample and standards were transferred to a set of tubes with flat bottoms. The tubes were placed in a rack equipped at the bottom with a reflecting surface, allowing light to pass through the solution. The colors of the samples and standards were compared by looking down through the solutions. Until recently, a modified form of this method was listed as a standard method for the analysis of ammonia in water and wastewater.[1]

Colorimetry, in which a sample absorbs visible light, is one example of a spectroscopic method of analysis. At the end of the nineteenth century, spectroscopy was limited to the absorption, emission, and scattering of visible, ultraviolet, and infrared electromagnetic radiation. During the twentieth century, spectroscopy has been extended to include other forms of electromagnetic radiation (photon spectroscopy), such as X-rays, microwaves, and radio waves, as well as energetic particles (particle spectroscopy), such as electrons and ions.[2]

10A Overview of Spectroscopy

The focus of this chapter is photon spectroscopy, using ultraviolet, visible, and infrared radiation. Because these techniques use a common set of optical devices for dispersing and focusing the radiation, they often are identified as optical spectroscopies. For convenience we will usually use the simpler term "spectroscopy" in place of photon spectroscopy or optical spectroscopy; however, it should be understood that we are considering only a limited part of a much broader area of analytical methods. Before we examine specific spectroscopic methods, however, we first review the properties of electromagnetic radiation.

10A.1 What Is Electromagnetic Radiation

Electromagnetic radiation, or light, is a form of energy whose behavior is described by the properties of both waves and particles. The optical properties of electromagnetic radiation, such as diffraction, are explained best by describing light as a wave. Many of the interactions between electromagnetic radiation and matter, such as absorption and emission, however, are better described by treating light as a particle, or photon. The exact nature of electromagnetic radiation remains unclear, as it has since the development of quantum mechanics in the first quarter of the twentieth century.[3] Nevertheless, the dual models of wave and particle behavior provide a useful description for electromagnetic radiation.

Wave Properties of Electromagnetic Radiation Electromagnetic radiation consists of oscillating electric and magnetic fields that propagate through space along a linear path and with a constant velocity (Figure 10.1). In a vacuum, electromagnetic radiation travels at the speed of light, c, which is 2.99792×10^8 m/s. Electromagnetic radiation moves through a medium other than a vacuum with a velocity, v, less than that of the speed of light in a vacuum. The difference between v and c is small enough (< 0.1%) that the speed of light to three significant figures, 3.00×10^8 m/s, is sufficiently accurate for most purposes.

Oscillations in the electric and magnetic fields are perpendicular to each other, and to the direction of the wave's propagation. Figure 10.1 shows an example of plane-polarized electromagnetic radiation consisting of an oscillating electric field and an oscillating magnetic field, each of which is constrained to a single plane. Normally, electromagnetic radiation is unpolarized, with oscillating electric and

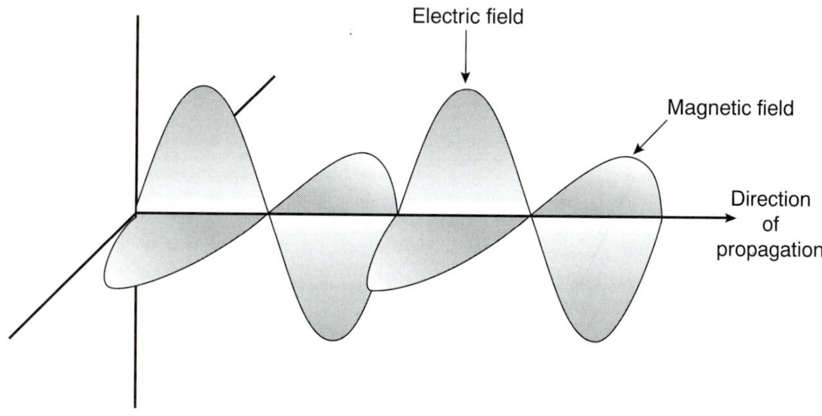

Figure 10.1
Plane-polarized electromagnetic radiation showing the electric field, the magnetic field, and the direction of propagation.

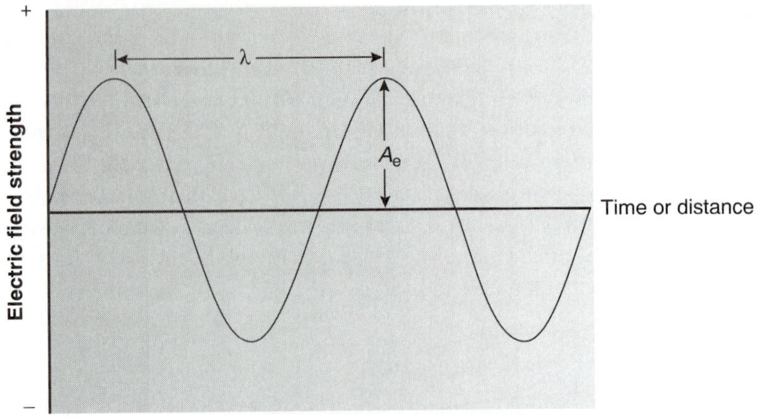

Figure 10.2
Electric field component of plane-polarized electromagnetic radiation.

magnetic fields in all possible planes oriented perpendicular to the direction of propagation.

The interaction of electromagnetic radiation with matter can be explained using either the electric field or the magnetic field. For this reason, only the electric field component is shown in Figure 10.2. The oscillating electric field is described by a sine wave of the form

$$E = A_e \sin(2\pi\nu t + \Phi)$$

where E is the magnitude of the electric field at time t, A_e is the electric field's maximum amplitude, ν is the **frequency,** or the number of oscillations in the electric field per unit time, and Φ is a phase angle accounting for the fact that the electric field's magnitude need not be zero at $t = 0$. An identical equation can be written for the magnetic field, M

frequency
The number of oscillations of an electromagnetic wave per second (ν).

$$M = A_m \sin(2\pi\nu t + \Phi)$$

where A_m is the magnetic field's maximum amplitude.

An electromagnetic wave, therefore, is characterized by several fundamental properties, including its velocity, amplitude, frequency, phase angle, polarization, and direction of propagation.[4] Other properties, which are based on these fundamental properties, also are useful for characterizing the wave behavior of electromagnetic radiation. The **wavelength** of an electromagnetic wave, λ, is defined as the distance between successive maxima, or successive minima (see Figure 10.2). For ultraviolet and visible electromagnetic radiation the wavelength is usually expressed in nanometers (nm, 10^{-9} m), and the wavelength for infrared radiation is given in microns (μm, 10^{-6} m). Unlike frequency, wavelength depends on the electromagnetic wave's velocity, where

wavelength
The distance between any two consecutive maxima or minima of an electromagnetic wave (λ).

$$\lambda = \frac{v}{\nu} = \frac{c}{\nu} \quad \text{(in vacuum)}$$

Thus, for electromagnetic radiation of frequency, ν, the wavelength in vacuum is longer than in other media. Another unit used to describe the wave properties of electromagnetic radiation is the **wavenumber,** $\bar{\nu}$, which is the reciprocal of wavelength

wavenumber
The reciprocal of wavelength ($\bar{\nu}$).

$$\bar{\nu} = \frac{1}{\lambda}$$

Wavenumbers are frequently used to characterize infrared radiation, with the units given in reciprocal centimeter (cm^{-1}).

Example

EXAMPLE 10.1

In 1817, Josef Fraunhofer (1787–1826) studied the spectrum of solar radiation, observing a continuous spectrum with numerous dark lines. Fraunhofer labeled the most prominent of the dark lines with letters. In 1859, Gustav Kirchhoff (1824–1887) showed that the "D" line in the solar spectrum was due to the absorption of solar radiation by sodium atoms. The wavelength of the sodium D line is 589 nm. What are the frequency and the wavenumber for this line?

SOLUTION

The frequency and wavenumber of the sodium D line are

$$\nu = \frac{c}{\lambda} = \frac{3.00 \times 10^8 \text{ m/s}}{589 \times 10^{-9} \text{ m}} = 5.09 \times 10^{14} \text{ s}^{-1}$$

$$\bar{\nu} = \frac{1}{\lambda} = \frac{1}{589 \times 10^{-9} \text{ m}} \times \frac{1 \text{ m}}{100 \text{ cm}} = 1.70 \times 10^4 \text{ cm}^{-1}$$

Two additional wave properties are **power,** *P,* and **intensity,** *I,* which give the flux of energy from a source of electromagnetic radiation.

power
The flux of energy per unit time (*P*).

intensity
The flux of energy per unit time per area (*I*).

Particle Properties of Electromagnetic Radiation When a sample absorbs electromagnetic radiation it undergoes a change in energy. The interaction between the sample and the electromagnetic radiation is easiest to understand if we assume that electromagnetic radiation consists of a beam of energetic particles called **photons.** When a photon is absorbed by a sample, it is "destroyed," and its energy acquired by the sample.[5] The energy of a photon, in joules, is related to its frequency, wavelength, or wavenumber by the following equations

photon
A particle of light carrying an amount of energy equal to $h\nu$.

$$E = h\nu$$

$$= \frac{hc}{\lambda}$$

$$= hc\bar{\nu}$$

where h is Planck's constant, which has a value of 6.626×10^{-34} J • s.

Example

EXAMPLE 10.2

What is the energy per photon of the sodium D line (λ = 589 nm)?

SOLUTION

The energy of the sodium D line is

$$E = \frac{hc}{\lambda} = \frac{(6.626 \times 10^{-34} \text{ J} \cdot \text{s})(3.00 \times 10^8 \text{ m/s})}{589 \times 10^{-9} \text{ m}} = 3.37 \times 10^{-19} \text{ J}$$

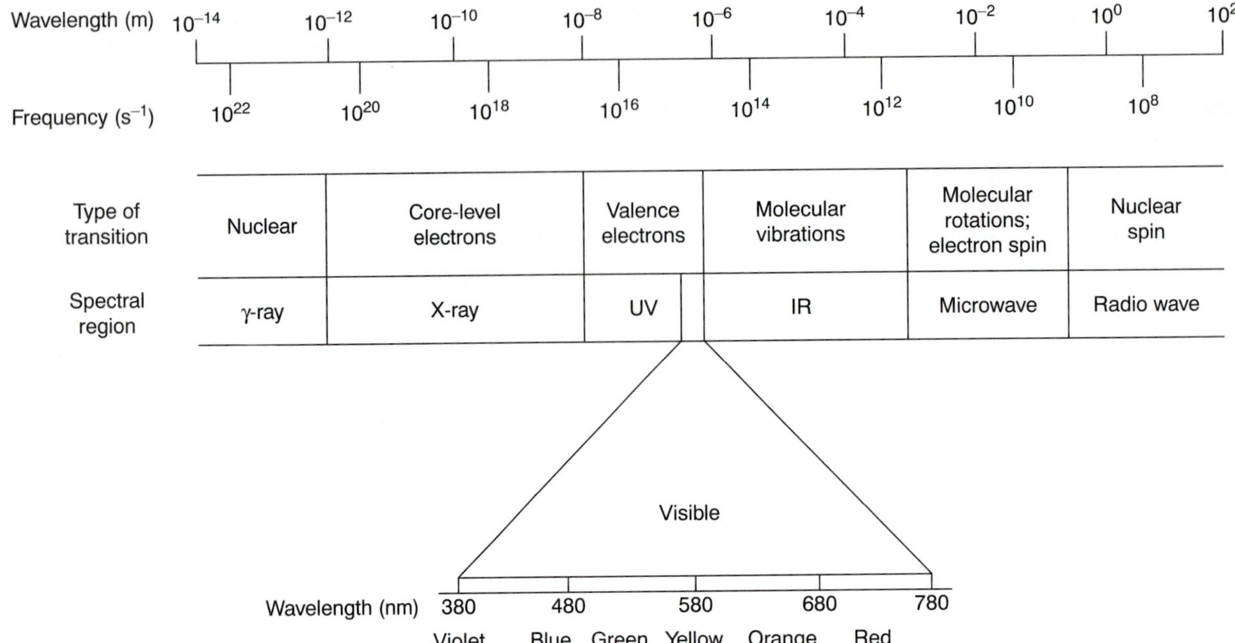

Figure 10.3
The electromagnetic spectrum showing the colors of the visible spectrum.

Colorplate 9 shows the spectrum of visible light.

The energy of a photon provides an additional characteristic property of electromagnetic radiation.

The Electromagnetic Spectrum The frequency and wavelength of electromagnetic radiation vary over many orders of magnitude. For convenience, electromagnetic radiation is divided into different regions based on the type of atomic or molecular transition that gives rise to the absorption or emission of photons (Figure 10.3). The boundaries describing the **electromagnetic spectrum** are not rigid, and an overlap between spectral regions is possible.

electromagnetic spectrum
The division of electromagnetic radiation on the basis of a photon's energy.

10A.2 Measuring Photons as a Signal

In the previous section we defined several characteristic properties of electromagnetic radiation, including its energy, velocity, amplitude, frequency, phase angle, polarization, and direction of propagation. Spectroscopy is possible only if the photon's interaction with the sample leads to a change in one or more of these characteristic properties.

Spectroscopy is conveniently divided into two broad classes. In one class of techniques, energy is transferred between a photon of electromagnetic radiation and the analyte (Table 10.1). In absorption spectroscopy the energy carried by a photon is absorbed by the analyte, promoting the analyte from a lower-energy state to a higher-energy, or excited, state (Figure 10.4). The source of the energetic state depends on the photon's energy. The electromagnetic spectrum in Figure 10.3, for example, shows that absorbing a photon of visible light causes a valence electron in the analyte to move to a higher-energy level. When an analyte absorbs infrared radiation, on the other hand, one of its chemical bonds experiences a change in vibrational energy.

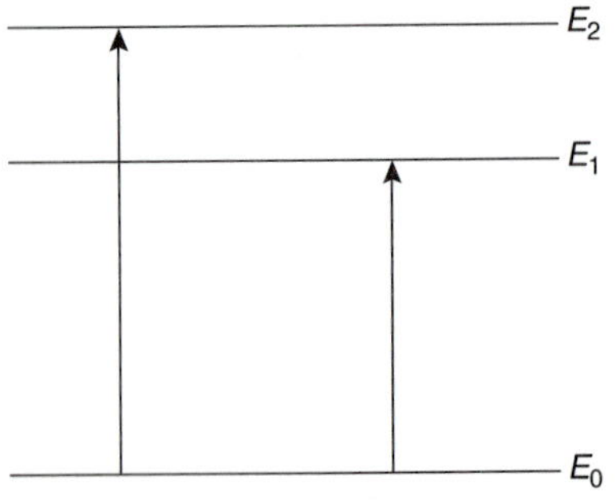

E_2

E_1

E_0

Figure 10.4
Simplified energy level diagram showing absorption of a photon.

Table 10.1 Representative Spectroscopies Involving an Exchange of Energy

Type of Energy Transfer	Region of the Electromagnetic Spectrum	Spectroscopic Technique
absorption	γ-ray	Mossbauer spectroscopy
	X-ray	X-ray absorption spectroscopy
	UV/Vis[a]	UV/Vis spectroscopy[b] atomic absorption spectroscopy[b]
	infrared	infrared spectroscopy[b] raman spectroscopy
	microwave	microwave spectroscopy electron spin resonance spectroscopy
	radio waves	nuclear magnetic resonance spectroscopy
emission (thermal excitation)	UV/Vis	atomic emission spectroscopy[b]
photoluminescence	X-ray	X-ray fluorescence
	UV/Vis	fluorescence spectroscopy[b] phosphorescence spectroscopy[b] atomic fluorescence spectroscopy

[a]UV/Vis: ultraviolet and visible ranges.
[b]Techniques discussed in this text.

The intensity of photons passing through a sample containing the analyte is attenuated because of absorption. The measurement of this attenuation, which we call **absorbance,** serves as our signal. Note that the energy levels in Figure 10.4 have well-defined values (i.e., they are quantized). Absorption only occurs when the photon's energy matches the difference in energy, ΔE, between two energy levels. A plot of absorbance as a function of the photon's energy is called an **absorbance spectrum** (Figure 10.5).

Emission of a photon occurs when an analyte in a higher-energy state returns to a lower-energy state (Figure 10.6). The higher-energy state can be achieved in several ways, including thermal energy, radiant energy from a photon, or by a

absorbance
The attenuation of photons as they pass through a sample (A).

absorbance spectrum
A graph of a sample's absorbance of electromagnetic radiation versus wavelength (or frequency or wavenumber).

emission
The release of a photon when an analyte returns to a lower-energy state from a higher-energy state.

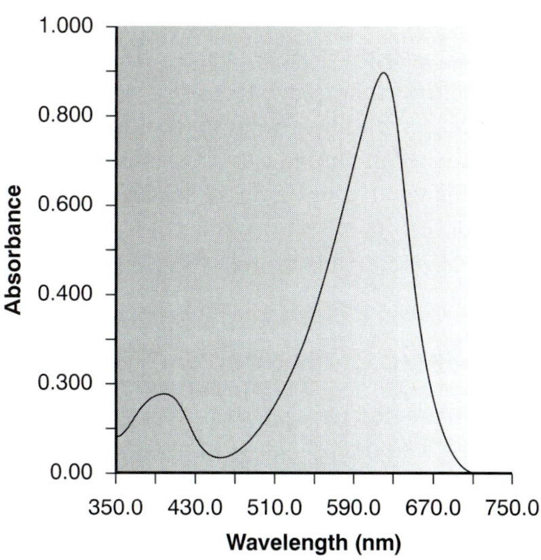

Figure 10.5
Ultraviolet/visible absorption spectrum for bromothymol blue.

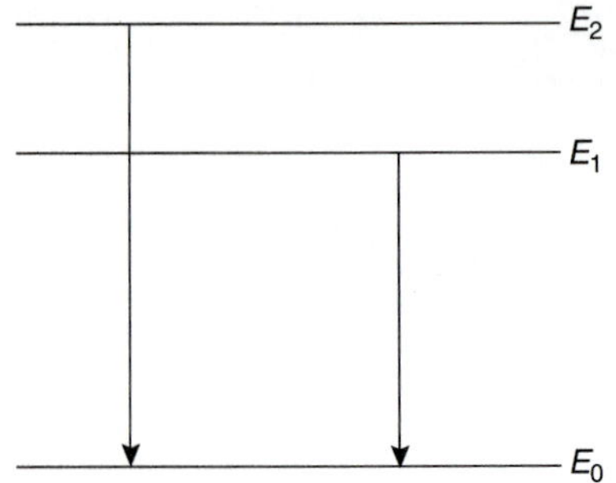

Figure 10.6
Simplified energy level diagram showing emission of a photon.

Table 10.2 Representative Spectroscopies That Do Not Involve an Exchange of Energy

Region of the Electromagnetic Spectrum	Type of Interaction	Spectroscopic Technique
X-ray	diffraction	X-ray diffraction
UV/Vis[a]	refraction	refractometry
	scattering	nephelometry[b] turbidimetry[b]
	dispersion	optical rotary dispersion

[a]UV/Vis: Ultraviolet and visible ranges.
[b]Techniques covered in this text.

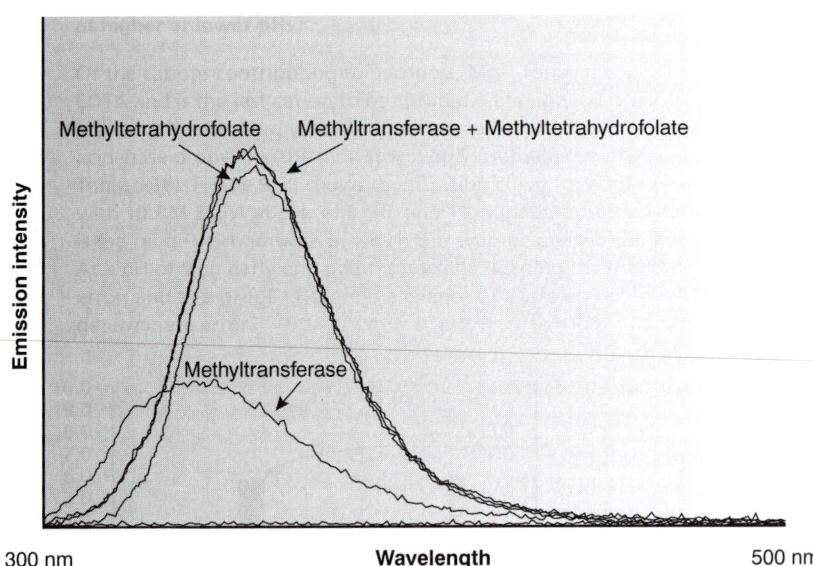

Figure 10.7
Photoluminescent spectra for methyltetrahydrofolate and the enzyme methyltransferase. When methyltetrahydrofolate and methyltransferase are mixed, the enzyme is no longer photoluminescent, but the photoluminescence of methyltetrahydrofolate is enhanced.
(Spectra courtesy of Dave Roberts, DePauw University.)

chemical reaction. Emission following the absorption of a photon is also called **photoluminescence,** and that following a chemical reaction is called **chemiluminescence.** A typical **emission spectrum** is shown in Figure 10.7.

photoluminescence
Emission following absorption of a photon.

chemiluminescence
Emission induced by a chemical reaction.

emission spectrum
A graph of emission intensity versus wavelength (or frequency or wavenumber).

In the second broad class of spectroscopy, the electromagnetic radiation undergoes a change in amplitude, phase angle, polarization, or direction of propagation as a result of its refraction, reflection, scattering, diffraction, or dispersion by the sample. Several representative spectroscopic techniques are listed in Table 10.2.

10B Basic Components of Spectroscopic Instrumentation

The instruments used in spectroscopy consist of several common components, including a source of energy that can be input to the sample, a means for isolating a narrow range of wavelengths, a detector for measuring the signal, and a signal processor to display the signal in a form convenient for the analyst. In this section we introduce the basic components used to construct spectroscopic in-

Table 10.3 Common Sources of Electromagnetic Radiation for Spectroscopy

Source	Wavelength Region	Useful for
H_2 and D_2 lamp	continuum source from 160–380 nm	UV molecular absorption
tungsten lamp	continuum source from 320–2400 nm	Vis molecular absorption
Xe arc lamp	continuum source from 200–1000 nm	molecular fluorescence
Nernst glower	continuum source from 0.4–20 μm	IR molecular absorption
globar	continuum source from 1–40 μm	IR molecular absorption
nichrome wire	continuum source from 0.75–20 μm	IR molecular absorption
hollow cathode lamp	line source in UV/Vis	atomic absorption
Hg vapor lamp	line source in UV/Vis	molecular fluorescence
laser	line source in UV/Vis	atomic and molecular absorption, fluorescence and scattering

Abbreviations: UV: ultraviolet; Vis: visible; IR: infrared.

struments. A more detailed discussion of these components can be found in the suggested end-of-chapter readings. Specific instrument designs are considered in later sections.

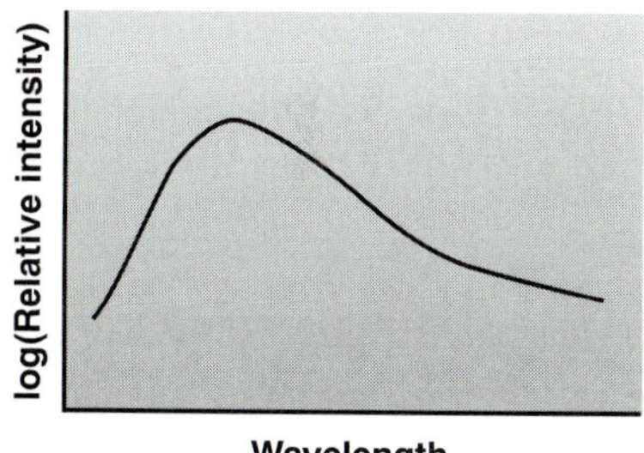

Figure 10.8
Emission spectrum from a typical continuum source.

10B.1 Sources of Energy

All forms of spectroscopy require a source of energy. In absorption and scattering spectroscopy this energy is supplied by photons. Emission and luminescence spectroscopy use thermal, radiant (photon), or chemical energy to promote the analyte to a less stable, higher energy state.

Sources of Electromagnetic Radiation A source of electromagnetic radiation must provide an output that is both intense and stable in the desired region of the electromagnetic spectrum. Sources of electromagnetic radiation are classified as either continuum or line sources. A **continuum source** emits radiation over a wide range of wavelengths, with a relatively smooth variation in intensity as a function of wavelength (Figure 10.8). **Line sources,** on the other hand, emit radiation at a few selected, narrow wavelength ranges (Figure 10.9). Table 10.3 provides a list of the most common sources of electromagnetic radiation.

continuum source
A source that emits radiation over a wide range of wavelengths.

line source
A source that emits radiation at only select wavelengths.

Sources of Thermal Energy The most common sources of thermal energy are flames and plasmas. Flame sources use the combustion of a fuel and an oxidant such as acetylene and air, to achieve temperatures of 2000–3400 K. Plasmas, which are hot, ionized gases, provide temperatures of 6000–10,000 K.

Chemical Sources of Energy Exothermic reactions also may serve as a source of energy. In chemiluminescence the analyte is raised to a higher-energy state by means of a chemical reaction, emitting characteristic radiation when it returns to a lower-energy state. When the chemical reaction results from a biological or enzymatic reaction, the emission of radiation is called bioluminescence. Commercially available "light sticks" and the flash of light from a firefly are examples of chemiluminescence and bioluminescence, respectively.

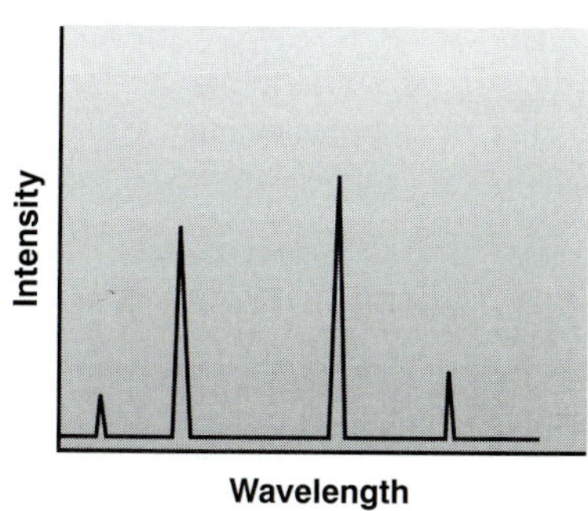

Figure 10.9
Emission spectrum from a typical line source.

10B.2 Wavelength Selection

In Nessler's original colorimetric method for ammonia, described at the beginning of the chapter, no attempt was made to narrow the wavelength range of visible light passing through the sample. If more than one component in the sample contributes to the absorption of radiation, however, then a quantitative analysis using Nessler's original method becomes impossible. For this reason we usually try to select a single wavelength where the analyte is the only absorbing species. Unfortunately, we cannot isolate a single wavelength of radiation from a continuum source. Instead, a wavelength selector passes a narrow band of radiation (Figure 10.10) characterized by a **nominal wavelength,** an **effective bandwidth,** and a maximum throughput of radiation. The effective bandwidth is defined as the width of the radiation at half the maximum throughput.

nominal wavelength
The wavelength which a wavelength selector is set to pass.

effective bandwidth
The width of the band of radiation passing through a wavelength selector measured at half the band's height.

The ideal wavelength selector has a high throughput of radiation and a narrow effective bandwidth. A high throughput is desirable because more photons pass through the wavelength selector, giving a stronger signal with less background noise. A narrow effective bandwidth provides a higher **resolution,** with spectral features separated by more than twice the effective bandwidth being resolved. Generally these two features of a wavelength selector are in opposition (Figure 10.11). Conditions favoring a higher throughput of radiation usually provide less resolution. Decreasing the effective bandwidth improves resolution, but at the cost of a noisier signal. For a qualitative analysis, resolution is generally more important than the throughput of radiation; thus, smaller effective bandwidths are desirable. In a quantitative analysis a higher throughput of radiation is usually desirable.[6]

resolution
In spectroscopy, the separation between two spectral features, such as absorption or emission lines.

Wavelength Selection Using Filters The simplest method for isolating a narrow band of radiation is to use an absorption or interference filter. Absorption **filters** work by selectively absorbing radiation from a narrow region of the electromagnetic spectrum. Interference filters use constructive and destructive interference to isolate a narrow range of wavelengths. A simple example of an absorption filter is a piece of colored glass. A purple filter, for example, removes the complementary color green from 500–560 nm. Commercially available absorption filters provide effective bandwidths from 30–250 nm. The maximum throughput for the smallest effective bandpasses, however, may be only 10% of the source's emission intensity over that range of wavelengths. Interference filters are more expensive than absorption filters, but have narrower effective bandwidths, typically 10–20 nm, with maximum throughputs of at least 40%.

filter
A wavelength selector that uses either absorption, or constructive and destructive interference to control the range of selected wavelengths.

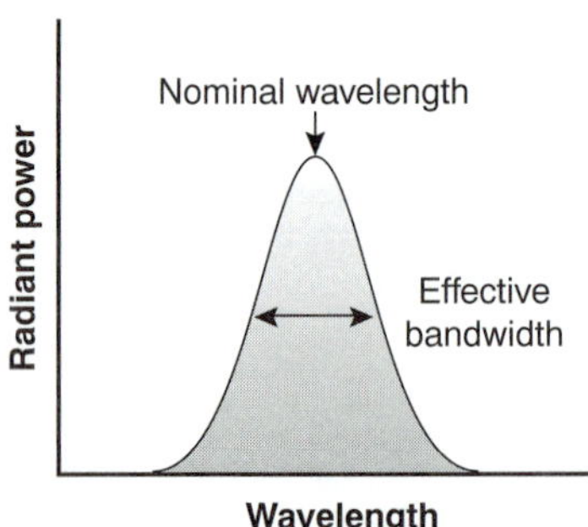

Figure 10.10
Band of radiation exiting wavelength selector showing the nominal wavelength and effective bandpass.

Wavelength Selection Using Monochromators One limitation of an absorption or interference filter is that they do not allow for a continuous selection of wavelength. If measurements need to be made at two wavelengths, then the filter must be changed in between measurements. A further limitation is that filters are available for only selected nominal ranges of wavelengths. An alternative approach to wavelength selection, which provides for a continuous variation of wavelength, is the **monochromator.**

monochromator
A wavelength selector that uses a diffraction grating or prism, and that allows for a continuous variation of the nominal wavelength.

The construction of a typical monochromator is shown in Figure 10.12. Radiation from the source enters the monochromator through an entrance slit. The radiation is collected by a collimating mirror, which reflects a parallel beam of radiation to a diffraction grating. The diffraction grating is an optically reflecting surface with

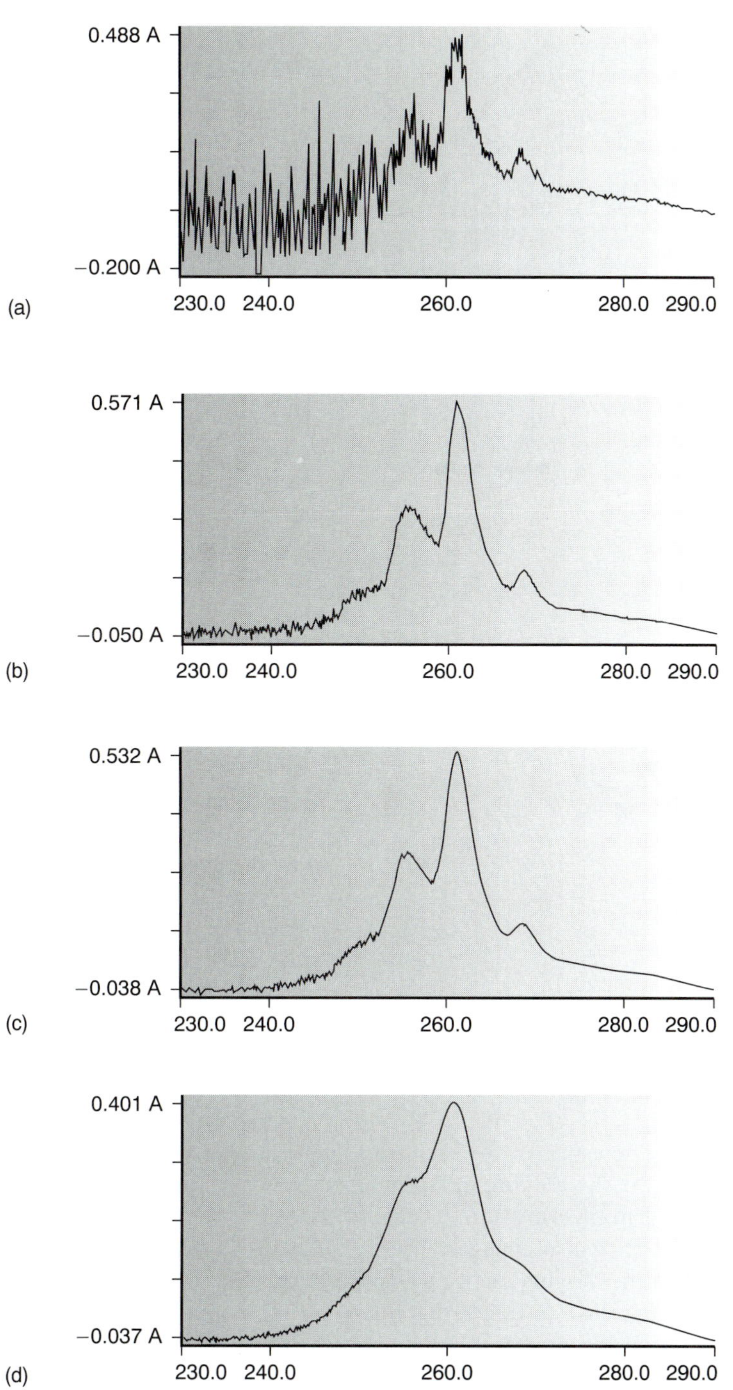

Figure 10.11
Effect of the monochromator's slit width on noise and resolution for the ultraviolet absorption spectrum of benzene. The slit width increases from spectrum (a) to spectrum (d) with effective bandpasses of 0.25 nm, 1.0 nm, 2.0 nm, and 4.0 nm.

a large number of parallel grooves (see inset to Figure 10.12). Diffraction by the grating disperses the radiation in space, where a second mirror focuses the radiation onto a planar surface containing an exit slit. In some monochromators a prism is used in place of the diffraction grating.

Radiation exits the monochromator and passes to the detector. As shown in Figure 10.12, a **polychromatic** source of radiation at the entrance slit is converted at the exit slit to a **monochromatic** source of finite effective bandwidth. The choice of

polychromatic
Electromagnetic radiation of more than one wavelength.

monochromatic
Electromagnetic radiation of a single wavelength.

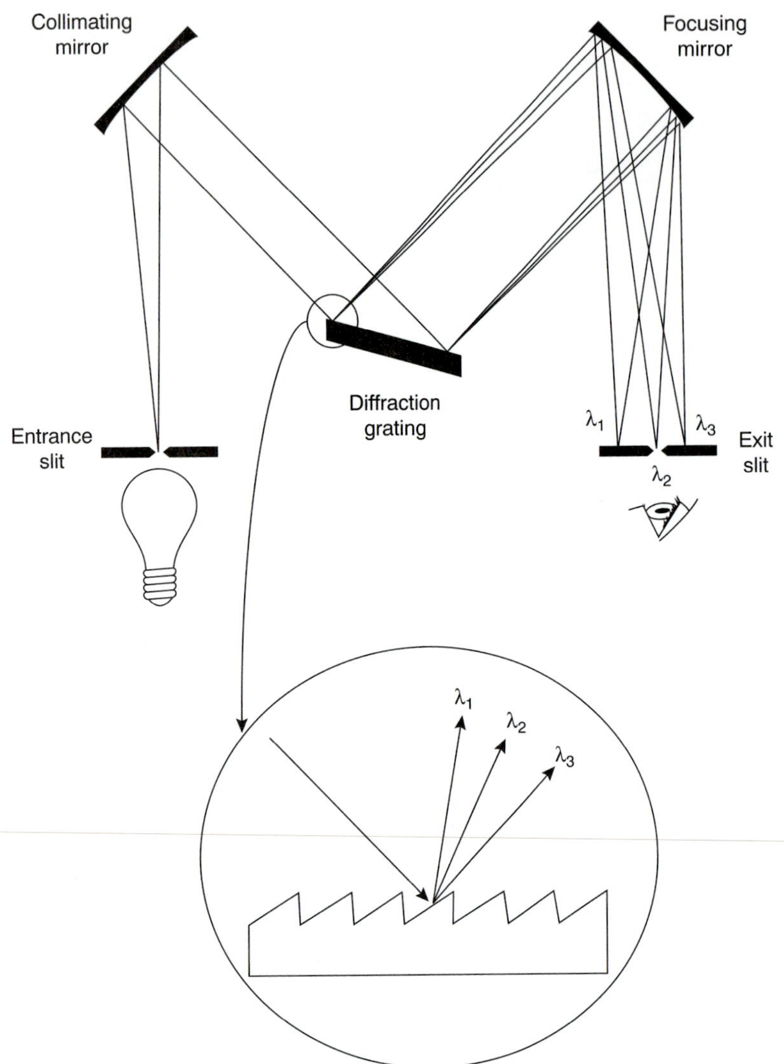

Figure 10.12
Typical grating monochromator with inset showing the dispersion of the radiation by the diffraction grating.

interferometer
A device that allows all wavelengths of light to be measured simultaneously, eliminating the need for a wavelength selector.

which wavelength exits the monochromator is determined by rotating the diffraction grating. A narrower exit slit provides a smaller effective bandwidth and better resolution, but allows a smaller throughput of radiation.

Monochromators are classified as either fixed-wavelength or scanning. In a fixed-wavelength monochromator, the wavelength is selected by manually rotating the grating. Normally, a fixed-wavelength monochromator is only used for quantitative analyses where measurements are made at one or two wavelengths. A scanning monochromator includes a drive mechanism that continuously rotates the grating, allowing successive wavelengths to exit from the monochromator. Scanning monochromators are used to acquire spectra and, when operated in a fixed-wavelength mode, for quantitative analysis.

Interferometers An **interferometer** provides an alternative approach for wavelength selection. Instead of filtering or dispersing the electromagnetic radiation, an interferometer simultaneously allows source radiation of all wavelengths to reach the detector (Figure 10.13). Radiation from the source is focused on a beam splitter that transmits half of the radiation to a fixed mirror, while reflecting the other half to a movable mirror. The radiation recombines at the beam splitter, where constructive and destructive interference determines, for each wavelength, the intensity of light reaching the detector. As the moving mirror changes position, the wavelengths of light experiencing maximum constructive interference and maximum destructive interference also changes. The signal at the detector shows intensity as a function of the moving mirror's position, expressed in units of distance or time. The result is called an interferogram, or a time domain spectrum. The time domain spectrum is converted mathematically, by a process called a Fourier transform, to the normal spectrum (also called a frequency domain spectrum) of intensity as a function of the radiation's energy. Further details about interferometers and the mathematics of the Fourier transform can be found in the suggested readings listed at the end of the chapter.

In comparison with a monochromator, interferometers provide two significant advantages. The first advantage, which is termed Jacquinot's advantage, results from the higher throughput of source radiation. Since an interferometer does not use slits and has fewer optical components from which radiation can be scattered and lost, the throughput of radiation reaching the detector is 80–200 times greater than that achieved with a monochromator. The result is an im-

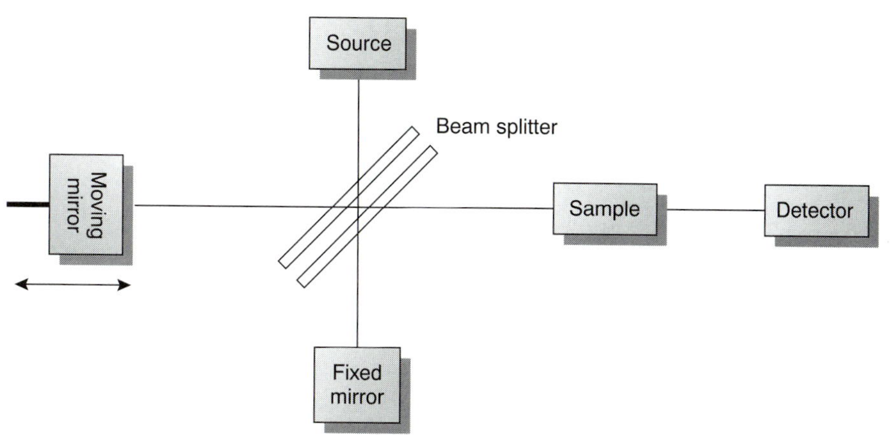

Figure 10.13
Block diagram of an interferometer.

proved **signal-to-noise ratio.** The second advantage, which is called Fellgett's advantage, reflects a savings in the time needed to obtain a spectrum. Since all frequencies are monitored simultaneously, an entire spectrum can be recorded in approximately 1 s, as compared to 10–15 min with a scanning monochromator.

signal-to-noise ratio
The ratio of the signal's intensity to the average intensity of the surrounding noise.

10B.3 Detectors

The first detector for optical spectroscopy was the human eye, which, of course, is limited both by its accuracy and its limited sensitivity to electromagnetic radiation. Modern detectors use a sensitive **transducer** to convert a signal consisting of photons into an easily measured electrical signal. Ideally the detector's signal, S, should be a linear function of the electromagnetic radiation's power, P,

$$S = kP + D$$

where k is the detector's sensitivity, and D is the detector's **dark current,** or the background electric current when all radiation from the source is blocked from the detector.

transducer
A device that converts a chemical or physical property, such as pH or photon intensity, to an easily measured electrical signal, such as a voltage or current.

dark current
The background current present in a photon detector in the absence of radiation from the source.

Photon Transducers Two general classes of transducers are used for optical spectroscopy, several examples of which are listed in Table 10.4. Phototubes and photomultipliers contain a photosensitive surface that absorbs radiation in the ultraviolet, visible, and near infrared (IR), producing an electric current proportional to the number of photons reaching the transducer. Other photon detectors use a semiconductor as the photosensitive surface. When the semiconductor absorbs photons, valence electrons move to the semiconductor's conduction band, producing a measurable current. One advantage of the Si photodiode is that it is easily miniaturized. Groups of photodiodes may be gathered together in a linear array containing from 64 to 4096 individual photodiodes. With a width of 25 μm per diode, for example, a linear array of 2048 photodiodes requires only 51.2 mm of linear space. By placing a **photodiode array** along the monochromator's focal plane, it is possible to monitor simultaneously an entire range of wavelengths.

photodiode array
A linear array of photodiodes providing the ability to detect simultaneously radiation at several wavelengths.

Thermal Transducers Infrared radiation generally does not have sufficient energy to produce a measurable current when using a photon transducer. A thermal transducer, therefore, is used for infrared spectroscopy. The absorption of infrared photons by a thermal transducer increases its temperature, changing one or more of its characteristic properties. The pneumatic transducer, for example,

Table 10.4 Characteristics of Transducers for Optical Spectroscopy

Detector	Class	Wavelength Range	Output Signal
phototube	photon	200–1000 nm	current
photomultiplier	photon	110–1000 nm	current
Si photodiode	photon	250–1100 nm	current
photoconductor	photon	750–6000 nm	change in resistance
photovoltaic cell	photon	400–5000 nm	current or voltage
thermocouple	thermal	0.8–40 μm	voltage
thermistor	thermal	0.8–40 μm	change in resistance
pneumatic	thermal	0.8–1000 μm	membrane displacement
pyroelectric	thermal	0.3–1000 μm	current

consists of a small tube filled with xenon gas equipped with an IR-transparent window at one end, and a flexible membrane at the other end. A blackened surface in the tube absorbs photons, increasing the temperature and, therefore, the pressure of the gas. The greater pressure in the tube causes the flexible membrane to move in and out, and this displacement is monitored to produce an electrical signal.

10B.4 Signal Processors

signal processor
A device, such as a meter or computer, that displays the signal from the transducer in a form that is easily interpreted by the analyst.

The electrical signal generated by the transducer is sent to a **signal processor** where it is displayed in a more convenient form for the analyst. Examples of signal processors include analog or digital meters, recorders, and computers equipped with digital acquisition boards. The signal processor also may be used to calibrate the detector's response, to amplify the signal from the detector, to remove noise by filtering, or to mathematically transform the signal.

10C Spectroscopy Based on Absorption

Historically, the first spectroscopic studies involved characterizing the emission of visible light from the sun, from flames, and from salts added to flames. Our survey of spectroscopy, however, begins with absorption because it is the more important technique in modern analytical spectroscopy.

10C.1 Absorbance of Electromagnetic Radiation

In absorption spectroscopy a beam of electromagnetic radiation passes through a sample. Much of the radiation is transmitted without a loss in intensity. At selected frequencies, however, the radiation's intensity is attenuated. This process of attenuation is called absorption. Two general requirements must be met if an analyte is to absorb electromagnetic radiation. The first requirement is that there must be a mechanism by which the radiation's electric field or magnetic field interacts with the analyte. For ultraviolet and visible radiation, this interaction involves the electronic energy of valence electrons. A chemical bond's vibrational energy is altered by the absorbance of infrared radiation. A more detailed treatment of this interaction, and its importance in determining the intensity of absorption, is found in the suggested readings listed at the end of the chapter.

The second requirement is that the energy of the electromagnetic radiation must exactly equal the difference in energy, ΔE, between two of the analytes quantized energy states. Figure 10.4 shows a simplified view of the absorption of a photon. The figure is useful because it emphasizes that the photon's energy must match the difference in energy between a lower-energy state and a higher-energy state. What is missing, however, is information about the types of energetic states involved, which transitions between states are likely to occur, and the appearance of the resulting spectrum.

We can use the energy level diagram in Figure 10.14 to explain an absorbance spectrum. The thick lines labeled E_0 and E_1 represent the analyte's ground (lowest) electronic state and its first electronic excited state. Superimposed on each electronic energy level is a series of lines representing vibrational energy levels.

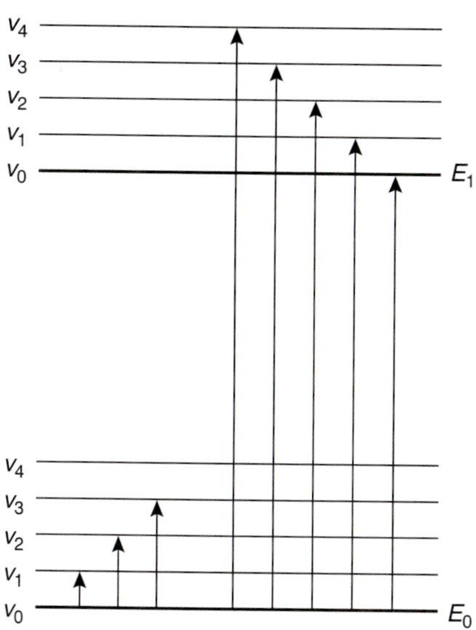

Figure 10.14
Energy level diagram showing difference between the absorption of infrared radiation (*left*) and ultraviolet–visible radiation (*right*).

Infrared Spectra for Molecules and Polyatomic Ions The energy of infrared radiation is sufficient to produce a change in the vibrational energy of a molecule or polyatomic ion (see Table 10.1). As shown in Figure 10.14, vibrational energy levels are quantized; that is, a molecule may have only certain, discrete vibrational energies. The energy for allowed vibrational modes, E_v, is

$$E_v = \left(v + \frac{1}{2}\right)h\nu_0$$

where v is the vibrational quantum number, which may take values of 0, 1, 2, . . ., and ν_0 is the bond's fundamental vibrational frequency. Values for ν_0 are determined by the bond's strength and the mass at each end of the bond and are characteristic of the type of bond. For example, a carbon–carbon single bond (C—C) absorbs infrared radiation at a lower energy than a carbon–carbon double bond (C=C) because a C—C bond is weaker than a C=C bond.

At room temperature most molecules are in their ground vibrational state ($v = 0$). A transition from the ground vibrational state to the first vibrational excited state ($v = 1$) requires the absorption of a photon with an energy of $h\nu_0$. Transitions in which Δv is ±1 give rise to the fundamental absorption lines. Weaker absorption lines, called overtones, are due to transitions in which Δv is ±2 or ±3. The number of possible normal vibrational modes for a linear molecule is $3N - 5$, and for a nonlinear molecule is $3N - 6$, where N is the number of atoms in the molecule. Not surprisingly, infrared spectra often show a considerable number of absorption bands. Even a relatively simple molecule, such as benzene (C_6H_6), for example, has 30 possible normal modes of vibration, although not all of these vibrational modes give rise to an absorption. A typical IR spectrum is shown in Figure 10.15.

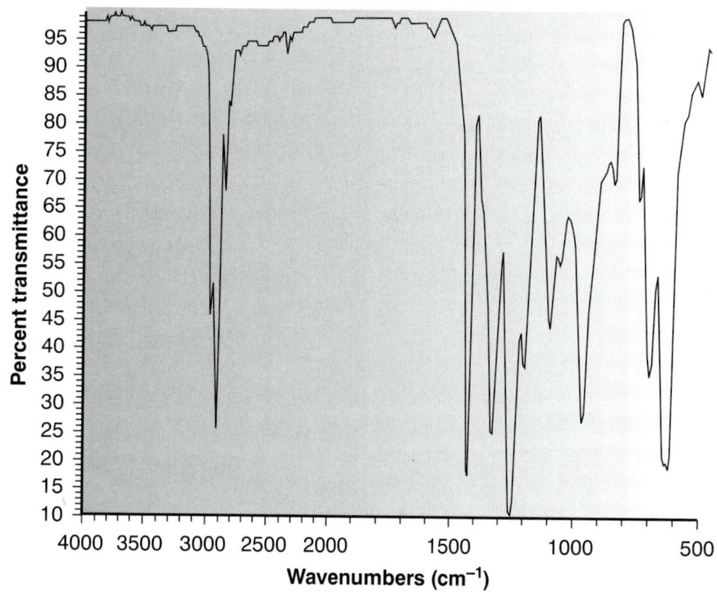

Figure 10.15
Fourier transform infrared (FT–IR) spectrum of polyvinylchloride.

Table 10.5 Electronic Transitions Involving n, σ, and π Molecular Orbitals

Transition	Wavelength Range (nm)	Examples
$\sigma \rightarrow \sigma^*$	< 200	C—C, C—H
$n \rightarrow \sigma^*$	160–260	H_2O, CH_3OH, CH_3Cl
$\pi \rightarrow \pi^*$	200–500	C=C, C=O, C=N, C≡C
$n \rightarrow \pi^*$	250–600	C=O, C=N, N=N, N=O

UV/Vis Spectra for Molecules and Ions When a molecule or ion absorbs ultraviolet or visible radiation it undergoes a change in its valence electron configuration. The valence electrons in organic molecules, and inorganic anions such as CO_3^{2-}, occupy quantized sigma bonding, σ, pi bonding, π, and nonbonding, n, molecular orbitals. Unoccupied sigma antibonding, σ^*, and pi antibonding, π^*, molecular orbitals often lie close enough in energy that the transition of an electron from an occupied to an unoccupied orbital is possible.

Four types of transitions between quantized energy levels account for molecular UV/Vis spectra. The approximate wavelength ranges for these absorptions, as well as a partial list of bonds, functional groups, or molecules that give rise to these transitions is shown in Table 10.5. Of these transitions, the most important are the $n \rightarrow \pi^*$ and $\pi \rightarrow \pi^*$, because they involve functional groups that are characteristic of the analyte and wavelengths that are easily accessible. The bonds and functional groups that give rise to the absorption of ultraviolet and visible radiation are called **chromophores.**

chromophore
The specific bonds or functional groups in a molecule responsible for the absorption of a particular wavelength of light.

Many transition metal ions, such as Cu^{2+} and Co^{2+}, form solutions that are colored because the metal ion absorbs visible light. The transitions giving rise to this absorption are due to valence electrons in the metal ion's *d*-orbitals. For a free metal ion, the five *d*-orbitals are of equal energy. In the presence of a complexing ligand or solvent molecule, however, the *d*-orbitals split into two or more groups that differ in energy. For example, in the octahedral complex $Cu(H_2O)_6^{2+}$ the six water molecules perturb the *d*-orbitals into two groups as shown in Figure 10.16. The resulting *d–d* transitions for transition metal ions are relatively weak.

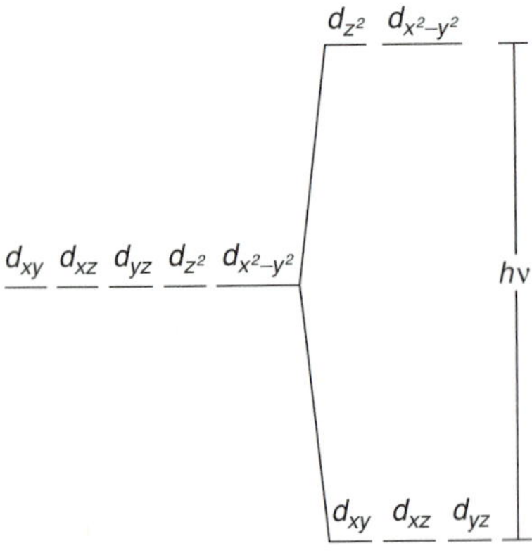

Figure 10.16
Splitting of *d*-orbitals in an octahedral field.

A more important source of UV/Vis absorption for inorganic metal–ligand complexes is charge transfer, in which absorbing a photon produces an excited state species that can be described in terms of the transfer of an electron from the metal, *M*, to the ligand, *L*.

$$M\text{—}L + h\nu \rightarrow M^{+}\text{—}L^{-}$$

Charge-transfer absorption is important because it produces very large absorbances, providing for a much more sensitive analytical method. One important example of a charge-transfer complex is that of *o*-phenanthroline with Fe^{2+}, the UV/Vis spectrum for which is shown in Figure 10.17. Charge-transfer absorption in which the electron moves from the ligand to the metal also is possible.

Comparing the IR spectrum in Figure 10.15 to the UV/Vis spectrum in Figure 10.17, we note that UV/Vis absorption bands are often significantly broader than those for IR absorption. Figure 10.14 shows why this is true. When a species

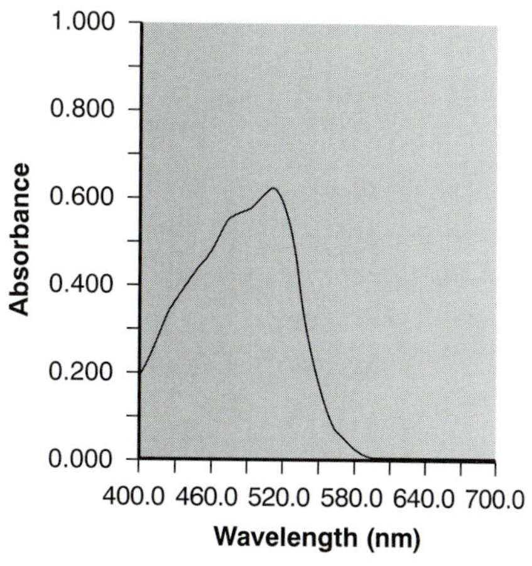

Figure 10.17
UV/Vis spectrum for Fe(*o*-phenanthroline)$_3^{2+}$.

absorbs UV/Vis radiation, the transition between electronic energy levels may also include a transition between vibrational energy levels. The result is a number of closely spaced absorption bands that merge together to form a single broad absorption band.

UV/Vis Spectra for Atoms As noted in Table 10.1, the energy of ultraviolet and visible electromagnetic radiation is sufficient to cause a change in an atom's valence electron configuration. Sodium, for example, with a valence shell electron configuration of [Ne] $3s^1$, has a single valence electron in its 3*s* atomic orbital. Unoccupied, higher energy atomic orbitals also exist. Figure 10.18 shows a partial energy level diagram for sodium's occupied and unoccupied valence shell atomic orbitals. This configuration of atomic orbitals, which shows a splitting of the *p* orbitals into two levels with slightly different energies, may differ from that encountered in earlier courses. The reasons for this splitting, however, are beyond the level of this text, and unimportant in this context.

Absorption of a photon is accompanied by the excitation of an electron from a lower-energy atomic orbital to an orbital of higher energy. Not all possible transitions between atomic orbitals are allowed. For sodium the only allowed transitions are those in which there is a change of ±1 in the orbital quantum number (ℓ); thus transitions from $s \rightarrow p$ orbitals are allowed, but transitions from $s \rightarrow d$ orbitals are forbidden. The wavelengths of electromagnetic radiation that must be absorbed to cause several allowed transitions are shown in Figure 10.18.

The atomic absorption spectrum for Na is shown in Figure 10.19 and is typical of that found for most atoms. The most obvious feature of this spectrum is that it consists of a few, discrete absorption lines corresponding to transitions between the ground state (the 3*s* atomic orbital) and the 3*p* and 4*p* atomic orbitals. Absorption from excited states, such as that from the 3*p* atomic orbital to the 4*s* or 3*d* atomic orbital, which are included in the energy level diagram in Figure 10.18, are too weak to detect. Since the

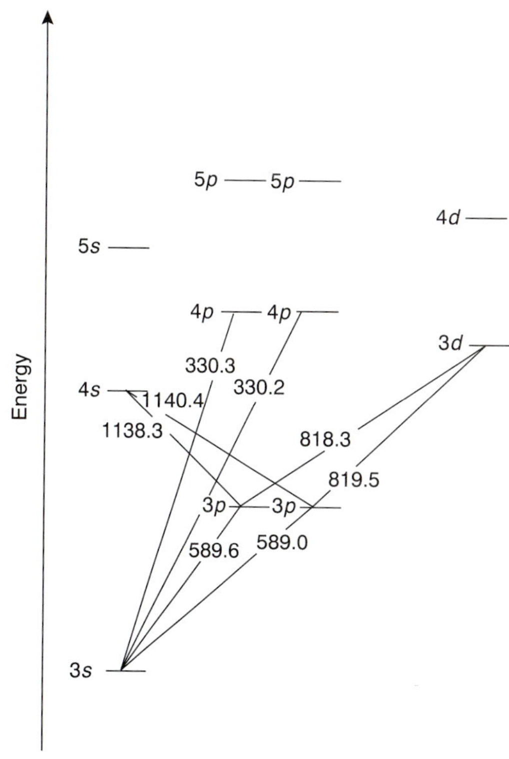

Figure 10.18
Valence shell energy diagram for sodium.

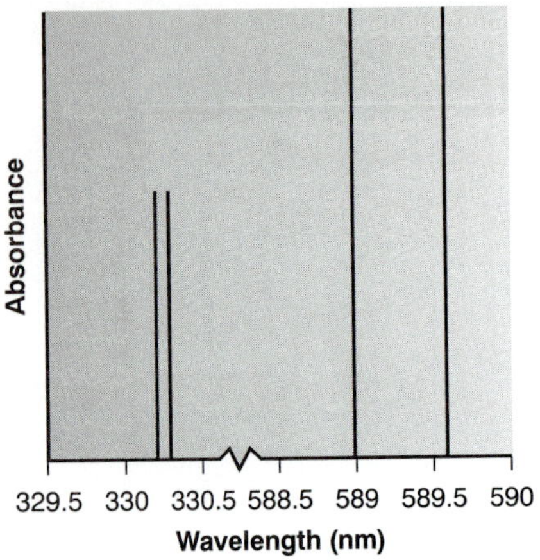

Figure 10.19
Atomic absorption spectrum for sodium.

lifetime of an excited state is short, typically 10^{-7}–10^{-8} s, an atom in the excited state is likely to return to the ground state before it has an opportunity to absorb a photon.

Another feature of the spectrum shown in Figure 10.19 is the narrow width of the absorption lines, which is a consequence of the fixed difference in energy between the ground and excited states. Natural line widths for atomic absorption, which are governed by the uncertainty principle, are approximately 10^{-5} nm. Other contributions to broadening increase this line width to approximately 10^{-3} nm.

10C.2 Transmittance and Absorbance

transmittance
The ratio of the radiant power passing through a sample to that from the radiation's source (T).

The attenuation of electromagnetic radiation as it passes through a sample is described quantitatively by two separate, but related terms: transmittance and absorbance. **Transmittance** is defined as the ratio of the electromagnetic radiation's power exiting the sample, P_T, to that incident on the sample from the source, P_0, (Figure 10.20a).

$$T = \frac{P_T}{P_0} \tag{10.1}$$

Multiplying the transmittance by 100 gives the percent transmittance (%T), which varies between 100% (no absorption) and 0% (complete absorption). All methods of detection, whether the human eye or a modern photoelectric transducer, measure the transmittance of electromagnetic radiation.

Attenuation of radiation as it passes through the sample leads to a transmittance of less than 1. As described, equation 10.1 does not distinguish between the different ways in which the attenuation of radiation occurs. Besides absorption by the analyte, several additional phenomena contribute to the net attenuation of radiation, including reflection and absorption by the sample container, absorption by components of the sample matrix other than the analyte, and the scattering of radiation. To compensate for this loss of the electromagnetic radiation's power, we use a method blank (Figure 10.20b). The radiation's power exiting from the method blank is taken to be P_0.

An alternative method for expressing the attenuation of electromagnetic radiation is absorbance, A, which is defined as

$$A = -\log T = -\log \frac{P_T}{P_0} = \log \frac{P_0}{P_T} \tag{10.2}$$

Absorbance is the more common unit for expressing the attenuation of radiation because, as shown in the next section, it is a linear function of the analyte's concentration.

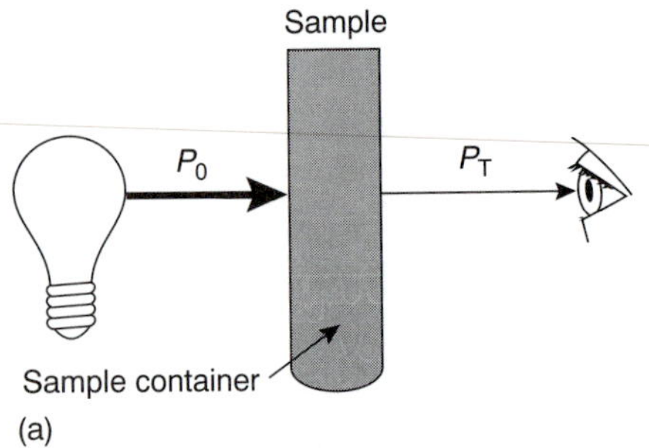

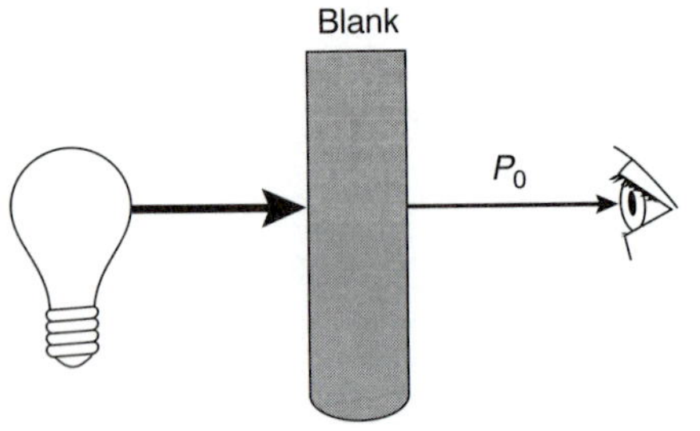

Figure 10.20
(a) Schematic diagram showing the attenuation of radiation passing through a sample; P_0 is the radiant power from the source and P_T is the radiant power transmitted by the sample. (b) Schematic diagram showing that P_0 is redefined as the radiant power transmitted by the blank, correcting the transmittance in (a) for any loss of radiation due to scattering, reflection or absorption by the cuvette, and absorption by the sample's matrix.

Example

EXAMPLE 10.3

A sample has a percent transmittance of 50.0%. What is its absorbance?

SOLUTION

With a percent transmittance of 50.0%, the transmittance of the sample is 0.500. Substituting into equation 10.2 gives

$$A = -\log T = -\log(0.500) = 0.301$$

Equation 10.1 has an important consequence for atomic absorption. Because of the narrow line width for atomic absorption, a continuum source of radiation cannot be used. Even with a high-quality monochromator, the effective bandwidth for a continuum source is 100–1000 times greater than that for an atomic absorption line. As a result, little of the radiation from a continuum source is absorbed ($P_o \approx P_T$), and the measured absorbance is effectively zero. For this reason, atomic absorption requires a line source.

10C.3 Absorbance and Concentration: Beer's Law

When monochromatic electromagnetic radiation passes through an infinitesimally thin layer of sample, of thickness dx, it experiences a decrease in power of dP (Figure 10.21). The fractional decrease in power is proportional to the sample's thickness and the analyte's concentration, C; thus

$$\frac{-dP}{P} = \alpha C dx \qquad \textbf{10.3}$$

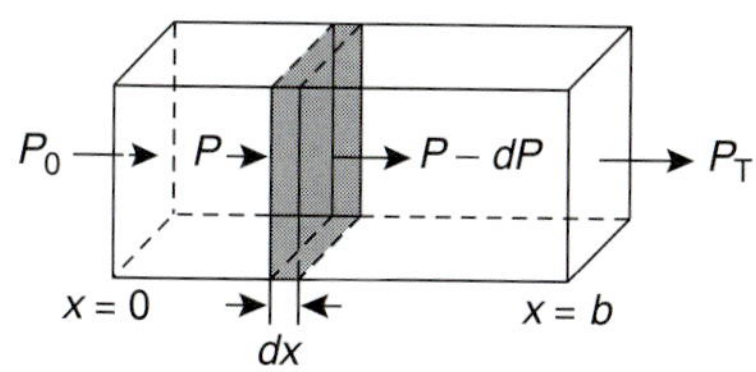

Figure 10.21
Factors used in deriving the Beer–Lambert law.

where P is the power incident on the thin layer of sample, and α is a proportionality constant. Integrating the left side of equation 10.3 from $P = P_0$ to $P = P_T$, and the right side from $x = 0$ to $x = b$, where b is the sample's overall thickness

$$-\int_{P=P_0}^{P=P_T} \frac{dP}{P} = \alpha C \int_{x=0}^{x=b} dx$$

gives

$$\ln\left(\frac{P_0}{P_T}\right) = \alpha b C$$

Converting from ln to log, and substituting equation 10.2, gives

$$A = abC \qquad \textbf{10.4}$$

where a is the analyte's absorptivity with units of cm^{-1} $conc^{-1}$. When concentration is expressed using molarity, the absorptivity is replaced by the molar absorptivity, ε (with units of cm^{-1} M^{-1})

$$A = \varepsilon b C \qquad \textbf{10.5}$$

The absorptivity and molar absorptivity give, in effect, the probability that the analyte will absorb a photon of given energy. As a result, values for both a and ε depend on the wavelength of electromagnetic radiation.

EXAMPLE 10.4

A 5.00×10^{-4} M solution of an analyte is placed in a sample cell that has a pathlength of 1.00 cm. When measured at a wavelength of 490 nm, the absorbance of the solution is found to be 0.338. What is the analyte's molar absorptivity at this wavelength?

SOLUTION

Solving equation 10.5 for ε and making appropriate substitutions gives

$$\varepsilon = \frac{A}{bC} = \frac{0.338}{(1.00\ \text{cm})(5.00 \times 10^{-4}\ \text{M})} = 676\ \text{cm}^{-1}\ \text{M}^{-1}$$

Beer's law
The relationship between a sample's absorbance and the concentration of the absorbing species ($A = \varepsilon bC$).

Equations 10.4 and 10.5, which establish the linear relationship between absorbance and concentration, are known as the Beer–Lambert law, or more commonly, as **Beer's law.** Calibration curves based on Beer's law are used routinely in quantitative analysis.

10C.4 Beer's Law and Multicomponent Samples

Beer's law can be extended to samples containing several absorbing components provided that there are no interactions between the components. Individual absorbances, A_i, are additive. For a two-component mixture of X and Y, the total absorbance, A_{tot}, is

$$A_{tot} = A_X + A_Y = \varepsilon_X bC_X + \varepsilon_Y bC_Y$$

Generalizing, the absorbance for a mixture of n components, A_m, is given as

$$A_m = \sum_{i=1}^{n} A_i = \sum_{i=1}^{n} \varepsilon_i bC_i \qquad \textbf{10.6}$$

10C.5 Limitations to Beer's Law

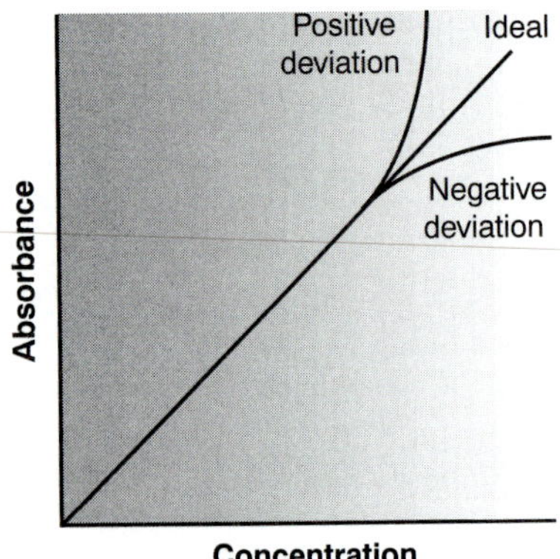

Figure 10.22
Calibration curves showing positive and negative deviations from Beer's law.

According to Beer's law, a calibration curve of absorbance versus the concentration of analyte in a series of standard solutions should be a straight line with an intercept of 0 and a slope of ab or εb. In many cases, however, calibration curves are found to be nonlinear (Figure 10.22). Deviations from linearity are divided into three categories: fundamental, chemical, and instrumental.

Fundamental Limitations to Beers Law Beer's law is a limiting law that is valid only for low concentrations of analyte. There are two contributions to this fundamental limitation to Beer's law. At higher concentrations the individual particles of analyte no longer behave independently of one another. The resulting interaction between particles of analyte may change the value of ε. A second contribution is that the absorptivity, a, and molar absorptivity, ε, depend on the sample's refractive index. Since the refractive index varies with the analyte's concentration, the values of a and ε will change. For sufficiently low concentrations of analyte, the refractive index remains essentially constant, and the calibration curve is linear.

Chemical Limitations to Beer's Law Chemical deviations from Beer's law can occur when the absorbing species is involved in an equilibrium reaction. Consider, as an example, an analysis for the weak acid, HA. To construct a Beer's law calibration curve, several standards containing known total concentrations of HA, C_{tot}, are prepared and the absorbance of each is measured at the same wavelength. Since HA is a weak acid, it exists in equilibrium with its conjugate weak base, A^-

$$HA + H_2O \rightleftharpoons H_3O^+ + A^-$$

If both HA and A^- absorb at the selected wavelength, then Beers law is written as

$$A = \varepsilon_{HA} bC_{HA} + \varepsilon_A bC_A \qquad \textbf{10.7}$$

where C_{HA} and C_A are the equilibrium concentrations of HA and A^-. Since the weak acid's total concentration, C_{tot}, is

$$C_{tot} = C_{HA} + C_A$$

the concentrations of HA and A^- can be written as

$$C_{HA} = \alpha_{HA} C_{tot} \qquad \textbf{10.8}$$

$$C_A = (1 - \alpha_{HA}) C_{tot} \qquad \textbf{10.9}$$

where α_{HA} is the fraction of weak acid present as HA. Substituting equations 10.8 and 10.9 into equation 10.7, and rearranging, gives

$$A = (\varepsilon_{HA}\alpha_{HA} + \varepsilon_A - \varepsilon_A\alpha_{HA}) b C_{tot} \qquad \textbf{10.10}$$

Because values of α_{HA} may depend on the concentration of HA, equation 10.10 may not be linear. A Beer's law calibration curve of A versus C_{tot} will be linear if one of two conditions is met. If the wavelength is chosen such that ε_{HA} and ε_A are equal, then equation 10.10 simplifies to

$$A = \varepsilon_A b C_{tot}$$

and a linear Beer's law calibration curve is realized. Alternatively, if α_{HA} is held constant for all standards, then equation 10.10 will be a straight line at all wavelengths. Because HA is a weak acid, values of α_{HA} change with pH. To maintain a constant value for α_{HA}, therefore, we need to buffer each standard solution to the same pH. Depending on the relative values of ε_{HA} and ε_A, the calibration curve will show a positive or negative deviation from Beer's law if the standards are not buffered to the same pH.

Instrumental Limitations to Beer's Law There are two principal instrumental limitations to Beer's law. The first limitation is that Beer's law is strictly valid for purely monochromatic radiation; that is, for radiation consisting of only one wavelength. As we learned in Section 10B.2, however, even the best wavelength selector passes radiation with a small, but finite effective bandwidth. Using polychromatic radiation always gives a negative deviation from Beer's law, but is minimized if the value of ε is essentially constant over the wavelength range passed by the wavelength selector. For this reason, as shown in Figure 10.23, it is preferable to make absorbance measurements at a broad absorption peak. In addition, deviations from Beer's law are less serious if the effective bandwidth from the source is less than one tenth of the natural bandwidth of the absorbing species.[7,8] When measurements must be made on a slope, linearity is improved by using a narrower effective bandwidth.

Stray radiation is the second contribution to instrumental deviations from Beer's law. Stray radiation arises from imperfections within the wavelength selector

stray radiation
Any radiation reaching the detector that does not follow the optical path from the source to the detector.

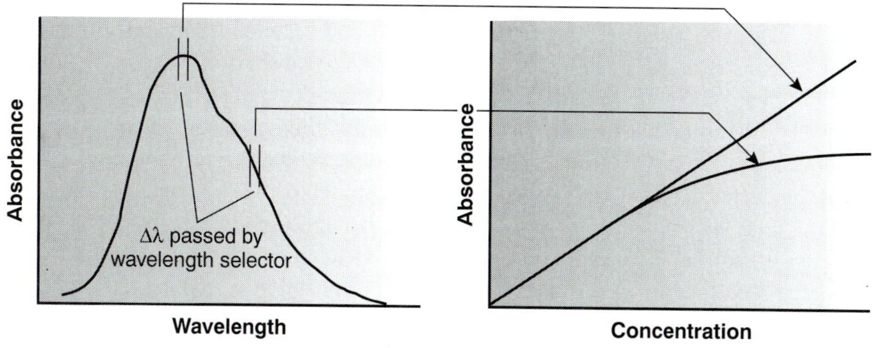

Figure 10.23
Effect of wavelength on the linearity of a Beer's law calibration curve.

that allows extraneous light to "leak" into the instrument. Stray radiation adds an additional contribution, P_{stray}, to the radiant power reaching the detector; thus

$$A = \log \frac{P_0 + P_{stray}}{P_T + P_{stray}}$$

For small concentrations of analyte, P_{stray} is significantly smaller than P_0 and P_T, and the absorbance is unaffected by the stray radiation. At higher concentrations of analyte, however, P_{stray} is no longer significantly smaller than P_T and the absorbance is smaller than expected. The result is a negative deviation from Beer's law.

10D Ultraviolet-Visible and Infrared Spectrophotometry

The earliest routine application of molecular absorption spectroscopy, which dates to the 1830s, was colorimetry, in which visible light was absorbed by a sample. The concentration of analyte was determined visually by comparing the sample's color to that of a set of standards using Nessler tubes (as described at the beginning of this chapter), or by using an instrument called a colorimeter. The development of visible absorption spectroscopy as a routine analytical technique was limited by the tedious nature of making visual color comparisons. Furthermore, although infrared radiation was discovered in 1800 and ultraviolet radiation in 1801, their use in optical molecular absorption spectroscopy was limited by the lack of a convenient means for detecting the radiation. During the 1930s and 1940s, advances in electronics resulted in the introduction of photoelectric transducers for ultraviolet and visible radiation, and thermocouples for infrared radiation. As a result, "modern" instrumentation for absorption spectroscopy routinely became available in the 1940s. Progress in these fields has been rapid ever since.

10D.1 Instrumentation

Frequently an analyst must select, from several instruments of different design, the one instrument best suited for a particular analysis. In this section we examine some of the different types of instruments used for molecular absorption spectroscopy, emphasizing their advantages and limitations. Methods of sample introduction are also covered in this section.

filter photometer
A simple instrument for measuring absorbance that uses absorption or interference filters to select the wavelength.

Instrument Designs for Molecular UV/Vis Absorption The simplest instrument currently used for molecular UV/Vis absorption is the **filter photometer** shown in Figure 10.24, which uses an absorption or interference filter to isolate a band of radiation. The filter is placed between the source and sample to prevent the sample from decomposing when exposed to high-energy radiation. A filter photometer has a single optical path between the source and detector and is called a single-beam instrument. The instrument is calibrated to 0% *T* while using a shutter to block the source radiation from the detector. After removing the shutter, the instrument is calibrated to 100% *T* using an appropriate blank. The blank is then replaced with the sample, and its transmittance is measured. Since the source's incident power and the sensitivity of the detector vary with wavelength, the photometer must be recalibrated whenever the filter is changed. In comparison with other spectroscopic instruments, photometers have the advantage of being relatively inexpensive, rugged, and easy to maintain. Another advantage of a photometer is its portability, making it a useful instrument for conducting spectroscopic analyses in the field. A disadvantage of a photometer is that it cannot be used to obtain an absorption spectrum.

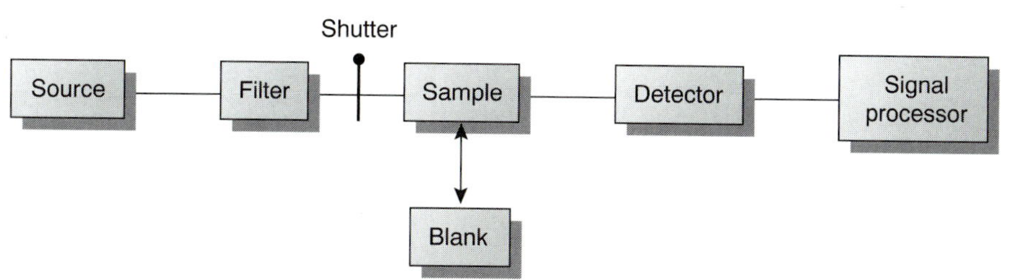

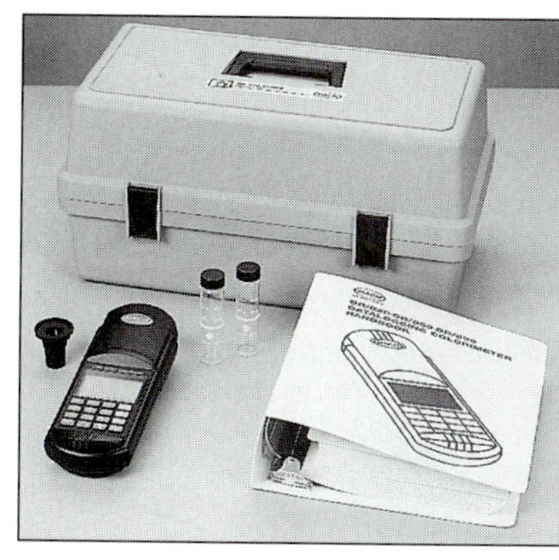

Figure 10.24
Block diagram for a filter photometer with photo showing a typical hand-held instrument suitable for field work.
Colorimeter™ is manufactured by Hach Company/photo courtesy of Hach Company.

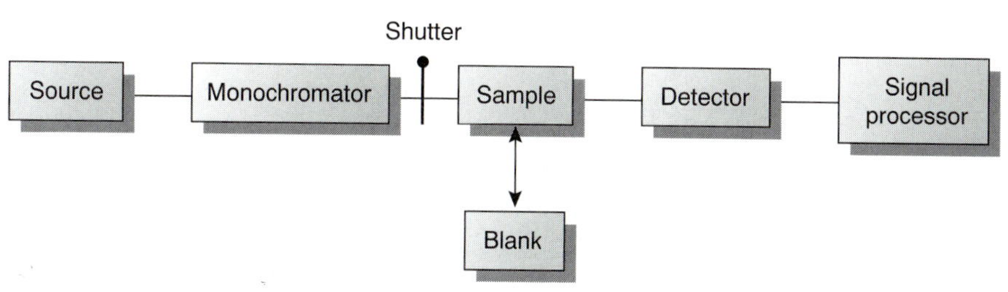

Figure 10.25
Block diagram for a single-beam fixed-wavelength spectrophotometer with photo of a typical instrument.
Photo courtesy of Fisher Scientific.

Instruments using monochromators for wavelength selection are called spectrometers. In absorbance spectroscopy, where the transmittance is a ratio of two radiant powers, the instrument is called a spectrophotometer. The simplest **spectrophotometer** is a single-beam instrument equipped with a fixed-wavelength monochromator, the block diagram for which is shown in Figure 10.25. Single-beam spectrophotometers are calibrated and used in the same manner as a photometer. One common example of a single-beam spectrophotometer is the Spectronic-20 manufactured by Milton-Roy. The Spectronic-20 can be used from 340 to 625 nm (950 nm with a red-sensitive detector), and has a fixed effective bandwidth of 20 nm. Because its effective bandwidth is fairly large, this instrument is more appropriate for a quantitative analysis than for a qualitative analysis. Battery-powered, hand-held single-beam spectrophotometers are available, which are easily transported and can be used for on-site analyses. Other single-beam spectrophotometers are available with effective bandwidths of 2–8 nm. Fixed-wavelength single-beam spectrophotometers are not practical for recording spectra since manually adjusting the wavelength and recalibrating the spectrophotometer is awkward and time-consuming. In addition, the accuracy of a single-beam spectrophotometer is limited by the stability of its source and detector over time.

spectrophotometer
An instrument for measuring absorbance that uses a monochromator to select the wavelength.

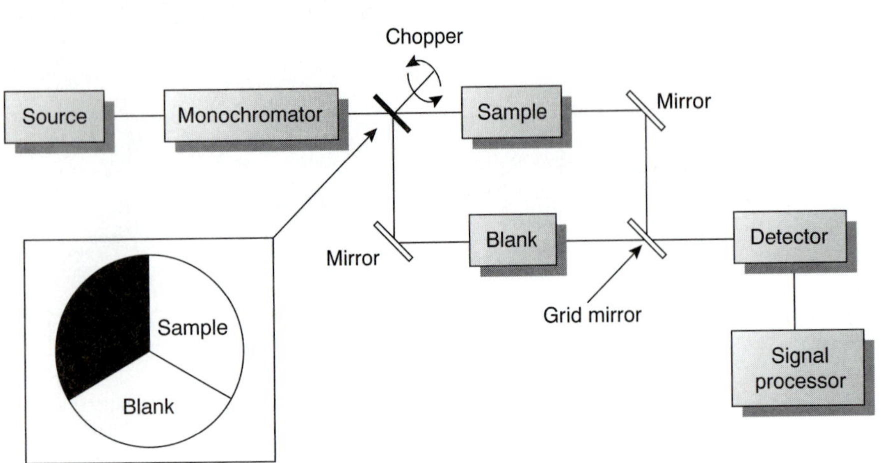

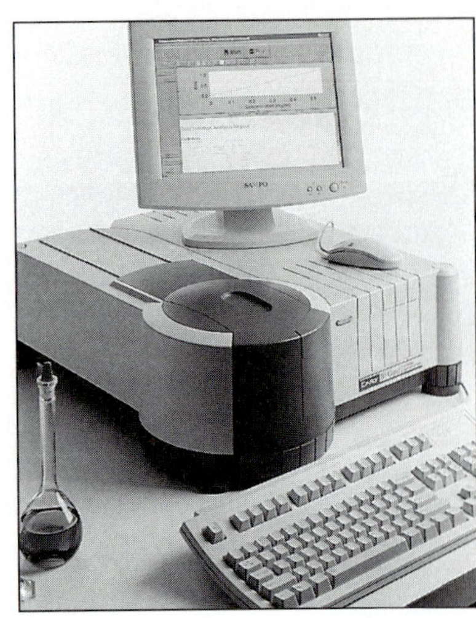

Figure 10.26
Block diagram for a double-beam in-time scanning spectrophotometer with photo of a typical instrument.
Photo courtesy of Varian, Inc.

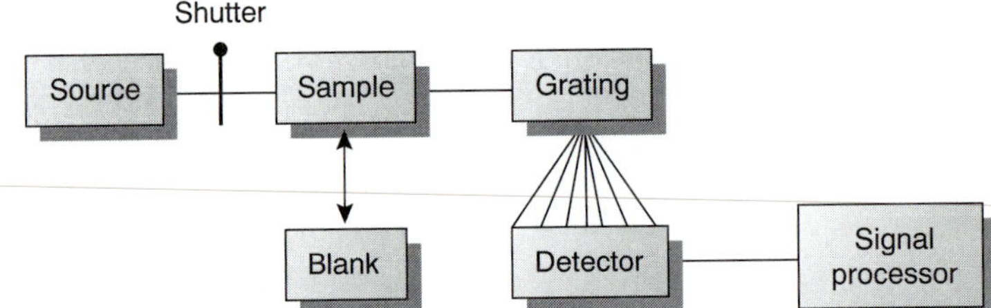

Figure 10.27
Block diagram for a diode array spectrophotometer.

The limitations of fixed-wavelength, single-beam spectrophotometers are minimized by using the double-beam in-time spectrophotometer as shown in Figure 10.26. A chopper, similar to that shown in the insert, controls the radiation's path, alternating it between the sample, the blank, and a shutter. The signal processor uses the chopper's known speed of rotation to resolve the signal reaching the detector into that due to the transmission of the blank (P_0) and the sample (P_T). By including an opaque surface as a shutter it is possible to continuously adjust the 0% T response of the detector. The effective bandwidth of a double-beam spectrophotometer is controlled by means of adjustable slits at the entrance and exit of the monochromator. Effective bandwidths of between 0.2 nm and 3.0 nm are common. A scanning monochromator allows for the automated recording of spectra. Double-beam instruments are more versatile than single-beam instruments, being useful for both quantitative and qualitative analyses; they are, however, more expensive.

The instrument designs considered thus far use a single detector and can only monitor one wavelength at a time. A linear photodiode array consists of multiple detectors, or channels, allowing an entire spectrum to be recorded in as little as 0.1 s. A block diagram for a typical multichannel spectrophotometer is shown in Figure 10.27. Source radiation passing through the sample is dispersed by a grating. The linear photodiode array is situated at the grating's focal plane, with each diode recording the radiant power over a narrow range of wavelengths.

One advantage of a linear photodiode array is the speed of data acquisition, which makes it possible to collect several spectra for a single sample. Individual spectra are added and averaged to obtain the final spectrum. This process of **signal averaging** improves a spectrum's signal-to-noise ratio. When a series of spectra is added, the sum of the signal at any point increases as (nS_x), where n is the number of spectra, and S_x is the signal for the spectrum's x-th point. The propagation of noise, which is a random event, increases as $(\sqrt{n}N_x)$, where N_x is the noise level for the spectrum's x-th point. The signal-to-noise ratio (S/N) at the x-th data point, therefore, increases by a factor of $\sqrt{n}$

signal averaging
The adding together of successive spectra to improve the signal-to-noise ratio.

$$\frac{S}{N} = \frac{nS_x}{\sqrt{n}N_x} = \sqrt{n}\,\frac{S_x}{N_x}$$

where (S_x/N_x) is the signal-to-noise ratio for a single scan. The effect of signal averaging is shown in Figure 10.28. The spectrum in Figure 10.28a shows the total signal for a single scan. Although there is an apparent peak near the center of the spectrum, the level of background noise makes it difficult to measure the peak's signal. Figures 10.28b and Figure 10.28c demonstrate the improvement in signal-to-noise ratio achieved by signal averaging. One disadvantage of a linear photodiode array is that the effective bandwidth per diode is roughly an order of magnitude larger than that obtainable with a high-quality monochromator.

The sample compartment for the instruments in Figures 10.24–10.27 provides a light-tight environment that prevents the loss of radiation, as well as the addition of stray radiation. Samples are normally in the liquid or solution state and are placed in cells constructed with UV/Vis-transparent materials, such as quartz, glass, and plastic (Figure 10.29). Quartz or fused-silica cells are required when working at wavelengths of less than 300 nm where other materials show a significant absorption. The most common cell has a pathlength of 1 cm, although cells with shorter (≥ 1 mm) and longer pathlengths (≤ 10 cm) are available. Cells with a longer pathlength are useful for the analysis of very dilute solutions or for gaseous samples. The highest quality cells are constructed in a rectangular shape, allowing the radiation to strike the cell at a 90° angle, where losses to reflection are minimal. These cells, which are usually available in matched pairs having identical optical properties, are the cells of choice for double-beam instruments. Cylindrical test tubes are often used as a sample cell for simple, single-beam instruments, although differences in the cell's pathlength and optical properties add an additional source of error to the analysis.

In some circumstances it is desirable to monitor a system without physically removing a sample for analysis. This is often the case, for example, with the on-line monitoring of industrial production lines or waste lines, for physiological monitoring, and for monitoring environmental systems. With the use of a fiber-optic probe it is possible to analyze samples in situ. A simple example of a remote-sensing, fiber-optic probe is shown in Figure 10.30a and consists of two bundles of fiber-optic cable. One bundle transmits radiation from the source to the sample cell, which is designed to allow for the easy flow of sample through the cell. Radiation from the source passes through the solution, where it is reflected back by a mirror. The second bundle of fiber-optic cable transmits the nonabsorbed radiation to the wavelength selector. In an alternative design (Figure 10.30b), the sample cell is a membrane containing a reagent phase capable of reacting with the analyte. When the analyte diffuses

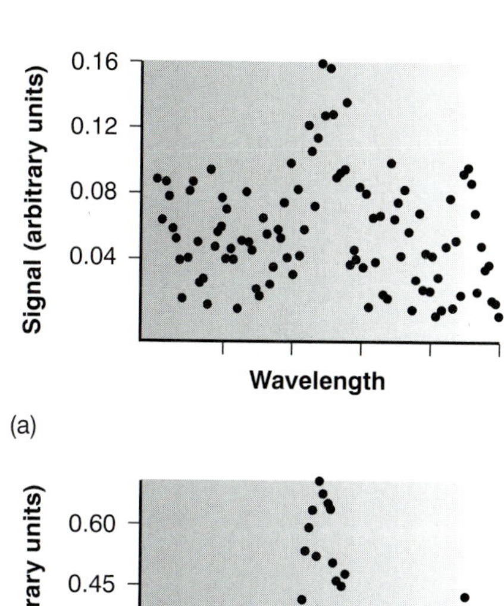

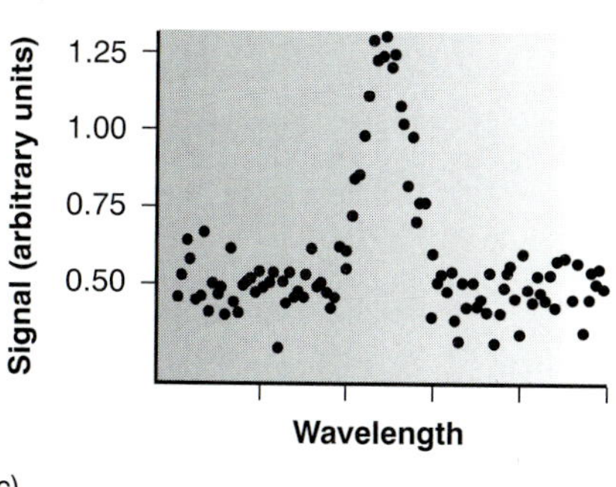

Figure 10.28
Effect of signal averaging on a spectrum's signal-to-noise ratio: (a) spectrum for a single scan; (b) spectrum after co-adding five spectra; (c) spectrum after co-adding ten spectra.

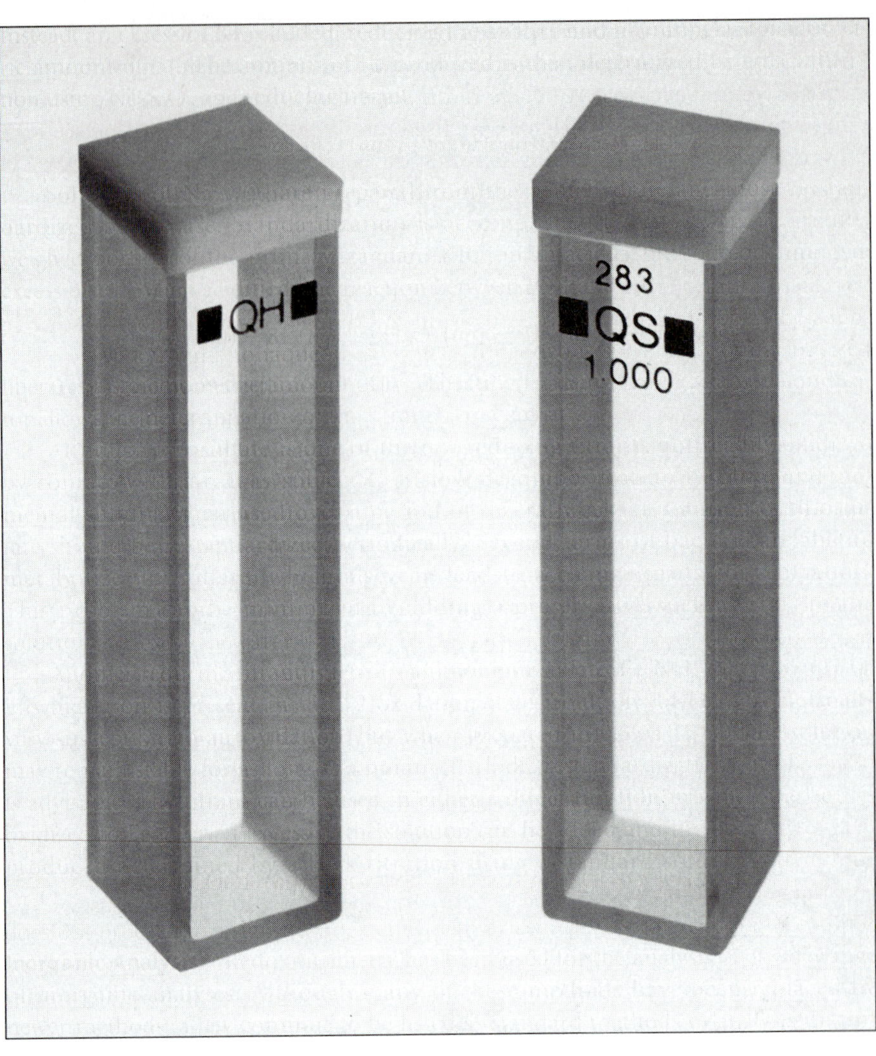

Figure 10.29
Typical cells used in UV/Vis spectroscopy. Courtesy of Fisher Scientific.

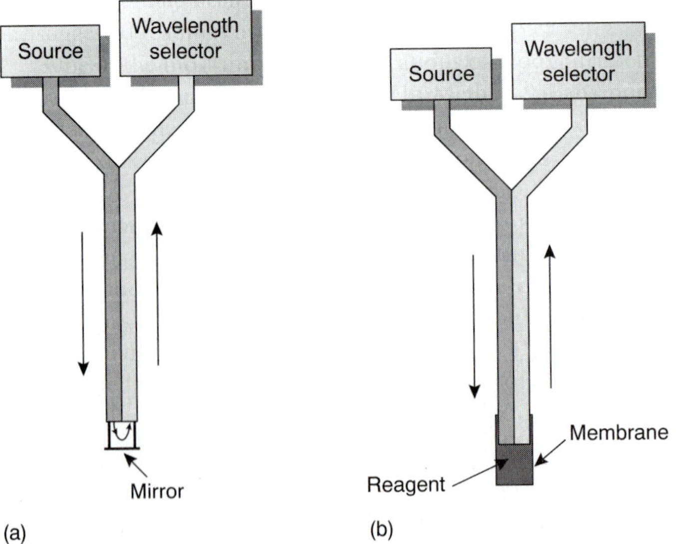

Figure 10.30
Example of fiber-optic probes.

across the membrane, it reacts with the reagent phase, producing a product that absorbs UV or visible radiation. Nonabsorbed radiation from the source is reflected or scattered back to the detector. Fiber-optic probes that show chemical selectivity are called optrodes.[9,10]

Instrument Designs for Infrared Absorption The simplest instrument for IR absorption spectroscopy is a filter photometer similar to that shown in Figure 10.24 for UV/Vis absorption. These instruments have the advantage of portability and typically are used as dedicated analyzers for gases such as HCN and CO.

Infrared instruments using a monochromator for wavelength selection are constructed using double-beam optics similar to that shown in Figure 10.26. Double-beam optics are preferred over single-beam optics because the sources and detectors for infrared radiation are less stable than that for UV/Vis radiation. In addition, it is easier to correct for the absorption of infrared radiation by atmospheric CO_2 and H_2O vapor when using double-beam optics. Resolutions of 1–3 cm^{-1} are typical for most instruments.

In a Fourier transform, infrared spectrometer, or FT–IR, the monochromator is replaced with an interferometer (see Figure 10.13). Because an FT–IR includes only a single optical path, it is necessary to collect a separate spectrum to compensate for the absorbance of atmospheric CO_2 and H_2O vapor. This is done by collecting a background spectrum without the sample and storing the result in the instrument's computer memory. The background spectrum is removed from the sample's spectrum by ratioing the two signals. In comparison to other IR instruments, an FT–IR provides for rapid data acquisition, allowing an enhancement in signal-to-noise ratio through signal averaging.

Infrared spectroscopy is routinely used for the analysis of samples in the gas, liquid, and solid states. Sample cells are made from materials, such as NaCl and KBr, that are transparent to infrared radiation. Gases are analyzed using a cell with a pathlength of approximately 10 cm. Longer pathlengths are obtained by using mirrors to pass the beam of radiation through the sample several times.

Liquid samples are analyzed in one of two ways. For nonvolatile liquids a suitable sample can be prepared by placing a drop of the liquid between two NaCl plates, forming a thin film that typically is less than 0.01 mm thick. Volatile liquids must be placed in a sealed cell to prevent their evaporation.

The analysis of solution samples is limited by the solvent's IR-absorbing properties, with CCl_4, CS_2, and $CHCl_3$ being the most common solvents. Solutions are placed in cells containing two NaCl windows separated by a Teflon spacer. By changing the Teflon spacer, pathlengths from 0.015 to 1.0 mm can be obtained. Sealed cells with fixed or variable pathlengths also are available.

The analysis of aqueous solutions is complicated by the solubility of the NaCl cell window in water. One approach to obtaining infrared spectra on aqueous solutions is to use attenuated total reflectance (ATR) instead of transmission. Figure 10.31 shows a diagram of a typical ATR sampler, consisting of an IR-transparent crystal of high-refractive index, such as ZnSe, surrounded by a sample of lower-refractive index. Radiation from the source enters the ATR crystal, where it undergoes a series of total internal reflections before exiting the crystal. During each reflection, the radiation penetrates into the sample to a depth of a few microns. The result is a selective attenuation of the radiation at those wavelengths at which the sample absorbs.

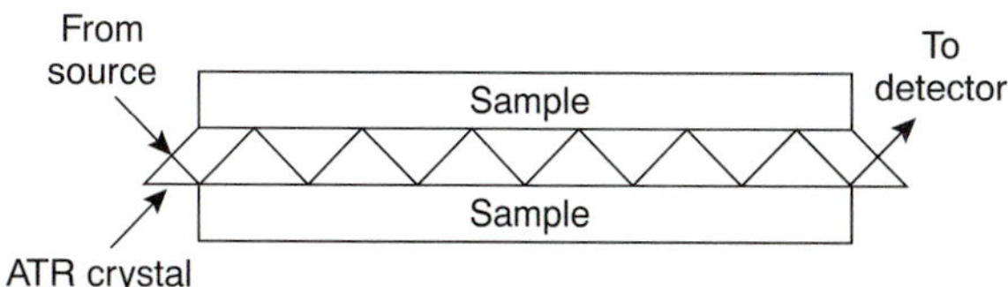

Figure 10.31
Attenuated total reflectance (ATR) cell for use in infrared spectroscopy.

ATR spectra are similar, but not identical, to those obtained by measuring the transmission of radiation.

Transparent solid samples can be analyzed directly by placing them in the IR beam. Most solid samples, however, are opaque and must be dispersed in a more transparent medium before recording a traditional transmission spectrum. If a suitable solvent is available, then the solid can be analyzed by preparing a solution and analyzing as described earlier. When a suitable solvent is not available, solid samples may be analyzed by preparing a mull of the finely powdered sample with a suitable oil. Alternatively, the powdered sample can be mixed with KBr and pressed into an optically transparent pellet.

Solid samples also can be analyzed by means of reflectance. The ATR sampler (see Figure 10.31) described for the analysis of aqueous solutions can be used for the analysis of solid samples, provided that the solid can be brought into contact with the ATR crystal. Examples of solids that have been analyzed by ATR include polymers, fibers, fabrics, powders, and biological tissue samples. Another reflectance method is diffuse reflectance, in which radiation is reflected from a rough surface, such as a powder. Powdered samples are mixed with a nonabsorbing material, such as powdered KBr, and the reflected light is collected and analyzed. As with ATR, the resulting spectrum is similar to that obtained by conventional transmission methods. Further details about these and other methods for preparing solids for infrared analysis can be found in the suggested readings listed at the end of the chapter.

10D.2 Quantitative Applications

The determination of an analyte's concentration based on its absorption of ultraviolet or visible radiation is one of the most frequently encountered quantitative analytical methods. One reason for its popularity is that many organic and inorganic compounds have strong absorption bands in the UV/Vis region of the electromagnetic spectrum. In addition, analytes that do not absorb UV/Vis radiation, or that absorb such radiation only weakly, frequently can be chemically coupled to a species that does. For example, nonabsorbing solutions of Pb^{2+} can be reacted with dithizone to form the red Pb–dithizonate complex. An additional advantage to UV/Vis absorption is that in most cases it is relatively easy to adjust experimental and instrumental conditions so that Beer's law is obeyed.

Quantitative analyses based on the absorption of infrared radiation, although important, are less frequently encountered than those for UV/Vis absorption. One reason is the greater tendency for instrumental deviations from Beer's law when using infrared radiation. Since infrared absorption bands are relatively narrow, deviations due to the lack of monochromatic radiation are more pronounced. In addition, infrared sources are less intense than sources of UV/Vis radiation, making stray radiation more of a problem. Differences in pathlength for samples and standards when using thin liquid films or KBr pellets are a problem, although an internal standard can be used to correct for any difference in pathlength. Finally, establishing a 100% T ($A = 0$) baseline is often difficult since the optical properties of NaCl sample cells may change significantly with wavelength due to contamination and degradation. This problem can be minimized by determining absorbance relative to a baseline established for the absorption band. Figure 10.32 shows how this is accomplished.

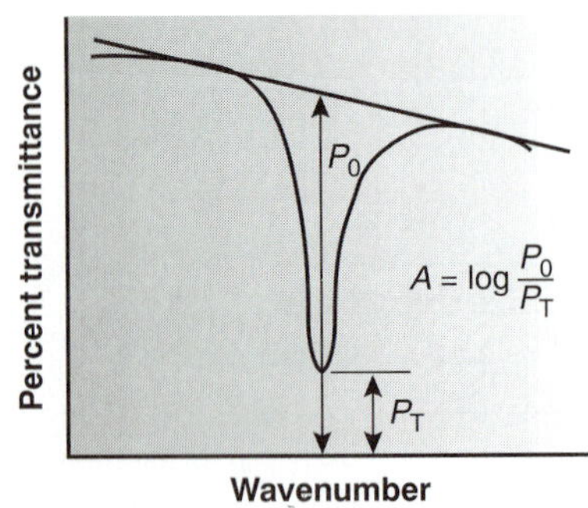

Figure 10.32
Method for determining absorbance from an IR transmission spectrum.

The applications of Beer's law for the quantitative analysis of samples in environmental chemistry, clinical chemistry, industrial chemistry and forensic chemistry are numerous. Examples from each of these fields follow.

Table 10.6 Selected Examples of the Application of UV/Vis Molecular Absorption to the Analysis of Waters and Wastewaters

Analyte	Method	λ (nm)
	Trace Metals	
aluminum	reaction with Eriochrome cyanide R dye at pH 6 produces red to pink complex	535
arsenic	reduce to AsH_3 using Zn and react with silver diethyldithiocarbamate to form red complex	535
cadmium	extraction into $CHCl_3$ containing dithizone from sample made basic with NaOH; pink to red complex	518
chromium	oxidize to Cr(VI) and react with diphenylcarbazide in acidic solution to give red-violet product	540
copper	react with neocuprine in neutral to slightly acid solution; extract into $CHCl_3/CH_3OH$ to give yellow solution	457
iron	react with *o*-phenanthroline in acidic solution to form orange-red complex	510
lead	extraction into $CHCl_3$ containing dithizone from sample made basic with ammoniacal buffer; cherry red complex	510
manganese	oxidize to MnO_4^- with persulfate	525
mercury	extraction into $CHCl_3$ containing dithizone from acidic sample; orange complex	492
zinc	reaction with zincon at pH 9 to form blue complex	620
	Inorganic Nonmetals	
ammonia	reaction with ammonia, hypochlorite, and phenol produces blue indophenol; catalyzed by manganous salt	630
cyanide	convert to CNCl by reaction with chloramine-T, followed by reaction with a pyridine-barbituric acid to form red-blue dye	578
fluoride	reaction with red Zr-SPADNS lake results in formation of ZrF_6^{2-} and decrease in concentration of the lake	570
chlorine (residual)	oxidation of leuco crystal violet to form product with a bluish color	592
nitrate	reduction to NO_2^- by Cd, colored azo dye formed by reaction with sulfanilamide and *N*-(1-naphthyl)-ethelyenediamine	543
phosphate	reaction with ammonium molybdate followed by reduction with stannous chloride to form molybdenum blue	690
	Organics	
phenol	reaction with 4-aminoantipyrine and $K_3Fe(CN)_6$ to form antipyrine dye	460
surfactants	formation of blue ion pair between anionic surfactant and the cationic dye methylene blue, which is extracted into $CHCl_3$	652

Environmental Applications Methods for the analysis of waters and wastewaters relying on the absorption of UV/Vis radiation are among some of the most frequently employed analytical methods. Many of these methods are outlined in Table 10.6, and a few are described later in more detail.

Although the quantitative analysis of metals in water and wastewater is accomplished primarily by atomic absorption or atomic emission spectroscopy, many metals also can be analyzed following the formation of a colored metal–ligand complex. One advantage to these spectroscopic methods is that they are easily adapted to the field analysis of samples using a filter photometer. One ligand used in the analysis of several metals is diphenylthiocarbazone, also known as dithizone. Dithizone is insoluble in water, but when a solution of dithizone in $CHCl_3$ is shaken with an aqueous solution containing an appropriate metal ion, a colored metal–dithizonate complex forms that is soluble in $CHCl_3$. The selectivity of dithizone is controlled by adjusting the pH of the aqueous sample. For example, Cd^{2+} is extracted from solutions that are made strongly basic with NaOH, Pb^{2+} from solutions that are made basic with an ammoniacal buffer, and Hg^{2+} from solutions that are slightly acidic.

When chlorine is added to water that portion available for disinfection is called the chlorine residual. Two forms of the chlorine residual are recognized. The free chlorine residual includes Cl_2, HOCl, and OCl^-. The combined chlorine residual, which forms from the reaction of NH_3 with HOCl, consists of monochloroamine, NH_2Cl, dichlororamine, $NHCl_2$, and trichloroamine, NCl_3. Since the free chlorine residual is more efficient at disinfection, analytical methods have been developed to determine the concentration of both forms of residual chlorine. One such method is the leuco crystal violet method. Free residual chlorine is determined by adding leuco crystal violet to the sample, which instantaneously oxidizes giving a bluish color that is monitored at 592 nm. Completing the analysis in less than 5 min prevents a possible interference from the combined chlorine residual. The total chlorine residual (free + combined) is determined by reacting a separate sample with iodide, which reacts with both chlorine residuals to form HOI. When the reaction is complete, leuco crystal violet is added and oxidized by HOI, giving the same bluish colored product. The combined chlorine residual is determined by difference.

The concentration of fluoride in drinking water may be determined indirectly by its ability to form a complex with zirconium. In the presence of the dye SPADNS,* solutions of zirconium form a reddish colored compound, called a "lake," that absorbs at 570 nm. When fluoride is added, the formation of the stable ZrF_6^{2-} complex causes a portion of the lake to dissociate, decreasing the absorbance. A plot of absorbance versus the concentration of fluoride, therefore, has a negative slope.

Spectroscopic methods also are used in determining organic constituents in water. For example, the combined concentrations of phenol, and ortho- and meta-substituted phenols are determined by using steam distillation to separate the phenols from nonvolatile impurities. The distillate is reacted with 4-aminoantipyrine at pH 7.9 ± 0.1 in the presence of $K_3Fe(CN)_6$, forming a colored antipyrine dye. The dye is extracted into $CHCl_3$, and the absorbance is monitored at 460 nm. A calibration curve is prepared using only the unsubstituted phenol, C_6H_5OH. Because the molar absorptivities of substituted phenols are generally less than that for phenol, the reported concentration represents the minimum concentration of phenolic compounds.

Molecular absorption also can be used for the analysis of environmentally significant airborne pollutants. In many cases the analysis is carried out by collecting the sample in water, converting the analyte to an aqueous form that can be analyzed by methods such as those described in Table 10.6. For example, the concentration of NO_2 can be determined by oxidizing NO_2 to NO_3^-. The concentration of NO_3^- is then determined by reducing to NO_2^- with Cd and reacting the NO_2^- with sulfanilamide and *N*-(1-naphthyl)-ethylenediamine to form a brightly colored azo dye. Another important application is the determination of SO_2, which is determined by collecting the sample in an aqueous solution of $HgCl_4^{2-}$ where it reacts to form $Hg(SO_3)_2^{2-}$. Addition of *p*-rosaniline and formaldehyde results in the formation of a bright purple complex that is monitored at 569 nm. Infrared absorption has proved useful for the analysis of organic vapors, including HCN, SO_2, nitrobenzene, methyl mercaptan, and vinyl chloride. Frequently, these analyses are accomplished using portable, dedicated infrared photometers.

*SPADNS is the acronym for the sodium salt of 2-(4-sulfophenylazo)-1,8-dihydroxy-3,6-naphthalenedisulfonic acid.

Table 10.7 Selected Examples of the Application of UV/Vis Molecular Absorption to the Analysis of Clinical Samples

Analyte	Method	λ (nm)
total serum protein	reaction with protein, NaOH, and Cu^{2+} produces blue-violet complex	540
serum cholesterol	reaction with Fe^{3+} in presence of isopropanol, acetic acid, and H_2SO_4 produces blue-violet complex	540
uric acid	reaction with phosphotungstic acid produces tungsten blue	710
serum barbiturates	barbiturates are extracted into $CHCl_3$, and then into 0.45 M NaOH	260
glucose	reaction with *o*-toludine at 100 °C produces blue-green complex	630
protein-bound iodine	decompose protein to release iodide; I^- determined by its catalytic effect on redox reaction between Ce^{4+} and As^{3+}	420

Clinical Applications UV/Vis molecular absorption is one of the most commonly employed techniques for the analysis of clinical samples, several examples of which are listed in Table 10.7.

The analysis of clinical samples is often complicated by the complexity of the sample matrix, which may contribute a significant background absorption at the desired wavelength. The determination of serum barbiturates provides one example of how this problem is overcome. The barbiturates are extracted from a sample of serum with $CHCl_3$, and extracted from the $CHCl_3$ into 0.45 M NaOH (pH ≈ 13). The absorbance of the aqueous extract is measured at 260 nm and includes contributions from the barbiturates as well as other components extracted from the serum sample. The pH of the sample is then lowered to approximately 10 by adding NH_4Cl, and the absorbance remeasured. Since the barbiturates do not absorb at this pH, the absorbance at pH 10 is used to correct the absorbance at pH 13; thus

$$A_{\text{barb}} = A_{\text{pH 13}} - \left(\frac{V_{\text{samp}} + V_{\text{NH}_4\text{Cl}}}{V_{\text{samp}}}\right) A_{\text{pH 10}}$$

Industrial Analysis UV/Vis molecular absorption is used for the analysis of a diverse array of industrial samples, including pharmaceuticals, food, paint, glass, and metals. In many cases the methods are similar to those described in Tables 10.6 and 10.7. For example, the iron content of food can be determined by bringing the iron into solution and analyzing using the *o*-phenanthroline method listed in Table 10.6.

Many pharmaceutical compounds contain chromophores that make them suitable for analysis by UV/Vis absorption. Products that have been analyzed in this fashion include antibiotics, hormones, vitamins, and analgesics. One example of the use of UV absorption is in determining the purity of aspirin tablets, for which the active ingredient is acetylsalicylic acid. Salicylic acid, which is produced by the hydrolysis of acetylsalicylic acid, is an undesirable impurity in aspirin tablets, and should not be present at more than 0.01% w/w. Samples can be screened for unacceptable levels of salicylic acid by monitoring the absorbance at a wavelength of

312 nm. Acetylsalicylic acid absorbs at 280 nm, but absorbs poorly at 312 nm. Conditions for preparing the sample are chosen such that an absorbance of greater than 0.02 signifies an unacceptable level of salicylic acid.

Forensic Applications UV/Vis molecular absorption is routinely used in the analysis of narcotics and for drug testing. One interesting forensic application is the determination of blood alcohol using the Breathalyzer test. In this test a 52.5-mL breath sample is bubbled through an acidified solution of $K_2Cr_2O_7$. Any ethanol present in the breath sample is oxidized by the dichromate, producing acetic acid and Cr^{3+} as products. The concentration of ethanol in the breath sample is determined from the decrease in absorbance at 440 nm where the dichromate ion absorbs. A blood alcohol content of 0.10%, which is the legal limit in most states, corresponds to 0.025 mg of ethanol in the breath sample.

Developing a Quantitative Method for a Single Component In developing a quantitative analytical procedure, the conditions under which Beer's law is obeyed must be established. First, the most appropriate wavelength for the analysis is determined from an absorption spectrum. In most cases the best wavelength corresponds to an absorption maximum because it provides greater sensitivity and is less susceptible to instrumental limitations to Beer's law due to the lack of monochromatic radiation. Second, if an instrument with adjustable slits is being used, then an appropriate slit width needs to be chosen. The absorption spectrum also aids in selecting a slit width. Generally the slit width should be as wide as possible to increase the throughput of radiation from the source, while being narrow enough to avoid instrumental limitations to Beer's law. Finally, a calibration curve is constructed to determine the range of concentrations for which Beer's law is valid. Additional considerations that are important in any quantitative method are the effect of potential interferents and establishing an appropriate blank.

Representative Methods

Method 10.1 Determination of Iron in Water and Wastewater[11]

Description of Method. Iron in the +2 oxidation state reacts with *o*-phenanthroline to form the orange-red $Fe(o\text{-phen})_3^{2+}$ complex (see Figure 10.17). The intensity of the complex's color is independent of solution acidity between a pH of 3 and 9. Due to a faster rate of complex formation in more acidic solutions, the reaction is usually carried out within the pH range of 3.0–3.5. Any iron present in the +3 oxidation state is reduced with hydroxylamine before adding *o*-phenanthroline. The most important interferents are strong oxidizing agents; polyphosphates; and metal ions, such as Cu^{2+}, Zn^{2+}, Ni^{2+}, and Cd^{2+}. Interferences from oxidizing agents are minimized by adding an excess of hydroxylamine, whereas the interference from polyphosphate is minimized by boiling the sample in the presence of acid. The absorbance of samples and standards is measured at a wavelength of 510 nm using a 1-cm cell (longer-pathlength cells may be used as well). Beer's law is obeyed for concentrations of iron within the range of 0.2–4.0 ppm.

N N

o-Phenanthroline

—Continued

Procedure. For samples containing less than 2 ppm Fe, directly transfer a 50-mL portion to a 125-mL Erlenmeyer flask. Samples containing more than 2 ppm Fe must be suitably diluted before acquiring the 50-mL portion. Add 2 mL of concentrated HCl and 1 mL of hydroxylamine to the sample in the Erlenmeyer flask. Heat the solution to boiling, and continue boiling until the solution's volume is reduced to between 15 and 20 mL. After cooling to room temperature, transfer the solution to a 50-mL volumetric flask, add 10 mL of an ammonium acetate buffer, 2 mL of a 1000 ppm solution of *o*-phenanthroline, and dilute to volume. Allow 10–15 min for color development before measuring the absorbance, using distilled water to set 100% *T*. Calibration standards, including a blank, are prepared by the same procedure using a stock solution containing a known concentration of Fe^{2+}.

Questions

1. **Explain why strong oxidizing agents interfere with this analysis and why an excess of hydroxylamine prevents such interferences from occurring.**

   To obtain an accurate result it is necessary that only the $Fe(o\text{-phen})_3^{2+}$ complex be present. Strong oxidizing agents interfere by oxidizing some of the Fe^{2+} to Fe^{3+}, producing the weakly absorbing $Fe(o\text{-phen})_3^{3+}$ complex. The excess hydroxylamine reacts with the oxidizing agents, removing them from the solution.

2. **The color intensity of the complex is stable between pH levels of 3 and 9. What are some possible complications at more acidic or more basic pHs?**

   As with EDTA, which we encountered in Chapter 9, *o*-phenanthroline is a ligand possessing acid–base properties. The formation of the $Fe(o\text{-phen})_3^{2+}$ complex, therefore, is less favorable at lower pH levels, where *o*-phenanthroline is protonated. The result is a decrease in absorbance. When the pH is greater than 9, competition for Fe^{2+} between OH^- and *o*-phenanthroline also leads to a decrease in absorbance. In addition, if the pH is sufficiently basic there is a risk that the iron will precipitate as $Fe(OH)_2$.

3. **Cadmium is an interferant because it forms a precipitate with *o*-phenanthroline. What effect would the formation of a Cd-*o*-phenanthroline precipitate have on the determination of the parts per million of Fe in a sample?**

   Since *o*-phenanthroline is present in large excess (2000 μg of *o*-phenanthroline for 100 μg of Fe^{2+}), it is not likely that the interference is due to an insufficient amount of *o*-phenanthroline being available to react with the Fe^{2+}. The presence of a precipitate in the sample cell results in the scattering of radiation and an apparent increase in absorbance. Since the measured absorbance is too high, the reported concentration also is too high.

4. **Even high-quality ammonium acetate contains a significant amount of iron. Why is this source of iron not a problem?**

   Since all samples and standards are prepared using the same volume of ammonium acetate buffer, the contribution of this source of iron is accounted for by the calibration curve's reagent blank.

Quantitative Analysis for a Single Analyte The concentration of a single analyte is determined by measuring the absorbance of the sample and applying Beer's law (equation 10.5) using any of the standardization methods described in Chapter 5. The most common methods are the normal calibration curve and the method of standard additions. Single-point standardizations also can be used, provided that the validity of Beer's law has been demonstrated.

Example

EXAMPLE 10.5

The determination of Fe in an industrial waste stream was carried out by the *o*-phenanthroline described in Method 10.1. Using the data shown in the following table, determine the concentration of Fe in the waste stream.

ppm Fe	Absorbance
0.00	0.000
1.00	0.183
2.00	0.364
3.00	0.546
4.00	0.727
unknown	0.269

SOLUTION

Linear regression of absorbance versus the concentration of Fe in the standards gives

$$A = 0.0006 + 0.1817 \times (\text{ppm Fe})$$

Substituting the unknown's absorbance into the calibration expression gives the concentration of Fe in the waste stream as 1.48 ppm.

Quantitative Analysis of Mixtures The analysis of two or more components in the same sample is straightforward if there are regions in the sample's spectrum in which each component is the only absorbing species. In this case each component can be analyzed as if it were the only species in solution. Unfortunately, UV/Vis absorption bands are so broad that it frequently is impossible to find appropriate wavelengths at which each component of a mixture absorbs separately. Earlier we learned that Beer's law is additive (equation 10.6); thus, for a two-component mixture of X and Y, the mixture's absorbance, A_m, is

$$(A_m)_{\lambda 1} = (\varepsilon_X)_{\lambda 1} bC_X + (\varepsilon_Y)_{\lambda 1} bC_Y \qquad \textbf{10.11}$$

where $\lambda 1$ is the wavelength at which the absorbance is measured. Since equation 10.11 includes terms for both the concentrations of X and Y, the absorbance at one wavelength does not provide sufficient information to determine either C_X or C_Y. If we measure the absorbance at a second wavelength, $\lambda 2$,

$$(A_m)_{\lambda 2} = (\varepsilon_X)_{\lambda 2} bC_X + (\varepsilon_Y)_{\lambda 2} bC_Y \qquad \textbf{10.12}$$

then C_X and C_Y can be determined by solving equations 10.11 and 10.12. Of course, it is necessary to determine values for ε for each component at both wavelengths. In general, for a mixture of n components, the absorbance must be measured at n different wavelengths.

Example

EXAMPLE 10.6

The concentrations of Fe^{3+} and Cu^{2+} in a mixture can be determined following their reaction with hexacyanoruthenate (II), $Ru(CN)_6^{4-}$, which forms a purple-blue complex with Fe^{3+} (λ_{max} = 550 nm), and a pale green complex with Cu^{2+} (λ_{max} = 396 nm).[12] The molar absorptivities (M^{-1} cm^{-1}) for the metal complexes at the two wavelengths are summarized in the following table.

	ε_{550}	ε_{396}
Fe^{3+}	9970	84
Cu^{2+}	34	856

When a sample containing Fe^{3+} and Cu^{2+} is analyzed in a cell with a pathlength of 1.00 cm, the absorbance at 550 nm is 0.183, and the absorbance at 396 nm is 0.109. What are the molar concentrations of Fe^{3+} and Cu^{2+} in the sample?

SOLUTION

Substituting known values into equations 10.11 and 10.12 gives

$$A_{550} = 0.183 = 9970C_{Fe} + 34C_{Cu}$$

$$A_{396} = 0.109 = 84C_{Fe} + 856C_{Cu}$$

To determine the C_{Fe} and C_{Cu} we solve the first equation for C_{Cu}

$$C_{Cu} = \frac{0.183 - 9970C_{Fe}}{34}$$

and substitute the result into the second equation.

$$0.109 = 84C_{Fe} + 856\left(\frac{0.183 - 9970C_{Fe}}{34}\right) = 4.607 - (2.51 \times 10^5)C_{Fe}$$

Solving for C_{Fe} gives the concentration of Fe^{3+} as 1.80×10^{-5} M. Substituting this concentration back into the equation for the mixture's absorbance at a wavelength of 396 nm gives the concentration of Cu^{2+} as 1.26×10^{-4} M.

To obtain results with good accuracy and precision the two wavelengths should be selected so that $\varepsilon_X > \varepsilon_Y$ at one wavelength and $\varepsilon_Y < \varepsilon_X$ at the other wavelength. The optimum precision is obtained when the difference in molar absorptivities is as large as possible. One method for locating the optimum wavelengths, therefore, is to plot $\varepsilon_X/\varepsilon_Y$ as a function of wavelength and determine the wavelengths at which $\varepsilon_X/\varepsilon_Y$ reaches maximum and minimum values.[13]

Two additional methods for determining the composition of a mixture deserve mention. In multiwavelength linear regression analysis (MLRA) the absorbance of a mixture is compared with that of standard solutions at several wavelengths.[14] If A_{SX} and A_{SY} are the absorbances of standard solutions of components X and Y at any wavelength, then

$$A_{SX} = \varepsilon_X b C_{SX} \qquad \textbf{10.13}$$

$$A_{SY} = \varepsilon_Y b C_{SY} \qquad \textbf{10.14}$$

where C_{SX} and C_{SY} are the known concentrations of X and Y in the standard solutions. Solving equations 10.13 and 10.14 for ε_X and ε_Y, substituting into equation 10.11 (the wavelength designation can be dropped), and rearranging gives

$$\frac{A_m}{A_{SX}} = \frac{C_X}{C_{SX}} + \left(\frac{C_Y}{C_{SY}}\right)\frac{A_{SY}}{A_{SX}}$$

To determine C_X and C_Y, the mixture's absorbance and the absorbances of the standard solutions are measured at several wavelengths. Plotting A_m/A_{SX} versus A_{SY}/A_{SX} gives a straight line with a slope of C_Y/C_{SY} and a y-intercept of C_X/C_{SX}.

The generalized standard addition method (GSAM) extends the analysis of mixtures to situations in which matrix effects prevent the determination of ε_X and ε_Y using external standards.[15] When adding a known concentration of analyte to a solution containing an unknown concentration of analyte, the concentrations usually are not additive (see question 9 in Chapter 5). Conservation of mass, however, is always obeyed. Equation 10.11 can be written in terms of moles, n, by using the relationship

$$C = \frac{n}{V} \tag{10.15}$$

where V is the total solution volume. Substituting equation 10.15 into 10.11 and 10.12 gives

$$Q_{\lambda 1} = V(A_m)_{\lambda 1} = (\varepsilon_X)_{\lambda 1} b n_X + (\varepsilon_Y)_{\lambda 1} b n_Y \tag{10.16}$$

$$Q_{\lambda 2} = V(A_m)_{\lambda 2} = (\varepsilon_X)_{\lambda 2} b n_X + (\varepsilon_Y)_{\lambda 2} b n_Y \tag{10.17}$$

where Q is the volume-corrected absorbance. If a standard is added to the sample, the moles of X and Y increase by the amount Δn_X and Δn_Y, and the new volume-corrected absorbances are

$$(Q_{\lambda 1})' = (\varepsilon_X)_{\lambda 1} b(n_X + \Delta n_X) + (\varepsilon_Y)_{\lambda 1} b(n_Y + \Delta n_Y) \tag{10.18}$$

$$(Q_{\lambda 2})' = (\varepsilon_X)_{\lambda 2} b(n_X + \Delta n_X) + (\varepsilon_Y)_{\lambda 2} b(n_Y + \Delta n_Y) \tag{10.19}$$

Subtracting equation 10.16 from 10.18, and 10.17 from 10.19 gives

$$\Delta Q_{\lambda 1} = (\varepsilon_X)_{\lambda 1} b\, \Delta n_X + (\varepsilon_Y)_{\lambda 1} b\, \Delta n_Y$$

$$\Delta Q_{\lambda 2} = (\varepsilon_X)_{\lambda 2} b\, \Delta n_X + (\varepsilon_Y)_{\lambda 2} b\, \Delta n_Y$$

Values for $(\varepsilon_X)_{\lambda 1}$, $(\varepsilon_Y)_{\lambda 1}$, $(\varepsilon_X)_{\lambda 2}$, and $(\varepsilon_Y)_{\lambda 2}$ are obtained by plotting $\Delta Q_{\lambda 1}$ versus Δn_X, $\Delta Q_{\lambda 1}$ versus Δn_Y, $\Delta Q_{\lambda 2}$ versus Δn_X, and $\Delta Q_{\lambda 2}$ versus Δn_Y and determining the slopes. Equations 10.16 and 10.17 can then be solved to determine n_X and n_Y.

10D.3 Qualitative Applications

As discussed earlier in Section 10C.1, ultraviolet, visible and infrared absorption bands result from the absorption of electromagnetic radiation by specific valence electrons or bonds. The energy at which the absorption occurs, as well as the intensity of the absorption, is determined by the chemical environment of the absorbing moiety. For example, benzene has several ultraviolet absorption bands due to $\pi \rightarrow \pi^*$ transitions. The position and intensity of two of these bands, 203.5 nm ($\varepsilon = 7400$) and 254 nm ($\varepsilon = 204$), are very sensitive to substitution. For benzoic acid, in which a carboxylic acid group replaces one of the aromatic hydrogens, the two bands shift to 230 nm (ε = 11,600) and 273 nm (ε = 970). Several rules have been developed to aid in correlating UV/Vis absorption bands to chemical structure. Similar correlations have been developed for determining structures using infrared absorption bands. For example the carbonyl, C=O, stretch is very sensitive to adjacent functional groups, occurring at 1650 cm^{-1} for acids, 1700 cm^{-1} for ketones, and 1800 cm^{-1} for acid chlorides. The qualitative manual interpretation of UV/Vis and IR spectra receives adequate coverage elsewhere in the chemistry curriculum, notably in organic chemistry and is therefore not considered further in this text.

With the availability of computerized data acquisition and storage it is possible to build database libraries of standard reference spectra. When a spectrum of an unknown compound is obtained, its identity can often be determined by searching through a library of reference spectra. This process is known as **spectral searching.** Comparisons are made by an algorithm that calculates the cumulative difference between the absorbances of the sample and reference spectra. For example, one simple algorithm uses the following equation

spectral searching
The matching of a spectrum for an unknown compound to a reference spectrum stored in a computer database.

$$D = \sum_{i=1}^{n} \left|(A_s)_i - (A_r)_i\right|$$

where D is the cumulative difference, A_s is the absorbance of the sample at wavelength or wavenumber i, A_r is the absorbance of the reference compound at the same wavelength or wavenumber, and n is the number of points for which the spectra were digitized. The cumulative difference is calculated for each reference spectrum. The reference compound with the smallest value of D provides the closest match to the unknown compound. The accuracy of spectral searching is limited by the number and type of compounds included in the library and by the effect of the sample's matrix on the spectrum.

Another advantage of computerized data acquisition is the ability to subtract one spectrum from another. When coupled with spectral searching it may be possible, by repeatedly searching and subtracting reference spectra, to determine the identity of several components in a sample without the need of a prior separation step. An example is shown in Figure 10.33 in which the composition of a two-component mixture consisting of mannitol and cocaine hydrochloride was identified by successive searching and subtraction. Figure 10.33a shows the spectrum of the mixture. A search of the spectral library selects mannitol (Figure 10.33b) as a likely component of the mixture. Subtracting mannitol's spectrum from the mixture's spectrum leaves a result (Figure 10.33c) that closely matches the spectrum of cocaine hydrochloride (Figure 10.33d) in the spectral library. Subtracting leaves only a small residual signal (Figure 10.33e).

10D.4 Characterization Applications

Molecular absorption, particularly in the UV/Vis range, has been used for a variety of different characterization studies, including determining the stoichiometry of metal–ligand complexes and determining equilibrium constants. Both of these examples are examined in this section.

Stoichiometry of a Metal–Ligand Complex The stoichiometry for a metal–ligand complexation reaction of the following general form

$$M + yL \rightleftharpoons ML_y$$

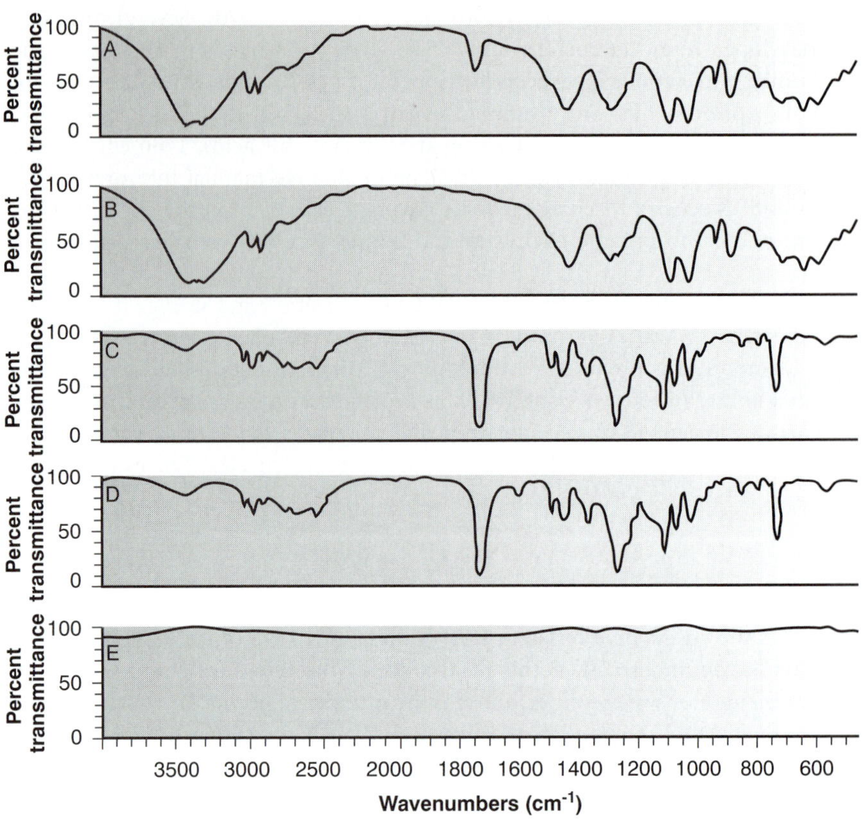

Figure 10.33
Spectral searching and stripping in the analysis of a mixture of mannitol and cocaine hydrochloride. (a) IR spectrum for the mixture; (b) Library IR spectrum of mannitol; (c) Result of subtracting mannitol's IR spectrum from that of the mixture; (d) Library IR spectrum for cocaine hydrochloride; (e) Result of subtracting the spectrum of cocaine hydrochloride from the mixture's IR spectrum.

can be determined by one of three methods: the method of continuous variations, the mole-ratio method, and the slope-ratio method.

method of continuous variations
A procedure for determining the stoichiometry between two reactants by preparing solutions containing different mole fractions of one reactant; also known as Job's method.

Of the three methods, the **method of continuous variations,** also called Job's method, is the most popular. In this method a series of solutions is prepared such that the total moles of metal and ligand, n_{tot}, in each solution is the same. Thus, if $(n_M)_i$ and $(n_L)_i$ are, respectively, the moles of metal and ligand in the i-th solution, then

$$n_{tot} = (n_M)_i + (n_L)_i$$

The relative amount of ligand and metal in each solution is expressed as the mole fraction of ligand, $(X_L)_i$, and the mole fraction of metal, $(X_M)_i$,

$$(X_L)_i = \frac{(n_L)_i}{n_{tot}}$$

$$(X_M)_i = 1 - (X_L)_i = \frac{(n_M)_i}{n_{tot}}$$

The concentration of the metal–ligand complex is determined by the limiting reagent, with the greatest concentration occurring when the metal and ligand are mixed stoichiometrically. If the reaction is monitored at a wavelength where only the metal–ligand complex absorbs, a plot of absorbance versus the mole fraction of ligand will show two linear branches: one when the ligand is the limiting reagent and a second when the metal is the limiting reagent. The intersection of these two branches occurs when a stoichiometric mixing of metal and ligand is reached. The mole fraction of ligand at this intersection is used to determine the value of y for the metal–ligand complex, ML_y.

$$y = \frac{n_L}{n_M} = \frac{X_L}{X_M} = \frac{X_L}{1 - X_L}$$

If there is no wavelength where only the metal–ligand complex absorbs, then the measured absorbances must be corrected for the absorbance that would be exhibited if the metal and ligand did not react to form ML_y.

$$A_{corr} = A_{meas} - \varepsilon_M b C_M - \varepsilon_L b C_L$$

In essence, the corrected absorbance gives the change in absorbance due to the formation of the metal–ligand complex. An example of the application of the method of continuous variations is shown in Example 10.7.

Example

EXAMPLE 10.7

To determine the formula for the complex between Fe^{2+} and *o*-phenanthroline, a series of solutions was prepared in which the total concentration of metal and ligand was held constant at 3.15×10^{-4} M. The absorbance of each solution was measured at a wavelength of 510 nm. Using the following data, determine the formula for the complex.

Colorplate 10 shows the solutions used to generate this data.

X_L	Absorbance	X_L	Absorbance
0.0	0.000	0.6	0.693
0.1	0.116	0.7	0.809
0.2	0.231	0.8	0.693
0.3	0.347	0.9	0.347
0.4	0.462	1.0	0.000
0.5	0.578		

SOLUTION

To determine the formula of the complex, we plot absorbance versus the mole fraction of ligand, obtaining the result shown in the graph.

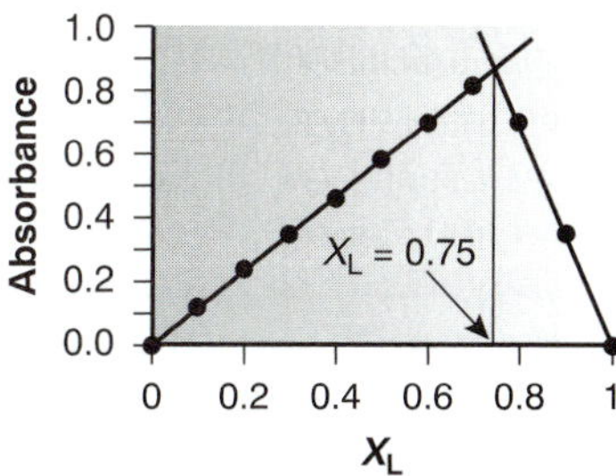

The maximum absorbance is determined by extrapolating the two linear portions of the plot. The intersection of the extrapolated lines corresponds to a mole fraction of ligand of 0.75. Solving for the value of y gives

$$y = \frac{X_L}{1 - X_L} = \frac{0.75}{1 - 0.75} = 3$$

and, the formula for the metal–ligand complex is $Fe(o\text{-phenanthroline})_3^{2+}$.

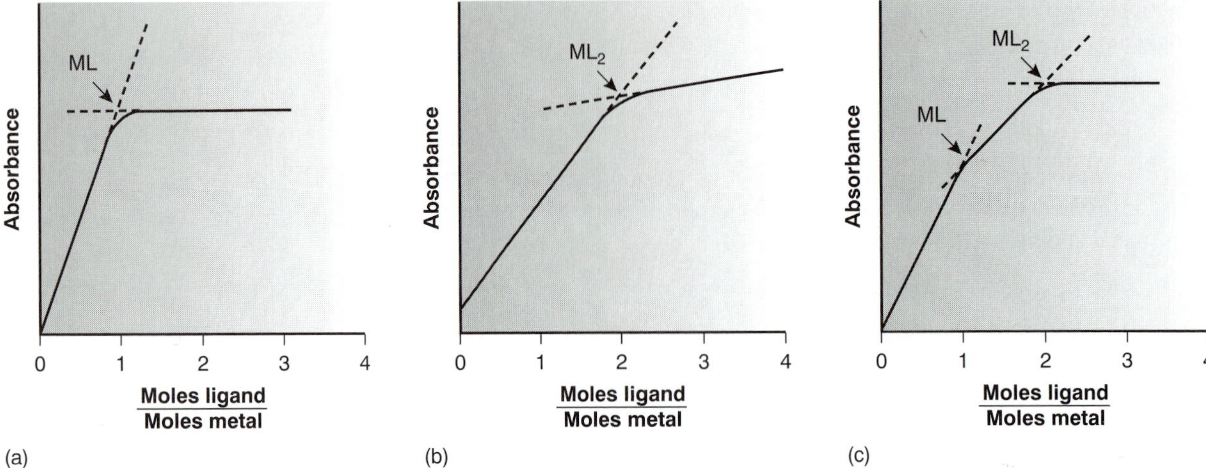

Figure 10.34
Mole-ratio plots used to determine the stoichiometry of a metal–ligand complexation reaction.

Several precautions are necessary when using the method of continuous variations. First, the method of continuous variations requires that a single metal–ligand complex be formed. To determine if this condition is true, plots of absorbance versus X_L should be constructed for several different wavelengths and for several different values of n_{tot}. If the maximum absorbance does not occur at the same value of X_L for each set of conditions, then more than one metal–ligand complex must be present. A second precaution is that the metal–ligand complex must obey Beer's law for the range of concentrations used in constructing the plot of absorbance versus X_L. Third, if the metal–ligand complex's formation constant is relatively small, the plot of absorbance versus X_L may show significant curvature. In this case it is often difficult to determine the stoichiometry by extrapolation. Finally, since the stability of the metal–ligand complex may be influenced by solution conditions, the composition of the solutions must be carefully controlled. When the ligand is a weak base, for example, the solutions must be buffered to the same pH.

mole-ratio method
A procedure for determining the stoichiometry between two reactants by preparing solutions containing different mole ratios of two reactants.

In the **mole-ratio method** the moles of one reactant, usually the metal, are held constant, while the moles of the other reactant are varied. The absorbance is monitored at a wavelength at which the metal–ligand complex absorbs. A plot of absorbance as a function of the ligand-to-metal mole ratio (n_L/n_M) has two linear branches that intersect at a mole ratio corresponding to the formula of the complex. Figure 10.34a shows a mole-ratio plot for the formation of a 1:1 complex in which the absorbance is monitored at a wavelength at which only the complex absorbs. Figure 10.34b shows a mole-ratio plot for a 1:2 complex in which the metal, the ligand, and the complex absorb at the selected wavelength. Unlike the method of continuous variations, the mole-ratio method can be used for complexation reactions that occur in a stepwise fashion, provided that the molar absorptivities of the metal–ligand complexes differ and the formation constants are sufficiently different. A typical mole-ratio plot for the stepwise formation of ML and ML_2 is shown in Figure 10.34c.

Both the method of continuous variations and the mole-ratio method rely on an extrapolation of absorbance data collected under conditions in which a linear relationship exists between absorbance and the relative amounts of metal and ligand. When a metal–ligand complex is very weak, a plot of absorbance versus X_L or n_L/n_M may be curved, making it impossible to determine the stoichiometry by extrapolation. In this case the slope ratio may be used.

In the **slope-ratio method** two sets of solutions are prepared. The first set consists of a constant amount of metal and a variable amount of ligand, chosen such that the total concentration of metal, C_M, is much greater than the total concentration of ligand, C_L. Under these conditions we may assume that essentially all the ligand is complexed. The concentration of a metal–ligand complex of the general form M_xL_y is

slope-ratio method
A procedure for determining the stoichiometry between two reactants by measuring the relative change in absorbance under conditions when each reactant is the limiting reagent.

$$[M_xL_y] = \frac{C_L}{y}$$

If absorbance is monitored at a wavelength where only M_xL_y absorbs, then

$$A = \varepsilon b[M_xL_y] = \frac{\varepsilon b C_L}{y}$$

and a plot of absorbance versus C_L will be linear with a slope, s_L, of

$$s_L = \frac{\varepsilon b}{y}$$

A second set of solutions is prepared with a fixed concentration of ligand that is much greater than the variable concentration of metal; thus

$$[M_xL_y] = \frac{C_M}{x}$$

$$A = \varepsilon b[M_xL_y] = \frac{\varepsilon b C_M}{x}$$

$$s_M = \frac{\varepsilon b}{x}$$

The mole ratio of ligand-to-metal is determined from the ratio of the two slopes.

$$\frac{s_M}{s_L} = \frac{\varepsilon b/x}{\varepsilon b/y} = \frac{y}{x}$$

An important assumption in the slope-ratio method is that the complexation reaction continues to completion in the presence of a sufficiently large excess of metal or ligand. The slope-ratio method also is limited to systems that obey Beer's law and in which only a single complex is formed.

Determination of Equilibrium Constants Another important application of molecular absorption is the determination of equilibrium constants. Let's consider, as a simple example, an acid–base reaction of the general form

$$HIn + H_2O \rightleftharpoons H_3O^+ + In^-$$

where HIn and In^- are the conjugate weak acid and weak base forms of a visual acid–base indicator. The equilibrium constant for this reaction is

$$K_a = \frac{[H_3O^+][In^-]}{[HIn]}$$

To determine the equilibrium constant's value, we prepare a solution in which the reaction exists in a state of equilibrium and determine the equilibrium concentration of H_3O^+, HIn, and In^-. The concentration of H_3O^+ is easily determined by measuring the solution's pH, whereas the concentration of HIn and In^- may be determined by measuring the solution's absorbance.

If both HIn and In^- absorb at the selected wavelength, then, from equation 10.6, we know that

$$A = \varepsilon_{HIn}b[HIn] + \varepsilon_{In}b[In^-] \qquad \textbf{10.20}$$

where ε_{HIn} and ε_{In} are the molar absorptivities for HIn and In^-. The total concentration of indicator, *C*, is given by a mass balance equation

$$C = [HIn] + [In^-] \qquad \textbf{10.21}$$

Solving equation 10.21 for [HIn] and substituting into equation 10.20 gives

$$A = \varepsilon_{HIn}b(C - [In^-]) + \varepsilon_{In}b[In^-]$$

which simplifies to

$$A = \varepsilon_{HIn}bC - \varepsilon_{HIn}b[In^-] + \varepsilon_{In}b[In^-]$$

$$A = A_{HIn} + b[In^-](\varepsilon_{In} - \varepsilon_{HIn}) \qquad \textbf{10.22}$$

where A_{HIn}, which is equal to $\varepsilon_{HIn}bC$, is the absorbance when the pH is acidic enough that essentially all the indicator is present as HIn. Solving equation 10.22 for the concentration of In^- gives

$$[In^-] = \frac{A - A_{HIn}}{b(\varepsilon_{In} - \varepsilon_{HIn})} \qquad \textbf{10.23}$$

Proceeding in the same fashion, we can derive a similar equation for the concentration of HIn; thus

$$[HIn] = \frac{A_{In} - A}{b(\varepsilon_{In} - \varepsilon_{HIn})} \qquad \textbf{10.24}$$

where A_{In}, which is equal to $\varepsilon_{In}bC$, is the absorbance when the pH is basic enough that only In^- contributes to the absorbance. Substituting equations 10.23 and 10.24 into the equilibrium constant expression for HIn gives

$$K_a = [H_3O^+]\left(\frac{A - A_{HIn}}{A_{In} - A}\right) \qquad \textbf{10.25}$$

Using equation 10.25, the value of K_a can be determined in one of two ways. The simplest approach is to prepare three solutions, each of which contains the same amount, *C*, of indicator. The pH of one solution is made acidic enough that [HIn] >> $[In^-]$. The absorbance of this solution gives A_{HIn}. The value of A_{In} is determined by adjusting the pH of the second solution such that $[In^-]$ >> [HIn]. Finally, the pH of the third solution is adjusted to an intermediate value, and the pH and absorbance, *A*, are recorded. The value of K_a can then be calculated by making appropriate substitutions into equation 10.25.

Example

EXAMPLE 10.8

The acidity constant for an acid–base indicator was determined by preparing three solutions, each of which has a total indicator concentration of 5.00×10^{-5} M. The first solution was made strongly acidic with HCl and has an absorbance of 0.250. The second solution was made strongly basic and has an absorbance of 1.40. The pH of the third solution was measured at 2.91, with an absorbance of 0.662. What is the value of K_a for the indicator?

SOLUTION

The value of K_a is determined by making appropriate substitutions into 10.25; thus

$$K_a = (1.23 \times 10^{-3})\left(\frac{0.662 - 0.250}{1.40 - 0.662}\right) = 6.87 \times 10^{-4}$$

A second approach is to prepare a series of solutions, each of which contains the same amount of indicator. Two solutions are used to determine values for A_{HIn} and A_{In}. Rewriting equation 10.25 in logarithmic form and rearranging

$$\log\left(\frac{A - A_{HIn}}{A_{In} - A}\right) = \text{pH} - \text{p}K_a$$

shows that a plot of log $[(A - A_{HIn})/(A_{In} - A)]$ versus pH is linear, with a slope of +1 and a *y*-intercept of $-\text{p}K_a$.

In developing this treatment for determining equilibrium constants, we have considered a relatively simple system in which the absorbance of HIn and In^- were easily measured, and for which it is easy to determine the concentration of H_3O^+. In addition to acid–base reactions, the same approach can be applied to any reaction of the general form

$$X + Y \rightleftharpoons Z$$

including metal–ligand complexation and redox reactions, provided that the concentration of the product, Z, and one of the reactants can be determined spectrophotometrically and the concentration of the other reactant can be determined by another method. With appropriate modifications, more-complicated systems, in which one or more of these parameters cannot be measured, also can be treated.[16]

10D.5 Evaluation

Scale of Operation Molecular UV/Vis absorption is routinely used for the analysis of trace analytes in macro and meso samples. Major and minor analytes can be determined by diluting samples before analysis, and concentrating a sample may allow for the analysis of ultratrace analytes. The scale of operations for infrared absorption is generally poorer than that for UV/Vis absorption.

Accuracy Under normal conditions relative errors of 1–5% are easily obtained with UV/Vis absorption. Accuracy is usually limited by the quality of the blank. Examples of the type of problems that may be encountered include the presence of particulates in a sample that scatter radiation and interferents that react with analytical reagents. In the latter case the interferant may react to form an absorbing species, giving rise to a positive determinate error. Interferents also may prevent the analyte from reacting, leading to a negative determinate error. With care, it may be possible to improve the accuracy of an analysis by as much as an order of magnitude.

Precision In absorption spectroscopy, precision is limited by indeterminate errors, or instrumental "noise," introduced when measuring absorbance. Precision is generally worse with very low absorbances due to the uncertainty of distinguishing a small difference between P_0 and P_T, and for very high absorbances when P_T approaches 0. We might expect, therefore, that precision will vary with transmittance.

Table 10.8 Effect of Indeterminate Instrumental Errors on Relative Uncertainty in Concentration

Category	Sources of Indeterminate Error	Equation for Relative Uncertainty in Concentration
$s_T = k_1$	%*T* readout resolution, noise in thermal detectors	$\frac{s_C}{C} = \frac{0.434k_1}{T \log T}$
$s_T = k_2\sqrt{T^2 + T}$	noise in photon detectors	$\frac{s_C}{C} = \frac{0.434k_2}{\log T}\sqrt{1 + \frac{1}{T}}$
$s_T = k_3T$	positioning of sample cell, fluctuations in source intensity	$\frac{s_C}{C} = \frac{0.434k_3}{\log T}$

We can derive an expression between precision and transmittance by applying the propagation of uncertainty as described in Chapter 4. To do so we write Beer's law as

$$C = -\frac{1}{\varepsilon b}\log T \qquad \textbf{10.26}$$

Using Table 4.9, the absolute uncertainty in the concentration, s_C, is given as

$$s_C = -\frac{0.434}{\varepsilon b} \times \frac{s_T}{T} \qquad \textbf{10.27}$$

where s_T is the absolute uncertainty for the transmittance. Dividing equation 10.27 by equation 10.26 gives the relative uncertainty in concentration, s_C/C, as

$$\frac{s_C}{C} = \frac{0.434s_T}{T\log T}$$

Thus, if s_T is known, the relative uncertainty in concentration can be determined for any transmittance.

Calculating the relative uncertainty in concentration is complicated by the fact that s_T may be a function of the transmittance. Three categories of indeterminate instrumental error have been observed.[17] Table 10.8 provides a summary of these categories. A constant s_T is observed for the uncertainty associated with the reading %*T* from a meter's analog or digital scale. Typical values are ±0.2–0.3% (k_1 of ±0.002–0.003) for an analog scale, and ±0.001% (k_1 of ±0.00001) for a digital scale. A constant s_T also is observed for the thermal transducers used in infrared spectrophotometers. The effect of a constant s_T on the relative uncertainty in concentration is shown by curve A in Figure 10.35. Note that the relative uncertainty is very large for both high and low absorbances, reaching a minimum when the absorbance is 0.434. This source of indeterminate error is important for infrared spectrophotometers and for inexpensive UV/Vis spectrophotometers. To obtain a relative uncertainty in concentration of ±1–2%, the absorbance must be kept between 0.1 and 1.

Values of s_T are a complex function of transmittance when indeterminate errors are dominated by the noise associated with photon transducers. Curve B in Figure 10.35 shows that the relative uncertainty in concentration is very large for low absorbances, but is less affected by higher absorbances. Although the relative uncertainty reaches a minimum when the absorbance is 0.96, there is little change in the relative uncertainty for absorbances between 0.5 and 2. This source of inde-

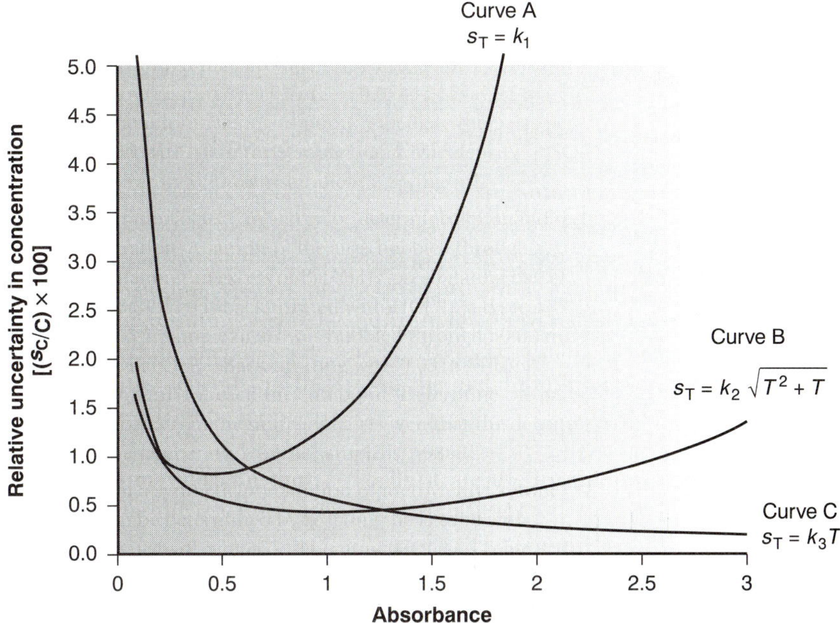

Figure 10.35
Relative uncertainties for absorption spectrophotometry as a function of absorbance for the three categories of indeterminate instrumental errors (see Table 10.8 for equations).

terminate error generally limits the precision of high-quality UV/Vis spectrophotometers for mid-to-high absorbances.

Finally, values of s_T are directly proportional to transmittance for indeterminate errors due to fluctuations in source intensity and for uncertainty in positioning the sample cell within the spectrometer. The latter is of particular importance since the optical properties of any sample cell are not uniform. As a result, repositioning the sample cell may lead to a change in the intensity of transmitted radiation. As shown by curve C in Figure 10.35, the effect of this source of indeterminate error is only important at low absorbances. This source of indeterminate errors is usually the limiting factor for high-quality UV/Vis spectrophotometers when the absorbance is relatively small.

When the relative uncertainty in concentration is limited by the $\%T$ readout resolution, the precision of the analysis can be improved by redefining the standards used to define 100% T and 0% T. Normally 100% T is established using a blank, and 0% T is established while using a shutter to prevent source radiation from reaching the detector. When the absorbance is too high, precision can be improved by resetting 100% T using a standard solution of analyte whose concentration is less than that of the sample (Figure 10.36a). For a sample whose absorbance is too low, precision can be improved by redefining 0% T, using a standard solution of analyte whose concentration is greater than that of the analyte (Figure 10.36b). In this case a calibration curve is required because a linear relationship between absorbance and concentration no longer exists. Precision can be further increased by combining these two methods (Figure 10.36c). Again, a calibration curve is necessary because the relationship between absorbance and concentration is no longer linear.

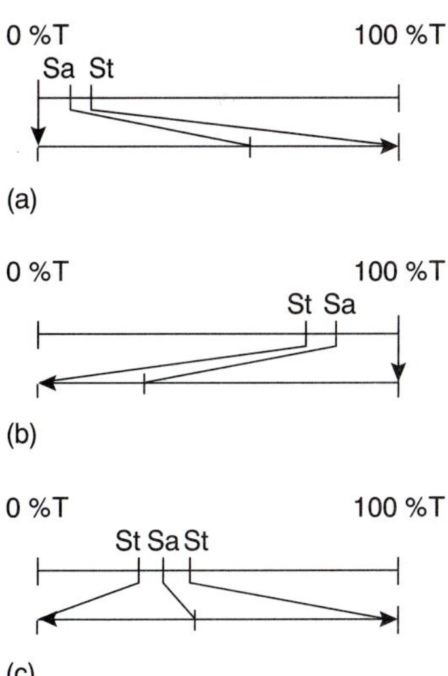

Figure 10.36
Methods for improving the precision of absorption measurements: (a) high-absorbance method; (b) low-absorbance method; and (c) maximum precision method. Abbreviations: Sa = sample; St = standard.

Sensitivity The sensitivity of a molecular absorption analysis is equivalent to the slope of a Beer's-law calibration curve and is determined by the product of the analyte's absorptivity and the pathlength of the sample cell. Sensitivity is improved by selecting a wavelength when absorbance is at a maximum or by increasing the pathlength.

Selectivity Selectivity is rarely a problem in molecular absorption spectrophotometry. In many cases it is possible to find a wavelength at which only the analyte absorbs or to use chemical reactions in a manner such that the analyte is the only species that absorbs at the chosen wavelength. When two or more species contribute to the measured absorbance, a multicomponent analysis is still possible, as shown in Example 10.6.

Time, Cost, and Equipment The analysis of a sample by molecular absorption spectroscopy is relatively rapid, although additional time may be required when it is necessary to use a chemical reaction to transform a nonabsorbing analyte into an absorbing form. The cost of UV/Vis instrumentation ranges from several hundred dollars for a simple, manually operated, single-beam instrument equipped with an inexpensive grating, to as much as $50,000 for a computer-controlled, high-resolution, double-beam instrument equipped with variable slits and operating over an extended range of wavelengths. Fourier transform infrared spectrometers can be obtained for as little as $15,000–$20,000, although more expensive models are available.

10E Atomic Absorption Spectroscopy

Atomic absorption, along with atomic emission, was first used by Guystav Kirchhoff and Robert Bunsen in 1859 and 1860, as a means for the qualitative identification of atoms. Although atomic emission continued to develop as an analytical technique, progress in atomic absorption languished for almost a century. Modern atomic absorption spectroscopy was introduced in 1955 as a result of the independent work of A. Walsh and C. T. J. Alkemade.[18] Commercial instruments were in place by the early 1960s, and the importance of atomic absorption as an analytical technique was soon evident.

10E.1 Instrumentation

Figure 10.37
Photo of a typical atomic absorption spectrophotometer.
Courtesy of Varian, Inc.

Atomic absorption spectrophotometers (Figure 10.37) are designed using either the single-beam or double-beam optics described earlier for molecular absorption spectrophotometers (see Figures 10.25 and 10.26). There are, however, several important differences that are considered in this section.

atomization
The process of converting an analyte into a free atom.

Atomization The most important difference between a spectrophotometer for atomic absorption and one for molecular absorption is the need to convert the analyte into a free atom. The process of converting an analyte in solid, liquid, or solution form to a free gaseous atom is called **atomization.** In most cases the sample containing the analyte undergoes some form of sample preparation that leaves the analyte in an organic or aqueous solution. For this reason, only the introduction of solution samples is considered in this text. Two general methods of atomization are used: flame atomization and electrothermal atomization. A few elements are atomized using other methods.

Flame Atomizers In flame atomization the sample is first converted into a fine mist consisting of small droplets of solution. This is accomplished using a nebulizer assembly similar to that shown in the inset to Figure 10.38. The sample is aspirated into a spray chamber by passing a high-pressure stream consisting of one or more combustion gases, past the end of a capillary tube immersed in the sample. The impact of the sample with the glass impact bead produces an aerosol mist. The aerosol

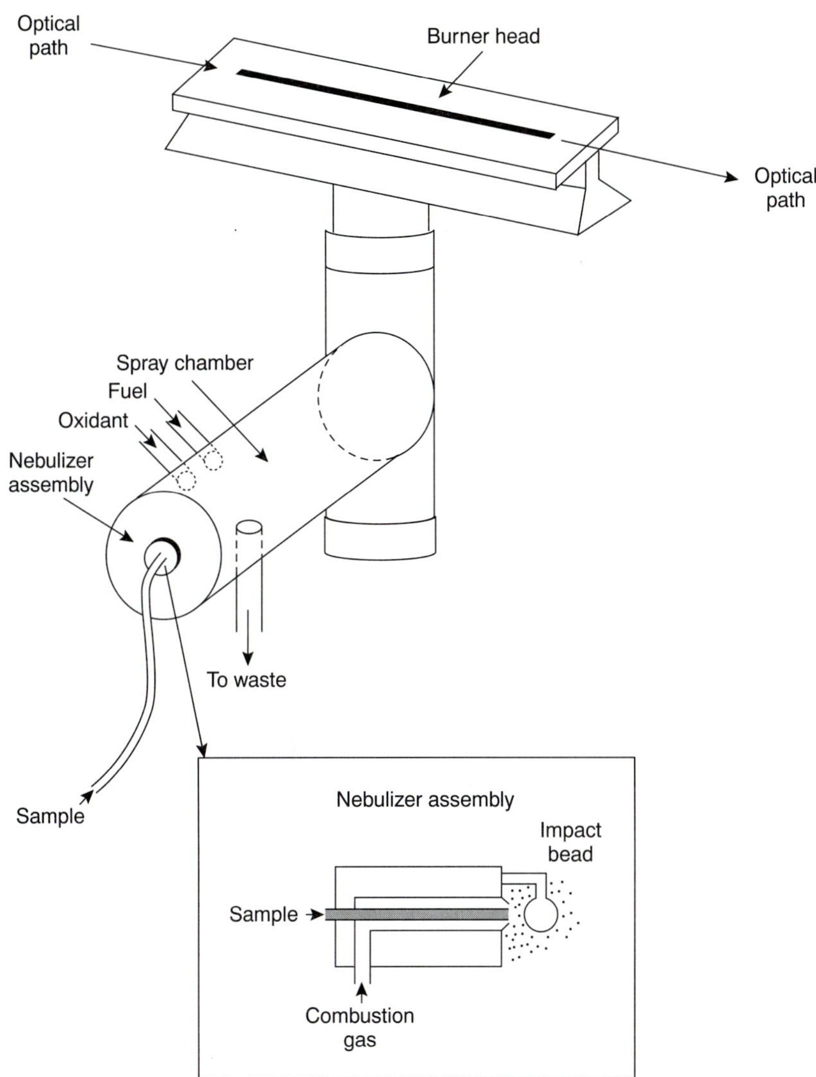

Figure 10.38
Flame atomization assembly equipped with spray chamber and slot burner. The inset shows the nebulizer assembly.

mist mixes with the combustion gases in the spray chamber before passing to the burner where the flame's thermal energy desolvates the aerosol mist to a dry aerosol of small, solid particles. Subsequently, thermal energy volatilizes the particles, producing a vapor consisting of molecular species, ionic species, and free atoms.

Thermal energy in flame atomization is provided by the combustion of a fuel–oxidant mixture. Common fuels and oxidants and their normal temperature ranges are listed in Table 10.9. Of these, the air–acetylene and nitrous oxide-acetylene flames are used most frequently. Normally, the fuel and oxidant are mixed in an approximately stoichiometric ratio; however, a fuel-rich mixture may be desirable for atoms that are easily oxidized. The most common design for the burner is the slot burner shown in Figure 10.38. This burner provides a long path length for monitoring absorbance and a stable flame.

The burner is mounted on an adjustable stage that allows the entire burner assembly to move horizontally and vertically. Horizontal adjustment is necessary to ensure that the flame is aligned with the instrument's optical path. Vertical adjustments are needed to adjust the height within the flame from which absorbance is

Table 10.9 Fuels and Oxidants Used for Flame Combustion

Fuel	Oxidant	Temperature Range (°C)
natural gas	air	1700–1900
hydrogen	air	2000–2100
acetylene	air	2100–2400
acetylene	nitrous oxide	2600–2800
acetylene	oxygen	3050–3150

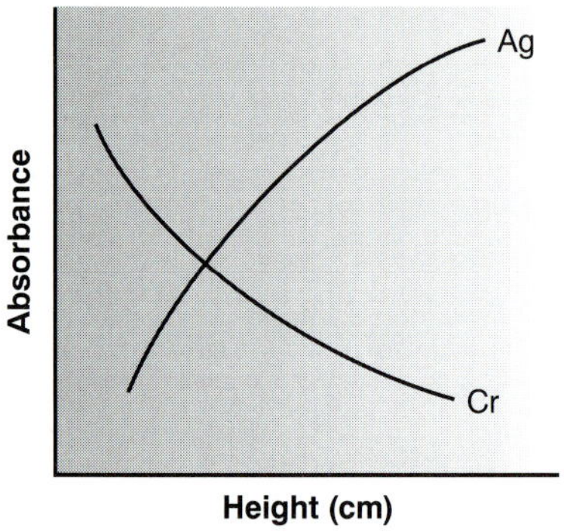

Figure 10.39
Absorbance profile for Ag and Cr in flame atomic absorption spectroscopy.

monitored. This is important because two competing processes affect the concentration of free atoms in the flame. An increased residence time in the flame results in a greater atomization efficiency; thus, the production of free atoms increases with height. On the other hand, longer residence times may lead to the formation of metal oxides that absorb at a wavelength different from that of the atom. For easily oxidized metals, such as Cr, the concentration of free atoms is greatest just above the burner head. For metals, such as Ag, which are difficult to oxidize, the concentration of free atoms increases steadily with height (Figure 10.39). Other atoms show concentration profiles that maximize at a characteristic height.

The most common means for introducing samples into a flame atomizer is continuous aspiration, in which the sample is continuously passed through the burner while monitoring the absorbance. Continuous aspiration is sample-intensive, typically requiring 2–5 mL of sample. Flame microsampling provides a means for introducing a discrete sample of fixed volume and is useful when the volume of sample is limited or when the sample's matrix is incompatible with the flame atomizer. For example, the continuous aspiration of a sample containing a high concentration of dissolved solids, such as sea water, may result in the build-up of solid deposits on the burner head. These deposits partially obstruct the flame, lowering the absorbance. Flame microsampling is accomplished using a micropipet to place 50–250 μL of sample in a Teflon funnel connected to the nebulizer, or by dipping the nebulizer tubing into the sample for a short time. Dip sampling is usually accomplished with an automatic sampler. The signal for flame microsampling is a transitory peak whose height or area is proportional to the amount of analyte that is injected.

The principal advantage of flame atomization is the reproducibility with which the sample is introduced into the spectrophotometer. A significant disadvantage to flame atomizers is that the efficiency of atomization may be quite poor. This may occur for two reasons. First, the majority of the aerosol mist produced during nebulization consists of droplets that are too large to be carried to the flame by the combustion gases. Consequently, as much as 95% of the sample never reaches the flame. A second reason for poor atomization efficiency is that the large volume of combustion gases significantly dilutes the sample. Together, these contributions to the efficiency of atomization reduce sensitivity since the analyte's concentration in the flame may be only 2.5×10^{-6} of that in solution.[19]

graphite furnace
An electrothermal atomizer that relies on resistive heating to atomize samples.

Electrothermal Atomizers A significant improvement in sensitivity is achieved by using resistive heating in place of a flame. A typical electrothermal atomizer, also known as a **graphite furnace,** consists of a cylindrical graphite tube approximately

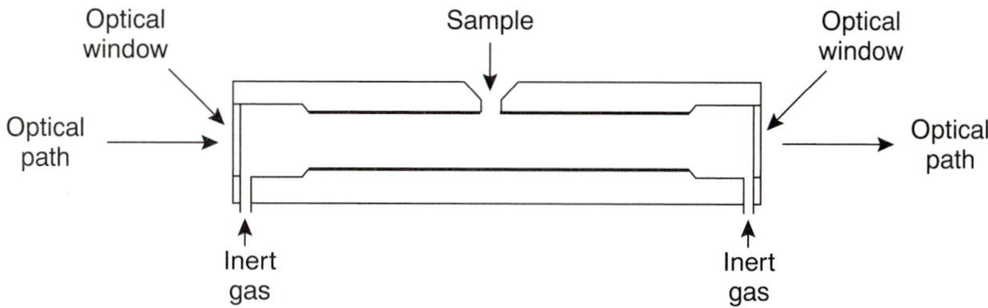

Figure 10.40
Diagram of an electrothermal analyzer.

1–3 cm in length, and 3–8 mm in diameter (Figure 10.40). The graphite tube is housed in an assembly that seals the ends of the tube with optically transparent windows. The assembly also allows for the passage of a continuous stream of inert gas, protecting the graphite tube from oxidation, and removing the gaseous products produced during atomization. A power supply is used to pass a current through the graphite tube, resulting in resistive heating.

Samples between 5 and 50 μL are injected into the graphite tube through a small-diameter hole located at the top of the tube. Atomization is achieved in three stages. In the first stage the sample is dried using a current that raises the temperature of the graphite tube to about 110 °C. Desolvation leaves the sample as a solid residue. In the second stage, which is called ashing, the temperature is increased to 350–1200 °C. At these temperatures, any organic material in the sample is converted to CO_2 and H_2O, and volatile inorganic materials are vaporized. These gases are removed by the inert gas flow. In the final stage the sample is atomized by rapidly increasing the temperature to 2000–3000 °C. The result is a transient absorbance peak whose height or area is proportional to the absolute amount of analyte injected into the graphite tube. The three stages are complete in approximately 45–90 s, with most of this time used for drying and ashing the sample.

Electrothermal atomization provides a significant improvement in sensitivity by trapping the gaseous analyte in the small volume of the graphite tube. The analyte's concentration in the resulting vapor phase may be as much as 1000 times greater than that produced by flame atomization.[20] The improvement in sensitivity, and the resulting improvement in detection limits, is offset by a significant decrease in precision. Atomization efficiency is strongly influenced by the sample's contact with the graphite tube, which is difficult to control reproducibly.

Miscellaneous Atomization Methods A few elements may be atomized by a chemical reaction that produces a volatile product. Elements such as As, Se, Sb, Bi, Ge, Sn, Te, and Pb form volatile hydrides when reacted with $NaBH_4$ in acid. An inert gas carries the volatile hydrides to either a flame or to a heated quartz observation tube situated in the optical path. Mercury is determined by the cold-vapor method in which it is reduced to elemental mercury with $SnCl_2$. The volatile Hg is carried by an inert gas to an unheated observation tube situated in the instrument's optical path.

10E.2 Quantitative Applications

Atomic absorption using either flame or electrothermal atomization is widely used for the analysis of trace metals in a variety of sample matrices. Using the atomic absorption analysis for zinc as an example, procedures have been developed for its determination in samples as diverse as water and wastewater, air, blood, urine, muscle tissue, hair, milk, breakfast cereals, shampoos, alloys, industrial plating baths, gasoline, oil, sediments, and rocks.

Developing a quantitative atomic absorption method requires several considerations, including choosing a method of atomization, selecting the wavelength and slit width, preparing the sample for analysis, minimizing spectral and chemical interferences, and selecting a method of standardization. Each of these topics is considered in this section.

Flame Versus Electrothermal Atomization The choice of atomization method is determined primarily by the analyte's concentration in the samples being analyzed. Because of its greater sensitivity, detection limits for most elements are significantly lower when using electrothermal atomization (Table 10.10). A better precision when using flame atomization makes it the method of choice when the analyte's concentration is significantly greater than the detection limit for flame atomization. In addition, flame atomization is subject to fewer interferences, allows for a greater throughput of samples, and requires less expertise from the operator. Electrothermal atomization is the method of choice when the analyte's concentration is lower than the detection limit for flame atomization. Electrothermal atomization is also useful when the volume of sample is limited.

Selecting the Wavelength and Slit Width The source for atomic absorption is a hollow cathode lamp consisting of a cathode and anode enclosed within a glass tube filled with a low pressure of Ne or Ar (Figure 10.41). When a potential is applied across the electrodes, the filler gas is ionized. The positively charged ions collide with the negatively charged cathode, dislodging, or "sputtering," atoms from the cathode's surface. Some of the sputtered atoms are in the excited state and emit radiation characteristic of the metal from which the cathode was manufactured. By fashioning the cathode from the metallic analyte, a hollow cathode lamp provides emission lines that correspond to the analyte's absorption spectrum.

characteristic concentration
The concentration of analyte giving an absorbance of 0.00436.

The sensitivity of an atomic absorption line is often described by its **characteristic concentration,** which is the concentration of analyte giving an absorbance of 0.00436 (corresponding to a percent transmittance of 99%). For example, Table 10.11 shows a list of wavelengths and characteristic concentrations for copper.

Usually the wavelength providing the best sensitivity is used, although a less sensitive wavelength may be more appropriate for a high concentration of analyte. A less sensitive wavelength also may be appropriate when significant interferences occur at the most sensitive wavelength. For example, atomizing a sample produces atoms of not only the analyte, but also of other components present in the sample's matrix. The presence of other atoms in the flame does not result in an interference unless the absorbance lines for the analyte and the potential interferant are within approximately 0.01 nm. When this is a problem, an interference may be

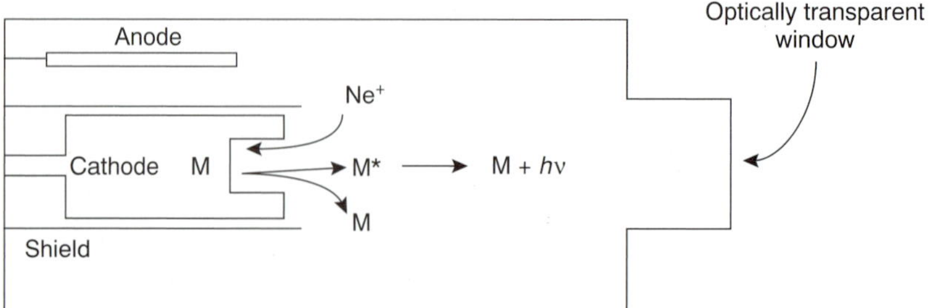

Figure 10.41
Schematic diagram of a hollow cathode lamp showing mechanism by which atomic emission is obtained.

Table 10.10 Atomic Absorption Detection Limits for Selected Elements

Element	Detection Limits (ppb) Flame Atomization	Electrothermal Atomization
Ag	0.9	0.001
Al	20	0.01
As[a]	20	0.08
Au	6	0.01
B	700	15
Ba	8	0.04
Be	1	0.003
Bi[a]	20	0.1
Ca	0.5	0.01
Cd	0.5	0.0002
Co	2	0.008
Cr	2	0.004
Cs	8	0.04
Fe	3	0.01
Ga	50	0.01
Ge	50	0.1
Hg[b]	200	0.2
K	1	0.004
Li	0.3	0.01
Mg	0.1	0.0002
Mn	0.8	0.0006
Mo	10	0.02
Na	0.2	0.004
Ni	2	0.05
Pb	10	0.007
Pd	10	0.05
Pt	40	0.2
Sb[a]	30	0.08
Se[a]	100	0.05
Si	20	0.005
Sn	10	0.03
Sr	2	0.01
Ti	10	0.3
V	20	0.1
Zn	0.8	0.0006

Source: Compiled from Parson, M. L.; Major, S.; Forster, A. R. *Appl. Spectrosc.* **1983,** *37,* 411–418; Weltz, B. *Atomic Absorption Spectrometry,* VCH: Deerfield Beach, FL, 1985.

[a]Detection limit by hydride vaporization method:
As 0.02 ppb
Bi 0.02 ppb
Sb 0.1 ppb
Se 0.02 ppb

[b]Detection limit by cold-vapor method:
Hg 0.001 ppb

Table 10.11 Absorption Lines and Characteristic Concentrations for Copper

Wavelength (nm)	Characteristic Concentration (ppm)
324.8	0.04
327.4	0.1
217.9	0.6
222.6	2.0
249.2	10
244.2	40

avoided by selecting another wavelength at which the analyte, but not the interferant, absorbs.

The emission spectrum from a hollow cathode lamp includes, besides emission lines for the analyte, additional emission lines for impurities present in the metallic cathode and the filler gas. These additional lines serve as a potential source of stray radiation that may lead to an instrumental deviation from Beer's law. Normally the monochromator's slit width is set as wide as possible, improving the throughput of radiation, while being narrow enough to eliminate this source of stray radiation.

Preparing the Sample Flame and electrothermal atomization require that the sample be in a liquid or solution form. Samples in solid form are prepared for analysis by dissolving in an appropriate solvent. When the sample is not soluble, it may be digested, either on a hot plate or by microwave, using HNO_3, H_2SO_4, or $HClO_4$. Alternatively, the analyte may be extracted via a Soxhlet extraction. Liquid samples may be analyzed directly or may be diluted or extracted if the matrix is incompatible with the method of atomization. Serum samples, for instance, may be difficult to aspirate when using flame atomization and may produce unacceptably high background absorbances when using electrothermal atomization. A liquid–liquid extraction using an organic solvent containing a chelating agent is frequently used to concentrate analytes. Dilute solutions of Cd^{2+}, Co^{2+}, Cu^{2+}, Fe^{3+}, Pb^{2+}, Ni^{2+}, and Zn^{2+}, for example, can be concentrated by extracting with a solution of ammonium pyrrolidine dithiocarbamate in methyl isobutyl ketone.

Minimizing Spectral Interference A spectral interference occurs when an analyte's absorption line overlaps with an interferant's absorption line or band. As noted previously, the overlap of two atomic absorption lines is seldom a problem. On the other hand, a molecule's broad absorption band or the scattering of source radiation is a potentially serious spectral interference.

An important question to consider when using a flame as an atomization source, is how to correct for the absorption of radiation by the flame. The products of combustion consist of molecular species that may exhibit broad-band absorption, as well as particulate material that may scatter radiation from the source. If this spectral interference is not corrected, then the intensity of the transmitted radiation decreases. The result is an apparent increase in the sample's absorbance. Fortunately, absorption and scattering of radiation by the flame are corrected by analyzing a blank.

Spectral interferences also occur when components of the sample's matrix react in the flame to form molecular species, such as oxides and hydroxides. Absorption and scattering due to components in the sample matrix other than the analyte constitute the sample's background and may present a significant problem, particularly at wavelengths below 300 nm, at which the scattering of radiation becomes more important. If the composition of the sample's matrix is known, then standards can be prepared with an identical matrix. In this case the background absorption is the same for both the samples and standards. Alternatively, if the background is due to a known matrix component, then that component can be added in excess to all samples and standards so that the contribution of the naturally occurring interferant is insignificant. Finally, many interferences due to the sample's matrix can be eliminated by adjusting the flame's composition. For example, by switching to a higher temperature flame it may be possible to prevent the formation of interfering oxides and hydroxides.

When the identity of the matrix interference is unknown, or when it is impossible to adjust the flame to eliminate the interference, then other means must be used to compensate for the background interference. Several methods have been developed to compensate for matrix interferences, and most atomic absorption spectrophotometers include one or more of these methods.

One of the most common methods for **background correction** is the use of a continuum source, such as a D_2 lamp. Since the D_2 lamp is a continuum source, the absorbance of its radiation by the analyte's narrow absorption line is negligible. Any absorbance of radiation from the D_2 lamp, therefore, is due to the background. Absorbance of radiation from the hollow cathode lamp, however, is due to both the analyte and the background. Subtracting the absorbance for the D_2 lamp from that for the hollow cathode lamp gives an absorbance that has been corrected for the background interference. Although this method of background correction may be quite effective, it assumes that the background absorbance is constant over the range of wavelengths passed by the monochromator. When this is untrue, subtracting the two absorbances may under- or over-correct for the background.

background correction
In atomic absorption spectroscopy, the correction of the net absorbance from that due to the sample matrix.

Other methods of background correction have been developed, including Zeeman effect background correction and Smith–Hieftje background correction, both of which are included in some commercially available atomic absorption spectrophotometers. Further details about these methods can be found in several of the suggested readings listed at the end of the chapter.

Minimizing Chemical Interferences The quantitative analysis of some elements is complicated by chemical interferences occurring during atomization. The two most common chemical interferences are the formation of nonvolatile compounds containing the analyte and ionization of the analyte. One example of a chemical interference due to the formation of nonvolatile compounds is observed when PO_4^{3-} or Al^{3+} is added to solutions of Ca^{2+}. In one study, for example, adding 100 ppm Al^{3+} to a solution of 5 ppm Ca^{2+} decreased the calcium ion's absorbance from 0.50 to 0.14, whereas adding 500 ppm PO_4^{3-} to a similar solution of Ca^{2+} decreased the absorbance from 0.50 to 0.38.[21] These interferences were attributed to the formation of refractory particles of $Ca_3(PO_4)_2$ and an Al–Ca–O oxide.

The formation of nonvolatile compounds often can be minimized by increasing the temperature of the flame, either by changing the fuel-to-oxidant ratio or by switching to a different combination of fuel and oxidant. Another approach is to add a releasing agent or protecting agent to solutions containing the analyte. A **releasing agent** is a species whose reaction with the interferent is more favorable than that of the analyte. Adding Sr^{2+} or La^{3+} to solutions of Ca^{2+}, for example, minimizes the effect of PO_4^{3-} and Al^{3+} by reacting in place of the analyte. Thus, adding 2000 ppm $SrCl_2$ to the Ca^{2+}/PO_4^{3-} and Ca^{2+}/Al^{3+} mixtures discussed in the preceding paragraph gave absorbances for each of 0.48, whereas a solution of 2000 ppm $SrCl_2$ and Ca^{2+} alone gave an absorbance of 0.49. **Protecting agents** react with the analyte to form a stable volatile complex. Adding 1% w/w EDTA to the Ca^{2+}/PO_4^{3-} solution discussed in the preceding paragraph gave an absorbance of 0.52, compared with an absorbance of 0.55 for just the Ca^{2+} and EDTA. On the other hand, EDTA does not serve as a protecting agent for solutions of Ca^{2+} and Al^{3+}.

releasing agent
A reagent whose reaction with an interferant is more favorable than the interferant's reaction with the analyte.

protecting agent
A reagent that reacts with the analyte, preventing it from transforming into a nonanalyzable form.

Ionization interferences occur when thermal energy from the flame or electrothermal atomizer is sufficient to ionize the analyte

$$M \rightleftharpoons M^+ + e^- \qquad \mathbf{10.28}$$

where M is the analyte in atomic form, and M^+ is the cation of the analyte formed by ionization. Since the absorption spectra for M and M^+ are different, the position of the equilibrium in reaction 10.28 affects absorbance at wavelengths where M absorbs. If another species is present that ionizes more easily than M, then the equilibrium in reaction 10.28 shifts to the left. Variations in the concentration of easily ionized species, therefore, may have a significant effect on a sample's absorbance, resulting in a determinate error. The effect of ionization can be minimized by adding a high concentration of an **ionization suppressor,** which is simply another species that ionizes more easily than the analyte. If the concentration of the ionization suppressor is sufficient, then the increased concentration of electrons in the flame pushes reaction 10.28 to the left, preventing the analyte's ionization. Potassium and cesium are frequently used as ionization suppressors because of their low ionization energy.

ionization suppressor
A reagent that is more easily ionizable than the analyte.

Standardizing the Method Because Beer's law also applies to atomic absorption, we might expect atomic absorption calibration curves to be linear. In practice, however, most atomic absorption calibration curves are nonlinear, or linear for only a limited range of concentrations. Nonlinearity in atomic absorption is a consequence of instrumental limitations, including stray radiation from the hollow cathode lamp and a nonconstant molar absorptivity due to the narrow width of the absorption line. Accurate quantitative work, therefore, often requires a suitable means for computing the calibration curve from a set of standards. Nonlinear calibration curves may be fit using quadratic and cubic equations, although neither works well over a broad range of concentrations. More accurate results may be obtained using some of the methods mentioned in Section 5C.5 in Chapter 5.

When possible, a quantitative analysis is best conducted using external standards. Unfortunately, matrix interferences are a frequent problem, particularly when using electrothermal atomization. For this reason the method of standard additions is often used. One limitation to this method of standardization, however, is the requirement that there be a linear relationship between absorbance and concentration.

Method 10.2 Determination of Cu and Zn in Tissue Samples[22]

Description of Method. Copper and zinc are isolated by digesting tissue samples after extracting any fatty tissue. The concentration of copper and zinc in the supernatant are determined by atomic absorption using an air–acetylene flame.

Procedure. Tissue samples are obtained by a muscle needle biopsy and are dried for 24–30 hours at 105 °C to remove all traces of moisture. The fatty tissue in the dried samples is removed by extracting overnight with anhydrous ether. After removing the ether, the sample is dried to obtain the fat-free dry tissue weight (FFDT). The sample is digested at 68 °C for 20–24 h using 3 mL of 0.75 M HNO_3. After centrifuging at 2500 rpm for 10 min, the supernatant is transferred to a 5-mL volumetric flask. The digestion is repeated two more times, for 2–4 h each, using 0.9-mL aliquots of 0.75 M HNO_3. These supernatants are added to the 5-mL volumetric flask, which is diluted to volume with 0.75 M HNO_3. The concentration of Cu and Zn in the diluted supernatant is determined by atomic absorption spectroscopy using an air–acetylene flame and external standards. Copper is analyzed at a wavelength of 324.8 nm with a slit width of 0.5 nm, and zinc is analyzed at 213.9 nm with a slit width of 1.0 nm. Background correction is used for zinc. Results are reported as micrograms of Cu or Zn per gram of FFDT.

Questions

1. **What is the proper matrix for the external standards and the blank?**

   The matrix for the standards and the blank should match that of the samples; thus, an appropriate matrix is 0.75 M HNO_3. Any interferences from other components of the sample matrix are minimized by background correction.

2. **Why is background correction necessary for the analysis of Zn, but not for the analysis of Cu?**

   Background correction is used to compensate for background absorption and scattering due to interferents in the sample. Such interferences are most severe for analytes, such as Zn, that absorb at wavelengths of less than 300 nm.

3. **The following absorbances were obtained for a set of Cu calibration standards**

ppm Cu	Absorbance
0.000	0.000
0.100	0.006
0.200	0.013
0.300	0.020
0.400	0.026
0.500	0.033
0.600	0.039
0.700	0.046
1.000	0.066

   What is the concentration of copper, in micrograms per gram FFDT, for a 11.23-mg FFDT tissue sample that yields an absorbance of 0.023?

   Linear regression of the calibration standards gives the relationship between absorbance and concentration as

$$A = -0.0002 + 0.0661(\text{ppm Cu})$$

   Substituting the sample's absorbance into the preceding equation gives the concentration of copper in solution as 0.351 ppm. The concentration in the tissue sample, therefore, is

$$\frac{(0.351\ \mu\text{g/mL})(5\ \text{mL})}{0.01123\ \text{g}} = 156\ \mu\text{g Cu/g FFDT}$$

10E.3 Evaluation

Scale of Operation Atomic absorption spectroscopy is ideally suited for the analysis of trace and ultratrace analytes, particularly when using electrothermal atomization. By diluting samples, atomic absorption also can be applied to minor and major analytes. Most analyses use macro or meso samples. The small volume requirement for electrothermal atomization or flame microsampling, however, allows the use of micro, or even ultramicro samples.

Accuracy When spectral and chemical interferences are minimized, accuracies of 0.5–5% are routinely possible. With nonlinear calibration curves, higher accuracy is obtained by using a pair of standards whose absorbances closely bracket the sample's absorbance and assuming that the change in absorbance is linear over the limited concentration range. Determinate errors for electrothermal atomization are frequently greater than that obtained with flame atomization due to more serious matrix interferences.

Precision For absorbances greater than 0.1–0.2, the relative standard deviation for atomic absorption is 0.3–1% for flame atomization, and 1–5% for electrothermal atomization. The principal limitation is the variation in the concentration of free-analyte atoms resulting from a nonuniform rate of aspiration, nebulization, and atomization in flame atomizers, and the consistency with which the sample is heated during electrothermal atomization.

Sensitivity The sensitivity of an atomic absorption analysis with flame atomization is influenced strongly by the flame's composition and the position in the flame from which absorption is monitored. Normally the sensitivity for an analysis is optimized by aspirating a standard and adjusting operating conditions, such as the fuel-to-oxidant ratio, the nebulizer flow rate, and the height of the burner, to give the greatest absorbance. With electrothermal atomization, sensitivity is influenced by the drying and ashing stages that precede atomization. The temperature and time used for each stage must be worked out for each type of sample.

Sensitivity is also influenced by the sample's matrix. We have already noted, for example, that sensitivity can be decreased by chemical interferences. An increase in sensitivity can often be realized by adding a low-molecular-weight alcohol, ester, or ketone to the solution or by using an organic solvent.

Selectivity Due to the narrow width of absorption lines, atomic absorption provides excellent selectivity. Atomic absorption can be used for the analysis of over 60 elements at concentrations at or below the level of parts per million.

Time, Cost, and Equipment The analysis time when using flame atomization is rapid, with sample throughputs of 250–350 determinations per hour when using a fully automated system. Electrothermal atomization requires substantially more time per analysis, with maximum sample throughputs of 20–30 determinations per hour. The cost of a new instrument ranges from \$10,000 to \$50,000 for flame atomization and \$18,000 to \$70,000 for electrothermal atomization. The more expensive instruments in each price range include double-beam optics and automatic samplers, are computer controlled, and can be programmed for multielemental analysis by allowing the wavelength and hollow cathode lamp to be changed automatically.

10F Spectroscopy Based on Emission

An analyte in an excited state possesses an energy, E_2, that is greater than that when it is in a lower energy state, E_1. When the analyte returns, or relaxes to a lower energy state the excess energy, ΔE,

$$\Delta E = E_2 - E_1$$

must be released. Figure 10.5 shows a simplified picture of this process.

The **lifetime** of an analyte in the excited state, A^*, is short; typically 10^{-5}–10^{-9} s for electronic excited states and 10^{-15} s for vibrational excited states. **Relaxation** occurs through collisions between A^* and other species in the sample, by photochemical reactions, and by the emission of photons. In the first process, which is called vibrational deactivation, or nonradiative relaxation, the excess energy is released as heat; thus

$$A^* \rightarrow A + \text{heat}$$

Relaxation by a photochemical reaction may involve a decomposition reaction in which A^* splits apart

$$A^* \rightarrow \text{X} + \text{Y}$$

or a reaction between A^* and another species

$$A^* + \text{Z} \rightarrow \text{X} + \text{Y}$$

In either case the excess energy is used up in the chemical reaction or released as heat.

In the third mechanism excess energy is released as a photon of electromagnetic radiation.

$$A^* \rightarrow A + h\nu$$

The release of a photon following thermal excitation is called emission, and that following the absorption of a photon is called photoluminescence. In chemiluminescence and bioluminescence, excitation results from a chemical or biochemical reaction, respectively. Spectroscopic methods based on photoluminescence are the subject of Section 10G, and atomic emission is covered in Section 10H.

lifetime
The length of time that an analyte stays in an excited state before returning to a lower-energy state.

relaxation
Any process by which an analyte returns to a lower-energy state from a higher-energy state.

10G Molecular Photoluminescence Spectroscopy

Photoluminescence is divided into two categories: fluorescence and phosphorescence. Absorption of an ultraviolet or visible photon promotes a valence electron from its ground state to an excited state with conservation of the electron's spin. For example, a pair of electrons occupying the same electronic ground state have opposite spins (Figure 10.42a) and are said to be in a singlet spin state. Absorbing a photon promotes one of the electrons to a **singlet excited state** (Figure 10.42b). Emission of a photon from a singlet excited state to a singlet ground state, or between any two energy levels with the same spin, is called **fluorescence.** The probability of a fluorescent transition is very high, and the average lifetime of the electron in the excited state is only 10^{-5}–10^{-8} s. Fluorescence, therefore, decays rapidly after the excitation source is removed. In some cases an electron in a singlet excited state is transformed to a **triplet excited state** (Figure 10.42c) in which its spin is no longer paired with that of the ground state. Emission between a triplet excited state and a singlet ground state, or between any two energy levels that differ in their respective

singlet excited state
An excited state in which all electron spins are paired.

fluorescence
Emission of a photon when the analyte returns to a lower-energy state with the same spin as the higher-energy state.

triplet excited state
An excited state in which unpaired electron spins occur.

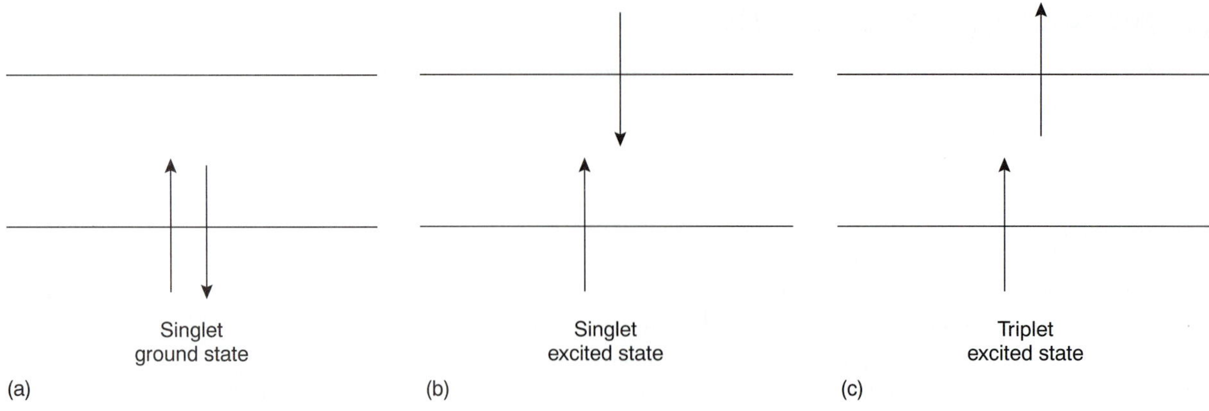

Figure 10.42
Difference between singlet and triplet states.

phosphorescence
Emission of a photon when the analyte returns to a lower-energy state with the opposite spin as the higher-energy state.

spin states, is called **phosphorescence.** Because the average lifetime for phosphorescence ranges from 10^{-4} to 10^{4} s, phosphorescence may continue for some time after removing the excitation source.

The use of molecular fluorescence for qualitative analysis and semiquantitative analysis can be traced to the early to mid-1800s, with more accurate quantitative methods appearing in the 1920s. Instrumentation for fluorescence spectroscopy using filters and monochromators for wavelength selection appeared in, respectively, the 1930s and 1950s. Although the discovery of phosphorescence preceded that of fluorescence by almost 200 years, qualitative and quantitative applications of molecular phosphorescence did not receive much attention until after the development of fluorescence instrumentation.

10G.1 Molecular Fluorescence and Phosphorescence Spectra

To appreciate the origin of molecular fluorescence and phosphorescence, we must consider what happens to a molecule following the absorption of a photon. Let's assume that the molecule initially occupies the lowest vibrational energy level of its electronic ground state. The ground state, which is shown in Figure 10.43, is a singlet state labeled S_0. Absorption of a photon of correct energy excites the molecule to one of several vibrational energy levels in the first excited electronic state, S_1, or the second electronic excited state, S_2, both of which are singlet states. Relaxation to the ground state from these excited states occurs by a number of mechanisms that are either radiationless, in that no photons are emitted, or involve the emission of a photon. These relaxation mechanisms are shown in Figure 10.43. The most likely pathway by which a molecule relaxes back to its ground state is that which gives the shortest lifetime for the excited state.

vibrational relaxation
A form of radiationless relaxation in which an analyte moves from a higher vibrational energy level to a lower vibrational energy level in the same electronic level.

Radiationless Deactivation One form of radiationless deactivation is **vibrational relaxation,** in which a molecule in an excited vibrational energy level loses energy as it moves to a lower vibrational energy level in the same electronic state. Vibrational relaxation is very rapid, with the molecule's average lifetime in an excited vibrational energy level being 10^{-12} s or less. As a consequence, molecules that are excited to different vibrational energy levels of the same excited electronic state quickly return to the lowest vibrational energy level of this excited state.

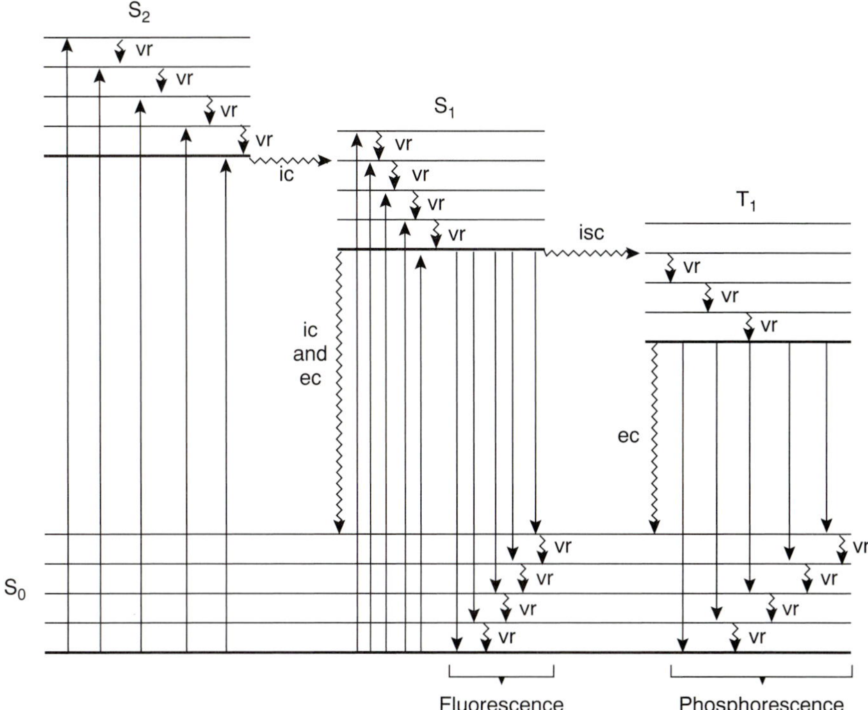

Figure 10.43
Energy level diagram for a molecule showing pathways for deactivation of an excited state: vr is vibrational relaxation; ic is internal conversion; ec is external conversion, and isc is intersystem crossing. The lowest vibrational energy level for each electronic state is indicated by the thicker line.

Another form of radiationless relaxation is **internal conversion,** in which a molecule in the ground vibrational level of an excited electronic state passes directly into a high vibrational energy level of a lower energy electronic state of the same spin state. By a combination of internal conversions and vibrational relaxations, a molecule in an excited electronic state may return to the ground electronic state without emitting a photon. A related form of radiationless relaxation is **external conversion** in which excess energy is transferred to the solvent or another component in the sample matrix.

A final form of radiationless relaxation is an **intersystem crossing** in which a molecule in the ground vibrational energy level of an excited electronic state passes into a high vibrational energy level of a lower energy electronic energy state with a different spin state. For example, an intersystem crossing is shown in Figure 10.43 between a singlet excited state, S_1, and a triplet excited state, T_1.

internal conversion
A form of radiationless relaxation in which the analyte moves from a higher electronic energy level to a lower electronic energy level.

external conversion
A form of radiationless relaxation in which energy is transferred to the solvent or sample matrix.

intersystem crossing
A form of radiationless relaxation in which the analyte moves from a higher electronic energy level to a lower electronic energy level with a different spin state.

Fluorescence Fluorescence occurs when a molecule in the lowest vibrational energy level of an excited electronic state returns to a lower energy electronic state by emitting a photon. Since molecules return to their ground state by the fastest mechanism, fluorescence is only observed if it is a more efficient means of relaxation than the combination of internal conversion and vibrational relaxation. A quantitative expression of the efficiency of fluorescence is the fluorescent **quantum yield, Φ_f,** which is the fraction of excited molecules returning to the ground state by fluorescence. Quantum yields range from 1, when every molecule in an excited state undergoes fluorescence, to 0 when fluorescence does not occur.

quantum yield
The fraction of absorbed photons that produce a desired event, such as fluorescence or phosphorescence (Φ).

The intensity of fluorescence, I_f, is proportional to the amount of the radiation from the excitation source that is absorbed and the quantum yield for fluorescence

$$I_f = k\Phi_f(P_0 - P_T) \qquad \textbf{10.29}$$

where k is a constant accounting for the efficiency of collecting and detecting the fluorescent emission. From Beer's law we know that

$$\frac{P_T}{P_0} = 10^{-\varepsilon bC} \tag{10.30}$$

where C is the concentration of the fluorescing species. Solving equation 10.30 for P_T and substituting into equation 10.29 gives, after simplifying

$$I_f = k\Phi_f P_0(1 - 10^{-\varepsilon bC}) \tag{10.31}$$

For low concentrations of the fluorescing species, where εbC is less than 0.01, this equation simplifies to

$$I_f = 2.303k\Phi_f P_0 \varepsilon bC \tag{10.32}$$

The intensity of fluorescence therefore, increases with an increase in quantum efficiency, incident power of the excitation source, and the molar absorptivity and concentration of the fluorescing species.

Fluorescence is generally observed with molecules where the lowest energy absorption is a $\pi \rightarrow \pi^*$ transition, although some $n \rightarrow \pi^*$ transitions show weak fluorescence. Most unsubstituted, nonheterocyclic aromatic compounds show favorable fluorescence quantum yields, although substitution to the aromatic ring can have a significant effect on Φ_f. For example, the presence of an electron-withdrawing group, such as $—NO_2$, decreases Φ_f, whereas adding an electron-donating group, such as —OH, increases Φ_f. Fluorescence also increases for aromatic ring systems and for aromatic molecules with rigid planar structures.

A molecule's fluorescence quantum yield is also influenced by external variables such as temperature and solvent. Increasing temperature generally decreases Φ_f because more frequent collisions between the molecule and the solvent increases external conversion. Decreasing the solvent's viscosity decreases Φ_f for similar reasons. For an analyte with acidic or basic functional groups, a change in pH may change the analyte's structure and, therefore, its fluorescent properties. Changes in both the wavelength and intensity of fluorescence may be affected.

As shown in Figure 10.43, fluorescence may return the molecule to any of several vibrational energy levels in the ground electronic state. Fluorescence, therefore, occurs over a range of wavelengths. Because the change in energy for fluorescent emission is generally less than that for absorption, a molecule's fluorescence spectrum is shifted to higher wavelengths than its absorption spectrum.

Phosphorescence A molecule in the lowest vibrational energy level of an excited triplet electronic state normally relaxes to the ground state by an intersystem crossing to a singlet state or by external conversion. Phosphorescence is observed when relaxation occurs by the emission of a photon. As shown in Figure 10.43, phosphorescence occurs over a range of wavelengths, all of which are at a lower energy than the molecule's absorption band. The intensity of phosphorescence, I_p, is given by an equation similar to equation 10.32 for fluorescence

$$I_p = 2.303k\Phi_p P_0 \varepsilon bC \tag{10.33}$$

where Φ_p is the quantum yield for phosphorescence.

Phosphorescence is most favorable for molecules that have $n \rightarrow \pi^*$ transitions, which have a higher probability for an intersystem crossing than do $\pi \rightarrow \pi^*$ transitions. For example, phosphorescence is observed with aromatic molecules containing carbonyl groups or heteroatoms. Aromatic compounds containing halide atoms also have a higher efficiency for phosphorescence. In general, an increase in phosphorescence corresponds to a decrease in fluorescence.

Since the average lifetime for phosphorescence is very long, ranging from 10^{-4} to 10^4 s, the quantum yield for phosphorescence is usually quite small. An improvement in Φ_p is realized by decreasing the efficiency of external conversion. This may be accomplished in several ways, including lowering the temperature, using a more viscous solvent, depositing the sample on a solid substrate, or trapping the molecule in solution.

Excitation Versus Emission Spectra Photoluminescence spectra are recorded by measuring the intensity of emitted radiation as a function of either the excitation wavelength or the emission wavelength. An **excitation spectrum** is obtained by monitoring emission at a fixed wavelength while varying the excitation wavelength. Figure 10.44 shows the excitation spectrum for the hypothetical system described by the energy level diagram in Figure 10.43. When corrected for variations in source intensity and detector response, a sample's excitation spectrum is nearly identical to its absorbance spectrum. The excitation spectrum provides a convenient means for selecting the best excitation wavelength for a quantitative or qualitative analysis.

excitation spectrum
A fluorescence or phosphorescence spectrum in which the emission intensity at a fixed wavelength is measured as a function of the wavelength used for excitation.

In an emission spectrum a fixed wavelength is used to excite the molecules, and the intensity of emitted radiation is monitored as a function of wavelength. Although a molecule has only a single excitation spectrum, it has two emission spectra, one for fluorescence and one for phosphorescence. The corresponding emission spectra for the hypothetical system in Figure 10.43 are shown in Figure 10.44.

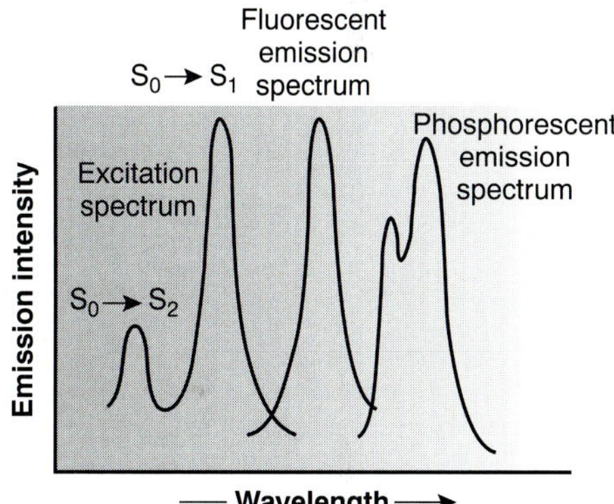

Figure 10.44
Example of molecular excitation and emission spectra.

10G.2 Instrumentation

The basic design of instrumentation for monitoring molecular fluorescence and molecular phosphorescence is similar to that found for other spectroscopies. The most significant differences are discussed in the following sections.

Molecular Fluorescence A typical instrumental block diagram for molecular fluorescence is shown in Figure 10.45. In contrast to instruments for absorption spectroscopy, the optical paths for the source and detector are usually positioned at an angle of 90°.

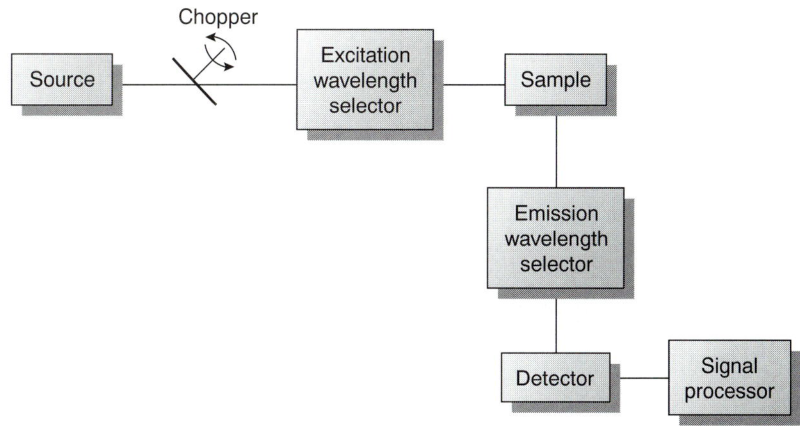

Figure 10.45
Block diagram for molecular fluorescence spectrometer.

fluorometer
An instrument for measuring fluorescence that uses filters to select the excitation and emission wavelengths.

spectrofluorometer
An instrument for measuring fluorescence that uses a monochromator to select the excitation and emission wavelengths.

Two basic instrumental designs are used for measuring molecular fluorescence. In a fluorometer the excitation and emission wavelengths are selected with absorption or interference filters. The excitation source for a **fluorometer** is usually a low-pressure mercury vapor lamp that provides intense emission lines distributed throughout the ultraviolet and visible region (254, 312, 365, 405, 436, 546, 577, 691, and 773 nm). When a monochromator is used to select the excitation and emission wavelengths, the instrument is called a **spectrofluorometer.** With a monochromator, the excitation source is usually a high-pressure Xe arc lamp, which has a continuum emission spectrum. Either instrumental design is appropriate for quantitative work, although only a spectrofluorometer can be used to record an excitation or emission spectrum.

The sample cells for molecular fluorescence are similar to those for optical molecular absorption. Remote sensing with fiber-optic probes (see Figure 10.30) also can be adapted for use with either a fluorometer or spectrofluorometer. An analyte that is fluorescent can be monitored directly. For analytes that are not fluorescent, a suitable fluorescent probe molecule can be incorporated into the tip of the fiber-optic probe. The analyte's reaction with the probe molecule leads to an increase or decrease in fluorescence.

Molecular Phosphorescence Instrumentation for molecular phosphorescence must discriminate between phosphorescence and fluorescence. Since the lifetime for fluorescence is much shorter than that for phosphorescence, discrimination is easily achieved by incorporating a delay between exciting and measuring phosphorescent emission. A typical instrumental design is shown in Figure 10.46. As shown

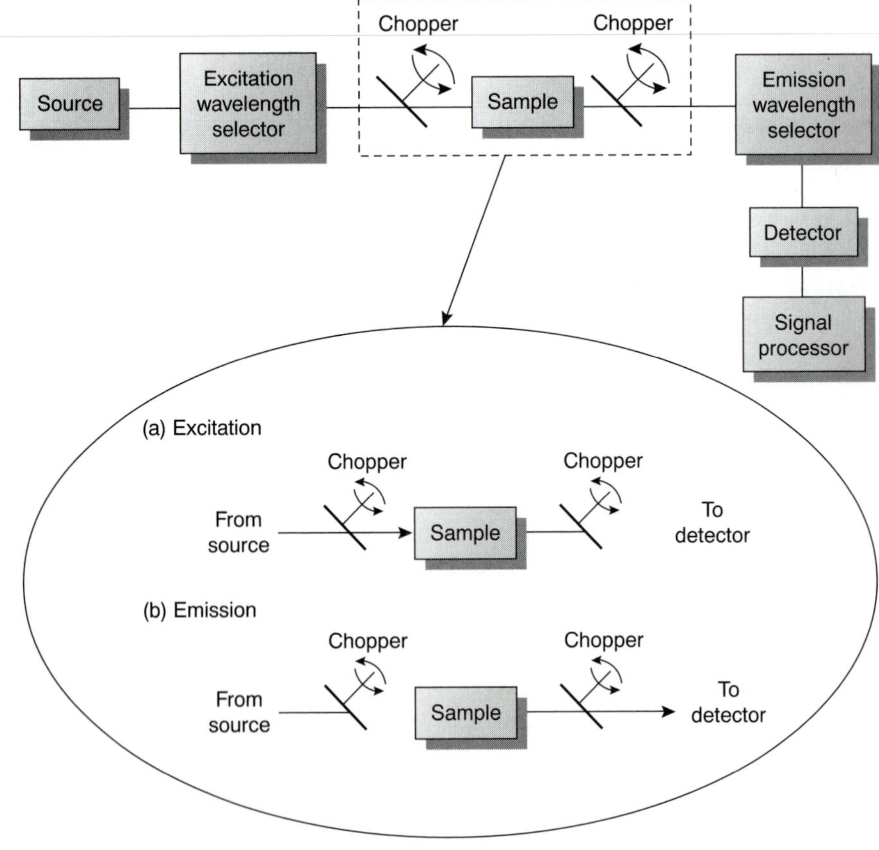

Figure 10.46
Block diagram for molecular phosphorescence spectrometer with inset showing how choppers are used to isolate excitation and emission.

in the inset, the two choppers are rotated out of phase, such that fluorescent emission is blocked from the detector when the excitation source is focused on the sample, and the excitation source is blocked from the sample when measuring the phosphorescent emission.

Because phosphorescence is such a slow process, provision must be made to prevent deactivation of the excited state by external conversion. Traditionally, this has been accomplished by dissolving the sample in a suitable organic solvent, usually a mixture of ethanol, isopentane, and diethyl ether. The resulting solution is frozen at liquid-N_2 temperatures, forming an optically clear solid. The solid matrix minimizes external conversion due to collisions between the analyte and the solvent. External conversion also is minimized by immobilizing the sample on a solid substrate, allowing the measurement of phosphorescence at room temperature. One approach is to place a drop of solution containing the analyte on a small filter paper disk mounted on a sample probe. After drying the sample under a heat lamp, the sample probe is placed in the spectrofluorometer for analysis. Other solid surfaces that have been used include silica gel, alumina, sodium acetate, and sucrose. This approach is particularly useful for the analysis of thin-layer chromatography plates.

10G.3 Quantitative Applications Using Molecular Luminescence

Molecular fluorescence and, to a lesser extent, phosphorescence have been used for the direct or indirect quantitative analysis of analytes in a variety of matrices. A direct quantitative analysis is feasible when the analyte's quantum yield for fluorescence or phosphorescence is favorable. When the analyte is not fluorescent or phosphorescent or when the quantum yield for fluorescence or phosphorescence is unfavorable, an indirect analysis may be feasible. One approach to an indirect analysis is to react the analyte with a reagent, forming a product with fluorescent properties. Another approach is to measure a decrease in fluorescence when the analyte is added to a solution containing a fluorescent molecule. A decrease in fluorescence is observed when the reaction between the analyte and the fluorescent species enhances radiationless deactivation, or produces a nonfluorescent product. The application of fluorescence and phosphorescence to inorganic and organic analytes is considered in this section.

Inorganic Analytes Except for a few metal ions, most notably UO_2^+, most inorganic ions are not sufficiently fluorescent to allow a direct analysis. Many metal ions can be determined indirectly by reacting with an organic ligand to form a fluorescent, or less commonly, a phosphorescent metal–ligand complex. One example of a chelating ligand is the sodium salt of 2,4,3′-trihydroxyazobenzene-5′-sulfonic acid, also known as alizarin garnet R, which forms a fluorescent metal–ligand complex with Al^{3+} (Figure 10.47). The analysis is carried out using an excitation wavelength of 470 nm, with fluorescence monitored at 500 nm. Other examples of chelating reagents that form fluorescent metal–ligand complexes with metal ions are listed in Table 10.12. A few inorganic nonmetals are determined by their ability to decrease, or quench, the fluorescence of another species. One example is the analysis for F^-, which is based on its ability to quench the fluorescence of the Al^{3+}–alizarin garnet R complex.

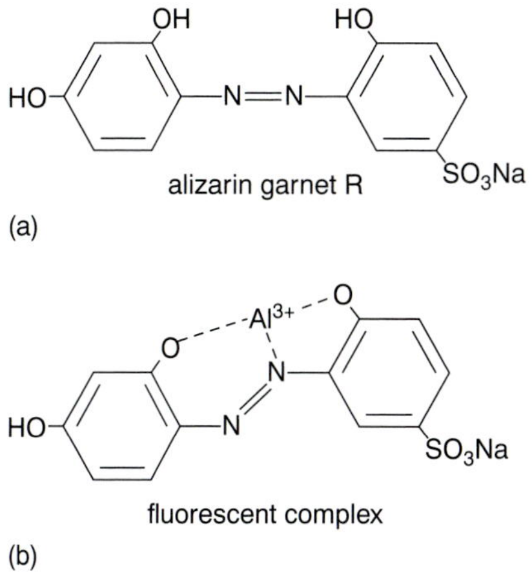

Figure 10.47
Structure of (a) alizarin garnet R, and (b) its metal–ligand complex with Al^{3+}.

Organic Analytes As noted earlier, organic compounds containing aromatic rings generally are fluorescent, but aromatic heterocycles are often phosphorescent. Many important biochemical, pharmaceutical, and environmental compounds are aromatic and, therefore, can be analyzed quantitatively by fluorometry

Table 10.12 Selected Chelating Agents for the Fluorometric Analysis of Inorganic Metal Ions

Chelating Agent	Metal Ions
8-hydroxyquinoline	Al^{3+}, Be^{2+}, Zn^{2+}, Li^{+}, Mg^{2+} (and others)
flavonal	Zr^{2+}, Sn^{4+}
benzoin	$B_4O_7^{2-}$, Zn^{2+}
2′,3,4′,5,7-pentahydroxyflavone	Be^{2+}
2-(*o*-hydroxyphenyl) benzoxazole	Cd^{2+}

Table 10.13 Selected Examples of Organic Compounds of Biochemical, Pharmaceutical, and Environmental Significance That Show Natural Fluorescence or Phosphorescence

Class	Compounds[a]
aromatic amino acids	phenylalanine (F) tyrosine (F) tryptophan (F, P)
vitamins	vitamin A (F) vitamin B_2 (F) vitamin B_6 (F) vitamin B_{12} (F) vitamin E (F) folic acid (F)
catecholamines	dopamine (F) norepinephrine (F)
pharmaceuticals and drugs	quinine (F) salicylic acid (F, P) morphine (F) barbiturates (F) LSD (F) codeine (P) caffeine (P) sulfanilamide (P)

environmental pollutants	polycyclic aromatic hydrocarbons:
	  pyrene (F)
	  benzo[*a*]pyrene (F)
	organothiophosphorous pesticides (F)
	carbamate insecticides (F)
	DDT (P)

[a]*Abbreviations:* F = fluorescence, P = phosphorescence, LSD = lysergic acid diethylamide, DDT = dichlorodiphenyltrichloroethane.

or phosphorometry. Several examples are listed in Table 10.13. When an organic analyte is not naturally fluorescent or phosphorescent, it may be possible to incorporate it into a chemical reaction that produces a fluorescent or phosphorescent product. For example, the enzyme creatine phosphokinase can be determined by using it to catalyze the formation of creatine from phosphocreatine. The creatine that is formed reacts with ninhydrin, producing a fluorescent product of unknown structure.

Standardizing the Method Equations 10.32 and 10.33 show that the intensity of fluorescent or phosphorescent emission is proportional to the concentration of the photoluminescent species, provided that the absorbance of radiation from the excitation source ($A = \varepsilon bC$) is less than approximately 0.01. Quantitative methods are usually standardized using a set of external standards. Calibration curves are linear over as much as four to six orders of magnitude for fluorescence and two to four orders of magnitude for phosphorescence. Calibration curves become nonlinear for high concentrations of the photoluminescent species at which the intensity of emission is given by equation 10.31. Nonlinearity also may be observed at low concentrations due to the presence of fluorescent or phosphorescent contaminants. As discussed earlier, the quantum efficiency for emission is sensitive to temperature and sample matrix, both of which must be controlled if external standards are to be used. In addition, emission intensity depends on the molar absorptivity of the photoluminescent species, which is sensitive to the sample matrix.

Representative Methods

Method 10.3 Determination of Quinine in Urine[23]

Description of Method. Quinine is an alkaloid used in treating malaria (it also is found in tonic water). It is a strongly fluorescent compound in dilute solutions of H_2SO_4 ($\Phi_f = 0.55$). The excitation spectrum of quinine shows two absorption bands at 250 nm and 350 nm, and the emission spectrum shows a single emission band at 450 nm. Quinine is rapidly excreted from the body in urine and is easily determined by fluorescence following its extraction from the urine sample.

Procedure. Transfer a 2.00-mL sample of urine to a 15-mL test tube, and adjust the pH to between 9 and 10 using 3.7 M NaOH. Add 4 mL of a 3:1 (v/v) mixture of chloroform and isopropanol, and shake the contents of the test tube for 1 min. Allow the organic and aqueous (urine) layers to separate, and transfer the organic phase to a clean test tube. Add 2.00 mL of 0.05 M H_2SO_4 to the organic phase, and shake the contents for 1 min. Allow the organic and aqueous layers to separate, and transfer the aqueous phase to the sample cell. Measure the fluorescent emission at 450 nm, using an excitation wavelength of 350 nm. Determine the concentration of quinine in the urine sample using a calibration curve prepared with a set of external standards in 0.05 M H_2SO_4 and a distilled water blank. The external standards are prepared from a 100.0-ppm solution of quinine in 0.05 M H_2SO_4.

Questions

1. **Chloride ion is known to quench the intensity of quinine's fluorescent emission. For example, the presence of 100 ppm NaCl (61 ppm Cl^-) gives an emission intensity that is only 83% of that without chloride, whereas the presence of 1000 ppm NaCl (610 ppm Cl^-) gives a fluorescent emission that is only 29% as intense. The concentration of chloride in urine typically ranges from 4600 to 6700 ppm Cl^-. How is an interference from chloride avoided in this procedure?**

   Extracting the aqueous urine sample with a mixture of chloroform and isopropanol separates the quinine and chloride, with the chloride remaining in the urine sample.

2. **Samples of urine that do not contain quinine still contain a small amount of fluorescent material after the extraction steps. How can the quantitative procedure described earlier be modified to take this into account?**

—Continued

Continued from page 431

One approach is to prepare a sample blank using urine known to be free of quinine. The fluorescent signal for the sample blank is subtracted from the urine sample's measured fluorescence.

3. **The fluorescent emission for quinine at 450 nm can be induced using an excitation frequency of either 250 nm or 350 nm. The fluorescent quantum efficiency is known to be the same for either excitation wavelength, and the UV absorption spectrum shows that ε_{250} is greater than ε_{350}. Nevertheless, fluorescent emission intensity is greater when using 350 nm as the excitation wavelength. Speculate on why this is the case.**

   From equation 10.32, I_f is a function of k, Φ_f, P_0, ε, b, and C. Since Φ_f, b, and C are the same for either excitation wavelength, and ε is larger for a wavelength of 250 nm, the greater I_f for an excitation wavelength of 350 nm must be due to larger values for P_0 or k at a wavelength of 350 nm. In fact, P_0 at 350 nm for a high-pressure Xe arc lamp is about 170% of that at 250 nm. In addition, the sensitivity of a typical photomultiplier transducer (which contributes to the value of k) at 350 nm is about 140% of that at 250 nm.

10G.4 Evaluation

Scale of Operation Molecular photoluminescence can be used for the routine analysis of trace and ultratrace analytes in macro and meso samples. Detection limits for fluorescence spectroscopy are strongly influenced by the analyte's quantum yield. For analytes with $\Phi_f > 0.5$, detection limits in the picomolar range are possible when using a high-quality spectrofluorometer. As an example, the detection limit for quinine sulfate, for which Φ_f is 0.55, is generally between 1 ppb and 1 ppTr (part per trillion). Detection limits for phosphorescence are somewhat poorer than those for fluorescence, with typical values in the nanomolar range for low-temperature phosphorometry and in the micromolar range for room-temperature phosphorometry using a solid substrate.

Accuracy The accuracy of a fluorescence method is generally 1–5% when spectral and chemical interferences are insignificant. Accuracy is limited by the same types of problems affecting other spectroscopic methods. In addition, accuracy is affected by interferences influencing the fluorescent quantum yield. The accuracy of phosphorescence is somewhat greater than that for fluorescence.

Precision When the analyte's concentration is well above the detection limit, the relative standard deviation for fluorescence is usually 0.5–2%. The limiting instrumental factor affecting precision is the stability of the excitation source. The precision for phosphorescence is often limited by reproducibility in preparing samples for analysis, with relative standard deviations of 5–10% being common.

Sensitivity From equations 10.32 and 10.33 we can see that the sensitivity of a fluorescent or phosphorescent method is influenced by a number of parameters. The importance of quantum yield and the effect of temperature and solution composition on Φ_f and Φ_p already have been considered. Besides quantum yield, the sensitivity of an analysis can be improved by using an excitation source that has a greater

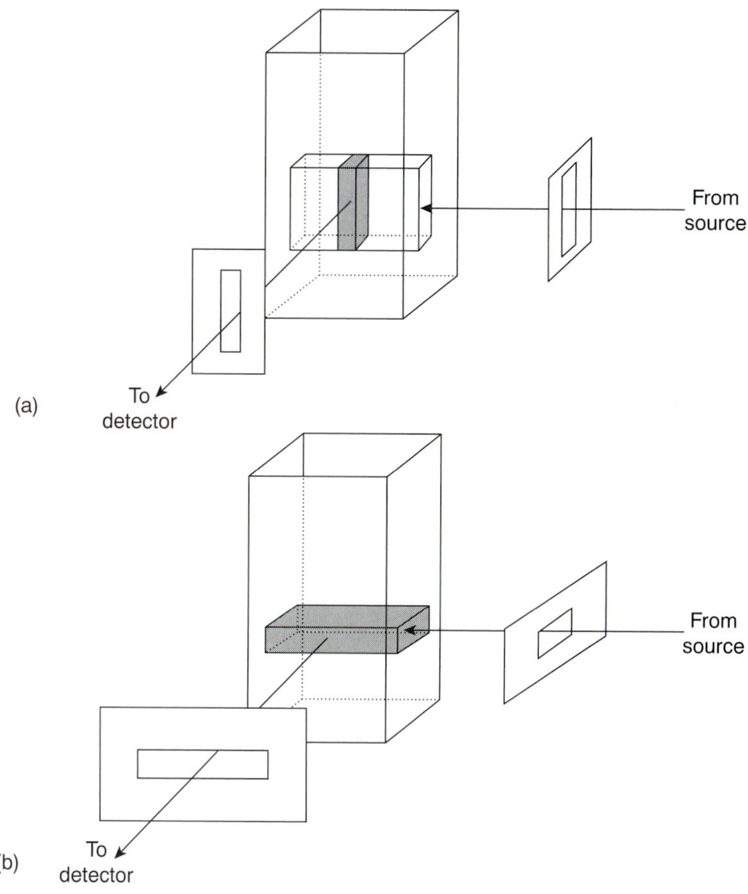

Figure 10.48
Use of slit orientation to change the volume from which fluorescence is measured: (a) normal; (b) rotated 90°.

emission intensity (P_0) at the desired wavelength and by selecting an excitation wavelength that corresponds to an absorption maximum (ε). Another approach that can be used to increase sensitivity is to increase the volume in the sample from which emission is monitored. Figure 10.48 shows how a 90° rotation of the slits used to focus the excitation source on the sample and to collect emission from the sample can produce a 5–30-fold increase in the signal.

Selectivity The selectivity of molecular fluorescence and phosphorescence is superior to that of absorption spectrophotometry for two reasons: first, not every compound that absorbs radiation is fluorescent or phosphorescent, and, second, selectivity between an analyte and an interferant is possible if there is a difference in either their excitation or emission spectra. In molecular luminescence the total emission intensity is a linear sum of that from each fluorescent or phosphorescent species. The analysis of a sample containing n components, therefore, can be accomplished by measuring the total emission intensity at n wavelengths.

Time, Cost, and Equipment As with other optical spectroscopic methods, fluorescent and phosphorescent methods provide a rapid means of analysis and are capable of automation. Fluorometers are relatively inexpensive, ranging from several hundred to several thousand dollars, and often are very satisfactory for quantitative work. Spectrofluorometers are more expensive, with some models costing as much as $50,000.

10H Atomic Emission Spectroscopy

In principle, emission spectroscopy can be applied to both atoms and molecules. Molecular infrared emission, or blackbody radiation played an important role in the early development of quantum mechanics and has been used for the analysis of hot gases generated by flames and rocket exhausts. Although the availability of FT–IR instrumentation extended the application of IR emission spectroscopy to a wider array of samples, its applications remain limited. For this reason IR emission is not considered further in this text. Molecular UV/Vis emission spectroscopy is of little importance since the thermal energies needed for excitation generally result in the sample's decomposition.

The focus of this section is the emission of ultraviolet and visible radiation following thermal or electrical excitation of atoms. Atomic emission spectroscopy has a long history. Qualitative applications based on the color of flames were used in the smelting of ores as early as 1550 and were more fully developed around 1830 with the observation of atomic spectra generated by flame emission and spark emission.[24] Quantitative applications based on the atomic emission from electrical sparks were developed by Norman Lockyer (1836–1920) in the early 1870s, and quantitative applications based on flame emission were pioneered by H. G. Lundegardh in 1930. Atomic emission based on emission from a plasma was introduced in 1964.

10H.1 Atomic Emission Spectra

Atomic emission occurs when a valence electron in a higher-energy atomic orbital returns to a lower-energy atomic orbital. Figure 10.23 shows a portion of the energy level diagram for sodium used earlier to discuss atomic absorption spectra. An atomic emission spectrum, therefore, consists of a series of discrete lines at wavelengths corresponding to the difference in energy between two atomic orbitals.

The intensity, I, of an emission line is proportional to the number of atoms, N^*, populating the excited state

$$I = kN^* \qquad \textbf{10.34}$$

where k is a constant related to the efficiency of the transition. For a system in thermal equilibrium, the population of the excited state is related to the total concentration of atoms, N, by the Boltzmann distribution. For many elements at temperatures of less than 5000 K the Boltzmann distribution for the ith excited state is approximated as

$$N^* = N\left(\frac{g_i}{g_0}\right)e^{-E_i/kT} \qquad \textbf{10.35}$$

where g_i and g_0 are statistical factors accounting for the number of equivalent energy levels for the excited state and ground state, E_i is the energy of the excited state relative to that of the ground state ($E_0 = 0$), k is Boltzmann's constant

(1.3807×10^{-23} J/K), and T is the temperature in kelvin. From equation 10.35 we can see that excited states with lower energies have larger populations and, therefore, the most intense emission lines. Furthermore, emission intensity increases with temperature.

10H.2 Equipment

Instrumentation for atomic emission spectroscopy is similar in design to that used for atomic absorption. In fact, most flame atomic absorption spectrometers are easily adapted for use as flame atomic emission spectrometers by turning off the hollow cathode lamp and monitoring the difference between the intensity of radiation emitted when aspirating the sample and that emitted when aspirating a blank. Many atomic emission spectrometers, however, are dedicated instruments designed to take advantage of features unique to atomic emission, including the use of plasmas, arcs, sparks, and lasers, as atomization and excitation sources and have an enhanced capability for multielemental analysis.

plasma
A hot, ionized gas containing an abundance of ions and electrons.

Atomization and Excitation Atomic emission requires a means for converting an analyte in solid, liquid, or solution form to a free gaseous atom. The same source of thermal energy usually serves as the excitation source. The most common methods are flames and plasmas, both of which are useful for liquid or solution samples. Solid samples may be analyzed by dissolving in solution and using a flame or plasma atomizer.

Flame Sources Atomization and excitation in flame atomic emission is accomplished using the same nebulization and spray chamber assembly used in atomic absorption (see Figure 10.38). The burner head consists of single or multiple slots or a Meker-style burner. Older atomic emission instruments often used a total consumption burner in which the sample is drawn through a capillary tube and injected directly into the flame.

Plasma Sources A **plasma** consists of a hot, partially ionized gas, containing an abundant concentration of cations and electrons that make the plasma a conductor. The plasmas used in atomic emission are formed by ionizing a flowing stream of argon, producing argon ions and electrons. The high temperatures in a plasma result from resistive heating that develops due to the movement of the electrons and argon ions. Because plasmas operate at much higher temperatures than flames, they provide better atomization and more highly populated excited states. Besides neutral atoms, the higher temperatures of a plasma also produce ions of the analyte.

A schematic diagram of the inductively coupled plasma (ICP) torch is shown in Figure 10.49. The ICP torch consists of three concentric quartz tubes, surrounded at the top by a radio-frequency induction coil. The sample is mixed with a stream of Ar using a spray chamber nebulizer similar to that used for flame emission and is carried to the plasma through the torch's central tube. Plasma formation is initiated by a spark from a Tesla coil. An alternating radio-frequency current in the induction coils creates a fluctuating magnetic field that induces the argon ions and electrons to move in a circular path. The resulting collisions with the abundant unionized gas give rise to resistive heating, provid-

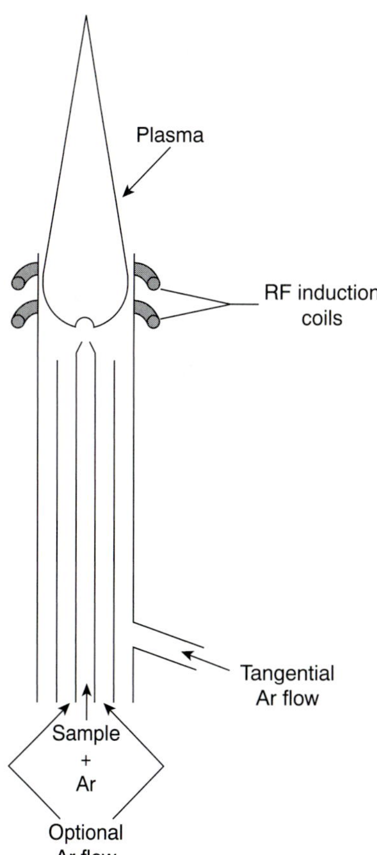

Figure 10.49
Diagram of an inductively coupled plasma torch.

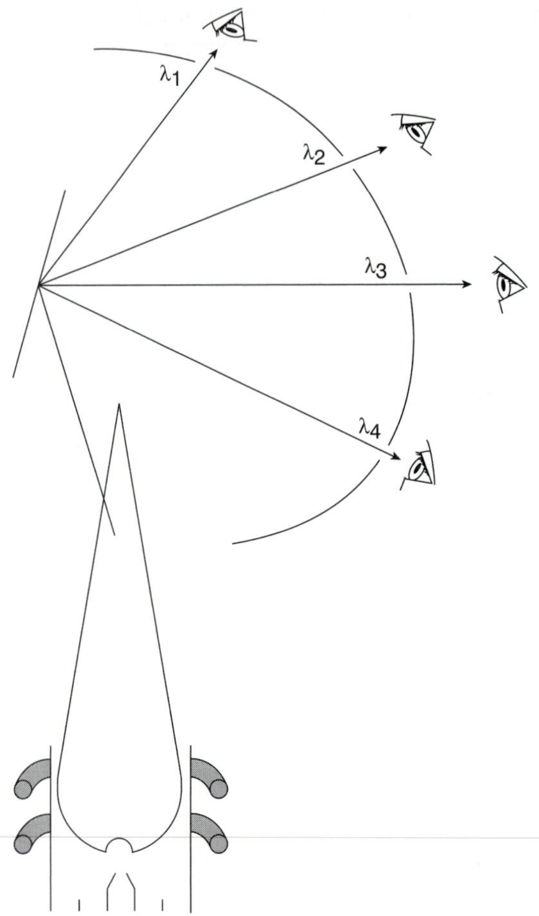

Figure 10.50
Schematic diagram of a multichannel atomic emission spectrometer, showing the arrangement of multiple exit slits and detectors for the simultaneous analysis of several elements.

ing temperatures as high as 10,000 K at the base of the plasma, and between 6000 and 8000 K at a height of 15–20 mm above the coil, where emission is usually measured. At these high temperatures the outer quartz tube must be thermally isolated from the plasma. This is accomplished by the tangential flow of argon shown in the schematic diagram.

Multielemental Analysis Atomic emission spectroscopy is ideally suited for multielemental analysis because all analytes in a sample are excited simultaneously. A scanning monochromator can be programmed to move rapidly to an analyte's desired wavelength, pausing to record its emission intensity before moving to the next analyte's wavelength. Proceeding in this fashion, it is possible to analyze three or four analytes per minute.

Another approach to multielemental analysis is to use a multichannel instrument that allows for the simultaneous monitoring of many analytes. A simple design for a multichannel spectrometer consists of a standard diffraction grating and 48–60 separate exit slits and detectors positioned in a semicircular array around the diffraction grating at positions corresponding to the desired wavelengths (Figure 10.50).

10H.3 Quantitative Applications

Atomic emission is used for the analysis of the same types of samples that may be analyzed by atomic absorption. The development of a quantitative atomic emission method requires several considerations, including choosing a source for atomization and excitation, selecting a wavelength and slit width, preparing the sample for analysis, minimizing spectral and chemical interferences, and selecting a method of standardization.

Choice of Atomization and Excitation Source Except for the alkali metals, detection limits when using an ICP are significantly better than those obtained with flame emission (Table 10.14). Plasmas also are subject to fewer spectral and chemical interferences. For these reasons a plasma emission source is usually the better choice.

Selecting the Wavelength and Slit Width The choice of wavelength is dictated by the need for sensitivity and freedom from interference due to unresolved emission lines from other constituents in the sample. Because an atomic emission spectrum usually has an abundance of emission lines, particularly when using a high-temperature plasma source, it is inevitable that some overlap will occur between emission lines. For example, an analysis for Ni using the atomic emission line at 349.30 nm is complicated by the atomic emission line for Fe at 349.06 nm. Narrower slit widths provide for better resolution. The easiest approach to selecting a wavelength is to obtain an emission spectrum for the sample and then to look for an emission line for the analyte that provides an intense signal and is resolved from other emission lines.

Preparing the Sample Flame and plasma sources are best suited for the analysis of samples in solution and liquid form. Although solids can be analyzed by direct insertion into the flame or plasma, they usually are first brought into solution by digestion or extraction.

Minimizing Spectral Interferences The most important spectral interference is a continuous source of background emission from the flame or plasma and emission bands from molecular species. This background emission is particularly severe for flames in which the temperature is insufficient to break down refractory compounds, such as oxides and hydroxides. Background corrections for flame emission are made by scanning over the emission line and drawing a baseline (Figure 10.51). Because the temperature of a plasma is

Table 10.14 Atomic Emission Detection Limits for Selected Elements

Element	Detection Limits (ppb)	
	Flame Emission	ICP
Ag	2	0.2
Al	3	0.2
As	2000	2
Au	500	0.9
B	50	0.1
Ba	1	0.01
Be	100	0.003
Bi	1000	10
Ca	0.1	0.0001
Cd	300	0.07
Co	5	0.1
Cr	1	0.08
Cs	0.02	—
Fe	10	0.09
Ga	5	0.6
Ge	400	0.5
Hg	150	1
K	0.01	30
Li	0.001	0.02
Mg	1	0.003
Mn	1	0.01
Mo	10	0.2
Na	0.01	0.1
Ni	10	0.2
Pb	0.2	1
Pd	40	2
Pt	2000	0.9
Sb	100	10
Se	—	1
Si	—	2
Sn	100	3
Sr	0.1	0.002
Ti	30	0.03
V	7	0.06
Zn	1000	0.1

Source: Compiled from Parsons, M. L.; Major, S.; Forster, A. R., *App. Spectrosc.* **1983,** *37,* 411–418.

self-absorption
In atomic emission, the decrease in emission intensity when light emitted by excited state atoms in the center of a flame or plasma is absorbed by atoms in the outer portion of the flame.

much higher, background interferences due to molecular emission are less problematic. Emission from the plasma's core is strong but is insignificant at a height of 10–30 mm above the core, where measurements normally are made.

Minimizing Chemical Interferences Flame emission is subject to the same types of chemical interferences as atomic absorption. These interferences are minimized by adjusting the flame composition and adding protecting agents, releasing agents, and ionization suppressors. An additional chemical interference results from **self-absorption.** Since the temperature of a flame is greatest at its center, the concentration of analyte atoms in an excited state is greater at the center than at the outer edges. If an excited state atom in the center of the flame emits radiation while returning to its ground state, then ground state atoms in the cooler, outer regions of the flame may absorb the radiation, thereby decreasing emission intensity. At high analyte concentrations a self-reversal may be seen in which the center of the emission band decreases (Figure 10.52).

Chemical interferences with plasma sources generally are insignificant. The higher temperature of the plasma limits the formation of nonvolatile species. For example, the presence of PO_4^{3-} in solutions being analyzed for Ca^{2+}, which is a significant interferant for flame emission, has a negligible effect when using a plasma source. In addition, the high concentration of electrons from the ionization of argon minimizes the effects of ionization interferences.

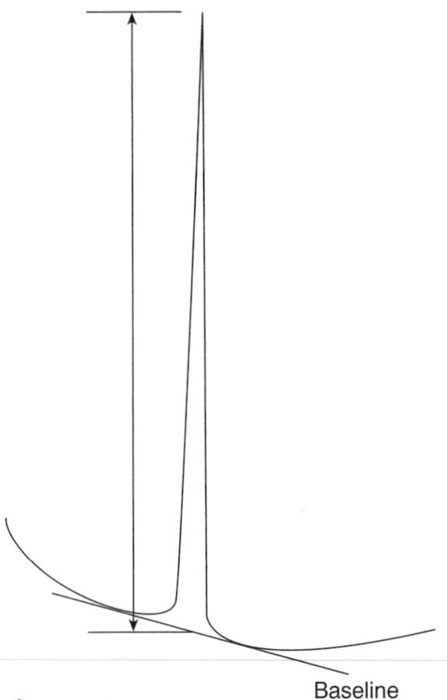

Figure 10.51
Method for background correction in flame atomic emission.

Standardizing the Method Equation 10.34 shows that emission intensity is proportional to the population of the excited state, N^*, from which the emission line originates. If the emission source is in thermal equilibrium, then the excited state population is proportional to the total population of analyte atoms, N, through the Boltzmann distribution (equation 10.35).

Calibration curves for flame emission are generally linear over two to three orders of magnitude, with chemical interferences due to ionization limiting linearity for lower concentrations of analyte, and self-absorption limiting linearity for higher concentrations of analyte. Plasma sources, which suffer from fewer chemical interferences, often yield calibration curves that are linear over four to five orders of magnitude and that are not affected significantly by changes in the matrix of the standards.

When possible, quantitative analyses are best conducted using external standards. Emission intensity, however, is affected significantly by many parameters, including the temperature of the excitation source and the efficiency of atomization. An increase in temperature of 10 K, for example, results in a 4% change in the fraction of Na atoms present in the 3*p* excited state. The method of internal standards can be used when variations in source parameters are difficult to control. In this case an internal standard is selected that has an emission line close to that of the analyte to compensate for changes in the temperature of the excitation source. In addition, the internal standard should be subject to the same chemical interferences to compensate for changes in atomization efficiency. To accurately compensate for these errors, the analyte and internal standard emission lines must be monitored simultaneously. The method of standard additions also can be used.

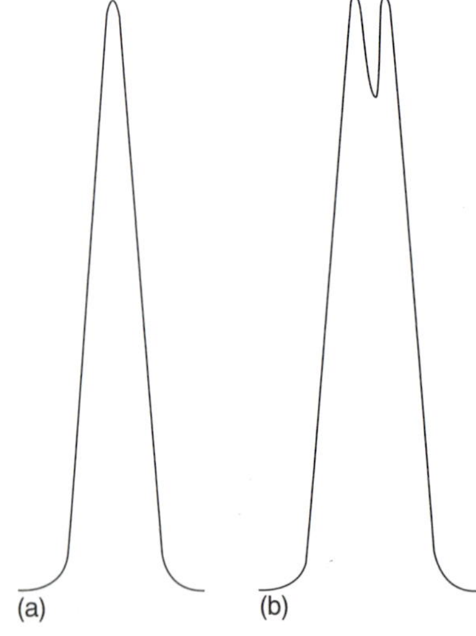

Figure 10.52
Atomic emission line at (a) low concentration of analyte, and (b) high concentration of analyte showing the effect of self-absorption.

Representative Methods

Method 10.4 Determination of Sodium in a Salt Substitute[25]

Description of Method. Salt substitutes, which are used in place of table salt for individuals on a low-sodium diet, contain KCl. Depending on the brand, fumaric acid, calcium hydrogen phosphate, or potassium tartrate also may be present. Typically, the concentration of sodium in a salt substitute is about 100 ppm. The concentration of sodium is easily determined by flame atomic emission. Because it is difficult to match the matrix of the standards to that of the sample, the analysis is accomplished by the method of standard additions.

Procedure. A sample is prepared by placing an approximately 10-g portion of the salt substitute in 10 mL of 3 M HCl and 100 mL of distilled water. After dissolving the sample, it is transferred to a 250-mL volumetric flask and diluted to volume with distilled water. A series of standard additions is prepared by placing 25-mL portions of the diluted sample into separate 50-mL volumetric flasks, spiking each with a known amount of an approximately 10-ppm standard solution of Na^+ and diluting to volume. After zeroing the instrument with an appropriate blank, the instrument is optimized at a wavelength of 589.0 nm while aspirating the standard solution of Na^+. The emission intensity is measured for each of the standard addition samples, and the concentration of sodium in the salt substitute is reported in parts per million.

Questions

1. **What type of chemical interference in the sample necessitates the use of the method of standard additions?**

   Potassium is more easily ionized than sodium. The high concentration of potassium in the sample suppresses the ionization of sodium, increasing its emission relative to that of a standard of equal concentration that does not contain potassium.

2. **Why is it inadvisable to match the matrix of the standards to the sample by adding KCl to each standard?**

   Sodium is a common contaminant found in many chemicals. Reagent grade KCl, for example, may contain 40–50 ppm sodium. This is a significant source of sodium, given that its concentration in the salt substitute is about 100 ppm.

3. **One problem with the analysis of salt samples is their tendency to clog the aspirator and burner assembly. What effect would this have on the analysis?**

   Clogging the aspirator and burner assembly decreases the rate of aspiration, decreasing the analyte's concentration in the flame. The result is a decrease in the signal and the introduction of a determinate error.

4. **The following results were obtained for the analysis of a 10.0077-g sample of salt substitute that was analyzed by the procedure described earlier.**

Concentration of Added Sodium (ppm)	Emission
0.000	1.79
0.420	2.63
1.051	3.54
2.102	4.94
3.153	6.18

—Continued

Continued from page 439

What is the concentration of sodium, in parts per million, in the salt substitute?

A standard addition calibration curve of emission versus the concentration of added sodium gives, by linear regression, an equation of

$$I = 1.97 + 1.37 \times (\text{ppm Na})$$

The concentration of sodium in the standard addition samples is determined from the absolute value of the *x*-intercept (see Figure 5.7b); thus, substituting zero for the emission intensity gives the concentration of sodium as 1.44 ppm. The concentration of sodium in the salt substitute, therefore, is

$$\frac{(1.44\ \mu\text{g Na}^+/\text{mL solution}) \times (50.00\ \text{mL}/25.00\ \text{mL}) \times 250.0\ \text{mL}}{10.0077\ \text{g}} = 71.9\ \text{ppm Na}^+$$

10H.4 Evaluation

Scale of Operation The scale of operations for atomic emission is ideal for the direct analysis of trace and ultratrace analytes in macro and meso samples. With appropriate dilutions, atomic emission also can be applied to major and minor analytes.

Accuracy When spectral and chemical interferences are insignificant, atomic emission is capable of producing quantitative results with accuracies of 1–5%. Accuracy in flame emission frequently is limited by chemical interferences. Because the higher temperature of a plasma source gives rise to more emission lines, accuracy when using plasma emission often is limited by stray radiation from overlapping emission lines.

Precision For samples and standards in which the concentration of analyte exceeds the detection limit by at least a factor of 50, the relative standard deviation for both flame and plasma emission is about 1–5%. Perhaps the most important factor affecting precision is the stability of the flame's or plasma's temperature. For example, in a 2500 K flame a temperature fluctuation of ±2.5 K gives a relative standard deviation of 1% in emission intensity. Significant improvements in precision may be realized when using internal standards.

Sensitivity Sensitivity in flame atomic emission is strongly influenced by the temperature of the excitation source and the composition of the sample matrix. Normally, sensitivity is optimized by aspirating a standard solution and adjusting the flame's composition and the height from which emission is monitored until the emission intensity is maximized. Chemical interferences, when present, decrease the sensitivity of the analysis. With plasma emission, sensitivity is less influenced by the sample matrix. In some cases, for example, a plasma calibration curve prepared using standards in a matrix of distilled water can be used for samples with more complex matrices.

Selectivity The selectivity of atomic emission is similar to that of atomic absorption. Atomic emission has the further advantage of rapid sequential or simultaneous analysis.

Time, Cost, and Equipment Sample throughput with atomic emission is very rapid when using automated systems capable of multielemental analysis. For example, sampling rates of 3000 determinations per hour have been achieved using an ICP with simultaneous analysis, and 300 determinations per hour with a sequential ICP. Flame emission is often accomplished using an atomic absorption spectrometer, which typically costs $10,000–50,000. Sequential ICPs range in price from $55,000 to $150,000, whereas an ICP capable of simultaneous multielemental analysis costs $80,000–200,000. Combination ICPs that are capable of both sequential and simultaneous analysis range in price from $150,000 to $300,000. The cost of Ar, which is consumed in significant quantities, cannot be overlooked when considering the expense of operating an ICP.

10I Spectroscopy Based on Scattering

The blue color of the sky during the day and the red color of the sun at sunset result from the scattering of light by small particles of dust, molecules of water, and other gases in the atmosphere. The efficiency with which light is scattered depends on its wavelength. The sky is blue because violet and blue light are scattered to a greater extent than other, longer wavelengths of light. For the same reason, the sun appears to be red when observed at sunset because red light is less efficiently scattered and, therefore, transmitted to a greater extent than other wavelengths of light. The scattering of radiation has been studied since the late 1800s, with applications beginning soon thereafter. The earliest quantitative applications of scattering, which date from the early 1900s, used the elastic scattering of light to determine the concentration of colloidal particles in a suspension.

10I.1 Origin of Scattering

A focused, monochromatic beam of radiation of wavelength λ, passing through a medium containing particles whose largest dimensions are less than $\frac{3}{2}\lambda$ is observed to scatter in all directions. For example, visible radiation of 500 nm is scattered by particles as large as 750 nm in the longest dimension. With larger particles, radiation also may be reflected or refracted. Two general categories of scattering are recognized. In elastic scattering, radiation is absorbed by the analyte and re-emitted without a change in the radiation's energy. When the radiation is re-emitted with a change in energy, the scattering is said to be inelastic. Only elastic scattering is considered in this text.

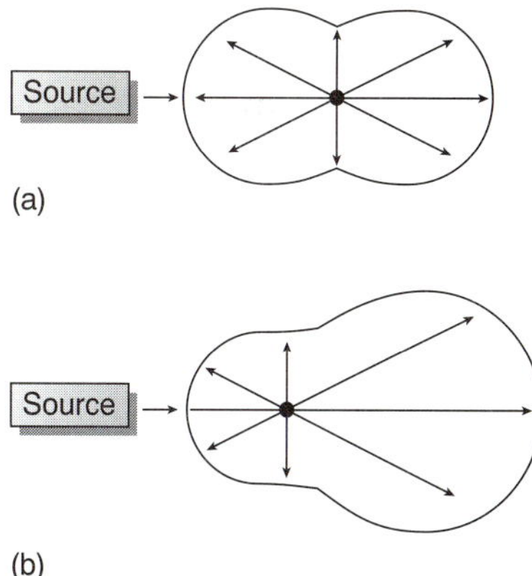

Figure 10.53
Distribution of radiation for (a) Rayleigh scattering and (b) large-particle scattering.

Elastic scattering is divided into two types: Rayleigh, or small-particle scattering, and large-particle scattering. Rayleigh scattering occurs when the scattering particles largest dimension is less than 5% of the radiation's wavelength. The intensity of the scattered radiation is symmetrically distributed (Figure 10.53a) and is proportional to its frequency to the fourth power (ν^4), accounting for the greater scattering of blue light compared with red light. For larger particles, the distribution of scattered light increases in the forward direction and decreases in the backward direction as the result of constructive and destructive interferences (Figure 10.53b).

10I.2 Turbidimetry and Nephelometry

Turbidimetry and nephelometry are two related techniques in which an incident source of radiation is elastically scattered by a suspension of colloidal particles. In **turbidimetry,** the detector is placed in line with the radiation source, and the

turbidimetry
A method in which the decrease in transmitted radiation due to scattering is measured.

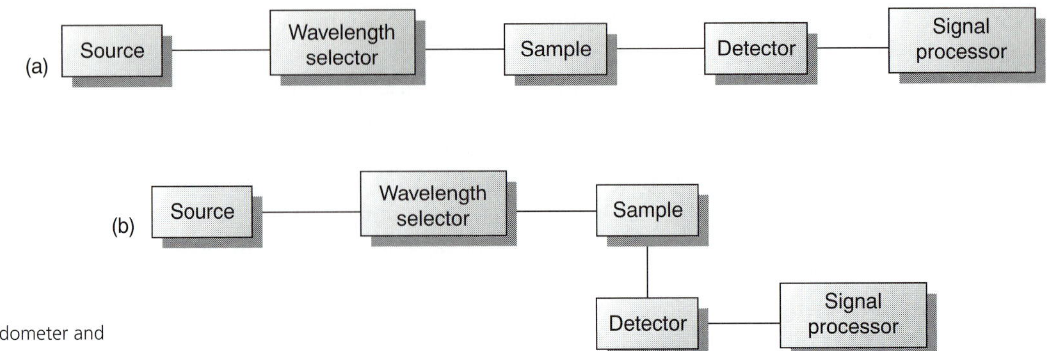

Figure 10.54
Block diagrams for (a) a turbidometer and (b) a nephelometer.

nephelometry
A method in which the intensity of scattered radiation is measured at an angle of 90° to the source.

decrease in the radiation's transmitted power is measured. In **nephelometry,** scattered radiation is measured at an angle of 90° to the radiation source. The similarity of the measurement of turbidimetry to absorbance, and of nephelometry to fluorescence, is evident in the block instrumental designs shown in Figure 10.54. In fact, turbidity can be measured using a UV/Vis spectrophotometer, such as a Spectronic-20, whereas a spectrofluorometer is suitable for nephelometry.

Turbidimetry Versus Nephelometry Choosing between turbidimetry and nephelometry is determined by two principal factors. The most important consideration is the intensity of the transmitted or scattered radiation relative to the intensity of radiation from the source. When the solution contains a small concentration of scattering particles, the intensity of the transmitted radiation, I_T, will be very similar to the intensity of the radiation source, I_0. As we learned earlier in the section on molecular absorption, determining a small difference between two intense signals is subject to a substantial uncertainty. Thus, nephelometry is a more appropriate choice for samples containing few scattering particles. On the other hand, turbidimetry is a better choice for samples containing a high concentration of scattering particles.

The second consideration in choosing between turbidimetry and nephelometry is the size of the scattering particles. For nephelometry, the intensity of scattered radiation at 90° will be greatest if the particles are small enough that Rayleigh scattering is in effect. For larger particles, as shown in Figure 10.37, scattering intensity is diminished at 90°. When using an ultraviolet or visible source of radiation, the optimum particle size is 0.1–1 μm. The size of the scattering particles is less important for turbidimetry, in which the signal is the relative decrease in transmitted radiation. In fact, turbidimetric measurements are still feasible even when the size of the scattering particles results in an increase in reflection and refraction (although a linear relationship between the signal and the concentration of scattering particles may no longer hold).

Determining Concentration by Turbidimetry In turbidimetry the measured transmittance, T, is the ratio of the transmitted intensity of the source radiation, I_T, to the intensity of source radiation transmitted by a blank, I_0.

$$T = \frac{I_T}{I_0}$$

The relationship between transmittance and the concentration of the scattering particles is similar to that given by Beer's law

$$-\log T = kbC \qquad \textbf{10.36}$$

where C is the concentration of the scattering particles in mass per unit volume (w/v), b is the pathlength, and k is a constant that depends on several factors, including the size and shape of the scattering particles and the wavelength of the source radiation. As with Beer's law, equation 10.36 may show appreciable deviations from linearity. The exact relationship is established by a calibration curve prepared using a series of standards of known concentration.

Determining Concentration by Nephelometry In nephelometry, the relationship between the intensity of scattered radiation, I_S, and the concentration (% w/v) of scattering particles is given as

$$I_S = k_S I_0 C \qquad \textbf{10.37}$$

where k_S is an empirical constant for the system, and I_0 is the intensity of the incident source radiation. The value of k_S is determined from a calibration curve prepared using a series of standards of known concentration.

Selecting a Wavelength for the Incident Radiation Selecting a wavelength for the incident radiation is based primarily on the need to minimize potential interferences. For turbidimetry, where the incident radiation is transmitted through the sample, it is necessary to avoid radiation that is absorbed by the sample. Since absorption is a common problem, the wavelength must be selected with some care, using a filter or monochromator for wavelength selection. For nephelometry, the absorption of incident radiation is not a problem unless it induces fluorescence from the sample. With a nonfluorescent sample there is no need for wavelength selection, and a source of white light may be used as the incident radiation. When using a filter or monochromator, other considerations include the dependence of scattering intensity, transducer sensitivity, and source intensity on the wavelength. For example, many common photon transducers are more sensitive to radiation at 400 nm than at 600 nm.

Preparing the Sample Although equations 10.36 and 10.37 relate scattering to the concentration of scattering particles, the intensity of scattered radiation is also influenced strongly by the particle's size and shape. For example, samples containing the same number of scattering particles may show significantly different values for $-\log T$ or I_s depending on the average diameter of the particles. For a quantitative analysis, therefore, it is necessary to maintain a uniform distribution of particle sizes throughout the sample and between samples and standards.

Most turbidimetric and nephelometric methods rely on the formation of the scattering particles by precipitation. As we learned in the discussion of precipitation gravimetry (Chapter 8), the properties of a precipitate are determined by the conditions used to effect the precipitation. To maintain a reproducible distribution of particle sizes between samples and standards, it is necessary to control parameters such as the concentration of reagents, order of adding reagents, pH, temperature, agitation or stirring rate, ionic strength, and time between the precipitate's initial formation and the measurement of transmittance or scattering. In many cases a

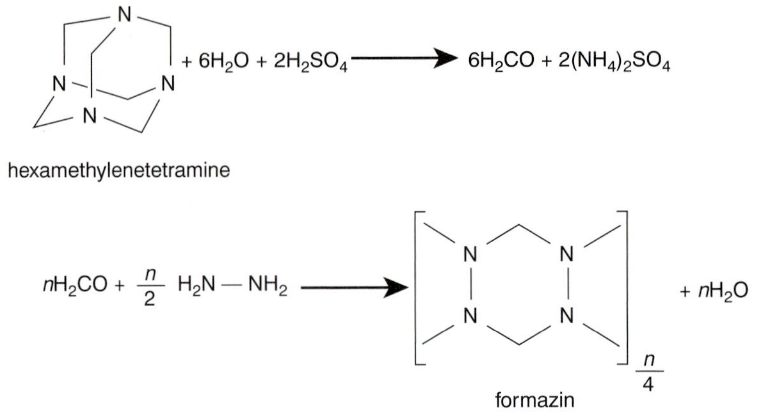

$$\text{hexamethylenetetramine} + 6H_2O + 2H_2SO_4 \longrightarrow 6H_2CO + 2(NH_4)_2SO_4$$

$$nH_2CO + \frac{n}{2}\,H_2N{-}NH_2 \longrightarrow [\text{formazin}]_{\frac{n}{4}} + nH_2O$$

Figure 10.55
Synthetic scheme for the preparation of formazin.

surface-active agent, such as glycerol, gelatin, or dextrin, is added to stabilize the precipitate in a colloidal state and to prevent the coagulation of the particles.

Applications Turbidimetry and nephelometry are widely used to determine the clarity of water, beverages, and food products. For example, the turbidity of water is determined using nephelometry by comparing the sample's scattering to that of a set of standards. The primary standard for measuring turbidity is formazin (Figure 10.55), which is an easily prepared, stable polymer suspension.[26] Formazin prepared by mixing a 1 g/100 mL solution of hydrazine sulfate, $N_2H_4{\cdot}H_2SO_4$, with a 10-g/100 mL solution of hexamethylenetetramine produces a suspension that is defined as 4000 nephelometric turbidity units (NTU). A set of standards with NTUs between 0 and 40 is prepared and used to construct a calibration curve. This method is readily adapted to the analysis of the clarity of orange juice, beer, and maple syrup.

A number of inorganic cations and anions can be determined by precipitating them under well-defined conditions and measuring the transmittance or scattering of radiation from the precipitated particles. The transmittance or scattering, as given by equation 10.36 or 10.37 is proportional to the concentration of the scattering particles, which, in turn, is related by the stoichiometry of the precipitation reaction to the analyte's concentration. Examples of analytes that have been determined in this way are listed in Table 10.15. The turbidimetric determination of SO_4^{2-} in water following its precipitation as $BaSO_4$ is described in Method 10.5.

Table 10.15 Selected Precipitates Used in Turbidimetric and Nephelometric Methods

Analyte	Precipitant	Precipitate
Ag^+	NaCl	AgCl
Ca^{2+}	$Na_2C_2O_4$	CaC_2O_4
Cl^-	$AgNO_3$	AgCl
CN^-	$AgNO_3$	AgCN
CO_3^{2-}	$BaCl_2$	$BaCO_3$
F^-	$CaCl_2$	CaF_2
SO_4^{2-}	$BaCl_2$	$BaSO_4$

Representative Methods

Method 10.5 Turbidimetric Determination of Sulfate in Water[27]

Description of Method. Adding $BaCl_2$ to an acidified sample precipitates SO_4^{2-} as $BaSO_4$. The concentration of SO_4^{2-} may be determined either by turbidimetry or nephelometry using an incident source of radiation of 420 nm. External standards containing known concentrations of SO_4^{2-} are used to standardize the method.

Procedure. A 100-mL sample is transferred to a 250-mL Erlenmeyer flask along with 5.00 mL of a conditioning reagent. The composition of the conditioning reagent is 50 mL glycerol, 30 mL concentrated HCl, 300 mL distilled water, 100 mL 95% ethyl or isopropyl alcohol, and 75 g NaCl. The sample and conditioning reagent are placed on a magnetic stirrer that is operated at the same speed for all samples and standards. A portion of crystalline $BaCl_2$ is added using a measuring spoon whose capacity is 0.2–0.3 mL, precipitating the SO_4^{2-} as $BaSO_4$. Timing begins when the $BaCl_2$ is added and the solution and precipitate are allowed to stir for exactly 1 min. At the end of the stirring period a portion of the suspension is poured into the turbidimeter or nephelometer cell, and the transmittance or scattering intensity measured at 30-s intervals for 4 min. The minimum transmittance or maximum scattering intensity recorded during this period is the analytical signal. Calibration curves over the range 0–40 ppm SO_4^{2-} are prepared by diluting a 100-ppm SO_4^{2-} standard. The standards and a reagent blank must be treated in exactly the same fashion as samples.

Questions

1. **What is the role of the conditioning reagent?**

   The conditioning reagent is used to stabilize the precipitate of $BaSO_4$. The high ionic strength and acidity, due to NaCl and HCl, prevent the formation of microcrystalline particles of $BaSO_4$, and glycerol and alcohol help stabilize the precipitate's suspension.

2. **Why is it important to use the same stirring rate and time for all samples and standards?**

   The stirring speed and time influence the size of the precipitate's particles, as well as their suspension in solution.

3. **The following results were obtained for the turbidimetric analysis of a set of standard SO_4^{2-} solutions using the procedure described earlier**

ppm SO_4^{2-}	Transmittance (T)
0.00	1.00
10.00	0.646
20.00	0.417
30.00	0.269
40.00	0.174

   When a sample of a surface water was carried through this procedure, the transmittance was found to be 0.538. What is the concentration of sulfate in the sample?

   A calibration curve of $-\log(T)$ versus concentration gives a standardization equation of

$$-\log(T) = -1.04 \times 10^{-5} + 0.0190(\text{ppm } SO_4^{2-})$$

   Substituting the sample's transmittance into the standardization equation gives the concentration of sulfate in the sample as 14.2 ppm.

10J KEY TERMS

absorbance (*p. 373*)
absorbance spectrum (*p. 373*)
atomization (*p. 412*)
background correction (*p. 419*)
Beer's law (*p. 386*)
characteristic concentration (*p. 416*)
chemiluminescence (*p. 374*)
chromophore (*p. 382*)
continuum source (*p. 375*)
dark current (*p. 379*)
effective bandwidth (*p. 376*)
electromagnetic spectrum (*p. 372*)
emission (*p. 373*)
emission spectrum (*p. 374*)
excitation spectrum (*p. 427*)
external conversion (*p. 425*)
filter (*p. 376*)
filter photometer (*p. 388*)
fluorescence (*p. 423*)
fluorometer (*p. 428*)
frequency (*p. 370*)
graphite furnace (*p. 414*)
intensity (*p. 371*)
interferometer (*p. 378*)
internal conversion (*p. 425*)
intersystem crossing (*p. 425*)
ionization suppressor (*p. 420*)
lifetime (*p. 423*)
line source (*p. 375*)
method of continuous variations (*p. 404*)
mole-ratio method (*p. 406*)
monochromatic (*p. 377*)
monochromator (*p. 376*)
nephelometry (*p. 442*)
nominal wavelength (*p. 376*)
phosphorescence (*p. 424*)
photodiode array (*p. 379*)
photoluminescence (*p. 374*)
photon (*p. 371*)
plasma (*p. 435*)
polychromatic (*p. 377*)
power (*p. 371*)
protecting agent (*p. 420*)
quantum yield (*p. 425*)
relaxation (*p. 423*)
releasing agent (*p. 420*)
resolution (*p. 376*)
self-absorption (*p. 438*)
signal averaging (*p. 391*)
signal processor (*p. 380*)
signal-to-noise ratio (*p. 379*)
singlet excited state (*p. 423*)
slope-ratio method (*p. 407*)
spectral searching (*p. 403*)
spectrofluorometer (*p. 428*)
spectrophotometer (*p. 389*)
stray radiation (*p. 387*)
transducer (*p. 379*)
transmittance (*p. 384*)
triplet excited state (*p. 423*)
turbidimetry (*p. 441*)
vibrational relaxation (*p. 424*)
wavelength (*p. 370*)
wavenumber (*p. 370*)

10K SUMMARY

Spectroscopic methods of analysis covered in this chapter include those based on the absorption, emission, or scattering of electromagnetic radiation. When a molecule absorbs UV/Vis radiation, it undergoes a change in its valence shell configuration, whereas a change in vibrational energy results for the absorption of IR radiation. Experimentally we measure the fraction of radiation transmitted (T) through the sample. Instrumentation for molecular absorption requires a source of electromagnetic radiation, a means for selecting the wavelength for which transmittance is measured, and a detector for measuring the transmittance. Beer's law relates absorbance to both transmittance and the concentration of the absorbing species ($A = -\log T = \varepsilon bC$).

In atomic absorption we measure the absorption of radiation by gaseous atoms. Samples are atomized using thermal energy from either a flame or a graphite furnace. Because the width of an atom's absorption band is so narrow, the continuum sources common for molecular absorption cannot be used. Instead, a hollow cathode lamp provides the necessary line source of radiation. Atomic absorption suffers from a number of spectral and chemical interferences. The absorption or scattering of radiation from the sample's matrix is an important spectral interference that may be minimized by background correction. Chemical interferences include the formation of nonvolatile forms of the analyte in the flame and ionization of the analyte. The former interference is minimized by using a releasing agent or a protecting agent, and an ionization suppressor helps minimize the latter interference.

When a molecule absorbs radiation, it moves from a lower-energy state to a higher-energy state. In returning to the lower-energy state the molecule may emit radiation. This process is called photoluminescence. One form of photoluminescence is fluorescence, in which the analyte emits a photon without undergoing a change in its spin state. In phosphorescence, emission occurs with a change in the analyte's spin state. For low concentrations of analyte, both fluorescent and phosphorescent emission intensities are a linear function of the analyte's concentration. Thermally excited atoms also emit radiation, forming the basis for atomic emission spectroscopy. Thermal excitation is achieved using either a flame or a plasma.

Spectroscopic measurements may also involve the scattering of light by a particulate form of the analyte. In turbidimetry, the decrease in the radiation's transmittance through the sample is measured and related to the analyte's concentration through Beer's law. In nephelometry we measure the intensity of scattered radiation, which varies linearly with the analyte's concentration.

10L *Suggested* EXPERIMENTS

Experiments

The following experiments may be used to illustrate the application of spectroscopy to quantitative or characterization problems. Experiments are divided into four groups: those using UV/Vis absorption, those using IR absorption, those using atomic absorption or atomic emission, and those using fluorescence. Other experiments are described in the highlighted method sections. A brief description is included with each experiment providing details such as the type of sample analyzed and the method of standardization.

The first group of experiments are for the analysis of samples using UV/Vis absorption.

Allen, H. C.; Brauers, T.; Finlayson-Pitts, B. J. "Illustrating Deviations in the Beer–Lambert Law in an Instrumental Analysis Laboratory: Measuring Atmospheric Pollutants by Differential Optical Absorption Spectrometry," *J. Chem. Educ.* **1997,** *74,* 1459–1462.

This experiment demonstrates the chemical limitations to Beer's law using the NO_2–N_2O_4 equilibrium as an example.

Blanco, M.; Iturriaga, H.; Maspoch, S.; et al. "A Simple Method for Spectrophotometric Determination of Two-Components with Overlapped Spectra," *J. Chem. Educ.* **1989,** *66,* 178–180.

This experiment describes the application of multiwavelength linear regression to the analysis of two-component mixtures. Directions are given for the analysis of permanganate–dichromate mixtures, Ti(IV)–V(V) mixtures and Cu(II)–Zn(II) mixtures.

Bruneau, E.; Lavabre, D.; Levy, G.; et al. "Quantitative Analysis of Continuous-Variation Plots with a Comparison of Several Methods," *J. Chem. Educ.* **1992,** *69,* 833–837.

In this experiment the method of continuous variations is used to determine the stoichiometry and equilibrium constant for the organic complex of 3-aminopyridine with picric acid in $CHCl_3$, and the inorganic complex of Fe^{3+} with salicylic acid.

Crisp, P. T.; Eckert, J. M.; Gibson, N. A. "The Determination of Anionic Surfactants in Natural and Waste Waters," *J. Chem. Educ.* **1983,** *60,* 236–238.

The concentration of anionic surfactants at the sub-ppm level in natural waters and industrial waters are determined spectrophotometrically. The anionic surfactants are extracted into a nonaqueous solvent following the formation of an ion association complex with a suitable cation.

Domínguez, A.; Fernández, A.; González, N.; et al. "Determination of Critical Micelle Concentration of Some Surfactants by Three Techniques," *J. Chem. Educ.* **1997,** *74,* 1227–1231.

Surfactants are long-chain compounds containing a hydrophobic tail and an ionic head. In polar solvents the surfactants arrange themselves in a "spherical" structure known as a micelle in which the hydrophobic tails form the core. The concentration of surfactant needed to form micelles is called the critical micelle concentration (CMC). This experiment describes both a UV absorption and a fluorescence method for determining the CMC. Directions also are given for determining the CMC by measuring electrical conductivity.

Hill, Z. D.; MacCarthy, P. "Novel Approach to Job's Method," *J. Chem. Educ.* **1986,** *63,* 162–167.

Data from the spectrophotometric titrations of Fe^{3+} with SCN^-, and of Cu^{2+} with EDTA are used to determine the stoichiometry of the resulting complexes using the method of continuous variations.

Long, J. R.; Drago, R. S. "The Rigorous Evaluation of Spectrophotometric Data to Obtain an Equilibrium Constant," *J. Chem. Educ.* **1982,** *59,* 1037–1039.

Although this experiment is written as a "dry-lab," it can be adapted to the laboratory. Details are given for the determination of the equilibrium constant for the binding of the Lewis base 1-methylimidazole to the Lewis acid cobalt(II)4-trifluoromethyl-*o*-phenylene-4,6-methoxysalicylideniminate in toluene. The equilibrium constant is found by a linear regression analysis of the absorbance data to a theoretical equilibrium model.

McDevitt, V. L.; Rodriquez, A.; Williams, K. R. "Analysis of Soft Drinks: UV Spectrophotometry, Liquid Chromatography, and Capillary Electrophoresis," **1998,** *75,* 625–629.

Procedures for determining the concentrations of caffeine, benzoic acid and aspartame in soda by these three methods are provided. In the example provided in this paper, the concentrations of caffeine and benzoic acid in Mello Yellow are determined spectrophotometrically.

Mehra, M. C.; Rioux, J. "An Analytical Chemistry Experiment in Simultaneous Spectrophotometric Determination of Fe(III) and Cu(II) with Hexacyanoruthenate(II) Reagent," *J. Chem. Educ.* **1982,** *59,* 688–689.

This experiment describes a standard multicomponent analysis for two analytes based on measuring the absorbance at two wavelengths. Hexacyanoruthenate(II) is used as a complexing agent, forming a purple-blue complex with Fe(III) and a pale green complex with Cu(II).

—Continued

Experiments

Continued from page 447

Örstan, A.; Wojcik, J. F. "Spectroscopic Determination of Protein-Ligand Binding Constants," *J. Chem. Educ.* **1987,** *64,* 814–816.

This experiment provides a nice example of the application of spectroscopy to biochemistry. After presenting the basic theory for the spectroscopic treatment of protein–ligand interactions, a procedure for characterizing the binding of methyl orange to bovine serum albumin is described.

Pandey, S.; Powell, J. R.; McHale, M. E. R.; et al. "Quantitative Determination of Cr(III) and Co(II) Using a Spectroscopic H-Point Standard Addition," *J. Chem. Educ.* **1997,** *74,* 848–850.

Another nice example of a multicomponent analysis based on current research is presented in this experiment. Although the H-point standard addition is not discussed in this text, this paper provides adequate theory and references to the original literature.

Raymond, M.; Jochum, C.; Kowalski, B. R. "Optimal Multicomponent Analysis Using the Generalized Standard Addition Method," *J. Chem. Educ.* **1983,** *60,* 1072–1073.

This experiment demonstrates the application of the generalized standard additions method for the analysis of mixtures of $K_2Cr_2O_7$ and $KMnO_4$.

Tucker, S.; Robinson, R.; Keane, C.; et al. "Colorimetric Determination of pH," *J. Chem. Educ.* **1989,** *66,* 769–771.

In this experiment mixtures of dyes are used to provide a means for determining spectrophotometrically a sample's pH.

Walmsley, F. "Aggregation in Dyes: A Spectrophotometric Study," *J. Chem. Educ.* **1992,** *69,* 583.

A dye used in the coloration of materials, such as fibers, must be present as the monomer if it is to adsorb to the material's surface. This experiment describes how spectrophotometry can be used to determine the equilibrium constant between a monomer and a dimer for the dye pinacyanol iodide in water.

Williams, K. R.; Cole, S. R.; Boyette, S. E.; et al. "The Use of Dristan Nasal Spray as the Unknown for Simultaneous Spectrophotometric Analysis of a Mixture," *J. Chem. Educ.* **1990,** *67,* 535.

The analysis of a pharmaceutical preparation for pheniramine maleate and phenylephrine hydrochloride is described in this experiment.

Yarnelle, M. K.; West, K. J. "Modification of an Ultraviolet Spectrophotometric Determination of the Active Ingredients in APC Tablets," *J. Chem. Educ.* **1989,** *66,* 601–602.

The analysis of APC tablets (a mixture of aspirin, phenacetin, and caffeine) has been a common undergraduate laboratory experiment. This experiment describes modifications to the standard analysis for APC tablets in which paracetamol (also known as acetaminophen) replaces phenacetin.

The second block of experiments provide examples of the application of infrared spectroscopy.

Frohlich, H. "Using Infrared Spectroscopy Measurements to Study Intermolecular Hydrogen Bonding," *J. Chem. Educ.* **1993,** *70,* A3–A6.

This experiment describes a characterization analysis in which the degree of association, equilibrium constant, and hydrogen bond energy are measured for benzyl alcohol and phenol in CCl_4.

Mathias, L. J.; Hankins, M. G.; Bertolucci, C. M.; et al. "Quantitative Analysis by FT-IR: Thin Films of Copolymers of Ethylene and Vinyl Acetate," *J. Chem. Educ.* **1992,** *69,* A217–A219.

The % w/w vinyl acetate in a copolymer is determined using a ratio of the peak at 1020 cm^{-1} due to C—O and the polyethylene C—H peak at 720 cm^{-1}.

Seasholtz, M. B.; Pence, L. E.; Moe Jr., O. A. "Determination of Carbon Monoxide in Automobile Exhaust by FT-IR Spectroscopy," *J. Chem. Educ.* **1988,** *65,* 820–823.

Samples of car exhaust are collected using a 4-L glass bottle evacuated to a level of less than 2 torr. A normal calibration curve using external standards of known P_{CO} is used to determine the P_{CO} in the exhaust samples.

The next set of experiments describe suitable applications of atomic absorption spectroscopy.

Gilles de Pelichy, L. D.; Adams, C.; Smith, E. T. "Analysis of the Essential Nutrient Strontium in Marine Aquariums by Atomic Absorption Spectroscopy," *J. Chem. Educ.* **1997,** *74,* 1192–1194.

The concentration of strontium in a sea water aquarium is determined by atomic absorption using the method of standard additions. The effect of adding La^{3+} as a releasing agent also is explored.

Hoskins, L. C.; Reichardt, P. B.; Stolzberg, R. J. "Determination of the Extraction Constant for Zinc Pyrrolidinecarbodithioate," *J. Chem. Educ.* **1981,** *58,* 580–581.

—Continued

Experiments

Aqueous solutions buffered to a pH of 5.2 and containing known total concentrations of Zn^{2+} are prepared. A solution containing ammonium pyrrolidinecarbodithioate (APCD) is added along with methyl isobutyl ketone (MIBK). The mixture is shaken briefly and then placed on a rotary shaker table for 30 min. At the end of the extraction period the aqueous and organic phases are separated and the concentration of zinc in the aqueous layer determined by atomic absorption. The concentration of zinc in the organic phase is determined by difference and the equilibrium constant for the extraction calculated.

Lehman, T. A.; Everett, W. W. "Solubility of Lead Sulfate in Water and in Sodium Sulfate Solutions," *J. Chem. Educ.* **1982,** *59,* 797.

Saturated solutions are prepared by adding $PbSO_4$ to solutions containing $0 - 2 \times 10^{-3}$ M Na_2SO_4. After allowing time for equilibration, the aqueous solution is isolated by decantation and analyzed by atomic absorption for the concentration of Pb^{2+}. The apparent K_{sp} for $PbSO_4$ is calculated, showing the effect of ionic strength on solubility products.

Markow, P. G. "Determining the Lead Content of Paint Chips," *J. Chem. Educ.* **1996,** *73,* 178–179.

The %w/w lead in a lead-based paint Standard Reference Material and in unknown paint chips is determined by atomic absorption using external standards.

Masina, M. R.; Nkosi, P. A.; Rasmussen, P. W.; et al. "Determination of Metal Ions in Pineapple Juice and Effluent of a Fruit Canning Industry," *J. Chem. Educ.* **1989,** *66,* 342–343.

The concentrations of iron, lead, tin, and aluminum are determined using the method of standard additions.

Quigley, M. N. "Determination of Calcium in Analgesic Tablets Using Atomic Absorption Spectrophotometry," *J. Chem. Educ.* **1994,** *71,* 800.

Analgesic tablets are ground into a fine powder, dissolved in HCl, and analyzed for calcium by atomic absorption. A releasing agent of La^{3+} is used to prevent an interference due to the formation of calcium pyrophosphate.

Quigley, M. N.; Vernon, F. "Determination of Trace Metal Ion Concentrations in Seawater," *J. Chem. Educ.* **1996,** *73,* 671–675.

Trace metals in sea water are preconcentrated either by coprecipitating with $Fe(OH)_3$ and recovering by dissolving the precipitate or by ion exchange. The concentrations of several trace metals are determined by standard additions using graphite furnace atomic absorption spectrometry.

Rheingold, A. L.; Hues, S.; Cohen, M. N. "Strontium and Zinc Content in Bones as an Indication of Diet," *J. Chem. Educ.,* **1983,** *60,* 233–234.

Samples of animal bones weighing approximately 3 g are ashed at 600 °C until the entire bone is ash-white. Samples are then crushed in a mortar and pestle. A portion of the sample is digested in HCl and diluted to a known volume. The concentrations of zinc and strontium are determined by atomic absorption. The analysis for strontium illustrates the use of a protecting agent as $La(NO_3)_3$ is added to prevent an interference due to the formation of refractory strontium phosphate.

Rocha, F. R. P.; Nóbrega, J. A. "Effects of Solution Physical Properties on Copper and Chromium Signals in Flame Atomic Absorption Spectrometry," *J. Chem. Educ.* **1996,** *73,* 982–984.

A nice experiment illustrating the importance of a sample's matrix. The effect on the absorbance of copper for solutions with different %v/v ethanol, and the effect on the absorbance of chromium for solutions with different concentrations of added surfactants are evaluated.

The last set of experiments illustrate the use of fluorescence spectroscopy.

Buccigross, J. M.; Bedell, C. M.; Suding-Moster, H. L. "Fluorescent Measurement of TNS Binding to Calmodulin," *J. Chem. Educ.* **1996,** *73,* 275–278.

Although intended for the biochemistry lab, this experiment provides analytical students with a practical characterization analysis. Of particular interest is the use of Job's method to determine the number of TNS (2-*p*-toludinylnaphthalene-6-sulfonate) binding sites on calmodulin. Fluorescence is measured at 475 nm using an excitation wavelength of 330 nm.

Henderleiter, J. A.; Hyslopo, R. M. "The Analysis of Riboflavin in Urine by Fluorescence," *J. Chem. Educ.* **1996,** *73,* 563–564.

Samples of urine are analyzed for riboflavin before and after taking a vitamin tablet containing riboflavin. Concentrations are determined using external standards or by the method of standard additions. Fluorescence is monitored at 525 nm using an excitation wavelength of 280 nm.

10M PROBLEMS

1. Provide the missing information in the following table

Wavelength (m)	Frequency (s^{-1})	Wavenumber (cm^{-1})	Energy (J/molecule)
4.50×10^{-9}			
	1.33×10^{15}		
		3215	
			7.20×10^{-19}

2. Provide the missing information in the following table

[Analyte] (M)	Absorb-ance	% Trans-mittance	Molor absorptivity ($M^{-1}cm^{-1}$)	Pathlength (cm)
1.40×10^{-4}			1120	1.00
	0.563		750	1.00
2.56×10^{-4}	0.225		440	
1.55×10^{-3}	0.167			5.00
		33.3	565	1.00
4.35×10^{-3}		21.2	1550	
1.20×10^{-4}		81.3		10.00

3. The transmittance of a solution is found to be 35.0%. What is the transmittance if the solution is diluted in half?

4. The transmittance of a solution is found to be 85.0% when measured in a cell whose pathlength is 1.00 cm. What is the percent transmittance if the pathlength is increased to 10.00 cm?

5. The accuracy of a spectrophotometer can be evaluated by preparing a solution of 60.06-ppm $K_2Cr_2O_7$ in 0.0050 M H_2SO_4 and measuring its absorbance at a wavelength of 350 nm using a cell with a pathlength of 1.00 cm. The absorbance should be 0.640. What is the molar absorptivity of $K_2Cr_2O_7$ at this wavelength?

6. Chemical deviations to Beer's law may occur when the concentration of an absorbing species is affected by the position of an equilibrium reaction. Consider a weak acid, HA, for which K_a is 2×10^{-5}. Construct Beer's law calibration curves of absorbance versus the total concentration of weak acid (C_{tot} = [HA] + [A^-]), using values for C_{tot} of 1.0×10^{-5}, 3.0×10^{-5}, 5.0×10^{-5}, 9.0×10^{-5}, 11×10^{-5}, and 13×10^{-5} M for the following sets of conditions: (a) $\varepsilon_{HA} = \varepsilon_{A^-} = 2000$, and solution is not buffered; (b) $\varepsilon_{HA} = 2000$ and $\varepsilon_{A^-} = 500$, and solution is not buffered; and, (c) $\varepsilon_{HA} = 2000$ and $\varepsilon_{A^-} = 500$, and solution is buffered to a pH of 4.50. Assume a constant pathlength of 1.00 cm for all samples. All values of ε have units of $M^{-1}cm^{-1}$.

7. One instrumental limitation to Beer's law is the use of polychromatic radiation instead of monochromatic radiation. Consider a radiation source that emits two wavelengths of radiation, λ′ and λ″. When treated separately, the absorbances at these wavelengths, A′ and A″, are

$$A' = \log \frac{P_0'}{P_T'} = \varepsilon' bC$$

$$A'' = \log \frac{P_0''}{P_T''} = \varepsilon'' bC$$

When both wavelengths are measured simultaneously, the absorbance is given as

$$A = \log \frac{(P_0' + P_0'')}{(P_T' + P_T'')}$$

(a) Show that when the molar absorptivity at λ′ and λ″ are the same (ε′ = ε″ = ε), the absorbance is equivalent to

$$A = \varepsilon bC$$

(b) Construct Beer's law calibration curves over the concentration range of zero to 1×10^{-4} M using ε′ = 1000 and ε″ = 1000, and ε′ = 1900 and ε″ = 100. Assume a value of 1.00 cm for the pathlength and that $P_0' = P_0''$. Explain the difference between the two curves.

8. A second instrumental limitation to Beer's law is stray radiation. The following data were obtained using a cell with a pathlength of 1.00 cm, when stray light is insignificant ($P_{stray} = 0$).

[Analyte] (M)	Absorbance
0.000	0.00
0.002	0.40
0.004	0.80
0.006	1.20
0.008	1.60
0.010	2.00

Calculate the absorbance of each solution when P_{stray} is 5% of P_0, and plot Beer's law calibration curves for both sets of data. Explain any differences between the two curves. (*Hint:* Assume that P_0 is 100).

9. In the process of performing a spectrophotometric determination of Fe, an analyst prepares a calibration curve using a single-beam spectrometer, such as a Spec-20. After preparing the calibration curve, the analyst drops the cuvette used for the method blank and the standards. The analyst acquires a new cuvette, measures the absorbance of the sample, and determines the %w/w Fe in the sample. Will the change in cuvette lead to a determinate error in the analysis? Explain.

10. One method for the analysis of Fe^{3+} that can be used with a variety of sample matrices, is to form the highly colored Fe^{3+}–thioglycolic acid complex. The complex absorbs strongly at 535 nm. Standardizing the method is accomplished using external standards. A 10.00 ppm Fe^{3+} working standard is prepared by transferring a 10-mL aliquot of a 100.0 ppm stock solution of Fe^{3+} to a 100-mL volumetric flask and diluting to volume. Calibration standards of 1.00, 2.00, 3.00, 4.00, and 5.00 ppm are then prepared by transferring appropriate amounts of the 100.0-ppm working solution into separate 50-mL volumetric flasks, each containing 5 mL of thioglycolic acid, 2 mL of 20% w/v ammonium citrate, and 5 mL of 0.22 M NH_3. After diluting to volume and mixing, the absorbances of the external standards are measured against an appropriate blank. Samples are prepared for analysis by taking a portion known to contain approximately 0.1 g of Fe^{3+}, dissolving in a minimum amount of HNO_3, and diluting to volume in a 1-L volumetric flask. A 1.00-mL aliquot of this solution is transferred to a 50-mL volumetric flask, along with 5 mL of thioglycolic acid, 2 mL of 20% w/v ammonium citrate, and 5 mL of 0.22 M NH_3 and diluted to volume. The absorbance of this solution is used to determine the concentration of Fe^{3+} in the sample.

(a) What is an appropriate blank for this procedure?

(b) Ammonium citrate is added to prevent the precipitation of Al^{3+}. What effect would the presence of trace amount of Fe^{3+} in the ammonium citrate have on the reported concentration of iron in the sample? (c) Why does the procedure call for taking an amount of sample containing approximately 0.1 grams of Fe^{3+}?

(d) Unbeknownst to the analyst, the 100-mL volumetric flask used to prepare the 10.00 ppm working standard of Fe^{3+} has a volume that is significantly smaller than 100.0mL. What effect will this have on the reported concentration of iron in the sample?

11. A spectrophotometric method for the analysis of iron has a linear calibration curve for standards of 0.00, 5.00, 10.00, 15.00, and 20.00 ppm. An iron ore sample with an expected iron content of 40–60% w/w is to be analyzed by this method. An approximately 0.5-g sample is taken, dissolved in a minimum of concentrated HCl, and diluted to 1 L in a volumetric flask using distilled water. A 5.00-mL aliquot is removed with a pipet. To what volume (10, 25, 50, 100, 250, 500, or 1000 mL) should it be diluted to minimize the uncertainty in the analysis? Explain.

12. In a recent paper, Lozano-Calero and colleagues describe a new method for the quantitative analysis of phosphorus in cola beverages.[28] The method is based on the formation of an intensely blue-colored phosphomolybdate complex, $(NH_4)_3[PO_4(MoO_3)_{12}]$. The complex is formed by adding $(NH_4)_6Mo_7O_{24}$ to the sample in the presence of a reducing agent, such as ascorbic acid. The concentration of the complex is determined spectrophotometrically at a wavelength of 830 nm, using a normal calibration curve as a method of standardization.

In a typical analysis, a set of standard solutions containing known amounts of phosphorus was prepared by placing appropriate volumes of a 4.00-ppm solution of P_2O_5 in a 5-mL volumetric flask, adding 2 mL of the ascorbic acid reducing solution, and diluting to volume with distilled water. Cola beverages were prepared for analysis by pouring a sample into a beaker and allowing it to stand for 24 h to expel the dissolved gases (CO_2). A 2.50-mL sample of the degassed sample was transferred to a 50-mL volumetric flask and diluted to volume. A 250-μL aliquot of the diluted sample was then transferred to a 5-mL volumetric flask, treated with 2 mL of the ascorbic acid reducing solution, and diluted to volume with distilled water.

(a) The authors note that this method only can be applied to noncolored cola beverages, such as Crystal Pepsi. Clearly explain why this is the case. (b) How might you modify this method so that it could be applied to any cola beverage? (c) Why is it necessary to ensure that the dissolved gases have been removed? (d) How would you prepare a blank for this method? (e) The calibration curve reported by the authors yields a regression line of

$$\text{Abs} = -0.02 + 0.72(\text{ppm } P_2O_5)$$

A sample of Crystal Pepsi, analyzed as described here, yields an absorbance of 0.565. What is the concentration of phosphorus, reported as parts per million of P, in the original sample of Crystal Pepsi?

13. EDTA forms colored complexes with a variety of metal ions that may serve as the basis for a quantitative spectrophotometric method of analysis. The molar absorptivities of the EDTA complexes of Cu^{2+}, Co^{2+}, and Ni^{2+} at three wavelengths are summarized in the following table (all values of ε are in M^{-1} cm^{-1})

Metal	$\varepsilon_{462.9}$	$\varepsilon_{732.0}$	$\varepsilon_{378.7}$
Co^{2+}	15.8	2.11	3.11
Cu^{2+}	2.32	95.2	7.73
Ni^{2+}	1.79	3.03	13.5

Using this information, determine (a) the concentration of Cu^{2+} in a solution that has an absorbance of 0.338 at a wavelength of 732.0 nm; (b) the concentrations of Cu^{2+} and Co^{2+} in a solution that has an absorbance of 0.453 at a wavelength of 732.0 nm and 0.107 at a wavelength of 462.9 nm; and (c) the concentrations of Cu^{2+}, Co^{2+}, and Ni^{2+} in a sample that has an absorbance of 0.423 at a wavelength of 732.0 nm, 0.184 at a wavelength of 462.9 nm, and 0.291 at a wavelength of 378.7 nm. The pathlength, *b*, for all measurements is 1.00 cm.

14. The concentration of phenol in a water sample is determined by separating the phenol from nonvolatile impurities by steam distillation, followed by reacting with 4-aminoantipyrine and $K_3Fe(CN)_6$ at pH 7.9 to form a colored antipyrine dye. A phenol standard with a concentration of

4.00 ppm has an absorbance of 0.424 at a wavelength of 460 nm using a 1.00-cm cell. A water sample is steam-distilled, and a 50.00-mL aliquot of the distillate is placed in a 100-mL volumetric flask and diluted to volume with distilled water. The absorbance of this solution is found to be 0.394. What is the concentration of phenol (in parts per million) in the water sample?

15. Saito described a quantitative spectrophotometric procedure for iron based on a solid-phase extraction using bathophenanthroline in a poly(vinyl chloride) membrane.[29] In the absence of Fe^{2+}, the membrane is colorless, but when immersed in a solution of Fe^{2+} and I^-, the membrane develops a red color as a result of the formation of a Fe^{2+}–bathophenanthroline complex. A calibration curve determined using a set of external standards with known molar concentrations of Fe^{2+} gave a standardization relationship of

$$A = (8.60 \times 10^3\ M^{-1})[Fe^{2+}]$$

What is the concentration of iron in parts per million for a sample with an absorbance of 0.100?

16. In the DPD colorimetric method for the free chlorine residual, which is reported as parts per million of Cl_2, the oxidizing power of free chlorine converts the colorless amine *N*,*N*-diethyl-*p*-phenylenediamine to a colored dye that absorbs strongly over the wavelength range of 440–580 nm. Analysis of a set of calibration standards gave the following results

ppm Cl_2	absorbance
0	0.000
0.50	0.270
1.00	0.543
1.50	0.813
2.00	1.084

A sample from a public water supply is analyzed to determine the free chlorine residual, giving an absorbance of 0.113. What is the free chlorine residual for the sample in parts per million Cl_2?

17. Brown and Lin reported a quantitative method for methanol based on its effect on the visible spectrum of methylene blue.[30] In the absence of methanol, the visible spectrum for methylene blue shows two prominent absorption bands centered at approximately 610 nm and 660 nm, corresponding to the monomer and dimer, respectively. In the presence of methanol, the intensity of the dimer's absorption band decreases, and that of the monomer increases. For concentrations of methanol between 0 and 30% v/v, the ratio of the absorbance at 663 nm, A_{663}, to that at 610 nm, A_{610}, is a linear function of the amount of methanol. Using the following standardization data, determine the %v/v methanol in a sample for which A_{610} is 0.75 and A_{663} is 1.07.

% Methanol (v/v)	$\frac{A_{663}}{A_{610}}$
0	1.21
5.0	1.29
10.0	1.42
15.0	1.52
20.0	1.62
25.0	1.74
30.0	1.84

18. The concentration of the barbiturate barbital in a blood sample was determined by extracting 3.00 mL of the blood with 15 mL of $CHCl_3$. The chloroform, which now contains the barbital, is then extracted with 10.0 mL of 0.045 M NaOH (pH ≈ 13). A 3.00-mL sample of the aqueous extract is placed in a 1.00-cm cell, and an absorbance of 0.115 is measured. The pH of the sample in the absorption cell is then adjusted to approximately 10 by adding 0.5 mL of 16% w/v NH_4Cl, giving an absorbance of 0.023. When 3.00 mL of a standard barbital solution with a concentration of 3.0 mg/100 mL is taken through the same procedure, the absorbance at pH 13 is 0.295, whereas that at a pH of 10 is 0.002. Report the concentration of barbital, (milligrams per 100 mL), in the sample.

19. Jones and Thatcher developed a spectrophotometric method for analyzing analgesic tablets containing aspirin, phenacetin, and caffeine.[31] The sample is dissolved in $CHCl_3$ and extracted with an aqueous solution of $NaHCO_3$ to remove the aspirin. After the extraction is complete, the chloroform is then transferred to a 250-mL volumetric flask and diluted to volume with $CHCl_3$. A 2.00-mL portion of this solution is diluted to volume in a 200-mL volumetric flask with $CHCl_3$. The absorbance of the final solution is measured at wavelengths of 250 nm and 275 nm, at which the absorptivities, in $ppm^{-1}\ cm^{-1}$, for caffeine and phenacetin are

$$\text{caffeine:}\quad a_{250} = 0.0131 \quad \text{and} \quad a_{275} = 0.0485$$

$$\text{phenacetin:}\quad a_{250} = 0.0702 \quad \text{and} \quad a_{275} = 0.0159$$

Aspirin is determined by neutralizing the $NaHCO_3$ in the aqueous solution and extracting the aspirin into $CHCl_3$. The combined extracts are diluted to 500 mL in a volumetric flask. A 20.00-mL portion of the solution is placed in a 100-mL volumetric flask and diluted to volume with $CHCl_3$. The absorbance of this solution is measured at 277 nm, where the absorptivity of aspirin is 0.00682 $ppm^{-1}\ cm^{-1}$. An analgesic tablet treated by this procedure is found to have absorbances of 0.466 at 250 nm, 0.164 at 275 nm, and 0.600 at 277 nm when using a cell with a 1.00-cm pathlength. Report the milligrams of aspirin, caffeine, and phenacetin in the analgesic tablet.

20. The concentration of SO_2 in a sample of air was determined by the *p*-rosaniline method. The SO_2 was collected in a 10.00-mL solution of $HgCl_4^{2-}$, where it forms $Hg(SO_3)_2^{2-}$, by pulling the air through the solution for 75 min at a rate of 1.6 L/min.

After adding *p*-rosaniline and formaldehyde, the colored solution was diluted to 25 mL in a volumetric flask. The absorbance was measured at 569 nm in a 1-cm cell, yielding a value of 0.485. A standard sample was prepared by substituting a 1.00-mL sample of a standard solution containing the equivalent of 15.00 ppm SO_2 for the air sample. The absorbance of the standard was found to be 0.181. Report the concentration of SO_2 in the air in parts per million. The density of air may be taken as 1.18 g/L.

21. Seasholtz and colleagues describe a method for the quantitative analysis of CO in automobile exhaust based on the measurement of infrared radiation at 2170 cm^{-1}.[32] A calibration curve was prepared by filling a 10-cm infrared gas cell with a known pressure of CO and measuring the absorbance using an FT–IR. The standardization relationship was found to be

$$A = -1.1 \times 10^{-4} + 9.9 \times 10^{-4}(P_{CO} \text{ in torr})$$

Samples were prepared by using a vacuum manifold to fill the gas cell. After measuring the total pressure, the absorbance of the sample at 2170 cm^{-1} was measured. Results are reported as %CO (P_{CO}/P_{tot}). Five exhaust samples were obtained from a 1973 coupe, yielding the following results

P_{tot} (torr)	Absorbance
595	0.1146
354	0.0642
332	0.0591
233	0.0412
143	0.0254

Determine the %CO for each sample, and report the mean value and the 95% confidence interval.

22. A recent innovation in IR sample preparation is the use of disposable sample cards made from thin sheets of either polyethylene (PE) or polytetrafluoroethylene (PTFE). Samples of analyte are dissolved in a suitable solvent and placed on the IR card. After the solvent evaporates, the sample's spectrum is obtained. Because the thickness of the PE or PTFE film is not uniform, the primary use for IR cards has been for qualitative analysis. Zhao and Malinowski showed how a quantitative analysis for polystyrene could be performed by adding an internal standard of KSCN to the sample.[33] Polystyrene was monitored at 1494 cm^{-1} and KSCN at 2064 cm^{-1}. Standard solutions were prepared by placing weighed portions of polystyrene in a 10-mL volumetric flask and diluting to volume with a solution of 10 g/L KSCN in methyl isobutyl ketone. Results obtained with a PE card are shown below

g polystyrene	0.1609	0.3290	0.4842	0.6402	0.8006
$A_{1494\ cm^{-1}}$	0.0452	0.1138	0.1820	0.3275	0.3195
$A_{2064\ cm^{-1}}$	0.1948	0.2274	0.2525	0.3580	0.2703

When a 0.8006-g sample of a poly(styrene/maleic anhydride) copolymer was prepared in the same manner, the following results were obtained

Replicate	$A_{1494\ cm^{-1}}$	$A_{2064\ cm^{-1}}$
1	0.2729	0.3582
2	0.2074	0.2820
3	0.2785	0.3642

What is the %w/w polystyrene in the copolymer? Given that the reported %w/w polystyrene is 67%, is there any evidence for a determinate error at $\alpha = 0.05$?

23. The following table lists the molar absorptivities for the Arsenazo complexes of copper and barium at selected wavelengths.[34] Determine the optimum wavelengths for the analysis of a mixture of copper and barium.

Wavelength (nm)	ε_{Cu}	ε_{Ba}
595	11900	7100
600	15500	7200
607	18300	7400
611	19300	6900
614	19300	7000
620	17800	7100
626	16300	8400
635	10900	9900
641	7500	10500
645	5300	10000
650	3500	8600
655	2200	6600
658	1900	6500
665	1500	3900
670	1500	2800
680	1800	1500

24. Blanco and co-workers[14] reported several examples of the application of multiwavelength linear regression analysis for the simultaneous determination of mixtures containing two components with overlapping spectra. For each of the following, determine the molar concentration of each analyte in the mixture.

(a) A mixture of MnO_4^- and $Cr_2O_7^{2-}$, and standards of 1.0×10^{-4} M $KMnO_4$ and 1.0×10^{-4} M $K_2Cr_2O_7$ gave the following results

Wavelength (nm)	Absorbances MnO_4^- standard	$Cr_2O_7^{2-}$ standard	Mixture
266	0.042	0.410	0.766
288	0.082	0.283	0.571
320	0.168	0.158	0.422
350	0.125	0.318	0.672
360	0.056	0.181	0.366

(b) Titanium and vanadium were determined by forming the H_2O_2 complexes. Results for a mixture of Ti(IV) and V(V) and standards of 63.1 ppm Ti(IV) and 96.4 ppm V(V) are listed in the following table

Wavelength (nm)	Absorbances Ti(IV) standard	V(V) standard	Mixture
390	0.895	0.326	0.651
430	0.884	0.497	0.743
450	0.694	0.528	0.665
470	0.481	0.512	0.547
510	0.173	0.374	0.314

(c) Copper and zinc were determined by forming colored complexes with 2-pyridyl-azo-resorcinol (PAR). The absorbances for PAR, a mixture of Cu^{2+} and Zn^{2+}, and standards of 1.0-ppm Cu^{2+} and 1.0-ppm Zn^{2+} are listed in the following table. Note that you must correct the absorbances for the metal for the absorbance due to the PAR.

Wavelength (nm)	Absorbances PAR	Cu^{2+} standard	Zn^{2+} standard	Mixture
480	0.211	0.698	0.971	0.656
496	0.137	0.732	1.018	0.668
510	0.100	0.732	0.891	0.627
526	0.072	0.602	0.672	0.498
540	0.056	0.387	0.306	0.290

25. Jochum and associates[15b] reported the following results for the analysis of Ni^{2+} and Cu^{2+} by the generalized standard addition method

Δn_{Ni} (mol)	Δn_{Cu} (mol)	$\Delta Q_{\lambda 1}$	$\Delta Q_{\lambda 2}$
1.00×10^{-5}	0	0.174	0.058
2.00×10^{-5}	0	0.345	0.093
3.00×10^{-5}	0	0.532	0.198
4.00×10^{-5}	0	0.732	0.266
4.00×10^{-5}	2.00×10^{-5}	1.11	1.81
4.00×10^{-5}	4.00×10^{-5}	1.49	3.28
4.00×10^{-5}	6.00×10^{-5}	1.89	4.88
4.00×10^{-5}	8.00×10^{-5}	2.31	6.44

Determine the moles of Ni^{2+} and Cu^{2+} in a sample if the volume corrected absorbances at λ1 and λ2 are, respectively, 0.442 and 0.869 before the standard additions.

26. The stoichiometry for a metal–ligand complex was determined by the method of continuous variations. A series of solutions was prepared in which the combined concentrations of the metal and ligand were held constant at 5.15×10^{-4} M. The absorbances of these solutions were measured at a wavelength at which only the metal–ligand complex absorbs. Using the following data, determine the formula of the complex.

Mole fraction M	Mole fraction L	Absorbance
1.0	0.0	0.001
0.9	0.1	0.126
0.8	0.2	0.260
0.7	0.3	0.389
0.6	0.4	0.515
0.5	0.5	0.642
0.4	0.6	0.775
0.3	0.7	0.771
0.2	0.8	0.513
0.1	0.9	0.253
0.0	1.0	0.000

27. The stoichiometry of a metal–ligand complex was determined by the mole-ratio method. A series of solutions was prepared in which the concentration of the metal was held constant at 3.65×10^{-4} M, and the concentration of the ligand varied from 1×10^{-4} M to 1×10^{-3} M. Using the following data, determine the stoichiometry of the metal–ligand complex.

[Ligand] (M)	Absorbance
1.0×10^{-4}	0.122
2.0×10^{-4}	0.251
3.0×10^{-4}	0.376
4.0×10^{-4}	0.496
5.0×10^{-4}	0.625
6.0×10^{-4}	0.752
7.0×10^{-4}	0.873
8.0×10^{-4}	0.937
9.0×10^{-4}	0.962
1.0×10^{-3}	1.001

28. The stoichiometry of a metal–ligand complex was determined by the slope-ratio method. Two sets of solutions were prepared. For the first set of solutions the concentration of the metal was held constant at 0.010 M, and the concentration of the ligand was varied. When the absorbance of these solutions was measured at a wavelength at which only the metal–ligand complex absorbs the following data were obtained.

[Ligand] (M)	Absorbance
1.0×10^{-5}	0.012
2.0×10^{-5}	0.029
3.0×10^{-5}	0.042
4.0×10^{-5}	0.055
5.0×10^{-5}	0.069

For the second set of solutions the concentration of the ligand was held constant at 0.010 M, and the concentration of the metal was varied, yielding the following absorbances.

[Metal] (M)	Absorbance
1.0×10^{-5}	0.040
2.0×10^{-5}	0.085
3.0×10^{-5}	0.125
4.0×10^{-5}	0.162
5.0×10^{-5}	0.206

Using this data, determine the stoichiometry of the metal–ligand complex.

29. Kawakami and Igarashi[35] developed a new spectrophotometric method for nitrite based on its reaction with 5,10,15,20-tetrakis(4-aminophenyl)porphrine (TAPP). As part of their study, they investigated the stoichiometry of the reaction between TAPP and NO_2^-; the following data are derived from a figure in their paper.

[TAPP] (M)	$[NO_2^-]$ (M)	Absorbance
8.0×10^{-7}	1.6×10^{-7}	0.227
8.0×10^{-7}	3.2×10^{-7}	0.192
8.0×10^{-7}	4.8×10^{-7}	0.158
8.0×10^{-7}	8.0×10^{-7}	0.126
8.0×10^{-7}	1.6×10^{-6}	0.065
8.0×10^{-7}	2.4×10^{-6}	0.047
8.0×10^{-7}	3.2×10^{-6}	0.042
8.0×10^{-7}	4.0×10^{-6}	0.042

What is the stoichiometry of the reaction?

30. The equilibrium constant for an acid–base indicator is determined by preparing three solutions, each of which has a total indicator concentration of 1.35×10^{-5} M. The pH of the first solution is adjusted until it is acidic enough to ensure that only the acid form of the indicator is present, yielding an absorbance of 0.673. The absorbance of the second solution, whose pH was adjusted to give only the base form of the indicator, was measured at 0.118. The pH of the third solution was adjusted to 4.17 and had an absorbance of 0.439. What is the acidity constant for the acid–base indicator?

31. The acidity constant for an organic weak acid was determined by measuring its absorbance as a function of pH while maintaining a constant total concentration of the acid. Using the data in the following table, determine the acidity constant for the organic weak acid.

pH	Absorbance
1.53	0.010
2.20	0.010
3.66	0.035
4.11	0.072
4.35	0.103
4.75	0.169
4.88	0.193
5.09	0.227
5.69	0.288
7.20	0.317
7.78	0.317

32. When using a spectrophotometer for which the precision of absorbance measurements is limited by the uncertainty of reading $\%T$, the analysis of highly absorbing solutions can lead to an unacceptable level of indeterminate errors. Consider the analysis of a sample for which the molar absorptivity is 1.0×10^4 M^{-1} cm^{-1}, and the pathlength is 1.00 cm. (a) What is the relative uncertainty in concentration for an analyte whose concentration is 2.0×10^{-4} M if s_T is ± 0.002? (b) What is the relative uncertainty in the concentration if the spectrophotometer is calibrated using a blank consisting of a 1.0×10^{-4} M solution of the analyte?

33. Hobbins reported the following calibration data for the flame atomic absorption analysis for phosphorus.[36]

ppm P	Absorbance
2130	0.048
4260	0.110
6400	0.173
8530	0.230

To determine the purity of a sample of Na_2HPO_4, a 2.469-g sample is dissolved and diluted to volume in a 100-mL volumetric flask. Analysis of the resulting solution gives an absorbance of 0.135. What is the purity of the Na_2HPO_4?

34. Bonert and Pohl reported results for the atomic absorption analysis of several metals in caustic suspensions produced during the manufacture of soda by the ammonia-soda process.[37] (a) The concentration of Cu was determined by acidifying a 200-mL sample of the caustic solution with 20 mL of concentrated HNO_3, adding 1 mL of 27% w/v H_2O_2 and boiling for 30 min. The resulting solution was diluted to 500 mL, filtered, and analyzed by flame atomic absorption using matrix-matched standards. The results for a typical analysis are shown in the following table.

Solution	ppm Cu	Absorbance
blank	0	0.007
standard 1	0.200	0.014
standard 2	0.500	0.036
standard 3	1.000	0.072
standard 4	2.000	0.146
sample		0.027

Determine the concentration of Cu in the caustic suspension. (b) The determination of Cr was accomplished by acidifying a 200-mL sample of the caustic solution with 20 mL of concentrated HNO_3, adding 0.2 g of Na_2SO_3 and boiling for 30 min. The Cr was isolated from the sample by adding 20 mL of NH_3, producing a precipitate that includes the chromium as well as other oxides. The precipitate was isolated by filtration, washed, and transferred with wash water to a beaker. After acidifying with 10 mL of HNO_3, the solution was evaporated to dryness. The residue was redissolved in a combination of HNO_3 and HCl and evaporated to dryness. Finally, the residue was dissolved in 5 mL of HCl, filtered, diluted to volume in a 50-mL volumetric flask, and analyzed by atomic absorption using the method of standard additions. The atomic absorption results are summarized in the following table.

Sample	ppm Cr_{add}	Absorbance
blank		0.001
sample		0.045
standard addition 1	0.200	0.083
standard addition 2	0.500	0.118
standard addition 3	1.000	0.192

Report the concentration of Cr in the caustic suspension.

35. Quigley and Vernon report results for the determination of trace metals in sea water using a graphite furnace atomic absorption spectrophotometer.[38] Calibration was achieved by the method of standard additions. The trace metals were first separated from their complex, high-salt matrix by coprecipitating with Fe^{3+}. In a typical analysis a 5.00-mL portion of 2000-ppm Fe^{3+} was added to 1.00 L of sea water. The pH was adjusted to 9 using NH_4OH, and the precipitate of $Fe(OH)_3$ allowed to stand overnight. After isolating and rinsing the precipitate, the $Fe(OH)_3$ and coprecipitated metals were dissolved in 2 mL of concentrated HNO_3 and diluted to volume in a 50-mL volumetric flask. To analyze for Mn^{2+}, a 1.00-mL sample of this solution was diluted to 100 mL in a volumetric flask. The following samples were injected into the graphite furnace and analyzed

	Absorbance
2.5-μL sample + 2.5 μL of 0-ppb Mn^{2+} standard	0.223
2.5-μL sample + 2.5 μL of 2.5-ppb Mn^{2+} standard	0.294
2.5-μL sample + 2.5 μL of 5.0-ppb Mn^{2+} standard	0.361

Report the parts per billion of Mn^{2+} in the sample of sea water.

36. The concentration of Na in plant materials may be determined by flame atomic emission. The material to be analyzed is prepared by grinding, homogenizing, and drying at 103 °C. A sample of approximately 4 g is transferred to a quartz crucible and heated on a hot plate to char the organic material. The sample is heated in a muffle furnace at 550 °C for several hours. After cooling to room temperature the residue is dissolved by adding 2 mL of 1:1 HNO_3 and evaporated to dryness. The residue is redissolved in 10 mL of 1:9 HNO_3, filtered, and diluted to 50 mL in a volumetric flask. The following data were obtained during a typical analysis for the concentration of Na in a 4.0264-g sample of oat bran

Sample	ppm Na	Emission (arbitrary units)
blank	0	0
standard 1	2.00	90.3
standard 2	4.00	181
standard 3	6.00	272
standard 4	8.00	363
standard 5	10.00	448
sample		238

Determine the parts per million Na in the sample of oat bran.

37. Gluodenis describes the use of ICP to analyze samples containing Pb and Ni in brass.[39] The analysis for Pb uses external standards prepared from brass samples containing known amounts of lead. Results are shown in the following table.

%w/w Pb	Emission Intensity
0.000	4.29×10^4
0.0100	1.87×10^5
0.0200	3.20×10^5
0.0650	1.28×10^6
0.350	6.22×10^6
0.700	1.26×10^7
1.04	1.77×10^7
2.24	3.88×10^7
3.15	5.61×10^7
9.25	1.64×10^8

What is the %w/w Pb in a sample of brass that gives an emission intensity of 9.25×10^4?

The analysis for Ni uses an internal standard. Results for a typical calibration are shown in the following table.

%w/w Ni	Ratio of Emission Intensity
0.000	0.00267
0.0140	0.00154
0.0330	0.00312
0.130	0.120
0.280	0.246
0.280	0.247
0.560	0.533
1.30	1.20
4.82	4.44

What is the %w/w Ni in a sample for which the ratio of emission intensity is 1.10×10^{-3}?

38. Yan and associates developed a method for the analysis of iron based on its formation of a fluorescent metal–ligand complex with the ligand 5-(4-methylphenylazo)-8-aminoquinoline.[40] In the presence of the surfactant cetyltrimethyl ammonium bromide the analysis is carried out using an excitation wavelength of 316 nm with emission monitored at 528 nm. Standardization with external standards gives

$$I = -0.03 + 1.594(\text{ppm Fe}^{3+})$$

A 0.5113-g sample of dry dog food was ashed to remove organic materials, and the residue dissolved in a small amount of HCl and diluted to volume in a 50-mL volumetric flask. Analysis of the resulting solution gave a fluorescent emission intensity of 5.72. Determine the parts per million of Fe in the sample of dog food.

39. A solution of 5.00×10^{-5} M 1,3-dihydroxynaphthelene in 2 M NaOH has a fluorescence intensity of 4.85 at a wavelength of 459 nm. What is the concentration of 1,3-dihydroxynaphthelene in a solution with a fluorescence intensity of 3.74 under identical conditions?

40. The following data was recorded for the phosphorescence intensity for several standard solutions of benzo[*a*]pyrene.

Molar Concentration benzo[*a*]pyrene	Intensity of Phosphorescent Emission
0	0
1.00×10^{-5}	0.98
3.00×10^{-5}	3.22
6.00×10^{-5}	6.25
1.00×10^{-4}	10.21

What is the concentration of benzo[*a*]pyrene in a sample yielding a phosphorescent emission intensity of 4.97?

41. The concentration of acetylsalicylic acid, $C_9H_8O_4$, in aspirin tablets can be determined by hydrolyzing to the salicylate ion, $C_7H_5O_2^-$, and determining the concentration of the salicylate ion spectrofluorometrically. A stock standard solution is prepared by weighing 0.0774 g of salicylic acid, $C_7H_6O_2$, into a 1-L volumetric flask and diluting to volume with distilled water. A set of calibration standards is prepared by pipeting 0, 2.00, 4.00, 6.00, 8.00, and 10.00 mL of the stock solution into separate 100-mL volumetric flasks containing 2.00 mL of 4 M NaOH and diluting to volume with distilled water. The fluorescence of the calibration standards was measured at an emission wavelength of 400 nm using an excitation wavelength of 310 nm; results are listed in the following table.

Milliliters of stock standard solution	Fluorescence emission intensity
0	0
2.00	3.02
4.00	5.98
6.00	9.18
8.00	12.13
10.00	14.96

Several aspirin tablets are ground to a fine powder in a mortar and pestle. A 0.1013-g portion of the powder is placed in a 1-L volumetric flask and diluted to volume with distilled water. A portion of this solution is filtered to remove insoluble binders, and a 10.00-mL aliquot transferred to a 100-mL volumetric flask containing 2.00 mL of 4 M NaOH. After diluting to volume the fluorescence of the resulting solution is found to be 8.69. What is the %w/w acetylsalicylic acid in the aspirin tablets?

42. Selenium (IV) in natural waters can be determined by complexing with ammonium pyrrolidine dithiocarbamate and extracting into $CHCl_3$. This step serves to concentrate the Se(IV) and to separate it from Se(VI). The Se(IV) is then extracted back into an aqueous matrix using HNO_3. After complexing with 2,3-diaminonaphthelene, the complex is extracted into cyclohexane. Fluorescence is measured at 520 nm following its excitation at 380 nm. Calibration is by adding known amounts of Se(IV) to the water sample before beginning the analysis. Given the following results, what is the concentration of Se(IV) in the sample?

Concentration of Se(IV) Added (nM)	Fluorescence Intensity
0	323
2.00	597
4.00	862
6.00	1123

43. Fibrinogen is a protein found in human plasma that is produced by the liver. Its concentration in plasma is clinically important to blood clotting. Many of the analytical methods used to determine the concentration of fibrinogen in plasma are based on light scattering following its precipitation. da Silva, and colleagues describe such a method in which fibrinogen precipitates in the presence of ammonium sulfate in a guanidine hydrochloride buffer.[41] Light scattering is measured nephelometrically at a wavelength of 340 nm. Analysis of a set of external calibration standards gives the following calibration equation.

$$I = -4.66 + 9907.63C_A$$

where I is the intensity of scattered light, and C_A is the concentration of fibrinogen in grams per liter. A 9.00-mL sample of plasma was collected from a patient and mixed with 1.00-mL of an anticoagulating agent. A 1.00-mL aliquot was then diluted to 250 mL in a volumetric flask. Analysis of the resulting solution gave a scattering intensity of 44.70. What is the concentration of fibrinogen (in grams per liter) in the plasma sample?

10N SUGGESTED READINGS

The history of spectroscopy is discussed in the following sources.

Laitenen, H. A.; Ewing, G. W., eds. *A History of Analytical Chemistry,* The Division of Analytical Chemistry of the American Chemical Society: Washington, D.C., 1977, pp. 103–243.

Thomas, N. C. "The Early History of Spectroscopy," *J. Chem. Educ.* **1991,** *68,* 631–633.

An explanation of the many units used to describe the energy of electromagnetic radiation is provided in the following paper.

Ball, D. W. "Units! Units! Units!" *Spectroscopy* **1995,** *10(8),* 44–47.

The following sources present a theoretical treatment of the interaction of electromagnetic radiation with matter.

Ingle, J. D.; Crouch, S. R. *Spectrochemical Analysis.* Prentice Hall, Englewood Cliffs, N.J., 1988.

Macomber, R. S. "A Unifying Approach to Absorption Spectroscopy at the Undergraduate Level," *J. Chem. Educ.* **1997,** *74,* 65–67.

Orchin, M.; Jaffe, H. H. *Symmetry, Orbitals and Spectra.* Wiley-Interscience: New York, 1971.

Two noncalculus-based approaches to discovering the Beer–Lambert law are found in the following papers.

Lykos, P. "The Beer–Lambert Law Revisited: A Development without Calculus," *J. Chem. Educ.* **1992,** *69,* 730–732.

Ricci, R. W.; Ditzler, M. A.; Nestor, L. P. "Discovering the Beer–Lambert Law," *J. Chem. Educ.* **1994,** *71,* 983–985.

The following sources provide further information on the optical characteristics of gratings and their method of production.

Grossman, W. E. L. "The Optical Characteristics and Production of Diffraction Gratings," *J. Chem. Educ.* **1993,** *70,* 741–748.

Palmer, C. "Diffraction Gratings," *Spectroscopy* **1995,** *10(2),* 14–15.

The evaluation of instrumentation for molecular UV/Vis spectroscopy is reviewed in the following pair of papers.

Altermose, I. R. "Evolution of Instrumentation for UV–Visible Spectrophotometry: Part I," *J. Chem. Educ.* **1986,** *63,* A216–A223.

Altermose, I. R. "Evolution of Instrumentation for UV–Visible Spectrophotometry: Part II," *J. Chem. Educ.* **1986,** *63,* A262–A266.

Listed below is a two-part series on the application of photodiode arrays in UV/Vis spectroscopy.

Jones, D. G. "Photodiode Array Detectors in UV–Vis Spectroscopy: Part I," *Anal. Chem.* **1985,** *57,* 1057A–1073A.

Jones, D. G. "Photodiode Array Detectors in UV-Vis Spectroscopy: Part II," *Anal. Chem.* **1985,** *11,* 1207A–1214A.

There is an abundant literature covering the Fourier transform and its application in spectroscopy, several examples of which are listed here.

Bracewell, R. N. "The Fourier Transform," *Sci. American* **1989,** *260(6),* 85–95.

Glasser, L. "Fourier Transforms for Chemists: Part I. Introduction to the Fourier Transform," *J. Chem. Educ.* **1987,** *64,* A228–A233.

Glasser, L. "Fourier Transforms for Chemists: Part II. Fourier Transforms in Chemistry and Spectroscopy," *J. Chem. Educ.* **1987,** *64,* A260–A266.

Glasser, L. "Fourier Transforms for Chemists: Part III. Fourier Transforms in Data Treatment," *J. Chem. Educ.* **1987,** *64,* A306–A313.

Graff, D. K. "Fourier and Hadamard: Transforms in Spectroscopy," *J. Chem. Educ.* **1995,** *72,* 304–309.

Griffiths, P. R. *Chemical Fourier Transform Spectroscopy.* Wiley-Interscience: New York, 1975.

Griffiths, P. R. ed. *Transform Techniques in Chemistry.* Plenum Press: New York, 1978.

Perkins, W. E. "Fourier Transform Infrared Spectroscopy: Part I. Instrumentation," *J. Chem. Educ.* **1986,** *63,* A5–A10.

Perkins, W. E. "Fourier Transform Infrared Spectroscopy: Part II. Advantages of FT–IR," *J. Chem. Educ.* **1987,** *64,* A269–A271.

Perkins, W. E. "Fourier Transform Infrared Spectroscopy: Part III. Applications," *J. Chem. Educ.* **1987,** *64,* A296–A305.

Strong III, F. C. "How the Fourier Transform Infrared Spectrophotometer Works," *J. Chem. Educ.* **1979,** *56,* 681–684.

Consult the following sources for more information about reflectance techniques for IR spectroscopy.

Leyden, D. E.; Shreedhara Murthy, R. S. "Surface-Selective Sampling Techniques in Fourier Transform Infrared Spectroscopy," *Spectroscopy* **1987,** *2(2),* 28–36.

Optical Spectroscopy: Sampling Techniques Manual. Harrick Scientific Corporation: Ossining, N.Y., 1987.

Porro, T. J.; Pattacini, S. C. "Sample Handling for Mid-Infrared Spectroscopy, Part I: Solid and Liquid Sampling," *Spectroscopy* **1993,** *8(7),* 40–47.

Porro, T. J.; Pattacini, S. C. "Sample Handling for Mid-Infrared Spectroscopy, Part II: Specialized Techniques," *Spectroscopy* **1993,** *8(8),* 39–44.

A thorough treatment of the multicomponent quantitative analysis of samples based on Beer's law, and the analysis of samples for which the pathlength is indeterminate is found in the following review article.

Brown, C. W.; Obremski, R. J. "Multicomponent Quantitative Analysis," *Appl. Spectrosc. Rev.* **1984,** *20,* 373–418.

For more information on optical luminescence spectroscopy, see the following sources.

Guilbault, G. G. *Practical Fluorescence.* Decker: New York, 1990.

Schenk, G. "Historical Overview of Fluorescence Analysis to 1980," *Spectroscopy* **1997,** *12,* 47–56.

Vo-Dinh, T. *Room-Temperature Phosphorimetry for Chemical Analysis.* Wiley-Interscience: New York, 1984.

Winefordner, J. D.; Schulman, S. G.; O'Haver, T. C. *Luminescence Spectroscopy in Analytical Chemistry.* Wiley-Interscience: New York, 1969.

The following sources provide additional information on atomic absorption and atomic emission.

Blades, M. W.; Weir, D. G. "Fundamental Studies of the Inductively Coupled Plasma," *Spectroscopy* **1994,** *9,* 14–21.

Hieftje, G. M. "Atomic Absorption Spectrometry—Has It Gone or Where Is It Going?" *J. Anal. At. Spectrom.* **1989,** *4,* 117–122.

Koirtyohann, S. R. "A History of Atomic Absorption Spectrometry from an Academic Perspective," *Anal. Chem.* **1991,** *63,* 1024A–1031A.

L'Vov, B. V. "Graphite Furnace Atomic Absorption Spectrometry," *Anal. Chem.* **1991,** *63,* 924A–931A.

Slavin, W. "A Comparison of Atomic Spectroscopic Analytical Techniques," *Spectroscopy* **1991,** *6,* 16–21.

Van Loon, J. C. *Analytical Atomic Absorption Spectroscopy.* Academic Press: New York, 1980.

Walsh, A. "The Development of Atomic Absorption Methods of Elemental Analysis 1952–1962," *Anal. Chem.* **1991,** *63,* 933A–941A.

Welz, B. *Atomic Absorption Spectroscopy.* VCH: Deerfield Beach, FL, 1985.

Several texts may be consulted for further examples of spectroscopic methods of analysis.

Christian, G. D.; Callis, J. B. eds. *Trace Analysis and Spectroscopic Methods for Molecules.* Wiley-Interscience: New York, 1986.

Skoog, D. A.; Holler, F. J.; Nieman, T. A. *Principles of Instrumental Analysis.* Saunders College Publishing: Philadelphia, 1998.

Vandecasteele, C.; Block, C. B. *Modern Methods for Trace Element Determination.* Wiley: Chichester, England, 1994.

Van Loon, J. C. *Selected Methods of Trace Metal Analysis: Biological and Environmental Samples.* Wiley-Interscience: New York, 1985.

10O REFERENCES

1. Method 417B in *Standard Methods for the Analysis of Water and Wastewater,* American Public Health Association: Washington, D.C.; 15 ed., 1981, pp. 356–360.
2. (a) Sheppard, N. In Andrews, D. L. Ed. *Perspectives in Modern Chemical Spectroscopy.* Springer-Verlag: Berlin, 1990, pp. 1–41; (b) Tyson, J. *Analysis: What Analytical Chemists Do.* Royal Society of Chemistry: Letchworth, England, 1988, pp. 49–52.
3. Home, D.; Gribbin, J. *New Scientist* **1991,** 2 November, 30–33.
4. Ball, D. W. *Spectroscopy* **1994,** *9(5),* 24–25.
5. Ball, D. W. *Spectroscopy* **1994,** *9(6),* 20–21.
6. Jiang, S; Parker, G. A. *Am. Lab.* **1981,** October, 38–43.
7. Strong, F. C., III *Anal. Chem.* **1976,** *48,* 2155–2161.
8. Gilbert, D. D. *J. Chem. Educ.* **1991,** *68,* A278–A281.
9. Seitz, W. R. *Anal. Chem.* **1984,** *56,* 16A–34A.
10. Angel, S. M. *Spectroscopy* **1987,** *2(2),* 38–48.
11. Method 3500-Fe-B as published in *Standard Methods for the Analysis of Water and Wastewater.* American Public Health Association: Washington, D.C., 19th ed., 1998, pp. 3–76 to 3–78.
12. DiTusa, M. R.; Schilt, A. A. *J. Chem. Educ.* **1985,** *62,* 541–542.
13. Mehra, M. C.; Rioux, J. *J. Chem. Educ.* **1982,** *59,* 688–689.
14. Blanco, M.; Iturriaga, H.; Maspoch, S.; et al. *Chem. Educ.* **1989,** *66,* 178–180.
15. (a) Saxberg, B. E. H.; Kowalski, B. R. *Anal. Chem.* **1979,** *51,* 1031–1038; (b) Jochum, C.; Jochum, P.; Kowalski, B. R. *Anal. Chem.* **1981,** *53,* 85–92; (c) Gerlach, R. W.; Kowalski, B. R. *Anal. Chim. Acta.* **1982,** *134,* 119; (d) Kalivas, J. H.; Kowalski, B. R. *Anal. Chem.* **1981,** *53,* 2207–2212; Raymond, M.; Jochum, C.; Kowalski, B. R. *J. Chem. Educ.* **1983,** *60,* 1072–1073.
16. Ramette, R. W. *Chemical Equilibrium and Analysis.* Addison-Wesley: Reading, MA, 1981, Chapter 13.
17. Rothman, L. D.; Crouch, S. R.; Ingle, J. D. Jr. *Anal. Chem.* **1975,** *47,* 1226–1233.
18. (a) Walsh, A. *Anal. Chem.* **1991,** *63,* 933A–941A; (b) Koirtyohann, S. R. *Anal. Chem.* **1991,** *63,* 1024A–1031A; (c) Slavin, W. *Anal. Chem.* **1991,** *63,* 1033A–1038A.
19. Ingle, J. D.; Crouch, S. R. *Spectrochemical Analysis.* Prentice-Hall: Englewood Cliffs, N.J., 1988; p. 275.
20. Parsons, M. L.; Major, S.; Forster, A. R. *Appl. Spectrosc.* **1983,** *37,* 411–418.
21. Hosking, J. W.; Snell, N. B.; Sturman, B. T. *J. Chem. Educ.* **1977,** *54,* 128–130.
22. (a) Bhattacharya, S. K.; Goodwin, T. G.; Crawford, A. J. *Anal. Lett.* **1984,** *17,* 1567–1593;(b) Crawford, A. J.; Bhattacharya, S. K. "Microanalysis of Copper and Zinc in Biopsy-Sized Tissue Specimens by Atomic Absorption Spectroscopy Using a Stoichiometric Air–Acetylene Flame"; Varian Instruments at Work, Number AA-46, April 1985.
23. (a) Mule, S. J.; Hushin, P. L. *Anal. Chem.* **1971,** *43,* 708–711; (b) O'Reilly, J. E. *J. Chem. Educ.* **1975,** *52,* 610–612.
24. Dawson, J. B. *J. Anal. At. Spectrosc.* **1991,** *6,* 93–98.
25. Goodney, D. E. *J. Chem. Educ.* **1982,** *59,* 875–876.

26. Hach, C. C.; Bryant, M. "Turbidity Standards"; Technical Information Series-Booklet No. 12, Hach Company: Loveland, CO, 1995.
27. Method 4500-SO_4^{2-}-E in *Standard Methods for the Analysis of Water and Wastewater,* American Public Health Association: Washington, D.C.; 20th ed., 1998, pp. 4–178 to 4–179.
28. Lozano-Calero, D.; Martín-Palomeque, P.; Madueño-Loriguillo, S. *J. Chem. Educ.* **1996,** *73,* 1173–1174.
29. Saito, T. *Anal. Chim. Acta.* **1992,** *268,* 351–355.
30. Lin, J.; Brown, C. W. *Spectroscopy* **1995,** *10(5),* 48–51.
31. Jones, M.; Thatcher, R. L. *Anal. Chem.* **1951,** *23,* 957–960.
32. Seaholtz, M. B.; Pence, L. E.; Moe, O. A. Jr. *J. Chem. Educ.* **1988,** *65,* 820–823.
33. Zhao, Z.; Malinowski, E. R. *Spectroscopy* **1996,** *11(7),* 44–49.
34. Grossman, O.; Turanov, A. N. *Anal. Chim. Acta.* **1992,** *257,* 195–202.
35. Kawakami, T.; Igarashi, S. *Anal. Chim. Acta.* **1996,** *333,* 175–180.
36. Hobbins, W. B. "Direct Determination of Phosphorus in Aqueous Matrices by Atomic Absorption"; Varian Instruments at Work, Number AA-19, February 1982.
37. Bonert, K.; Pohl, B. "The Determination of Cd, Cr, Cu, Ni, and Pb in Concentrated $CaCl_2$/NaCl Solutions by AAS"; AA Instruments at Work (Varian), Number 98, November 1990.
38. Quigley, M. N.; Vernon, F. *J. Chem. Educ.* **1996,** *73,* 671–673.
39. Gluodenis, Jr., T. J. *Am. Lab.* November **1998,** 245–275.
40. Yan, G.; Shi, G; Liu, Y. *Anal. Chim. Acta.* **1992,** *264,* 121–124.
41. da Silva, M. P.; Fernández-Romero, J. M.; Luque de Castro, M. D. *Anal. Chim. Acta.* **1996,** *327,* 101–106.

Chapter 11

Electrochemical Methods of Analysis

In Chapter 10 we examined several analytical methods based on the interaction of electromagnetic radiation with matter. In this chapter we turn our attention to analytical methods in which a measurement of potential, current, or charge in an electrochemical cell serves as the analytical signal.

11A Classification of Electrochemical Methods

Although there are only three principal sources for the analytical signal—potential, current, and charge—a wide variety of experimental designs are possible; too many, in fact, to cover adequately in an introductory textbook. The simplest division is between bulk methods, which measure properties of the whole solution, and interfacial methods, in which the signal is a function of phenomena occurring at the interface between an electrode and the solution in contact with the electrode. The measurement of a solution's conductivity, which is proportional to the total concentration of dissolved ions, is one example of a bulk electrochemical method. A determination of pH using a pH electrode is one example of an interfacial electrochemical method. Only interfacial electrochemical methods receive further consideration in this text.

11A.1 Interfacial Electrochemical Methods

The diversity of interfacial electrochemical methods is evident from the partial family tree shown in Figure 11.1. At the first level, interfacial electrochemical methods are divided into static methods and dynamic methods. In static methods no current passes between the electrodes, and the concentrations of species in the electrochemical cell remain unchanged, or static. Potentiometry, in which the potential of an electrochemical cell is measured under static conditions, is one of the most important quantitative electrochemical methods, and is discussed in detail in Section 11B.

The largest division of interfacial electrochemical methods is the group of dynamic methods, in which current flows and concentrations change as the result of a redox reaction. Dynamic methods are further subdivided by whether we choose to control the current or the potential. In controlled-current coulometry, which is covered in Section 11C, we completely oxidize or reduce the analyte by passing a fixed current through the analytical solution. Controlled-potential methods are subdivided further into controlled-potential coulometry and amperometry, in which a constant potential is applied during the analysis, and voltammetry, in which the potential is systematically varied. Controlled-potential coulometry is discussed in Section 11C, and amperometry and voltammetry are discussed in Section 11D.

11A.2 Controlling and Measuring Current and Potential

Electrochemical measurements are made in an electrochemical cell, consisting of two or more electrodes and associated electronics for controlling and measuring the current and potential. In this section the basic components of electrochemical instrumentation are introduced. Specific experimental designs are considered in greater detail in the sections that follow.

The simplest electrochemical cell uses two electrodes. The potential of one of the electrodes is sensitive to the analyte's concentration and is called the working, or **indicator electrode.** The second electrode, which is called the **counter electrode,** serves to complete the electric circuit and provides a reference potential against which the working electrode's potential is measured. Ideally the counter electrode's potential remains constant so that any change in the overall cell potential is attributed to the working electrode. In a dynamic method, where the passage of current changes the concentration of species in the electrochemical cell, the potential of the counter electrode may change over time. This problem is eliminated by replacing the counter electrode with two electrodes: a **reference electrode,** through which no

indicator electrode
The electrode whose potential is a function of the analyte's concentration (also known as the working electrode).

counter electrode
The second electrode in a two-electrode cell that completes the circuit.

reference electrode
An electrode whose potential remains constant and against which other potentials can be measured.

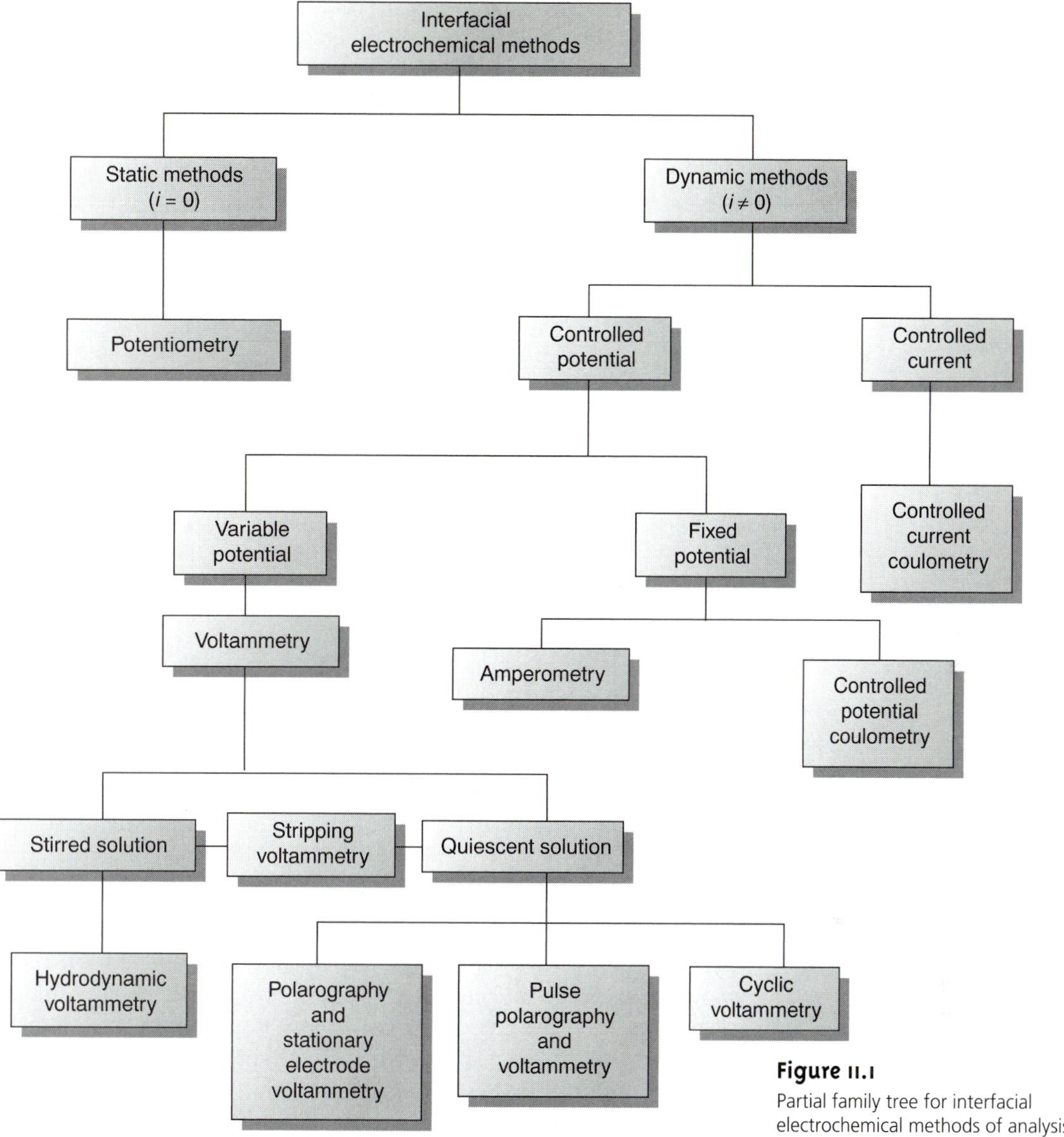

Figure 11.1
Partial family tree for interfacial electrochemical methods of analysis.

current flows and whose potential remains constant; and an **auxiliary electrode** that completes the electric circuit and through which current is allowed to flow.

auxiliary electrode
The third electrode in a three-electrode cell that completes the circuit.

Although many different electrochemical methods of analysis are possible (Figure 11.1) there are only three basic experimental designs: (1) measuring the potential under static conditions of no current flow; (2) measuring the potential while controlling the current; and (3) measuring the current while controlling the potential. Each of these experimental designs, however, is based on **Ohm's law** that a current, i, passing through an electric circuit of resistance, R, generates a potential, E; thus

$$E = iR$$

Ohm's law
The statement that the current moving through a circuit is proportional to the applied potential and inversely proportional to the circuit's resistance ($E = iR$).

Each of these experimental designs also uses a different type of instrument. To aid in understanding how they control and measure current and potential, these instruments are described as if they were operated manually. To do so the analyst observes a change in current or potential and manually adjusts the instrument's settings to maintain the desired experimental conditions. It is important to understand that modern electrochemical instruments provide an automated, electronic means of controlling and measuring current and potential. They do so by using very different electronic circuitry than that shown here. Further details about such instruments can be found in the suggested readings listed at the end of the chapter.

potentiometer
A device for measuring the potential of an electrochemical cell without drawing a current or altering the cell's composition.

Potentiometers Measuring the potential of an electrochemical cell under conditions of zero current is accomplished using a **potentiometer.** A schematic diagram of a manual potentiometer is shown in Figure 11.2. The current in the upper half of the circuit is

$$i_{up} = \frac{E_{PS}}{R_{ab}}$$

where E_{PS} is the power supply's potential, and R_{ab} is the resistance between points *a* and *b* of the slide-wire resistor. In a similar manner, the current in the lower half of the circuit is

$$i_{low} = \frac{E_{cell}}{R_{cb}}$$

where E_{cell} is the potential difference between the working electrode and the counter electrode, and R_{cb} is the resistance between the points *c* and *b* of the slide-wire resistor. When

$$i_{up} = i_{low} = 0$$

no current flows through the galvanometer and the cell potential is given by

$$E_{cell} = \frac{R_{cb}}{R_{ab}} \times E_{PS}$$

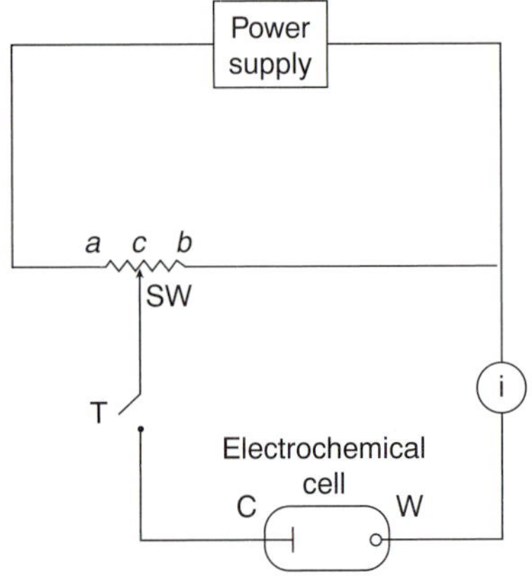

Figure 11.2
Schematic diagram of a manual potentiostat: C = counter electrode; W = working electrode; SW = slide-wire resistor; T = tap key; i = galvanometer.

To make a measurement the tap key is pressed momentarily, and the current is noted at the galvanometer. If a nonzero current is registered, then the slide wire is adjusted and the current remeasured. This process is continued until the galvanometer registers a current of zero. Using the tap key minimizes the total amount of current allowed to flow through the cell. Provided that the total current is negligible, the change in the analyte's concentration is insignificant. For example, a current of 10^{-9} A drawn for 1 s consumes only about 10^{-14} mol of analyte. Modern potentiometers use operational amplifiers to create a high-impedance voltmeter capable of measuring potentials while drawing currents of less than 10^{-9} A.

galvanostat
A device used to control the current in an electrochemical cell.

Galvanostats A **galvanostat** is used for dynamic methods, such as constant-current coulometry, in which it is necessary to control the current flowing through an electrochemical cell. A schematic diagram of a manual constant-current galvanostat is shown in Figure 11.3. If the resistance, *R*, of the galvanostat is significantly larger than the resistance of the electrochemical cell, and the applied voltage from the power supply is much greater than the cell potential, then the current between the auxiliary and working electrodes is equal to

$$i = \frac{E_{PS}}{R}$$

The potential of the working electrode, which changes as the composition of the electrochemical cell changes, is monitored by including a reference electrode and a high-impedance potentiometer.

Potentiostats A **potentiostat** is used for dynamic methods when it is necessary to control the potential of the working electrode. Figure 11.4 shows a schematic diagram for a manual potentiostat that can be used to maintain a constant cell potential. The potential of the working electrode is monitored by a reference electrode connected to the working electrode through a high-impedance potentiometer. The desired potential is achieved by adjusting the slide-wire resistor connected to the auxiliary electrode. If the working electrode's potential begins to drift from the desired value, then the slide-wire resistor is manually readjusted, returning the potential to its initial value. The current flowing between the auxiliary and working electrodes is measured with a galvanostat. Modern potentiostats include waveform generators allowing a time-dependent potential profile, such as a series of potential pulses, to be applied to the working electrode.

potentiostat
A device used to control the potential in an electrochemical cell.

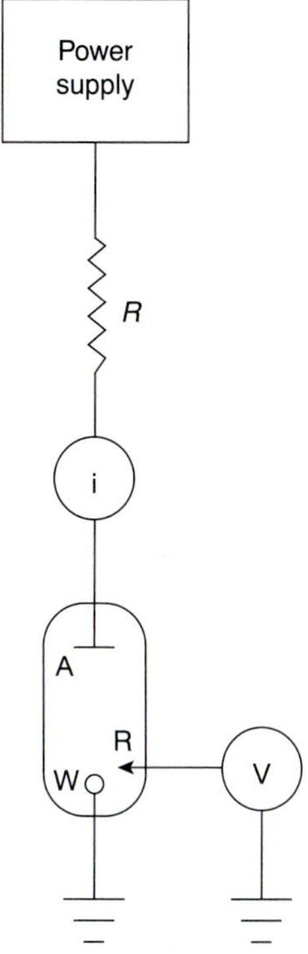

Figure 11.3
Schematic diagram of a galvanostat: *R* = resistor; i = galvanometer; A = auxiliary electrode; W = working electrode; R = reference electrode; V = voltmeter or potentiometer (optional).

11B Potentiometric Methods of Analysis

In potentiometry the potential of an electrochemical cell is measured under static conditions. Because no current, or only a negligible current, flows while measuring a solution's potential, its composition remains unchanged. For this reason, potentiometry is a useful quantitative method. The first quantitative potentiometric applications appeared soon after the formulation, in 1889, of the Nernst equation relating an electrochemical cell's potential to the concentration of electroactive species in the cell.[1]

When first developed, potentiometry was restricted to redox equilibria at metallic electrodes, limiting its application to a few ions. In 1906, Cremer discovered that a potential difference exists between the two sides of a thin glass membrane when opposite sides of the membrane are in contact with solutions containing different concentrations of H_3O^+. This discovery led to the development of the glass pH electrode in 1909. Other types of membranes also yield useful potentials. Kolthoff and Sanders, for example, showed in 1937 that pellets made from AgCl could be used to determine the concentration of Ag^+. Electrodes based on membrane potentials are called ion-selective electrodes, and their continued development has extended potentiometry to a diverse array of analytes.

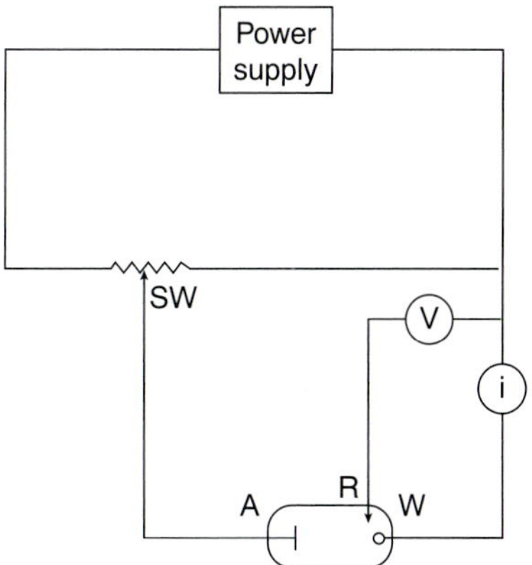

Figure 11.4
Schematic diagram of a manual potentiostat: SW = slide-wire resistor; A = auxiliary electrode; R = reference electrode; W = working electrode; V = voltmeter or potentiometer; i = galvanometer.

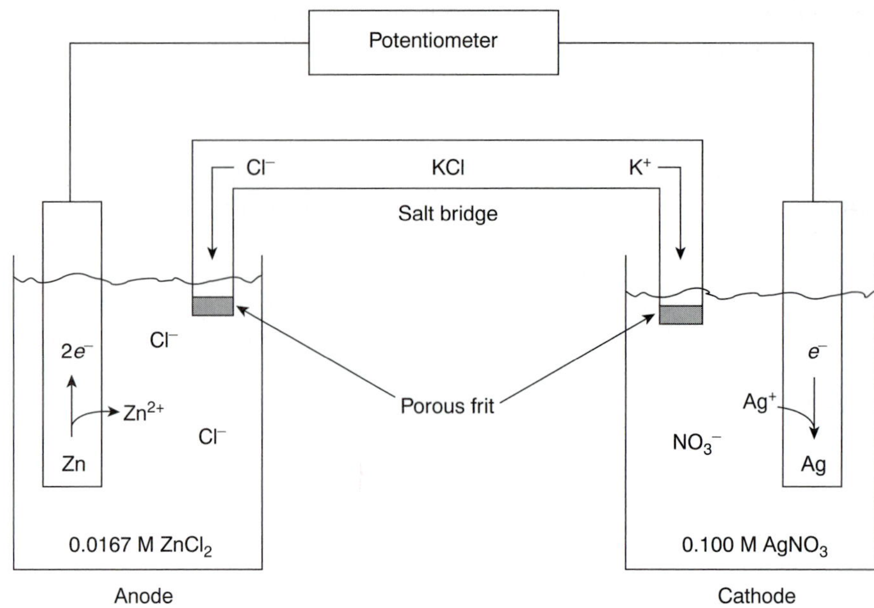

Figure 11.5
Electrochemical cell for potentiometry.

11B.1 Potentiometric Measurements

Potentiometric measurements are made using a potentiometer to determine the difference in potential between a working or, indicator, electrode and a counter electrode (see Figure 11.2). Since no significant current flows in potentiometry, the role of the counter electrode is reduced to that of supplying a reference potential; thus, the counter electrode is usually called the reference electrode. In this section we introduce the conventions used in describing potentiometric electrochemical cells and the relationship between the measured potential and concentration.

Potentiometric Electrochemical Cells A schematic diagram of a typical potentiometric electrochemical cell is shown in Figure 11.5. Note that the electrochemical cell is divided into two half-cells, each containing an electrode immersed in a solution containing ions whose concentrations determine the electrode's potential. This separation of electrodes is necessary to prevent the redox reaction from occurring spontaneously on the surface of one of the electrodes, short-circuiting the electrochemical cell and making the measurement of cell potential impossible. A **salt bridge** containing an inert electrolyte, such as KCl, connects the two half-cells. The ends of the salt bridge are fixed with porous frits, allowing ions to move freely between the half-cells and the salt bridge, while preventing the contents of the salt bridge from draining into the half-cells. This movement of ions in the salt bridge completes the electric circuit.

salt bridge
A connection between two solutions that allows the movement of current in the form of ionic charge.

By convention, the electrode on the left is considered to be the **anode,** where oxidation occurs

anode
The electrode where oxidation occurs.

$$Zn(s) \rightleftharpoons Zn^{2+}(aq) + 2e^-$$

and the electrode on the right is the **cathode,** where reduction occurs

cathode
The electrode where reduction occurs.

$$Ag^+(aq) + e^- \rightleftharpoons Ag(s)$$

The electrochemical cell's potential, therefore, is for the reaction

$$Zn(s) + 2Ag^+(aq) \rightleftharpoons 2Ag(s) + Zn^{2+}(aq)$$

Also, by convention, potentiometric electrochemical cells are defined such that the indicator electrode is the cathode (right half-cell) and the reference electrode is the anode (left half-cell).

Shorthand Notation for Electrochemical Cells Although Figure 11.5 provides a useful picture of an electrochemical cell, it does not provide a convenient representation. A more useful representation is a shorthand, or schematic, notation that uses symbols to indicate the different phases present in the electrochemical cell, as well as the composition of each phase. A vertical slash (|) indicates a phase boundary where a potential develops, and a comma (,) separates species in the same phase, or two phases where no potential develops. Shorthand cell notations begin with the anode and continue to the cathode. The electrochemical cell in Figure 11.5, for example, is described in shorthand notation as

$$\mathrm{Zn}(s) \mid \mathrm{ZnCl_2}\ (aq,\ 0.0167\ \mathrm{M}) \parallel \mathrm{AgNO_3}\ (aq,\ 0.100\ \mathrm{M}) \mid \mathrm{Ag}(s)$$

The double vertical slash (||) indicates the salt bridge, the contents of which are normally not indicated. Note that the double vertical slash implies that there is a potential difference between the salt bridge and each half-cell.

Example

EXAMPLE 11.1

What are the anodic, cathodic, and overall reactions responsible for the potential in the electrochemical cell shown here? Write the shorthand notation for the electrochemical cell.

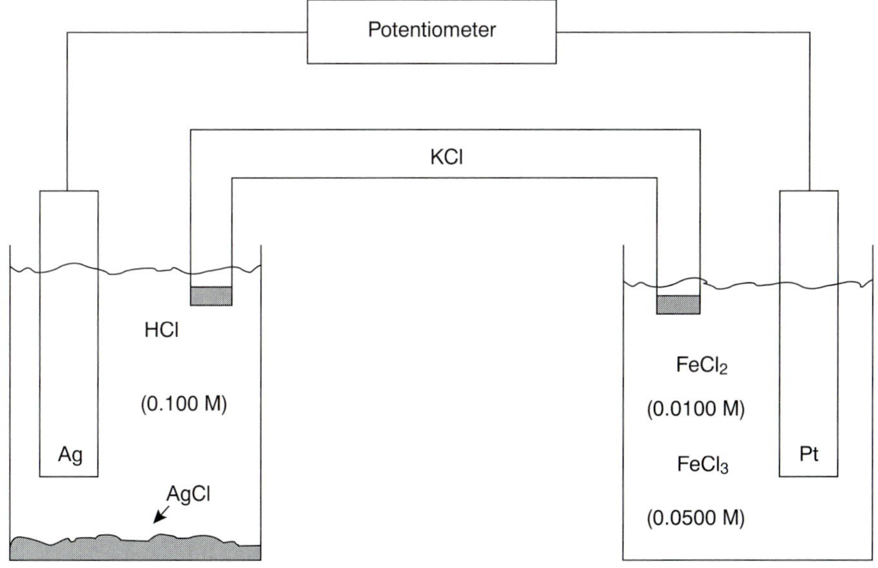

SOLUTION

The oxidation of Ag to Ag^+ occurs at the anode (the left-hand cell). Since the solution contains a source of Cl^-, the anodic reaction is

$$\mathrm{Ag}(s) + \mathrm{Cl^-}(aq) \rightleftharpoons \mathrm{AgCl}(s) + e^-$$

The cathodic reaction (the right-hand cell) is the reduction of Fe^{3+} to Fe^{2+}

$$Fe^{3+}(aq) + e^- \rightleftharpoons Fe^{2+}(aq)$$

The overall cell reaction, therefore, is

$$Ag(s) + Fe^{3+}(aq) + Cl^-(aq) \rightleftharpoons AgCl(s) + Fe^{2+}(aq)$$

The electrochemical cell's shorthand notation is

$$Ag(s) \mid HCl\ (aq, 0.100\ M), AgCl\ (sat'd) \parallel FeCl_2\ (aq, 0.0100\ M), FeCl_3\ (aq, 0.0500\ M) \mid Pt$$

Note that the Pt cathode is an inert electrode that carries electrons to the reduction half-reaction. The electrode itself does not undergo oxidation or reduction.

Potential and Concentration—The Nernst Equation The potential of a potentiometric electrochemical cell is given as

$$E_{cell} = E_c - E_a \qquad 11.1$$

where E_c and E_a are reduction potentials for the reactions occurring at the cathode and anode. These reduction potentials are a function of the concentrations of those species responsible for the electrode potentials, as given by the Nernst equation

$$E = E^\circ - \frac{RT}{nF}\ln Q$$

where E° is the standard-state reduction potential, R is the gas constant, T is the temperature in Kelvins, n is the number of electrons involved in the reduction reaction, F is Faraday's constant, and Q is the reaction quotient.* Under typical laboratory conditions (temperature of 25 °C or 298 K) the Nernst equation becomes

$$E = E^\circ - \frac{0.05916}{n}\log Q \qquad 11.2$$

where E is given in volts.

Using equation 11.2 the potential of the anode and cathode in Figure 11.5 are

$$E_a = E^\circ_{Zn^{2+}/Zn} - \frac{0.05916}{2}\log\frac{1}{[Zn^{2+}]}$$

$$E_c = E^\circ_{Ag^+/Ag} - 0.05916\log\frac{1}{[Ag^+]}$$

Note, again, that the Nernst equations for both E_c and E_a are written for reduction reactions. The cell potential, therefore, is

$$E_{cell} = \left(E^\circ_{Ag^+/Ag} - 0.05916\ \log\frac{1}{[Ag^+]}\right) - \left(E^\circ_{Zn^{2+}/Zn} - \frac{0.05916}{2}\log\frac{1}{[Zn^{2+}]}\right) \qquad 11.3$$

*See Chapter 6 for a review of the Nernst equation.

Substituting known values for the standard-state reduction potentials (see Appendix 3D) and the concentrations of Ag^+ and Zn^{2+}, gives a potential for the electrochemical cell in Figure 11.5 of

$$E_{cell} = \left(+0.7996 - 0.05916 \log \frac{1}{0.100}\right) - \left(-0.7618 - \frac{0.05916}{2} \log \frac{1}{0.0167}\right)$$

$$= +1.555 \text{ V}$$

Example

EXAMPLE 11.2

What is the potential of the electrochemical cell shown in Example 11.1?

SOLUTION

The potential for the electrochemical cell is

$$E_{cell} = \left(E^\circ_{Fe^{3+}/Fe^{2+}} - 0.05916 \log \frac{[Fe^{2+}]}{[Fe^{3+}]}\right) - (E^\circ_{AgCl/Ag} - 0.05916 \log [Cl^-])$$

$$= \left(+0.771 - 0.05916 \log \frac{0.0100}{0.0500}\right) - [+0.2223 - 0.05916 \log (0.100)]$$

$$= +0.531 \text{ V}$$

In potentiometry, the concentration of analyte in the cathodic half-cell is generally unknown, and the measured cell potential is used to determine its concentration. Thus, if the potential for the cell in Figure 11.5 is measured at +1.50 V, and the concentration of Zn^{2+} remains at 0.0167 M, then the concentration of Ag^+ is determined by making appropriate substitutions to equation 11.3

$$E_{cell} = +1.50 \text{ V}$$

$$= \left(+0.7996 - 0.05916 \log \frac{1}{[Ag^+]}\right) - \left(-0.7618 - \frac{0.05916}{2} \log \frac{1}{0.0167}\right)$$

Solving for $[Ag^+]$ gives its concentration as 0.0118 M.

Example

EXAMPLE 11.3

What is the concentration of Fe^{3+} in an electrochemical cell similar to that shown in Example 11.1 if the concentration of HCl in the left-hand cell is 1.0 M, the concentration of $FeCl_2$ in the right-hand cell is 0.0151 M and the measured potential is +0.546 V?

SOLUTION

Making appropriate substitutions into the Nernst equation for the electrochemical cell (see Example 11.2)

$$E_{cell} = +0.546 = \left(+0.771 - 0.05916 \log \frac{0.0151}{[Fe^{3+}]}\right) - [+0.2223 - 0.05916 \log(1.0)]$$

and solving for $[Fe^{3+}]$ gives its concentration as 0.0136 M.

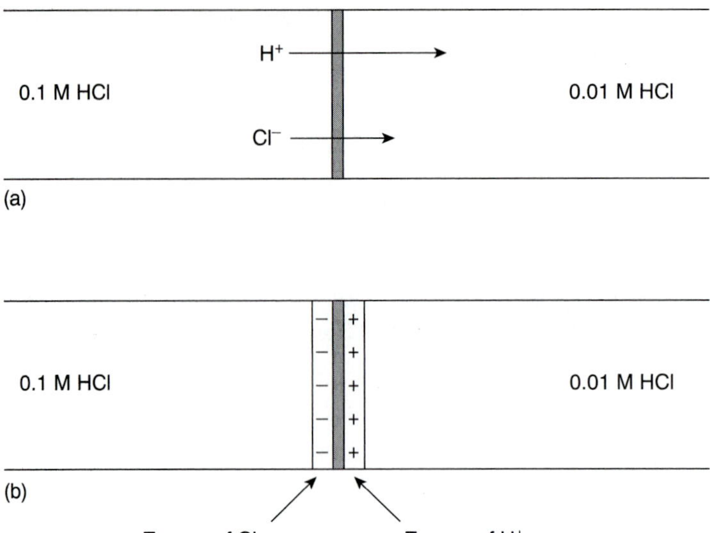

Figure 11.6
Origin of liquid junction potential between solutions of 0.1 M HCl and 0.01 M HCl.

Despite the apparent ease of determining an analyte's concentration using the Nernst equation, several problems make this approach impractical. One problem is that standard-state potentials are temperature-dependent, and most values listed in reference tables are for a temperature of 25 °C. This difficulty can be overcome by maintaining the electrochemical cell at a temperature of 25 °C or by measuring the standard-state potential at the desired temperature.

Another problem is that the Nernst equation is a function of activities, not concentrations.* As a result, cell potentials may show significant matrix effects. This problem is compounded when the analyte participates in additional equilibria. For example, the standard-state potential for the Fe^{3+}/Fe^{2+} redox couple is +0.767 V in 1 M $HClO_4$, +0.70 V in 1 M HCl, and +0.53 in 10 M HCl. The shift toward more negative potentials with an increasing concentration of HCl is due to chloride's ability to form stronger complexes with Fe^{3+} than with Fe^{2+}. This problem can be minimized by replacing the standard-state potential with a matrix-dependent formal potential. Most tables of standard-state potentials also include a list of selected formal potentials (see Appendix 3D).

A more serious problem is the presence of additional potentials in the electrochemical cell, not accounted for by equation 11.1. In writing the shorthand notation for the electrochemical cell in Figure 11.5, for example, we use a double slash (||) for the salt bridge, indicating that a potential difference exists at the interface between each end of the salt bridge and the solution in which it is immersed. The origin of this potential, which is called a liquid junction potential, and its significance are discussed in the following section.

liquid junction potential
A potential that develops at the interface between two ionic solutions that differ in composition, because of a difference in the mobilities of the ions (E_{lj}).

Liquid Junction Potentials A **liquid junction potential** develops at the interface between any two ionic solutions that differ in composition and for which the mobility of the ions differs. Consider, for example, solutions of 0.1 M HCl and 0.01 M HCl separated by a porous membrane (Figure 11.6a). Since the concentration of HCl on the left side of the membrane is greater than that on the right side of the membrane, there is a net diffusion of H^+ and Cl^- in the direction of the arrows. The mobility of H^+, however, is greater than that for Cl^-, as shown by the difference in the

* See Chapter 6 for a review of activity.

lengths of their respective arrows. As a result, the solution on the right side of the membrane develops an excess of H^+ and has a positive charge (Figure 11.6b). Simultaneously, the solution on the left side of the membrane develops a negative charge due to the greater concentration of Cl^-. The difference in potential across the membrane is called a liquid junction potential, E_{lj}.

The magnitude of the liquid junction potential is determined by the ionic composition of the solutions on the two sides of the interface and may be as large as 30–40 mV. For example, a liquid junction potential of 33.09 mV has been measured at the interface between solutions of 0.1 M HCl and 0.1 M NaCl.[2] The magnitude of a salt bridge's liquid junction potential is minimized by using a salt, such as KCl, for which the mobilities of the cation and anion are approximately equal. The magnitude of the liquid junction potential also is minimized by incorporating a high concentration of the salt in the salt bridge. For this reason salt bridges are frequently constructed using solutions that are saturated with KCl. Nevertheless, a small liquid junction potential, generally of unknown magnitude, is always present.

When the potential of an electrochemical cell is measured, the contribution of the liquid junction potential must be included. Thus, equation 11.1 is rewritten as

$$E_{cell} = E_c - E_a + E_{lj}$$

Since the junction potential is usually of unknown value, it is normally impossible to directly calculate the analyte's concentration using the Nernst equation. Quantitative analytical work is possible, however, using the standardization methods discussed in Chapter 5.

11B.2 Reference Electrodes

Potentiometric electrochemical cells are constructed such that one of the half-cells provides a known reference potential, and the potential of the other half-cell indicates the analyte's concentration. By convention, the reference electrode is taken to be the anode; thus, the shorthand notation for a potentiometric electrochemical cell is

$$\text{Reference} \parallel \text{Indicator}$$

and the cell potential is

$$E_{cell} = E_{ind} - E_{ref} + E_{lj}$$

The ideal reference electrode must provide a stable potential so that any change in E_{cell} is attributed to the indicator electrode, and, therefore, to a change in the analyte's concentration. In addition, the ideal reference electrode should be easy to make and to use. Three common reference electrodes are discussed in this section.

Standard Hydrogen Electrode The **standard hydrogen electrode** (SHE) is rarely used for routine analytical work, but is important because it is the reference electrode used to establish standard-state potentials for other half-reactions. The SHE consists of a Pt electrode immersed in a solution in which the hydrogen ion activity is 1.00 and in which H_2 gas is bubbled at a pressure of 1 atm (Figure 11.7). A conventional salt bridge connects the SHE to the indicator half-cell. The shorthand notation for the standard hydrogen electrode is

$$\text{Pt}(s), \text{H}_2\ (g, 1\text{ atm}) \mid \text{H}^+\ (aq, a = 1.00) \parallel$$

and the standard-state potential for the reaction

$$2\text{H}^+(aq) + e^- \rightleftharpoons \text{H}_2(g)$$

standard hydrogen electrode
Reference electrode based on the reduction of $H^+(aq)$ to $H_2(g)$ at a Pt electrode; that is, $H^+(aq) + e^- \rightleftharpoons \frac{1}{2}H_2(g)$.

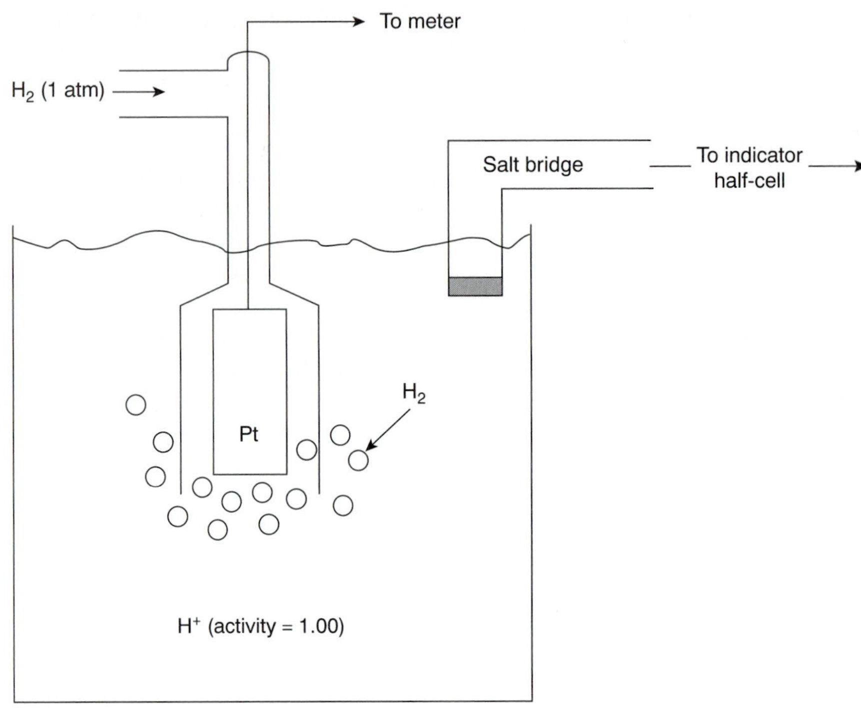

Figure 11.7
Schematic diagram of the standard hydrogen electrode (SHE).

is, by definition, 0.00 V for all temperatures. Despite its importance as the fundamental reference electrode against which all other potentials are measured, the SHE is rarely used because it is difficult to prepare and inconvenient to use.

Calomel Electrodes Calomel reference electrodes are based on the redox couple between Hg_2Cl_2 and Hg (calomel is a common name for Hg_2Cl_2).

$$Hg_2Cl_2(s) + 2e^- \rightleftharpoons 2Hg(\ell) + 2Cl^-(aq)$$

The Nernst equation for the calomel electrode is

$$E = E^\circ_{Hg_2Cl_2/Hg} - \frac{0.05916}{2}\log[Cl^-]^2 = +0.2682 - \frac{0.05916}{2}\log[Cl^-]^2$$

The potential of a calomel electrode, therefore, is determined by the concentration of Cl^-.

saturated calomel electrode
Reference electrode based on the reduction of Hg_2Cl_2 to Hg in an aqueous solution saturated with KCl; that is, $Hg_2Cl_2(s) + 2e^- \rightleftharpoons 2Hg(\ell) + 2Cl^-(aq)$.

The **saturated calomel electrode** (SCE), which is constructed using an aqueous solution saturated with KCl, has a potential at 25 °C of +0.2444 V. A typical SCE is shown in Figure 11.8 and consists of an inner tube, packed with a paste of Hg, Hg_2Cl_2, and saturated KCl, situated within a second tube filled with a saturated solution of KCl. A small hole connects the two tubes, and an asbestos fiber serves as a salt bridge to the solution in which the SCE is immersed. The stopper in the outer tube may be removed when additional saturated KCl is needed. The shorthand notation for this cell is

$$Hg(\ell) \mid Hg_2Cl_2\ (\text{sat'd}),\ KCl\ (aq,\ \text{sat'd}) \parallel$$

The SCE has the advantage that the concentration of Cl^-, and, therefore, the potential of the electrode, remains constant even if the KCl solution partially evaporates. On the other hand, a significant disadvantage of the SCE is that the solubility of KCl is sensitive to a change in temperature. At higher temperatures the concentration of Cl^- increases, and the electrode's potential decreases. For example, the potential of the SCE at 35 °C is +0.2376 V. Electrodes containing unsaturated solutions of KCl have potentials that are less temperature-dependent, but experience a change in potential if the concentration of KCl increases due to evaporation. Another disadvantage to calomel electrodes is that they cannot be used at temperatures above 80 °C.

Silver/Silver Chloride Electrodes Another common reference electrode is the **silver/silver chloride electrode,** which is based on the redox couple between AgCl and Ag.

silver/silver chloride electrode
Reference electrode based on the reduction of AgCl to Ag; that is, $AgCl(s) + e^- \rightleftharpoons Ag(s) + Cl^-(aq)$.

$$AgCl(s) + e^- \rightleftharpoons Ag(s) + Cl^-(aq)$$

As with the saturated calomel electrode, the potential of the Ag/AgCl electrode is determined by the concentration of Cl^- used in its preparation.

$$E + E^\circ_{AgCl/Ag} - 0.05916 \log [Cl^-] = +0.2223 - 0.05916 \log [Cl^-]$$

When prepared using a saturated solution of KCl, the Ag/AgCl electrode has a potential of +0.197 V at 25 °C. Another common Ag/AgCl electrode uses a solution of 3.5 M KCl and has a potential of +0.205 at 25 °C. The Ag/AgCl electrode prepared with saturated KCl, of course, is more temperature-sensitive than one prepared with an unsaturated solution of KCl.

A typical Ag/AgCl electrode is shown in Figure 11.9 and consists of a silver wire, the end of which is coated with a thin film of AgCl. The wire is immersed in a solution that contains the desired concentration of KCl and that is saturated with AgCl. A porous plug serves as the salt bridge. The shorthand notation for the cell is

$$Ag(s) \mid AgCl \text{ (sat'd)}, KCl \ (x \text{ M}) \parallel$$

where x is the concentration of KCl.

In comparison to the SCE the Ag/AgCl electrode has the advantage of being useful at higher temperatures. On the other hand, the Ag/AgCl electrode is more prone to reacting with solutions to form insoluble silver complexes that may plug the salt bridge between the electrode and the solution.

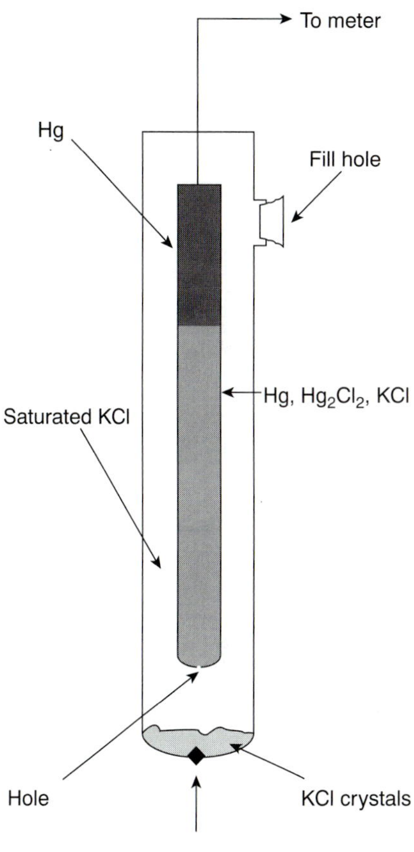

Figure 11.8
Schematic diagram of the saturated calomel electrode (SCE).

11B.3 Metallic Indicator Electrodes

The potential of the indicator electrode in a potentiometric electrochemical cell is proportional to the concentration of analyte. Two classes of indicator electrodes are used in potentiometry: metallic electrodes, which are the subject of this section, and ion-selective electrodes, which are covered in the next section.

The potential of a metallic electrode is determined by the position of a redox reaction at the electrode–solution interface. Three types of metallic electrodes are commonly used in potentiometry, each of which is considered in the following discussion.

Electrodes of the First Kind When a copper electrode is immersed in a solution containing Cu^{2+}, the potential of the electrode due to the reaction

$$Cu^{2+}(aq) + 2e^- \rightleftharpoons Cu(s)$$

is determined by the concentration of copper ion.

$$E = E^\circ_{Cu^{2+}/Cu} - \frac{0.05916}{2} \log \frac{1}{[Cu^{2+}]} = +0.3419 - \frac{0.05916}{2} \log \frac{1}{[Cu^{2+}]}$$

If the copper electrode is the indicator electrode in a potentiometric electrochemical cell that also includes a saturated calomel reference electrode

$$\text{SCE} \,||\, \text{Cu}^{2+}\,(\text{unk}) \,|\, \text{Cu}(s)$$

then the cell potential can be used to determine an unknown concentration of Cu^{2+} in the indicator half-cell

$$E_{\text{cell}} = E_{\text{ind}} - E_{\text{ref}} + E_{\text{lj}} = +0.3419 - \frac{0.05916}{2}\log\frac{1}{[\text{Cu}^{2+}]} - +0.2444 + E_{\text{lj}}$$

electrode of the first kind
A metallic electrode whose potential is a function of the concentration of M^{n+} in an M^{n+}/M redox half-reaction.

Metallic indicator electrodes in which a metal is in contact with a solution containing its ion are called **electrodes of the first kind.** In general, for a metal M, in a solution of M^{n+}, the cell potential is given as

$$E_{\text{cell}} = K - \frac{0.05916}{n}\log\frac{1}{[\text{M}^{n+}]} = K + \frac{0.05916}{n}\log[\text{M}^{n+}]$$

where K is a constant that includes the standard-state potential for the M^{n+}/M redox couple, the potential of the reference electrode, and the junction potential. For a variety of reasons, including slow kinetics for electron transfer, the existence of surface oxides and interfering reactions, electrodes of the first kind are limited to Ag, Bi, Cd, Cu, Hg, Pb, Sn, Tl, and Zn. Many of these electrodes, such as Zn, cannot be used in acidic solutions where they are easily oxidized by H^+.

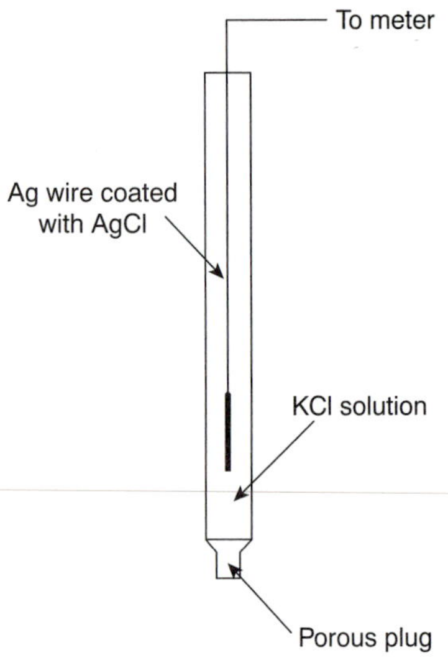

Figure 11.9
Schematic diagram of a Ag/AgCl electrode.

Electrodes of the Second Kind An electrode of the first kind involving an M^{n+}/M redox couple will respond to the concentration of another species if that species is in equilibrium with M^{n+}. For example, the potential of a silver electrode in a solution of Ag^+ is given by

$$E = E^{\circ}_{\text{Ag}^+/\text{Ag}} - 0.05916\log\frac{1}{[\text{Ag}^+]} = +0.7996 - 0.05916\log\frac{1}{[\text{Ag}^+]} \qquad \textbf{11.4}$$

If the solution is saturated with AgI, then the solubility reaction

$$\text{AgI}(s) \rightleftharpoons \text{Ag}^+(aq) + \text{I}^-(aq)$$

determines the concentration of Ag^+; thus

$$[\text{Ag}^+] = \frac{K_{\text{sp,AgI}}}{[\text{I}^-]} \qquad \textbf{11.5}$$

where $K_{sp,AgI}$ is the solubility product for AgI. Substituting equation 11.5 into 11.4

$$E = +0.7996 - 0.05916\log\frac{[\text{I}^-]}{K_{\text{sp,AgI}}}$$

shows that the potential of the silver electrode is a function of the concentration of I^-. When this electrode is incorporated into a potentiometric electrochemical cell

$$\text{REF} \,||\, \text{AgI (sat'd)}, \text{I}^-\,(\text{unk}) \,|\, \text{Ag}(s)$$

the cell potential is

$$E_{\text{cell}} = K - 0.05916\log\,[\text{I}^-]$$

where K is a constant that includes the standard-state potential for the Ag^+/Ag redox couple, the solubility product for AgI, the potential of the reference electrode, and the junction potential.

When the potential of an electrode of the first kind responds to the potential of another ion that is in equilibrium with M^{n+}, it is called an **electrode of the second kind.** Two common electrodes of the second kind are the calomel and silver/silver chloride reference electrodes. Electrodes of the second kind also can be based on complexation reactions. For example, an electrode for EDTA is constructed by coupling a Hg^{2+}/Hg electrode of the first kind to EDTA by taking advantage of its formation of a stable complex with Hg^{2+}.

electrode of the second kind
A metallic electrode whose potential is a function of the concentration of X in an MX_n/M redox half-reaction.

Redox Electrodes Electrodes of the first and second kind develop a potential as the result of a redox reaction in which the metallic electrode undergoes a change in its oxidation state. Metallic electrodes also can serve simply as a source of, or a sink for, electrons in other redox reactions. Such electrodes are called **redox electrodes.** The Pt cathode in Example 11.1 is an example of a redox electrode because its potential is determined by the concentrations of Fe^{2+} and Fe^{3+} in the indicator half-cell. Note that the potential of a redox electrode generally responds to the concentration of more than one ion, limiting their usefulness for direct potentiometry.

redox electrode
An inert electrode that serves as a source or sink for electrons for a redox half-reaction.

11B.4 Membrane Electrodes

If metallic electrodes were the only useful class of indicator electrodes, potentiometry would be of limited applicability. The discovery, in 1906, that a thin glass membrane develops a potential, called a **membrane potential,** when opposite sides of the membrane are in contact with solutions of different pH led to the eventual development of a whole new class of indicator electrodes called ion-selective electrodes (ISEs). Following the discovery of the glass pH electrode, **ion-selective electrodes** have been developed for a wide range of ions. Membrane electrodes also have been developed that respond to the concentration of molecular analytes by using a chemical reaction to generate an ion that can be monitored with an ion-selective electrode. The development of new membrane electrodes continues to be an active area of research.

membrane potential
A potential developing across a conductive membrane whose opposite sides are in contact with solutions of different composition.

ion-selective electrode
An electrode in which the membrane potential is a function of the concentration of a particular ion in solution.

Membrane Potentials Ion-selective electrodes, such as the glass pH electrode, function by using a membrane that reacts selectively with a single ion. Figure 11.10 shows a generic diagram for a potentiometric electrochemical cell equipped with an ion-selective electrode. The shorthand notation for this cell is

$$\mathrm{Ref(samp)} \parallel [\mathrm{A}]_{\mathrm{samp}} \mid [\mathrm{A}]_{\mathrm{int}} \parallel \mathrm{Ref(int)}$$

where the membrane is represented by the vertical slash (|) separating the two solutions containing analyte. Two reference electrodes are used; one positioned within the internal solution, and one in the sample solution. The cell potential, therefore, is

$$E_{\mathrm{cell}} = E_{\mathrm{Ref(int)}} - E_{\mathrm{Ref(samp)}} + E_{\mathrm{mem}} + E_{\mathrm{lj}} \qquad \mathbf{11.6}$$

where E_{mem} is the potential across the membrane. Since the liquid junction potential and reference electrode potentials are constant, any change in the cell's potential is attributed to the membrane potential.

Interaction of the analyte with the membrane results in a membrane potential if there is a difference in the analyte's concentration on opposite sides of the membrane. One side of the membrane is in contact with an internal solution containing a fixed concentration of analyte, while the other side of the membrane is in contact with the sample. Current is carried through the membrane by the movement of either the analyte or an ion already present in the membrane's matrix. The membrane potential is given by a Nernst-like equation

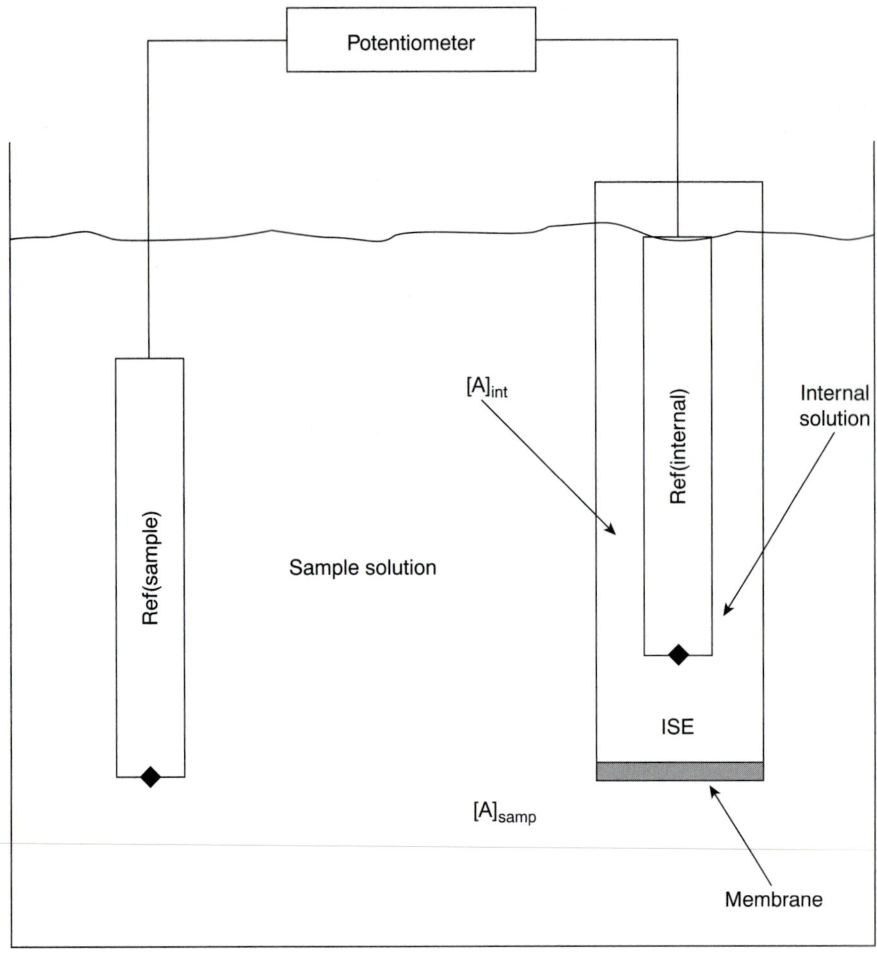

Figure 11.10
Electrochemical cell for potentiometry with an ion-selective membrane electrode.

$$E_{mem} = E_{asym} - \frac{RT}{zF}\ln\frac{[A]_{int}}{[A]_{samp}} \qquad 11.7$$

where $[A]_{samp}$ and $[A]_{int}$ are the concentrations of analyte in the sample and the internal solution, respectively, and z is the analyte's charge. Ideally, E_{mem} should be zero when the concentrations of analyte on both sides of the membrane are equal. The term E_{asym}, which is called an **asymmetry potential,** accounts for the fact that the membrane potential is usually not zero under these conditions.

asymmetry potential
The membrane potential when opposite sides of the membrane are in contact with identical solutions yet a nonzero potential is observed.

Substituting equation 11.7 into equation 11.6, assuming a temperature of 25 °C and rearranging gives

$$E_{cell} = K + \frac{0.05916}{z}\log[A]_{samp} \qquad 11.8$$

where K is a constant accounting for the potentials of the reference electrodes, any liquid junction potentials, the asymmetry potential, and the concentration of analyte in the internal solution. Equation 11.8 is a general equation, and applies to all types of ion-selective electrodes.

Selectivity of Membranes Membrane potentials result from a chemical interaction between the analyte and active sites on the membrane's surface. Because the signal depends on a chemical process, most membranes are not selective toward a single analyte. Instead, the membrane potential is proportional to the concentration of all ions in the sample solution capable of interacting at the membrane's active sites. Equation 11.8 can be generalized to include the contribution of an interferent, *I*,

$$E_{\text{cell}} = K + \frac{0.05916}{z_{\text{A}}} \log([\text{A}] + K_{\text{A,I}}[\text{I}]^{z_{\text{A}}/z_{\text{I}}})$$

where z_A and z_I are the charges of the analyte and interferent, and $K_{A,I}$ is a selectivity coefficient accounting for the relative response of the interferent.* The selectivity coefficient is defined as

$$K_{\text{A,I}} = \frac{[\text{A}]_E}{[\text{I}]_E^{z_{\text{A}}/z_{\text{I}}}}$$

where $[\text{A}]_E$ and $[\text{I}]_E$ are the concentrations of analyte and interferent yielding identical cell potentials. When the selectivity coefficient is 1.00, the membrane responds equally to the analyte and interferent. A membrane shows good selectivity for the analyte when $K_{A,I}$ is significantly less than 1.00.

Selectivity coefficients for most commercially available ion-selective electrodes are provided by the manufacturer. If the selectivity coefficient is unknown, it can be determined experimentally. The easiest method for determining $K_{A,I}$ is to prepare a series of solutions, each of which contains the same concentration of interferent, $[\text{I}]_{\text{add}}$, but a different concentration of analyte. A plot of cell potential versus the log of the analyte's concentration has two distinct linear regions (Figure 11.11). When the analyte's concentration is significantly larger than $K_{A,I}[\text{I}]_{\text{add}}$, the potential is a linear function of log [A], as given by equation 11.8. If $K_{A,I}[\text{I}]_{\text{add}}$ is significantly larger than the analyte's concentration, however, the cell potential remains constant. The concentration of analyte and interferent at the intersection of these two linear regions is used to calculate $K_{A,I}$.

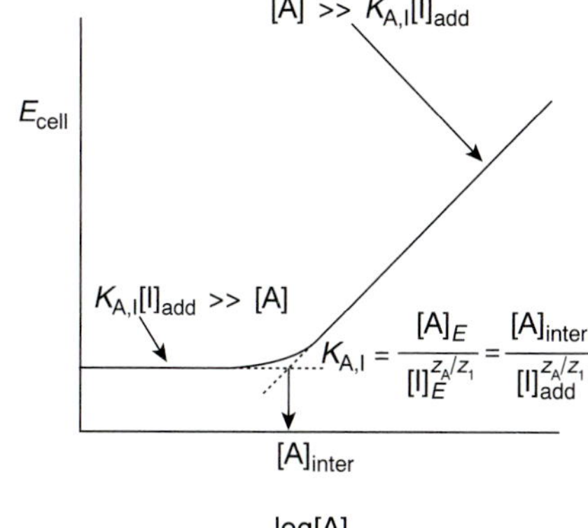

Figure 11.11
Plot of cell potential versus the log of the analyte's concentration in the presence of a fixed concentration of interferent, showing the determination of the selectivity coefficient.

glass electrode
An ion-selective electrode based on a glass membrane in which the potential develops from an ion-exchange reaction on the membrane's surface.

Glass Ion-Selective Electrodes The first commercial **glass electrodes** were manufactured using Corning 015, a glass with a composition of approximately 22% Na_2O, 6% CaO, and 72% SiO_2. When immersed in an aqueous solution, the outer approximately 10 nm of the membrane becomes hydrated over the course of several hours. Hydration of the glass membrane results in the formation of negatively charged sites, G^-, that are part of the glass membrane's silica framework. Sodium ions, which are able to move through the hydrated layer, serve as the counterions. Hydrogen ions from solution diffuse into the membrane and, since they bind more strongly to the glass than does Na^+, displace the sodium ions

$$\text{H}^+(aq) + \text{G}^-\text{–Na}^+(s) \rightleftharpoons \text{G}^-\text{–H}^+(s) + \text{Na}^+(aq)$$

giving rise to the membrane's selectivity for H^+. The transport of charge across the membrane is carried by the Na^+ ions. The potential of glass electrodes using Corning 015 obeys the equation

$$E_{\text{cell}} = K + 0.05916 \log [\text{H}^+] \qquad \textbf{11.9}$$

over a pH range of approximately 0.5–9. Above a pH of 9–10, the glass membrane may become more responsive to other cations, such as Na^+ and K^+.

*Note that this treatment of sensitivity is similar to that introduced in Chapter 3.

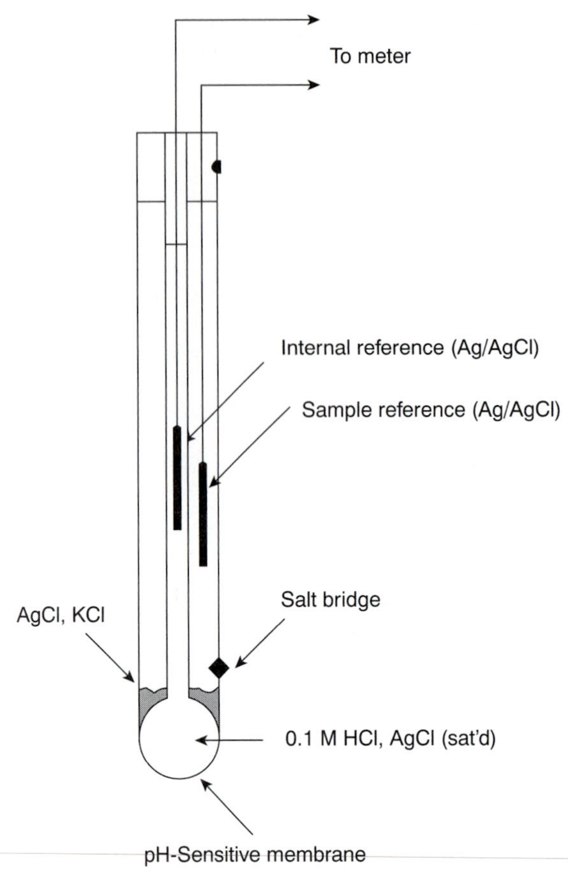

Figure 11.12
Schematic diagram of a combination glass electrode for measuring pH.

Example

EXAMPLE 11.4

The selectivity coefficient K_{H^+/Na^+} for Corning 015 is approximately 10^{-11}. What error in pH is expected for a solution of 0.05 M NaOH.

SOLUTION

A solution of 0.05 M NaOH has an actual H^+, $[H^+]_{act}$, concentration of 2×10^{-13} M and a pH of 12.7. The electrode responds, however, to both H^+ and Na^+, with the apparent concentration of H^+, $[H^+]_{app}$, given by

$$[H^+]_{app} = [H^+]_{act} + K_{H^+/Na^+}[Na^+] = 2 \times 10^{-13} + (10^{-11})(0.05)$$

$$= 7 \times 10^{-13}$$

The apparent concentration of H^+ corresponds to a pH of 12.2, giving an error of –0.5 pH units.

Replacing Na_2O and CaO with Li_2O and BaO extends the useful pH range of glass membrane electrodes to pH levels greater than 12.

Glass membrane pH electrodes are often available in a combination form that includes both the indicator and the reference electrode. The use of a single electrode greatly simplifies the measurement of pH. An example of a typical combination electrode is shown in Figure 11.12.

Table 11.1 Representative Examples of Glass Membrane Ion-Selective Electrodes

Analyte	Membrane Composition	Selectivity Coefficients[a]
Na^+	11% Na_2O, 18% Al_2O_3, 71% SiO_2	$K_{Na^+/H^+} = 1000$ $K_{Na^+/K^+} = K_{Na^+/Li^+} = 10^{-3}$
Li^+	15% Li_2O, 25% Al_2O_3, 60% SiO_2	$K_{Li^+/Na^+} = 0.3$ $K_{Li^+/K^+} = 10^{-3}$
K^+	27% Na_2O, 5% Al_2O_3, 68% SiO_2	$K_{K^+/Na^+} = 0.05$

[a]Selectivity constants are approximate, and those found experimentally may vary substantially from the listed values.[3]

The response of the Corning 015 glass membrane to monovalent cations other than H^+ at high pH led to the development of glass membranes possessing a greater selectivity for other cations. For example, a glass membrane with a composition of 11% Na_2O, 18% Al_2O_3, and 71% SiO_2 is used as a Na^+ ion-selective electrode. Other glass electrodes have been developed for the analysis of Li^+, K^+, Rb^+, Cs^+, NH_4^+, Ag^+, and Tl^+. Several representative examples of glass membrane electrodes are listed in Table 11.1.

Since the typical thickness of the glass membrane in an ion-selective electrode is about 50 μm, they must be handled carefully to prevent the formation of cracks or breakage. Before a glass electrode can be used it must be conditioned by soaking for several hours in a solution containing the analyte. Glass electrodes should not be allowed to dry out, as this destroys the membrane's hydrated layer. If a glass electrode has been allowed to dry out, it must be reconditioned before it can be used. The composition of a glass membrane changes over time, affecting the electrode's performance. The average lifetime for a glass electrode is several years.

Crystalline Solid-State Ion-Selective Electrodes **Solid-state ion-selective electrodes** use membranes fashioned from polycrystalline or single-crystal inorganic salts. Polycrystalline ion-selective electrodes are made by forming a thin pellet of Ag_2S, or a mixture of Ag_2S and either a second silver salt or another metal sulfide. The pellet, which is 1–2 mm in thickness, is sealed into the end of a nonconducting plastic cylinder, and an internal solution containing the analyte and a reference electrode are placed in the cylinder. Charge is carried across the membrane by Ag^+ ions.

solid-state ion-selective electrode
An ion-selective electrode based on a sparingly soluble inorganic crystalline material.

The membrane potential for a Ag_2S pellet develops as the result of a difference in the equilibrium position of the solubility reaction

$$Ag_2S(s) \rightleftharpoons 2Ag^+(aq) + S^{2-}(aq)$$

on the two sides of the membrane. When used to monitor the concentration of Ag^+ ions, the cell potential is

$$E_{cell} = K + 0.05916 \log [Ag^+]$$

The membrane also responds to the concentration of S^{2-}, with the cell potential given as

$$E_{cell} = K - \frac{0.05916}{2} \log [S^{2-}]$$

If a mixture of an insoluble silver salt and Ag_2S is used to make the membrane, then the membrane potential also responds to the concentration of the anion of the added silver salt. Thus, pellets made from a mixture of Ag_2S and AgCl can serve as a Cl^- ion-selective electrode, with a cell potential of

$$E_{cell} = K - 0.05916 \log [Cl^-]$$

Membranes fashioned from a mixture of Ag_2S with CdS, CuS, or PbS are used to make ion-selective electrodes that respond to the concentration of Cd^{2+}, Cu^{2+}, or Pb^{2+}. In this case the cell potential is

$$E_{cell} = K + \frac{0.05916}{2} \log [M^{2+}]$$

where $[M^{2+}]$ is the concentration of the appropriate metal ion.

Several examples of polycrystalline, Ag_2S-based ion-selective electrodes are listed in Table 11.2. The selectivity of these ion-selective electrodes is determined by solubility. Thus, a Cl^- ion-selective electrode constructed using a Ag_2S/AgCl membrane is more selective for Br^- ($K_{Cl^-/Br^-} = 10^2$) and I^- ($K_{Cl^-/I^-} = 10^6$) since AgBr and AgI are less soluble than AgCl. If the concentration of Br^- is sufficiently high, the AgCl at the membrane–solution interface is replaced by AgBr, and the electrode's response to Cl^- decreases substantially. Most of the ion-selective electrodes listed in Table 11.2 can be used over an extended range of pH levels. The equilibrium between S^{2-} and HS^- limits the analysis for S^{2-} to a pH range of 13–14. Solutions of CN^-, on the other hand, must be kept basic to avoid the release of HCN.

The membrane of a F^- ion-selective electrode is fashioned from a single crystal of LaF_3 that is usually doped with a small amount of EuF_2 to enhance the membrane's conductivity. Since EuF_2 provides only two F^- ions, compared with three for LaF_3, each EuF_2 produces a vacancy in the crystal lattice. Fluoride ions move through the membrane by moving into adjacent vacancies. The LaF_3 membrane is sealed into the end of a nonconducting plastic tube, with a standard solution of F^-, typically 0.1 M NaF, and a Ag/AgCl reference electrode.

The membrane potential for a F^- ion-selective electrode results from a difference in the solubility of LaF_3 on opposite sides of the membrane, with the potential given by

$$E_{cell} = K - 0.05916 \log [F^-]$$

One advantage of the F^- ion-selective electrode is its freedom from interference. The only significant exception is OH^- ($K_{F^-/OH^-} = 0.1$), which imposes a maximum pH limit for a successful analysis.

Example

EXAMPLE 11.5

What is the maximum pH that can be tolerated in the analysis of a 1×10^{-5} M solution of F^- if the error is to be less than 1%?

SOLUTION

To achieve an error of less than 1%, the difference between the actual, $[F^-]_{act}$, and apparent, $[F^-]_{app}$, concentrations of F^- must be less than 1%; thus

$$[F^-]_{app} - [F^-]_{act} \le 0.01[F^-]_{act}$$

$$([F^-]_{act} + K_{F^-/OH^-}[OH^-]) - [F^-]act \le 0.01[F^-]_{act}$$

$$K_{F^-/OH^-}[OH^-] \le 0.01[F^-]_{act}$$

Solving for $[OH^-]$, and making appropriate substitutions gives

$$[OH^-] \le \frac{0.01[F^-]_{act}}{K_{F^-/OH^-}} = \frac{(0.01)(1 \times 10^{-5})}{0.1} = 1 \times 10^{-6}$$

corresponding to a pH of less than 8.

Table 11.2 Representative Examples of Polycrystalline Ion-Selective Electrodes

Analyte	Membrane Composition	Selectivity Coefficients[a]
Ag^+	Ag_2S	$K_{Ag^+/Cu^{2+}} = 10^{-6}$ $K_{Ag^+/Pb^{2+}} = 10^{-10}$ Hg^{2+} interferes
Cd^{2+}	CdS/Ag_2S	$K_{Cd^{2+}/Fe^{2+}} = 200$ $K_{Cd^{2+}/Pb^{2+}} = 6$ Ag^+, Hg^{2+}, Cu^{2+} must be absent
Cu^{2+}	CuS/Ag_2S	$K_{Cu^{2+}/Fe^{3+}} = 10$ $K_{Cu^{2+}/Cu^+} = 1$ Ag^+, Hg^{2+} must be absent
Pb^{2+}	PbS/Ag_2S	$K_{Pb^{2+}/Fe^{3+}} = 1$ $K_{Pb^{2+}/Cd^{2+}} = 1$ Ag^+, Hg^{2+}, Cu^{2+} must be absent
Br^-	$AgBr/Ag_2S$	$K_{Br^-/I^-} = 5000$ $K_{Br^-/CN^-} = 100$ $K_{Br^-/Cl^-} = 5 \times 10^{-3}$ $K_{Br^-/OH^-} = 1 \times 10^{-5}$ S^{2-} must be absent
Cl^-	$AgCl/Ag_2S$	$K_{Cl^-/I^-} = 1 \times 10^6$ $K_{Cl^-/CN^-} = 1 \times 10^4$ $K_{Cl^-/Br^-} = 100$ $K_{Cl^-/OH^-} = 0.01$ S^{2-} must be absent
CN^-	AgI/Ag_2S	$K_{CN^-/I^-} = 100$ $K_{CN^-/Br^-} = 1 \times 10^{-4}$ $K_{CN^-/Cl^-} = 1 \times 10^{-6}$ $K_{CN^-/OH^-} = 1 \times 10^{-8}$ S^{2-} must be absent
I^-	AgI/Ag_2S	$K_{I^-/S^{2-}} = 30$ $K_{I^-/CN^-} = 0.01$ $K_{I^-/Br^-} = 1 \times 10^{-4}$ $K_{I^-/Cl^-} = 1 \times 10^{-6}$ $K_{I^-/OH^-} = 1 \times 10^{-7}$
SCN^-	$AgSCN/Ag_2S$	$K_{SCN^-/I^-} = 1000$ $K_{SCN^-/Br^-} = 100$ $K_{SCN^-/CN^-} = 100$ $K_{SCN^-/Cl^-} = 0.1$ $K_{SCN^-/OH^-} = 0.01$ S^{2-} must be absent
S^{2-}	Ag_2S	Hg^{2+} interferes

[a]Selectivity constants are approximate, and those found experimentally may vary substantially from the listed values.[3]

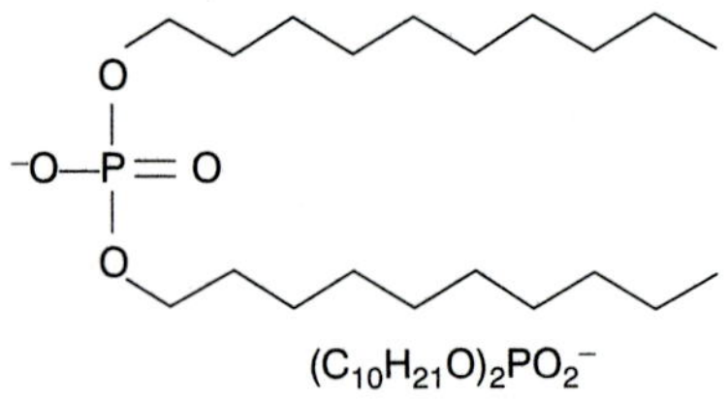

$(C_{10}H_{21}O)_2PO_2^-$

Figure 11.13
Structure and formula of di-(*n*-decyl) phosphate.

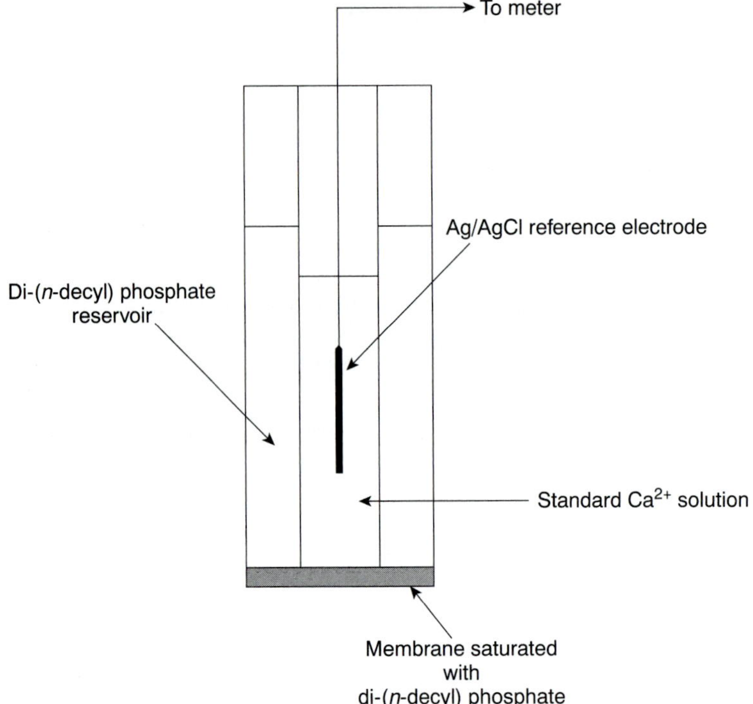

Figure 11.14
Schematic diagram of a Ca^{2+} liquid-based ion-selective electrode.

Below a pH of 4 the predominate form of fluoride in solution is HF, which, unlike F^-, does not contribute to the membrane potential. For this reason, an analysis for total fluoride must be carried out at a pH greater than 4.

Unlike ion-selective electrodes using glass membranes, crystalline solid-state ion-selective electrodes do not need to be conditioned before use and may be stored dry. The surface of the electrode is subject to poisoning, as described earlier for a Cl^- ISE in contact with an excessive concentration of Br^-. When this happens, the electrode can be returned to its original condition by sanding and polishing the crystalline membrane.

Liquid-Based Ion-Selective Electrodes Another approach to constructing an ion-selective electrode is to use a hydrophobic membrane containing a selective, liquid organic complexing agent. Three types of organic liquids have been used: cation exchangers, anion exchangers, and neutral **ionophores.** When the analyte's concentration on the two sides of the membrane is different, a membrane potential is the result. Current is carried through the membrane by the analyte.

One example of a **liquid-based ion-selective electrode** is that for Ca^{2+}, which uses a porous plastic membrane saturated with di-(*n*-decyl) phosphate (Figure 11.13). As shown in Figure 11.14, the membrane is placed at the end of a nonconducting cylindrical tube and is in contact with two reservoirs. The outer reservoir contains di-(*n*-decyl) phosphate in di-*n*-octylphenylphosphonate, which soaks into the porous membrane. The inner reservoir contains a standard aqueous solution of Ca^{2+} and a Ag/AgCl reference electrode. Calcium ion-selective electrodes are also available in which the di-(*n*-decyl) phosphate is immobilized in a polyvinyl chloride

ionophore
A neutral ligand whose exterior is hydrophobic and whose interior is hydrophilic.

liquid-based ion-selective electrode
An ion-selective electrode in which a chelating agent is incorporated into a hydrophobic membrane.

Table 11.3 Representative Examples of Liquid-Based Ion-Selective Electrodes

Analyte	Membrane Composition	Selectivity Coefficients[a]
Ca^{2+}	di-(*n*-decyl) phosphate in PVC	$K_{Ca^{2+}/Zn^{2+}} = 1–5$ $K_{Ca^{2+}/Al^{3+}} = 0.90$ $K_{Ca^{2+}/Mn^{2+}} = 0.38$ $K_{Ca^{2+}/Cu^{2+}} = 0.070$ $K_{Ca^{2+}/Mg^{2+}} = 0.032$
K^+	Valinomycin in PVC	$K_{K^+/Rb^+} = 1.9$ $K_{K^+/Cs^+} = 0.38$ $K_{K^+/Li^+} = 1 \times 10^{-4}$ $K_{K^+/Na^+} = 1 \times 10^{-5}$
Li^+	ETH 149 in PVC	$K_{Li^+/H^+} = 1$ $K_{Li^+/Na^+} = 0.05$ $K_{Li^+/K^+} = 7 \times 10^{-3}$
NH_4^+	Nonactin and monactin in PVC	$K_{NH_4^+/K^+} = 0.12$ $K_{NH_4^+/H^+} = 0.016$ $K_{NH_4^+/Li^+} = 4.2 \times 10^{-3}$ $K_{NH_4^+/Na^+} = 2 \times 10^{-3}$
ClO_4^-	$Fe(o\text{-phen})_3^{3+}$ in *p*-nitrocymene with porous membrane	$K_{ClO_4^-/OH^-} = 1$ $K_{ClO_4^-/I^-} = 0.012$ $K_{ClO_4^-/NO_3^-} = 1.5 \times 10^{-3}$ $K_{ClO_4^-/Br^-} = 5.6 \times 10^{-4}$ $K_{ClO_4^-/Cl^-} = 2.2 \times 10^{-4}$
NO_3^-	tetradodecyl ammonium nitrate in PVC	$K_{NO_3^-/Cl^-} = 6 \times 10^{-3}$ $K_{NO_3^-/F^-} = 9 \times 10^{-4}$

[a]Selectivity constants are approximate, and those found experimentally may vary substantially from the listed values.[3]

(PVC) membrane, eliminating the need for a reservoir containing di-(*n*-decyl) phosphate.

A membrane potential develops as the result of a difference in the equilibrium position of the complexation reaction

$$Ca^{2+}(aq) + 2(C_{10}H_{21}O)_2PO_2^-\,(m) \rightleftharpoons Ca[(C_{10}H_{21}O)_2PO_2]_2(m)$$

on the two sides of the membrane, where (m) indicates that the species is present in the membrane. The cell potential for the Ca^{2+} ion-selective electrode is

$$E_{cell} = K + \frac{0.05916}{2}\log[Ca^{2+}]$$

The selectivity of the electrode for Ca^{2+} is very good, with only Zn^{2+} showing greater selectivity.

The properties of several representative liquid-based ion-selective electrodes are presented in Table 11.3. An electrode using a liquid reservoir can be stored in a dilute solution of analyte and needs no additional conditioning before use. The lifetime of an electrode with a PVC membrane, however, is proportional to its exposure to aqueous solutions. For this reason these electrodes are best stored by covering the membrane with a cap containing a small amount of wetted gauze to

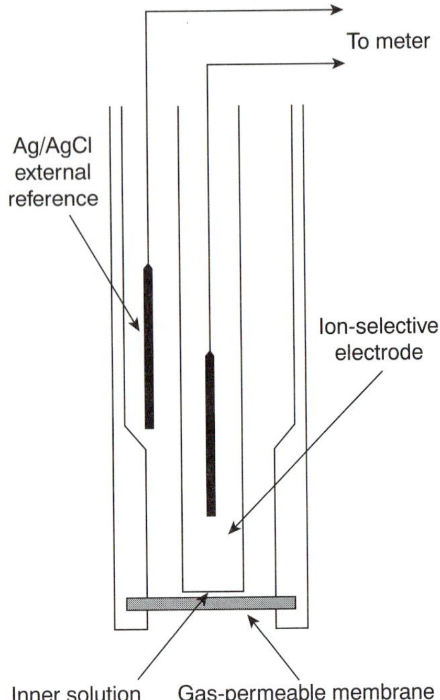

Figure 11.15
Schematic diagram of a gas-sensing membrane electrode.

maintain a humid environment. The electrode must then be conditioned before use by soaking in a solution of analyte for 30–60 min.

Gas-Sensing Electrodes A number of membrane electrodes have been developed that respond to the concentration of dissolved gases. The basic design of these electrodes is shown in Figure 11.15 and consists of a thin membrane separating the sample from an inner solution containing an ion-selective electrode. The membrane is permeable to the gaseous analyte, but is not permeable to nonvolatile components in the sample matrix. Once the gaseous analyte passes through the membrane, it reacts in the inner solution, producing a species whose concentration can be monitored by an appropriate ion-selective electrode. For example, in the CO_2 electrode, CO_2 reacts in the inner solution to produce H_3O^+.

$$CO_2(aq) + 2H_2O(\ell) \rightleftharpoons HCO_3^-(aq) + H_3O^+(aq) \qquad \textbf{11.10}$$

The change in the concentration of H_3O^+ is monitored with a pH ion-selective electrode, for which the cell potential is given by equation 11.9. The relationship between the concentration of H_3O^+ and CO_2 is given by rearranging the equilibrium constant expression for reaction 11.10; thus

$$[H_3O^+] = K\frac{[CO_2]}{[HCO_3^-]} \qquad \textbf{11.11}$$

where K is the equilibrium constant. If the amount of HCO_3^- in the internal solution is sufficiently large, then its concentration is unaffected by the presence of CO_2 and remains constant. Substituting equation 11.11 into equation 11.9 gives

$$E_{cell} = K' + 0.05916 \log [CO_2]$$

where K' is a constant that includes the constant for the pH ion-selective electrode, the equilibrium constant for reaction 11.10, and the concentration of HCO_3^-.

Gas-sensing electrodes have been developed for a variety of gases, the characteristics for which are listed in Table 11.4. The composition of the inner solution changes with use, and both it and the membrane must be replaced periodically. Gas-sensing electrodes are stored in a solution similar to the internal solution to minimize their exposure to atmospheric gases.

Potentiometric Biosensors Potentiometric electrodes for the analysis of molecules of biochemical importance can be constructed in a fashion similar to that used for gas-sensing electrodes. The most common class of potentiometric biosensors are the so-called **enzyme electrodes,** in which an enzyme is trapped or immobilized at the surface of an ion-selective electrode. Reaction of the analyte with the enzyme produces a product whose concentration is monitored by the ion-selective electrode. Potentiometric biosensors have also been designed around other biologically active species, including antibodies, bacterial particles, tissue, and hormone receptors.

enzyme electrodes
An electrode that responds to the concentration of a substrate by reacting the substrate with an immobilized enzyme, producing an ion that can be monitored with an ion-selective electrode.

One example of an enzyme electrode is the urea electrode, which is based on the catalytic hydrolysis of urea by urease

$$CO(NH_2)_2(aq) + 2H_2O(\ell) \rightleftharpoons 2NH_4^+(aq) + CO_3^{2-}(aq)$$

In one version of the urea electrode, shown in Figure 11.16, an NH_3 electrode is modified by adding a dialysis membrane that physically traps a pH 7.0 buffered solution of urease between the dialysis membrane and the gas-permeable

Table 11.4 Characteristics of Gas-Sensing Membrane Electrodes

Analyte	Reaction in Inner Solution	Inner Solution	Ion-Selective Electrode
CO_2	$CO_2 + 2H_2O \rightleftharpoons HCO_3^- + H_3O^+$	0.01 M $NaHCO_3$, 0.01 M NaCl	glass pH electrode
HCN	$HCN + H_2O \rightleftharpoons CN^- + H_3O^+$	0.01 M $KAg(CN)_2$	Ag_2S membrane electrode
HF	$HF + H_2O \rightleftharpoons F^- + H_3O^+$	1 M H_3O^+	F^- electrode
H_2S	$H_2S + H_2O \rightleftharpoons HS^- + H_3O^+$	pH 5 citrate buffer	Ag_2S membrane electrode
NH_3	$NH_3 + H_2O \rightleftharpoons NH_4^+ + OH^-$	0.01 M NH_4Cl, 0.1 M KNO_3	glass pH electrode
NO_2	$2NO_2 + 3H_2O \rightleftharpoons NO_3^- + NO_2^- + 2H_3O^+$	0.02 M $NaNO_2$, 0.1 M KNO_3	glass pH electrode
SO_2	$SO_2 + 2H_2O \rightleftharpoons HSO_3^- + H_3O^+$	0.001 M $NaHSO_3$, pH 5	glass pH electrode

Source: Data compiled from Cammann, K. *Working with Ion-Selective Electrodes.* Springer-Verlag: Berlin, 1977.

membrane.[4] When immersed in the sample, urea diffuses through the dialysis membrane, where it reacts with the enzyme urease. The NH_4^+ that is produced is in equilibrium with NH_3

$$NH_4^+(aq) + H_2O(\ell) \rightleftharpoons H_3O^+(aq) + NH_3(aq)$$

which, in turn, diffuses through the gas-permeable membrane, where it is detected by a pH electrode. The response of the electrode to the concentration of urea is given by

$$E_{cell} = K - 0.05916 \log [\text{urea}] \qquad \textbf{11.12}$$

Another version of the urea electrode (Figure 11.17) immobilizes the enzyme in a polymer membrane formed directly on the tip of a glass pH electrode.[5] In this case, the electrode's response is

$$\text{pH} = K[\text{urea}] \qquad \textbf{11.13}$$

Few potentiometric biosensors are commercially available. As shown in Figures 11.16 and 11.17, however, available ion-selective and gas-sensing electrodes may be easily converted into biosensors. Several representative examples are described in Table 11.5, and additional examples can be found in several reviews listed in the suggested readings at the end of the chapter.

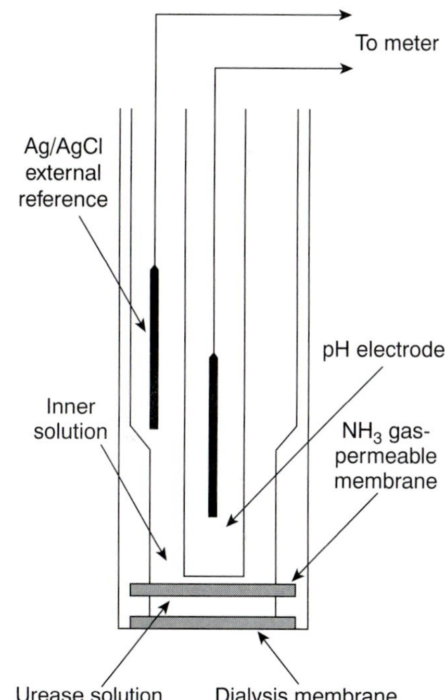

Figure 11.16
Schematic diagram of an enzyme-based potentiometric biosensor for urea in which urease is trapped between two membranes.

11B.5 Quantitative Applications

The potentiometric determination of an analyte's concentration is one of the most common quantitative analytical techniques. Perhaps the most frequently employed, routine quantitative measurement is the potentiometric determination of a solution's pH, a technique considered in more detail in the following discussion. Other areas in which potentiometric applications are important include clinical chemistry, environmental chemistry, and potentiometric titrations. Before considering these applications, however, we must first examine more closely the relationship between cell potential and the analyte's concentration, as well as methods for standardizing potentiometric measurements.

Activity Versus Concentration In describing metallic and membrane indicator electrodes, the Nernst equation relates the measured cell potential to the concentration of analyte. In writing the Nernst equation, we often ignore an important detail—the

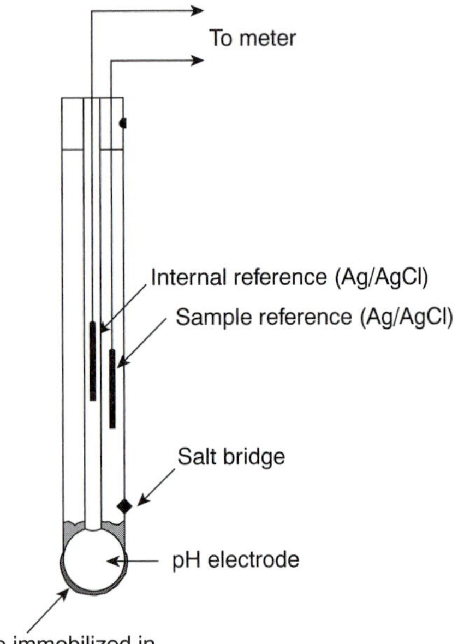

Figure 11.17
Schematic diagrams of a second enzyme-based potentiometric biosensor for urea in which urease is immobilized in a polymer matrix.

potential of an electrochemical cell is a function of activity, not concentration. Thus, the Nernst equation for a metallic electrode of the first kind is more appropriately written as

$$E_{\text{cell}} = K - \frac{0.05916}{n}\log\frac{1}{a_{\text{M}^{n+}}} \qquad \textbf{11.14}$$

where $a_{\text{M}^{n+}}$ is the activity of the metal ion. As described in Chapter 6, the activity of an ion is equal to the product of its concentration, $[\text{M}^{n+}]$, and a matrix-dependent activity coefficient, $\gamma_{\text{M}^{n+}}$.

$$a_{\text{M}^{n+}} = [\text{M}^{n+}]\gamma_{\text{M}^{n+}} \qquad \textbf{11.15}$$

Substituting equation 11.15 into equation 11.14 and rearranging gives

$$E_{\text{cell}} = K - \frac{0.05916}{n}\log\frac{1}{\gamma_{\text{M}^{n+}}} - \frac{0.05916}{n}\log\frac{1}{[\text{M}^{n+}]} \qquad \textbf{11.16}$$

Equation 11.16 can be solved for the metal ion's concentration if its activity coefficient is known. This presents a serious complication since the activity coefficient may be difficult to determine. If, however, the standards and samples have an identical matrix, then $\gamma_{\text{M}^{n+}}$ remains constant, and equation 11.16 simplifies to

$$E_{\text{cell}} = K' - \frac{0.05916}{n}\log\frac{1}{[\text{M}^{n+}]}$$

where K' includes the activity coefficient.

Quantitative Analysis Using External Standards To determine the concentration of analyte in a sample, it is necessary to standardize the electrode. If the electrode's response obeys the Nernst equation,

Table 11.5 Representative Examples of Potentiometric Biosensors

Analyte	Biologically Active Phase[a]	Substance Determined
5′-adenosinemonophosphate (5′-AMP)	AMP-deaminase (E)	NH_3
L-arginine	arginase + urease (E)	NH_3
asparagine	asparaginase (E)	NH_4^+
L-cysteine	*Proteus morganii* (B)	H_2S
L-glutamate	yellow squash (T)	CO_2
L-glutamine	*Sarcina flava* (B)	NH_3
oxalate	oxalate decarboxylase (E)	CO_2
penicillin	penicillinase (E)	H_3O^+
L-phenylalanine	L-amino acid oxidase and horseradish peroxidase (E)	I^-
sugars	bacteria from human dental plaque (B)	H_3O^+
urea	urease (E)	NH_3 or H_3O^+

Source: Compiled from Cammann, K. *Working with Ion-Selective Electrodes.* Springer-Verlag: Berlin, 1977; and Lunte, C. E.; Heineman, W. R. "Electrochemical Techniques in Bioanalysis." In Steckham, E., ed. *Topics in Current Chemistry,* Vol. 143, Springer-Verlag: Berlin, 1988, p. 8.[3,6]

[a]*Abbreviations:* E = enzyme; B = bacterial particle; T = tissue.

then only the constant K need be determined, and standardizing with a single external standard is possible. Since small deviations from the ideal "Nerstian" slope of $\pm RT/nF$ or $\pm RT/zF$ are frequently observed, standardization is usually accomplished using two or more external standards.

In most quantitative analyses we are interested in determining the concentration, not the activity, of the analyte. As noted earlier, however, the electrode's response is a function of the analyte's activity. In the absence of interferents, a calibration curve of potential versus activity is a straight line. A plot of potential versus concentration, however, may be curved at higher concentrations of analyte due to changes in the analyte's activity coefficient. A curved calibration curve may still be used to determine the analyte's concentration if the standard's matrix matches that of the sample. When the exact composition of the sample matrix is unknown, which often is the case, matrix matching becomes impossible.

Another approach to matrix matching, which does not rely on knowing the exact composition of the sample's matrix, is to add a high concentration of inert electrolyte to all samples and standards. If the concentration of added electrolyte is sufficient, any difference between the sample's matrix and that of the standards becomes trivial, and the activity coefficient remains essentially constant. The solution of inert electrolyte added to the sample and standards is called a **total ionic strength adjustment buffer** (TISAB).

total ionic strength adjustment buffer
A solution containing a relatively high concentration of inert electrolytes such that its composition fixes the ionic concentration of all solutions to which it is added.

Example

EXAMPLE 11.6

The concentration of Ca^{2+} in a water sample was determined by the method of external standards. The ionic strength of the samples and standards was maintained at a nearly constant level by making each solution 0.5 M in KNO_3. The measured cell potentials for the external standards are shown in the following table.

$[Ca^{2+}]$ (M)	E_{cell} (V)
1.00×10^{-5}	–0.125
5.00×10^{-5}	–0.103
1.00×10^{-4}	–0.093
5.00×10^{-4}	–0.072
1.00×10^{-3}	–0.065
5.00×10^{-3}	–0.043
1.00×10^{-2}	–0.033

What is the concentration of Ca^{2+} in a water sample if its cell potential is found to be –0.084 V?

SOLUTION

Linear regression gives the equation for the calibration curve as

$$E_{cell} = 0.027 + 0.0303 \log [Ca^{2+}]$$

Substituting the cell potential for the sample gives the concentration of Ca^{2+} as 2.17×10^{-4} M. Note that the slope of the calibration curve is slightly different from the ideal value of $0.05916/2 = 0.02958$.

Quantitative Analysis Using the Method of Standard Additions Because of the difficulty of maintaining a constant matrix for samples and standards, many quantitative potentiometric methods use the method of standard additions. A sample of volume, V_X, and analyte concentration, C_X, is transferred to a sample cell, and the potential, $(E_{cell})_X$, measured. A standard addition is made by adding a small volume, V_S, of a standard containing a known concentration of analyte, C_S, to the sample, and the potential, $(E_{cell})_S$, measured. Provided that V_S is significantly smaller than V_X, the change in sample matrix is ignored, and the analyte's activity coefficient remains constant. Example 11.7 shows how a one-point standard addition can be used to determine the concentration of an analyte.

Example

EXAMPLE 11.7

The concentration of Ca^{2+} in a sample of sea water is determined using a Ca ion-selective electrode and a one-point standard addition. A 10.00-mL sample is transferred to a 100-mL volumetric flask and diluted to volume. A 50.00-mL aliquot of sample is placed in a beaker with the Ca ion-selective electrode and a reference electrode, and the potential is measured as –0.05290 V. A 1.00-mL aliquot of a 5.00×10^{-2} M standard solution of Ca^{2+} is added, and a potential of –0.04417 V is measured. What is the concentration of Ca^{2+} in the sample of sea water?

SOLUTION

To begin, we write Nernst equations for the two measured cell potentials. The cell potential for the sample is

$$(E_{cell})_X = K + \frac{0.05916}{2}\log C_X$$

and that following the standard addition is

$$(E_{cell})_S = K + \frac{0.05916}{2}\log\left(\frac{V_X}{V_T}C_X + \frac{V_S}{V_T}C_S\right)$$

where V_T is the total volume $(V_S + V_X)$ after the standard addition. Subtracting the first equation from the second equation gives

$$(E_{cell})_S - (E_{cell})_X = \frac{0.05916}{2}\log\left(\frac{V_X}{V_T}C_X + \frac{V_S}{V_T}C_S\right) - \frac{0.05916}{2}\log C_X$$

Replacing $(E_{cell})_S - (E_{cell})_X$ with ΔE and rearranging yields

$$\frac{2\Delta E}{0.05916} = \log\left[\frac{(V_X/V_T)C_X + (V_S/V_T)C_S}{C_X}\right] = \log\left(\frac{V_X}{V_T} + \frac{V_S C_S}{V_T C_X}\right)$$

Substituting known values for ΔE, V_X, V_S, V_T, and C_S,

$$\frac{2(0.00873)}{0.05916} = \log\left[\frac{50.00\text{ mL}}{51.00\text{ mL}} + \frac{(1.00\text{ mL})(5.00 \times 10^{-2}\text{ M})}{(51.00\text{ mL})C_X}\right]$$

$$0.2951 = \log\left(0.9804 + \frac{9.804 \times 10^{-4}}{C_X}\right)$$

and taking the inverse log of both sides gives

$$1.973 = 0.9804 + \frac{9.804 \times 10^{-4}}{C_X}$$

Finally, solving for C_X gives the concentration of Ca^{2+} as 9.88×10^{-4} M. Since the original sample of sea water was diluted by a factor of 10, the concentration of Ca^{2+} in the sea water sample is 9.88×10^{-3} M.

Free Ions Versus Complexed Ions In discussing the F^- ion-selective electrode, we noted that the membrane potential is influenced by the concentration of F^-, but not the concentration of HF. An analysis for fluoride, therefore, is pH-dependent. Below a pH of approximately 4, fluoride is present predominantly as HF, and a quantitative analysis for total fluoride is impossible. If the pH is increased to greater than 4, however, the equilibrium

$$HF(aq) + H_2O(\ell) \rightleftharpoons H_3O^+(aq) + F^-(aq)$$

shifts to the right, and a quantitative analysis for total fluoride is possible.

Most potentiometric electrodes are selective for only the free, uncomplexed analyte and do not respond to complexed forms of the analyte. Solution conditions, therefore, must be carefully controlled if the purpose of the analysis is to determine the analyte's total concentration. On the other hand, this selectivity provides a significant advantage over other quantitative methods of analysis when it is necessary to determine the concentration of free ions. For example, calcium is present in urine both as free Ca^{2+} ions and as protein-bound Ca^{2+} ions. If a urine sample is analyzed by atomic absorption spectroscopy, the signal is proportional to the total concentration of Ca^{2+}, since both free and bound calcium are atomized. Analysis with a Ca^{2+} ISE, however, gives a signal that is a function of only free Ca^{2+} ions since the protein-bound ions cannot interact with the electrode's membrane.

Representative Method Ion-selective electrodes find application in numerous quantitative analyses, each of which has its own unique considerations. The following procedure for the analysis of fluoride in toothpaste provides an instructive example.

Representative Methods

Method 11.1 Determination of Fluoride in Toothpaste[7]

Description of the Method. The concentration of fluoride in toothpastes containing soluble F^- may be determined with a F^- ion-selective electrode, using a calibration curve prepared with external standards. Although the F^- ISE is very selective (only OH^- with K_{F^-/OH^-} of 0.1 is a significant interferent), Fe^{3+} and Al^{3+} interfere with the analysis by forming soluble fluoride complexes that do not interact with the ion-selective electrode's membrane. This interference is minimized by reacting any Fe^{3+} and Al^{3+} with a suitable complexing agent.

Procedure. Prepare 1 L of a standard solution of 1.00% w/v SnF_2, and transfer to a plastic bottle for storage. Using this solution, prepare 100 mL each of standards containing 0.32%, 0.36%, 0.40%, 0.44%, and 0.48% w/v SnF_2, adding

—Continued

Continued from page 489

400 mg of malic acid to each solution as a stabilizer. Transfer the standards to plastic bottles for storage. Prepare a total ionic strength adjustment buffer (TISAB) by mixing 500 mL of water, 57 mL of glacial acetic acid, 58 g of NaCl, and 4 g of the disodium salt of DCTA (*trans*-1,2-cyclohexanetetraacetic acid) in a 1-L beaker, stirring until dissolved. Cool the beaker in a water bath, and add 5 M NaOH until the pH is between 5 and 5.5. Transfer the contents of the beaker to a 1-L volumetric flask, and dilute to volume. Standards are prepared by placing approximately 1 g of a fluoride-free toothpaste, 30 mL of distilled water, and 1.00 mL of the standard into a 50-mL plastic beaker and stirring vigorously for 2 min with a stir bar. The resulting suspension is quantitatively transferred to a 100-mL volumetric flask along with 50 mL of TISAB and diluted to volume with distilled water. The entire standard solution is then transferred to a 250-mL plastic beaker until its potential is measured. Samples of toothpaste are prepared for analysis by using approximately 1-g portions and treating in the same manner as the standards. The cell potential for the standards and samples are measured using a F^- ion-selective electrode and an appropriate reference electrode. The solution is stirred during the measurement, and 2–3 min is allowed for equilibrium to be reached. The concentration of F^- in the toothpaste is reported as %w/w SnF_2.

Questions

1. **The total ionic strength adjustment buffer serves several purposes in this procedure. Identify these purposes.**

   The composition of the TISAB accomplishes three things: (1) The high concentration of NaCl (approximately 1 M) ensures that the ionic strength of the standards and samples are essentially identical. Since the activity coefficient for fluoride will be the same in all solutions, the Nernst equation can be written using the concentration of F^- in place of its activity; (2) Glacial acetic acid and NaOH are used to prepare an acetic acid/acetate buffer of pH 5–5.5. The pH of this buffer is high enough to ensure that the predominant form of fluoride is F^- instead of HF; and (3) DCTA is added as a complexing agent for any Fe^{3+} or Al^{3+} that might be present, preventing the formation of FeF_6^{3-} or AlF_6^{3-}.

2. **Why is a fluoride-free toothpaste added to the standard solutions?**

   Fluoride-free toothpaste is added as a precaution against any matrix effects that might influence the ion-selective electrode's response. This assumes, of course, that the matrices of the two toothpastes are otherwise similar.

3. **The procedure specifies that the standard and sample solutions should be stored in plastic containers. Why is it not a good idea to store the solutions in glass containers?**

   The fluoride ion is capable of reacting with glass to form SiF_4.

4. **The slope of the calibration curve is found to be –57.98 mV per tenfold change in the concentration of F^-, compared with the expected slope of –59.16 mV per tenfold change in concentration. What effect does this have on the quantitative analysis for %w/w SnF_2 in the toothpaste samples?**

   No effect at all—this is the reason for preparing a calibration curve with multiple standards.

Measurement of pH With the availability of inexpensive glass pH electrodes and pH meters, the determination of pH has become one of the most frequent quantitative analytical measurements. The potentiometric determination of pH, however, is not without complications, several of which are discussed in this section.

One complication is the meaning of pH.[8,9] The conventional definition of pH as presented in most introductory texts is

$$\text{pH} = -\log [\text{H}^+] \tag{11.17}$$

The pH of a solution, however, is defined by the response of an electrode to the H^+ ion and, therefore, is a measure of its activity.

$$\text{pH} = -\log(a_{\text{H}^+}) \tag{11.18}$$

Calculating the pH of a solution using equation 11.17 only approximates the true pH. Thus, a solution of 0.1 M HCl has a calculated pH of 1.00 using equation 11.17, but an actual pH of 1.1 as defined by equation 11.18.[8] The difference between the two values occurs because the activity coefficient for H^+ is not unity in a matrix of 0.1 M HCl. Obviously the true pH of a solution is affected by the composition of its matrix. As an extreme example, the pH of 0.01 M HCl in 5 *m* LiCl is 0.8, a value that is more acidic than that of 0.1 M HCl![8]

A second complication in measuring pH results from uncertainties in the relationship between potential and activity. For a glass membrane electrode, the cell potential, E_X, for a solution of unknown pH is given as

$$E_X = K - \frac{RT}{F}\ln\frac{1}{a_{\text{H}^+}} = K - \frac{2.303RT}{F}\text{pH}_X \tag{11.19}$$

where K includes the potential of the reference electrode, the asymmetry potential of the glass membrane and any liquid junction potentials in the electrochemical cell. All the contributions to K are subject to uncertainty and may change from day to day, as well as between electrodes. For this reason a pH electrode must be calibrated using a standard buffer of known pH. The cell potential for the standard, E_S, is

$$E_S = K - \frac{2.303RT}{F}\text{pH}_S \tag{11.20}$$

where pH_S is the pH of the standard. Subtracting equation 11.20 from equation 11.19 and solving for pH gives

$$\text{pH}_X = \text{pH}_S - \frac{(E_X - E_S)F}{2.303RT} \tag{11.21}$$

which is the operational definition of pH adopted by the International Union of Pure and Applied Chemistry.*

*Equations 11.19–11.21 are defined for a potentiometric electrochemical cell in which the pH electrode is the cathode. In this case an increase in pH decreases the cell potential. Many pH meters are designed with the pH electrode as the anode so that an increase in pH increases the cell potential. The operational definition of pH then becomes

$$\text{pH}_X = \text{pH}_S - \frac{(E_X - E_S)F}{2.303RT}$$

This difference, however, does not affect the operation of a pH meter.

Table 11.6 pH Values for Selected NIST Primary Standard Buffers[a]

Temperature (°C)	Saturated (25 °C) $KHC_4H_4O_6$ (tartrate)	0.05 *m* $KH_2C_6H_5O_7$ (citrate)	0.05 *m* $KHC_8H_4O_4$ (phthalate)	0.025 *m* KH_2PO_4, 0.025 *m* Na_2HPO_4	0.008695 *m* KH_2PO_4, 0.03043 *m* Na_2HPO_4	0.01 *m* $Na_4B_4O_7$ (borax)	0.025 *m* $NaHCO_3$, 0.025 *m* Na_2CO_3
0	—	3.863	4.003	6.984	7.534	9.464	10.317
5	—	3.840	3.999	6.951	7.500	9.395	10.245
10	—	3.820	3.998	6.923	7.472	9.332	10.179
15	—	3.802	3.999	6.900	7.448	9.276	10.118
20	—	3.788	4.002	6.881	7.429	9.225	10.062
25	3.557	3.776	4.008	6.865	7.413	9.180	10.012
30	3.552	3.766	4.015	6.854	7.400	9.139	9.966
35	3.549	3.759	4.024	6.844	7.389	9.102	9.925
40	3.547	3.753	4.035	6.838	7.380	9.068	9.889
45	3.547	3.750	4.047	6.834	7.373	9.038	9.856
50	3.549	3.749	4.060	6.833	7.367	9.011	9.828

Source: Values taken from Bates, R. G. *Determination of pH: Theory and Practice,* 2nd ed. Wiley: New York, 1973.[10]
[a]Concentrations are given in molality (moles solute per kilograms solvent).

Calibrating the electrode presents a third complication since a standard with an accurately known activity for H^+ needs to be used. Unfortunately, it is not possible to calculate rigorously the activity of a single ion. For this reason pH electrodes are calibrated using a standard buffer whose composition is chosen such that the defined pH is as close as possible to that given by equation 11.18. Table 11.6 gives pH values for several primary standard buffer solutions accepted by the National Institute of Standards and Technology.

A pH electrode is normally standardized using two buffers: one near a pH of 7 and one that is more acidic or basic depending on the sample's expected pH. The pH electrode is immersed in the first buffer, and the "standardize" or "calibrate" control is adjusted until the meter reads the correct pH. The electrode is placed in the second buffer, and the "slope" or "temperature" control is adjusted to the-buffer's pH. Some pH meters are equipped with a temperature compensation feature, allowing the pH meter to correct the measured pH for any change in temperature. In this case a thermistor is placed in the sample and connected to the pH meter. The "temperature" control is set to the solution's temperature, and the pH meter is calibrated using the "calibrate" and "slope" controls. If a change in the sample's temperature is indicated by the thermistor, the pH meter adjusts the slope of the calibration based on an assumed Nerstian response of $2.303RT/F$.

Clinical Applications Perhaps the area in which ion-selective electrodes receive the widest use is in clinical analysis, where their selectivity for the analyte in a complex matrix provides a significant advantage over many other analytical methods. The most common analytes are electrolytes, such as Na^+, K^+, Ca^{2+}, H^+, and Cl^-, and dissolved gases, such as CO_2. For extracellular fluids, such as blood and urine, the analysis can be made in vitro with conventional electrodes, provided that sufficient sample is available. Some clinical analyzers place a series of ion-selective electrodes in a flow

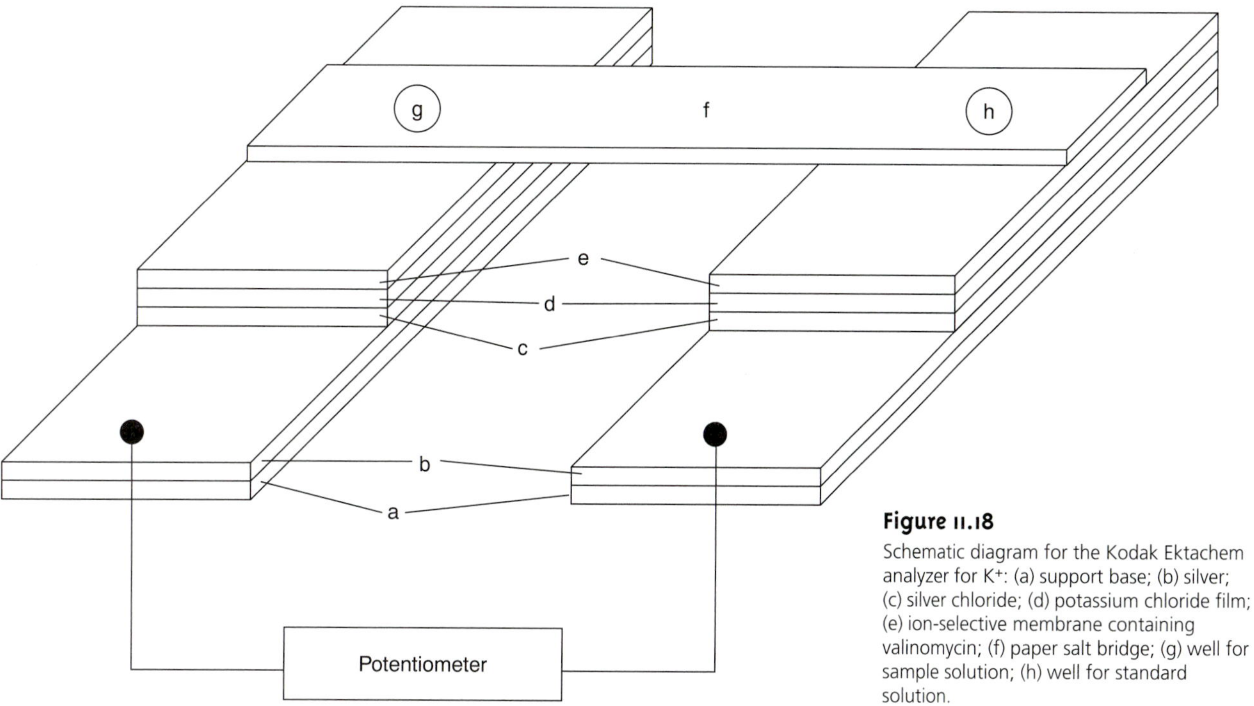

Figure 11.18
Schematic diagram for the Kodak Ektachem analyzer for K^+: (a) support base; (b) silver; (c) silver chloride; (d) potassium chloride film; (e) ion-selective membrane containing valinomycin; (f) paper salt bridge; (g) well for sample solution; (h) well for standard solution.

cell, allowing several analytes to be monitored simultaneously. Standards, samples, and rinse solutions are pumped through the flow cell and across the surface of the electrodes. For smaller volumes of sample the analysis can be conducted using disposable ion-selective systems, such as the Kodak Ektachem analyzer for K^+ shown in Figure 11.18. The analyzer consists of separate electrodes for the sample and reference solutions. Each electrode is constructed from several thin films, consisting of a Ag/AgCl reference electrode, a salt bridge and an ion-selective membrane, deposited on a support base. The two electrodes are connected by a paper salt bridge saturated with the sample and reference solutions. The overall dimensions of the analyzer are 2.8 cm × 2.4 cm with a thickness of 150 μm and require only 10 μL each of sample and reference solution. Similar analyzers are available for the determination of Na^+, Cl^-, and CO_2.

The analysis of intercellular fluids requires an ion-selective electrode that can be inserted directly into the desired cell. Liquid-based membrane microelectrodes with tip diameters of less than 1 μm are constructed by heating and drawing out a hard-glass capillary tube with an initial diameter of approximately 1–2 mm (Figure 11.19). The tip of the microelectrode is made hydrophobic by dipping in dichlorodimethyl silane. An inner solution appropriate for the desired analyte and a Ag/AgCl wire reference electrode are placed within the microelectrode. The tip of the microelectrode is then dipped into a solution containing the liquid complexing agent. The small volume of liquid complexing agent entering the microelectrode is retained within the tip by capillary action, eliminating the need for a solid membrane. Potentiometric microelectrodes have been developed for a number of clinically important analytes, including H^+, K^+, Na^+, Ca^{2+}, Cl^-, and I^-.

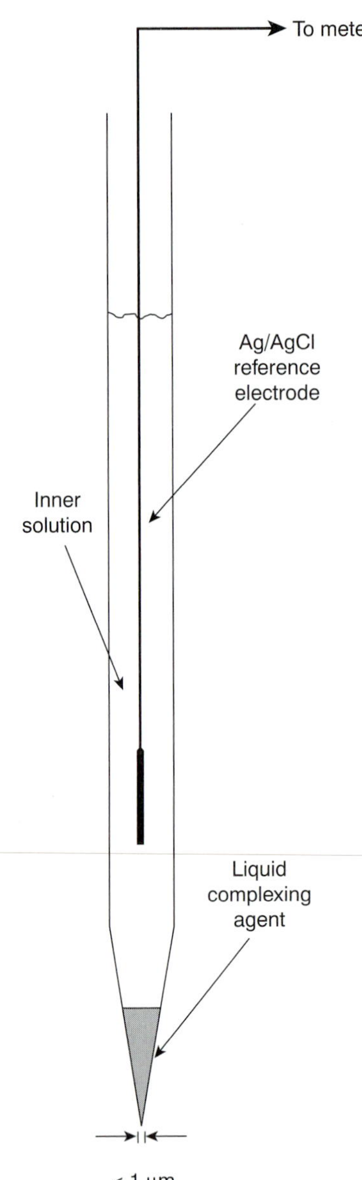

Figure 11.19
Schematic diagram of a liquid-based ion-selective microelectrode.

Environmental Applications Although ion-selective electrodes find use in environmental analysis, their application is not as widespread as in clinical analysis. Standard methods have been developed for the analysis of CN^-, F^-, NH_3, and NO_3^- in water and wastewater. Except for F^-, however, other analytical methods are considered superior. By incorporating the ion-selective electrode into a flow cell, the continuous monitoring of wastewater streams and other flow systems is possible. Such applications are limited, however, by the electrode's response to the analyte's activity, rather than its concentration. Considerable interest has been shown in the development of biosensors for the field screening and monitoring of environmental samples for a number of priority pollutants.[11]

Potentiometric Titrations In Chapter 9 we noted that one method for determining the equivalence point of an acid–base titration is to follow the change in pH with a pH electrode. The potentiometric determination of equivalence points is feasible for acid–base, complexation, redox, and precipitation titrations, as well as for titrations in aqueous and nonaqueous solvents. Acid–base, complexation, and precipitation potentiometric titrations are usually monitored with an ion-selective electrode that is selective for the analyte, although an electrode that is selective for the titrant or a reaction product also can be used. A redox electrode, such as a Pt wire, and a reference electrode are used for potentiometric redox titrations. More details about potentiometric titrations are found in Chapter 9.

11B.6 Evaluation

Scale of Operation The working range for most ion-selective electrodes is from a maximum concentration of 0.1–1 M to a minimum concentration of $10^{-5} – 10^{-10}$ M. This broad working range extends from major to ultratrace analytes, and is significantly greater than many other analytical methods. For conventional ion-selective electrodes, macro-sized samples with minimum volumes of 0.05–10 mL are necessary. Microelectrodes and specially designed analyzers, such as the Kodak Ektachem analyzer for K^+ shown in Figure 11.18, may be used with ultramicro-sized samples provided that the sample taken for analysis is sufficiently large to be representative of the original sample.

Accuracy The accuracy of a potentiometric analysis is limited by the measurement error for the cell's potential. Several factors contribute to this measurement error, including the contribution to the potential from interfering ions, the finite current drawn through the cell while measuring the potential, differences in the analyte's activity coefficient in the sample and standard solutions, and liquid junction potentials. Errors in accuracy due to interfering ions often can be eliminated by including a separation step before the potentiometric analysis. Modern high-impedance potentiometers minimize errors due to the passage of current through the electrochemical cell. Errors due to activity coefficients and liquid junction potentials are minimized by matching the matrix of the standards to that of the sample. Even in the best circumstances, however, a difference in potential of approximately ±1 mV is observed for samples and standards at equal concentration.

The effect of an uncertainty in potential on the accuracy of a potentiometric method of analysis is evaluated using a propagation of uncertainty. For a membrane ion-selective electrode the general expression for potential is given as

$$E_{\text{cell}} = K + \frac{RT}{zF}\ln[\text{A}]$$

where z is the charge of the analyte. From Table 4.9 in Chapter 4, the error in the cell potential, ΔE_{cell} is

$$\Delta E_{\text{cell}} = \frac{RT}{zF}\frac{\Delta[\text{A}]}{[\text{A}]}$$

Rearranging and multiplying through by 100 gives the percent relative error in concentration as

$$\%\ \text{relative error} = \frac{\Delta[\text{A}]}{[\text{A}]} \times 100 = \frac{\Delta E_{\text{cell}}}{RT/zF} \times 100 \qquad \mathbf{11.22}$$

The relative measurement error in concentration, therefore, is determined by the magnitude of the error in measuring the cell's potential and by the charge of the analyte. Representative values are shown in Table 11.7 for ions with charges of ±1 and ±2, at a temperature of 25 °C. Accuracies of 1–5% for monovalent ions and 2–10% for divalent ions are typical. Although equation 11.22 was developed for membrane electrodes, it also applies to metallic electrodes of the first and second kind when z is replaced by n.

Precision The precision of a potentiometric measurement is limited by variations in temperature and the sensitivity of the potentiometer. Under most conditions, and with simple, general-purpose potentiometers, the potential can be measured with a repeatability of ±0.1 mV. From Table 11.7 this result corresponds to an uncertainty of ±0.4% for monovalent analytes, and ±0.8% for divalent analytes. The reproducibility of potentiometric measurements is about a factor of 10 poorer.

Sensitivity The sensitivity of a potentiometric analysis is determined by the term RT/nF or RT/zF in the Nernst equation. Sensitivity is best for smaller values of n or z.

Table 11.7 Relationship Between Measurement Error in Potential and Relative Error in Concentration

Error in Potential (±mV)	Relative Error in Concentration (%)	
	$z = 1$	$z = 2$
0.1	±0.4	±0.8
0.5	±1.9	±3.9
1.0	±3.9	±7.8
1.5	±5.8	±11.1
2.0	±7.8	±15.6

Selectivity As described earlier, most ion-selective electrodes respond to more than one analyte. For many ion-selective electrodes, however, the selectivity for the analyte is significantly greater than for most interfering ions. Published selectivity coefficients for ion-selective electrodes (representative values are found in Tables 11.1 through 11.3) provide a useful guide in helping the analyst determine whether a potentiometric analysis is feasible for a given sample.

Time, Cost, and Equipment In comparison with competing methods, potentiometry provides a rapid, relatively low-cost means for analyzing samples. Commercial instruments for measuring pH or potential are available in a variety of price ranges and include portable models for use in the field.

11C Coulometric Methods of Analysis

In potentiometry, the potential of an electrochemical cell under static conditions is used to determine an analyte's concentration. As seen in the preceding section, potentiometry is an important and frequently used quantitative method of analysis. Dynamic electrochemical methods, such as **coulometry,** voltammetry, and amperometry, in which current passes through the electrochemical cell, also are important analytical techniques. In this section we consider coulometric methods of analysis. Voltammetry and amperometry are covered in Section 11D.

coulometry
An electrochemical method in which the current required to exhaustively oxidize or reduce the analyte is measured.

Coulometric methods of analysis are based on an exhaustive electrolysis of the analyte. By exhaustive we mean that the analyte is quantitatively oxidized or reduced at the working electrode or reacts quantitatively with a reagent generated at the working electrode. There are two forms of coulometry: controlled-potential coulometry, in which a constant potential is applied to the electrochemical cell, and controlled-current coulometry, in which a constant current is passed through the electrochemical cell.

The total charge, Q, in coulombs, passed during an electrolysis is related to the absolute amount of analyte by **Faraday's law**

Faraday's law
The current or charge passed in a redox reaction is proportional to the moles of the reaction's reactants and products.

$$Q = nFN \qquad \textbf{11.23}$$

where n is the number of electrons transferred per mole of analyte, F is Faraday's constant (96487 C mol^{-1}), and N is the moles of analyte. A coulomb is also equivalent to an A·s; thus, for a constant current, i, the charge is given as

$$Q = it_e \qquad \textbf{11.24}$$

where t_e is the electrolysis time. If current varies with time, as it does in controlled-potential coulometry, then the total charge is given by

$$Q = \int_{t=0}^{t=t_e} i(t)\,dt \qquad \textbf{11.25}$$

In coulometry, current and time are measured, and equation 11.24 or equation 11.25 is used to calculate Q. Equation 11.23 is then used to determine the moles of analyte. To obtain an accurate value for N, therefore, all the current must result in the analyte's oxidation or reduction. In other words, coulometry requires 100% **current efficiency** (or an accurately measured current efficiency established using a standard), a factor that must be considered in designing a coulometric method of analysis.

current efficiency
The percentage of current that actually leads to the analyte's oxidation or reduction.

11C.1 Controlled-Potential Coulometry

The easiest method for ensuring 100% current efficiency is to maintain the working electrode at a constant potential that allows for the analyte's quantitative oxidation or reduction, without simultaneously oxidizing or reducing an interfering species. The current flowing through an electrochemical cell under a constant potential is proportional to the analyte's concentration. As electrolysis progresses the analyte's concentration decreases, as does the current. The resulting current-versus-time profile for controlled-potential coulometry, which also is known as potentiostatic coulometry, is shown in Figure 11.20. Integrating the area under the curve (equation 11.25), from $t = 0$ until $t = t_e$, gives the total charge. In this section we consider the experimental parameters and instrumentation needed to develop a controlled-potential coulometric method of analysis.

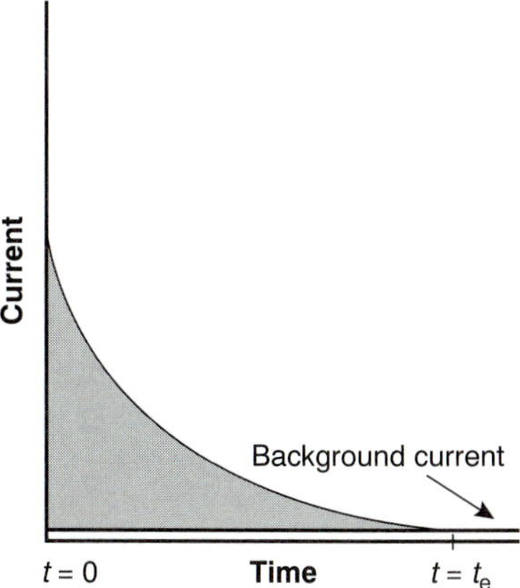

Figure 11.20
Current–time curve for controlled-potential coulometry.

Selecting a Constant Potential In controlled-potential coulometry, the potential is selected so that the desired oxidation or reduction reaction goes to completion without interference from redox reactions involving other components of the sample matrix. To see how an appropriate potential for the working electrode is selected, let's develop a constant-potential coulometric method for Cu^{2+} based on its reduction to copper metal at a Pt cathode working electrode.

$$Cu^{2+}(aq) + 2e^- \rightleftharpoons Cu(s) \qquad \mathbf{11.26}$$

A ladder diagram for a solution of Cu^{2+} (Figure 11.21) provides a useful means for evaluating the solution's redox properties. From the ladder diagram we can see that reaction 11.26 is favored when the working electrode's potential is more negative than +0.342 V versus the SHE (+0.093 V versus the SCE). To maintain a 100% current efficiency, however, the potential must be selected so that the reduction of H_3O^+ to H_2 does not contribute significantly to the total charge passed at the electrode.

The potential needed for a quantitative reduction of Cu^{2+} can be calculated using the Nernst equation

$$E = E^{\circ}_{Cu^{2+}/Cu} - \frac{0.05916}{2}\log\frac{1}{[Cu^{2+}]} \qquad \mathbf{11.27}$$

If we define a quantitative reduction as one in which 99.99% of the Cu^{2+} is reduced to Cu, then the concentration of Cu^{2+} at the end of the electrolysis must be

$$[Cu^{2+}] \le 10^{-4}[Cu^{2+}]_0 \qquad \mathbf{11.28}$$

where $[Cu^{2+}]_0$ is the initial concentration of Cu^{2+} in the sample. Substituting equation 11.28 into equation 11.27 gives the desired potential electrode as

$$E = E^{\circ}_{Cu^{2+}/Cu} - \frac{0.05916}{2}\log\frac{1}{10^{-4}[Cu^{2+}]_0}$$

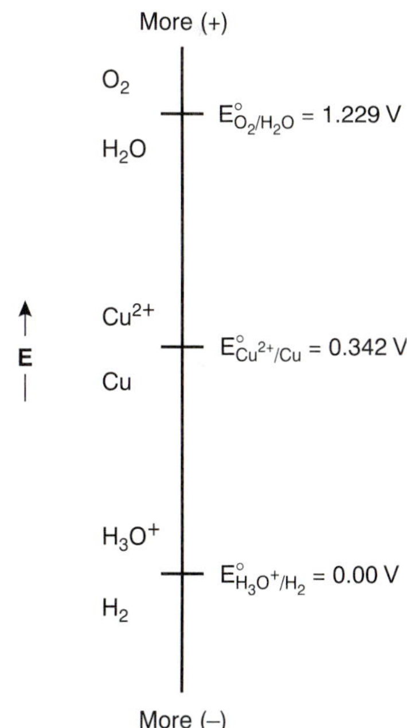

Figure 11.21
Ladder diagram for aqueous solutions of Cu^{2+}.

If the initial concentration of Cu^{2+} is 1.00×10^{-4} M, for example, then the cathode's potential must be more negative than +0.105 V versus the SHE (–0.139 V versus the SCE) to achieve a quantitative reduction of Cu^{2+} to Cu. Note that at this potential H_3O^+ is not reduced to H_2, maintaining a 100% current efficiency. Many of the published procedures for the controlled-potential coulometric analysis of Cu^{2+} call for potentials that are more negative than that shown for the reduction of H_3O^+ in Figure 11.21.[12] Such potentials can be used, however, because the slow kinetics for reducing H_3O^+ results in a significant **overpotential** that shifts the potential of the H_3O^+/H_2 redox couple to more negative potentials.

overpotential
The difference between the potential actually required to initiate an oxidation or reduction reaction, and the potential predicted by the Nernst equation.

Minimizing Electrolysis Time The current-time curve for controlled-potential coulometry in Figure 11.20 shows that the current decreases continuously throughout electrolysis. An exhaustive electrolysis, therefore, may require a long time. Since time is an important consideration in choosing and designing analytical methods, the factors that determine the analysis time need to be considered.

The change in current as a function of time in controlled-potential coulometry is approximated by an exponential decay; thus, the current at time t is

$$i = i_0 e^{-kt} \qquad \textbf{11.29}$$

where i_0 is the initial current, and k is a constant that is directly proportional to the area of the working electrode and the rate of stirring and inversely proportional to the volume of the solution. For an exhaustive electrolysis in which 99.99% of the analyte is oxidized or reduced, the current at the end of the analysis, t_e, may be approximated as

$$i \leq (10^{-4}) i_0 \qquad \textbf{11.30}$$

Substituting equation 11.30 into equation 11.29 and solving for t_e gives the minimum time for an exhaustive electrolysis as

$$t_e = -\frac{1}{k}\ln(10^{-4}) = \frac{9.21}{k}$$

From this equation we see that increasing k leads to a shorter analysis time. For this reason controlled-potential coulometry is carried out in small-volume electrochemical cells, using electrodes with large surface areas and with high stirring rates. A quantitative electrolysis typically requires approximately 30–60 min, although shorter or longer times are possible.

Instrumentation The potential in controlled-potential coulometry is set using a three-electrode potentiostat. Two types of working electrodes are commonly used: a Pt electrode manufactured from platinum-gauze and fashioned into a cylindrical tube, and an Hg pool electrode. The large overpotential for reducing H_3O^+ at mercury makes it the electrode of choice for analytes requiring negative potentials. For example, potentials more negative than –1 V versus the SCE are feasible at an Hg electrode (but not at a Pt electrode), even in very acidic solutions. The ease with which mercury is oxidized, however, prevents its use at potentials that are positive with respect to the SHE. Platinum working electrodes are used when positive potentials are required. The auxiliary electrode, which is often a Pt wire, is separated by a salt bridge from the solution containing the analyte. This is necessary to prevent electrolysis products generated at the auxiliary electrode from reacting with the analyte and interfering in the analysis. A saturated calomel or Ag/AgCl electrode serves as the reference electrode.

The other essential feature of instrumentation for controlled-potential coulometry is a means of determining the total charge passed during electrolysis. One method is to monitor the current as a function of time and determine the area under the curve (see Figure 11.20). Modern instruments, however, use electronic integration to monitor charge as a function of time. The total charge at the end of the electrolysis then can be read directly from a digital readout or from a plot of charge versus time (Figure 11.22).

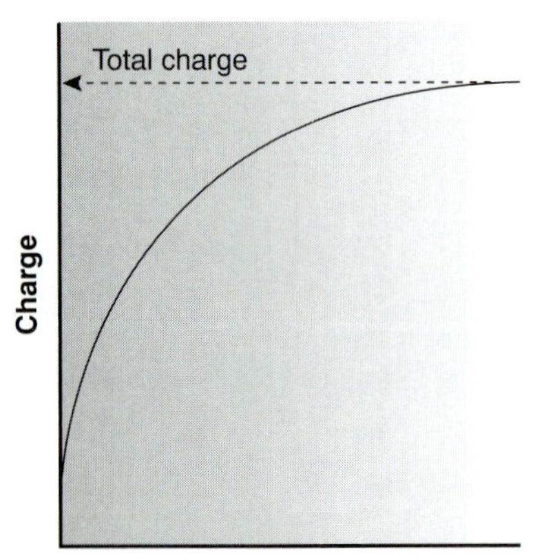

Figure 11.22
Charge–time curve obtained by integrating the current–time curve in Figure 11.20.

11C.2 Controlled-Current Coulometry

A second approach to coulometry is to use a constant current in place of a constant potential (Figure 11.23). Controlled-current coulometry, also known as amperostatic coulometry or coulometric titrimetry, has two advantages over controlled-potential coulometry. First, using a constant current makes for a more rapid analysis since the current does not decrease over time. Thus, a typical analysis time for controlled-current coulometry is less than 10 min, as opposed to approximately 30–60 min for controlled-potential coulometry. Second, with a constant current the total charge is simply the product of current and time (equation 11.24). A method for integrating the current–time curve, therefore, is not necessary.

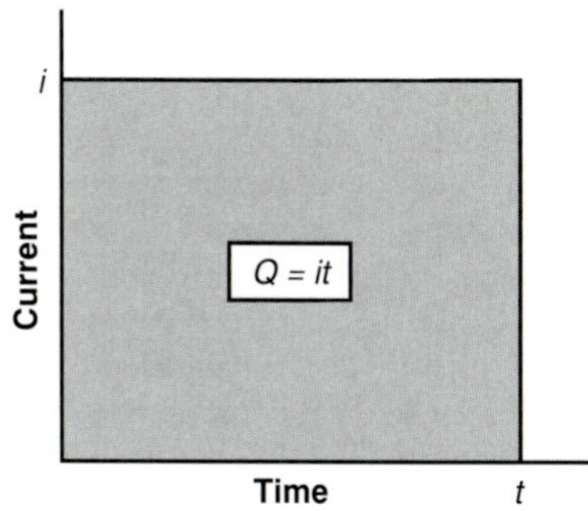

Figure 11.23
Current–time curve for controlled-current coulometry.

Using a constant current does present two important experimental problems that must be solved if accurate results are to be obtained. First, as electrolysis occurs the analyte's concentration and, therefore, the current due to its oxidation or reduction steadily decreases. To maintain a constant current the cell potential must change until another oxidation or reduction reaction can occur at the working electrode. Unless the system is carefully designed, these secondary reactions will produce a current efficiency of less than 100%. The second problem is the need for a method of determining when the analyte has been exhaustively electrolyzed. In controlled-potential coulometry this is signaled by a decrease in the current to a constant background or residual current (see Figure 11.20). In controlled-current coulometry, however, a constant current continues to flow even when the analyte has been completely oxidized or reduced. A suitable means of determining the end-point of the reaction, t_e, is needed.

Maintaining Current Efficiency To illustrate why changing the working electrode's potential can lead to less than 100% current efficiency, let's consider the coulometric analysis for Fe^{2+} based on its oxidation to Fe^{3+} at a Pt working electrode in 1 M H_2SO_4.

$$Fe^{2+}(aq) \rightleftharpoons Fe^{3+}(aq) + e^-$$

The ladder diagram for this system is shown in Figure 11.24a. Initially the potential of the working electrode remains nearly constant at a level near the standard-state potential for the Fe^{3+}/Fe^{2+} redox couple. As the concentration of Fe^{2+} decreases, however, the potential of the working electrode shifts toward more positive values until another oxidation reaction can provide the necessary current. Thus, in this case the potential eventually increases to a level at which the oxidation of H_2O occurs.

$$6H_2O(\ell) \rightleftharpoons O_2(g) + 4H_3O^+(aq) + 4e^-$$

Since the current due to the oxidation of H_3O^+ does not contribute to the oxidation of Fe^{2+}, the current efficiency of the analysis is less than 100%. To maintain a 100% current efficiency the products of any competing oxidation reactions must react both rapidly and quantitatively with the remaining Fe^{2+}. This may be accomplished, for example, by adding an excess of Ce^{3+} to the analytical solution (Figure 11.24b). When the potential of the working electrode shifts to a more positive potential, the first species to be oxidized is Ce^{3+}.

$$Ce^{3+}(aq) \rightleftharpoons Ce^{4+}(aq) + e^-$$

The Ce^{4+} produced at the working electrode rapidly mixes with the solution, where it reacts with any available Fe^{2+}.

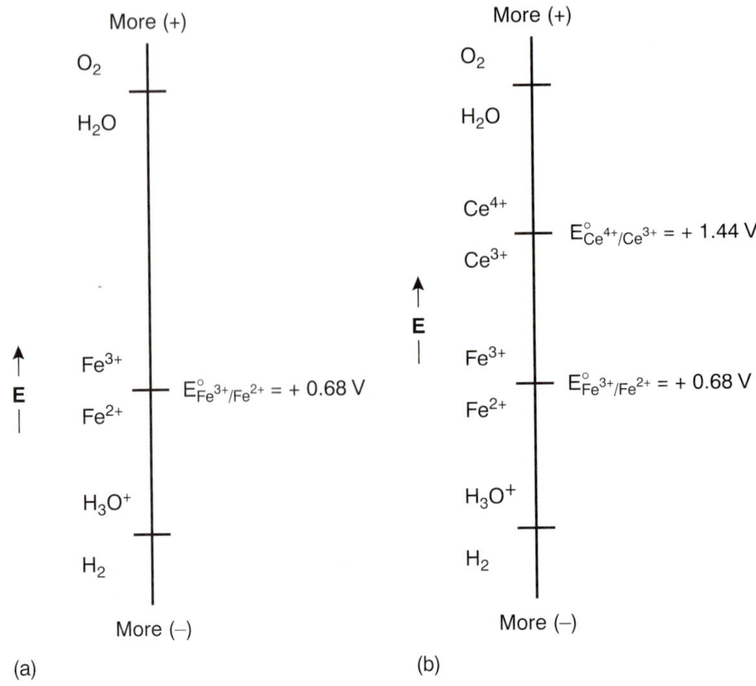

Figure 11.24
Ladder diagrams for the controlled-current coulometric analysis of Fe^{2+} (a) without the addition of Ce^{3+}, and (b) with the addition of Ce^{3+}. The matrix is 1 M H_2SO_4 in both cases.

$$Ce^{4+}(aq) + Fe^{2+}(aq) \rightleftharpoons Fe^{3+}(aq) + Ce^{3+}(aq) \qquad 11.31$$

Combining these reactions gives the desired overall reaction of

$$Fe^{2+}(aq) \rightleftharpoons Fe^{3+}(aq) + e^-$$

In this manner, a current efficiency of 100% is maintained. Furthermore, since the concentration of Ce^{3+} remains at its initial level, the potential of the working electrode remains constant as long as any Fe^{2+} is present. This prevents other oxidation reactions, such as that for H_2O, from interfering with the analysis. A species, such as Ce^{3+}, which is used to maintain 100% current efficiency, is called a **mediator.**

mediator
A species that transfers electrons from the electrode to the analyte.

End Point Determination Adding a mediator solves the problem of maintaining 100% current efficiency, but does not solve the problem of determining when the analyte's electrolysis is complete. Using the same example, once all the Fe^{2+} has been oxidized current continues to flow as a result of the oxidation of Ce^{3+} and, eventually, the oxidation of H_2O. What is needed is a means of indicating when the oxidation of Fe^{2+} is complete. In this respect it is convenient to treat a controlled-current coulometric analysis as if electrolysis of the analyte occurs only as a result of its reaction with the mediator. A reaction between an analyte and a mediator, such as that shown in reaction 11.31, is identical to that encountered in a redox titration. Thus, the same end points that are used in redox titrimetry (see Chapter 9), such as visual indicators, and potentiometric and conductometric measurements, may be used to signal the end point of a controlled-current coulometric analysis. For example, ferroin may be used to provide a visual end point for the Ce^{3+}-mediated coulometric analysis for Fe^{2+}.

Instrumentation Controlled-current coulometry normally is carried out using a galvanostat and an electrochemical cell consisting of a working electrode and a counterelectrode. The working electrode, which often is constructed from Pt, is also called the generator electrode since it is where the mediator reacts to generate the species reacting with the analyte. The counterelectrode is isolated from the analytical solution by a salt bridge or porous frit to prevent its electrolysis products from reacting with the analyte. Alternatively, oxidizing or reducing the mediator can be carried out externally, and the appropriate products flushed into the analytical solution. Figure 11.25 shows one simple method by which oxidizing and reducing agents can be generated externally. A solution containing the mediator flows under the influence of gravity into a small-volume electrochemical cell. The products generated at the anode and cathode pass through separate tubes, and the appropriate oxidizing or reducing reagent can be selectively delivered to the analytical solution. For example, external generation of Ce^{4+} can be obtained using an aqueous solution of Ce^{3+} and the products generated at the anode.

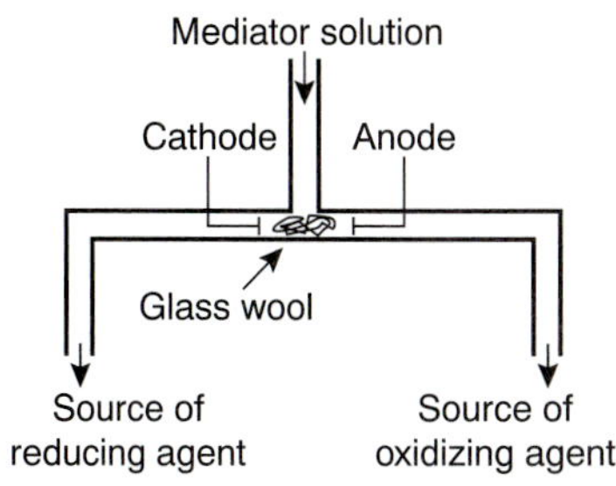

Figure 11.25
Method for the external generation of oxidizing and reducing agents in coulometric titrations.

The other necessary instrumental component for controlled-current coulometry is an accurate clock for measuring the electrolysis time, t_e, and a switch for starting and stopping the electrolysis. Analog clocks can read time to the nearest ±0.01 s, but the need to frequently stop and start the electrolysis near the end point leads to a net uncertainty of ±0.1 s. Digital clocks provide a more accurate measurement of time, with errors of ±1 ms being possible. The switch must control the flow of current and the clock, so that an accurate determination of the electrolysis time is possible.

Coulometric Titrations Controlled-current coulometric methods commonly are called **coulometric titrations** because of their similarity to conventional titrations. We already have noted, in discussing the controlled-current coulometric determination of Fe^{2+}, that the oxidation of Fe^{2+} by Ce^{4+} is identical to the reaction used in a redox titration. Other similarities between the two techniques also exist. Combining equations 11.23 and 11.24 and solving for the moles of analyte gives

coulometric titrations
A titration in which the equivalence point is the time required for a constant current to completely oxidize or reduce the analyte.

$$N = \left(\frac{i}{nF}\right)(t_e) \qquad 11.32$$

Compare this equation with the relationship between the moles of strong acid, N, titrated with a strong base of known concentration.

$$N = (M_{\text{base}})(V_{\text{base}})$$

The titrant in a conventional titration is replaced in a coulometric titration by a constant-current source whose current is analogous to the titrant's molarity. The time needed for an exhaustive electrolysis takes the place of the volume of titrant, and the switch for starting and stopping the electrolysis serves the same function as a buret's stopcock.

11C.3 Quantitative Applications

Coulometry may be used for the quantitative analysis of both inorganic and organic compounds. Examples of controlled-potential and controlled-current coulometric methods are discussed in the following sections.

Controlled-Potential Coulometry The majority of controlled-potential coulometric analyses involve the determination of inorganic cations and anions, including trace metals and halides. Table 11.8 provides a summary of several of these methods.

The ability to control selectivity by carefully selecting the working electrode's potential, makes controlled-potential coulometry particularly useful for the analysis of alloys. For example, the composition of an alloy containing Ag, Bi, Cd, and Sb

Table 11.8 Representative Examples for the Controlled-Potential Coulometric Analysis of Inorganic Ions

Analyte	Electrolytic Reaction[a]	Electrode
Antimony	$Sb(III) + 3e^- \rightleftharpoons Sb$	Pt
Arsenic	$As(III) \rightleftharpoons As(V) + 2e^-$	Pt
Cadmium	$Cd(II) + 2e^- \rightleftharpoons Cd$	Pt or Hg
Cobalt	$Co(II) + 2e^- \rightleftharpoons Co$	Pt or Hg
Copper	$Cu(II) + 2e^- \rightleftharpoons Cu$	Pt or Hg
Halides	$Ag + X^- \rightleftharpoons AgX + e^-$	Ag
Iron	$Fe(II) \rightleftharpoons Fe(III) + e^-$	Pt
Lead	$Pb(II) + 2e^- \rightleftharpoons Pb$	Pt or Hg
Nickel	$Ni(II) + 2e^- \rightleftharpoons Ni$	Pt or Hg
Plutonium	$Pu(III) \rightleftharpoons Pu(IV) + e^-$	Pt
Silver	$Ag(I) + e^- \rightleftharpoons Ag$	Pt
Tin	$Sn(II) + 2e^- \rightleftharpoons Sn$	Pt
Uranium	$U(VI) + 2e^- \rightleftharpoons U(IV)$	Pt or Hg
Zinc	$Zn(II) + 2e^- \rightleftharpoons Zn$	Pt or Hg

Source: Compiled from Rechnitz, G. A. *Controlled-Potential Analysis.* Macmillan: New York, 1963, p. 49.
[a]Electrolytic reactions are written in terms of the change in oxidation state. The actual species in solution depend on the composition of the sample matrix.

can be determined by dissolving the sample and placing it in a matrix of 0.2 M H_2SO_4. A platinum working electrode is immersed in the solution and held at a constant potential of +0.40 V versus the SCE. At this potential Ag(I) deposits on the Pt electrode as Ag, and the other metal ions remain in solution. When electrolysis is complete, the total charge is used to determine the amount of silver in the alloy. The potential of the platinum electrode is then shifted to –0.08 V versus the SCE, depositing Bi on the working electrode. When the coulometric analysis for bismuth is complete, antimony is determined by shifting the working electrode's potential to –0.33 V versus the SCE, depositing Sb. Finally, cadmium is determined following its electrodeposition on the Pt electrode at a potential of –0.80 V versus the SCE.

Another area where controlled-potential coulometry has found application is in nuclear chemistry, in which elements such as uranium and polonium can be determined at trace levels. For example, microgram quantities of uranium in a medium of H_2SO_4 can be determined by reducing U(VI) to U(IV) at a mercury working electrode.

Controlled-potential coulometry also can be applied to the quantitative analysis of organic compounds, although the number of applications is significantly less than that for inorganic analytes. One example is the six-electron reduction of a nitro group, $-NO_2$, to a primary amine, $-NH_2$, at a mercury electrode. Solutions of picric acid, for instance, can be analyzed by reducing to triaminophenol.

$$\text{(OH)C}_6\text{H}_2(\text{NO}_2)_3 + 18H_3O^+ + 18e^- \longrightarrow \text{(OH)C}_6\text{H}_2(\text{NH}_2)_3 + 24H_2O$$

Table 11.9 Representative Examples of Coulometric Redox Titrations

Mediator	Electrochemically Generated Reagent	Generator-Electrode Reaction	Representative Application[a]
Ag^+	Ag^{2+}	$Ag^+ \rightleftharpoons Ag^{2+} + e^-$	$\underline{H_2C_2O_4} + 2Ag^{2+} + 2H_2O \rightleftharpoons 2CO_2 + 2Ag^+ + 2H_3O^+$
Br^-	Br_2	$2Br^- \rightleftharpoons Br_2 + 2e^-$	$\underline{H_2S} + Br_2 + 2H_2O \rightleftharpoons S + 2Br^- + 2H_3O^+$
Ce^{3+}	Ce^{4+}	$Ce^{3+} \rightleftharpoons Ce^{4+} + e^-$	$\underline{Fe(CN)_6{}^{4-}} + Ce^{4+} \rightleftharpoons Fe(CN)_6{}^{3-} + Ce^{3+}$
Cl^-	Cl_2	$2Cl^- \rightleftharpoons Cl_2 + 2e^-$	$\underline{Tl(I)} + Cl_2 \rightleftharpoons Tl(III) + 2Cl^-$
Fe^{3+}	Fe^{2+}	$Fe^{3+} + e^- \rightleftharpoons Fe^{2+}$	$\underline{Cr_2O_7{}^{2-}} + 6Fe^{2+} + 14H_3O^+ \rightleftharpoons 2Cr^{3+} + 6Fe^{3+} + 21H_2O$
I^-	I_3^-	$3I^- \rightleftharpoons I_3^- + 2e^-$	$\underline{2S_2O_3{}^{2-}} + I_3^- \rightleftharpoons S_4O_6{}^{2-} + 3I^-$
Mn^{2+}	Mn^{3+}	$Mn^{2+} \rightleftharpoons Mn^{3+} + e^-$	$\underline{As(III)} + 2Mn^{3+} \rightleftharpoons As(V) + 2Mn^{2+}$

[a]The analyte is the underlined species in each reaction.

Another example is the successive reduction of trichloroacetate to dichloroacetate, and of dichloroacetate to monochloroacetate

$$Cl_3CCOO^-(aq) + H_3O^+(aq) + 2e^- \rightleftharpoons Cl_2HCCOO^-(aq) + Cl^-(aq) + H_2O(\ell)$$

$$Cl_2HCCOO^-(aq) + H_3O^+(aq) + 2e^- \rightleftharpoons ClH_2CCOO^-(aq) + Cl^-(aq) + H_2O(\ell)$$

Mixtures of trichloroacetate and dichloroacetate are analyzed by selecting an initial potential at which only the more easily reduced trichloroacetate is reduced. When its electrolysis is complete, the potential is switched to a more negative potential at which dichloroacetate is reduced. The total charge for the first electrolysis is used to determine the amount of trichloroacetate, and the difference in total charge between the first and second electrolyses gives the amount of dichloroacetate.

Controlled-Current Coulometry The use of a mediator makes controlled-current coulometry a more versatile analytical method than controlled-potential coulometry. For example, the direct oxidation or reduction of a protein at the working electrode in controlled-potential coulometry is difficult if the protein's active redox site lies deep within its structure. The controlled-current coulometric analysis of the protein is made possible, however, by coupling its oxidation or reduction to a mediator that is reduced or oxidized at the working electrode. Controlled-current coulometric methods have been developed for many of the same analytes that may be determined by conventional redox titrimetry. These methods, several of which are summarized in Table 11.9, also are called coulometric redox titrations.

Coupling the mediator's oxidation or reduction to an acid–base, precipitation, or complexation reaction involving the analyte allows for the coulometric titration of analytes that are not easily oxidized or reduced. For example, when using H_2O as a mediator, oxidation at the anode produces H_3O^+

$$6H_2O(\ell) \rightleftharpoons 4H_3O^+(aq) + O_2(g) + 4e^-$$

while reduction at the cathode produces OH^-.

$$2H_2O(\ell) + 2e^- \rightleftharpoons 2OH^-(aq) + H_2(g)$$

If the oxidation or reduction of H_2O is carried out externally using the generator cell shown in Figure 11.25, then H_3O^+ or OH^- can be dispensed selectively into a solution containing a basic or acidic analyte. The resulting reaction is identical to that in an acid–base titration. Coulometric acid–base titrations have been used for

Table 11.10 Representative Examples of Coulometric Titrations Using Acid–Base, Complexation, and Precipitation Reactions

Type of Reaction	Mediator	Electrochemically Generated Reagent	Generator-Electrode Reaction	Representative Reaction[a]
Acid–Base	H_2O	H_3O^+	$6H_2O \rightleftharpoons 4H_3O^+ + O_2 + 4e^-$	$\underline{OH^-} + H_3O^+ \rightleftharpoons 2H_2O$
	H_2O	OH^-	$2H_2O + 2e^- \rightleftharpoons 2OH^- + H_2$	$\underline{H_3O^+} + OH^- \rightleftharpoons 2H_2O$
Complexation	$HgNH_3Y^{2-}$	EDTA (HY^{3-})	$HgNH_3Y^{2-} + NH_4^+ + 2e^- \rightleftharpoons HY^{3-} + Hg + 2NH_3$	$\underline{Ca^{2+}} + HY^{3-} + H_2O \rightleftharpoons CaY^{2-} + H_3O^+$
Precipitation	Ag	Ag^+	$Ag \rightleftharpoons Ag^+ + e^-$	$\underline{I^-} + Ag^+ \rightleftharpoons AgI$
	Hg	Hg_2^{2+}	$2Hg \rightleftharpoons Hg_2^{2+} + 2e^-$	$2\underline{Cl^-} + Hg_2^{2+} \rightleftharpoons Hg_2Cl_2$
	$Fe(CN)_6^{3-}$	$Fe(CN)_6^{4-}$	$Fe(CN)_6^{3-} + e^- \rightleftharpoons Fe(CN)_6^{4-}$	$3\underline{Zn^{2+}} + 2K^+ + 2Fe(CN)_6^{4-} \rightleftharpoons K_2Zn_3[Fe(CN)_6]_2$

[a]The analyte is the underlined species in each reaction.

the analysis of strong and weak acids and bases, in both aqueous and nonaqueous matrices. Examples of coulometric titrations involving acid–base, complexation, and precipitation reactions are summarized in Table 11.10.

In comparison with conventional titrimetry, there are several advantages to the coulometric titrations listed in Tables 11.9 and 11.10. One advantage is that the electrochemical generation of a "titrant" that reacts immediately with the analyte allows the use of reagents whose instability prevents their preparation and storage as a standard solution. Thus, highly reactive reagents such as Ag^{2+} and Mn^{3+} can be used in coulometric titrations. Because it is relatively easy to measure small quantities of charge, coulometric titrations can be used to determine small quantities of analyte that cannot be measured accurately by a conventional titration.

Quantitative Calculations The absolute amount of analyte in a coulometric analysis is determined by applying Faraday's law (equation 11.23) with the total charge during the electrolysis given by equation 11.24 or equation 11.25. Example 11.8 shows the calculations for a typical coulometric analysis.

Example

EXAMPLE 11.8

The purity of a sample of $Na_2S_2O_3$ was determined by a coulometric redox titration using I^- as a mediator, and I_3^- as the "titrant." A sample weighing 0.1342 g is transferred to a 100-mL volumetric flask and diluted to volume with distilled water. A 10.00-mL portion is transferred to an electrochemical cell along with 25 mL of 1 M KI, 75 mL of a pH 7.0 phosphate buffer, and several drops of a starch indicator solution. Electrolysis at a constant current of 36.45 mA required 221.8 s to reach the starch indicator end point. Determine the purity of the sample.

SOLUTION

From Table 11.9 we see that the coulometric titration of $S_2O_3^{2-}$ with I_3^- is

$$2S_2O_3^{2-}(aq) + I_3^-(aq) \rightleftharpoons S_4O_6^{2-}(aq) + 3I^-(aq)$$

Oxidizing $S_2O_3^{2-}$ to $S_4O_6^{2-}$ requires one electron per $S_2O_3^{2-}$ ($n = 1$). Combining equations 11.23 and 11.24, and making an appropriate substitution for moles of $Na_2S_2O_3$ gives

$$\frac{nF(\text{g Na}_2\text{S}_2\text{O}_3)}{\text{FW Na}_2\text{S}_2\text{O}_3} = it_e$$

Solving for the grams of $Na_2S_2O_3$ gives

$$\text{g Na}_2\text{S}_2\text{O}_3 = \frac{it_e(\text{FW Na}_2\text{S}_2\text{O}_3)}{nF}$$

$$= \frac{(0.03645\ \text{A})(221.8\ \text{s})(158.1\ \text{g/mol})}{(1\ \text{mol}\ e^-)(96487\ \text{C/mol}\ e^-)} = 0.01325\ \text{g Na}_2\text{S}_2\text{O}_3$$

This represents the amount of $Na_2S_2O_3$ in a 10.00-mL portion of a 100-mL sample, thus 0.1325 g of $Na_2S_2O_3$ is present in the original sample. The purity of the sample, therefore, is

$$\frac{0.1325\ \text{g Na}_2\text{S}_2\text{O}_3}{0.1342\text{-g sample}} \times 100 = 98.73\%\ \text{w/w Na}_2\text{S}_2\text{O}_3$$

Note that the calculation is worked as if $S_2O_3^{2-}$ is oxidized directly at the working electrode instead of in solution.

Representative Method Every controlled-potential or controlled-current coulometric method has its own unique considerations. Nevertheless, the following procedure for the determination of dichromate by a coulometric redox titration provides an instructive example.

Representative Methods

Method 11.2 Determination of Dichromate by a Coulometric Redox Titration[13]

Description of the Method. The concentration of $Cr_2O_7^{2-}$ in a sample is determined by a coulometric redox titration using Fe^{3+} as a mediator and electrogenerated Fe^{2+} as the "titrant." The end point of the coulometric redox titration is determined potentiometrically.

Procedure. The electrochemical cell consists of a Pt working electrode and a Pt counterelectrode that is maintained in a separate compartment connected to the analytical solution by a porous glass disk. The counter electrode's compartment is filled with 0.2 M Na_2SO_4, the level of which must always be maintained above that of the solution in the electrochemical cell. Platinum and tungsten electrodes connected to a potentiometer are used to follow the change in potential during the analysis. A solution of approximately 0.3 M $NH_4Fe(SO_4)_2$ is prepared for use as the mediator. The coulometric redox titration is carried out by adding 5.00 mL of the sample solution, 2 mL of 9 M H_2SO_4 and 10–25 mL of the $NH_4Fe(SO_4)_2$ mediator solution. Distilled water is added as needed until the electrodes are covered. Pure N_2 is bubbled through the solution for 15 min to remove traces of O_2. A flow of N_2 is maintained during the electrolysis, but is turned off momentarily when measuring the potential. A magnetic stir bar is used to stir the solution. Adjust the current to 15–50 mA, and begin the titration. Periodically stop the titration, and measure the potential of the solution. In this manner a titration

—Continued

Continued from page 505

curve of potential versus time is recorded. The time needed to reach the equivalence point is read from the titration curve.

Questions

1. **Does the platinum working electrode serve as the cathode or the anode in this analysis?**

   Reduction of Fe^{3+} to Fe^{2+} occurs at the working electrode, making it the cathode in the electrochemical cell.

2. **Pure N_2 is bubbled through the solution to remove any trace of dissolved O_2. Why is this necessary?**

   If O_2 is present, some of the electrogenerated Fe^{2+} may be oxidized back to Fe^{3+} by the reaction

   $$4Fe^{2+}(aq) + O_2(g) + 4H_3O^+(aq) \rightleftharpoons 4Fe^{3+}(aq) + 6H_2O(\ell)$$

   Any Fe^{2+} lost in this fashion must be replaced by the additional reduction of Fe^{3+}, reducing the current efficiency and increasing the time needed to reach the titration's end point. The net result is that the reported concentration of $Cr_2O_7^{2-}$ is too large.

3. **What is the effect on the analysis if the $NH_4Fe(SO_4)_2$ used to prepare the mediator solution is contaminated with trace amounts of Fe^{2+}?**

   The Fe^{2+} introduced when the mediator solution is first added will react with the $Cr_2O_7^{2-}$ before the electrolysis begins. As a result, the amount of Fe^{2+} that must be generated is less than expected, decreasing the time needed to reach the end point of the titration. The current efficiency, therefore, is greater than 100%, and the reported concentration of $Cr_2O_7^{2-}$ is too small. Trace amount of Fe^{2+} can be removed from the mediator solution by adding H_2O_2 and heating at 50–70 °C until the evolution of O_2 ceases. Alternatively, a blank titration can be used to correct for any Fe^{2+} initially present in the mediator.

11C.4 Characterization Applications

Studies aimed at characterizing the mechanisms of electrode reactions often make use of coulometry for determining the number of electrons involved in the reaction. To make such measurements a known amount of a pure compound is subject to a controlled-potential electrolysis. The coulombs of charge needed to complete the electrolysis are used to determine the value of *n* using Faraday's law (equation 11.23).

Example

EXAMPLE 11.9

A 0.3619-g sample of tetrachloropicolinic acid, $C_6HNO_2Cl_4$, is dissolved in distilled water, transferred to a 1000-mL volumetric flask, and diluted to volume. An exhaustive controlled-potential electrolysis of a 10.00-mL portion of this solution at a spongy silver cathode requires 5.374 C of charge. What is the value of *n* for this reduction reaction?

SOLUTION

The 10.00-mL portion of sample contains 3.619 mg, or 1.39×10^{-5} mol of tetrachloropicolinic acid. Solving equation 11.23 for n and making appropriate substitutions gives

$$n = \frac{Q}{FN} = \frac{5.374 \text{ C}}{(96478 \text{ C/mol } e^{-})(1.39 \times 10^{-5} \text{ mol } C_6HNO_2Cl_4)} = 4.01$$

Thus, reducing a molecule of tetrachloropicolinic acid requires four electrons. The overall reaction, which results in the selective formation of 3,6-dichloropicolinic acid, is

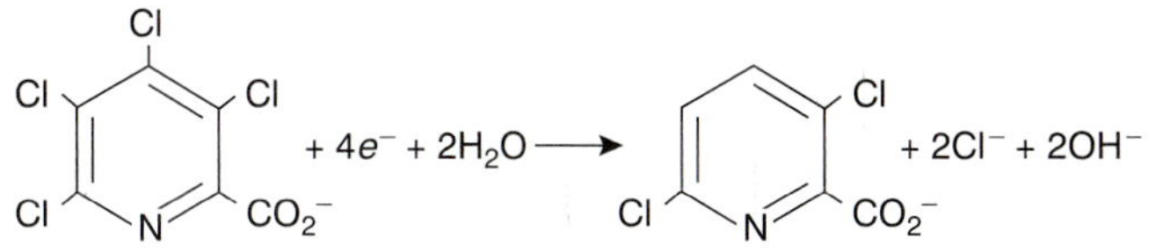

Cl
Cl Cl
Cl N CO_2^- $+ 4e^- + 2H_2O \longrightarrow$ Cl
Cl N CO_2^- $+ 2Cl^- + 2OH^-$

11C.5 Evaluation

Scale of Operation Coulometric methods of analysis can be used to analyze small absolute amounts of analyte. In controlled-current coulometry, for example, the moles of analyte consumed during an exhaustive electrolysis is given by equation 11.32. An electrolysis carried out with a constant current of 100 μA for 100 s, therefore, consumes only 1×10^{-7} mol of analyte if $n = 1$. For an analyte with a molecular weight of 100 g/mol, 1×10^{-7} mol corresponds to only 10 μg. The concentration of analyte in the electrochemical cell, however, must be sufficient to allow an accurate determination of the end point. When using visual end points, coulometric titrations require solution concentrations greater than 10^{-4} M and, as with conventional titrations, are limited to major and minor analytes. A coulometric titration to a preset potentiometric end point is feasible even with solution concentrations of 10^{-7} M, making possible the analysis of trace analytes.[14]

Accuracy The accuracy of a controlled-current coulometric method of analysis is determined by the current efficiency, the accuracy with which current and time can be measured, and the accuracy of the end point. With modern instrumentation the maximum measurement error for current is about ±0.01%, and that for time is approximately ±0.1%. The maximum end point error for a coulometric titration is at least as good as that for conventional titrations and is often better when using small quantities of reagents. Taken together, these measurement errors suggest that accuracies of 0.1–0.3% are feasible. The limiting factor in many analyses, therefore, is current efficiency. Fortunately current efficiencies of greater than 99.5% are obtained routinely and often exceed 99.9%.

In controlled-potential coulometry, accuracy is determined by current efficiency and the determination of charge. Provided that no interferents are present that are easier to oxidize or reduce than the analyte, current efficiencies of greater than 99.9% are easily obtained. When interferents are present, however, they can often be eliminated by applying a potential such that the exhaustive electrolysis of the interferents is possible without the simultaneous electrolysis of the analyte. Once the interferents have been removed the potential can be switched to a level at which electrolysis of the analyte is feasible. The limiting factor in the accuracy of many controlled-potential coulometric methods of analysis is the determination of charge. With modern electronic integrators, the total charge can be determined with an accuracy of better than 0.5%.

So what is to be done when an acceptable current efficiency is not feasible? If the analyte's oxidation or reduction leads to its deposition on the working electrode, it may be possible to determine the analyte's mass. In this case the working electrode is weighed before beginning the electrolysis and reweighed when electrolysis of the analyte is complete. The difference in the electrode's weight gives the analyte's mass. This technique is known as electrogravimetry. More details about this technique are provided in the suggested reading listed at the end of this chapter.

Precision Precision is determined by the uncertainties of measuring current, time, and the end point in controlled-current coulometry and of measuring charge in controlled-potential coulometry. Precisions of ±0.1–0.3% are routinely obtained for coulometric titrations, and precisions of ±0.5% are typical for controlled-potential coulometry.

Sensitivity For a coulometric method of analysis, the calibration sensitivity is equivalent to nF in equation 11.25. In general, coulometric methods in which the analyte's oxidation or reduction involves a larger value of n show a greater sensitivity.

Selectivity Selectivity in controlled-potential and controlled-current coulometry is improved by carefully adjusting solution conditions and by properly selecting the electrolysis potential. In controlled-potential coulometry the potential is fixed by the potentiostat, whereas in controlled-current coulometry the potential is determined by the redox reaction involving the mediator. In either case, the ability to control the potential at which electrolysis occurs affords some measure of selectivity. By adjusting pH or adding a complexing agent, it may be possible to shift the potential at which an analyte or interferent undergoes oxidation or reduction. For example, the standard-state reduction potential for Zn^{2+} is –0.762 V versus the SHE, but shifts to –1.04 for $Zn(NH_3)_4^{2+}$. This provides an additional means for controlling selectivity when an analyte and interferent undergo electrolysis at similar potentials.

Time, Cost, and Equipment Controlled-potential coulometry is a relatively time-consuming analysis, with a typical analysis requiring 30–60 min. Coulometric titrations, on the other hand, require only a few minutes and are easily adapted for automated analysis. Commercial instrumentation for both controlled-potential and controlled-current coulometry is available and is relatively inexpensive. Low-cost potentiostats and constant-current sources are available for less than $1000.

IID Voltammetric Methods of Analysis

In **voltammetry** a time-dependent potential is applied to an electrochemical cell, and the current flowing through the cell is measured as a function of that potential. A plot of current as a function of applied potential is called a **voltammogram** and is the electrochemical equivalent of a spectrum in spectroscopy, providing quantitative and qualitative information about the species involved in the oxidation or reduction reaction.[15] The earliest voltammetric technique to be introduced was polarography, which was developed by Jaroslav Heyrovsky

voltammetry
An electrochemical method in which we measure current as a function of the applied potential.

voltammogram
A plot of current as a function of applied potential.

(1890–1967) in the early 1920s, for which he was awarded the Nobel Prize in chemistry in 1959. Since then, many different forms of voltammetry have been developed, a few of which are noted in Figure 11.1. Before examining these techniques and their applications in more detail, however, we must first consider the basic experimental design for making voltammetric measurements and the factors influencing the shape of the resulting voltammogram.

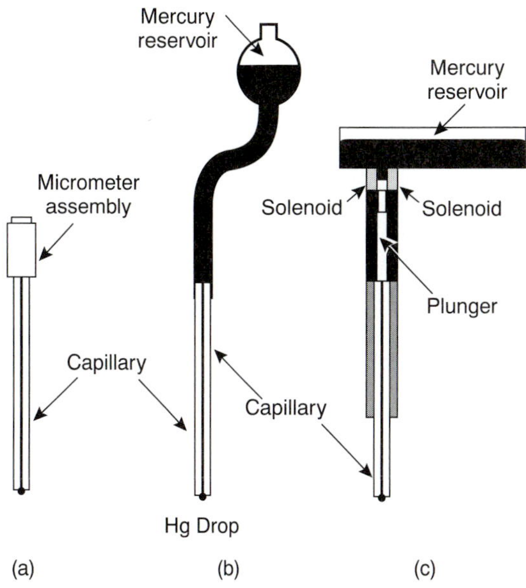

Figure 11.26
Mercury electrodes: (a) hanging mercury drop electrode; (b) dropping mercury electrode; (c) static mercury drop electrode.

11D.1 Voltammetric Measurements

Although early voltammetric methods relied on the use of only two electrodes, modern voltammetry makes use of a three-electrode potentiostat, such as that shown in Figure 11.4. A time-dependent potential excitation signal is applied to the working electrode, changing its potential relative to the fixed potential of the reference electrode. The resulting current between the working and auxiliary electrodes is measured. The auxiliary electrode is generally a platinum wire, and the SCE and Ag/AgCl electrode are common reference electrodes.

Several different materials have been used as working electrodes, including mercury, platinum, gold, silver, and carbon. The earliest voltammetric techniques, including polarography, used mercury for the working electrode. Since mercury is a liquid, the working electrode often consists of a drop suspended from the end of a capillary tube (Figure 11.26). In the **hanging mercury drop electrode,** or HMDE, a drop of the desired size is formed by the action of a micrometer screw that pushes the mercury through a narrow capillary tube. In the **dropping mercury electrode,** or DME, mercury drops form at the end of the capillary tube as a result of gravity. Unlike the HMDE, the mercury drop of a DME grows continuously and has a finite lifetime of several seconds. At the end of its lifetime the mercury drop is dislodged, either manually or by gravity, and replaced by a new drop. The **static mercury drop electrode,** or SMDE, uses a solenoid-driven plunger to control the flow of mercury. The SMDE can be used as either a hanging mercury drop electrode or as a dropping mercury electrode. A single activation of the solenoid momentarily lifts the plunger, allowing enough mercury to flow through the capillary to form a single drop. To obtain a dropping mercury electrode the solenoid is activated repeatedly. A mercury film electrode consists of a thin layer of mercury deposited on the surface of a solid carbon, platinum, or gold electrode. The solid electrode is placed in a solution of Hg^{2+} and held at a potential at which the reduction of Hg^{2+} to Hg is favorable, forming a thin mercury film.

hanging mercury drop electrode
An electrode in which a drop of Hg is suspended from a capillary tube.

dropping mercury electrode
An electrode in which successive drops of Hg form at the end of a capillary tube as a result of gravity, with each drop providing a fresh electrode surface.

static mercury drop electrode
An electrode in which successive drops of Hg form at the end of a capillary tube as the result of a mechanical plunger, with each drop providing a fresh electrode surface.

Mercury has several advantages as a working electrode. Perhaps its most important advantage is its high overpotential for the reduction of H_3O^+ to H_2, which allows for the application of potentials as negative as –1 V versus the SCE in acidic solutions, and –2 V versus the SCE in basic solutions. A species such as Zn^{2+}, which is difficult to reduce at other electrodes without simultaneously reducing H_3O^+, is easily reduced at a mercury working electrode. Other advantages include the ability of metals to dissolve in the mercury, resulting in the formation of an **amalgam,** and the ability to easily renew the surface of the electrode by extruding a new drop. One limitation to its use as a working electrode is the ease with which Hg is oxidized. For this reason, mercury electrodes cannot be used as at potentials more positive than –0.3 V to +0.4 V versus the SCE, depending on the composition of the solution.

amalgam
A metallic solution of mercury with another metal.

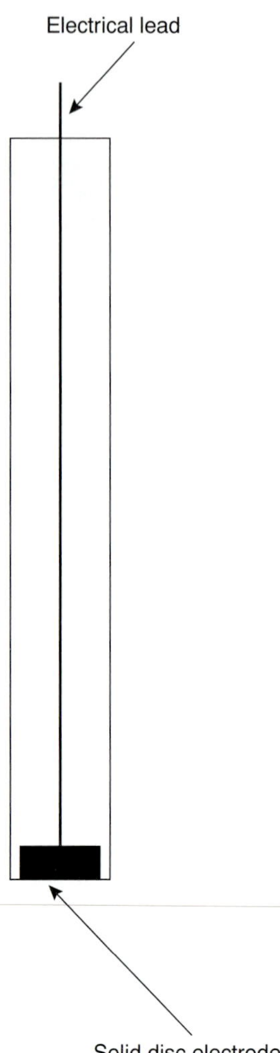

Figure 11.27
Schematic diagram of a solid electrode.

faradaic current
Any current in an electrochemical cell due to an oxidation or reduction reaction.

cathodic current
A faradaic current due to a reduction reaction.

anodic current
A faradaic current due to an oxidation reaction.

Solid electrodes constructed using platinum, gold, silver, or carbon may be used over a range of potentials, including potentials that are negative and positive with respect to the SCE. For example, the potential range for a Pt electrode extends from approximately +1.2 V to –0.2 V versus the SCE in acidic solutions and from +0.7 V to –1 V versus the SCE in basic solutions. Solid electrodes, therefore, can be used in place of mercury for many voltammetric analyses requiring negative potentials and for voltammetric analyses at positive potentials at which mercury electrodes cannot be used. Except for the carbon paste electrode, solid electrodes are fashioned into disks that are sealed into the end of an inert support and are in contact with an electrical lead (Figure 11.27). The carbon paste electrode is made by filling the cavity at the end of the inert support with a paste consisting of carbon particles and a viscous oil. Solid electrodes are not without problems, the most important of which is the ease with which the electrode's surface may be altered by the adsorption of solution species or the formation of oxide layers. For this reason solid electrodes need frequent reconditioning, either by applying an appropriate potential or by polishing.

A typical arrangement for a voltammetric electrochemical cell is shown in Figure 11.28. Besides the working, reference, and auxiliary electrodes, the cell also includes a N_2 purge line for removing dissolved O_2 and an optional stir bar. Electrochemical cells are available in a variety of sizes, allowing for the analysis of solution volumes ranging from more than 100 mL to as small as 50 μL.

11D.2 Current in Voltammetry

When an analyte is oxidized at the working electrode, a current passes electrons through the external electric circuitry to the auxiliary electrode, where reduction of the solvent or other components of the solution matrix occurs. Reducing an analyte at the working electrode requires a source of electrons, generating a current that flows from the auxiliary electrode to the cathode. In either case, a current resulting from redox reactions at the working and auxiliary electrodes is called a **faradaic current.** In this section we consider the factors affecting the magnitude of this faradaic current, as well as the source of any nonfaradaic currents.

Sign Conventions Since the reaction of interest occurs at the working electrode, the classification of current is based on this reaction. A current due to the analyte's reduction is called a **cathodic current** and, by convention, is considered positive. **Anodic currents** are due to oxidation reactions and carry a negative value.

Influence of Applied Potential on the Faradaic Current As an example, let's consider the faradaic current when a solution of $Fe(CN)_6^{3-}$ is reduced to $Fe(CN)_6^{4-}$ at the working electrode. The relationship between the concentrations of $Fe(CN)_6^{3-}$, $Fe(CN)_6^{4-}$, and the potential of the working electrode is given by the Nernst equation; thus

$$E = +0.356 - 0.05916 \log \frac{[Fe(CN)_6^{4-}]_{x=0}}{[Fe(CN)_6^{3-}]_{x=0}}$$

where +0.356 is the standard-state potential for the $Fe(CN)_6^{3-}/Fe(CN)_6^{4-}$ redox couple, and $x = 0$ indicates that the concentrations of $Fe(CN)_6^{3-}$ and $Fe(CN)_6^{4-}$ are those at the surface of the electrode. Surface concentrations are used instead of bulk concentrations since the equilibrium position for the redox reaction

$$Fe(CN)_6^{3-}(aq) + e^- \rightleftharpoons Fe(CN)_6^{4-}(aq)$$

can only be established electrochemically at the electrode's surface.

Let's assume that we have a solution for which the concentration of $Fe(CN)_6^{3-}$ is 1.0 mM and in which $Fe(CN)_6^{4-}$ is absent. A ladder diagram for this redox example is shown in Figure 11.29. If a potential of +0.530 V is applied to the working electrode, the concentrations of $Fe(CN)_6^{3-}$ and $Fe(CN)_6^{4-}$ at the surface of the electrode are unaffected, and no faradaic current is observed. Switching the potential to +0.356 V, however, requires that

$$[Fe(CN)_6^{4-}]_{x=0} = [Fe(CN)_6^{3-}]_{x=0} = 0.50 \text{ mM}$$

which is only possible if a portion of the $Fe(CN)_6^{3-}$ at the electrode surface is reduced to $Fe(CN)_6^{4-}$. If this was all that occurred after the potential was applied, the result would be a brief surge of faradaic current that would quickly return to zero. However, although the concentration of $Fe(CN)_6^{4-}$ at the electrode surface is 0.50 mM, its concentration in the bulk of solution is zero. As a result, a concentration gradient exists between the solution at the electrode surface and the bulk solution. This concentration gradient creates a driving force that transports $Fe(CN)_6^{4-}$ away from the electrode surface (Figure 11.30). The subsequent decrease in the concentration of $Fe(CN)_6^{4-}$ at the electrode surface requires the further reduction of $Fe(CN)_6^{3-}$, as well as its transport from bulk solution to the electrode surface. Thus, a faradaic current continues to flow until there is no difference between the concentrations of $Fe(CN)_6^{3-}$ and $Fe(CN)_6^{4-}$ at the electrode surface and their concentrations in the bulk of solution.*

Although the applied potential at the working electrode determines if a faradaic current flows, the magnitude of the current is determined by the rate of the resulting oxidation or reduction reaction at the electrode surface. Two factors contribute to the rate of the electrochemical reaction: the rate at which the reactants and products are transported to and from the surface of the electrode, and the rate at which electrons pass between the electrode and the reactants and products in solution.

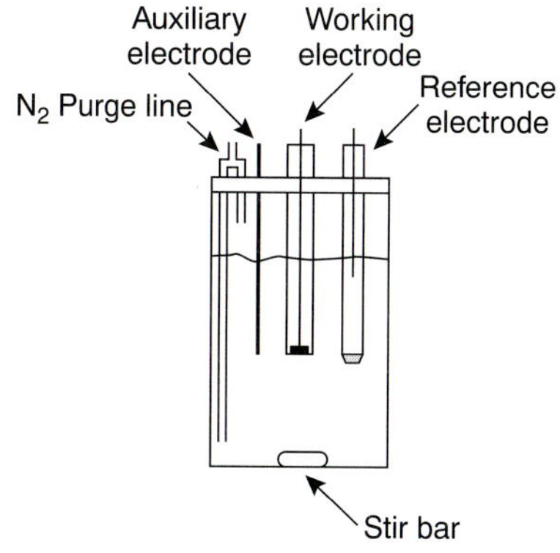

Figure 11.28
Typical electrochemical cell for use in voltammetry.

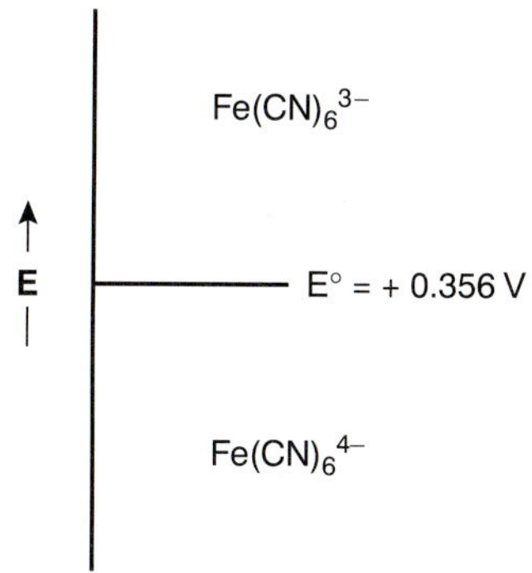

Figure 11.29
Ladder diagram for the $Fe(CN)_6^{3-}/Fe(CN)_6^{4-}$ redox half-reaction.

Influence of Mass Transport on the Faradaic Current There are three modes of **mass transport** that influence the rate at which reactants and products are transported to and from the electrode surface: diffusion, migration, and convection. **Diffusion** from a region of high concentration to a region of low concentration occurs whenever the concentration of an ion or molecule at the surface of the electrode is different from that in bulk solution. When the potential applied to the working electrode is sufficient to reduce or oxidize the analyte at the electrode surface, a concentration gradient similar to that shown in Figure 11.31 is established. The volume of solution in which the concentration gradient exists is called the **diffusion layer.** Without other modes of mass transport, the width of the diffusion layer, δ, increases with time as the concentration of reactants near the electrode surface decreases. The contribution of diffusion to the rate of mass transport, therefore, is time-dependent.

mass transport
The movement of material toward or away from the electrode surface.

diffusion
The movement of material in response to a concentration gradient.

diffusion layer
The layer of solution adjacent to the electrode in which diffusion is the only means of mass transport.

*In voltammetry the working electrode's surface area is significantly smaller than that used in coulometry. Consequently, very little analyte undergoes electrolysis, and the analyte's concentration in bulk solution remains essentially unchanged.

convection
The movement of material in response to a mechanical force, such as stirring a solution.

migration
The movement of a cation or anion in response to an applied potential.

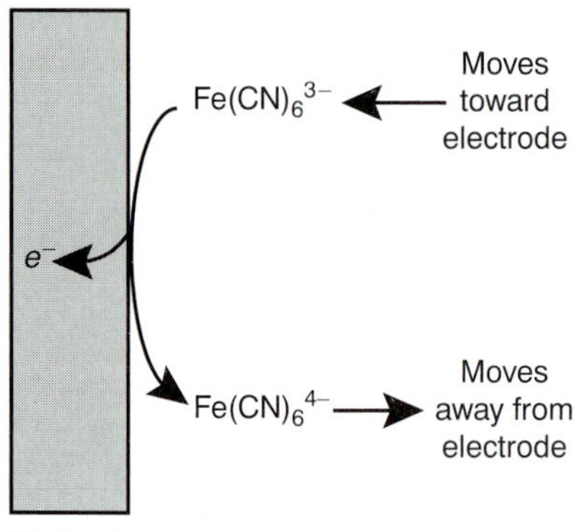

Figure 11.30
Schematic showing transport of $Fe(CN)_6^{3-}$ toward the electrode and $Fe(CN)_6^{4-}$ away from the electrode following the reduction of $Fe(CN)_6^{3-}$.

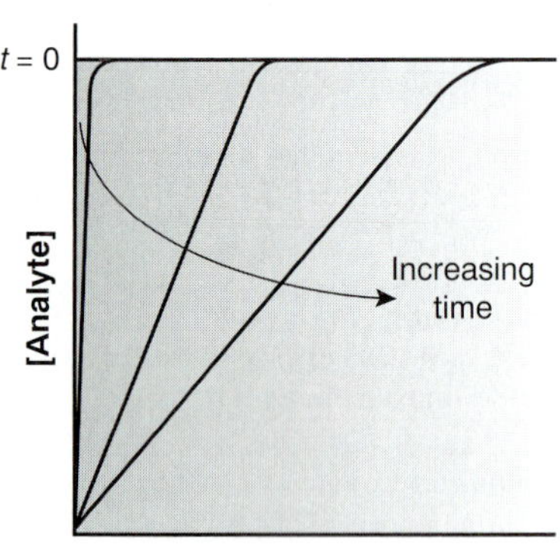

Figure 11.31
Concentration gradients for the analyte in the absence of convection, showing the time-dependent change in diffusion as a method of mass transport.

nonfaradaic current
A current in an electrochemical cell that is not the result of a redox reaction.

Convection occurs when a mechanical means is used to carry reactants toward the electrode and to remove products from the electrode. The most common means of convection is to stir the solution using a stir bar. Other methods include rotating the electrode and incorporating the electrode into a flow cell.

The final mode of mass transport is **migration,** which occurs when charged particles in solution are attracted or repelled from an electrode that has a positive or negative surface charge. Thus, when the electrode is positively charged, negatively charged particles move toward the electrode, while positively charged particles move toward the bulk solution. Unlike diffusion and convection, migration only affects the mass transport of charged particles.

The flux of material to and from the electrode surface is a complex function of all three modes of mass transport. In the limit in which diffusion is the only significant means for the mass transport of the reactants and products, the current in a voltammetric cell is given by

$$i = \frac{nFAD(C_{\text{bulk}} - C_{x=0})}{\delta} \qquad \textbf{11.33}$$

where n is the number of electrons transferred in the redox reaction, F is Faraday's constant, A is the area of the electrode, D is the diffusion coefficient for the reactant or product, C_{bulk} and $C_{x=0}$ are the concentration of the analyte in bulk solution and at the electrode surface, and δ is the thickness of the diffusion layer.

For equation 11.33 to be valid, convection and migration must not interfere with the formation of a diffusion layer between the electrode and the bulk of solution. Migration is eliminated by adding a high concentration of an inert supporting electrolyte to the analytical solution. Ions of similar charge are equally attracted or repelled from the surface of the electrode and, therefore, have an equal probability of undergoing migration. The large excess of inert ions, however, ensures that few reactant and product ions will move as a result of migration. Although convection may be easily eliminated by not physically agitating the solution, in some situations it is desirable either to stir the solution or to push the solution through an electrochemical flow cell. Fortunately, the dynamics of a fluid moving past an electrode results in a small diffusion layer, typically of 0.001 – 0.01-cm thickness, in which the rate of mass transport by convection drops to zero (Figure 11.32).

Influence of the Kinetics of Electron Transfer on the Faradaic Current The rate of mass transport is one factor influencing the current in a voltammetric experiment. The ease with which electrons are transferred between the electrode and the reactants and products in solution also affects the current. When electron transfer kinetics are fast, the redox reaction is at equilibrium, and the concentrations of reactants and products at the electrode are those specified by the Nernst equation. Such systems are considered electrochemically reversible. In other systems, when electron transfer kinetics are sufficiently slow, the concentration of reactants and products at the electrode surface, and thus the current, differ from that predicted by the Nernst equation. In this case the system is electrochemically irreversible.

Nonfaradaic Currents Faradaic currents result from a redox reaction at the electrode surface. Other currents may also exist in an electrochemical cell that are unrelated to any redox reaction. These currents are called **nonfaradaic currents** and must be accounted for if the faradaic component of the measured current is to be determined.

The most important example of a nonfaradaic current occurs whenever the electrode's potential is changed. In discussing migration as a means of mass transport, we noted that negatively charged particles in solution migrate toward a positively charged electrode, and positively charged particles move away from the same electrode. When an inert electrolyte is responsible for migration, the result is a structured electrode–surface interface called the **electrical double layer,** or EDL, the exact structure of which is of no concern in the context of this text. The movement of charged particles in solution, however, gives rise to a short-lived, nonfaradaic **charging current.** Changing the potential of an electrode causes a change in the structure of the EDL, producing a small charging current.

Residual Current Even in the absence of analyte, a small current inevitably flows through an electrochemical cell. This current, which is called the **residual current,** consists of two components: a faradaic current due to the oxidation or reduction of trace impurities, and the charging current. Methods for discriminating between the faradaic current due to the analyte and the residual current are discussed later in this chapter.

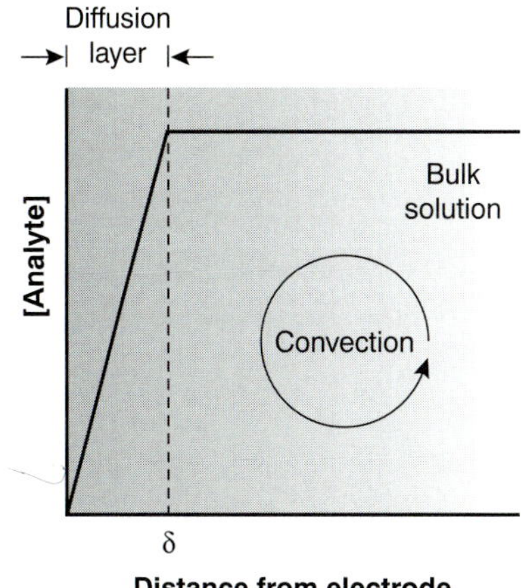

Figure 11.32
Concentration gradient for the analyte showing the effects of diffusion and convection as methods of mass transport.

electrical double layer
The interface between a positively or negatively charged electrode and the negatively or positively charged layer of solution in contact with the electrode.

charging current
A current in an electrochemical cell due to the electrical double layer's formation.

residual current
The current in an electrochemical cell that is present even in the absence of the analyte.

hydrodynamic voltammetry
A form of voltammetry in which the solution is mechanically stirred.

11D.3 Shape of Voltammograms

The shape of a voltammogram is determined by several experimental factors, the most important of which are how the current is measured and whether convection is included as a means of mass transport. Despite an abundance of different voltammetric techniques, several of which are discussed in this chapter, only three shapes are common for voltammograms (Figure 11.33).

The voltammogram in Figure 11.33a is characterized by a current that increases from the background residual current to a limiting current at potentials at which the analyte is oxidized or reduced. Since the magnitude of a faradaic current is inversely proportional to δ (equation 11.33), a limiting current implies that the thickness of the diffusion layer remains constant. The simplest method for obtaining a limiting current is to stir the solution (Figure 11.32), which can be accomplished with a magnetic stir bar, or by rotating the electrode. Voltammetric techniques that include convection by stirring are called **hydrodynamic voltammetry.** When convection is absent, the thickness of the diffusion layer increases with time (Figure 11.31), resulting in a peak current in place of a limiting current (Figure 11.33b).

In the voltammograms in Figures 11.33a and 11.33b, the current is monitored as a function of the applied potential. Alternatively, the change in current following

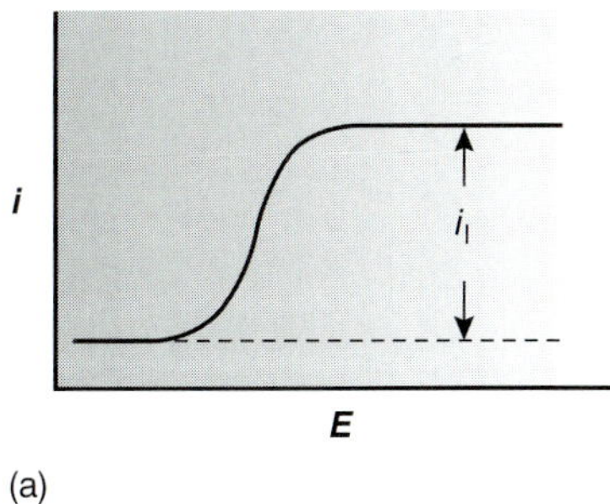

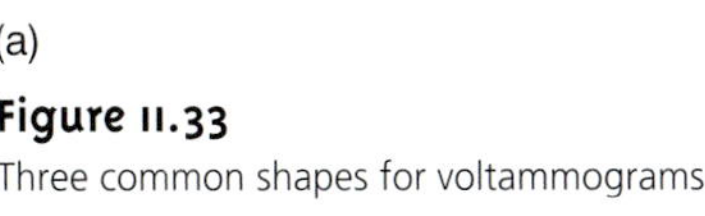

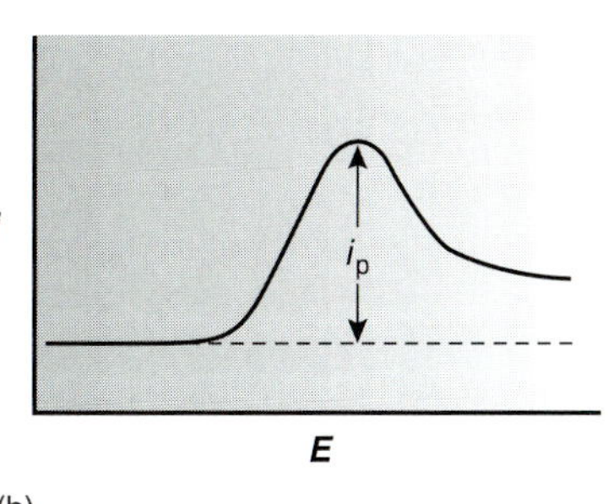

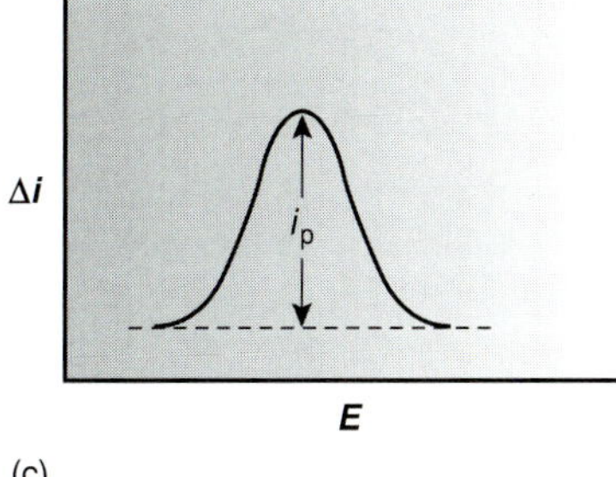

(a) (b) (c)

Figure 11.33
Three common shapes for voltammograms.

a change in potential may be measured. The resulting voltammogram, which is shown in Figure 11.33c, also is characterized by a peak current.

11D.4 Quantitative and Qualitative Aspects of Voltammetry

Earlier we described a voltammogram as the electrochemical equivalent of a spectrum in spectroscopy. In this section we consider how quantitative and qualitative information may be extracted from a voltammogram. Quantitative information is obtained by relating current to the concentration of analyte in the bulk solution. Qualitative information is obtained from the voltammogram by extracting the standard-state potential for the redox reaction. For simplicity we only consider voltammograms similar to that shown in Figure 11.33a.

Determining Concentration Let's assume that the redox reaction at the working electrode is

$$O + ne^- \rightleftharpoons R \tag{11.34}$$

and that initially only O is present in the bulk solution. The current is determined by the rate at which O diffuses through the fixed diffusion layer (see Figure 11.32), and is given by equation 11.33, or

$$i = K_O([O]_{bulk} - [O]_{x=0}) \tag{11.35}$$

where K_O is a constant equal to $nFAD_O/\delta$. When the limiting current is reached, the concentration of O at the electrode surface is zero, and this equation simplifies to

$$i_{lim} = K_O[O]_{bulk} \tag{11.36}$$

Thus, the limiting current, i_{lim}, is a linear function of the concentration of O in bulk solution, and a quantitative analysis is possible using any of the standardization methods discussed in Chapter 5. Equations similar to equation 11.35 can be developed for other forms of voltammetry, in which peak currents are related to the analyte's concentration in bulk solution.

Determining the Standard-State Potential To extract the standard-state potential, or formal potential, for reaction 11.34 from a voltammogram, it is necessary to rewrite the Nernst equation

$$E = E^\circ_{O/R} - \frac{0.05916}{n}\log\frac{[R]_{x=0}}{[O]_{x=0}} \tag{11.37}$$

in terms of current instead of the concentration of O and R. Substituting equation 11.36 into equation 11.35 and rearranging gives

$$[O]_{x=0} = \frac{i_{lim} - i}{K_O} \tag{11.38}$$

To derive a similar equation for the concentration of R at the electrode surface we note that

$$i = K_R([R]_{x=0} - [R]_{bulk})$$

Since the concentration of R in bulk solution is zero, this equation simplifies to

$$i = K_R[R]_{x=0}$$

and

$$[R]_{x=0} = \frac{i}{K_R} \qquad \textbf{11.39}$$

Substituting equations 11.39 and 11.38 into equation 11.37 and rearranging gives

$$E = E^{\circ}_{O/R} - \frac{0.05916}{n}\log\frac{K_O}{K_R} - \frac{0.05916}{n}\log\frac{i}{i_{lim} - i} \qquad \textbf{11.40}$$

When the current is half that of the limiting current

$$i = \frac{1}{2}i_{lim}$$

equation 11.40 simplifies to

$$E_{1/2} = E^{\circ}_{O/R} - \frac{0.05916}{n}\log\frac{K_O}{K_R} \qquad \textbf{11.41}$$

where $E_{1/2}$ is the half-wave potential (Figure 11.34). If K_O is approximately equal to K_R, which is often the case, then the half-wave potential is equal to the standard-state potential. Note that equation 11.41 is only valid if the redox reaction is electrochemically reversible. Voltammetric techniques giving peak potentials also can be used to determine a redox reaction's standard-state potential.

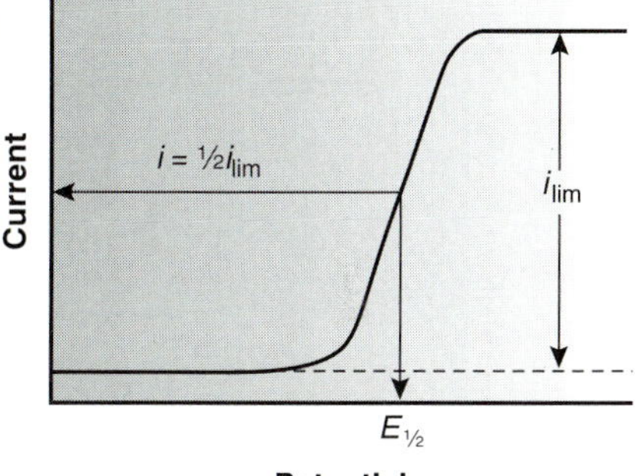

Figure 11.34
Determination of limiting current and half-wave potential in linear scan hydrodynamic voltammetry.

11D.5 Voltammetric Techniques

A number of voltammetric experiments are routinely used in quantitative and qualitative analyses. Several of these methods are briefly described in this section.

polarography
A form of voltammetry using a dropping mercury electrode or a static mercury drop electrode.

Polarography The earliest voltammetric experiment was normal **polarography** at a dropping mercury electrode. In normal polarography the potential is linearly scanned, producing voltammograms such as that shown in Figure 11.35. Although polarography takes place in an unstirred solution, a limiting current is obtained because the falling Hg drops mix the solution. Each new Hg drop, therefore, grows in a solution whose composition is identical to that of the initial bulk solution. Oscillations in the current are due to the growth of the Hg drop, which leads to a

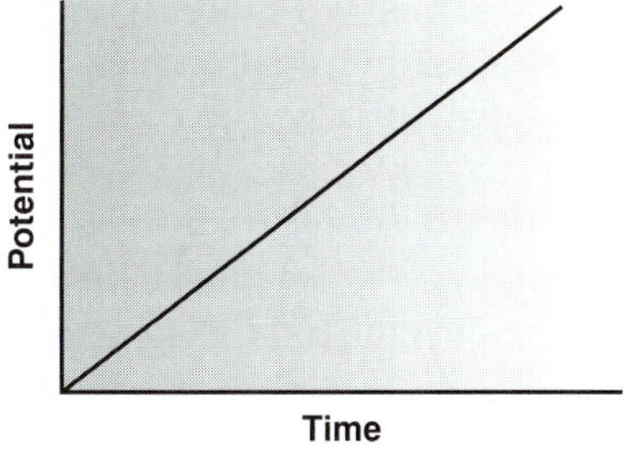

(a)

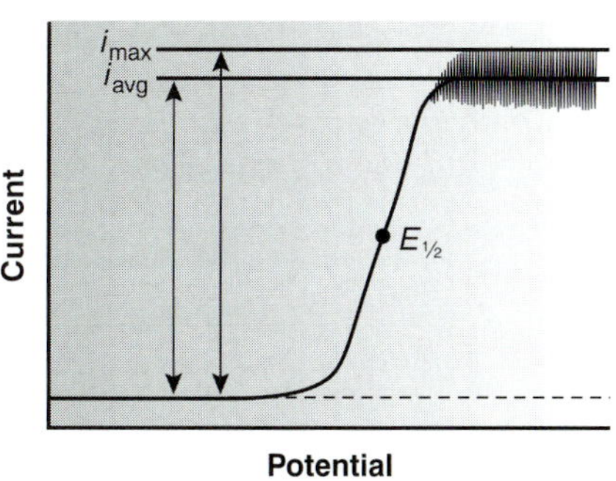

(b)

Figure 11.35
Potential-excitation signal and voltammogram for normal polarography.

time-dependent change in the area of the working electrode. The limiting current, which is also called the diffusion current, may be measured from the maximum current, i_{max}, or from the average current, i_{avg}. The relationship between the concentration of analyte, C_A, and the limiting current is given by the Ilikovic equation

$$(i_{lim})_{max} = 706nD^{1/2}m^{2/3}t^{1/6}C_A$$

$$(i_{lim})_{avg} = 607nD^{1/2}m^{2/3}t^{1/6}C_A$$

where n is the number of electrons transferred in the redox reaction, D is the analyte's diffusion coefficient, m is the flow rate of the Hg, and t is the drop time. The half-wave potential, $E_{1/2}$, provides qualitative information about the redox reaction (see Appendix 3E for a list of selected polarographic half-wave potentials).

Normal polarography has been replaced by various forms of pulse polarography,[16] several examples of which are shown in Figure 11.36. Differential pulse polarography (Figure 11.36b), for example, uses a series of potential pulses characterized by a cycle of time τ, a pulse time of t_p, a potential pulse of ΔE_p, and a potential step per cycle of ΔE_s. Typical experimental conditions for differential pulse polarography are $\tau \approx 1$ s, $t_p \approx 50$ ms, $\Delta E_p \approx 50$ mV, $\Delta E_s \approx 2$ mV. The current is measured twice, for approximately 17 ms before the forward pulse and for approximately 17 ms before the reverse pulse. The difference in the two currents gives rise to a peak-shaped voltammogram. Other forms of pulse polarography include normal pulse polarography (Figure 11.36a), staircase polarography (Figure 11.36c), and square-wave polarography (Figure 11.36d). Limiting and peak currents are directly proportional to the concentration of analyte, and half-wave and peak potentials can be used for qualitative purposes. The popularity of pulse polarography is due to a substantial improvement in sensitivity and detection limits from those in normal polarography.

Polarography is used extensively for the analysis of metal ions and inorganic anions, such as IO_3^- and NO_3^-. Organic compounds containing easily reducible or oxidizable functional groups also can be studied polarographically. Functional groups that have been used include carbonyls, carboxylic acids, and carbon–carbon double bonds.

Hydrodynamic Voltammetry In polarography a limiting current is obtained because each falling drop of mercury returns the solution near the electrode to its initial composition. As noted earlier, a limiting current is also obtained whenever the solution is stirred during the analysis. The simplest means of stirring the solution is with a magnetic stir bar. More commonly, however, stirring is achieved by rotating the electrode.

In hydrodynamic voltammetry current is measured as a function of the potential applied to a solid working electrode. The same potential profiles used for polarography, such as a linear scan or a differential pulse, are used in hydrodynamic voltammetry. The resulting voltammograms are identical to those for polarography, except for the lack of current oscillations resulting from the growth of the mercury drops. Because hydrodynamic voltammetry is not limited to Hg electrodes, it is useful for the analysis of analytes that are reduced or oxidized at more positive potentials.

stripping voltammetry
A form of voltammetry in which the analyte is first deposited on the electrode and then removed, or "stripped," electrochemically while monitoring the current as a function of the applied potential.

Stripping Voltammetry One of the most important quantitative voltammetric techniques is **stripping voltammetry,** which is composed of three related techniques: anodic, cathodic, and adsorptive stripping voltammetry. Since anodic strip-

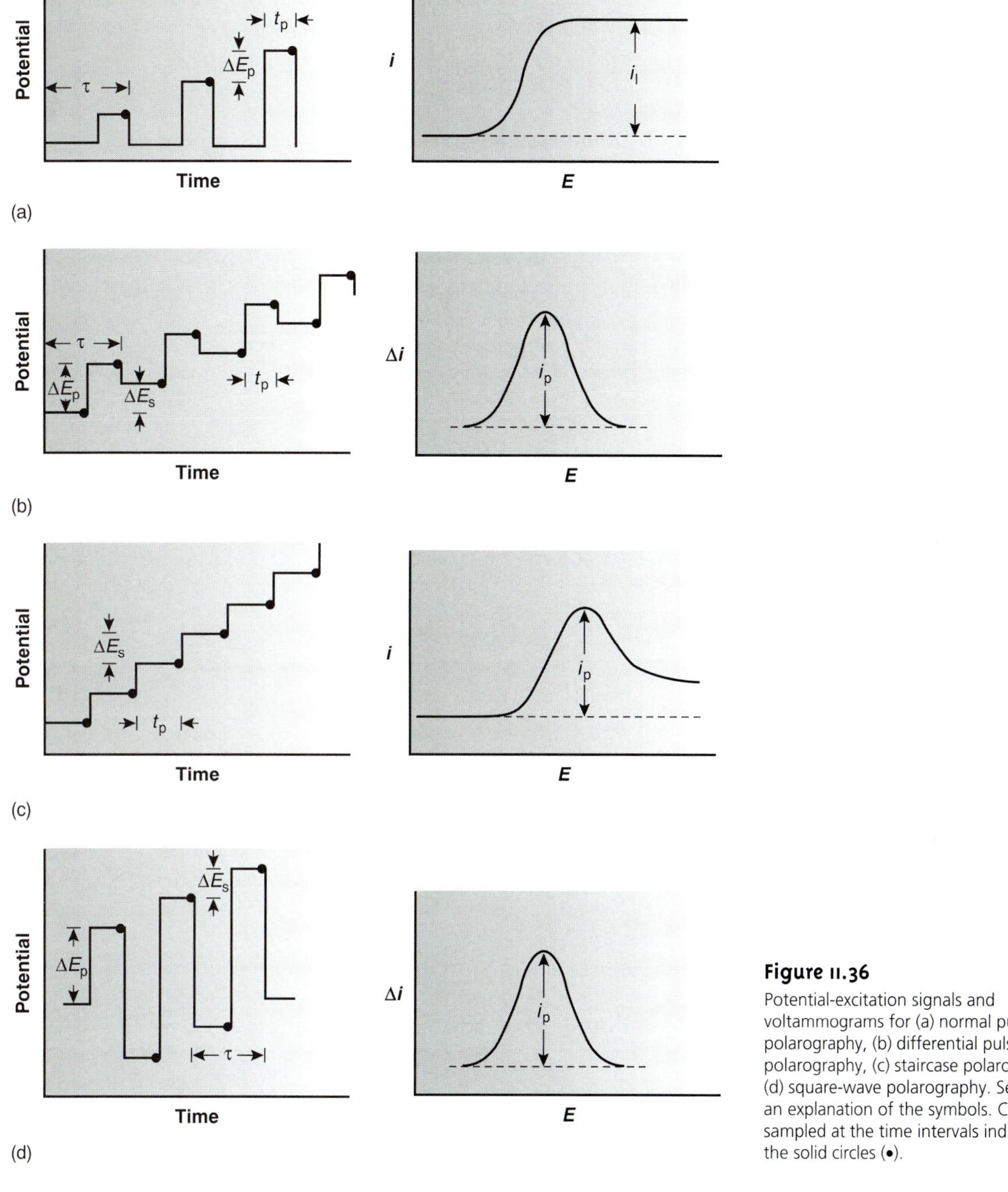

Figure 11.36
Potential-excitation signals and voltammograms for (a) normal pulse polarography, (b) differential pulse polarography, (c) staircase polarography, and (d) square-wave polarography. See text for an explanation of the symbols. Current is sampled at the time intervals indicated by the solid circles (•).

ping voltammetry has found the widest application, we consider it in the greatest detail.

Anodic stripping voltammetry consists of two steps (Figure 11.37). The first is a controlled potential electrolysis in which the working electrode, usually a hanging mercury drop or mercury film, is held at a cathodic potential sufficient to deposit the metal ion on the electrode. For example, with Cu^{2+} the deposition reaction is

$$Cu^{2+}(aq) + 2e^- \rightleftharpoons Cu(Hg)$$

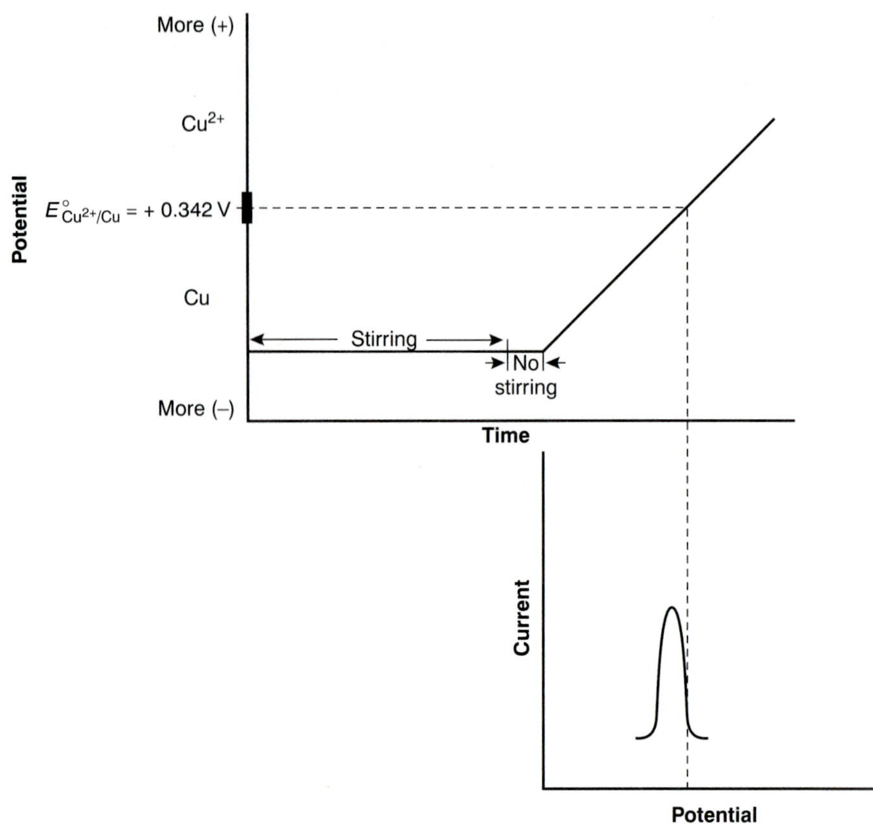

Figure 11.37
Potential-excitation signal and voltammogram for anodic stripping voltammetry at a hanging mercury drop electrode.

where Cu(Hg) indicates that the copper is amalgamated with the mercury. This step essentially serves as a means of preconcentrating the analyte from the larger volume of the solution to the smaller volume of the electrode. The solution is stirred during electrolysis to increase the rate of deposition. Near the end of the deposition time stirring is stopped, eliminating convection as a mode of mass transport. Deposition times of 1–30 min are common, with longer times being used for analytes at lower concentrations.

In the second step, the potential is scanned anodically toward more positive potentials. When the potential of the working electrode is sufficiently positive the analyte is stripped from the electrode, returning to solution as its oxidized form

$$\mathrm{Cu(Hg)} \rightleftharpoons \mathrm{Cu^{2+}}(aq) + 2e^-$$

The current during the stripping step is monitored as a function of potential, giving rise to peak-shaped voltammograms similar to that shown in Figure 11.37. The peak current is proportional to the analyte's concentration in the solution.

Anodic stripping voltammetry is very sensitive to experimental conditions, which must be carefully controlled if results are to be accurate and precise. Key variables include the area of the mercury film electrode or the size of the Hg drop when using a hanging mercury drop electrode, the deposition time, the rest time, the rate of stirring, and the scan rate during the stripping step. Anodic stripping voltammetry is best used for metals that form amalgams with mercury, several examples of which are listed in Table 11.11.

The experimental design for cathodic stripping voltammetry is similar to that for anodic stripping voltammetry with two exceptions. First, the deposition step in-

Table 11.11 Representative Examples of Analytes Determined by Stripping Voltammetry

Anodic Stripping Voltammetry	Cathodic Stripping Voltammetry	Absorptive Stripping Voltammetry
bismuth	bromide	bilirubin
cadmium	chloride	codeine
copper	iodide	cocaine
gallium	mercaptans (RSH)	digitoxin
indium	sulfide	dopamine
lead	thiocyanate	heme
thallium		monensin
tin		testosterone
zinc		

Source: Compiled from Peterson, W. M.; Wong, R. V. *Am. Lab.* November 1981, 116–128; and Wang, J. *Am. Lab.* May 1985, 41–50.[17,18]

volves the oxidation of the Hg electrode to Hg_2^{2+}, which then reacts with the analyte to form an insoluble film at the surface of the electrode. For example, when Cl^- is the analyte the deposition step is

$$2Hg(\ell) + 2Cl^-(aq) \rightleftharpoons Hg_2Cl_2(s) + 2e^-$$

Second, stripping is accomplished by scanning cathodically toward a more negative potential, reducing Hg_2^{2+} back to Hg and returning the analyte to solution.

$$Hg_2Cl_2(s) + 2e^- \rightleftharpoons 2Hg(\ell) + 2Cl^-(aq)$$

Table 11.11 lists several analytes that have been analyzed successfully by cathodic stripping voltammetry.

In adsorptive stripping voltammetry the deposition step occurs without electrolysis. Instead, the analyte adsorbs to the electrode's surface. During deposition the electrode is maintained at a potential that enhances adsorption. For example, adsorption of a neutral molecule on a Hg drop is enhanced if the electrode is held at –0.4 V versus the SCE, a potential at which the surface charge of mercury is approximately zero. When deposition is complete the potential is scanned in an anodic or cathodic direction depending on whether we wish to oxidize or reduce the analyte. Examples of compounds that have been analyzed by absorptive stripping voltammetry also are listed in Table 11.11.

Amperometry The final voltammetric technique to be considered is **amperometry,** in which a constant potential is applied to the working electrode, and current is measured as a function of time. Since the potential is not scanned, amperometry does not lead to a voltammogram.

amperometry
A form of voltammetry in which we measure current as a function of time while maintaining a constant potential.

One important application of amperometry is in the construction of chemical sensors. One of the first amperometric sensors to be developed was for dissolved O_2 in blood, which was developed in 1956 by L. C. Clark. The design of the amperometric sensor is shown in Figure 11.38 and is similar to potentiometric membrane electrodes. A gas-permeable membrane is stretched across the end of the sensor and is separated from the working and counter electrodes by a thin solution of KCl. The working electrode is a Pt disk cathode, and an Ag ring anode is the

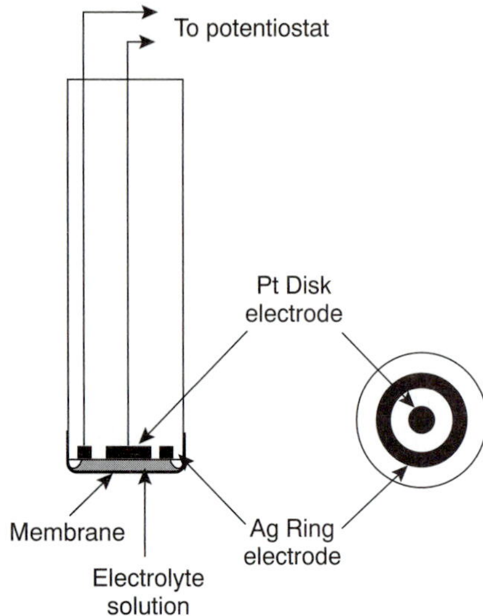

Figure 11.38
Clark amperometric sensor for the determination of dissolved O_2.

counter electrode. Although several gases can diffuse across the membrane, including O_2, N_2, and CO_2, only oxygen is reduced at the cathode.

$$O_2(aq) + 4H_3O^+(aq) + 4e^- \rightleftharpoons 6H_2O(\ell)$$

Another example of an amperometric sensor is the glucose sensor. In this case the single membrane in Figure 11.38 is replaced with three membranes. The outermost membrane is of polycarbonate, which is permeable to glucose and O_2. The second membrane contains an immobilized preparation of glucose oxidase that catalyzes the oxidation of glucose to gluconolactone and hydrogen peroxide.

$$\beta\text{-D-glucose}(aq) + O_2(g) + H_2O(\ell) \rightleftharpoons \text{gluconolactone}(aq) + H_2O_2(aq)$$

The hydrogen peroxide then diffuses through the innermost membrane of cellulose acetate, where it is oxidized at a Pt anode.

$$H_2O_2(aq) + 2OH^-(aq) \rightleftharpoons O_2(g) + 2H_2O(\ell) + 2e^-$$

Figure 11.39 summarizes the reactions taking place in this amperometric sensor. FAD is the oxidized form of flavin adenine nucleotide (the active site of the enzyme glucose oxidase), and $FADH_2$ is the active site's reduced form. Note that O_2 serves as a mediator, carrying electrons to the electrode. Other mediators, such as $Fe(CN)_6^{3-}$, can be used in place of O_2.

By changing the enzyme and mediator, the amperometric sensor in Figure 11.39 is easily extended to the analysis of other substrates. Other bioselective materials may be incorporated into amperometric sensors. For example, a CO_2 sensor has been developed using an amperometric O_2 sensor with a two-layer membrane, one of which contains an immobilized preparation of autotrophic bacteria.[19] As CO_2 diffuses through the membranes, it is converted to O_2 by the bacteria, increasing the concentration of O_2 at the Pt cathode.

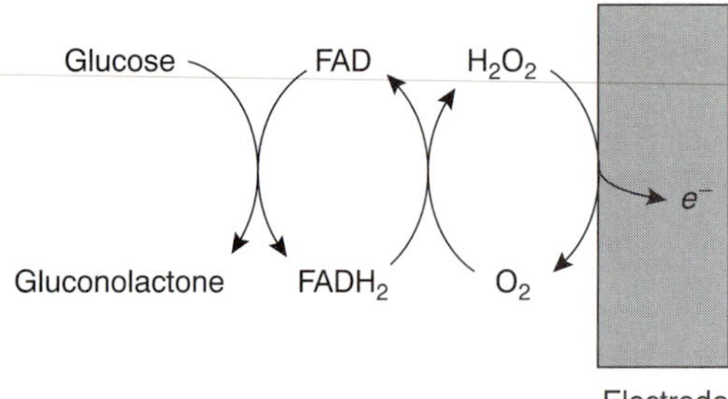

Figure 11.39
Schematic showing the reactions by which an amperometric biosensor responds to glucose.

11D.6 Quantitative Applications

Quantitative voltammetry has been applied to a wide variety of sample types, including environmental samples, clinical samples, pharmaceutical formulations, steels, gasoline, and oil.

Selecting the Voltammetric Technique The choice of which voltammetric technique to use depends on the sample's characteristics, including the analyte's expected concentration and the location of the sample. Amperometry is best suited for use as a detector in flow systems or as a selective sensor for the rapid analysis of a single analyte. The portability of amperometric sensors, which are similar to potentiometric sensors, make them ideal for field studies.

Pulse polarography and stripping voltammetry can frequently be used interchangeably, although each has its advantages and disadvantages. Pulse polarography is better for analyzing a wider range of inorganic and organic analytes because the need to preconcentrate the analyte at the electrode surface restricts the application of anodic and cathodic stripping voltammetry.

When either pulse polarography or anodic stripping voltammetry can be used, the selection is often based on the analyte's expected concentration and the desired accuracy and precision. Detection limits for normal pulse polarography generally are on the order of 10^{-6}–10^{-7} M, whereas those for differential pulse polarography, staircase, and square-wave polarography are between 10^{-7} M and 10^{-8} M. Preconcentrating the analyte in stripping voltammetry lowers the detection limit for many analytes to as little as 10^{-10} M. On the other hand, the current in stripping voltammetry is much more sensitive than pulse polarography to changes in experimental conditions, which may lead to poorer precision and accuracy.

Anodic stripping voltammetry also suffers from occasional interferences when two metals, such as Cu and Zn, combine to form an intermetallic compound in the mercury amalgam. The deposition potential for Zn^{2+} is sufficiently negative that any Cu^{2+} present in the sample is also deposited. After deposition, intermetallic compounds such as CuZn and $CuZn_2$ form within the mercury amalgam. During the stripping step, the zinc in the intermetallic compounds strips at potentials near that of copper, decreasing the current for zinc and increasing the current for copper. This problem can often be overcome by adding a third element that forms a stronger intermetallic compound with the interfering metal. Thus, adding Ga^{3+} minimizes this problem by forming an intermetallic compound of Cu and Ga.

Correcting for Residual Current In any quantitative analysis the signal due to the analyte must be corrected for signals arising from other sources. The total measured current in any voltammetric experiment, i_{tot}, consists of two parts: that due to the analyte's oxidation or reduction, i_a, and a background, or residual, current, i_r.

$$i_{tot} = i_a + i_r$$

The residual current, in turn, has two sources. One source is a faradaic current due to the oxidation or reduction of trace impurities in the sample, i_i. The other source is the charging current, i_{ch}, that is present whenever the working electrode's potential changes.

$$i_r = i_i + i_{ch}$$

Faradaic currents due to impurities can usually be minimized by carefully preparing the sample. For example, one important impurity is dissolved O_2, which is reduced first to H_2O_2 and then to H_2O. Dissolved O_2 is removed by bubbling an inert gas such as N_2 through the sample before the analysis.

Two methods are commonly used to correct for the residual current. One method is to extrapolate the total measured current when the analyte's faradaic current is zero. This is the method shown in the voltammograms included in this chapter. The advantage of this method is that it does not require any additional data. On the other hand, extrapolation assumes that changes in the residual current with potential are predictable, which often is not the case. A second, and more rigorous, approach is to obtain a voltammogram for an appropriate blank. The blank's residual current is then subtracted from the total current obtained with the sample.

Analysis for Single Components The analysis of samples containing only a single electroactive analyte is straightforward. Any of the standardization methods discussed in Chapter 5 can be used to establish the relationship between current and the concentration of analyte.

Example

EXAMPLE 11.10

The concentration of As(III) in water can be determined by differential pulse polarography in 1 M HCl. The initial potential is set to –0.1 V versus the SCE, and is scanned toward more negative potentials at a rate of 5 mV/s. Reduction of As(III) to As(0) occurs at a potential of approximately –0.44 V versus the SCE. The peak currents, corrected for the residual current, for a set of standard solutions are shown in the following table.

[As(III)] (M)	i_p (μA)
1.00×10^{-6}	0.298
3.00×10^{-6}	0.947
6.00×10^{-6}	1.83
9.00×10^{-6}	2.72

What is the concentration of As(III) in a sample of water if the peak current under the same conditions is 1.37 μA?

SOLUTION

Linear regression gives the equation for the calibration curve as

$$i_p(\mu A) = 0.0176 + 3.01 \times 10^5[\text{As(III)}]$$

Substituting the sample's peak current into the regression equation gives the concentration of As(III) as 4.49×10^{-6} M.

Example

EXAMPLE 11.11

The concentration of copper in a sample of sea water is determined by anodic stripping voltammetry using the method of standard additions. When a 50.0-mL sample is analyzed, the peak current is 0.886 μA. A 5.00-μL spike of 10.0-ppm Cu^{2+} is added, giving a peak current of 2.52 μA. Calculate the parts per million of copper in the sample of sea water.

SOLUTION

Peak currents in anodic stripping voltammetry are a linear function of concentration

$$i_p = k(\text{ppm Cu}^{2+})$$

where k is a constant. Thus, for the sample we write

$$0.886 = k(\text{ppm Cu}^{2+})$$

and for the standard addition

$$2.52 = k\left[\frac{0.0500 \text{ L}}{0.0500 \text{ L} + 5.00 \times 10^{-6} \text{ L}}(\text{ppm Cu}^{2+}) + \frac{5.00 \times 10^{-6} \text{ L}}{0.0500 \text{ L} + 5.00 \times 10^{-6} \text{ L}}(10.0 \text{ ppm})\right]$$

Solving the first equation for *k*, substituting into the second equation, and simplifying gives

$$2.52 = 0.8859 + \frac{(8.859 \times 10^{-5})(10.0 \text{ ppm})}{(\text{ppm } Cu^{2+})}$$

Solving gives the concentration of Cu^{2+} as 5.42×10^{-4} ppm, or 0.542 ppb.

Multicomponent Analysis One advantage of voltammetry as a quantitative method of analysis is its capability for analyzing two or more analytes in a single sample. As long as the components behave independently, the resulting voltammogram for a multicomponent mixture is a summation of their respective individual voltammograms. If the separation between the half-wave potentials or peak potentials is sufficient, each component can be determined independently as if it were the only component in the sample (Figure 11.40). The minimum separation between the half-wave potentials or peak potentials for the independent analysis of two components depends on several factors, including the type of electrode and the potential-excitation signal. For normal polarography the separation must be at least ±0.2–0.3 V, and for differential pulse voltammetry a minimum separation of ±0.04–0.05 V is needed.

When the overlap between the voltammograms for two components prevents their independent analysis, a simultaneous analysis similar to that used in spectrophotometry may be possible. An example of this approach is outlined in Example 11.12.

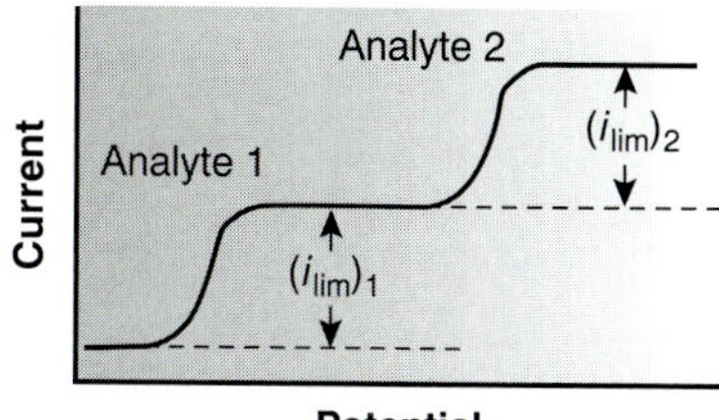

(a)

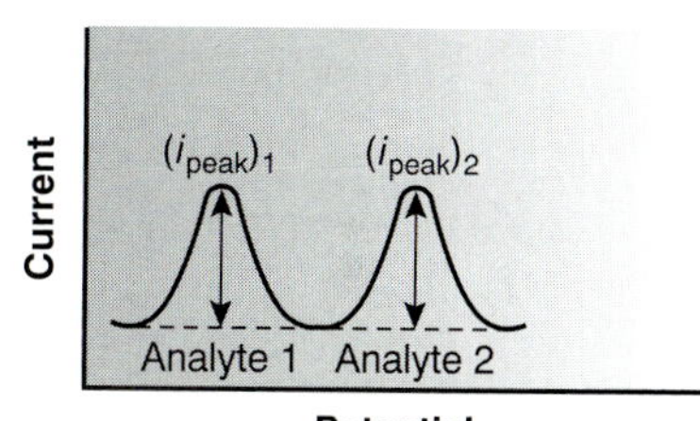

(b)

Figure 11.40
Voltammograms showing the independent analysis of two components.

Example

EXAMPLE 11.12

The differential pulse polarographic analysis of mixtures of indium and cadmium in 0.1 M HCl is complicated by the overlap of their respective voltammograms.[20] The peak potential for indium is at –0.557 V, and that for cadmium occurs at a potential of –0.597 V. When a 0.800-ppm indium standard is analyzed, the peak current (in arbitrary units) is found to be 200.5 at –0.557 V and 87.5 at –0.597 V. A standard solution of 0.793-ppm cadmium gives peak currents of 58.5 at –0.557 V and 128.5 at –0.597 V. What is the concentration of indium and cadmium in a sample if the peak current is 167.0 at a potential of –0.557 V and 99.5 at a potential of –0.597 V?

SOLUTION

Peak currents in differential pulse polarography are a linear function of the concentration of analyte; thus

$$i_p = k(\text{ppm analyte})$$

where *k* is a constant that depends on the analyte and the applied potential. To determine the concentrations of indium and cadmium in the sample, we must first find the values of *k* for each analyte at each potential. For simplicity we will identify the potential of –0.557 V as *E*1, and that for –0.597 V as *E*2. The values of *k* are

$$k_{In,E1} = \frac{200.5}{0.800 \text{ ppm}} = 250.6 \text{ ppm}^{-1} \qquad k_{In,E2} = \frac{87.5}{0.800 \text{ ppm}} = 109.4 \text{ ppm}^{-1}$$

$$k_{\text{Cd},E1} = \frac{58.5}{0.793 \text{ ppm}} = 73.8 \text{ ppm}^{-1} \qquad k_{\text{Cd},E2} = \frac{128.5}{0.793 \text{ ppm}} = 162.0 \text{ ppm}^{-1}$$

Next, we write simultaneous equations for the current at the two potentials

$$i_{E1} = 250.6 \text{ ppm}^{-1} \text{ (ppm In)} + 73.8 \text{ ppm}^{-1} \text{ (ppm Cd)} = 167.0$$

$$i_{E2} = 109.4 \text{ ppm}^{-1} \text{ (ppm In)} + 162.0 \text{ ppm}^{-1} \text{ (ppm Cd)} = 99.5$$

Solving the simultaneous equations, which is left as an exercise, gives the concentration of indium as 0.606 ppm and the concentration of cadmium as 0.205 ppm.

Environmental Samples One area in which quantitative voltammetry has had a significant effect is in the analysis of trace metals in environmental samples. The most common samples are natural waters, including groundwater, lakes, rivers and streams, sea water, rain, and snow. Concentrations of trace metals at the parts-per-billion level can be determined using differential pulse polarography, whereas with anodic stripping voltammetry the determination of trace metals at the pptr (parts-per-trillion) level is possible. The combination of low detection limits and the capability for the simultaneous analysis of several analytes makes differential pulse polarography and anodic stripping voltammetry ideally suited for such samples.

One interesting application of anodic stripping voltammetry to the analysis of natural waters is the determination of the speciation, or chemical form, of the trace metals. The speciation of a trace metal is important because its bioavailability, toxicity, and ease of transport through the environment often depend on its chemical form. For example, trace metals strongly bound to colloidal particles are generally not available to aquatic lifeforms and, therefore, are not toxic. Unfortunately, anodic stripping voltammetry cannot distinguish the exact chemical form of a trace metal as closely related species, such as Pb^{2+} and $PbCl^{+}$, yield only a single stripping peak. Instead, trace metals are divided into several "operationally defined" categories that have environmental significance.

Several speciation schemes have been developed,[21] but we consider only the speciation scheme proposed by Batley and Florence,[22] which uses a combination of anodic stripping voltammetry with ion-exchange and UV irradiation to divide soluble trace metals into seven classes. Anodic stripping voltammetry in a pH 4.8 acetic acid buffer is used to distinguish labile metals present as hydrated ions, weakly bound complexes, or weakly adsorbed on colloidal surfaces from those metals that are bound in stronger complexes or strongly adsorbed. Only those metals that are hydrated, weakly bound, or weakly adsorbed deposit at the electrode. Ion exchange and UV irradiation are used to further subdivide the trace metals. A Chelex-100 ion-exchange resin is used to distinguish between ionic metals and strongly bound metals, whereas UV radiation is used to separate metals bound to organic and inorganic phases. Table 11.12 shows how trace metals are divided into seven classes using these three experimental techniques. The analysis of sea water samples, for example, showed that cadmium, copper, and lead were primarily present as labile organic complexes or as labile adsorbates on organic colloids (see group II, Table 11.12).

Differential pulse polarography and stripping voltammetry have been applied to the analysis of trace metals in airborne particulates, incinerator fly ash, rocks,

Table 11.12 Speciation of Soluble Trace Metals According to the Scheme of Batley and Florence[22]

Method	**Speciation of Soluble Metals**						
ASV	**Labile Metals**			**Nonlabile or Bound Metals**			
Ion-Exchange	**Removed**	**Not Removed**		**Removed**		**Not Removed**	
UV Irradiation		**Released**	**Not Released**	**Released**	**Not Released**	**Released**	**Not Released**
Description	free metal ions, labile organic complexes, and labile inorganic complexes	labile organic complexes and labile metals absorbed on organics	labile inorganic complexes and labile metals absorbed on inorganics	nonlabile organic complexes	nonlabile inorganic complexes	nonlabile organic complexes and nonlabile metals absorbed on organics	nonlabile inorganic complexes and nonlabile metals absorbed on inorganics
Group	I	II	III	IV	V	VI	VII

minerals, and sediments. The trace metals, of course, must be brought into solution by digesting or extracting before the voltammetric analysis.

Amperometric sensors also are used to analyze environmental samples. For example, the dissolved O_2 sensor described earlier is routinely used for the determination of dissolved oxygen and biochemical oxygen demand, or BOD, in waters and wastewaters. The latter test, which is a measure of the amount of oxygen required by aquatic bacteria during the decomposition of organic matter, is of importance in evaluating the efficiency of wastewater treatment plants and in monitoring organic pollution in natural waters. A high BOD corresponds to a high concentration of organic material that may seriously deplete the level of dissolved oxygen in the water. Other amperometric sensors have been developed to monitor anionic surfactants in water, and CO_2, H_2SO_4, and NH_3 in atmospheric gases.[19]

Clinical Samples Differential pulse polarography and stripping voltammetry have been used to determine the concentration of trace metals in a variety of matrices, including blood, urine, and tissue samples. The determination of lead in blood is of considerable interest due to concerns about lead poisoning. Because the concentration of lead in blood is so small, anodic stripping voltammetry frequently is the method of choice. The analysis is complicated, however, by the presence of proteins that may adsorb at the surface of the mercury electrode, inhibiting either the deposition or stripping of lead. In addition, proteins may prevent the electrodeposition of lead through the formation of stable, nonlabile complexes. For these reasons samples of whole blood must be digested or ashed before the analysis. Differential pulse polarography is one of the few techniques that can be used for the routine quantitative analysis of drugs in biological fluids at concentrations of less than 10^{-6} M.[23] Amperometric sensors based on enzyme catalysts also have wide applicability.[24] Table 11.13 provides a partial list of enzymatic amperometric sensors.

Miscellaneous Samples Besides environmental and clinical samples, differential pulse polarography and stripping voltammetry have been used for the analysis of trace metals in other samples, including food, steels and other alloys, gasoline, gunpowder residues, and pharmaceuticals. Voltammetry is also an important tool for

Table 11.13 Representative Examples of Amperometric Biosensors

Analyte	Enzyme	Species Detected
choline	choline oxidase	H_2O_2
ethanol	alcohol oxidase	H_2O_2
formaldehyde	formaldehyde dehydrogenase	NADH[a]
glucose	glucose oxidase	H_2O_2
glutamine	glutaminase, glutamate oxidase	H_2O_2
glycerol	glycerol dehydrogenase	NADH, O_2
lactate	lactate oxidase	H_2O_2
phenol	polyphenol oxidase	quinone
inorganic P	nucleoside phosphorylase	O_2

Source: Compiled from Cammann, K.; Lemke, U.; Rohen, A.; et al. *Angew. Chem. Int. Ed. Engl.* **1991,** *30,* 516–539.

[a]NADH is the reduced form of nicotinamide adenine dinucleotide.

the quantitative analysis of organics, particularly in the pharmaceutical industry, in which it is used to determine the concentration of drugs and vitamins in formulations.[23] For example, voltammetric methods have been developed for the quantitative analysis of vitamin A, niacinamide, and riboflavin. When the compound of interest is not electroactive, it often can be derivatized to an electroactive form. One example is the differential pulse polarographic determination of sulfanilamide, in which it is converted into an electroactive azo dye by coupling with sulfamic acid and 1-naphthol.

Representative Methods

Method 11.3 Determination of Chlorpromazine[25]

Description of Method. The amount of chlorpromazine in a pharmaceutical formulation is determined voltammetrically at a graphite working electrode in a nonstirred solution. Calibration is achieved using the method of standard additions.

Procedure. Place 10.00 mL of a solution consisting of 0.01 M HCl and 0.1 M KCl in the electrochemical cell. Place a graphite working electrode, a Pt auxiliary electrode, and a SCE reference electrode in the cell, and record the voltammogram from +0.2 V to 2.0 V at a scan rate of 50 mV/s. Weigh out an appropriate amount of the pharmaceutical formulation, and dissolve it in a small amount of the electrolyte. Transfer the solution to a 100-mL volumetric flask, and dilute to volume with the electrolyte. Filter a small amount of the diluted solution, and transfer 1.00 mL of the filtrate to the voltammetric cell. Mix the contents of the voltammetric cell, and allow the solution to sit for 10 s before recording the voltammogram. Return the potential to +0.2 V, add 1.00 mL of a chlorpromazine standard, and record the voltammogram. Report the %w/w chlorpromazine in the formulation.

Questions

1. **Is chlorpromazine oxidized or reduced at the graphite working electrode?**

   Since the direction of the scan is toward more positive potentials, the chlorpromazine is oxidized at the graphite working electrode.

—Continued

2. **Why is it not necessary to remove dissolved O_2 from the solution before recording the voltammogram?**

   Dissolved O_2 is a problem when scanning toward more negative potentials where its reduction can lead to a significant cathodic current. In this procedure we are scanning toward more positive potentials and generating anodic currents; thus, dissolved O_2 is not an interferent.

3. **What is the purpose of waiting 10 s after mixing the contents of the electrochemical cell before recording the voltammogram?**

   Mixing the solution leads to convection, but the delay allows convection to cease.

4. **In preparing the sample for analysis the initial solution is filtered. Why is it not necessary to collect the entire filtrate before proceeding?**

   This analysis is an example of a concentration technique. Once the original sample is brought to volume in the 100-mL volumetric flask, any portion of the sample solution, even that obtained on filtering, may be used for the analysis.

11D.7 Characterization Applications

In the previous section we saw how voltammetry can be used to determine the concentration of an analyte. Voltammetry also can be used to obtain additional information, including verifying electrochemical reversibility, determining the number of electrons transferred in a redox reaction, and determining equilibrium constants for coupled chemical reactions. Our discussion of these applications is limited to the use of voltammetric techniques that give limiting currents, although other voltammetric techniques also can be used to obtain the same information.

Electrochemical Reversibility and Determination of *n* In deriving a relationship between $E_{1/2}$ and the standard-state potential for a redox couple (11.41), we noted that the redox reaction must be reversible. How can we tell if a redox reaction is reversible from its voltammogram? For a reversible reaction, equation 11.40 describes the voltammogram.

$$E = E^{\circ}_{O/R} - \frac{0.05916}{n}\log\frac{K_O}{K_R} - \frac{0.05916}{n}\log\frac{i}{i_{lim} - i}$$

A plot of E versus $\log(i/i_{lim} - i)$ for a reversible reaction, therefore, should be a straight line with a slope of $-0.05916/n$. In addition, the slope should yield an integer value for n.

Example

EXAMPLE 11.13

The following data were obtained from the linear scan hydrodynamic voltammogram of a reversible reduction reaction

Potential (V vs. SCE)	Current (μA)
-0.358	0.37
-0.372	0.95
-0.382	1.71
-0.400	3.48
-0.410	4.20
-0.435	4.97

The limiting current was 5.15 μA. Show that the reduction reaction is reversible, and determine values for n and $E_{1/2}$.

SOLUTION

If the reaction is reversible, then a plot of E versus $\log(i/i_{lim} - i)$ will be a straight line with a slope that yields an integer value for n. As shown in Figure 11.41, plotting the data in this manner gives a straight line. Linear regression gives the equation for the straight line as

$$E = -0.391 - 0.02999 \log \frac{i}{i_{lim} - i}$$

Knowing that

$$\text{Slope} = -0.02999 = -\frac{0.05916}{n}$$

gives a value for n of 1.97, or 2 electrons. Combining equations 11.40 and 11.41 shows that the y-intercept for a plot of E versus $\log(i/i_{lim} - i)$ is equal to the half-wave potential. Thus, $E_{1/2}$ for the reduction reaction is –0.391 V versus the SCE.

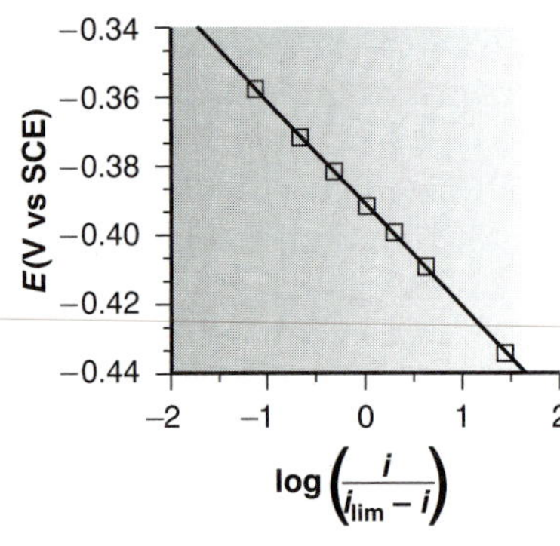

Figure 11.41
Graphical determination of electrochemical reversibility, n, and half-wave potential in linear scan hydrodynamic voltammetry.

Determining Equilibrium Constants for Coupled Chemical Reactions Another important application of voltammetry is the determination of equilibrium constants for solution reactions that are coupled to a redox reaction occurring at the electrode. The presence of the solution reaction affects the ease of electron transfer, shifting the potential to more negative or more positive potentials. Consider, for example, the reduction of O to R

$$O + ne^- \rightleftharpoons R$$

the voltammogram for which is shown in Figure 11.42. If a ligand, L, capable of forming a strong complex with O, is present, then the reaction

$$O + pL \rightleftharpoons OL_p$$

also must be considered. The overall reaction, therefore, is

$$OL_p + ne^- \rightleftharpoons R + pL$$

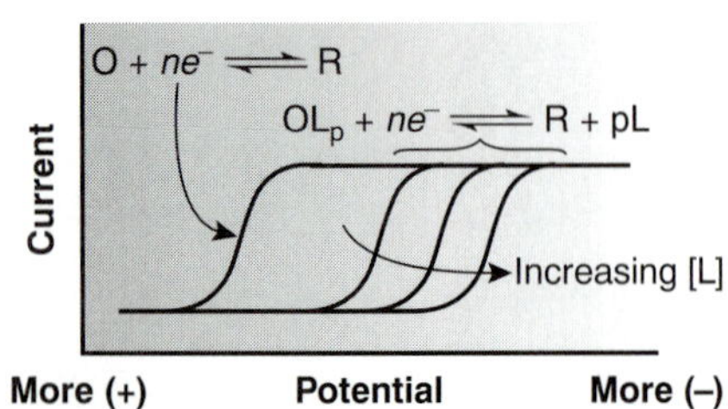

Figure 11.42
Effect of complexation on voltammograms.

Because of its stability, reduction of the OL_p complex is less favorable than the reduction of O. As shown in Figure 11.42, the voltammograms for OL_p occur at potentials more negative than those for O. Furthermore, the shift in the voltammogram depends on the ligand's concentration.

The shift in the voltammogram for a metal ion in the presence of a ligand may be used to determine both the metal–ligand complex's stoichiometry and its formation constant. To derive a relationship between the relevant variables we begin with two equations: the Nernst equation for the reduction of O

$$E = E^{\circ}_{O/R} - \frac{0.05916}{n}\log\frac{[R]_{x=0}}{[O]_{x=0}}$$

and the stability constant for the complex at the electrode surface

$$\beta_p = \frac{[OL_p]_{x=0}}{[O]_{x=0}[L]^p_{x=0}} \qquad \textbf{11.42}$$

In the absence of ligand the half-wave potential occurs when $[R]_{x=0}$ and $[O]_{x=0}$ are equal; thus, from the Nernst equation we have

$$(E_{1/2})_{nc} = E^{\circ}_{O/R} \qquad \textbf{11.43}$$

where the subscript "nc" signifies that no complex is present.

When ligand is present we must account for its effect on the concentration of O. Solving equation 11.42 for $[O]_{x=0}$ and substituting into the Nernst equation gives

$$E = E^{\circ}_{O/R} - \frac{0.05916}{n}\log\frac{[R]_{x=0}[L]^p_{x=0}\beta_p}{[OL_p]_{x=0}} \qquad \textbf{11.44}$$

If the ligand is present in excess and the formation constant is sufficiently large, such that all of O is present as the complex, then $[R]_{x=0}$ and $[OL_p]_{x=0}$ are equal at the half-wave potential, and equation 11.44 simplifies to

$$(E_{1/2})_c = E^{\circ}_{O/R} - \frac{0.05916}{n}\log[L]^p_{x=0}\beta_p \qquad \textbf{11.45}$$

where the subscript "c" indicates that the complex is present. Defining $\Delta E_{1/2}$ as

$$\Delta E_{1/2} = (E_{1/2})_c - (E_{1/2})_{nc} \qquad \textbf{11.46}$$

and substituting equations 11.44 and 11.45 gives, after rearranging

$$\Delta E_{1/2} = -\frac{0.05916}{n}\log\beta_p - \frac{0.05916p}{n}\log[L]$$

A plot of $\Delta E_{1/2}$ versus log [L], therefore, gives a straight line. The slope of the line is used to determine the stoichiometric coefficient, p, and the y-intercept gives the value for the formation constant β_p.

Example

EXAMPLE 11.14

A voltammogram for the two-electron reduction of M has a half-wave potential of –0.226 V versus the SCE. In the presence of an excess of the ligand L, the following half-wave potentials are recorded

[L] (M)	$(E_{1/2})_c$ (V vs. SCE)
0.020	–0.494
0.040	–0.512
0.060	–0.523
0.080	–0.530
0.100	–0.536

Determine values for the stoichiometry of the complex and the formation constant of the complex.

SOLUTION

First we must calculate values of $\Delta E_{1/2}$ using equation 11.46; thus

[L] (M)	$\Delta E_{1/2}$ (V vs. SCE)
0.020	–0.268
0.040	–0.286
0.060	–0.297
0.080	–0.304
0.100	–0.310

The plot of $\Delta E_{1/2}$ as a function of the log of the ligand concentration is shown in Figure 11.43. Linear regression gives the equation for the straight line as

$$\Delta E_{1/2} = -0.370 - 0.0601 \log [\mathrm{L}]$$

From the slope and intercept we know that

$$-0.0601 = -\frac{0.05916p}{2}$$

and

$$-0.371 = -\frac{0.05916}{2} \log \beta_p$$

Solving these equations gives the stoichiometry of the complex as ML_2 ($p = 2.03$), with a formation constant of $\beta_2 = 3.5 \times 10^{12}$.

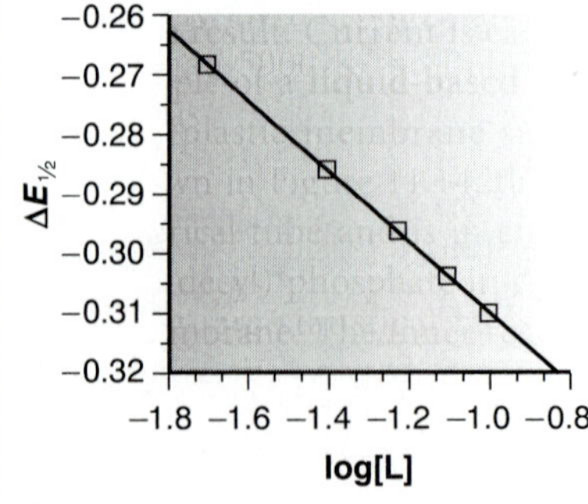

Figure 11.43
Graphical determination of the stoichiometry and formation constant for a complexation reaction.

11D.8 Evaluation

Scale of Operation Voltammetry is routinely used to analyze samples at the parts-per-million level and, in some cases, can be used to detect analytes at the parts-per-billion or parts-per-trillion level. Most analyses are carried out in conventional electrochemical cells using macro samples; however, microcells are available that require as little as 50 μL of sample. Microelectrodes, with diameters as small as 2 μm, allow voltammetric measurements to be made on even smaller samples. For example, the concentration of glucose in 200-μm pond snail neurons has been successfully monitored using a 2-μm amperometric glucose electrode.[26]

Accuracy The accuracy of a voltammetric analysis often is limited by the ability to correct for residual currents, particularly those due to charging. For analytes at the parts-per-million level, accuracies of ±1–3% are easily obtained. As expected, a decrease in accuracy is experienced when analyzing samples with significantly smaller concentrations of analyte.

Precision Precision is generally limited by the uncertainty in measuring the limiting or peak current. Under most experimental conditions, precisions of ±1–3% can be reasonably expected. One exception is the analysis of ultratrace analytes in complex matrices by stripping voltammetry, for which precisions as poor as ±25% are possible.

Sensitivity In many voltammetric experiments, sensitivity can be improved by adjusting the experimental conditions. For example, in stripping voltammetry, sensitivity is improved by increasing the deposition time, by increasing the rate of the linear potential scan, or by using a differential-pulse technique. One reason for the popularity of potential pulse techniques is an increase in current relative to that obtained with a linear potential scan.

Selectivity Selectivity in voltammetry is determined by the difference between half-wave potentials or peak potentials, with minimum differences of ±0.2–0.3 V required for a linear potential scan, and ±0.04–0.05 V for differential pulse voltammetry. Selectivity can be improved by adjusting solution conditions. As we have seen, the presence of a complexing ligand can substantially shift the potential at which an analyte is oxidized or reduced. Other solution parameters, such as pH, also can be used to improve selectivity.

Time, Cost, and Equipment Commercial instrumentation for voltammetry ranges from less than $1000 for simple instruments to as much as $20,000 for more sophisticated instruments. In general, less expensive instrumentation is limited to linear potential scans, and the more expensive instruments allow for more complex potential-excitation signals using potential pulses. Except for stripping voltammetry, which uses long deposition times, voltammetric analyses are relatively rapid.

11E KEY TERMS

amalgam (p. 509)
amperometry (p. 519)
anode (p. 466)
anodic current (p. 510)
asymmetry potential (p. 476)
auxiliary electrode (p. 463)
cathode (p. 466)
cathodic current (p. 510)
charging current (p. 513)
convection (p. 512)
coulometric titrations (p. 501)
coulometry (p. 496)
counter electrode (p. 462)
current efficiency (p. 496)
diffusion (p. 511)
diffusion layer (p. 511)
dropping mercury electrode (p. 509)
electrical double layer (p. 513)
electrode of the first kind (p. 474)
electrode of the second kind (p. 475)
enzyme electrodes (p. 484)
faradaic current (p. 510)
Faraday's law (p. 496)
galvanostat (p. 464)
glass electrode (p. 477)
hanging mercury drop electrode (p. 509)
hydrodynamic voltammetry (p. 513)
indicator electrode (p. 462)
ionophore (p. 482)
ion-selective electrode (p. 475)
liquid-based ion-selective electrode (p. 482)
liquid junction potential (p. 470)
mass transport (p. 511)
mediator (p. 500)
membrane potential (p. 475)
migration (p. 512)
nonfaradaic current (p. 512)
Ohm's law (p. 463)
overpotential (p. 497)
polarography (p. 515)
potentiometer (p. 464)
potentiostat (p. 465)
redox electrode (p. 475)
reference electrode (p. 462)
residual current (p. 513)
salt bridge (p. 466)
saturated calomel electrode (p. 472)
silver/silver chloride electrode (p. 473)
solid-state ion-selective electrode (p. 479)
standard hydrogen electrode (p. 471)
static mercury drop electrode (p. 509)
stripping voltammetry (p. 516)
total ionic strength adjustment buffer (p. 487)
voltammetry (p. 508)
voltammogram (p. 508)

11F SUMMARY

Electrochemical methods covered in this chapter include potentiometry, coulometry, and voltammetry. Potentiometric methods are based on the measurement of an electrochemical cell's potential when only a negligible current is allowed to flow. In principle the Nernst equation can be used to calculate the concentration of species in the electrochemical cell by measuring its potential and solving the Nernst equation; the presence of liquid junction potentials, however, necessitates the use of an external standardization or the use of standard additions.

Potentiometric electrodes are divided into two classes: metallic electrodes and membrane electrodes. The smaller of these classes are the metallic electrodes. Electrodes of the first kind respond to the concentration of their cation in solution; thus the potential of an Ag wire is determined by the concentration of Ag^+ in solution. When another species is present in solution and in equilibrium with the metal ion, then the electrode's potential will respond to the concentration of that ion. For example, an Ag wire in contact with a solution of Cl^- will respond to the concentration of Cl^- since the relative concentrations of Ag^+ and Cl^- are fixed by the solubility product for AgCl. Such electrodes are called electrodes of the second kind.

The potential of a membrane electrode is determined by a difference in the composition of the solution on either side of the membrane. Electrodes using a glass membrane respond to ions that bind to negatively charged sites on the glass membrane's surface. A pH electrode is one example of a glass membrane electrode. Other kinds of membrane electrodes include those using insoluble crystalline solids and liquid ion exchangers incorporated in a hydrophobic membrane. The F^- ion-selective electrode, which uses a single crystal of LaF_3 as the ion-selective membrane, is an example of a solid-state electrode. The Ca^{2+} ion-selective electrode, in which a chelating ligand such as di-(*n*-decyl)phosphate is immobilized in a PVC membrane, is an example of a liquid-based ion-selective electrode.

Potentiometric electrodes also can be designed to respond to molecules by incorporating a reaction producing an ion whose concentration can be determined using a traditional ion-selective electrode. Gas-sensing electrodes, for example, include a gas-permeable membrane that isolates the ion-selective electrode from the solution containing the analyte. Diffusion of a dissolved gas across the membrane alters the composition of the inner solution in a manner that can be followed with an ion-selective electrode. Enzyme electrodes operate in the same way.

Coulometric methods are based on Faraday's law that the total charge or current passed during an electrolysis is proportional to the amount of reactants and products in the redox reaction. If the electrolysis is 100% efficient, in that only the analyte is oxidized or reduced, then the total charge or current can be used to determine the amount of analyte in a sample. In controlled-potential coulometry, a constant potential is applied and the current is measured as a function of time, whereas in controlled-current coulometry, the current is held constant and the time required to completely oxidize or reduce the analyte is measured.

In voltammetry we measure the current in an electrochemical cell as a function of the applied potential. Individual voltammetric methods differ in terms of the type of electrode used, how the applied potential is changed, and whether the transport of material to the electrode's surface is enhanced by stirring.

Polarography is a voltammetric experiment conducted at a mercury electrode under conditions in which the solution is not stirred. Normal polarography uses a dropping mercury electrode (or a static mercury drop electrode) and a linear potential scan in an unstirred solution. Other forms of polarography include normal pulse polarography, differential pulse polarography, staircase polarography, and square-wave polarography, all of which apply a series of potential pulses to the mercury electrode.

In hydrodynamic voltammetry the solution is stirred either by using a magnetic stir bar or by rotating the electrode. Because the solution is stirred, a dropping mercury electrode cannot be used and is replaced with a solid electrode. Both linear potential scans or potential pulses can be applied.

In stripping voltammetry the analyte is first deposited on the electrode, usually as the result of an oxidation or reduction reaction. The potential is then scanned, either linearly or by using potential pulses, in a direction that removes the analyte by a reduction or oxidation reaction.

Amperometry is a voltammetric method in which a constant potential is applied to the electrode and the resulting current is measured. Amperometry is most often used in the construction of chemical sensors that, as with potentiometric sensors, are used for the quantitative analysis of single analytes. One important example, for instance, is the Clark O_2 electrode, which responds to the concentration of dissolved O_2 in solutions such as blood and water.

11G *Suggested* EXPERIMENTS

Experiments

The following set of suggested experiments describes the preparation of solid-state and liquid ion-exchange ion-selective electrodes, as well as potentiometric biosensors.

Chan, W. H.; Wong, M. S.; Yip, C. W. "Ion-Selective Electrode in Organic Analysis: A Salicylate Electrode," *J. Chem. Educ.* **1986,** *63,* 915–916.

The preparation of an ion-selective electrode for salicylate is described. The electrode incorporates an ion-pair of crystal violet and salicylate in a PVC matrix as the ion-selective membrane. Its use for the determination of acetylsalicylic acid in aspirin tablets is described. A similar experiment is described by Creager, S. E.; Lawrence, K. D.; Tibbets, C. R. in "An Easily Constructed Salicylate-Ion-Selective Electrode for Use in the Instructional Laboratory," *J. Chem. Educ.* **1995,** *72,* 274–276.

Christopoulos, T. K.; Diamandis, E. P. "Use of a Sintered Glass Crucible for Easy Construction of Liquid-Membrane Ion-Selective Electrodes," *J. Chem. Educ.* **1988,** *65,* 648.

Liquid-membrane ion-selective electrodes are prepared by removing the bottom below the glass filter, cleaning the filter, and making it hydrophobic using trimethylchlorosilane. Suitable liquid ion exchangers are added and allowed to penetrate the hydrophobic glass membrane. Adding an appropriate internal reference solution and reference electrode completes the ion-selective electrode.

Fricke, G. H.; Kuntz, M. J. "Inexpensive Solid-State Ion-Selective Electrodes for Student Use," *J. Chem. Educ.* **1977,** *54,* 517–520.

Instructions are provided in this experiment for the preparation of solid-state electrodes for the quantitative analysis of Ag^+, Cl^-, Cu^{2+}, and Pb^{2+}. All electrodes are based on AgCl, AgCl/Ag_2S, CuS/Ag_2S, and PbS/Ag_2S preparations. Electrodes are prepared by (a) mixing the solid with silicon grease and coating on an Ag wire; (b) mixing the solid with silicone grease and packing into the end of a glass tube; or (c) preparing a pressed pellet.

Lloyd, B. W.; O'Brien, F. L.; Wilson, W. D. "Student Preparation and Analysis of Chloride and Calcium Ion Selective Electrodes," *J. Chem. Educ.* **1976,** *53,* 328–330.

This experiment describes the preparation of liquid ion-exchange electrodes for Cl^- and Ca^{2+}. The liquid ion-exchange solutions are incorporated into PVC membranes and fixed to the end of glass tubing. The internal solutions are either NaCl or $CaCl_2$, and a Ag/AgCl reference electrode is situated in the internal solution.

Martínez-Fàbregas, E.; Alegret, S. "A Practical Approach to Chemical Sensors through Potentiometric Transducers: Determination of Urea in Serum by Means of a Biosensor," *J. Chem. Educ.* **1994,** *71,* A67–A70.

Directions are provided for constructing and characterizing an ammonium ion-selective electrode. The electrode is then modified to respond to urea by adding a few milligrams of urease and covering with a section of dialysis membrane. Directions for determining urea in serum also are provided.

—*Continued*

Experiments

Continued from page 533

Mifflin, T. E.; Andriano, K. M.; Robbins, W. B. "Determination of Penicillin Using an Immobilized Enzyme Electrode," *J. Chem. Educ.* **1984,** *61,* 638–639.

Directions for preparing a potentiometric biosensor for penicillin are provided in this experiment. The enzyme penicillinase is immobilized in a polyacrylamide polymer formed on the surface of a glass pH electrode. The electrode shows a linear response to penicillin G over a concentration range of 10^{-5} M to 10^{-3} M.

Palanivel, A.; Riyazuddin, P. "Fabrication of an Inexpensive Ion-Selective Electrode," *J. Chem. Educ.* **1984,** *61,* 290.

This experiment describes the preparation of a Ag_2S/CuS electrode using the graphite rod from a dry cell D-type battery. After obtaining the graphite rod, it is inserted and sealed in a polyethylene tube, leaving a 2-cm portion exposed. The protruding portion of the graphite rod, is placed in a saturated solution of Na_2S for 1 h, and in a saturated solution of CuS for an additional hour. In a related experiment (Riyazuddin, P.; Devika, D. "Potentiometric Acid–Base Titrations with Activated Graphite Electrodes," *J. Chem. Educ.* **1997,** *74,* 1198–1199.), activated graphite electrodes are used for the potentiometric titration of HCl with NaOH.

Ramaley, L; Wedge, P. J.; Crain, S. M. "Inexpensive Instrumental Analysis: Part 1. Ion-Selective Electrodes," *J. Chem. Educ.* **1994,** *71,* 164–167.

This experiment describes the preparation and evaluation of two liquid-membrane Na^+ ion-selective electrodes, using either the sodium salt of monensin or a hemisodium ionophore as ion exchangers incorporated into a PVC matrix. Electrodes prepared using monensin performed poorly, but those prepared using hemisodium showed a linear response over a range of 0.1 M to 3×10^{-5} M Na^+ with slopes close to the theoretical value.

The following set of experiments illustrate several applications of potentiometric electrodes.

Harris, T. M. "Potentiometric Measurement in a Freshwater Aquarium," *J. Chem. Educ.* **1993,** *70,* 340–341.

This experiment describes a semester-long project in which the concentration of several ions in a fresh water aquarium are monitored. Ions that are monitored using potentiometric electrodes include H^+ (pH electrode), Cl^- (chloride electrode), HCO_3^- (CO_2 electrode), NH_4^+ (NH_3 electrode), and NO_3^- (NH_3 electrode). Nitrate concentrations were determined following its conversion to ammonia.

Radic, N.; Komijenovic, J. "Potentiometric Determination of an Overall Formation Constant Using an Ion-Selective Membrane Electrode," *J. Chem. Educ.* **1993,** *70,* 509–511.

The titration of Al^{3+} with F^- in a water–acetonitrile mixture is followed potentiometrically using a fluoride ion-selective electrode. The cumulative formation constant for AlF_6^{3-} is determined from the titration curve.

Selig, W. S. "Potentiometric Titrations Using Pencil and Graphite Sensors," *J. Chem. Educ.* **1984,** *61,* 80–81.

This experiment describes the use of coated graphite electrodes for the potentiometric monitoring of precipitation, acid–base, complexation, and redox titrations.

The following four experiments describe useful and accessible applications of coulometry.

Bertotti, M.; Vaz, J. M.; Telles, R. "Ascorbic Acid Determination in Natural Orange Juice," *J. Chem. Educ.* **1995,** *72,* 445–447.

The titration of ascorbic acid using coulometrically generated I_2 and Br_2 is described in this experiment. Details are also given for the polarographic analysis of ascorbic acid.

Kalbus, G. E.; Lieu, V. T. "Dietary Fat and Health: An Experiment on the Determination of Iodine Number of Fats and Oils by Coulometric Titration," *J. Chem. Educ.* **1991,** *68,* 64–65.

The iodine number of fats and oils provides a quantitative measurement of the degree of unsaturation. A solution containing a 100% excess of ICl is added to the sample, reacting across the double-bonded sites of unsaturation. The excess ICl is converted to I_2 by adding KI. The resulting I_2 is reacted with a known excess of $Na_2S_2O_3$. To complete the analysis the excess $S_2O_3^{2-}$ is back titrated with coulometrically generated I_2.

Lötz, A. "A Variety of Electrochemical Methods in a Coulometric Titration Experiment," *J. Chem. Educ.* **1998,** *75,* 775–777.

This experiment describes three coulometric titrations for the quantitative analysis of mixtures containing KI and HCl. Iodide is determined by titrating with coulometrically generated Br_2. Chloride is determined indirectly by the coulometric titration of H^+. Finally, iodide and chloride are determined together by precipitating AgI and AgCl following the anodic dissolution of an Ag wire.

Swim, J.; Earps, E.; Reed, L. M.; et al. "Constant-Current Coulometric Titration of Hydrochloric Acid," *J. Chem. Educ.* **1996,** *73,* 679–683.

Directions are provided for constructing an inexpensive constant-current source and demonstrate its use in determining the concentration of HCl coulometrically.

—Continued

Experiments

The final set of experiments provide examples of voltammetric and amperometric analyses.

García-Armada, P.; Losada, J.; de Vicente-Pérez, S. "Cation Analysis Scheme by Differential Pulse Polarography," *J. Chem. Educ.* **1996,** *73,* 544–547.

Directions are provided for the simultaneous qualitative and quantitative analysis of Bi(III), Cd(II), Co(II), Cu(II), Cr(III), Mn(II), Ni(II), Pb(II), Sb(III), and Zn(II). The analysis is accomplished using differential pulse polarography in three separate supporting electrolytes.

Herrera-Melián, J. A.; Doña-Rodríguez, J. M.; Hernández-Brito, J.; et al. "Voltammetric Determination of Ni and Co in Water Samples," *J. Chem. Educ.* **1997,** *74,* 1444–1445.

This experiment describes the application of cathodic stripping voltammetry to the analysis of Ni and Co in fresh water and sea water.

Marin, D.; Mendicuti, F. "Polarographic Determination of Composition and Thermodynamic Stability Constant of a Complex Metal Ion," *J. Chem. Educ.* **1988,** *65,* 916–918.

This experiment describes the determination of the stability (cumulative formation) constant for the formation of $Pb(OH)_3^-$ by measuring the shift in the half-wave potential for the reduction of Pb^{2+} as a function of the concentration of OH^-. The influence of ionic strength is also considered, and results are extrapolated to zero ionic strength to determine the thermodynamic formation constant.

Sittampalam, G.; Wilson, G. S. "Amperometric Determination of Glucose at Parts Per Million Levels with Immobilized Glucose Oxidase," *J. Chem. Educ.* **1982,** *59,* 70–73.

This experiment describes the use of a commercially available amperometric biosensor for glucose that utilizes the enzyme glucose oxidase. The concentration of glucose in artificial unknowns, serum, or plasma can be determined using a normal calibration curve.

Town, J. L.; MacLaren, F.; Dewald, H. D. "Rotating Disk Voltammetry Experiment," *J. Chem. Educ.* **1991,** *68,* 352–354.

This experiment introduces hydrodynamic voltammetry using a rotating working electrode. Its application for the quantitative analysis of $K_4Fe(CN)_6$ is demonstrated.

Wang, J. "Sensitive Electroanalysis Using Solid Electrodes," *J. Chem. Educ.* **1982,** *59,* 691–692.

The technique of hydrodynamic modulation voltammetry (HMV), in which the rate of stirring is pulsed between high and low values, is demonstrated in this experiment. The application of HMV for the quantitative analysis of ascorbic acid in vitamin C tablets using the method of standard additions also is outlined.

Wang, J. "Anodic Stripping Voltammetry," *J. Chem. Educ.* **1983,** *60,* 1074–1075.

The application of anodic stripping voltammetry for the quantitative analysis of Cd, Pb, and Cu in natural waters is described in this experiment.

Wang, J.; Maccà, C. "Use of Blood-Glucose Test Strips for Introducing Enzyme Electrodes and Modern Biosensors," *J. Chem. Educ.* **1996,** *73,* 797–800.

Commercially available kits for monitoring blood-glucose use an amperometric biosensor incorporating the enzyme glucose oxidase. This experiment describes how such monitors can be adapted to the quantitative analysis of glucose in beverages.

11H PROBLEMS

1. Identify the anode and cathode for the following electrochemical cells, and write the oxidation or reduction reaction occurring at each electrode.
(a) Pt | $FeCl_2$ (0.015 M), $FeCl_3$ (0.045 M) || $AgNO_3$ (0.1 M) | Ag
(b) Ag | AgBr(s) | NaBr (1.0 M) || $CdCl_2$ (0.05 M) | Cd
(c) Pb, $PbSO_4$ | H_2SO_4 (1.5 M) || H_2SO_4 (2.0 M) | $PbSO_4$, PbO_2

2. Calculate the potential for the electrochemical cells in problem 1.

3. Calculate the molar concentration for the underlined component in the following cell if the cell potential is measured at +0.294 V

Ag | AgCl (sat'd) | NaCl (0.1 M) || <u>KI</u> (*x* M) | I_2(s) | Pt

4. What reaction prevents Zn from being used as an electrode of the first kind in acidic solution? Which other electrodes of the first kind would you expect to behave in the same manner as Zn when immersed in acidic solutions?

5. Creager and colleagues designed a salicylate ion-selective electrode using a PVC membrane impregnated with tetraalkylammonium salicylate.[27] To determine the ion-selective electrode's selectivity coefficient for benzoate, a set of salicylate calibration standards was prepared in which the concentration of benzoate was held constant at 0.10 M. Using the following data, determine the value of the selectivity coefficient.

[Salicylate] (M)	Potential (mV)
1.0	20.2
1.0×10^{-1}	73.5
1.0×10^{-2}	126
1.0×10^{-3}	168
1.0×10^{-4}	182
1.0×10^{-5}	182
1.0×10^{-6}	177

What is the maximum acceptable concentration of benzoate if this ion-selective electrode is to be used for analyzing samples containing as little as 10^{-5} M salicylate with an accuracy of better than 1%?

6. Watanabe and co-workers described a new membrane electrode for the determination of cocaine, which is a weak base alkaloid with a pK_a of 8.64.[28] The response of the electrode for a fixed concentration of cocaine was found to be independent of pH in the range of 1–8, but decreased sharply above a pH of 8. Offer an explanation for the source of this pH dependency.

7. Show that equation 11.12 for the urea electrode is correct.

8. The pH-based urea electrode has the response shown in equation 11.13. Rewrite the equation to show the relationship between potential, *E*, and the concentration of urea. Comment on your result.

9. Mifflin and associates described a membrane electrode for the quantitative analysis of penicillin in which the enzyme penicillinase is immobilized in a polyacrylamide gel that is coated on a glass pH electrode.[29] The following data were collected for a series of penicillin standards.

[Penicillin] (M)	Potential (mV)
1.0×10^{-2}	220
2.0×10^{-3}	204
1.0×10^{-3}	190
2.0×10^{-4}	153
1.0×10^{-4}	135
1.0×10^{-5}	96
1.0×10^{-6}	80

Construct a calibration curve for the electrode, and report (a) the range of concentrations in which a linear response is observed, (b) the equation for the calibration curve in this range, and (c) the concentration of penicillin in a sample that yields a potential of 142 mV.

10. Ion-selective electrodes can be incorporated in flow cells to monitor the concentration of an analyte in standards and samples that are pumped through the flow cell. As the analyte passes through the cell, a potential spike is recorded instead of a steady-state potential. The concentration of K^+ in serum has been determined in this fashion, using standards prepared in a matrix of 0.014 M NaCl.[30]

$[K^+]$ (mM)	Potential (arb. units)
0.1	25.5
0.2	37.2
0.4	50.8
0.6	58.7
0.8	64.0
1.0	66.8

A 1.00-mL sample of serum is diluted to volume in a 10-mL volumetric flask and analyzed, giving a potential of 51.1. Report the concentration of K^+ in the sample of serum.

11. Wang and Taha described an interesting application of potentiometry called batch injection.[31] As shown in the following figure, an ion-selective electrode is placed in an inverted position in a large-volume tank, and a fixed volume of a sample or standard solution is injected toward the electrode's surface using a micropipet.

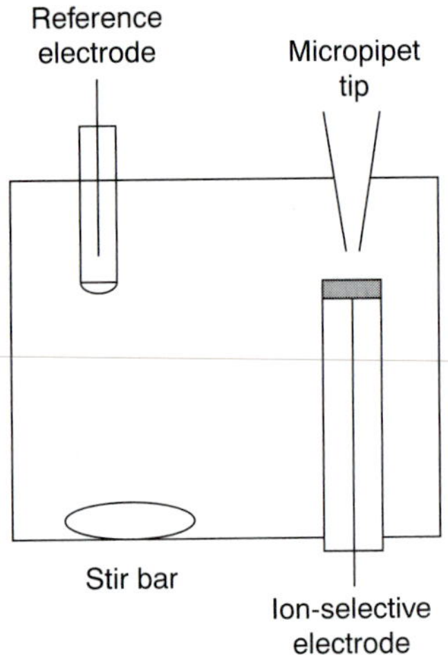

The response of the electrode is a peak whose height is proportional to the analyte's concentration. The following data were collected using a pH electrode

pH	Potential (mV)
2	+300
3	+240
4	+168
5	+81
6	+35
8	–92
9	–168
10	–235
11	–279

Determine the pH for each of the following samples, given the recorded peak potential: (a) tomato juice, +167 mV; (b) tap water, –27 mV; (c) coffee, 122 mV.

12. The concentration of NO_3^- in a water sample is determined by a one-point standard addition using an NO_3^- ion-selective electrode. A 25.00-mL sample is placed in a beaker, and a potential of +0.102 V is measured. A 1.00-mL aliquot of a 200.0 ppm standard solution of NO_3^- is added, after which the potential is found to be +0.089 V. Report the concentration of NO_3^- in parts per million.

13. The following data were collected for the analysis of fluoride in tap water and in toothpaste. (a) For the analysis of tap water, three 25.0-mL samples were each mixed with 25.0 mL of TISAB, and the potential was measured with an F^- ISE relative to a saturated calomel electrode. Five 1.00-mL additions of a standard solution of 100.0-ppm F^- were added to each, measuring the potential following each addition.

Milliliters of standard F^- added	Potential (mV) Trial 1	Trial 2	Trial 3
0.00	−79	−82	−81
1.00	−119	−119	−118
2.00	−133	−133	−133
3.00	−142	−142	−142
4.00	−149	−148	−148
5.00	−154	−153	−153

Determine the parts per million of F^- in the tap water. (b) For the analysis of toothpaste a 0.3619-g sample was transferred to a 100-mL volumetric flask along with 50.0 mL of TISAB and diluted to volume with distilled water. Three 20.0-mL aliquots were removed, and the potential was measured with an F^- ion-selective electrode using a saturated calomel electrode as a reference. Five separate 1.00-mL additions of a 100.0-ppm solution of F^- were added to each, measuring the potential following each addition.

Milliliters of standard F^- added	Potential (mV) Trial 1	Trial 2	Trial 3
0.00	−55	−54	−55
1.00	−82	−82	−83
2.00	−94	−94	−94
3.00	−102	−103	−102
4.00	−108	−108	−109
5.00	−112	−112	−113

Report the parts per million of F^- in the sample of toothpaste.

14. You are responsible for determining the amount of KI in iodized salt and decide to use an I^- ion-selective electrode. Describe how you would perform this analysis using (a) external standards and (b) the method of standard additions.

15. The purity of a sample of picric acid, $C_6H_3N_3O_7$, is determined by controlled-potential coulometry, converting the picric acid to triaminophenol, $C_6H_9N_3O$. A 0.2917-g sample of picric acid is placed in a 1000-mL volumetric flask and diluted to volume. A 10.00-mL portion of this solution is transferred to a coulometric cell and diluted till the Pt cathode is immersed. The exhaustive electrolysis of the sample requires 21.67 C of charge. Report the purity of the picric acid.

16. The concentration of H_2S in the drainage from an abandoned mine can be determined by a coulometric titration using KI as a mediator and I_3^- as the "titrant."

$$H_2S(aq) + I_3^-(aq) + 2H_2O(\ell) \rightleftharpoons 2H_3O^+(aq) + 3I^-(aq) + S(s)$$

A 50.00-mL sample of water is placed in a coulometric cell, along with an excess of KI and a small amount of starch as an indicator. Electrolysis is carried out at a constant current of 84.6 mA, requiring 386 s to reach the starch end point. Report the concentration of H_2S in the sample in parts per million.

17. One method for the determination of H_3AsO_3 is by a coulometric titration using I_3^- as a "titrant." The relevant reactions and standard-state potentials are summarized as follows.

$$H_3AsO_4(aq) + 2H_3O^+(aq) + 2e^- \rightleftharpoons H_3AsO_3(aq) + 3H_2O(\ell) \qquad E^\circ = +0.559$$

$$I_3^-(aq) + 2e^- \rightleftharpoons 3I^- \qquad E^\circ = +0.536$$

Explain why the coulometric titration must be done in neutral solutions (pH ≈ 7), instead of in strongly acidic solutions (pH < 0).

18. The reduction of acrylonitrile, $CH_2{=}CHCN$, to adiponitrile, $NC(CH_2)_4CN$, is an important industrial process. A 0.594-g sample of acrylonitrile is placed in a 1-L volumetric flask and diluted to volume. An exhaustive controlled-potential electrolysis of a 1.00-mL portion of the diluted acrylonitrile requires 1.080 C of charge. What is the value of *n* for this reduction?

19. A linear-potential scan hydrodynamic voltammogram for a mixture of Fe^{2+} and Fe^{3+} is shown in the figure, where $i_{l,a}$ and $i_{l,c}$ are the anodic and cathodic limiting currents. (a) Show that the potential is given by

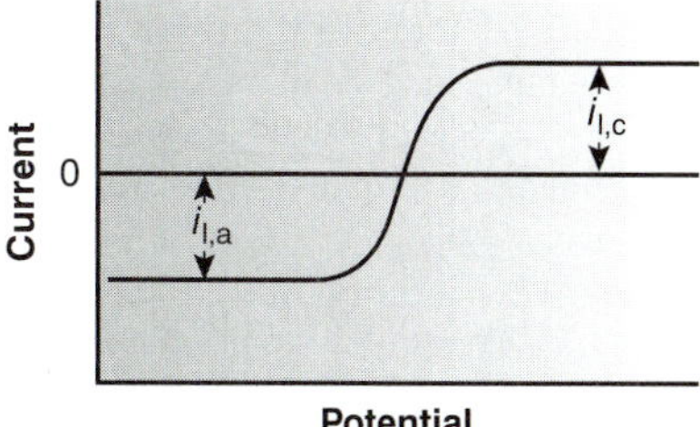

$$E = E^\circ_{Fe^{3+}/Fe^{2+}} - 0.05916 \log \frac{K_{Fe^{3+}}}{K_{Fe^{2+}}} - 0.05916 \log \frac{i - i_{l,a}}{i_{l,c} - i}$$

(b) What is the potential when $i = 0$ for a solution that is 0.1 mM Fe^{3+} and 0.05 mM Fe^{2+}? You may assume that $K_{Fe^{3+}} \approx K_{Fe^{2+}}$.

20. The amount of sulfur in aromatic monomers can be determined by differential pulse polarography. Standard solutions are prepared for analysis by dissolving 1.000 mL of the purified monomer in 25.00 mL of an electrolytic solvent, adding a known amount of S, deaerating, and measuring the peak current. The following results were obtained for a set of calibration standards

Micrograms of added S	Peak current (μA)
0	0.14
28	0.70
56	1.23
112	2.41
168	3.42

Analysis of a 1.000-mL sample, treated in the same manner as the standards, gives a peak current of 1.77 μA. Report the amount of sulfur present in the sample in milligrams per milliliter.

21. The purity of a sample of $K_3Fe(CN)_6$ was determined using linear-potential scan hydrodynamic voltammetry at a glassy carbon electrode using the method of external standards. The following data were obtained for a set of calibration standards.

$[K_3Fe(CN)_6]$ (mM)	Limiting current (μA)
2.0	127
4.0	252
6.0	376
8.0	500
10.0	624

A sample was prepared for analysis by diluting a 0.246-g sample to volume in a 100-mL volumetric flask. The limiting current for the sample was found to be 444 μA. Report the purity of this sample of $K_3Fe(CN)_6$.

22. Anodic stripping voltammetry at a mercury film electrode can be used to determine whether an individual has recently fired a gun by looking for traces of antimony in residue collected from the individual's hands. In a typical analysis a sample is collected with a cotton-tipped swab that had been wetted with 5% v/v HNO_3. When returned to the lab, the swab is placed in a vial containing 5.00 mL of 4 M HCl that is 0.02 M in hydrazine sulfate. After allowing the swab to soak overnight, 4.00 mL of the solution is transferred to an electrochemical cell, and 100 μL of 1.00×10^{-2} M $HgCl_2$ is added. Anodic stripping voltammetry of the sample gives a peak current of 0.38 μA. After a standard addition of 100 μL of 5.00×10^{-2} ppb Sb was added, anodic stripping voltammetry is repeated, giving a peak current of 1.14 μA. How many nanograms of Sb is collected from the individual's hand?

23. Zinc can be used as an internal standard in the analysis of thallium by differential pulse polarography. A standard solution containing 5.00×10^{-5} M Zn^{2+} and 2.50×10^{-5} M Tl^{+} gave peak currents of 5.71 μA and 3.19 μA, respectively. An 8.713-g sample of an alloy known to be free of zinc was dissolved in acid, transferred to a 500-mL volumetric flask, and diluted to volume. A 25.0-mL portion of this solution was mixed with 25.0 mL of a 5.00×10^{-4} M solution of Zn^{2+}. Analysis of this solution gave a peak current for Zn^{2+} of 12.3 μA, and for Tl^{+} of 20.2 μA. Report the Tl% w/w in the alloy.

24. Differential pulse voltammetry at a carbon working electrode can be used to determine the concentrations of ascorbic acid and caffeine in drug formulations.[32] In a typical analysis, a 0.9183-g tablet is crushed and ground into a fine powder. A 0.5630-g sample of this powder is transferred to a 100-mL volumetric flask, brought into solution, and diluted to volume. A 0.500-mL portion is then transferred to a voltammetric cell containing 20.00 mL of a suitable supporting electrolyte. The resulting voltammogram gives peak currents of 1.40 μA and 3.88 μA for ascorbic acid and caffeine, respectively. A 0.500-mL aliquot of a standard solution containing 250.0 ppm ascorbic acid and 200.0 ppm caffeine is then added. A voltammogram of this solution gives peak currents of 2.80 μA and 8.02 μA for ascorbic acid and caffeine, respectively. Report the number of milligrams of ascorbic acid and of caffeine in the tablet.

25. Ratana-ohpas and co-workers describe a stripping analysis method for determining levels of tin in canned fruit juices.[33] Standards containing 50.0 ppb Sn^{4+}, 100.0 ppb Sn^{4+}, and 150.0 ppb Sn^{4+} were analyzed, giving signals of 83.0, 171.6, and 260.2, respectively. A 2.00-mL sample of lychee juice was mixed with 20.00 mL of 1:1 HCl/HNO_3. A 0.500-mL portion of this mixture was added to 10 mL of 6 M HCl and the volume adjusted to 30.00 mL. Analysis of this diluted sample gave a signal of 128.2. Report the parts per million of Sn^{4+} in the original sample of lychee juice.

26. Sittampalam and Wilson described the preparation and use of an amperometric sensor for glucose.[34] The sensor is calibrated by measuring the steady-state current when it is immersed in standard solutions of glucose. A typical set of calibration data is shown in the following table.

Milligrams of glucose/100 mL	Current (arb. units)
2.0	17.2
4.0	32.9
6.0	52.1
8.0	68.0
10.0	85.8

A 2.00-mL sample of a solution containing an unknown amount of glucose is diluted to 10 mL in a volumetric flask, and a steady-state current of 23.6 is measured. What is the concentration of glucose in the sample in milligrams per 100 mL?

27. Differential pulse polarography is used to determine the concentrations of lead, thallium, and indium in a mixture.[20]

Since the polarographic peaks for lead and thallium, and for thallium and indium overlap, a simultaneous analysis is necessary. Peak currents (in arbitrary units) at potentials of –0.385 V, –0.455 V, and –0.557 V were measured for a single standard solution and for a sample, giving the results shown in the following table.

Analyte	Concentration (μg/mL)	i_{peak} at –0.385 V	i_{peak} at –0.455V	i_{peak} at –0.557 V
Pb^{2+}	1.0	26.1	2.9	0
Tl^{+}	2.0	7.8	23.5	3.2
In^{3+}	0.4	0	0	22.9
sample		60.6	28.8	54.1

Report the concentrations of Pb^{2+}, Tl^{+}, and In^{3+} in the sample.

28. Abass and colleagues developed an amperometric biosensor for NH_4^+ that uses the enzyme glutamate dehydrogenase to catalyze the following reaction.

$$\text{2-oxyglutarate}(aq) + NH_4^+(aq) + NADH(aq) \rightarrow \text{glutamate}(aq) + NAD^+(aq) + H_2O(\ell)$$

where NADH is the reduced form of nicotinamide adenine dinucleotide.[35] The biosensor actually responds to the concentration of NADH; however, the rate of the reaction depends on the concentration of NH_4^+. If the initial concentrations of 2-oxyglutarate and NADH are the same for all samples and standards, then the signal is proportional to the concentration of NH_4^+. As shown in the following table, the sensitivity of the method is dependent on pH.

pH	Sensitivity (nA s^{-1} M^{-1})
6.2	1.67×10^3
6.75	5.00×10^3
7.3	9.33×10^3
7.7	1.03×10^4
8.3	1.27×10^4
9.3	2.67×10^3

Two possible explanations for the effect of pH on the sensitivity of this analysis are the acid–base chemistry of NH_4^+, and, the acid–base chemistry of the enzyme. Given that the pK_a for NH_4^+ is 9.244, explain the source of this pH-dependent sensitivity.

29. The speciation scheme of Batley and Florence[22] requires eight measurements on four samples. After removing insoluble particulates by filtration, the solution is analyzed for the concentration of anodic stripping voltammetry (ASV) labile metal and the total concentration of metal. A portion of the filtered solution is passed through an ion-exchange column, and the concentrations of ASV metal and total metal are determined. A second portion of the filtered solution is irradiated with UV light, and the concentrations of ASV metal and total metal are measured. Finally, a third portion of the filtered solution is irradiated with UV light, passed through an ion-exchange column, and the concentrations of ASV-labile metal and total metal are again determined. The groups that are included in each measurement are summarized in the following table (see Table 11.12).

Separation Step	Groups Remaining in Solution	Groups Contributing to Concentration of ASV- Labile Metal	Groups Contributing to Concentration of Total Metal
1. filtration	I, II, III, IV, V, VI, VII	I, II, III	I, II, III, IV, V, VI, VII
2. ion-exchange	II, III, VI, VII	II, III	II, III, VI, VII
3. UV irradiation	I, II, III, IV, V, VI, VII	I, II, III, IV, VI	I, II, III, IV, V, VI, VII
4. UV irradiation/ ion-exchange	III, VII	III	III, VII

(a) Show how these eight measurements can be used to determine the concentration of metal present in each group.

(b) Batley and Florence[22a] report the following results for the speciation of cadmium, lead, and copper in a sample of sea water

Measurement	ppb Cd^{2+}	ppb Pb^{2+}	ppb Cu^{2+}
filtration: ASV-labile	0.24	0.39	0.26
filtration: total	0.28	0.50	0.40
ion-exchange: ASV-labile	0.21	0.33	0.17
ion-exchange: total	0.26	0.43	0.24
UV: ASV-labile	0.26	0.37	0.33
UV: total	0.28	0.50	0.43
ion-exchange/UV: labile	0	0	0
ion-exchange/UV: total	0.02	0.12	0.10

Determine the speciation of each metal in this sample of sea water.

30. The polarographic half-wave potentials (versus the SCE) for Pb^{2+} and Tl^{+} in 1 M HCl are, respectively, –0.44 V and –0.45 V. In an electrolyte of 1 M NaOH, however, the half-wave potentials are –0.76 V for Pb^{2+} and –0.48 V for Tl^{+}. Why does the change in electrolyte have such a great effect on the half-wave potential for Pb^{2+}, but not on the half-wave potential for Tl^{+}?

31. The following data were collected for the reduction of Pb^{2+} by normal pulse polarography

Potential (V vs. SCE)	Current (μA)
–0.345	0.16
–0.370	0.98
–0.383	2.05
–0.393	3.13
–0.409	4.62
–0.420	5.16

The limiting current was 5.67 μA. Verify that the reduction reaction is reversible, and determine values for n and $E_{1/2}$. The half-wave potentials for the normal pulse polarograms of Pb^{2+} in the presence of several different concentrations of OH^- are shown in the following table.

$[OH^-]$ (M)	$E_{1/2}$ (V vs. SCE)
0.050	-0.646
0.100	-0.673
0.150	-0.689
0.300	-0.715

Determine the stoichiometry of the Pb–hydroxide complex and its formation constant.

32. The concentration of Cu^{2+} in sea water may be determined by anodic stripping voltammetry at a hanging mercury drop electrode. To determine the total amount of Cu^{2+} in a sample of sea water, it is necessary to free any copper bound in organic complexes. To a 20.00-mL sample of sea water is added 1 mL of 0.05 M HNO_3 and 1 mL of 0.1% H_2O_2. The sample is irradiated with UV light for 8 h and then diluted to volume in a 25-mL volumetric flask. Deposition of Cu^{2+} takes place at –0.3 V for 10 min, producing a peak current of 26.1 (arbitrary units). A second 20.00-mL sample of the sea water is treated identically, except that 0.1 mL of a 5.00 μM solution of Cu^{2+} is added, producing a peak current of 38.4. Report the concentration of Cu^{2+} in the sea water in parts per billion.

33. Thioamide drugs, such as that shown here, can be determined by cathodic stripping analysis. Deposition occurs at +0.05 V versus the SCE. During the stripping step, the potential is scanned cathodically, and a stripping peak is observed at –0.55 V. In a typical application a 2.00-mL sample of urine was mixed with 2.00 mL of a pH 4.78 buffer. Following a 2-minute deposition, a peak current of 0.562 μA was measured. A 0.10-mL addition of a 5.00 μM solution of the thioamide drug was added to the same solution. A peak current of 0.837 μA was recorded following the same 2-min deposition time. Report the drug's molar concentration in the urine sample.

CH_3

N

$CSNH_2$

34. The concentration of vanadium (V) in sea water can be determined by adsorptive stripping voltammetry after forming a complex with catechol.[36] The catechol-V(V) complex is deposited on a hanging mercury drop electrode at a potential of –0.1 V (vs. Ag/AgCl). A cathodic potential scan gives a stripping peak that is proportional to the concentration of V(V) in the sample of sea water. The following standard additions were used to analyze a sample of sea water.

$[V(V)]_{add}$ (M)	i_{peak} (nA)
2.0×10^{-8}	24
4.0×10^{-8}	33
8.0×10^{-8}	52
1.2×10^{-7}	69
1.8×10^{-7}	97
2.8×10^{-7}	140

Determine the molar concentration of V (V) in the sample of sea water, assuming that the standard additions result in a negligible change in the sample's volume.

11I SUGGESTED READINGS

The following general references provide a broad introduction to electrochemical methods of analysis.

Bard, A. J.; Faulkner, L. R. *Electrochemical Methods.* Wiley: New York, 1980.

Faulkner, L. R. "Electrochemical Characterization of Chemical Systems." In Kuwana, T. E., ed., *Physical Methods in Modern Chemical Analysis,* Vol. 3. Academic Press: New York, 1983, pp. 137–248.

Kissinger, P. T.; Heineman, W. R. *Laboratory Techniques in Electroanalytical Chemistry.* 2nd ed. Marcel Dekker: New York, 1996.

Lingane, J. J. *Electroanalytical Chemistry.* 2nd ed. Interscience: New York, 1958.

Sawyer, D. T.; Roberts, J. L., Jr. *Experimental Electrochemistry for Chemists.* Wiley-Interscience: New York, 1974.

Vassos, B. H.; Ewing, G. W. *Electroanalytical Chemistry.* Wiley-Interscience: New York, 1983.

Three short articles providing a good understanding of important general principles of electrochemistry follow.

Faulkner, L. R. "Understanding Electrochemistry: Some Distinctive Concepts," *J. Chem. Educ.* **1983,** *60,* 262–264.

Maloy, J. T. "Factors Affecting the Shape of Current-Potential Curves," *J. Chem. Educ.* **1983,** *60,* 285–289.

Zoski, C. G. "Charging Current Discrimination in Analytical Voltammetry," *J. Chem. Educ.* **1986,** *63,* 910–914.

Additional information on potentiometry and ion-selective electrodes can be found in the following sources.

Bates, R. G. *Determination of pH: Theory and Practice.* 2nd ed. Wiley: New York, 1973.

Buck, R. P. "Potentiometry: pH Measurements and Ion Selective Electrodes." In Weissberger, A., ed., *Physical Methods of Organic Chemistry,* Vol. 1, Part IIA. Wiley: New York, 1971, pp. 61–162.

Cammann, K. *Working with Ion-Selective Electrodes.* Springer-Verlag: Berlin, 1977.

Evans, A. *Potentiometry and Ion-Selective Electrodes.* Wiley: New York, 1987.

Frant, M. S. "Where Did Ion Selective Electrodes Come From?" *J. Chem. Educ.* **1997,** *74,* 159–166.

Light, T. S. "Industrial Use and Application of Ion-Selective Electrodes," *J. Chem. Educ.* **1997,** *74,* 171–177.

Rechnitz, G. A. "Ion and Bio-Selective Membrane Electrodes," *J. Chem. Educ.* **1983,** *60,* 282–284.

Ruzicka, J. "The Seventies—Golden Age for Ion-Selective Electrodes," *J. Chem. Educ.* **1997,** *74,* 167–170.

Young, C. C. "Evolution of Blood Chemistry Analyzers Based on Ion Selective Electrodes," *J. Chem. Educ.* **1997,** *74,* 177–182.

The following sources provide additional information on electrochemical biosensors.

Alvarez-Icasa, M.; Bilitewski, U. "Mass Production of Biosensors," *Anal. Chem.* **1993,** *65,* 525A–533A.

Meyerhoff, M. E.; Fu, B.; Bakker, E.; et al. "Polyion-Sensitive Membrane Electrodes for Biomedical Analysis," *Anal. Chem.* **1996,** *68,* 168A–175A.

Nicolini, C.; Adami, M; Antolini, F.; et al. "Biosensors: A Step to Bioelectronics," *Phys. World,* May 1992, 30–34.

Schultz, J. S. "Biosensors," *Sci. Am.* August 1991, 64–69.

Thompson, M.; Krull, U. "Biosensors and the Transduction of Molecular Recognition," *Anal. Chem.* **1991,** *63,* 393A–405A.

Vadgama, P. "Designing Biosensors," *Chem. Brit.* **1992,** *28,* 249–252.

The following is a good source covering the clinical application of electrochemistry.

Wang, J. *Electroanalytical Techniques in Clinical Chemistry and Laboratory Medicine.* VCH: New York, 1998.

Coulometry is covered in the following texts.

Milner, G. W. C.; Phillips, G. *Coulometry in Analytical Chemistry.* Pergamon: New York, 1967.

Rechnitz, G. A. *Controlled-Potential Analysis.* Macmillan: New York, 1963.

For a description of electrogravimetry, see the following resource.

Tanaka, N. "Electrodeposition," In Kolthoff, I. M.; Elving, P. J., eds. *Treatise on Analytical Chemistry,* Part I: *Theory and Practice,* Vol. 4. Interscience: New York, 1963.

The following sources provide additional information on polarography and pulse polarography.

Flato, J. B. "The Renaissance in Polarographic and Voltammetric Analysis," *Anal. Chem.* **1972,** *44(11),* 75A–87A.

Kolthoff, I. M.; Lingane, J. J. *Polarography.* Interscience: New York, 1952.

Osteryoung, J. "Pulse Voltammetry," *J. Chem. Educ.* **1983,** *60,* 296–298.

Additional information on stripping voltammetry is available in the following text.

Wang, J. *Stripping Analysis.* VCH Publishers: Deerfield Beach, FL, 1985, pp. 116–124.

11J REFERENCES

1. Stork, J. T. *Anal. Chem.* **1993,** *65,* 344A–351A.
2. Sawyer, D. T.; Roberts, J. L., Jr. *Experimental Electrochemistry for Chemists.* Wiley-Interscience: New York, 1974, p. 22.
3. Cammann, K. *Working with Ion-Selective Electrodes.* Springer-Verlag: Berlin, 1977.
4. (a) Papastathopoulos, D. S.; Rechnitz, G. A. *Anal. Chim. Acta* **1975,** *79,* 17; (b) Riechel, T. L. *J. Chem. Educ.* **1984,** *61,* 640–642.
5. Tor, R.; Freeman, A. *Anal. Chem.* **1986,** *58,* 1042–1046.
6. Lunte, C. E.; Heineman, W. R. "Electrochemical Techniques in Bioanalysis" in Steckhan, E., ed. *Topics in Current Chemistry.* Vol. 143. Springer-Verlag: Berlin, 1988, p. 8.
7. Procedure adapted from Kennedy, J. H. *Analytical Chemistry—Practice.* Harcourt Brace Jovanovich: San Diego, 1984, p. 117–118.
8. Hawkes, S. J. *J. Chem. Educ.* **1994,** *71,* 747–749.
9. Kristensen, H. B.; Saloman, A.; Kokholm, G. *Anal. Chem.* **1991,** *63,* 885A–891A.
10. Bates, R. G. *Determination of pH: Theory and Practice.* 2nd ed. Wiley: New York, 1973.
11. Rogers, K. R.; Williams, L. R. *Trends Anal. Chem.* **1995,** *14,* 289–294.
12. Rechnitz, G. A. *Controlled-Potential Analysis.* Macmillan: New York, 1963, p. 49.
13. Procedure adapted from Bassett, J.; Denney, R. C.; Jeffery, G. H.; et al. *Vogel's Textbook of Quantitative Inorganic Analysis.* Longman: London, 1978, pp. 559–560.
14. Curran, D. J. "Constant-Current Coulometry." Chapter 20 in Kissinger, P. T.; Heineman, W. R., eds. *Laboratory Techniques in Electroanalytical Chemistry.* Marcel Dekker, Inc.: New York, 1984, pp. 539–568.
15. Maloy, J. T. *J. Chem. Educ.* **1983,** *60,* 285–289.
16. Osteryoung, J. *J. Chem. Educ.* **1983,** *60,* 296–298.
17. Peterson, W. M.; Wong, R. V. *Am. Lab.* November 1981, 116–128.
18. Wang, J. *Am. Lab.* May 1985, 41–50.
19. Karube, I.; Nomura, Y.; Arikawa, Y. *Trends in Anal. Chem.* **1995,** *14,* 295–299.
20. Lanza, P. *J. Chem. Educ.* **1990,** *67,* 704–705.
21. Wang, J. *Stripping Analysis.* VCH Publishers: Deerfield Beach, FL, 1985, pp. 116–124.
22. (a) Batley, G. E.; Florence, T. M. *Anal. Lett.* **1976,** *9,* 379–388; (b) Batley, G. E.; Florence, T. M. *Talanta* **1977,** *24,* 151; (c) Batley, G. E.; Florence, T. M. *Anal. Chem.* **1980,** *52,* 1962–1963; (d) Florence, T. M.; Batley, G. E. *CRC Crit. Rev. Anal. Chem.* **1980,** *9,* 219–296.
23. Brooks, M. A. "Application of Electrochemistry to Pharmaceutical Analysis" Chapter 21. In Kissinger, P. T.; Heineman, W. R., eds.

Laboratory Techniques in Electroanalytical Chemistry. Marcel Dekker, Inc.: New York, 1984, pp. 569–607.

24. Cammann, K.; Lemke, U.; Rohen, A.; et al. *Angew. Chem. Int. Ed. Engl.* **1991,** *30,* 516–539.
25. Procedure adapted from Pungor, E. *A Practical Guide to Instrumental Analysis.* CRC Press: Boca Raton, FL, 1995, pp. 34–37.
26. Abe, T.; Lau, L. L.; Ewing, A. G. *J. Am. Chem. Soc.* **1991,** *113,* 7421–7423.
27. Creager, S. E.; Lawrence, K. D.; Tibbets, C. R. *J. Chem. Educ.* **1995,** *72,* 274–276.
28. Watanabe, K.; Okada, K.; Oda, H.; et al. *Anal. Chim. Acta* **1995,** *316,* 371–375.
29. Mifflin, T. E.; Andriano, K. M.; Robbins, W. B. *J. Chem. Educ.* **1984,** *61,* 638–639.
30. Meyerhoff, M. E.; Kovach, P. M. *J. Chem. Educ.* **1983,** *9,* 766–768.
31. Wang, J.; Taha, Z. *Anal. Chim. Acta* **1991,** *252,* 215–221.
32. Lau, O.; Luk, S.; Cheung, Y. *Analyst,* **1989,** *114,* 1047–1051.
33. Ratana-ohpas, R; Kanatharana, P.; Ratana-ohpas, W.; et al. *Anal. Chim. Acta* **1996,** *333,* 115–118.
34. Sittampalam, G; Wilson, G. S. *J. Chem. Educ.* **1982,** *59,* 70–73.
35. Abass, A. K.; Hart, J. P.; Cowell, D. C.; et al. *Anal. Chim. Acta* **1998,** *373,* 1–8.
36. van der Berg, C. M. G.; Huang, Z. Q. *Anal. Chem.* **1984,** *56,* 2383–2386.

Chapter 12

Chromatographic and Electrophoretic Methods

Drawing from an arsenal of analytical techniques, many of which were the subject of the preceding four chapters, analytical chemists have designed methods for the analysis of analytes at increasingly lower concentrations and in increasingly more complex matrices. Despite the power of these techniques, they often suffer from a lack of selectivity. For this reason, many analytical procedures include a step to separate the analyte from potential interferents. Several separation methods, such as liquid–liquid extractions and solid-phase microextractions, were discussed in Chapter 7. In this chapter we consider two additional separation methods that combine separation and analysis: chromatography and electrophoresis.

12A Overview of Analytical Separations

In Chapter 7 we examined several methods for separating an analyte from potential interferents. For example, in a liquid–liquid extraction the analyte and interferent are initially present in a single liquid phase. A second, immiscible liquid phase is introduced, and the two phases are thoroughly mixed by shaking. During this process the analyte and interferents partition themselves between the two phases to different extents, affecting their separation. Despite the power of these separation techniques, there are some significant limitations.

12A.1 The Problem with Simple Separations

Suppose we have a sample containing an analyte in a matrix that is incompatible with our analytical method. To determine the analyte's concentration we first separate it from the matrix using, for example, a liquid–liquid extraction. If there are additional analytes, we may need to use additional extractions to isolate them from the analyte's matrix. For a complex mixture of analytes this quickly becomes a tedious process.

Furthermore, the extent to which we can effect a separation depends on the distribution ratio of each species in the sample. To separate an analyte from its matrix, its distribution ratio must be significantly greater than that for all other components in the matrix. When the analyte's distribution ratio is similar to that of another species, then a separation becomes impossible. For example, let's assume that an analyte, A, and a matrix interferent, I, have distribution ratios of 5 and 0.5, respectively. In an attempt to separate the analyte from its matrix, a simple liquid–liquid extraction is carried out using equal volumes of sample and a suitable extraction solvent. Following the treatment outlined in Chapter 7, it is easy to show that a single extraction removes approximately 83% of the analyte and 33% of the interferent. Although it is possible to remove 99% of A with three extractions, 70% of I is also removed. In fact, there is no practical combination of number of extractions or volume ratio of sample and extracting phases that produce an acceptable separation of the analyte and interferent by a simple liquid–liquid extraction.

12A.2 A Better Way to Separate Mixtures

The problem with a simple extraction is that the separation only occurs in one direction. In a liquid–liquid extraction, for example, we extract a solute from its initial phase into the extracting phase. Consider, again, the separation of an analyte and a matrix interferent with distribution ratios of 5 and 0.5, respectively. A single liquid–liquid extraction transfers 83% of the analyte and 33% of the interferent to the extracting phase (Figure 12.1). If the concentrations of A and I in the sample were identical, then their concentration ratio in the extracting phase after one extraction is

$$\frac{[A]}{[I]} = \frac{0.83}{0.33} = 2.5$$

Thus, a single extraction improves the separation of the solutes by a factor of 2.5. As shown in Figure 12.1, a second extraction actually leads to a poorer separation. After combining the two portions of the extracting phase, the concentration ratio decreases to

$$\frac{[A]}{[I]} = \frac{0.97}{0.55} = 1.8$$

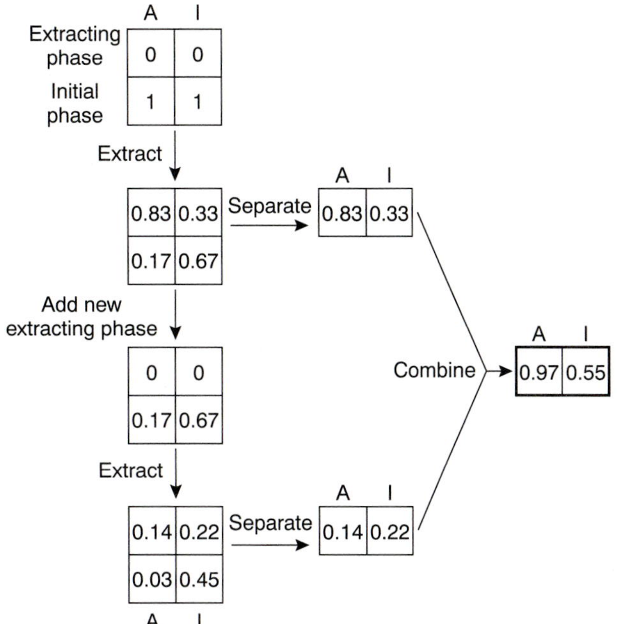

Figure 12.1
Progress of a liquid–liquid extraction using two identical extractions of a sample (initial phase) with fresh portions of the extracting phase. All numbers are fractions of solute in the phases; A = analyte, I = interferent.

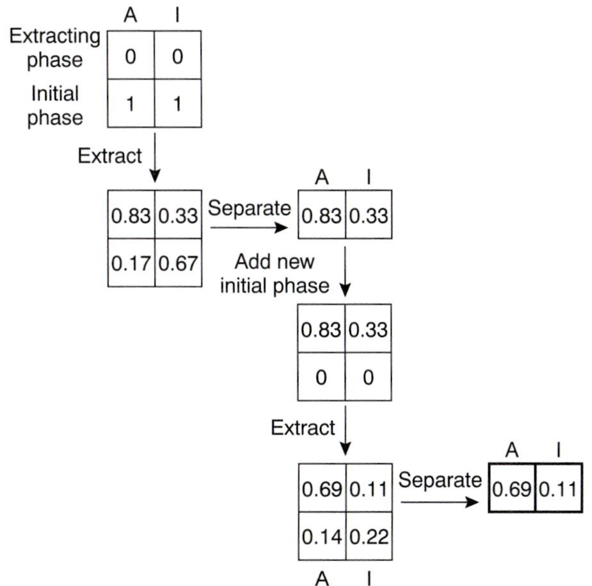

Figure 12.2
Progress of a liquid–liquid extraction in which the solutes are first extracted into the extracting phase and then extracted back into a fresh portion of the initial phase. All numbers are fractions of solute in the phases; A = analyte, I = interferent.

We can improve the separation by first extracting the solutes into the extracting phase, and then extracting them back into a fresh portion of the initial phase (Figure 12.2). Because solute A has the larger distribution ratio, it is extracted to a greater extent during the first extraction and to a lesser extent during the second extraction. In this case the final concentration ratio of

$$\frac{[A]}{[I]} = \frac{0.69}{0.11} = 6.3$$

countercurrent extraction
A liquid–liquid extraction in which solutes are extracted back and forth between fresh portions of two extracting phases.

mobile phase
In chromatography, the extracting phase that moves through the system.

stationary phase
In chromatography, the extracting phase that remains in a fixed position.

in the extracting phase is significantly greater. The process of extracting the solutes back and forth between fresh portions of the two phases, which is called a **countercurrent extraction,** was developed by Craig in the 1940s.[1*] The same phenomenon forms the basis of modern chromatography.

Chromatographic separations are accomplished by continuously passing one sample-free phase, called a **mobile phase,** over a second sample-free phase that remains fixed, or stationary. The sample is injected, or placed, into the mobile phase. As it moves with the mobile phase, the sample's components partition themselves between the mobile and **stationary phases.** Those components whose distribution ratio favors the stationary phase require a longer time to pass through the system. Given sufficient time, and sufficient stationary and mobile phase, solutes with similar distribution ratios can be separated.

chromatography
A separation in which solutes partition between a mobile and stationary phase.

The history of modern chromatography can be traced to the turn of the century when the Russian botanist Mikhail Tswett (1872–1919) used a column packed with a stationary phase of calcium carbonate to separate colored pigments from plant extracts. The sample was placed at the top of the column and carried through the stationary phase using a mobile phase of petroleum ether. As the sample moved through the column, the pigments in the plant extract separated into individual colored bands. Once the pigments were adequately separated, the calcium carbonate was removed from the column, sectioned, and the pigments recovered by extraction. Tswett named the technique **chromatography,** combining the Greek words for "color" and "to write." There was little interest in Tswett's technique until 1931 when chromatography was reintroduced as an analytical technique for biochemical separations. Pioneering work by Martin and Synge in 1941[2] established the importance of liquid–liquid partition chromatography and led to the development of a theory for chromatographic separations; they were awarded the 1952 Nobel Prize in chemistry for this work. Since then, chromatography in its many forms has become the most important and widely used separation technique. Other separation methods, such as electrophoresis, effect a separation without the use of a stationary phase.

12A.3 Classifying Analytical Separations

Analytical separations may be classified in three ways: by the physical state of the mobile phase and stationary phase; by the method of contact between the mobile phase and stationary phase; or by the chemical or physical mechanism responsible for separating the sample's constituents. The mobile phase is usually a liquid or a gas, and the stationary phase, when present, is a solid or a liquid film coated on a solid surface. Chromatographic techniques are often named by listing the type of mobile phase, followed by the type of stationary phase. Thus, in gas–liquid chromatography the mobile phase is a gas and the stationary phase is a liquid. If only one phase is indicated, as in gas chromatography, it is assumed to be the mobile phase.

column chromatography
A form of chromatography in which the stationary phase is retained in a column.

planar chromatography
A form of chromatography in which the stationary phase is immobilized on a flat surface.

Two common approaches are used to bring the mobile phase and stationary phase into contact. In **column chromatography,** the stationary phase is placed in a narrow column through which the mobile phase moves under the influence of gravity or pressure. The stationary phase is either a solid or a thin, liquid film coating on a solid particulate packing material or the column's walls. In **planar chromatography** the stationary phase coats a flat glass, metal, or plastic plate

*The theory behind countercurrent extractions is outlined in Appendix 6.

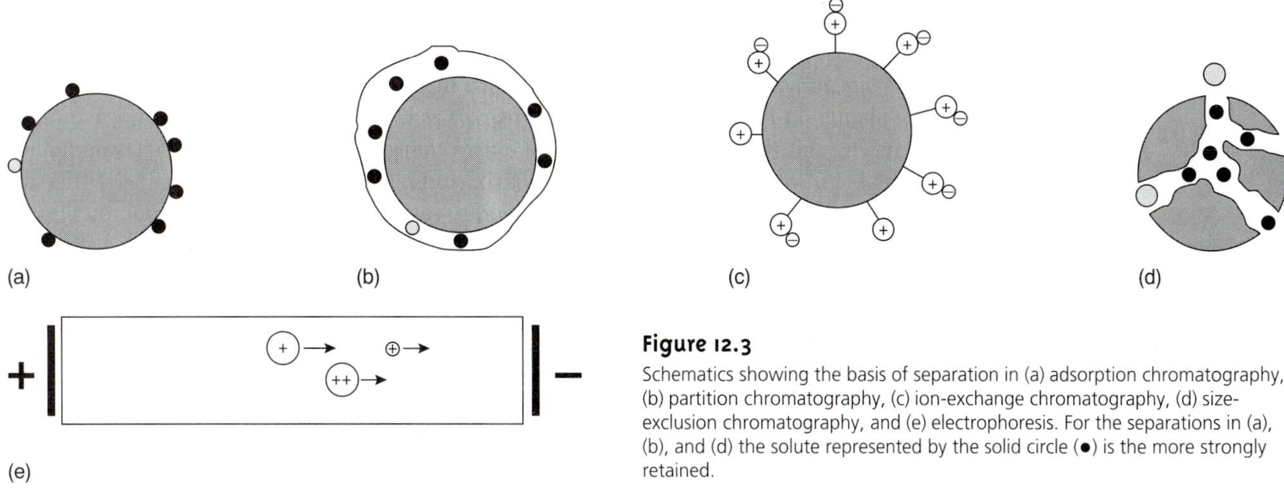

Figure 12.3
Schematics showing the basis of separation in (a) adsorption chromatography, (b) partition chromatography, (c) ion-exchange chromatography, (d) size-exclusion chromatography, and (e) electrophoresis. For the separations in (a), (b), and (d) the solute represented by the solid circle (•) is the more strongly retained.

and is placed in a developing chamber. A reservoir containing the mobile phase is placed in contact with the stationary phase, and the mobile phase moves by capillary action.

The mechanism by which solutes separate provides a third means for characterizing a separation (Figure 12.3). In adsorption chromatography, solutes separate based on their ability to adsorb to a solid stationary phase. In partition chromatography, a thin liquid film coating a solid support serves as the stationary phase. Separation is based on a difference in the equilibrium partitioning of solutes between the liquid stationary phase and the mobile phase. Stationary phases consisting of a solid support with covalently attached anionic (e.g., $-SO_3^-$) or cationic (e.g., $-N(CH_3)_3^+$) functional groups are used in ion-exchange chromatography. Ionic solutes are attracted to the stationary phase by electrostatic forces. Porous gels are used as stationary phases in size-exclusion chromatography, in which separation is due to differences in the size of the solutes. Large solutes are unable to penetrate into the porous stationary phase and so quickly pass through the column. Smaller solutes enter into the porous stationary phase, increasing the time spent on the column. Not all separation methods require a stationary phase. In an electrophoretic separation, for example, charged solutes migrate under the influence of an applied potential field. Differences in the mobility of the ions account for their separation.

12B General Theory of Column Chromatography

Of the two methods for bringing the stationary and mobile phases into contact, the more important is column chromatography. In this section we develop a general theory that we may apply to any form of column chromatography. With appropriate modifications, this theory also can be applied to planar chromatography.

A typical column chromatography experiment is outlined in Figure 12.4. Although the figure depicts a liquid–solid chromatographic experiment similar to that first used by Tswett, the design of the column and the physical state of the

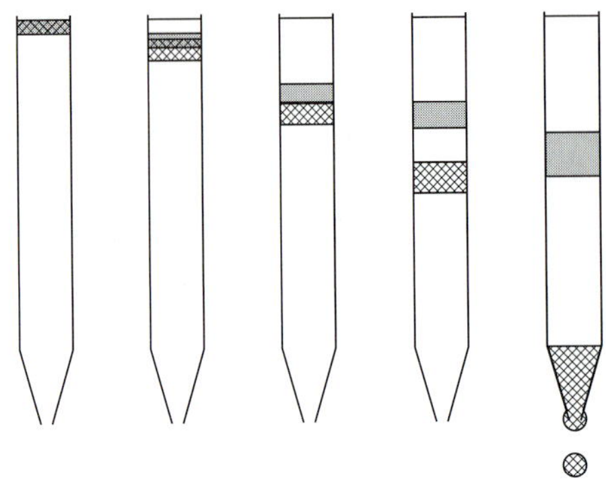

Figure 12.4
Progress of a column chromatographic separation showing the separation of two solute bands.

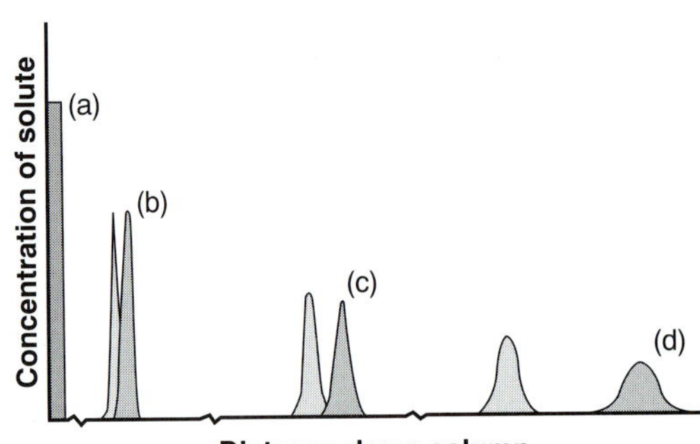

Figure 12.5
Another view of the progress of a column chromatographic separation showing the separation of two solute bands.

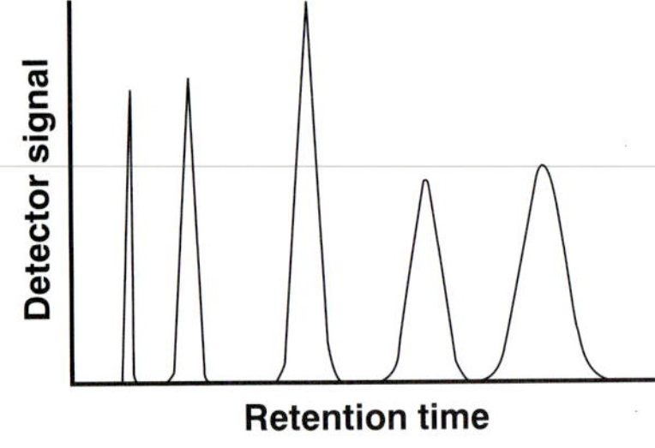

Figure 12.6
Typical chromatogram of detector response as a function of retention time.

chromatogram
A plot of the detector's signal as function of elution time or volume.

retention time
The time a solute takes to move from the point of injection to the detector (t_r).

retention volume
The volume of mobile phase needed to move a solute from its point of injection to the detector (V_r).

baseline width
The width of a solute's chromatographic band measured at the baseline (w).

stationary and mobile phases may vary. The sample is introduced at the top of the column as a narrow band. Ideally, the solute's initial concentration profile is rectangular (Figure 12.5a). As the sample moves down the column the solutes begin to separate, and the individual solute bands begin to broaden and develop a Gaussian profile (Figures 12.5b,c). If the strength of each solute's interaction with the stationary phase is sufficiently different, then the solutes separate into individual bands (Figure 12.5d). The progress of a chromatographic separation is monitored with a suitable detector situated at the end of the column. A plot of the detector's signal as a function of time or volume of eluted mobile phase is known as a **chromatogram** (Figure 12.6) and consists of a peak for each of the separated solute bands.

A chromatographic peak may be characterized in many ways, two of which are shown in Figure 12.7. The **retention time,** t_r, is the elapsed time from the introduction of the solute to the peak maximum. The retention time also can be measured indirectly as the volume of mobile phase eluting between the solute's introduction and the appearance of the solute's peak maximum. This is known as the **retention volume,** V_r. Dividing the retention volume by the mobile phase's flow rate, u, gives the retention time.

The second important parameter is the chromatographic peak's width at the baseline, w. As shown in Figure 12.7, **baseline width** is determined by the intersection with the baseline of tangent lines drawn through the inflection points on either side of the chromatographic peak. Baseline width is measured in units of time or volume, depending on whether the retention time or retention volume is of interest.

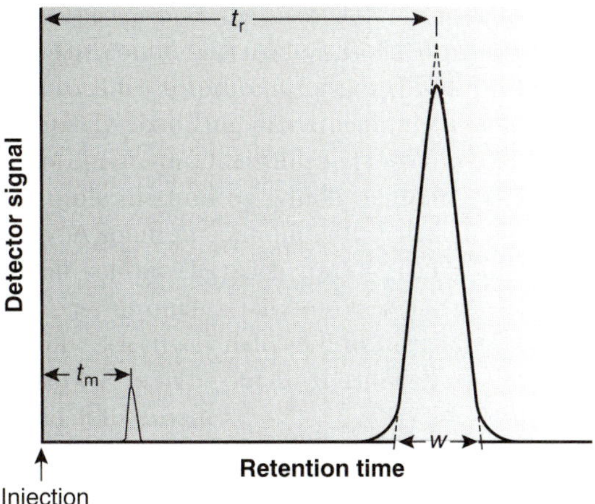

Figure 12.7
Measurement of the column's void time, t_m, and the retention time, t_r, and baseline width, w, for a solute.

Besides the solute peak, Figure 12.7 also shows a small peak eluted soon after the sample is injected into the mobile phase. This peak results from solutes that move through the column at the same rate as the mobile phase. Since these solutes do not interact with the stationary phase, they are considered nonretained. The time or volume of mobile phase required to elute nonretained components is called the column's **void time,** t_m, or **void volume.**

void time
The time required for unretained solutes to move from the point of injection to the detector (t_m).

void volume
The volume of mobile phase needed to move an unretained solute from the point of injection to the detector.

12B.1 Chromatographic Resolution

The goal of chromatography is to separate a sample into a series of chromatographic peaks, each representing a single component of the sample. **Resolution** is a quantitative measure of the degree of separation between two chromatographic peaks, A and B, and is defined as

resolution
The separation between two chromatographic bands (R).

$$R = \frac{t_{r,B} - t_{r,A}}{0.5(w_B + w_A)} = \frac{2\Delta t_r}{w_B + w_A} \qquad 12.1$$

As shown in Figure 12.8, the degree of separation between two chromatographic peaks improves with an increase in R. For two peaks of equal size, a resolution of 1.5 corresponds to an overlap in area of only 0.13%. Because resolution is a quantitative measure of a separation's success, it provides a useful way to determine if a change in experimental conditions leads to a better separation.

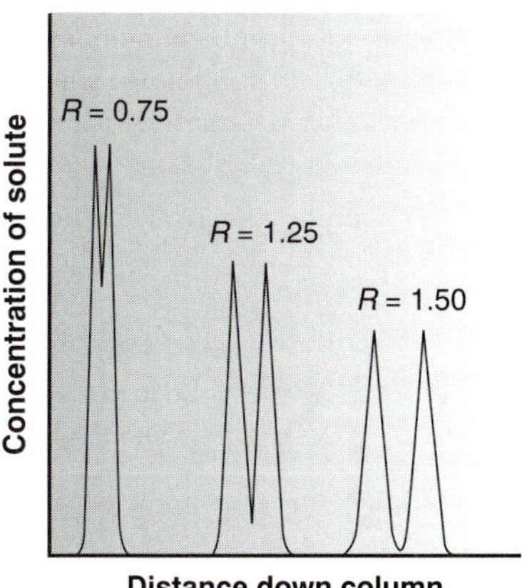

Figure 12.8
Three examples of chromatographic resolution.

Example

EXAMPLE 12.1

In a chromatographic analysis of lemon oil a peak for limonene has a retention time of 8.36 min with a baseline width of 0.96 min. γ-Terpinene elutes at 9.54 min, with a baseline width of 0.64 min. What is the resolution between the two peaks?

SOLUTION

Using equation 12.1, we find that the resolution is

$$R = \frac{2\Delta t_r}{w_B + w_A} = \frac{2(9.54 - 8.36)}{0.64 + 0.96} = 1.48$$

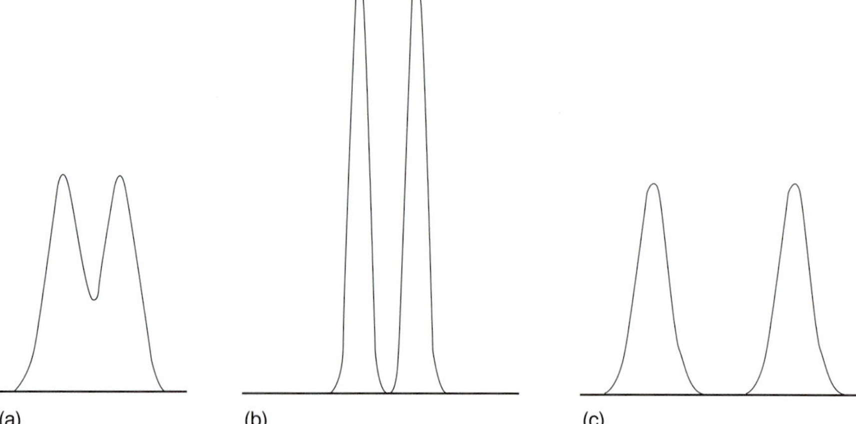

Figure 12.9
Two methods for improving chromatographic resolution: (a) Original separation showing a pair of poorly resolved solutes; (b) Improvement in resolution due to an increase in column efficiency; (c) Improvement in resolution due to a change in column selectivity.

From equation 12.1 it is clear that resolution may be improved either by increasing Δt_r or by decreasing w_A or w_B (Figure 12.9). We can increase Δt_r by enhancing the interaction of the solutes with the column or by increasing the column's selectivity for one of the solutes. Peak width is a kinetic effect associated with the solute's movement within and between the mobile phase and stationary phase. The effect is governed by several factors that are collectively called column efficiency. Each of these factors is considered in more detail in the following sections.

12B.2 Capacity Factor

The distribution of a solute, S, between the mobile phase and stationary phase can be represented by an equilibrium reaction

$$S_m \rightleftharpoons S_s$$

and its associated partition coefficient, K_D, and distribution ratio, D,

$$K_D = \frac{[S_s]}{[S_m]}$$

$$D = \frac{[S_s]_{tot}}{[S_m]_{tot}} \qquad 12.2$$

where the subscripts m and s refer to the mobile phase and stationary phase, respectively. As long as the solute is not involved in any additional equilibria in either the mobile phase or stationary phase, the equilibrium partition coefficient and the distribution ratio will be the same.

Conservation of mass requires that the total moles of solute remain constant throughout the separation, thus

$$(\text{moles S})_{tot} = (\text{moles S})_m + (\text{moles S})_s \qquad 12.3$$

Solving equation 12.3 for the moles of solute in the stationary phase and substituting into equation 12.2 gives

$$D = \frac{\{(\text{moles S})_{\text{tot}} - (\text{moles S})_{\text{m}}\}/V_{\text{s}}}{(\text{moles S})_{\text{m}}/V_{\text{m}}} = \frac{(\text{moles S})_{\text{tot}} V_{\text{m}} - (\text{moles S})_{\text{m}} V_{\text{m}}}{(\text{moles S})_{\text{m}} V_{\text{s}}}$$

where V_m and V_s are the volumes of the mobile and stationary phases. Rearranging and solving for the fraction of solute in the mobile phase, f_m, gives

$$f_{\text{m}} = \frac{(\text{moles S})_{\text{m}}}{(\text{moles S})_{\text{tot}}} = \frac{V_{\text{m}}}{V_{\text{m}} + DV_{\text{s}}} \qquad 12.4$$

Note that this equation is identical to that describing the extraction of a solute in a liquid–liquid extraction (equation 7.25 in Chapter 7). Since the volumes of the stationary and mobile phase may not be known, equation 12.4 is simplified by dividing both the numerator and denominator by V_m; thus

$$f_{\text{m}} = \frac{1}{1 + D(V_{\text{s}}/V_{\text{m}})} = \frac{1}{1 + k'} \qquad 12.5$$

where

$$k' = D\frac{V_{\text{s}}}{V_{\text{m}}} \qquad 12.6$$

is the solute's **capacity factor.**

capacity factor
A measure of how strongly a solute is retained by the stationary phase (k').

A solute's capacity factor can be determined from a chromatogram by measuring the column's void time, t_m, and the solute's retention time, t_r (see Figure 12.7). The mobile phase's average linear velocity, u, is equal to the length of the column, L, divided by the time required to elute a nonretained solute.

$$u = \frac{L}{t_{\text{m}}} \qquad 12.7$$

By the same reasoning, the solute's average linear velocity, v, is

$$v = \frac{L}{t_{\text{r}}} \qquad 12.8$$

The solute can only move through the column when it is in the mobile phase. Its average linear velocity, therefore, is simply the product of the mobile phase's average linear velocity and the fraction of solute present in the mobile phase.

$$v = uf_{\text{m}} \qquad 12.9$$

Substituting equations 12.5, 12.7, and 12.8 into equation 12.9 gives

$$\frac{L}{t_{\text{r}}} = \frac{L}{t_{\text{m}}}\left(\frac{1}{1 + k'}\right)$$

Finally, solving this equation for k' gives

$$k' = \frac{t_{\text{r}} - t_{\text{m}}}{t_{\text{m}}} = \frac{t_{\text{r}}'}{t_{\text{m}}} \qquad 12.10$$

where t_r' is known as the **adjusted retention time.**

adjusted retention time
The difference between a solute's retention time and column's void time (t_r').

Example

EXAMPLE 12.2

In a chromatographic analysis of low-molecular-weight acids, butyric acid elutes with a retention time of 7.63 min. The column's void time is 0.31 min. Calculate the capacity factor for butyric acid.

SOLUTION

$$k' = \frac{t_r - t_m}{t_m} = \frac{7.63 \text{ min} - 0.31 \text{ min}}{0.31 \text{ min}} = 23.6$$

12B.3 Column Selectivity

selectivity factor
The ratio of capacity factors for two solutes showing the column's selectivity for one of the solutes (α).

The relative selectivity of a chromatographic column for a pair of solutes is given by the **selectivity factor,** α, which is defined as

$$\alpha = \frac{k'_B}{k'_A} = \frac{t_{r,B} - t_m}{t_{r,A} - t_m} \qquad \textbf{12.11}$$

The identities of the solutes are defined such that solute A always has the smaller retention time. Accordingly, the selectivity factor is equal to 1 when the solutes elute with identical retention times, and is greater than 1 when $t_{r,B}$ is greater than $t_{r,A}$.

Example

EXAMPLE 12.3

In the same chromatographic analysis for low-molecular-weight acids considered in Example 12.2, the retention time for isobutyric acid is 5.98 min. What is the selectivity factor for isobutyric acid and butyric acid?

SOLUTION

First we must calculate the capacity factor for isobutyric acid. Using the void time from Example 12.2, this is

$$k' = \frac{t_r - t_m}{t_m} = \frac{5.98 \text{ min} - 0.31 \text{ min}}{0.31 \text{ min}} = 18.3$$

The selectivity factor, therefore, is

$$\alpha = \frac{k'_{\text{buty}}}{k'_{\text{iso}}} = \frac{23.6}{18.3} = 1.29$$

12B.4 Column Efficiency

At the beginning of a chromatographic separation the solute occupies a narrow band of finite width. As the solute passes through the column, the width of its band continually increases in a process called **band broadening.** Column efficiency provides a quantitative measure of the extent of band broadening.

In their original theoretical model of chromatography, Martin and Synge[2] treated the chromatographic column as though it consists of discrete sections at which partitioning of the solute between the stationary and mobile phases occurs. They called each section a **theoretical plate** and defined column efficiency in terms of the number of theoretical plates, *N*, or the height of a theoretical plate, *H*; where

$$N = \frac{L}{H} \qquad \textbf{12.12}$$

A column's efficiency improves with an increase in the number of theoretical plates or a decrease in the height of a theoretical plate.

Assuming a Gaussian profile, the extent of band broadening is measured by the variance or standard deviation of a chromatographic peak. The height of a theoretical plate is defined as the variance per unit length of the column

$$H = \frac{\sigma^2}{L} \qquad \textbf{12.13}$$

where the variance, σ^2, has units of distance squared. Because retention time and peak width are usually measured in seconds or minutes, it is more convenient to express the standard deviation in units of time, τ, by dividing σ by the mobile phase's average linear velocity.

$$\tau = \frac{\sigma}{u} = \frac{\sigma t_r}{L} \qquad \textbf{12.14}$$

When a chromatographic peak has a Gaussian shape, its width at the baseline, *w*, is four times its standard deviation, τ.

$$w = 4\tau \qquad \textbf{12.15}$$

Combining equations 12.13 through 12.15 gives the height of a theoretical plate in terms of the easily measured chromatographic parameters t_r and *w*.

$$H = \frac{Lw^2}{16t_r^2} \qquad \textbf{12.16}$$

The number of theoretical plates in a chromatographic column is obtained by combining equations 12.12 and 12.16.

$$N = 16\left(\frac{t_r}{w}\right)^2 \qquad \textbf{12.17}$$

Alternatively, the number of theoretical plates can be approximated as

$$N = 5.545\left(\frac{t_r}{w_{1/2}}\right)^2$$

where $w_{1/2}$ is the width of the chromatographic peak at half its height.

band broadening
The increase in a solute's baseline width as it moves from the point of injection to the detector.

theoretical plate
A quantitative means of evaluating column efficiency that treats the column as though it consists of a series of small zones, or plates, in which partitioning between the mobile and stationary phases occurs.

Example

EXAMPLE 12.4

A chromatographic analysis for the chlorinated pesticide Dieldrin gives a peak with a retention time of 8.68 min and a baseline width of 0.29 min. How many theoretical plates are involved in this separation? Given that the column used in this analysis is 2.0 meters long, what is the height of a theoretical plate?

SOLUTION

Using equation 12.17, the number of theoretical plates is

$$N = 16\left(\frac{t_r}{w}\right)^2 = 16\left(\frac{8.68 \text{ min}}{0.29 \text{ min}}\right)^2 = 14,300 \text{ plates}$$

Solving equation 12.12 for H gives the average height of a theoretical plate as

$$H = \frac{L}{N} = \frac{(2.0 \text{ m})(1000 \text{ mm/m})}{14,300 \text{ plates}} = 0.14 \text{ mm/plate}$$

It is important to remember that a theoretical plate is an artificial construct and that no such plates exist in a chromatographic column. In fact, the number of theoretical plates depends on both the properties of the column and the solute. As a result, the number of theoretical plates for a column is not fixed and may vary from solute to solute.

12B.5 Peak Capacity

peak capacity
The maximum number of solutes that can be resolved on a particular column (n_c).

Another important consideration is the number of solutes that can be baseline resolved on a given column. An estimate of a column's **peak capacity,** n_c, is

$$n_c = 1 + \frac{\sqrt{N}}{4}\ln\frac{V_{max}}{V_{min}} \qquad 12.18$$

where V_{min} and V_{max} are the smallest and largest volumes of mobile phase in which a solute can be eluted and detected.[3] A column with 10,000 theoretical plates, for example, can resolve no more than

$$n_c = 1 + \frac{\sqrt{10,000}}{4}\ln\frac{30 \text{ mL}}{1 \text{ mL}} = 86 \text{ solutes}$$

if the minimum and maximum volumes of mobile phase in which the solutes can elute are 1 mL and 30 mL. This estimate provides an upper bound on the number of solutes that might be separated and may help to exclude from consideration columns that do not have enough theoretical plates to separate a complex mixture. Just because a column's theoretical peak capacity is larger than the number of solutes to be separated, however, does not mean that the separation will be feasible. In most situations the peak capacity obtained is less

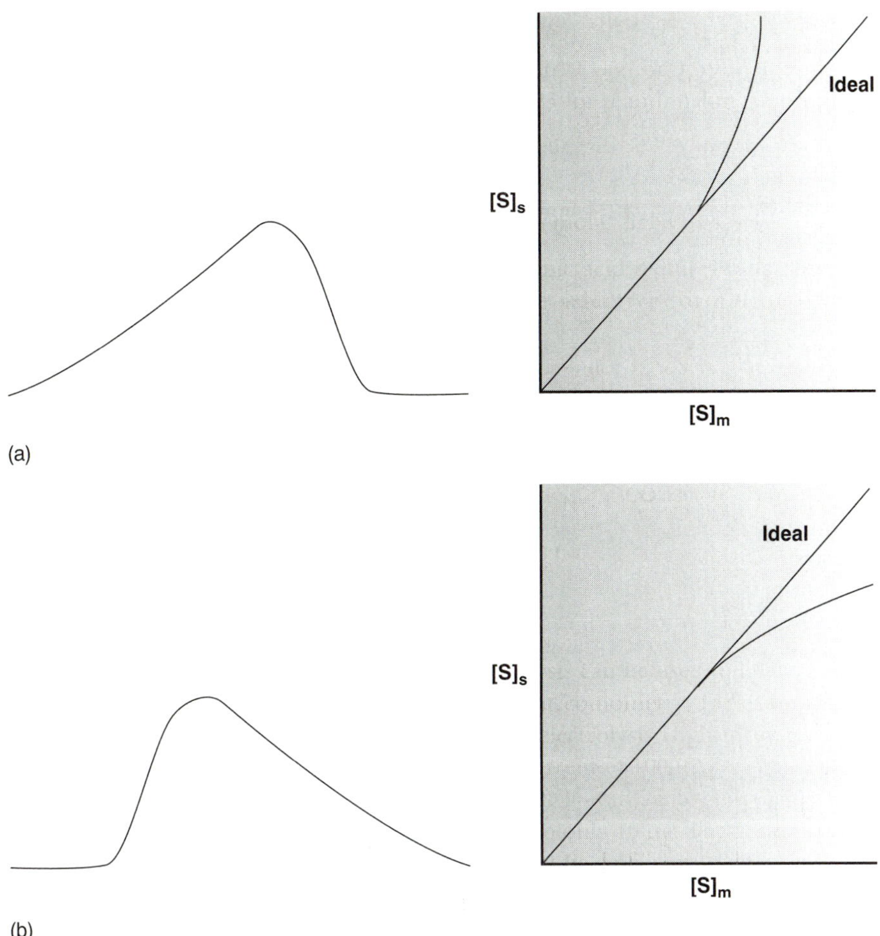

Figure 12.10
Nonideal asymmetrical chromatographic bands showing (a) fronting and (b) tailing. Also depicted are the corresponding sorption isotherms showing the relationship between the concentration of solute in the stationary phase as a function of its concentration in the mobile phase.

than the estimated value because the retention characteristics of some solutes are too similar to effect their separation. Nevertheless, columns with more theoretical plates, or a greater range of possible elution volumes, are more likely to separate a complex mixture.

12B.6 Nonideal Behavior

The treatment of chromatography outlined in Section 12B assumes that a solute elutes as a symmetrical band, such as that shown in Figure 12.7. This ideal behavior occurs when the solute's partition coefficient, K_D, is constant for all concentrations of solute. In some situations, chromatographic peaks show nonideal behavior, leading to asymmetrical peaks, similar to those shown in Figure 12.10. The chromatographic peak in Figure 12.10a is an example of "**fronting**" and is most often the result of overloading the column with sample. Figure 12.10b, which is an example of "**tailing,**" occurs when some sites on the stationary phase retain the solutes more strongly than other sites.

fronting
A tail at the beginning of a chromatographic peak, usually due to injecting too much sample.

tailing
A tail at the end of a chromatographic peak, usually due to the presence of highly active sites in the stationary phase.

12C Optimizing Chromatographic Separations

Now that we have defined capacity factor, selectivity, and column efficiency we consider their relationship to chromatographic resolution. Since we are only interested in the resolution between solutes eluting with similar retention times, it is safe to assume that the peak widths for the two solutes are approximately the same. Equation 12.1, therefore, is written as

$$R = \frac{t_{r,B} - t_{r,A}}{w_B} \qquad 12.19$$

Solving equation 12.17 for w_B and substituting into equation 12.19 gives

$$R = \frac{1}{4}\sqrt{N_B}\left(\frac{t_{r,B} - t_{r,A}}{t_{r,B}}\right) \qquad 12.20$$

The retention times for solutes A and B are replaced with their respective capacity factors by rearranging equation 12.10

$$t_r = k' t_m + t_m$$

and substituting into equation 12.20.

$$R = \frac{1}{4}\sqrt{N_B}\left(\frac{k'_B - k'_A}{1 + k'_B}\right)$$

Finally, solute A's capacity factor is eliminated using equation 12.11. After rearranging, the equation for the resolution between the chromatographic peaks for solutes A and B is

$$R = \frac{1}{4}\sqrt{N_B}\left(\frac{\alpha - 1}{\alpha}\right)\left(\frac{k'_B}{1 + k'_B}\right) \qquad 12.21$$

Besides resolution, another important factor in chromatography is the amount of time required to elute a pair of solutes. The time needed to elute solute B is

$$t_{r,B} = \frac{16R^2 H}{u}\left(\frac{\alpha}{\alpha - 1}\right)^2 \frac{(1 + k'_B)^3}{(k'_B)^2} \qquad 12.22$$

Equations 12.21 and 12.22 contain terms corresponding to column efficiency, column selectivity, and capacity factor. These terms can be varied, more or less independently, to obtain the desired resolution and analysis time for a pair of solutes. The first term, which is a function of the number of theoretical plates or the height of a theoretical plate, accounts for the effect of column efficiency. The second term is a function of α and accounts for the influence of column selectivity. Finally, the third term in both equations is a function of k'_B, and accounts for the effect of solute B's capacity factor. Manipulating these parameters to improve resolution is the subject of the remainder of this section.

12C.1 Using the Capacity Factor to Optimize Resolution

One of the simplest ways to improve resolution is to adjust the capacity factor for solute B. If all other terms in equation 12.21 remain constant, increasing k'_B improves resolution. As shown in Figure 12.11, however, the effect is greatest when the

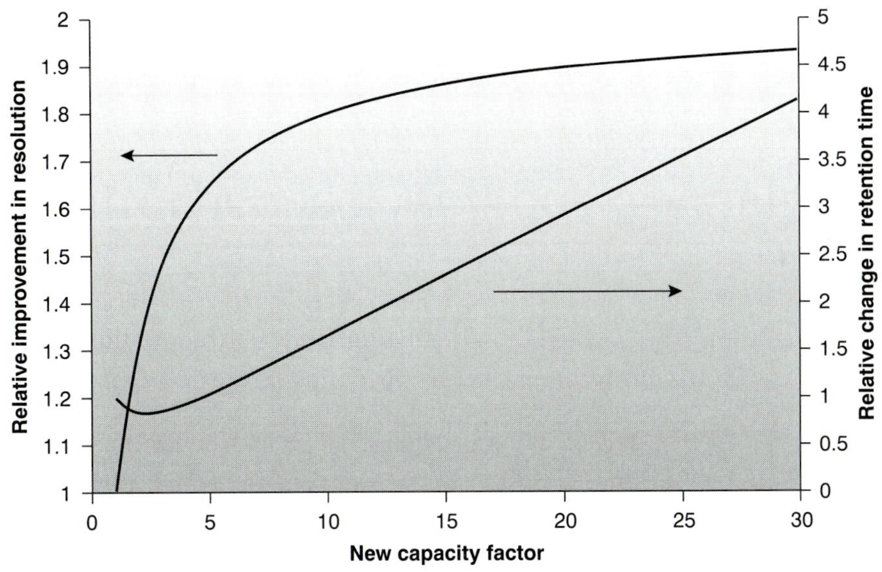

Figure 12.11
Effect of a change in k'_B on resolution and retention time. The original value of k'_B is assumed to be 1.

original capacity factor is small. Furthermore, large increases in k'_B do not lead to proportionally larger increases in resolution. For example, when the original value of k'_B is 1, increasing its value to 10 gives an 82% improvement in resolution; a further increase to 15 provides a net improvement in resolution of only 87.5%.

Any improvement in resolution obtained by increasing k'_B generally comes at the expense of a longer analysis time. This is also indicated in Figure 12.11, which shows the relative change in retention time as a function of the new capacity factor. Note that a minimum in the retention time curve occurs when k'_B is equal to 2, and that retention time increases in either direction. Increasing k'_B from 2 to 10, for example, approximately doubles solute B's retention time.

The relationship between capacity factor and analysis time can be advantageous when a separation produces an acceptable resolution with a large k'_B. In this case it may be possible to decrease k'_B with little loss in resolution while significantly shortening the analysis time.

A solute's capacity factor is directly proportional to its distribution ratio (equation 12.6), which, in turn, is proportional to the solute's equilibrium distribution constant. To increase k'_B without significantly changing α, which also is a function of k'_B, it is necessary to alter chromatographic conditions in a way that leads to a general, nonselective increase in the capacity factor for both solutes. In gas chromatography, this is usually accomplished by decreasing the column's temperature. At a lower temperature a solute's vapor pressure decreases, ensuring that it spends more time in the stationary phase increasing its capacity factor. In liquid chromatography, changing the mobile phase's solvent strength is the easiest way to change a solute's capacity factor. When the mobile phase has a low solvent strength, solutes spend proportionally more time in the stationary phase, thereby increasing their capacity factors. Additionally, equation 12.6 shows that the capacity factor is proportional to the volume of stationary phase. Increasing the volume of stationary phase, therefore, also leads to an increase in k'_B.

Adjusting the capacity factor to improve resolution between one pair of solutes may lead to an unacceptably long retention time for other solutes. For example, improving resolution for solutes with short retention times by increasing

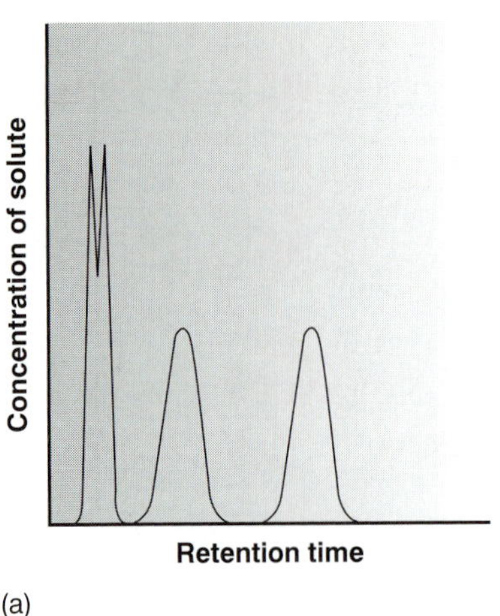

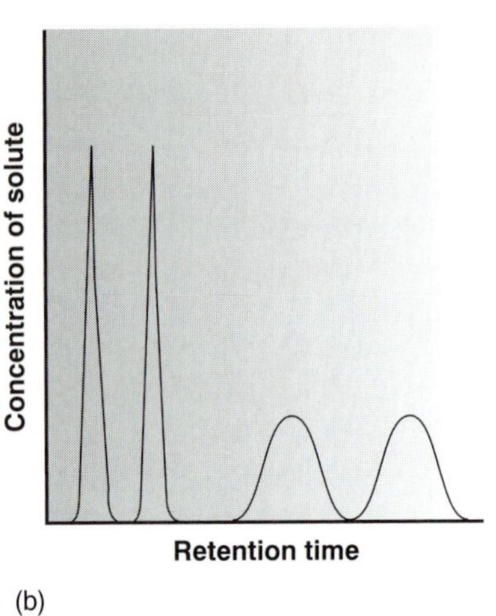

Figure 12.12
The general elution problem in chromatography. Improving the resolution of the overlapping bands in chromatogram (a) results in a longer analysis time for chromatogram (b).

k'_B may substantially increase the retention times for later eluting solutes. On the other hand, decreasing k'_B as a means of shortening the overall analysis time may lead to a loss of resolution for solutes eluting with shorter retention times. This difficulty is encountered so frequently that it is known as the general elution problem (Figure 12.12). One solution to the general elution problem is to make incremental adjustments to the capacity factor over time. Thus, initial chromatographic conditions are adjusted to enhance the resolution for solutes with short retention times. As the separation progresses, chromatographic conditions are changed in a manner that increases the elution rate (decreases the retention time) for later eluting solutes. In gas chromatography this is accomplished by **temperature programming.** The column's initial temperature is selected such that the first solutes to elute are fully resolved. The temperature is then increased, either continuously or in steps, to bring off later eluting components with both an acceptable resolution and a reasonable analysis time. In liquid chromatography the same effect can be obtained by increasing the solvent's eluting strength. This is known as a **gradient elution.**

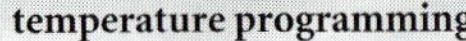

temperature programming
The process of changing the column's temperature to enhance the separation of both early and late eluting solutes.

gradient elution
The process of changing the mobile phase's solvent strength to enhance the separation of both early and late eluting solutes.

12C.2 Using Column Selectivity to Optimize Resolution

A second approach to improving resolution is to adjust alpha, α. In fact, when α is nearly 1, it usually is not possible to improve resolution by adjusting k'_B or N. Changes in α often have a more dramatic effect on resolution than k'_B. For example, changing α from 1.1 to 1.5 improves resolution by 267%.

A change in α is possible if chromatographic conditions are altered in a manner that is more selective for one of the solutes. If a solute participates in a secondary equilibrium reaction in either the stationary or mobile phase, then it may be possible to alter that phase in a way that selectively changes the solute's capacity factor. For example, Figure 12.13a shows how the pH of an aqueous mobile phase can be used to control the retention times, and thus the capacity factors, for two substituted benzoic acids. The resulting change in α is shown in Figure 12.13b. In gas chromatography, adjustments in α are usually accomplished by changing the stationary phase, whereas changing the composition of the mobile phase is used in liquid chromatography.

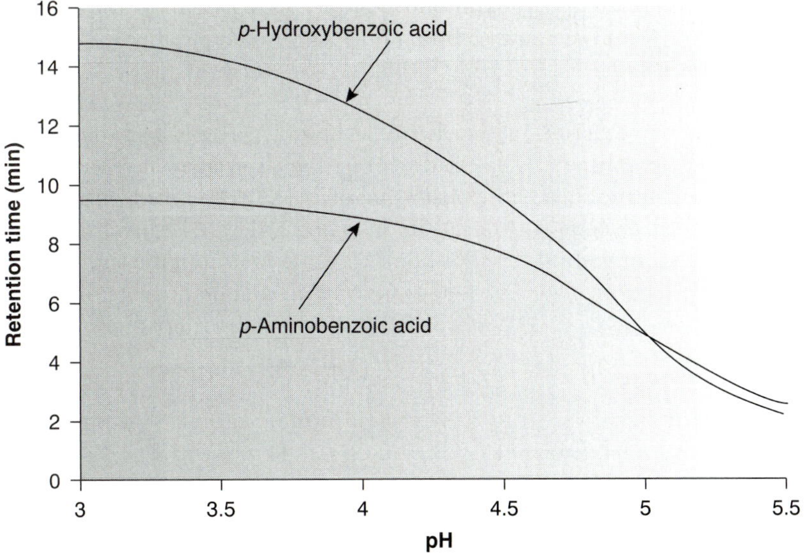

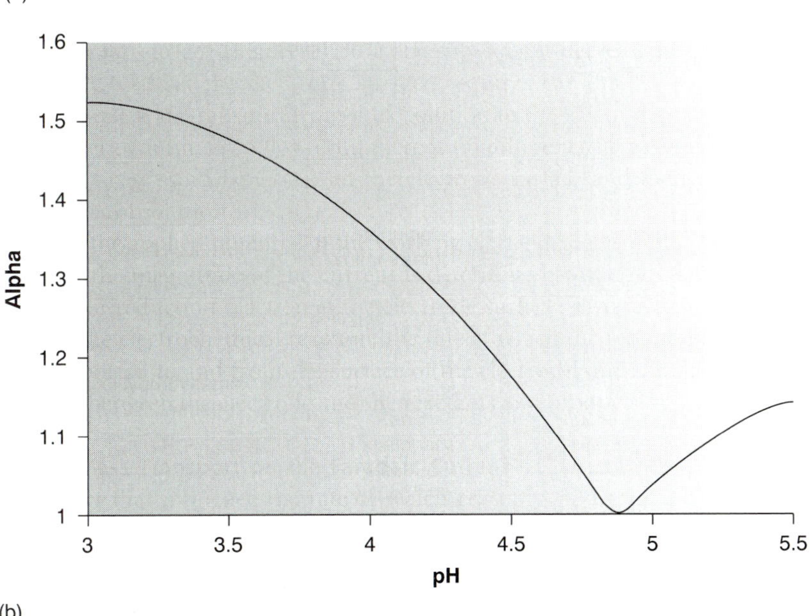

Figure 12.13
Use of column selectivity to improve chromatographic resolution showing: (a) the variation in retention time with mobile phase pH, and (b) the resulting change in alpha with mobile phase pH.

12C.3 Using Column Efficiency to Optimize Resolution

If the capacity factor and α are known, then equation 12.21 can be used to calculate the number of theoretical plates needed to achieve a desired resolution (Table 12.1). For example, given $\alpha = 1.05$ and $k'_B = 2.0$, a resolution of 1.25 requires approximately 24,800 theoretical plates. If the column only provides 12,400 plates, half of what is needed, then the separation is not possible. How can the number of theoretical plates be doubled? The easiest way is to double the length of the column; however, this also requires a doubling of the analysis time. A more desirable approach is to cut the height of a theoretical plate in half, providing the desired resolution without changing the analysis time. Even better, if H can be decreased by more than

Table 12.1 Number of Theoretical Plates Needed to Achieve Desired Resolution for Selected Values of k'_B and α

	R = 1.00		R = 1.25		R = 1.50	
k'_B	α = 1.05	α = 1.10	α = 1.05	α = 1.10	α = 1.05	α = 1.10
0.5	63,500	17,400	99,200	27,200	143,000	39,200
1.0	28,200	7,740	44,100	12,100	63,500	17,400
1.5	19,600	5,380	30,600	8,400	44,100	12,100
2.0	15,900	4,360	24,800	6,810	35,700	9,800
3.0	12,500	3,440	19,600	5,380	28,200	7,740
5.0	10,200	2,790	15,900	4,360	22,900	6,270
10.0	8,540	2,340	13,300	3,660	19,200	5,270

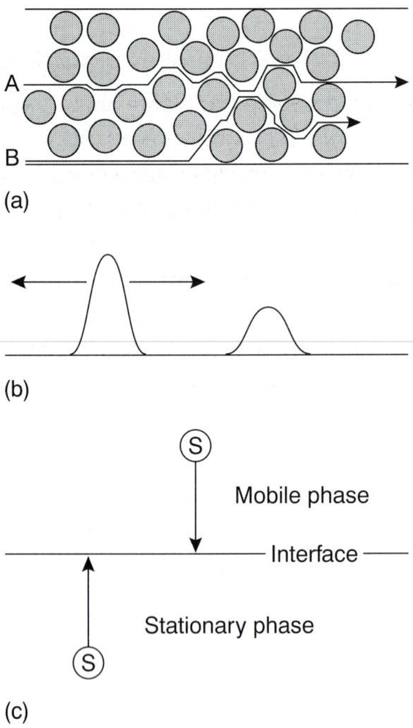

Figure 12.14
Schematics illustrating the contributions to band broadening due to (a) multiple paths, (b) longitudinal diffusion, and (c) mass transfer.

longitudinal diffusion
One contribution to band broadening in which solutes diffuse from areas of high concentration to areas of low concentration.

50%, it also may be possible to achieve the desired resolution with an even shorter analysis time by decreasing k'_B or α.

To determine how the height of a theoretical plate can be decreased, it is necessary to understand the experimental factors contributing to the broadening of a solute's chromatographic band. Several theoretical treatments of band broadening have been proposed. We will consider one approach in which the height of a theoretical plate is determined by four contributions: multiple paths, longitudinal diffusion, mass transfer in the stationary phase, and mass transfer in the mobile phase.

Multiple Paths Solute molecules passing through a chromatographic column travel separate paths that may differ in length. Because of these differences in path length, solute molecules injected simultaneously elute at different times. The principal factor contributing to this variation in path length is a nonhomogeneous packing of the stationary phase in the column. Differences in particle size and packing consistency cause solute molecules to travel paths of different length. Some solute molecules follow relatively straight paths through the column, but others follow a longer, more tortuous path (Figure 12.14a). The contribution of multiple paths to the height of a theoretical plate, H_p, is

$$H_p = 2\lambda d_p \tag{12.23}$$

where d_p is the average diameter of the particulate packing material, and λ is a constant accounting for the consistency of the packing. A smaller range of particle sizes and a more consistent packing produce a smaller value for λ. Note that for an open tubular column, which does not contain packing material, H_p is 0.

Longitudinal Diffusion The second contribution to band broadening is the result of the solute's **longitudinal diffusion** in the mobile phase. Even if the mobile phase velocity is 0, solute molecules are constantly in motion, diffusing through the mobile phase. Since the concentration of solute is greatest at the center of a chromatographic band, more solute diffuses toward the band's forward and rear edges than diffuses toward the band's center. The net result is an increase in the band's width (Figure 12.14b). The contribution of longitudinal diffusion to the height of a theoretical plate, H_d, is

$$H_d = \frac{2\gamma D_m}{u} \qquad 12.24$$

where D_m is the solute's diffusion coefficient in the mobile phase, u is the mobile phase velocity, and γ is a constant related to the column packing. The effect of H_d on the height of a theoretical plate is minimized by a high mobile-phase velocity. Because a solute's diffusion coefficient is larger in a gaseous mobile phase than in a liquid mobile phase, longitudinal diffusion is a more serious problem in gas chromatography.

Mass Transfer The final two contributions to band broadening result from the finite time required for a solute molecule to diffuse through the stationary phase and mobile phase. A chromatographic separation occurs because solutes move between the stationary and mobile phases. For a solute to move from one phase to the other, it must first diffuse to the interface between the two phases (Figure 12.14c)—a process called **mass transfer.** A contribution to band broadening occurs whenever the solute's movement to the interface is not fast enough to maintain a true equilibrium distribution of solute between the two phases. Thus, solute molecules in the mobile phase move farther down the column than expected before passing into the stationary phase. Solute molecules in the stationary phase, on the other hand, take longer than expected to cross into the mobile phase. The contributions of mass transfer in the stationary phase, H_s, and mass transfer in the mobile phase, H_m, are given by

mass transfer
One contribution to band broadening due to the time required for a solute to move from the mobile phase or the stationary phase to the interface between the two phases.

$$H_s = \frac{qk'd_f^2}{(1+k')^2 D_s} u \qquad 12.25$$

$$H_m = \frac{fn(d_p^2, d_c^2)}{D_m} u \qquad 12.26$$

where d_f is the thickness of the stationary phase, d_c is the column's diameter, D_s is the solute's diffusion coefficient in the stationary phase, q is a constant related to the column packing material, and the remaining terms are as previously defined. As indicated in equation 12.26, the exact form of H_m is unknown, although it is a function of particle size and column diameter. The contribution of mass transfer to the height of a theoretical plate is smallest for slow mobile-phase velocities, smaller diameter packing materials, and thinner films of stationary phase.

Putting It All Together The net height of a theoretical plate is a summation of the contributions from each of the terms in equations 12.23–12.26; thus,

$$H = H_p + H_d + H_s + H_m \qquad 12.27$$

An alternative form of this equation is the **van Deemter equation**

van Deemter equation
An equation showing the effect of the mobile phase's flow rate on the height of a theoretical plate.

$$H = A + \frac{B}{u} + Cu \qquad 12.28$$

which emphasizes the importance of the mobile phase's flow rate. In the van Deemter equation, A accounts for multiple paths (H_p), B/u for longitudinal diffusion (H_d), and Cu for the solute's mass transfer in the stationary and mobile phases (H_s and H_m).

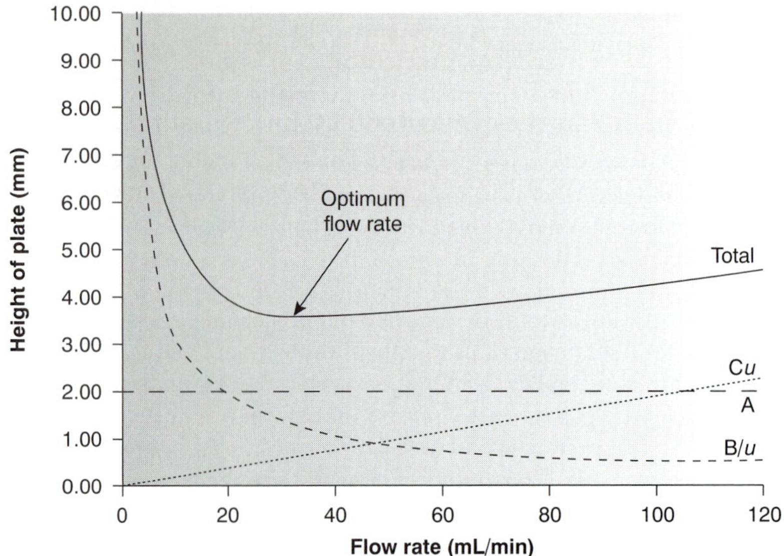

Figure 12.15
Plot of the height of a theoretical plate as a function of mobile-phase velocity using the van Deemter equation. The contributions to the terms *A*, *B/u*, and *Cu* also are shown.

There is some disagreement on the correct equation for describing the relationship between plate height and mobile-phase velocity.[4] In addition to the van Deemter equation (equation 12.28), another equation is that proposed by Hawkes

$$H = \frac{B}{u} + (C_s + C_m)u$$

where C_s and C_m are the mass transfer terms for the stationary and mobile phases respectively. A third equation was devised by Knox.

$$H = Au^{1/3} + \frac{B}{u} + Cu$$

All three equations, and others, have been used to characterize chromatographic systems, with no single equation providing the best explanation in every case.[5]

To increase the number of theoretical plates without increasing the length of the column, it is necessary to decrease one or more of the terms in equation 12.27 or equation 12.28. The easiest way to accomplish this is by adjusting the velocity of the mobile phase. At a low mobile-phase velocity, column efficiency is limited by longitudinal diffusion, whereas at higher velocities efficiency is limited by the two mass transfer terms. As shown in Figure 12.15 (which is interpreted in terms of equation 12.28), the optimum mobile-phase velocity corresponds to a minimum in a plot of H as a function of u.

The remaining parameters affecting the height of a theoretical plate are determined by the construction of the column and suggest how the column's design may be used to improve efficiency. For example, both H_p and H_m are a function of the size of the particles used for the packing material. Decreasing particle size, therefore, is one approach to improving efficiency. A decrease in particle size is limited, however, by the need for a greater pressure to push the mobile phase through the column.

capillary column
A narrow bored column that usually does not contain a particulate packing material.

One of the most important advances in column construction has been the development of open tubular, or **capillary columns** that contain no packing material ($d_p = 0$). Instead, the interior wall of a capillary column is coated with a thin film of the stationary phase. The absence of packing material means that the mobile phase

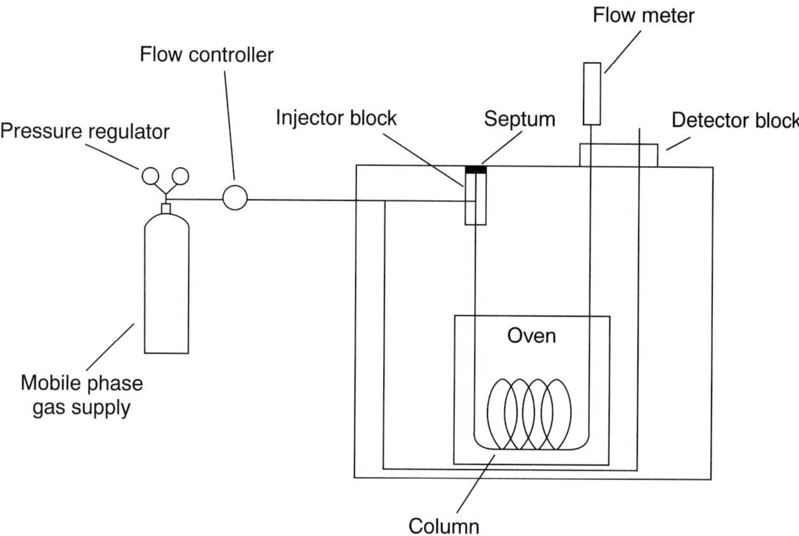

Figure 12.16
Schematic diagram for a typical gas chromatograph.

can move through the column with substantially less pressure. As a result, capillary columns can be manufactured with much greater lengths than is possible with a packed column. Furthermore, plate height is reduced because the H_p term in equation 12.27 disappears and the H_m term becomes smaller. The combination of a smaller height for a theoretical plate and a longer column leads to an approximate 100-fold increase in the number of theoretical plates. Capillary columns are not without disadvantages. Because capillary columns are much narrower than packed columns, they require a significantly smaller amount of sample. Difficulties with reproducibly injecting small samples complicates the use of capillary chromatography for quantitative work.

Another approach to improving resolution is to use thin films of stationary phase. Capillary columns used in gas chromatography and the bonded phases commonly used in HPLC provide a significant decrease in plate height due to the reduction of the H_s term in equation 12.27.

12D Gas Chromatography

In **gas chromatography** (GC) the sample, which may be a gas or liquid, is injected into a stream of an inert gaseous mobile phase (often called the carrier gas). The sample is carried through a packed or capillary column where the sample's components separate based on their ability to distribute themselves between the mobile and stationary phases. A schematic diagram of a typical gas chromatograph is shown in Figure 12.16.

gas chromatography
A chromatographic technique in which the mobile phase is a gas.

12D.1 Mobile Phase

The most common mobile phases for GC are He, Ar, and N_2, which have the advantage of being chemically inert toward both the sample and the stationary phase. The choice of which carrier gas to use is often determined by the instrument's detector. With packed columns the mobile-phase velocity is usually within the range of 25–150 mL/min, whereas flow rates for capillary columns are 1–25 mL/min. Actual flow rates are determined with a flow meter placed at the column outlet.

12D.2 Chromatographic Columns

A chromatographic column provides a location for physically retaining the stationary phase. The column's construction also influences the amount of sample that can be handled, the efficiency of the separation, the number of analytes that can be easily separated, and the amount of time required for the separation. Both packed and capillary columns are used in gas chromatography.

packed column
A wide-bore column containing a particulate packing material.

Packed Columns A **packed column** is constructed from glass, stainless steel, copper or aluminum and is typically 2–6 m in length, with an internal diameter of 2–4 mm. The column is filled with a particulate solid support, with particle diameters ranging from 37–44 μm to 250–354 μm.

The most widely used particulate support is diatomaceous earth, which is composed of the silica skeletons of diatoms. These particles are quite porous, with surface areas of 0.5–7.5 m^2/g, which provides ample contact between the mobile phase and stationary phase. When hydrolyzed, the surface of a diatomaceous earth contains silanol groups (–SiOH), providing active sites that absorb solute molecules in gas–solid chromatography.

gas–liquid chromatography
A chromatographic technique in which the mobile phase is a gas and the stationary phase is a liquid coated either on a solid packing material or on the column's walls.

In **gas–liquid chromatography** (GLC), separation is based on the partitioning of solutes between a gaseous mobile phase and a liquid stationary phase coated on the solid packing material. To avoid the adsorption of solute molecules on exposed packing material, which degrades the quality of the separation, surface silanols are deactivated by silanizing with dimethyldichlorosilane and washing with an alcohol (typically methanol) before coating with stationary phase.

$$\text{—Si—OH} \xrightarrow{Si(CH_3)_2Cl_2} \text{—Si—O—}Si(CH_3)_2Cl + HCl \xrightarrow{ROH} \text{—Si—O—}Si(CH_3)_2OR + HCl$$

More recently, solid supports made from glass beads or fluorocarbon polymers have been introduced. These supports have the advantage of being more inert than diatomaceous earth.

To minimize the multiple path and mass transfer contributions to plate height (equations 12.23 and 12.26), the packing material should be of as small a diameter as is practical and loaded with a thin film of stationary phase (equation 12.25). Compared with capillary columns, which are discussed in the next section, packed columns can handle larger amounts of sample. Samples of 0.1–10 μL are routinely analyzed with a packed column. Column efficiencies are typically several hundred to 2000 plates/m, providing columns with 3000–10,000 theoretical plates. Assuming V_{max}/V_{min} is approximately 50,[3] a packed column with 10,000 theoretical plates has a peak capacity (equation 12.18) of

$$n_c = 1 + \frac{\sqrt{10,000}}{4}\ln(50) \approx 100$$

open tubular column
A capillary column that does not contain a particulate packing material.

Capillary Columns Capillary, or **open tubular columns** are constructed from fused silica coated with a protective polymer. Columns may be up to 100 m in length with an internal diameter of approximately 150–300 μm (Figure 12.17). Larger bore columns of 530 μm, called megabore columns, also are available.

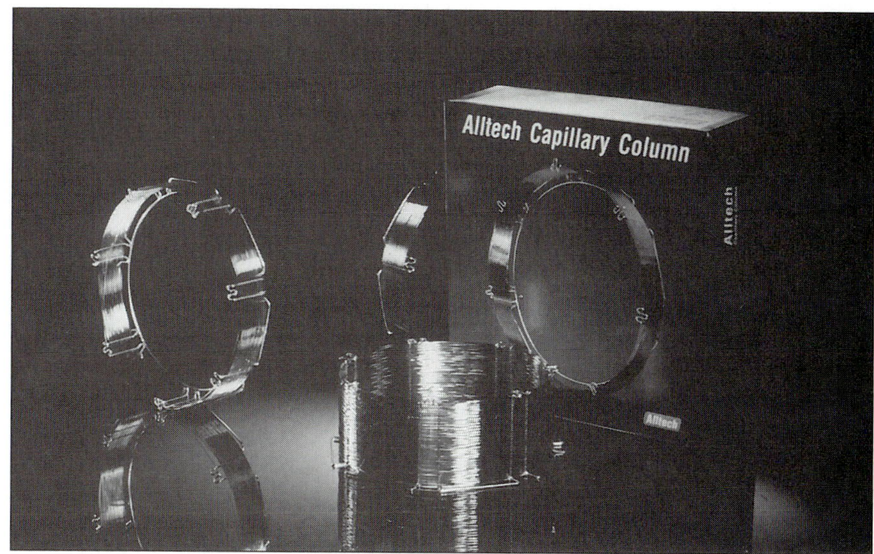

Figure 12.17
Photo of a capillary column.
Courtesy of Alltech Associates, Inc., Deerfield, Illinois.

Capillary columns are of two principal types. **Wall-coated open tubular columns** (WCOT) contain a thin layer of stationary phase, typically 0.25 μm thick, coated on the capillary's inner wall. In **support-coated open tubular columns** (SCOT), a thin layer of a solid support, such as a diatomaceous earth, coated with a liquid stationary phase is attached to the capillary's inner wall.

wall-coated open tubular column
An open tubular column in which the stationary phase is coated on the column's walls.

support-coated open tubular column
An open tubular column in which the stationary phase is coated on a solid support that is attached to the column's walls.

Capillary columns provide a significant improvement in separation efficiency. The pressure needed to move the mobile phase through a packed column limits its length. The absence of packing material allows a capillary column to be longer than a packed column. Although most capillary columns contain more theoretical plates per meter than a packed column, the more important contribution to their greater efficiency is the ability to fashion longer columns. For example, a 50-m capillary column with 3000 plates/m has 150,000 theoretical plates and, assuming V_{max}/V_{min} is approximately 50,[3] a peak capacity of almost 380. On the other hand, packed columns can handle larger samples. Due to its smaller diameter, capillary columns require smaller samples; typically less than 10^{-2} μL.

12D.3 Stationary Phases

Selectivity in gas chromatography is influenced by the choice of stationary phase. Elution order in GLC is determined primarily by the solute's boiling point and, to a lesser degree, by the solute's interaction with the stationary phase. Solutes with significantly different boiling points are easily separated. On the other hand, two solutes with similar boiling points can be separated only if the stationary phase selectively interacts with one of the solutes. In general, nonpolar solutes are more easily separated with a nonpolar stationary phase, and polar solutes are easier to separate using a polar stationary phase.

The main criteria for selecting a stationary phase are that it should be chemically inert, thermally stable, of low volatility, and of an appropriate polarity for the solutes being separated. Although hundreds of stationary phases have been developed, many of which are commercially available, the majority of GLC separations are accomplished with perhaps five to ten common stationary phases. Several of

Table 12.2 Selected Stationary Phases for Gas–Liquid Chromatography

Stationary Phase	Polarity	Trade Names	Temperature Limit (°C)	Applications
squalane	nonpolar	squalane	150	low-boiling aliphatic hydrocarbons
Apezion L	nonpolar	Apezion L	300	amides fatty acid methyl esters high-boiling aliphatic hydrocarbons terpenoids
polydimethyl siloxane	slightly polar	SE-30	300–350	alkaloids amino acid derivatives drugs pesticides phenols steroids
50% methyl-50% phenyl polysiloxane	moderately polar	OV-17	375	alkaloids drugs pesticides polyaromatic hydrocarbons polychlorinated biphenyls
50% trifluoropropyl-50% methyl polysiloxane	moderately polar	OV-210	275	alkaloids amino acid derivatives drugs halogenated compounds ketones phenols
50% cyanopropyl-50% phenylmethyl polysiloxane	polar	OV-225	275	nitriles pesticides steroids
polyethylene glycol	polar	Carbowax 20M	225	aldehydes esters ethers phenols

these are listed in Table 12.2, in order of increasing polarity, along with their physical properties and typical applications.

Many stationary phases have the general structure shown in Figure 12.18a. A stationary phase of polydimethyl siloxane, in which all the –R groups are methyl groups ($-CH_3$), is nonpolar and often makes a good first choice for a new separation. The order of elution when using polydimethyl siloxane usually follows the boiling points of the solutes, with lower boiling solutes eluting first. Replacing some of the methyl groups with other substituents increases the stationary phase's polarity, providing greater selectivity. Thus, in 50% methyl-50% phenyl polysiloxane, 50% of the –R groups are phenyl groups ($-C_6H_5$), producing a slightly polar stationary phase. Increasing polarity is provided by substituting trifluoropropyl ($-C_3H_6CF_3$) and cyanopropyl ($-C_3H_6CN$) functional groups or using a stationary phase based on polyethylene glycol (Figure 12.18b).

bleed
The tendency of a stationary phase to elute from the column.

An important problem with all liquid stationary phases is their tendency to "**bleed**" from the column. The temperature limits listed in Table 12.2 are those that minimize the loss of stationary phase. When operated above these limits, a column's useful lifetime is significantly shortened. Capillary columns with bonded or cross-linked stationary phases provide superior stability. Bonded stationary phases are attached to the capillary's silica surface. Cross-linking, which is done after the stationary phase is placed in the capillary column, links together separate polymer chains, thereby providing greater stability.

Another important characteristic of a gas chromatographic column is the thickness of the stationary phase. As shown in equation 12.25, separation efficiency improves with thinner films. The most common film thickness is 0.25 μm. Thicker films are used for highly volatile solutes, such as gases, because they have a greater capacity for retaining such solutes. Thinner films are used when separating solutes of low volatility, such as steroids.

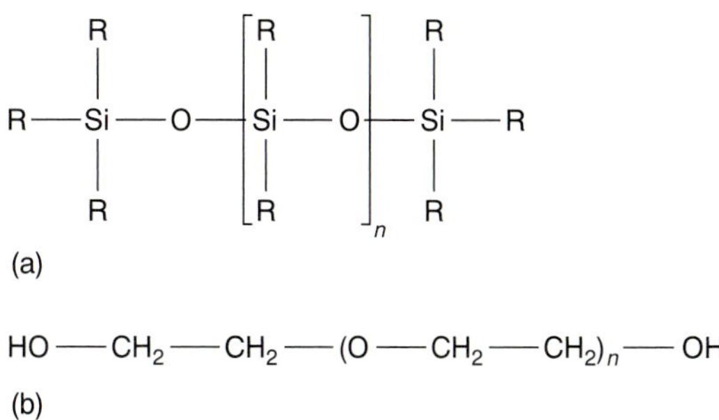

Figure 12.18
General structures of common stationary phases for gas chromatography.

A few GLC stationary phases rely on chemical selectivity. The most notable are stationary phases containing chiral functional groups, which can be used for separating enantiomers.[6]

12D.4 Sample Introduction

Three considerations determine how samples are introduced to the gas chromatograph. First, all constituents injected into the GC must be volatile. Second, the analytes must be present at an appropriate concentration. Finally, injecting the sample must not degrade the separation.

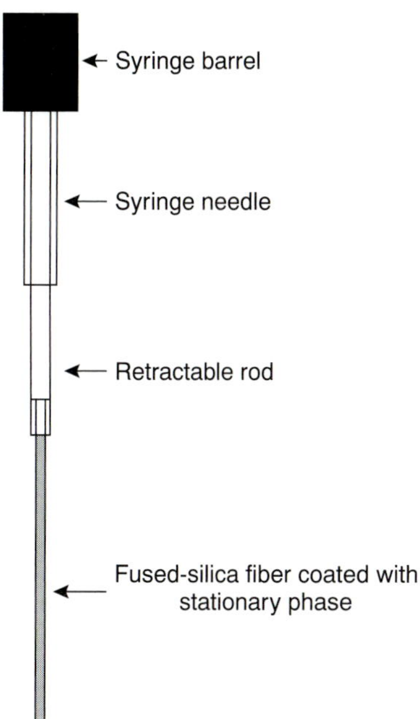

Figure 12.19
Schematic diagram of a device for solid-phase microextractions.

Preparing a Volatile Sample Gas chromatography can be used to separate analytes in complex matrices. Not every sample that can potentially be analyzed by GC, however, can be injected directly into the instrument. To move through the column, the sample's constituents must be volatile. Solutes of low volatility may be retained by the column and continue to elute during the analysis of subsequent samples. Nonvolatile solutes condense on the column, degrading the column's performance.

Volatile analytes can be separated from a nonvolatile matrix using any of the extraction techniques described in Chapter 7. Liquid–liquid extractions, in which analytes are extracted from an aqueous matrix into methylene chloride or other organic solvent, are commonly used. Solid-phase extractions also are used to remove unwanted matrix constituents.

An attractive approach to isolating analytes is a **solid-phase microextraction** (SPME). In one approach, which is illustrated in Figure 12.19, a fused silica fiber is placed inside a syringe needle. The fiber, which is coated with a thin organic film, such as polydimethyl siloxane, is lowered into the sample by depressing a plunger and is exposed to the sample for a predetermined time. The fiber is then withdrawn into the needle and transferred to the gas chromatograph for analysis.

Volatile analytes also can be separated from a liquid matrix using a purge and trap or by headspace sampling. In a purge and trap (see Figure 7.19 in Chapter 7), an inert gas, such as He or N_2, is bubbled through the sample, purging the volatile compounds. These compounds are swept through a trap packed with an absorbent material, such as Tenax, where they are collected. Heating the trap and back flushing with carrier gas transfers the volatile compounds to the gas chromatograph. In **headspace sampling** the sample is placed in a closed vial with an overlying air space. After allowing time for the volatile analytes to equilibrate between the sample and the overlying air, a portion of the vapor phase is sampled by syringe and injected into the gas chromatograph.

solid-phase microextraction
A solid-phase extraction in which the solid adsorbent is coated on a fused-silica fiber held within a syringe needle.

headspace sampling
The sampling of the vapor phase overlying a liquid phase.

Thermal desorption is used to release volatile analytes from solids. A portion of the solid is placed in a glass-lined, stainless steel tube and held in place with plugs of glass wool. After purging with carrier gas to remove O_2 (which could lead to oxidation reactions when heating the sample), the sample is heated. Volatile analytes are swept from the tube by the carrier gas and carried to the GC. To maintain efficiency the solutes often are concentrated at the top of the column by cooling the column inlet below room temperature, a process known as **cryogenic focusing.**

cryogenic focusing
The process of concentrating volatile solutes by cooling the column's inlet below room temperature.

Nonvolatile analytes must be chemically converted to a volatile derivative before analysis. For example, amino acids are not sufficiently volatile to analyze directly by gas chromatography. Reacting an amino acid with 1-butanol and acetyl chloride produces an esterfied amino acid. Subsequent treatment with trifluoroacetic acid gives the amino acid's volatile *N*-trifluoroacetyl-*n*-butyl ester derivative.

Adjusting the Analyte's Concentration Analytes present at concentrations too small to give an adequate signal need to be concentrated before analyzing. A side benefit of many of the extraction methods outlined earlier is that they often concentrate the analytes. Volatile organic materials isolated from aqueous samples by a purge and trap, for example, can be concentrated by as much as 1000-fold.

When an analyte is too concentrated, it is easy to overload the column, thereby seriously degrading the separation. In addition, the analyte may be present at a concentration level that exceeds the detector's linear response. Dissolving the sample in a volatile solvent, such as methylene chloride, makes its analysis feasible.

Injecting the Sample To avoid any precolumn loss in resolution due to band broadening, a sample of sufficient size must be introduced in a small volume of mobile phase. An example of a simple injection port for a packed column is shown in Figure 12.20. Injections are made through a rubber septum using a microliter syringe. The injector block is heated to a temperature that is at least 50 °C above the sample component with the highest boiling point. In this way rapid vaporization of the entire sample is ensured.

Capillary columns require the use of a special injector to avoid overloading the column with sample. Several capillary injectors are available, the most common of which is a split/splitless injector.[7] When used for a **split injection** only about 0.1–1% of the sample enters the column, with the remainder carried off as waste. In a **splitless injection,** which is useful for trace analysis, the column temperature is held 20–25 °C below the solvent's boiling point. As the solvent enters the column, it condenses, forming a barrier that traps the solutes. After allowing time for the solutes to concentrate, the column's temperature is increased, and the separation begins. A splitless injection allows a much higher percentage of the solutes to enter the chromatographic column.

split injection
A technique for injecting samples onto a capillary column in which only a small portion of the sample enters the column.

splitless injection
A technique for injecting a sample onto a capillary column that allows a higher percentage of the sample to enter the column.

For samples that decompose easily, an **on-column injection** may be necessary. In this method the sample is injected on the column without heating. The column temperature is then increased, volatilizing the sample with as low a temperature as is practical.

on-column injection
The direct injection of thermally unstable samples onto a capillary column.

12D.5 Temperature Control

As noted earlier, control of the column's temperature is critical to attaining a good separation in gas chromatography. For this reason the column is located inside a thermostated oven. In an isothermal separation the column is maintained at a constant temperature, the choice of which is dictated by the solutes. Normally, the temperature is set slightly below that for the lowest boiling solute so as to increase the solute's interaction with the stationary phase.

One difficulty with an isothermal separation is that a temperature favoring the separation of low-boiling solutes may cause unacceptably long retention times for higher boiling solutes. Ovens capable of temperature programming provide a solution to this problem. The initial temperature is set below that for the lowest boiling solute. As the separation progresses, the temperature is slowly increased at either a uniform rate or in a series of steps.

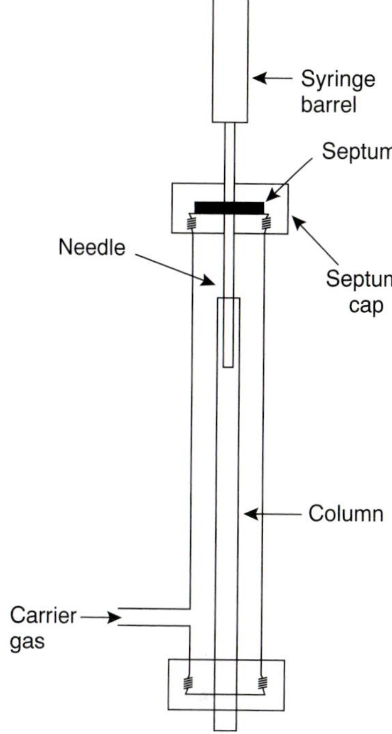

Figure 12.20
Schematic diagram of an injector for packed column gas chromatography.

12D.6 Detectors for Gas Chromatography

The final part of a gas chromatograph is the detector. The ideal detector has several desirable features, including low detection limits, a linear response over a wide range of solute concentrations (which makes quantitative work easier), responsiveness to all solutes or selectivity for a specific class of solutes, and an insensitivity to changes in flow rate or temperature.

Thermal Conductivity Detector One of the earliest gas chromatography detectors, which is still widely used, is based on the mobile phase's thermal conductivity (Figure 12.21). As the mobile phase exits the column, it passes over a tungsten–rhenium wire filament. The filament's electrical resistance depends on its temperature, which, in turn, depends on the thermal conductivity of the mobile phase. Because of its high thermal conductivity, helium is the mobile phase of choice when using a **thermal conductivity detector** (TCD).

thermal conductivity detector
A universal GC detector in which the signal is a change in the thermal conductivity of the mobile phase.

When a solute elutes from the column, the thermal conductivity of the mobile phase decreases and the temperature of the wire filament, and thus its resistance, increases. A reference cell, through which only the mobile phase passes, corrects for any time-dependent variations in flow rate, pressure, or electrical power, all of which may lead to a change in the filament's resistance.

A TCD detector has the advantage of universality, since it gives a signal for any solute whose thermal conductivity differs from that of helium. Another advantage is that it gives a linear response for solute concentrations over a range of 10^4–10^5 orders of magnitude. The detector also is nondestructive, making it possible to isolate solutes with a postdetector cold trap. Unfortunately, the thermal

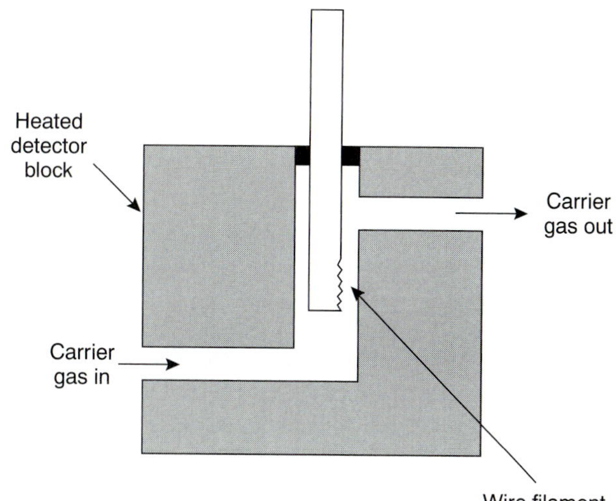

Figure 12.21
Schematic diagram of a thermal conductivity detector for gas chromatography.

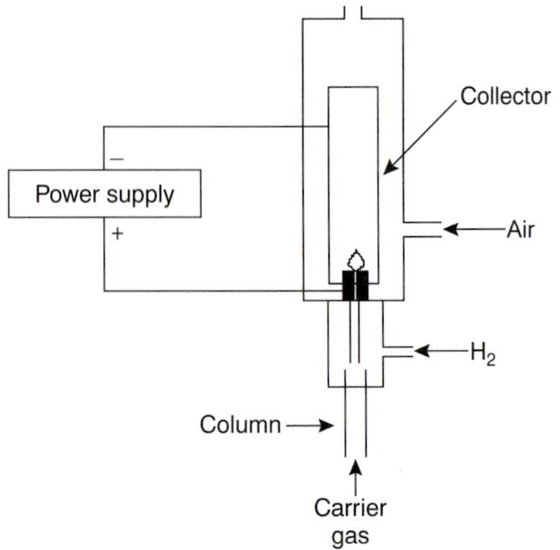

Figure 12.22
Schematic diagram of a flame ionization detector for gas chromatography.

conductivity detector's detection limit is poor in comparison with other popular detectors.

Flame Ionization Detector Combustion of an organic compound in an H_2/air flame results in a flame rich in electrons and ions. If a potential of approximately 300 V is applied across the flame, a small current of roughly 10^{-9}–10^{-12} A develops. When amplified, this current provides a useful analytical signal. This is the basis of the popular **flame ionization detector** (FID), a schematic of which is shown in Figure 12.22.

Most carbon atoms, except those in carbonyl and carboxylic groups, generate a signal, making the FID an almost universal detector for organic compounds. Most inorganic compounds and many gases, such as H_2O and CO_2, cannot be detected, making the FID detector ideal for the analysis of atmospheric and aqueous environmental samples. Advantages of the FID include a detection limit that is approximately two to three orders of magnitude smaller than that for a thermal conductivity detector and a linear response over 10^6–10^7 orders of magnitude in the amount of analyte injected. The sample, of course, is destroyed when using a flame ionization detector.

flame ionization detector
A nearly universal GC detector in which the solutes are combusted in an H_2/air flame, producing a measurable current.

electron capture detector
A detector for GC that provides selectivity for solutes with halogen and nitro functional groups.

Electron Capture Detector The **electron capture detector** is an example of a selective detector. The detector consists of a beta emitter (a beta particle is an electron) such as ^{63}Ni. The emitted electrons ionize the mobile phase, which is usually N_2, resulting in the production of additional electrons that give rise to an electric current between a pair of electrodes (Figure 12.23). When a solute with a high cross section for the capture of electrons elutes from the column, the electric current decreases. This decrease in electric current serves as the signal. The ECD is highly selective toward solutes with electronegative functional groups, such as halogens, and nitro groups and is relatively insensitive to amines, alcohols, and hydrocarbons. Although its detection limit is excellent, its linear range extends over only about two orders of magnitude.

Other Detectors Two additional detectors are similar in design to a flame ionization detector. In the flame photometric detector optical emission from phosphorus and sulfur provides a detector selective for compounds containing these elements. The thermionic detector responds to compounds containing nitrogen or phosphorus.

Two common detectors, which also are independent instruments, are Fourier transform infrared spectrophotometers (FT–IR) and mass spectrometers (MS). In GC–FT–IR, effluent from the column flows through an optical cell constructed

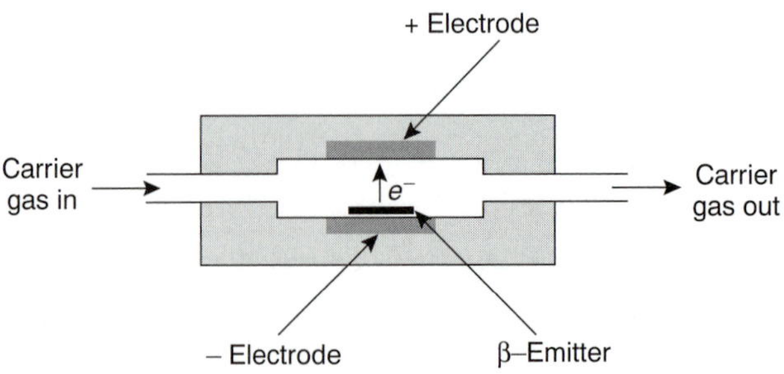

Figure 12.23
Schematic diagram of an electron capture detector for gas chromatography.

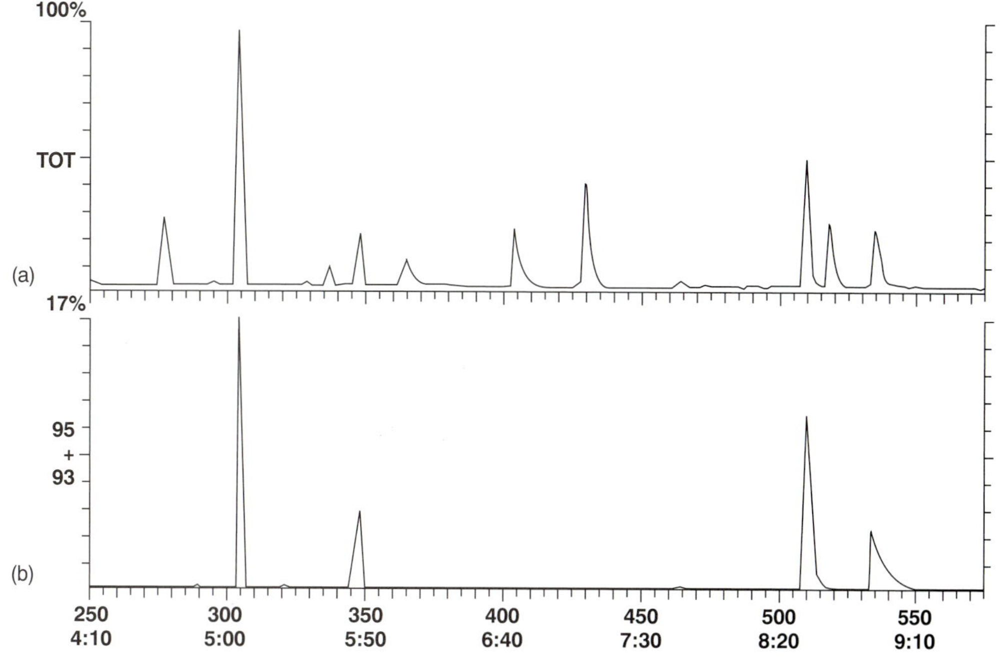

Figure 12.24
(a) Total ion chromatogram for a ten-component mixture; (b) Chromatogram recorded using selective ion monitoring for mass-to-charge ratios of 93 and 95, which are characteristic ions for the monoterpenes α-pinene (t_r = 5.08 min), β-pinene (t_r = 5.81 min), camphor (t_r = 8.51 min), and menthol (t_r = 8.93 min). (Chromatograms courtesy of Bryan Hanson and Sara Peters, DePauw University).

from a 10–40-cm Pyrex tube with an internal diameter of 1–3 mm. The cell's interior surface is coated with a reflecting layer of gold. Multiple reflections of the source radiation as it is transmitted through the cell increase the optical path length through the sample.

In GC–MS effluent from the column is introduced directly into the mass spectrometer's ionization chamber in a manner that eliminates the majority of the carrier gas. In the ionization chamber all molecules (remaining carrier gas, solvent, and solutes) are ionized, and the ions are separated by their mass-to-charge ratio. Because each solute undergoes a characteristic fragmentation into smaller ions, its **mass spectrum** of ion intensity as a function of mass-to-charge ratio provides qualitative information that can be used to identify the solute.

mass spectrum
A plot of ion intensity as a function of the ion's mass-to-charge ratios.

As a GC detector, the total ion current for all ions reaching the detector is usually used to obtain the chromatogram (Figure 12.24a). Selectivity can be achieved by monitoring only specific mass-to-charge ratios (Figure 12.24b), a process called selective ion monitoring. A mass spectrometer provides excellent detection limits, typically 25 fg to 100 pg, with a linear range spanning five orders of magnitude.

12D.7 Quantitative Applications

Gas chromatography is widely used for the analysis of a diverse array of samples in environmental, clinical, pharmaceutical, biochemical, forensic, food science, and petrochemical laboratories. Examples of these applications are discussed in the following sections.

Environmental Analysis One of the most important environmental applications of gas chromatography is for the analysis of numerous organic pollutants in air, water, and wastewater. The analysis of volatile organics in drinking water, for example, is accomplished by a purge and trap, followed by their separation on a capillary column with a nonpolar stationary phase. A flame ionization, electron capture, or

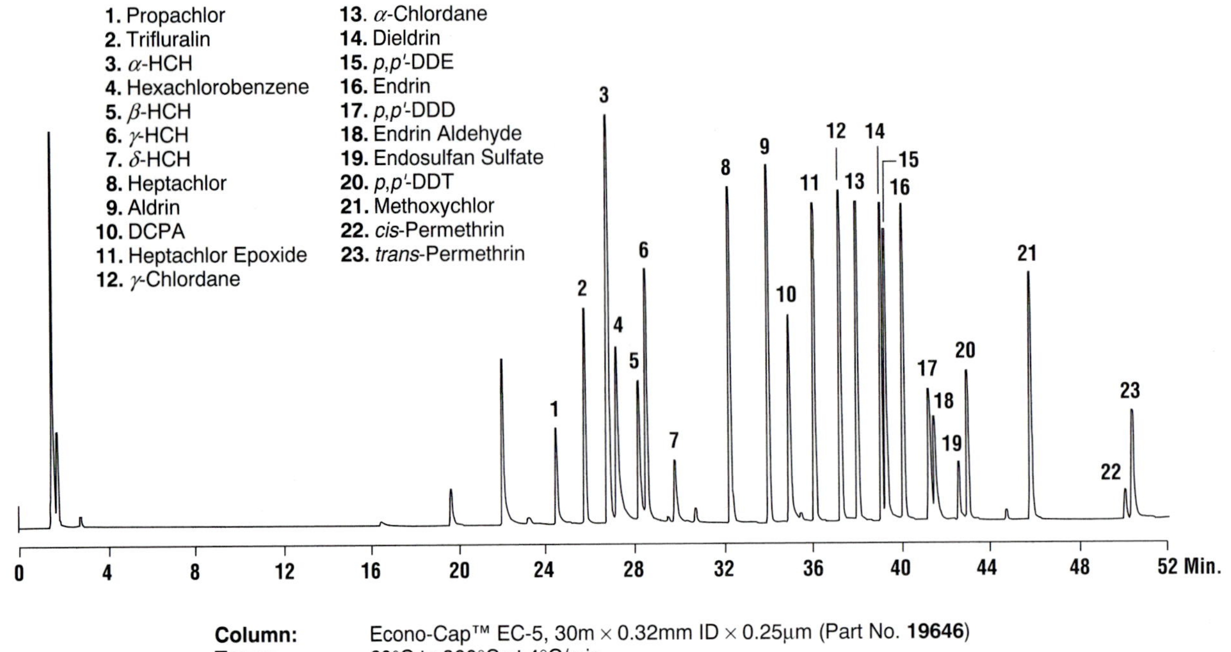

Figure 12.25
Examples of the application of gas chromatography to the analysis of (a) chlorinated pesticides in water, (b) blood alcohols, (c) Scotch whiskey, and (d) unleaded gasoline. (Chromatograms courtesy of Alltech Associates, Inc. Deerfield, IL).

mass spectrometer can be used as a detector. Figure 12.25a shows a typical chromatogram for the analysis of chlorinated pesticides in water.

Clinical Analysis Clinical, pharmaceutical, and forensic labs make frequent use of gas chromatography for the analysis of drugs. Because the sample's matrix is often incompatible with the GC column, analytes generally must be isolated by extraction. Figure 12.25b shows how gas chromatography can be used in monitoring blood alcohol levels.

Consumer Goods Many flavors, spices, and fragrances are readily analyzed by GC, using headspace analysis or thermal desorption. Foods and beverages are analyzed either directly or following a suitable extraction. Volatile materials, such as those found in spices and fragrances, often can be obtained by headspace sampling. Figure 12.25c shows a typical analysis of a sample of Scotch whiskey.

Petroleum Industry Gas chromatography is ideally suited for the analysis of petroleum products, including gasoline, diesel fuel, and oil. A typical chromatogram for the analysis of unleaded gasoline is shown in Figure 12.25d.

Quantitative Calculations In a quantitative analysis, the height or area of an analyte's chromatographic peak is used to determine its concentration. Although peak height is easy to measure, its utility is limited by the inverse relationship between the height and width of a chromatographic peak. Unless chromatographic conditions are carefully controlled to maintain a constant column efficiency, variations in

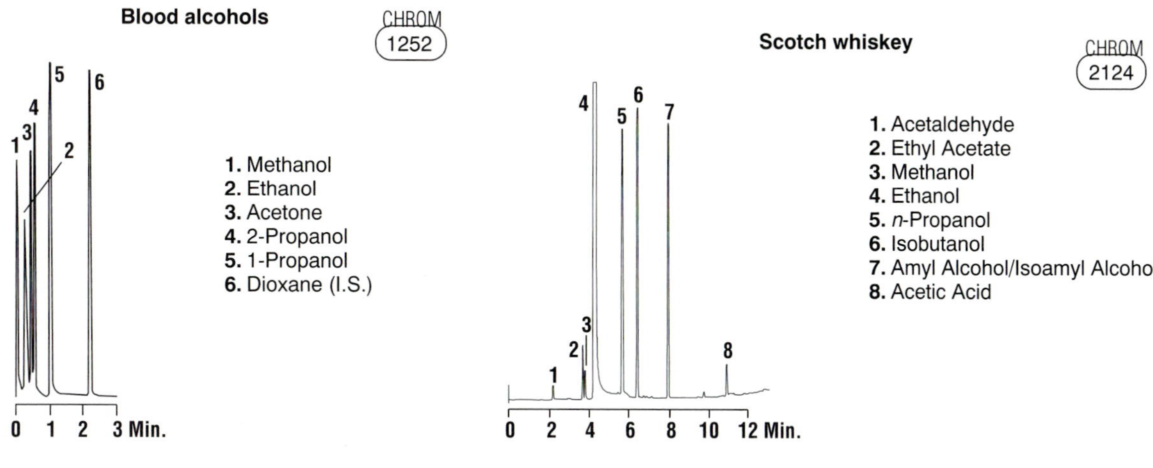

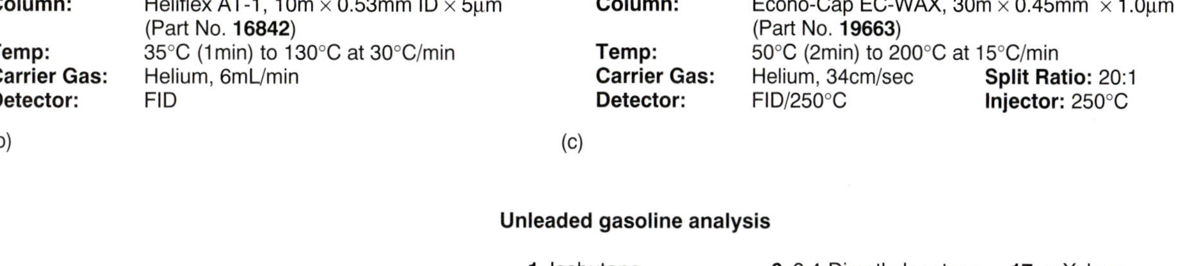

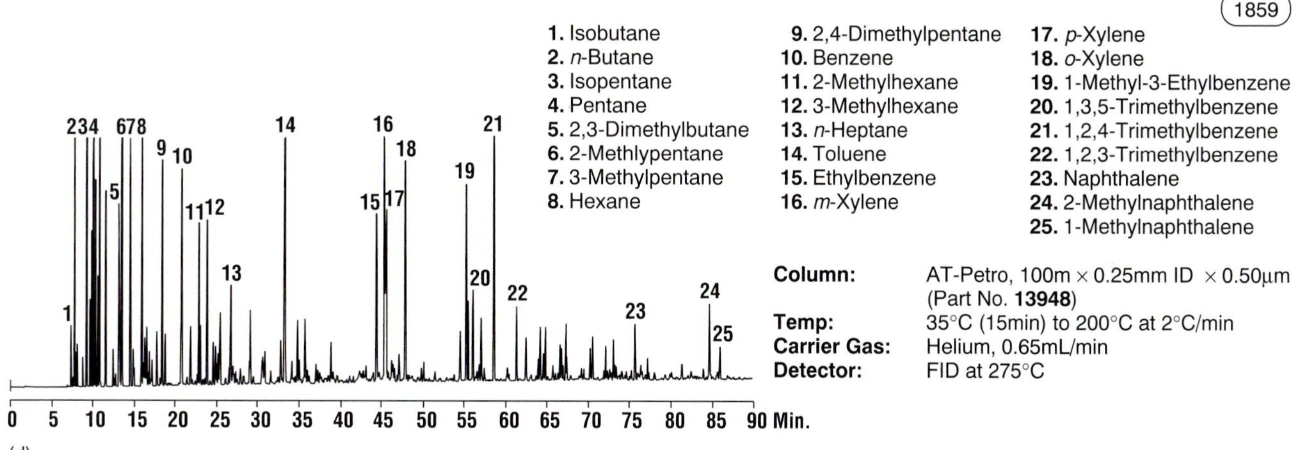

Figure 12.25—*Continued*

peak height may decrease the accuracy and precision of the quantitative analysis. A better choice is to measure the area under the chromatographic peak with an integrating recorder. Since peak area is directly proportional to the amount of analyte that was injected, changes in column efficiency will not affect the accuracy or precision of the analysis.

Calibration curves are usually constructed by analyzing a series of external standards and plotting the detector's signal as a function of their known concentrations. As long as the injection volume is identical for every standard and sample, calibration curves prepared in this fashion give both accurate and precise results. Unfortunately, even under the best of conditions, replicate injections may have volumes that differ by as much as 5% and often may be substantially worse. For this reason, quantitative work requiring high accuracy and precision is accomplished using an internal standard.

Example

EXAMPLE 12.5

Marriott and Carpenter[8] report the following data for five replicate injections of a mixture of 1% v/v methylisobutylketone (peak 1) and 1% v/v *p*-xylene (peak 2).

Injection	Peak	Peak Area
A	1	49075
	2	78112
B	1	85829
	2	135404
C	1	84136
	2	132332
D	1	71681
	2	112889
E	1	58054
	2	91287

Assume that *p*-xylene is the analyte and that methylisobutylketone is the internal standard. Determine the 95% confidence interval for a single-point standardization, with and without using the internal standard.

SOLUTION

For a single-point external standard (omitting the internal standard) the relationship between peak area, A_2, and the concentration, C_2, of *p*-xylene is

$$A_2 = kC_2$$

Substituting the known concentration for the *p*-xylene standard and the appropriate peak areas, gives the following values for the constant *k*.

78112 135404 132332 112889 91287

The average value for *k* is 110,000, with a standard deviation of 25,100 (a relative standard deviation of 22.8%). The 95% confidence interval is 110,000 ± 31,200.

For an internal standardization, the relationship between the peak areas for the analyte, A_2, and the internal standard, A_1, and their respective concentrations, C_1 and C_2, is

$$\frac{A_2}{A_1} = k\frac{C_2}{C_1}$$

Substituting the known concentrations and the appropriate peak areas gives the following values for the constant *k*.

1.5917 1.5776 1.5728 1.5749 1.5724

The average value for *k* is 1.5779, with a standard deviation of 0.0080 (a relative standard deviation of 0.507%). The 95% confidence interval is 1.5779 ± 0.0099.

As this example clearly shows, the variation in individual peak areas between injections is substantial. The use of an internal standard, however, corrects for these variations, providing a means for accurate and precise calibration.

12D.8 Qualitative Applications

Gas chromatography also can be used for qualitative purposes. When using an FT–IR or a mass spectrometer as the detector, the available spectral information often can be used to identify individual solutes.

With conventional nonspectroscopic detectors, other methods must be used to identify the solutes. One approach is to spike the sample by adding an aliquot of a suspected analyte and looking for an increase in peak height. Retention times also can be compared with values measured for standards, provided that the operating conditions are identical. Because of the difficulty of exactly matching such conditions, tables of retention times are of limited utility.

Kovat's retention index provides one solution to the matching of retention times. Under isothermal conditions, the adjusted retention times of normal alkanes increase logarithmically. Kovat defined the retention index, *I*, for a normal alkane as 100 times the number of carbon atoms; thus, the retention index is 400 for butane and 500 for pentane. To determine the retention index for another compound, its adjusted retention time is measured relative to that for the normal alkanes eluting just before and after. For example, a compound eluting between butane and pentane has a retention index between 400 and 500. The exact value for the compound's retention index, I_{cpd}, is given as

Kovat's retention index
A means for normalizing retention times by comparing a solute's retention time with those for normal alkanes.

$$I_{\mathrm{cpd}} = 100\left[\frac{(\log t_r')_{\mathrm{cpd}} - (\log t_r')_x}{(\log t_r')_{x+1} - (\log t_r')_x}\right] + I_x \qquad 12.29$$

where x is the normal alkane eluting before the compound, and x + 1 is the normal alkane eluting just after the compound.

Example

EXAMPLE 12.6

In a separation of a mixture of hydrocarbons, the following adjusted retention times were measured.

propane	2.23 min
isobutane	5.71 min
butane	6.67 min

What is the Kovat's retention index for each of these hydrocarbons?

SOLUTION

Kovat's retention index for a normal alkane is 100 times the number of carbons; thus

$$I_{\mathrm{propane}} = 100 \times 3 = 300$$

$$I_{\mathrm{butane}} = 100 \times 4 = 400$$

To find Kovat's retention index for isobutane, we use equation 12.29.

$$I_{\mathrm{isobutane}} = 100\left[\frac{(\log t_r')_{\mathrm{isobutane}} - (\log t_r')_{\mathrm{propane}}}{(\log t_r')_{\mathrm{butane}} - (\log t_r')_{\mathrm{propane}}}\right] + I_{\mathrm{propane}}$$

$$= 100\left[\frac{\log(5.71) - \log(2.23)}{\log(6.67) - \log(2.23)}\right] + 300 = 386$$

12D.9 Representative Method

Although each gas chromatographic method has its own unique considerations, the following description of the determination of trihalomethanes in drinking water provides an instructive example of a typical procedure.

Representative Methods

Method 12.1 Determination of Trihalomethanes in Drinking Water[9]

Description of Method. Trihalomethanes, such as chloroform ($CHCl_3$) and bromoform ($CHBr_3$), are found in most chlorinated waters. Since chloroform is a suspected carcinogen, the determination of trihalomethanes in public drinking water supplies is of considerable importance. In this method the trihalomethanes $CHCl_3$, $CHBrCl_2$, $CHBr_2Cl$, and $CHBr_3$ are isolated by a liquid–liquid extraction with pentane and determined by gas chromatography using an electron capture detector. Because of its volatility and ubiquitous presence in most labs, chloroform from other sources is a significant interferent.

Procedure. Samples are collected in 40-mL vials with screw-caps lined with a Teflon septum. Fill the vial to overflowing, ensuring that there are no air bubbles. Add a reducing agent of ascorbic acid (25 mg/40 mL) to quench the further production of trihalomethanes, and seal the vial. Store samples at 4 °C, and analyze within 14 days.

Prepare a standard stock solution for each trihalomethane by placing 9.8 mL of methanol in a 10-mL volumetric flask. Let the volumetric flask stand for 10 min, or until all surfaces wetted with methanol are dry. Weigh the volumetric flask to the nearest ±0.1 mg. Using a 100-μL syringe, add 2 or 3 drops of the trihalomethane to the volumetric flask, allowing it to drop directly into the methanol. Reweigh the flask before diluting to volume and mixing. Transfer to a 15-mL screw-cap vial with Teflon liner, and report the concentration in micrograms per milliliter. Standard stock solutions are stable for 4 weeks when stored at 4 °C.

Prepare a single multicomponent working standard from the stock standards by making appropriate dilutions with methanol. Concentrations in the working standards should be at such a level that a 20-μL sample added to 100 mL of water gives a calibration standard whose response for each trihalomethane is within ±25% of that for the samples to be analyzed.

Samples and calibration standards are prepared for analysis using a 10-mL syringe. Add 10.00 mL of each sample and standard to separate 14-mL screw-cap vials containing 2.00 mL of pentane. Shake vigorously for 1 min to effect the separation. Wait 60 s for the phases to separate. Inject 3.0-μL aliquots of the pentane layer into a GC equipped with a 2-mm internal diameter, 2-m long glass column packed with a stationary phase of 10% squalane on a packing material of 80/100 mesh Chromosorb WAW. Operate the column at 67 °C and a flow rate of 25 mL/min.

Questions

1. **A simple liquid–liquid extraction rarely extracts 100% of the analyte. How does this method account for incomplete extractions?**

   Both the samples and standards are treated identically so their relative concentrations are unaffected by an incomplete extraction.

2. **This method uses a short, packed column that generally produces a poor resolution of chromatographic peaks. The liquid–liquid extraction used to extract the trihalomethanes is nonselective. Besides the trihalomethanes, a wide range of nonpolar and polar organic constituents, such as benzene and**

—Continued

phenol, also are extracted. Why does the presence of these other compounds not interfere with this analysis?

An electron capture detector is relatively insensitive to nonhalogenated compounds, providing the additional selectivity.

3. **Although chloroform is an analyte, it also can be interferent. Due to its volatility, chloroform present in the laboratory air may diffuse through the sample vial's Teflon septum, contaminating the samples. How can we determine whether samples have been contaminated in this manner?**

   A sample blank of trihalomethane-free water can be kept with the samples at all times. If the sample blank shows no evidence for chloroform, then we can safely assume that the samples also are free from contamination.

4. **Why is it necessary to collect samples such that there is no headspace (layer of air overlying the liquid) in the sample vial?**

   Due to the volatility of trihalomethanes, the presence of a headspace allows for the possible loss of analyte.

12D.10 Evaluation

Scale of Operation Analytes present at levels from major to ultratrace components have been successfully determined by gas chromatography. Depending on the choice of detector, samples with major and minor analytes may need to be diluted before analysis. The thermal conductivity and flame ionization detectors can handle larger amounts of analyte; other detectors, such as the electron capture detector or a mass spectrometer, require substantially smaller amounts of analyte. Although the volume of sample injected is quite small (often less than a microliter), the amount of available material from which the injection volume is taken must be sufficient to be a representative sample. For trace analytes, the actual amount of analyte injected is often in the picogram range. Using the trihalomethane analysis described in Method 12.1 as an example, a 3.0-μL injection of a water sample containing 1 μg/L of $CHCl_3$ corresponds to 15 pg of $CHCl_3$ (assuming a complete extraction of $CHCl_3$).

Accuracy The accuracy of a gas chromatographic method varies substantially from sample to sample. For routine samples, accuracies of 1–5% are common. For analytes present at very low concentration levels, for samples with complex matrices, or for samples requiring significant processing before analysis, accuracy may be substantially poorer. In the analysis for trihalomethanes described in Method 12.1, for example, determinate errors as large as ±25% are possible.[9]

Precision The precision of a gas chromatographic analysis includes contributions from sampling, sample preparation, and the instrument. The relative standard deviation due to the gas chromatographic portion of the analysis is typically 1–5%, although it can be significantly higher. The principal limitations to precision are detector noise and the reproducibility of injection volumes. In quantitative work, the use of an internal standard compensates for any variability in injection volumes.

Sensitivity In a gas chromatographic analysis, sensitivity (the slope of a calibration curve) is determined by the detector's characteristics. Of greater interest for quantitative work is the detector's linear range; that is, the range of concentrations over which a calibration curve is linear. Detectors with a wide linear range, such as a thermal conductivity detector and flame ionization detector, can be used to analyze samples of varying concentration without adjusting operating conditions. Other detectors, such as the electron capture detector, have a much narrower linear range.

Selectivity Because it combines separation with analysis, gas chromatography provides excellent selectivity. By adjusting conditions it is usually possible to design a separation such that the analytes elute by themselves. Additional selectivity can be provided by using a detector, such as the electron capture detector, that does not respond to all compounds.

Time, Cost, and Equipment Analysis time can vary from several minutes for samples containing only a few constituents to more than an hour for more complex samples. Preliminary sample preparation may substantially increase the analysis time. Instrumentation for gas chromatography ranges in price from inexpensive (a few thousand dollars) to expensive (more than $50,000). The more expensive models are equipped for capillary columns and include a variety of injection options and more sophisticated detectors, such as a mass spectrometer. Packed columns typically cost $50–$200, and the cost of a capillary column is typically $200–$1000.

12E High-Performance Liquid Chromatography

Although gas chromatography is widely used, it is limited to samples that are thermally stable and easily volatilized. Nonvolatile samples, such as peptides and carbohydrates, can be analyzed by GC, but only after they have been made more volatile by a suitable chemical derivatization. For this reason, the various techniques included within the general scope of liquid chromatography are among the most commonly used separation techniques. Although simple column chromatography, first introduced by Tswett, is still used in large-scale preparative work, the focus of this section is on **high-performance liquid chromatography** (HPLC).

high-performance liquid chromatography
A chromatographic technique in which the mobile phase is a liquid.

In HPLC, a liquid sample, or a solid sample dissolved in a suitable solvent, is carried through a chromatographic column by a liquid mobile phase. Separation is determined by solute/stationary-phase interactions, including liquid–solid adsorption, liquid–liquid partitioning, ion exchange and size exclusion, and by solute/mobile-phase interactions. In each case, however, the basic instrumentation is essentially the same. A schematic diagram of a typical HPLC instrument is shown in Figure 12.26. The remainder of this section deals exclusively with HPLC separations based on liquid–liquid partitioning. Other forms of liquid chromatography receive consideration later in this chapter.

12E.1 HPLC Columns

An HPLC typically includes two columns: an analytical column responsible for the separation and a guard column. The guard column is placed before the analytical column, protecting it from contamination.

Analytical Columns The most commonly used columns for HPLC are constructed from stainless steel with internal diameters between 2.1 mm and 4.6 mm, and

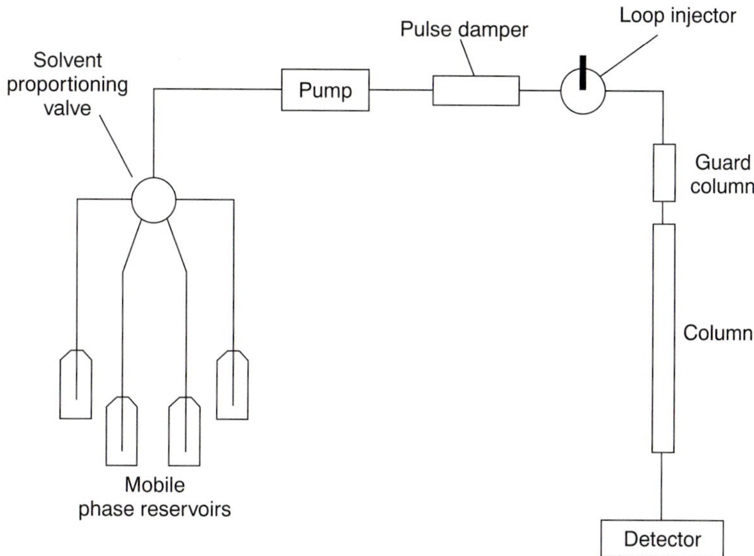

Figure 12.26
Schematic diagram of a high-performance liquid chromatograph.

lengths ranging from approximately 30 mm to 300 mm. These columns are packed with 3–10 μm porous silica particles that may have an irregular or spherical shape. Typical column efficiencies are 40,000–60,000 theoretical plates/m. Assuming V_{max}/V_{min} is approximately 50,[3] a 25-cm column with 50,000 plates/m has 12,500 theoretical plates and a peak capacity (equation 12.18) of 110.

Microcolumns use less solvent and, because the sample is diluted to a lesser extent, produce larger signals at the detector. These columns are made from fused silica capillaries with internal diameters of 44–200 μm and lengths of up to several meters. Microcolumns packed with 3–5-μm particles have been prepared with column efficiencies of up to 250,000 theoretical plates.[10]

Open tubular microcolumns also have been developed, with internal diameters of 1–50 μm and lengths of approximately 1 m. These columns, which contain no packing material, may be capable of obtaining column efficiencies of up to 1 million theoretical plates.[11] The development of open tubular columns, however, has been limited by the difficulty of preparing columns with internal diameters less than 10 μm.

Guard Columns Two problems tend to shorten the lifetime of an analytical column. First, solutes binding irreversibly to the stationary phase degrade the column's performance by decreasing the available stationary phase. Second, particulate material injected with the sample may clog the analytical column. To minimize these problems, a **guard column** is placed before the analytical column. Guard columns usually contain the same particulate packing material and stationary phase as the analytical column, but are significantly shorter and less expensive; a length of 7.5 mm and a cost one-tenth of that for the corresponding analytical column is typical. Because they are intended to be sacrificial, guard columns are replaced regularly.

guard column
An inexpensive column used to protect a more expensive analytical column.

12E.2 Stationary Phases

In liquid–liquid chromatography the stationary phase is a liquid film coated on a packing material consisting of 3–10 μm porous silica particles. The stationary phase may be partially soluble in the mobile phase, causing it to "bleed" from the column

bonded stationary phase
A liquid stationary phase that is chemically bonded to a particulate packing material.

over time. To prevent this loss of stationary phase, it is covalently bound to the silica particles. **Bonded stationary phases** are attached by reacting the silica particles with an organochlorosilane of the general form $Si(CH_3)_2RCl$, where R is an alkyl or substituted alkyl group.

$$\begin{array}{c} \diagdown \\ O \\ \diagup \\ -Si-OH \\ \diagdown \\ O \\ \diagup \end{array} \xrightarrow{Si(CH_3)_2RCl} \begin{array}{c} \diagdown \\ O \\ \diagup \\ -Si-O-Si(CH_3)_2R \\ \diagdown \\ O \\ \diagup \end{array} + HCl$$

To prevent unwanted interactions between the solutes and any unreacted –SiOH groups, the silica frequently is "capped" by reacting it with $Si(CH_3)_3Cl$; such columns are designated as end-capped.

The properties of a stationary phase are determined by the nature of the organosilane's alkyl group. If R is a polar functional group, then the stationary phase will be polar. Examples of polar stationary phases include those for which R contains a cyano ($-C_2H_4CN$), diol ($-C_3H_6OCH_2CHOHCH_2OH$), or amino ($-C_3H_6NH_2$) functional group. Since the stationary phase is polar, the mobile phase is a nonpolar or moderately polar solvent. The combination of a polar stationary phase and a nonpolar mobile phase is called **normal-phase chromatography.**

normal-phase chromatography
Liquid chromatography using a polar stationary phase and a nonpolar mobile phase.

reverse-phase chromatography
Liquid chromatography using a nonpolar stationary phase and a polar mobile phase.

In **reverse-phase chromatography,** which is the more commonly encountered form of HPLC, the stationary phase is nonpolar and the mobile phase is polar. The most common nonpolar stationary phases use an organochlorosilane for which the R group is an *n*-octyl (C_8) or *n*-octyldecyl (C_{18}) hydrocarbon chain. Most reverse-phase separations are carried out using a buffered aqueous solution as a polar mobile phase. Because the silica substrate is subject to hydrolysis in basic solutions, the pH of the mobile phase must be less than 7.5.

12E.3 Mobile Phases

The elution order of solutes in HPLC is governed by polarity. In a normal-phase separation the least polar solute spends proportionally less time in the polar stationary phase and is the first solute to elute from the column. Retention times are controlled by selecting the mobile phase, with a less polar mobile phase leading to longer retention times. If, for example, a separation is poor because the solutes are eluting too quickly, switching to a less polar mobile phase leads to longer retention times and more opportunity for an acceptable separation. When two solutes are adequately resolved, switching to a more polar mobile phase may provide an acceptable separation with a shorter analysis time. In a reverse-phase separation the order of elution is reversed, with the most polar solute being the first to elute. Increasing the polarity of the mobile phase leads to longer retention times, whereas shorter retention times require a mobile phase of lower polarity.

polarity index
A quantitative measure of a solvent's polarity.

Choosing a Mobile Phase Several indices have been developed to assist in selecting a mobile phase, the most useful of which is the **polarity index.**[12] Table 12.3 provides values for the polarity index, P', of several commonly used mobile phases, in which larger values of P' correspond to more polar solvents. Mobile phases of intermediate polarity can be fashioned by mixing together two or more of the mobile phases in Table 12.3. For example, a binary mobile phase made by combining solvents A and B has a polarity index, P'_{AB}, of

Table 12.3 Properties of HPLC Mobile Phases

Mobile Phase	Polarity Index (P')	UV Cutoff (nm)
cyclohexane	0.04	210
n-hexane	0.1	210
carbon tetrachloride	1.6	265
i-propyl ether	2.4	220
toluene	2.4	286
diethyl ether	2.8	218
tetrahydrofuran	4.0	220
ethanol	4.3	210
ethyl acetate	4.4	255
dioxane	4.8	215
methanol	5.1	210
acetonitrile	5.8	190
water	10.2	—

$$P'_{AB} = \phi_A P'_A + \phi_B P'_B \qquad \textbf{12.30}$$

where P'_A and P'_B are the polarity indexes for solvents A and B, and ϕ_A and ϕ_B are the volume fractions of the two solvents.

Example

EXAMPLE 12.7

A reverse-phase HPLC separation is carried out using a mobile-phase mixture of 60% v/v water and 40% v/v methanol. What is the mobile phase's polarity index?

SOLUTION

From Table 12.3 we find that the polarity index is 10.2 for water and 5.1 for methanol. Using equation 12.30, the polarity index for a 60:40 water–methanol mixture is

$$P'_{AB} = (0.60)(10.2) + (0.40)(5.1) = 8.2$$

A useful guide when using the polarity index is that a change in its value of 2 units corresponds to an approximate tenfold change in a solute's capacity factor. Thus, if k' is 22 for the reverse-phase separation of a solute when using a mobile phase of water ($P' = 10.2$), then switching to a 60:40 water–methanol mobile phase ($P' = 8.2$) will decrease k' to approximately 2.2. Note that the capacity factor decreases because we are switching from a more polar to a less polar mobile phase in a reverse-phase separation.

Changing the mobile phase's polarity index, by changing the relative amounts of two solvents, provides a means of changing a solute's capacity factor. Such

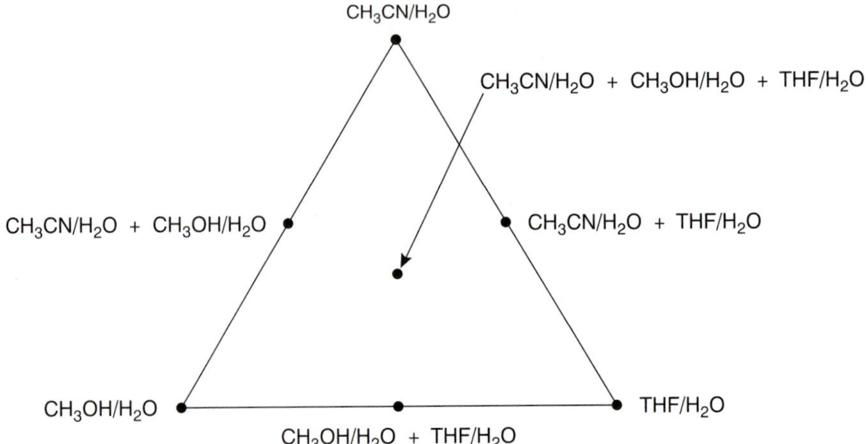

Figure 12.27
Solvent triangle for optimizing reverse-phase HPLC separations. Binary and ternary mixtures contain equal volumes of each of the aqueous mobile phases making up the vertices of the triangle.

changes, however, are not very selective; thus, two solutes that significantly overlap may continue to be poorly resolved even after making a significant change in the mobile phase's polarity.

To effect a better separation between two solutes it is often necessary to improve the selectivity factor, α. Two approaches are commonly used to accomplish this improvement. When a solute is a weak acid or a weak base, adjusting the pH of the aqueous mobile phase can lead to significant changes in the solute's retention time. This is shown in Figure 12.13a for the reverse-phase separation of *p*-aminobenzoic acid and *p*-hydroxybenzoic acid on a nonpolar C_{18} column. At more acidic pH levels, both weak acids are present as neutral molecules. Because they partition favorably into the stationary phase, the retention times for the solutes are fairly long. When the pH is made more basic, the solutes, which are now present as their conjugate weak base anions, are less soluble in the stationary phase and elute more quickly. Similar effects can be achieved by taking advantage of metal–ligand complexation and other equilibrium reactions.[13]

A second approach to changing the selectivity factor for a pair of solutes is to change one or more of the mobile-phase solvents. In a reverse-phase separation, for example, this is accomplished by changing the solvent mixed with water. Besides methanol, other common solvents for adjusting retention times are acetonitrile and tetrahydrofuran (THF). A common strategy for finding the best mobile phase is to use the solvent triangle shown in Figure 12.27. The separation is first optimized using an aqueous mobile phase of acetonitrile to produce the best separation within the desired analysis time (methanol or THF also could be chosen first). Table 12.4 is used to estimate the composition of methanol/H_2O and THF/H_2O mobile phases that will produce similar analysis times. These mobile phases are then adjusted, if necessary, establishing the three points of the solvent triangle. Four additional mobile phases are prepared using the binary and ternary mobile phases indicated in Figure 12.27. From these seven mobile phases it is possible to estimate how a change in the mobile-phase composition might affect the separation.

Isocratic Versus Gradient Elution When a separation uses a single mobile phase of fixed composition it is called an **isocratic elution.** It is often difficult, however, to find a single mobile-phase composition that is suitable for all solutes. Recalling the general elution problem, a mobile phase that is suitable for early eluting solutes may lead to unacceptably long retention times for later eluting solutes. Optimizing con-

isocratic elution
The use of a mobile phase whose composition remains constant throughout the separation.

Table 12.4 Organic Solvent/H_2O Mobile-Phase Compositions Having Approximately Equal Solvent Strength

%v/v CH_3OH	%v/v CH_3CN	%v/v THF
0	0	0
10	6	4
20	14	10
30	22	16
40	32	24
50	40	30
60	50	36
70	60	44
80	72	52
90	87	62
100	99	71

ditions for late eluting solutes, on the other hand, may provide an inadequate separation of early eluting solutes. Changing the composition of the mobile phase with time provides a solution to this problem. For a reverse-phase separation the initial mobile-phase composition is relatively polar. As the separation progresses, the mobile phase's composition is made less polar. Such separations are called gradient elutions.

12E.4 HPLC Plumbing

An important feature of HPLC instrumentation (see Figure 12.26) is the presence of several solvent reservoirs. As discussed in the previous section, controlling the mobile phase's polarity plays an important role in improving a liquid chromatographic separation. The availability of several solvent reservoirs allows the mobile phase's composition to be quickly and easily varied. This is essential when using a gradient elution, for which the mobile-phase composition is systematically changed from a weaker solvent to a stronger solvent.

Before they are used, mobile-phase solvents must be treated to remove dissolved gases, such as N_2 and O_2, and small particulate matter, such as dust. Dissolved gases often lead to the formation of gas bubbles when the mobile phase enters the detector, resulting in a distortion of the detector's signal. Degassing is accomplished in several ways, but the most common are the use of a vacuum pump or sparging with an inert gas, such as He, which has a low solubility in the mobile phase. Particulate material capable of clogging the HPLC tubing or column is removed by filtering. If the instrument is not designed to do so, degassing and filtering can be completed before the solvents are placed in their reservoirs.

The mobile-phase solvents are pulled from their reservoirs by the action of a pump. Most HPLC instruments use a reciprocating pump consisting of a piston whose back-and-forth movement is capable both of maintaining a constant flow rate of up to several milliliters per minute and of obtaining the high output pressure needed to push the mobile phase through the chromatographic column. A solvent proportioning valve controls the mobile phase's composition, making possible the

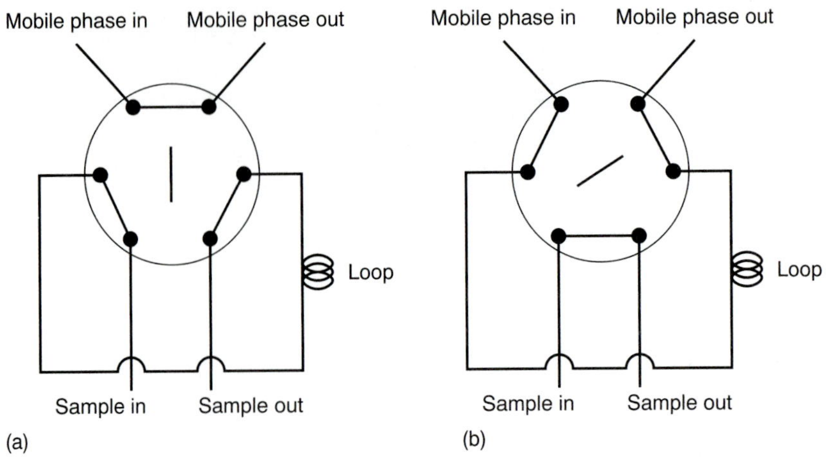

Mobile phase in Mobile phase out Mobile phase in Mobile phase out
Loop Loop
Sample in Sample out Sample in Sample out
(a) (b)

Figure 12.28
Schematic diagram of a loop injector in the (a) load and (b) inject positions.

necessary change in the mobile phase's composition when using a gradient elution. The back and forth movement of a reciprocating pump results in a pulsed flow that contributes noise to the chromatogram. To eliminate this problem a pulse damper is placed at the outlet of the pump.

12E.5 Sample Introduction

The typical operating pressure of an HPLC is sufficiently high that it is impossible to inject the sample in the same manner as in gas chromatography. Instead, the sample is introduced using a **loop injector** (Figure 12.28). Sampling loops are interchangeable, and available with volumes ranging from 0.5 μL to 2 mL.

loop injector
A means for injecting samples in which the sample is loaded into a short section of tubing and injected onto the column by redirecting the mobile phase through the loop.

In the load position the sampling loop is isolated from the mobile phase and is open to the atmosphere. A syringe with a capacity several times that of the sampling loop is used to place the sample in the loop. Any extra sample beyond that needed to fill the sample loop exits through the waste line. After loading the sample, the injector is turned to the inject position. In this position the mobile phase is directed through the sampling loop, and the sample is swept onto the column.

12E.6 Detectors for HPLC

As with gas chromatography, numerous detectors have been developed for use in monitoring HPLC separations.[14] To date, the majority of HPLC detectors are not unique to the method, but are either stand-alone instruments or modified versions of the same.

Colorplate 12 shows a photo of an HPLC equipped with a diode array detector.

Spectroscopic Detectors The most popular HPLC detectors are based on spectroscopic measurements, including UV/Vis absorption, and fluorescence. These detectors range from simple designs, in which the analytical wavelength is selected using appropriate filters, to essentially a modified spectrophotometer equipped with a flow cell. When using a UV/Vis detector, the resulting chromatogram is a plot of absorbance as a function of elution time. Instruments utilizing a diode array spectrophotometer record entire spectra, giving a three-dimensional chromatogram showing absorbance as a function of wavelength and elution time. Figure 12.29a shows a typical flow cell for HPLC when using a UV/Vis spectrophotometer as a detector. The flow cell has a volume of 1–10 μL and a path length of 0.2–1 cm. One limitation to using absorbance is that the mobile phase must not absorb strongly at

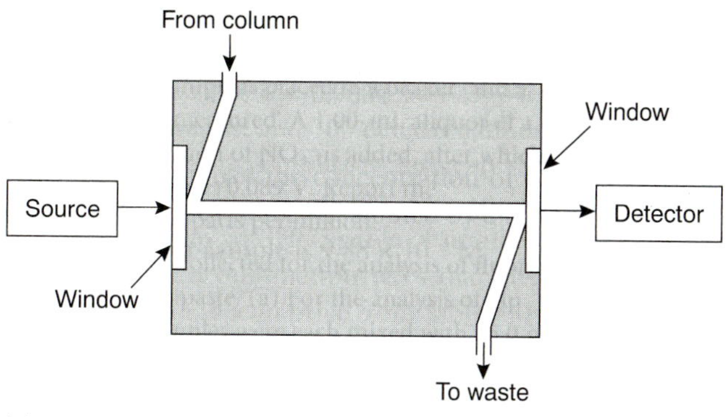

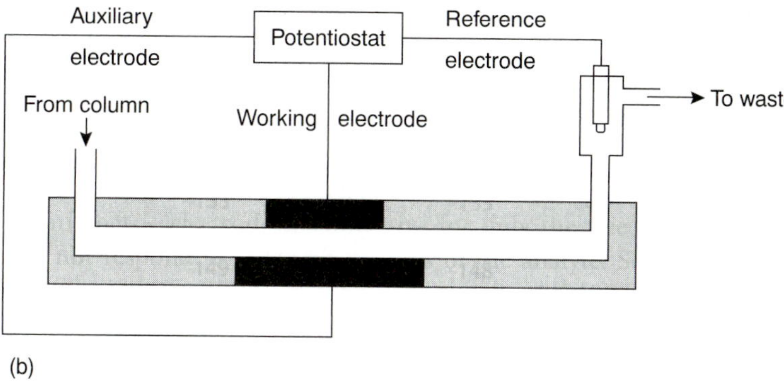

Figure 12.29
Schematic diagrams of flow cell detectors for HPLC using (a) UV/Vis absorption spectrophotometry and (b) amperometry.

the chosen wavelength. Table 12.3 lists the wavelengths below which UV/Vis absorbance cannot be used for different mobile phases. Detectors based on absorbance provide detection limits of as little as 100 pg–1 ng of injected analyte. Fluorescence detectors provide additional selectivity since fewer solutes are capable of fluorescing. The resulting chromatogram is a plot of fluorescence intensity as a function of time. Detection limits are as little as 1–10 pg of injected analyte.

Electrochemical Detectors Another common group of HPLC detectors are those based on electrochemical measurements such as amperometry, voltammetry, coulometry, and conductivity. Figure 12.29b, for example, shows an amperometric flow cell. Effluent from the column passes over the working electrode, which is held at a potential favorable for oxidizing or reducing the analytes. The potential is held constant relative to a downstream reference electrode, and the current flowing between the working and auxiliary electrodes is measured. Detection limits for amperometric electrochemical detection are 10 pg–1 ng of injected analyte.

Other Detectors Several other detectors have been used in HPLC. Measuring a change in the mobile phase's refractive index is analogous to monitoring the mobile phase's thermal conductivity in gas chromatography. A refractive index detector is nearly universal, responding to almost all compounds, but has a poorer detection limit of 100 ng–1 μg of injected analyte. Furthermore, a refractive index detector is not useful for a gradient elution unless the mobile-phase components have identical refractive indexes. Another useful detector is a mass spectrometer. The advantages of using a mass spectrometer in HPLC are the same as for gas chromatography. Detection limits are quite good, typically 100 pg–1 ng of injected analyte, with values as low as 1–10 pg in some situations. In addition, a mass spectrometer provides qualitative, structural information that can help identify the analytes. The interface between the HPLC and mass spectrometer is technically more difficult than that in a GC–MS because of the incompatibility of a liquid mobile phase with the mass spectrometer's high vacuum requirement. Recent developments in mass spectrometry, however, have led to a growing interest in LC–MS.

12E.7 Quantitative Applications

HPLC is routinely used for both qualitative and quantitative analyses of environmental, pharmaceutical, industrial, forensic, clinical, and consumer product samples. Figure 12.30 shows several representative examples.

Preparing Samples for Analysis Samples in liquid form can be analyzed directly, after a suitable clean-up to remove any particulate materials or after a suitable extraction to remove matrix interferents. In determining polyaromatic hydrocarbons (PAH) in wastewater, for example, an initial extraction with CH_2Cl_2 serves the dual purpose of concentrating the analytes and isolating them from matrix interferents. Solid samples must first be dissolved in a suitable solvent, or the analytes of interest must be brought into solution by extraction. For example, an HPLC analysis for the active ingredients and degradation products in a pharmaceutical tablet often begins by extracting the powdered tablet with a portion of mobile phase. Gases are collected by bubbling through a trap containing a suitable solvent. Organic isocyanates in industrial atmospheres can be determined in this manner by bubbling the air through a solution of 1-(2-methoxyphenyl)piperazine in toluene. Reacting the isocyanates with 1-(2-methoxyphenyl)piperazine serves the dual purposes of stabilizing them against degradation before the HPLC analysis while also forming a derivative that can be monitored by UV absorption.

Quantitative Calculations Quantitative analyses are often easier to conduct with HPLC than GC because injections are made with a fixed-volume injection loop instead of a syringe. As a result, variations in the amount of injected sample are minimized, and quantitative measurements can be made using external standards and a normal calibration curve.

Example

EXAMPLE 12.8

The concentration of PAHs in soil can be determined by first extracting the PAHs with methylene chloride. The extract is then diluted, if necessary, and the PAHs are separated by HPLC using a UV/Vis or fluorescence detector. Calibration is achieved using one or more external standards. In a typical analysis, a 2.013-g sample of dried soil is extracted with 20.00 mL of methylene chloride. After filtering to remove the soil, a 1-mL portion of the extract is removed and diluted to 10 mL with acetonitrile. Injecting 5 μL of the diluted extract into an HPLC gives a signal of 0.217 (arbitrary units) for the PAH fluoranthene. When 5 μL of a 20.0-ppm fluoranthene standard is analyzed using the same conditions, a signal of 0.258 is measured. Report the parts per million of fluoranthene in the soil.

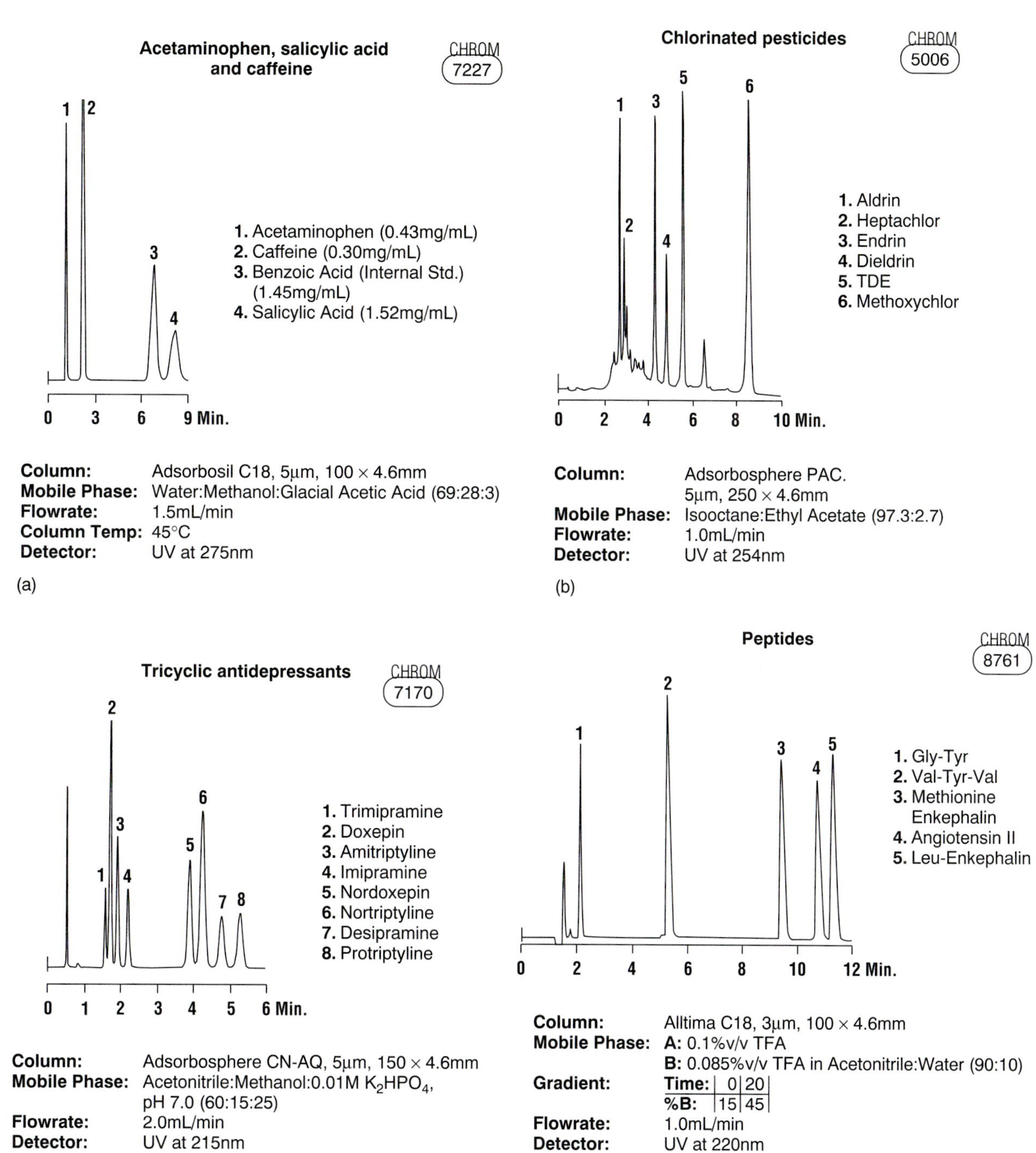

Figure 12.30

Examples of the application of HPLC to the analysis of (a) acetaminophen, salicylic acid, and caffeine; (b) chlorinated pesticides; (c) tricyclic antidepressants; and (d) peptides. (Chromatograms courtesy of Alltech Associates, Inc. Deerfield, IL).

SOLUTION

For a single-point external standard, the relationship between the signal, S, and the concentration, C, of fluoranthene is

$$S = kC$$

Substituting the known values for the standard's signal and concentration gives the value of k as

$$k = \frac{S}{C} = \frac{0.258}{20.0 \text{ ppm}} = 0.0129 \text{ ppm}^{-1}$$

Using this value for k and the signal obtained for the extracted and diluted soil sample gives a fluoranthene concentration of

$$C = \frac{S}{k} = \frac{0.217}{0.0129 \text{ ppm}^{-1}} = 16.8 \text{ ppm}$$

This, of course, is the concentration of fluoranthene in the sample injected into the HPLC. The concentration of fluoranthene in the soil is

$$\frac{(16.8\ \mu\text{g/mL}) \times (10 \text{ mL}/1 \text{ mL}) \times 20 \text{ mL}}{2.013 \text{ g}} = 1670 \text{ ppm}$$

12E.8 Representative Method

Although each HPLC method has its own unique considerations, the following description of the determination of the fluoxetine in serum provides an instructive example of a typical procedure.

Representative Methods

Method 12.2 Determination of Fluoxetine in Serum[15]

Description of Method. Fluoxetine, whose structure is shown in Figure 12.31a, is another name for the antidepressant drug Prozac. The determination of fluoxetine and its metabolite norfluoxetine, Figure 12.31b, in serum is an important part of monitoring its therapeutic use. The analysis is complicated by the complex matrix of serum samples. A solid-phase extraction followed by an HPLC analysis using a fluorescence detector provides the necessary selectivity and detection limits.

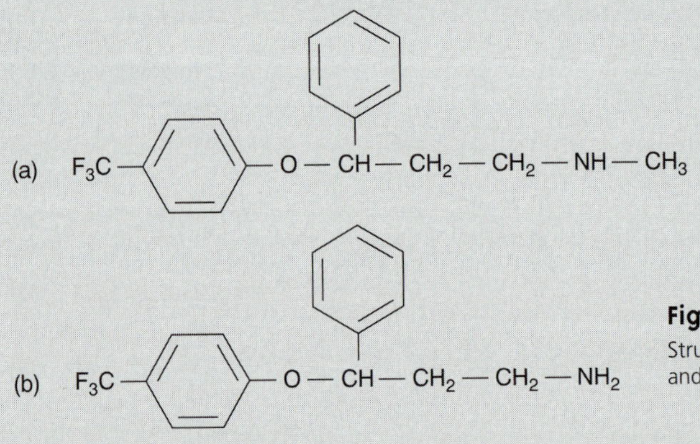

(a) $F_3C-C_6H_4-O-CH(C_6H_5)-CH_2-CH_2-NH-CH_3$

(b) $F_3C-C_6H_4-O-CH(C_6H_5)-CH_2-CH_2-NH_2$

Figure 12.31 Structures of (a) fluoxetine and (b) norfluoxetine.

—Continued

Procedure. A known amount of the antidepressant protriptyline is added to a serum sample, serving as an internal standard. A 0.5-mL aliquot of the serum is passed through a solid-phase extraction cartridge containing silica particles with a bonded C_{18} phase. After washing to remove interfering constituents from the sample matrix, the remaining constituents, including both analytes and the internal standard, are removed by washing the cartridge with 0.25 mL of a 25:75 v/v mixture of 0.1 M $HClO_4$ and acetonitrile. A 20-μL aliquot is injected onto a 15-cm × 4.6-mm column packed with a 5-μm C_8-bonded stationary phase. An isocratic mobile-phase mixture of 37.5:62.5 v/v acetonitrile and water (containing 1.5 g of tetramethylammonium perchlorate and 0.1 mL of 70% v/v $HClO_4$) is used to elute the sample. Detection is with a fluorescence detector using an excitation wavelength of 235 nm and an emission wavelength of 310 nm.

Questions

1. **What is the purpose of including an initial solid-phase extraction?**

   A direct injection of serum is not advisable since the presence of particulate material in the sample matrix may clog the column. In addition, some of the sample's constituents may absorb too strongly to the stationary phase, thus degrading the column's performance. Finally, although an HPLC is capable of separating and analyzing complex mixtures, an analysis may still be difficult if there are too many constituents to provide adequate resolution of the analytes. The solid-phase extraction serves the purpose of cleaning up the sample before using the HPLC.

2. **One advantage of an HPLC analysis is that a loop injector often eliminates the need for an internal standard. Why is an internal standard used in this analysis? What assumption(s) must we make about the internal standard?**

   An internal standard is used because of difficulties introduced in the solid-phase extraction. For example, the volume of serum taken for the solid-phase extraction and the volume of solvent used to remove the analyte and internal standard are quite small (0.5 mL and 0.25 mL, respectively). The precision and accuracy with which these volumes can be measured are not as good as when using larger volumes. Using an internal standard compensates for any variations in these volumes. To be useful, the analytes and internal standard must be assumed to be retained to the same extent during the initial loading and washing of the cartridge, and they must be assumed to be extracted to the same extent during the final elution.

3. **If the peaks for fluoxetine and protriptyline are insufficiently resolved, how might you alter the mobile phase to improve their separation?**

   Decreasing the amount of acetonitrile and increasing the amount of water in the mobile will increase retention times, thereby providing a better resolution.

12E.9 Evaluation

When compared with gas chromatography, HPLC has only a few differences in the scale of operation; accuracy; precision; sensitivity; selectivity; and time, cost, and equipment necessary. Injection volumes in HPLC are usually significantly larger than in GC because of the greater capacity of HPLC columns. Precision in HPLC is often better due to the routine use of loop injectors. Because HPLC is not limited to volatile analytes, the range of compounds that can be analyzed is somewhat larger than for GC. Capillary GC columns, on the other hand, have more theoretical plates, providing greater resolving power for complex mixtures.

12F Liquid–Solid Adsorption Chromatography

liquid–solid adsorption chromatography
A form of liquid chromatography in which the stationary phase is a solid adsorbent.

In **liquid–solid adsorption chromatography** (LSC) the column packing also serves as the stationary phase. In Tswett's original work the stationary phase was finely divided $CaCO_3$, but modern columns employ porous 3–10-μm particles of silica or alumina. Since the stationary phase is polar, the mobile phase is usually a nonpolar or moderately polar solvent. Typical mobile phases include hexane, isooctane, and methylene chloride. The usual order of elution, from shorter to longer retention times, is

olefins < aromatic hydrocarbons < ethers < esters, aldehydes, ketones
< alcohols, amines < amides < carboxylic acids

For most samples liquid–solid chromatography does not offer any special advantages over liquid–liquid chromatography (LLC). One exception is for the analysis of isomers, where LLC excels. Figure 12.32 shows a typical LSC separation of two amphetamines on a silica column using an 80:20 mixture of methylene chloride and methanol containing 1% NH_4OH as a mobile phase. Nonpolar stationary phases, such as charcoal-based absorbents, also may be used.

12G Ion-Exchange Chromatography

ion-exchange chromatography
A form of liquid chromatography in which the stationary phase is an ion-exchange resin.

In **ion-exchange chromatography** (IEC) the stationary phase is a cross-linked polymer resin, usually divinylbenzene cross-linked polystyrene, with covalently attached ionic functional groups (Figure 12.33). The counterions to these fixed charges are mobile and can be displaced by ions that compete more favorably for the exchange sites. Ion-exchange resins are divided into four categories: strong acid cation exchangers; weak acid cation exchangers; strong base anion exchangers; and weak base anion exchangers. Table 12.5 provides a list of several common ion-exchange resins.

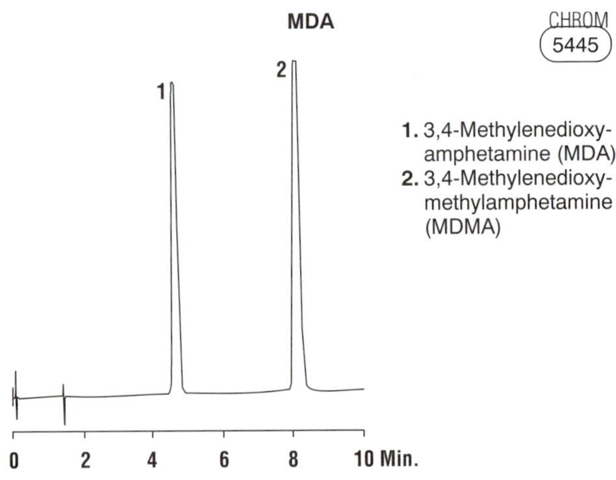

Figure 12.32
Example of the application of liquid–solid chromatography to the analysis of amphetamines. (Chromatogram courtesy of Alltech Associates, Inc. Deerfield, IL).

Strong acid cation exchangers include a sulfonic acid functional group that retains its anionic form, and thus its capacity for ion-exchange, in strongly acidic solutions. The functional groups for a weak acid cation exchanger, however, are fully protonated at pH levels less then 4, thereby losing their exchange capacity. The strong base anion exchangers are fashioned using a quaternary amine, therefore retaining a positive charge even in strongly basic solutions. Weak base anion exchangers, however, remain protonated only at pH levels that are moderately basic. Under more basic conditions, a weak base anion exchanger loses its positive charge and, therefore, its exchange capacity.

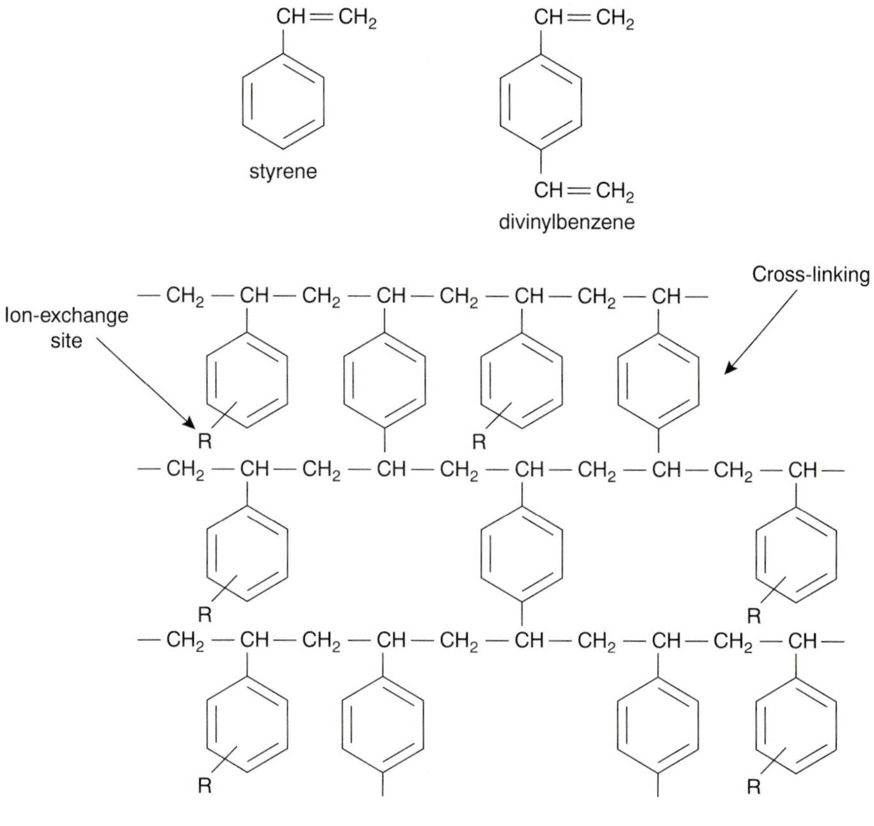

Figure 12.33
Structures of styrene, divinylbenzene, and a styrene-divinylbenzene co-polymer modified for use as an ion-exchange resin. The ion-exchange sites, indicated by R, are mostly in the *para* position and are not necessarily bound to all styrene units.

Table 12.5 Examples of Common Ion-Exchange Resins

Type	Functional Group	Examples
strong acid cation exchanger	sulfonic acid	$-SO_3^-$ $-CH_2CH_2SO_3^-$
weak acid cation exchanger	carboxylic acid	$-COO^-$ $-CH_2COO^-$
strong base anion exchanger	quaternary amine	$-CH_2N(CH_3)_3^+$ $-CH_2CH_2N(CH_2CH_3)_3^+$
weak base anion exchanger	amine	$-NH_3^+$ $-CH_2CH_2NH(CH_2CH_3)_2^+$

The ion-exchange reaction of a monovalent cation, M^+, at a strong acid exchange site is

$$-SO_3^-{-}H^+(s) + M^+(aq) \rightleftharpoons -SO_3^-{-}M^+(s) + H^+(aq)$$

The equilibrium constant for this ion-exchange reaction, which is also called the selectivity coefficient, is

$$K = \frac{\{-SO_3^- -M^+\}[H^+]}{\{-SO_3^- -H^+\}[M^+]} \qquad \mathbf{12.31}$$

where the brackets { } indicate a surface concentration. Rearranging equation 12.31 shows that the distribution ratio for the exchange reaction

$$D = \frac{\text{amount of } M^+ \text{ in stationary phase}}{\text{amount of } M^+ \text{ in mobile phase}} = \frac{\{-SO_3^- -M^+\}}{[M^+]} = K\frac{\{-SO_3^- -H^+\}}{[H^+]}$$

is a function of the concentration of H^+ and, therefore, the pH of the mobile phase.

Ion-exchange resins are incorporated into HPLC columns either as micron-sized porous polymer beads or by coating the resin on porous silica particles. Selectivity is somewhat dependent on whether the resin includes a strong or weak exchange site and on the extent of cross-linking. The latter is particularly important because it controls the resin's permeability and, therefore, the accessibility of the exchange sites. An approximate order of selectivity for a typical strong acid cation exchange resin, in order of decreasing *D*, is

$$Al^{3+} > Ba^{2+} > Pb^{2+} > Ca^{2+} > Ni^{2+} > Cd^{2+} > Cu^{2+} > Co^{2+} > Zn^{2+} > Mg^{2+}$$
$$> Ag^+ > K^+ > NH_4^+ > Na^+ > H^+ > Li^+$$

Note that highly charged ions bind more strongly than ions of lower charge. Within a group of ions of similar charge, those ions with a smaller hydrated radius (Table 6.1 in Chapter 6) or those that are more polarizable bind more strongly. For a strong base anion exchanger the general order is

$$SO_4^{2-} > I^- > HSO_4^- > NO_3^- > Br^- > NO_2^- > Cl^- > HCO_3^- > CH_3COO^- > OH^- > F^-$$

Again, ions of higher charge and smaller hydrated radius bind more strongly than ions with a lower charge and a larger hydrated radius.

The mobile phase in IEC is usually an aqueous buffer, the pH and ionic composition of which determines a solute's retention time. Gradient elutions are possible in which the ionic strength or pH of the mobile phase is changed with time. For example, an IEC separation of cations might use a dilute solution of HCl as the mobile phase. Increasing the concentration of HCl speeds the elution rate for more strongly retained cations, since the higher concentration of H^+ allows it to compete more successfully for the ion-exchange sites.

Ion-exchange columns can be substituted into the general HPLC instrument shown in Figure 12.26. The most common detector measures the conductivity of the mobile phase as it elutes from the column. The high concentration of electrolyte in the mobile phase is a problem, however, because the mobile-phase ions dominate the conductivity. For example, if a dilute solution of HCl is used as the mobile phase, the presence of large concentrations of H_3O^+ and Cl^- produces a background conductivity that may prevent the detection of analytes eluting from the column.

ion-suppressor column
A column used to minimize the conductivity of the mobile phase in ion-exchange chromatography.

To minimize the mobile phase's contribution to conductivity, an **ion-suppressor column** is placed between the analytical column and the detector. This column selectively removes mobile-phase electrolyte ions without removing solute ions. For example, in cation ion-exchange chromatography using a dilute solution of HCl as the mobile phase, the suppressor column contains an anion-exchange resin. The exchange reaction

$$H^+(aq) + Cl^-(aq) + Resin^+–OH^- \rightleftharpoons Resin^+–Cl^- + H_2O(\ell)$$

replaces the ionic HCl with H_2O. Analyte cations elute as hydroxide salts instead of as chloride salts. A similar process is used in anion ion-exchange chromatography in which a cation ion-exchange resin is placed in the suppressor column. If the mobile phase contains Na_2CO_3, the exchange reaction

$$2Na^+(aq) + CO_3^{2-}(aq) + 2Resin^-–H^+ \rightleftharpoons 2Resin^-–Na^+ + H_2CO_3(aq)$$

replaces a strong electrolyte, Na_2CO_3, with a weak electrolyte, H_2CO_3.

Ion suppression is necessary when using a mobile phase containing a high concentration of ions. **Single-column ion chromatography,** in which an ion-suppressor column is not needed, is possible if the concentration of ions in the mobile phase can be minimized. Typically this is done by using a stationary phase resin with a low capacity for ion exchange and a mobile phase with a small concentration of ions. Because the background conductivity due to the mobile phase is sufficiently small, it is possible to monitor a change in conductivity as the analytes elute from the column.

single-column ion chromatography
Ion-exchange chromatography in which conditions are adjusted so that an ion-suppressor column is not needed.

A UV/Vis absorbance detector can also be used if the solute ions absorb ultraviolet or visible radiation. Alternatively, solutions that do not absorb in the UV/Vis range can be detected indirectly if the mobile phase contains a UV/Vis-absorbing species. In this case, when a solute band passes through the detector, a decrease in absorbance is measured at the detector.

Ion-exchange chromatography has found important applications in water analysis and in biochemistry. For example, Figure 12.34a shows how ion-exchange chromatography can be used for the simultaneous analysis of seven common anions in approximately 12 min. Before IEC, a complete analysis of the same set of anions required 1–2 days. Ion-exchange chromatography also has been used for the analysis of proteins, amino acids, sugars, nucleotides, pharmaceuticals, consumer products, and clinical samples. Several examples are shown in Figure 12.34.

12H Size-Exclusion Chromatography

Two classes of micron-sized stationary phases have been encountered in this section: silica particles and cross-linked polymer resin beads. Both materials are porous, with pore sizes ranging from approximately 50 to 4000 Å for silica particles and from 50 to 1,000,000 Å for divinylbenzene cross-linked polystyrene resins. In **size-exclusion chromatography,** also called molecular-exclusion or gel-permeation chromatography, separation is based on the solute's ability to enter into the pores of the column packing. Smaller solutes spend proportionally more time within the pores and, consequently, take longer to elute from the column.

size-exclusion chromatography
A form of liquid chromatography in which the stationary phase is a porous material and in which separations are based on the size of the solutes.

The size selectivity of a particular packing is not infinite, but is limited to a moderate range. All solutes significantly smaller than the pores move through the column's entire volume and elute simultaneously, with a retention volume, V_r, of

$$V_r = V_i + V_o \qquad \textbf{12.32}$$

where V_i is the volume of mobile phase occupying the packing material's pore space, and V_o is volume of mobile phase in the remainder of the column. The maximum size for which equation 12.32 holds is the packing material's **inclusion limit,**

inclusion limit
In size-exclusion chromatography, the smallest solute that can be separated from other solutes; all smaller solutes elute together.

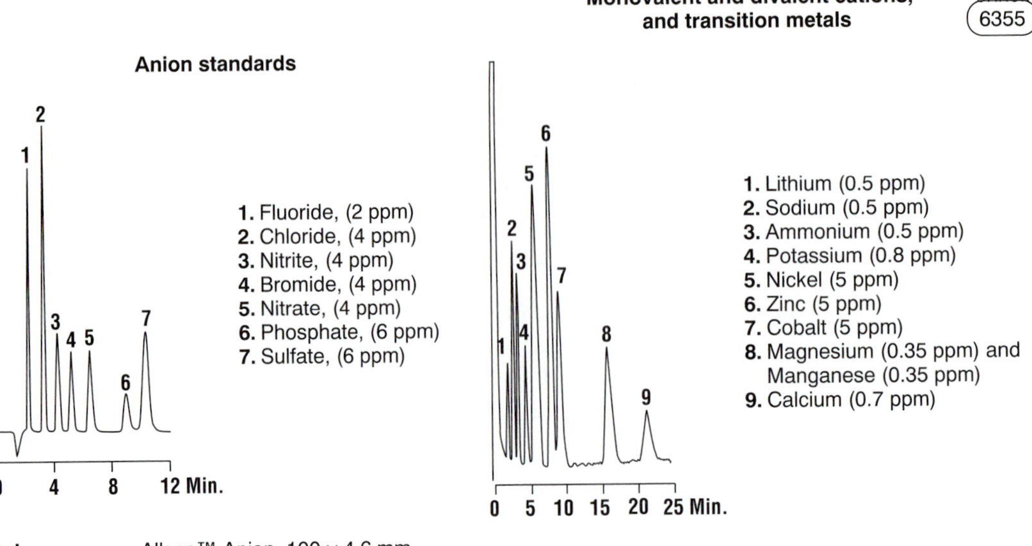

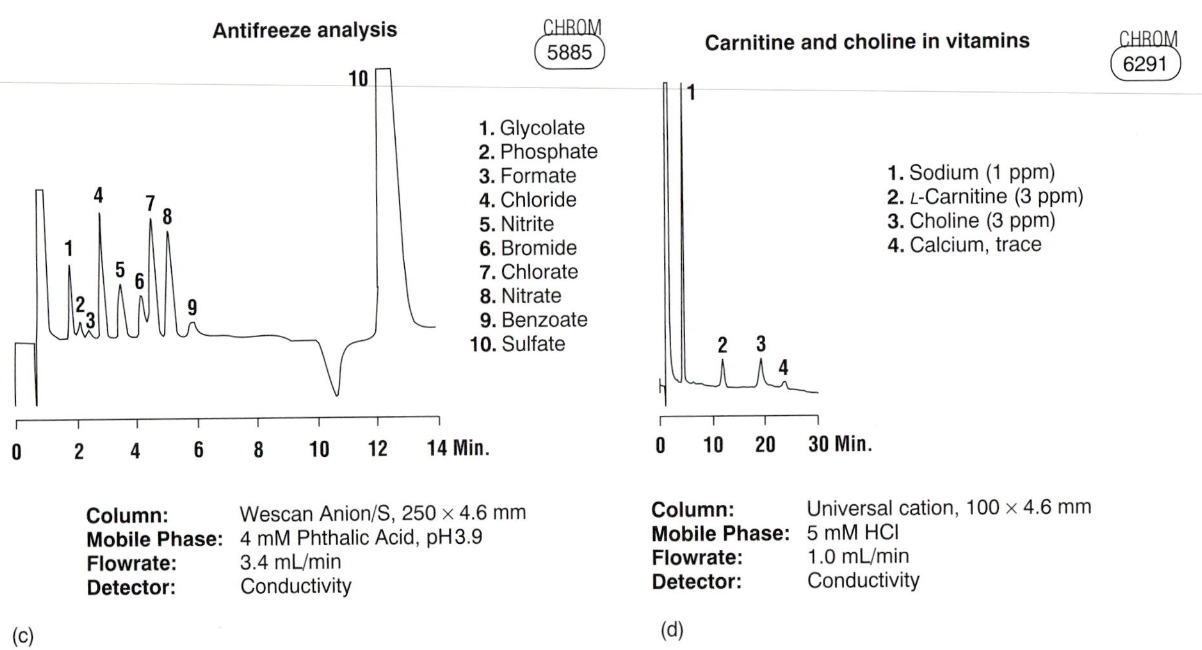

Figure 12.34

Examples of the application of ion-exchange chromatography to the analysis of (a) inorganic anions, (b) inorganic cations, (c) antifreeze, and (d) vitamins. (Chromatograms courtesy of Alltech Associates, Inc. Deerfield, IL).

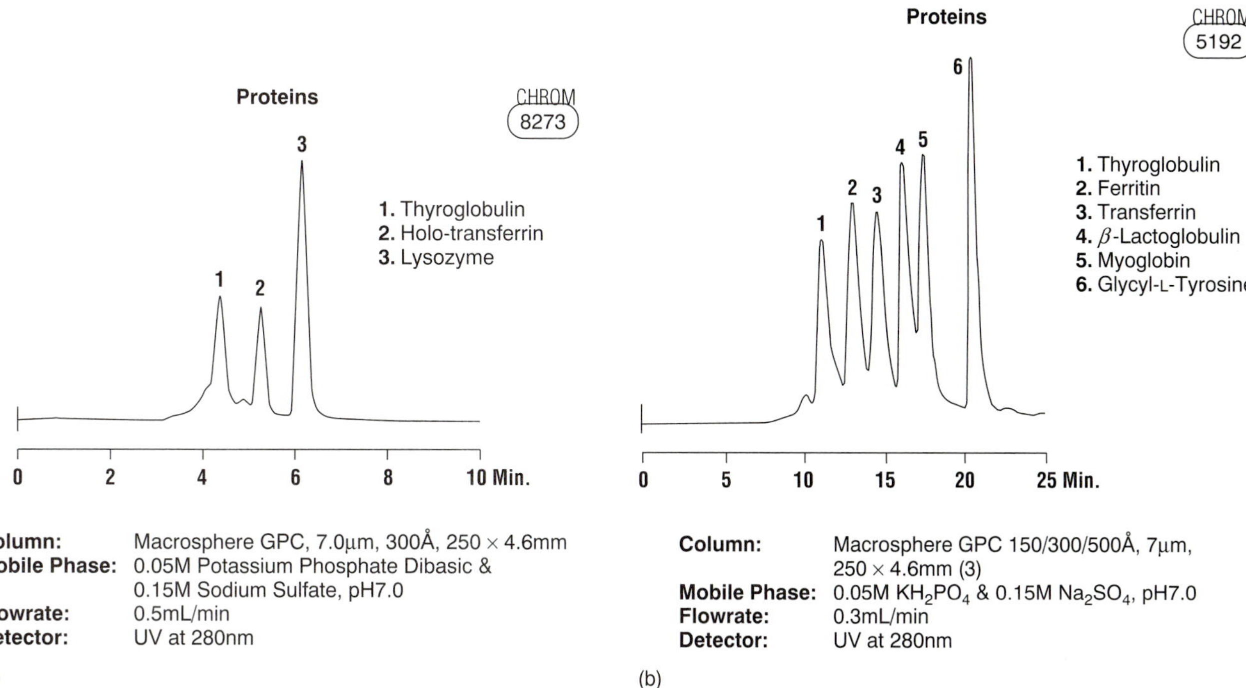

Figure 12.35
Examples of the application of size-exclusion chromatography to the analysis of proteins. The separation in (a) uses a single column; that in (b) uses three columns, providing a wider range of size selectivity. (Chromatograms courtesy of Alltech Associates, Inc. Deerfield, IL).

or permeation limit. All solutes too large to enter the pores elute simultaneously with a retention volume of

$$V_r = V_o \qquad \textbf{12.33}$$

Equation 12.33 defines the packing material's **exclusion limit.**

exclusion limit
In size-exclusion chromatography, the largest solute that can be separated from other solutes; all larger solutes elute together.

In between the inclusion limit and the exclusion limit, each solute spends an amount of time in the pore space proportional to its size. The retention volume for a solute is

$$V_r = V_o + DV_i \qquad \textbf{12.34}$$

where D is the solute's distribution ratio, which ranges from 0 at the exclusion limit to 1 at the inclusion limit. The validity of equation 12.34 requires that size exclusion be the only interaction between the solute and the stationary phase responsible for the separation. To this end, silica particles used for size exclusion are deactivated as described earlier, and polymer resins are synthesized without exchange sites.

Size-exclusion chromatography provides a rapid means for separating larger molecules, including polymers and biomolecules. Figure 12.35 shows the application of size-exclusion chromatography for the analysis of protein mixtures. In Figure 12.35a, a column packing with 300 Å pores, with an inclusion limit of 7500 g/mol and an exclusion limit of 1.2×10^6 g/mol, is used to separate a mixture of three proteins. Mixtures containing a wider range of formula weights can be separated by joining together several columns in series. Figure 12.35b shows an example spanning an inclusion limit of 4000 g/mol and an exclusion limit of 7.5×10^6 g/mol.

Another important application is for the determination of formula weights. Calibration curves of log(formula weight) versus V_r are prepared between the exclusion limit and inclusion limit (Figure 12.36). Since the retention volume is, to some

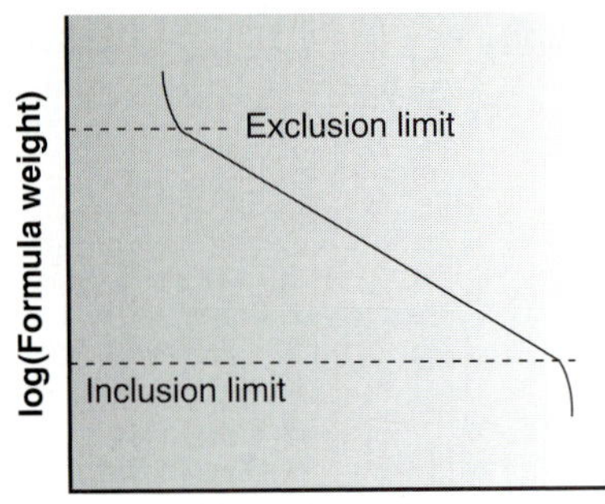

Figure 12.36
Calibration curve for the determination of formula weight by size-exclusion chromatography.

degree, a function of a solute's size and shape, reasonably accurate determinations of formula weight are possible only if the standards are carefully chosen to minimize the effect of shape.

Size-exclusion chromatography can be carried out using conventional HPLC instrumentation, replacing the HPLC column with an appropriate size-exclusion column. A UV/Vis detector is the most common means for obtaining the chromatogram.

12I Supercritical Fluid Chromatography

Despite their importance, gas chromatography and liquid chromatography cannot be used to separate and analyze all types of samples. Gas chromatography, particularly when using capillary columns, provides for rapid separations with excellent resolution. Its application, however, is limited to volatile analytes or those analytes that can be made volatile by a suitable derivatization. Liquid chromatography can be used to separate a wider array of solutes; however, the most commonly used detectors (UV, fluorescence, and electrochemical) do not respond as universally as the flame ionization detector commonly used in gas chromatography.

supercritical fluid chromatography
A separation technique in which the mobile phase is a supercritical fluid.

Supercritical fluid chromatography (SFC) provides a useful alternative to gas chromatography and liquid chromatography for some samples. The mobile phase in supercritical fluid chromatography is a gas held at a temperature and pressure exceeding its critical point (Figure 12.37). Under these conditions the mobile phase is neither a gas nor a liquid. Instead, the mobile phase is a supercritical fluid whose properties are intermediate between those of a gas and a liquid (Table 12.6). Specifically, supercritical fluids have viscosities that are similar to those of gases, which means that they can move through either capillary or packed columns without the need for the high pressures encountered in HPLC. Analysis time and resolution, although not as good as in GC, are usually better than that obtainable with conventional HPLC. The density of a supercritical fluid, however, is much closer to that of a liquid, accounting for its ability to function as a solvent. The mobile phase in SFC, therefore, behaves more like the liquid mobile phase in HPLC than the gaseous mobile phase in GC.

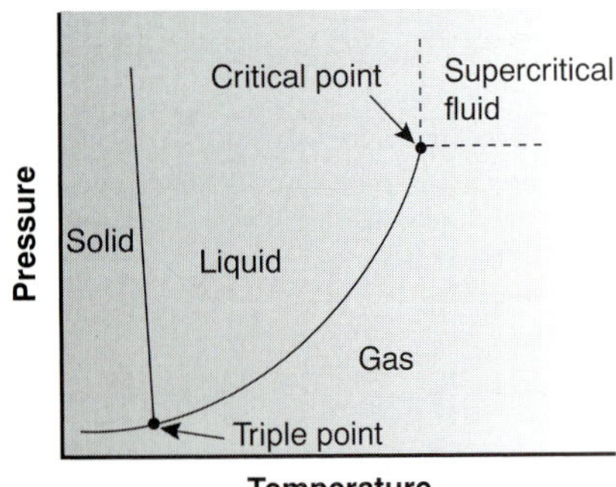

Figure 12.37
Phase diagram for a supercritical fluid.

The most common mobile phase for supercritical fluid chromatography is CO_2. Its low critical temperature, 31 °C, and critical pressure, 72.9 atm, are relatively easy to achieve and maintain. Although supercritical CO_2 is a good solvent for nonpolar organics, it is less useful for polar solutes. The addition of an organic modifier, such as methanol, improves the mobile phase's elution strength. Other common mobile phases and their critical temperatures and pressures are listed in Table 12.7.

Colorplate 11 shows the phase transition of liquid CO_2 to supercritical CO_2.

Table 12.6 Typical Properties of Gases, Liquids, and Supercritical Fluids[a]

Phase	Density (g cm^{-3})	Viscosity (g cm^{-1} s^{-1})	Diffusion coefficient (cm^2 s^{-1})
gas	$\approx 10^{-3}$	$\approx 10^{-4}$	$\approx 10^{-1}$
supercritical fluid	$\approx$ 0.1–1	$\approx 10^{-4}$–10^{-3}	$\approx 10^{-4}$–10^{-3}
liquid	$\approx$ 1	$\approx 10^{-2}$	$< 10^{-5}$

[a] Values are reported to the nearest factor of 10.

The instrumentation necessary for supercritical fluid chromatography is essentially the same as that for a standard GC or HPLC. The only important addition is the need for a pressure restrictor to maintain the critical pressure. Gradient elutions, similar to those in HPLC, are accomplished by changing the applied pressure over time. The resulting change in the density of the mobile phase affects its solvent strength. Detection can be accomplished using standard GC detectors or HPLC detectors.

Supercritical fluid chromatography has found many applications in the analysis of polymers, fossil fuels, waxes, drugs, and food products. Its application in the analysis of triglycerides is shown in Figure 12.38.

Table 12.7 Critical Point Properties for Selected Supercritical Fluids

Compound	Critical Temperature (°C)	Critical Pressure (atm)
carbon dioxide	31.3	72.9
ethane	32.4	48.3
nitrous oxide	36.5	71.4
ammonia	132.3	111.3
diethyl ether	193.6	36.3
isopropanol	235.3	47.0
methanol	240.5	78.9
ethanol	243.4	63.0
water	374.4	226.8

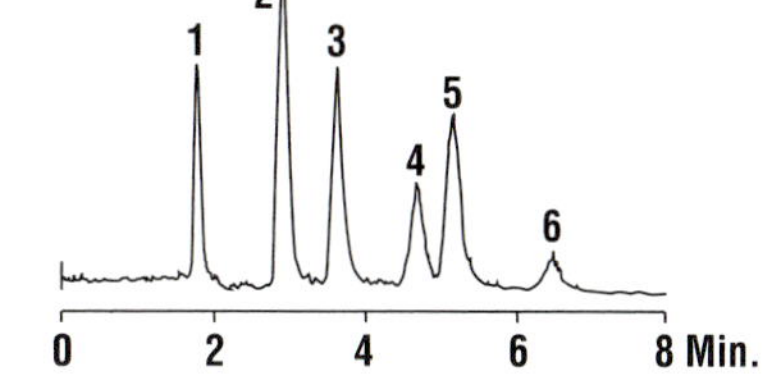

Figure 12.38
Example of the application of supercritical fluid chromatography to the analysis of triglycerides. (Chromatogram courtesy of Alltech Associates, Inc. Deerfield, IL).

12J Electrophoresis

Thus far all the separations we have considered involve a mobile phase and a stationary phase. Separation of a complex mixture of analytes occurs because each analyte has a different ability to partition between the two phases. An analyte whose distribution ratio favors the stationary phase is retained on the column for a longer time, thereby eluting with a longer retention time. Although the methods described in the preceding sections involve different types of stationary and mobile phases, all are forms of chromatography.

Electrophoresis is another class of separation techniques in which analytes are separated based on their ability to move through a conductive medium, usually an aqueous buffer, in response to an applied electric field. In the absence of other effects, cations migrate toward the electric field's negatively charged cathode, and anions migrate toward the positively charged anode. More highly charged ions and ions of smaller size, which means they have a higher charge-to-size ratio, migrate at a faster rate than larger ions, or ions of lower charge. Neutral species do not experience the electric field and remain stationary. As we will see shortly, under normal conditions even neutral species and anions migrate toward the cathode. In either case, differences in their rate of migration allow for the separation of complex mixtures of analytes.

electrophoresis
A separation technique based on a solute's ability to move through a conductive medium under the influence of an electric field.

There are several forms of electrophoresis. In slab gel electrophoresis the conducting buffer is retained within a porous gel of agarose or polyacrylamide. Slabs are formed by pouring the gel between two glass plates separated by spacers. Typical thicknesses are 0.25–1 mm. Gel electrophoresis is an important technique in biochemistry, in which it is frequently used for DNA sequencing. Although it is a powerful tool for the qualitative analysis of complex mixtures, it is less useful for quantitative work.

In **capillary electrophoresis** the conducting buffer is retained within a capillary tube whose inner diameter is typically 25–75 μm. Samples are injected into one end of the capillary tube. As the sample migrates through the capillary, its components separate and elute from the column at different times. The resulting **electropherogram** looks similar to the chromatograms obtained in GC or HPLC and provides

capillary electrophoresis
Electrophoresis taking place in a capillary tube.

electropherogram
The equivalent of a chromatogram in electrophoresis.

both qualitative and quantitative information. Only capillary electrophoretic methods receive further consideration in this text.

12J.1 Theory of Capillary Electrophoresis

In capillary electrophoresis the sample is injected into a buffered solution retained within a capillary tube. When an electric field is applied to the capillary tube, the sample's components migrate as the result of two types of mobility: electrophoretic mobility and electroosmotic mobility. **Electrophoretic mobility** is the solute's response to the applied electric field. As described earlier, cations move toward the negatively charged cathode, anions move toward the positively charged anode, and neutral species, which do not respond to the electric field, remain stationary. The other contribution to a solute's migration is **electroosmotic flow,** which occurs when the buffer solution moves through the capillary in response to the applied electric field. Under normal conditions the buffer solution moves toward the cathode, sweeping most solutes, even anions, toward the negatively charged cathode.

electrophoretic mobility
A measure of a solute's ability to move through a conductive medium in response to an applied electric field (μ_{ep}).

electroosmotic flow
The movement of the conductive medium in response to an applied electric field.

Electrophoretic Mobility The velocity with which a solute moves in response to the applied electric field is called its **electrophoretic velocity,** ν_{ep}; it is defined as

electrophoretic velocity
The velocity with which a solute moves through the conductive medium due to its electrophoretic mobility (ν_{ep}).

$$\nu_{ep} = \mu_{ep}E \qquad \textbf{12.35}$$

where μ_{ep} is the solute's electrophoretic mobility, and E is the magnitude of the applied electric field. A solute's electrophoretic mobility is defined as

$$\mu_{ep} = \frac{q}{6\pi\eta r} \qquad \textbf{12.36}$$

where q is the solute's charge, η is the buffer solvent's viscosity, and r is the solute's radius. Using equations 12.35 and 12.36, we can make several important conclusions about a solute's electrophoretic velocity. Electrophoretic mobility, and, therefore, electrophoretic velocity, is largest for more highly charged solutes and solutes of smaller size. Since q is positive for cations and negative for anions, these species migrate in opposite directions. Neutral species, for which q is 0, have an electrophoretic velocity of 0.

Electroosmotic Mobility When an electric field is applied to a capillary filled with an aqueous buffer, we expect the buffer's ions to migrate in response to their electrophoretic mobility. Because the solvent, H_2O, is neutral, we might reasonably expect it to remain stationary. What is observed under normal conditions, however, is that the buffer solution moves toward the cathode. This phenomenon is called the electroosmotic flow.

Electroosmosis occurs because the walls of the capillary tubing are electrically charged. The surface of a silica capillary contains large numbers of silanol groups (Si–OH). At pH levels greater than approximately 2 or 3, the silanol groups ionize to form negatively charged silanate ions ($Si–O^-$). Cations from the buffer are attracted to the silanate ions. As shown in Figure 12.39, some of these cations bind tightly to the silanate ions, forming an inner, or fixed, layer. Other cations are more loosely bound, forming an outer, or mobile, layer. Together these two layers are called the double layer. Cations in the outer layer migrate toward the cathode. Because these cations are solvated, the solution is also pulled along, producing the electroosmotic flow.

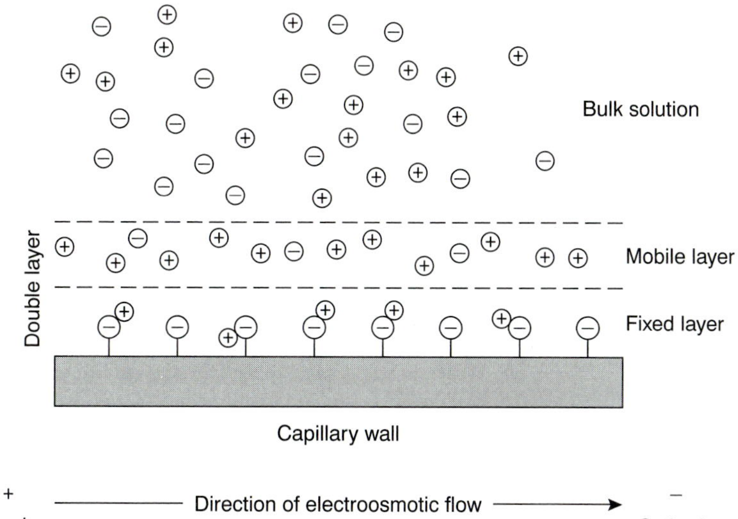

Figure 12.39
Schematic diagram showing the origin of electroosmotic flow.

Electroosmotic flow velocity, ν_{eof}, is a function of the magnitude of the applied electric field and the buffer solution's electroosmotic mobility, μ_{eof}.

$$\nu_{eof} = \mu_{eof}E \qquad 12.37$$

electroosmotic flow velocity
The velocity with which the solute moves through the capillary due to the electroosmotic flow (ν_{eof}).

Electroosmotic mobility is defined as

$$\mu_{eof} = \frac{\varepsilon\zeta}{4\pi\eta} \qquad 12.38$$

where ε is the buffer solution's dielectric constant, ζ is the zeta potential, and η is the buffer solution's viscosity.

Examining equations 12.37 and 12.38 shows that the **zeta potential** plays an important role in determining the electroosmotic flow velocity. Two factors determine the zeta potential and, therefore, the electroosmotic velocity. First, the zeta potential is directly proportional to the charge on the capillary walls, with a greater density of silanate ions corresponding to a larger zeta potential. Below a pH of 2, for example, there are few silanate ions; thus, the zeta potential and electroosmotic flow velocity are 0. As the pH level is increased, both the zeta potential and the electroosmotic flow velocity increase. Second, the zeta potential is proportional to the thickness of the double layer. Increasing the buffer solution's ionic strength provides a higher concentration of cations, decreasing the thickness of the double layer.

zeta potential
The change in potential across a double layer (ζ).

The electroosmotic flow profile is very different from that for a phase moving under forced pressure. Figure 12.40 compares the flow profile for electroosmosis with that for hydrodynamic pressure. The uniform, flat profile for electroosmosis helps to minimize band broadening in capillary electrophoresis, thus improving separation efficiency.

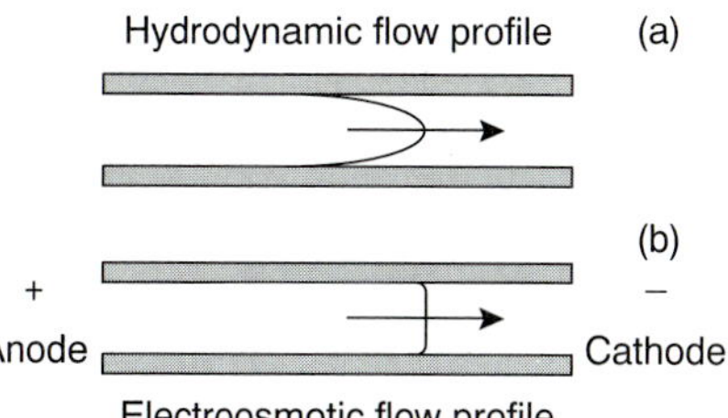

Figure 12.40
Schematic showing a comparison of the flow profiles for (a) GC and HPLC, and (b) electrophoresis.

Total Mobility A solute's net, or total velocity, ν_{tot}, is the sum of its electrophoretic velocity and the electroosmotic flow velocity; thus,

$$\nu_{tot} = \nu_{ep} + \nu_{eof}$$

and

$$\mu_{tot} = \mu_{ep} + \mu_{eof}$$

Under normal conditions the following relationships hold

$$(v_{tot})_{cations} > v_{eof}$$

$$(v_{tot})_{anions} < v_{eof}$$

$$(v_{tot})_{neutrals} = v_{eof}$$

Thus, cations elute first in an order corresponding to their electrophoretic mobilities, with small, highly charged cations eluting before larger cations of lower charge. Neutral species elute as a single band, with an elution rate corresponding to the electroosmotic flow velocity. Finally, anions are the last components to elute, with smaller, highly charged anions having the longest elution time.

Migration Time A solute's total velocity is given by

$$v_{tot} = \frac{l}{t_m}$$

where l is the distance the solute travels between its point of injection and the detector, and t_m is the migration time. Since

$$v_{tot} = \mu_{tot}E = (\mu_{ep} + \mu_{eof})E$$

we have, after rearranging,

$$t_m = \frac{l}{(\mu_{ep} + \mu_{eof})E} \qquad 12.39$$

Finally, the magnitude of the electric field is

$$E = \frac{V}{L} \qquad 12.40$$

where V is the applied potential, and L is the length of the capillary tube. Substituting equation 12.40 into equation 12.39 gives

$$t_m = \frac{lL}{(\mu_{ep} + \mu_{eof})V} \qquad 12.41$$

Examining equation 12.41 shows that we can decrease a solute's migration time (and thus the total analysis time) by applying a higher voltage or by using a shorter capillary tube. Increasing the electroosmotic flow also shortens the analysis time, but, as we will see shortly, at the expense of resolution.

Efficiency The efficiency of capillary electrophoresis is characterized by the number of theoretical plates, N, just as it is in GC or HPLC. In capillary electrophoresis, the number of theoretic plates is determined by

$$N = \frac{(\mu_{ep} + \mu_{eof})V}{2D} \qquad 12.42$$

where D is the solute's diffusion coefficient. From equation 12.42 it is easy to see that the efficiency of a capillary electrophoretic separation increases with higher voltages. Again, increasing the electroosmotic flow velocity improves efficiency, but at the expense of resolution. Two additional observations deserve comment.

First, solutes with larger electrophoretic mobilities (in the same direction as the electroosmotic flow) have greater efficiencies; thus, smaller, more highly charged solutes are not only the first solutes to elute, but do so with greater efficiency. Second, efficiency in capillary electrophoresis is independent of the capillary's length. Typical theoretical plate counts are approximately 100,000–200,000 for capillary electrophoresis.

Selectivity In chromatography, selectivity is defined as the ratio of the capacity factors for two solutes (equation 12.11). In capillary electrophoresis, the analogous expression for selectivity is

$$\alpha = \frac{\mu_{ep,1}}{\mu_{ep,2}}$$

where $\mu_{ep,1}$ and $\mu_{ep,2}$ are the electrophoretic mobilities for solutes 1 and 2, respectively, chosen such that $\alpha \geq 1$. Selectivity often can be improved by adjusting the pH of the buffer solution. For example, NH_4^+ is a weak acid with a pK_a of 9.24. At a pH of 9.24 the concentrations of NH_4^+ and NH_3 are equal. Decreasing the pH below 9.24 increases its electrophoretic mobility because a greater fraction of the solute is present as the cation NH_4^+. On the other hand, raising the pH above 9.24 increases the proportion of the neutral NH_3, decreasing its electrophoretic mobility.

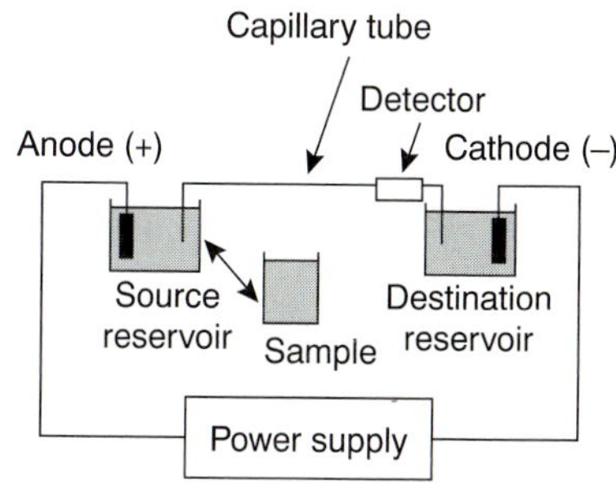

Figure 12.41
Schematic diagram for capillary electrophoresis. The sample and source reservoir are switched when making injections.

Resolution The resolution between two solutes is

$$R = \frac{0.177(\mu_{ep,2} - \mu_{ep,1})V^{1/2}}{\sqrt{(\mu_{avg} + \mu_{eof})D}} \qquad 12.43$$

where μ_{avg} is the average electrophoretic mobility for the two solutes. Examining equation 12.43 shows that increasing the applied voltage and decreasing the electroosmotic flow velocity improves resolution. The latter effect is particularly important because increasing electroosmotic flow improves analysis time and efficiency while decreasing resolution.

12J.2 Instrumentation

The basic instrumentation for capillary electrophoresis is shown in Figure 12.41 and includes a power supply for applying the electric field, anode and cathode compartments containing reservoirs of the buffer solution, a sample vial containing the sample, the capillary tube, and a detector. Each part of the instrument receives further consideration in this section.

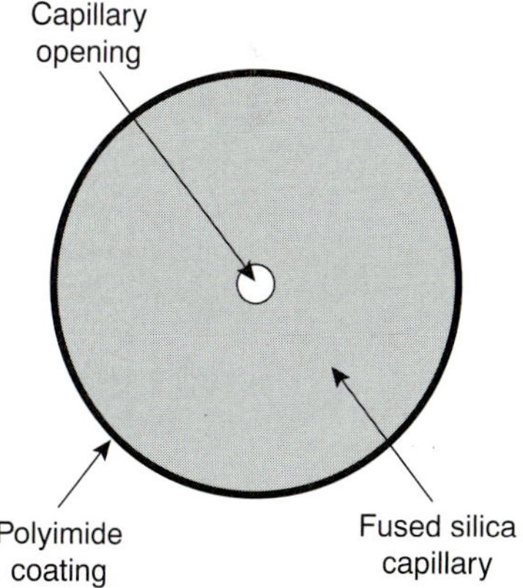

Figure 12.42
Schematic diagram showing a cross section of a capillary column for capillary electrophoresis.

Capillary Tubes Figure 12.42 shows a cross section of a typical capillary tube. Most capillary tubes are made from fused silica coated with a 20–35-μm layer of polyimide to give it mechanical strength. The inner diameter is typically 25–75 μm, which is smaller than that for a capillary GC column, with an outer diameter of 200–375 μm.

The narrow bore of the capillary column and the relative thickness of the capillary's walls are important. When an electric field is applied to a capillary containing a conductive medium, such as a buffer solution, current flows through the capillary. This current leads to **Joule heating,** the extent of which is proportional to the capillary's radius and the magnitude of the electric field. Joule heating is a problem because it changes the buffer solution's viscosity, with the solution at the center of the

Joule heating
The heating of a conductive solution due to the passage of an electric current through the solution.

capillary being less viscous than that near the capillary walls. Since the solute's electrophoretic mobility depends on the buffer's viscosity (see equation 12.36), solutes in the center of the capillary migrate at a faster rate than solutes near the capillary walls. The result is additional band broadening that degrades the separation. Capillaries with smaller inner diameters generate less Joule heating, and those with larger outer diameters are more effective at dissipating the heat. Capillary tubes may be placed inside a thermostated jacket to control heating, in which case smaller outer diameters allow a more rapid dissipation of thermal energy.

Injecting the Sample The mechanism by which samples are introduced in capillary electrophoresis is quite different from that used in GC or HPLC. Two types of injection are commonly used: hydrodynamic injection and electrokinetic injection. In both cases the capillary tube is filled with buffer solution. One end of the capillary tube is placed in the destination reservoir, and the other is placed in the sample vial.

hydrodynamic injection
An injection technique in capillary electrophoresis in which pressure is used to inject sample into the capillary column.

Hydrodynamic injection uses pressure to force a small portion of the sample into the capillary tubing. To inject a sample hydrodynamically a difference in pressure is applied across the capillary by either pressurizing the sample vial or by applying a vacuum to the destination reservoir. The volume of sample injected, in liters, is given by the following equation

$$V_{\text{inj}} = \frac{\Delta P d^4 \pi t}{128 \eta L} \times 10^3 \qquad \textbf{12.44}$$

where ΔP is the pressure difference across the capillary in pascals, d is the capillary's inner diameter in meters, t is the amount of time that the pressure is applied in seconds, η is the buffer solution's viscosity in kilograms per meter per second (kg m^{-1} s^{-1}), and L is the length of the capillary tubing in meters. The factor of 10^3 changes the units from cubic meters to liters.

Example

EXAMPLE 12.9

A hydrodynamic injection is made by applying a pressure difference of 2.5×10^3 Pa (approximately 0.02 atm) for 2 s to a 75-cm long capillary tube with an internal diameter of 50 μm. Assuming that the buffer solution's viscosity is 10^{-3} kg m^{-1} s^{-1}, what volume of sample is injected?

SOLUTION

Making appropriate substitutions into equation 12.44 gives the volume of injected sample as

$$V_{\text{inj}} = \frac{(2.5 \times 10^3 \text{ Pa})(50 \times 10^{-6} \text{ m})^4 (3.14)(2 \text{ s})}{(128)(0.001 \text{ kg m}^{-1} \text{ s}^{-1})(0.75 \text{ m})} \times 10^3 = 1 \times 10^{-9} \text{ L} = 1 \text{ nL}$$

Since the injected sample plug is cylindrical, its length, l_{plug}, is easily calculated using the equation for the volume of a cylinder.

$$V = \pi r^2 l_{\text{plug}}$$

Thus,

$$l_{\text{plug}} = \frac{V}{\pi r^2} = \frac{(1 \times 10^{-9} \text{ L})(10^{-3} \text{ m}^3/L)}{(3.14)(25 \times 10^{-6} \text{ m})^2} = 5 \times 10^{-4} \text{ m} = 0.5 \text{ mm}$$

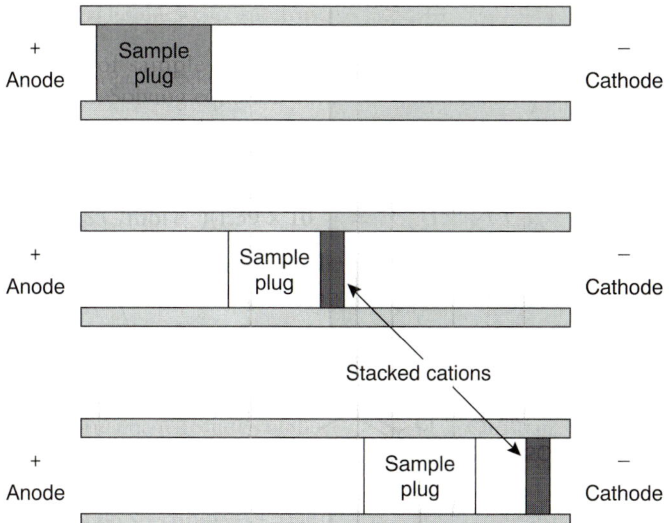

Figure 12.43
Schematic diagram demonstrating stacking.

electrokinetic injection
An injection technique in capillary electrophoresis in which an electric field is used to inject sample into the capillary column.

Electrokinetic injections are made by placing both the capillary and the anode into the sample vial and briefly applying an electric field. The moles of solute injected into the capillary, n_{inj}, are determined using

$$n_{inj} = \pi C t r^2 (\mu_{ep} + \mu_{eof}) E \frac{\kappa_{buf}}{\kappa_{samp}} \qquad \textbf{12.45}$$

where C is the solute's concentration in the sample, t is the amount of time that the electric field is applied, r is the capillary's radius, μ_{ep} is the solute's electrophoretic mobility, μ_{eof} is the electroosmotic mobility, E is the applied electric field, and κ_{buf} and κ_{samp} are the conductivities of the buffer solution and sample, respectively. An important consequence of equation 12.45 is that it is inherently biased toward sampling solutes with larger electrophoretic mobilities. Those solutes with the largest electrophoretic mobilities (smaller, more positively charged ions) are injected in greater numbers than those with the smallest electrophoretic mobilities (smaller, more negatively charged ions).

When a solute's concentration in the sample is too small to reliably analyze, it may be possible to inject the solute in a manner that increases its concentration in the capillary tube. This method of injection is called **stacking.** Stacking is accomplished by placing the sample in a solution whose ionic strength is significantly less than that of the buffering solution. Because the sample plug has a lower concentration of ions than the buffering solution, its resistance is greater. Since the electric current passing through the capillary is fixed, we know from Ohm's law

stacking
A means of concentrating solutes in capillary electrophoresis after their injection onto the capillary column.

$$E = iR$$

that the electric field in the sample plug is greater than that in the buffering solution. Electrophoretic velocity is directly proportional to the electric field (see equation 12.35); thus, ions in the sample plug migrate with a greater velocity. When the solutes reach the boundary between the sample plug and the buffering solution, the electric field decreases and their electrophoretic velocity slows down, "stacking" together in a smaller sampling zone (Figure 12.43).

Applying the Electric Field Migration in electrophoresis occurs in response to the applied electric field. The ability to apply a large electric field is important because

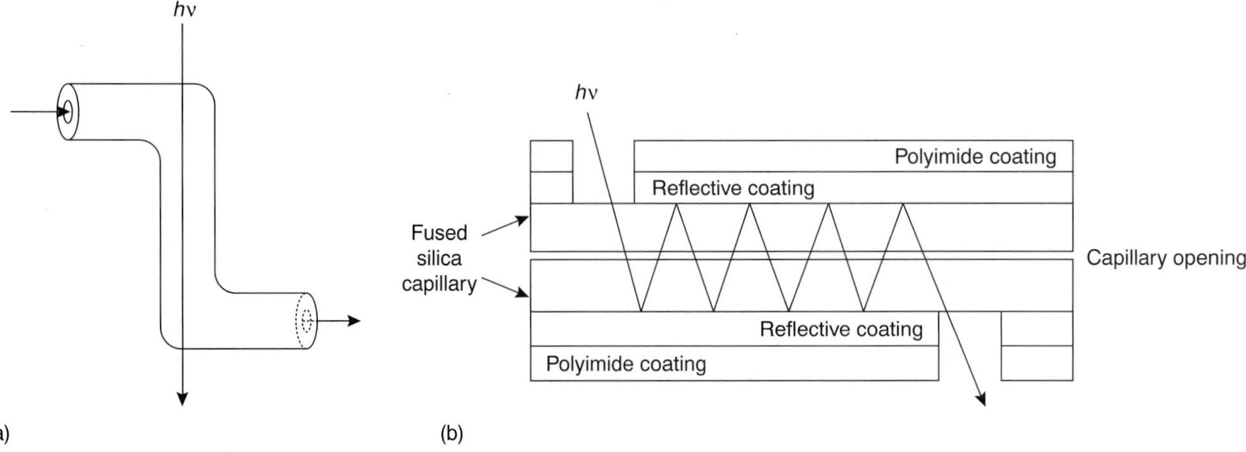

Figure 12.44
Schematic diagrams of two approaches to on-column detection using UV/Vis absorption spectroscopy.

higher voltages lead to shorter analysis times (see equation 12.41), more efficient separations (see equation 12.42), and better resolution (see equation 12.43). Because narrow-bore capillary tubes dissipate Joule heating so efficiently, voltages of up to 40 kV can be applied.

Detectors Most of the detectors used in HPLC also find use in capillary electrophoresis. Among the more common detectors are those based on the absorption of UV/Vis radiation, fluorescence, conductivity, amperometry, and mass spectrometry. Whenever possible, detection is done "on-column" before the solutes elute from the capillary tube and additional band broadening occurs.

UV/Vis detectors are among the most popular. Because absorbance is directly proportional to path length, the capillary tubing's small diameter leads to signals that are smaller than those obtained in HPLC. Several approaches have been used to increase the path length, including a Z-shaped sample cell or multiple reflections (Figure 12.44). Detection limits are about 10^{-7} M.

Better detection limits are obtained using fluorescence, particularly when using a laser as an excitation source. When using fluorescence detection, a small portion of the capillary's protective coating is removed and the laser beam is focused on the inner portion of the capillary tubing. Emission is measured at an angle of 90° to the laser. Because the laser provides an intense source of radiation that can be focused to a narrow spot, detection limits are as low as 10^{-16} M.

Solutes that do not absorb UV/Vis radiation or undergo fluorescence can be detected by other detectors. Table 12.8 provides a list of detectors used in capillary electrophoresis along with some of their important characteristics.

12J.3 Capillary Electrophoresis Methods

There are several different forms of capillary electrophoresis, each of which has its particular advantages. Several of these methods are briefly described in this section.

capillary zone electrophoresis
A form of capillary electrophoresis in which separations are based on differences in the solutes' electrophoretic mobilities.

Capillary Zone Electrophoresis The simplest form of capillary electrophoresis is **capillary zone electrophoresis** (CZE). In CZE the capillary tube is filled with a buffer solution and, after loading the sample, the ends of the capillary tube are placed in reservoirs containing additional buffer solution. Under normal conditions, the end of the capillary containing the sample is the anode, and solutes migrate toward

Table 12.8 Characteristics of Selected Detectors for Capillary Electrophoresis

Detector	Selectivity	Detection Limit Moles Injected	Detection Limit Molarity[a]	On-Column Detection?
UV/Vis absorbance	solute must have UV/Vis absorbing chromophore	10^{-13}–10^{-16}	10^{-5}–10^{-7}	yes
indirect absorbance	universal	10^{-12}–10^{-15}	10^{-4}–10^{-6}	yes
fluorescence	solute must have favorable fluorescent quantum efficiency	10^{-15}–10^{-17}	10^{-7}–10^{-9}	yes
laser fluorescence	solute must have favorable fluorescent quantum efficiency	10^{-18}–10^{-20}	10^{-13}–10^{-16}	yes
mass spectrometer	universal when monitoring all ions; selective when monitoring single ion	10^{-16}–10^{-17}	10^{-8}–10^{-10}	no
amperometry	solute must undergo oxidation or reduction	10^{-18}–10^{-19}	10^{-7}–10^{-10}	no
conductivity	universal	10^{-15}–10^{-16}	10^{-7}–10^{-9}	no
radiometric	solutes must be radioactive	10^{-17}–10^{-19}	10^{-10}–10^{-12}	yes

Source: Adapted from Baker, D. R. *Capillary Electrophoresis.* Wiley-Interscience: New York, 1995.[16]
[a]Concentration depends on the volume of sample injected.

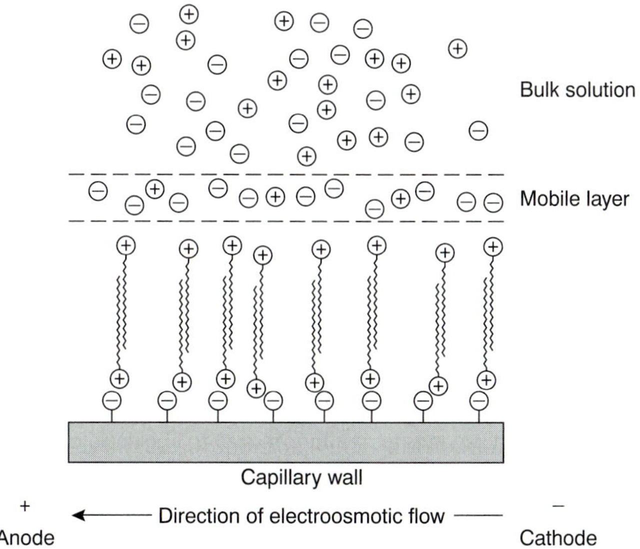

Figure 12.45
Schematic diagram showing the reversal of electroosmotic flow.

the cathode at a velocity determined by their electrophoretic mobility and the electroosmotic flow. Cations elute first, with smaller, more highly charged cations eluting before larger cations with smaller charges. Neutral species elute as a single band. Finally, anions are the last species to elute, with smaller, more negatively charged anions being the last to elute.

The direction of electroosmotic flow and, therefore, the order of elution in CZE can be reversed. This is accomplished by adding an alkylammonium salt to the buffer solution. As shown in Figure 12.45, the positively charged end of the alkylammonium ion binds to the negatively charged silanate ions on the capillary's walls. The alkylammonium ion's "tail" is hydrophobic and associates with the tail of another alkylammonium ion. The result is a layer of positive charges to which anions in the buffer solution are attracted. The migration of these solvated anions toward the anode reverses the electroosmotic flow's direction. The order of elution in this case is exactly the opposite of that observed under normal conditions.

Capillary zone electrophoresis also can be accomplished without an electroosmotic flow by coating the capillary's walls with a nonionic reagent. In the absence of electroosmotic flow only cations migrate from the anode to the cathode. Anions elute into the source reservoir while neutral species remain stationary.

Capillary zone electrophoresis provides effective separations of any charged species, including inorganic anions and cations, organic acids and amines, and large biomolecules such as proteins. For example, CZE has been used to separate a mixture of 36 inorganic and organic ions in less than 3 minutes.[17] Neutral species, of course, cannot be separated.

micellar electrokinetic capillary chromatography
A form of capillary electrophoresis in which neutral solutes are separated based on their ability to partition into a charged micelle.

micelle
An agglomeration of molecules containing ionic "heads" and hydrophobic "tails," which form into a structure with a hydrophobic interior and a hydrophilic exterior.

Micellar Electrokinetic Capillary Chromatography One limitation to CZE is its inability to separate neutral species. **Micellar electrokinetic chromatography** (MEKC) overcomes this limitation by adding a surfactant, such as sodium dodecylsulfate (Figure 12.46a) to the buffer solution. Sodium dodecylsulfate, (SDS) has a long-chain hydrophobic "tail" and an ionic functional group, providing a negatively charged "head." When the concentration of SDS is sufficiently large, a micelle forms. A **micelle** consists of an agglomeration of 40–100 surfactant molecules in which the hydrocarbon tails point inward, and the negatively charged heads point outward (Figure 12.46b).

Because micelles are negatively charged, they migrate toward the cathode with a velocity less than the electroosmotic flow velocity. Neutral species partition themselves between the micelles and the buffer solution in much the same manner as they do in HPLC. Because there is a partitioning between two phases, the term "chromatography" is used. Note that in MEKC both phases are "mobile."

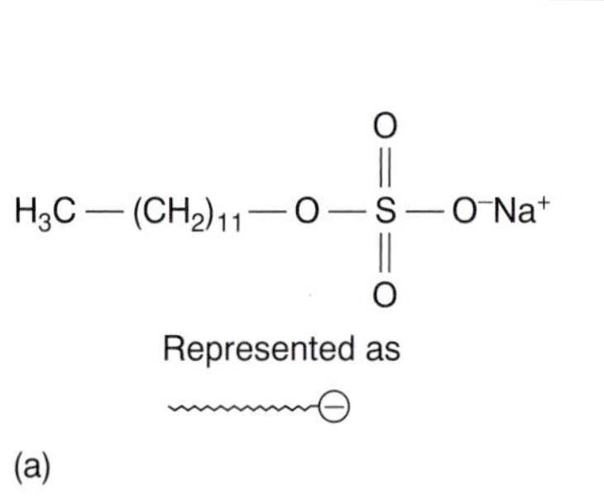

(b)

Figure 12.46
(a) Structure of sodium dodecylsulfate; (b) structure of a micelle.

The elution order for neutral species in MEKC depends on the extent to which they partition into the micelles. Hydrophilic neutrals are insoluble in the micelle's hydrophobic inner environment and elute as a single band as they would in CZE. Neutral solutes that are extremely hydrophobic are completely soluble in the micelle, eluting with the micelles as a single band. Those neutral species that exist in a partition equilibrium between the buffer solution and the micelles elute between the completely hydrophilic and completely hydrophobic neutrals. Those neutral species favoring the buffer solution elute before those favoring the micelles. Micellar electrokinetic chromatography has been used to separate a wide variety of samples, including mixtures of pharmaceutical compounds, vitamins, and explosives.

capillary gel electrophoresis
A form of capillary electrophoresis in which the capillary column contains a gel enabling separations based on size.

Capillary Gel Electrophoresis In **capillary gel electrophoresis** (CGE) the capillary tubing is filled with a polymeric gel. Because the gel is porous, solutes migrate through the gel with a velocity determined both by their electrophoretic mobility and their size. The ability to effect a separation based on size is useful when the solutes have similar electrophoretic mobilities. For example, fragments of DNA of varying length have similar charge-to-size ratios, making their separation by CZE difficult. Since the DNA fragments are of different size, a CGE separation is possible.

The capillary used for CGE is usually treated to eliminate electroosmotic flow, thus preventing the gel's extrusion from the capillary tubing. Samples are injected electrokinetically because the gel provides too much resistance for hydrodynamic sampling. The primary application of CGE is the separation of large biomolecules, including DNA fragments, proteins, and oligonucleotides.

Capillary Electrochromatography Another approach to separating neutral species is **capillary electrochromatography** (CEC). In this technique the capillary tubing is packed with 1.5–3-μm silica particles coated with a bonded, nonpolar stationary phase. Neutral species separate based on their ability to partition between the stationary phase and the buffer solution (which, due to electroosmotic flow, is the mobile phase). Separations are similar to the analogous HPLC separation, but without the need for high-pressure pumps. Furthermore, efficiency in CEC is better than in HPLC, with shorter analysis times.

capillary electrochromatography
A form of capillary electrophoresis in which a stationary phase is included within the capillary column.

12J.4 Representative Method

Although each capillary electrophoretic method has its own unique considerations, the following description of the determination of a vitamin B complex provides an instructive example of a typical procedure.

Representative Methods

Method 12.3 Determination of a Vitamin B Complex by CZE or MEKC[18]

Description of Method. The water-soluble vitamins B_1 (thiamine hydrochloride), B_2 (riboflavin), B_3 (niacinamide), and B_6 (pyridoxine hydrochloride) may be determined by CZE using a pH 9 sodium tetraborate/sodium dihydrogen phosphate buffer or by MEKC using the same buffer with the addition of sodium dodecylsulfate. Detection is by UV absorption at 200 nm. An internal standard of o-ethoxybenzamide is used to standardize the method.

Procedure. A vitamin B complex tablet is crushed and placed in a beaker with 20.00 mL of a 50% v/v methanol solution that is 20 mM in sodium tetraborate and contains 100.0 ppm of o-ethoxybenzamide. After mixing for 2 min to ensure that the B vitamins are dissolved, a 5.00-mL portion is passed through a 0.45-μm filter to remove insoluble binders. An approximately 4-nL sample is loaded into a 50-μm internal diameter capillary column. For CZE the capillary column contains a 20 mM pH 9 sodium tetraborate/sodium dihydrogen phosphate buffer. For MEKC the buffer is also 150 mM in sodium dodecylsulfate. A 40-kV/m electric field is used to effect both the CZE and MEKC separations.

Questions

1. **Methanol, which elutes at 4.69 min, is included as a neutral species to indicate the electroosmotic flow. When using standard solutions of each vitamin, CZE peaks are found at 3.41 min, 4.69 min, 6.31 min, and 8.31 min. Examine the structures and pK_a information in Figure 12.47, and determine the order in which the four B vitamins elute.**

   Vitamin B_1 is a cation and must, therefore, elute before the neutral species methanol; thus it elutes first at 3.41 min. Vitamin B_3 is a neutral species and should elute with methanol at 4.69 min. The remaining two B vitamins are weak acids that partially ionize in the pH 9 buffer. Of the two, vitamin B_6 is the stronger acid and is ionized (as the anion) to a greater extent. Vitamin B_6, therefore, is the last of the vitamins to elute.

—Continued

Continued from page 607

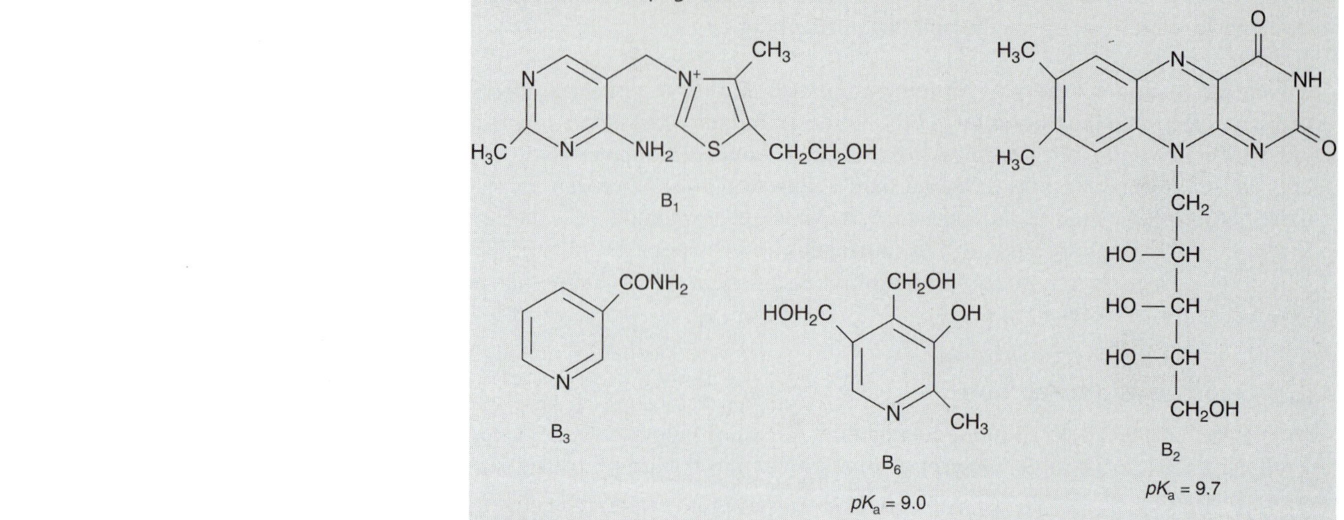

B_1

B_3

B_6

$pK_a = 9.0$

B_2

$pK_a = 9.7$

Figure 12.47
Structures of the vitamins B_1, B_2, B_3, and B_6.

2. **The order of elution when using MEKC is vitamin B_3 (5.58 min), vitamin B_6 (6.59 min), vitamin B_2 (8.81 min), and vitamin B_1 (11.21 min). What conclusions can you make about the solubility of the B vitamins in the sodium dodecylsulfate micelles?**

   The elution time for vitamin B_1 shows the greatest change, increasing from 3.41 min to 11.21 min. Clearly vitamin B_1 has the greatest solubility in the micelles. Vitamins B_2 and B_3 have a more limited solubility in the micelles, showing slightly longer elution times. Interestingly, the elution time for vitamin B_6 decreases in the presence of the micelles.

3. **A quantitative analysis for vitamin B_1 was carried out using this procedure. When a solution of 100.0 ppm B_1 and 100.0 ppm *o*-ethoxybenzamide was analyzed, the peak area for vitamin B_1 was 71% of that for the internal standard. The analysis of a 0.125-g vitamin B complex tablet gave a peak area for vitamin B_1 that was 1.82 times as great as that for the internal standard. How many milligrams of vitamin B_1 are in the tablet?**

   For an internal standardization the relevant equation is

$$\frac{S_A}{S_{IS}} = k\frac{C_A}{C_{IS}}$$

   where S_A and S_{IS} are, respectively, the signals for the analyte and internal standard, and C_A and C_{IS} are their respective concentrations. Making appropriate substitutions for the standard solution

$$\frac{71}{100} = k \times \frac{100.0 \text{ ppm}}{100.0 \text{ ppm}}$$

   gives k as 0.71. Substituting values for the sample

$$\frac{1.82}{1} = 0.71 \times \frac{C_A}{100.0 \text{ ppm}}$$

   gives the concentration of vitamin B_1 as 256 ppm. This is the concentration in the sample as injected. To determine the number of milligrams of vitamin B_1, we must account for the sample's dissolution; thus

$$\frac{256 \text{ mg}}{\text{L}} \times 0.0200 \text{ L} = 5.1 \text{ mg vitamin } B_1$$

12J.5 Evaluation

When compared with GC and HPLC, capillary electrophoresis provides similar levels of accuracy, precision, and sensitivity and a comparable degree of selectivity. The amount of material injected into a capillary electrophoretic column is significantly smaller than that for GC and HPLC; typically 1 nL versus 0.1 μL for capillary GC and 1–100 μL for HPLC. Detection limits for capillary electrophoresis, however, are 100–1000 times poorer than those for GC and HPLC. The most significant advantages of capillary electrophoresis are improvements in separation efficiency, time, and cost. Capillary electrophoretic columns contain substantially more theoretical plates ($\approx 10^6$ plates/m) than that found in HPLC ($\approx 10^5$ plates/m) and capillary GC columns ($\approx 10^3$ plates/m), providing unparalleled resolution and peak capacity. Separations in capillary electrophoresis are fast and efficient. Furthermore, the capillary column's small volume means that a capillary electrophoresis separation requires only a few microliters of buffer solution, compared with 20–30 mL of mobile phase for a typical HPLC separation.

12K KEY TERMS

adjusted retention time (*p. 551*)
band broadening (*p. 553*)
baseline width (*p. 548*)
bleed (*p. 566*)
bonded stationary phase (*p. 580*)
capacity factor (*p. 551*)
capillary column (*p. 562*)
capillary electrochromatography (*p. 607*)
capillary electrophoresis (*p. 597*)
capillary gel electrophoresis (*p. 606*)
capillary zone electrophoresis (*p. 604*)
chromatogram (*p. 548*)
chromatography (*p. 546*)
column chromatography (*p. 546*)
countercurrent extraction (*p. 546*)
cryogenic focusing (*p. 568*)
electrokinetic injection (*p. 603*)
electron capture detector (*p. 570*)
electroosmotic flow (*p. 598*)
electroosmotic flow velocity (*p. 599*)
electropherogram (*p. 597*)
electrophoresis (*p. 597*)
electrophoretic mobility (*p. 598*)
electrophoretic velocity (*p. 598*)
exclusion limit (*p. 595*)
flame ionization detector (*p. 570*)
fronting (*p. 555*)
gas chromatography (*p. 563*)
gas–liquid chromatography (*p. 564*)
gradient elution (*p. 558*)
guard column (*p. 579*)
headspace sampling (*p. 567*)
high-performance liquid chromatography (*p. 578*)
hydrodynamic injection (*p. 602*)
inclusion limit (*p. 593*)
ion-exchange chromatography (*p. 590*)
ion-suppressor column (*p. 592*)
isocratic elution (*p. 582*)
Joule heating (*p. 601*)
Kovat's retention index (*p. 575*)
liquid–solid adsorption chromatography (*p. 590*)
longitudinal diffusion (*p. 560*)
loop injector (*p. 584*)
mass spectrum (*p. 571*)
mass transfer (*p. 561*)
micellar electrokinetic capillary chromatography (*p. 606*)
micelle (*p. 606*)
mobile phase (*p. 546*)
normal-phase chromatography (*p. 580*)
on-column injection (*p. 568*)
open tubular column (*p. 564*)
packed column (*p. 564*)
peak capacity (*p. 554*)
planar chromatography (*p. 546*)
polarity index (*p. 580*)
resolution (*p. 549*)
retention time (*p. 548*)
retention volume (*p. 548*)
reverse-phase chromatography (*p. 580*)
selectivity factor (*p. 552*)
single-column ion chromatography (*p. 593*)
size-exclusion chromatography (*p. 593*)
solid-phase microextraction (*p. 567*)
split injection (*p. 568*)
splitless injection (*p. 568*)
stacking (*p. 603*)
stationary phase (*p. 546*)
supercritical fluid chromatography (*p. 596*)
support-coated open tubular column (*p. 565*)
tailing (*p. 555*)
temperature programming (*p. 558*)
theoretical plate (*p. 553*)
thermal conductivity detector (*p. 569*)
van Deemter equation (*p. 561*)
void time (*p. 549*)
void volume (*p. 549*)
wall-coated open tubular column (*p. 565*)
zeta potential (*p. 599*)

12L SUMMARY

Chromatography and electrophoresis are powerful analytical techniques that can separate a sample into its components while providing a means for determining their concentration. Chromatographic separations utilize the selective partitioning of the sample's components between a stationary phase that is immobilized within a column and a mobile phase that passes through the column.

The effectiveness of a separation is described by the resolution between the chromatographic bands for two components and is a function of the component's capacity factor, the column's efficiency, and the column's selectivity. A component's capacity factor is a measure of the degree to which it successfully partitions into the stationary phase, with larger capacity factors corresponding to more strongly retained components. The column's selectivity for two components is the ratio of the component's capacity factors, providing a relative measure of the column's ability to retain the two components. Column efficiency accounts for those factors that cause a component's chromatographic band to increase in width during the separation. Column efficiency is defined in terms of the number of theoretical plates and the height of a theoretical plate, the latter of which is a function of a number of parameters, most notably the mobile phase's flow rate. Chromatographic separations are optimized by increasing the number of theoretical plates, increasing the column's selectivity, or increasing the component's capacity factors.

In gas chromatography (GC) the mobile phase is an inert gas, and the stationary phase is a nonpolar or polar organic liquid that is either coated on a particulate material and packed into a wide-bore column or coated on the walls of a narrow-bore capillary column. Gas chromatography is useful for the analysis of volatile components.

In high-performance liquid chromatography (HPLC) the mobile phase is either a nonpolar solvent (normal phase) or a polar solvent (reverse phase). A stationary phase of opposite polarity, which is bonded to a particulate material, is packed into a wide-bore column. HPLC can be applied to a wider range of samples than GC; however, the separation efficiency for HPLC is not as good as that for GC.

Together, GC and HPLC account for the largest number of chromatographic separations. Other separation techniques, however, find specialized applications. Of particular importance are: ion-exchange chromatography, which is useful for separating anions and cations; size-exclusion chromatography, which is useful for separating large molecules; and supercritical fluid chromatography, which combines many of the advantages of GC and HPLC for the analysis of materials that are not easily analyzed by either of these methods.

In capillary zone electrophoresis a sample's components are separated based on their ability to move through a conductive medium under the influence of an applied electric field. Because of the effect of electroosmotic flow, positively charged solutes elute first, with smaller, more highly charged cationic solutes eluting before larger cations of lower charge. Neutral species elute without undergoing further separation. Finally, anions elute last, with smaller, more negatively charged anions being the last to elute. By adding a surfactant, neutral species also can be separated by micellar electrokinetic capillary chromatography. Electrophoretic separations also can take advantage of the ability of polymeric gels to separate solutes by size (capillary gel electrophoresis) and the ability of solutes to partition into a stationary phase (capillary electrochromatography). In comparison to GC and HPLC, capillary electrophoresis provides faster and more efficient separations.

12M *Suggested* EXPERIMENTS

Experiments

The following experiments may be used to illustrate the application of chromatography and electrophoresis to a number of different types of samples. Experiments are grouped by the type of technique, and each is briefly annotated.

The first set of experiments describes the application of gas chromatography. These experiments encompass a variety of different types of samples, columns, and detectors. Most experiments may be easily modified to use available equipment and detectors.

Elderd, D. M.; Kildahl, N. K.; Berka, L. H. "Experiments for Modern Introductory Chemistry: Identification of Arson Accelerants by Gas Chromatography," *J. Chem. Educ.* **1996**, *73*, 675–677.

Although aimed at the introductory class, this simple experiment provides a nice demonstration of the use of GC for a qualitative analysis. Students obtain chromatograms for several possible accelerants using headspace sampling and then analyze the headspace over a sealed sample of charred wood to determine the accelerant used in burning the wood. Separations are carried out using a wide-bore capillary column with a stationary phase of methyl 50% phenyl silicone and a flame ionization detector.

—Continued

Experiments

Graham, R. C.; Robertson, J. K. "Analysis of Trihalomethanes in Soft Drinks," *J. Chem. Educ.* **1988,** *65,* 735–737.

Trihalomethanes are extracted from soft drinks using a liquid–liquid extraction with pentane. Samples are analyzed using a packed column containing 20% OV-101 on 80/100 mesh Gaschrom Q equipped with an electron capture detector.

Kegley, S. E.; Hansen, K. J.; Cunningham, K. L. "Determination of Polychlorinated Biphenyls (PCBs) in River and Bay Sediments," *J. Chem. Educ.* **1996,** *73,* 558–562.

This somewhat lengthy experiment provides a thorough introduction to the use of GC for the analysis of trace-level environmental pollutants. Sediment samples are extracted by sonicating with 3 × 100-mL portions of 1:1 acetone:hexane. The extracts are then filtered and concentrated before bringing to a final volume of 10 mL. Samples are analyzed with a capillary column using a stationary phase of 5% phenylmethyl silicone, a splitless injection, and an ECD detector.

Quach, D. T.; Ciszkowski, N. A.; Finlayson-Pitts, B. J. "A New GC-MS Experiment for the Undergraduate Instrumental Analysis Laboratory in Environmental Chemistry: Methyl-*t*-butyl Ether and Benzene in Gasoline," *J. Chem. Educ.* **1998,** *75,* 1595–1598.

This experiment describes the determination of methyl-*t*-butyl ether and benzene in gasoline using the method of standard additions. Two compounds naturally present at high concentration (*o*-xylene and toluene) are used as internal standards to correct for variations in the amount of sample injected into the GC. Because of the complexity of gasoline, single-ion monitoring is used to determine the signals for the analytes and internal standards. Separations are carried out using a capillary column with a stationary phase of 5% diphenyl/95% dimethylsiloxane.

Rice, G. W. "Determination of Impurities in Whiskey Using Internal Standard Techniques," *J. Chem. Educ.* **1987,** *64,* 1055–1056.

An internal standard of 1-butanol is used to determine the concentrations of one or more of the following impurities commonly found in whiskey: acetaldehyde, methanol, ethyl acetate, 1-propanol, 2-methyl-1-propanol, acetic acid, 2-methyl-1-butanol and 3-methyl-1-butanol. A packed column using 5% Carbowax 20m on 80/120 Carbopak B and an FID detector were used.

Rubinson, J. F.; Neyer-Hilvert, J. "Integration of GC-MS Instrumentation into the Undergraduate Laboratory: Separation and Identification of Fatty Acids in Commercial Fats and Oils," *J. Chem. Educ.* **1997,** *74,* 1106–1108.

Fatty acids from commercial fats and oils, such as peanut oil, are extracted with methanolic NaOH and made volatile by derivatizing with a solution of methanol/BF_3. Separations are carried out using a capillary 5% phenylmethyl silicone column with MS detection. By searching the associated spectral library students are able to identify the fatty acids present in their sample. Quantitative analysis is by external standards.

Rudzinski, W. E.; Beu, S. "Gas Chromatographic Determination of Environmentally Significant Pesticides," *J. Chem. Educ.* **1982,** *59,* 614–615.

Students analyze samples of orange juice that have been spiked with diazinon, malathion, and ethion. Samples are extracted with acetonitrile and then extracted with pet ether. The pesticide residues are then purified using an activated magnesium silicate (Florisil) column, eluting the pesticides with mixtures of pet ether and ethyl ether. After removing most of the solvent, samples are analyzed by GC using a packed glass column containing 1.5% SP-2250/1.95% SP-2401 on 100/120 Supelcoport. Both electron capture and flame ionization detection are used.

Welch, W. C.; Greco, T. G. "An Experiment in Manual Multiple Headspace Extraction for Gas Chromatography," *J. Chem. Educ.* **1993,** *70,* 333–335.

The principle of headspace sampling is introduced in this experiment using a mixture of methanol, chloroform, 1,2-dichloroethane, 1,1,1-trichloroethane, benzene, toluene, and *p*-xylene. Directions are given for evaluating the distribution coefficient for the partitioning of a volatile species between the liquid and vapor phase and for its quantitative analysis in the liquid phase. Both packed (OV-101) and capillary (5% phenyl silicone) columns were used. The GC is equipped with a flame ionization detector.

Another experiment with the same focus is Ramachandran, B. R.; Allen, J. M.; Halpern, A. M. "Air-Water Partitioning of Environmentally Important Organic Compounds," *J. Chem. Educ.* **1996,** *73,* 1058–1061.

This experiment provides an alternative approach to measuring the partition coefficient (Henry's law constant) for volatile organic compounds in water. A OV-101 packed column and flame ionization detector are used.

Williams, K. R.; Pierce, R. E. "The Analysis of Orange Oil and the Aqueous Solubility of *d*-Limonene," *J. Chem. Educ.* **1998,** *75,* 223–226.

Two experiments are described in this paper. In the first experiment students determine the %w/w orange oil in a prepared sample by analyzing for *d*-limonene using anisole as an internal standard. Separations are accomplished using

—Continued

Experiments

Continued from page 611

a megabore open tubular column with a 5% phenylmethyl silicone bonded stationary phase and a thermal conductivity detector. In the second experiment the solubility of *d*-limonene is determined by equilibrating different volumes of *d*-limonene with water and measuring the amount of *d*-limonene in the overlying vapor phase using headspace sampling.

Wong, J. W.; Ngim, K. K.; Shibamoto, T.; et al. "Determination of Formaldehyde in Cigarette Smoke," *J. Chem. Educ.* **1997,** *74,* 1100–1103.

Formaldehyde from cigarette smoke is collected by trapping the smoke in a 1-L separatory funnel and extracting into an aqueous solution. To aid in its detection, cysteamine is included in the aqueous extracting solution, leading to the formation of a thiazolidine derivative. Samples are analyzed using a DB-1 capillary column with a thermionic or flame photometric detector. Directions also are given for using an HPLC. Formaldehyde is derivatized using 2,4-dinitrophenylhydrazine, and samples are analyzed using a C_{18} column with a UV detector set to 365 nm.

Yang, M. J.; Orton, M. L., Pawliszyn, J. "Quantitative Determination of Caffeine in Beverages Using a Combined SPME-GC/MS Method," *J. Chem. Educ.* **1997,** *74,* 1130–1132.

Caffeine in coffee, tea, and soda is determined by a solid-phase microextraction using an uncoated silica fiber, followed by a GC analysis using a capillary SPB-5 column with an MS detector. Standard solutions are spiked with $^{13}C_3$ caffeine as an internal standard.

The second set of experiments describes the application of high-performance liquid chromatography. These experiments encompass a variety of different types of samples and a variety of common detectors.

Bidlingmeyer, B. A.; Schmitz, S. "The Analysis of Artificial Sweeteners and Additives in Beverages by HPLC," *J. Chem. Educ.* **1991,** *68,* A195–A200.

The concentrations of benzoic acid, aspartame, caffeine, and saccharin in a variety of beverages are determined in this experiment. A C_{18} column and a mobile phase of 80% v/v acetic acid (pH = 4.2) and 20% v/v methanol are used to effect the separation. A UV detector set to 254 nm is used to measure the eluent's absorbance. The ability to adjust retention times by changing the mobile phase's pH is also explored.

DiNunzio, J. E. "Determination of Caffeine in Beverages by High Performance Liquid Chromatography," *J. Chem. Educ.* **1985,** *62,* 446–447.

The concentration of caffeine in a typical serving of coffee and soda is determined in this experiment. Separations are achieved using a C_{18} column with a mobile phase of 30% v/v methanol in water, with UV detection at a wavelength of 254 nm.

Ferguson, G. K. "Quantitative HPLC Analysis of an Analgesic/Caffeine Formulation: Determination of Caffeine," *J. Chem. Educ.* **1998,** *75,* 467–469.

The %w/w caffeine in an analgesic formulation is determined in this experiment. The separation uses a C_{18} column with a mobile phase of 94.1% v/v water, 5.5% v/v acetonitrile, 0.2% v/v triethylamine, and 0.2% v/v acetic acid. A UV detector is set to 254 nm.

Ferguson, G. K. "Quantitative HPLC Analysis of a Psychotherapeutic Medication: Simultaneous Determination of Amitriptyline Hydrochloride and Perphenazine," *J. Chem. Educ.* **1998,** *75,* 1615–1618.

This experiment describes a quantitative analysis for the active ingredients in a prescription antipsychotic medication. The separation makes use of a cyanopropyl derivatized column and a mobile phase of 70% v/v acetonitrile, 5% v/v methanol, and 25% v/v 0.1 M aqueous KH_2PO_4. A UV detector set to 215 nm is used to measure the eluent's absorbance.

Haddad, P.; Hutchins, S.; Tuffy, M. "High Performance Liquid Chromatography of Some Analgesic Compounds," *J. Chem. Educ.* **1983,** *60,* 166–168.

This experiment focuses on developing an HPLC separation capable of distinguishing acetylsalicylic acid, paracetamol, salicylamide, caffeine, and phenacetin. A C_{18} column and UV detection are used to obtain chromatograms. Solvent parameters used to optimize the separation include the pH of the buffered aqueous mobile phase, the %v/v methanol added to the aqueous mobile phase, and the use of tetrabutylammonium phosphate as an ion-pairing reagent.

Mueller, B. L.; Potts, L. W. "HPLC Analysis of an Asthma Medication," *J. Chem. Educ.* **1988,** *65,* 905–906.

This experiment describes the quantitative analysis of the asthma medication Quadrinal for the active ingredients theophylline, salicylic acid, phenobarbital, ephedrine HCl, and potassium iodide. Separations are carried out using a C_{18} column with a mobile phase of 19% v/v acetonitrile, 80% v/v water, and 1% acetic acid. A small amount of triethylamine (0.03% v/v) is included to ensure the elution of ephedrine HCl. A UV detector set to 254 nm is used to record the chromatogram.

—Continued

Experiments

Remcho, V. T.; McNair, H. M.; Rasmussen, H. T. "HPLC Method Development with the Photodiode Array Detector," *J. Chem. Educ.* **1992,** *69,* A117–A119.

A mixture of methyl paraben, ethyl paraben, propyl paraben, diethyl phthalate, and butyl paraben is separated by HPLC. This experiment emphasizes the development of a mobile-phase composition capable of separating the mixture. A photodiode array detector demonstrates the coelution of the two compounds.

Siturmorang, M.; Lee, M. T. B.; Witzeman, L. K.; et al. "Liquid Chromatography with Electrochemical Detection (LC-EC): An Experiment Using 4-Aminophenol," *J. Chem. Educ.* **1998,** *75,* 1035–1038.

The use of an amperometric detector is emphasized in this experiment. Hydrodynamic voltammetry (see Chapter 11) is first performed to identify a potential for the oxidation of 4-aminophenol without an appreciable background current due to the oxidation of the mobile phase. The separation is then carried out using a C_{18} column and a mobile phase of 50% v/v pH 5, 20 mM acetate buffer with 0.02 M $MgCl_2$, and 50% v/v methanol. The analysis is easily extended to a mixture of 4-aminophenol, ascorbic acid, and catechol, and to the use of a UV detector.

Tran, C. D.; Dotlich, M. "Enantiomeric Separation of Beta-Blockers by High Performance Liquid Chromatography," *J. Chem. Educ.* **1995,** *72,* 71–73.

This experiment introduces the use of a chiral column (a β-cyclodextrin-bonded C_{18} column) to separate the beta-blocker drugs Inderal LA (*S*-propranolol and *R*-propranolol), Tenormim (DL-atenolol) and Lopressor (DL-metaprolol). The mobile phase was 90:10 (v/v) acetonitrile and water. A UV detector set to 254 nm is used to obtain the chromatogram.

Van Arman, S. A.; Thomsen, M. W. "HPLC for Undergraduate Introductory Laboratories," *J. Chem. Educ.* **1997,** *74,* 49–50.

In this experiment students analyze an artificial RNA digest consisting of cytidine, uridine, thymidine, guanosine, and adenosine using a C_{18} column and a mobile phase of 0.4% v/v triethylammonium acetate, 5% v/v methanol, and 94.6% v/v water. The chromatogram is recorded using a UV detector at a wavelength of 254 nm.

Wingen, L. M.; Low, J. C.; Finlayson-Pitts, B. J. "Chromatography, Absorption, and Fluorescence: A New Instrumental Analysis Experiment on the Measurement of Polycyclic Aromatic Hydrocarbons in Cigarette Smoke," *J. Chem. Educ.* **1998,** *75,* 1599–1603.

The analysis of cigarette smoke for 16 different polyaromatic hydrocarbons is described in this experiment. Separations are carried out using a polymeric bonded silica column with a mobile phase of 50% v/v water, 40% v/v acetonitrile, and 10% v/v tetrahydrofuran. A notable feature of this experiment is the evaluation of two means of detection. The ability to improve sensitivity by selecting the optimum excitation and emission wavelengths when using a fluorescence detector is demonstrated. A comparison of fluorescence detection with absorbance detection shows that better detection limits are obtained when using fluorescence.

The third set of experiments provides a few representative applications of ion chromatography.

Bello, M. A.; Gustavo González, A. "Determination of Phosphate in Cola Beverages Using Nonsuppressed Ion Chromatography," *J. Chem. Educ.* **1996,** *73,* 1174–1176.

In this experiment phosphate is determined by single-column, or nonsuppressed, ion chromatography using an anionic column and a conductivity detector. The mobile phase is a mixture of *n*-butanol, acetonitrile, and water (containing sodium gluconate, boric acid, and sodium tetraborate).

Kieber, R. J.; Jones, S. B. "An Undergraduate Laboratory for the Determination of Sodium, Potassium, and Chloride," *J. Chem. Educ.* **1994,** *71,* A218–A222.

Three techniques, one of which is ion chromatography, are used to determine the concentrations of three ions in solution. The combined concentrations of Na^+ and K^+ are determined by an ion exchange with H^+, the concentration of which is subsequently determined by an acid–base titration using NaOH. Flame atomic absorption is used to measure the concentration of Na^+, and K^+ is determined by difference. The concentration of Cl^- is determined by ion-exchange chromatography on an anionic column using a mobile phase of HCO_3^- and CO_3^{2-} with ion suppression. A conductivity detector is used to record the chromatogram.

Koubek, E.; Stewart, A. E. "The Analysis of Sulfur in Coal," *J. Chem. Educ.* **1992,** *69,* A146–A148.

Sulfur in coal is converted into a soluble sulfate by heating to 800 °C in the presence of MgO and Na_2CO_3. After dissolving in water, the concentration of sulfate is determined by single-column ion chromatography, using an anionic column and a mobile phase of 1 mM potassium hydrogen phthalate. A conductivity detector is used to record the chromatogram.

Luo, P.; Luo, M. A.; Baldwin, R. P. "Determination of Sugars in Food Products," *J. Chem. Educ.,* **1993,** *70,* 679–681.

—Continued

Experiments

Continued from page 613

The concentrations of nine sugars (fucose, methylglucose, arabinose, glucose, fructose, lactose, sucrose, cellobiose, and maltose) in beer, milk, and soda are determined using an anionic column and a mobile phase of 0.1 M NaOH. Detection is by amperometry at a Cu working electrode.

The last set of experiments provides examples of the application of capillary electrophoresis. These experiments encompass a variety of different types of samples and include examples of capillary zone electrophoresis and micellar electrokinetic chromatography.

Conradi, S.; Vogt, C.; Rohde, E. "Separation of Enatiomeric Barbiturates by Capillary Electrophoresis Using a Cyclodextrin-Containing Run Buffer," *J. Chem. Educ.* **1997,** *74,* 1122–1125.

In this experiment the enantiomers of cyclobarbital and thiopental, and phenobarbital are separated using MEKC with cyclodextran as a chiral selector. By adjusting the pH of the buffer solution and the concentration and type of cyclodextran, students are able to find conditions in which the enantiomers of cyclobarbital and thiopental are resolved.

Conte, E. D.; Barry, E. F.; Rubinstein, H. "Determination of Caffeine in Beverages by Capillary Zone Electrophoresis," *J. Chem. Educ.* **1996,** *73,* 1169–1170.

Caffeine in tea and coffee is determined by CZE using nicotine as an internal standard. The buffer solution is 50 mM sodium borate adjusted to pH 8.5 with H_3PO_4. A UV detector set to 214 nm is used to record the electropherograms.

Hage, D. S.; Chattopadhyay, A.; Wolfe, C. A. C.; et al. "Determination of Nitrate and Nitrite in Water by Capillary Electrophoresis," *J. Chem. Educ.* **1998,** *75,* 1588–1590.

In this experiment the concentrations of NO_2^- and NO_3^- are determined by CZE using IO_4^- as an internal standard. The buffer solution is 0.60 M sodium acetate buffer adjusted to a pH of 4.0. A UV detector set to 222 nm is used to record the electropherogram.

Janusa, M. A.; Andermann, L. J.; Kliebert, N. M.; et al. "Determination of Chloride Concentration Using Capillary Electrophoresis," *J. Chem. Educ.* **1998,** *75,* 1463–1465.

Directions are provided for the determination of chloride in samples using CZE. The buffer solution includes pyromellitic acid which allows the indirect determination of chloride by monitoring absorbance at 250 nm.

McDevitt, V. L.; Rodríguez, A.; Williams, K. R. "Analysis of Soft Drinks: UV Spectrophotometry, Liquid Chromatography, and Capillary Electrophoresis," *J. Chem. Educ.* **1998,** *75,* 625–629.

Caffeine, benzoic acid, and aspartame in soft drinks are analyzed by three methods. Using several methods to analyze the same sample provides students with the opportunity to compare results with respect to accuracy, volume of sample required, ease of performance, sample throughput, and detection limit.

Thompson, L.; Veening, H.; Strain, T. G. "Capillary Electrophoresis in the Undergraduate Instrumental Analysis Laboratory: Determination of Common Analgesic Formulations," *J. Chem. Educ.* **1997,** *74,* 1117–1121.

Students determine the concentrations of caffeine, acetaminophen, acetylsalicylic acid, and salicylic acid in several analgesic preparations using both CZE (70 mM borate buffer solution, UV detection at 210 nm) and HPLC (C_{18} column with 3% v/v acetic acid mixed with methanol as a mobile phase, UV detection at 254 nm).

Vogt, C.; Conradi, S.; Rhode, E. "Determination of Caffeine and Other Purine Compounds in Food and Pharmaceuticals by Micellar Electrokinetic Chromatography," *J. Chem. Educ.* **1997,** *74,* 1126–1130.

This experiment describes a quantitative analysis for caffeine, theobromine, and theophylline in tea, pain killers, and cocoa. Separations are accomplished by MEKC using a pH 8.25 borate–phosphate buffer with added SDS. A UV detector set to 214 nm is used to record the electropherogram. An internal standard of phenobarbital is included for quantitative work.

Weber, P. L.; Buck, D. R. "Capillary Electrophoresis: A Fast and Simple Method for the Determination of the Amino Acid Composition of Proteins," *J. Chem. Educ.* **1994,** *71,* 609–612.

This experiment describes a method for determining the amino acid composition of cyctochrome c and lysozyme. The proteins are hydrolyzed in acid, and an internal standard of α-aminoadipic acid is added. Derivatization with naphthalene-2,3-dicarboxaldehyde gives derivatives that absorb at 420 nm. Separation is by MEKC using a buffer solution of 50 mM SDS in 20 mM sodium borate.

12N PROBLEMS

1. The following data were obtained for four compounds separated on a 20-m capillary column.

Compound	t_r (min)	w (min)
A	8.04	0.15
B	8.26	0.15
C	8.43	0.16

(a) Calculate the number of theoretical plates for each compound and the average number of theoretical plates for the column. (b) Calculate the average height of a theoretical plate. (c) Explain why it is possible for each compound to have a different number of theoretical plates.

2. Using the data from Problem 1, calculate the resolution and selectivity factors for each pair of adjacent compounds. For resolution, use both equations 12.1 and 12.21, and compare your results. Discuss how you might improve the resolution between compounds B and C. The retention time for an unretained solute is 1.19 min.

3. Using the chromatogram shown here, which was obtained on a 2-m column, determine values for t_r, w, t_r', k', N, and H.

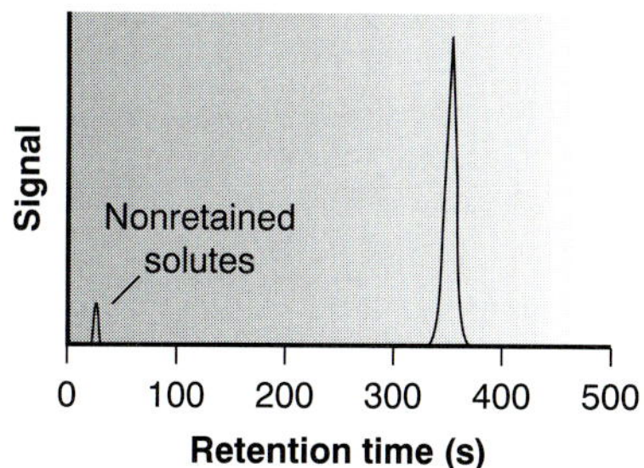

4. Using the partial chromatogram shown here, determine the resolution between the two solute bands.

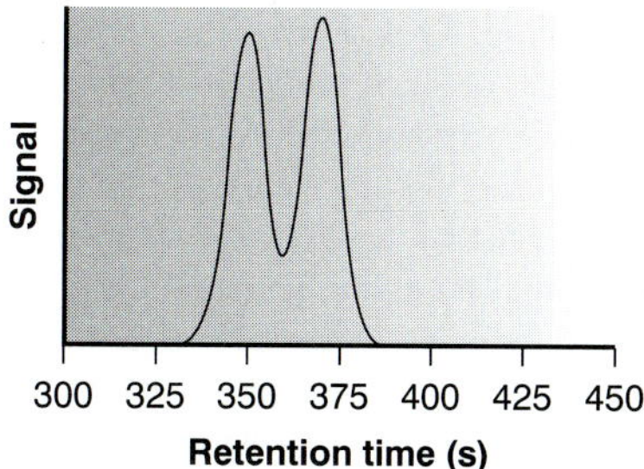

5. The chromatogram in Problem 4 was obtained on a 2-m column with a column dead time of 50 s. How long a column is needed to achieve a resolution of 1.5? What height of a theoretical plate is needed to achieve a resolution of 1.5 without increasing the length of the column?

6. Complete the following table.

N_B	α	k_B'	R
100,000	1.05	0.50	
10,000	1.10		1.50
10,000		4.0	1.00
	1.05	3.0	1.75

7. Moody[19] studied the efficiency of a GC separation of 2-butanone on a dinonyl phthalate column. Evaluating the plate height as a function of flow rate gave a van Deemter equation for which A is 1.65 mm, B is 25.8 mm · mL min^{-1}, and C is 0.0236 mm · min mL^{-1}. (a) Prepare a graph of H versus u for flow rates in the range of 5–120 mL/min. (b) For what range of flow rates does each term in the van Deemter equation have the greatest effect? (c) What are the optimum flow rate and the height of a theoretical plate at that flow rate? (d) For open tubular columns the A term is no longer needed. If the B and C terms remain unchanged, what are the optimum flow rate and the height of a theoretical plate at that flow rate? (e) How many more theoretical plates will there be in the open tubular column compared with the packed column? (f) Equation 12.28 is written in terms of the linear velocity (centimeters per second), yet we have evaluated it in this problem using the flow rate (milliliters per minute). Why can we do this?

8. Hsieh and Jorgenson[20] prepared 12–33-μm HPLC columns packed with 5.44-μm spherical stationary phase particles. To evaluate these columns they measured reduced plate height, h,

$$h = \frac{H}{d_p}$$

as a function of reduced flow rate, v,

$$v = \frac{ud_p}{D_m}$$

where d_p is the particle diameter, and D_m is the solute's diffusion coefficient in the mobile phase. The data were analyzed using van Deemter plots, with a portion of their results summarized in the following table for the solute norepinephrine.

Column Internal Diameter (μm)	A	B	C
33	0.63	1.32	0.10
33	0.67	1.30	0.08
23	0.40	1.34	0.09
23	0.58	1.11	0.09
17	0.31	1.47	0.09
17	0.40	1.41	0.11
12	0.22	1.53	0.11
12	0.19	1.27	0.1[illegible]

(a) Construct separate van Deemter plots using the data in the first and last rows for flow rates in the range 0.7–15. Determine the optimum flow rate and plate height for each case, given $d_p = 5.44$ μm and $D_m = 6.23 \times 10^{-6}$ cm^2 s^{-1}. (b) The *A* term in the van Deemter equation appears to be strongly correlated with the column's inner diameter, with smaller diameter columns providing smaller values of *A*. Explain why this effect is seen (*Hint:* Consider how many particles can fit across a capillary of each diameter).

9. Method 12.1 describes the analysis of the trihalomethanes $CHCl_3$, $CHBr_3$, $CHCl_2Br$, and $CHClBr_2$ in drinking water using a packed column with a nonpolar stationary phase. Predict the order in which these four trihalomethanes will elute.

10. A mixture of *n*-heptane, tetrahydrofuran, 2-butanone, and *n*-propanol elutes in this order when using a polar stationary phase such as Carbowax. The elution order is exactly the opposite when using a nonpolar stationary phase such as polydimethyl siloxane. Explain the order of elution in each case.

11. The analysis of trihalomethanes in drinking water is described in Method 12.1. A single standard gives the following results when carried through the described procedure.

Trihalomethane	Concentration (ppb)	Peak Area
$CHCl_3$	1.30	1.35×10^4
$CHCl_2Br$	0.90	6.12×10^4
$CHClBr_2$	4.00	1.71×10^4
$CHBr_3$	1.20	1.52×10^4

Analysis of water from a drinking fountain gives areas of 1.56×10^4, 5.13×10^4, 1.49×10^4, and 1.76×10^4 for $CHCl_3$, $CHCl_2Br$, $CHClBr_2$, and $CHBr_3$, respectively. Determine the concentration of each of the trihalomethanes in the sample of water.

12. Zhou and colleagues determined the %w/w H_2O in methanol by GC, using a capillary column coated with a nonpolar stationary phase and a thermal conductivity detector.[21] A series of calibration standards gave the following results.

% w/w H_2O	Peak Height (arb. units)
0.00	1.15
0.0145	2.74
0.0472	6.33
0.0951	11.58
0.1757	20.43
0.2901	32.97

(a) What is the %w/w H_2O in a sample giving a peak height of 8.63? (b) The %w/w H_2O in a freeze-dried antibiotic is determined in the following manner. A 0.175-g sample is placed in a vial along with 4.489 g of methanol. Water in the vial extracts into the methanol. Analysis of the sample gave a peak height of 13.66. What is the %w/w H_2O in the antibiotic?

13. Loconto and co-workers describe a method for determining trace levels of water in soil.[22] The method takes advantage of the reaction of water with calcium carbide, CaC_2, to produce acetylene gas, C_2H_2. By carrying out the reaction in a sealed vial, the amount of acetylene produced may be determined by sampling the headspace. In a typical analysis a sample of soil is placed in a sealed vial with CaC_2. Analysis of the headspace gave a blank-corrected signal of 2.70×10^5. A second sample is prepared in the same manner except that a standard addition of 5.0 mg H_2O/g solid is added, giving a blank-corrected signal of 1.06×10^6. Determine the number of milligrams of H_2O/g soil in the soil sample.

14. Van Atta and Van Atta used gas chromatography to determine the %v/v methyl salicylate in rubbing alcohol.[23] A set of standard additions was prepared by transferring 20.00 mL of rubbing alcohol to separate 25-mL volumetric flasks and pipeting 0.00 mL, 0.20 mL, and 0.50 mL of methyl salicylate to the flasks. All three flasks were then diluted to volume using isopropanol. Analysis of the three samples gave peak heights for methyl salicylate of 57.00 mm, 88.5 mm, and 132.5 mm, respectively. Determine the %v/v methyl salicylate in the rubbing alcohol.

15. The amount of camphor in an analgesic ointment can be determined by GC using the method of internal standards.[24] A standard sample was prepared by placing 45.2 mg of camphor and 2.00 mL of a 6.00 mg/mL internal standard solution of terpene hydrate in a 25-mL volumetric flask and diluting to volume with CCl_4. When an approximately 2-μL sample of the standard was injected, the FID signals for the two components were measured (in arbitrary units) as 67.3 for camphor and 19.8 for terpene hydrate. A 53.6-mg sample of an analgesic ointment was prepared for analysis by placing it in a 50-mL Erlenmeyer flask along with 10 mL of CCl_4. After heating to 50 °C in a water bath, the sample was cooled to below room temperature and filtered. The residue was washed with two 5-mL portions of CCl_4, and the combined filtrates were collected in a 25-mL volumetric flask. After adding 2.00 mL of the internal standard solution, the contents of the flask were diluted to volume with CCl_4. Analysis of an approximately 2-μL sample gave FID signals of 13.5 for the terpene hydrate and 24.9 for the camphor. Report the %w/w camphor in the analgesic ointment.

16. The concentration of pesticide residues on agricultural products, such as oranges, may be determined by GC-MS.[25] Pesticide residues are extracted from the sample using methylene chloride, and the concentrations of the extracted pesticides are concentrated by evaporating the methylene chloride to a smaller volume. Calibration is accomplished using anthracene-d_{10} as an internal standard. In a study to determine the parts per billion of heptachlor epoxide on oranges, a 50.0-g sample of orange rinds was chopped and extracted with 50.00 mL of methylene chloride. After removing any insoluble material by filtration, the methylene chloride was reduced in volume, spiked with a known amount of the internal standard, and diluted to 10 mL in a volumetric flask. Analysis of the sample gives a peak–area ratio (A_{anal}/$A_{int\ stan}$) of 0.108. A series of calibration standards, each containing the same amount of anthracene-d_{10} as the sample, give the following results.

ppb Heptachlor Epoxide	$A_{anal}/A_{int\ stan}$
20.0	0.065
60.0	0.153
200.0	0.637
500.0	1.554
1000.0	3.198

Report the concentration of heptachlor epoxide residue (in nanograms per gram) on the oranges.

17. The adjusted retention times for octane, toluene, and nonane on a particular GC column are 15.98 min, 17.73 min, and 20.42 min, respectively. What is the retention index for all three compounds?

18. The following data were collected for a series of normal alkanes using a stationary phase of Carbowax 20M.

Alkane	t_r' (min)
pentane	0.79
hexane	1.99
heptane	4.47
octane	14.12
nonane	33.11

What is the retention index for a compound whose adjusted retention time is 9.36 min?

19. The following data have been reported for the gas chromatographic analysis of *p*-xylene and methylisobutylketone (MIBK) on a capillary column.[8]

Injection Mode	Compound	t_r (min)	Peak Area	Peak Width (min)
split	MIBK	1.878	54285	0.028
	p-xylene	5.234	123483	0.044
splitless	MIBK	3.420	2493005	1.057
	p-xylene	5.795	3396656	1.051

Explain the difference in the retention times, the peak areas, and the peak widths when switching from a split injection to a splitless injection.

20. Otto and Wegscheider report the following capacity factors for the reverse phase separation of 2-aminobenzoic acid on a C_{18} column when using 10% v/v methanol as a mobile phase.[26]

pH	k'
2.0	10.5
3.0	16.7
4.0	15.8
5.0	8.0
6.0	2.2
7.0	1.8

Explain the changes in capacity factor.

21. Haddad and associates report the following capacity factors for the reverse-phase separation of salicylamide (k'_{sal}) and caffeine (K'_{caff}).[27]

%v/v methanol	30%	35%	40%	45%	50%	55%
k'_{sal}	2.4	1.6	1.6	1.0	0.7	0.7
k'_{caff}	4.3	2.8	2.3	1.4	1.1	0.9

Explain the changes in capacity factor. What is the advantage of using a mobile phase with a smaller %v/v methanol? Are there any disadvantages?

22. Suppose that you are to separate a mixture of benzoic acid, aspartame, and caffeine in a diet soda. The following information is available to you.

	t_r in Aqueous Mobile Phase Buffered to a pH of			
Compound	3.0	3.5	4.0	4.5
benzoic acid	7.4	7.0	6.9	4.4
aspartame	5.9	6.0	7.1	8.1
caffeine	3.6	3.7	4.1	4.4

(a) Explain the change in retention time for each compound. (b) Plot retention time versus pH for each compound on the same graph, and identify a pH level that will yield an acceptable separation.

23. The composition of a multivitamin tablet is conveniently determined using an HPLC with a diode array UV/Vis detector. A 5-μL standard sample containing 170 ppm vitamin C, 130 ppm niacin, 120 ppm niacinamide, 150 ppm pyridoxine, 60 ppm thiamine, 15 ppm folic acid, and 10 ppm riboflavin is injected into the HPLC, giving signals (in arbitrary units) of, respectively, 0.22, 1.35, 0.90, 1.37, 0.82, 0.36, and 0.29. The multivitamin tablet is prepared for analysis by grinding into a powder and transferring to a 125-mL Erlenmeyer flask containing 10 mL of 1% v/v NH_3 in dimethyl sulfoxide. After sonicating in an ultrasonic bath for 2 min, 90 mL of 2% acetic acid is added, and the mixture is stirred for 1 min and sonicated at 40 °C for 5 min. The extract is then filtered through a 0.45-μm membrane filter. Injection of a 5-μL sample into the HPLC gives signals of 0.87 for vitamin C, 0.00 for niacin, 1.40 for niacinamide, 0.22 for pyridoxine, 0.19 for thiamine, 0.11 for folic acid, and 0.44 for riboflavin. Report the number of milligrams of each vitamin present in the tablet.

24. The amount of caffeine in an analgesic tablet was determined by HPLC using a normal calibration curve. Standard solutions of caffeine were prepared and analyzed using a 10-μL fixed-volume injection loop. Results for the standards are summarized in the following table.

Concentration of Standards (ppm)	Signal (arbitrary units)
50.0	8354
100.0	16925
150.0	25218
200.0	33584
250.0	42002

The sample was prepared by placing a single analgesic tablet in a small beaker and adding 10 mL of methanol. After allowing the sample to dissolve, the contents of the beaker, including the insoluble binder, were quantitatively transferred to a 25-mL volumetric flask and diluted to volume with methanol. The sample was then filtered, and a 1.00-mL aliquot was transferred to a 10-mL volumetric flask and diluted to volume with methanol. When analyzed by HPLC, the signal for the caffeine was found to be 21469. Report the number of milligrams of caffeine in the analgesic tablet.

25. Kagel and Farwell report a reverse-phase HPLC method for determining the concentration of acetylsalicylic acid (ASA) and caffeine (CAF) in analgesic tablets using salicylic acid (SA) as an internal standard.[28] A series of standards was prepared by adding known amounts of acetylsalicylic acid and caffeine to 250-mL Erlenmeyer flasks and adding 100 mL of methanol. A 10.00-mL aliquot of a standard solution of salicylic acid was then added to each. The following results are obtained for a typical set of standard solutions.

Standard	Milligrams ASA	Milligrams CAF	Peak Height Ratio ASA/SA	Peak Height Ratio CAF/SA
1	200.0	20.0	20.5	10.6
2	250.0	40.0	25.1	23.0
3	300.0	60.0	30.9	36.8

A sample of an analgesic tablet was placed in a 250-mL Erlenmeyer flask and dissolved in 100 mL of methanol. After adding a 10.00-mL portion of the internal standard, the solution was filtered. Analysis of the sample gave a peak height ratio of 23.2 for ASA and 17.9 for CAF. (a) Determine the number of milligrams ASA and CAF in the tablet. (b) Why was it necessary to filter the sample? (c) The directions indicate that approximately 100 mL of methanol is used to dissolve the standards and samples. Why is it not necessary to measure this volume more precisely? (d) In the presence of moisture, ASA decomposes to SA and acetic acid. What complication might this present for this analysis? How might you evaluate whether this is a problem?

26. Bohman and colleagues described a reverse-phase HPLC method for the quantitative analysis of vitamin A in food using the method of standard additions.[29] In a typical example, a 10.067-g sample of cereal is placed in a 250-mL Erlenmeyer flask along with 1 g of sodium ascorbate, 40 mL of ethanol, and 10 mL of 50% w/v KOH. After refluxing for 30 min, 60 mL of ethanol is added, and the solution is cooled to room temperature. Vitamin A is extracted using three 100-mL portions of hexane. The combined portions of hexane are evaporated, and the residue containing vitamin A is transferred to a 5-mL volumetric flask and diluted to volume with methanol. A standard addition is prepared in a similar manner using a 10.093-g sample of the cereal and spiking it with 0.0200 mg of vitamin A. Injecting the sample and standard addition into the HPLC gives peak areas of 6.77×10^3 and 1.32×10^4, respectively. Report the vitamin A content of the sample in milligrams/100 g cereal.

27. Ohta and Tanaka reported a method for the simultaneous analysis of several inorganic anions and the cations Mg^{2+} and Ca^{2+} in water by ion-exchange chromatography.[30] The mobile phase includes 1,2,4-benzenetricarboxylate, which absorbs strongly at 270 nm. Indirect detection of the analytes is possible because their presence in the detector leads to a decrease in absorbance. Unfortunately, Ca^{2+} and Mg^{2+}, which are present at high concentrations in many environmental waters, form stable complexes with 1,2,4-benzenetricarboxylate that interfere with the analysis. (a) Adding EDTA to the mobile phase eliminates the interference caused by Ca^{2+} and Mg^{2+}; explain why. (b) A standard solution containing 1.0 M $NaHCO_3$, 0.20 mM $NaNO_2$, 0.20 mM $MgSO_4$, 0.10 mM $CaCl_2$, and 0.10 mM $Ca(NO_3)_2$ gives the following typical peak areas (arbitrary units).

Ion	HCO_3^-	Cl^-	NO_2^-	NO_3^-
Peak Area	373.5	322.5	264.8	262.7
Ion	Ca^{2+}	Mg^{2+}	SO_4^{2-}	
Peak Area	458.9	352.0	341.3	

Analysis of a river water sample (pH of 7.49) gives the following results.

Ion	HCO_3^-	Cl^-	NO_2^-	NO_3^-
Peak Area	310.0	403.1	3.97	157.6
Ion	Ca^{2+}	Mg^{2+}	SO_4^{2-}	
Peak Area	734.3	193.6	324.3	

Determine the concentration of each ion in the sample of rain water. (c) The detection of HCO_3^- actually gives the total concentration of carbonate in solution ($[CO_3^{2-}] + [HCO_3^-] + [H_2CO_3]$). Given that the pH of the water is 7.49, what is the actual concentration of HCO_3^-? (d) An independent analysis gives the following additional concentrations.

$$[Na^+] = 0.60 \text{ mM} \qquad [NH_4^+] = 0.014 \text{ mM} \qquad [K^+] = 0.046 \text{ mM}$$

A solution's ionic balance is defined as the ratio of the total cation charge to the total anion charge. Determine the ion balance for this sample of water, and comment on whether the result is reasonable.

28. The concentrations of Cl^-, NO_3^-, and SO_4^{2-} may be determined by ion chromatography. A 50-μL standard sample of 10.0-ppm Cl^-, 2.00-ppm NO_3^-, and 5.00-ppm SO_4^{2-} gave signals (in arbitrary units) of 59.3, 16.1, and 6.08, respectively. A sample of effluent from a wastewater treatment plant was diluted tenfold, and a 50-μL portion gave signals of 44.2 for

Cl^-, 2.73 for NO_3^-, and 5.04 for SO_4^{2-}. Report the parts per million of each anion in the effluent sample.

29. A series of polyvinylpyridine standards of different molecular weight were analyzed by size-exclusion chromatography, yielding the following results.

Formula Weight	Retention Volume (mL)
600,000	6.42
100,000	7.98
20,000	9.30
3,000	10.94

When a preparation of polyvinylpyridine of unknown formula weight was analyzed the retention volume was found to be 8.45. Report the average formula weight for the preparation.

30. Diet soft drinks contain appreciable quantities of aspartame, benzoic acid, and caffeine. What is the expected order of elution for these compounds in a capillary zone electrophoresis separation using a pH 9.4 buffer solution, given that aspartame has pK_a values of 2.964 and 7.37, benzoic acid's pK_a is 4.2, and the pK_a for caffeine is less than 0.

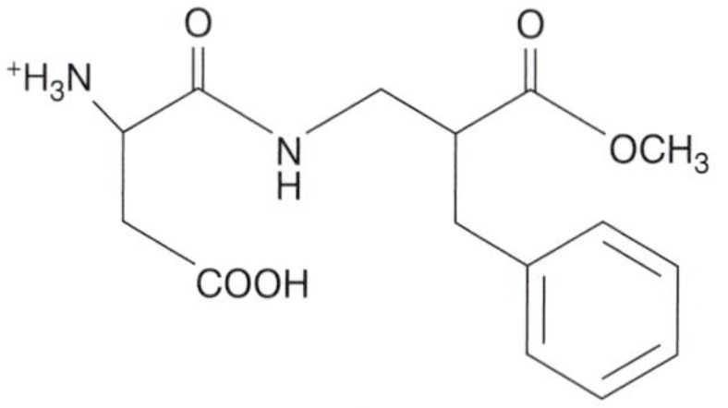

aspartame

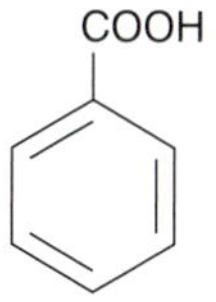

benzoic acid

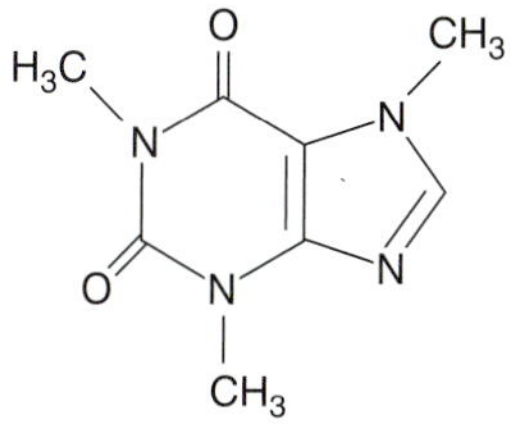

caffeine

31. Janusa and co-workers report the determination of chloride by CZE.[31] Analysis of a series of external standards gives the following calibration curve.

$$\text{Area} = -883 + 5590(\text{ppm Cl}^-)$$

A standard sample of 57.22% w/w Cl^- was analyzed by placing 0.1011 g in a 100-mL volumetric flask and diluting to volume. Three unknowns were prepared by pipeting 0.250 mL, 0.500 mL, and 0.750 mL of the bulk unknown into separate 50-mL volumetric flasks and diluting to volume. Analysis of the three unknowns gave areas of 15310, 31546, and 47582, respectively. Evaluate the accuracy of this analysis.

32. The analysis of NO_3^- in aquarium water was carried out by CZE using IO_4^- as an internal standard. A standard solution of 15.0-ppm NO_3^- and 10.0-ppm IO_4^- gives peak heights (arbitrary units) of 95.0 and 100.1, respectively. A sample of water from an aquarium is diluted 1:100, and sufficient internal standard added to make its concentration 10.0 ppm. Analysis gives signals of 29.2 and 105.8 for NO_3^- and IO_4^-, respectively. Report the parts per million of NO_3^- in the sample of aquarium water.

33. Suggest conditions for separating a mixture of 2-aminobenzoic acid (pK_{a1} = 2.08, pK_{a2} = 4.96), benzylamine (pK_a = 9.35), and 4-methylphenol (pK_a = 10.26) by capillary zone electrophoresis.

34. McKillop and associates have examined the electrophoretic separation of alkylpyridines by CZE.[32] Separations were carried out using either 50-μm or 75-μm inner diameter capillaries, with a total length of 57 cm and a length of 50 cm from the point of injection to the detector. The run buffer was a pH 2.5 lithium phosphate buffer. Separations were achieved using an applied voltage of 15 kV. The electroosmotic flow velocity, as measured using a neutral marker, was found to be 6.398×10^{-5} cm² V⁻¹ s⁻¹. The diffusion coefficient, *D*, for the alkylpyridines may be taken to be 1.0×10^{-5} cm² s⁻¹.(a) Calculate the electrophoretic mobility for 2-ethylpyridine, given that its elution time is 8.20 min. (b) How many theoretical plates are there for 2-ethylpyridine? (c) The electrophoretic mobilities for 3-ethylpyridine and 4-ethylpyridine are 3.366×10^{-4} cm² V⁻¹ s⁻¹ and 3.397×10^{-4} cm² V⁻¹ s⁻¹, respectively. What is the expected resolution between these two alkylpyridines? (d) Explain the trends in electrophoretic mobility shown in the following table.

Alkylpyridine	μ_{ep} (cm² V⁻¹ s⁻¹)
2-methylpyridine	3.581×10^{-4}
2-ethylpyridine	3.222×10^{-4}
2-propylpyridine	2.923×10^{-4}
2-pentylpyridine	2.534×10^{-4}
2-hexylpyridine	2.391×10^{-4}

(e) Explain the trends in electrophoretic mobility shown in the following table.

Alkylpyridine	μ_{ep} (cm² V⁻¹ s⁻¹)
2-ethylpyridine	3.222×10^{-4}
3-ethylpyridine	3.366×10^{-4}
4-ethylpyridine	3.397×10^{-4}

(f) The pK_a for pyridine is 5.229. At a pH of 2.5 the electrophoretic mobility of pyridine is 4.176×10^{-4} cm² V⁻¹ s⁻¹. What is the expected electrophoretic mobility if the run buffer's pH is 7.5?

12O SUGGESTED READINGS

The following texts provide a good introduction to the broad field of separations, including chromatography and electrophoresis.

Giddings, J. C. *Unified Separation Science.* Wiley-Interscience: New York, 1991.

Karger, B. L.; Snyder, L. R.; Harvath, C. *An Introduction to Separation Science.* Wiley-Interscience: New York, 1973.

Miller, J. M. *Separation Methods in Chemical Analysis.* Wiley-Interscience: New York, 1975.

A more recent discussion of peak capacity is presented in the following paper.

Shen, Y.; Lee, M. "General Equation for Peak Capacity in Column Chromatography," *Anal. Chem.* **1998,** *70,* 3853–3856.

The following references may be consulted for more information on gas chromatography.

Grob, R. L., ed. *Modern Practice of Gas Chromatography.* Wiley-Interscience: New York, 1972.

Hinshaw, J. V. "A Compendium of GC Terms and Techniques," *LC•GC* **1992,** *10,* 516–522.

Ioffe, B. V.; Vitenberg, A. G. *Head-Space Analysis and Related Methods in Gas Chromatography.* Wiley-Interscience: New York, 1982.

Kitson, F. G.; Larsen, B. S.; McEwen, C. N. *Gas Chromatography and Mass Spectrometry: A Practical Guide.* Academic Press: San Diego, 1996.

The following references may be consulted for more information on high-performance liquid chromatography.

Dorschel, C. A.; Ekmanis, J. L.; Oberholtzer, J. E.; et al. "LC Detectors," *Anal. Chem.* **1989,** *61,* 951A–968A.

Simpson, C. F., ed. *Techniques in Liquid Chromatography.* Wiley-Hayden: Chichester, England, 1982.

Snyder, L. R.; Glajch, J. L.; Kirkland, J. J. *Practical HPLC Method Development.* Wiley-Interscience: New York, 1988.

The following references may be consulted for more information on ion chromatography.

Shpigun, O. A.; Zolotov, Y. A. *Ion Chromatography in Water Analysis.* Ellis Horwood: Chichester, England, 1988.

Smith, F. C. Jr.; Chang, R. C. *The Practice of Ion Chromatography.* Wiley-Interscience: New York, 1983.

The following references may be consulted for more information on supercritical fluid chromatography.

Palmieri, M. D. "An Introduction to Supercritical Fluid Chromatography. Part I: Principles and Applications," *J. Chem. Educ.* **1988,** *65,* A254–A259.

Palmieri, M. D. "An Introduction to Supercritical Fluid Chromatography. Part II: Applications and Future Trends," *J. Chem. Educ.* **1989,** *66,* A141–A147.

The following references may be consulted for more information on capillary electrophoresis.

Baker, D. R. *Capillary Electrophoresis.* Wiley-Interscience: New York, 1995.

Copper, C. L. "Capillary Electrophoresis: Part I. Theoretical and Experimental Background," *J. Chem. Educ.* **1998,** *75,* 343–347.

Copper, C. L.; Whitaker, K. W. "Capillary Electrophoresis: Part II. Applications," *J. Chem. Educ.* **1998,** *75,* 347–351.

The application of spreadsheets and computer programs for modeling chromatography is described in the following papers.

Abbay, G. N.; Barry, E. F.; Leepipatpiboon, S.; et al. "Practical Applications of Computer Simulation for Gas Chromatography Method Development," *LC•GC* **1991,** *9,* 100–114.

Drouen, A.; Dolan, J. W.; Snyder, L. R.; et al. "Software for Chromatographic Method Development," *LC•GC* **1991,** *9,* 714–724.

Kevra, S. A.; Bergman, D. L.; Maloy, J. T. "A Computational Introduction to Chromatographic Bandshape Analysis," *J. Chem. Educ.* **1994,** *71,* 1023–1028.

Sundheim, B. R. "Column Operations: A Spreadsheet Model." *J. Chem. Educ.* **1992,** *69,* 1003–1005.

12P REFERENCES

1. Craig, L. C. *J. Biol. Chem.* **1944,** *155,* 519.
2. Martin, A. J. P.; Synge, R. L. M. *Biochem. J.* **1941,** *35,* 1358.
3. Giddings, J. C. *Unified Separation Science.* Wiley-Interscience: New York, 1991.
4. Hawkes, S. J. *J. Chem. Educ.* **1983,** *60,* 393–398.
5. Kennedy, R. T.; Jorgenson, J. W. *Anal. Chem.* **1989,** *61,* 1128–1135.
6. Hinshaw, J. V. *LC•GC,* **1993,** *11,* 644–648.
7. Grob, K. *Anal. Chem.* **1994,** *66,* 1009A–1019A.
8. Marriott, P. J.; Carpenter, P. D. *J. Chem. Educ.* **1996,** *73,* 96–99.
9. Method 6232B as published in *Standard Methods for the Examination of Water and Wastewater,* 18th ed. American Public Health Association: Washington, D.C., 1992.
10. Novotny, M. *Science* **1989,** *246,* 51–57.
11. (a) Jorgenson, J. W.; Guthrie, E. J. *J. Chromatog.* **1983,** *255,* 335; (b) Kennedy, R. T.; Oates, M. D.; Cooper, B. R.; et al. *Science* **1989,** *246,* 57–63.

12. Snyder, L. R.; Glajch, J. L.; Kirkland, J. J. *Practical HPLC Method Development.* Wiley-Interscience: New York, 1988.
13. Foley, J. P. *Chromatography* **1987,** *2(6),* 43–51.
14. Yeung, E. S. *LC•GC* **1989,** *7,* 118–128.
15. Smyth, W. F. *Analytical Chemistry of Complex Matrices.* Wiley Teubner: Chichester, England, 1996, pp. 187–189.
16. Baker, D. R. *Capillary Electrophoresis.* Wiley-Interscience: New York, 1995.
17. Jones, W. R.; Jandik, P. *J. Chromatog.* **1992,** *608,* 385–393.
18. Smyth, W. F. *Analytical Chemistry of Complex Matrices.* Wiley Teubner: Chichester, England, 1996, pp. 154–156.
19. Moody, H. W. *J. Chem. Educ.* **1982,** *59,* 218–219.
20. Hsieh, S.; Jorgenson, J. W. *Anal. Chem.* **1996,** *68,* 1212–1217.
21. Zhou, X.; Hines, P. A.; White, K. C.; et al. *Anal. Chem.* **1998,** *70,* 390–394.
22. Loconto, P. R.; Pan, Y. L.; Voice, T. C. *LC•GC* **1996,** *14,* 128–132.
23. Van Atta, R. E.; Van Atta, R. L. *J. Chem. Educ.* **1980,** *57,* 230–231.
24. Pant, S. K.; Gupta, P. N.; Thomas, K. M.; et al. *LC•GC* **1990,** *8,* 322–325.
25. Feigel, C. *Varian GC/MS Application Note,* Number 22.
26. Otto, M.; Wegscheider, W. *J. Chromatogr.* **1983,** *258,* 11–22.
27. Haddad, P.; Hutchins, S.; Tuffy, M. *J. Chem. Educ.* **1983,** *60,* 166–168.
28. Kagel, R. A.; Farwell, S. O. *J. Chem. Educ.* **1983,** *60,* 163–166.
29. Bohman, O.; Engdahl, K. A.; Johnsson, H. *J. Chem. Educ.* **1982,** *59,* 251–252.
30. Ohta, K.; Tanaka, K. *Anal. Chim. Acta* **1998,** *373,* 189–195.
31. Janusa, M. A.; Andermann, L. J.; Kliebert, N. M.; et al. *J. Chem. Educ.* **1998,** *75,* 1463–1465.
32. McKillop, A. G.; Smith, R. M.; Rowe, R. C.; et al. *Anal. Chem.* **1999,** *71,* 497–503.

Chapter 13

Kinetic Methods of Analysis

A system under thermodynamic control is in a state of equilibrium, and its signal has a constant, or steady-state value (Figure 13.1a). When a system is under kinetic control, however, its signal changes with time (Figure 13.1b) until equilibrium is established. Thus far, the techniques we have considered have involved measurements made when the system is at equilibrium.

By changing the time at which measurements are made, an analysis can be carried out under either thermodynamic control or kinetic control. For example, one method for determining the concentration of NO_2^- in groundwater involves the diazotization reaction shown in Figure 13.2.[1] The final product, which is a reddish-purple azo dye, absorbs visible light at a wavelength of 543 nm. Since the concentration of dye is determined by the amount of NO_2^- in the original sample, the solution's absorbance can be used to determine the concentration of NO_2^-. The reaction in the second step, however, is not instantaneous. To achieve a steady-state signal, such as that in Figure 13.1a, the absorbance is measured following a 10-min delay. By measuring the signal during the 10-min development period, information about the rate of the reaction is obtained. If the reaction's rate is a function of the concentration of NO_2^-, then the rate also can be used to determine its concentration in the sample.[2]

There are many potential advantages to kinetic methods of analysis, perhaps the most important of which is the ability to use chemical reactions that are slow to reach equilibrium. In this chapter we examine three techniques that rely on measurements made while the analytical system is under kinetic rather than thermodynamic control: chemical kinetic techniques, in which the rate of a chemical reaction is measured; radiochemical techniques, in which a radioactive element's rate of nuclear decay is measured; and flow injection analysis, in which the analyte is injected into a continuously flowing carrier stream, where its mixing and reaction with reagents in the stream are controlled by the kinetic processes of convection and diffusion.

13A Methods Based on Chemical Kinetics

The earliest examples of analytical methods based on chemical kinetics, which date from the late nineteenth century, took advantage of the catalytic activity of enzymes. Typically, the enzyme was added to a solution containing a suitable substrate, and the reaction between the two was monitored for a fixed time. The enzyme's activity was determined by measuring the amount of substrate that had reacted. Enzymes also were used in procedures for the quantitative analysis of hydrogen peroxide and carbohydrates. The application of catalytic reactions continued in the first half of the twentieth century, and developments included the use of nonenzymatic catalysts, noncatalytic reactions, and differences in reaction rates when analyzing samples with several analytes.

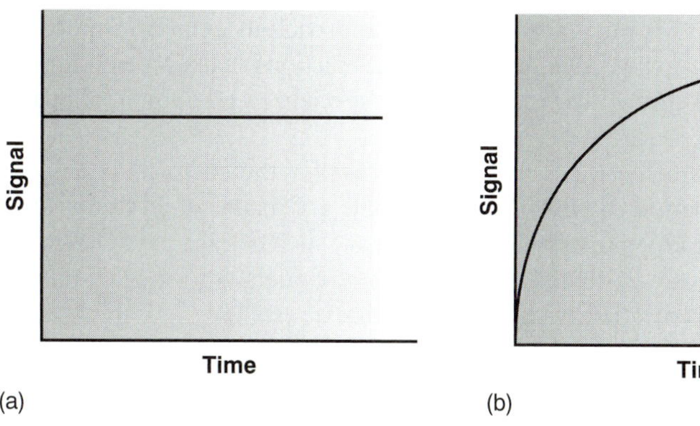

Figure 13.1
Plot of signal versus time for an analytical system that is under (a) thermodynamic control; and (b) under kinetic control.

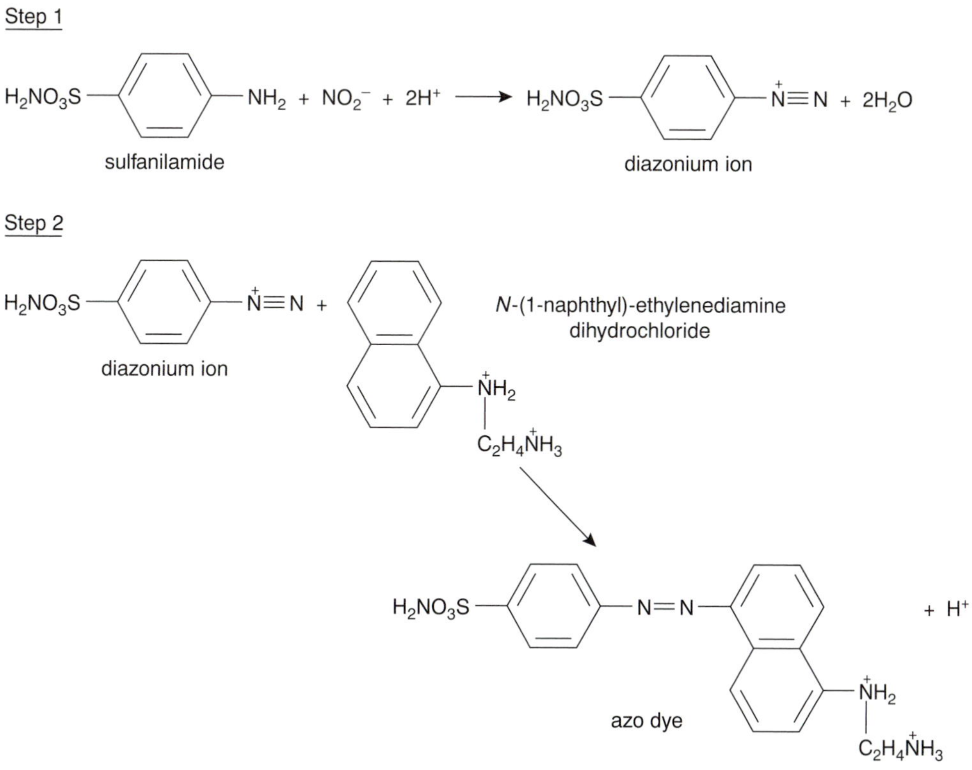

Step 1

H_2NO_3S–(sulfanilamide)–NH_2 + NO_2^- + $2H^+$ ⟶ H_2NO_3S–(diazonium ion)–$\overset{+}{N}{\equiv}N$ + $2H_2O$

Step 2

H_2NO_3S–(diazonium ion)–$\overset{+}{N}{\equiv}N$ + *N*-(1-naphthyl)-ethylenediamine dihydrochloride (–$\overset{+}{N}H_2$–$C_2H_4\overset{+}{N}H_3$) ⟶ H_2NO_3S–(azo dye)–$N{=}N$–(naphthyl)–$\overset{+}{N}H_2$–$C_2H_4\overset{+}{N}H_3$ + H^+

Figure 13.2
Analytical scheme for the analysis of NO_2^- in groundwater.

Despite the variety of methods that had been developed, by 1960 kinetic methods were no longer in common use. The principal limitation to a broader acceptance of chemical kinetic methods was their greater susceptibility to errors from uncontrolled or poorly controlled variables, such as temperature and pH, and the presence of interferents that activate or inhibit catalytic reactions. Many of these limitations, however, were overcome during the 1960s, 1970s, and 1980s with the development of improved instrumentation and data analysis methods compensating for these errors.[3]

13A.1 Theory and Practice

rate
The change in a property's value per unit change in time; the rate of a reaction is a change in concentration per unit change in time.

rate law
An equation relating a reaction's rate at a given time to the concentrations of species affecting the rate.

rate constant
In a rate law, the proportionality constant between a reaction's rate and the concentrations of species affecting the rate (k).

Every chemical reaction occurs at a finite rate and, therefore, can potentially serve as the basis for a chemical kinetic method of analysis. To be effective, however, the chemical reaction must meet three conditions. First, the **rate** of the chemical reaction must be fast enough that the analysis can be conducted in a reasonable time, but slow enough that the reaction does not approach its equilibrium position while the reagents are mixing. As a practical limit, reactions reaching equilibrium within 1 s are not easily studied without the aid of specialized equipment allowing for the rapid mixing of reactants.

A second requirement is that the **rate law** for the chemical reaction must be known for the period in which measurements are made. In addition, the rate law should allow the kinetic parameters of interest, such as rate constants and concentrations, to be easily estimated. For example, the rate law for a reaction that is first order in the concentration of the analyte, A, is expressed as

$$\text{rate} = -\frac{d[\text{A}]}{dt} = k[\text{A}] \qquad \textbf{13.1}$$

where k is the reaction's **rate constant.** As shown in Appendix 5,* the integrated form of this rate law

$$\ln\,[\text{A}]_t = \ln\,[\text{A}]_0 - kt \qquad \text{or} \qquad [\text{A}]_t = [\text{A}]_0 e^{-kt} \qquad \textbf{13.2}$$

provides a simple mathematical relationship between the rate constant, the reaction's elapsed time, t, the initial concentration of analyte, $[\text{A}]_0$, and the analyte's concentration at time t, $[\text{A}]_t$.

Unfortunately, most reactions of analytical interest do not follow the simple rate laws shown in equations 13.1 and 13.2. Consider, for example, the following reaction between an analyte, A, and a reagent, R, to form a product, P

$$\text{A} + \text{R} \underset{k_b}{\overset{k_f}{\rightleftharpoons}} \text{P}$$

where k_f is the rate constant for the forward reaction, and k_b is the rate constant for the reverse reaction. If the forward and reverse reactions occur in single steps, then the rate law is

$$\text{Rate} = k_f[\text{A}][\text{R}] - k_b[\text{P}] \qquad \textbf{13.3}$$

Although the rate law for the reaction is known, there is no simple integrated form. We can simplify the rate law for the reaction by restricting measurements to the

*Appendix 5 provides a general review of kinetics.

beginning of the reaction when the product's concentration is negligible. Under these conditions, the second term in equation 13.3 can be ignored; thus

$$\text{Rate} = k_f[\text{A}][\text{R}] \qquad \textbf{13.4}$$

The integrated form of the rate law for equation 13.4, however, is still too complicated to be analytically useful. We can simplify the kinetics, however, by carefully adjusting the reaction conditions.[4] For example, pseudo-first-order kinetics can be achieved by using a large excess of R (i.e. $[\text{R}]_0 >> [\text{A}]_0$), such that its concentration remains essentially constant. Under these conditions

$$\text{Rate} = -\frac{d[\text{A}]}{dt} = k[\text{R}]_0[\text{A}] = k'[\text{A}] \qquad \textbf{13.5}$$

$$\ln[\text{A}]_t = \ln[\text{A}]_0 - k't \qquad \text{or} \qquad [\text{A}]_t = [\text{A}]_0 e^{-k't} \qquad \textbf{13.6}$$

It may even be possible to adjust conditions such that measurements are made under pseudo-zero-order conditions where

$$\text{Rate} = -\frac{d[\text{A}]}{dt} = k[\text{A}]_0[\text{R}]_0 = k'' \qquad \textbf{13.7}$$

$$[\text{A}]_t = [\text{A}]_0 - k''t \qquad \textbf{13.8}$$

A final requirement for a chemical kinetic method of analysis is that it must be possible to monitor the reaction's progress by following the change in concentration for one of the reactants or products as a function of time. Which species is used is not important; thus, in a quantitative analysis the rate can be measured by monitoring the analyte, a reagent reacting with the analyte, or a product. For example, the concentration of phosphate can be determined by monitoring its reaction with Mo(VI) to form 12-molybdophosphoric acid (12-MPA).

$$H_3PO_4 + 6\text{Mo(VI)} + 9H_2O \rightarrow \text{12-MPA} + 9H_3O^+ \qquad \textbf{13.9}$$

We can monitor the progress of this reaction by coupling it to a second reaction in which 12-MPA is reduced to form heteropolyphosphomolybdenum blue, PMB,

$$\text{12-MPA} + n\text{Red} \rightarrow \text{PMB} + n\text{Ox}$$

where Red is a suitable reducing agent, and Ox is its conjugate form.[5,6] The rate of formation of PMB is measured spectrophotometrically and is proportional to the concentration of 12-MPA. The concentration of 12-MPA, in turn, is proportional to the concentration of phosphate. Reaction 13.9 also can be followed spectrophotometrically by monitoring the formation of 12-MPA.[6,7]

Classifying Chemical Kinetic Methods A useful scheme for classifying chemical kinetic methods of analysis is shown in Figure 13.3.[3] Methods are divided into two main categories. For those methods identified as direct-computation methods, the concentration of analyte, $[\text{A}]_0$, is calculated using the appropriate rate law. Thus, for a first-order reaction in A, equation 13.2 is used to determine $[\text{A}]_0$, provided that values for *k*, *t*, and $[\text{A}]_t$ are known. With a curve-fitting method, regression is used to find the best fit between the data (e.g., $[\text{A}]_t$ as a function of time) and the known mathematical model for the rate law. In this case, kinetic parameters, such as *k* and $[\text{A}]_0$, are adjusted to find the best fit. Both categories are further subdivided into rate methods and integral methods.

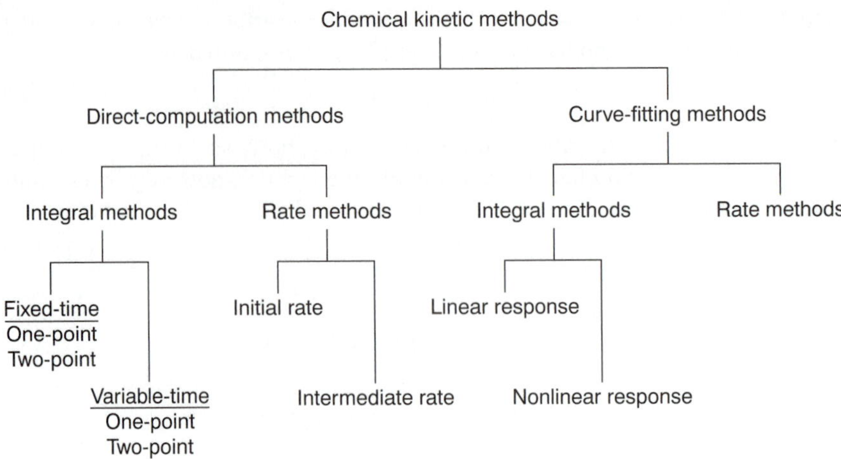

Figure 13.3
Classification of chemical kinetic methods of analysis.

Direct-Computation Integral Methods Integral methods for analyzing kinetic data make use of the integrated form of the rate law. In the one-point fixed-time integral method, the concentration of analyte is determined at a single time. The initial concentration of analyte, $[A]_0$, is calculated using equation 13.2, 13.6, or 13.8, depending on whether the reaction follows first-order, pseudo-first-order, or pseudo-zero-order kinetics. The rate constant for the reaction is determined in a separate experiment using a standard solution of analyte. Alternatively, the analyte's initial concentration can be determined using a calibration curve consisting of a plot of $[A]_t$ for several standard solutions of known $[A]_0$.

Example

EXAMPLE 13.1

The concentration of nitromethane, CH_3NO_2, can be determined from the kinetics of its decomposition in basic solution. In the presence of excess base the reaction is pseudo-first-order in nitromethane. For a standard solution of 0.0100 M nitromethane, the concentration of nitromethane after 2.00 s was found to be 4.24×10^{-4} M. When a sample containing an unknown amount of nitromethane was analyzed, the concentration remaining after 2.00 s was found to be 5.35×10^{-4} M. What is the initial concentration of nitromethane in the sample?

SOLUTION

The value for the pseudo-first-order rate constant is determined by solving equation 13.6 for k' and making appropriate substitutions; thus

$$k' = \frac{\ln[A]_0 - \ln[A]_t}{t} = \frac{\ln(0.0100) - \ln(4.24 \times 10^{-4})}{2.00 \text{ s}} = 1.58 \text{ s}^{-1}$$

Equation 13.6 can then be solved for the initial concentration of nitromethane. This is easiest to do using the exponential form of equation 13.6.

$$[A]_t = [A]_0 e^{-k't}$$

$$[A]_0 = \frac{[A]_t}{e^{-k't}} = \frac{5.35 \times 10^{-4} \text{ M}}{e^{-(1.58 \text{ s}^{-1})(2.00 \text{ s})}} = 0.0126 \text{ M}$$

In Example 13.1 the initial concentration of analyte is determined by measuring the amount of unreacted analyte at a fixed time. Sometimes it is more convenient to measure the concentration of a reagent reacting with the analyte or the concentration of one of the reaction's products. The one-point fixed-time integral method can still be applied if the stoichiometry is known between the analyte and the species being monitored. For example, if the concentration of the product in the reaction

$$A + R \rightarrow P$$

is monitored, then the concentration of the analyte at time t is

$$[A]_t = [A]_0 - [P]_t \qquad \textbf{13.10}$$

since the stoichiometry between the analyte and product is 1:1. Substituting equation 13.10 into equation 13.6 gives

$$\ln([A]_0 - [P]_t) = \ln [A]_0 - k't \qquad \textbf{13.11}$$

which is simplified by writing in exponential form

$$[A]_0 - [P]_t = [A]_0 e^{-k't}$$

and solving for $[A]_0$.

$$[A]_0 = \frac{[P]_t}{1 - e^{-k't}} \qquad \textbf{13.12}$$

Example

EXAMPLE 13.2

The concentration of thiocyanate, SCN^-, can be determined from the pseudo-first-order kinetics of its reaction with excess Fe^{3+} to form a reddish colored complex of $Fe(SCN)^{2+}$. The reaction's progress is monitored by measuring the absorbance of $Fe(SCN)^{2+}$ at a wavelength of 480 nm. When a standard solution of 0.100 M SCN^- is used, the concentration of $Fe(SCN)^{2+}$ after 10.0 s is found to 0.0516 M. The analysis of a sample containing an unknown amount of SCN^- results in a concentration of $Fe(SCN)^{2+}$ of 0.0420 M after 10.0 s. What is the initial concentration of SCN^- in the sample?

SOLUTION

The pseudo-first-order rate constant is determined by solving equation 13.11 for k' and making appropriate substitutions

$$k' = \frac{\ln[A]_0 - \ln([A]_0 - [P]_t)}{t} = \frac{\ln(0.1) - \ln(0.1 - 0.0516)}{10.0 \text{ s}} = 0.0726 \text{ s}^{-1}$$

Equation 13.12 then can be used to determine the initial concentration of SCN^-.

$$[A]_0 = \frac{[P]_t}{1 - e^{-k't}} = \frac{0.0420 \text{ M}}{1 - e^{-(0.0726 \text{ s}^{-1})(10.0 \text{ s})}} = 0.0814 \text{ M}$$

The one-point fixed-time integral method has the advantage of simplicity since only a single measurement is needed to determine the analyte's initial concentration. As with any method relying on a single determination, however, a one-point fixed-time integral method cannot compensate for constant sources of determinate error. Such corrections can be made by making measurements at two points in time and using the difference between the measurements to determine the analyte's initial concentration. Constant sources of error affect both measurements equally, thus, the difference between the measurements is independent of these errors. For a two-point fixed-time integral method, in which the concentration of analyte for a pseudo-first-order reaction is measured at times t_1 and t_2, we can write

$$[A]_{t_1} = [A]_0 e^{-k't_1} \qquad \textbf{13.13}$$

$$[A]_{t_2} = [A]_0 e^{-k't_2}$$

Subtracting the second equation from the first equation and solving for $[A]_0$ gives

$$[A]_0 = \frac{[A]_{t_1} - [A]_{t_2}}{e^{-k't_1} - e^{-k't_2}} \qquad \textbf{13.14}$$

The rate constant for the reaction can be calculated from equation 13.14 by measuring $[A]_{t_1}$ and $[A]_{t_2}$ for a standard solution of analyte. The analyte's initial concentration also can be found using a calibration curve consisting of a plot of $([A]_{t_1} - [A]_{t_2})$ versus $[A]_0$.

Fixed-time integral methods are advantageous for systems in which the signal is a linear function of concentration. In this case it is not necessary to determine the concentration of the analyte or product at times t_1 or t_2, because the relevant concentration terms can be replaced by the appropriate signal. For example, when a pseudo-first-order reaction is followed spectrophotometrically, when Beer's law

$$(\text{Abs})_t = \varepsilon b[A]_t$$

is valid, equations 13.6 and 13.14 can be rewritten as

$$(\text{Abs})_t = [A]_0(e^{-k't})\varepsilon b = c[A]_0$$

$$[A]_0 = \frac{(\text{Abs})_{t_1} - (\text{Abs})_{t_2}}{e^{-k't_1} - e^{-k't_2}} \times \frac{1}{\varepsilon b} = c[(\text{Abs})_{t_1} - (\text{Abs})_{t_2}]$$

where $(\text{Abs})_t$ is the absorbance at time t, and c is a constant.

An alternative to a fixed-time method is a variable-time method, in which we measure the time required for a reaction to proceed by a fixed amount. In this case the analyte's initial concentration is determined by the elapsed time, Δt, with a higher concentration of analyte producing a smaller Δt. For this reason variable-time integral methods are appropriate when the relationship between the detector's response and the concentration of analyte is not linear or is unknown. In the one-point variable-time integral method, the time needed to cause a desired change in concentration is measured from the start of the reaction. With the two-point variable-time integral method, the time required to effect a change in concentration is measured.

One important application of the variable-time integral method is the quantitative analysis of catalysts, which is based on the catalyst's ability to increase the rate of a reaction. As the initial concentration of catalyst is increased, the time needed to reach the desired extent of reaction decreases. For many catalytic systems the relationship between the elapsed time, Δt, and the initial concentration of analyte is

$$\frac{1}{\Delta t} = F_{\text{cat}}[A]_0 + F_{\text{uncat}}$$

where F_{cat} and F_{uncat} are constants that are functions of the rate constants for the catalyzed and uncatalyzed reactions, and the extent of the reaction during the time span Δt.[8]

Example

EXAMPLE 13.3

Sandell and Kolthoff[9] developed a quantitative method for iodide based on its catalytic effect on the following redox reaction.

$$As^{3+} + 2Ce^{4+} \rightarrow As^{5+} + 2Ce^{3+}$$

Standards were prepared by adding a known amount of KI to fixed amounts of As^{3+} and Ce^{4+} and measuring the time for all the Ce^{4+} to be reduced. The following results were obtained:

Micrograms I^-	Δt (min)
5.0	0.9
2.5	1.8
1.0	4.5

How many micrograms of I^- are in a sample for which Δt is found to be 3.2 min?

SOLUTION

The relationship between the concentration of I^- and Δt is shown by the calibration curve in Figure 13.4, for which

$$\frac{1}{\Delta t} = -8.67 \times 10^{-9} + 0.222(\mu g\ I^-)$$

Substituting 3.2 min for Δt in the preceding equation gives 1.4 μg as the amount of I^- originally present in the sample.

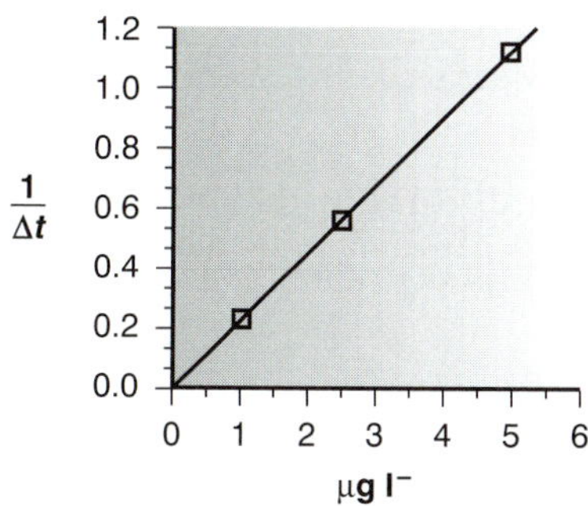

Figure 13.4
Calibration curve for the variable-time integral determination of I^-.

Direct-Computation Rate Methods Rate methods for analyzing kinetic data are based on the differential form of the rate law. The rate of a reaction at time t, $(\text{rate})_t$, is determined from the slope of a curve showing the change in concentration for a reactant or product as a function of time (Figure 13.5). For a reaction that is first-order, or pseudo-first-order in analyte, the rate at time t is given as

$$(\text{rate})_t = k[\text{A}]_t$$

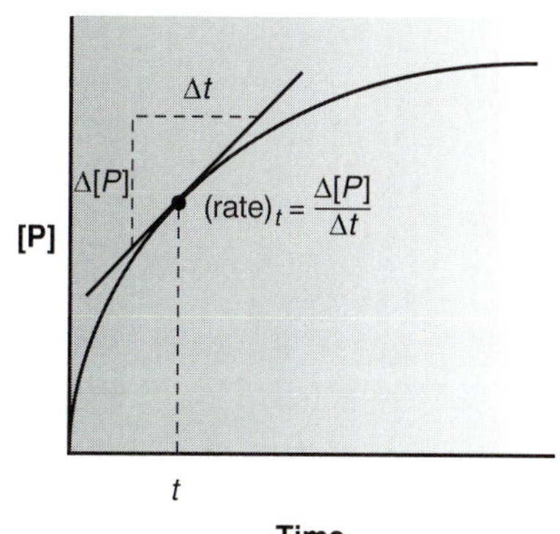

Figure 13.5
Determination of reaction rate from a tangent line at time t.

Substituting an equation similar to 13.13 into the preceding equation gives the following relationship between the rate at time t and the analyte's initial concentration.

$$(\text{rate})_t = k[\text{A}]_0 e^{-kt}$$

If the rate is measured at a fixed time, then both k and e^{-kt} are constant, and a calibration curve of $(\text{rate})_t$ versus $[\text{A}]_0$ can be used for the quantitative analysis of the analyte.

The use of the initial rate ($t = 0$) has the advantage that the rate is at its maximum, providing an improvement in sensitivity. Furthermore, the initial rate is measured under pseudo-zero-order conditions, in which the change in concentration with time is effectively linear, making the determination of slope easier. Finally, when using the initial rate, complications due to competing reactions are avoided. One disadvantage of the initial rate method is that there may be insufficient time for a complete mixing of the reactants. This problem is avoided by using a rate measured at an intermediate time ($t > 0$).

Example

EXAMPLE 13.4

The concentration of aluminum in serum can be determined by adding 2-hydroxy-1-naphthaldehyde *p*-methoxybenzoyl-hydrazone and measuring the initial rate of the resulting complexation reaction under pseudo-first-order conditions.[10] The rate of reaction is monitored by the fluorescence of the metal–ligand complex. Initial rates, with units of emission intensity per second, were measured for a set of standard solutions, yielding the following results

$[Al^{3+}]$ (μM)	0.300	0.500	1.00	3.00
$(\text{rate})_{t=0}$	0.261	0.599	1.44	4.82

A serum sample treated in the same way as the standards has an initial rate of 0.313 emission intensity/s. What is the concentration of aluminum in the serum sample?

SOLUTION

A calibration curve of emission intensity per second versus the concentration of Al^{3+} (Figure 13.6) is a straight line, where

$$(\text{rate})_{t=0} = 1.69 \times [\text{Al}^{3+} (\mu\text{M})] - 0.246$$

Substituting the sample's initial rate into the calibration equation gives an aluminum concentration of 0.331 μM.

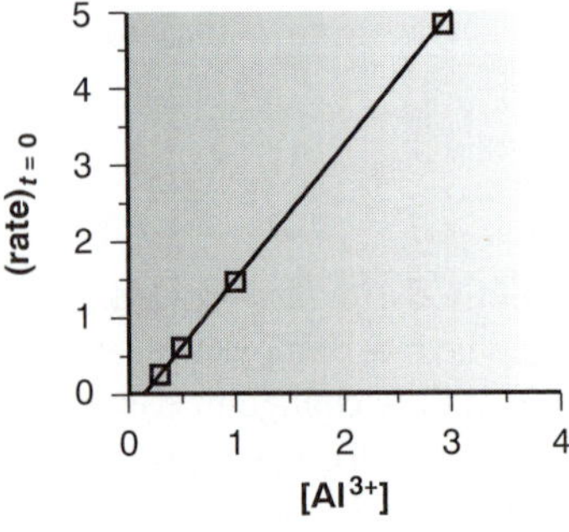

Figure 13.6
Result of curve-fitting for the kinetic data in Example 13.4.

Curve-Fitting Methods In the direct-computation methods discussed earlier, the analyte's concentration is determined by solving the appropriate rate equation at one or two discrete times. The relationship between the analyte's concentration and the measured response is a function of the rate constant, which must be measured in a separate experiment. This may be accomplished using a single external standard (as in Example 13.2) or with a calibration curve (as in Example 13.4).

In a curve-fitting method the concentration of a reactant or product is monitored continuously as a function of time, and a regression analysis is used to fit an appropriate differential or integral rate equation to the data. For example, the initial concentration of analyte for a pseudo-first-order reaction, in which the concentration of a product is followed as a function of time, can be determined by fitting a rearranged form of equation 13.12

$$[\mathrm{P}]_t = [\mathrm{A}]_0(1 - e^{-k't})$$

to the kinetic data using both $[\mathrm{A}]_0$ and k' as adjustable parameters. By using data from more than one or two discrete times, curve-fitting methods are capable of producing more reliable results. Although curve-fitting methods are computationally more demanding, the calculations are easily handled by computer.

Example

EXAMPLE 13.5

The data shown in the following table were collected for a reaction known to follow pseudo-zero-order kinetics during the time in which the reaction was monitored.

Time (s)	$[\mathrm{A}]_t$ (mM)
3	0.0731
4	0.0728
5	0.0681
6	0.0582
7	0.0511
8	0.0448
9	0.0404
10	0.0339
11	0.0217
12	0.0143

What are the rate constant and the initial concentration of analyte in the sample?

SOLUTION

For a pseudo-zero-order reaction a plot of $[\mathrm{A}]_t$ versus time should be linear with a slope of $-k$, and a y-intercept of $[\mathrm{A}]_0$ (equation 13.8). A plot of the kinetic data is shown in Figure 13.7. Linear regression gives an equation of

$$[\mathrm{A}]_t = 0.0986 - 0.00677t$$

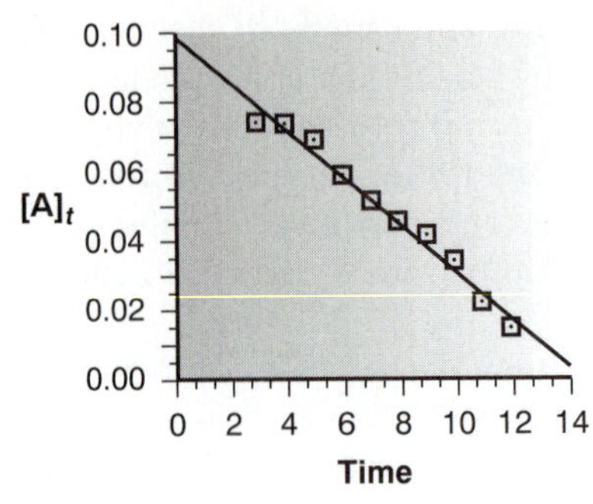

Figure 13.7
Result of curve-fitting for the kinetic data in Example 13.5.

Thus, the rate constant is 0.00677 s^{-1}, and the initial concentration of analyte is 0.0986 mM.

Miscellaneous Methods At the beginning of this section we noted that kinetic methods are susceptible to significant errors when experimental variables affecting the reaction's rate are difficult to control. Many variables, such as temperature, can be controlled with proper instrumentation. Other variables, such as interferents in the sample matrix, are more difficult to control and may lead to significant errors. Although not discussed in this text, direct-computation and curve-fitting methods have been developed that compensate for these sources of error.[3]

Representative Method Although each chemical kinetic method has its own unique considerations, the determination of creatinine in urine based on the kinetics of its reaction with picrate provides an instructive example of a typical procedure.

Representative Methods

Method 13.1 Determination of Creatinine in Urine[11]

Description of Method. Creatine is an organic acid found in muscle tissue that supplies energy for muscle contractions. One of its metabolic products is creatinine, which is excreted in urine. Because the concentration of creatinine in urine and serum is an important indication of renal function, rapid methods for its analysis are clinically important. In this method the rate of reaction between creatinine and picrate in an alkaline medium is used to determine the concentration of creatinine in urine. Under the conditions of the analysis, the reaction is first-order in picrate, creatinine, and hydroxide.

$$\text{Rate} = k[\text{picrate}][\text{creatinine}][\text{OH}^-]$$

The rate of reaction is monitored using a picrate ion-selective electrode.

Procedure. Prepare a set of external standards containing 0.5 g/L to 3.0 g/L creatinine (in 5 mM H_2SO_4) using a stock solution of 10.00 g/L creatinine in 5 mM H_2SO_4. In addition, prepare a solution of 1.00×10^{-2} M sodium picrate. Pipet 25.00 mL of 0.20 M NaOH, adjusted to an ionic strength of 1.00 M using Na_2SO_4, into a thermostated reaction cell at 25 °C. Add 0.500 mL of the 1.00×10^{-2} M picrate solution to the reaction cell. Suspend a picrate ion-selective electrode in the solution, and monitor the potential until it stabilizes. When the potential is stable, add 2.00 mL of a

—*Continued*

creatinine external standard, and record the potential as a function of time. Repeat the procedure using the remaining external standards. Construct a calibration curve of rate ($\Delta E/\Delta t$) versus [creatinine]. Samples of urine (2.00 mL) are analyzed in a similar manner, with the concentration of creatinine determined from the calibration curve.

Questions

1. **The analysis is carried out under conditions in which the reaction's kinetics are pseudo-first-order in picrate. Show that under these conditions, a plot of potential as a function of time will be linear.**

   The response of the picrate ion-selective electrode is

   $$E = K - \frac{RT}{F} \ln [\text{picrate}]$$

   We know from equation 13.6 that for a pseudo-first-order reaction, the concentration of picrate at time *t* is

   $$\ln [\text{picrate}]_t = \ln [\text{picrate}]_0 - k't$$

   where k' is the pseudo-first-order rate constant. Substituting the kinetic expression into the equation for the ion-selective electrode's potential leaves us with

   $$E_t = K - \frac{RT}{F} (\ln [\text{picrate}]_0 - k't)$$

   $$E_t = K - \frac{RT}{F} \ln [\text{picrate}]_0 + \frac{RTk'}{F} t$$

   Since both *K* and $(RT/F) \ln [\text{picrate}]_0$ are constants, a plot of E_t versus *t* will be a straight line whose slope, RTk'/F, is the reaction's rate ($\Delta E/\Delta t$).

2. **As carried out the rate of the reaction is pseudo-first-order in picrate and pseudo-zero-order in creatinine and OH^-. Explain why it is possible to prepare a calibration curve of rate versus [creatinine].**

   Since the reaction is carried out under conditions in which it is pseudo-zero-order in creatinine and OH^-, the rate constant, k', is

   $$k' = k[\text{creatinine}][OH^-]$$

   where *k* is the reaction's true rate constant. The rate, therefore, is

   $$\text{Rate} = \frac{RTk[\text{creatinine}][OH^-]}{F} = c[\text{creatinine}]$$

   where *c* is a constant.

3. **Why is it necessary to use a thermostat in the reaction cell?**

   The rate of a reaction is temperature-dependent. To avoid a determinate error resulting from a systematic change in temperature or to minimize indeterminate errors due to fluctuations in temperature, the reaction cell must have a thermostat to maintain a constant temperature.

4. **Why is it necessary to prepare the NaOH solution so that it has an ionic strength of 1.00 M?**

   The potential of the ion-selective electrode actually responds to the activity of picrate in solution. By adjusting the NaOH solution to a high ionic strength, we maintain a constant ionic strength in all standards and samples. Because the relationship between activity and concentration is a function of ionic strength (see Chapter 6), the use of a constant ionic strength allows us to treat the potential as though it were a function of the concentration of picrate.

13A.2 Instrumentation

Quantitative information about a chemical reaction can be made using any of the techniques described in the preceding chapters. For reactions that are kinetically slow, an analysis may be performed without worrying about the possibility that significant changes in concentration occur while measuring the signal. When the reaction's rate is too fast, which is usually the case, significant errors may be introduced if changes in concentration are ignored. One solution to this problem is to stop, or **quench,** the reaction by suitably adjusting experimental conditions. For example, many reactions involving enzymes show a strong pH dependency and may be quenched by adding a strong acid or strong base. Once the reaction is stopped, the concentration of the desired species can be determined at the analyst's convenience. Another approach is to use a visual indicator that changes color after the reaction occurs to a fixed extent. You may recall that this variable-time method is the basis of the so-called "clock reactions" commonly used to demonstrate kinetics in the general chemistry classroom and laboratory. Finally, reactions with fast kinetics may be monitored continuously using the same types of spectroscopic and electrochemical detectors found in chromatographic instrumentation.

quench
To stop a reaction by suddenly changing the reaction conditions.

Two additional problems for chemical kinetic methods of analysis are the need to control the mixing of the sample and reagents in a rapid and reproducible fashion and the need to control the acquisition and analysis of the signal. Many kinetic determinations are made early in the reaction when pseudo-zero-order or pseudo-first-order conditions are in effect. Depending on the rate of reaction, measurements are typically made within a span of a few milliseconds or seconds. This is both an advantage and a disadvantage. The disadvantage is that transferring the sample and reagent to a reaction vessel and their subsequent mixing must be automated if a reaction with rapid kinetics is to be practical. This usually requires a dedicated instrument, thereby adding an additional expense to the analysis. The advantage is that a rapid, automated analysis allows for a high throughput of samples. For example, an instrument for the automated kinetic analysis of phosphate, based on reaction 13.9, has achieved sampling rates of 3000 determinations per hour.[6]

A variety of designs have been developed to automate kinetic analyses.[6] The **stopped-flow** apparatus, which is shown schematically in Figure 13.8, has found use in kinetic determinations involving very fast reactions. Sample and reagents are loaded into separate syringes, and precisely measured volumes are dispensed by the action of a syringe drive. The two solutions are rapidly mixed in the mixing chamber before flowing through an observation cell. The flow of sample and reagents is stopped by applying back pressure with the stopping syringe. The back pressure completes the mixing, after which the reaction's progress is monitored spectrophotometrically. With a stopped-flow apparatus, it is possible to complete the mixing of sample and reagent and initiate the kinetic measurements within approximately 0.5 ms. The stopped-flow apparatus shown in Figure 13.8 can be modified by attaching an automatic sampler to the sample syringe, thereby allowing the sequential analysis of multiple samples. In this way the stopped-flow apparatus can be used for the routine analysis of several hundred samples per hour.

stopped flow
A kinetic method of analysis designed to rapidly mix samples and reagents when using reactions with very fast kinetics.

Another automated approach to kinetic analyses is the centrifugal analyzer, a partial cross section of which is shown in Figure 13.9. In this technique the sample and reagents are placed in separate wells oriented radially around a circular transfer disk attached to the rotor of a centrifuge. As the centrifuge spins, the

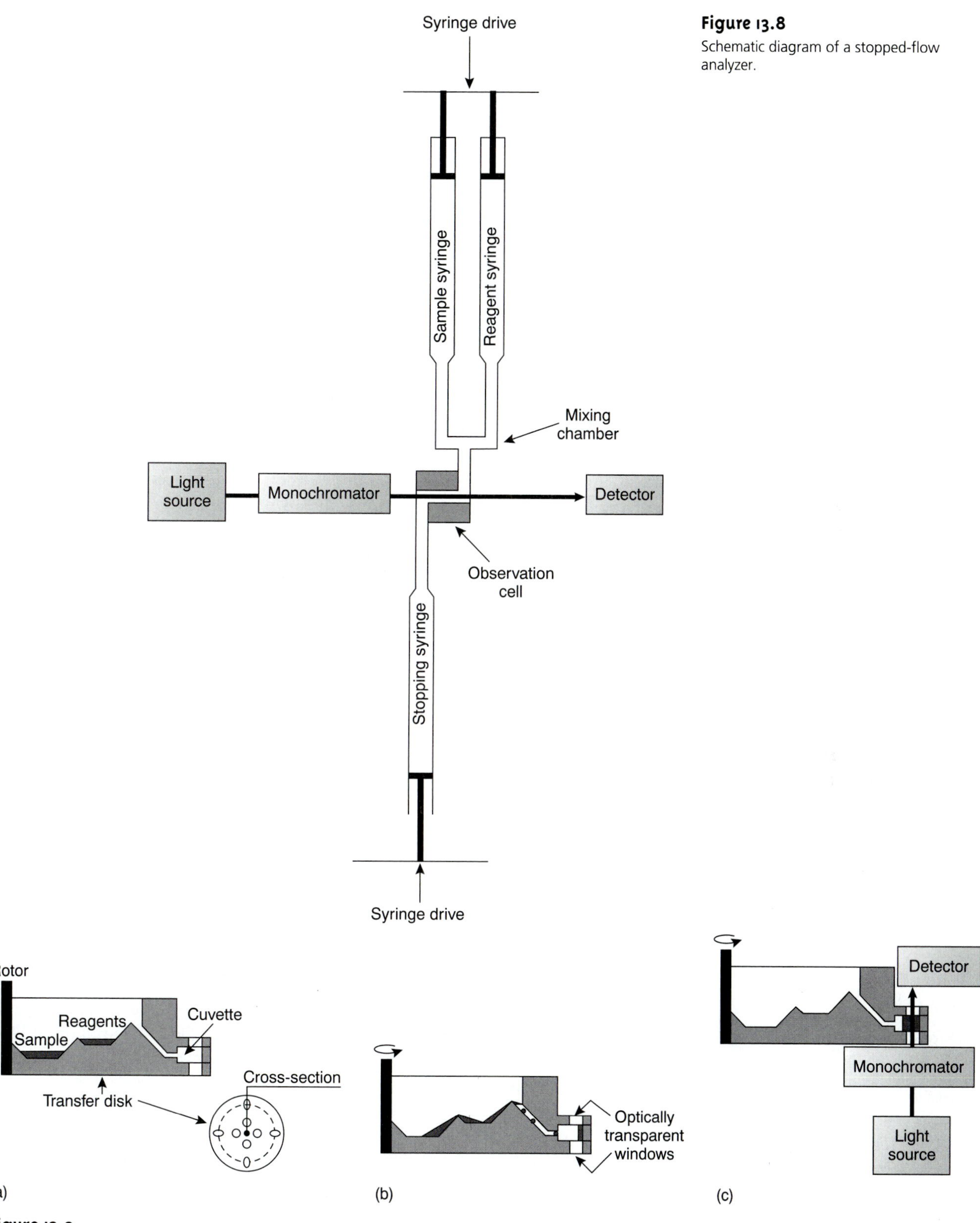

Figure 13.8
Schematic diagram of a stopped-flow analyzer.

Figure 13.9
Schematic diagram of a centrifugal analyzer showing (a) the wells for holding the sample and reagent; (b) mixing of the sample and reagent; and (c) the configuration of the spectrophotometric detector.

sample and reagents are pulled by the centrifugal force to the cuvette, where mixing occurs. A single optical source and detector, located above and below the transfer disk's outer edge, allows the absorbance of the reaction mixture to be measured as it passes through the optical beam. The centrifugal analyzer allows a number of samples to be analyzed simultaneously. For example, if a transfer plate contains 30 cuvettes and rotates with a speed of 600 rpm, it is possible to collect 10 data points per sample for each second of rotation.

The ability to collect kinetic data for several hundred samples per hour is of little consequence if the analysis of the data must be accomplished manually. Besides time, the manual analysis of kinetic data is limited by noise in the detector's signal and the accuracy with which the analyst can determine reaction rates from tangents drawn to differential rate curves. Not surprisingly, the development of automated kinetic analyzers was paralleled by the development of analog and digital circuitry, as well as computer software for the smoothing, on-line integration and differentiation, and analysis of kinetic signals.[12]

13A.3 Quantitative Applications

Chemical kinetic methods of analysis continue to find use for the analysis of a variety of analytes, most notably in clinical laboratories, where automated methods aid in handling a large volume of samples. In this section several general quantitative applications are considered.

enzyme
A protein that catalyzes biochemical reactions.

substrate
The specific molecule for which an enzyme serves as a catalyst.

Enzyme-Catalyzed Reactions **Enzymes** are highly specific catalysts for biochemical reactions, with each enzyme showing a selectivity for a single reactant, or **substrate.** For example, acetylcholinesterase is an enzyme that catalyzes the decomposition of the neurotransmitter acetylcholine to choline and acetic acid. Many enzyme–substrate reactions follow a simple mechanism consisting of the initial formation of an enzyme–substrate complex, ES, which subsequently decomposes to form product, releasing the enzyme to react again.

$$E + S \underset{k_{-1}}{\overset{k_1}{\rightleftharpoons}} ES \underset{k_{-2}}{\overset{k_2}{\rightleftharpoons}} E + P \qquad 13.15$$

If measurements are made early in the reaction, the product's concentration is negligible, and the step described by the rate constant k_{-2} can be ignored. Under these conditions the rate of the reaction is

$$\frac{d[P]}{dt} = k_2[ES] \qquad 13.16$$

steady-state approximation
In a kinetic process, the assumption that a compound formed during the reaction reaches a concentration that remains constant until the reaction is nearly complete.

To be analytically useful equation 13.16 needs to be written in terms of the concentrations of enzyme and substrate. This is accomplished by applying the **"steady-state" approximation,** in which we assume that the concentration of ES is essentially constant. After an initial period in which the enzyme–substrate complex first forms, the rate of formation of ES

$$\left(\frac{d[ES]}{dt}\right)_f = k_1[E][S] = k_1([E]_0 - [ES])[S]$$

and its rate of disappearance

$$\left(\frac{-d[ES]}{dt}\right)_d = k_{-1}[ES] + k_2[ES]$$

are equal. Combining these equations gives

$$k_1([\mathrm{E}]_0 - [\mathrm{ES}])[\mathrm{S}] = k_{-1}[\mathrm{ES}] + k_2[\mathrm{ES}]$$

which is solved for the concentration of the enzyme–substrate complex

$$[\mathrm{ES}] = \frac{[\mathrm{E}]_0[\mathrm{S}]}{\{(k_{-1} + k_2)/k_1\} + [\mathrm{S}]} = \frac{[\mathrm{E}]_0[\mathrm{S}]}{K_\mathrm{m} + [\mathrm{S}]} \qquad 13.17$$

where K_m is called the **Michaelis constant.** Substituting equation 13.17 into equation 13.16 leads to the final rate equation

Michaelis constant
A combination of several rate constants affecting the rate of an enzyme-substrate reaction.

$$\frac{d[\mathrm{P}]}{dt} = \frac{k_2[\mathrm{E}]_0[\mathrm{S}]}{K_\mathrm{m} + [\mathrm{S}]} \qquad 13.18$$

A plot of equation 13.18, shown in Figure 13.10, is instructive for defining conditions under which the rate of an enzymatic reaction can be used for the quantitative analysis of enzymes and substrates. For high substrate concentrations, where $[\mathrm{S}] >> K_\mathrm{m}$, equation 13.18 simplifies to

$$\frac{d[\mathrm{P}]}{dt} = k_2[\mathrm{E}]_0 = V_\mathrm{max} \qquad 13.19$$

where V_max is the maximum rate for the catalyzed reaction. Under these conditions the rate of the reaction is pseudo-zero-order in substrate, and the maximum rate can be used to calculate the enzyme's concentration. Typically, this determination is made by a variable-time method. At lower substrate concentrations, where $[\mathrm{S}] << K_\mathrm{m}$, equation 13.18 becomes

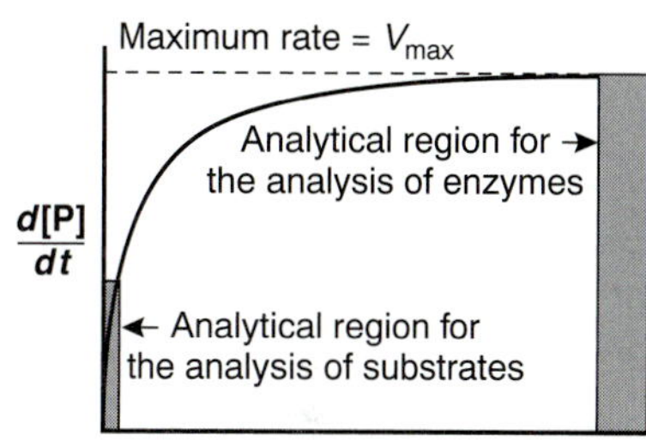

Figure 13.10
Plot of equation 13.18 showing limits for which a chemical kinetic method of analysis can be used to determine the concentration of a catalyst or a substrate.

$$\frac{d[\mathrm{P}]}{dt} = \frac{k_2[\mathrm{E}]_0[\mathrm{S}]}{K_\mathrm{m}} = \frac{V_\mathrm{max}[\mathrm{S}]}{K_\mathrm{m}} \qquad 13.20$$

The reaction is now first-order in substrate, and the rate of the reaction can be used to determine the substrate's concentration by a fixed-time method.

Chemical kinetic methods have been applied to the quantitative analysis of a number of enzymes and substrates.[13] One example, is the determination of glucose based on its oxidation by the enzyme glucose oxidase.[6]

$$\text{Glucose} + \mathrm{H_2O} + \mathrm{O_2} \xrightarrow{\text{glucose oxidase}} \text{gluconolactone} + \mathrm{H_2O_2}$$

Conditions are controlled, such that equation 13.20 is valid. The reaction is monitored by following the rate of change in the concentration of dissolved O_2 using an appropriate voltammetric technique.

Nonenzyme-Catalyzed Reactions The variable-time method has also been used to determine the concentration of nonenzymatic catalysts. Because a trace amount of catalyst can substantially enhance a reaction's rate, a kinetic determination of a catalyst's concentration is capable of providing an excellent detection limit. One of the most commonly used reactions is the reduction of H_2O_2 by reducing agents, such as thiosulfate, iodide, and hydroquinone. These reactions are catalyzed by trace levels of selected metal ions. For example the reduction of H_2O_2 by I^-

$$2\mathrm{I^-} + \mathrm{H_2O_2} + 2\mathrm{H_3O^+} \rightleftharpoons 2\mathrm{H_2O} + \mathrm{I_2}$$

is catalyzed by Mo(VI), W(VI), and Zr(IV). A variable-time analysis is conducted by adding a small, fixed amount of ascorbic acid to each solution. As I_2 is produced, it rapidly oxidizes the ascorbic acid and is, itself, reduced back to I^-. Once all the ascorbic acid is consumed, the presence of excess I_2 provides a visual end point.

Noncatalytic Reactions Chemical kinetic methods are not as common for the quantitative analysis of analytes in noncatalytic reactions. Because they lack the enhancement of reaction rate obtained when using a catalyst, noncatalytic methods generally are not used for the determination of analytes at low concentrations.[4] Noncatalytic methods for analyzing inorganic analytes are usually based on a complexation reaction. One example was outlined in Example 13.4, in which the concentration of aluminum in serum was determined by the initial rate of formation of its complex with 2-hydroxy-1-naphthaldehyde *p*-methoxybenzoyl-hydrazone.[10] The greatest number of noncatalytic methods, however, are for the quantitative analysis of organic analytes. For example, the insecticide methyl parathion has been determined by measuring its rate of hydrolysis in alkaline solutions.[14]

13A.4 Characterization Applications

Chemical kinetic methods also find use in determining rate constants and elucidating reaction mechanisms. These applications are illustrated by two examples from the chemical kinetic analysis of enzymes.

Determining V_{max} and K_m for Enzyme-Catalyzed Reactions The value of V_{max} and K_m for an enzymatic reaction are of significant interest in the study of cellular chemistry.[15] From equation 13.19 we see that V_{max} provides a means for determining the rate constant k_2. For enzymes that follow the mechanism shown in reaction 13.15, k_2 is equivalent to the enzyme's turnover number, k_{cat}. The turnover number is the maximum number of substrate molecules converted to product by a single active site on the enzyme, per unit time. Thus, the turnover number provides a direct indication of the catalytic efficiency of an enzyme's active site. The Michaelis constant, K_m, is significant because it provides an estimate of the substrate's intracellular concentration.[15b]

Values of V_{max} and K_m for reactions obeying the mechanism shown in reaction 13.15 can be determined using equation 13.18 by measuring the rate of reaction as a function of the substrate's concentration. The curved nature of the relationship between rate and the concentration of substrate (see Figure 13.10), however, is inconvenient for this purpose. Equation 13.18 can be rewritten in a linear form by taking its reciprocal

$$\frac{1}{d[\mathrm{P}]/dt} = \frac{1}{v} = \left(\frac{K_m}{V_{max}}\right)\frac{1}{[\mathrm{S}]} + \frac{1}{V_{max}}$$

Lineweaver–Burk plot
A graphical means for evaluating enzyme kinetics.

A plot of $1/v$ versus $1/[S]$, which is called a double reciprocal, or **Lineweaver–Burk plot,** is a straight line with a slope of K_m/V_{max}, a *y*-intercept of $1/V_{max}$, and an *x*-intercept of $-1/K_m$ (Figure 13.11).

Elucidating Mechanisms for the Inhibition of Enzyme Catalysis An inhibitor interacts with an enzyme in a manner that decreases the enzyme's catalytic efficiency. Examples of inhibitors include some drugs and poisons. Irreversible inhibitors covalently bind to the enzyme's active site, producing a permanent loss in catalytic efficiency even when the inhibitor's concentration is decreased. Reversible inhibitors form noncovalent complexes with the enzyme, thereby causing a temporary de-

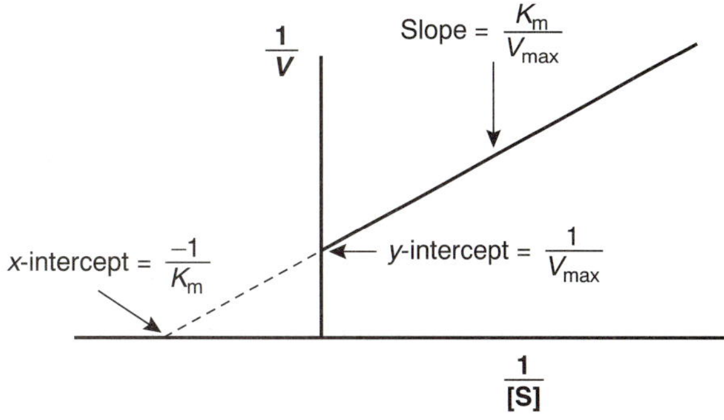

Figure 13.11
Lineweaver–Burk plot of equation 13.18.

crease in catalytic efficiency. If the inhibitor is removed, the enzyme's catalytic efficiency returns to its normal level.

The reversible binding of an inhibitor to an enzyme can occur through several pathways, as shown in Figure 13.12. In competitive inhibition (Figure 13.12a), the substrate and the inhibitor compete for the same active site on the enzyme. Because the substrate cannot bind to an enzyme that is already bound to an inhibitor, the enzyme's catalytic efficiency for the substrate is decreased. With noncompetitive inhibition (Figure 13.12b) the substrate and inhibitor bind to separate active sites on the enzyme, forming an enzyme–substrate–inhibitor, or ESI complex. The presence of the ESI complex, however, decreases catalytic efficiency since only the enzyme–substrate complex can react to form product. Finally, in uncompetitive inhibition (Figure 13.12c) the inhibitor cannot bind to the free enzyme, but can bind to the enzyme–substrate complex. Again, the result is the formation of an inactive ESI complex.

The three reversible mechanisms for enzyme inhibition are distinguished by observing how changing the inhibitor's concentration affects the relationship between the rate of reaction and the concentration of substrate. As shown in Figure 13.13, when kinetic data are displayed as a Lineweaver–Burk plot, it is possible to determine which mechanism is in effect.

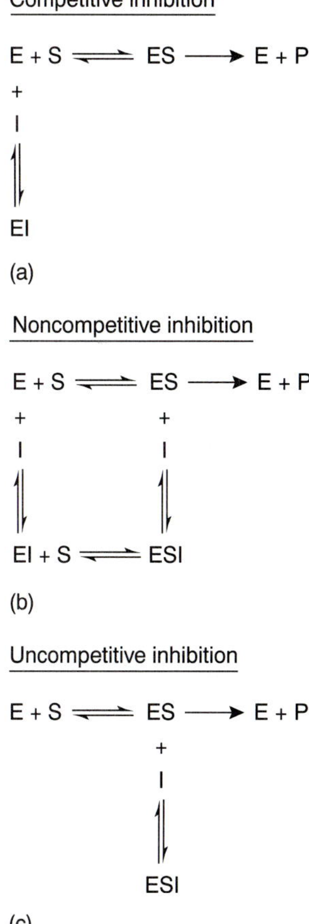

Figure 13.12
Mechanisms for the inhibition of enzyme catalysis.

13A.5 Evaluation of Chemical Kinetic Methods

Scale of Operation The detection limit for chemical kinetic methods ranges from minor components to ultratrace components (see Figure 3.6 in Chapter 3) and is principally determined by two factors: the rate of the reaction and the instrumental method used for monitoring the rate. Because the signal is directly proportional to the reaction rate, faster reactions generally result in lower detection limits. All other considerations being equal, the detection limit is smaller for catalytic reactions than for noncatalytic reactions. Not surprisingly, chemical kinetic methods based on catalysis were among the earliest techniques for trace-level analysis. Ultratrace analysis is also possible when using catalytic reactions. For example, ultratrace levels of Cu ($<$1 ppb) can be determined by measuring its catalytic effect on the redox reaction between hydroquinone and H_2O_2. Without a catalyst, most chemical kinetic methods for organic compounds involve reactions with relatively slow rates, limiting the method to minor and higher concentration trace analytes. Noncatalytic chemical kinetic methods for inorganic compounds involving metal–ligand complexation may be fast or slow, with detection limits ranging from trace to minor levels of analyte.

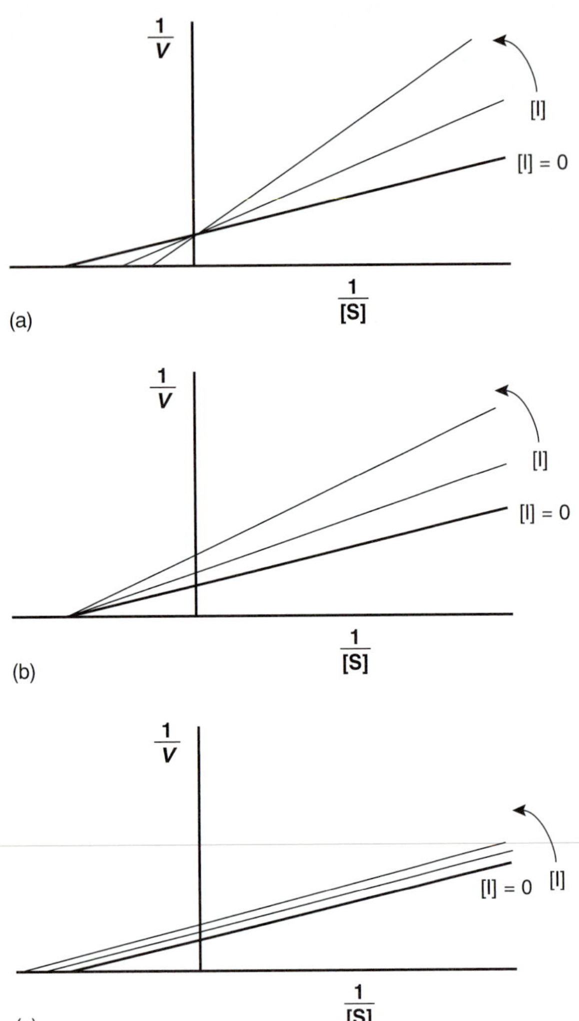

Figure 13.13
Effect of the concentration of inhibitor on the Lineweaver–Burk plots for (a) competitive inhibition, (b) noncompetitive inhibition, and (c) uncompetitive inhibition. The inhibitor's concentration increases in the direction shown by the arrows.

The second factor influencing detection limits is the instrumental method used to monitor the reaction's progress. Most reactions are monitored spectrophotometrically or electrochemically. The scale of operation for these methods was discussed in Chapters 10 and 11 and, therefore, is not discussed here.

Accuracy As noted earlier, chemical kinetic methods are potentially subject to larger errors than equilibrium methods due to the effect of uncontrolled or poorly controlled variables, such as temperature and solution pH. Although the direct-computation chemical kinetic methods described in this chapter can yield results with moderate accuracy (1–5%), reaction systems are encountered in which accuracy is quite poor. An improvement in accuracy may be realized by using curve-fitting methods. In one study, for example, accuracy was improved by two orders of magnitude (500–5%) by replacing a direct-computation analysis with a curve-fitting analysis.[16] Although not discussed in this chapter, data analysis methods that include the ability to compensate for experimental error[3,17–19] can lead to a significant improvement in accuracy.

Precision The precision of a chemical kinetic method is limited by the signal-to-noise ratio of the instrumental method used to monitor the reaction's progress. With integral methods, precisions of 1–2% are routinely possible. The precision for differential methods may be somewhat poorer, particularly for noisy signals, due to the difficulty in measuring the slope of a noisy rate curve.[19] It may be possible to improve the precision in this case by using a combination of signal averaging and smoothing of the data before its analysis.

Sensitivity The sensitivity for a one-point fixed-time integral method of analysis is improved by making measurements under conditions in which the concentration of the monitored species is larger rather than smaller. When the analyte's concentration, or the concentration of any other reactant, is monitored, measurements are best made early in the reaction before its concentration has substantially decreased. On the other hand, when a product is used to monitor the reaction, measurements are more appropriately made at longer times. For a two-point fixed-time integral method, sensitivity is improved by increasing the difference between times t_1 and t_2. As discussed earlier, the sensitivity of a rate method improves when using the initial rate.

Selectivity The analysis of closely related compounds, as we have seen in earlier chapters, is often complicated by their tendency to interfere with one another. To overcome this problem, the analyte and interferent must first be separated. An advantage of chemical kinetic methods is that conditions can often be adjusted so that the analyte and interferent have different reaction rates. If the difference in rates is large enough, one species may react completely before the other species has a chance to react. For example, many enzymes selectively catalyze a single substrate, allowing the quantitative analysis of that substrate in the presence of similar substrates.

The conditions necessary to ensure that a faster-reacting species can be kinetically separated from a more slowly reacting species can be determined from the appropriate integrated rate laws. As an example, let's consider a system consisting of an analyte, A, and an interferent, B, both of which show first-order kinetics with a common reagent. To avoid an interference, the relative magnitudes of their rate constants must be sufficiently different. The fractions, *f*, of A and B remaining at any point in time, *t*, are given as

$$(f_A)_t = \frac{[A]_t}{[A]_0} \qquad \textbf{13.21}$$

$$(f_B)_t = \frac{[B]_t}{[B]_0} \qquad \textbf{13.22}$$

For a first-order reaction we can write, from equations 13.2, 13.21, and 13.22

$$\ln(f_A)_t = -k_A t$$

$$\ln(f_B)_t = -k_B t$$

Taking the ratio of these two equations gives

$$\frac{k_A}{k_B} = \frac{\ln(f_A)_t}{\ln(f_B)_t}$$

Thus, for example, if 99% of A is to react before 1% of B has reacted

$$\frac{k_A}{k_B} = \frac{\ln(0.01)}{\ln(0.99)} = 458$$

then the rate constant for A must be 458 times larger than that for B. Under these conditions the analyte's concentration can be determined before the interferent begins to react. If the analyte has the slower reaction, then it can be determined after the interferent's reaction is complete.

The method described here is impractical when the simultaneous analysis of both A and B is desired. The difficulty in this case is that conditions favoring the analysis of A generally do not favor the analysis of B. For example, if conditions are adjusted such that 99% of A reacts in 5 s, then B must reach 99% completion in either 0.01 s if it has the faster kinetics or 2300 s if it has the slower kinetics.

Several additional approaches for analyzing mixtures have been developed that do not require such a large difference in rate constants.[3,4] Because both A and B react at the same time, the integrated form of the first-order rate law becomes

$$C_t = [A]_t + [B]_t = [A]_0 e^{-k_A t} + [B]_0 e^{-k_B t} \qquad \textbf{13.23}$$

where C_t is the total concentration of A and B. If C_t is measured at times t_1 and t_2, the resulting pair of simultaneous equations can be solved to give $[A]_0$ and $[B]_0$. The rate constants k_A and k_B must be determined in separate experiments using standard solutions of A and B. Alternatively, if A and B react to form a common product, P, equation 13.23 can be written as

$$P_t = [A]_0(1 - e^{-k_A t}) + [B]_0(1 - e^{-k_B t})$$

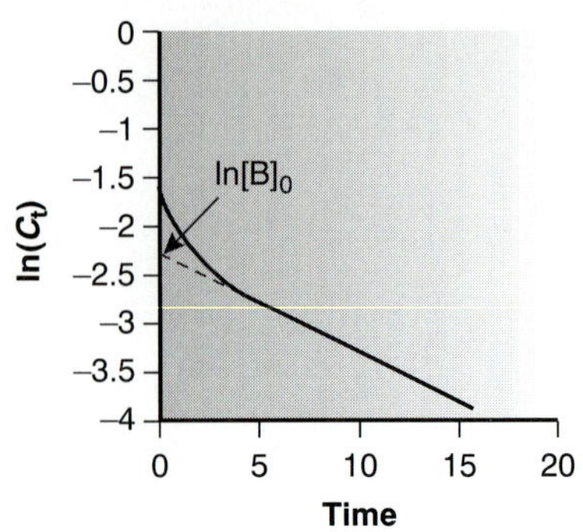

Figure 13.14
Determination of the concentration of a slowly reacting analyte, B, in the presence of a faster reacting analyte, A.

Again, a pair of simultaneous equations at times t_1 and t_2 can be solved for $[A]_0$ and $[B]_0$.

Equation 13.23 can also be used as the basis for a curve-fitting method. As shown in Figure 13.14, a plot of $\ln(C_t)$ as a function of time consists of two regions. At short times the plot is curved since A and B are reacting simultaneously. At later times, however, the concentration of the faster-reacting component, A, decreases to 0, and equation 13.23 simplifies to

$$C_t \approx [B]_t = [B]_0 e^{-k_B t}$$

Under these conditions a plot of $\ln(C_t)$ versus time is linear. Extrapolating the linear portion to time 0 gives $[B]_0$, and $[A]_0$ is determined by difference.

Time, Cost, and Equipment Automated chemical kinetic methods of analysis provide a rapid means for analyzing samples, with throughputs ranging from several hundred to several thousand determinations per hour. The initial start-up costs, however, may be fairly high because an automated analysis requires a dedicated instrument designed to meet the specific needs of the analysis. When handled manually, chemical kinetic methods can be accomplished using equipment and instrumentation routinely available in most laboratories. Sample throughput, however, is much lower than with automated methods.

13B Radiochemical Methods of Analysis

Atoms with the same number of protons but a different number of neutrons are called **isotopes.** To identify an isotope we use the symbol ${}^{A}_{Z}E$, where E is the element's atomic symbol, Z is the element's atomic number (which is the number of protons), and A is the element's atomic mass number (which is the sum of the number of protons and neutrons). Although isotopes of a given element have the same chemical properties, their nuclear properties are different. The most important difference between isotopes is their stability. The nuclear configuration of a stable isotope remains constant with time. Unstable isotopes, however, spontaneously disintegrate, emitting radioactive particles as they transform into a more stable form.

isotopes
Atoms with the same number of protons but different numbers of neutrons are called isotopes.

The most important types of radioactive particles are alpha particles, beta particles, gamma rays, and X-rays. An **alpha particle,** which is symbolized as α, is equivalent to a helium nucleus, ${}^{4}_{2}He$. Thus, emission of an alpha particle results in a new isotope whose atomic number and atomic mass number are, respectively, 2 and 4 less than that for the unstable parent isotope.

alpha particle
A positively charged subatomic particle equivalent to a helium nucleus (α).

$${}^{238}_{92}U \rightarrow {}^{234}_{90}Th + \alpha$$

Beta particles, β, come in two forms. A **negatron,** ${}^{0}_{-1}\beta$, is equivalent to an electron, and is produced when a neutron is converted to a proton, increasing the atomic number by 1.

beta particle
A charged subatomic particle produced when a neutron converts to a proton, or a proton converts to a neutron (β).

negatron
The beta particle formed when a neutron converts to a proton; equivalent to an electron (${}^{0}_{-1}\beta$).

$${}^{214}_{82}Pb \rightarrow {}^{214}_{83}Bi + {}^{0}_{-1}\beta$$

Converting a proton to a neutron results in the emission of a **positron,** ${}^{0}_{1}\beta$.

positron
The beta particle formed when a proton converts to a neutron (${}^{0}_{1}\beta$).

$${}^{30}_{15}P \rightarrow {}^{30}_{14}Si + {}^{0}_{1}\beta$$

Emission of an alpha or beta particle often produces an isotope in an unstable, high-energy state. This excess energy is released as a **gamma ray,** γ, or an X-ray. Gamma ray and X-ray emission may also occur without the release of alpha or beta particles.

gamma ray
High-energy radiation emitted by nuclei (γ).

Although similar to chemical kinetic methods of analysis, radiochemical methods are best classified as nuclear kinetic methods. In this section we review the kinetics of radioactive decay and examine several quantitative and characterization applications.

13B.1 Theory and Practice

The rate of decay, or activity, for a radioactive isotope follows first-order kinetics

$$A = -\frac{dN}{dt} = \lambda N \qquad \textbf{13.24}$$

where A is the activity, N is the number of radioactive atoms present in the sample at time t, and λ is the radioisotope's decay constant. Activity is given in units of disintegrations per unit time, which is equivalent to the number of atoms undergoing radioactive decay per unit time.

As with any first-order process, equation 13.24 can be expressed in an integrated form.

$$N = N_0 e^{-\lambda t} \qquad \textbf{13.25}$$

Substituting equation 13.25 into equation 13.24 gives

$$A = \lambda N_0 e^{-\lambda t} = A_0 e^{-\lambda t} \qquad \textbf{13.26}$$

By measuring the activity at time t, therefore, we can determine the initial activity, A_0, or the number of radioactive atoms originally present in the sample, N_0.

An important characteristic property of a radioactive isotope is its **half-life,** $t_{1/2}$, which is the amount of time required for half of the radioactive atoms to disintegrate. For first-order kinetics the half-life is independent of concentration and is given as

half-life
The time required for half of the initial number of a radioactive isotope's atoms to disintegrate ($t_{1/2}$).

$$t_{1/2} = \frac{0.693}{\lambda} \qquad \textbf{13.27}$$

Since the half-life is independent of the number of radioactive atoms, it remains constant throughout the decay process. Thus, 50% of the radioactive atoms disintegrate in one half-life, 75% in two half-lives, and 87.5% in three half-lives.

Kinetic information about radioactive isotopes is usually given in terms of the half-life because it provides a more intuitive sense of the isotope's stability. Knowing, for example, that the decay constant for ${}^{90}_{38}Sr$ is 0.0247 yr^{-1} does not give an immediate sense of how fast it disintegrates. On the other hand, knowing that the half-life for ${}^{90}_{38}Sr$ is 28.1 years makes it clear that the concentration of ${}^{90}_{38}Sr$ in a sample remains essentially constant over a short period of time.

13B.2 Instrumentation

Alpha particles, beta particles, gamma rays, and X-rays are measured using the particle's energy to produce an amplified pulse of electric current in a detector. These pulses are counted to give the rate of disintegration. Three types of detectors commonly are encountered: gas-filled detectors, scintillation counters, and semiconductor detectors. The gas-filled detector consists of a tube filled with an inert gas, such as Ar. When radioactive particles enter the tube, they ionize the inert gas, producing a large number of Ar^+/e^- ion pairs. Movement of the electrons toward an anode and the Ar^+ toward a cathode generates a measurable electric current. A **Geiger counter** is one example of a gas-filled detector. A **scintillation counter** uses a fluorescent

Geiger counter
An instrument for counting radioactive particles based on their ability to ionize an inert gas such as Ar.

scintillation counter
An instrument for counting radioactive particles based on their ability to initiate fluorescence in another substance.

material to convert radioactive particles into easily measurable photons. For example, one solid-state scintillation counter, consisting of a crystal of NaI with 0.2% TlI, produces several thousand photons for each radioactive particle. Finally, in a semiconductor detector the adsorption of a radioactive particle promotes thousands of electrons to the semiconductor's conduction band, leading to an increase in its conductivity. Further details about radiation detectors, and the signal processors used to count particles, can be found in the suggested readings listed at the end of this chapter.

13B.3 Quantitative Applications

Three common quantitative applications of radiochemical methods of analysis are considered in this section: the direct analysis of radioactive isotopes by measuring their rate of disintegration, neutron activation, and the use of radioactive isotopes as tracers in isotope dilution.

Direct Analysis of Radioactive Analytes The concentration of a long-lived radioactive isotope is essentially constant during the period of analysis. As shown in Example 13.6, the sample's activity can be used to calculate the number of radioactive particles that are present.

Example

EXAMPLE 13.6

The activity in a 10.00-mL sample of radioactive wastewater containing ${}^{90}_{38}Sr$ was found to be 9.07×10^6 disintegrations/s. What is the molar concentration of ${}^{90}_{38}Sr$ in the sample? The half-life for ${}^{90}_{38}Sr$ is 28.1 years.

SOLUTION

Substituting equation 13.27 into equation 13.24 and solving for N gives

$$N = \frac{A \times t_{1/2}}{0.693}$$

Before the number of atoms of ${}^{90}_{38}Sr$ can be determined, it is necessary to express the activity and the half-life in the same units. Converting the half-life for ${}^{90}_{38}Sr$ to seconds gives $t_{1/2}$ as 8.86×10^8 s. Substituting known values gives the number of atoms of ${}^{90}_{38}Sr$ as

$$\frac{(9.07 \times 10^6 \text{ disintegrations/s})(8.86 \times 10^8 \text{ s})}{0.693} = 1.16 \times 10^{16} \text{ atoms of } {}^{90}_{38}\text{Sr}$$

Thus, the concentration of ${}^{90}_{38}Sr$ in the sample is

$$\frac{1.16 \times 10^{16} \text{ atoms}}{(6.022 \times 10^{23} \text{ atoms/mol})(0.01000 \text{ L})} = 1.93 \times 10^{-6} \text{ M } {}^{90}_{38}\text{Sr}$$

The direct analysis of short-lived radioactive isotopes using the method outlined in Example 13.6 is less useful since it provides only a transient measure of the isotope's concentration. The concentration of the isotope at a particular moment can be determined by measuring its activity after an elapsed time, t, and using equation 13.26 to calculate N_0.

Neutron Activation Analysis Few samples of interest are naturally radioactive. For many elements, however, radioactivity may be induced by irradiating the sample with neutrons in a process called neutron activation analysis (NAA). The radioactive element formed by **neutron activation** decays to a stable isotope by emitting gamma rays and, if necessary, other nuclear particles. The rate of gamma-ray emission is proportional to the analyte's initial concentration in the sample. For example, when a sample containing nonradioactive $^{27}_{13}Al$ is placed in a nuclear reactor and irradiated with neutrons, the following nuclear reaction results.

neutron activation
A means of inducing radioactivity in a nonradioactive sample by bombarding the sample with neutrons.

$$^{27}_{13}\text{Al} + ^{1}_{0}n \rightarrow ^{28}_{13}\text{Al}$$

The radioactive isotope of $^{28}_{13}Al$ has a characteristic decay process that includes the release of a beta particle and a gamma ray.

$$^{28}_{13}\text{Al} \rightarrow ^{28}_{14}\text{Si} + ^{0}_{-1}\beta + \gamma$$

When irradiation is complete, the sample is removed from the nuclear reactor, allowed to cool while any short-lived interferences that might be present decay to the background, and the rate of gamma-ray emission is measured.

The initial activity at the end of irradiation depends on the number of $^{28}_{13}Al$ atoms that are present. This, in turn, is equal to the difference between the rate of formation for $^{28}_{13}Al$ and its rate of disintegration,

$$\frac{d(N_{^{28}_{13}\text{Al}})}{dt} = \Phi\sigma(N_{^{27}_{13}\text{Al}}) - \lambda N_{^{28}_{13}\text{Al}} \qquad \mathbf{13.28}$$

where Φ is the neutron flux, and σ is the reaction cross-section, or probability for the capture of a neutron by the $^{27}_{13}Al$ nucleus. Integrating equation 13.28 over the time of irradiation, t_i, and multiplying by λ gives the initial activity, A_0, at the end of irradiation as

$$A_0 = \lambda(N_{^{28}_{13}\text{Al}}) = \Phi\sigma(N_{^{28}_{13}\text{Al}})(1 - e^{-\lambda t_i})$$

Thus, the number of atoms of $^{27}_{13}Al$ initially present in the sample can be calculated from A_0 if values for Φ, σ, λ, and t_i are known.

A simpler approach for analyzing neutron activation data is to use one or more external standards. Letting $(A_0)_x$ and $(A_0)_s$ represent the initial activity for the analyte in an unknown and a single external standard, and letting w_x and w_s represent the weight of analyte in the unknown and the external standard, gives a pair of equations

$$(A_0)_x = kw_x \qquad \mathbf{13.29}$$

$$(A_0)_s = kw_s \qquad \mathbf{13.30}$$

that can be solved to determine the mass of analyte in the sample.

As noted earlier, gamma-ray emission is measured following a "cooling" period in which short-lived interferents are allowed to decay away. The initial activity therefore, is determined by extrapolating a curve of activity versus time back to $t = 0$ (Figure 13.15). Alternatively, if the samples and standards are irradiated simultaneously, and the activities are measured at the same time, then these activities may be used in place of $(A_0)_x$ and $(A_0)_s$ in the preceding equations.

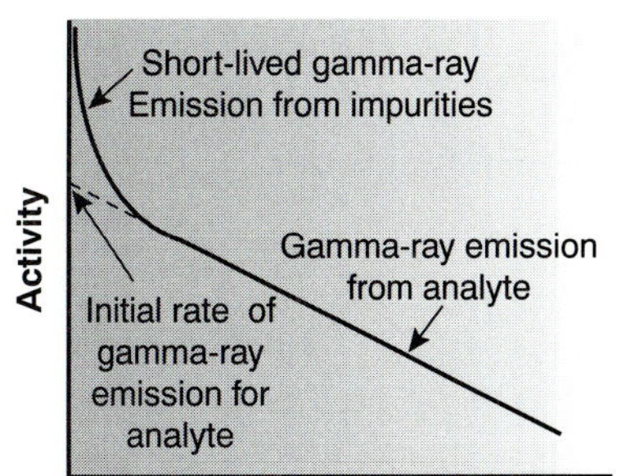

Figure 13.15
Plot of gamma-ray emission as a function of time showing how the analyte's initial activity is determined experimentally.

Example

EXAMPLE 13.7

The concentration of Mn in steel can be determined by a neutron activation analysis using the method of external standards. A 1.000-g sample of an unknown steel sample and a 0.950-g sample of a standard steel known to contain 0.463% w/w Mn, are irradiated with neutrons in a nuclear reactor for 10 h. After a 40-min "cooling" period, the activities for gamma-ray emission were found to be 2542 cpm (counts per minute) for the unknown and 1984 cpm for the standard. What is the %w/w Mn in the unknown steel sample?

SOLUTION

Combining equations 13.29 and 13.30 gives

$$w_x = \frac{A_x}{A_s} \times w_s$$

The weight of Mn in the standard sample is

$$w_s = \frac{0.00463 \text{ g Mn}}{\text{g steel}} \times 0.950 \text{ g steel} = 0.00440 \text{ g Mn}$$

Substituting into the previous equation gives

$$w_x = \frac{2542 \text{ cpm}}{1984 \text{ cpm}} \times 0.00440 \text{ g Mn} = 0.00564 \text{ g Mn}$$

Since the original mass of steel is 1.000 g, the %w/w Mn is 0.564%.

One of the important advantages of NAA is its applicability to almost all elements in the periodic table. Another advantage of neutron activation is that it is nondestructive. Consequently, NAA is an important technique for analyzing archaeological and forensic samples, as well as works of art.

isotope dilution
A form of internal standardization in which a radioactive form of the analyte serves as the internal standard.

tracer
A radioactive species used as an internal standard.

Isotope Dilution Another important quantitative radiochemical method is **isotope dilution.** In this method of analysis a sample of analyte, called a **tracer,** is prepared in a radioactive form with a known activity, A_T, for its radioactive decay. A measured mass of the tracer, w_T, is added to a sample containing an unknown mass, w_x, of a nonradioactive analyte, and the material is homogenized. The sample is then processed to isolate w_A grams of purified analyte, containing both radioactive and nonradioactive materials. The activity of the isolated sample, A_A, is measured. If all the analyte, both radioactive and nonradioactive, is recovered, then A_A and A_T will be equal. Normally, some of the analyte is lost during isolation and purification. In this case A_A is less than A_T, and

$$A_A = A_T\left(\frac{w_A}{w_x + w_T}\right) \qquad \textbf{13.31}$$

The ratio of weights in equation 13.31 accounts for the "dilution" of the activity due to a failure to recover all the analyte. Solving equation 13.31 for w_x gives

$$w_x = \frac{A_T}{A_A} w_A - w_T \qquad \textbf{13.32}$$

Example

EXAMPLE 13.8

The concentration of insulin in a production vat is determined by the method of isotope dilution. A 1.00-mg sample of insulin labeled with ^{14}C, with an activity of 549 cpm, is added to a 10.0-mL sample taken from the production vat. After homogenizing the sample, a portion of the insulin is separated and purified, yielding 18.3 mg of pure insulin. The activity for the isolated insulin is measured at 148 cpm. How many milligrams of insulin are in the original sample?

SOLUTION

Substituting known values into equation 13.32 gives

$$w_x = \frac{549 \text{ cpm}}{148 \text{ cpm}} \times 18.3 \text{ mg} - 1.00 \text{ mg} = 66.9 \text{ mg insulin}$$

Equations 13.31 and 13.32 are only valid if the radioactive element in the tracer has a half-life that is considerably longer than the time needed to conduct the analysis. If this is not the case, then the decrease in activity is due both to the effect of dilution and the natural decrease in the isotope's activity. Some common radioactive isotopes for use in isotope dilution are listed in Table 13.1.

An important feature of isotope dilution is that it is not necessary to recover all the analyte to determine the amount of analyte present in the original sample. Isotope dilution, therefore, is useful for the analysis of samples with complex matrices, when a complete recovery of the analyte is difficult.

13B.4 Characterization Applications

One example of a characterization application is the determination of a sample's age based on the kinetics for the decay of a radioactive isotope present in the sample. The most common example is carbon-14 dating, which is used to determine the age of natural organic materials.

As cosmic rays pass through the upper atmosphere, some of the $^{14}_{7}N$ present is converted to $^{14}_{6}C$ by the capture of high-energy neutrons. The $^{14}_{6}C$ then migrates into

Table 13.1 Common Isotopes for Use as Tracers

Isotope	Half-Life
^{3}H	12.5 years
^{14}C	5730 years
^{32}P	14.3 days
^{35}S	87.1 days
^{45}Ca	152 days
^{55}Fe	2.91 years
^{60}Co	5.3 years
^{131}I	8 days

the lower atmosphere, where it is oxidized to produce radioactive CO_2, which is subsequently incorporated into living organisms. As a result, all living plants and animals have approximately the same ratio of ${}^{14}_{6}C$ to ${}^{12}_{6}C$ in their tissues. When an organism dies, however, the radioactive decay of ${}^{14}_{6}C$ to ${}^{14}_{7}N$ by β^- emission ($t_{1/2}$ = 5730 years) results in a gradual reduction in the ratio of ${}^{14}_{6}C$ to ${}^{12}_{6}C$. The change in this ratio can be used to date samples that are as much as 30,000 years old, although the precision of the analysis is best when the sample's age is less than 7000 years. The accuracy of carbon-14 data is limited by the assumption that the ratio of ${}^{14}_{6}C$ to ${}^{12}_{6}C$ has remained constant over time. Some variation in the ratio has occurred as the result of the increased consumption of fossil fuel, and the production of ${}^{14}_{6}C$ during the testing of nuclear weapons. Correction factors have been developed that increase the accuracy of carbon-14 dating.

Example

EXAMPLE 13.9

A sample of fabric was dated by measuring the relative ratio of ${}^{14}_{6}C$ to ${}^{12}_{6}C$, with the result being 80.9% of that found in modern fibers. How old is the fabric?

SOLUTION

Equations 13.26 and 13.27 provide us with a way to convert a change in the ratio of ${}^{14}_{6}C$ to ${}^{12}_{6}C$ to the fabric's age. Letting A_0 be the ratio of ${}^{14}_{6}C$ to ${}^{12}_{6}C$ in modern fibers, we assign it a value of 1.00. The ratio of ${}^{14}_{6}C$ to ${}^{12}_{6}C$ in the sample, A, is 0.809. Solving gives

$$t = \ln\left(\frac{A_0}{A}\right) \times \frac{t_{1/2}}{0.693} = \ln\left(\frac{1.00}{0.809}\right) \times \left(\frac{5730 \text{ years}}{0.693}\right) = 1750 \text{ years}$$

Other isotopes can be used to determine the age of samples. The age of rocks, for example, has been determined from the ratio of the number of radioactive ${}^{238}_{92}U$ atoms to the number of stable ${}^{206}_{82}Pb$ atoms produced by radioactive decay. For rocks that do not contain uranium, dating is accomplished by comparing the ratio of radioactive ${}^{40}_{19}K$ to the stable ${}^{40}_{18}Ar$. Another example is the dating of sediments collected from lakes by measuring the amount of ${}^{210}_{82}Pb$ present.

13B.5 Evaluation

Radiochemical methods are routinely used for the analysis of trace analytes in macro- and mesosamples. The accuracy and precision of radiochemical analyses are generally within the range of 1–5%. Precision is limited by the random nature of radioactive decay and is improved by counting the emission of radioactive particles over as long a time as is practical. If the number of counts, M, is reasonably large ($M \geq 100$), and the counting period is significantly less than the isotopes half-life, then the percent relative standard deviation for the activity, $(\sigma_A)_{rel}$, is estimated as

$$(\sigma_A)_{rel} = \frac{1}{\sqrt{M}} \times 100$$

For example, when the activity is determined by counting 10,000 radioactive particles, the relative standard deviation is 1%. The analytical sensitivity of a radiochemical method is inversely proportional to the standard deviation of the measured activity and, therefore, is improved by increasing the number of particles that are counted.

Selectivity rarely is of concern with radiochemical methods because most samples contain only a single radioactive isotope. When several radioactive isotopes are present, differences in the energies of their respective radioactive particles can be used to determine each isotope's activity.

In comparison with most other analytical techniques, radiochemical methods are usually more expensive and require more time to complete an analysis. Radiochemical methods also are subject to significant safety concerns due to the analyst's potential exposure to high-energy radiation and the need to safely dispose of radioactive waste.

13C Flow Injection Analysis

The focus of this chapter is on methods in which the signal is time-dependent. Methods based on the kinetics of chemical and nuclear reactions were presented in the previous two sections. In this section we consider the technique of flow injection analysis. In this technique the sample is injected into a flowing carrier stream and swept through the detector, giving rise to a transient signal. The shape of this signal depends on the physical and chemical kinetic processes occurring in the carrier stream during the time between injection and detection.

13C.1 Theory and Practice

Flow injection analysis (FIA) was developed in the mid-1970s as a highly efficient technique for the automated analyses of samples.[20,21] Unlike the centrifugal analyzer described earlier in this chapter, in which samples are simultaneously analyzed in batches of limited size, FIA allows for the rapid, sequential analysis of an unlimited number of samples. FIA is one member of a class of techniques called continuous-flow analyzers, in which samples are introduced sequentially at regular intervals into a liquid carrier stream that transports the samples to the detector.[22]

flow injection analysis
An analytical technique in which samples are injected into a carrier stream of reagents, or in which the sample merges with other streams carrying reagents before passing through a detector.

Figure 13.16 is a schematic diagram detailing the basic components of a flow injection analyzer. The reagent serving as the carrier is stored in a reservoir, and a propelling unit maintains a constant flow of the carrier through the system of tubing comprising the instrument. The sample is injected directly into the flowing carrier stream, where it travels through a mixing and reaction zone before passing through the detector's flow-cell. Figure 13.16 is the simplest design for a flow injection analyzer, consisting of a single channel with one reagent reservoir. Multiple-channel instruments, in which reagents contained in separate reservoirs are combined

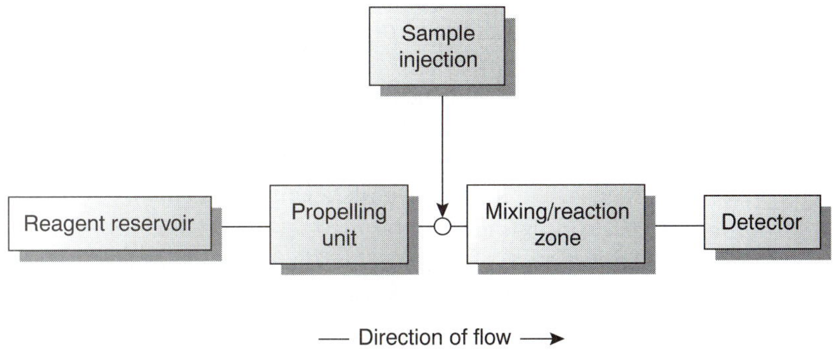

Figure 13.16
Schematic diagram of a simple flow injection analyzer showing the principal units.

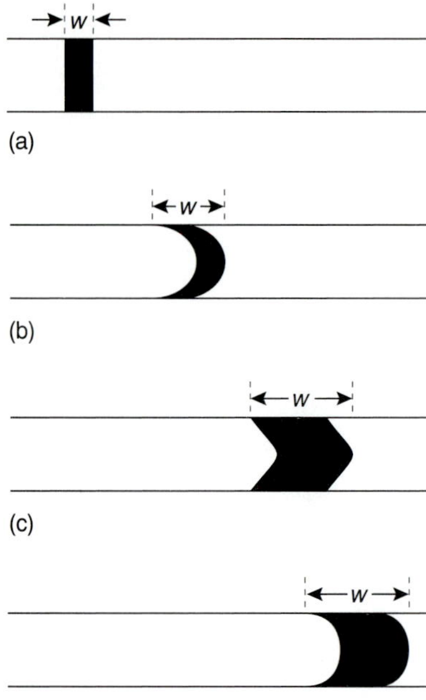

Figure 13.17
Effect of dispersion on a sample's flow profile at different times during a flow injection analysis: (a) at injection; and when the dispersion is due to (b) convection; (c) convection and diffusion; and (d) diffusion. The widths of the flow profiles also are shown.

fiagram
A graph showing the detector's response as a function of time in a flow injection analysis.

by merging channels, also are possible. A more detailed discussion of FIA instrumentation is found in the next section.

When a sample is injected into the carrier stream it has the rectangular flow profile (of width w) shown in Figure 13.17a. As the sample is carried through the mixing and reaction zone, the width of the flow profile increases as the sample disperses into the carrier stream. Dispersion results from two processes: convection due to the flow of the carrier stream and diffusion due to a concentration gradient between the sample and the carrier stream. Convection of the sample occurs by laminar flow, in which the linear velocity of the sample at the tube's walls is zero, while the sample at the center of the tube moves with a linear velocity twice that of the carrier stream. The result is the parabolic flow profile shown in Figure 13.7b. Convection is the primary means of dispersion in the first 100 ms following the sample's injection.

The second contribution to the sample's dispersion is diffusion due to the concentration gradient between the sample and the carrier stream. Diffusion occurs parallel (axial) and perpendicular (radial) to the flow of the carrier stream, with only the latter contribution being important. Radial diffusion decreases the linear velocity of the sample at the center of the tubing, but the sample at the edge of the tubing experiences an increase in its linear velocity. Diffusion helps to maintain the integrity of the sample's flow profile (Figure 13.17c), preventing samples in the carrier stream from dispersing into one another. Both convection and diffusion make significant contributions to dispersion from approximately 3–20 s after the sample's injection. This is the normal time scale for a flow injection analysis. After approximately 25 s, diffusion becomes the only significant contributor to dispersion, resulting in a flow profile similar to that shown in Figure 13.17d.

An FIA curve, or **"fiagram,"** is a plot of the detector's signal as a function of time. Figure 13.18 shows a typical fiagram for conditions in which both convection and diffusion contribute to the sample's dispersion. Also shown on the figure are several parameters used to characterize the fiagram. Two parameters are used to define the time required for the sample to move from the injector to the detector. The travel time, t_a, is the elapsed time from the sample's injection to the arrival of the leading edge of its flow profile at the detector. Residence time, T, on the other hand, is the time required to obtain the maximum signal. The difference between the residence time and travel time is given as t'. The value for t' approaches 0 when convection is the primary means of dispersion and increases in value as the contribution from diffusion becomes more important.

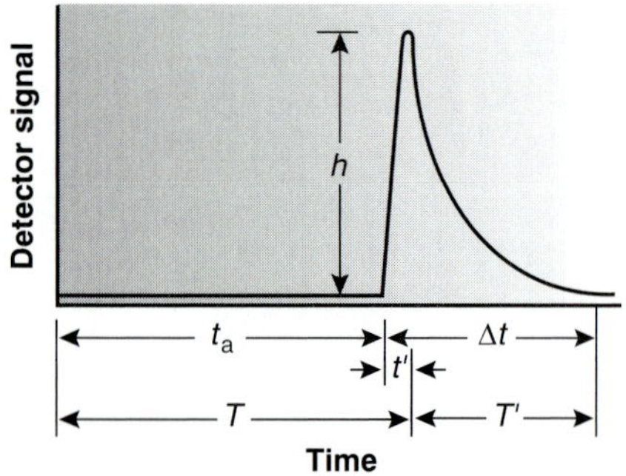

Figure 13.18
Typical FIA curve showing characteristic parameters.

The time required for the sample to pass through the detector's flow cell, and for the signal to return to the baseline, is also described by two parameters. The baseline-to-baseline time, Δt, is the elapsed time between the arrival of the leading edge of the sample's flow profile to the departure of its trailing edge. The elapsed time between the maximum signal and its return to the baseline is called the return time, T'. The final characteristic parameter of a fiagram is the peak height, h, which is equivalent to the difference between the maximum signal and the signal at the baseline.

Of the six parameters shown in Figure 13.18, the most important are peak height and return time. The peak height is related, directly or indirectly, to the analyte's concentration and is used for quantitative work. The sensitivity of the method, therefore, is also determined by the peak height. The return time determines the frequency with which samples may be injected.[22] Figure 13.19 shows that when a second sample is injected at a time T' after injecting the first sample,

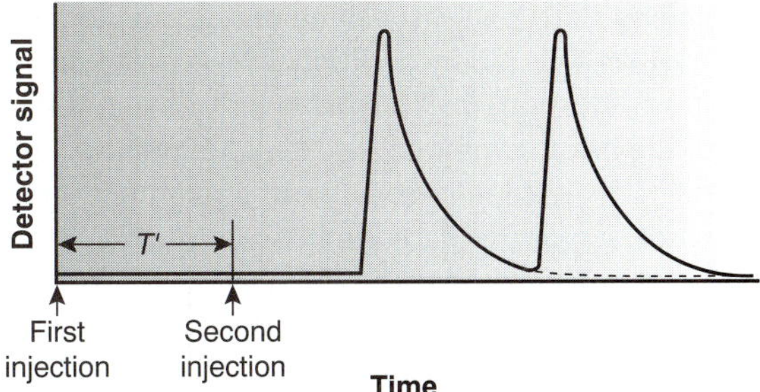

Figure 13.19
Effect of return time on sampling frequency.

the overlap of the two FIA curves is minimal. By injecting samples at intervals of T', the maximum sampling rate is realized.

Peak heights and return times are influenced by the dispersion of the sample's flow profile and are influenced by the physical and chemical properties of the flow injection system. Physical parameters affecting the peak height and return time include the volume of sample injected; the flow rate; the length, diameter, and geometry of the mixing and reaction zone; and the presence of mixing points where separate channels merge together. The kinetics of any chemical reactions involving the sample and reagents in the carrier stream also influences the peak height and return time.

Unfortunately, there is no good theory that can be used to consistently predict the peak height and return time for a given set of physical and chemical parameters. The design of a flow injection analyzer for a particular analytical problem still occurs largely by a process of experimentation. Nevertheless, some general observations about the effects of physical and chemical parameters can be made. In the absence of chemical effects, sensitivity (larger peak height) is improved by injecting larger samples, increasing the flow rate, decreasing the length and diameter of the tubing in the mixing and reaction zone, and merging separate channels before the point where the sample is injected. Except for sample volume, an improvement in the sampling rate (smaller return time) is achieved by the same combination of physical parameters. Larger sample volumes, however, lead to longer return times and a decrease in sample throughput. The effect of chemical reactivity depends on whether the species monitored by the detector is a reactant or a product. For example, when the monitored species is a reactant, sensitivity is improved by selecting a combination of physical parameters that enables the sample to reach the detector more quickly. Adjusting the chemical composition of the carrier stream in a manner that decreases the rate of the reaction also improves sensitivity in this case.

13C.2 Instrumentation

The basic components of a flow injection analyzer are shown in Figure 13.16 and include a unit for propelling the carrier stream, a means for injecting the sample into the carrier stream, and a detector for monitoring the composition of the carrier stream. These units are connected by a transport system that provides a means for bringing together separate channels and that enables an appropriate mixing of the sample with the carrier stream. Separations modules also may be incorporated in the flow injection analyzer. Each of these components is considered in greater detail in this section.

peristaltic pump
A device for propelling liquids through flexible tubing.

Propelling Unit The propelling unit is used to move the carrier stream in the main channel, as well as any additional reagent streams in secondary channels, through the flow injection analyzer. Although several different propelling units have been used, the most common is a peristaltic pump. A **peristaltic pump** consists of a set of rollers attached to the outside of a rotating drum (Figure 13.20). Tubing from the reagent reservoirs is placed in between the rollers and a fixed plate. As the drum rotates the rollers squeeze the tubing, forcing the contents of the tubing to move in the direction of the rotation. Peristaltic pumps are capable of providing a constant flow rate, which is controlled by the drum's speed of rotation and the inner diameter of the tubing. Flow rates from 0.0005–40 mL/min are possible, which is more than adequate to meet the needs of FIA, for which flow rates of 0.5–2.5 mL/min are common. One limitation to peristaltic pumps is that they produce a pulsed flow, particularly at higher flow rates, which may lead to oscillations in the signal.

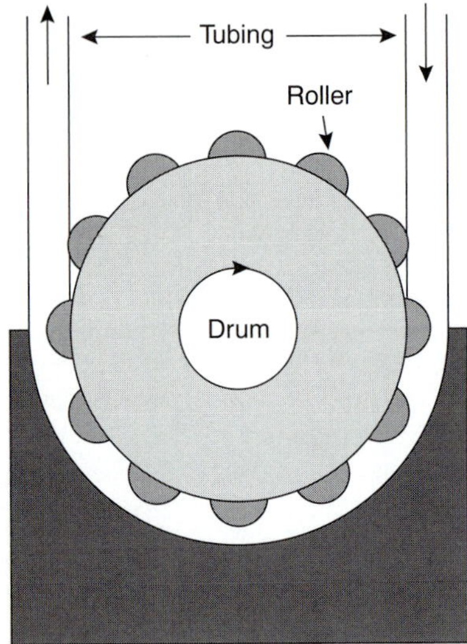

Figure 13.20
Schematic diagram of a peristaltic pump.

Injector The sample, typically 5–200 μL, is placed in the carrier stream by injection. Although syringe injections through a rubber septum are used, a more common means of injection is the rotary, or loop, injector used in HPLC and shown in Figure 12.28 of Chapter 12. This type of injector provides reproducible injection volumes and is easily adaptable to automation, a feature that is particularly important when high sampling rates are desired.

Detector Detection in FIA may be accomplished using many of the electrochemical and optical detectors used in HPLC. These detectors were discussed in Chapter 12 and are not considered further in this section. In addition, FIA detectors also have been designed around the use of ion-selective electrodes and atomic absorption spectroscopy.

Transport System The heart of a flow injection analyzer is the transport system used to bring together the carrier stream, the sample, and any reagents that must react with the sample to generate the desired signal. Each reagent reservoir connected to the flow injection analyzer is considered a separate channel. All channels must merge before the carrier stream reaches the detector, with the merging points determined by the chemistry involved in the method. The completed assembly of channels is called a **manifold.**

manifold
The complete system of tubing for merging together samples and reagents in a flow injection analysis.

The simplest manifold includes only a single channel, the basic outline of which is shown in Figure 13.21. This type of manifold is commonly used for direct analyses that do not require a chemical reaction. In this case the carrier stream only serves as a means for transporting the sample to the detector rapidly and reproducibly. For example, this manifold design has been used as a means of sample introduction in atomic absorption spectroscopy, achieving sampling rates as high as 700 samples/h. This manifold also is used for determining a sample's pH or the concentration of metal ions using ion-selective electrodes.

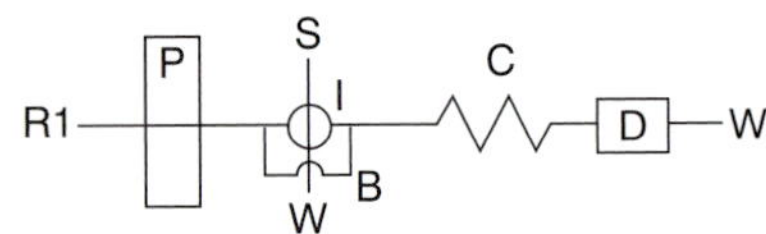

Figure 13.21
Example of a single-channel manifold for use in flow injection analysis where R1 is a reagent reservoir; P is the pump; S is the sample; I is the injector; B is a bypass loop; W is waste; C is the mixing and reaction coil; and D is the detector.

A single-channel manifold also can be used for systems in which a chemical reaction generates the species responsible for the analytical signal. In this case the carrier stream both transports the sample to the detector and reacts with the sample. Because the sample must mix with the carrier stream, flow rates are lower than when no chemical reaction is involved. One example is the determination of chloride in water, which is based on the following sequence of reactions.[23]

$$Hg(SCN)_2(aq) + 2Cl^-(aq) \rightleftharpoons HgCl_2(aq) + 2SCN^-(aq)$$

$$Fe^{3+}(aq) + SCN^-(aq) \rightleftharpoons Fe(SCN)^{2+}(aq)$$

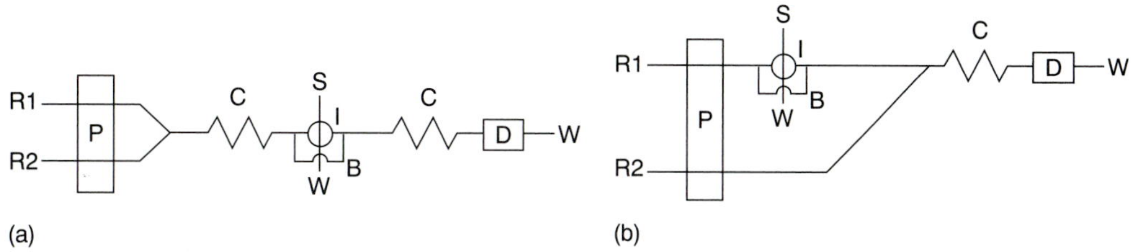

Figure 13.22
Two examples of dual-channel manifolds for use in flow injection analysis where R1 and R2 are reagent reservoirs; P is the pump; S is the sample; I is the injector; B is a bypass loop; W is waste; C is the mixing and reaction coil; and D is the detector.

The carrier stream consists of an acidic solution of $Hg(SCN)_2$ and Fe^{3+}. When a sample containing chloride is injected into the carrier stream, the chloride displaces the thiocyanate from $Hg(SCN)_2$. The displaced thiocyanate then reacts with Fe^{3+} to form the reddish colored $Fe(SCN)^{2+}$ complex, the absorbance of which is monitored at a wavelength of 480 nm. Sampling rates of approximately 120 samples/h have been achieved with this system.

Most flow injection analyses requiring a chemical reaction use a manifold containing more than one channel. This provides more control over the mixing of reagents and the interaction of the reagents with the sample. Two configurations are possible for dual-channel systems. The dual-channel manifold shown in Figure 13.22a is used when a mixture of the reagents is unstable. For example, in acidic solutions phosphate reacts with molybdate to form the heteropoly acid $H_3P(Mo_{12}O_{40})$. In the presence of ascorbic acid the molybdenum in the heteropoly acid is reduced from Mo(VI) to Mo(V), forming a blue-colored complex that is monitored spectrophotometrically at 660 nm.[23] Solutions of molybdate and ascorbic acid are maintained in separate reservoirs because they are not stable when mixed together. The two reagent channels are merged and mixed just before the point where the sample is injected.

A dual-channel manifold can also be used to mix a reagent in a secondary channel with a sample that has been injected into the primary channel (Figure 13.22b). This style of manifold has been used for the quantitative analysis of many analytes, including the determination of chemical oxygen demand (COD) in wastewater.[24] The COD is a measure of the amount of oxygen required to completely oxidize the organic matter in the water sample and is an indicator of the level of organic pollution. In the conventional method of analysis, COD is determined by refluxing the sample for 2 h in the presence of acid and a strong oxidizing agent, such as $K_2Cr_2O_7$ or $KMnO_4$. When refluxing is complete, the amount of oxidant consumed in the reaction is determined by a redox titration. In the flow injection version of this analysis, the sample is injected into a carrier stream of aqueous H_2SO_4, which merges with a solution of the oxidant from a secondary channel. The oxidation reaction is kinetically slow. As a result, the mixing and reaction coils are very long, typically 40 m, and are submerged in a thermostated bath. The sampling rate is lower than that for most flow injection analyses, but at 10–30 samples/h it is substantially greater than that for the conventional method.

More complex manifolds involving three or more channels are common,[22] but the possible combination of designs is too numerous to discuss in this text. One example of a four-channel manifold is shown in Figure 13.23.

Separation Modules Incorporating a separation module in the flow injection manifold allows separations, such as dialysis, gaseous diffusion, and liquid–liquid extraction, to be included in a flow injection analysis. Such separations are never complete, but are reproducible if the operating conditions are carefully controlled.

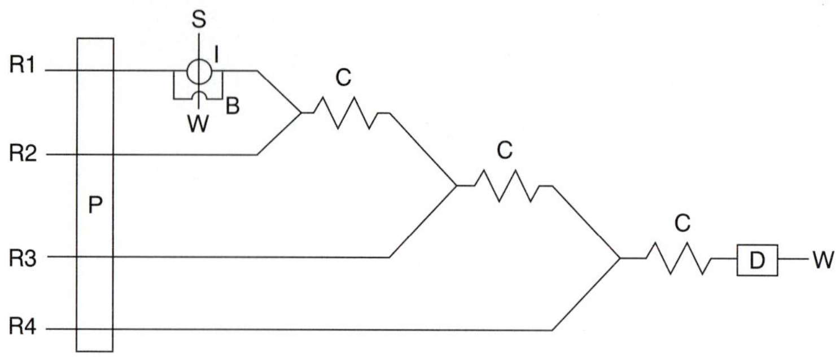

Figure 13.23
Example of a four-channel manifold for a flow injection analysis where R1 through R4 are reagent reservoirs; P is the pump; S is the sample; I is the injector; B is a bypass loop; W is waste; C is a mixing and reaction coil; and D is the detector.

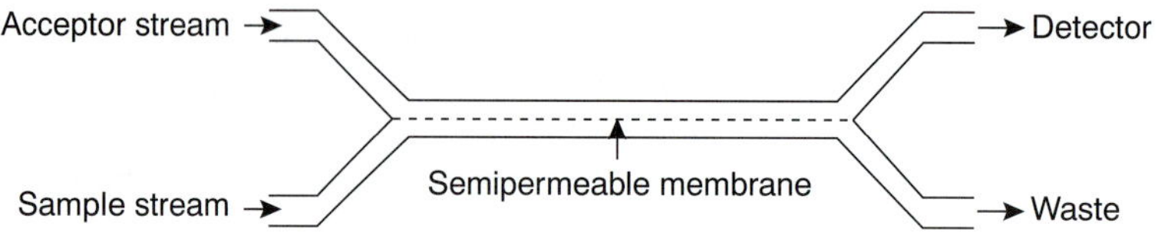

Figure 13.24
Separation module for a flow injection analysis using a semipermeable membrane for dialysis and gaseous diffusion.

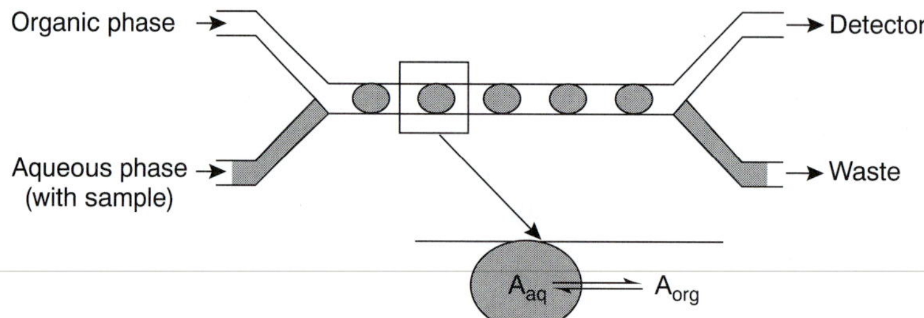

Figure 13.25
Separation module for a flow injection analysis using a liquid–liquid extraction (inset shows the equilibrium reaction).

Dialysis and gaseous diffusion are accomplished by placing a semipermeable membrane between the carrier stream containing the sample and an acceptor stream (Figure 13.24). As the sample stream passes through the separation module, a portion of those species capable of crossing the semipermeable membrane do so, entering the acceptor stream. This type of separation module is common in the analysis of clinical samples, such as serum and urine, for which dialysis separates the analyte from its complex matrix. Semipermeable gaseous diffusion membranes have been used for the determination of ammonia and carbon dioxide in blood. For example, ammonia is determined by injecting the sample into a carrier stream of aqueous NaOH. Ammonia diffuses across the semipermeable membrane into an acceptor stream containing an acid–base indicator. The resulting acid–base reaction between ammonia and the indicator is monitored spectrophotometrically.

Liquid–liquid extractions are accomplished by merging together two immiscible fluids, each carried in a separate channel. The result is a segmented flow through the separation module, consisting of alternating portions of the two phases. At the outlet of the separation module, the two fluids are separated by taking advantage of the difference in their densities. Figure 13.25 shows a typical configuration for a separation module in which the sample is injected into an aqueous phase and extracted into a less dense organic phase that passes through the detector.

Table 13.2 Selected Examples of the Application of FIA to the Analysis of Environmental Samples

Analtye	Sample	Volume Sample (μL)	Concentration Range	Sampling Frequency (h^{-1})
Ca^{2+}	fresh water	20	0.8–7.2 ppm	80
Cu^{2+}	groundwater	70–700	100–400 ppb	20
Pb^{2+}	groundwater	70–700	0–40 ppb	20
Zn^{2+}	sea water	1000	1–100 ppb	30–60
NH_4^+	sea water	60	0.18–18.1 ppm	288
NO_3^-	rain water	1000	1–10 ppm	40
SO_4^{2-}	fresh water	400	4–140 ppm	180
CN^-	industrial	10	0.3–100 ppm	40

Source: Adapted from Valcárcel, M.; Luque de Castro, M. D. *Flow-Injection Analysis: Principles and Applications.* Ellis Horwood: Chichester, England, 1987.

13C.3 Quantitative Applications

In a quantitative flow injection analysis a calibration curve is determined by injecting standard samples containing known concentrations of analyte. The format of the calibration curve, such as absorbance versus concentration, is determined by the method of detection. Calibration curves for standard spectroscopic and electrochemical methods were discussed in Chapters 10 and 11 and are not considered further in this chapter.

Flow injection analysis has been applied to a wide variety of samples, including environmental, clinical, agricultural, industrial, and pharmaceutical samples. The majority of analyses to date involve environmental and clinical samples, which is the focus of this section.

Quantitative analytical methods using FIA have been developed for cationic, anionic, and molecular pollutants in wastewater, fresh waters, groundwaters, and marine waters, several examples of which were described in the previous section. Table 13.2 provides a partial listing of other analytes that have been determined using FIA, many of which are modifications of conventional standard spectrophotometric and potentiometric methods. An additional advantage of FIA for environmental analysis is its ability to provide for the continuous, in situ monitoring of pollutants in the field.[25]

As noted in Chapter 9, several standard methods for the analysis of water involve acid–base, complexation, or redox titrations. Flow injection analysis also can be used for rapidly conducting such titrations using a single-channel manifold similar to that shown in Figure 13.21.[26] A solution, consisting of a titrant (the concentration of which is stoichiometrically less than that of the analyte) and an indicator, is placed in the reservoir and continuously pumped through the system. When a sample is injected, it thoroughly mixes with the titrant in the carrier stream. The reaction between the analyte, which is in excess, and the titrant produces a relatively broad flow profile for the sample. As the sample moves toward the detector, additional mixing occurs. Due to the continued reaction between titrant and analyte, the analyte's concentration decreases, and the width of its flow profile also decreases. As the sample passes through the detector, the width of its flow profile, Δt, is determined by monitoring the indicator's absorbance (Figure 13.26). Solutions with higher initial concentrations of analyte have a greater Δt. Calibration curves of Δt versus log [analyte] are prepared using standard solutions of analyte.

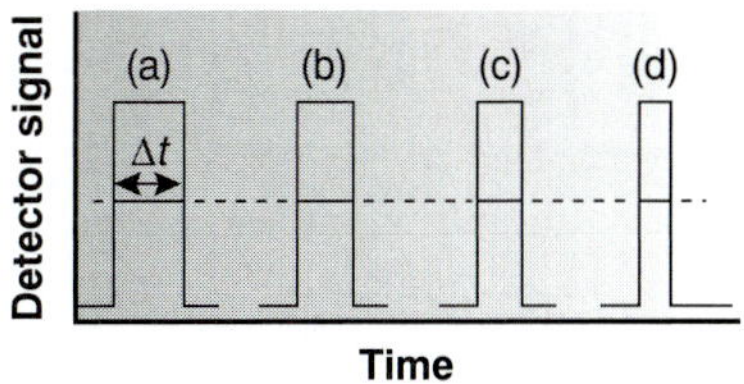

Figure 13.26
Use of FIA for titration experiments showing the determination of Δt. The concentration of the analyte decreases from (a) to (d).

Table 13.3 Selected Examples of the Application of FIA to the Analysis of Clinical Samples

Analyte	Sample	Volume Sample (μL)	Concentration Range	Sampling Frequency (h^{-1})
Nonenzymatic Methods				
Cu^{2+}	serum	20	0.7–1.5 ppm	70
Cl^-	serum	60	50–150 meq/L	125
PO_4^{3-}	serum	200	10–60 ppm	130
total CO_2	blood plasma	50	10–50 mM	70
chloropromazine	urine	200	1.5–9 μM	24
Enzymatic Methods				
glucose	blood serum	26.5	0.5–15 mM	60
urea	blood serum	30	4–20 mM	60
ethanol	blood	30	5–30 ppm	50

Source: Adapted from Valcárcel, M.; Luque de Castro, M. D. *Flow-Injection Analysis: Principles and Applications.* Ellis Horwood: Chichester, England, 1987.

Flow injection analysis has also found numerous applications in the analysis of clinical samples, using both enzymatic and nonenzymatic methods. A list of selected examples is given in Table 13.3.

Representative Method Although each FIA method has its own unique considerations, the determination of phosphate described in the following method provides an instructive example of a typical procedure.

Representative Methods

Method 13.2 Determination of Phosphate by FIA[27]

Description of Method. The FIA determination of phosphate is an adaptation of a standard spectrophotometric analysis for phosphate. In the presence of acid, phosphate reacts with molybdate to form a yellow-colored complex in which molybdenum is present as Mo(VI).

$$H_3PO_4(aq) + 12H_2MoO_4(aq) \rightleftharpoons H_3P(Mo_{12}O_{40})(aq) + 12H_2O(\ell)$$

In the presence of a reducing agent, such as ascorbic acid, the yellow-colored complex is reduced to a blue-colored complex of Mo(V).

Procedure. Prepare reagent solutions of 0.005 M ammonium molybdate in 0.40 M HNO_3, and 0.7% w/v ascorbic acid in 1% v/v glycerin. Using a stock solution of 100.0-ppm phosphate, prepare a set of external standards with phosphate concentrations of 10, 20, 30, 40, 50, and 60 ppm. Use a manifold similar to that shown in Figure 13.22a, with a 50-cm mixing coil and a 50-cm reaction coil. Set the flow rate to 0.5 mL/min. Prepare a calibration curve by injecting 50 μL of each standard, measuring the absorbance at 650 nm. Samples are analyzed in the same manner.

—Continued

Questions

1. **As noted, this procedure is adapted from a standard spectrophotometric method. The instructions for the spectrophotometric method indicate that the absorbance should be measured 5–10 min after adding the ascorbic acid. Why is this waiting period necessary in the spectrophotometric method, but not necessary in the FIA method?**

   The reduction of the yellow-colored Mo(VI) complex to the blue-colored Mo(V) complex is a slow reaction. In the standard spectrophotometric method, it is difficult to reproducibly control the amount of time that reagents are allowed to react before measuring the absorbance. To achieve good precision, therefore, the reaction is allowed sufficient time to proceed to completion before measuring the absorbance. In the FIA method, the flow rate and the dimensions of the reaction coil determine the elapsed time between sample introduction and the measurement of absorbance (about 30 s in this configuration). Since this time is precisely controlled, the reaction time is the same for all standards and samples.

2. **The spectrophotometric method recommends using standard solutions of phosphate in the range of 2–10 ppm, whereas the FIA method recommends standards in the range of 10–60 ppm. Explain why the methods use a different range of standards.**

   In the FIA method we measure the absorbance before the reduction of the yellow-colored Mo(VI) complex is complete. For this reason, the absorbance for any standard solution of phosphate will always be smaller when using the FIA method. This means that the FIA method is less sensitive, and higher concentrations of phosphate are necessary.

3. **How would you incorporate a reagent blank into the FIA analysis?**

   A reagent blank can be obtained by injecting a sample of distilled water in place of the external standard or sample. The reagent blank's absorbance is subtracted from the absorbances obtained for the standards and samples.

4. **The following data were obtained for a set of external phosphate standards. All absorbances have been corrected for a reagent blank.**

Parts per Million PO_4^{3-}	Absorbance
10.00	0.079
20.00	0.160
30.00	0.233
40.00	0.316
60.00	0.482

   What is the concentration of phosphate in the sample if it gives an absorbance of 0.287?

   A calibration curve of absorbance versus parts per million PO_4^{3-} gives the following calibration relationship.

$$\text{Absorbance} = -0.0030 + 8.04 \times 10^{-3}(\text{ppm } PO_4^{3-})$$

   Substituting the sample's absorbance gives its concentration of phosphate as 36.07 ppm.

13C.4 Evaluation

The majority of FIA applications are modifications of conventional titrimetric, spectrophotometric, and electrochemical methods of analysis. For this reason it is appropriate to evaluate FIA in relation to these conventional methods. The scale of operations for FIA allows for the routine analysis of minor and trace analytes and for macro-, meso-, and microsamples. The ability to work with microliter injection volumes is useful when the sample is scarce. Conventional methods of analysis, however, may allow the determination of smaller concentrations of analyte.

The accuracy and precision of FIA are comparable to that obtained by conventional methods of analysis. The precision of a flow injection analysis is influenced by variables that are not encountered in conventional methods, including the stability of the flow rate and the reproducibility of the sample's injection. In addition, results from FIA may be more susceptible to temperature variations. These variables, therefore, must be carefully controlled.

In general, the sensitivity of FIA is less than that for conventional methods of analysis for two principal reasons. First, as with chemical kinetic methods, measurements in FIA are made under nonequilibrium conditions when the signal has yet to reach its maximum value. Second, dispersion of the sample as it progresses through the system results in its dilution. As discussed earlier, however, the variables that influence sensitivity are known. As a result the FIA manifold can be designed to optimize the sensitivity of the analysis.

Selectivity in FIA is often better than that for conventional methods of analysis. In many cases this is due to the kinetic nature of the measurement process, in which potential interferents may react more slowly than the analyte. Contamination from external sources also is less of a problem since reagents are stored in closed reservoirs and are pumped through a system of transport tubing that, except for waste lines, is closed to the environment.

Finally, FIA is an attractive technique with respect to demands on time, cost, and equipment. When employed for automated analyses, FIA provides for very high sampling rates. Most analyses can be operated with sampling rates of 20–120 samples/h, but rates as high as 1700 samples/h have been realized.[22] Because the volume of the flow injection manifold is small, typically less than 2 mL, consumption of reagents is substantially less than with conventional methods. This can lead to a significant decrease in the cost per analysis. Flow injection analysis requires additional equipment, beyond that used for similar conventional methods of analysis, which adds to the expense of the analysis. On the other hand, flow injection analyzers can be assembled from equipment already available in many laboratories.

13D KEY TERMS

alpha particle (*p. 642*)
beta particle (*p. 642*)
enzyme (*p. 636*)
fiagram (*p. 650*)
flow injection analysis (*p. 649*)
gamma ray (*p. 642*)
Geiger counter (*p. 643*)
half-life (*p. 643*)
isotope dilution (*p. 646*)
isotopes (*p. 642*)
Lineweaver–Burk plot (*p. 638*)
manifold (*p. 652*)
Michaelis constant (*p. 637*)
negatron (*p. 642*)
neutron activation (*p. 645*)
peristaltic pump (*p. 652*)
positron (*p. 642*)
quench (*p. 634*)
rate (*p. 624*)
rate constant (*p. 624*)
rate law (*p. 624*)
scintillation counter (*p. 643*)
steady-state approximation (*p. 636*)
stopped flow (*p. 634*)
substrate (*p. 636*)
tracer (*p. 646*)

13E SUMMARY

Kinetic methods of analysis are based on the rate at which a chemical or physical process involving the analyte occurs. Three types of kinetic methods are discussed in this chapter: chemical kinetic methods, radiochemical methods, and flow injection analysis.

Chemical kinetic methods are based on the rate at which a chemical reaction involving the analyte proceeds. Either the integrated or differential form of the rate law may be used. When using an integral method, the concentration of analyte, or a reactant or product stoichiometrically related to the analyte, is determined at one or more points in time following the reactions initiation. The initial concentration of analyte is then determined using the integral form of the reaction's rate law. Alternatively, the time required to effect a given change in concentration may be measured. In a differential kinetic method the rate of the reaction is measured at a time, t, and the differential form of the rate law used to determine the analyte's concentration.

Chemical kinetic methods are particularly useful for reactions that are too slow for a convenient analysis by other analytical methods. In addition, chemical kinetic methods are often easily adapted to an automated analysis. For reactions with fast kinetics, automation allows hundreds (or more) of samples to be analyzed per hour. Another important application of chemical kinetic methods is for the quantitative analysis of enzymes and their substrates and for the characterization of enzyme catalysis.

Radiochemical methods of analysis take advantage of the decay of radioactive isotopes. A direct measurement of the rate at which a radioactive isotope decays may be used to determine its concentration in a sample. For analytes that are not naturally radioactive, neutron activation often can be used to induce radioactivity. Isotope dilution, in which a radioactively labeled form of an analyte is spiked into the sample, can be used as an internal standard for quantitative work.

In a flow injection analysis the sample is injected into a flowing carrier stream that often is merged with additional streams carrying reagents. As the sample moves with the carrier stream, it both reacts with the contents of the carrier stream and any additional reagent streams, and undergoes dispersion. The resulting fiagram of signal versus time bears some resemblance to a chromatogram. Unlike chromatography, however, flow injection analysis is not a separation technique. Because all components in a sample move with the carrier stream's flow rate, it is possible to introduce a second sample before the first sample reaches the detector. As a result, flow injection analysis is ideally suited for the rapid throughput of samples.

13F *Suggested* EXPERIMENTS

Experiments

The following experiments may be used to illustrate the application of kinetic methods of analysis. Experiments are divided into two groups: those based on chemical kinetics and those using flow injection analysis. Each suggested experiment includes a brief description.

The following experiments introduce the application of chemical kinetics, including enzyme kinetics.

Bateman, Jr. R. C.; Evans, J. A. "Using the Glucose Oxidase/Peroxidase Systems in Enzyme Kinetics," *J. Chem. Educ.* **1995,** *72,* A240–A241.

Glucose oxidase catalyzes the oxidation of glucose to gluconolactone and hydrogen peroxide. In the presence of horseradish peroxidase, the hydrogen peroxide is reduced by a water-soluble dye that forms a blue-green color when oxidized. The kinetics of the enzymatic reactions are followed by monitoring absorbance at 725 nm or 414 nm. Procedures are given for determining the activity of glucose oxidase and for studying the effect of glucose concentration on the reaction rate. The adaptation of the method for other sugars is also discussed.

Diamandis, E. P.; Koupparis, M. A.; Hadjiionnou, T. P. "Kinetic Studies with Ion-Selective Electrodes: Determination of Creatinine in Urine with a Picrate Ion-Selective Electrode," *J. Chem. Educ.* **1983,** *60,* 74–76.

This experiment includes instructions for preparing a picrate ion-selective electrode. The application of the electrode in determining the concentration of creatinine in urine (which is further described in Method 13.1) also is outlined.

Pandey, S.; McHale, M. E. R.; Horton, A. M.; et al. "Kinetics-Based Indirect Spectrophotometric Method for the Simultaneous Determination of MnO_4^- and $Cr_2O_7^{2-}$, *J. Chem. Educ.* **1998,** *75,* 450–452.

This experiment describes a kinetic method for determining the concentrations of MnO_4^- and $Cr_2O_7^{2-}$ in a mixture. The analysis is based on a difference in the rate at

—Continued

Experiments

Continued from page 659

which the analytes oxidize a solution of pyrogallol red; oxidation of pyrogallol red by MnO_4^- is faster than that for $Cr_2O_7^{2-}$. The reaction is followed by measuring the absorbance of pyrogallol at a wavelength of 488 nm at two reaction times. Calibration curves prepared using external standards provide the necessary information to determine the concentrations of MnO_4^- and $Cr_2O_7^{2-}$ in an unknown. Results can be compared with more traditional spectrophotometric methods for analyzing mixtures.

The following set of experiments provide examples of the application of flow injection analysis or the characterization of the behavior of a flow injection analysis system.

Carroll, M. K.; Tyson, J. F. "An Experiment Using Time-Based Detection in Flow Injection Analysis," *J. Chem. Educ.* **1993,** *70,* A210–A216.

Most flow injection analyses use peak height as the analytical signal. When there is insufficient time for reagents to merge with the sample, the result is a split-peak, or "doublet," due to reaction at the sample's leading and trailing edges. This experiment describes how the difference between the peak times can be used for quantitative work.

Hansen, E. H.; Ruzicka, J. "The Principles of Flow Injection Analysis as Demonstrated by Three Lab Exercises," *J. Chem. Educ.* **1979,** *56,* 677–680.

Directions are provided for the quantitative analyses of Cl^- and PO_4^{3-}, and for conducting titrations of strong acids and strong bases.

McKelvie, I. D.; Cardwell, T. J.; Cattrall, R. W. "A Microconduit Flow Injection Analysis Demonstration Using a 35-mm Slide Projector," *J. Chem. Educ.* **1990,** *67,* 262–263.

Directions are provided for constructing a small-scale FIA system that can be used to demonstrate the features of flow injection analysis. For another example see Grudpan, K.; Thanasarn, T. "Overhead Projector Injection Analysis," *Anal. Proc.* **1993,** *30,* 10–12.

Meyerhoff, M. E.; Kovach, P. M. "An Ion-Selective Electrode/Flow Injection Analysis Experiment: Determination of Potassium in Serum," *J. Chem. Educ.* **1983,** *60,* 766–768.

This experiment describes the preparation of a flow-through potentiometric electrode assembly incorporating a valinomycin–PVC membrane in the transport tubing. The application of this detector to the analysis of K^+ in serum is outlined.

Ríos, A.; Luque de Castro, M.; Valcárcel, M. "Determination of Reaction Stoichiometries by Flow Injection Analysis," *J. Chem. Educ.* **1986,** *63,* 552–553.

This experiment describes the use of FIA for determining the stoichiometry of the Fe^{2+}–*o*-phenanthroline complex using the method of continuous variations and the mole-ratio method. Directions are also provided for determining the stoichiometry of the oxidation of ascorbic acid by dichromate and for determining the rate constant for the reaction at different pH levels and different concentration ratios of the reactants.

Stults, C. L. M.; Wade, A. P.; Crouch, S. R. "Investigation of Temperature Effects on Dispersion in a Flow Injection Analyzer," *J. Chem. Educ.* **1988,** *65,* 645–647.

Many factors affect a sample's dispersion in an FIA analysis. In this experiment students study the effect of temperature on dispersion.

Wolfe, C. A. C.; Oates, M. R.; Hage, D. S. "Automated Protein Assay Using Flow Injection Analysis," *J. Chem. Educ.* **1998,** *75,* 1025–1028.

This experiment describes the adaptation of the bicinchoninic acid (BCA) protein assay to a flow injection analysis. The assay is based on the reduction of Cu^{2+} to Cu^+ by the protein, followed by the reaction of Cu^+ with bicinchoninic acid to form a purple complex that absorbs at 562 nm. Directions are provided for the analysis of bovine serum albumin and rabbit immunoglobulin G, and suggestions are provided for additional analyses.

13G PROBLEMS

1 Equation 13.14 shows how $[A]_0$ is determined for a two-point fixed-time integral method in which the concentration of A for the pseudo-first-order reaction

$$A + R \rightarrow P$$

is measured at times t_1 and t_2. Derive a similar equation for the case when the product is monitored under pseudo-first-order conditions.

2. The concentration of phenylacetate can be determined from the kinetics of its pseudo-first-order hydrolysis reaction in an ethylamine buffer. When a standard solution of 0.55 mM phenylacetate is analyzed, the concentration of phenylacetate after 60 s is found to be 0.17 mM. When an unknown is analyzed, the concentration of phenylacetate remaining after 60 s is found to be 0.23 mM. What is the initial concentration of phenylacetate in the unknown?

3. In the presence of acid, solutions of iodide are oxidized by hydrogen peroxide

$$2I^- + H_2O_2 + 2H_3O^+ \rightarrow 4H_2O + I_2$$

When I^- and H_3O^+ are present in excess, the kinetics of the reaction are pseudo-first-order in H_2O_2, and can be used to determine the concentration of H_2O_2 by following the production of I_2 with time. In one analysis the absorbance of the solution was measured after 240 s at 348 nm (where Beer's law holds for I_2). When a set of standard solutions of H_2O_2 was analyzed, the following results were obtained

$[H_2O_2]$ (μM)	Absorbance
100.0	0.236
200.0	0.471
400.0	0.933
800.0	1.872

What is the concentration of H_2O_2 in a sample that has an absorbance of 0.669 after 240 s?

4. The concentration of chromic acid can be determined from its reduction by alcohols under conditions when the kinetics are pseudo-first-order in analyte. One approach is to monitor the absorbance of the solution at a wavelength of 355 nm. A standard solution of 5.1×10^{-4} M chromic acid yields absorbances of 0.855 and 0.709 at, 100 s and 300 s, respectively, after the reaction's initiation. When a sample with an unknown amount of chromic acid is analyzed under identical conditions, absorbances of 0.883 and 0.706 are obtained. What is the concentration of chromic acid in this sample?

5. Malmstadt and Pardue developed a variable-time method for the determination of glucose based on its oxidation by the enzyme glucose oxidase.[28] To monitor the reaction's progress, iodide is added to the samples and standards. The H_2O_2 produced by the oxidation of glucose reacts with the I^-, giving I_2 as a product. The time required to produce a fixed amount of I_2 is determined spectrophotometrically. The following data were reported for a set of calibration standards

Glucose (ppm)		Time (s)	
5.0	146.5	150.0	149.6
10.0	69.2	67.1	66.0
20.0	34.8	35.0	34.0
30.0	22.3	22.7	22.6
40.0	16.7	16.5	17.0
50.0	13.3	13.3	13.8

To verify the method a 1.00-mL aliquot of a standard solution of 40.0-ppm glucose was added to 1.00 mL of the combined reagents, requiring 34.6 s to produce the same extent of reaction. Determine the calculated concentration of glucose in the standard and the percent error for the analysis.

6. Deming and Pardue studied the kinetics for the hydrolysis of *p*-nitrophenyl phosphate by the enzyme alkaline phosphatase.[29] The progress of the reaction was monitored by measuring the absorbance due to *p*-nitrophenol, which is one of the products of the reaction. A plot of the rate of the reaction (with units of $\mu mol\ mL^{-1}\ s^{-1}$) versus the volume, V, (in milliliters) of a serum calibration standard containing the enzyme yielded a straight line with the following equation

$$\text{Rate (in } \mu\text{mol mL}^{-1}\text{ s}^{-1}) = 2.7 \times 10^{-7} + 3.485 \times 10^{-5} \times V$$

A 10.00-mL sample of serum is analyzed, yielding a rate of 6.84×10^{-5} $\mu mol\ mL^{-1}\ s^{-1}$. How much more dilute is the enzyme in the serum sample than in the serum calibration standard?

7. The following data were collected for a reaction known to be pseudo-first-order in analyte, A, during the time in which the reaction is monitored.

Time (s)	$[A]_t$ (mM)
2	1.36
4	1.24
6	1.12
8	1.02
10	0.924
12	0.838
14	0.760
16	0.690
18	0.626
20	0.568

What are the rate constant and the initial concentration of analyte in the sample?

8. The enzyme acetylcholinesterase catalyzes the decomposition of acetylcholine to choline and acetic acid. Under a given set of conditions the enzyme has a K_m of 9×10^{-5} M, and a k_2 of 1.4×10^4 s^{-1}. What is the concentration of acetylcholine in a sample for which the rate of reaction in the presence of 6.61×10^{-7} M enzyme is 12.33 μM s^{-1}? You may assume that the concentration of acetylcholine is significantly smaller than K_m.

9. The enzyme fumarase catalyzes the stereospecific addition of water to fumarate to form L-malate. A standard solution of fumarase, with a concentration of 0.150 μM, gave a rate of reaction of 2.00 μM min^{-1} under conditions in which the concentration of the substrate was significantly greater than K_m. The rate of reaction for a sample, under identical conditions, was found to be 1.15 μM min^{-1}. What is the concentration of fumarase in the sample?

10. The enzyme urease catalyzes the hydrolysis of urea. The rate of this reaction was determined for a series of solutions in which the concentration of urea was changed while maintaining a fixed urease concentration of 5.0 μM. The following data were obtained.

[urea] (M)	Rate (Ms^{-1})
1.00×10^{-7}	6.25×10^{-6}
2.00×10^{-7}	1.25×10^{-5}
3.00×10^{-7}	1.88×10^{-5}
4.00×10^{-7}	2.50×10^{-5}
5.00×10^{-7}	3.12×10^{-5}
6.00×10^{-7}	3.75×10^{-5}
7.00×10^{-7}	4.37×10^{-5}
8.00×10^{-7}	5.00×10^{-5}
9.00×10^{-7}	5.62×10^{-5}
1.00×10^{-6}	6.25×10^{-5}

Determine the values of V_{max}, k_2, and K_m for urease.

11. In a study of the effect of an enzyme inhibitor, values for V_{max} and K_m were measured for several concentrations of inhibitor. As the concentration of the inhibitor was increased, V_{max} was found to remain essentially constant while the value of K_m increased. Which mechanism for enzyme inhibition is in effect?

12. In the case of competitive inhibition, the equilibrium between the enzyme, E, the inhibitor, I, and the enzyme–inhibitor complex, EI, is described by the equilibrium constant K_I. Show that for competitive inhibition the equation for the rate of reaction is

$$\frac{d[P]}{dt} = \frac{V_{max}[S]}{K_m\{1 + ([I]/K_I)\} + [S]}$$

where K_I is the dissociation constant for the EI complex

$$E + I \rightleftharpoons EI$$

You may assume that $k_2 << k_{-1}$.

13. Analytes A and B react with a common reagent R with first-order kinetics. If 99.9% of A must react before 0.1% of B has reacted, what is the minimum acceptable ratio for their respective rate constants?

14. A mixture of two analytes, A and B, is analyzed simultaneously by monitoring their combined concentration, C = [A] + [B], as a function of time when they react with a common reagent. Both A and B are known to follow first-order kinetics with the reagent, and A is known to react faster than B. Given the data in the following table,

Time (min)	C (mM)
1	0.313
6	0.200
11	0.136
16	0.098
21	0.074
26	0.058
31	0.047
36	0.038
41	0.032
46	0.027
51	0.023
56	0.019
61	0.016
66	0.014
71	0.012

determine the initial concentrations of A and B, and the first-order rate constants, k_A and k_B.

15. Table 13.1 provides a list of several isotopes commonly used as tracers. The half-lives for these isotopes also are listed. What is the rate constant for the radioactive decay of each isotope?

16. ^{60}Co is a long-lived isotope ($t_{1/2}$ = 5.3 years) that frequently is used as a radiotracer. The activity in a 5.00-mL sample of a solution of ^{60}Co was found to be 2.1×10^7 disintegrations/s. What is the molar concentration of ^{60}Co in the sample?

17. The concentration of Ni in a new alloy is determined by a neutron activation analysis using the method of external standards. A 0.500-g sample of the alloy and a 1.000-g sample of a standard alloy known to contain 5.93% w/w Ni are irradiated with neutrons in a nuclear reactor. When irradiation is complete, the sample and standard are allowed to cool, and the gamma-ray activities are measured. Given that the activity is 1020 cpm for the sample and 3540 cpm for the standard, determine the %w/w Ni in the alloy.

18. The vitamin B_{12} content of a multivitamin tablet is determined by dissolving ten tablets in water. The dissolved tablets are transferred to a 100-mL volumetric flask and diluted to volume. A 50.00-mL portion is removed and treated with 0.500 mg of radioactive vitamin B_{12} having an activity of 572 cpm. After homogenization, the vitamin B_{12} in the sample is isolated and purified, producing 18.6 mg with an activity of 361 cpm. Calculate the average concentration of vitamin B_{12} in the tablet (in milligrams per tablet).

19. The oldest sample that can be dated by carbon-14 is approximately 30,000 years. What percentage of the ^{14}C still remains after this time span?

20. Potassium–argon dating is based on the nuclear decay of ^{40}K ($t_{1/2} = 1.3 \times 10^9$ years) to ^{40}Ar. Assuming that there is no ^{40}Ar originally present in the rock and that the ^{40}Ar cannot escape to the atmosphere, then the relative amounts of ^{40}K and ^{40}Ar can be used to determine the age of the rock. When 100.0 mg of a rock sample was analyzed, it was found to contain 4.63×10^{-6} mol of ^{40}K and 2.09×10^{-6} mol ^{40}Ar. How old is the rock sample?

21. The steady-state activity for ^{14}C is 15 cpm/g of carbon. What mass of carbon is needed to give a percent relative standard deviation of 1.0% for the activity of a sample if counting is limited to 1 h? How long must the radioactive decay from a 0.50-g sample of carbon be monitored to give a percent relative standard deviation of 1% for the activity?

22. Shown here is a fiagram obtained for a solution of 100.0-ppm PO_4^{3-}. Determine h, t_a, T, t', Δt, and T'. What is the sensitivity of this FIA method (assuming a linear relationship between absorbance and concentration)? How many samples can be analyzed per hour?

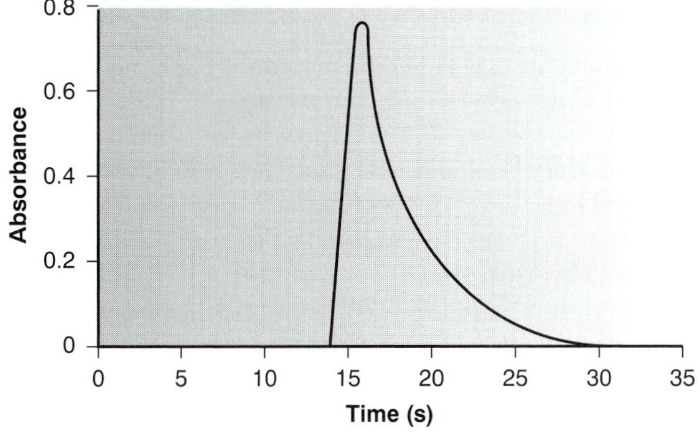

23. A sensitive method for the flow injection analysis of Cu^{2+} is based on its ability to catalyze the oxidation of di-2-pyridyl ketone hydrazone (DPKH) by atmospheric oxygen.[30] The product of the reaction is fluorescent and can be used to generate a signal when using a fluorometer as a detector. The yield of the reaction is at a maximum when the solution is made basic with NaOH. The fluorescence, however, is greatest in the presence of HCl. Sketch an FIA manifold that will be appropriate for this analysis.

24. The concentration of chloride in sea water can be determined by a flow injection analysis. The analysis of a set of calibration standards gives the results in the following table.

Cl^- (ppm)	Absorbance
5.00	0.057
10.00	0.099
20.00	0.230
30.00	0.354
40.00	0.478
50.00	0.594
75.00	0.840

A 1.00-mL sample of sea water is placed in a 500-mL volumetric flask and diluted to volume with distilled water. When injected into the flow injection analyzer, an absorbance of 0.317 is measured. What is the concentration of Cl^- in the sample?

25. Ramsing and colleagues[26] developed an FIA method for acid–base titrations using a carrier stream mixture of 2.0×10^{-3} M NaOH and the acid–base indicator bromthymol blue. Standard solutions of HCl were injected, and the following values of Δt were measured from the resulting fiagrams.

[HCl] (M)	Δt (s)
0.008	3.13
0.010	3.59
0.020	5.11
0.040	6.39
0.060	7.06
0.080	7.71
0.100	8.13
0.200	9.27
0.400	10.45
0.600	11.40

A sample with an unknown concentration of HCl was analyzed five times, giving values for Δt of 7.43, 7.28, 7.41, 7.37, and 7.33 s. Determine the concentration of HCl in the sample.

26. Milardovíc and co-workers used a flow injection analysis method with an amperometric biosensor to determine the concentration of glucose in blood.[31] Given that a blood sample that is 6.93 mM in glucose has a signal of 7.13 nA, what is the concentration of glucose in a sample of blood giving a signal of 11.50 nA?

27. Fernández-Abedul and Costa-García developed an FIA method for detecting cocaine ($C_{17}H_{21}NO_4$) in samples.[32] The following data were collected for 12 replicate injections of a sample with a cocaine concentration of 6.2×10^{-6} M.

Replicate	Signal (arbitrary units)
1	24.5
2	24.1
3	24.1
4	23.8
5	23.9
6	25.1
7	23.9
8	24.8
9	23.7
10	23.3
11	23.2
12	23.2

(a) What is the relative standard deviation for the analysis of this sample? (b) Given that these replicate analyses required 10.5 min, what is the expected sample throughput in samples per hour? (c) The following calibration data are available.

[cocaine] (M)	Signal (arbitrary units)
1.8×10^{-7}	0.8
3.6×10^{-7}	2.1
6.0×10^{-7}	2.4
8.1×10^{-7}	3.2
1.0×10^{-6}	4.5
2.0×10^{-6}	8.1
4.0×10^{-6}	14.4
6.0×10^{-6}	21.6
8.0×10^{-6}	27.1
1.0×10^{-5}	32.9

In a typical analysis a 10.0-mg sample is dissolved in water and diluted to volume in a 25-mL volumetric flask. A 125-μL aliquot is transferred to a 25-mL volumetric flask and diluted to volume with a pH 9 buffer. When injected into the flow injection apparatus a signal of 21.4 is obtained. What is the %w/w cocaine in the sample?

28. Holman and associates described an FIA method for determining the concentration of H_2SO_4 in nonaqueous solvents.[33] Agarose beads (22 – 45-μm diameter) with a bonded acid–base indicator are soaked in NaOH and immobilized in the detector's flow cell. Solutions of H_2SO_4 in *n*-butanol are injected into the carrier stream. As the sample passes through the flow cell, the reaction of H_2SO_4 and NaOH takes place. The end point of this neutralization reaction is signaled by a change in the bound indicator's color and is detected spectrophotometrically. The elution volume needed to reach the titration's end point is inversely proportional to the concentration of H_2SO_4; thus, a plot of end point volume versus $[H_2SO_4]^{-1}$ should be linear. The following data are typical of that obtained using a set of external standards.

$[H_2SO_4]$ (M)	End Point Volume (mL)
3.58×10^{-4}	0.266
4.36×10^{-4}	0.227
5.60×10^{-4}	0.176
7.52×10^{-4}	0.136
1.38×10^{-3}	0.075
2.98×10^{-3}	0.037
5.62×10^{-3}	0.017

What is the concentration of H_2SO_4 in a sample whose end point volume is 0.157 mL?

13H SUGGESTED READINGS

The following source provides a general review of the importance of kinetics in analytical chemistry.

Mottola, H. A. "Some Kinetic Aspects Relevant to Contemporary Analytical Chemistry." *J. Chem. Ed.* **1981,** *58,* 399–403.

A brief history of chemical kinetic methods of analysis is found in the following text.

Laitinen, H. A.; Ewing, G. W., eds. *A History of Analytical Chemistry,* The Division of Analytical Chemistry of the American Chemical Society: Washington, D.C., 1977, pp. 97–102.

The following are useful resources for further information regarding chemical kinetic methods of analysis.

Malmstadt, H. V.; Delaney, C. J.; Cordos, E. A. "Reaction-Rate Methods of Chemical Analysis," *Crit. Rev. Anal. Chem.* **1972,** *2,* 559–619.

Mark, H. B.; Rechnitz, G. A. *Kinetics in Analytical Chemistry.* Wiley: New York, 1968.

Mottola, H. A. "Catalytic and Differential Reaction-Rate Methods of Chemical Analysis," *Crit. Rev. Anal. Chem.* **1974,** *4,* 229–280.

Mottola, H. A. *Kinetic Aspects of Analytical Chemistry.* Wiley: New York, 1988.

Pardue, H. L. "Kinetic Aspects of Analytical Chemistry," *Anal. Chim. Acta* **1989,** *216,* 69–107.

Perez-Bendito, D.; Silva, M. *Kinetic Methods in Analytical Chemistry.* Ellis Horwood: Chichester, England, 1988.

Additional information on the kinetics of enzyme catalyzed reactions may be found in the following texts.

Bergmyer, H. U.; Grassl, M. *Methods of Enzymatic Analysis,* 3rd ed. Verlag Chemie: Deerfield Beach, FL, 1983.

Pisakiewicz, D. *Kinetics of Chemical and Enzyme-Catalyzed Reactions.* Oxford University Press: New York, 1977.

The following instrumental analysis textbooks may be consulted for further information on the detectors and signal analyzers used in radiochemical methods of analysis.

Skoog, D. A.; Leary, J. J. *Principles of Instrumental Analysis,* 4th ed., Saunders College Publishing/Harcourt Brace and Co.: Philadelphia, 1998, Chapter 32.

Strobel, H. A.; Heineman, W. R. *Chemical Instrumentation: A Systematic Approach,* 3rd ed., Wiley-Interscience: New York, 1989.

The following resources provide additional information on the theory and application of flow injection analysis.

Betteridge, D. "Flow Injection Analysis," *Anal. Chem.* **1978,** *50,* 832A–846A.

Kowalski, B. R.; Ruzicka, J. Christian, G. D. "Flow Chemography—The Future of Chemical Education," *Trends Anal. Chem.* **1990,** *9,* 8–13.

Mottola, H. A. "Continuous Flow Analysis Revisited," *Anal. Chem.* **1981,** *53,* 1312A–1316A.

Ruzicka, J. "Flow Injection Analysis: From Test Tube to Integrated Microconduits," *Anal. Chem.* **1983,** *55,* 1040A–1053A.

Ruzicka, J.; Hansen, E. H. "Flow-Injection Analysis," Wiley-Interscience: New York, 1989.

Stewart, K. K. "Flow Injection Analysis: New Tools for Old Assays, New Approaches to Analytical Measurements," *Anal. Chem.* **1983,** *55,* 931A–940A.

Valcárcel, M.; Luque de Castro, M. D. *Flow-Injection Analysis: Principles and Applications,* Ellis Horwood: Chichester, England, 1987.

The use of flow injection as a means of introducing samples in atomic absorption spectroscopy is covered in the following paper.

Tyson, J. F. "Atomic Spectrometry and Flow Injection Analysis: A Synergic Combination," *Anal. Chim. Acta* **1988,** *214,* 57–75.

13I REFERENCES

1. Method 4500-NO_2^- B in *Standard Methods for the Analysis of Waters and Wastewaters,* 20th ed. American Public Health Association: Washington, D.C., 1998, pp. 4-112–4-114.
2. Karayannis, M. I.; Piperaki, E. A.; Maniadake, M. M. *Anal. Lett.* **1986,** *19,* 13–23.
3. Pardue, H. L. *Anal. Chim. Acta* **1989,** *216,* 69–107.
4. Mottola, H. A. *Anal. Chim. Acta* **1993,** *280,* 279–287.
5. Crouch, S. R.; Malmstadt, H. V.; *Anal. Chem.* **1967,** *39,* 1084–1089, 1090–1093.
6. Malmstadt, H. V.; Cordos, E. A.; Delaney, C. J. *Anal. Chem.* **1972,** *44(12),* 26A–41A.
7. Javier, A. C.; Crouch, S. R.; Malmstadt, H. V. *Anal. Chem.* **1969,** *41,* 239–243.
8. Mark, H. B.; Rechnitz, G. A. *Kinetics in Analytical Chemistry.* Interscience: New York, 1968.
9. Sandell, E. B.; Kolthoff, I. M. *J. Am. Chem. Soc.* **1934,** *56,* 1426.
10. Ioannou, P. C.; Piperaki, E. A. *Clin. Chem.* (Winston-Salem, NC) **1986,** *32,* 1481–1483.
11. Diamandis, E. P.; Koupparis, M. A.; Hadjiioannou, T. P. *J. Chem. Educ.* **1983,** *60,* 74–76.
12. Malmstadt, H. V.; Delaney, C. J.; Cordos, E. A. *Anal. Chem.* **1972,** *44(12),* 79A–89A.
13. Guilbault, G. G. *Enzymatic Methods of Analysis,* Pergamon: New York, 1970.
14. Cruces Blanco, C.; Garcia Sanchez, F. *Int. J. Environ. Anal. Chem.* **1990,** *38,* 513.
15. (a) Northrop, D. B. *J. Chem. Educ.* **1998,** *75,* 1153–1157; (b) Zubzy, G. *Biochemistry,* 2nd ed. Macmillan Publishing Co.: New York, 1988, p. 269.
16. Pausch, J. B.; Margerum, D. W. *Anal. Chem.* **1969,** *41,* 226–232.
17. Holler, F. J.; Calhoun, R. K.; McClanahan, S. F. *Anal. Chem.* **1982,** *54,* 755–761.
18. Wentzell, P. D.; Crouch, S. R. *Anal. Chem.* **1986,** *58,* 2851–2855.
19. Wentzell, P. D.; Crouch, S. R. *Anal. Chem.* **1986,** *58,* 2855–2858.
20. Ruzicka, J.; Hansen, E. H. *Anal. Chim. Acta* **1975,** *78,* 145–157.
21. Stewart, K. K.; Beecher, G. R.; Hare, P. E. *Anal. Biochem.* **1976,** *70,* 167–173.
22. Valcárcel, M.; Luque de Castro, M. D. *Flow-Injection Analysis: Principles and Applications,* Ellis Horwood: Chichester, England, 1987.
23. Hansen, E. H.; Ruzicka, J. *J. Chem. Educ.* **1979,** *56,* 677–680.
24. Korenaga, T.; Ikatsu, H. *Anal. Chim. Acta* **1982,** *141,* 301–309.
25. Andrew, K. N.; Blundell, N. J.; Price, D.; et al. *Anal. Chem.* **1994,** *66,* 916A–922A.
26. Ramsing, A. U.; Ruzicka, J.; Hansen, E. H. *Anal. Chim. Acta* **1981,** *129,* 1–17.
27. Guy, R. D.; Ramaley, L.; Wentzell, P. D. *J. Chem. Educ.* **1998,** *75,* 1028–1033.
28. Malmstadt, H. V.; Pardue, H. L. *Anal. Chem.* **1961,** *33,* 1040–1047.
29. Deming, S. N.; Pardue, H. L. *Anal. Chem.* **1971,** *43,* 192–200.
30. Lazaro, F.; Luque de Castro, M. D.; Valcárcel, M. *Analyst* **1984,** *109,* 333.
31. Milardovíc, S.; Kruhak, I.; Ivekovíc, D.; et al. *Anal. Chim. Acta* **1997,** *350,* 91–96.
32. Fernández-Abedul, M; Costa-García, A. *Anal. Chim. Acta* **1996,** *328,* 67–71.
33. Holman, D. A.; Christian, G. D.; Ruzicka, J. *Anal. Chem.* **1997,** *69,* 1763–1765.

Chapter 14

Developing a Standard Method

In Chapter 1 we made a distinction between analytical chemistry and chemical analysis. The goals of analytical chemistry are to improve established methods of analysis, to extend existing methods of analysis to new types of samples, and to develop new analytical methods. Once a method has been developed and tested, its application is best described as chemical analysis. We recognize the status of such methods by calling them standard methods. A standard method may be unique to a particular laboratory, which developed the method for their specific purpose, or it may be a widely accepted method used by many laboratories.

Numerous examples of standard methods have been presented and discussed in the preceding six chapters. What we have yet to consider, however, is what constitutes a standard method. In this chapter we consider how a standard method is developed, including optimizing the experimental procedure, verifying that the method produces acceptable precision and accuracy in the hands of a single analyst, and validating the method for general use.

14A Optimizing the Experimental Procedure

In the presence of H_2O_2 and H_2SO_4, solutions of vanadium form a reddish brown color that is believed to be a compound with the general formula $(VO)_2(SO_4)_3$. Since the intensity of the color depends on the concentration of vanadium, the absorbance of the solution at a wavelength of 450 nm can be used for the quantitative analysis of vanadium. The intensity of the color, however, also depends on the amount of H_2O_2 and H_2SO_4 present. In particular, a large excess of H_2O_2 is known to decrease the solution's absorbance as it changes from a reddish brown to a yellowish color.[1]

Developing a **standard method** for vanadium based on its reaction with H_2O_2 and H_2SO_4 requires that their respective concentrations be optimized to give a maximum absorbance. Using terminology adapted by statisticians, the absorbance of the solution is called the **response.** Hydrogen peroxide and sulfuric acid are **factors** whose concentrations, or **factor levels,** determine the system's response. Optimization involves finding the best combination of factor levels. Usually we desire a maximum response, such as maximum absorbance in the quantitative analysis for vanadium as $(VO)_2(SO_4)_3$. In other situations, such as minimizing percent error, we seek a minimum response.

standard method
A method that has been identified as providing acceptable results.

response
The property of a system that is measured (R).

factor
A property of a system that is experimentally varied and that may affect the response.

factor level
A factor's value.

14A.1 Response Surfaces

One of the most effective ways to think about optimization is to visualize how a system's response changes when we increase or decrease the levels of one or more of its factors. A plot of the system's response as a function of the factor levels is called a **response surface.** The simplest response surface is for a system with only one factor. In this case the response surface is a straight or curved line in two dimensions. A calibration curve, such as that shown in Figure 14.1, is an example of a one-factor response surface in which the response (absorbance) is plotted on the y-axis versus the factor level (concentration of analyte) on the x-axis. Response surfaces can also be expressed mathematically. The response surface in Figure 14.1, for example, is

$$A = 0.008 + 0.0896C_A$$

where A is the absorbance, and C_A is the analyte's concentration in parts per million.

response surface
A graph showing how a system's response changes as a function of its factors.

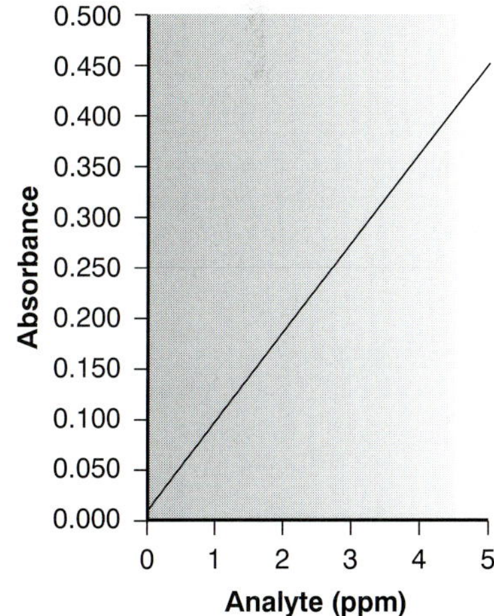

Figure 14.1
Example of a one-factor response surface.

For a two-factor system, such as the quantitative analysis for vanadium described earlier, the response surface is a flat or curved plane plotted in three dimensions. For example, Figure 14.2a shows the response surface for a system obeying the equation

$$R = 3.0 - 0.30A + 0.020AB$$

where R is the response, and A and B are the factor levels. Alternatively, we may represent a two-factor response surface as a contour plot, in which contour lines indicate the magnitude of the response (Figure 14.2b).

The response surfaces in Figure 14.2 are plotted for a limited range of factor levels ($0 \le A \le 10$, $0 \le B \le 10$), but can be extended toward more positive or more negative values. This is an example of an unconstrained response surface. Most response surfaces of interest to analytical chemists, however, are naturally constrained by the nature of the factors or the response or are constrained by practical limits set by the analyst. The response surface in Figure 14.1, for example, has a natural constraint on its factor since the smallest possible concentration for the analyte is zero. Furthermore, an upper limit exists because it is usually undesirable to extrapolate a calibration curve beyond the highest concentration standard.

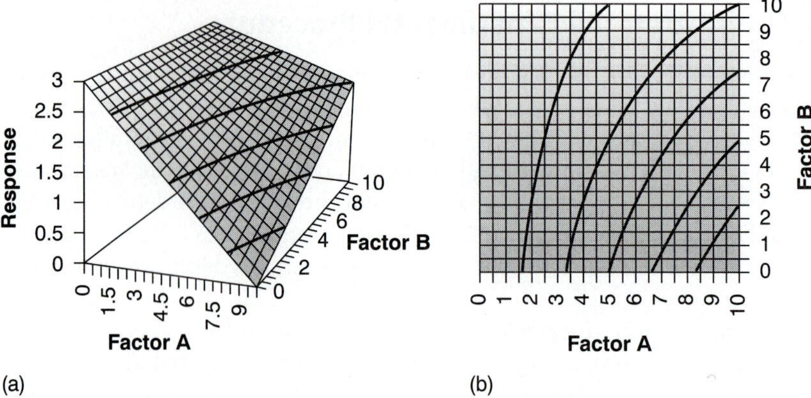

Figure 14.2
Example of a two-factor response surface displayed as (a) a pseudo-three-dimensional graph and (b) a contour plot. Contour lines are shown for intervals of 0.5 response units.

If the equation for the response surface is known, then the optimum response is easy to locate. Unfortunately, the response surface is usually unknown; instead, its shape, and the location of the optimum response must be determined experimentally. The focus of this section is on useful experimental designs for optimizing analytical methods. These experimental designs are divided into two broad categories: searching methods, in which an algorithm guides a systematic search for the optimum response; and modeling methods, in which we use a theoretical or empirical model of the response surface to predict the optimum response.

14A.2 Searching Algorithms for Response Surfaces

Imagine that you wish to climb to the top of a mountain. Because the mountain is covered with trees that obscure its shape, the shortest path to the summit is unknown. Nevertheless, you can reach the summit by always walking in a direction that moves you to a higher elevation. The route followed (Figure 14.3) is the result of a systematic search for the summit. Of course, many routes are possible leading from the initial starting point to the summit. The route taken, therefore, is determined by the set of rules (the algorithm) used to determine the direction of each step. For example, one algorithm for climbing a mountain is to always move in the direction that has the steepest slope.

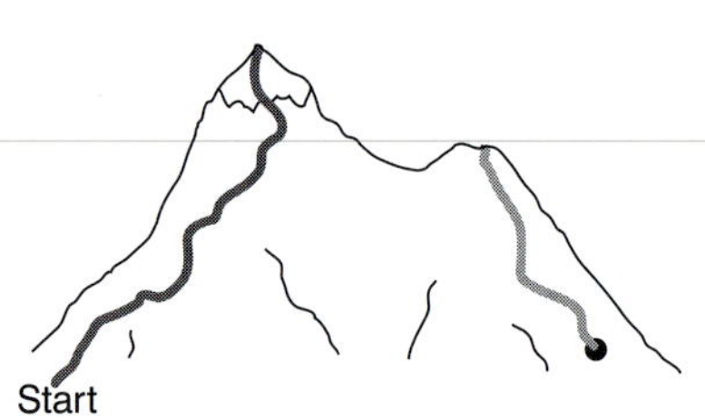

Figure 14.3
Mountain-climbing analogy to using a searching algorithm to find the optimum response for a response surface. The path on the left leads to the global optimum, and the path on the right leads to a local optimum.

A systematic searching algorithm can also be used to locate the optimum response for an analytical method. To find the optimum response, we select an initial set of factor levels and measure the response. We then apply the rules of the searching algorithm to determine the next set of factor levels. This process is repeated until the algorithm indicates that we have reached the optimum response. Two common searching algorithms are described in this section. First, however, we must consider how to evaluate a searching algorithm.

Effectiveness and Efficiency A searching algorithm is characterized by its effectiveness and its efficiency. To be effective, the experimentally determined optimum should closely coincide with the system's true optimum. A searching algorithm may fail to find the true optimum for several reasons, including a poorly designed algorithm, uncertainty in measuring the response, and the presence of local optima. A poorly designed algorithm may prematurely end the search. For example, an algorithm for climbing a mountain that allows movement to the north, south, east, or west will fail to find a summit that can only be reached by moving to the northwest.

When measuring the response is subject to relatively large random errors, or noise, a spuriously high response may produce a false optimum from which the searching algorithm cannot move. When climbing a mountain, boulders encountered along the way are examples of "noise" that must be avoided if the true optimum is to be found. The effect of noise can be minimized by increasing the size of the individual steps such that the change in response is larger than the noise.

Finally, a response surface may contain several local optima, only one of which is the system's true, or global, optimum. This is a problem because a set of initial conditions near a local optimum may be unable to move toward the global optimum. The mountain shown in Figure 14.3, for example, contains two peaks, with the peak on the left being the true summit. A search for the summit beginning at the position identified by the dot may find the local peak instead of the true summit. Ideally, a searching algorithm should reach the global optimum regardless of the initial set of factor levels. One way to evaluate a searching algorithm's effectiveness, therefore, is to use several sets of initial factor levels, finding the optimum response for each, and comparing the results.

A second desirable characteristic for a searching algorithm is efficiency. An efficient algorithm moves from the initial set of factor levels to the optimum response in as few steps as possible. The rate at which the optimum is approached can be increased by taking larger steps. If the step size is too large, however, the difference between the experimental optimum and the true optimum may be unacceptably large. One solution is to adjust the step size during the search, using larger steps at the beginning, and smaller steps as the optimum response is approached.

One-Factor-at-a-Time Optimization One approach to optimizing the quantitative method for vanadium described earlier is to select initial concentrations for H_2O_2 and H_2SO_4 and measure the absorbance. We then increase or decrease the concentration of one reagent in steps, while the second reagent's concentration remains constant, until the absorbance decreases in value. The concentration of the second reagent is then adjusted until a decrease in absorbance is again observed. This process can be stopped after one cycle or repeated until the absorbance reaches a maximum value or exceeds an acceptable threshold value.

Table 14.1 Example of Two Independent Factors

Factor A	Factor B	Response
A_1	B_1	40
A_2	B_1	80
A_1	B_2	60
A_2	B_2	100

A one-factor-at-a-time optimization is consistent with a commonly held belief that to determine the influence of one factor it is necessary to hold constant all other factors. This is an effective, although not necessarily an efficient, experimental design when the factors are independent.[2] Two factors are considered independent when changing the level of one factor does not influence the effect of changing the other factor's level. Table 14.1 provides an example of two independent factors. When factor B is held at level B_1, changing factor A from level A_1 to level A_2 increases the response from 40 to 80; thus, the change in response, ΔR, is

$$\Delta R = 80 - 40 = 40$$

In the same manner, when factor B is at level B_2, we find that

$$\Delta R = 100 - 60 = 40$$

when the level of factor A changes from A_1 to A_2. We can see this independence graphically by plotting the response versus the factor levels for factor A (Figure 14.4). The parallel lines show that the level of factor B does not influence the effect on the response of changing factor A. In the same manner, the effect of changing factor B's level is independent of the level of factor A.

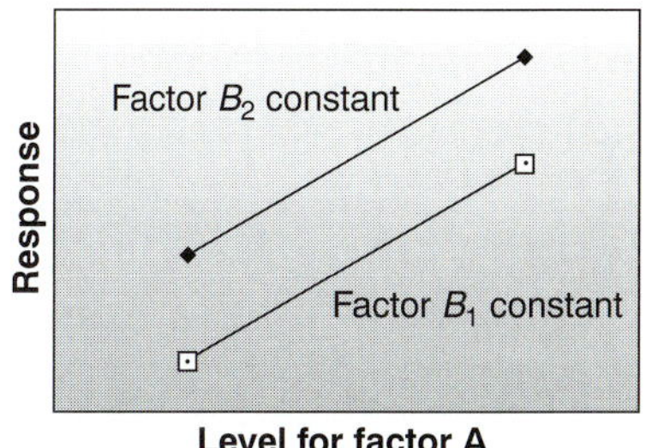

Figure 14.4
Factor effect plot for two independent factors.

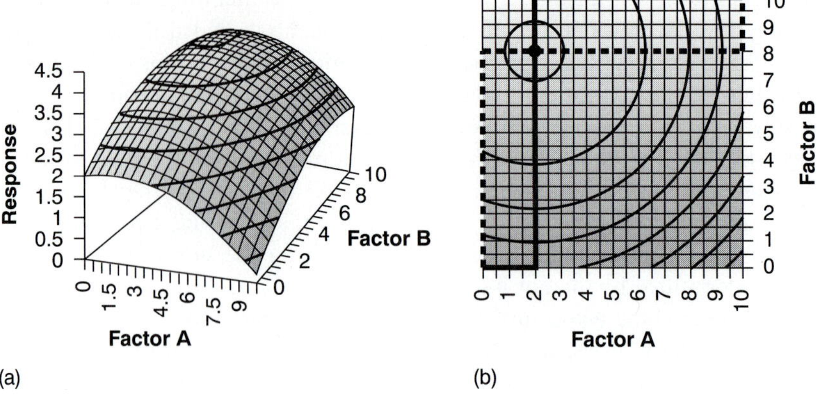

Figure 14.5
Two views of a two-factor response surface for which the factors are independent. The optimum response in (b) is indicated by the • at (2, 8). Contour lines are shown for intervals of 0.5 response units.

Mathematically, two factors are independent if they do not appear in the same term in the algebraic equation describing the response surface. For example, factors A and B are independent when the response, R, is given as

$$R = 2.0 + 0.12A + 0.48B - 0.03A^2 - 0.03B^2 \qquad \textbf{14.1}$$

The resulting response surface for equation 14.1 is shown in Figure 14.5.

The progress of a searching algorithm is best followed by mapping its path on a contour plot of the response surface. Positions on the response surface are identified as (a, b) where a and b are the levels for factors A and B. Four examples of a one-factor-at-a-time optimization of the response surface for equation 14.1 are shown in Figure 14.5b. For those paths indicated by a solid line, factor A is optimized first, followed by factor B. The order of optimization is reversed for paths marked by a dashed line. The effectiveness of this algorithm for optimizing independent factors is shown by the fact that the optimum response at (2, 8) is reached in a single cycle from any set of initial factor levels. Furthermore, it does not matter which factor is optimized first. Although this algorithm is effective at locating the optimum response, its efficiency is limited by requiring that only a single factor can be changed at a time.

Table 14.2 Example of Two Dependent Factors

Factor A	Factor B	Response
A_1	B_1	20
A_2	B_1	80
A_1	B_2	60
A_2	B_2	80

Unfortunately, it is more common to find that two factors are not independent. In Table 14.2, for instance, changing the level of factor B from level B_1 to level B_2 has a significant effect on the response when factor A is at level A_1

$$\Delta R = 60 - 20 = 40$$

but has no effect when factor A is at level A_2.

$$\Delta R = 80 - 80 = 0$$

This effect is seen graphically in Figure 14.6. Factors that are not independent are said to interact. In this case the equation for the response includes an interaction term in which both factors A and B are present. Equation 14.2, for example, contains a final term accounting for the interaction between the factors A and B.

$$R = 5.5 + 1.5A + 0.6B - 0.15A^2 - 0.0245B^2 - 0.0857AB \qquad \textbf{14.2}$$

The resulting response surface for equation 14.2 is shown in Figure 14.7a.

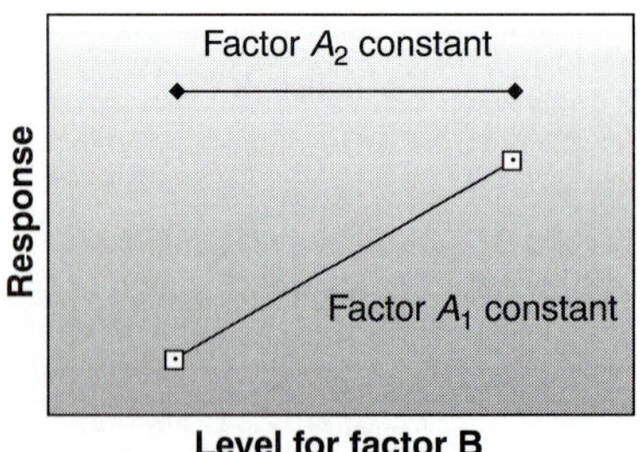

Figure 14.6
Factor effect plot for two interacting factors.

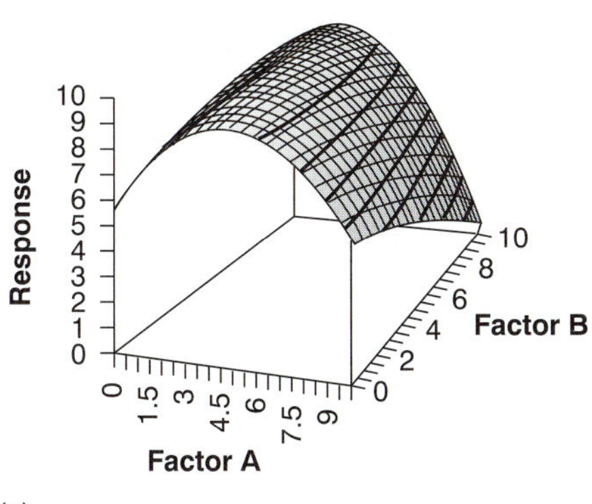

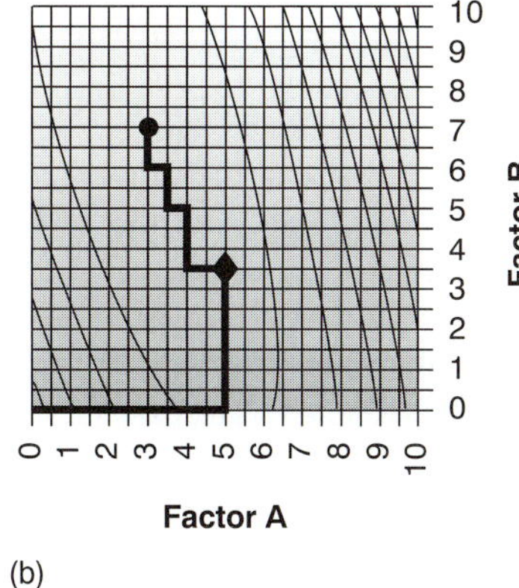

Figure 14.7
Two views of a two-factor response surface for which the factors interact. The optimum response in (b) is indicated by the • at (3, 7). The response at the end of the first cycle is shown in (b) by the ♦. Contour lines are shown for intervals of 1.0 response units.

The progress of a one-factor-at-a-time optimization for the response surface given by equation 14.2 is shown in Figure 14.7b. In this case the optimum response of (3, 7) is not reached in a single cycle. If we start at (0, 0), for example, optimizing factor A follows the solid line to the point (5, 0). Optimizing factor B completes the first cycle at the point (5, 3.5). If our algorithm allows for only a single cycle, then the optimum response is not found. The optimum response usually can be reached by continuing the search through additional cycles, as shown in Figure 14.7b. The efficiency of a one-factor-at-a-time searching algorithm is significantly less when the factors interact. An additional complication with interacting factors is the possibility that the search will end prematurely on a ridge of the response surface, where a change in level for any single factor results in a smaller response (Figure 14.8).

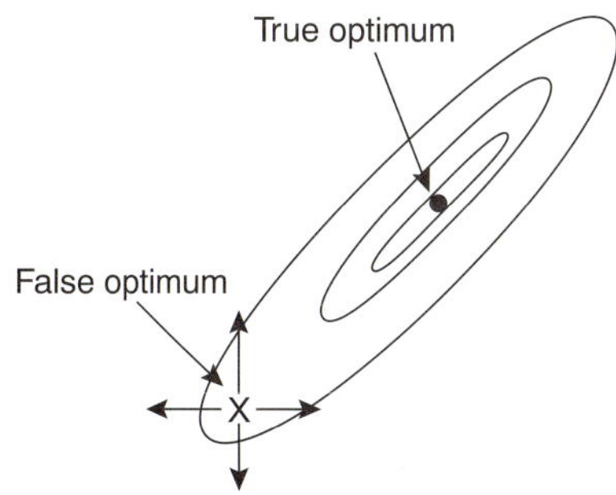

Figure 14.8
Example of a false optimum for a one-factor-at-a-time searching algorithm.

Simplex Optimization The efficiency of a searching algorithm is improved by allowing more than one factor to be changed at a time. A convenient way to accomplish this with two factors is to select three sets of initial factor levels, positioned as the vertices of a triangle (Figure 14.9), and to measure the response for each. The set of factor levels giving the smallest response is rejected and replaced with a new set of factor levels using a set of rules. This process is continued until no further optimization is possible. The set of factor levels is called a simplex. In general, for k factors a simplex is a $(k + 1)$-dimensional geometric figure.[3,4]

The initial simplex is determined by choosing a starting point on the response surface and selecting step sizes for each factor. Ideally the step sizes for each factor should produce an approximately equal change in the response. For two factors a convenient set of factor levels is (a, b), $(a + s_A, b)$, and $(a + 0.5s_A, b + 0.87s_B)$, where s_A and s_B are the step sizes for factors A and B.[5] Optimization is achieved using the following set of rules:

Rule 1. Rank the response for each vertex of the simplex from best to worst.

Rule 2. Reject the worst vertex, and replace it with a new vertex generated by reflecting the worst vertex through the midpoint of the remaining vertices. The factor levels for the new vertex are twice the average factor levels for the retained vertices minus the factor levels for worst vertex.

Rule 3. If the new vertex has the worst response, then reject the vertex with the second-worst response, and calculate the new vertex using rule 2. This rule ensures that the simplex does not return to the previous simplex.

Rule 4. Boundary conditions are a useful way to limit the range of possible factor levels. For example, it may be necessary to limit the concentration of a factor

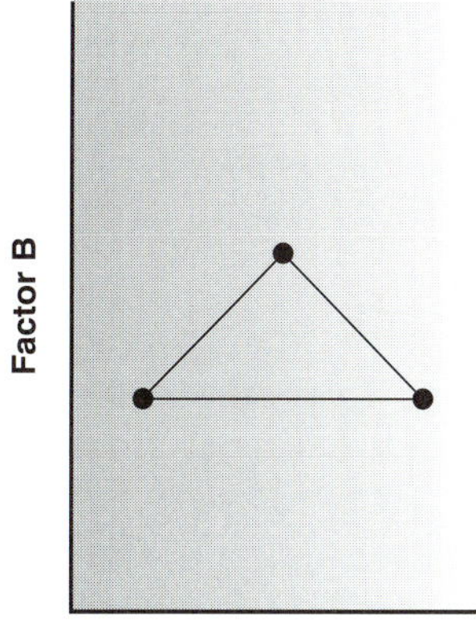

Figure 14.9
Simplex for two factors.

for solubility reasons or to limit temperature due to a reagent's thermal stability. If the new vertex exceeds a boundary condition, then assign it a response lower than all other responses, and follow rule 3.

Because the size of the simplex remains constant during the search, this algorithm is called a fixed-sized **simplex optimization.** Example 14.1 illustrates the application of these rules.

simplex optimization
An efficient optimization method that allows several factors to be optimized at the same time.

Example

EXAMPLE 14.1

Find the optimum response for the response surface in Figure 14.7 using the fixed-sized simplex searching algorithm. Use (0, 0) for the initial factor levels, and set the step size for each factor to 1.0.

SOLUTION

Letting $a = 0$, $b = 0$, $s_a = 1$ and $s_b = 1$ gives the vertices for the initial simplex as

$$\text{Vertex 1: } (a, b) = (0, 0)$$

$$\text{Vertex 2: } (a + s_A, b) = (1, 0)$$

$$\text{Vertex 3: } (a + 0.5s_A, b + 0.87s_B) = (0.5, 0.87)$$

The responses (calculated using equation 14.2) for the three vertices are shown in the following table

Vertex	Factor A	Factor B	Response
V_1	0	0	5.50
V_2	1.00	0	6.85
V_3	0.50	0.87	6.68

with V_1 giving the worst response and V_3 the best response (rule 1). We reject V_1 and replace it with a new vertex whose factor levels are calculated using rule 2; thus

$$\text{New } a = 2\times\left(\frac{a \text{ for } V_2 + a \text{ for } V_3}{2}\right) - (a \text{ for } V_1) = 2\times\frac{1.00 + 0.50}{2} - 0 = 1.50$$

$$\text{New } b = 2\times\left(\frac{b \text{ for } V_2 + b \text{ for } V_3}{2}\right) - (b \text{ for } V_1) = 2\times\frac{0 + 0.87}{2} - 0 = 0.87$$

The new simplex, therefore, is

Vertex	Factor A	Factor B	Response
V_2	1.00	0	6.85
V_3	0.50	0.87	6.68
V_4	1.50	0.87	7.80

The worst response is for vertex 3, which we replace with the following new vertex

$$a = 2\times\frac{1.00 + 1.50}{2} - 0.5 = 2.00 \qquad b = 2\times\frac{0 + 0.87}{2} - 0.87 = 0$$

The resulting simplex now consists of the following vertices

Vertex	Factor A	Factor B	Response
V_2	1.00	0	6.85
V_4	1.50	0.87	7.80
V_5	2.00	0	7.90

The calculation of the remaining vertices is left as an exercise. The progress of the completed optimization is shown in Table 14.3 and in Figure 14.10. The optimum response of (3, 7) first appears in the twenty-fourth simplex, but a total of 29 steps is needed to verify that the optimum has been found.

Table 14.3 Progress of Fixed-Sized Simplex Optimization for Response Surface in Figure 14.10

Simplex	Vertices	Notes
1	1, 2, 3	
2	2, 3, 4	
3	2, 4, 5	
4	4, 5, 6	
5	5, 6, 7	
6	6, 7, 8	
7	7, 8, 9	
8	8, 9, 10	
9	8, 10, 11	
10	10, 11, 12	
11	11, 12, 13	
12	12, 13, 14	follow rule 3
13	13, 14, 15	
14	13, 15, 16	
15	13, 16, 17	follow rule 3
16	16, 17, 18	
17	16, 18, 19	
18	16, 19, 20	follow rule 3
19	19, 20, 21	
20	19, 21, 22	follow rule 3
21	21, 22, 23	
22	21, 23, 24	follow rule 3
23	23, 24, 25	
24	23, 25, 26	
25	23, 26, 27	follow rule 3
26	26, 27, 28	follow rule 3
27	26, 28, 29	
28	26, 29, 30	follow rule 3
29	26, 30, 31	vertex 31 same as vertex 25

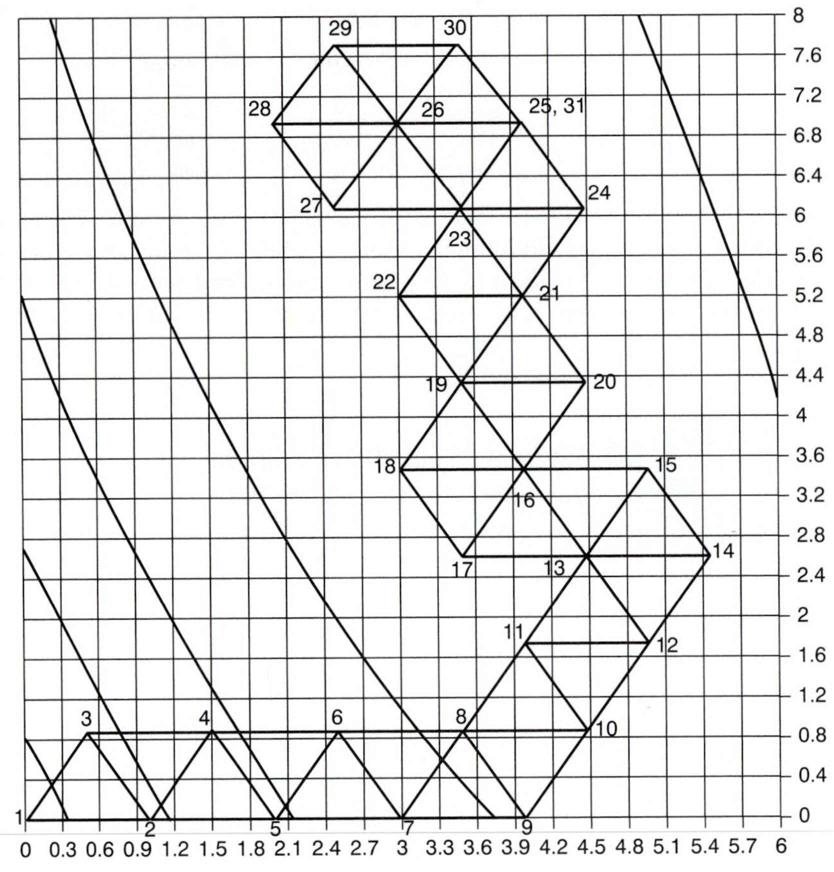

Figure 14.10
Progress of a fixed-sized simplex optimization for the response surface of Example 14.1. The optimum response at (3, 7) corresponds to vertex 26.

The fixed-size simplex searching algorithm is effective at locating the optimum response for both independent and interacting factors. Its efficiency, however, is limited by the simplex's size. We can increase its efficiency by allowing the size of the simplex to expand or contract in response to the rate at which the optimum is being approached.[3,6] Although the algorithm for a variable-sized simplex is not presented here, an example of its increased efficiency is shown Figure 14.11. The references and suggested readings may be consulted for further details.

14A.3 Mathematical Models of Response Surfaces

Earlier we noted that a response surface can be described mathematically by an equation relating the response to its factors. If a series of experiments is carried out in which we measure the response for several combinations of factor levels, then linear regression can be used to fit an equation describing the response surface to the data. The calculations for a linear regression when the system is first-order in one factor (a straight line) were described in Chapter 5. A complete mathematical treatment of linear regression for systems that are second-order or that contain more than one factor is beyond the scope of this text. Nevertheless, the computations for

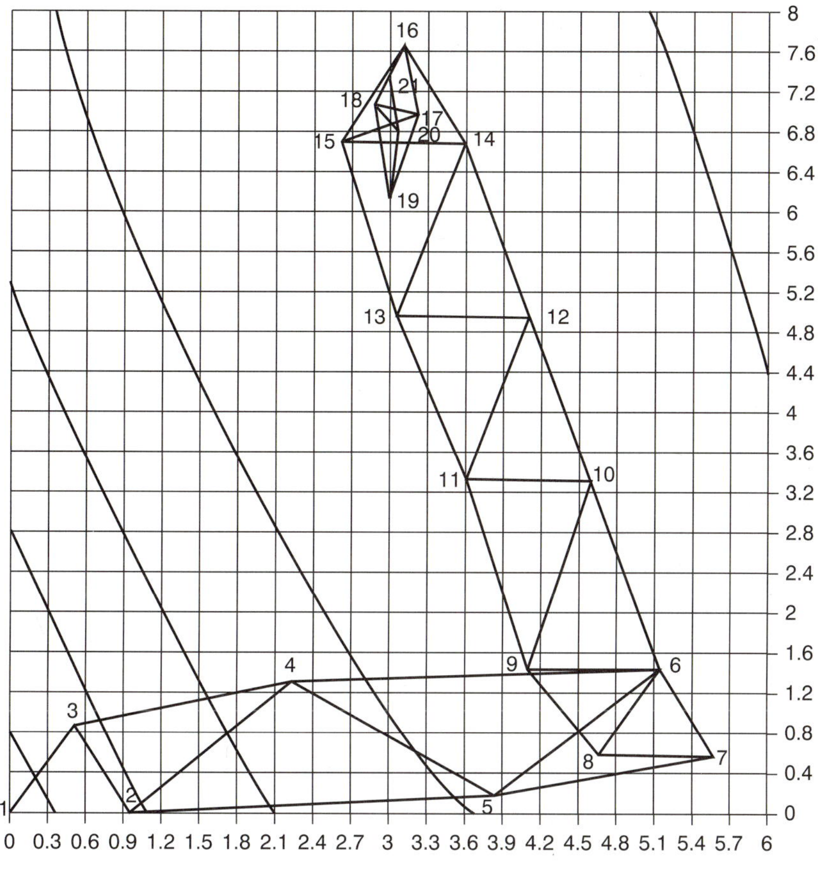

Figure 14.11
Progress of a variable-sized simplex optimization for the response surface of Example 14.1. The optimum response is at (3, 7).

a few special cases are straightforward and are considered in this section. A more comprehensive treatment of linear regression can be found in several of the suggested readings listed at the end of this chapter.

Theoretical Models of the Response Surface Mathematical models for response surfaces are divided into two categories: those based on theory and those that are empirical. **Theoretical models** are derived from known chemical and physical relationships between the response and the factors. In spectrophotometry, for example, Beer's law is a theoretical model relating a substance's absorbance, A, to its concentration, C_A

$$A = \varepsilon b C_A$$

where ε is the molar absorptivity, and b is the pathlength of the electromagnetic radiation through the sample. A Beer's law calibration curve, therefore, is a theoretical model of a response surface.

theoretical model
A model describing a system's response that has a theoretical basis and can be derived from theoretical principles.

Empirical Models of the Response Surface In many cases the underlying theoretical relationship between the response and its factors is unknown, making impossible a theoretical model of the response surface. A model can still be developed if we make some reasonable assumptions about the equation describing the response surface. For example, a response surface for two factors, A and B, might be represented by an equation that is first-order in both factors

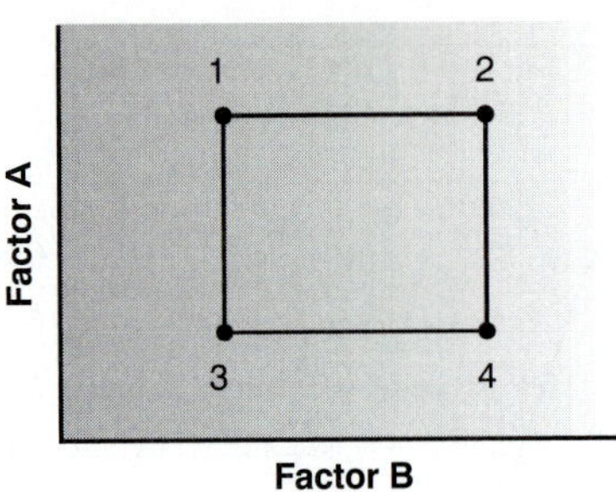

Runs	Factor levels A	Factor levels B
1	H_A	L_B
2	H_A	H_B
3	L_A	L_B
4	L_A	H_B

(a)

Factor A
Factor B
Factor C

Runs	Factor levels A	Factor levels B	Factor levels C
1	H_A	L_B	L_C
2	H_A	L_B	H_C
3	H_A	H_B	H_C
4	H_A	H_B	L_C
5	L_A	L_B	L_C
6	L_A	L_B	H_C
7	L_A	H_B	H_C
8	L_A	H_B	L_C

(b)

Figure 14.12
2^k factorial designs for (a) $k = 2$ and (b) $k = 3$.

$$R = \beta_0 + \beta_a A + \beta_b B$$

first-order in both factors with an additional interaction term

$$R = \beta_0 + \beta_a A + \beta_b B + \beta_{ab} AB$$

or a second-order polynomial equation

$$R = \beta_0 + \beta_a A + \beta_b B + \beta_{aa} A^2 + \beta_{bb} B^2$$

The terms β_0, β_a, β_b, β_{ab}, β_{aa}, and β_{bb} are adjustable parameters whose values are determined by using linear regression to fit the data to the equation. Such equations are **empirical models** of the response surface because they have no basis in a theoretical understanding of the relationship between the response and its factors. An empirical model may provide an excellent description of the response surface over a wide range of factor levels. It is more common, however, to find that an empirical model only applies to the range of factor levels for which data have been collected.

empirical model
A model describing a system's response that is not derived from theoretical principles.

To develop an empirical model for a response surface, it is necessary to collect the right data using an appropriate experimental design. Two popular experimental designs are considered in the following sections.

Factorial Designs To determine a factor's effect on the response, it is necessary to measure the response for at least two factor levels. For convenience these levels are labeled high, H_f, and low, L_f, where f is the factor; thus H_A is the high level for factor A, and L_B is the low level for factor B. When more than one factor is included in the empirical model, then each factor's high level should be paired with both the high and low levels for all other factors. In the same way, the low level for each factor should be paired with the high and low levels for all other factors (Figure 14.12). All together, a minimum of 2^k experiments is necessary, where k is the number of factors. This experimental design is known as a 2^k factorial design.

The linear regression calculations for a 2^k factorial design are straightforward and can be done without the aid of a sophisticated statistical software package. To simplify the computations, factor levels are coded as +1 for the high level, and –1 for the low level. The relationship between a factor's coded level, x_f^*, and its actual value, x_f, is given as

$$x_f = c_f + d_f x_f^* \qquad 14.3$$

where c_f is the factor's average level, and d_f is the absolute difference between the factor's average level and its high and low values. Equation 14.3 is used to transform coded results back to their actual values.

Example

EXAMPLE 14.2

The factor A has coded levels of +1 and –1 with an average factor level of 100, and d_A equal to 5. What are the actual factor levels?

SOLUTION

The actual factor levels are

$$H_A = 100 + (1)(5) = 105 \qquad L_A = 100 + (-1)(5) = 95$$

Let's start by considering a simple example involving two factors, A and B, to which we wish to fit the following empirical model.

$$R = \beta_0 + \beta_a A + \beta_b B + \beta_{ab} AB \qquad 14.4$$

A 2^k factorial design with two factors requires four runs, or sets of experimental conditions, for which the uncoded levels, coded levels, and responses are shown in Table 14.4. The terms β_0, β_a, β_b, and β_{ab} in equation 14.4 account for, respectively, the mean effect (which is the average response), first-order effects due to factors A and B, and the interaction between the two factors. Estimates for these parameters are given by the following equations

$$\beta_0 \approx b_0 = \frac{1}{n} \Sigma R_i \qquad 14.5$$

$$\beta_a \approx b_a = \frac{1}{n} \Sigma A_i^* R_i \qquad 14.6$$

$$\beta_b \approx b_b = \frac{1}{n} \Sigma B_i^* R_i \qquad 14.7$$

Table 14.4 Example of Uncoded and Coded Factor Levels and Responses for a 2^2 Factorial Design

Run	*A*	*B*	*A**	*B**	Response
1	15	30	+1	+1	22.5
2	15	10	+1	–1	11.5
3	5	30	–1	+1	17.5
4	5	10	–1	–1	8.5

$$\beta_{ab} \approx b_{ab} = \frac{1}{n}\sum A_i^* B_i^* R_i \qquad \mathbf{14.8}$$

where n is the number of runs, and A_i^* and B_i^* are the coded factor levels for the ith run. Solving for the estimated parameters using the data in Table 14.4

$$b_0 = \frac{1}{4}(22.5 + 11.5 + 17.5 + 8.5) = 15.0$$

$$b_a = \frac{1}{4}(22.5 + 11.5 - 17.5 - 8.5) = 2.0$$

$$b_b = \frac{1}{4}(22.5 - 11.5 + 17.5 - 8.5) = 5.0$$

$$b_{ab} = \frac{1}{4}(22.5 - 11.5 - 17.5 + 8.5) = 0.5$$

leaves us with the following empirical model for the response surface

$$R = 15.0 + 2.0A^* + 5.0B^* + 0.5A^*B^* \qquad \mathbf{14.9}$$

The suitability of this model can be evaluated by substituting values for A^* and B^* from Table 14.4 and comparing the calculated response to the known response. Using the values for the first run as an example gives

$$R = 15.0 + (2.0)(+1) + (5.0)(+1) + (0.5)(+1)(+1) = 22.5$$

which agrees with the known response.

Example

EXAMPLE 14.3

Equation 14.9 gives the empirical model of the response surface for the data in Table 14.4 when the factors are in coded form. Convert the equation to its uncoded form.

SOLUTION

To convert the equation to its uncoded form, it is necessary to solve equation 14.3 for each factor. Values for c_f and d_f are determined from the high and low levels for each factor; thus

$$c_A = \frac{H_A + L_A}{2} = \frac{15 + 5}{2} = 10 \qquad d_A = H_A - c_A = 15 - 10 = 5$$

$$c_B = \frac{H_B + L_B}{2} = \frac{30 + 10}{2} = 20 \qquad d_B = H_B - c_B = 30 - 20 = 10$$

Substituting known values into equation 14.3,

$$A = 10 + 5A^* \qquad B = 20 + 10B^*$$

and rearranging, gives

$$A^* = 0.2A - 2 \qquad B^* = 0.1B - 2$$

Substituting these equations into equation 14.9 gives, after simplifying, the uncoded equation for the response surface.

$$R = 15.0 + 2(0.2A - 2) + 5(0.1B - 2) + 0.5(0.2A - 2)(0.1B - 2)$$

$$= 15.0 + 0.4A - 4 + 0.5B - 10 + 0.01AB - 0.2A - 0.1B + 2$$

$$= 3.0 + 0.2A + 0.4B + 0.01AB$$

We can verify this equation by substituting values for A and B from Table 14.4 and solving for the response. Using values for the first run, for example, gives

$$R = 3.0 + (0.2)(15) + (0.4)(30) + (0.01)(15)(30) = 22.5$$

which agrees with the expected value.

The computation just outlined is easily extended to any number of factors. For a system with three factors, for example, a 2^3 factorial design can be used to determine the parameters for the empirical model described by the following equation

$$R = \beta_0 + \beta_a A + \beta_b B + \beta_c C + \beta_{ab} AB + \beta_{ac} AC + \beta_{bc} BC + \beta_{abc} ABC \qquad \textbf{14.10}$$

where A, B, and C are the factors. The terms β_0, β_a, β_b, and β_{ab} are estimated using equations 14.6–14.9. The remaining parameters are estimated using the following equations.

$$\beta_c \approx b_c = \frac{1}{n} \Sigma C_i^* R_i \qquad \textbf{14.11}$$

$$\beta_{ac} \approx b_{ac} = \frac{1}{n} \Sigma A_i^* C_i^* R_i \qquad \textbf{14.12}$$

$$\beta_{bc} \approx b_{bc} = \frac{1}{n} \Sigma B_i^* C_i^* R_i \qquad \textbf{14.13}$$

$$\beta_{abc} \approx b_{abc} = \frac{1}{n} \Sigma A_i^* B_i^* C_i^* R_i \qquad \textbf{14.14}$$

Example

EXAMPLE 14.4

Table 14.5 lists the uncoded factor levels, coded factor levels, and responses for a 2^3 factorial design. Determine the coded and uncoded empirical model for the response surface based on equation 14.10.

SOLUTION

We begin by calculating the estimated parameters using equations 14.6–14.9 and 14.11–14.14.

$$b_0 = \frac{1}{8}(137.25 + 54.75 + 73.75 + 30.25 + 61.75 + 30.25 + 41.25 + 18.75) = 56.0$$

$$b_a = \frac{1}{8}(137.25 + 54.75 + 73.75 + 30.25 - 61.75 - 30.25 - 41.25 - 18.75) = 18.0$$

$$b_b = \frac{1}{8}(137.25 + 54.75 - 73.75 - 30.25 + 61.75 + 30.25 - 41.25 - 18.75) = 15.0$$

Table 14.5 Uncoded and Coded Factor Levels and Responses for the 2^3 Factorial Design of Example 14.4

Run	*A*	*B*	*C*	*A**	*B**	*C**	Response
1	15	30	45	+1	+1	+1	137.25
2	15	30	15	+1	+1	−1	54.75
3	15	10	45	+1	−1	+1	73.75
4	15	10	15	+1	−1	−1	30.25
5	5	30	45	−1	+1	+1	61.75
6	5	30	15	−1	+1	−1	30.25
7	5	10	45	−1	−1	+1	41.25
8	5	10	15	−1	−1	−1	18.75

$$b_c = \frac{1}{8}(137.25 - 54.75 + 73.75 - 30.25 + 61.75 - 30.25 + 41.25 - 18.75) = 22.5$$

$$b_{ab} = \frac{1}{8}(137.25 + 54.75 - 73.75 - 30.25 - 61.75 - 30.25 + 41.25 + 18.75) = 7.0$$

$$b_{ac} = \frac{1}{8}(137.25 - 54.75 + 73.75 - 30.25 - 61.75 + 30.25 - 41.25 + 18.75) = 9.0$$

$$b_{bc} = \frac{1}{8}(137.25 - 54.75 - 73.75 + 30.25 + 61.75 - 30.25 - 41.25 + 18.75) = 6.0$$

$$b_{abc} = \frac{1}{8}(137.25 - 54.75 - 73.75 + 30.25 - 61.75 + 30.25 + 41.25 - 18.75) = 3.75$$

The coded empirical model, therefore, is

$$R = 56 + 18A^* + 15B^* + 22.5C^* + 7A^*B^* + 9A^*C^* + 6B^*C^* + 3.75A^*B^*C^*$$

To check the result we substitute the coded factor levels for the first run into the coded empirical model, giving

$$R = 56 + (18)(+1) + (15)(+1) + (22.5)(+1) + (7)(+1)(+1) + (9)(+1)(+1) + (6)(+1)(+1) + (3.75)(+1)(+1)(+1) = 137.25$$

which agrees with the measured response.

To transform the coded empirical model into its uncoded form, it is necessary to replace A^*, B^*, and C^* with the following relationships

$$A^* = 0.2A - 2 \qquad B^* = 0.1B - 2 \qquad C^* = \frac{C}{15} - 2$$

the derivations of which are left as an exercise. Substituting these relationships into the coded empirical model and simplifying (which also is left as an exercise) gives the following result for the uncoded empirical model

$$R = 3 + 0.2A + 0.4B + 0.5C - 0.01AB + 0.02AC - 0.01BC + 0.005ABC$$

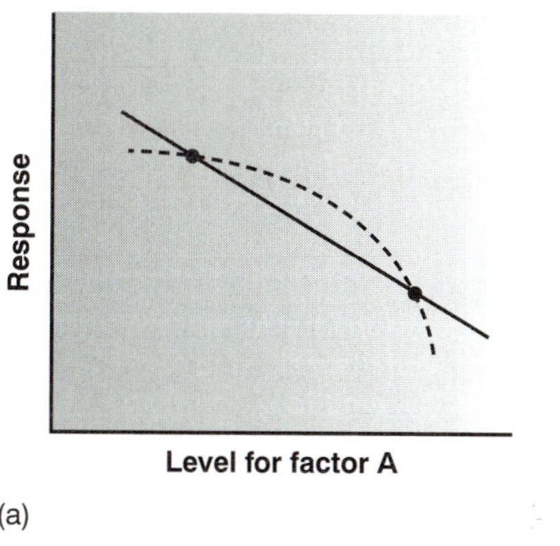

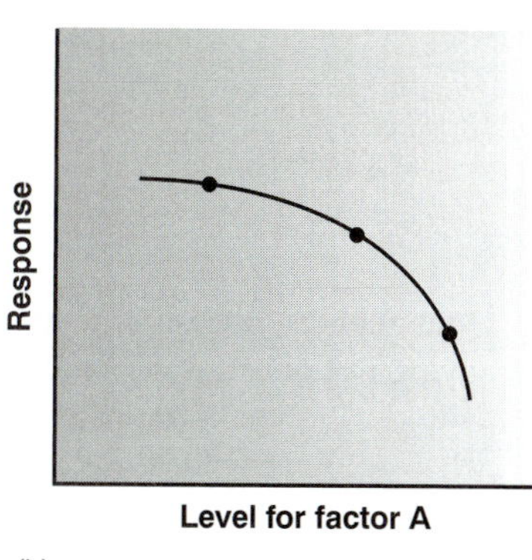

Figure 14.13
Curved one-factor response surface showing (a) the limitation of a 2^k factorial design for modeling second-order effects; and (b) the application of a 3^k factorial design for modeling second-order effects.

A 2^k factorial design is limited to models that include only a factor's first-order effects on the response. Thus, for a 2^2 factorial design, it is possible to determine the first-order effect for each factor (β_a and β_b), as well as the interaction between the factors (β_{ab}). There is insufficient information in the factorial design, however, to determine any higher order effects or interactions. This limitation is a consequence of having only two levels for each factor. Consider, for example, a system in which the response is a function of a single factor. Figure 14.13a shows the experimentally measured response for a 2^1 factorial design in which only two levels of the factor are used. The only empirical model that can be fit to the data is that for a straight line.

$$R = \beta_0 + \beta_a A$$

If the actual response is that represented by the dashed curve, then the empirical model is in error. To fit an empirical model that includes curvature, a minimum of three levels must be included for each factor. The 3^1 factorial design shown in Figure 14.13b, for example, can be fit to an empirical model that includes second-order effects for the factor.

$$R = \beta_0 + \beta_a A + \beta_{aa} A^2$$

In general, an n-level factorial design can include single-factor and interaction terms up to the $(n - 1)$th order.

The effectiveness of a first-order empirical model can be judged by measuring the response at the center of the factorial design. If there are no higher order effects, the average response of the runs in a 2^k factorial design should be equal to the measured response at the center of the factorial design. The influence of random error can be accounted for by making several determinations of the response at the center of the factorial design and establishing a suitable confidence interval. If the difference between the two responses is significant, then a first-order empirical model is probably not appropriate.

EXAMPLE 14.5

At the beginning of this section we noted that the concentration of vanadium can be determined spectrophotometrically by making the solution acidic with H_2SO_4 and reacting with H_2O_2 to form a reddish brown compound with the general formula $(VO)_2(SO_4)_3$. Palasota and Deming[7] studied the effect on the absorbance of the relative amounts of H_2SO_4 and H_2O_2, reporting the following results for a 2^2 factorial design.

H_2SO_4	H_2O_2	Response
+1	+1	0.330
+1	−1	0.359
−1	+1	0.293
−1	−1	0.420

Four replicate measurements were made at the center of the factorial design, giving responses of 0.334, 0.336, 0.346, and 0.323. Determine if a first-order empirical model is appropriate for this system. Use a 90% confidence interval when accounting for the effect of random error.

SOLUTION

We begin by determining the confidence interval for the response at the center of the factorial design. The mean response is 0.335, with a standard deviation of 0.0094. The 90% confidence interval, therefore, is

$$\mu = \overline{X} \pm \frac{ts}{\sqrt{n}} = 0.335 \pm \frac{(2.35)(0.0094)}{\sqrt{4}} = 0.335 \pm 0.011$$

The average response, $\overline{R}$, from the factorial design is

$$\overline{R} = \frac{1}{n}\sum R_i = \frac{1}{4}(0.330 + 0.359 + 0.293 + 0.420) = 0.350$$

Because $\overline{R}$ exceeds the confidence interval's upper limit of 0.346, there is reason to believe that a 2^2 factorial design and a first-order empirical model are inappropriate for this system. A complete empirical model for this system is presented in problem 10 in the end-of-chapter problem set.

Many systems that cannot be represented by a first-order empirical model can be described by a full second-order polynomial equation, such as that for two factors.

$$R = \beta_0 + \beta_a A + \beta_b B + \beta_{aa} A^2 + \beta_{bb} B^2 + \beta_{ab} AB$$

Because each factor must be measured for at least three levels, a convenient experimental design is a 3^k factorial design. A 3^2 factorial design for two factors, for example, is shown in Figure 14.14. The computations for 3^k factorial designs are not as easily generalized as those for a 2^k factorial design and are not considered in this text. Further details about the calculations are found in the suggested readings at the end of the chapter.

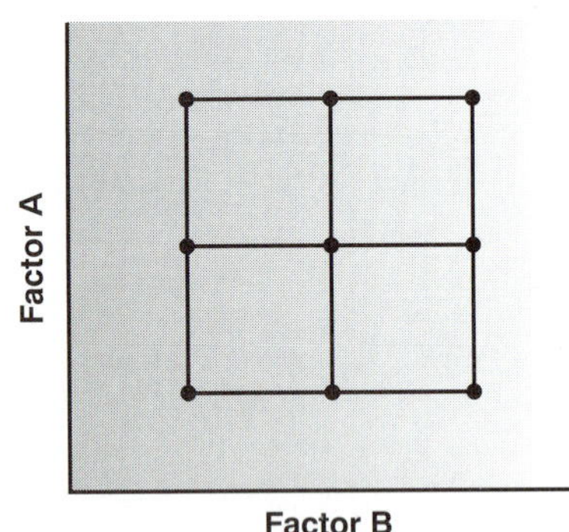

Figure 14.14
3^k factorial design for $k = 2$.

Central Composite Designs One limitation to a 3^k factorial design is the number of trials that must be run. For two factors, as shown in Figure 14.14, a total of nine trials is needed. This number increases to 27 for three factors and 81 for four factors. A more efficient experimental design for systems containing more than two factors is the central composite design, two examples of which are shown in Figure 14.15. The central composite design consists of a 2^k factorial design, which provides data for estimating the first-order effects for each factor and interactions between the factors, and a "star" design consisting of $2k + 1$ points, which provides data for estimating second-order effect. Although a central composite design for two factors requires the same number of trials, 9, as a 3^2 factorial design, it requires only 15 trials and 25 trials, respectively, for systems in-

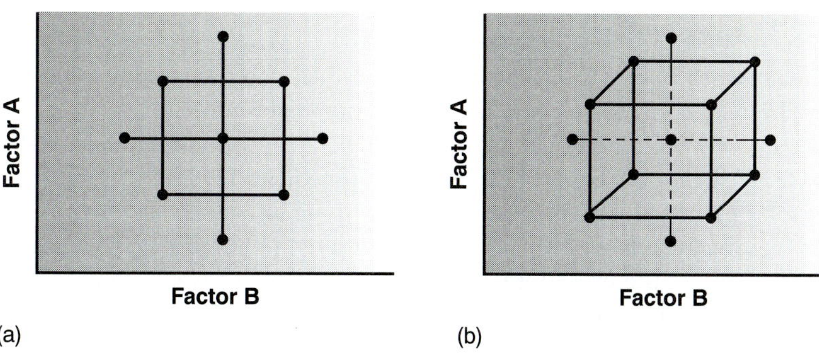

Figure 14.15
Central composite designs for (a) $k = 2$ and (b) $k = 3$.

volving three or four factors. A discussion of central composite designs, including computational considerations, can be found in the suggested readings at the end of the chapter.

14B Verifying the Method

After developing and optimizing a method, it is necessary to determine the quality of results that can reasonably be expected when the method is used by a single analyst. Generally, three steps are included in the process of verifying a method: determining single-operator characteristics, the blind analysis of standards, and determining the method's ruggedness. In addition, if an alternative standard method exists, both the standard method and the new method can be used to analyze the same sample, and the results compared. If the quality of the results is unacceptable, the method is not suitable for consideration as a standard method.

14B.1 Single-Operator Characteristics

The first step in verifying a method is to determine the precision, accuracy, and detection limit when a single analyst uses the method to analyze a standard sample of known composition. The detection limit, which was discussed in Chapter 4, is determined by analyzing a reagent blank for each type of sample matrix for which the method will be used. Precision is determined by analyzing replicate portions, preferably more than ten, of a standard sample. Finding the method's accuracy is evaluated by a t-test, as described in Chapter 4. Precision and accuracy should be evaluated for several different concentration levels of analyte, including at least one concentration near the detection limit, and for each type of sample matrix that will be encountered. The analysis of several concentrations allows for the detection of constant sources of determinate error and establishes the range of concentrations for which the method is applicable.

14B.2 Blind Analysis of Standard Samples

Single-operator characteristics are determined by analyzing a sample whose concentration of analyte is known to the analyst. The second step in verifying a method is the **blind analysis** of standard samples where the analyte's concentration remains unknown to the analyst. The standard sample is analyzed several times, and the average concentration of the analyte is determined. This value should be within three, and preferably two standard deviations (as determined from the single-operator characteristics) of the analyte's known concentration.

blind analysis
The analysis of a standard sample whose composition is unknown to the analyst.

14B.3 Ruggedness Testing

In many cases an optimized method may produce excellent results in the laboratory developing the method, but poor results in other laboratories. This is not surprising since a method is often optimized by a single analyst under an ideal set of conditions, in which the sources of reagents, equipment, and instrumentation remain the same for each trial. The procedure might also be influenced by environmental factors, such as the temperature or relative humidity in the laboratory, whose levels are not specified in the procedure and which may differ between laboratories. Finally, when optimizing a method the analyst usually takes particular care to perform the analysis in exactly the same way during every trial.

An important step in developing a standard method is to determine which factors have a pronounced effect on the quality of the analytical method's result. The procedure can then be written to specify the degree to which these factors must be controlled. A procedure that, when carefully followed, produces high-quality results in different laboratories is considered rugged. The method by which the critical factors are discovered is called **ruggedness testing.**[8]

ruggedness testing
The process of evaluating a method to determine those factors for which a small change in value has a significant effect on the method's results.

Ruggedness testing is often performed by the laboratory developing the standard method. Potential factors are identified and their effects evaluated by performing the analysis while the factors are held at two levels. Normally one level for each factor is that given in the procedure, and the other is a level likely to be encountered when the procedure is used by other laboratories.

This approach to ruggedness testing can be time-consuming. If seven potential factors are identified, for example, ruggedness testing can be accomplished with a 2^7 factorial design. This requires a total of 128 trials, which is a prohibitively large amount of work. A simpler experimental design is shown in Table 14.6, in which the two factor levels are identified by upper case and lower case letters. This design, which is similar to that for the 2^3 factorial design discussed in the previous section, is called a fractional factorial design and provides information about the first-order effect of each factor. It does not, however, provide sufficient information to evaluate higher order effects or potential interactions between factors, both of which are assumed to be of less importance than first-order effects.

The experimental design for ruggedness testing is balanced in that each factor level is paired an equal number of times with the upper case and lower case levels

Table 14.6 Experimental Design for a Ruggedness Test Involving Seven Factors

	Factor							
Run	A	B	C	D	E	F	G	Response
1	A	B	C	D	E	F	G	R_1
2	A	B	c	D	e	f	g	R_2
3	A	b	C	d	E	f	g	R_3
4	A	b	c	d	e	F	G	R_4
5	a	B	C	d	e	F	g	R_5
6	a	B	c	d	E	f	G	R_6
7	a	b	C	D	e	f	G	R_7
8	a	b	c	D	E	F	g	R_8

for every other factor. The effect of changing the level for any one factor, E_f, is determined by subtracting the average response when the factor is at its upper case level from the average value when it is at its lower case level

$$E_f = \frac{\left(\sum R_i\right)_{\text{upper case}}}{4} - \frac{\left(\sum R_i\right)_{\text{lower case}}}{4} \qquad \mathbf{14.15}$$

Because the design is balanced, the levels for the remaining factors appear an equal number of times in both summation terms, and, consequently, their effect on the response is canceled. For example, the effect of changing the level for factor A is determined by averaging the responses from runs 1 through 4 and subtracting the average response from runs 5 through 8. Factor B does not affect E_f because its upper case levels in runs 1 and 2 are canceled by the upper case levels in runs 5 and 6, and its lower case levels in runs 3 and 4 are canceled by the lower case levels in runs 7 and 8. Similar reasoning shows that the remaining factors are also canceled.

After each of the effects is calculated, they are ranked from largest to smallest, without regard to sign, and those factors whose effects are substantially larger than the other factors are identified. The estimated standard deviation for the analysis is given by

$$s = \sqrt{\frac{2}{7}\sum E_i^2} \qquad \mathbf{4.16}$$

This provides the best estimate of the expected standard deviation for results obtained by several laboratories, due to the effects of small changes in uncontrolled or poorly controlled factors. If this standard deviation is unacceptably large, then the procedure may be modified to bring under greater control those factors whose levels have the greatest effect on the response.

Example

EXAMPLE 14.6

The concentration of trace metals in sediment samples collected from rivers and lakes can be determined by extracting with acid and analyzing the extract by atomic absorption spectrophotometry. One procedure calls for an overnight extraction with dilute HCl or HNO_3. The samples are placed in plastic bottles with 25 mL of acid and extracted on a shaking table at a moderate speed and ambient temperature. To determine the ruggedness of the method, the effect of a change in level for the following factors was studied using the experimental design shown in Table 14.6.

Factor A – extraction time	A = 24 h	a = 12 h
Factor B – shaking speed	B = medium	b = high
Factor C – acid type	C = HCl	c = HNO_3
Factor D – acid concentration	D = 0.1 M	d = 0.05 M
Factor E – volume of acid	E = 25 mL	e = 35 mL
Factor F – type of container	F = plastic	f = glass
Factor G – temperature	G = ambient	g = 25 °C

A standard sample containing a known amount of analyte was carried through the procedure. The percentage of analyte actually found in the eight trials were found to be

$$R_1 = 98.9 \qquad R_2 = 99.0 \qquad R_3 = 97.5 \qquad R_4 = 97.7$$

$$R_5 = 97.4 \qquad R_6 = 97.3 \qquad R_7 = 98.6 \qquad R_8 = 98.6$$

Determine which factors, if any, appear to have a significant effect on the response, and estimate the method's expected standard deviation.

SOLUTION

The effect of a change in level for each factor is calculated using equation 14.15

$$E_A = \frac{R_1 + R_2 + R_3 + R_4}{4} - \frac{R_5 + R_6 + R_7 + R_8}{4} = 0.30$$

$$E_B = \frac{R_1 + R_2 + R_5 + R_6}{4} - \frac{R_3 + R_4 + R_7 + R_8}{4} = 0.05$$

$$E_C = \frac{R_1 + R_3 + R_5 + R_7}{4} - \frac{R_2 + R_4 + R_6 + R_8}{4} = -0.05$$

$$E_D = \frac{R_1 + R_2 + R_7 + R_8}{4} - \frac{R_3 + R_4 + R_5 + R_6}{4} = 1.30$$

$$E_E = \frac{R_1 + R_3 + R_6 + R_8}{4} - \frac{R_2 + R_4 + R_5 + R_7}{4} = -0.10$$

$$E_F = \frac{R_1 + R_4 + R_5 + R_8}{4} - \frac{R_2 + R_3 + R_6 + R_7}{4} = 0.05$$

$$E_G = \frac{R_1 + R_4 + R_6 + R_7}{4} - \frac{R_2 + R_3 + R_5 + R_8}{4} = 0.00$$

Ordering the factors by their absolute values

Factor D	1.30
Factor A	0.35
Factor E	−0.10
Factor B	0.05
Factor C	−0.05
Factor F	0.05
Factor G	0.00

shows that the concentration of acid (factor D) has a substantial effect on the response, with a concentration of 0.05 M providing a much lower percent recovery. The extraction time (factor A) also appears to be significant, but its effect is not as important as that for the acid's concentration. All other factors appear to be insignificant.

The method's estimated standard deviation is

$$s = \sqrt{\frac{2}{7}\{0.30^2 + 0.05^2 + (-0.05)^2 + 1.30^2 + (-0.10)^2 + 0.05^2 + 0.00^2\}} = 0.72$$

which, for an average recovery of 98.1% gives a relative standard deviation of approximately 0.7%. If the acid's concentration is controlled such that its effect approaches that for factors B, C, and F, then the relative standard deviation becomes 0.18, or approximately 0.2%.

14B.4 Equivalency Testing

If a standard method is available, the performance of a new method can be evaluated by comparing results with those obtained with an approved standard method. The comparison should be done at a minimum of three concentrations to evaluate the applicability of the new method for different amounts of analyte. Alternatively, we can plot the results obtained by the new method against those obtained by the approved standard method. A linear regression analysis should give a slope of 1 and a y-intercept of 0 if the results of the two methods are equivalent.

14C Validating the Method as a Standard Method

For an analytical method to be of use, it must be capable of producing results with acceptable accuracy and precision. The process of verifying a method as described in the previous section determines whether the method meets this goal for a single analyst. A further requirement for a standard method is that an analysis should not be affected by a change in the analyst performing the work, the laboratory in which the work is performed, or the time when the analysis is conducted. The process by which a method is approved for general use is known as validation and involves a collaborative test of the method by analysts in several laboratories. Collaborative testing is used routinely by regulatory agencies and professional organizations, such as the U.S. Environmental Protection Agency, the American Society for Testing and Materials, the Association of Official Analytical Chemists, and the American Public Health Association, in establishing their standard methods of analysis.

When an analyst performs a single analysis on a sample, the difference between the experimentally determined value and the expected value is influenced by three sources of error: random error, systematic errors inherent to the method, and systematic errors unique to the analyst. If enough replicate analyses are performed, a distribution of results can be plotted (Figure 14.16a). The width of this distribution is described by the standard deviation and can be used to determine the effect of random error on the analysis. The position of the distribution relative to the sample's true value, μ, is determined both by systematic errors inherent to the method and those systematic errors unique to the analyst. For a single analyst there is no way to separate the total systematic error into its component parts.

The goal of a **collaborative test** is to determine the expected magnitude of all three sources of error when a method is placed into general practice. When several analysts each analyze the same sample one time, the variation in their collective results (Figure 14.16b) includes contributions from random errors and those systematic errors (biases) unique to the analysts. Without additional information, the standard deviation for the pooled data cannot be used to separate the precision of the analysis from the systematic errors of the analysts. The position of the distribution, however, can be used to detect the presence of a systematic error in the method.

collaborative testing
A validation method used to evaluate the sources of random and systematic errors affecting an analytical method.

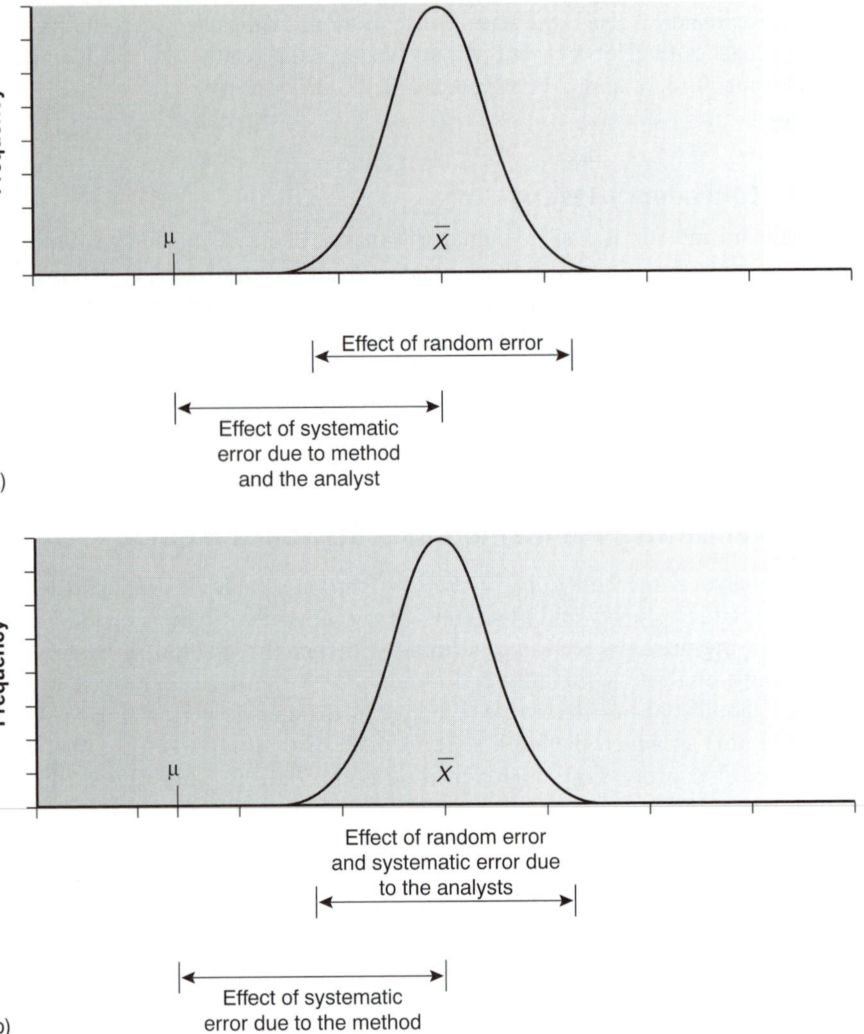

Figure 14.16
Partitioning of random error, systematic errors due to the analyst, and systematic error due to the method for (a) replicate analyses performed by a single analyst and (b) single determinations performed by several analysts.

14C.1 Two-Sample Collaborative Testing

The design of a collaborative test must provide the additional information needed to separate the effect of random error from that due to systematic errors introduced by the analysts. One simple approach, which is accepted by the Association of Official Analytical Chemists, is to have each analyst analyze two samples, X and Y, that are similar in both matrix and concentration of analyte. The results obtained by each analyst are plotted as a single point on a two-sample chart, using the result for one sample as the *x*-coordinate and the value for the other sample as the *y*-coordinate.[8,9]

A two-sample chart is divided into four quadrants, identified as (+, +), (–, +), (–, –), and (+, –), depending on whether the points in the quadrant have values for the two samples that are larger or smaller than the mean values for samples X and Y. Thus, the quadrant (+, –) contains all points for which the result for sample X is larger than the mean for sample X, and for which the result for sample Y is less than the mean for sample Y. If the variation in results is dominated by random errors,

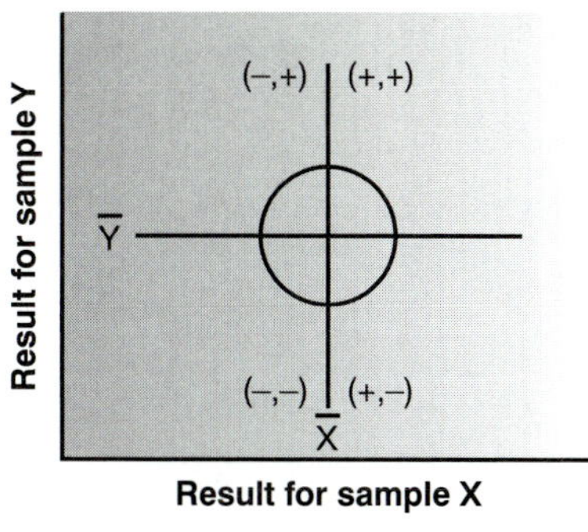

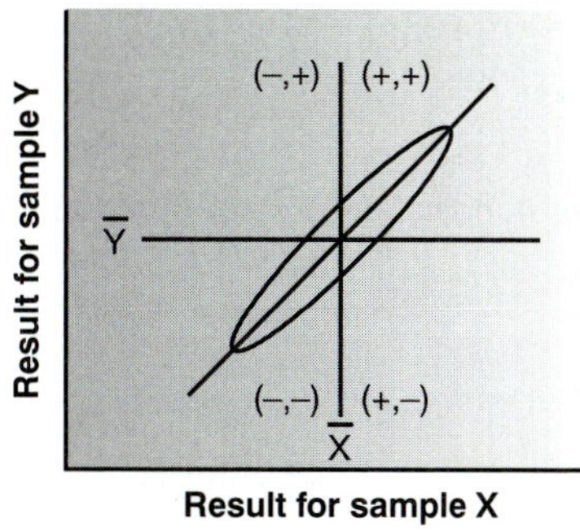

Figure 14.17
Typical two-sample plot when (a) random errors are larger than systematic errors due to the analysts and (b) systematic errors due to the analysts are larger than the random errors.

then the points should be distributed randomly in all four quadrants, with an equal number of points in each quadrant. Furthermore, the points will be clustered in a circular pattern whose center is the mean values for the two samples (Figure 14.17a). When systematic errors are significantly larger than random errors, then the points will be found primarily in the (+, +) and (–, –) quadrants and will be clustered in an elliptical pattern around a line bisecting these quadrants at a 45° angle (Figure 14.17b).

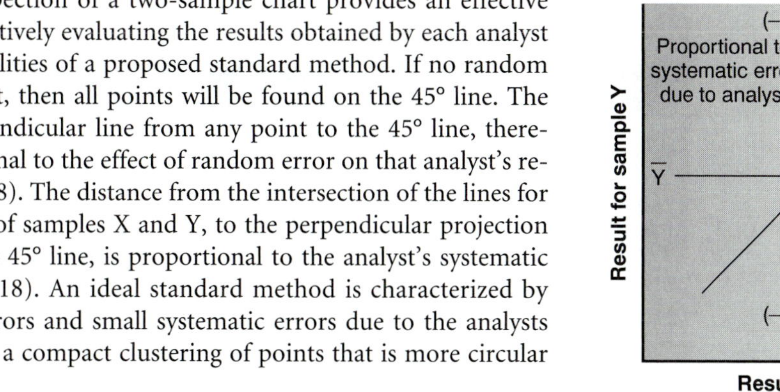

A visual inspection of a two-sample chart provides an effective means for qualitatively evaluating the results obtained by each analyst and of the capabilities of a proposed standard method. If no random errors are present, then all points will be found on the 45° line. The length of a perpendicular line from any point to the 45° line, therefore, is proportional to the effect of random error on that analyst's results (Figure 14.18). The distance from the intersection of the lines for the mean values of samples X and Y, to the perpendicular projection of a point on the 45° line, is proportional to the analyst's systematic error (Figure 14.18). An ideal standard method is characterized by small random errors and small systematic errors due to the analysts and should show a compact clustering of points that is more circular than elliptical.

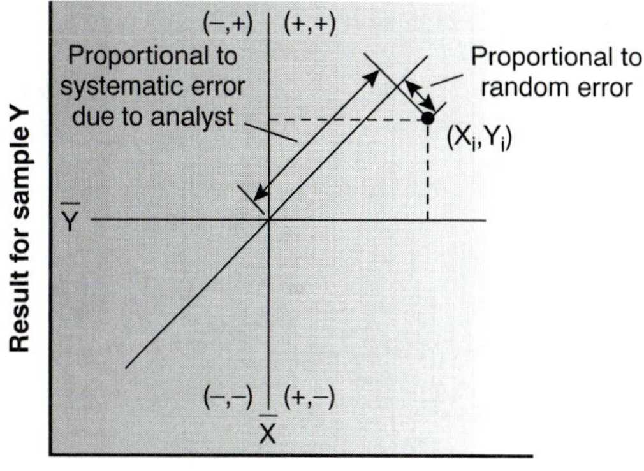

Figure 14.18
Relationship between point in a two-sample plot and the random error and systematic error due to the analyst.

The data used to construct a two-sample chart can also be used to separate the total variation of the data, σ_{tot}, into contributions from random error, σ_{rand}, and systematic errors due to the analysts, σ_{sys}.[9] Since an analyst's systematic errors should be present at the same level in the analysis of samples X and Y, the difference, D, between the results for the two samples

$$D_i = X_i - Y_i$$

is influenced only by random error. The contribution from random error, therefore, is estimated by the standard deviation of the differences for each analyst

$$s_D = \sqrt{\frac{\sum (D_i - \overline{D})^2}{2(n-1)}} = s_{rand} \approx \sigma_{rand} \qquad \textbf{14.17}$$

where n is the number of analysts. The inclusion of a factor of 2 in the denominator of equation 14.17 is the result of using two values to determine D_i. The total, T, of each analyst's results for the two samples

$$T_i = X_i + Y_i$$

contains contributions from both random error and twice the analyst's systematic error; thus

$$\sigma_{tot}^2 = \sigma_{rand}^2 + 2\sigma_{sys}^2 \qquad 14.18$$

The standard deviation of the totals provides an estimate for σ_{tot}.

$$s_T = \sqrt{\frac{\sum(T_i - \overline{T})^2}{2(n-1)}} = s_{tot} \approx \sigma_{tot} \qquad 14.19$$

Again, the factor of 2 in the denominator is the result of using two values to determine T_i.

If systematic errors due to the analysts are significantly larger than random errors, then s_T should be larger than s_D. This can be tested statistically using a one-tailed F-test

$$F = \frac{s_T^2}{s_D^2}$$

where the degrees of freedom for both the numerator and the denominator are $n - 1$. If s_T is significantly larger than s_D, then we can use equation 14.18 to separate the data's total variation into that due to random error and that due to the systematic errors of the analysts.

Example

EXAMPLE 14.7

As part of a collaborative study of a new method for determining the amount of total cholesterol in blood, two samples were sent to ten analysts with instructions to analyze each sample one time. The following results, in milligrams of total cholesterol per 100 mL of serum, were obtained

Analyst	Tablet 1	Tablet 2
1	245.0	229.4
2	247.4	249.7
3	246.0	240.4
4	244.9	235.5
5	255.7	261.7
6	248.0	239.4
7	249.2	255.5
8	225.1	224.3
9	255.0	246.3
10	243.1	253.1

Using this data estimate the values for σ_{rand} and σ_{sys} for the method assuming $\alpha = 0.05$.

SOLUTION

A two-sample plot of the data is shown in Figure 14.19, with the average value for sample 1 shown by the vertical line at 245.9, and the average value for sample 2 shown by the horizontal line at 243.5. To estimate σ_{rand} and σ_{sys}, it first is necessary to calculate values for D_i and T_i; thus

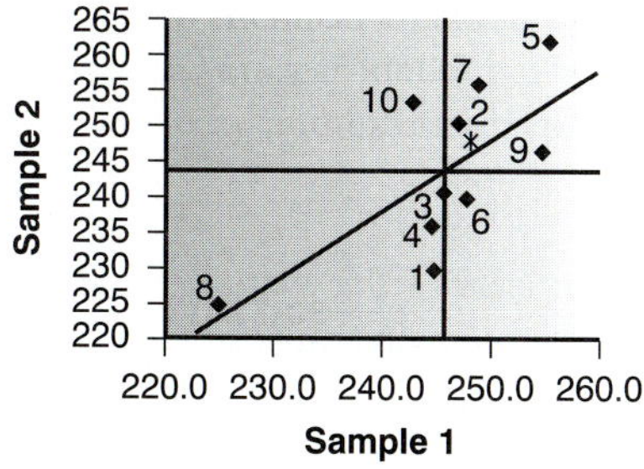

Figure 14.19
Two-sample plot for data in Example 14.7. The analyst responsible for each data point is indicated by the associated number. The true values for the two samples are indicated by the *.

Analyst	D_i	T_i
1	15.6	474.4
2	–2.3	497.1
3	5.6	486.4
4	9.4	480.4
5	–6.0	517.4
6	8.6	487.4
7	–6.3	504.7
8	0.8	449.4
9	8.7	501.3
10	–10.0	496.2

The standard deviations for the differences, s_D, and the totals, s_T, are calculated* using equations 14.17 and 14.19, giving

$$s_D = 5.95 \qquad s_T = 13.3$$

To determine if the systematic errors of the analysts are significant, an F-test is performed using s_T and s_D

$$F = \frac{s_T^2}{s_D^2} = \frac{(13.3)^2}{(5.95)^2} = 5.00$$

Because the F-ratio is larger than $F(0.05, 9, 9)$, which is 3.179, we conclude that the systematic errors of the analysts are significant at the 95% confidence levels. The estimated precision for a single analyst is

$$\sigma_{rand} \approx s_D = 5.95$$

The estimated standard deviation due to systematic errors between analysts is calculated from equation 14.18.

$$\sigma_{sys} = \sqrt{\frac{\sigma_{tot}^2 - \sigma_{rand}^2}{2}} \approx \sqrt{\frac{s_T^2 - s_D^2}{2}} = \sqrt{\frac{(13.3)^2 - (5.95)^2}{2}} = 8.41$$

*Here is a short-cut that simplifies the calculation of s_D and s_T. Enter the values for D_i into your calculator, and use its built-in functions to find the standard deviation. Divide this result by $\sqrt{2}$ to obtain s_D. You can use the same approach to calculate s_T.

When the true values for the two samples are known, it is possible to test for the presence of systematic errors in the method. If no systematic method errors occur, then the sum of the true values for the samples, μ_{tot}

$$\mu_{tot} = \mu_{samp\ X} + \mu_{samp\ Y}$$

will fall within the confidence interval around the average value, $\overline{T}$. A two-tailed t-test, as described in Chapter 4, with the following null and alternative hypotheses

$$H_0: \overline{T} = \mu_{tot}$$

$$H_A: \overline{T} \neq \mu_{tot}$$

is used to determine whether there is any evidence for systematic method errors. The test statistic, t_{exp}, is given as

$$t_{exp} = \frac{|\overline{T} - \mu_{tot}|\sqrt{n}}{s_T\sqrt{2}} \qquad \textbf{14.20}$$

and has $n - 1$ degrees of freedom. The $\sqrt{2}$ in the denominator is included because s_T, as given in equation 14.19, underestimates the standard deviation when comparing $\overline{T}$ to μ_{tot}.

Example

EXAMPLE 14.8

The two samples analyzed in Example 14.7 are known to contain the following concentrations of cholesterol.

$$\mu_{samp\ 1} = 248.3 \text{ mg/100 mL} \qquad \mu_{samp\ 2} = 247.6 \text{ mg/100 mL}$$

Determine if there is any evidence for a systematic error in the method at the 95% confidence level.

SOLUTION

Using the data from Example 14.7 and the true values for the samples, we know that s_T is 13.3, and

$$\overline{T} = \overline{X}_{samp\ 1} + \overline{X}_{samp\ 2} = 245.9 + 243.5 = 489.4$$

$$\mu_{tot} = \mu_{samp\ 1} + \mu_{samp\ 2} = 248.3 + 247.6 = 495.9$$

Substituting these values into equation 14.20 gives

$$t_{exp} = \frac{|489.4 - 495.9|\sqrt{10}}{13.3\sqrt{2}} = 1.09$$

This value for t_{exp} is smaller than the critical value of 2.26 for $t(0.05, 9)$. Thus, there is no evidence for a systematic error in the method at the 95% confidence level.

In Examples 14.7 and 14.8 we have seen how a collaborative test using a pair of closely related samples can be used to evaluate a method. Ideally, a collaborative test should involve several pairs of samples, whose concentrations of analyte span the method's anticipated useful range. In this way the method can be evaluated for constant sources of error, and the expected relative standard deviation and bias for different levels of analyte can be determined.

14C.2 Collaborative Testing and Analysis of Variance

In the two-sample collaborative test, each analyst performs a single determination on two separate samples. The resulting data are reduced to a set of differences, *D*, and a set of totals, *T*, each characterized by a mean value and a standard deviation. Extracting values for random errors affecting precision and systematic differences between analysts is relatively straightforward for this experimental design.

An alternative approach for collaborative testing is to have each analyst perform several replicate determinations on a single, common sample. This approach generates a separate data set for each analyst, requiring a different statistical treatment to arrive at estimates for σ_{rand} and σ_{sys}.

A variety of statistical methods may be used to compare three or more sets of data. The most commonly used method is an **analysis of variance** (ANOVA). In its simplest form, a one-way ANOVA allows the importance of a single variable, such as the identity of the analyst, to be determined. The importance of this variable is evaluated by comparing its variance with the variance explained by indeterminate sources of error inherent to the analytical method.

analysis of variance
A statistical method for comparing three or more sets of data.

Variance was introduced in Chapter 4 as one measure of a data set's spread around its central tendency. In the context of an analysis of variance, it is useful to see that variance is simply a ratio of the sum of squares for the differences between individual values and their mean, to the degrees of freedom. For example, the variance, s^2, of a data set consisting of n measurements is given as

$$s^2 = \frac{\text{sum of squares}}{\text{degrees of freedom}} = \frac{\sum_{i=1}^{n}(X_i - \overline{X})^2}{n-1}$$

where X_i is the value of a single measurement, and $\overline{X}$ is the mean. As we will see, the ability to partition the variance into separate terms for the sum of squares and the degrees of freedom greatly simplifies the calculations in a one-way ANOVA.

Let's use a simple example to develop the rationale behind a one-way ANOVA calculation. The data in Table 14.7 show the results obtained by several analysts in determining the purity of a single pharmaceutical preparation of sulfanilamide. Each column in this table lists the results obtained by an individual analyst. For convenience, entries in the table are represented by the symbol X_{ij}, where i identifies the analyst and j indicates the replicate number; thus $X_{3,5}$ is the fifth replicate for the third analyst (and is equal to 94.24%). The variability in the results shown in Table 14.7 arises from two sources: indeterminate errors associated with the analytical procedure that are experienced equally by all analysts, and systematic or determinate errors introduced by the analysts.

One way to view the data in Table 14.7 is to treat it as a single system, characterized by a global mean, $\overline{\overline{X}}$, and a global variance, $\overline{\overline{s^2}}$. These parameters are calculated using the following equations.

$$\overline{\overline{X}} = \frac{\sum_{i=1}^{h}\sum_{j=1}^{n_i} X_{ij}}{N}$$

Table 14.7 Results of Four Analysts for the %Purity of a Preparation of Sulfanilamide

Replicate	Analyst A	Analyst B	Analyst C	Analyst D
1	94.09	99.55	95.14	93.88
2	94.64	98.24	94.62	94.23
3	95.08	101.1	95.28	96.05
4	94.54	100.4	94.59	93.89
5	95.38	100.1	94.24	94.95
6	93.62			95.49

$$\overline{\overline{s^2}} = \frac{\sum_{i=1}^{h}\sum_{j=1}^{n_i}\left(X_{ij} - \overline{\overline{X}}\right)^2}{N-1} \qquad \textbf{14.21}$$

where h is the total number of samples (in this case the number of analysts), n_i is the number of replicates for the ith sample (in this case the ith analyst), and N is the total number of data points in the system. The global variance provides a measure of the combined influence of indeterminate and systematic errors.

A second way to work with the data in Table 14.7 is to treat the results for each analyst separately. Because the repeatability for any analyst is influenced by indeterminate errors, the variance, s^2, of the data in each column provides an estimate of σ^2_{rand}. A better estimate is obtained by pooling the individual variances. The result, which is called the within-sample variance (s_{w}^2), is calculated by summing the squares of the differences between the replicates for each sample and that sample's mean, and dividing by the degrees of freedom.

$$s_{\text{w}}^2 = \sigma^2_{\text{rand}} = \frac{\sum_{i=1}^{h}\sum_{j=1}^{n}(X_{ij} - \overline{X}_i)^2}{N-h} \qquad \textbf{14.22}$$

Finally, the data for each analyst can be reduced to separate mean values, $\overline{X}_i$. The variance of the individual means about the global mean is called the between-sample variance, s_{b}^2, and is calculated as

$$s_{\text{b}}^2 = \frac{\sum_{i=1}^{h} n_i\left(\overline{X}_i - \overline{\overline{X}}\right)^2}{h-1} \qquad \textbf{14.23}$$

where n_i is the number of replicates for the ith sample. The between-sample variance includes contributions from both random and systematic errors and, therefore, provides an estimate for both σ^2_{rand} and σ^2_{sys}

$$s_{\text{b}}^2 = \sigma^2_{\text{rand}} + \overline{n}\sigma^2_{\text{sys}} \qquad \textbf{14.24}$$

where $\overline{n}$ is approximated as the average number of replicates per analyst.

$$\overline{n} = \frac{\sum_{i=1}^{h} n_i}{h}$$

Table 14.8 Summary of Calculations for a One-Way Analysis of Variance

Source	Sum of Squares	Degrees of Freedom	Variance	Expected Variance	*F*-ratio
Between sample	$SS_b = \sum_{i=1}^{h} n_i(\overline{X}_i - \overline{\overline{X}})^2$	$h - 1$	$s_b^2 = \dfrac{SS_b}{h-1}$	$s_b^2 = \sigma_{rand}^2 + \bar{n}\sigma_{sys}^2$	$F = \dfrac{s_b^2}{s_w^2}$
Within sample	$SS_w = SS_t - SS_b$	$N - h$	$s_w^2 = \dfrac{SS_w}{N-h}$	$s_w^2 = \sigma_{rand}^2$	
Total	$SS_t = \sum_{i=1}^{h}\sum_{j=1}^{n}(X_{ij} - \overline{\overline{X}})^2 = \overline{\overline{s}}^2(N-1)$				

In a one-way ANOVA of the data in Table 14.7, the null hypothesis is that no significant differences exist between the mean values for each analyst. The alternative hypothesis is that at least one of the means is significantly different. If the null hypothesis is true, then σ_{sys}^2 must be zero. Thus, from equations 14.22 and 14.24, both s_w^2 and s_b^2 are predictors of σ_{rand}^2 and should have similar values. If s_b^2 is significantly greater than s_w^2, then σ_{sys}^2 is greater than zero. In this case the alternative hypothesis must be accepted, and a significant difference between the means for the analysts has been demonstrated. The test statistic is the *F*-ratio

$$F = \frac{s_b^2}{s_w^2}$$

which is compared with the critical value $F(\alpha, h - 1, N - h)$. This is a one-tailed significance test because we are only interested in whether s_b^2 is significantly greater than s_w^2.

Both s_b^2 and s_w^2 are easy to determine for small data sets. For larger data sets, however, calculating s_w^2 becomes tedious.* Its calculation is simplified by taking advantage of the relationship between the sum-of-squares terms for the global variance, the within-sample variance, and the between-sample variance. The numerator of equation 14.21, which also is known as the total sum of squares, SS_t, can be split into two terms

$$SS_t = SS_w + SS_b$$

where the sum of squares for the variation within the sample, SS_w, is the numerator of equation 14.22, and the sum of squares between the sample, SS_b, is the numerator of equation 14.23. Calculating SS_t and SS_b gives SS_w by difference. Dividing SS_w and SS_b by their respective degrees of freedom gives s_w^2 and s_b^2. Table 14.8 summarizes all the necessary equations for a one-way ANOVA calculation. The application of a one-way ANOVA is outlined in Example 14.9.

Example

EXAMPLE 14.9

Table 14.7 shows the results obtained by four analysts determining the purity of a pharmaceutical preparation of sulfanilamide. Determine if the difference in their results is significant at $\alpha = 0.05$. If such a difference exists, estimate values for σ_{rand}^2 and σ_{sys}^2.

*A number of statistics and spreadsheet software packages are available to perform ANOVA calculations.

SOLUTION

To begin, we calculate the global mean and variance and the means for each analyst. These results are

$\overline{\overline{X}} = 95.87 \quad \overline{\overline{s^2}} = 5.506 \quad \overline{X}_A = 94.56 \quad \overline{X}_B = 99.88 \quad \overline{X}_C = 94.77 \quad \overline{X}_D = 94.75$

Using these values, we then calculate the total sum of squares

$$SS_t = \overline{\overline{s^2}}\,(N - 1) = (5.506)(22 - 1) = 115.63$$

the between-sample sum of squares

$$SS_b = \sum_{i=1}^{4} n_i\left(\overline{X}_i - \overline{\overline{X}}\right)^2 = 6(94.56 - 95.87)^2 + 5(99.88 - 95.87)^2$$
$$+ 5(94.77 - 95.87)^2 + 6(94.75 - 95.87)^2 = 104.27$$

and the within-sample sum of squares

$$SS_w = SS_t - SS_b = 115.63 - 104.27 = 11.36$$

The remainder of the necessary calculations are summarized in the following table.

Source	Sum of Squares	Degrees of Freedom	Variance
between	104.27	$h - 1 = 4 - 1 = 3$	34.76
within	11.36	$N - h = 22 - 4 = 18$	0.631

The value of the test statistic is

$$F_{exp} = \frac{s_b^2}{s_w^2} = \frac{34.76}{0.631} = 55.08$$

for which the critical value of $F(0.05, 3, 18)$ is 3.16. Because F_{exp} is greater than $F(0.05, 3, 18)$, the null hypothesis is rejected, and the results for at least one analyst are significantly different from those for the other analysts. The value for σ_{rand}^2 is estimated by the within-sample variance

$$\sigma_{rand}^2 \approx s_w^2 = 0.631$$

whereas, from equation 14.24, σ_{sys}^2 is

$$\sigma_{sys}^2 \approx \frac{s_b^2 - s_w^2}{\overline{n}} = \frac{34.76 - 0.631}{22/4} = 6.205$$

In this example the variance due to systematic differences between the analysts is almost an order of magnitude greater than the variance due to the method's precision.

Fisher's least significant difference
A modified form of the *t*-test for comparing several sets of data.

Once a significant difference has been demonstrated by an analysis of variance, a modified version of the *t*-test, known as **Fisher's least significant difference,** can be used to determine which analyst or analysts are responsible for the difference. The test statistic for comparing the mean values $\overline{X}_1$ and $\overline{X}_2$ is the *t*-test described in Chapter 4, except that s_{pool} is replaced by the square root of the within-sample variance obtained from an analysis of variance.

$$t_{\text{exp}} = \frac{|\overline{X}_1 - \overline{X}_2|}{\sqrt{s_{\text{w}}^2[(1/n_1) + (1/n_2)]}} \qquad \textbf{14.25}$$

This value of t_{exp} is compared with the critical value for $t(\alpha, \nu)$, where the significance level is the same as that used in the ANOVA calculation, and the degrees of freedom is the same as that for the within-sample variance. Because we are interested in whether the larger of the two means is significantly greater than the other mean, the value of $t(\alpha, \nu)$ is that for a one-tail significance test.

Example

EXAMPLE 14.10

Determine the source of the significant difference for the data in Example 14.9.

SOLUTION

Individual comparisons using Fisher's least significant difference test are based on the following null hypothesis and one-tailed alternative hypothesis

$$H_0: \overline{X}_i = \overline{X}_j \qquad H_A: \overline{X}_i > \overline{X}_j \quad \text{or} \quad \overline{X}_i < \overline{X}_j$$

Using equation 14.25, we can calculate values of t_{exp} for each possible comparison. These values can then be compared with the one-tailed critical value of 1.73 for $t(0.05, 18)$, as found in Appendix 1B. For example, t_{exp} when comparing the results for analysts A and B is

$$(t_{\text{exp}})_{\text{AB}} = \frac{|\overline{X}_{\text{A}} - \overline{X}_{\text{B}}|}{\sqrt{s_{\text{w}}^2[(1/n_{\text{A}}) + (1/n_{\text{B}})]}} = \frac{|94.56 - 99.88|}{\sqrt{(0.631)[(1/6) + (1/5)]}} = 11.06$$

Because $(t_{\text{exp}})_{\text{AB}}$ is greater than $t(0.05, 18)$, we reject the null hypothesis and accept the alternative hypothesis that the results for analyst B are significantly greater than those for analyst A. Working in the same fashion, it is easy to show that

$(t_{\text{exp}})_{\text{AC}} = 0.437$ H_0 is retained

$(t_{\text{exp}})_{\text{AD}} = 0.414$ H_0 is retained

$(t_{\text{exp}})_{\text{BC}} = 10.17$ H_0 is rejected and H_A is accepted

$(t_{\text{exp}})_{\text{BD}} = 10.67$ H_0 is rejected and H_A is accepted

$(t_{\text{exp}})_{\text{CD}} = 0.04$ H_0 is accepted

Taken together, these results suggest that there is a significant systematic difference between the work of analyst B and that of the other analysts. There is no way to decide, however, whether any of the four analysts has done accurate work.

An analysis of variance can be extended to systems involving more than a single variable. For example, a two-way ANOVA can be used in a collaborative study to determine the importance to an analytical method of both the analyst and the instrumentation used. The treatment of multivariable ANOVA is beyond the scope of this text, but is covered in several of the texts listed as suggested readings at the end of the chapter.

14C.3 What Is a Reasonable Result for a Collaborative Study?

Collaborative testing provides a means for estimating the variability (or reproducibility) among analysts in different labs. If the variability is significant, we can determine that portion due to random errors traceable to the method (σ_{rand}) and that due to systematic differences between the analysts (σ_{sys}). In the previous two sections we saw how a two-sample collaborative test, or an analysis of variance can be used to estimate σ_{rand} and σ_{sys} (or σ^2_{rand} and σ^2_{sys}). We have not considered, however, what is a reasonable value for a method's reproducibility.

An analysis of nearly 10,000 collaborative studies suggests that a reasonable estimate for a method's reproducibility is

$$R = 2^{(1-0.5 \log C)} \qquad \textbf{14.26}$$

where R is the percent relative standard deviation for the results included in the collaborative study, and C is the fractional amount of analyte in the sample on a weight-to-weight basis.[10] For example, when the sample used in a collaborative study is 1 ppm (μg/g) in analyte, C has a value of 10^{-6}. The estimated percent relative standard deviation, therefore, is

$$R = 2^{[1-0.5 \log(10^{-6})]} = 16\%$$

Equation 14.26 appears to be independent of the type of analyte, the type of matrix, and the method of analysis.

Example

EXAMPLE 14.11

What is the estimated relative standard deviation for the results of a collaborative study in which the sample is pure analyte (100% w/w analyte)? Repeat for the case in which the analyte's concentration is 0.1% w/w.

SOLUTION

When the sample is 100% w/w analyte ($C = 1$), the estimated percent relative standard deviation is

$$R = 2^{[1-0.5 \log(1)]} = 2\%$$

Thus, we expect that approximately 67% of the participants in the collaborative study ($\pm 1\sigma$) will report the analyte's concentration within the range of 98% w/w to 102% w/w. When the analyte's concentration is 0.1% w/w ($C = 0.001$), the estimated percent relative standard deviation is

$$R = 2^{[1 - 0.5 \log(0.001)]} = 5.7\%$$

and we expect that 67% of the analysts will report the analyte's concentration within the range of 0.094% w/w to 0.106% w/w.

Of course, equation 14.26 only provides an estimate of the variability in results submitted by those participating in a collaborative study. A relative standard deviation that is within one-half to twice the estimated value suggests that the method performs acceptably when used by analysts in different laboratories. The percent relative standard deviation for a single analyst should be one-half to two-thirds that for the variability between analysts.

14D KEY TERMS

analysis of variance (*p. 693*)
blind analysis (*p. 683*)
collaborative testing (*p. 687*)
empirical model (*p. 676*)
factor (*p. 667*)
factor level (*p. 667*)
Fisher's least significant difference (*p. 696*)
response (*p. 667*)
response surface (*p. 667*)
ruggedness testing (*p. 684*)
simplex optimization (*p.672*)
standard method (*p. 667*)
theoretical model (*p. 675*)

14E SUMMARY

One of the goals of analytical chemistry is to develop new analytical methods that are recognized or accepted as being standard methods. In this chapter we considered how a standard method is developed, including finding the optimum experimental conditions, verifying that the method produces acceptable precision and accuracy, and validating the method for general use.

In optimizing a method, we seek to find the combination of experimental parameters producing the best result or response. We can visualize this process as being similar to finding the highest point on a mountain, in which the mountain's topography, called a response surface, is a plot of the system's response as a function of the factors under our control.

One approach to finding the optimum response is to use a searching algorithm. In a one-factor-at-a-time optimization, one factor is varied while holding constant all other factors, until there is no further improvement in the response. The process then continues with the next factor until no further improvement is found when changing any of the factors. This approach to finding the optimum response is often effective, but not efficient. Another searching algorithm, that is both effective and efficient, is a simplex optimization, the rules of which allow us to change the values of all factors simultaneously.

Another approach to optimizing a method is to develop a mathematical model of the response surface. Such models can be theoretical, in that they are derived from a known chemical and physical relationship between the response and its factors. Alternatively, an empirical model, which does not have a firm theoretical basis, can be developed by fitting an empirical equation to experimental data. One approach is to use a 2^k factorial design in which each factor is tested at both a high and a low level, and against every other factor at each of their levels.

After optimizing a method, it is necessary to demonstrate that it can produce acceptable results. Verifying a method usually includes establishing single-operator characteristics, performing a blind analysis of standard samples, and determining the method's ruggedness. Single-operator characteristics include the method's precision, accuracy, and detection limit when used by a single analyst. To test against possible bias on the part of the analyst, the analyst then analyzes a set of blind samples in which the analyst does not know the concentration of analyte in the samples. Finally, ruggedness testing is used to determine which experimental factors must be carefully controlled to avoid unexpectedly large determinate or indeterminate sources of error.

The last step in establishing a standard method is to validate its transferability to other laboratories. An important step in the process of validating a method is collaborative testing, in which a common set of samples is analyzed by different laboratories. In a well-designed collaborative test, it is possible to establish limits for the method's precision and accuracy.

14F *Suggested* EXPERIMENTS

Experiments

The following set of experiments provides practical examples of the optimization of experimental conditions. Examples include simplex optimization, factorial designs used to develop empirical models of response surfaces, and the fitting of experimental data to theoretical models of the response surface.

Harvey, D. T.; Byerly, S.; Bowman, A.; et al. "Optimization of HPLC and GC Separations Using Response Surfaces," *J. Chem. Educ.* **1991,** *68,* 162–168.

In this experiment a theoretical model is used to optimize the HPLC separation of substituted benzoic acids by adjusting the pH of the mobile phase. An empirical model is then used to find the optimum combination of organic modifiers (methanol, acetonitrile, and tetrahydrofuran) to shorten the analysis time. A fixed-size simplex optimization of a GC separation using column temperature and carrier gas flow rate is also presented.

—Continued

Experiments

Continued from page 699

Leggett, D. L. "Instrumental Simplex Optimization," *J. Chem. Educ.* **1983,** *60,* 707–710.

A variable-size simplex optimization of a gas chromatographic separation using oven temperature and carrier gas flow rate as factors is described in this experiment.

Oles, P. J. "Fractional Factorial Experimental Design as a Teaching Tool for Quantitative Analysis," *J. Chem. Educ.* **1998,** *75,* 357–359.

This experiment describes the use of a fractional factorial design to examine the effects of volume of HNO_3, molarity of $AgNO_3$, volume of $AgNO_3$, digestion temperature, and composition of wash water on the gravimetric analysis for chloride.

Sangsila, S.; Labinaz, G.; Poland, J. S.; et al. "An Experiment on Sequential Simplex Optimization of an Atomic Absorption Analysis Procedure," *J. Chem. Educ.* **1989,** *66,* 351–353.

This experiment describes a fixed-size simplex optimization of a system involving four factors. The goal of the optimization is to maximize the absorbance of As by hydride generation atomic absorption spectroscopy using the concentration of HCl, the N_2 flow rate, the mass of $NaBH_4$, and reaction time as factors.

Shavers, C. L.; Parsons, M. L.; Deming, S. N. "Simplex Optimization of Chemical Systems," *J. Chem Educ.* **1979,** *56,* 307–309.

This experiment describes a variable-size simplex optimization of the quantitative analysis of vanadium as $(VO)_2(SO_4)_3$ using the amount of H_2O_2 and H_2SO_4 as factors. A related experiment [Palasota, J. A.; Deming, S. N. "Central Composite Experimental Design," *J. Chem. Educ.* **1992,** *69,* 560–561] examines this same analytical method using a central composite experimental design.

Stieg, S. "A Low-Noise Simplex Optimization Experiment," *J. Chem. Educ.* **1986,** *63,* 547–548.

In this experiment the goal is to mix solutions of 1 M HCl and 20-ppm methyl violet to give the maximum absorbance at a wavelength of 425 nm (corresponding to a maximum concentration for the acid form of methyl violet). A variable-size simplex optimization is used to find the optimum mixture.

Stolzberg, R. J. "Screening and Sequential Experimentation: Simulations and Flame Atomic Absorption Spectrometry Experiments," *J. Chem. Educ.* **1997,** *74,* 216–220.

This experiment describes a fractional factorial design used to examine the effects of flame height, flame stoichiometry, acetic acid, lamp current, wavelength, and slit width on the flame atomic absorbance obtained using a solution of 2.00-ppm Ag^+.

Van Ryswyk, H.; Van Hecke, G. R. "Attaining Optimal Conditions," *J. Chem. Educ.* **1991,** *66,* 878–882.

This experiment examines the effect of reaction time, temperature, and mole ratio of reactants on the synthetic yield of acetylferrocene by a Friedel–Crafts acylation of ferrocene. A central composite experimental design is used to find the optimum conditions, but the experiment could be modified to use a factorial design.

14G PROBLEMS

1. For each of the following equations, determine the optimum response, using the one-factor-at-a-time searching algorithm. Begin the search at (0, 0) with factor A, and use a step size of 1 for both factors. The boundary conditions for each response surface are $0 \le A \le 10$ and $0 \le B \le 10$. Continue the search through as many cycles as necessary until the optimum response is found. Compare your optimum response for each equation with the true optimum.

(a) $R = 1.68 + 0.24A + 0.56B - 0.04A^2 - 0.04B^2$ $\quad \mu_{opt} = (3, 7)$

(b) $R = 4.0 - 0.4A + 0.08AB$ $\quad \mu_{opt} = (10, 10)$

(c) $R = 3.264 + 1.537A + 0.5664B - 0.1505A^2 - 0.02734B^2 - 0.05785AB$ $\quad \mu_{opt} = (3.91, 6.22)$

Note: These equations are from Deming, S. N.; Morgan, S. L. *Experimental Design: A Chemometric Approach.* Elsevier: Amsterdam, 1987, and pseudo-three-dimensional plots of the response surfaces can be found in their Figures 11.4, 11.5, and 11.14. The response surface for problem (a) also is shown in Color Plate 13.

2. Determine the optimum response for the equation in problem 1c, using the fixed-sized simplex searching algorithm. Compare your optimum response with the true optimum.

3. A 2^k factorial design was used to determine the equation for the response surface in problem 1b. The uncoded levels, coded levels, and the responses are shown in the following table.

A	B	A*	B*	Response
8	8	+1	+1	5.92
8	2	+1	−1	2.08
2	8	−1	+1	4.48
2	2	−1	−1	3.52

Determine the coded and uncoded equation for the response surface.

4. Koscielniak and Parczewski investigated the influence of Al on the determination of Ca by atomic absorption spectrophotometry using the 2^k factorial design shown in the following table.[11]

Ca^{2+} (ppm)	Al^{3+} (ppm)	Ca*	Al*	Response
10	160	+1	+1	54.29
10	0	+1	−1	98.44
4	160	−1	+1	19.18
4	0	−1	−1	38.53

(a) Determine the coded equation for the response surface. (b) If you wish to analyze a sample for which the concentration of Ca^{2+} is 6.0 ppm, what is the maximum concentration of Al^{3+} that can be present if the error in the response must be less than 5.0%?

5. Strange reports the following information for a 2^3 factorial design used to investigate the yield of a chemical process.[12]

Factor	High (+1) Level	Low (−1) Level
X: Temperature	140 °C	120 °C
Y: Catalyst	Type B	Type A
Z: [Reactant]	0.50 M	0.25 M

Run	X*	Y*	Z*	% Yield
1	−1	−1	−1	28
2	1	−1	−1	17
3	−1	1	−1	41
4	1	1	−1	34
5	−1	−1	1	56
6	1	−1	1	51
7	−1	1	1	42
8	1	1	1	36

(a) Determine the coded equation for this data. (b) If β terms of less than ± 1 are insignificant, what main effect and interaction terms in the coded equation are important? Write down this simpler form for the coded equation. (c) Explain why the coded equation for these data cannot be transformed into an uncoded form. (d) Which is the better catalyst, A or B? (e) What is the yield using this catalyst if the temperature is set to 125 °C and the concentration of the reactant is 0.45 M?

6. Pharmaceutical tablets coated with lactose often develop a brown discoloration. The factors that primarily affect the discoloration are temperature, relative humidity, and the presence of a base acting as a catalyst. The following data for a 2^3 factorial design have been reported.[13]

Factor	High (+1) Level	Low (−1) Level
X: benzocaine	present	absent
Y: temperature	40 °C	25 °C
Z: relative humidity	75%	50%

Run	X*	Y*	Z*	Color (arb. units)
1	−1	−1	−1	1.55
2	1	−1	−1	5.40
3	−1	1	−1	3.50
4	1	1	−1	6.75
5	−1	−1	1	2.45
6	1	−1	1	3.60
7	−1	1	1	3.05
8	1	1	1	7.10

(a) Determine the coded equation for these data. (b) If β terms of less than 0.5 are insignificant, what main effect and interaction terms in the coded equation are important? Write down this simpler form for the coded equation.

7. The following data for a 2^3 factorial design were collected during a study of the effect of temperature, pressure, and residence time on the %yield of a reaction.[14]

Factor	High (+1) Level	Low (−1) Level
X: temperature	200 °C	100 °C
Y: pressure	0.6 MPa	0.2 MPa
Z: residence time	20 min	10 min

Run	X*	Y*	Z*	Yield
1	−1	−1	−1	2
2	1	−1	−1	6
3	−1	1	−1	4
4	1	1	−1	8
5	−1	−1	1	10
6	1	−1	1	18
7	−1	1	1	8
8	1	1	1	12

(a) Determine the coded equation for these data. (b) If β terms of less than ± 1 are insignificant, what main effect and interaction terms in the coded equation are important? Write down this simpler form for the coded equation. (c) Three runs were made at the center of the factorial design (temperature = 150 °C, pressure = 0.4 MPa, residence time = 15 min), giving yields of 8, 9, and 8.8. Determine if a first-order empirical model is appropriate for this system at $\alpha = 0.05$.

8. Duarte and colleagues used a factorial design to optimize a flow injection analysis method for determining penicillin potentiometrically.[15] Three factors were studied—reactor length, carrier flow rate, and sample volume, with the high and low values summarized in the following table.

Factor	Low (–1) Level	High (+1) Level
X: reactor length (cm)	1.5	2.0
Y: carrier flow rate (mL/min)	1.6	2.2
Z: sample volume (μL)	100	150

An optimum response was defined as the greatest sensitivity, as determined by the measured potential for a standard solution of penicillin, and the largest sampling rate. The results of the optimization studies are shown in the following table.

Run	X^*	Y^*	Z^*	ΔE (mV)	Samples/h
1	–1	–1	–1	37.45	21.5
2	1	–1	–1	31.70	26.0
3	–1	1	–1	32.10	30.0
4	1	1	–1	27.20	33.0
5	–1	–1	1	39.85	21.0
6	1	–1	1	32.85	19.5
7	–1	1	1	35.00	30.0
8	1	1	1	32.15	34.0

(a) Determine the coded equation for the response surface where ΔE is the response. (b) Determine the coded equation for the response surface where samples/h is the response. (c) Based on the coded equations, do the conditions favoring sensitivity also improve the sampling rate? (d) What conditions would you choose if your goal is to optimize both sensitivity and sampling rate?

9. Here is a challenge! McMinn and co-workers investigated the effect of five factors for optimizing an H_2-atmosphere flame ionization detector using a 2^5 factorial design.[16] The factors and their levels were

Factor	High (+1) Level	Low (–1) Level
A: H_2 flow rate (mL/min)	1460	1382
B: SiH_4 (ppm)	20.0	12.2
C: $O_2 + N_2$ flow rate (mL/min)	255	210
D: O_2/N_2	1.36	1.19
E: electrode height	75	55

The coded ("+" = +1, "–" = –1) factor levels and responses, R, for the 32 experiments are shown in the following table

Run	A^*	B^*	C^*	D^*	E^*	R	Run	A^*	B^*	C^*	D^*	E^*	R
1	–	–	–	–	–	0.36	17	–	–	–	–	+	0.39
2	+	–	–	–	–	0.51	18	+	–	–	–	+	0.45
3	–	+	–	–	–	0.15	19	–	+	–	–	+	0.32
4	+	+	–	–	–	0.39	20	+	+	–	–	+	0.25
5	–	–	+	–	–	0.79	21	–	–	+	–	+	0.18
6	+	–	+	–	–	0.83	22	+	–	+	–	+	0.29
7	–	+	+	–	–	0.74	23	–	+	+	–	+	0.07
8	+	+	+	–	–	0.69	24	+	+	+	–	+	0.19
9	–	–	–	+	–	0.60	25	–	–	–	+	+	0.53
10	+	–	–	+	–	0.82	26	+	–	–	+	+	0.60
11	–	+	–	+	–	0.42	27	–	+	–	+	+	0.36
12	+	+	–	+	–	0.59	28	+	+	–	+	+	0.43
13	–	–	+	+	–	0.96	29	–	–	+	+	+	0.23
14	+	–	+	+	–	0.87	30	+	–	+	+	+	0.51
15	–	+	+	+	–	0.76	31	–	+	+	+	+	0.13
16	+	+	+	+	–	0.74	32	+	+	+	+	+	0.43

(a) Determine the coded equation for this response surface, ignoring β terms less than ±0.03. (b) A simplex optimization of this system finds optimal values for the factors of A = 2278 mL/min, B = 9.90 ppm, C = 260.6 mL/min, and D = 1.71. The value for E was maintained at its high level. Are these values consistent with your analysis of the factorial design?

10. A good empirical model provides an accurate picture of the response surface for factor levels included within the experimental design. The same model, however, may yield an inaccurate prediction for the response at other factor levels. An empirical model, therefore, should be tested before extrapolating to conditions other than those used to determine the model. For example, Palasota and Deming[7] studied the effect of the relative amounts of H_2SO_4 and H_2O_2 on the absorbance of solutions of vanadium using the following central composite design.

Run	Drops 1% H_2O_2	Drops 20% H_2SO_4
1	15	22
2	10	20
3	20	20
4	8	15
5	15	15
6	15	15
7	15	15
8	15	15
9	22	15
10	10	10
11	20	10
12	15	8

The reaction of H_2O_2 and H_2SO_4 generates a reddish brown solution whose absorbance is measured at a wavelength of 450 nm. A regression analysis on their data yielded the following uncoded equation for the response (Absorbance × 1000).

$$R = 835.90 - 36.82X_1 - 21.34X_2 + 0.52(X_1)^2 + 0.15(X_2)^2 + 0.98X_1X_2$$

where X_1 is the drops of H_2O_2, and X_2 is the drops of H_2SO_4. Calculate the predicted absorbances for 10 drops of H_2O_2 and 0 drops of H_2SO_4, 0 drops of H_2O_2 and 10 drops of H_2SO_4, and for 0 drops of each reagent. Are these results reasonable? Explain. What does your answer tell you about this empirical model?

11. A newly proposed method is to be tested for its single-operator characteristics. To be competitive with the standard method, the new method must have a relative standard deviation of less than 10%, with a bias of less than 10%. To test the method, an analyst performs ten replicate analyses on a standard sample known to contain 1.30 ppm of the analyte. The results for the ten trials are

1.25 1.26 1.29 1.56 1.46 1.23 1.49 1.27 1.31 1.43

Are the single-operator characteristics for this method acceptable?

12. A proposed gravimetric method was evaluated for its ruggedness by varying the following factors

Factor A—sample size	$A = 1$ g	$a = 1.1$ g
Factor B—pH	$B = 6.5$	$b = 6.0$
Factor C—digestion time	$C = 3$ h	$c = 1$ h
Factor D—number of rinses	$D = 3$ rinses	$d = 5$ rinses
Factor E—precipitant	$E =$ reagent 1	$e =$ reagent 2
Factor F—digestion temperature	$F = 50$ °C	$f = 60$ °C
Factor G—drying temperature	$G = 110$ °C	$g = 140$ °C

A standard sample containing a known amount of analyte was carried through the procedure using the experimental design in Table 14.6. The percentage of the known amount of analyte actually found in the eight trials were found to be

$R_1 = 98.9$ $R_2 = 98.5$ $R_3 = 97.7$ $R_4 = 97.0$

$R_5 = 98.8$ $R_6 = 98.5$ $R_7 = 97.7$ $R_8 = 97.3$

Determine which factors, if any, appear to have a significant effect on the response, and estimate the expected standard deviation for the method.

13. The two-sample plot for the data in Example 14.7 is shown in Figure 14.19. Identify the analyst whose work is (a) the most accurate; (b) the most precise; (c) the least accurate; and (d) the least precise.

14. Chichilo reports the following data for the determination of the %w/w Al in two samples of limestone.[17]

Analyst	Sample 1	Sample 2
1	1.35	1.57
2	1.35	1.33
3	1.34	1.47
4	1.50	1.60
5	1.52	1.62
6	1.39	1.52
7	1.30	1.36
8	1.32	1.53

Construct a two-sample plot for these data, and estimate values for σ_{rand} and σ_{sys} assuming $\alpha = 0.05$.

15. The importance of between-laboratory variability on the results of an analytical method can be determined by having several laboratories analyze the same sample. In one such study seven laboratories analyzed a sample of homogenized milk for a selected alfatoxin.[18] The results, in parts per billion, are summarized in the following table.

Replicate	Lab A	Lab B	Lab C	Lab D	Lab E	Lab F	Lab G
1	1.6	4.6	1.2	1.5	6.0	6.2	3.3
2	2.9	2.8	1.9	2.7	3.9	3.8	3.8
3	3.5	3.0	2.9	3.4	4.3	5.5	5.5
4	1.8	4.5	1.1	2.0	5.8	4.2	4.9
5	2.2	3.1	2.9	3.4	4.0	5.3	4.5

(a) Determine if the between-laboratory variability is significantly greater than the within-laboratory variability at $\alpha = 0.05$. If the between-laboratory variability is significant, then determine the source of that variability. (b) Estimate values for σ^2_{rand} and σ^2_{sys}.

16. A class of analytical students is asked to analyze a steel sample to determine the %w/w Mn. (a) Given that the steel sample is 0.26% w/w Mn, estimate the expected relative standard deviation for the class' results. (b) The actual results obtained by the students are

0.26% 0.28% 0.27% 0.24% 0.26% 0.25%

0.26% 0.28% 0.25% 0.24% 0.26% 0.25%

0.29% 0.24% 0.27% 0.23% 0.26% 0.24%

Are these results consistent with the estimated relative standard deviation?

14H SUGGESTED READINGS

The following texts and articles provide an excellent discussion of optimization methods based on searching algorithms and mathematical modeling, including a discussion of the relevant calculations.

Bayne, C. K.; Rubin, I. B. *Practical Experimental Designs and Optimization Methods for Chemists.* VCH Publishers: Deerfield Beach, FL, 1986.

Deming, S. N.; Morgan, S. L. *Experimental Design: A Chemometric Approach.* Elsevier: Amsterdam, 1987.

Hendrix, C. D. "What Every Technologist Should Know About Experimental Design," *Chemtech* **1979,** *9,* 167–174.

Hendrix, C. D. "Through the Response Surface with Test Tube and Pipe Wrench," *Chemtech* **1980,** *10,* 488–497.

Morgan, E. *Chemometrics: Experimental Design.* John Wiley and Sons: Chichester, England, 1991.

The following texts provide additional information about ANOVA calculations, including discussions of two-way analysis of variance.

Graham, R. C. *Data Analysis for the Chemical Sciences.* VCH Publishers: New York, 1993.

Miller, J. C.; Miller, J. N. *Statistics for Analytical Chemistry,* 3rd ed. Ellis Horwood Limited: Chichester, England, 1993.

14I REFERENCES

1. *Vogel's Textbook of Quantitative Inorganic Analysis.* Longman: London, 1978; p. 752.
2. Sharaf, M. A.; Illman, D. L.; Kowalski, B. R. *Chemometrics.* Wiley-Interscience: New York, 1986.
3. Spendley, W.; Hext, G. R.; Himsworth, F. R. *Technometrics* **1962,** *4,* 441–461.
4. Deming, S. N.; Parker, L. R. *CRC Crit. Rev. Anal. Chem.* **1978,** *7(3),* 187–202.
5. Long, D. E. *Anal. Chim. Acta* **1969,** *46,* 193–206.
6. Nelder, J. A.; Mead, R. *Computer J.* **1965,** *7,* 308–313.
7. Palasota, J. A.; Deming, S. N. *J. Chem. Educ.* **1992,** *62,* 560–563.
8. Youden, W. J. *Anal. Chem.* **1960,** *32(13),* 23A–37A.
9. Youden, W. J. "Statistical Techniques for Collaborative Tests" in *Statistical Manual of the Association of Official Analytical Chemists.* Association of Official Analytical Chemists: Washington, D.C., 1975.
10. (a) Horwitz, W. *Anal. Chem.* **1982,** *54,* 67A–76A; (b) Hall, P.; Selinger, B. *Anal. Chem.* **1989,** *61,* 1465–1466; (c) Albert, R.; Horwitz, W. *Anal. Chem.* **1997,** *69,* 789–790.
11. Koscielniak, P.; Parczewski, A. *Anal. Chim. Acta* **1983,** *153,* 111–119.
12. Strange, R. S. *J. Chem. Educ.* **1990,** *67,* 113–115.
13. Armstrong, N. A.; James, K. C. *Pharmaceutical Experimental Design and Interpretation.* Taylor and Francis: London, 1996 as cited in González, A. G. *Anal. Chim. Acta* **1998,** *360,* 227–241.
14. Akhnazarova, S.; Kafarov, V. *Experimental Optimization in Chemistry and Chemical Engineering.* MIR Publishers: Moscow, 1982 as cited in González, A. G. *Anal. Chim. Acta* **1998,** *360,* 227–241.
15. Duarte, M. M. M. B.; de O. Netro, G.; Kubota, L. T.; et al. *Anal. Chim. Acta* **1997,** *350,* 353–357.
16. McMinn, D. G.; Eatherton, R. L.; Hill, H. H. *Anal. Chem.* **1984,** *56,* 1293–1298.
17. Chichilo, P. *J. Assoc. Offc. Agr. Chemists* **1964**, *47,* 1019 as reported in Youden, W. J. "Statistical Techniques for Collaborative Tests," in *Statistical Manual of the Association of Official Analytical Chemists.* Association of Official Analytical Chemists: Washington, D.C., 1975.
18. Massart, D. L.; Vandeginste, B. G. M.; Deming, S. N.; et al. *Chemometrics: A Textbook.* Elsevier: Amsterdam, 1988.

Chapter 15

Quality Assurance

In Chapter 1 we noted that each field of chemistry brings a unique perspective to the broader discipline of chemistry. For analytical chemistry this perspective was identified as an approach to solving problems, which was presented as a five-step process: (1) Identify and define the problem; (2) Design the experimental procedure; (3) Conduct an experiment and gather data; (4) Analyze the experimental data; and (5) Propose a solution to the problem. The analytical approach, as presented thus far, appears to be a straightforward process of moving from problem-to-solution. Unfortunately (or perhaps fortunately for those who consider themselves to be analytical chemists!), an analysis is seldom routine. Even a well-established procedure, carefully followed, can yield poor data of little use.

An important feature of the analytical approach, which we have neglected thus far, is the presence of a "feedback loop" involving steps 2, 3, and 4. As a result, the outcome of one step may lead to a reevaluation of the other two steps. For example, after standardizing a spectrophotometric method for the analysis of iron we may find that its sensitivity does not meet the original design criteria. Considering this information we might choose to select a different method, to change the original design criteria, or to improve the sensitivity.

The "feedback loop" in the analytical approach is maintained by a **quality assurance** program (Figure 15.1), whose objective is to control systematic and random sources of error.[1–5] The underlying assumption of a quality assurance program is that results obtained when an analytical system is in statistical control are free of bias and are characterized by well-defined confidence intervals. When used properly, a quality assurance program identifies the practices necessary to bring a system into statistical control, allows us to determine if the system remains in statistical control, and suggests a course of corrective action when the system has fallen out of statistical control.

The focus of this chapter is on the two principal components of a quality assurance program: quality control and quality assessment. In addition, considerable attention is given to the use of control charts for routinely monitoring the quality of analytical data.

quality assurance
The steps taken during an analysis to ensure that the analysis is under control and that it is properly monitored.

quality control
Those steps taken to ensure that an analysis is under statistical control.

good laboratory practices
Those general laboratory procedures that, when followed, help ensure the quality of analytical work.

good measurement practices
Those instructions outlining how to properly use equipment and instrumentation to ensure the quality of measurements.

15A Quality Control

Quality control encompasses all activities used to bring a system into statistical control. The most important facet of quality control is a set of written directives describing all relevant laboratory-specific, technique-specific, sample-specific, method-specific, and protocol-specific operations.[1,3,6] **Good laboratory practices** (GLPs) describe the general laboratory operations that need to be followed in any analysis. These practices include properly recording data and maintaining records, using chain-of-custody forms for samples that are submitted for analysis, specifying and purifying chemical reagents, preparing commonly used reagents, cleaning and calibrating glassware, training laboratory personnel, and maintaining the laboratory facilities and general laboratory equipment.

Good measurement practices (GMPs) describe operations specific to a technique. In general, GMPs provide instructions for maintaining, calibrating, and using the equipment and instrumentation that form the basis for a specific technique. For example, a GMP for a titration describes how to calibrate a buret (if nec-

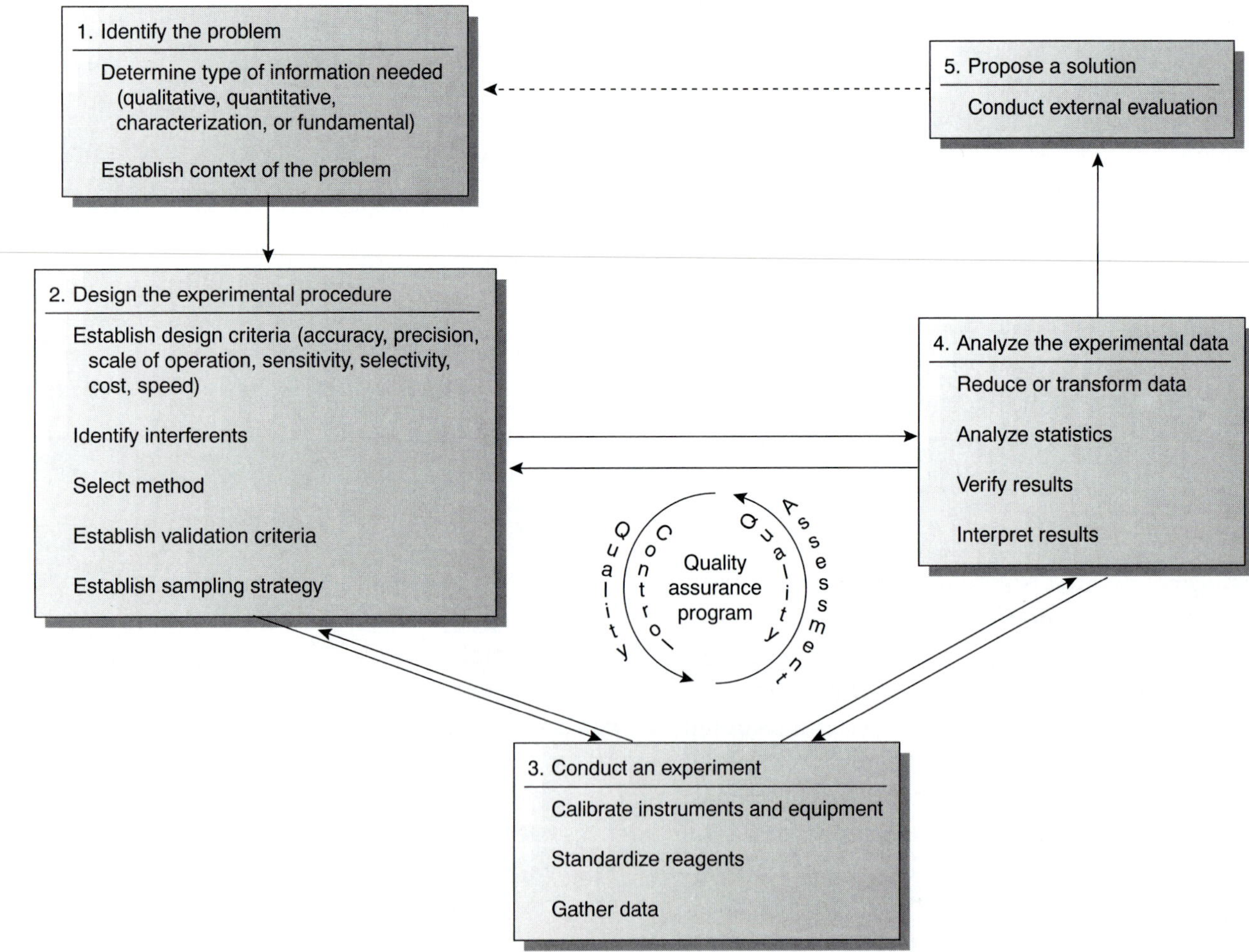

Figure 15.1
Schematic diagram of the analytical approach to problem solving, showing the role of the quality assurance program.

essary), how to fill a buret with the titrant, the correct way to read the volume of titrant in the buret, and the correct way to dispense the titrant.

The operations that need to be performed when analyzing a specific analyte in a specific matrix are defined by a **standard operations procedure** (SOP). The SOP describes all steps taken during the analysis, including: how the sample is processed in the laboratory, the analyte's separation from potential interferents, how the method is standardized, how the analytical signal is measured, how the data are transformed into the desired result, and the quality assessment tools that will be used to maintain quality control. If the laboratory is responsible for sampling, then the SOP will also state how the sample is to be collected and preserved and the nature of any prelaboratory processing. A SOP may be developed and used by a single laboratory, or it may be a standard procedure approved by an organization such as the American Society for Testing and Materials or the Federal Food and Drug Administration. A typical SOP is provided in the following example.

standard operations procedure
The procedure followed in collecting and analyzing samples and in interpreting the results of an analysis.

Example

EXAMPLE 15.1

Provide an SOP for the determination of cadmium in lake sediments by atomic absorption spectrophotometry using a normal calibration curve.

SOLUTION

Sediment samples should be collected using a bottom grab sampler and stored at 4 °C in acid-washed polyethylene bottles during transportation to the laboratory. Samples should be dried to constant weight at 105 °C and ground to a uniform particle size. The cadmium in a 1-g sample of the sediment is extracted by adding the sediment and 25 mL of 0.5 M HCl to an acid-washed 100-mL polyethylene bottle and shaking for 24 h. After filtering, the sample is analyzed by atomic absorption spectrophotometry using an air–acetylene flame, a wavelength of 228.8 nm, and a slit width of 0.5 nm. A normal calibration curve is prepared using five standards with nominal concentrations of 0.20, 0.50, 1.00, 2.00, and 3.00 ppm. The accuracy of the calibration curve is checked periodically by analyzing the 1.00-ppm standard. An accuracy of ±10% is considered acceptable.

Although an SOP provides a written procedure, it is not necessary to follow the procedure exactly as long as any modifications are identified. On the other hand, a **protocol for a specific purpose** (PSP), which is the most detailed of the written quality control directives, must be followed exactly if the results of the analysis are to be accepted. In many cases the required elements of a PSP are established by the agency sponsoring the analysis. For example, labs working under contract with the Environmental Protection Agency must develop a PSP that addresses such items as sampling and sample custody, frequency of calibration, schedules for the preventive maintenance of equipment and instrumentation, and management of the quality assurance program.

protocol for a specific purpose
A precisely written protocol for an analysis that must be followed exactly.

Two additional aspects of a quality control program deserve mention. The first is the physical inspection of samples, measurements and results by the individuals responsible for collecting and analyzing the samples.[1] For example, sediment samples might be screened during collection, and samples containing "foreign objects," such as pieces of metal, be discarded without being analyzed. Samples that are discarded can then be replaced with additional samples. When a sudden change in the performance of an instrument is observed, the analyst may choose to repeat those measurements that might be adversely influenced. The analyst may also decide to reject a result and reanalyze the sample when the result is clearly unreasonable. By identifying samples, measurements, and results that may be subject to gross errors, inspection helps control the quality of an analysis.

A final component of a quality control program is the certification of an analyst's competence to perform the analysis for which he or she is responsible.[7] Before an analyst is allowed to perform a new analytical method, he or she may be required to successfully analyze an independent check sample with acceptable accuracy and precision. The check sample should be similar in composition to samples that the analyst will routinely encounter, with a concentration that is 5 to 50 times that of the method's detection limit.

15B Quality Assessment

The written directives of a quality control program are a necessary, but not a sufficient, condition for obtaining and maintaining an analysis in a state of statistical control. Although quality control directives explain how an analysis should be properly conducted, they do not indicate whether the system is under statistical control. This is the role of **quality assessment,** which is the second component of a quality assurance program.

quality assessment
The steps taken to evaluate whether an analysis is under statistical control.

The goals of quality assessment are to determine when a system has reached a state of statistical control; to detect when the system has moved out of statistical control; and, if possible, to suggest why a loss of statistical control has occurred so that corrective actions can be taken. For convenience, the methods of quality assessment are divided into two categories: internal methods that are coordinated within the laboratory and external methods for which an outside agency or individual is responsible. The incorporation of these methods into a quality assurance program is covered in Section 15C.

15B.1 Internal Methods of Quality Assessment

The most useful methods for quality assessment are those that are coordinated by the laboratory and that provide the analyst with immediate feedback about the system's state of statistical control. Internal methods of quality assessment included in this section are the analysis of duplicate samples, the analysis of blanks, the analysis of standard samples, and spike recoveries.

duplicate samples
Two samples taken from a single gross sample and used to evaluate an analytical method's precision.

Analysis of Duplicate Samples An effective method for determining the precision of an analysis is to analyze **duplicate samples.** In most cases the duplicate samples are taken from a single gross sample (also called a split sample), although in some cases the duplicates must be independently collected gross samples. The results from the duplicate samples, X_1 and X_2, are evaluated by determining the difference, d, or the relative difference, $(d)_r$, between the samples

$$d = X_1 - X_2$$

$$(d)_r = \frac{d}{(X_1 + X_2)/2} \times 100$$

and comparing the results with accepted values, such as those shown in Table 15.1 for the analysis of waters and wastewaters.[7] Alternatively, the results for a set of n duplicates are combined to estimate the standard deviation for the analysis

Table 15.1 Selected Quality Assessment Limits for the Analysis of Waters and Wastewaters

Analyte	Limits for Spike Recovery (%)	$(d)_r$ When [Analyte] < 20 × MDL (±%)	$(d)_r$ When [Analyte] > 20 × MDL (±%)
acids	60–140	40	20
anions	80–120	25	10
bases or neutrals	70–130	40	20
carbamate pesticides	50–150	40	20
herbicides	40–160	40	20
metals	80–120	25	10
other inorganics	80–120	25	10
volatile organics	70–130	40	20

Abbreviation: MDL = method's detection limit.

$$s = \sqrt{\frac{\sum d_i^2}{2n}}$$

where d_i is the difference between the ith pair of duplicates. The degrees of freedom for the standard deviation is the same as the number of duplicate samples. If duplicate samples from several sources are combined, then the precision of the measurement process must be approximately the same for each. The precision obtained is then compared with the precision needed to accept the results of the analysis.

Example

EXAMPLE 15.2

To evaluate the precision for the determination of potassium in blood serum, duplicate analyses were performed on six samples, yielding the following results.

Duplicate	X_1	X_2
1	160	147
2	196	202
3	207	196
4	185	193
5	172	188
6	133	119

Calculate the standard deviation for the analysis.

SOLUTION

The standard deviation is determined as follows.

Duplicate	$d = X_1 - X_2$	d^2
1	13	169
2	−6	36
3	11	121
4	−8	64
5	−16	256
6	14	196

$$s = \sqrt{\frac{169 + 36 + 121 + 64 + 256 + 196}{(2)(6)}} = \sqrt{\frac{842}{12}} = 8.4$$

The Analysis of Blanks The use of a blank was introduced in Chapter 3 as a means of correcting the measured signal for contributions from sources other than the analyte. The most common blank is a method, or reagent blank, in which an analyte-free sample, usually distilled water, is carried through the analysis using the same reagents, glassware, and instrumentation. Method blanks are used to identify and correct systematic errors due to impurities in the reagents and contamination in the glassware and instrumentation. At a minimum, method blanks should be analyzed whenever new reagents are used, although a more frequent analysis provides an ongoing monitoring of the purity of the reagents. A new method blank should also be run whenever a sample with a high concentration of the analyte is analyzed, because any residual carryover of the analyte may contaminate the glassware or instrumentation.

field blank
A blank sample collected in the field.

trip blank
A blank prepared in the laboratory that accompanies a set of sample containers in the field and laboratory.

When samples are collected in the field, the method blank may be augmented with field and trip blanks.[8] A **field blank** is an analyte-free sample carried from the laboratory to the sampling site. At the sampling site the blank is transferred to a clean sample container, exposing it to the local environment, preserved, and transported back to the laboratory for analysis. Field blanks are used to identify and correct systematic errors due to sampling, transport, and analysis. **Trip blanks** are analyte-free samples carried from the laboratory to the sampling site and returned to the laboratory without being opened. A trip blank is used to identify and correct systematic errors due to cross-contamination of volatile organic compounds during transport, handling, storage, and analysis.

Analysis of Standards The analysis of a standard containing a known concentration of analyte also can be used to monitor a system's state of statistical control. Ideally, a standard reference material (SRM) should be used, provided that the matrix of the SRM is similar to that of the samples being analyzed. A variety of appropriate SRMs are available from the National Institute of Standards and Technology (NIST). If a suitable SRM is not available, then an independently prepared synthetic sample can be used if it is prepared from reagents of known purity. At a minimum, a standardization of the method is verified by periodically analyzing one of the calibration standards. In all cases, the analyte's experimentally determined concentration in the standard must fall within predetermined limits if the system is to be considered under statistical control.

spike recovery
An analysis of a sample after spiking with a known amount of analyte.

Spike Recoveries One of the most important quality assessment tools is the recovery of a known addition, or spike, of analyte to a method blank, field blank, or sample. To determine a **spike recovery,** the blank or sample is split into two portions, and a known amount of a standard solution of the analyte is added to one portion. The concentration of the analyte is determined for both the spiked, F, and unspiked portions, I, and the percent recovery, $\%R$, is calculated as

$$\%R = \frac{F - I}{A} \times 100$$

where A is the concentration of the analyte added to the spiked portion.

Example

EXAMPLE 15.3

A spike recovery for the analysis of chloride in well water was performed by adding 5.00 mL of a 25,000-ppm solution of Cl^- to a 500-mL volumetric flask and diluting to volume with the sample. Analysis of the sample and the spiked sample resulted in chloride concentrations of 183 ppm and 409 ppm, respectively. Determine the percent recovery of the spike.

SOLUTION

The concentration of the added spike is calculated by taking into account the effect of dilution.

$$A = 25{,}000 \text{ ppm} \times \frac{5.00 \text{ mL}}{500.0 \text{ mL}} = 250 \text{ ppm}$$

Thus, the spike recovery is

$$\%R = \frac{409 - 183}{250} \times 100 = 90.4\%$$

Spike recoveries on method blanks and field blanks are used to evaluate the general performance of an analytical procedure. The concentration of analyte added to the blank should be between 5 and 50 times the method's detection limit. Systematic errors occurring during sampling and transport will result in an unacceptable recovery for the field blank, but not for the method blank. Systematic errors occurring in the laboratory, however, will affect the recoveries for both the field and method blanks.

Spike recoveries for samples are used to detect systematic errors due to the sample matrix or the stability of the sample after its collection. Ideally, samples should be spiked in the field at a concentration between 1 and 10 times the expected concentration of the analyte or 5 to 50 times the method's detection limit, whichever is larger. If the recovery for a field spike is unacceptable, then a sample is spiked in the laboratory and analyzed immediately. If the recovery for the laboratory spike is acceptable, then the poor recovery for the field spike may be due to the sample's deterioration during storage. When the recovery for the laboratory spike also is unacceptable, the most probable cause is a matrix-dependent relationship between the analytical signal and the concentration of the analyte. In this case the samples should be analyzed by the method of standard additions. Typical limits for acceptable spike recoveries for the analysis of waters and wastewaters are shown in Table 15.1.[7]

15B.2 External Methods of Quality Assessment

Internal methods of quality assessment should always be viewed with some level of skepticism because of the potential for bias in their execution and interpretation. For this reason, external methods of quality assessment also play an important role in quality assurance programs. One external method of quality assessment is the certification of a laboratory by a sponsoring agency. Certification is based on the successful analysis of a set of **proficiency standards** prepared by the sponsoring agency. For example, laboratories involved in environmental analyses may be required to analyze standard samples prepared by the Environmental Protection

proficiency standard
A standard sample provided by an external agency as part of certifying the quality of a laboratory's work.

Agency. A second example of an external method of quality assessment is the voluntary participation of the laboratory in a collaborative test (Chapter 14) sponsored by a professional organization such as the Association of Official Analytical Chemists. Finally, individuals contracting with a laboratory can perform their own external quality assessment by submitting blind duplicate samples and blind standard samples to the laboratory for analysis. If the results for the quality assessment samples are unacceptable, then there is good reason to consider the results suspect for other samples provided by the laboratory.

15C Evaluating Quality Assurance Data

In the previous section we described several internal methods of quality assessment that provide quantitative estimates of the systematic and random errors present in an analytical system. Now we turn our attention to how this numerical information is incorporated into the written directives of a complete quality assurance program. Two approaches to developing quality assurance programs have been described[9]: a prescriptive approach, in which an exact method of quality assessment is prescribed; and a performance-based approach, in which any form of quality assessment is acceptable, provided that an acceptable level of statistical control can be demonstrated.

15C.1 Prescriptive Approach

With a prescriptive approach to quality assessment, duplicate samples, blanks, standards, and spike recoveries are measured following a specific protocol. The result for each analysis is then compared with a single predetermined limit. If this limit is exceeded, an appropriate corrective action is taken. Prescriptive approaches to quality assurance are common for programs and laboratories subject to federal regulation. For example, the Food and Drug Administration (FDA) specifies quality assurance practices that must be followed by laboratories analyzing products regulated by the FDA.

A good example of a prescriptive approach to quality assessment is the protocol outlined in Figure 15.2, published by the Environmental Protection Agency (EPA) for laboratories involved in monitoring studies of water and wastewater.[10] Independent samples A and B are collected simultaneously at the sample site. Sample A is split into two equal-volume samples, and labeled A_1 and A_2. Sample B is also split into two equal-volume samples, one of which, B_{SF}, is spiked with a known amount of analyte. A field blank, D_F, also is spiked with the same amount of analyte. All five samples (A_1, A_2, B, B_{SF}, and D_F) are preserved if necessary and transported to the laboratory for analysis.

The first sample to be analyzed is the field blank. If its spike recovery is unacceptable, indicating that a systematic error is present, then a laboratory method blank, D_L, is prepared and analyzed. If the spike recovery for the method blank is also unsatisfactory, then the systematic error originated in the laboratory. An acceptable spike recovery for the method blank, however, indicates that the systematic error occurred in the field or during transport to the laboratory. Systematic errors in the laboratory can be corrected, and the analysis continued. Any systematic errors occurring in the field, however, cast uncertainty on the quality of the samples, making it necessary to collect new samples.

If the field blank is satisfactory, then sample B is analyzed. If the result for B is above the method's detection limit, or if it is within the range of 0.1 to 10 times the amount of analyte spiked into B_{SF}, then a spike recovery for B_{SF} is determined. An

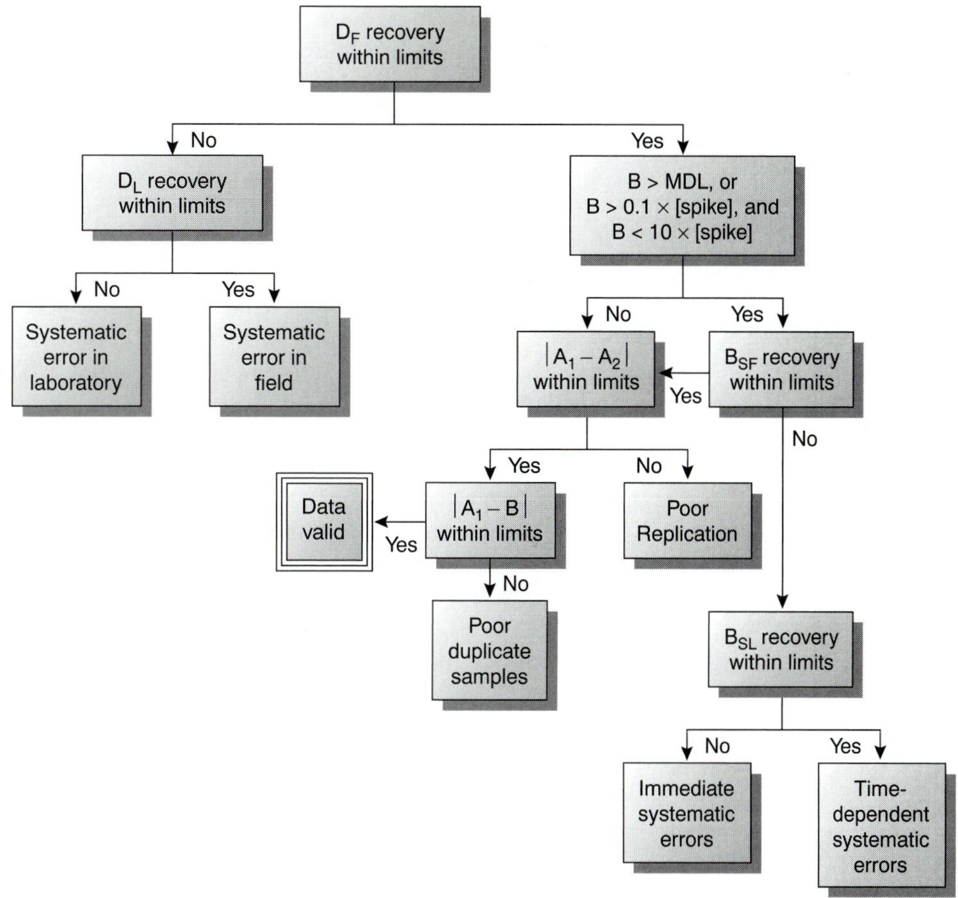

Figure 15.2
Example of a prescriptive approach to quality assurance. Adapted from Environmental Monitoring and Support Laboratory, U.S. Environmental Protection Agency, "Handbook for Analytical Quality Control in Water and Wastewater Laboratories," March 1979.

unacceptable spike recovery for B_{SF} indicates the presence of a systematic error involving the sample. To determine the source of the systematic error, a laboratory spike, B_{SL}, is prepared using sample B and analyzed. If the spike recovery for B_{SL} is acceptable, then the systematic error requires a long time to have a noticeable effect on the spike recovery. One possible explanation is that the analyte has not been properly preserved or has been held beyond the acceptable holding time. An unacceptable spike recovery for B_{SL} suggests an immediate systematic error, such as that due to the influence of the sample's matrix. In either case, the systematic errors are fatal and must be corrected before the sample is reanalyzed.

If the spike recovery for B_{SF} is acceptable, or if the result for sample B is below the method's detection limit or outside the range of 0.1 to 10 times the amount of analyte spiked in B_{SF}, then the duplicate samples A_1 and A_2 are analyzed. The results for A_1 and A_2 are discarded if the difference between their values is excessive. If the difference between the results for A_1 and A_2 is within the accepted limits, then the results for samples A_1 and B are compared. Since samples collected from the same sampling site at the same time should be identical in composition, the results are discarded if the difference between their values is unsatisfactory, and accepted if the difference is satisfactory.

This protocol requires four to five evaluations of quality assessment data before the result for a single sample can be accepted; a process that must be repeated for each analyte and for each sample. Other prescriptive protocols are equally demanding. For example, Figure 3.7 in Chapter 3 shows a portion of the quality assurance protocol used for the graphite furnace atomic absorption analysis of trace metals in aqueous solutions. This protocol involves the analysis of an initial calibration verification standard and an initial calibration blank, followed by the analysis of samples in groups of ten. Each group of samples is preceded and followed by continuing calibration verification (CCV) and continuing calibration blank (CCB) quality assessment samples. Results for each group of ten samples can be accepted only if both sets of CCV and CCB quality assessment samples are acceptable.

The advantage to a prescriptive approach to quality assurance is that a single consistent set of guidelines is used by all laboratories to control the quality of analytical results. A significant disadvantage, however, is that the ability of a laboratory to produce quality results is not taken into account when determining the frequency of collecting and analyzing quality assessment data. Laboratories with a record of producing high-quality results are forced to spend more time and money on quality assessment than is perhaps necessary. At the same time, the frequency of quality assessment may be insufficient for laboratories with a history of producing results of poor quality.

15C.2 Performance-Based Approach

In a performance-based approach to quality assurance, a laboratory is free to use its experience to determine the best way to gather and monitor quality assessment data. The quality assessment methods remain the same (duplicate samples, blanks, standards, and spike recoveries) since they provide the necessary information about precision and bias. What the laboratory can control, however, is the frequency with which quality assessment samples are analyzed, and the conditions indicating when an analytical system is no longer in a state of statistical control. Furthermore, a performance-based approach to quality assessment allows a laboratory to determine if an analytical system is in danger of drifting out of statistical control. Corrective measures are then taken before further problems develop.

control chart
A graph showing the time-dependent change in the results of an analysis that is used to monitor whether an analysis is in a state of statistical control.

The principal tool for performance-based quality assessment is the **control chart.** In a control chart the results from the analysis of quality assessment samples are plotted in the order in which they are collected, providing a continuous record of the statistical state of the analytical system. Quality assessment data collected over time can be summarized by a mean value and a standard deviation. The fundamental assumption behind the use of a control chart is that quality assessment data will show only random variations around the mean value when the analytical system is in statistical control. When an analytical system moves out of statistical control, the quality assessment data is influenced by additional sources of error, increasing the standard deviation or changing the mean value.

Control charts were originally developed in the 1920s as a quality assurance tool for the control of manufactured products.[11] Two types of control charts are commonly used in quality assurance: a property control chart in which results for single measurements, or the means for several replicate measurements, are plotted sequentially; and a precision control chart in which ranges or standard deviations are plotted sequentially. In either case, the control chart consists of a line representing the mean value for the measured property or the precision, and two or more boundary lines whose positions are determined by the precision of the measurement process. The position of the data points about the boundary lines determines whether the system is in statistical control.

Construction of Property Control Charts The simplest form for a property control chart is a sequence of points, each of which represents a single determination of the property being monitored. To construct the control chart, it is first necessary to determine the mean value of the property and the standard deviation for its measurement. These statistical values are determined using a minimum of 7 to 15 samples (although 30 or more samples are desirable), obtained while the system is known to be under statistical control. The center line (CL) of the control chart is determined by the average of these n points

$$CL = \overline{X} = \frac{\sum X_i}{n}$$

The positions of the boundary lines are determined by the standard deviation, S, of the points used to determine the central line

$$S = \sqrt{\frac{\sum (X_i - \overline{X})^2}{n - 1}}$$

with the upper and lower warning limits (UWL and LWL), and the upper and lower control limits (UCL and LCL) given by

$$UWL = CL + 2S$$

$$LWL = CL - 2S$$

$$UCL = CL + 3S$$

$$LCL = CL - 3S$$

Example

EXAMPLE 15.4

Construct a property control chart for the following spike recovery data (all values are for percentage of spike recovered).

Sample:	1	2	3	4	5
Result:	97.3	98.1	100.3	99.4	100.9
Sample:	6	7	8	9	10
Result:	98.6	96.9	99.6	101.1	100.4
Sample:	11	12	13	14	15
Result:	100.0	95.9	98.3	99.2	102.1
Sample:	16	17	18	19	20
Result:	98.5	101.7	100.4	99.1	100.3

SOLUTION

The mean and the standard deviation for the 20 data points are 99.4 and 1.6, respectively, giving the UCL as 104.2, the UWL as 102.6, the LWL as 96.2 and the LCL as 94.6. The resulting property control chart is shown in Figure 15.3.

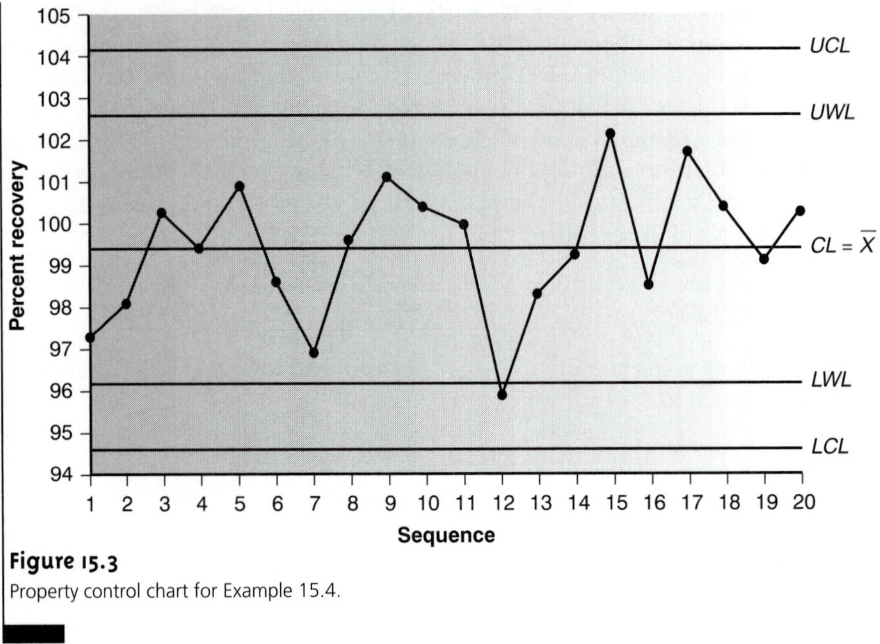

Figure 15.3
Property control chart for Example 15.4.

Property control charts can also be constructed using points that are the mean value, $\overline{X}_i$, for a set of r replicate determinations on a single sample. The mean for the ith sample is given by

$$\overline{X}_i = \frac{\sum_{j=1}^{r} X_{ij}}{j}$$

where X_{ij} is the jth replicate. The center line for the control chart, therefore, is

$$CL = \overline{X} = \frac{\sum \overline{X}_i}{n}$$

To determine the standard deviation for the warning and control limits, it is necessary to calculate the variance for each sample, s_i^2.

$$s_i^2 = \frac{\sum_{j=1}^{r} (X_{ij} - \overline{X}_i)^2}{r - 1}$$

The overall standard deviation, S, is the square root of the average variance for the samples used to establish the control plot.

$$S = \sqrt{\frac{\sum s_i^2}{n}}$$

Finally, the resulting warning and control limits are

$$UWL = CL + \frac{2S}{\sqrt{r}}$$

$$LWL = CL - \frac{2S}{\sqrt{r}}$$

$$UCL = CL + \frac{3S}{\sqrt{r}}$$

Table 15.2 Statistical Factors for the Upper Warning Limit and Upper Control Limit

Replicates	f_{UWL}	f_{UCL}
2	2.512	3.267
3	2.050	2.575
4	1.855	2.282
5	1.743	2.115
6	1.669	2.004

$$LCL = CL - \frac{3S}{\sqrt{r}}$$

Constructing a Precision Control Chart The most common measure of precision used in constructing a precision control chart is the range, R, between the largest and smallest results for a set of j replicate analyses on a sample.

$$R = X_{\text{large}} - X_{\text{small}}$$

To construct the control chart, ranges for a minimum of 15–20 samples (preferably 30 or more samples) are obtained while the system is known to be in statistical control. The line for the average range, $\overline{R}$, is determined by the mean of these n samples

$$\overline{R} = \frac{\sum R_i}{n}$$

The upper control line and the upper warning line are given by

$$UCL = f_{\text{UCL}} \times \overline{R}$$

$$UWL = f_{\text{UWL}} \times \overline{R}$$

where f_{UCL} and f_{UWL} (Table 15.2) are statistical factors determined by the number of replicates used to determine the range. Because the range always is greater than or equal to zero, there is no lower control limit or lower warning limit.

Example

EXAMPLE 15.5

Construct a precision control chart using the following 20 ranges, each determined from a duplicate analysis of a 10-ppm calibration standard

Sample:	1	2	3	4	5
Result:	0.36	0.09	0.11	0.06	0.25
Sample:	6	7	8	9	10
Result:	0.15	0.28	0.27	0.03	0.28
Sample:	11	12	13	14	15
Result:	0.21	0.19	0.06	0.13	0.37
Sample:	16	17	18	19	20
Result:	0.01	0.19	0.39	0.05	0.05

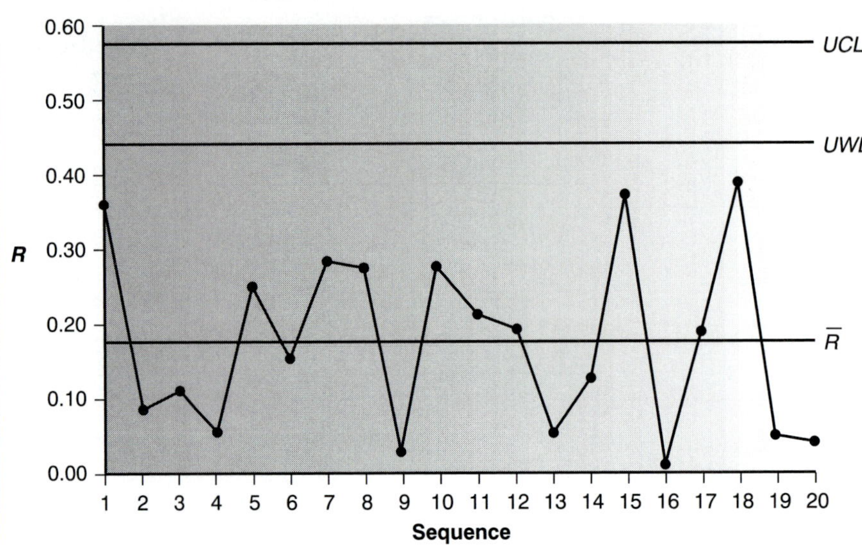

Figure 15.4
Precision control chart for Example 15.5.

SOLUTION

The average range for the 20 duplicate samples is 0.177. Because two replicates were used for each point, the *UWL* and *UCL* are

$$UWL = (2.512)(0.177) = 0.44$$

$$UCL = (3.267)(0.177) = 0.58$$

The complete control chart is shown in Figure 15.4.

The precision control chart is strictly valid only for the replicate analysis of identical samples, such as a calibration standard or a standard reference material. Its use for the analysis of nonidentical samples, such as a series of clinical or environmental samples, is complicated by the fact that the range usually is not independent of the magnitude of X_{large} and X_{small}. For example, Table 15.3 shows the relationship between $\overline{R}$ and the concentration of chromium in water.[10] Clearly the significant difference in the average range for these concentrations of Cr makes a single precision control chart impossible. One solution to this problem is to prepare separate precision control charts, each of which covers a range of concentrations for which $\overline{R}$ is approximately constant (Figure 15.5).

Interpreting Control Charts The purpose of a control chart is to determine if a system is in statistical control. This determination is made by examining the location of individual points in relation to the warning limits and the control limits, and the distribution of the points around the central line. If we assume that the data are normally distributed, then the probability of finding a point at any distance from the mean value can be determined from the normal distribution curve. The upper and lower control limits for a property control chart, for example, are set to $\pm 3S$, which, if S is a good approximation for σ, includes 99.74% of the data. The probability that a point will fall outside the *UCL* or *LCL*, therefore, is only 0.26%. The

Table 15.3 Average Range for Duplicate Samples for Different Concentrations of Chromium in Water

Cr (ppb)	Number of Duplicate Samples	$\bar{R}$
5 to <10	32	0.32
10 to <25	15	0.57
25 to <50	16	1.12
50 to <150	15	3.80
150 to <500	8	5.25
>500	5	76.0

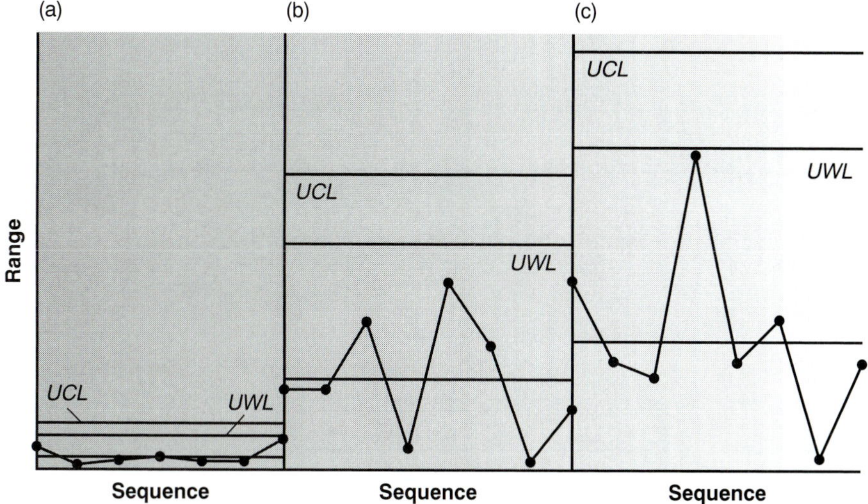

Figure 15.5
Example of the use of subrange precision control charts for samples that span a range of analyte concentrations. The precision control charts are used for (a) low concentrations of analyte; (b) intermediate concentrations of analyte; and (c) high concentrations of analyte.

most likely explanation when a point exceeds a control limit is that a systematic error has occurred or that the precision of the measurement process has deteriorated. In either case the system is assumed to be out of statistical control.

Rule 1. A system is considered to be out of statistical control if any single point exceeds either the *UCL* or the *LCL*.

The upper and lower warning limits, which are located at $\pm 2S$, should only be exceeded by 5% of the data; thus

Rule 2. A system is considered to be out of statistical control if two out of three consecutive points are between the *UWL* and *UCL* or between the *LWL* and *LCL*.

When a system is in statistical control, the data points should be randomly distributed about the center line. The presence of an unlikely pattern in the data is another indication that a system is no longer in statistical control.[4,12] Thus,

Rule 3. A system is considered to be out of statistical control if a run of seven consecutive points is completely above or completely below the center line (Figure 15.6a).

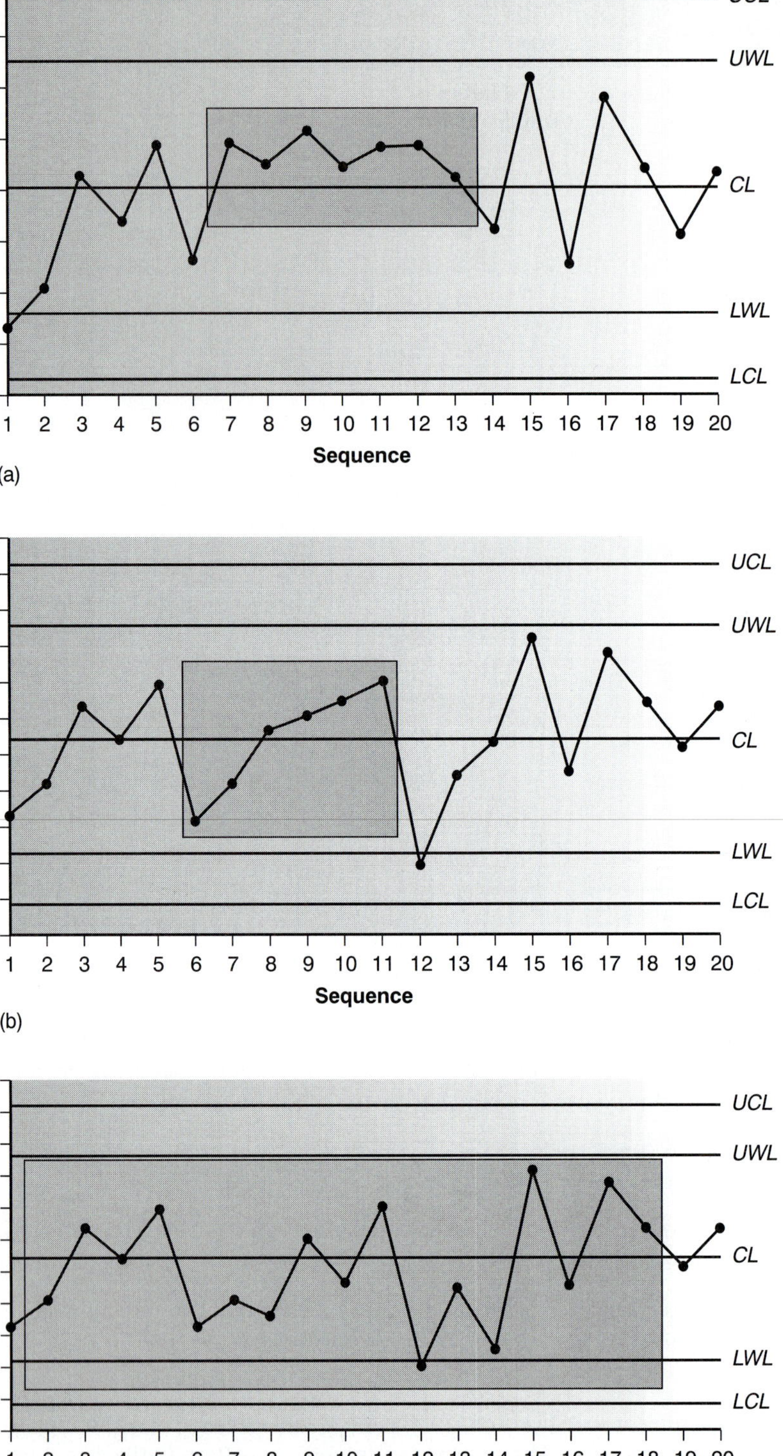

Figure 15.6

Examples of property control charts that show a run of data (highlighted in box) indicating that the system is out of statistical control.

Rule 4. A system is considered to be out of statistical control if six consecutive points are all increasing in value or all decreasing in value (Figure 15.6b). The points may be on either side of the center line.

Rule 5. A system is considered to be out of statistical control if 14 consecutive points alternate up and down in value (Figure 15.6c). The points may be on either side of the center line.

Rule 6. A system is considered to be out of statistical control if any obvious "nonrandom" pattern is observed.

The same rules apply to precision control charts with the exception that there are no lower warning and lower control limits.

Using Control Charts for Quality Assurance Control charts play an important role in a performance-based program of quality assurance because they provide an easily interpreted picture of the statistical state of an analytical system. Quality assessment samples such as blanks, standards, and spike recoveries can be monitored with property control charts. A precision control chart can be used to monitor duplicate samples.

The first step in using a control chart for quality assurance is to determine the mean value and the standard deviation (except when using the range) for the quality assessment data while the system is under statistical control. These values must be established under the same conditions that will be present during the normal use of the control chart. Thus, preliminary data should be randomly collected throughout the day, as well as over several days, to account for short-term and long-term variability. The preliminary data are used to construct an initial control chart, and discrepant points are determined using the rules discussed in the previous section. Questionable points are dropped, and the control chart is replotted. As the control chart is used, it may become apparent that the original limits need adjusting. Control limits can be recalculated if the number of new data points is at least equivalent to the amount of data used to construct the original control chart. For example, if 15 points were initially used, the limits can be reevaluated after 15 additional points are collected. The 30 points are pooled together to calculate the new limits. A second modification can be made after a further 30 points have been collected. Another indication that a control chart needs to be modified is when points rarely exceed the warning limits. In this case the new limits can be recalculated using the last 20 points.

Once a control chart is in use, new quality assessment data should be added at a rate sufficient to ensure that the system remains in statistical control. As with prescriptive approaches to quality assurance, when a quality assessment sample is found to be out of statistical control, all samples analyzed since the last successful verification of statistical control must be reanalyzed. The advantage of a performance-based approach to quality assurance is that a laboratory may use its experience, guided by control charts, to determine the frequency for collecting quality assessment samples. When the system is stable, quality assessment samples can be acquired less frequently.

15D KEY TERMS

control chart *(p. 714)*
duplicate samples *(p. 708)*
field blank *(p. 710)*
good laboratory practices *(p. 706)*
good measurement practices *(p. 706)*
proficiency standard *(p. 711)*
protocol for a specific purpose *(p. 707)*
quality assessment *(p. 708)*
quality assurance *(p. 706)*
quality control *(p. 706)*
spike recovery *(p. 710)*
standard operations procedure *(p. 707)*
trip blank *(p. 710)*

15E SUMMARY

Few analyses are so straightforward that high-quality results are easily obtained. Good analytical work requires careful planning and an attention to detail. Creating and maintaining a quality assurance program is one way to help ensure the quality of analytical results. Quality assurance programs usually include elements of quality control and quality assessment.

Quality control encompasses all activities used to bring a system into statistical control. The most important facet of quality control is written documentation, including statements of good laboratory practices, good measurement practices, standard operating procedures, and protocols for a specific purpose.

Quality assessment includes the statistical tools used to determine whether an analysis is in a state of statistical control and, if possible, to suggest why an analysis has drifted out of statistical control. Among the tools included in quality assessment are the analysis of duplicate samples, the analysis of blanks, the analysis of standards, and the analysis of spike recoveries.

Another important quality assessment tool, which provides an ongoing evaluation of an analysis, is a control chart. A control chart plots a property, such as a spike recovery, as a function of time. Results exceeding warning and control limits, or unusual patterns of data indicate that an analysis is no longer under statistical control.

15F Suggested EXPERIMENTS

Experiments

The following three experiments introduce aspects of quality assurance and quality control.

Bell, S. C.; Moore, J. "Integration of Quality Assurance/Quality Control into Quantitative Analysis," *J. Chem. Educ.* **1998,** *75,* 874–877.

The use of several QA/QC methods is described in this article, including control charts for monitoring the concentration of solutions of thiosulfate that have been prepared and stored with and without proper preservation; the use of method blanks and standard samples to determine the presence of determinate error and to establish single-operator characteristics; and the use of spiked samples and recoveries to identify the presence of determinate errors associated with collecting and analyzing samples.

Laquer, F. C. "Quality Control Charts in the Quantitative Analysis Laboratory Using Conductance Measurement," *J. Chem. Educ.* **1990,** *67,* 900–902.

The conductivities of a standard solution of KCl, laboratory distilled water, and synthetic-process samples are monitored weekly and evaluated using a control chart.

Marcos, J.; Ríos, A.; Valcárcel, M. "Practicing Quality Control in a Bioanalytical Experiment," *J. Chem. Educ.* **1995,** *72,* 947–949.

This experiment demonstrates how control charts and an analysis of variance can be used to evaluate the quality of results in a quantitative analysis for chlorophyll a and b in plant material.

15G PROBLEMS

1. Make a list of good laboratory practices for the lab accompanying this course (or another lab if this course does not have an associated laboratory). Explain the rationale for each item on your list.
2. Write directives outlining good measurement practices for (a) a buret, (b) a pH meter, and (c) a spectrophotometer.
3. A method for the analysis of lead in industrial wastewater has a method detection limit of 10 ppb. The relationship between the analytical signal and the concentration of lead, as determined from a calibration curve is

$$S_{meas} = 0.349 \times (\text{ppm Pb})$$

Analysis of a sample in duplicate gives S_{meas} as 0.554 and 0.516. Is the precision between these two duplicates acceptable based on the limits shown in Table 15.1?

4. The following data were obtained for the duplicate analysis of a 5.00-ppm NO_3^- standard

Sample	X_1 (ppm)	X_2 (ppm)
1	5.02	4.90
2	5.10	5.18
3	5.07	4.95
4	4.96	5.01
5	4.88	4.98
6	5.04	4.97

Calculate the standard deviation for the analysis of these duplicate samples. If the maximum limit for the relative standard deviation is 1.5%, are these results acceptable?

5. Gonzalez and colleagues developed a voltammetric procedure for the determination of *tert*-butylhydroxyanisole (BHA) in chewing gum.[13] Analysis of a commercial chewing gum gave results of 0.20 mg/g. To evaluate the accuracy of their results, they performed five spike recoveries, adding an amount of BHA equivalent to 0.135 mg/g to each sample. The experimentally determined concentrations of BHA in these samples were reported as 0.342, 0.340, 0.340, 0.324, and 0.322 mg/g. Determine the % recovery for each sample and the average % recovery.

6. A sample is to be analyzed following the protocol shown in Figure 15.2, using a method with a detection limit of 0.05 ppm. The relationship between the analytical signal and the concentration of the analyte, as determined from a calibration curve is

$$S_{meas} = 0.273 \times (\text{ppm analyte})$$

Answer the following questions if the limits for a successful spike recovery are ±10%.

(a) A field blank is spiked with the analyte to a concentration of 2.00 ppm and returned to the lab. Analysis of the spiked field blank gives a signal of 0.573. Is the spike recovery for the field blank acceptable?

(b) The analysis of a spiked field blank is found to be unacceptable. To determine the source of the problem, a spiked method blank is prepared by spiking distilled water with the analyte to a concentration of 2.00 ppm. Analysis of the spiked method blank gives a signal of 0.464. Is the source of the problem in the laboratory or in the field?

(c) The analysis for a spiked field sample, B_{SF}, was found to be unacceptable. To determine the source of the problem, the sample was spiked in the laboratory by adding sufficient analyte to increase the concentration by 2.00 ppm. Analysis of the sample before and after the spike gave signals of 0.456 for B and 1.03 for sample B_{SL}. Considering these data, what is the most likely source of the systematic error?

7. The following data were obtained for the repetitive analysis of a stable standard.[14]

Sample	X_i (ppm)	Sample	X_i (ppm)	Sample	X_i (ppm)
1	35.1	10	35.0	18	36.4
2	33.2	11	31.4	19	32.1
3	33.7	12	35.6	20	38.2
4	35.9	13	30.2	21	33.1
5	33.5	14	32.7	22	34.9
6	34.5	15	31.1	23	36.2
7	34.4	16	34.8	24	34.0
8	34.3	17	34.3	25	33.8
9	31.8				

Construct a property control chart for these data, and evaluate the state of statistical control.

8. The following data were obtained for the repetitive spike recoveries of field samples.[15]

Sample	% Recovery	Sample	% Recovery	Sample	% Recovery
1	94.6	10	104.6	18	104.6
2	93.1	11	123.8	19	91.5
3	100.0	12	93.8	20	83.1
4	122.3	13	80.0	21	100.8
5	120.8	14	99.2	22	123.1
6	93.1	15	101.5	23	96.2
7	117.7	16	74.6	24	96.9
8	96.2	17	108.5	25	102.3
9	73.8				

Construct a property control chart for these data, and evaluate the state of statistical control.

9. The following data were obtained for the duplicate analysis of a stable standard.[14]

Sample	X_1 (ppm)	X_2 (ppm)	Sample	X_1 (ppm)	X_2 (ppm)
1	50	46	14	36	36
2	37	36	15	47	45
3	22	19	16	16	20
4	17	20	17	18	21
5	32	34	18	26	22
6	46	46	19	35	36
7	26	28	20	26	25
8	26	30	21	49	51
9	61	58	22	33	32
10	44	45	23	40	38
11	40	44	24	16	13
12	36	35	25	39	42
13	29	31			

Construct a precision control chart for these data, and evaluate the state of statistical control.

15H SUGGESTED READINGS

The following texts and articles may be consulted for an additional discussion of the various aspects of quality assurance and quality control.

Amore, F. "Good Analytical Practices," *Anal. Chem.* **1979,** *51,* 1105A–1110A.

Barnard, Jr. A. J.; Mitchell, R. M.; Wolf, G. E. "Good Analytical Practices in Quality Control," *Anal. Chem.* **1978,** *50,* 1079A–1086A.

Cairns, T.; Rogers, W. M. "Acceptable Analytical Data for Trace Analysis," *Anal. Chem.* **1993,** *55,* 54A–57A.

Taylor, J. K. *Quality Assurance of Chemical Measurements.* Lewis Publishers: Chelsea, MI, 1987.

Additional information about the construction and use of control charts may be found in the following sources.

Miller, J. C.; Miller, J. N. *Statistics for Analytical Chemistry,* 3rd ed. Ellis Horwood Limited: Chichester, England, 1993.

Ouchi, G. I. "Creating Control Charts with a Spreadsheet Program," *LC•GC* **1993,** *11,* 416–423 and *LC•GC* **1997,** *15,* 336–344.

Simpson, J. M. "Spreadsheet Statistics," *J. Chem. Educ.* **1994,** *71,* A88–A89.

15I REFERENCES

1. Taylor, J. K. *Anal. Chem.* **1981,** *53,* 1588A–1596A
2. Taylor, J. K. *Anal. Chem.* **1983,** *55,* 600A–608A.
3. Taylor, J. K. *Am. Lab.* October **1985,** 67–75.
4. Nadkarni, R. A. *Anal. Chem.* **1991,** *63,* 675A–682A.
5. Valcárcel, M.; Ríos, A. *Trends Anal. Chem.* **1994,** *13,* 17–23.
6. ACS Committee for Environmental Improvement, "Principles of Environmental Analysis," *Anal. Chem.* **1983,** *55,* 2210–2218.
7. American Public Health Association, *Standard Methods for the Analysis of Water and Wastewater,* 18th ed. Washington, D.C., 1992.
8. Keith, L. H. *Environmental Sampling and Analysis: A Practical Guide.* Lewis Publishers, Chelsea, MI, 1991.
9. Poppiti, J. *Environ. Sci. Technol.* **1994,** *28,* 151A–152A.
10. Environmental Monitoring and Support Laboratory, U.S. Environmental Protection Agency, "Handbook for Analytical Quality Control in Water and Wastewater Laboratories," March 1979.
11. Shewhart, W. A. *Economic Control of the Quality of Manufactured Products.* Macmillan: London, 1931.
12. Mullins, E. *Analyst* **1994,** *119,* 369–375.
13. Gonzalez, A.; Ruiz, M. A.; Yanez-Sedeno, P.; et al. *Anal. Chim. Acta* **1994,** *285,* 63–71.
14. American Public Health Association, *Standard Methods for the Analysis of Water and Wastewater,* 18th ed. Washington, D.C., 1992. Data from Table 1030:I on page 1–10.
15. American Public Health Association, *Standard Methods for the Analysis of Water and Wastewater,* 18th ed. Washington, D.C., 1992. Data adapted from Table 1030:II on page 1–10.

Appendix 1

Appendix 1A

Single-Sided Normal Distribution[a]

z	0.00	0.01	0.02	0.03	0.04	0.05	0.06	0.07	0.08	0.09
0.0	0.5000	0.4960	0.4920	0.4880	0.4840	0.4801	0.4761	0.4721	0.4681	0.4641
0.1	0.4602	0.4562	0.4522	0.4483	0.4443	0.4404	0.4365	0.4325	0.4286	0.4247
0.2	0.4207	0.4168	0.4129	0.4090	0.4052	0.4013	0.3974	0.3936	0.3897	0.3859
0.3	0.3821	0.3783	0.3745	0.3707	0.3669	0.3632	0.3594	0.3557	0.3520	0.3483
0.4	0.3446	0.3409	0.3372	0.3336	0.3300	0.3264	0.3228	0.3192	0.3156	0.3121
0.5	0.3085	0.3050	0.3015	0.2981	0.2946	0.2912	0.2877	0.2843	0.2810	0.2776
0.6	0.2743	0.2709	0.2676	0.2643	0.2611	0.2578	0.2546	0.2514	0.2483	0.2451
0.7	0.2420	0.2389	0.2358	0.2327	0.2296	0.2266	0.2236	0.2206	0.2177	0.2148
0.8	0.2119	0.2090	0.2061	0.2033	0.2005	0.1977	0.1949	0.1922	0.1894	0.1867
0.9	0.1841	0.1814	0.1788	0.1762	0.1736	0.1711	0.1685	0.1660	0.1635	0.1611
1.0	0.1587	0.1562	0.1539	0.1515	0.1492	0.1469	0.1446	0.1423	0.1401	0.1379
1.1	0.1357	0.1335	0.1314	0.1292	0.1271	0.1251	0.1230	0.1210	0.1190	0.1170
1.2	0.1151	0.1131	0.1112	0.1093	0.1075	0.1056	0.1038	0.1020	0.1003	0.0985
1.3	0.0968	0.0951	0.0934	0.0918	0.0901	0.0885	0.0869	0.0853	0.0838	0.0823
1.4	0.0808	0.0793	0.0778	0.0764	0.0749	0.0735	0.0721	0.0708	0.0694	0.0681
1.5	0.0668	0.0655	0.0643	0.0630	0.0618	0.0606	0.0594	0.0582	0.0571	0.0559
1.6	0.0548	0.0537	0.0526	0.0516	0.0505	0.0495	0.0485	0.0475	0.0465	0.0455
1.7	0.0446	0.0436	0.0427	0.0418	0.0409	0.0401	0.0392	0.0384	0.0375	0.0367
1.8	0.0359	0.0351	0.0344	0.0336	0.0329	0.0322	0.0314	0.0307	0.0301	0.0294
1.9	0.0287	0.0281	0.0274	0.0268	0.0262	0.0256	0.0250	0.0244	0.0239	0.0253
2.0	0.0228	0.0222	0.0217	0.0212	0.0207	0.0202	0.0197	0.0192	0.0188	0.0183
2.1	0.0179	0.0174	0.0170	0.0166	0.0162	0.0158	0.0154	0.0150	0.0146	0.0143
2.2	0.0139	0.0136	0.0132	0.0129	0.0125	0.0122	0.0119	0.0116	0.0113	0.0110
2.3	0.0107	0.0104	0.0102		0.00964		0.00914		0.00866	
2.4	0.00820		0.00776		0.00734		0.00695		0.00657	
2.5	0.00621		0.00587		0.00554		0.00523		0.00494	
2.6	0.00466		0.00440		0.00415		0.00391		0.00368	
2.7	0.00347		0.00326		0.00307		0.00289		0.00272	
2.8	0.00256		0.00240		0.00226		0.00212		0.00199	
2.9	0.00187		0.00175		0.00164		0.00154		0.00144	
3.0	0.00135									
3.1	0.000968									
3.2	0.000687									

continued

Single-Sided Normal Distribution[a]—*continued*

u	0.00	0.01	0.02	0.03	0.04	0.05	0.06	0.07	0.08	0.09
3.3	0.000483									
3.4	0.000337									
3.5	0.000233									
3.6	0.000159									
3.7	0.000108									
3.8	0.0000723									
3.9	0.0000481									
4.0	0.0000317									
4.1	0.0000207									
4.2	0.0000133									
4.3	0.00000854									
4.4	0.00000541									
4.5	0.00000340									
4.6	0.00000211									
4.7	0.00000130									
4.8	0.000000793									
4.9	0.000000479									
5.0	0.000000287									

[a]This table gives the proportion, *P*, of the area under a normal distribution curve that lies to the right of the deviation *z*, where *z* is defined as

$$z = (X - \mu)/\sigma$$

For example, the proportion of the area under a normal distribution curve that lies to the right of a deviation of 0.04 is 0.4840, or 48.40%. The area to the left of the deviation is given as 1 – *P*. Thus, 51.60% of the area under the normal distribution curve lies to the left of a deviation of 0.04. When the deviation is negative, the values in the table give the proportion of the area under the normal distribution curve that lies to the left of *z;* therefore, 48.40% of the area lies to the left, and 51.60% of the area lies to the right of a deviation of –0.04.

Appendix 1B

t-Table[a]

Value of *t* for confidence interval of:	90%	95%	98%	99%
Critical value of $\lvert t \rvert$ for α values of:	0.10	0.05	0.02	0.01
Degrees of Freedom				
1	6.31	12.71	31.82	63.66
2	2.92	4.30	6.96	9.92
3	2.35	3.18	4.54	5.84
4	2.13	2.78	3.75	4.60
5	2.02	2.57	3.36	4.03
6	1.94	2.45	3.14	3.71
7	1.89	2.36	3.00	3.50
8	1.86	2.31	2.90	3.36
9	1.83	2.26	2.82	3.25
10	1.81	2.23	2.76	3.17
12	1.78	2.18	2.68	3.05
14	1.76	2.14	2.62	2.98
16	1.75	2.12	2.58	2.92
18	1.73	2.10	2.55	2.88
20	1.72	2.09	2.53	2.85
30	1.70	2.04	2.46	2.75
50	1.68	2.01	2.40	2.68
∞	1.64	1.96	2.33	2.58

[a]The *t*-values in this table are for a two-tailed test. For a one-tailed test, the α values for each column are half of the stated value. For example, the first column for a one-tailed test is for the 95% confidence level, $\alpha = 0.05$.

Appendix 1C.1

F-Table for One-Tailed Test at $\alpha = 0.05$ (95% Confidence Level)

ν_2/ν_1[a]	1	2	3	4	5	6	7	8	9	10	15	20	∞
1	161.4	199.5	215.7	224.6	230.2	234.0	236.8	238.9	240.5	241.9	245.9	248.0	254.3
2	18.51	19.00	19.16	19.25	19.30	19.33	19.35	19.37	19.38	19.40	19.43	19.45	19.50
3	10.13	9.552	9.277	9.117	9.013	8.941	8.887	8.845	8.812	8.786	8.703	8.660	8.526
4	7.709	6.944	6.591	6.388	6.256	6.163	6.094	6.041	5.999	5.964	5.858	5.803	5.628
5	6.608	5.786	5.409	5.192	5.050	4.950	4.876	4.818	4.772	4.735	4.619	4.558	4.365
6	5.987	5.143	4.757	4.534	4.387	4.284	4.207	4.147	4.099	4.060	3.938	3.874	3.669
7	5.591	4.737	4.347	4.120	3.972	3.866	3.787	3.726	3.677	3.637	3.511	3.445	3.230
8	5.318	4.459	4.066	3.838	3.687	3.581	3.500	3.438	3.388	3.347	3.218	3.150	2.928
9	5.117	4.256	3.863	3.633	3.482	3.374	3.293	3.230	3.179	3.137	3.006	2.936	2.707
10	4.965	4.103	3.708	3.478	3.326	3.217	3.135	3.072	3.020	2.978	2.845	2.774	2.538
11	4.844	3.982	3.587	3.357	3.204	3.095	3.012	2.948	2.896	2.854	2.719	2.646	2.404
12	4.747	3.885	3.490	3.259	3.106	2.996	2.913	2.849	2.796	2.753	2.617	2.544	2.296
13	4.667	3.806	3.411	3.179	3.025	2.915	2.832	2.767	2.714	2.671	2.533	2.459	2.206
14	4.600	3.739	3.344	3.112	2.958	2.848	2.764	2.699	2.646	2.602	2.463	2.388	2.131
15	4.534	3.682	3.287	3.056	2.901	2.790	2.707	2.641	2.588	2.544	2.403	2.328	2.066
16	4.494	3.634	3.239	3.007	2.852	2.741	2.657	2.591	2.538	2.494	2.352	2.276	2.010
17	4.451	3.592	3.197	2.965	2.810	2.699	3.614	2.548	2.494	2.450	2.308	2.230	1.960
18	4.414	3.555	3.160	2.928	2.773	2.661	2.577	2.510	2.456	2.412	2.269	2.191	1.917
19	4.381	3.552	3.127	2.895	2.740	2.628	2.544	2.477	2.423	2.378	2.234	2.155	1.878
20	4.351	3.493	3.098	2.866	2.711	2.599	2.514	2.447	2.393	2.348	2.203	2.124	1.843
∞	3.842	2.996	2.605	2.372	2.214	2.099	2.010	1.938	1.880	1.831	1.666	1.570	1.000

[a] ν_1 = degrees of freedom in numerator; ν_2 = degrees of freedom in denominator.

Appendix 1C.2

F-Table for Two-Tailed Test at $\alpha = 0.05$ (95% Confidence Level)

ν_2/ν_1[a]	1	2	3	4	5	6	7	8	9	10	15	20	∞
1	647.8	799.5	864.2	899.6	921.8	937.1	948.2	956.7	963.3	968.6	984.9	993.1	1018
2	38.51	39.00	39.17	39.25	39.30	39.33	39.36	39.37	39.39	39.40	39.43	39.45	39.498
3	17.44	16.04	15.44	15.10	14.88	14.73	14.62	14.54	14.47	14.42	14.25	14.17	13.902
4	12.22	10.65	9.979	9.605	9.364	9.197	9.074	8.980	8.905	8.844	8.657	8.560	8.257
5	10.01	8.434	7.764	7.388	7.146	6.978	6.853	6.757	6.681	6.619	6.428	6.329	6.015
6	8.813	7.260	6.599	6.227	5.988	5.820	5.695	5.600	5.523	5.461	5.269	5.168	4.849
7	8.073	6.542	5.890	5.523	5.285	5.119	4.995	4.899	4.823	4.761	4.568	4.467	4.142
8	7.571	6.059	5.416	5.053	4.817	4.652	4.529	4.433	4.357	4.295	4.101	3.999	3.670
9	7.209	5.715	5.078	4.718	4.484	4.320	4.197	4.102	4.026	3.964	3.769	3.667	3.333
10	6.937	5.456	4.826	4.468	4.236	4.072	3.950	3.855	3.779	3.717	3.522	3.419	3.080
11	6.724	5.256	4.630	4.275	4.044	3.881	3.759	3.664	3.588	3.526	3.330	3.226	2.883
12	6.544	5.096	4.474	4.121	3.891	3.728	3.607	3.512	3.436	3.374	3.177	3.073	2.725
13	6.414	4.965	4.347	3.996	3.767	3.604	3.483	3.388	3.312	3.250	3.053	2.948	2.596
14	6.298	4.857	4.242	3.892	3.663	3.501	3.380	3.285	3.209	3.147	2.949	2.844	2.487
15	6.200	4.765	4.153	3.804	3.576	3.415	3.293	3.199	3.123	3.060	2.862	2.756	2.395

continued

F-Table for Two-Tailed Test at $\alpha = 0.05$ (95% Confidence Level)—continued

ν_2/ν_1[a]	1	2	3	4	5	6	7	8	9	10	15	20	∞
16	6.115	4.687	4.077	3.729	3.502	3.341	3.219	3.125	3.049	2.986	2.788	2.681	2.316
17	6.042	4.619	4.011	3.665	3.438	3.277	3.156	3.061	2.985	2.922	2.723	2.616	2.247
18	5.978	4.560	3.954	3.608	3.382	3.221	3.100	3.005	2.929	2.866	2.667	2.559	2.187
19	5.922	4.508	3.903	3.559	3.333	3.172	3.051	2.956	2.880	2.817	2.617	2.509	2.133
20	5.871	4.461	3.859	3.515	3.289	3.128	3.007	2.913	2.837	2.774	2.573	2.464	2.085
∞	5.024	3.689	3.116	2.786	2.567	2.408	2.288	2.192	2.114	2.048	1.833	1.708	1.000

[a] ν_1 = degrees of freedom in numerator; ν_2 = degrees of freedom in denominator.

Appendix 1D

Critical Values for Q-Test

N/α	0.1	0.05	0.04	0.02	0.01
3	0.941	0.970	0.976	0.988	0.994
4	0.765	0.829	0.846	0.889	0.926
5	0.642	0.710	0.729	0.780	0.821
6	0.560	0.625	0.644	0.698	0.740
7	0.507	0.568	0.586	0.637	0.680
8	0.468	0.526	0.543	0.590	0.634
9	0.437	0.493	0.510	0.555	0.598
10	0.412	0.466	0.483	0.527	0.568

Appendix 1E

Random Number Table

23733	43499	92848	39382	65423	52443	41812	90777	42345	39906
57880	74874	94181	90599	55012	18321	18766	31656	22117	31932
16446	72008	81701	90740	56193	15918	11383	19009	12891	79288
29317	89360	22245	19021	13613	48029	18412	13461	76692	22631
33897	18660	58252	10512	64982	11378	20874	42643	79780	57597
42119	71919	34890	28512	46025	26436	90134	80172	88083	24811
99173	19823	55086	63276	26295	96233	89540	87300	37651	72632
26509	83658	57363	20297	43957	99727	24619	88851	46397	75952
99527	34791	35965	39549	31292	76368	64897	10940	31625	37920
51477	86644	97853	27286	52256	61036	13485	39191	86568	36054
77050	97664	52009	79924	46681	87091	91046	58698	96379	64325
36167	85529	43159	29083	10326	99195	25138	73863	62097	26824
58567	27030	52337	18779	79381	22082	50045	93486	70336	87715
86573	94804	42825	54559	97987	41256	60800	70200	31662	40669
31893	90234	41062	62755	33545	59069	17986	99255	98675	92860
88094	84469	50846	88467	50297	31536	36189	84433	48281	52258
45251	13773	60075	82857	31205	66074	68602	37893	33277	57724
66147	61078	91044	25890	85664	66375	21245	67183	69835	19025
82454	79002	50734	75464	26500	78228	77829	14984	42235	10653
12791	38550	18383	56867	77215	50166	97525	16877	67902	87906

Appendix 2

Recommended Reagents for Preparing Primary Standards

All compounds should be of the highest available purity. Metals should be cleaned with dilute acid to remove surface impurities and rinsed with distilled water. Unless otherwise indicated, compounds should be dried to constant weight at 110 °C. Most compounds can be dissolved in dilute acid (1:1 HCl or 1:1 HNO_3), with heating if necessary; some of the compounds are water-soluble.

Element	Compound	FW (g/mol)	Comments
aluminum	Al metal	26.982	
antimony	Sb metal	121.760	
	$KSbOC_4H_4O_6$	324.92	compound prepared by drying $KSbOC_4H_4O_6 \cdot 1/2H_2O$ at 110 °C and storing in desiccator
arsenic	As metal	74.922	
	As_2O_3	197.84	toxic
barium	$BaCO_3$	197.35	dry at 200 °C for 4 h
bismuth	Bi metal	208.980	
	Bi_2O_3	465.96	not considered a primary standard
boron	H_3BO_3	61.83	do not dry
bromine	KBr	119.01	
cadmium	Cd metal	112.411	
	CdO	128.40	not considered a primary standard
calcium	$CaCO_3$	100.09	
cerium	Ce metal	140.116	
	$(NH_4)_2Ce(NO_3)_6$	548.23	not considered a primary standard
cesium	Cs_2CO_3	325.82	
	Cs_2SO_4	361.87	
chlorine	NaCl	58.44	
chromium	Cr metal	51.996	
	$K_2Cr_2O_7$	294.19	
cobalt	Co metal	58.933	
copper	Cu metal	63.546	
	CuO	79.54	
fluorine	NaF	41.99	do not store in glass container
iodine	KI	166.00	
	KIO_3	214.00	
iron	Fe metal	55.845	
lead	Pb metal	207.2	
	$Pb(NO_3)_2$	331.20	approaches primary standard
lithium	Li_2CO_3	73.89	

continued

Element	Compound	FW (g/mol)	Comments
magnesium	Mg metal	24.305	
	MgO	40.31	not considered a primary standard
manganese	Mn metal	54.938	
	$MnSO_4 \cdot H_2O$	169.01	not considered a primary standard; may be dried at 110 °C without loss of hydrated water
mercury	Hg metal	200.59	
molybdenum	Mo metal	95.94	
nickel	Ni metal	58.693	
phosphorus	KH_2PO_4	136.09	not considered a primary standard
potassium	KCl	74.56	
	K_2CO_3	138.21	
	$K_2Cr_2O_7$	294.19	
	$KHC_8H_4O_4$	204.23	
silicon	Si metal	28.085	
	SiO_2	60.08	
silver	Ag metal	107.868	
	$AgNO_3$	169.87	approaches primary standard
sodium	NaCl	58.44	
	Na_2CO_3	106.00	
	$Na_2C_2O_4$	134.00	
strontium	$SrCO_3$	147.63	
sulfur	elemental S	32.066	
	K_2SO_4	174.27	
	Na_2SO_4	142.04	
tin	Sn metal	118.710	
titanium	Ti metal	47.867	
tungsten	W metal	183.84	
uranium	U metal	238.029	
	U_3O_8	842.09	
vanadium	V metal	50.942	
	V_2O_5	181.88	not considered a primary standard
zinc	Zn metal	65.39	
	ZnO	81.37	not considered a primary standard

Source: Information compiled from Moody, J. R.; Greenberg, R. R.; Pratt, K. W.; et al. *Anal. Chem.* **1988,** *60,* 1203A–1218A; and Smith, B. W.; Parsons, M. L. *J. Chem. Educ.* **1973,** *50,* 679–681.

Appendix 3

Appendix 3A

Solubility Products

Bromide (Br^-)	pK_{sp}	K_{sp}
CuBr	8.3	5×10^{-9}
AgBr	12.30	5.0×10^{-13}
Hg_2Br_2	22.25	5.6×10^{-23}
$HgBr_2$(μ = 0.5 M)	18.9	1.3×10^{-19}
$PbBr_2$(μ = 4.0 M)	5.68	2.1×10^{-6}

Carbonate (CO_3^{2-})	pK_{sp}	K_{sp}
$MgCO_3$	7.46	3.5×10^{-8}
$CaCO_3$ (calcite)	8.35	4.5×10^{-9}
$CaCO_3$ (aragonite)	8.22	6.0×10^{-9}
$SrCO_3$	9.03	9.3×10^{-10}
$BaCO_3$	8.30	5.0×10^{-9}
$MnCO_3$	9.30	5.0×10^{-10}
$FeCO_3$	10.68	2.1×10^{-11}
$CoCO_3$	9.98	1.0×10^{-10}
$NiCO_3$	6.87	1.3×10^{-7}
Ag_2CO_3	11.09	8.1×10^{-12}
Hg_2CO_3	16.05	8.9×10^{-17}
$ZnCO_3$	10.00	1.0×10^{-10}
$CdCO_3$	13.74	1.8×10^{-14}
$PbCO_3$	13.13	7.4×10^{-14}

Chloride (Cl^-)	pK_{sp}	K_{sp}
CuCl	6.73	1.9×10^{-7}
AgCl	9.74	1.8×10^{-10}
Hg_2Cl_2	17.91	1.2×10^{-18}
$PbCl_2$	4.78	1.7×10^{-5}

Chromate (CrO_4^{2-})	pK_{sp}	K_{sp}
$BaCrO_4$	9.67	2.1×10^{-10}
$CuCrO_4$	5.44	3.6×10^{-6}
Ag_2CrO_4	11.92	1.2×10^{-12}
Hg_2CrO_4	8.70	2.0×10^{-9}

Cyanide (CN^-)	pK_{sp}	K_{sp}
AgCN	15.66	2.2×10^{-16}
$Zn(CN)_2$ (μ = 3.0 M)	15.5	3×10^{-16}
$Hg_2(CN)_2$	39.3	5×10^{-40}

Ferrocyanide ($[Fe(CN)_6^{4-}]$)	pK_{sp}	K_{sp}
$Zn_2[Fe(CN)_6]$	15.68	2.1×10^{-16}
$Cd_2[Fe(CN)_6]$	17.38	4.2×10^{-18}
$Pb_2[Fe(CN)_6]$	18.02	9.5×10^{-19}

Fluoride (F^-)	pK_{sp}	K_{sp}
MgF_2	8.18	6.6×10^{-9}
CaF_2	10.41	3.9×10^{-11}
SrF_2	8.54	2.9×10^{-9}
BaF_2	5.76	1.7×10^{-6}
PbF_2	7.44	3.6×10^{-8}

Hydroxide (OH^-)	pK_{sp}	K_{sp}
$Mg(OH)_2$	11.15	7.1×10^{-12}
$Ca(OH)_2$	5.19	6.5×10^{-6}
$Ba(OH)_2 \cdot 8H_2O$	3.6	3×10^{-4}
$La(OH)_3$	20.7	2×10^{-21}
$Mn(OH)_2$	12.8	1.6×10^{-13}
$Fe(OH)_2$	15.1	8×10^{-16}
$Co(OH)_2$	14.9	1.3×10^{-15}
$Ni(OH)_2$	15.2	6×10^{-16}
$Cu(OH)_2$	19.32	4.8×10^{-20}
$Fe(OH)_3$	38.8	1.6×10^{-39}
$Co(OH)_3$ (T = 19 °C)	44.5	3×10^{-45}
Ag_2O ($+H_2O \rightleftharpoons 2Ag^+ + 2OH^-$)	15.42	3.8×10^{-16}
Cu_2O ($+ H_2O \rightleftharpoons 2Cu^+ + 2OH^-$)	29.4	4×10^{-30}
$Zn(OH)_2$ (amorphous)	15.52	3.0×10^{-16}
$Cd(OH)_2$ (β)	14.35	4.5×10^{-15}

continued

Solubility Products—*continued*

Hydroxide (OH^-)	pK_{sp}	K_{sp}
HgO (red) (+ $H_2O \rightleftharpoons Hg^{2+} + 2OH^-$)	25.44	3.6×10^{-26}
SnO (+ $H_2O \rightleftharpoons Sn^{2+} + 2OH^-$)	26.2	6×10^{-27}
PbO (yellow) (+ $H_2O \rightleftharpoons Pb^{2+} + 2OH^-$)	15.1	8×10^{-16}
$Al(OH)_3$ (α)	33.5	3×10^{-34}

Iodate (IO_3^-)	pK_{sp}	K_{sp}
$Ca(IO_3)_2$	6.15	7.1×10^{-7}
$Ba(IO_3)_2$	8.81	1.5×10^{-9}
$AgIO_3$	7.51	3.1×10^{-8}
$Hg_2(IO_3)_2$	17.89	1.3×10^{-18}
$Zn(IO_3)_2$	5.41	3.9×10^{-6}
$Cd(IO_3)_2$	7.64	2.3×10^{-8}
$Pb(IO_3)_2$	12.61	2.5×10^{-13}

Iodide (I^-)	pK_{sp}	K_{sp}
AgI	16.08	8.3×10^{-17}
Hg_2I_2	28.33	4.7×10^{-29}
HgI_2 (μ = 0.5 M)	27.95	1.1×10^{-28}
PbI_2	8.10	7.9×10^{-9}

Oxalate ($C_2O_4^{2-}$)	pK_{sp}	K_{sp}
CaC_2O_4 (μ = 0.1 M, T = 20 °C)	7.9	1.3×10^{-8}
BaC_2O_4 (μ = 0.1 M, T = 20 °C)	6.0	1×10^{-6}
SrC_2O_4 (μ = 0.1 M, T = 20 °C)	6.4	4×10^{-7}

Phosphate (PO_4^{3-})	pK_{sp}	K_{sp}
$Fe_3(PO_4)_2 \cdot 8H_2O$	36.0	1×10^{-36}
$Zn_3(PO_4)_2 \cdot 4H_2O$	35.3	5×10^{-36}
Ag_3PO_4	17.55	2.8×10^{-18}
$Pb_3(PO_4)_2$ (T = 38 °C)	43.53	3.0×10^{-44}

Sulfate (SO_4^{2-})	pK_{sp}	K_{sp}
$CaSO_4$	4.62	2.4×10^{-5}
$SrSO_4$	6.50	3.2×10^{-7}
$BaSO_4$	9.96	1.1×10^{-10}
Ag_2SO_4	4.83	1.5×10^{-5}
Hg_2SO_4	6.13	7.4×10^{-7}
$PbSO_4$	7.79	1.6×10^{-8}

Sulfide (S^{2-})	pK_{sp}	K_{sp}
MnS (green)	13.5	3×10^{-14}
FeS	18.1	8×10^{-19}
CoS (β)	25.6	3×10^{-26}
NiS (γ)	26.6	3×10^{-27}
CuS	36.1	8×10^{-37}
Cu_2S	48.5	3×10^{-49}
Ag_2S	50.1	8×10^{-51}
ZnS (α)	24.7	2×10^{-25}
CdS	27.0	1×10^{-27}
Hg_2S (red)	53.3	5×10^{-54}
PbS	27.5	3×10^{-28}

Thiocyanate (SCN^-)	pK_{sp}	K_{sp}
CuSCN (μ = 5.0 M)	13.40	4.0×10^{-14}
AgSCN	11.97	1.1×10^{-12}
$Hg_2(SCN)_2$	19.52	3.0×10^{-20}
$Hg(SCN)_2$ (μ = 1.0 M)	19.56	2.8×10^{-20}

Source: All values are from Martell, A. E.; Smith, R. M. *Critical Stability Constants,* Vol. 4. Plenum Press: New York, 1976. Unless otherwise stated, values are for 25 °C and zero ionic strength.

Appendix 3B

Acid Dissociation Constants

Compound	Conjugate Acid	pK_a	K_a
acetic acid	CH_3COOH	4.757	1.75×10^{-5}
adipic acid	$HOOC(CH_2)_4COOH$	4.42	3.8×10^{-5}
		5.42	3.8×10^{-6}
alanine	COOH–$CHCH_3$–NH_3^+	2.348 (COOH)	4.49×10^{-3}
		9.867 (NH_3)	1.36×10^{-10}
aminobenzene	$C_6H_5NH_3^+$	4.601	2.51×10^{-5}

Compound	Conjugate Acid	pK_a	K_a
4-aminobenzene sulfonic acid	^-O_3S—C_6H_4—NH_3^+	3.232	5.86×10^{-4}
2-aminobenzoic acid	C_6H_4(COOH)(NH_3^+)	2.08 (COOH) 4.96 (NH_3)	8.3×10^{-3} 1.1×10^{-5}
2-aminophenol	C_6H_4(OH)(NH_3^+)	4.78 (NH_3); (T = 20 °C) 9.97 (OH); (T = 20 °C)	1.7×10^{-5} 1.05×10^{-10}
ammonia	NH_4^+	9.244	5.70×10^{-10}
arginine	COOH—$CHCH_2CH_2CH_2NHC(=NH_2^+)NH_2$—$NH_3^+$	1.823 (COOH) 8.991 (NH_3) (12.48) (NH_2)	1.50×10^{-2} 1.02×10^{-9} 3.3×10^{-13}
arsenic acid	H_3AsO_4	2.24 6.96 11.50	5.8×10^{-3} 1.1×10^{-7} 3.2×10^{-12}
asparagine	COOH—$CHCH_2C(=O)NH_2$—NH_3^+	2.14 (COOH); (μ = 0.1 M) 8.72 (NH_3); (μ = 0.1 M)	7.2×10^{-3} 1.9×10^{-9}
aspartic acid	COOH—$CHCH_2COOH$—NH_3^+	1.990 (α-COOH) 3.900 (β-COOH) 10.002 (NH_3)	1.02×10^{-2} 1.26×10^{-4} 9.95×10^{-11}
benzoic acid	C_6H_5—COOH	4.202	6.28×10^{-5}
benzylamine	C_6H_5—$CH_2NH_3^+$	9.35	4.5×10^{-10}
boric acid	H_3BO_3	9.236 (12.74); (T = 20 °C) (13.80); (T = 20 °C)	5.81×10^{-10} 1.82×10^{-13} 1.58×10^{-14}
carbonic acid	H_2CO_3	6.352 10.329	4.45×10^{-7} 4.69×10^{-11}
catechol	C_6H_4(OH)(OH)	9.40 12.8	4.0×10^{-10} 1.6×10^{-13}

continued

Acid Dissociation Constants—*continued*

Compound	Conjugate Acid	pK_a	K_a
chloroacetic acid	$ClCH_2COOH$	2.865	1.36×10^{-3}
chromic acid	H_2CrO_4	–0.2; (T = 20 °C) 6.51	1.6 3.1×10^{-7}
citric acid	COOH $HOOCH_2C$—C—CH_2COOH OH	3.128 (COOH) 4.761 (COOH) 6.396 (COOH)	7.45×10^{-4} 1.73×10^{-5} 4.02×10^{-7}
cupferron	NO C_6H_5—N OH	4.16; (μ = 0.1 M)	6.9×10^{-5}
cysteine	COOH $CHCH_2SH$ NH_3^+	(1.71) (COOH) 8.36 (SH) 10.77 (NH_3)	1.9×10^{-2} 4.4×10^{-9} 1.7×10^{-11}
dichloroacetic acid	$Cl_2CHCOOH$	1.30	5.0×10^{-2}
diethylamine	$(CH_3CH_2)_2NH_2^+$	10.933	1.17×10^{-11}
dimethylamine	$(CH_3)_2NH_2^+$	10.774	1.68×10^{-11}
dimethylgloxime	HON NOH H_3C CH_3	10.66 12.0	2.2×10^{-11} 1×10^{-12}
ethylamine	$CH_3CH_2NH_3^+$	10.636	2.31×10^{-11}
ethylenediamine	$^+H_3NCH_2CH_2NH_3^+$	6.848 9.928	1.42×10^{-7} 1.18×10^{-10}
ethylenediaminetetraacetic acid (EDTA)	$HOOCH_2C$ CH_2COOH $^+HNCH_2CH_2NH^+$ $HOOCH_2C$ CH_2COOH	0.0 (COOH); (μ = 1.0 M) 1.5 (COOH); (μ = 0.1 M) 2.0 (COOH); (μ = 0.1 M) 2.68 (COOH); (μ = 0.1 M) 6.11 (NH); (μ = 0.1 M) 10.17 (NH); (μ = 0.1 M)	1.0 3.2×10^{-2} 1.0×10^{-2} 2.1×10^{-3} 7.8×10^{-7} 6.8×10^{-11}
formic acid	HCOOH	3.745	1.80×10^{-4}
fumaric acid	COOH HOOC	3.053 4.494	8.85×10^{-4} 3.21×10^{-5}
glutamic acid	COOH $CHCH_2CH_2COOH$ NH_3^+	2.23 (α-COOH) 4.42 (λ-COOH) 9.95 (NH_3)	5.9×10^{-3} 3.8×10^{-5} 1.12×10^{-10}

Compound	Conjugate Acid	pK_a	K_a
glutamine	$HOOC-CH(NH_3^+)-CH_2CH_2C(=O)NH_2$	2.17 (COOH); (μ = 0.1 M) 9.01 (NH_3); (μ = 0.1 M)	6.8×10^{-3} 9.8×10^{-10}
glycine	$^+H_3NCH_2COOH$	2.350 (COOH) 9.778 (NH_3)	4.47×10^{-3} 1.67×10^{-10}
glycolic acid	$HOCH_2COOH$	3.831 (COOH)	1.48×10^{-4}
histidine	$HOOC-CH(NH_3^+)-CH_2-$(imidazolium ring, N–H and N^+–H)	1.7 (COOH); (μ = 0.1 M) 6.02 (NH); (μ = 0.1 M) 9.08 (NH_3); (μ = 0.1 M)	2×10^{-2} 9.5×10^{-7} 8.3×10^{-10}
hydrogen cyanide	HCN	9.21	6.2×10^{-10}
hydrogen fluoride	HF	3.17	6.8×10^{-4}
hydrogen peroxide	H_2O_2	11.65	2.2×10^{-12}
hydrogen sulfide	H_2S	7.02 13.9	9.5×10^{-8} 1.3×10^{-14}
hydrogen thiocyanate	HSCN	0.9	1.3×10^{-1}
8-hydroxyquinoline	(quinolinium ring, N^+–H, with OH at position 8)	4.91 (NH) 9.81 (OH)	1.23×10^{-5} 1.55×10^{-10}
hydroxylamine	$HONH_3^+$	5.96	1.1×10^{-6}
hypobromous	HOBr	8.63	2.3×10^{-9}
hypochlorous	HOCl	7.53	3.0×10^{-8}
hypoiodous	HOI	10.64	2.3×10^{-11}
iodic acid	HIO_3	0.77	1.7×10^{-1}
isoleucine	$HOOC-CH(NH_3^+)-CH(CH_3)CH_2CH_3$	2.319 (COOH) 9.754 (NH_3)	4.80×10^{-3} 1.76×10^{-10}
leucine	$HOOC-CH(NH_3^+)-CH_2CH(CH_3)_2$	2.329 (COOH) 9.747 (NH_3)	4.69×10^{-3} 1.79×10^{-10}
lysine	$HOOC-CH(NH_3^+)-CH_2CH_2CH_2CH_2NH_3^+$	2.04 (COOH); (μ = 0.1 M) 9.08 (α-NH_3); (μ = 0.1 M) 10.69 (ε-NH_3); (μ = 0.1 M)	9.1×10^{-3} 8.3×10^{-10} 2.0×10^{-11}

continued

Acid Dissociation Constants—*continued*

Compound	Conjugate Acid	pK_a	K_a
maleic acid	HOOC–CH=CH–COOH (cis)	1.910 6.332	1.23×10^{-2} 4.66×10^{-7}
malic acid	$HOOCH_2C$—CH(OH)COOH	3.459 (COOH) 5.097 (COOH)	3.48×10^{-4} 8.00×10^{-6}
malonic acid	$HOOCCH_2COOH$	2.847 5.696	1.42×10^{-3} 2.01×10^{-6}
methionine	$HOOC-CH(NH_3^+)CH_2CH_2SCH_3$	2.20 (COOH); (μ = 0.1 M) 9.05 (NH_3); (μ = 0.1 M)	6.3×10^{-3} 8.9×10^{-10}
methylamine	$CH_3NH_3^+$	10.64	2.3×10^{-11}
2-methylaniline	benzene ring with CH_3 and NH_3^+ (ortho)	4.447	3.57×10^{-5}
4-methylaniline	H_3C–C_6H_4–NH_3^+ (para)	5.084	8.24×10^{-6}
2-methylphenol	benzene ring with CH_3 and OH (ortho)	10.28	5.2×10^{-11}
4-methylphenol	H_3C–C_6H_4–OH (para)	10.26	5.5×10^{-11}
nitrilotriacetic acid	$^+HN(CH_2COOH)_3$	1.1 (COOH); (T = 20 °C , μ = 1.0 M) 1.650 (COOH); (T = 20 °C) 2.940 (COOH); (T = 20 °C) 10.334 (NH_3); (T = 20 °C)	8×10^{-2} 2.24×10^{-2} 1.15×10^{-3} 4.63×10^{-11}
2-nitrobenzoic acid	benzene ring with COOH and NO_2 (ortho)	2.179	6.62×10^{-3}
3-nitrobenzoic acid	benzene ring with COOH and NO_2 (meta)	3.449	3.56×10^{-4}

Compound	Conjugate Acid	pK_a	K_a
4-nitrobenzoic acid	$O_2N-C_6H_4-COOH$	3.442	3.61×10^{-4}
2-nitrophenol	OH, NO_2 (ortho)	7.21	6.2×10^{-8}
3-nitrophenol	OH, NO_2 (meta)	8.39	4.1×10^{-9}
4-nitrophenol	$O_2N-C_6H_4-OH$	7.15	7.1×10^{-8}
nitrous acid	HNO_2	3.15	7.1×10^{-4}
oxalic acid	$H_2C_2O_4$	1.252	5.60×10^{-2}
		4.266	5.42×10^{-5}
1,10-phenanthroline	NH^+ N	4.86	1.38×10^{-5}
phenol	C_6H_5-OH	9.98	1.05×10^{-10}
phenylalanine	COOH–$CHCH_2C_6H_5$–NH_3^+	2.20 (COOH)	6.3×10^{-3}
		9.31 (NH_3)	4.9×10^{-10}
phosphoric acid	H_3PO_4	2.148	7.11×10^{-3}
		7.199	6.32×10^{-8}
		12.35	4.5×10^{-13}
phthalic acid	COOH, COOH (ortho)	2.950	1.12×10^{-3}
		5.408	3.91×10^{-6}
piperidine	NH_2^+	11.123	7.53×10^{-12}
proline	COOH, N H_2^+	1.952 (COOH)	1.12×10^{-2}
		10.640 (NH)	2.29×10^{-11}

continued

Acid Dissociation Constants—*continued*

Compound	Conjugate Acid	pK_a	K_a
propanoic acid	CH_3CH_2COOH	4.874	1.34×10^{-5}
propylamine	$CH_3CH_2CH_2NH_3^+$	10.566	2.72×10^{-11}
pyridine	NH^+ (pyridinium ring)	5.229	5.90×10^{-6}
resorcinol	OH, OH (benzene ring)	9.30	5.0×10^{-10}
		11.06	8.7×10^{-12}
salicylic acid	COOH, OH (benzene ring)	2.97 (COOH)	1.07×10^{-3}
		13.74 (OH)	1.8×10^{-14}
serine	$COOH—CHCH_2OH—NH_3^+$	2.187 (COOH)	6.50×10^{-3}
		9.209 (NH_3)	6.18×10^{-10}
succinic acid	$HOOCCH_2CH_2COOH$	4.207	6.21×10^{-5}
		5.636	2.31×10^{-6}
sulfuric acid	H_2SO_4	strong	strong
		1.99	1.0×10^{-2}
sulfurous acid	H_2SO_3	1.91	1.2×10^{-2}
		7.18	6.6×10^{-8}
D-tartaric acid	HOOC—CH(OH)—CH(OH)—COOH	3.036 (COOH)	9.20×10^{-4}
		4.366 (COOH)	4.31×10^{-5}
threonine	$COOH—CHCHOHCH_3—NH_3^+$	2.088 (COOH)	8.17×10^{-3}
		9.100 (NH_3)	7.94×10^{-10}
thiosulfuric acid	$H_2S_2O_3$	0.6	3×10^{-1}
		1.6	3×10^{-2}
trichloroacetic acid	Cl_3CCOOH	0.66; (μ = 0.1 M)	2.2×10^{-1}
triethanolamine	$(HOCH_2CH_2)_3NH^+$	7.762	1.73×10^{-8}
triethylamine	$(CH_3CH_2)_3NH^+$	10.715	1.93×10^{-11}

Compound	Conjugate Acid	pK_a	K_a
trimethylamine	$(CH_3)_3NH^+$	9.800	1.58×10^{-10}
tris(hydroxymethyl)-aminomethane (TRIS or THAM)	$(HOCH_2)_3CNH_3^+$	8.075	8.41×10^{-9}
tryptophan	COOH–CHCH$_2$–(indol-3-yl, N–H)–NH$_3^+$	2.35 (COOH); (μ = 0.1 M) 9.33 (NH_3); (μ = 0.1 M)	4.5×10^{-3} 4.7×10^{-10}
tyrosine	COOH–CHCH$_2$–(C$_6$H$_4$)–OH–NH$_3^+$	2.17 (COOH); (μ = 0.1 M) 9.19 (NH_3) 10.47 (OH)	6.8×10^{-3} 6.5×10^{-10} 3.4×10^{-11}
valine	COOH–CHCH$(CH_3)_2$–NH$_3^+$	2.286 (COOH) 9.718 (OH)	5.18×10^{-3} 1.91×10^{-10}

Source: All values are from Martell, A. E.; Smith, R. M. *Critical Stability Constants,* Vol. 1–4. Plenum Press: New York, 1976. Unless otherwise stated, values are for 25 °C and zero ionic strength. Values in parentheses are considered less reliable.

Appendix 3C

Metal–Ligand Formation Constants

Acetate CH_3COO^-	**log K_1**	**log K_2**	**log K_3**	**log K_4**	**log K_5**	**log K_6**
Mg^{2+}	1.27					
Ca^{2+}	1.18					
Ba^{2+}	1.07					
Mn^{2+}	1.40					
Fe^{2+}	1.40					
Co^{2+}	1.46					
Ni^{2+}	1.43					
Cu^{2+}	2.22	1.41				
Ag^+	0.73	–0.09				
Zn^{2+}	1.57					
Cd^{2+}	1.93	1.22	–0.89			
Pb^{2+}	2.68	1.40				

continued

Metal–Ligand Formation Constants—*continued*

Ammonia NH_3	log K_1	log K_2	log K_3	log K_4	log K_5	log K_6
Ag^+	3.31	3.91				
Co^{2+} (T = 20 °C)	1.99	1.51	0.93	0.64	0.06	–0.74
Ni^{2+}	2.72	2.17	1.66	1.12	0.67	–0.03
Cu^{2+}	4.04	3.43	2.80	1.48		
Zn^{2+}	2.21	2.29	2.36	2.03		
Cd^{2+}	2.55	2.01	1.34	0.84		

Chloride Cl^-	log K_1	log K_2	log K_3	log K_4	log K_5	log K_6
Cu^{2+}	0.40					
Fe^{3+}	1.48	0.65				
Ag^+ (μ = 5.0 M)	3.70	1.92	0.78	–0.3		
Zn^{2+}	0.43	0.18	–0.11	–0.3		
Cd^{2+}	1.98	1.62	–0.2	–0.7		
Pb^{2+}	1.59	0.21	–0.1	–0.3		

Cyanide CN^-	log K_1	log K_2	log K_3	log K_4	log K_5	log K_6
Fe^{2+}						35.4 (β_6)
Fe^{3+}						43.6 (β_6)
Ag^+		20.48 (β_2)	0.92			
Zn^{2+}		11.07 (β_2)	4.98	3.57		
Cd^{2+}	6.01	5.11	4.53	2.27		
Hg^{2+}	17.00	15.75	3.56	2.66		
Ni^{2+}				30.22 (β_4)		

Ethylenediamine $H_2NCH_2CH_2NH_2$	log K_1	log K_2	log K_3	log K_4	log K_5	log K_6
Ni^{2+}	7.38	6.18	4.11			
Cu^{2+}	10.48	9.07				
Ag^+ (T = 20 °C, μ = 0.1 M)	4.70	3.00				
Zn^{2+}	5.66	4.98	3.25			
Cd^{2+}	5.41	4.50	2.78			

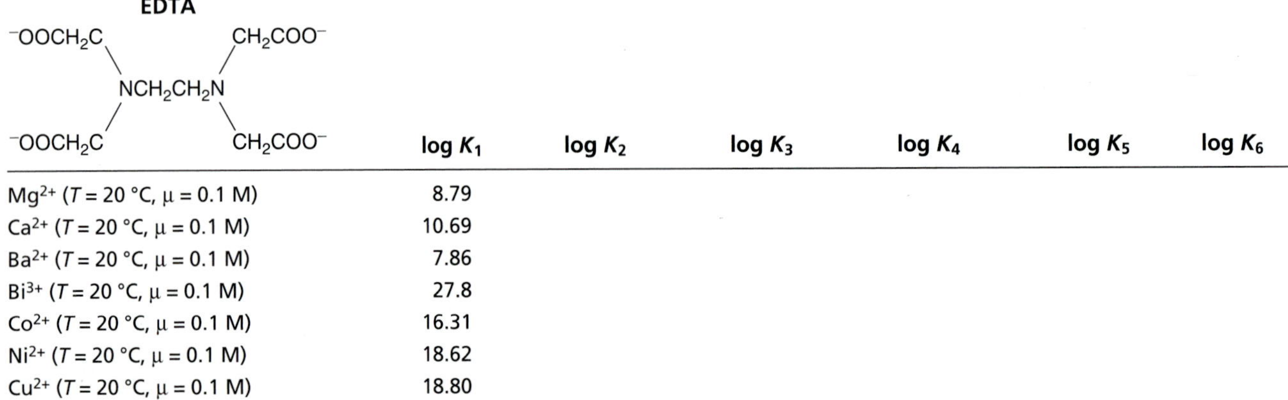

EDTA	log K_1	log K_2	log K_3	log K_4	log K_5	log K_6
Mg^{2+} (T = 20 °C, μ = 0.1 M)	8.79					
Ca^{2+} (T = 20 °C, μ = 0.1 M)	10.69					
Ba^{2+} (T = 20 °C, μ = 0.1 M)	7.86					
Bi^{3+} (T = 20 °C, μ = 0.1 M)	27.8					
Co^{2+} (T = 20 °C, μ = 0.1 M)	16.31					
Ni^{2+} (T = 20 °C, μ = 0.1 M)	18.62					
Cu^{2+} (T = 20 °C, μ = 0.1 M)	18.80					

EDTA

```
⁻OOCH₂C           CH₂COO⁻
       \         /
        NCH₂CH₂N
       /         \
⁻OOCH₂C           CH₂COO⁻
```

EDTA	log K_1	log K_2	log K_3	log K_4	log K_5	log K_6
Cr^{3+} (T = 20 °C, μ = 0.1 M)	(23.4)					
Fe^{3+} (T = 20 °C, μ = 0.1 M)	25.1					
Ag^{+} (T = 20 °C, μ = 0.1 M)	7.32					
Zn^{2+} (T = 20 °C, μ = 0.1 M)	16.50					
Cd^{2+} (T = 20 °C, μ = 0.1 M)	16.46					
Hg^{2+} (T = 20 °C, μ = 0.1 M)	21.7					
Pb^{2+} (T = 20 °C, μ = 0.1 M)	18.04					
Al^{3+} (T = 20 °C, μ = 0.1 M)	16.3					

Fluoride

F^-	log K_1	log K_2	log K_3	log K_4	log K_5	log K_6
Al^{3+} (μ = 0.5 M)	6.11	5.01	3.88	3.0	1.4	0.4

Hydroxide

OH^-	log K_1	log K_2	log K_3	log K_4	log K_5	log K_6
Al^{3+}	9.01	(9.69)	(8.3)	6.0		
Co^{2+}	4.3	4.1	1.3	0.5		
Fe^{2+}	4.5	(2.9)	2.6	−0.4		
Fe^{3+}	11.81	10.5	12.1			
Ni^{2+}	4.1	3.9	3			
Pb^{2+}	6.3	4.6	3.0			
Zn^{2+}	5.0	(6.1)	2.5	(1.2)		

Iodide

I^-	log K_1	log K_2	log K_3	log K_4	log K_5	log K_6
Ag^{+} (T = 18 °C)	6.58	(5.12)	(1.4)			
Cd^{2+}	2.28	1.64	1.08	1.0		
Pb^{2+}	1.92	1.28	0.7	0.6		

Nitriloacetate

```
CH₂COO⁻
|
N — CH₂COO⁻
|
CH₂COO⁻
```

Nitriloacetate	log K_1	log K_2	log K_3	logK_4	log K_5	log K_6
Mg^{2+} (T = 20 °C, μ = 0.1 M)	5.41					
Ca^{2+} (T = 20 °C, μ = 0.1 M)	6.41					
Ba^{2+} (T = 20 °C, μ = 0.1 M)	4.82					
Mn^{2+} (T = 20 °C, μ = 0.1 M)	7.44					
Fe^{2+} (T = 20 °C, μ = 0.1 M)	8.33					
Co^{2+} (T = 20 °C, μ = 0.1 M)	10.38					
Ni^{2+} (T = 20 °C, μ = 0.1 M)	11.53					
Cu^{2+} (T = 20 °C, μ = 0.1 M)	12.96					
Fe^{3+} (T = 20 °C, μ = 0.1 M)	15.9					
Zn^{2+} (T = 20 °C, μ = 0.1 M)	10.67					
Cd^{2+} (T = 20 °C, μ = 0.1 M)	9.83					
Pb^{2+} (T = 20 °C, μ = 0.1 M)	11.39					

continued

Metal–Ligand Formation Constants—*continued*

Oxalate $C_2O_4^{2-}$	log K_1	log K_2	log K_3	logK_4	log K_5	log K_6
Ca^{2+} (μ = 1 M)	1.66	1.03				
Fe^{2+} (μ = 1 M)	3.05	2.10				
Co^{2+}	4.72	2.28				
Ni^{2+}	5.16					
Cu^{2+}	6.23	4.04				
Fe^{3+} (μ = 0.5 M)	7.53	6.11	4.85			
Zn^{2+}	4.87	2.78				

1,10-Phenanthroline

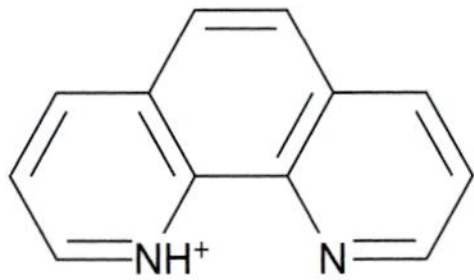

1,10-Phenanthroline	log K_1	log K_2	log K_3	logK_4	log K_5	log K_6
Fe^{2+}			20.7 (β_3)			
Mn^{2+} (μ = 0.1 M)	4.0	3.3	3.0			
Co^{2+} (μ = 0.1 M)	7.08	6.64	6.08			
Ni^{2+} (μ = 0.1 M)	8.6	8.1	7.6			
Fe^{3+}			13.8 (β_3)			
Ag^{+} (μ = 0.1 M)	5.02	7.04				
Zn^{2+}	6.2	(5.9)	(5.2)			

Thiosulfate $S_2O_3^{2-}$	log K_1	log K_2	log K_3	logK_4	log K_5	log K_6
Ag^{+} (*T* = 20 °C)	8.82	4.85	0.53			

Thiocyanate SCN^{-}	log K_1	log K_2	log K_3	logK_4	log K_5	log K_6
Mn^{2+}	1.23					
Fe^{2+}	1.31					
Co^{2+}	1.72					
Ni^{2+}	1.76					
Cu^{2+}	2.33					
Fe^{3+}	3.02					
Ag^{+}	4.8	3.43	1.27	0.2		
Zn^{2+}	1.33	0.58	0.09	–0.4		
Cd^{2+}	1.89	0.89	0.02	–0.5		
Hg^{2+}		17.26 (β_2)	2.71	1.83		

Source: All values are from Martell, A. E.; Smith, R. M. *Critical Stability Constants,* Vol. 1–4, Plenum Press: New York, 1976. Unless otherwise stated, values are for 25 °C and zero ionic strength. Values in parentheses are considered less reliable.

Appendix 3D

Standard Reduction Potentials[a]

Aluminum	E° (V)	$E^{\circ\prime}$ (V)
$Al^{3+} + 3e^- \rightleftharpoons Al(s)$	−1.676	
$Al(OH)_4^- + 3e^- \rightleftharpoons Al(s) + 4OH^-$	−2.310	
$AlF_6^{3-} + 3e^- \rightleftharpoons Al(s) + 6F^-$	−2.07	

Antimony	E° (V)	$E^{\circ\prime}$ (V)
$Sb(s) + 3H^+ + 3e^- \rightleftharpoons SbH_3(g)$	−0.510	
$Sb_2O_5(s) + 6H^+ + 4e^- \rightleftharpoons 2SbO^+ + 3H_2O$	0.605	
$SbO^+ + 2H^+ + 3e^- \rightleftharpoons Sb(s) + H_2O$	0.212	

Arsenic	E° (V)	$E^{\circ\prime}$ (V)
$As(s) + 3H^+ + 3e^- \rightleftharpoons AsH_3(g)$	−0.225	
$H_3AsO_4 + 2H^+ + 2e^- \rightleftharpoons HAsO_2 + 2H_2O$	0.560	
$HAsO_2 + 3H^+ + 3e^- \rightleftharpoons As(s) + 2H_2O$	0.240	

Barium	E° (V)	$E^{\circ\prime}$ (V)
$Ba^{2+} + 2e^- \rightleftharpoons Ba(s)$	−2.92	
$BaO(s) + 2H^+ + 2e^- \rightleftharpoons Ba(s) + H_2O$	2.365	

Beryllium	E° (V)	$E^{\circ\prime}$ (V)
$Be^{2+} + 2e^- \rightleftharpoons Be(s)$	−1.99	

Bismuth	E° (V)	$E^{\circ\prime}$ (V)
$Bi^{3+} + 3e^- \rightleftharpoons Bi(s)$	0.317	
$BiCl_4^- + 3e^- \rightleftharpoons Bi(s) + 4Cl^-$	0.199	

Boron	E° (V)	$E^{\circ\prime}$ (V)
$B(OH)_3 + 3H^+ + 3e^- \rightleftharpoons B(s) + 3H_2O$	−0.890	
$B(OH)_4^- + 3e^- \rightleftharpoons B(s) + 4OH^-$	−1.811	

Bromine	E° (V)	$E^{\circ\prime}$ (V)
$Br_2 + 2e^- \rightleftharpoons 2Br^-$	1.087	
$HOBr + H^+ + 2e^- \rightleftharpoons Br^- + H_2O$	1.341	
$HOBr + H^+ + e^- \rightleftharpoons \frac{1}{2}Br_2(\ell) + H_2O$	1.604	
$BrO^- + H_2O + 2e^- \rightleftharpoons Br^- + 2OH^-$		0.76 1 M NaOH
$BrO_3^- + 6H^+ + 5e^- \rightleftharpoons \frac{1}{2}Br_2(\ell) + 3H_2O$	1.5	
$BrO_3^- + 6H^+ + 6e^- \rightleftharpoons Br^- + 3H_2O$	1.478	

Cadmium	E° (V)	$E^{\circ\prime}$ (V)
$Cd^{2+} + 2e^- \rightleftharpoons Cd(s)$	−0.4030	
$Cd(CN)_4^{2-} + 2e^- \rightleftharpoons Cd(s) + 4CN^-$	−0.943	
$Cd(NH_3)_4^{2+} + 2e^- \rightleftharpoons Cd(s) + 4NH_3$	−0.622	

Calcium	E° (V)	$E^{\circ\prime}$ (V)
$Ca^{2+} + 2e^- \rightleftharpoons Ca(s)$	−2.84	

Carbon	E° (V)	$E^{\circ\prime}$ (V)
$CO_2(g) + 2H^+ + 2e^- \rightleftharpoons CO(g) + H_2O$	−0.106	
$CO_2(g) + 2H^+ + 2e^- \rightleftharpoons HCO_2H$	−0.20	
$2CO_2(g) + 2H^+ + 2e^- \rightleftharpoons H_2C_2O_4$	−0.481	
$HCHO + 2H^+ + 2e^- \rightleftharpoons CH_3OH$	0.2323	

Cerium	E° (V)	$E^{\circ\prime}$ (V)
$Ce^{3+} + 3e^- \rightleftharpoons Ce(s)$	−2.336	
$Ce^{4+} + e^- \rightleftharpoons Ce^{3+}$	1.72	1.70 1 M $HClO_4$
		1.44 1 M H_2SO_4
		1.61 1 M HNO_3
		1.28 1 M HCl

Chlorine	E° (V)	$E^{\circ\prime}$ (V)
$Cl_2(g) + 2e^- \rightleftharpoons 2Cl^-$	1.396	
$ClO^- + H_2O + e^- \rightleftharpoons \frac{1}{2}Cl_2(g) + 2OH^-$		0.421 1 M NaOH
$ClO^- + H_2O + 2e^- \rightleftharpoons Cl^- + 2OH^-$		0.890 1 M NaOH
$HClO_2 + 2H^+ + 2e^- \rightleftharpoons HOCl + H_2O$	1.64	
$ClO_3^- + 2H^+ + e^- \rightleftharpoons ClO_2(g) + H_2O$	1.175	
$ClO_3^- + 3H^+ + 2e^- \rightleftharpoons HClO_2 + H_2O$	1.181	
$ClO_4^- + 2H^+ + 2e^- \rightleftharpoons ClO_3^- + H_2O$	1.201	

continued

Standard Reduction Potentials[a]—*continued*

Chromium	E° (V)	$E^{\circ\prime}$ (V)	
$Cr^{3+} + e^- \rightleftharpoons Cr^{2+}$	−0.424		
$Cr^{2+} + 2e^- \rightleftharpoons Cr(s)$	−0.90		
$Cr_2O_7^{2-} + 14H^+ + 6e^- \rightleftharpoons 2Cr^{3+} + 7H_2O$	1.36		
$CrO_4^{2-} + 4H_2O + 3e^- \rightleftharpoons Cr(OH)_4^- + 4OH^-$		−0.13	1 M NaOH

Cobalt	E° (V)	$E^{\circ\prime}$ (V)
$Co^{2+} + 2e^- \rightleftharpoons Co(s)$	−0.277	
$Co^{3+} + e^- \rightleftharpoons Co^{2+}$	1.92	
$Co(NH_3)_6^{3+} + e^- \rightleftharpoons Co(NH_3)_6^{2+}$	0.1	
$Co(OH)_3(s) + e^- \rightleftharpoons Co(OH)_2(s) + OH^-$	0.17	
$Co(OH)_2(s) + 2e^- \rightleftharpoons Co(s) + 2OH^-$	−0.746	

Copper	E° (V)	$E^{\circ\prime}$ (V)
$Cu^+ + e^- \rightleftharpoons Cu(s)$	0.520	
$Cu^{2+} + e^- \rightleftharpoons Cu^+$	0.159	
$Cu^{2+} + 2e^- \rightleftharpoons Cu(s)$	0.3419	
$Cu^{2+} + I^- + e^- \rightleftharpoons CuI(s)$	0.86	
$Cu^{2+} + Cl^- + e^- \rightleftharpoons CuCl(s)$	0.559	

Fluorine	E° (V)	$E^{\circ\prime}$ (V)
$F_2(g) + 2H^+ + 2e^- \rightleftharpoons 2HF$	3.053	
$F_2(g) + 2e^- \rightleftharpoons 2F^-$	2.87	

Gallium	E° (V)	$E^{\circ\prime}$ (V)
$Ga^{3+} + 3e^- \rightleftharpoons Ga(s)$	−0.529	

Gold	E° (V)	$E^{\circ\prime}$ (V)
$Au^+ + e^- \rightleftharpoons Au(s)$	1.83	
$Au^{3+} + 2e^- \rightleftharpoons Au^+$	1.36	
$Au^{3+} + 3e^- \rightleftharpoons Au(s)$	1.52	
$AuCl_4^- + 3e^- \rightleftharpoons Au(s) + 4Cl^-$	1.002	

Hydrogen	E° (V)	$E^{\circ\prime}$ (V)
$2H^+ + 2e^- \rightleftharpoons H_2(g)$	0.00000	
$H_2O + e^- \rightleftharpoons \frac{1}{2}H_2(g) + OH^-$	−0.828	

Iodine	E° (V)	$E^{\circ\prime}$ (V)
$I_2(s) + 2e^- \rightleftharpoons 2I^-$	0.5355	
$I_3^- + 2e^- \rightleftharpoons 3I^-$	0.536	
$HIO + H^+ + 2e^- \rightleftharpoons I^- + H_2O$	0.985	
$IO_3^- + 6H^+ + 5e^- \rightleftharpoons \frac{1}{2}I_2(s) + 3H_2O$	1.195	
$IO_3^- + 3H_2O + 6e^- \rightleftharpoons I^- + 6OH^-$	0.257	

Iron	E° (V)	$E^{\circ\prime}$ (V)	
$Fe^{2+} + 2e^- \rightleftharpoons Fe(s)$	−0.44		
$Fe^{3+} + 3e^- \rightleftharpoons Fe(s)$	−0.037		
$Fe^{3+} + e^- \rightleftharpoons Fe^{2+}$	0.771	0.70	1 M HCl
		0.767	1 M $HClO_4$
		0.746	1 M HNO_3
		0.68	1 M H_2SO_4
		0.44	0.3 M H_3PO_4
$Fe(CN)_6^{3-} + e^- \rightleftharpoons Fe(CN)_6^{4-}$	0.356	0.71	1 M HCl
$Fe(phen)_3^{3+} + e^- \rightleftharpoons Fe(phen)_3^{2+}$	1.147		
$Fe(CN)_6^{3-} + e^- \rightleftharpoons Fe(CN)_6^{4-}$	0.356		

lanthanum	E° (V)	$E^{\circ\prime}$ (V)
$La^{3+} + 3e^- \rightleftharpoons La(s)$	−2.38	

Lead	E° (V)	$E^{\circ\prime}$ (V)
$Pb^{2+} + 2e^- \rightleftharpoons Pb(s)$	−0.126	
$PbO_2(s) + 4H^+ + 2e^- \rightleftharpoons Pb^{2+} + 2H_2O$	1.46	
$PbO_2(s) + SO_4^{2-} + 4H^+ + 2e^- \rightleftharpoons PbSO_4(s) + 2H_2O$	1.690	
$PbSO_4(s) + 2e^- \rightleftharpoons Pb(s) + SO_4^{2-}$	−0.356	

Lithium	E° (V)	$E^{\circ\prime}$ (V)
$Li^+ + e^- \rightleftharpoons Li(s)$	−3.040	

Magnesium	E° (V)	$E^{\circ\prime}$ (V)
$Mg^{2+} + 2e^- \rightleftharpoons Mg(s)$	−2.356	
$Mg(OH)_2(s) + 2e^- \rightleftharpoons Mg(s) + 2OH^-$	−2.687	

Manganese	E° (V)	$E^{\circ\prime}$ (V)
$Mn^{2+} + 2e^- \rightleftharpoons Mn(s)$	−1.17	
$M^{3+} + e^- \rightleftharpoons Mn^{2+}$	1.5	
$MnO_2(s) + 4H^+ + 2e^- \rightleftharpoons Mn^{2+} + 2H_2O$	1.23	
$MnO_4^- + 4H^+ + 3e^- \rightleftharpoons MnO_2(s) + 2H_2O$	1.70	
$MnO_4^- + 8H^+ + 5e^- \rightleftharpoons Mn^{2+} + 4H_2O$	1.51	
$MnO_4^- + 2H_2O + 3e^- \rightleftharpoons MnO_2(s) + 4OH^-$	0.60	

Mercury	E° (V)	$E^{\circ\prime}$ (V)
$Hg^{2+} + 2e^- \rightleftharpoons Hg(\ell)$	0.8535	
$2Hg^{2+} + 2e^- \rightleftharpoons Hg_2^{2+}$	0.911	
$Hg_2^{2+} + 2e^- \rightleftharpoons 2Hg(\ell)$	0.7960	
$Hg_2Cl_2(s) + 2e^- \rightleftharpoons 2Hg(\ell) + 2Cl^-$	0.2682	
$HgO(s) + 2H^+ + 2e^- \rightleftharpoons Hg(\ell) + H_2O$	0.926	
$Hg_2Br_2(s) + 2e^- \rightleftharpoons 2Hg(\ell) + 2Br^-$	0.1392	
$Hg_2I_2(s) + 2e^- \rightleftharpoons 2Hg(\ell) + 2I^-$	−0.0405	

Molybdenum	E° (V)	$E^{\circ\prime}$ (V)
$Mo^{3+} + 3e^- \rightleftharpoons Mo(s)$	–0.2	
$MoO_2(s) + 4H^+ + 4e^- \rightleftharpoons Mo(s) + 2H_2O$	–0.152	
$MoO_4^{2-} + 4H_2O + 6e^- \rightleftharpoons Mo(s) + 8OH^-$	–0.913	

Nickel	E° (V)	$E^{\circ\prime}$ (V)
$Ni^{2+} + 2e^- \rightleftharpoons Ni(s)$	–0.257	
$Ni(OH)_2(s) + 2e^- \rightleftharpoons Ni(s) + 2OH^-$	–0.72	
$Ni(NH_3)_6^{2+} + 2e^- \rightleftharpoons Ni(s) + 6NH_3$	–0.49	

Nitrogen	E° (V)	$E^{\circ\prime}$ (V)
$N_2(g) + 5H^+ + 4e^- \rightleftharpoons N_2H_5^+$	–0.23	
$N_2O(g) + 2H^+ + 2e^- \rightleftharpoons N_2(g) + H_2O$	1.77	
$2NO(g) + 2H^+ + 2e^- \rightleftharpoons N_2O(g) + H_2O$	1.59	
$HNO_2 + H^+ + e^- \rightleftharpoons NO(g) + H_2O$	0.996	
$2HNO_2 + 4H^+ + 4e^- \rightleftharpoons N_2O(g) + 3H_2O$	1.297	
$NO_3^- + 3H^+ + 2e^- \rightleftharpoons HNO_2 + H_2O$	0.94	

Oxygen	E° (V)	$E^{\circ\prime}$ (V)
$O_2(g) + 2H^+ + 2e^- \rightleftharpoons H_2O_2$	0.695	
$O_2(g) + 4H^+ + 4e^- \rightleftharpoons 2H_2O$	1.229	
$H_2O_2 + 2H^+ + 2e^- \rightleftharpoons 2H_2O$	1.763	
$O_2(g) + 2H_2O + 4e^- \rightleftharpoons 4OH^-$	0.401	
$O_3(g) + 2H^+ + 2e^- \rightleftharpoons O_2(g) + H_2O$	2.07	

Phosphorus	E° (V)	$E^{\circ\prime}$ (V)
$P(s, \text{white}) + 3H^+ + 3e^- \rightleftharpoons PH_3(g)$	–0.063	
$H_3PO_3 + 2H^+ + 2e^- \rightleftharpoons H_3PO_2 + H_2O$	–0.499	
$H_3PO_4 + 2H^+ + 2e^- \rightleftharpoons H_3PO_3 + H_2O$	–0.276	

Platinum	E° (V)	$E^{\circ\prime}$ (V)
$Pt^{2+} + 2e^- \rightleftharpoons Pt(s)$	1.188	
$PtCl_4^{2-} + 2e^- \rightleftharpoons Pt(s) + 4Cl^-$	0.758	

Potassium	E° (V)	$E^{\circ\prime}$ (V)
$K^+ + e^- \rightleftharpoons K(s)$	–2.924	

Ruthenium	E° (V)	$E^{\circ\prime}$ (V)
$Ru^{3+} + e^- \rightleftharpoons Ru^{2+}$	0.249	
$RuO_2(s) + 4H^+ + 4e^- \rightleftharpoons Ru(s) + 2H_2O$	0.68	
$Ru(NH_3)_6^{3+} + e^- \rightleftharpoons Ru(NH_3)_6^{2+}$	0.10	
$Ru(CN)_6^{3-} + e^- \rightleftharpoons Ru(CN)_6^{4-}$	0.86	

Selenium	E° (V)	$E^{\circ\prime}$ (V)
$Se(s) + 2e^- \rightleftharpoons Se^{2-}$		–0.670 1 M NaOH
$Se(s) + 2H^+ + 2e^- \rightleftharpoons H_2Se(g)$	–0.115	
$H_2SeO_3 + 4H^+ + 4e^- \rightleftharpoons Se(s) + 3H_2O$	0.74	
$SeO_4^{3-} + 4H^+ + e^- \rightleftharpoons H_2SeO_3 + H_2O$	1.151	

Silicon	E° (V)	$E^{\circ\prime}$ (V)
$SiF_6^{2-} + 4e^- \rightleftharpoons Si(s) + 6F^-$	–1.37	
$SiO_2(s) + 4H^+ + 4e^- \rightleftharpoons Si(s) + 2H_2O$	–0.909	
$SiO_2(s) + 8H^+ + 8e^- \rightleftharpoons SiH_4(g) + 2H_2O$	–0.516	

Silver	E° (V)	$E^{\circ\prime}$ (V)
$Ag^+ + e^- \rightleftharpoons Ag(s)$	0.7996	
$AgBr(s) + e^- \rightleftharpoons Ag(s) + Br^-$	0.071	
$Ag_2C_2O_4(s) + 2e^- \rightleftharpoons 2Ag(s) + C_2O_4^{2-}$	0.47	
$AgCl(s) + e^- \rightleftharpoons Ag(s) + Cl^-$	0.2223	
$AgI(s) + e^- \rightleftharpoons Ag(s) + I^-$	–0.152	
$Ag_2S(s) + 2e^- \rightleftharpoons 2Ag(s) + S^{2-}$	–0.71	
$Ag(NH_3)_2^+ + e^- \rightleftharpoons Ag(s) + 2NH_3$	0.373	

Sodium	E° (V)	$E^{\circ\prime}$ (V)
$Na^+ + e^- \rightleftharpoons Na(s)$	–2.713	

Strontium	E° (V)	$E^{\circ\prime}$ (V)
$Sr^{2+} + 2e^- \rightleftharpoons Sr(s)$	–2.89	

Sulfur	E° (V)	$E^{\circ\prime}$ (V)
$S(s) + 2e^- \rightleftharpoons S^{2-}$	–0.407	
$S(s) + 2H^+ + 2e^- \rightleftharpoons H_2S$	0.144	
$S_2O_6^{2-} + 4H^+ + 2e^- \rightleftharpoons 2H_2SO_3$	0.569	
$S_2O_8^{2-} + 2e^- \rightleftharpoons 2SO_4^{2-}$	1.96	
$S_4O_6^{2-} + 2e^- \rightleftharpoons 2S_2O_3^{2-}$	0.080	
$2SO_3^{2-} + 2H_2O + 2e^- \rightleftharpoons S_2O_4^{2-} + 4OH^-$	–1.13	
$2SO_3^{2-} + 3H_2O + 4e^- \rightleftharpoons S_2O_3^{2-} + 6OH^-$		–0.576 1 M NaOH
$2SO_4^{2-} + 4H^+ + 2e^- \rightleftharpoons S_2O_6^{2-} + 2H_2O$	–0.25	
$SO_4^{2-} + H_2O + 2e^- \rightleftharpoons SO_3^{2-} + 2OH^-$	–0.936	
$SO_4^{2-} + 4H^+ + 2e^- \rightleftharpoons H_2SO_3 + H_2O$	+0.172	

Thallium	E° (V)	$E^{\circ\prime}$ (V)
$Tl^{3+} + 2e^- \rightleftharpoons Tl^+$		1.25 1 M $HClO_4$
		0.77 1 M HCL
$Tl^3 + 3e^- \rightleftharpoons Tl(s)$	0.742	

continued

Standard Reduction Potentials[a]—*continued*

Tin	E° (V)	$E^{\circ\prime}$ (V)	
$Sn^{2+} + 2e^- \rightleftharpoons Sn(s)$		−0.19	1 M HCl
$Sn^{4+} + 2e^- \rightleftharpoons Sn^{2+}$	0.154	0.139	1 M HCl

Uranium	E° (V)	$E^{\circ\prime}$ (V)
$U^{3+} + 3e^- \rightleftharpoons U(s)$	−1.66	
$U^{4+} + e^- \rightleftharpoons U^{3+}$	−0.52	
$UO_2^{+} + 4H^+ + e^- \rightleftharpoons U^{4+} + 2H_2O$	0.27	
$UO_2^{2+} + e^- \rightleftharpoons UO_2^{+}$	0.16	
$UO_2^{2+} + 4H^+ + 2e^- \rightleftharpoons U^{4+} + 2H_2O$	0.327	

Titanium	E° (V)	$E^{\circ\prime}$ (V)
$Ti^{2+} + 2e^- \rightleftharpoons Ti(s)$	−1.63	
$Ti^{3+} + e^- \rightleftharpoons Ti^{2+}$	−0.37	

Vanadium	E° (V)	$E^{\circ\prime}$ (V)
$V^{2+} + 2e^- \rightleftharpoons V(s)$	−1.13	
$V^{3+} + e^- \rightleftharpoons V^{2+}$	−0.255	
$VO^{2+} + 2H^+ + e^- \rightleftharpoons V^{3+} + H_2O$	0.337	
$VO_2^{+} + 2H^+ + e^- \rightleftharpoons VO^{2+} + H_2O$	1.000	

Tungsten	E° (V)	$E^{\circ\prime}$ (V)
$WO_2(s) + 4H^+ + 4e^- \rightleftharpoons W(s) + 2H_2O$	−0.119	
$WO_3(s) + 6H^+ + 6e^- \rightleftharpoons W(s) + 3H_2O$	−0.090	

Zinc	E° (V)	$E^{\circ\prime}$ (V)
$Zn^{2+} + 2e^- \rightleftharpoons Zn(s)$	−0.7618	
$Zn(OH)_4^{2-} + 2e^- \rightleftharpoons Zn(s) + 4OH^-$	−1.285	
$Zn(NH_3)_4^{2+} + 2e^- \rightleftharpoons Zn(s) + 4NH_3$	−1.04	
$Zn(CN)_4^{2-} + 2e^- \rightleftharpoons Zn(s) + 4CN^-$	−1.34	

Source: Values are compiled from the following sources: Bard, A. J.; Parsons, R.; Jordon, J., eds. *Standard Potentials in Aqueous Solutions.* Dekker: New York, 1985; Milazzo, G.; Caroli, S.; Sharma, V. K. *Tables of Standard Electrode Potentials.* Wiley: London, 1978; Swift, E. H.; Butler, E. A. *Quantitative Measurements and Chemical Equilibria.* Freeman: New York, 1972.

[a]Solids, gases, and liquids are identified; all other species are aqueous. Reduction reactions in acidic solution are written using H^+ instead of H_3O^+. Reactions may be rewritten by replacing H^+ with H_3O^+ and adding one molecule of H_2O to the opposite side of the reaction for each H^+; thus

$$H_3AsO_4 + 2H^+ + 2e^- \rightleftharpoons HAsO_2 + 2H_2O$$

becomes

$$H_3AsO_4 + 2H_3O^+ + 2e^- \rightleftharpoons HAsO_2 + 4H_2O$$

Conditions for formal potentials ($E^{\circ\prime}$) are listed next to the potential.

Appendix 3E

Selected Polarographic Half-Wave Potentials[a]

Element	$E_{1/2}$ (volts vs. SCE)	Matrix
$Al^{3+} + 3e^- \rightleftharpoons Al(s)$	−0.5	0.2 M acetate (pH 4.5–4.7)
$Cd^{2+} + 2e^- \rightleftharpoons Cd(s)$	−0.60	0.1 M KCl
		0.05 M H_2SO_4
		1 M HNO_3
	−0.81	1 M NH_4Cl plus 1 M NH_3
$Cr^{+3} + 3e^- \rightleftharpoons Cr(s)$	−0.35 (+3 → +2)	1 M NH_4^+/NH_3 buffer (pH 8–9)
	−1.70 (+2 → 0)	
$Co^{3+} + 3e^- \rightleftharpoons Co(s)$	−0.5 (+3 → +2)	1 M NH_4Cl plus 1 M NH_3
	−1.3 (+2 → 0)	
$Co^{2+} + 2e^- \rightleftharpoons Co(s)$	−1.03	1 M KSCN
$Cu^{2+} + 2e^- \rightleftharpoons Cu(s)$	0.04	0.1 M KNO_3
		0.1 M NH_4ClO_4
		1 M Na_2SO_4
	−0.22	0.5 M potassium citrate (pH 7.5)
$Fe^{3+} + 3e^- \rightleftharpoons Fe(s)$	−0.17 (+3 → +2)	0.5 M sodium tartrate (pH 5.8)
	−1.52 (+2 → 0)	
$Fe^{3+} + e^- \rightleftharpoons Fe^{2+}$	−0.27	0.2 M $Na_2C_2O_4$ (pH $<$ 7.9)
$Pb^{2+} + 2e^- \rightleftharpoons Pb(s)$	−0.405	1 M HNO_3
	−0.435	1 M KCl
$Mn^{2+} + 2e^- \rightleftharpoons Mn(s)$	−1.65	1 M NH_4Cl plus 1 M NH_3
$Ni^{2+} + 2e^- \rightleftharpoons Ni(s)$	−0.70	1 M KSCN
	−1.09	1 M NH_4Cl plus 1 M NH_3
$Zn^{2+} + 2e^- \rightleftharpoons Zn(s)$	−0.995	0.1 M KCl
	−1.33	1 M NH_4Cl plus 1 M NH_3

Source: All values from Dean, J. A. *Analytical Chemistry Handbook.* McGraw-Hill: New York, 1995.

[a]In some cases two polarographic waves are seen as the redox reaction occurs in steps. Half-wave potentials are given for each wave, with the corresponding change in oxidation states shown in parentheses.

Appendix 4

Balancing Redox Reactions

Balancing a redox reaction is often more challenging than balancing other types of reactions, because we must balance both electrons and elements. Perhaps the simplest way to balance a redox reaction is by the half-reaction method, which consists of the following steps.

1. Identify the oxidizing and reducing agents and the appropriate conjugate reducing and oxidizing agents.
2. Balance all atoms in each half-reaction *except for O and H.*
3. Balance the oxygen in each half-reaction by adding appropriate amounts of H_2O.
4. For acidic solutions, balance the hydrogen in each half-reaction by adding H_3O^+ and H_2O to opposite sides of the reaction; for basic solutions, add OH^- and H_2O to opposite sides of the reaction.
5. Balance the net charge in each half-reaction by adding electrons; the electrons should be a reactant for the reduction half-reaction and a product for the oxidation half-reaction.
6. Adjust the coefficients of each half-reaction so that both half-reactions involve the same number of electrons.
7. Combine the two half-reactions, and simplify the stoichiometry.

The application of the half-reaction method is shown in Example A4.1.

Example

EXAMPLE A4.1

Balance the following redox reactions.

(a) $Cr_2O_7^{2-}(aq) + Zn(s) \rightleftharpoons Cr^{3+}(aq) + Zn^{2+}(aq)$ (acidic solution)

(b) $MnO_4^-(aq) + S^{2-}(aq) \rightleftharpoons MnO_2(s) + S(s)$ (basic solution)

SOLUTION

We begin by writing unbalanced equations for the oxidation and reduction half-reactions in part (a).

$$Cr_2O_7^{2-}(aq) \rightleftharpoons Cr^{3+}(aq) \qquad Zn(s) \rightleftharpoons Zn^{2+}(aq)$$

The atoms in each half-reaction are then balanced. The zinc half-reaction is already balanced in this respect, so we begin by balancing chromium in the dichromate half-reaction.

$$Cr_2O_7^{2-}(aq) \rightleftharpoons 2Cr^{3+}(aq)$$

Oxygen is then balanced by adding H_2O.

$$Cr_2O_7^{2-}(aq) \rightleftharpoons 2Cr^{3+}(aq) + 7H_2O(\ell)$$

and hydrogen is balanced by adding H_3O^+ and sufficient additional water.

$$14H_3O^+(aq) + Cr_2O_7^{2-}(aq) \rightleftharpoons 2Cr^{3+}(aq) + 21H_2O(\ell)$$

Both half-reactions are balanced with respect to charge by adding electrons as needed.

$$14H_3O^+(aq) + Cr_2O_7^{2-}(aq) + 6e^- \rightleftharpoons 2Cr^{3+}(aq) + 21H_2O(\ell)$$

$$Zn(s) \rightleftharpoons Zn^{2+}(aq) + 2e^-$$

The stoichiometric coefficients for the zinc half-reaction are multiplied by 3 so that both half-reactions involve 6 electrons.

$$14H_3O^+(aq) + Cr_2O_7^{2-}(aq) + 6e^- \rightleftharpoons 2Cr^{3+}(aq) + 21H_2O(\ell)$$

$$3Zn(s) \rightleftharpoons 3Zn^{2+}(aq) + 6e^-$$

Adding together the two half-reactions and simplifying gives us the final balanced redox reaction.

$$14H_3O^+(aq) + Cr_2O_7^{2-}(aq) + 3Zn(s) \rightleftharpoons 2Cr^{3+}(aq) + 21H_2O(\ell) + 3Zn^{2+}(aq)$$

Balancing the reaction in part (b) is accomplished in the same manner, except that H_2O and OH^- are used to balance hydrogen. Balancing the reaction is left as an exercise; the resulting balanced reaction is

$$2MnO_4^-(aq) + 4H_2O(\ell) + 3S^{2-}(aq) \rightleftharpoons 2MnO_2(s) + 8OH^-(aq) + 3S(s)$$

Appendix 5

Review of Chemical Kinetics

A reaction's equilibrium position defines the extent to which the reaction can occur. For example, we expect a reaction with a large equilibrium constant, such as the dissociation of HCl in water

$$HCl(aq) + H_2O(\ell) \rightleftharpoons H_3O^+(aq) + Cl^-(aq)$$

to proceed nearly to completion. The magnitude of an equilibrium constant, however, does not guarantee that a reaction will reach its equilibrium position. Many reactions with large equilibrium constants, such as the reduction of MnO_4^- by H_2O

$$4MnO_4^-(aq) + 2H_2O(\ell) \rightleftharpoons 4MnO_2(s) + 3O_2(g) + 4OH^-(aq)$$

do not occur to an appreciable extent. The study of the rate at which a chemical reaction approaches its equilibrium position is called kinetics.

A5.1 Chemical Reaction Rates

A study of the kinetics of a chemical reaction begins with the measurement of its reaction rate. Consider, for example, the general reaction shown in the following equation, involving the aqueous solutes A, B, C, and D, with stoichiometries of *a*, *b*, *c*, and *d*.

$$aA + bB \rightleftharpoons cC + dD \qquad \textbf{A5.1}$$

The rate, or velocity, at which this reaction approaches its equilibrium position can be determined by following the change in concentration of a reactant or a product as a function of time. For example, if we monitor the concentration of reactant A, we express the rate as

$$R = -\frac{d[A]}{dt} \qquad \textbf{A5.2}$$

where R is the measured rate expressed as a change in concentration as a function of time. Because a reactant's concentration decreases with time, we include a negative sign so that the rate has a positive value.

The rate also can be determined by following the change in concentration of a product as a function of time.

$$R' = \frac{d[C]}{dt} \qquad \textbf{A5.3}$$

Rates determined by monitoring different species in a chemical reaction need not have the same value. The rate R in equation A5.2 and the rate R' in equation A5.3 will have the same value only if the stoichiometric coefficients of A and C in reaction A5.1 are the same. In general, the relationship between the rates R and R' is

$$R = \frac{a}{c} \times R'$$

A5.2 The Rate Law

A rate law describes how the rate of a reaction is affected by the concentration of each species present in the reaction mixture. The rate law for reaction A5.1 takes the general form of

$$R = k[\mathrm{A}]^{\alpha}[\mathrm{B}]^{\beta}[\mathrm{C}]^{\gamma}[\mathrm{D}]^{\delta}[\mathrm{E}]^{\varepsilon} \ldots \qquad \textbf{A5.4}$$

where k is the rate constant, and α, β, γ, δ and ε are the orders of the reaction with respect to the species present in the reaction.

Several important points about the rate law are shown in equation A5.4. First, the rate of a reaction may depend on the concentrations of both reactants and products, as well as the concentrations of species that do not appear in the reaction's overall stoichiometry. Species E in equation A5.4, for example, may represent a catalyst. Second, the reaction order for a given species is not necessarily the same as its stoichiometry in the chemical reaction. Reaction orders may be positive, negative, or zero and may take integer or noninteger values. Finally, the overall reaction order is the sum of the individual reaction orders. Thus, the overall reaction order for equation A5.4 is $\alpha + \beta + \gamma + \delta + \varepsilon$.

A5.3 Kinetic Analysis of Selected Reactions

In this section we review the application of kinetics to several simple chemical reactions, focusing on how the integrated form of the rate law can be used to determine reaction orders. In addition, we consider how rate laws for more complex systems can be determined.

First-Order Reactions The simplest case is a first-order reaction in which the rate depends on the concentration of only one species. The best example of a first-order reaction is an irreversible thermal decomposition, which we can represent as

$$\mathrm{A} \rightarrow \text{Products} \qquad \textbf{A5.5}$$

with a rate law of

$$R = -\frac{d[\mathrm{A}]}{dt} = k[\mathrm{A}] \qquad \textbf{A5.6}$$

The simplest way to demonstrate that a reaction is first-order in A, is to double the concentration of A and note the effect on the reaction's rate. If the observed rate doubles, then the reaction must be first-order in A. Alternatively, we can derive a relationship between the [A] and time by rearranging equation A5.6

$$\frac{d[\mathrm{A}]}{[\mathrm{A}]} = -kdt$$

and integrating.

$$\int_{[\mathrm{A}]_0}^{[\mathrm{A}]_t} \frac{d[\mathrm{A}]}{[\mathrm{A}]} = -k\int_0^t dt \qquad \mathbf{A5.7}$$

Evaluating the integrals in equation A5.7,

$$\ln \frac{[\mathrm{A}]_t}{[\mathrm{A}]_0} = -kt \qquad \mathbf{A5.8}$$

and rearranging,

$$\ln\,[\mathrm{A}]_t = -kt + \ln\,[\mathrm{A}]_0 \qquad \mathbf{A5.9}$$

shows that for a first-order reaction, a plot of ln $[\mathrm{A}]_t$ versus time is linear with a slope of $-k$ and an intercept of ln $[\mathrm{A}]_0$. Equations A5.8 and A5.9 are known as integrated forms of the rate law.

Reaction A5.5 is not the only possible form of a first-order reaction. For example, the reaction

$$\mathrm{A} + \mathrm{B} \rightarrow \text{Products} \qquad \mathbf{A5.10}$$

shows first-order kinetics if the concentration of B does not affect the reaction rate. This may happen if the reaction's mechanism involves at least two steps. Imagine that in the first step, A is slowly converted to an intermediate species C. The intermediate and the remaining reactant, B, then react rapidly, in one or more steps, to form the products.

$$\mathrm{A} \rightarrow \mathrm{C} \qquad \text{(slow)}$$

$$\mathrm{B} + \mathrm{C} \rightarrow \text{Products} \qquad \text{(fast)}$$

Since the rate of a chemical reaction only depends on the slowest, or rate-determining step, and any preceding steps, species B will not show up in the rate law.

Second-Order Reactions The simplest overall reaction demonstrating second-order behavior is

$$2\mathrm{A} \rightarrow \text{Products}$$

for which the rate law is

$$R = -\frac{d[\mathrm{A}]}{dt} = k[\mathrm{A}]^2$$

Proceeding in the same manner as for a first-order reaction, the integrated form of the rate law is derived as follows

$$\frac{d[\mathrm{A}]}{[\mathrm{A}]^2} = -k\,dt$$

$$\int_{[\mathrm{A}]_0}^{[\mathrm{A}]_t} \frac{d[\mathrm{A}]}{[\mathrm{A}]^2} = -k\int_0^t dt$$

$$\frac{1}{[\mathrm{A}]_t} = kt + \frac{1}{[\mathrm{A}]_0}$$

Thus, for a second-order reaction, a plot of $[A]_t^{-1}$ versus t is linear, with a slope of k and an intercept of $[A]_0^{-1}$. Alternatively, a reaction can be shown to be second-order in A by observing the effect on the rate of changing the concentration of A. In this case, doubling the concentration of A produces a fourfold increase in the reaction's rate.

Example

EXAMPLE A5.1

The following data were obtained during a kinetic study of the hydration of *p*-methoxyphenylacetylene by measuring the relative amounts of reactants and products by nuclear magnetic resonance (NMR).[1]

Time (min)	*p*-Methoxyphenylacetylene (%)
67	85.9
161	70.0
241	57.6
381	40.7
479	32.4
545	27.7
604	24

Using appropriate graphs, determine whether this reaction is first- or second-order in *p*-methoxyphenylacetylene.

SOLUTION

To determine the reaction order we plot ln(%*p*-methoxyphenylacetylene) versus time for a first-order reaction, and $(\%p\text{-methoxyphenylacetylene})^{-1}$ versus time for a second-order reaction (Figure A5.1). Because the straight-line for the first-order plot fits the data nicely, we conclude that the reaction is first-order in *p*-methoxyphenylacetylene. Note that when plotted using the equation for a second-order reaction, the data show curvature that does not fit the straight-line model.

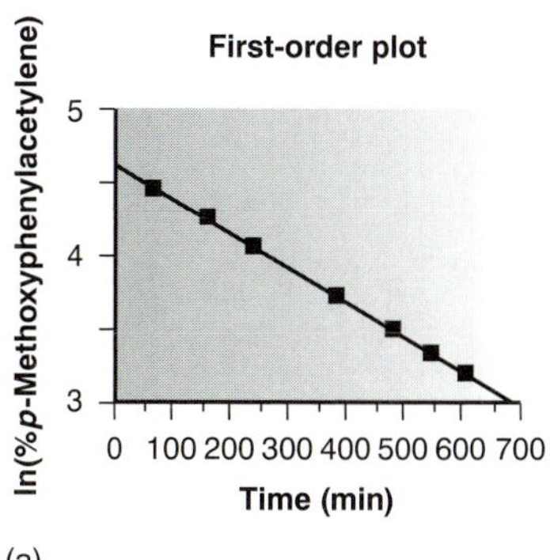

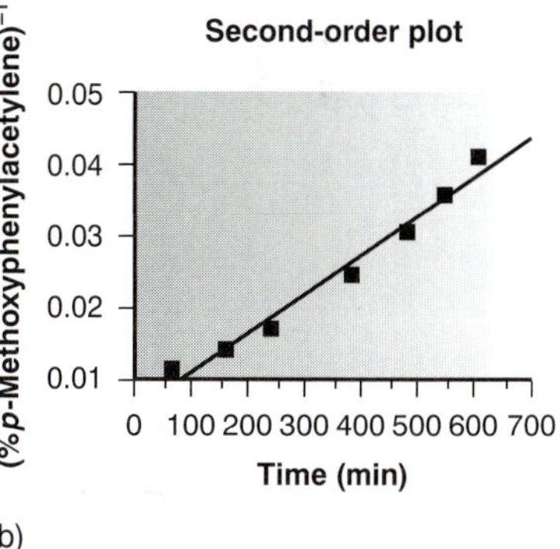

Figure A5.1
Graphs used to determine the reaction order for the data in Example A5.1.

Pseudo-Order Reactions and the Method of Initial Rates Unfortunately, most reactions of importance in analytical chemistry do not follow these simple first-order and second-order rate laws. We are more likely to encounter the second-order rate law given in equation A5.11 than that in equation A5.10.

$$R = k[A][B] \qquad \textbf{A5.11}$$

Demonstrating that a reaction obeys the rate law in equation A5.11 is complicated by the lack of a simple integrated form of the rate law. The kinetics can be simplified, however, by carrying out the analysis under conditions in which the concentrations of all species but one are so large that their concentrations are effectively constant during the reaction. For example, if the concentration of B is selected such that $[B] >> [A]$, then equation A5.11 simplifies to

$$R = k'[A]$$

where the rate constant k' is equal to $k[B]$. Under these conditions, the reaction appears to follow first-order kinetics in A and is termed a pseudo-first-order reaction in A. Verifying the reaction order for A can then be carried out by either using the integrated rate law or by noting the effect on the reaction rate of changing the concentration of A. The process can be repeated to determine the reaction order for B by making [A] >> [B].

A variation on the use of pseudo-ordered reactions is the initial rate method. In this approach to determining a reaction's rate law, a series of experiments is conducted in which the concentration of those species expected to affect the reaction's rate are changed one at a time. The initial rate of the reaction is determined for each set of conditions. Comparing the reaction's initial rate for two experiments in which the concentration of only a single species has been changed allows the reaction order for that species to be determined. The application of this method is outlined in the following example.

Example

EXAMPLE A5.2

The following data were collected during a kinetic study of the iodination of acetone by measuring the concentration of unreacted I_2 in solution.[2]

Experiment Number	$[C_3H_6O]$ (M)	$[H^+]$ (M)	$[I_2]$ (M)	Rate ($M\ s^{-1}$)
1	1.33	0.0404	6.65×10^{-3}	1.78×10^{-6}
2	1.33	0.0809	6.65×10^{-3}	3.89×10^{-6}
3	1.33	0.162	6.65×10^{-3}	8.11×10^{-6}
4	1.33	0.323	6.65×10^{-3}	16.6×10^{-6}
5	0.167	0.323	6.65×10^{-3}	1.64×10^{-6}
6	0.333	0.323	6.65×10^{-3}	3.76×10^{-6}
7	0.667	0.323	6.65×10^{-3}	7.55×10^{-6}
8	0.333	0.323	3.32×10^{-3}	3.57×10^{-6}

Determine the rate law for this reaction and the value of the rate constant.

SOLUTION

The order of the rate law with respect to the three reactants can be determined by comparing the rates of two experiments in which the concentration of only one of the reactants is changed. For example, in experiment 2 the $[H^+]$ and the rate are approximately twice as large as in experiment 1, indicating that the reaction is first-order in $[H^+]$. Working in the same manner, experiments 6 and 7 show that the reaction is also first-order with respect to $[C_3H_6O]$, and experiments 6 and 8 show that the rate of the reaction is independent of the $[I_2]$. Thus, the rate law is

$$R = k[C_3H_6O][H^+]$$

The value of the rate constant can be determined by substituting the rate, the $[C_3H_6O]$, and the $[H^+]$ for an experiment into the rate law and solving for k. Using the data from experiment 1, for example, gives a rate constant of $3.31 \times 10^{-5}\ M^{-1}\ s^{-1}$. The average rate constant for the eight experiments is $3.49 \times 10^{-5}\ M^{-1}\ s^{-1}$.

REFERENCES

1. Kaufman, D.; Sterner, C.; Masek, B.; et al. *J. Chem. Educ.* **1982,** *59,* 885–886.

2. Birk, J. P.; Walters, D. L. *J. Chem. Educ.* **1992,** *69,* 585–587.

Appendix 6

Countercurrent Separations

A solution to the problem of separating solutes with similar distribution ratios was introduced by Craig in the 1940s.[1] The technique, which is known as a countercurrent liquid–liquid extraction, is outlined in Figure A6.1 and discussed in detail in this appendix. In contrast to a simple liquid–liquid extraction, in which the phase containing the sample is extracted sequentially, a countercurrent extraction is based on a serial extraction of both the sample and extracting phases. Countercurrent separations are no longer in common use, having been replaced by chromatographic separations. Nevertheless, the theory of countercurrent extractions is instructive because it provides a useful introduction to the theory of chromatographic separations.

To keep track of the progress of a countercurrent liquid–liquid extraction, it is necessary to adopt a labeling convention. Each step in a countercurrent extraction includes the transferring of the upper phase followed by an extraction. Steps are labeled sequentially beginning with zero. Extractions take place in a series of tubes that also are labeled sequentially, starting with zero. The upper and lower phases in each tube are identified by a letter and number. The letters U and L represent, respectively, the upper phase and the lower phase, and the number indicates the step in the countercurrent extraction in which the phase was first introduced. Thus U_0 is the upper phase introduced at step 0 (the first extraction), and L_2 is the lower phase introduced at step 2 (the third extraction). Finally, partitioning of the solute in any extraction tube results in a fraction p remaining in the upper phase and a fraction q remaining in the lower phase. Values of q are calculated using equation A6.1

$$(q_{\text{aq}})_1 = \frac{(\text{moles aq})_1}{(\text{moles aq})_0} = \frac{V_{\text{aq}}}{DV_{\text{org}} + V_{\text{aq}}} \qquad \textbf{A6.1}$$

and p is equal to $1 - q$.

Let's assume that the solute to be separated is present in an aqueous phase of 1 M HCl and that the organic phase is benzene. Because benzene has the smaller density, it is the upper phase, and 1 M HCl is the lower phase. To begin the countercurrent extraction the aqueous sample containing the solute is placed in tube 0 along with a portion of benzene. As shown in Figure A6.1a, initially all the solute is present in phase L_0. After extracting (Figure A6.1b), a fraction p of the solute is present in phase U_0, and a fraction q is in phase L_0. This completes step 0 of the countercurrent extraction. Thus far there is no difference between a simple liquid–liquid extraction and a countercurrent extraction.

At the completion of step 0 the phase U_0 is removed, and a fresh portion of benzene, U_1, is added to tube 0 (Figure A6.1c). This, too, is just the same as in a

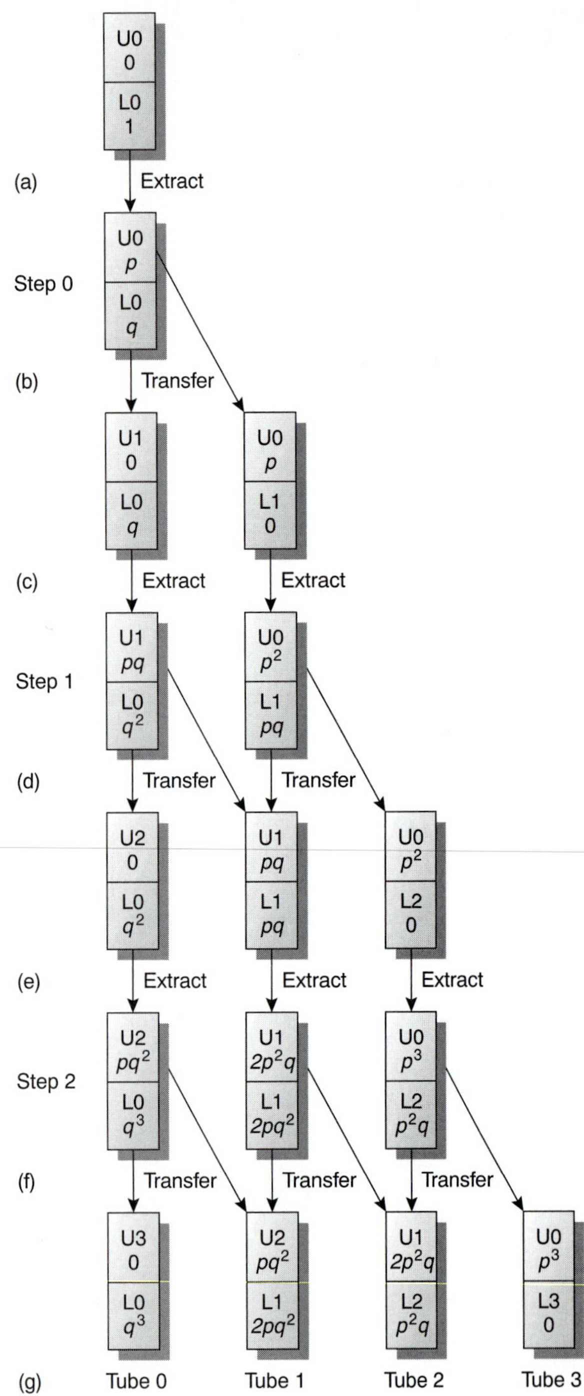

Figure A6.1
Scheme for a countercurrent extraction.

simple liquid–liquid extraction. The difference between the two extraction techniques, however, occurs at this point when phase U_0 is placed in tube 1 along with a portion of solute-free aqueous 1 M HCl as phase L1 (Figure A6.1c). Tube 0 now contains a fraction q of the solute, and tube 1 contains a fraction p of the solute. Carrying out the extraction in tube 0 results in a fraction p of its contents remaining in the upper phase and a fraction q remaining in the lower phase. Thus, phases U_1 and L_0 now contain, respectively, fractions pq and q^2 of the original amount of solute. In the same way it is easy to show that the phases U_0 and L_1 in tube 1 contain, respectively, fractions p^2 and pq of the original amount of solute. This completes step 1 of the extraction (Figure A6.1d). As shown in the remainder of Figure A6.1, the countercurrent extraction continues in this cycle of phase transfers and extractions.

In a countercurrent liquid–liquid extraction the lower phase in each tube remains in place, and the upper phase moves from tube 0 to higher numbered tubes. This difference in the movement of the phases is indicated by referring to the lower phase as a stationary phase and the upper phase as a mobile phase. With each transfer some of the solute in tube r is moved to tube $r + 1$, and a portion of the solute in tube $r - 1$ is moved to tube r. As a result, a solute introduced at tube 0 moves with the mobile phase. The solute, however, does not move at the same rate as the mobile phase since, at each step, a portion of the solute is extracted into the stationary phase. A solute that is preferentially extracted into the stationary phase spends proportionally less time in the mobile phase and moves at a slower rate. As the number of steps increases, solutes with different values of q separate into completely different sets of extraction tubes.

The effectiveness of a countercurrent extraction can be judged from a histogram indicating the fraction of solute present in each tube. To determine the total amount of solute in an extraction tube, we add together the fraction of solute present in the tube's upper and lower phases following each transfer. For example, at the beginning of step 3 (Figure A6.1g) the upper and lower phases of tube 1 contain fractions pq^2 and $2pq^2$ of the solute, respectively; thus, the total fraction of solute in the tube is $3pq^2$. Table A6.1 summarizes this for the steps outlined in Figure A6.1. A typical histogram, calculated assuming distribution ratios of 5.0 for solute A and 0.5 for solute B, is shown in Figure A6.2. Although four steps is not enough to separate the solutes in this instance, it is easy to see that by extending the countercurrent extraction the solutes will eventually be separated.

Figure A6.1 and Table A6.1 show how a solute's distribution changes during the first four steps of a countercurrent extraction. Now we consider how these results can be generalized to give the distribution of a solute in any tube, at any step during the extraction. You may recognize the pattern of entries in Table A6.1 as following the binomial distribution

Table A6.1 Fraction of Solute Remaining in Tube *r* After Extraction Step *n* for a Countercurrent Extraction

n/*r*	0	1	2	3
0	1			
1	q	p		
2	q^2	$2pq$	p^2	
3	q^3	$3pq^2$	$3p^2q$	p^3

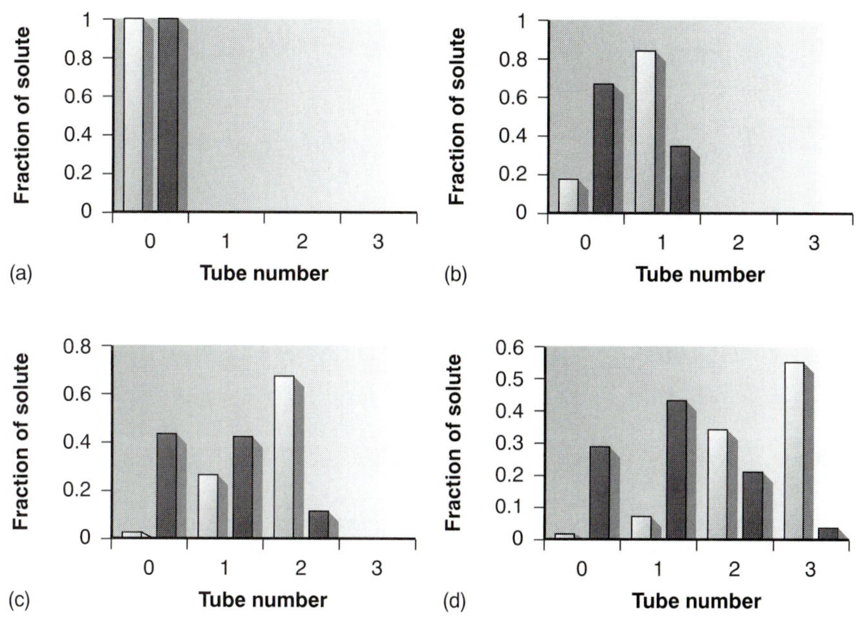

Figure A6.2
Progress of a countercurrent extraction after (a) step 0, (b) step 1, (c) step 2, and (d) step 3. The distribution ratio, D, is 5.0 for solute A and 0.5 for solute B.

$$f(r, n) = \frac{n!}{(n - r)!\, r!} p^r q^{n-r} \qquad \text{A6.2}$$

where $f(r, n)$ is the fraction of solute present in tube r at step n of the countercurrent extraction. After extracting, therefore, the upper phase contains a fraction $p \times f(r, n)$, and the lower phase contains a fraction $q \times f(r, n)$ of the solute.

EXAMPLE A6.1

The countercurrent extraction shown in Figure A6.2 is carried out through step 30. Calculate the fraction of solutes A and B in tubes 5, 10, 15, 20, 25, and 30.

SOLUTION

The fraction, q, of each solute that remains in the lower phase is calculated using equation A6.1. Since the volumes of the lower and upper phases are equal, we get

$$q_A = \frac{1}{D_A + 1} = \frac{1}{5 + 1} = 0.167$$

and

$$q_B = \frac{1}{D_B + 1} = \frac{1}{0.5 + 1} = 0.667$$

Thus, p_A is 0.833 and p_B is 0.333. For solute A, the fraction present in tubes 5, 10, 15, 20, 25, and 30 after step 30 are

$$f(5, 30) = \frac{30!}{(30 - 5)!\,5!}(0.833)^5(0.167)^{30-5} = 2.1 \times 10^{-15} \approx 0$$

$$f(10, 30) = \frac{30!}{(30 - 10)!\,10!}(0.833)^{10}(0.167)^{30-10} = 1.4 \times 10^{-9} \approx 0$$

$$f(15, 30) = \frac{30!}{(30 - 15)!\,15!}(0.833)^{15}(0.167)^{30-15} = 2.2 \times 10^{-5} \approx 0$$

$$f(20, 30) = \frac{30!}{(30 - 20)!\,20!}(0.833)^{20}(0.167)^{30-20} = 0.013$$

$$f(25, 30) = \frac{30!}{(30 - 25)!\,25!}(0.833)^{25}(0.167)^{30-25} = 0.192$$

$$f(30, 30) = \frac{30!}{(30 - 30)!\,30!}(0.833)^{30}(0.167)^{30-30} = 0.004$$

The fraction of solute B in tubes 5, 10, 15, 20, 25, and 30 are calculated in the same way, yielding respective values of 0.023, 0.153, 0.025, 0, 0, and 0. A complete histogram of the distribution of solutes A and B is given in Figure A6.3 and shows that the two solutes have been successfully separated.

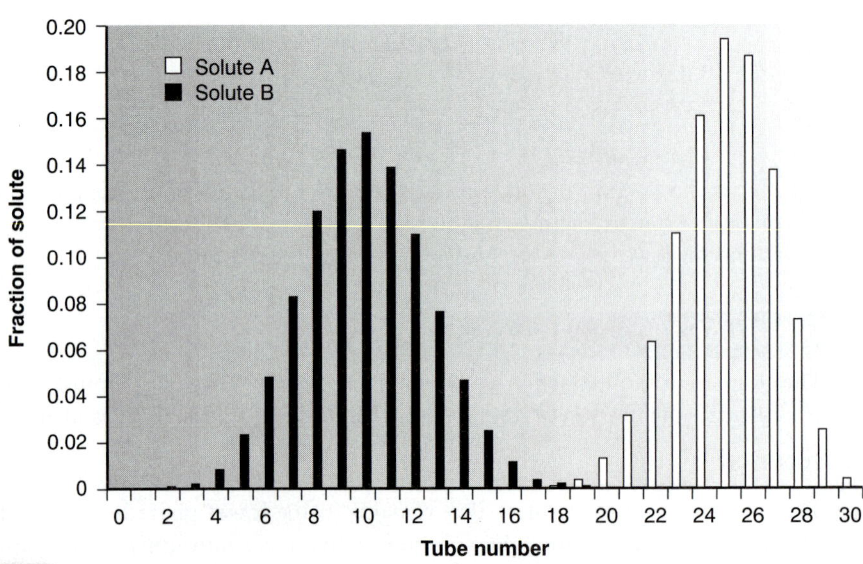

Figure A6.3
Progress of a countercurrent extraction for Example A6.1 after 30 steps.

Constructing a histogram using equation A6.2 is tedious, particularly when the number of steps is large. The fraction of solute in most tubes will be approximately zero. Consequently, the histogram's construction can be simplified by only solving equation A6.2 for those tubes containing an amount of solute exceeding a threshold value. Since the location of the solute obeys a binomial distribution, we can use the distribution's mean and standard deviation to determine which tubes contain a significant fraction of solute. The properties of a binomial distribution were covered in Chapter 4, with the mean, μ, and standard deviation, σ, given as

$$\mu = np$$

$$\sigma = \sqrt{np(1-p)} = \sqrt{npq}$$

Furthermore, when both np and nq are greater than 5, the binomial distribution is closely approximated by the normal distribution,[2] and the probability tables in Appendix 1A can be used to determine the location of the solute and its recovery.

Example

EXAMPLE A6.2

Two solutes, A and B, with distribution ratios of 9 and 4, respectively, are to be separated by a countercurrent extraction in which the volumes of the upper and lower phases are equal. After 100 steps, determine the 99% confidence interval for the location of each solute.

SOLUTION

The fraction, q, of each solute remaining in the lower phase is calculated using equation A6.1. Since the volumes of the lower and upper phases are equal, we get

$$q_A = \frac{1}{D_A + 1} = \frac{1}{9+1} = 0.10$$

$$q_B = \frac{1}{D_B + 1} = \frac{1}{4+1} = 0.20$$

and, consequently, p_A is 0.90 and p_B is 0.80. The mean and standard deviation for the distribution of solutes A and B after 100 steps of the countercurrent extraction are

$$\mu_A = np_A = (100)(0.90) = 90 \qquad \sigma_A = \sqrt{np_Aq_A} = \sqrt{(100)(0.90)(0.10)} = 3$$

$$\mu_B = np_B = (100)(0.80) = 80 \qquad \sigma_B = \sqrt{np_Bq_B} = \sqrt{(100)(0.80)(0.20)} = 4$$

The confidence interval for a normally distributed solute is

$$r = \mu \pm z\sigma$$

where r is the number of the tube, and the value of z is determined by the chosen significance level. For a 99% confidence interval, the value of z is 2.58 (Appendix 1B); thus,

$$r_A = 90 \pm (2.58)(3) = 90 \pm 8$$

$$r_B = 80 \pm (2.58)(4) = 80 \pm 10$$

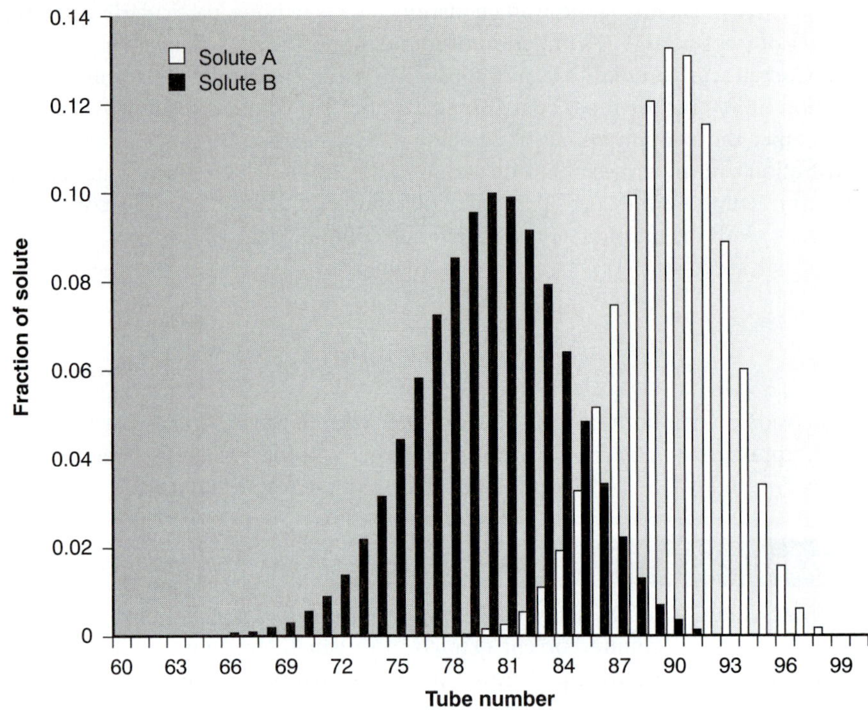

Figure A6.4
Progress of a countercurrent extraction for Example A6.2 after 100 steps.

Since the two confidence intervals overlap, a complete separation of the two solutes cannot be achieved in a 100-step countercurrent extraction. The complete distribution of the solutes is shown in Figure A6.4.

Example

EXAMPLE A6.3

For the countercurrent extraction in Example A6.2, calculate the recovery and separation factor for solute A if the contents of tubes 85–99 are pooled together.

SOLUTION

From Example A6.2 we know that after 100 steps of the countercurrent extraction, solute A is normally distributed about tube 90 with a standard deviation of 3. To determine the fraction of solute in tubes 85–99, we use the single-sided normal distribution in Appendix 1A to determine the fraction of solute in tubes 0–84 and in tube 100. The fraction of solute A in tube 100 is determined by calculating the deviation z (see Chapter 4)

$$z = \frac{r - \mu}{\sigma} = \frac{99 - 90}{3} = 3$$

and using the table in Appendix 1A to determine the corresponding fraction. For $z = 3$ this corresponds to 0.135% of solute A. To determine the fraction of solute A in tubes 0–84, we again calculate the deviation

$$z = \frac{r - \mu}{\sigma} = \frac{85 - 90}{3} = -1.67$$

From Appendix 1A we find that 4.75% of solute A is present in tubes 0–84. Solute A's recovery, therefore, is

$$100\% - 4.75\% - 0.135\% \approx 95\%$$

To calculate the separation factor, we must determine the recovery of solute B in tubes 85–99. This is determined by calculating the fraction of solute B in tubes 85–100 and subtracting the fraction of solute B in tube 100. By calculating z and using Appendix 1A, we find that approximately 10.6% of solute B is in tubes 85–100, and that essentially no solute B is in tube 100. The separation factor, $S_{B,A}$ therefore, is

$$S_{B,A} = \frac{R_B}{R_A} = \frac{10.6}{95} = 0.112$$

REFERENCES

1. Craig, L. C. *J. Biol. Chem.* **1944,** *155,* 519.
2. Mark, H.; Workman, J. *Spectroscopy,* **1990,** *5(3),* 55–56.

Appendix 7

Answers to Selected Problems

Chapter 1

1. (a) quantitative analysis; (b) qualitative or quantitative analysis; (c) qualitative analysis; (d) characterization analysis; (e) fundamental study; (f) quantitative analysis.

Chapter 2

1. (a) 3; (b) 3; (c) 5; (d) 3; (e) 4; (f) 3
2. (a) 0.894; (b) 0.893; (c) 0.894; (d) 0.900; (e) 0.0891
3. (a) 12.01 g/mol; (b) 16.0 g/mol; (c) 6.022×10^{23} mol^{-1}; (d) 9.65×10^4 C/mol
4. (a) 71.9; (b) 39.8; (c) 6.1×10^3; (d) 55; (e) 2.57×10^{-2}; (f) –4.185; (g) 7.2×10^{-8}; (h) 5.30×10^{-13}
5. 1.230% w/w Ni
6. 288.91 g/mol
7. 16.6 mL
8. (a) 0.20 N; (b) 0.10 N; (c) 0.20 N
9. 1.4×10^{-6} M
10. (a) 12.0 M; (b) 31.1 g, 26.3 mL
11. 2.4 mL
12. 1.80×10^{-5}% w/v, 0.180 ppm, 1.80×10^2 ppb
13. 12 mg NaF/gal
14. 5.160, 1.19×10^{-9} M
15. (a) moles Mg^{2+} = 2 × moles $Mg_2P_2O_7$; (b) moles HCl = 2 × moles $CaCO_3$; (c) 2 × moles AgCl = moles NH_3; (d) moles Fe^{2+} = 6 × moles $Cr_2O_7^{2-}$
16. 0.329 M $K_2Cr_2O_7$
17.

	Mass (g) Needed to Prepare		
Compound	0.10 M K^+	1.0×10^2 ppm K^+	1.0% w/v K^+
KCl	7.5	0.19	19
K_2SO_4	8.7	0.22	22
$K_3Fe(CN)_6$	11	0.28	28

18. Solution A: 4.00×10^{-3} M; Solution B: 1.00×10^{-3} M; Solution C: 4.00×10^{-5} M
19. Solution 1: 0.66 M NaCl; Solution 2: 0.1312 M NaCl
20. 0.061 M NO_3^-; pNO_3 = 1.21
21. 0.069 M Cl^-; pCl = 1.16
22. 8.44 M ethanol

Chapter 3

2. (a) 5 ng
3. (a) 1.6 ppm^{-1}; (b) 0.55 ppm^{-1}; (c) 0.35; (d) analyte; (e) $C_I/C_A < 0.029$
4. 2.01 ppm
5. –1.0
6. With ascorbic acid, $K_{A,I} = 30$; with methionine, $K_{A,I} = -2.7 \times 10^{-3}$
7. (a) 4.4; (b) ascorbic acid; (c) 4.4×10^{-3} M
8. (a) 66.5 A/M; (b) -7.9×10^{-3}; (c) hypoxanthine; (d) 1.42×10^{-6} M
9. C_C = 0.35 ppm, C_D = 0.17 ppm

Chapter 4

1. mean = 5.583 g, median = 5.552 g, range = 0.148 g, standard deviation = 0.056 g, variance = 3.1×10^{-3}
2. (a) mean = 243.5 mg, median = 243.4 mg, range = 37.4 mg, standard deviation = 11.9 mg, variance = 141; (b) 29.3%
3. (a) For 100-mg tablets: mean = 95.56 mg, standard deviation = 2.16 g; (b) For 100-mg tablets: 1.97%
4. 98.2%
5. (a) 16.0 g/mol ± 0.4 g/mol; (b) volume in liters and mass in grams
6. (a) 50.0 ppm ± 0.3 ppm; (b) no
7. 0.1175 M ± 0.0001 M
9. 100 mg
10. dilution (b) use 25-mL pipet four times
14. 90 carbons, 36.6%
15. (a) 22.4%; (b) 3.6%

16. mean = 16.883, standard deviation = 0.0794, 95% confidence interval = 16.883 ± 0.066% w/w Cr
17. (a) t_{exp} = 2.82 and $t(0.05, 8)$ = 2.31, difference is significant; (b) IUPAC: 2.81 ppTr, LOI: 11.42 ppTr, LOQ: 12.16 ppTr
18. t_{exp} = 3.23 and $t(0.01, 6)$ = 3.71, no evidence that difference is significant
19. t_{exp} = 1.43 and $t(0.05, 11)$ = 2.205, no evidence that difference is significant
20. t_{exp} = 1.07 and $t(0.05, 3)$ = 3.18, no evidence that difference is significant
21. t_{exp} = 4.28 and $t(0.05, 9)$ = 2.26, difference is significant
23. t_{exp} = 0.40 and $t(0.05, 12)$ = 2.18, no evidence that difference is significant
24. t_{exp} = 0.942 and $t(0.05, 4)$ = 2.78, no evidence that difference is significant
25. t_{exp} = 1.91 and $t(0.10, 8)$ = 1.86, difference is significant
26. t_{exp} = 2.17 and $t(0.05, 8)$ = 2.31, no evidence that difference is significant
27. t_{exp} = 1.19 and $t(0.05, 11)$ = 2.205, no evidence that difference is significant
28. t_{exp} = 0.09 and $t(0.05, 7)$ = 2.36, no evidence that difference is significant
29. t_{exp} = 3.69 and $t(0.05, 5)$ = 2.57, difference is significant
30. no evidence for any outliers

Chapter 5

1. (a) without correction: 10.011 mL, with correction 10.021 mL; (b) –0.010 mL, –0.10%
2. (a) without correction: 0.2500 g, with correction 0.2501 g; (b) –0.0001 g, –0.040%
3. proportional error
4. 4.94 g/cm^3
5.

Molarity	Serial Dilution Uncertainty	Single Dilution Uncertainty
1.00×10^{-2}	2.94×10^{-5}	2.94×10^{-5}
1.00×10^{-3}	3.64×10^{-6}	6.37×10^{-6}
1.00×10^{-4}	4.23×10^{-7}	6.37×10^{-7}
1.00×10^{-5}	4.75×10^{-8}	1.02×10^{-7}

6. 7.33 ppm
7. 2.20×10^{-3}% w/w
8. 0.191 ppm
10. 20.8 ppm
11. (a) right; (b) left; (c) left
12. (a) $S_{meas} = -2.69 + 0.477[Cd^{2+}\ (nM)]$, slope = 0.477 ± 0.036, y-intercept = –2.69 ± 2.02; (c) slope = 1.43, approximately three times more sensitive; (d) 49.9 ± 5.9 nM Cd^{2+}
13. 12.3 ± 0.2 ppb
14. (a) $S_A/S_{IS} = 0.3037 \pm 0.5576(C_A/C_{IS})$, slope = 0.5776 ± 0.0999, y-intercept = 0.3037 ± 0.2484
15. Plotting measured versus accepted gives t_{exp} = 0.679 for the y-intercept and t_{exp} = 0.345 for the slope. Because $t(0.05, 3)$ is 3.18, there is no evidence for a significant difference between the values.
16. μA = 2.61 μA + (14.43 μA ppm^{-1}) × (ppm Tl added)

Chapter 6

1. (a) $K = \dfrac{[NH_4^+][Cl^-]}{[NH_3][HCl]} = 1.75 \times 10^9$

   (c) $K = \dfrac{[Cd(CN)_4^{2-}][Y^{4-}]}{[CdY^{2-}][CN^-]^4} = 28.9$

   (e) $K = \dfrac{[Ba^{2+}][H_2CO_3]}{[H_3O^+]} = 2.4 \times 10^8$
3. 0.626 V
4. (a) $2MnO_4^-(aq) + 5H_2SO_3(aq) + H_2O(\ell) \rightleftharpoons 2Mn^{2+}(aq) + 5SO_4^{2-}(aq) + 4H_3O^+$
   (b) $IO_3^-(aq) + 5I^-(aq) + 6H_3O^+(aq) \rightleftharpoons 3I_2(s) + 9H_2O$
   (c) $3ClO^-(aq) + I^-(aq) \rightleftharpoons IO_3^-(aq) + 3Cl^-$
5. (a) solubility increases; (b) solubility decreases; (c) solubility decreases
6. (a) charge balance: $[Na^+] + [H_3O^+] = [OH^-] + [Cl^-]$
   mass balance: $0.1\ M = [Na^+] = [Cl^-]$
   (c) charge balance: $[H_3O^+] = [OH^-] + [F^-]$
   mass balance: $0.1\ M = [HF] + [F^-]$
   (e) charge balance: $2[Mg^{2+}] + [H_3O^+] = [OH^-] + [HCO_3^-] + 2[CO_3^{2-}]$
   mass balance: $[Mg^{2+}] = [H_2CO_3] + [HCO_3^-] + [CO_3^{2-}]$
   (g) charge balance: $[Na^+] + [H_3O^+] = [OH^-] + [Cl^-] + [NO_2^-]$
   mass balance: $0.1\ M = [Cl^-]$
   mass balance: $0.050\ M = [Na^+] = [HNO_2] + [NO_2^-]$
7. (b) 6.79; (d) 2.90; (f) 8.61
8. (a) 1.53, 4.15, 9.67
9. (a) 6.7×10^{-7} M; (b) 3.5×10^{-9} M; (c) 4.8×10^{-16} M
10. 2.1×10^{-4} M, 1.3×10^{-3} M
11. 710 M (very soluble), 1.2×10^{-4} M
12. 1.2×10^{-4} M
13. $[Ag^+] = 1.3 \times 10^{-5}$ M, $[Cl^-] = 1.3 \times 10^{-5}$ M, $[AgCl(aq)] = 9.0 \times 10^{-7}$ M, $[AgCl_2^-] = 9.7 \times 10^{-10}$ M
14. (a) 0.050 M; (b) 0.075 M; (c) 0.30 M
15. (a) 6.7×10^{-7} M; (b) 7.2×10^{-9} M; (c) 1.7×10^{-15} M
17. pH levels > 12.35
19. (a) 3.52; (b) 9.10; (c) 10.23
20. (a) 3.27; (b) 8.94; (c) 10.22
21. (a) 3.74; (b) 9.24; (c) 10.24

22. pM = 8.18 –log [ML]/[L], 8.18, 8.00
23. 0.761 V, 0.769 V

Chapter 7

1. (a) overall variance = 0.9144, method variance = 0.0330, sampling variance = 0.8814; (b) 96.4%
2. 74, 12, 56, 46, 40, 83, 88, 45, 29, 92
3. For a daily cycle, collect samples at least every 12 h, and samples should be collected at least every six months for a yearly cycle; more frequent sampling would be even better.
9. (a) 5.9%; (b) 8 samples
10. 25.7 g, 6.4 g
13. (a) 15 samples; (b) 3.50 g
14. (a) ±0.176; (b) ±0.518; (c) ±0.216
15. sampling plan (a), sampling plan (c)
16. $t_{exp} = 1.23$ and $t(0.05, 7) = 2.36$, no evidence for a significant difference
18. (a) 106.3%; (b) 38.6%
19. (a) $R_{Co} = 0.991$, $R_{Fe} = 0.019$; (b) 0.0196; (c) 8.72%; (d) –0.73%; (e) 0.57%
20. (a) $R_{Ca} = 0.932$, $R_{Mg} = 0.073$; (b) 54.7%
22. Cyanide, CN^-, becomes HCN at lower pH levels
24. (a) 88.2% extracted; (b) 95.7% extracted; (c) 98.5% extracted; (d) 99.0% extracted
25. (a) 6600 mL; (b) 408 mL; (c) 123.2 mL; (d) 99.4 mL
26. When using 50.0 mL of the organic solvent, D must be 99.0, but D must be 18.0 when using 25.0 mL of the organic solvent.
27. pH < 4.3
28. three extractions
29. pH > 10.1
30. (a) 4.8% of HA extracted, 99.6% of HB extracted; (b) aqueous; (c) $R_{HA} = 95.2\%$, $R_{HB} = 0.4\%$; (d) 4.2×10^{-3}; (e) –2.81%
31. (a) lower concentrations of I^- increase extraction efficiency; (b) $D = K_D/(1 + K_f[I^-]_{aq})$
32. (a) $D = K_D\beta_2[L^-]^2_{aq}/(1 + \beta_2[L^-]^2_{aq})$; (b) 82.1% extracted
34. 99.7% extracted

Chapter 8

2. $$S_{AgCl} = \frac{K_{sp}}{[Ag^+]} + K_1K_{sp} + \frac{2\beta_2K^2_{sp}}{[Ag^+]} + \frac{3\beta_3K^3_{sp}}{[Ag^+]^2}$$
4. (a) pH > 4.27; (b) pH > 6.51; (c) pH > 1.99; (d) pH > 10.33; (e) pH > 13.9
5. Occluded MnO_4^- in experiment 1 imparts color to $KClO_4$
6. RSS decreases from experiment 1 to experiment 3
7. (b) %w/w Al_2O_3 reported too high; (d) better sensitivity
8. (b) %w/w Ca reported too low
9. (a) %w/w Fe reported too low; (b) prevent peptization
11. 1.3-g samples
12. 4.17% w/w As_2O_3
13. 97.65% pure
14. 0.542 g $FeSO \cdot 7H_2O$/tablet
15. 2.44% w/w Fe_2O_3, 39.6% w/w CaO, 1.66% w/w MgO
16. three ethoxy groups per molecule
17. 22.22% w/w K_2SO_4
18. 22.05% w/w Mn, 62.22% w/w Fe
19. 30.72% $NaNO_3$
20. 54.931 g/mol
21. A is $AgNO_3$, B is $Pb(NO_3)_2$, C is KI, D is Na_2SO_4, and E is $BaCl_2$
23. direct method is more sensitive
24. 0.29-g samples
25. 32 mL
26. (a) 1.560 g $PbCrO_4$; (b) 1.005 mol CrO_4^{2-}/mol Pb^{2+}; (c) %w/w Pb reported too high
28. $2KO_3(s) \rightarrow 2KO_2(s) + O_2(g)$, 89.3% w/w KO_3
29. 37.65% w/w H_2O
31. (a) 30.01% w/w Fe; (b) 64.56% w/w C, 5.42% w/w H; (c) $FeC_{10}H_{10}$
32. (a) Polymer A: mean 30.07%, standard deviation 0.0208, Polymer B: mean 29.79%, standard deviation 0.0200; (b) $t_{exp} = 16.81$ and $t(0.05, 4) = 2.78$, difference is significant
33. 20 μmol/m^2
35. 14 mg/m^3, 1.4×10^{-2} mg/L
36. (a) 21.28% w/w fat; (b) $t_{exp} = 5.85$ and $t(0.05, 4) = 2.78$, difference is significant
37.

Average Depth (cm)	% w/w Organic Matter
1	14.0
3	16.8
5	34.0
7	26.9
9	40.8
11	54.8
13	65.1
15	9.04
17	not present

38. (a) macro- or mesosample, trace-level analyte; (b) $\Delta f = 7.97 + 2.18 \times 10^8$[thiourea (M)]; (c) 7.71×10^{-7} M; (d) $\pm 1.22 \times 10^{-7}$

Chapter 9

7. normal, first-derivative and second-derivative titration curves do not show distinct equivalence points, Gran plot shows equivalence point at approximately 5.5 mL; expected equivalence point is 5.07 mL

10. (a) aniline is too weak of a base to titrate in water; (b) report too many moles of aniline; (c) so that no more than one buret-full of titrant is needed

13. titrating three protons, 3 × moles H_4Y = moles NaOH

15. OH^- and HCO_3^- react to produce CO_3^{2-} and H_2O

16. (a) only OH^-, 1740 ppm OH^-; (b) CO_3^{2-} and HCO_3^-, 1630 ppm CO_3^{2-}, 2860 ppm HCO_3^-; (c) HCO_3^-, 4180 ppm HCO_3^-; (d) CO_3^{2-}, 4920 ppm CO_3^{2-}; (e) OH^- and CO_3^{2-}, 1240 ppm CO_3^{2-}, 1390 ppm OH^-

17. (a) 0.0585 M HPO_4^{2-} and 0.0553 M PO_4^{3-}; (b) 0.0474 M H_3PO_4; (c) 0.1091 M NaOH; (d) 0.0838 M H_3PO_4 and 0.0214 M $H_2PO_4^-$

18. 15.59% w/w protein

19. 36.7 ppm SO_2

20. 2480 ppm CO_2

21. 97.47% pure

22. 207

23. 138 g/equivalent

24. taurine

25. $pK_a = 4.83$

26. ultramicro sample; major or minor analyte

29. (a) systematic—molarity reported too low; (b) systematic—molarity reported too high; (c) systematic—molarity reported too low; (d) random; (e) systematic—molarity reported too high; (f) no effect; (g) no effect

35. EGTA

36. 12.8 mg Ca^{2+}/100 mL

37. 98.08% w/w $CaCO_3$

38. 54,300 ppm NaCN

39. 53.90% w/w Cd^{2+}

40. (b) 0.02483 M Fe^{3+}, 0.03536 M Al^{3+}

41. (c) 0.05032 M

45. (a) to remove any Sn^{2+} which will interfere with the titration (b) no; (c) no

46. (a) "50-mL portion of the resulting solution" and "50 mL of a standard solution of Fe^{2+}"; (b) report too much Cr^{3+}; (c) report too much Cr^{3+}

47. (a) report too much H_2O_2; (b) use a reagent blank

48. 98.42% w/w Fe_2O_3

49. 18.69% w/w Mn^{2+}

50. 23.5% w/w U

51. 6.80×10^{-5} cm

52. 280 ppm CO

53. 6.19 ppm O_2

54. (b) negative determinate error; (c) remove AgCl by filtering; (d) no

58. 26.19% w/w kBr

59. 29.86% $BaCl_2$

60. 90.9% pure

Chapter 10

1.

Wavelength (m)	Frequency (s^{-1})	Wavenumber (cm^{-1})	Energy (J)
4.50×10^{-9}	6.67×10^{16}	2.22×10^{6}	4.42×10^{-17}
2.26×10^{-7}	1.33×10^{15}	4.42×10^{4}	8.81×10^{-19}
3.11×10^{-6}	9.65×10^{13}	3.21×10^{3}	6.39×10^{-20}
2.76×10^{-7}	1.09×10^{15}	3.62×10^{4}	7.20×10^{-19}

2.

[Analyte] (M)	Absorbance	%*T*	ε	*b*
1.40×10^{-4}	0.157	69.7	1120	1.00
7.51×10^{-4}	0.563	27.4	750	1.00
2.56×10^{-4}	0.225	59.6	440	2.00
1.55×10^{-3}	0.167	68.1	21.5	5.00
8.46×10^{-4}	0.478	33.3	565	1.00
4.35×10^{-3}	0.674	21.2	1550	0.100
1.20×10^{-4}	0.0899	81.3	74.9	10.00

3. 59.2% *T*

4. 19.7% *T*

5. 3134 cm^{-1} M^{-1}

6.

Concentration (M)	Absorbance (a)	Absorbance (b)	Absorbance (c)
1.0×10^{-5}	0.020	0.009	0.014
3.0×10^{-5}	0.060	0.035	0.043
5.0×10^{-5}	0.100	0.065	0.071
9.0×10^{-5}	0.180	0.130	0.128
11×10^{-5}	0.220	0.163	0.156
13×10^{-5}	0.260	0.197	0.185

7. (b)

Concentration (M)	Absorbance for $\varepsilon' = \varepsilon'' = 1000$	Absorbance for $\varepsilon' = 1900$, $\varepsilon'' = 100$
0	0	0
2×10^{-5}	0.020	0.020
4×10^{-5}	0.040	0.039
6×10^{-5}	0.060	0.057
8×10^{-5}	0.080	0.074
1×10^{-4}	0.100	0.091

8.

[Analyte] (M)	Absorbance
0.000	0.00
0.002	0.37
0.004	0.70
0.006	0.97
0.008	1.15
0.010	1.24

9. yes
10. (a) 5 mL thioglycolic acid, 2 mL 20% w/v ammonium citrate and 5 mL 0.22 M NH_3 diluted to 50 mL; (b) no effect if blank is used; (c) to keep absorbance within range of calibration curve; (d) reported results for Fe^{3+} are too low
11. dilute to 100 mL
12. (e) 142 ppm P
13. (a) 3.55×10^{-3} M Cu^{2+}; (b) 6.24×10^{-3} M Co^{2+}, 4.62×10^{-3} M Cu^{2+}; (c) 3.69×10^{-3} M Cu^{2+}, 1.73×10^{-2} M Ni^{2+}, 9.14×10^{-3} M Co^{2+}
14. 7.44 ppm phenol
15. 0.648 ppm Fe^{2+}
16. 0.209 ppm Cl_2
17. 10.6% v/v methanol
18. 0.90 mg barbital/100 mL
19. 220 mg aspirin, 160 mg phenacetin, 32 mg caffeine
20. 0.28 ppm SO_2
21. mean = 18.4% CO; ± 0.83% CO
22. mean = 64.03% w/w polystyrene; $t_{exp} = 3.73$ and $t(0.05, 2) = 4.30$, no evidence for a determinate error
23. 613 nm and 658 nm
24. (a) $[MnO_4^-] = 8.2 \times 10^{-5}$ M, $[Cr_2O_7^{2-}] = 1.8 \times 10^{-4}$ M; (b) $C_V = 58.5$ ppm, $C_{Ti} = 31.5$ ppm; (c) 0.26 ppm Zn^{2+}, 0.51 ppm Cu^{2+}
25. 1.0×10^{-5} moles Cu^{2+}, 1.30×10^{-5} moles Ni^{2+}
26. ML_2
27. ML_2
28. ML_3
29. 1:1 stoichiometry
30. $pK_a = 4.31$
31. $pK_a = 4.72$
32. (a) 4%; (b) 0.9%
33. 95.5% pure
34. (a) 0.865 ppm Cu; (b) 0.0828 ppm Cr
35. 40.6 ppb Mn^{2+}
36. 65.5 ppm Na
37. 0.0110% w/w Pb, 7.61×10^{-3}% w/w Ni
38. 353 ppm Fe^{3+}
39. 3.86×10^{-5} M
40. 4.82×10^{-5} M benzo[a]pyrene
41. 64.9% w/w acetylsalicylic acid
42. 2.45 nM Se(IV)
43. 1.38 g fibrinogen/L

Chapter 11

1. (a) anode: Pt, $Fe^{2+} \rightleftharpoons Fe^{3+} + e^-$, cathode: Ag, $Ag^+ + e^- \rightleftharpoons Ag(s)$
   (b) anode: Ag, $Ag(s) + Br^- \rightleftharpoons AgBr(s) + e^-$, cathode: Cd, $Cd^{2+} + 2e^- \rightleftharpoons Cd(s)$
   (c) anode: Pb, $Pb(s) + SO_4^{2-} \rightleftharpoons PbSO_4(s) + 2e^-$ cathode: PbO_2, $PbO_2(s) + SO_4^{2-} + 2e^- + 4H^+ \rightleftharpoons PbSO_4(s) + 2H_2O$
2. (a) –0.059 V; (b) –0.512 V; (c) 2.10 V
3. 0.211 M I^-
4. $Zn(s) + 2H^+(aq) \rightleftharpoons Zn^{2+}(aq) + H_2(g)$
5. 7.11×10^{-3}; 1.3×10^{-5} M benzoate
6. the membrane responds only to the protonated form of cocaine
9. (a) 2×10^{-3} M to 1×10^{-5} M; (b) E (mV) = 331.4 + 47.76 log [penicillin]; (c) 1.1×10^{-4} M
10. 4.1 mM K^+
11. (a) pH = 3.98; (b) pH = 6.9; (c) pH = 4.67
12. 11.03 ppm NO_3^-
13. (a) 1.20 ppm F^-; (b) 0.0593% w/w F^-
15. 98.0% pure
16. 115 ppm H_2S
18. $n = 1$
19. (b) 0.789 V
20. 82.5 μg S/mL
21. 95.0% pure
22. 30.2 ng Sb
23. 0.863% w/w Tl
24. 29.2 mg caffeine/tablet, 38.8 mg ascorbic acid/tablet
25. 49.8 ppm Sn^{4+}
26. 14.0 mg glucose/100 mL
27. 2.0 μg Tl^+/mL, 2.0 μg Pb^{2+}/mL, 0.89 μg In^{3+}/mL
29.

Group	ppb Cd	ppb Pb	ppb Cu
I	0.03	0.06	0.09
II	0.21	0.33	0.17
III	0	0	0
IV	–0.01	0	0.10
V	0	+0.01	–0.015
VI	0.03	–0.02	–0.03
VII	0.02	0.12	0.10

31. $n = 2$, $E_{1/2} = -0.390$ V, $Pb(OH)_3^-$, $\beta_3 = 3.71 \times 10^{12}$

32. 3.37 ppb Cu^{2+}

33. 0.47 μM

34. 3.5×10^{-8} M

Chapter 12

1. (a) $N_A = 46{,}000$, $N_B = 48{,}500$, $N_c = 44{,}400$, average = 46,300; (b) 0.43 mm

2. using equation 12.1: $R_{AB} = 1.5$, $R_{BC} = 1.1$; using equation 12.21: $R_{AB} = 1.5$, $R_{BC} = 1.1$

3. $t_r = 350$ s, $w = 22$ s, $t_r' = 325$ s, $k' = 13$, N = 4050, H = 0.49 mm

4. $R = 1.0$

5. 4.5 m long, 0.167 mm

6.

N_B	α	k_B'	R
100,000	1.05	0.5	1.25
10,000	1.10	1.9	1.50
10,000	1.05	4.0	1.00
38,400	1.05	3.0	1.75

7. (b) *B* term 5–16 mL/min, *A* term 16–71 mL/min, *C* term 71–120 mL/min; (c) 33 mL/min, H = 3.20 mm; (d) 33 mL/min, H = 1.56 mm; (e) 2.1 times as many

8. (a) row 1: $v = 3.6$, $u = 4.1 \times 10^{-2}$ cm/s, $h = 1.36$, H = 7.4×10^{-6} m; row 2: $v = 3.3$, $u = 3.8 \times 10^{-2}$ cm/s, $h = 0.97$, H = 5.3×10^{-6} m

9. $CHCl_3$, $CHCl_2Br$, $CHClBr_2$, $CHBr_3$

11. 1.50 ppb $CHCl_3$, 0.754 ppb $CHCl_2Br$, 3.49 ppb $CHClBr_2$, 1.39 ppb $CHBr_3$

12. (a) 0.0682% w/w H_2O; (b) 2.96% w/w H_2O

13. 1.7 mg H_2O/g soil

14. 1.9% v/v methyl salicylate

15. 45.8% w/w camphor

16. 7.98 ng heptachlor epoxide/g orange rind

17. $I_{octane} = 800$, $I_{nonane} = 900$, $I_{toluene} = 842$

18. 765

23. 67 mg vitamin C, 19 mg niacinamide, 2.4 mg pyridoxine, 1.4 mg thiamin, 0.46 mg folic acid, 1.5 mg riboflavin, niacin is absent

24. 31.9 mg

25. (a) 228 mg acetylsalicylic acid, 31.5 mg caffeine

26. 0.21 mg vitamin A/100 g

27. (b) 0.83 M HCO_3^-, 0.25 mM Cl^-, 0.0030 mM NO_2^-, 0.12 mM NO_3^-, 0.32 mM Ca^{2+}, 0.11 mM Mg^{2+}, 0.19 mM SO_4^{2-}; (c) 0.77 M HCO_3^-; (d) 1.00

28. 74.5 ppm Cl^-, 3.39 ppm NO_3^-, 41.4 ppm SO_4^{2-}

29. 56,000

30. order is caffeine, aspartame, benzoic acid

32. 436 ppm NO_3^-

34. (a) 3.22×10^{-4} cm^2 V^{-1} s^{-1}; (b) approximately 290,000 theoretical plates; (c) $R = 1.1$; (f) $\mu_{ep} = 0$.

Chapter 13

1. $[A]_0 = \dfrac{[P]_{t2} - [P]_{t1}}{e^{-k't1} - e^{-k't2}}$

2. 0.74 mM phenylacetate

3. 286 μM H_2O_2

4. 6.2×10^{-4} M chromic acid

5. 39.4 ppm glucose, –1.5% error

6. approximately 5 times more dilute

7. $k = 0.0486$ s^{-1}, $[A]_0 = 1.50$ mM

8. 1.2×10^{-7} M acetylcholine

9. 0.0863 μM fumarase

10. $V_{max} = 0.406$ M s^{-1}, $K_m = 6.49 \times 10^{-3}$ M, $k_2 = 8.1 \times 10^4$ s^{-1}

11. competitive inhibition

13. $k_A/k_B = 6900$

14. $[A]_0 = 0.220$ mM, $[B]_0 = 0.125$ mM, $k_A = 0.135$ min^{-1}, $k_B = 0.0332$ min^{-1}

15.

Isotope	Rate Constant
^3H	5.54×10^{-2}/year
^{14}C	1.21×10^{-4}/year
^{32}P	4.85×10^{-2}/day
^{35}S	7.96×10^{-3}/day
^{45}Ca	4.56×10^{-3}/day
^{55}Fe	2.38×10^{-1}/year
^{60}Co	1.31×10^{-1}/year
^{131}I	8.66×10^{-2}/day

16. 1.7×10^{-6} M ^{60}Co

17. 3.42% w/w Ni

18. 5.79 mg vitamin B_{12} per tablet

19. 2.7%

20. 7.0×10^8 years

21. 11 g C, 1300 min

22. $h = 0.762$ abs units, $t_a = 14.1$ s, $T = 15.8$ s, $t' = 1.7$ s, $\Delta t = 15.2$ s, $T' = 13.5$ s, sensitivity = 7.62×10^{-3} ppm^{-1}, about 260–270 samples/h

24. 13,700 ppm Cl^-

25. 0.071 M HCl

26. 11.2 mM glucose

27. (a) 2.52%; (b) 68 samples/h; (c) 94.8% w/w cocaine

28. 6.27×10^{-4} M H_2SO_4

Chapter 14

3. $R = 4.00 + 1.20B^* + 0.72A^*B^*$, $R = 4.00 - 0.40A + 0.08AB$
4. (a) $R = 52.610 + 23.755\text{Ca}^* - 15.875\text{Al}^* - 6.200\text{Ca}^*\text{Al}^*$; (b) 16.9 ppm Al
5. (a) $R = 38.1 - 3.6X^* + 0.1Y^* + 8.1Z^* + 0.4X^*Y^* + 0.9X^*Z^* - 7.4Y^*Z^* - 0.6X^*Y^*Z^*$; (b) $R = 38.1 - 3.6X^* + 8.1Z^* - 7.4Y^*Z^*$; (d) catalyst A for high concentrations, catalyst B for low concentrations; (e) 49.2%
6. (a) $R = 4.18 + 1.54X^* + 0.92Y^* - 0.12Z^* + 0.29X^*Y^* - 0.24X^*Z^* + 0.10Y^*Z^* + 0.44X^*Y^*Z^*$; (b) $R = 4.18 + 1.54X^* + 0.92Y^*$
7. (a) $R = 8.5 + 2.5X^* - 0.5Y^* + 3.5Z^* - 0.5X^*Y^* + 0.5X^*Z^* - 1.5Y^*Z^* - 0.5X^*Y^*Z^*$; (b) $R = 8.5 + 2.5X^* + 3.5Z^* - 1.5Y^*Z^*$; (c) no evidence for curvature
8. (a) $R_{\Delta E} = 33.54 - 2.56X^* - 1.92Y^* + 1.42Z^* + 0.62X^*Y^* + 0.10X^*Z^* + 0.54Y^*Z^* + 0.41X^*Y^*Z^*$; $R_{samp} = 26.9 + 1.2X^* + 4.9Y^* - 0.8Z^* + 0.5X^*Y^* - 0.6X^*Z^* + Y^*Z^* + 0.9X^*Y^*Z^*$; (b) no
9. (a) $R = 0.9 + 0.05^* - 0.071B^* + 0.039C^* + 0.074D^* - 0.15E^* - 0.12C^*E^* + 0.049A^*C^*E^*$; (b) yes
11. yes
12. pH and digestion time, 0.72
13. (a) 2; (b) 8; (c) 8; (d) 1 and 10
14. $\sigma_{rand} \approx 0.055$, $\sigma_{syst} \approx 0.078$
15. (a) variability between labs is significant; (b) $\sigma_{rand} \approx 0.80$, $\sigma_{syst} \approx 1.3$
16. (a) 4.9%; (b) yes

Chapter 15

3. $(d)_r = 7.2\%$, yes
4. $s = 0.066$; relative standard deviation = 1.3%, precision acceptable
5. average recovery is 99%
6. (a) yes; (b) lab

Glossary

A

absorbance the attenuation of photons as they pass through a sample (A). *(p. 373)*

absorbance spectrum a graph of a sample's absorbance of electromagnetic radiation versus wavelength (or frequency or wavenumber). *(p. 373)*

accuracy a measure of the agreement between an experimental result and its expected value. *(p. 38)*

acid a proton donor. *(p. 140)*

acid–base titration a titration in which the reaction between the analyte and titrant is an acid–base reaction. *(p. 278)*

acid dissociation constant the equilibrium constant for a reaction in which an acid donates a proton to the solvent (K_a). *(p. 140)*

acidity a measure of a water's ability to neutralize base. *(p. 301)*

activity true thermodynamic constants use a species activity in place of its molar concentration (a). *(p. 172)*

activity coefficient the number that when multiplied by a species' concentration gives that species' activity (γ). *(p. 172)*

adjusted retention time the difference between a solute's retention time and column's void time (t_r'). *(p. 551)*

adsorbate a coprecipitated impurity that adsorbs to the surface of a precipitate. *(p. 239)*

aliquot a portion of a solution. *(p. 111)*

alkalinity a measure of a water's ability to neutralize acid. *(p. 300)*

alpha particle a positively charged subatomic particle equivalent to a helium nucleus (α). *(p. 642)*

alternative hypothesis a statement that the difference between two values is too great to be explained by indeterminate error; accepted if the significance test shows that null hypothesis should be rejected (H_A). *(p. 83)*

amalgam a metallic solution of mercury with another metal. *(p. 509)*

amperometry a form of voltammetry in which we measure current as a function of time while maintaining a constant potential. *(p. 519)*

amphiprotic a species capable of acting as both an acid and a base. *(p. 142)*

analysis a process that provides chemical or physical information about the constituents in the sample or the sample itself. *(p. 36)*

analysis of variance a statistical method for comparing three or more sets of data. *(p. 693)*

analytes the constituents of interest in a sample. *(p. 36)*

anode the electrode where oxidation occurs. *(p. 466)*

anodic current a faradaic current due to an oxidation reaction. *(p. 510)*

argentometric titration a precipitation titration in which Ag^+ is the titrant. *(p. 355)*

asymmetry potential the membrane potential when opposite sides of the membrane are in contact with identical solutions yet a nonzero potential is observed. *(p. 476)*

atomization the process of converting an analyte into a free atom. *(p. 412)*

auxiliary complexing agent a second ligand in a complexation titration that initially binds with the analyte but is displaced by the titrant. *(p. 316)*

auxiliary electrode the third electrode in a three-electrode cell that completes the circuit. *(p. 463)*

auxiliary oxidizing agent a reagent used to oxidize the analyte before its analysis by a redox titration. *(p. 341)*

auxiliary reducing agent a reagent used to reduce the analyte before its analysis by a redox titration. *(p. 341)*

B

background correction in atomic absorption spectroscopy, the correction of the net absorbance from that due to the sample matrix. *(p. 419)*

back titration a titration in which a reagent is added to a solution containing the analyte, and the excess reagent remaining after its reaction with the analyte is determined by a titration. *(p. 275)*

balance an apparatus used to measure mass. *(p. 25)*

band broadening the increase in a solute's baseline width as it moves from the point of injection to the detector. (*p. 553*)

base a proton acceptor. (*p. 140*)

base dissociation constant the equilibrium constant for a reaction in which a base accepts a proton from the solvent (K_b). (*p. 141*)

baseline width the width of a solute's chromatographic band measured at the baseline (w). (*p. 548*)

Beer's law the relationship between a sample's absorbance and the concentration of the absorbing species ($A = \varepsilon bC$). (*p. 386*)

beta particle a charged subatomic particle produced when a neutron converts to a proton, or a proton converts to a neutron (β). (*p. 642*)

binomial distribution probability distribution showing chance of obtaining one of two specific outcomes in a fixed number of trials. (*p. 72*)

bleed the tendency of a stationary phase to elute from the column. (*p. 566*)

blind analysis the analysis of a standard sample whose composition is unknown to the analyst. (*p. 683*)

bonded stationary phase a liquid stationary phase that is chemically bonded to a particulate packing material. (*p. 580*)

breakthrough volume the volume of sample that can be passed through a solid sorbent before the analytes are no longer retained. (*p. 196*)

buffer a solution containing a conjugate weak acid/weak base pair that is resistant to a change in pH when a strong acid or strong base is added. (*p. 167*)

buret volumetric glassware used to deliver variable, but known volumes of solution. (*p. 277*)

C

calibration the process of ensuring that the signal measured by a piece of equipment or an instrument is correct. (*p. 47*)

calibration curve the result of a standardization showing graphically how a method's signal changes with respect to the amount of analyte. (*p. 47*)

capacity factor a measure of how strongly a solute is retained by the stationary phase (k'). (*p. 551*)

capillary column a narrow bored column that usually does not contain a particulate packing material. (*p. 562*)

capillary electrochromatography a form of capillary electrophoresis in which a stationary phase is included within the capillary column. (*p. 607*)

capillary electrophoresis electrophoresis taking place in a capillary tube. (*p. 597*)

capillary gel electrophoresis a form of capillary electrophoresis in which the capillary column contains a gel enabling separations based on size. (*p. 606*)

capillary zone electrophoresis a form of capillary electrophoresis in which separations are based on differences in the solutes' electrophoretic mobilities. (*p. 604*)

cathode the electrode where reduction occurs. (*p. 466*)

cathodic current a faradaic current due to a reduction reaction. (*p. 510*)

central limit theorem the distribution of measurements subject to indeterminate errors is often a normal distribution. (*p. 79*)

characteristic concentration the concentration of analyte giving an absorbance of 0.00436. (*p. 416*)

characterization analysis an analysis in which we evaluate a sample's chemical or physical properties. (*p. 9*)

charge balance equation an equation stating that the total concentration of positive charge in a solution must equal the total concentration of negative charge. (*p. 159*)

charging current a current in an electrochemical cell due to the electrical double layer's formation. (*p. 513*)

chemiluminescence emission induced by a chemical reaction. (*p. 374*)

chromatogram a plot of the detector's signal as function of elution time or volume. (*p. 548*)

chromatography a separation in which solutes partition between a mobile and stationary phase. (*p. 546*)

chromophore the specific bonds or functional groups in a molecule responsible for the absorption of a particular wavelength of light. (*p. 382*)

coagulation the process of smaller particles of precipitate clumping together to form larger particles. (*p. 242*)

collaborative testing a validation method used to evaluate the sources of random and systematic errors affecting an analytical method. (*p. 687*)

column chromatography a form of chromatography in which the stationary phase is retained in a column. (*p. 546*)

common ion effect the solubility of an insoluble salt decreases when it is placed in a solution already containing one of the salt's ions. (*p. 158*)

complexation titration a titration in which the reaction between the analyte and titrant is a complexation reaction. (*p. 314*)

composite sample several grab samples combined to form a single sample. (*p. 186*)

concentration an expression stating the relative amount of solute per unit volume or unit mass of solution. (*p. 15*)

concentration techniques a technique in which the signal is proportional to the analyte's concentration; also called "instrumental" techniques. (*p. 38*)

conditional formation constant the equilibrium formation constant for a metal–ligand complex for a specific set of solution conditions, such as pH. (*p. 316*)

confidence interval range of results around a mean value that could be explained by random error. (*p. 75*)

coning and quartering a process for reducing the size of a gross sample. (*p. 199*)

constant determinate error a determinate error whose value is the same for all samples. (*p. 60*)

continuum source a source that emits radiation over a wide range of wavelengths. (*p. 375*)

control chart a graph showing the time-dependent change in the results of an analysis that is used to monitor whether an analysis is in a state of statistical control. *(p. 714)*

convection the movement of material in response to a mechanical force, such as stirring a solution. *(p. 512)*

convenience sampling a sampling plan in which samples are collected because they are easily obtained. *(p. 185)*

coulometric titrations a titration in which the equivalence point is the time required for a constant current to completely oxidize or reduce the analyte. *(p. 501)*

coulometry an electrochemical method in which the current required to exhaustively oxidize or reduce the analyte is measured. *(p. 496)*

countercurrent extraction a liquid–liquid extraction in which solutes are extracted back and forth between fresh portions of two extracting phases. *(p. 546)*

counter electrode the second electrode in a two-electrode cell that completes the circuit. *(p. 462)*

cryogenic focusing the process of concentrating volatile solutes by cooling the column's inlet below room temperature. *(p. 568)*

cumulative formation constant the formation constant for a metal–ligand complex in which two or more ligands are simultaneously added to a metal ion or to a metal–ligand complex (β_i). *(p. 144)*

current efficiency the percentage of current that actually leads to the analyte's oxidation or reduction. *(p. 496)*

D

dark current the background current present in a photon detector in the absence of radiation from the source. *(p. 379)*

degrees of freedom the number of independent values on which a result is based (ν). *(p. 80)*

desiccant a drying agent. *(p. 29)*

desiccator a closed container containing a desiccant; used to store samples in a moisture-free environment. *(p. 29)*

detection limit a statistical statement about the smallest amount of analyte that can be determined with confidence. *(p. 39)*

determinate error any systematic error that causes a measurement or result to always be too high or too small; can be traced to an identifiable source. *(p. 58)*

determination an analysis of a sample to find the identity, concentration, or properties of the analyte. *(p. 36)*

dialysis a method of separation that uses a semi-permeable membrane. *(p. 206)*

diffusion the movement of material in response to a concentration gradient. *(p. 511)*

diffusion layer the layer of solution adjacent to the electrode in which diffusion is the only means of mass transport. *(p. 511)*

digestion the process by which a precipitate is given time to form larger, purer particles. *(p. 239)*

dilution the process of preparing a less concentrated solution from a more concentrated solution. *(p. 31)*

displacement titration a titration in which the analyte displaces a species, usually from a complex, and the amount of the displaced species is determined by a titration. *(p. 275)*

dissociation constant the equilibrium constant for a reaction in which a metal–ligand complex dissociates to form uncomplexed metal ion and ligand (K_d). *(p. 144)*

distribution ratio a ratio expressing the total concentration of solute in one phase relative to a second phase; all forms of the solute are considered in defining the distribution ratio (D). *(p. 216)*

Dixon's *Q*-test statistical test for deciding if an outlier can be removed from a set of data. *(p. 93)*

dropping mercury electrode an electrode in which successive drops of Hg form at the end of a capillary tube as a result of gravity, with each drop providing a fresh electrode surface. *(p. 509)*

duplicate samples two samples taken from a single gross sample and used to evaluate an analytical method's precision. *(p. 708)*

E

effective bandwidth the width of the band of radiation passing through a wavelength selector measured at half the band's height. *(p. 376)*

electrical double layer the interface between a positively or negatively charged electrode and the negatively or positively charged layer of solution in contact with the electrode. *(p. 513)*

electrode of the first kind a metallic electrode whose potential is a function of the concentration of M^{n+} in an M^{n+}/M redox half-reaction. *(p. 474)*

electrode of the second kind a metallic electrode whose potential is a function of the concentration of X in an MX_n/M redox half-reaction. *(p. 475)*

electrogravimetry a gravimetric method in which the signal is the mass of an electrodeposit on the cathode or anode in an electrochemical cell. *(p. 234)*

electrokinetic injection an injection technique in capillary electrophoresis in which an electric field is used to inject sample into the capillary column. *(p. 603)*

electromagnetic spectrum the division of electromagnetic radiation on the basis of a photon's energy. *(p. 372)*

electron capture detector a detector for GC that provides selectivity for solutes with halogen and nitro functional groups. *(p. 570)*

electroosmotic flow the movement of the conductive medium in response to an applied electric field. *(p. 598)*

electroosmotic flow velocity the velocity with which the solute moves through the capillary due to the electroosmotic flow (ν_{eof}). *(p. 599)*

electropherogram the equivalent of a chromatogram in electrophoresis. *(p. 597)*

electrophoresis a separation technique based on a solute's ability to move through a conductive medium under the influence of an electric field. *(p. 597)*

electrophoretic mobility a measure of a solute's ability to move through a conductive medium in response to an applied electric field (μ_{ep}). *(p. 598)*

electrophoretic velocity the velocity with which a solute moves through the conductive medium due to its electrophoretic mobility (ν_{ep}). *(p. 598)*

emission the release of a photon when an analyte returns to a lower-energy state from a higher-energy state. *(p. 373)*

emission spectrum a graph of emission intensity versus wavelength (or frequency or wavenumber). *(p. 374)*

empirical model a model describing a system's response that is not derived from theoretical principles. *(p. 676)*

end point the point in a titration where we stop adding titrant. *(p. 274)*

enthalpy a change in enthalpy indicates the heat absorbed or released during a chemical reaction at constant pressure. *(p. 137)*

entropy a measure of disorder. *(p. 137)*

enzyme a protein that catalyzes biochemical reactions. *(p. 636)*

enzyme electrodes an electrode that responds to the concentration of a substrate by reacting the substrate with an immobilized enzyme, producing an ion that can be monitored with an ion-selective electrode. *(p. 484)*

equilibrium a system is at equilibrium when the concentrations of reactants and products remain constant. *(p. 136)*

equilibrium constant for a reaction at equilibrium, the equilibrium constant determines the relative concentrations of products and reactants. *(p. 138)*

equivalence point the point in a titration where stoichiometrically equivalent amounts of analyte and titrant react. *(p. 274)*

equivalent the moles of a species that can donate one reaction unit. *(p. 17)*

equivalent weight the mass of a compound containing one equivalent (EW). *(p. 17)*

error a measure of bias in a result or measurement. *(p. 64)*

excitation spectrum a fluorescence or phosphorescence spectrum in which the emission intensity at a fixed wavelength is measured as a function of the wavelength used for excitation. *(p. 427)*

exclusion limit in size-exclusion chromatography, the largest solute that can be separated from other solutes; all larger solutes elute together. *(p. 595)*

external conversion a form of radiationless relaxation in which energy is transferred to the solvent or sample matrix. *(p. 425)*

external standard a standard solution containing a known amount of analyte, prepared separately from samples containing the analyte. *(p. 109)*

extraction the process by which a solute is transferred from one phase to a new phase. *(p. 212)*

F

factor a property of a system that is experimentally varied and that may affect the response. *(p. 667)*

factor level a factor's value. *(p. 667)*

faradaic current any current in an electrochemical cell due to an oxidation or reduction reaction. *(p. 510)*

Faraday's law the current or charge passed in a redox reaction is proportional to the moles of the reaction's reactants and products. *(p. 496)*

fiagram a graph showing the detector's response as a function of time in a flow injection analysis. *(p. 650)*

field blank a blank sample collected in the field. *(p. 710)*

filter a wavelength selector that uses either absorption, or constructive and destructive interference to control the range of selected wavelengths. *(p. 376)*

filter photometer a simple instrument for measuring absorbance that uses absorption or interference filters to select the wavelength. *(p. 388)*

Fisher's least significant difference a modified form of the *t*-test for comparing several sets of data. *(p. 696)*

flame ionization detector a nearly universal GC detector in which the solutes are combusted in an H_2/air flame, producing a measurable current. *(p. 570)*

flow injection analysis an analytical technique in which samples are injected into a carrier stream of reagents, or in which the sample merges with other streams carrying reagents before passing through a detector. *(p. 649)*

fluorescence emission of a photon when the analyte returns to a lower-energy state with the same spin as the higher-energy state. *(p. 423)*

fluorometer an instrument for measuring fluorescence that uses filters to select the excitation and emission wavelengths. *(p. 428)*

formality the number of moles of solute, regardless of chemical form, per liter of solution (F). *(p. 15)*

formal potential the potential of a redox reaction for a specific set of solution conditions, such as pH and ionic composition. *(p. 332)*

formation constant the equilibrium constant for a reaction in which a metal and a ligand bind to form a metal–ligand complex (K_f). *(p. 144)*

formula weight the mass of a compound containing one mole (FW). *(p. 17)*

frequency the number of oscillations of an electromagnetic wave per second (ν). *(p. 370)*

fronting a tail at the beginning of a chromatographic peak, usually due to injecting too much sample. *(p. 555)*

***F*-test** statistical test for comparing two variances to see if their difference is too large to be explained by indeterminate error. *(p. 87)*

fundamental analysis an analysis whose purpose is to improve an analytical method's capabilities. *(p. 9)*

G

galvanostat a device used to control the current in an electrochemical cell. *(p. 464)*

gamma ray high-energy radiation emitted by nuclei (γ). *(p. 642)*

gas chromatography a chromatographic technique in which the mobile phase is a gas. *(p. 563)*

gas–liquid chromatography a chromatographic technique in which the mobile phase is a gas and the stationary phase is a liquid coated either on a solid packing material or on the column's walls. *(p. 564)*

Geiger counter an instrument for counting radioactive particles based on their ability to ionize an inert gas such as Ar. *(p. 643)*

Gibb's free energy a thermodynamic function for systems at constant temperature and pressure that indicates whether or not a reaction is favorable ($\Delta G < 0$), unfavorable ($\Delta G > 0$), or at equilibrium ($\Delta G = 0$). *(p. 137)*

glass electrode an ion-selective electrode based on a glass membrane in which the potential develops from an ion-exchange reaction on the membrane's surface. *(p. 477)*

good laboratory practices those general laboratory procedures that, when followed, help ensure the quality of analytical work. *(p. 706)*

good measurement practices those instructions outlining how to properly use equipment and instrumentation to ensure the quality of measurements. *(p. 706)*

grab sample a single sample removed from the target population. *(p. 185)*

gradient elution the process of changing the mobile phase's solvent strength to enhance the separation of both early and late eluting solutes. *(p. 558)*

Gran plot a linearized form of a titration curve. *(p. 293)*

graphite furnace an electrothermal atomizer that relies on resistive heating to atomize samples. *(p. 414)*

gravimetry any method in which the signal is a mass or change in mass. *(p. 233)*

gross sample the initial sample, collected from the target population without any processing. *(p. 193)*

guard column an inexpensive column used to protect a more expensive analytical column. *(p. 579)*

H

half-life the time required for half of the initial number of a radioactive isotope's atoms to disintegrate ($t_{1/2}$). *(p. 643)*

hanging mercury drop electrode an electrode in which a drop of Hg is suspended from a capillary tube. *(p. 509)*

headspace sampling the sampling of the vapor phase overlying a liquid phase. *(p. 567)*

Henderson–Hasselbalch equation equation showing the relationship between a buffer's pH and the relative amounts of the buffer's conjugate weak acid and weak base. *(p. 169)*

heterogeneous not uniform in composition. *(p. 58)*

high-performance liquid chromatography a chromatographic technique in which the mobile phase is a liquid. *(p. 578)*

histogram a plot showing the number of times an observation occurs as a function of the range of observed values. *(p. 77)*

homogeneous uniform in composition. *(p. 72)*

homogeneous precipitation a precipitation in which the precipitant is generated in situ by a chemical reaction. *(p. 241)*

hydrodynamic injection an injection technique in capillary electrophoresis in which pressure is used to inject sample into the capillary column. *(p. 602)*

hydrodynamic voltammetry a form of voltammetry in which the solution is mechanically stirred. *(p. 513)*

I

inclusion a coprecipitated impurity in which the interfering ion occupies a lattice site in the precipitate. *(p. 238)*

inclusion limit in size-exclusion chromatography, the smallest solute that can be separated from other solutes; all smaller solutes elute together. *(p. 593)*

indeterminate error any random error that causes some measurements or results to be too high while others are too low. *(p. 62)*

indicator a colored compound whose change in color signals the end point of a titration. *(p. 274)*

indicator electrode the electrode whose potential is a function of the analyte's concentration (also known as the working electrode). *(p. 462)*

in situ sampling sampling done within the population without physically removing the sample. *(p. 186)*

intensity the flux of energy per unit time per area (I). *(p. 371)*

interferometer a device that allows all wavelengths of light to be measured simultaneously, eliminating the need for a wavelength selector. *(p. 378)*

internal conversion a form of radiationless relaxation in which the analyte moves from a higher electronic energy level to a lower electronic energy level. *(p. 425)*

internal standard a standard, whose identity is different from the analyte's, that is added to all samples and standards containing the analyte. *(p. 116)*

intersystem crossing a form of radiationless relaxation in which the analyte moves from a higher electronic energy level to a lower electronic energy level with a different spin state. *(p. 425)*

ion-exchange chromatography a form of liquid chromatography in which the stationary phase is an ion-exchange resin. (*p. 590*)

ionic strength a quantitative method for reporting the ionic composition of a solution that takes into account the greater effect of more highly charged ions (μ). (*p. 172*)

ionization suppressor a reagent that is more easily ionizable than the analyte. (*p. 420*)

ionophore a neutral ligand whose exterior is hydrophobic and whose interior is hydrophilic. (*p. 482*)

ion-selective electrode an electrode in which the membrane potential is a function of the concentration of a particular ion in solution. (*p. 475*)

ion-suppressor column a column used to minimize the conductivity of the mobile phase in ion-exchange chromatography. (*p. 592*)

isocratic elution the use of a mobile phase whose composition remains constant throughout the separation. (*p. 582*)

isotope dilution a form of internal standardization in which a radioactive form of the analyte serves as the internal standard. (*p. 646*)

isotopes atoms with the same number of protons but different numbers of neutrons are called isotopes. (*p. 642*)

J

Jones reductor a reduction column using a Zn amalgam as a reducing agent. (*p. 341*)

Joule heating the heating of a conductive solution due to the passage of an electric current through the solution. (*p. 601*)

judgmental sampling samples collected from the target population using available information about the analyte's distribution within the population. (*p. 184*)

K

Kjeldahl analysis an acid–base titrimetric method for determining the amount of nitrogen in organic compounds. (*p. 302*)

Kovat's retention index a means for normalizing retention times by comparing a solute's retention time with those for normal alkanes. (*p. 575*)

L

laboratory sample sample taken into the lab for analysis after processing the gross sample. (*p. 199*)

ladder diagram a visual tool for evaluating systems at equilibrium. (*p. 150*)

Le Châtelier's principle when stressed, a system that was at equilibrium returns to its equilibrium state by reacting in a manner that relieves the stress. (*p. 148*)

leveling acids that are better proton donors than the solvent are leveled to the acid strength of the protonated solvent; bases that are better proton acceptors than the solvent are leveled to the base strength of the deprotonated solvent. (*p. 296*)

lifetime the length of time that an analyte stays in an excited state before returning to a lower-energy state. (*p. 423*)

ligand a Lewis base that binds with a metal ion. (*p. 144*)

limit of identification the smallest concentration or absolute amount of analyte such that the probability of type 1 and type 2 errors are equal (LOI). (*p. 95*)

limit of quantitation the smallest concentration or absolute amount of analyte that can be reliably determined (LOQ). (*p. 96*)

linear regression a mathematical technique for fitting an equation, such as that for a straight line, to experimental data. (*p. 118*)

line source a source that emits radiation at only select wavelengths. (*p. 375*)

Lineweaver–Burk plot a graphical means for evaluating enzyme kinetics. (*p. 638*)

liquid-based ion-selective electrode an ion-selective electrode in which a chelating agent is incorporated into a hydrophobic membrane. (*p. 482*)

liquid junction potential a potential that develops at the interface between two ionic solutions that differ in composition, because of a difference in the mobilities of the ions (E_{lj}). (*p. 470*)

liquid–solid adsorption chromatography a form of liquid chromatography in which the stationary phase is a solid adsorbent. (*p. 590*)

longitudinal diffusion one contribution to band broadening in which solutes diffuse from areas of high concentration to areas of low concentration. (*p. 560*)

loop injector a means for injecting samples in which the sample is loaded into a short section of tubing and injected onto the column by redirecting the mobile phase through the loop. (*p. 584*)

M

manifold the complete system of tubing for merging together samples and reagents in a flow injection analysis. (*p. 652*)

masking a pseudo-separation method in which a species is prevented from participating in a chemical reaction by binding it with a masking agent in an unreactive complex. (*p. 207*)

masking agent the reagent used to bind the species to be masked in an unreactive complex. (*p. 208*)

mass balance equation an equation stating that matter is conserved, and that the total amount of a species added to a solution must equal the sum of the amount of each of its possible forms present in solution. (*p. 159*)

mass spectrum a plot of ion intensity as a function of the ion's mass-to-charge ratios. (*p. 571*)

mass transfer one contribution to band broadening due to the time required for a solute to move from the mobile phase or the stationary phase to the interface between the two phases. *(p. 561)*

mass transport the movement of material toward or away from the electrode surface. *(p. 511)*

matrix all other constituents in a sample except for the analytes. *(p. 36)*

matrix matching adjusting the matrix of an external standard so that it is the same as the matrix of the samples to be analyzed. *(p. 110)*

mean the average value of a set of data ($\overline{X}$). *(p. 54)*

measurement an experimental determination of an analyte's chemical or physical properties. *(p. 36)*

measurement error an error due to limitations in the equipment and instruments used to make measurements. *(p. 58)*

median that value for a set of ordered data, for which half of the data is larger in value and half is smaller in value ($\widetilde{X}_{med}$). *(p. 55)*

mediator a species that transfers electrons from the electrode to the analyte. *(p. 500)*

membrane potential a potential developing across a conductive membrane whose opposite sides are in contact with solutions of different composition. *(p. 475)*

meniscus the curved surface of a liquid contained in a tube. *(p. 29)*

metallochromic indicator a visual indicator used to signal the end point in a complexation titration. *(p. 323)*

method a means for analyzing a sample for a specific analyte in a specific matrix. *(p. 36)*

method blank a sample that contains all components of the matrix except the analyte. *(p. 45)*

method error an error due to limitations in the analytical method used to analyze a sample. *(p. 58)*

method of continuous variations a procedure for determining the stoichiometry between two reactants by preparing solutions containing different mole fractions of one reactant; also known as Job's method. *(p. 404)*

method of standard additions a standardization in which aliquots of a standard solution are added to the sample. *(p. 110)*

micellar electrokinetic capillary chromatography a form of capillary electrophoresis in which neutral solutes are separated based on their ability to partition into a charged micelle. *(p. 606)*

micelle an agglomeration of molecules containing ionic "heads" and hydrophobic "tails," which form into a structure with a hydrophobic interior and a hydrophilic exterior. *(p. 606)*

Michaelis constant a combination of several rate constants affecting the rate of an enzyme–substrate reaction. *(p. 637)*

migration the movement of a cation or anion in response to an applied potential. *(p. 512)*

mobile phase in chromatography, the extracting phase that moves through the system. *(p. 546)*

molality the number of moles of solute per kilogram of solvent (m). *(p. 18)*

molarity the number of moles of solute per liter of solution (M). *(p. 15)*

mole-ratio method a procedure for determining the stoichiometry between two reactants by preparing solutions containing different mole ratios of two reactants. *(p. 406)*

monochromatic electromagnetic radiation of a single wavelength. *(p. 377)*

monochromator a wavelength selector that uses a diffraction grating or prism, and that allows for a continuous variation of the nominal wavelength. *(p. 376)*

multiple-point standardization any standardization using two or more standards containing known amounts of analyte. *(p. 109)*

N

negatron the beta particle formed when a neutron converts to a proton; equivalent to an electron ($^{0}_{-1}\beta$). *(p. 642)*

nephelometry a method in which the intensity of scattered radiation is measured at an angle of 90° to the source. *(p. 442)*

Nernst equation an equation relating electrochemical potential to the concentrations of products and reactants. *(p. 146)*

neutron activation a means of inducing radioactivity in a nonradioactive sample by bombarding the sample with neutrons. *(p. 645)*

nominal wavelength the wavelength which a wavelength selector is set to pass. *(p. 376)*

nonfaradaic current a current in an electrochemical cell that is not the result of a redox reaction. *(p. 512)*

normal calibration curve a calibration curve prepared using several external standards. *(p. 109)*

normal distribution "bell-shaped" probability distribution curve for measurements and results showing the effect of random error. *(p. 73)*

normality the number of equivalents of solute per liter of solution (N). *(p. 16)*

normal-phase chromatography liquid chromatography using a polar stationary phase and a nonpolar mobile phase. *(p. 580)*

null hypothesis a statement that the difference between two values can be explained by indeterminate error; retained if the significance test does not fail (H_0). *(p. 83)*

Nyquist theorem statement that a periodic signal must be sampled at least twice each period to avoid a determinate error in measuring its frequency. *(p. 184)*

O

occlusion a coprecipitated impurity trapped within a precipitate as it forms. *(p. 239)*

Ohm's law the statement that the current moving through a circuit is proportional to the applied potential and inversely proportional to the circuit's resistance ($E = iR$). *(p. 463)*

on-column injection the direct injection of thermally unstable samples onto a capillary column. *(p. 568)*

one-tailed significance test significance test in which the null hypothesis is rejected for values at only one end of the normal distribution. *(p. 84)*

open tubular column a capillary column that does not contain a particulate packing material. *(p. 564)*

outlier data point whose value is much larger or smaller than the remaining data. *(p. 93)*

overpotential the difference between the potential actually required to initiate an oxidation or reduction reaction, and the potential predicted by the Nernst equation. *(p. 497)*

oxidation a loss of electrons. *(p. 146)*

oxidizing agent a species that accepts electrons from another species. *(p. 146)*

P

packed column a wide-bore column containing a particulate packing material. *(p. 564)*

paired data two sets of data consisting of results obtained using several samples drawn from different sources. *(p. 88)*

paired *t*-test statistical test for comparing paired data to determine if their difference is too large to be explained by indeterminate error. *(p. 92)*

particulate gravimetry a gravimetric method in which the mass of a particulate analyte is determined following its separation from its matrix. *(p. 234)*

partition coefficient an equilibrium constant describing the distribution of a solute between two phases; only one form of the solute is used in defining the partition coefficient (K_D). *(p. 211)*

parts per billion nanograms of solute per gram of solution; for aqueous solutions the units are often expressed as micrograms of solute per liter of solution (ppb). *(p. 18)*

parts per million micrograms of solute per gram of solution; for aqueous solutions the units are often expressed as milligrams of solute per liter of solution (ppm). *(p. 18)*

peak capacity the maximum number of solutes that can be resolved on a particular column (n_c). *(p. 554)*

peptization the reverse of coagulation in which a coagulated precipitate reverts to smaller particles. *(p. 245)*

peristaltic pump a device for propelling liquids through flexible tubing. *(p. 652)*

personal error an error due to biases introduced by the analyst. *(p. 60)*

p-function a function of the form pX, where pX = –log(X). *(p. 19)*

pH defined as pH = $-\log[H_3O^+]$. *(p. 142)*

phosphorescence emission of a photon when the analyte returns to a lower-energy state with the opposite spin as the higher-energy state. *(p. 424)*

photodiode array a linear array of photodiodes providing the ability to detect simultaneously radiation at several wavelengths. *(p. 379)*

photoluminescence emission following absorption of a photon. *(p. 374)*

photon a particle of light carrying an amount of energy equal to $h\nu$. *(p. 371)*

pipet glassware designed to deliver a specific volume of solution when filled to its calibration mark. *(p. 27)*

planar chromatography a form of chromatography in which the stationary phase is immobilized on a flat surface. *(p. 546)*

plasma a hot, ionized gas containing an abundance of ions and electrons. *(p. 435)*

polarity index a quantitative measure of a solvent's polarity. *(p. 580)*

polarography a form of voltammetry using a dropping mercury electrode or a static mercury drop electrode. *(p. 515)*

polychromatic electromagnetic radiation of more than one wavelength. *(p. 377)*

population all members of a system. *(p. 71)*

positron the beta particle formed when a proton converts to a neutron ($^{0}_{-1}\beta$). *(p. 642)*

potentiometer a device for measuring the potential of an electrochemical cell without drawing a current or altering the cell's composition. *(p. 464)*

potentiostat a device used to control the potential in an electrochemical cell. *(p. 465)*

power the flux of energy per unit time (P). *(p. 371)*

precipitant a reagent that causes the precipitation of a soluble species. *(p. 235)*

precipitate an insoluble solid that forms when two or more soluble reagents are combined. *(p. 139)*

precipitation gravimetry a gravimetric method in which the signal is the mass of a precipitate. *(p. 234)*

precipitation titration a titration in which the reaction between the analyte and titrant involves a precipitation. *(p. 350)*

precision an indication of the reproducibility of a measurement or result. *(p. 39)*

preconcentration the process of increasing an analyte's concentration before its analysis. *(p. 223)*

primary reagent a reagent of known purity that can be used to make a solution of known concentration. *(p. 106)*

probability distribution plot showing frequency of occurrence for members of a population. *(p. 71)*

procedure written directions outlining how to analyze a sample. *(p. 36)*

proficiency standard a standard sample provided by an external agency as part of certifying the quality of a laboratory's work. (*p. 711*)

proportional determinate error a determinate error whose value depends on the amount of sample analyzed. (*p. 61*)

protecting agent a reagent that reacts with the analyte, preventing it from transforming into a nonanalyzable form. (*p. 420*)

protocol a set of written guidelines for analyzing a sample specified by an agency. (*p. 37*)

protocol for a specific purpose a precisely written protocol for an analysis that must be followed exactly. (*p. 707*)

purge and trap a technique for separating volatile analytes from liquid samples in which the analytes are subsequently trapped on a solid adsorbent. (*p. 214*)

Q

qualitative analysis an analysis in which we determine the identity of the constituent species in a sample. (*p. 8*)

quality assessment the steps taken to evaluate whether an analysis is under statistical control. (*p. 708*)

quality assurance the steps taken during an analysis to ensure that the analysis is under control and that it is properly monitored. (*p. 706*)

quality assurance and quality control those steps taken to ensure that the work conducted in an analytical lab is capable of producing acceptable results; also known as QA/QC. (*p. 48*)

quality control those steps taken to ensure that an analysis is under statistical control. (*p. 706*)

quantitative analysis an analysis in which we determine how much of a constituent species is present in a sample. (*p. 9*)

quantitative transfer the process of moving a sample from one container to another in a manner that ensures all material is transferred. (*p. 30*)

quantum yield the fraction of absorbed photons that produce a desired event, such as fluorescence or phosphorescence (Φ). (*p. 425*)

quench to stop a reaction by suddenly changing the reaction conditions. (*p. 634*)

R

random sample a sample collected at random from the target population. (*p. 183*)

range the numerical difference between the largest and smallest values in a data set (w). (*p. 56*)

rate the change in a property's value per unit change in time; the rate of a reaction is a change in concentration per unit change in time. (*p. 624*)

rate constant in a rate law, the proportionality constant between a reaction's rate and the concentrations of species affecting the rate (k). (*p. 624*)

rate law an equation relating a reaction's rate at a given time to the concentrations of species affecting the rate. (*p. 624*)

reagent grade reagents conforming to standards set by the American Chemical Society. (*p. 107*)

recovery the fraction of analyte or interferent remaining after a separation (R). (*p. 202*)

redox electrode an inert electrode that serves as a source or sink for electrons for a redox half-reaction. (*p. 475*)

redox indicator a visual indicator used to signal the end point in a redox titration. (*p. 339*)

redox reaction an electron-transfer reaction. (*p. 145*)

redox titration a titration in which the reaction between the analyte and titrant is an oxidation/reduction reaction. (*p. 331*)

reducing agent a species that donates electrons to another species. (*p. 146*)

reduction a gain of electrons. (*p. 146*)

reference electrode an electrode whose potential remains constant and against which other potentials can be measured. (*p. 462*)

relative supersaturation a measure of the extent to which a solution, or a localized region of solution, contains more dissolved solute than that expected at equilibrium (RSS). (*p. 241*)

relaxation any process by which an analyte returns to a lower-energy state from a higher-energy state. (*p. 423*)

releasing agent a reagent whose reaction with an interferant is more favorable than the interferant's reaction with the analyte. (*p. 420*)

repeatability the precision for an analysis in which the only source of variability is the analysis of replicate samples. (*p. 62*)

reproducibility the precision when comparing results for several samples, for several analysts or several methods. (*p. 62*)

residual current the current in an electrochemical cell that is present even in the absence of the analyte. (*p. 513*)

residual error the difference between an experimental value and the value predicted by a regression equation. (*p. 118*)

resolution in spectroscopy, the separation between two spectral features, such as absorption or emission lines. (*p. 376*)

response the property of a system that is measured (R). (*p. 667*)

response surface a graph showing how a system's response changes as a function of its factors. (*p. 667*)

retention time the time a solute takes to move from the point of injection to the detector (t_r). (*p. 548*)

retention volume the volume of mobile phase needed to move a solute from its point of injection to the detector (V_r). (*p. 548*)

reverse-phase chromatography liquid chromatography using a nonpolar stationary phase and a polar mobile phase. *(p. 580)*

robust a method that can be applied to analytes in a wide variety of matrices is considered robust. *(p. 42)*

rugged a method that is insensitive to changes in experimental conditions is considered rugged. *(p. 42)*

ruggedness testing the process of evaluating a method to determine those factors for which a small change in value has a significant effect on the method's results. *(p. 684)*

S

salt bridge a connection between two solutions that allows the movement of current in the form of ionic charge. *(p. 466)*

sample those members of a population that we actually collect and analyze. *(p. 71)*

sampling error an error introduced during the process of collecting a sample for analysis. *(p. 58)*

sampling plan a plan that ensures that a representative sample is collected. *(p. 182)*

saturated calomel electrode reference electrode based on the reduction of Hg_2Cl_2 to Hg in an aqueous solution saturated with KCl; that is, $Hg_2Cl_2(s) + 2e^- \rightleftharpoons 2Hg(\ell) + 2Cl^-(aq)$. *(p. 472)*

scientific notation a shorthand method for expressing very large or very small numbers by indicating powers of ten; for example, 1000 is 1×10^3. *(p. 12)*

scintillation counter an instrument for counting radioactive particles based on their ability to initiate fluorescence in another substance. *(p. 643)*

secondary reagent a reagent whose purity must be established relative to a primary reagent. *(p. 107)*

selectivity a measure of a method's freedom from interferences as defined by the method's selectivity coefficient. *(p. 40)*

selectivity coefficient a measure of a method's sensitivity for an interferent relative to that for the analyte ($K_{A,I}$). *(p. 40)*

selectivity factor the ratio of capacity factors for two solutes showing the column's selectivity for one of the solutes (α). *(p. 552)*

self-absorption in atomic emission, the decrease in emission intensity when light emitted by excited state atoms in the center of a flame or plasma is absorbed by atoms in the outer portion of the flame. *(p. 438)*

sensitivity a measure of a method's ability to distinguish between two samples; reported as the change in signal per unit change in the amount of analyte (k). *(p. 39)*

separation factor a measure of the effectiveness of a separation at separating an analyte from an interferent ($S_{I,A}$). *(p. 203)*

signal an experimental measurement that is proportional to the amount of analyte (S). *(p. 37)*

signal averaging the adding together of successive spectra to improve the signal-to-noise ratio. *(p. 391)*

signal processor a device, such as a meter or computer, that displays the signal from the transducer in a form that is easily interpreted by the analyst. *(p. 380)*

signal-to-noise ratio the ratio of the signal's intensity to the average intensity of the surrounding noise. *(p. 379)*

significance test a statistical test to determine if the difference between two values is significant. *(p. 83)*

significant figures the digits in a measured quantity, including all digits known exactly and one digit (the last) whose quantity is uncertain. *(p. 13)*

silver/silver chloride electrode reference electrode based on the reduction of AgCl to Ag; that is, $AgCl(s) + e^- \rightleftharpoons Ag(s) + Cl^-(aq)$. *(p. 473)*

simplex optimization an efficient optimization method that allows several factors to be optimized at the same time. *(p. 672)*

single-column ion chromatography ion-exchange chromatography in which conditions are adjusted so that an ion-suppressor column is not needed. *(p. 593)*

single-point standardization any standardization using a single standard containing a known amount of analyte. *(p. 108)*

singlet excited state an excited state in which all electron spins are paired. *(p. 423)*

SI units stands for *Système International d'Unités.* These are the internationally agreed on units for measurements. *(p. 12)*

size-exclusion chromatography a separation method in which a mixture passes through a bed of porous particles, with smaller particles taking longer to pass through the bed due to their ability to move into the porous structure. *(p. 206)*

slope-ratio method a procedure for determining the stoichiometry between two reactants by measuring the relative change in absorbance under conditions when each reactant is the limiting reagent. *(p. 407)*

solid-phase microextraction a solid-phase extraction in which the solid adsorbent is coated on a fused-silica fiber held within a syringe needle. *(p. 567)*

solid-state ion-selective electrode an ion-selective electrode based on a sparingly soluble inorganic crystalline material. *(p. 479)*

solubility product the equilibrium constant for a reaction in which a solid dissociates into its ions (K_{sp}). *(p. 140)*

spectral searching the matching of a spectrum for an unknown compound to a reference spectrum stored in a computer database. *(p. 403)*

spectrofluorometer an instrument for measuring fluorescence that uses a monochromator to select the excitation and emission wavelengths. *(p. 428)*

spectrophotometer an instrument for measuring absorbance that uses a monochromator to select the wavelength. *(p. 389)*

spike recovery an analysis of a sample after spiking with a known amount of analyte. (*p. 710*)

split injection a technique for injecting samples onto a capillary column in which only a small portion of the sample enters the column. (*p. 568*)

splitless injection a technique for injecting a sample onto a capillary column that allows a higher percentage of the sample to enter the column. (*p. 568*)

stacking a means of concentrating solutes in capillary electrophoresis after their injection onto the capillary column. (*p. 603*)

standard deviation a statistical measure of the "average" deviation of data from the data's mean value (s). (*p. 56*)

standard deviation about the regression the uncertainty in a regression analysis due to indeterminate error (s_r). (*p. 121*)

standard hydrogen electrode reference electrode based on the reduction of $H^+(aq)$ to $H_2(g)$ at a Pt electrode; that is, $H^+(aq) + e^- \rightleftharpoons \frac{1}{2}H_2(g)$. (*p. 471*)

standardization the process of establishing the relationship between the amount of analtye and a method's signal. (*p. 47*)

standard method a method that has been identified as providing acceptable results. (*p. 667*)

standard operations procedure the procedure followed in collecting and analyzing samples and in interpreting the results of an analysis. (*p. 707*)

standard reference material a material available from the National Institute of Standards and Technology certified to contain known concentrations of analytes. (*p. 61*)

standard state condition in which solids and liquids are in pure form, gases have partial pressures of 1 atm, solutes have concentrations of 1 M, and the temperature is 298 K. (*p. 137*)

static mercury drop electrode an electrode in which successive drops of Hg form at the end of a capillary tube as the result of a mechanical plunger, with each drop providing a fresh electrode surface. (*p. 509*)

stationary phase in chromatography, the extracting phase that remains in a fixed position. (*p. 546*)

steady-state approximation in a kinetic process, the assumption that a compound formed during the reaction reaches a concentration that remains constant until the reaction is nearly complete. (*p. 636*)

stepwise formation constant the formation constant for a metal–ligand complex in which only one ligand is added to the metal ion or to a metal–ligand complex (K_i). (*p. 144*)

stock solution a solution of known concentration from which other solutions are prepared. (*p. 30*)

stopped flow a kinetic method of analysis designed to rapidly mix samples and reagents when using reactions with very fast kinetics. (*p. 634*)

stratified sampling a sampling plan that divides the population into distinct strata from which random samples are collected. (*p. 185*)

stray radiation any radiation reaching the detector that does not follow the optical path from the source to the detector. (*p. 387*)

stripping voltammetry a form of voltammetry in which the analyte is first deposited on the electrode and then removed, or "stripped," electrochemically while monitoring the current as a function of the applied potential. (*p. 516*)

substrate the specific molecule for which an enzyme serves as a catalyst. (*p. 636*)

supercritical fluid a state of matter where a substance is held at a temperature and pressure that exceeds its critical temperature and pressure. (*p. 215*)

supercritical fluid chromatography a separation technique in which the mobile phase is a supercritical fluid. (*p. 596*)

supernatant the solution that remains after a precipitate forms. (*p. 244*)

support-coated open tubular column an open tubular column in which the stationary phase is coated on a solid support that is attached to the column's walls. (*p. 565*)

systematic–judgmental sampling a sampling plan that combines judgmental sampling with systematic sampling. (*p. 184*)

systematic sampling samples collected from the target population at regular intervals in time or space. (*p. 184*)

T

tailing a tail at the end of a chromatographic peak, usually due to the presence of highly active sites in the stationary phase. (*p. 555*)

technique a chemical or physical principle that can be used to analyze a sample. (*p. 36*)

temperature programming the process of changing the column's temperature to enhance the separation of both early and late eluting solutes. (*p. 558*)

theoretical model a model describing a system's response that has a theoretical basis and can be derived from theoretical principles. (*p. 675*)

theoretical plate a quantitative means of evaluating column efficiency that treats the column as though it consists of a series of small zones, or plates, in which partitioning between the mobile and stationary phases occurs. (*p. 553*)

thermal conductivity detector a universal GC detector in which the signal is a change in the thermal conductivity of the mobile phase. (*p. 569*)

thermogram a graph showing change in mass as a function of applied temperature. (*p. 256*)

thermogravimetry a form of volatilization gravimetry in which the change in a sample's mass is monitored while it is heated. (*p. 255*)

titrant the reagent added to a solution containing the analyte and whose volume is the signal. (*p. 274*)

titration curve a graph showing the progress of a titration as a function of the volume of titrant added. *(p. 276)*

titration error the determinate error in a titration due to the difference between the end point and the equivalence point. *(p. 274)*

titrimetry any method in which volume is the signal. *(p. 274)*

tolerance the maximum determinate measurement error for equipment or instrument as reported by the manufacturer. *(p. 58)*

total analysis techniques a technique in which the signal is proportional to the absolute amount of analyte; also called "classical" techniques. *(p. 38)*

total ionic strength adjustment buffer a solution containing a relatively high concentration of inert electrolytes such that its composition fixes the ionic concentration of all solutions to which it is added. *(p. 487)*

total Youden blank a blank that corrects the signal for analyte–matrix interactions. *(p. 129)*

tracer a radioactive species used as an internal standard. *(p. 646)*

transducer a device that converts a chemical or physical property, such as pH or photon intensity, to an easily measured electrical signal, such as a voltage or current. *(p. 379)*

transmittance the ratio of the radiant power passing through a sample to that from the radiation's source (T). *(p. 384)*

trip blank a blank prepared in the laboratory that accompanies a set of sample containers in the field and laboratory. *(p. 710)*

triplet excited state an excited state in which unpaired electron spins occur. *(p. 423)*

***t*-test** statistical test for comparing two mean values to see if their difference is too large to be explained by indeterminate error. *(p. 85)*

turbidimetry a method in which the decrease in transmitted radiation due to scattering is measured. *(p. 441)*

two-tailed significance test significance test in which the null hypothesis is rejected for values at either end of the normal distribution. *(p. 84)*

type 1 error the risk of falsely rejecting the null hypothesis (α). *(p. 84)*

type 2 error the risk of falsely retaining the null hypothesis (β). *(p. 84)*

U

uncertainty the range of possible values for a measurement. *(p. 64)*

unpaired data two sets of data consisting of results obtained using several samples drawn from a single source. *(p. 88)*

V

validation the process of verifying that a procedure yields acceptable results. *(p. 47)*

van Deemter equation an equation showing the effect of the mobile phase's flow rate on the height of a theoretical plate. *(p. 561)*

variance the square of the standard deviation (s^2). *(p. 57)*

vibrational relaxation a form of radiationless relaxation in which an analyte moves from a higher vibrational energy level to a lower vibrational energy level in the same electronic level. *(p. 424)*

void time the time required for unretained solutes to move from the point of injection to the detector (t_m). *(p. 549)*

void volume the volume of mobile phase needed to move an unretained solute from the point of injection to the detector. *(p. 549)*

volatilization gravimetry a gravimetric method in which the loss of a volatile species gives rise to the signal. *(p. 234)*

voltammetry an electrochemical method in which we measure current as a function of the applied potential. *(p. 508)*

voltammogram a plot of current as a function of applied potential. *(p. 508)*

volume percent milliliters of solute per 100 mL of solution (% v/v). *(p. 18)*

volumetric flask glassware designed to contain a specific volume of solution when filled to its calibration mark. *(p. 26)*

W

Walden reductor a reduction column using granular Ag as a reducing agent. *(p. 341)*

wall-coated open tubular column an open tubular column in which the stationary phase is coated on the column's walls. *(p. 565)*

wavelength the distance between any two consecutive maxima or minima of an electromagnetic wave (λ). *(p. 370)*

wavenumber the reciprocal of wavelength ($\bar{\nu}$). *(p. 370)*

weight percent grams of solute per 100 g of solution. (% w/w). *(p. 18)*

weight-to-volume percent grams of solute per 100 mL of solution (% w/v). *(p. 18)*

Z

zeta potential the change in potential across a double layer (ζ). *(p. 599)*

Index

Page numbers followed by italic *f* and *t* refer to figures and tables, respectively.

A

B

C

F

G

H

I

J

K

L

M

N

O

P

T

W

X

Y

Z